Stem Cell Biology and Regenerative Medicine

Volume 79

Our understanding of stem cells has grown rapidly over the last decade. While the apparently tremendous therapeutic potential of stem cells has not yet been realized, their routine use in regeneration and restoration of tissue and organ function is greatly anticipated. To this end, many investigators continue to push the boundaries in areas such as the reprogramming, the stem cell niche, nanotechnology, biomimetics and 3D bioprinting, to name just a few. The objective of the volumes in the Stem Cell Biology and Regenerative Medicine series is to capture and consolidate these developments in a timely way. Each volume is thought-provoking in identifying problems, offering solutions, and providing ideas to excite further innovation in the stem cell and regenerative medicine fields.

Francesco De Francesco • Kursad Turksen
Editors

Management and Strategies for Wound Healing

Springer

Editors
Francesco De Francesco
Department of Reconstructive Surgery and Hand Surgery
Azienda Ospedaliera Universitaria delle Marche
Ancona, Italy

Kursad Turksen
Regenerative Medicine
Ottawa Hospital Research Institute (retired)
Ottawa, ON, Canada

ISSN 2196-8985 ISSN 2196-8993 (electronic)
Stem Cell Biology and Regenerative Medicine
ISBN 978-3-032-12441-8 ISBN 978-3-032-12442-5 (eBook)
https://doi.org/10.1007/978-3-032-12442-5

This Springer imprint is published by the registered company Springer Nature Switzerland AG
The registered company address is: Gewerbestrasse 11, 6330 Cham, Switzerland

Preface

Wound healing represents one of the most dynamic, multifaceted, and vital areas of clinical, basic, and translational research. The ability to restore tissues and organs to their original structure and function following injury remains a central challenge—and an enduring goal—in regenerative medicine. This volume seeks to illuminate the complexities and innovations shaping this field.

To capture the breadth and depth of current advances, we have curated a collection of reviews that explore the clinical realities, fundamental biology, and emerging translational strategies in wound healing. From cellular mechanisms and molecular pathways to innovative biomaterials and therapeutic frontiers, this work aims to serve as both a comprehensive reference and a source of inspiration for continued research.

We extend our heartfelt thanks to Merry Stuber, Senior Editor of the Cell Biology & Biomedical Engineering series, whose support has been invaluable throughout this endeavor.

Our sincere gratitude also goes to the production team for their exceptional dedication in bringing this volume to life.

Finally, we deeply thank all the contributors for their commitment, insight, and scholarly excellence. Their willingness to share not only their scientific progress but also the unresolved challenges in the field has been instrumental in shaping this volume. We are confident that readers will find their contributions as thought-provoking and impactful as we have.

Ancona, Italy — Francesco De Francesco
Ottawa, ON, Canada — Kursad Turksen

Contents

1 **The History of Wound Care** . . . 1
Emma N. Winslow and Joseph A. Molnar

2 **Anatomy, Histology and Embryonic Development of the Skin** . . . 21
Marika Berardini, Luisa Gesualdi, Francesco De Francesco, and Giulia Ricci

3 **Pathophysiology of Wound Healing** . . . 45
Ali Golchin, Khan Sharun, and Aynaz Mihanfar

4 **Clinical Assessment of Hard-to-Heal Wounds: Tools, Evidence, and Multidisciplinary Strategies** . . . 61
Francesco De Francesco and Michele Riccio

5 **Cellular and Molecular Mechanisms in Wound Healing** . . . 85
Md Sariful Islam Howlader and Hiranmoy Das

6 **Mitochondrial Orchestration of Wound Healing: From Cellular Energetics to Therapeutic Innovation** . . . 125
Meghnad G. Joshi, Mrunal N. Damle, and Rakesh Kumar Sharma

7 **Cutaneous Wound Epithelialization: Role of Cadherin- and Integrin-Mediated Adhesions** . . . 181
Katheryn E. Rothenberg and Martine Dunnwald

8 **Cellular Activation of Surface G-Protein Coupled Receptors as Drug Targets for Skin Wound Healing** . . . 199
Apiwan Arinno and Pawin Pongkorpsakol

9 **Growth Factors and Cytokines: Unravelling Molecular Mechanisms for Therapeutic Innovations in Wound Care** . . . 219
Madhan Jeyaraman, Naveen Jeyaraman, Sathish Muthu, Shrideavi Murugan, and Sharun Khan

10 **Inflammation, Fibrosis, and Innate Immune Response in Diabetic Foot Ulcer** 237
Gurkaranjot Singh, Zuhair Chaudhry, Sankalp Nigam, Dominick Shoha, and Vikrant Rai

11 **Preclinical Models for Wound Healing and Repair Studies** 259
Rochelle Bitolas and David A. Brown

12 **Preclinical Insights into Wound Healing: Models, Mechanisms, and Translational Strategies** 287
Naveen Jeyaraman, Mainak Roy, Sathish Muthu, Sharun Khan, and Madhan Jeyaraman

13 **Extracellular Vesicles as Functional Mediators of Wound Healing** . . 311
Heidi Yuan and Luisa A. DiPietro

14 **Stem Cells and Extracellular Vesicles for the Treatment of Chronic Wounds and Peripheral Vascular Diseases** 353
Barbara Zavan and Letizia Ferroni

15 **The Role of Bone Marrow-Derived Cells (BMDCs) During Wound Healing** 383
Junko Okano and Hideto Kojima

16 **The Role of Adipose-Derived Stem Cells in Accelerating and Refining Wound Healing** 409
Francesco De Francesco

17 **Wound Dressings: Principles and Modern Applications** 439
Apirag Chuangsuwanich and Pitawan Rachata

18 **Current Biomaterials and Strategies in Tissue Engineering for Wound Healing** 463
F. Re, E. Borsani, M. Bonetti, and F. Bonomini

19 **Biomaterials as Skin Substitutes** 495
Eylül Odabaş-Kansız, Seyithan Kansız, Ayşe Eser Elçin, and Yaşar Murat Elçin

20 **Innovative Wound Dressing** 543
Apirag Chuangsuwanich and Pitawan Rachata

21 **Innovative Treatments for Accelerated Wound Healing** 573
David S. Salven and David A. Brown

22 **Regenerative Therapy for Burns** 597
Ahmed T. El-Serafi, Raghad Alghazali, Iman M. Talaat, Folke Sjöberg, Moustafa Elmasry, Maha Saber-Ayad, and Ibrahim El-Serafi

23 Cultured Epithelial Autografts Based on a Fibrin Matrix System: Optimising Engraftment in Extensive Burns 627
Khong-Yik Chew, Alvin W. C. Chua, Keya Rani Roy, Seng-Teik Lee, Yann Barrandon, and Bien-Keem Tan

24 Healing Beyond the Surface: Regenerative Strategies in Complex Burn Treatment 641
Frederik Schlottmann, Frederik J. Kloss, Lisa Lorbeer, Moritz R. Milewski, and Peter M. Vogt

25 Wound Healing and Regenerative Medicine in Diabetes 665
Martin Kauke-Navarro, Felix J. Klimitz, and Henry C. Hsia

26 Innovative Frontiers in Wound Healing: Integrating Bioengineering, Nanotechnology, and Regenerative Medicine 695
Sathish Muthu, Naveen Jeyaraman, A. S. P. V. S. Saketh, Sharun Khan, and Madhan Jeyaraman

27 Nanobiology of Cutaneous Wound Healing 713
Mohini Moulick, Aparajita Nandy, Pramod Bhasme, Shayani Mukherjee, Anita Yadav, and Subhadip Ghatak

28 Adjunctive Therapy for Wound Healing 747
Mohamed Abdelhakim and Rei Ogawa

29 Nutrition and Wound Healing: Mechanisms and Clinical Integration 765
Scott A. Sylvester, Abigail Peoples, Samuel Pendergraft, and Joseph A. Molnar

30 Scar Problems After Wound Healing 785
Mohamed Abdelhakim and Rei Ogawa

31 Therapeutic Approaches to Treat Hypertrophic Scars 805
Vanessa C. Smith, Steve Smith, and Tugba Ozdemir

32 AI and Personalized Medicine in Wound Care 833
Necip Gurler, Raphael A. Yaakov, H. Samet Bulbul, Patrick Cheng, Kyle Wu, and Ozgur Guler

Index 867

Contributors

Mohamed Abdelhakim Department of Plastic, Reconstructive, and Aesthetic Surgery, Nippon Medical School Hospital, Tokyo, Japan

Raghad Alghazali Department of Hand Surgery, Plastic Surgery and Burns, Linköping University Hospital, Linköping, Sweden

Department of Biomedical and Clinical Sciences, Linköping University, Linköping, Sweden

Apiwan Arinno Department of Science, Technology and Innovation, Faculty of Science, Chulabhorn Royal Academy, Bangkok, Thailand

Yann Barrandon Department of Plastic, Reconstructive and Aesthetic Surgery, Singapore General Hospital, Musculoskeletal Sciences Academic Clinical Programmes, Duke-NUS Medical School, Singapore, Singapore

Marika Berardini Department of Anatomy, Histology, Forensic-Medicine and Orthopedics, Section of Histology and Embryology, "Sapienza" University of Rome, Rome, Italy

Pramod Bhasme McGowan Institute for Regenerative Medicine, Department of Surgery, University of Pittsburgh, Pittsburgh, PA, USA

Rochelle Bitolas Duke University School of Medicine, Durham, NC, USA

M. Bonetti Division of Anatomy and Physiopathology, Department of Clinical and Experimental Sciences, University of Brescia, Brescia, Italy

F. Bonomini Division of Anatomy and Physiopathology, Department of Clinical and Experimental Sciences, University of Brescia, Brescia, Italy

E. Borsani Division of Anatomy and Physiopathology, Department of Clinical and Experimental Sciences, University of Brescia, Brescia, Italy

David A. Brown Duke Department of Surgery, Division of Plastic, Maxillofacial, and Oral Surgery, Duke University Medical Center, Durham, NC, USA

Duke University School of Medicine, Durham, NC, USA

H. Samet Bulbul eKare Inc, Fairfax, VA, USA

Zuhair Chaudhry College of Osteopathic Medicine of the Pacific, Western University of Health Sciences, Pomona, CA, USA

Patrick Cheng eKare Inc, Fairfax, VA, USA

Khong-Yik Chew Department of Plastic, Reconstructive and Aesthetic Surgery, Singapore General Hospital, Musculoskeletal Sciences Academic Clinical Programmes, Duke-NUS Medical School, Singapore, Singapore

Alvin W. C. Chua Department of Plastic, Reconstructive and Aesthetic Surgery, Singapore General Hospital, Musculoskeletal Sciences Academic Clinical Programmes, Duke-NUS Medical School, Singapore, Singapore

Apirag Chuangsuwanich Division of Plastic Surgery, Department of Surgery, Faculty of Medicine Siriraj Hospital, Mahidol University, Bangkok, Thailand

Mae Fah Luang Medical Center Hospital, Mae Fah Luang University, Chiang Rai, Thailand

Mrunal N. Damle Department of Stem Cells and Regenerative Medicine, D.Y. Patil Education Society (Deemed University), D. Y. Patil Vidyanagar, Kolhapur, Maharashtra, India

Stem Plus Biotech Pvt. Ltd, Sangli, Maharashtra, India

Hiranmoy Das Department of Pharmaceutical Sciences, Jerry H. Hodge School of Pharmacy, Texas Tech University Health Sciences Center, Amarillo, TX, USA

Francesco De Francesco Department of Reconstructive Surgery and Hand Surgery, Azienda Ospedaliera Universitaria delle Marche, Ancona, Italy

Luisa A. DiPietro Center for Wound Healing and Tissue Regeneration, University of Illinois Chicago, Chicago, IL, USA

Martine Dunnwald Department of Anatomy and Cell Biology, University of Iowa Carver College of Medicine, Iowa City, IA, USA

Ayşe Eser Elçin Tissue Engineering, Biomaterials and Nanobiotechnology Laboratory, Faculty of Science and Stem Cell Institute, Ankara University, Ankara, Türkiye

Yaşar Murat Elçin Tissue Engineering, Biomaterials and Nanobiotechnology Laboratory, Faculty of Science and Stem Cell Institute, Ankara University, Ankara, Türkiye

Biovalda Health Technologies, Inc., Ankara, Türkiye

Moustafa Elmasry Department of Hand Surgery, Plastic Surgery and Burns, Linköping University Hospital, Linköping, Sweden

Ahmed T. El-Serafi Department of Medical Laboratory Sciences, College of Health Sciences, University of Sharjah, Sharjah, United Arab Emirates

Department of Hand Surgery, Plastic Surgery and Burns, Linköping University Hospital, Linköping, Sweden

Department of Biomedical and Clinical Sciences, Linköping University, Linköping, Sweden

Ibrahim El-Serafi Basic Medical Sciences Department, College of Medicine, Ajman University, Ajman, United Arab Emirates

Letizia Ferroni Maria Cecilia Hospital, GVM Care & Research, Ravenna, Italy

Luisa Gesualdi Department of Anatomy, Histology, Forensic-Medicine and Orthopedics, Section of Histology and Embryology, "Sapienza" University of Rome, Rome, Italy

Subhadip Ghatak McGowan Institute for Regenerative Medicine, Department of Surgery, University of Pittsburgh, Pittsburgh, PA, USA

Ali Golchin Iranian Tissue Engineering and Regenerative Medicine Society (ITERMS), Tehran, Iran

Regenerative Medicine Group (REMED), Universal Scientific Education and Research Network (USERN), Tehran, Iran

Ozgur Guler eKare Inc, Fairfax, VA, USA

Necip Gurler eKare Inc, Fairfax, VA, USA

Md Sariful Islam Howlader Department of Pharmaceutical Sciences, Jerry H. Hodge School of Pharmacy, Texas Tech University Health Sciences Center, Amarillo, TX, USA

Henry C. Hsia Department of Surgery, Division of Plastic Surgery, Yale School of Medicine, New Haven, CT, USA

Madhan Jeyaraman Department of Orthopaedics, ACS Medical College and Hospital, Dr MGR Educational and Research Institute, Chennai, Tamil Nadu, India

Orthopaedic Research Group, Coimbatore, Tamil Nadu, India

Naveen Jeyaraman Department of Orthopaedics, ACS Medical College and Hospital, Dr MGR Educational and Research Institute, Chennai, Tamil Nadu, India

Orthopaedic Research Group, Coimbatore, Tamil Nadu, India

Meghnad G. Joshi Department of Stem Cells and Regenerative Medicine, D.Y. Patil Education Society (Deemed University), D. Y. Patil Vidyanagar, Kolhapur, Maharashtra, India

Stem Plus Biotech Pvt. Ltd, Sangli, Maharashtra, India

Seyithan Kansız Tissue Engineering, Biomaterials and Nanobiotechnology Laboratory, Faculty of Science and Stem Cell Institute, Ankara University, Ankara, Türkiye

Martin Kauke-Navarro Department of Surgery, Division of Plastic Surgery, Yale School of Medicine, New Haven, CT, USA

Sharun Khan Centre for Regenerative Nanomedicine, Northwestern University, Chicago, IL, USA

Graduate Institute of Medicine, Yuan Ze University, Taoyuan, Taiwan

Felix J. Klimitz Department of Surgery, Division of Plastic Surgery, Yale School of Medicine, New Haven, CT, USA

Frederik J. Kloss Department of Plastic, Aesthetic, Hand and Reconstructive Surgery, Hannover Medical School, Hannover, Germany

Hideto Kojima Department of Biocommunication Development, Shiga University of Medical Science, Otsu, Japan

Seng-Teik Lee Department of Plastic, Reconstructive and Aesthetic Surgery, Singapore General Hospital, Musculoskeletal Sciences Academic Clinical Programmes, Duke-NUS Medical School, Singapore, Singapore

C. D. Lokhande Center for Interdisciplinary Research, D Y Patil Education Society (Deemed to be University), Kolhapur, Maharashtra, India

Lisa Lorbeer Department of Plastic, Aesthetic, Hand and Reconstructive Surgery, Hannover Medical School, Hannover, Germany

Aynaz Mihanfar Iranian Tissue Engineering and Regenerative Medicine Society (ITERMS), Tehran, Iran

Department of Clinical Biochemistry, School of Medicine, Urmia University of Medical Sciences, Urmia, Iran

Moritz R. Milewski Department of Plastic, Aesthetic, Hand and Reconstructive Surgery, Hannover Medical School, Hannover, Germany

Joseph A. Molnar Department of Plastic and Reconstructive Surgery, Atrium Health Wake Forest Baptist Hospital, Medical Center Boulevard, Winston-Salem, NC, USA

Wake Forest Institute for Regenerative Medicine, Wake Forest University School of Medicine, Winston-Salem, NC, USA

Mohini Moulick McGowan Institute for Regenerative Medicine, Department of Surgery, University of Pittsburgh, Pittsburgh, PA, USA

Shayani Mukherjee McGowan Institute for Regenerative Medicine, Department of Surgery, University of Pittsburgh, Pittsburgh, PA, USA

Shrideavi Murugan Department of Orthopaedics, Tirunelveli Medical College and Hospital, Tirunelveli, Tamil Nadu, India

Sathish Muthu Orthopaedic Research Group, Coimbatore, Tamil Nadu, India

Central Research Laboratory, Meenakshi Medical College Hospital and Research Institute, Meenakshi Academy of Higher Education and Research, Chennai, Tamil Nadu, India

Aparajita Nandy McGowan Institute for Regenerative Medicine, Department of Surgery, University of Pittsburgh, Pittsburgh, PA, USA

Sankalp Nigam College of Osteopathic Medicine of the Pacific, Western University of Health Sciences, Pomona, CA, USA

Eylül Odabaş-Kansız Tissue Engineering, Biomaterials and Nanobiotechnology Laboratory, Faculty of Science and Stem Cell Institute, Ankara University, Ankara, Türkiye

Rei Ogawa Department of Plastic, Reconstructive, and Aesthetic Surgery, Nippon Medical School Hospital, Tokyo, Japan

Junko Okano Department of Plastic and Reconstructive Surgery, Shiga University of Medical Science, Otsu, Japan

Tugba Ozdemir Department of Nanoscience & Biomedical Engineering, South Dakota School of Mines and Technology, Rapid City, SD, USA

Samuel Pendergraft Wake Forest Institute for Regenerative Medicine, Wake Forest University School of Medicine, Winston-Salem, NC, USA

Abigail Peoples Department of Plastic and Reconstructive Surgery, Atrium Health Wake Forest Baptist Hospital, Medical Center Boulevard, Winston-Salem, NC, USA

Pawin Pongkorpsakol Princess Srisavangavadhana Faculty of Medicine, Chulabhorn Royal Academy, Bangkok, Thailand

Pitawan Rachata Division of Plastic Surgery, Department of Surgery, School of Medicine, Mae Fah Luang University, Chiang Rai, Thailand

Mae Fah Luang Medical Center Hospital, Mae Fah Luang University, Chiang Rai, Thailand

Vikrant Rai Department of Translational Research, Western University of Health Sciences, Pomona, CA, USA

F. Re Department of Clinical and Experimental Sciences, University of Brescia, Bone Marrow Transplant Unit, ASST Spedali Civili, Brescia, Italy

Centro di Ricerca Emato-Oncologica AIL (CREA), ASST Spedali Civili, Brescia, Italy

Giulia Ricci Department of Experimental Medicine, University of Campania "Luigi Vanvitelli", Naples, Italy

Michele Riccio Department of Reconstructive Surgery and Hand Surgery, Azienda Ospedaliera Universitaria delle Marche, Ancona, Italy

Katheryn E. Rothenberg Department of Biology, University of Iowa College of Liberal Arts and Sciences, Iowa City, IA, USA

Keya Rani Roy Department of Plastic, Reconstructive and Aesthetic Surgery, Singapore General Hospital, Musculoskeletal Sciences Academic Clinical Programmes, Duke-NUS Medical School, Singapore, Singapore

Mainak Roy Department of Orthopaedics, All India Institute of Medical Sciences, New Delhi, India

Maha Saber-Ayad Clinical Sciences Department, College of Medicine, University of Sharjah, Sharjah, United Arab Emirates

A. S. P. V. S. Saketh Department of Orthopaedics & Sports Medicine, Yashoda Hospitals, Malakpet, Hyderabad, Telangana, India

David S. Salven Division of Plastic, Maxillofacial, and Oral Surgery, Duke University School of Medicine, Durham, NC, USA

Frederik Schlottmann Department of Plastic, Aesthetic, Hand and Reconstructive Surgery, Hannover Medical School, Hannover, Germany

Rakesh Kumar Sharma Department of Obstetrics and Gynecology, D.Y. Patil Education Society Deemed University, D.Y. Patil Medical College, Kolhapur, Maharashtra, India

Khan Sharun Graduate Institute of Medicine, Yuan Ze University, Taoyuan, Taiwan

Regenerative Medicine Group, Department of Health Science and Technology, Aalborg University, Gistrup, Denmark

Dominick Shoha College of Osteopathic Medicine of the Pacific, Western University of Health Sciences, Pomona, CA, USA

Gurkaranjot Singh College of Osteopathic Medicine of the Pacific, Western University of Health Sciences, Pomona, CA, USA

Folke Sjöberg Department of Biomedical and Clinical Sciences, Linköping University, Linköping, Sweden

Steve Smith Department of Nanoscience & Biomedical Engineering, South Dakota School of Mines and Technology, Rapid City, SD, USA

Vanessa C. Smith Department of Nanoscience & Biomedical Engineering, South Dakota School of Mines and Technology, Rapid City, SD, USA

Scott A. Sylvester Department of Plastic and Reconstructive Surgery, Atrium Health Wake Forest Baptist Hospital, Medical Center Boulevard, Winston-Salem, NC, USA

Iman M. Talaat Clinical Sciences Department, College of Medicine, University of Sharjah, Sharjah, United Arab Emirates

Bien-Keem Tan Department of Plastic, Reconstructive and Aesthetic Surgery, Singapore General Hospital, Musculoskeletal Sciences Academic Clinical Programmes, Duke-NUS Medical School, Singapore, Singapore

Peter M. Vogt Department of Plastic, Aesthetic, Hand and Reconstructive Surgery, Hannover Medical School, Hannover, Germany

Emma N. Winslow Wake Forest University, Fredericksburg, VA, USA

Kyle Wu eKare Inc, Fairfax, VA, USA

Raphael A. Yaakov eKare Inc, Fairfax, VA, USA

Anita Yadav McGowan Institute for Regenerative Medicine, Department of Surgery, University of Pittsburgh, Pittsburgh, PA, USA

Heidi Yuan Center for Wound Healing and Tissue Regeneration, University of Illinois Chicago, Chicago, IL, USA

Barbara Zavan Translational Medicine Department, University of Ferrara, Ferrara, Italy

Chapter 1
The History of Wound Care

Emma N. Winslow and Joseph A. Molnar

Abstract As the future is a continuum of the past, it is important to review our predecessors' discoveries, viewpoints, and mistakes. This chapter discusses the history of wound care from prehistoric times to the modern era in several sections. Part 1 speculates on wound care in our early ancestors through examples in our closest relatives. Part 2 discusses wound care practices documented in early medical writings in Mesopotamia. Part 3 highlights wound cases from ancient Egypt. Part 4 explores medical philosophies and treatments of ancient Greece. Part 5 reviews the evolution of wound care in ancient Rome as the empire expanded. Part 6 examines ancient Chinese medicine with respect to wound care and the practice of wound care in the late Zhou to early Han Dynasties. Part 7 reveals the development of plastic surgery practices in ancient India. Part 8 describes the advancement of surgery and amputation medicine in medieval and early modern France. Part 9 describes the medical discoveries of early modern Great Britain. Part 10 reviews the development of twentieth-century advancements, including antibiotics, compression wraps, negative pressure wound therapy, skin grafts, and skin substitutes. Part 11 looks into the future of wound healing.

Keywords Wound care · Global history · Modern history · Medical texts · Ancient civilizations

Abbreviation

V.A.C Vacuum assisted closure

E. N. Winslow (✉) · J. A. Molnar
Wake Forest University, Fredericksburg, VA, USA

F. De Francesco, K. Turksen (eds.), *Management and Strategies for Wound Healing*, Stem Cell Biology and Regenerative Medicine 79,
https://doi.org/10.1007/978-3-032-12442-5_1

Introduction

Much of our knowledge on the history of wound care comes from texts written across the world (Fig. 1.1), but it is widely accepted that wound care in various forms existed long before recorded history. While there is no definitive proof, there are notable examples in our closest relatives that provide fuel for speculation.

Wolfgang Köhler, a German Gestalt psychologist born in 1887, during his years of primate research observed wound care in the animal kingdom. One of his chimpanzees developed a large abscess on its lower jaw. He wrote, "When it became noticeable through the extent of the inflamed surface and secretion of pus, another of the apes would not stir from the patient's side, but pressed and kneaded the injured jaw, until the pus was removed, revealing a raw, gaping wound. The animal thus treated made no objections. As apes like using diverse objects in all eager manipulations, the operator worked with a large piece of old rag in his hand. Yet, wonderful to relate, the wound—itself, probably originally caused by a skin treatment with filthy hands—healed rapidly and completely" [21]. Substituting sterile gauze for the old rag, one can see the similarity between the chimpanzee's work and a modern incision and drainage.

Another example comes from the remains of a Neanderthal man, fondly nicknamed "Nandy," dated to the Late Pleistocene. Based on his remains, this close relative to early humans appeared to have suffered a series of severe injuries from a young age. These included a crushing blow to his head, resulting in damage to his left eye, unilateral hearing loss, and paralysis of his contralateral arm and leg. He later lost his arm altogether, amputated above the elbow. Despite this, he survived beyond the average lifespan of a Neanderthal. The locations of his injuries all showed signs of healing, indicating that these injuries were not the cause of his death. It is thought that he died when his roof caved in and was later buried by friends and family. Given the great obstacles he faced, it is highly likely that he survived with the help of friends and family, which would have involved some form of medical attention and wound care [33].

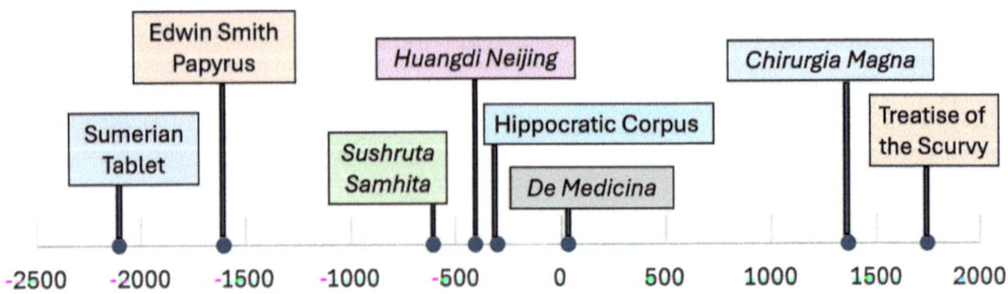

Fig. 1.1 Timeline of historic medical texts related to wound care (own work)

Ancient Mesopotamia

The first known written evidence of wound care lies in the earliest known medical text, the famous clay Sumerian tablet dated back to 2100 BC. Written in cuneiform, it roughly translates to "three healing gestures": washing the wound with hot water and beer, using plasters, and bandaging it. The tablet goes on to review fifteen prescriptions, including eight different types of plasters, which were likely meant to treat wounds and other local maladies. Three of which are as follows:

> *Pound together: dried wine dregs, juniper and prunes; pour beer on the mixture. Then rub with oil, and bind on.*
> *Pass through a sieve and then knead together: turtle shell, naga-si plant, salt, and mustard. Then wash the diseased part with beer of good quality and hot water, and rub with the mixture. Then, friction and rub again with oil, and put on pounded pine.*
> *Take some river sediment and pound it, knead it with water; then rub the diseased part with mineral oil, and bind on as a plaster.* [21]

While the individual components of these prescriptions may look different today, the essence of the three healing gestures has remarkable parallels to the modern wound care process.

The Sumerians brewed at least 19 different types of beer. While the alcohol content of beer is generally too low to have any significant antibacterial effect, ancient or otherwise, it may inherently have some antibacterial properties due to its compounds. Hops have been shown to inhibit the growth of certain strains of gram-positive bacteria, and the relatively low pH of beer is also known to inhibit bacterial growth [31]. There is also evidence behind their use of hot water specifically, as boiling water for even a few minutes can kill most of the bacteria, protozoa, and parasites living in it. Considering what would have been available at the time, washing the wounds with beer and hot water would be an excellent first step in cleaning the wounds.

Plasters are the next step in Sumerian wound healing. They commonly used oils, animal fats, sediments, resins, and plant material. Several of these dressings surely had their merits. Much like today's petroleum jelly, animal fats and oils are soothing and help maintain a moist wound bed. Compared to dry healing environments, moisture results in improved quality and speed of epithelialization [34]. The oils and fats would also prevent mechanical damage by keeping bandages from sticking to and ripping the healing tissues. Furthermore, they could prevent infection. Fats can last for days to months before turning rancid, and in preliminary experiments by D. Kekessy in Geneva, staphylococci plated in sesame oil died rapidly [21]. As for considering the usefulness of their other dressings, a major problem lies in linguistics. Even with expert interpreters, it remains difficult to discern which specific plant species and mineral types are represented by the various cuneiform symbols. We are still able to speculate on the use of several materials that may have been used. Among them are pine resin, cedar resin, cassia, frankincense, myrrh, and turpentine (resin of the Cyprus turpentine plant, not to be confused with the distillate used for paint thinner today). Many of these are native to the region and others would have been imported by traders. Several of these materials are antimicrobial to an extent, and their powerful fragrances would help ameliorate putrid odors [21].

One can also glimpse into the common wound care practices of ancient times through the Bible. In Luke 10:34, the parable of the good Samaritan reads: "He went to him and bandaged his wounds, pouring on oil and wine. Then he put the man on his own donkey, took him to an inn and took care of him." Isaiah 1:6 reads, "... only wounds and welts and open sores, not cleansed or bandaged or soothed with olive oil" [23]. This scripture supports the prevalence and enduring nature of the "three healing gestures."

Ancient Egypt

Arguably, the most famous medical text from ancient Egypt is the Edwin Smith Papyrus (Fig. 1.2), written 300 years later, around 1600 B.C. Of note, this papyrus contains evidence that it was written by two scribes as a copy of an even older document, and so the actual date of origin remains unknown. The document walks the reader through a series of medical cases, including wounds. For each case and for

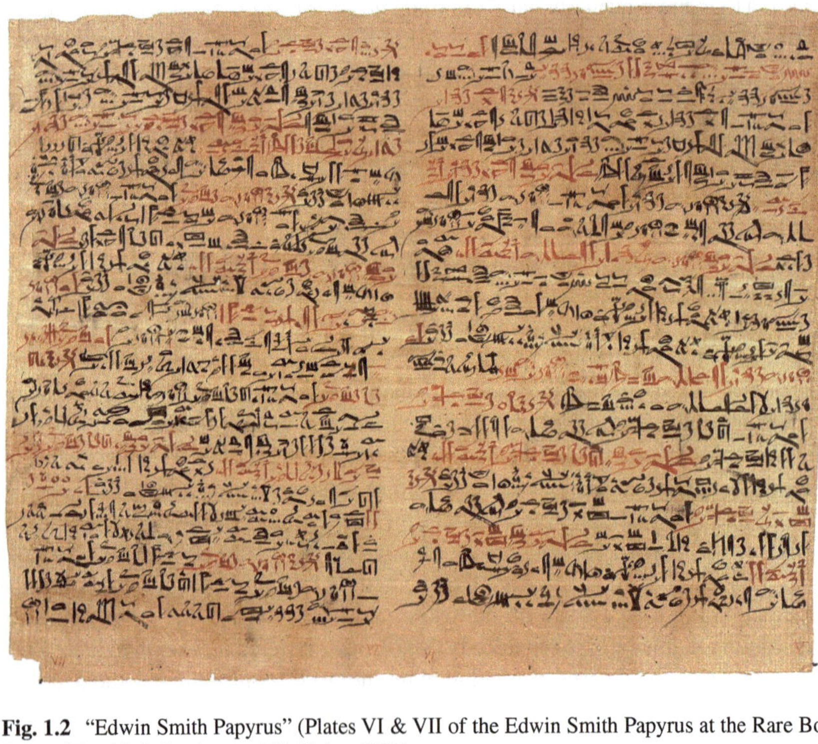

Fig. 1.2 "Edwin Smith Papyrus" (Plates VI & VII of the Edwin Smith Papyrus at the Rare Book Room, New York Academy of Medicine, CC0)

any complications, it runs through a well-recognized flow of examination, diagnosis, and treatment.

These are best illustrated through an example:

Case 47: Instructions concerning a gaping wound in his shoulder

First examination: If thou examinest a man having a gaping wound in his shoulder, its flesh being laid back and its sides separated, while he suffers with swelling (in) his shoulder blade, thou shouldst palpate his wound. Shouldst thou find its gash separated from its sides in his wound, as a roll of linen is unrolled, (and) it is painful when he raises his arm on account of it, thou shouldst draw together for him his gash with stitching.

Though shouldst say concerning him: 'One having a gaping wound in his shoulder, its flesh being laid back and its sides separated, while he suffers with swelling in his shoulder blade. An ailment which I will treat.'

Thou shouldst bind it with fresh meat the first day.

Second examination: If thou findest that wound open and its stitching loose, thou shouldst draw together for him its gash with two strips of linen over that gash; thou shouldst treat it afterward (with) grease, honey, (and) lint every day until he recovers. If thou finest a wound, its flesh laid back, its sides separated, in any member of a man, thou shouldst treat it according to these directions.

Third examination: If, however, thou findest that his flesh has developed inflammation from that wound which is in his shoulder, while that wound is inflamed, open, and its stitching loose, thou shouldst lay thy hand upon it. Shouldst thou find inflammation issuing from the mouth of his wound at thy touch, and secretions discharging therefrom are cool like wenesh-juice, thou should say concerning him: 'One having a gaping wound in his shoulder, it being inflamed, and he continues to have fever from it. An ailment with which I will contend.'

Fourth examination: If then, thou finest that man continuing to have fever, while that wound is inflamed, thou shall not bind it; thou shalt moor (him) at his mooring stakes, until the period of his injury passes by.

Fifth examination: If, however, his fever abates and the inflammation in the mouth of his wound dissipates entirely, thou shouldst treat him afterward (with) grease, honey (and) lint every day, until he recovers. [21]

With each condition, the Edwin Smith Papyrus's physician chose to convey one of three statements: "an ailment which I will treat," "an ailment with which I will contend," or "an ailment not to be treated." Reading similarly to an assessment in today's documentation, it is an indicator of prognosis, and of when it might be wise to resort to palliative measures for the patient's sake. This may be a form of proof that the ethical concept of nonmaleficence, or "do no harm," has been understood and respected for thousands of years.

These writings also emphasize the importance of physical examination: "lay thy hand upon it." In this specific case, the physician detects inflammation and purulent drainage. Notice here that with these signs of potential infection the prognosis became more guarded. "An ailment which I will treat" became "an ailment with which I will contend."

In all fairness, the grease, honey, and lint used over the wound site are a fair choice to promote healing and prevent infection even by modern standards. Lint acts as a form of packing gauze, the grease, as earlier mentioned, has a soothing and protective effect, and honey has a myriad of healing and antibacterial benefits being investigated still today. The combination of lint, grease, and honey was used in twenty-two out of forty-eight wound cases in the Edwin Smith Papyrus—an obvious favorite [21]. Another popular dressing of this era was “powder of green pigment,” referring to copper carbonate/malachite. While toxic, it has strong antibacterial properties [12]. Ground into a stunning blue-green pigment, it was used for eye makeup in both men and women. Now, the use of copper is seeing a modern resurgence from its Egyptian roots. A systematic review published in 2022 explored the healing benefits of copper nanoparticles in dressings. They noted their strong antibacterial properties, as well as their ability to promote angiogenesis and skin regeneration, with the cytotoxicity quelled by their lower concentrations used [30].

Binding the wound with fresh meat on the first day was an overarching theme throughout the Edwin Smith Papyrus. There may have been benefits from the hemostatic properties of muscle or crushed tissue, but ultimately the reasoning behind this practice is not entirely known. The ancient Egyptians also used fresh meat in the ceremony of the Opening of the Mouth. Here, the bandages over the mouth of the celebrated mummy were cut open to breathe life back into the deceased. A freshly amputated foreleg from a young bull would be offered as food for the spirit. The idea of fresh meat on a wound bed may have been thought akin to providing nourishment to flesh being claimed by death. Magic did play a role in the healing arts of ancient Egypt. The Ebers Papyrus, a 110-page scroll dating around 1550 B.C., lists a variety of incantations to be recited during medical treatments [21].

Ancient Greece

Few, if any, medical professionals would be able to say that their practice is untouched by ancient Greece. From the origination of countless medical words used worldwide, to the Hippocratic Oath and its associated ethical principles, the influence of Greek medicine and philosophy cannot be overstated. By far, the most influential collection of literature from this era was the Hippocratic Corpus. This assortment of medical teachings and bits of cases was written from circa 450 B.C. to 200 B.C. by various authors, likely including Hippocrates himself (Fig. 1.3), and associated with his medical school on the island of Kos. The Hippocratic Corpus not only gives a fantastic glimpse into the philosophies and standard of care of that time but also deepens our understanding of today’s medicine.

At the center of the ancient Greek concept of wellness are the four humors: blood, phlegm, yellow bile, and black bile. In health, these four exist in harmony, and imbalance is the root of disease. Another enemy was the cold. With several well-described examples of tetanus, catching cold was seen as being on a continuum, from chills to spasms to opisthotonus to death [21]. In wound care, physicians noted that warmth fostered bleeding and that bleeding fostered healing. Additionally, stagnated blood was associated with infection. It is true, of course, that good circulation is critical for tissue growth, that stirring up light bleeding during debridement helps refresh the healing cascade, and that stagnant blood is prone to infection. By the logic that opening vessels near the wound would let out bad humors, venesection became a common practice. One of the most essential aspects of wound care

Fig. 1.3 Statue of Hippocrates of Kos on the Facade of the Oxford University Museum of Natural History, England, UK. (Photography by Doina Gavrilov, CC0)

and surgery is the ability to control bleeding. In ancient Greece, their philosophy on the treatment of hemorrhage unfortunately involved additional hemorrhage. In these cases, a towel would be dipped in cold water and placed on the body near the site of bleeding, and a towel dipped in hot water would be placed on a site farther away, thus supposedly drawing the blood back into the body towards the warm towel and away from the cold bleeding vessel. A plug made from wool and fig tree juice would also be used to help with clotting. If these methods failed, a different vein was cut, ultimately placing the patient at risk of hemorrhagic shock [21]. The tourniquet was introduced in the Hippocratic writings but failed to gain traction, likely due to the risk of gangrene with prolonged use [14].

The ancient Greeks also remarked on different types of pus, and classified them as "good pus," and "bad pus." The "good pus" may have reflected exudate, fluid produced from the wound in the normal healing process, as opposed to true purulence. The absence of pus was also noted and considered worrisome, now recognized as likely reflecting a poor immune response in the patient. The Hippocratic writings advised, "the wound should be made to suppurate as quickly as possible; for, thus the parts surrounding the wound would be the least disposed to inflammation, and would become the soonest clean; for the flesh which has been chopped and bruised by the blow, must necessarily suppurate and slough away" [21]. While these writings assert that suppuration was critical and emphasized the importance of promoting this process through various salves, Hippocrates himself was believed to avoid this in favor of simply washing the wounds with wine or vinegar [12].

A common dressing of the time was a pad of wool dipped in a particular combination of ingredients, many of which can be recognized from the dressings from ancient Mesopotamia and ancient Egypt: sweet wine, copper acetate/verdigris, copper oxide, lead oxide, alum, myrrh, frankincense, gall nuts, vine flowers, and grease of wool. Like beer, wine is also bactericidal, yet not necessarily due to the alcohol content. It was found that wine with a nearly ten percent alcohol concentration was more bactericidal than a ten percent alcohol solution in water. This is due to the compounds malvoside and oenoside, which can kill *Vibrio cholera*, *Escherichia coli*, and *Salmonella typhi* in less than four hours [21].

They had also first documented the management of auto-amputation:

> *With regard to the gangrene of fleshy parts ... when the forearm and leg drop off, the patients readily recover ... Those parts of the body which are below the boundaries of the blackening are to be removed at the joint, as soon as they are completely dead and lost their sensibility; care being taken not to wound any living part.* [21].

Like in ancient Egypt, magic continued to play a role in ancient Greek medicine. When people were unable to be cured by more traditional medicine, they would seek aid at the temple of Asklepios, the patron god of physicians. Here, they would stay in the temple overnight and wait for Asklepios to visit them in their dreams. Food offerings were involved, as well as the temple dogs and snakes (Aesculapian snakes- constrictors harmless to humans, and the same as seen in the Rod of Aesculapius, the symbol of medicine). These treatments were evidently popular, as there were over 200 temples to Asklepsios built throughout the Greek empire. In the ruins of one of the temples in Epidauros was a series of successful case histories, chiseled around 350 B.C. (Originally written simply, word separations and punctuation are added to this quote for ease of reading).

> *Gorgias of Herkakleia, with pus. In a battle he had been wounded by an arrow in the lung and for a year and a half he had suppurated so badly that he filled sixty-seven basins with pus. While sleeping in the temple he saw a vision—it seemed to him the god extracted the arrowhead from his lung. When day came he walked out well, holding the arrowhead in his hands.* [21]

The Hippocratic Corpus showed no disdain for the ritual, and one could assume that it held a respected place in Greek medicine. If the authors felt otherwise, it is likely that they would have admonished it in the same way they did regarding the ornate bandage wrapping often performed by physicians in their day.

> *Leave aside theatrical bandages that serve no purpose; this is miserable and fit for charlatans, and often hurts the patient.* [14]

Ancient Rome

After the era of Hippocrates, the Roman Empire came to rise in the year 27 B.C. Until then, there was very little in the way of medicine and wound care. In 219 B.C., a physician from Greece—a wound care specialist—migrated over and within only a short time, and was met with animosity. Cato the Elder, a Roman historian of the time, even wrote to his son the following about this physician:

> *The Greeks are a worthless people, and an intractable one, and you must consider my words prophetic: when that race gives us its literature it will corrupt all things, and even all the more if it sends hither its physicians. They have conspired together to murder all foreigners with their physic, but this very thing they do for a fee, to gain credit and destroy us easily ... I have forbidden you to have dealings with physicians.* [21]

Greek medicine, with its surgeries and starvation-and-purging regimens, would have been considered radical and unpleasant to the Romans, who relied on crushed cabbage leaves as their mainstay of wound care. In time, they grew to accept Greek medicine, and soon another Greek physician arrived and fared much better. Granted, there was a bit of compromise that surely aided his acceptance, as he rejected venesection and celebrated good food, wine, massage, and music therapy as treatments [37]. These gentler treatments were a vast improvement due to their tolerability and lower risk of complications. There is also strong evidence today for his treatment measures, as lower psychological stress levels have consistently been shown to improve wound healing rates [38].

As the Roman Empire grew, so did the casualties. They developed military hospitals across their expanding territory, and Roman medicine evolved beyond cabbage leaves and the occasional Greek physician. One of the richest texts dedicated to Roman medicine was *De Medicina*, a section of an encyclopedia written around 47 A.D. by Aulus Cornelius Celsus. The four cardinal signs of acute inflammation, referenced commonly today, "Rubor et tumor cum galore et dolore," or "Redness and swelling with heat and pain," are found in this text. Within *De Medicina*, Celsus dedicated an entire sub-section to the treatment of wounds and ulcers. Wound dressings in the Roman period were similar to those of the Greeks. Familiar materials included copper acetate, lead oxide, alum, and pine resin [21]. While the dressings were not particularly groundbreaking for that era, Celsus's writings show that the Romans had made some substantial advancements.

For one, Celsus documented venous ligation, which was a huge stride from the counterproductive hemorrhage control strategies of the ancient Greeks.

> *The veins that are pouring out blood are to be seized, and round the wounded spot they are to be tied in two places, and cut across in between, so that each end may retract on itself, and yet have its orifice closed.* [21]

Celsus also made observations on how a patient's medical history, lifestyle, and risk factors impact the course of wound healing.

> *Again, both age and constitution and mode of life and the season have also some influence; for a boy or young adult heals more readily than does an old man; one who is strong than a weak man; a man who is not too thin or too fat than one who is either of these; one of sound habit than of unsound; one who takes exercise than a sluggard; one who is sober and temperate than one addicted to wine and venery. And the most opportune time for healing is the spring, or at any rate when the weather is neither too cold nor hot, for wounds are harmed by excessive heat and excessive cold, but most of all by variations of these; hence autumn is the most pernicious season.* [14]

While autumn has yet to be determined as a risk factor for wound healing in modern society, there is absolute certainty that general health and lifestyle play a key role in wound healing. A patient evaluation would be incomplete without ascertaining a patient's medical history and habits and considering these factors in prognosis and treatment.

Celsus also believed in the usefulness of wounds as a means to study anatomy. While others of the time dabbled in vivisection, he wrote:

> *I believe that medicine should be rational... But to open the bodies of men still alive is as cruel as it is needless. To open the dead is necessary to those who learn, who must know the positions and relations, which the corpse shows better than does a living and wounded man. As for all the rest that can be learned only from the living, actual practice will demonstrate it in the course of treating the wounded, in a somewhat slower, but milder way.* [21]

Much like in ancient India (more on this later), this method emphasized the importance of anatomy in medicine while simultaneously avoiding unnecessary suffering, which holds true to the ethics of medical research today.

Ancient China

The impact of ancient Chinese medicine was so powerful that it is still used even today all over the world. Like in Ancient Greece, ancient Chinese medicine also had its basis in philosophic principles. Rather than the four humors, it was based on the concepts of Yin and Yang. These are two inseparable powers that exist in harmony: Yang representing light, heat, dryness, and masculine energy, and Yin representing darkness, cold, moisture, and feminine energy. Imbalances of the two are thought to be the root of the disease state. Chi, or Qi, refers to vital energy, which flows through the universe and all living beings, and is affected by the forces of Yin and Yang. These principles are woven through the *Huangdi Neijing*, also known as the Yellow Emperor's Classic of Internal Medicine, written roughly between 480 and 300 B.C. It discussed the fundamental principles of Chinese medicine, the treatment of many medical conditions, as well as health and life in general [18]. The well-known practice of acupuncture was first described here as a way of altering the flow of the forces and restoring balance to the patient, thus restoring health. Ancient Chinese medicine took a more preventative approach to health and emphasized living according to the seasons to stave off disease.

Through other texts, we can also obtain greater insight into the medicine of that time, and of wound care specifically. *The Rites of Zhou*, written as a description of the ideal state during the late Zhou to early Han Dynasty, lists a great number of palace administrators, including but not limited to turtle-catchers, superintendent of wines, director of silk, officer of feathers, and five services of physicians. The imperial physicians were trained formally, going through examinations and clinical practice to determine rank. The groups include the chief of physicians (more on this in a moment), physicians for the animals (veterinarians), physicians for simple diseases, food physicians (essentially, nutritionists who arrange the preparation of foods and drinks in keeping with the seasons), and, finally, ulcer physicians. Somewhat like a combination between a medical board and a chief medical director today, the chief of physicians would ensure that the practice of the other physicians meets the standard of care. He would review cases and scrutinize death notices and those in which patients failed to recover from their illnesses, noting any mistakes by the physician. He could adjust the physician's level of appointment, or, in more severe cases, revoke their medical license. He would also oversee that appropriate drugs are available. This office comprises thirty people: two first-class medical graduates, four third-class medical graduates, two warehouse wardens, two scribes, and twenty apprentices [21].

The ulcer physicians were a set of eight third-class medical graduates in charge of managing all sorts of wounds, which they would do with drugs. Unfortunately, there was only one sentence on the matter: "In general, to heal the ulcers, they attack them with the five poisonous substances." It is believed that these refer to bile stone, mercuric sulphide, sulphur, magnetic stone, and arsenic oxide [21]. Although this concoction would be poisonous indeed, it would certainly be bactericidal. Herbs and tea leaves were also known to be used in wound care [1]. Tannic acid, a compound in tea leaves, was shown in recent years to have both antibacterial and anti-inflammatory effects and was incorporated into hydrogels and silver nanoparticle dressings to enhance healing [24].

Ancient India

Plastic surgery is an essential aspect of wound care today. Flap surgery and skin grafting can be life-changing operations for those with severe hidradenitis suppurativa, recalcitrant pressure sores, and others in need of tissue reconstruction. Named Sushruta, the father of plastic surgery, lived in Kashi, India, around 600 B.C. His thesis, known as the *Sushruta Samhita* (Fig. 1.4), is the oldest foundational surgical text.

The *Sushruta Samhita* described the first flap surgery—a reconstruction of a nose out of a pedicle flap of adjacent forehead tissue. Two small tubes placed in the nostrils maintained an open passageway for air, which were removed when the healing process neared an end. Preoperatively, patients were fed a light meal and were made drunk on wine to lessen the pain and decrease the likelihood of fainting. This operation was born from need—nose amputations were a common form of legal punishment in that time for a variety of misdeeds and continued even up to a more modern era. The nasal flap surgery became a way to improve comfort and restore dignity to those chastised. Large earrings were also common in that time (and around the world today), and so cheek and ear flap surgeries were also developed to mend overstretched or torn earlobes [10]. Even predating the full understanding of germ theory, the Sushruta emphasized the importance of cleanliness in surgeries, as well as high-quality surgical equipment. Instruments were to be made from iron, copper, or gold, and were to be cleaned thoroughly between patients. Substances like salt, mustard, and clarified butter were used in cleaning [10].

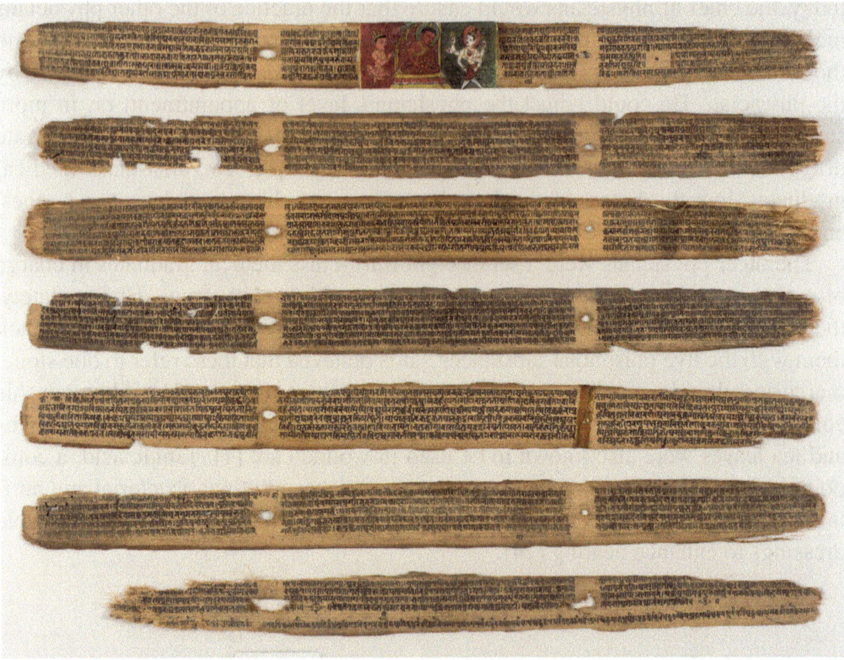

Fig. 1.4 "The Susruta-Samhita or Sahottara-Tantra (A Treatise on Ayurvedic Medicine)" (Los Angeles County Museum of Art, gift of Emeritus Professor and Mrs. Thomas O. Ballinger, Nepal, Text: twelfth–thirteenth century; Images: eighteenth–nineteenth century, CC0)

Regarding anatomical knowledge, it is thought that the people of ancient India relied on observations of their patients who sustained various injuries. It was likely because of this that Ayurvedic medicine developed the concept of Marmas, or "vital points." These spots, which are scattered across the body, are found in areas that, when wounded, lead to significant functional impairments or death [21]. For example, if an arrow injury to the *Lohitaksha* point in the groin could lead to leg paralysis or rapid hemorrhage and death, a surgeon of ancient India would wisely learn to avoid cutting into that area.

Medieval and Early Modern France

France made considerable advancements in wound care through the medieval and early modern period, particularly in relation to acute wounds and wartime injuries.

In 1363, Guy de Chauliac—a French surgeon and physician to Pope Clement VI—wrote the *Chirurgia Magna*, a guide to surgery. Ultimately becoming widely translated and one of the most influential surgical texts of its time, it not only commented on surgery alone but also on wound management. It described five basic

Fig. 1.5 "Bust of Ambroise Paré" (Smithsonian Institution, gift of Sarah Cooper Hewitt, 1819, CC0)

principles of acute wound treatment: removal of foreign bodies, reapproximation of tissues, maintenance of their apposition, conservation of substance, and the treatment of complications [12].

Ambroise Paré (1510–1590) (Fig. 1.5) was the royal surgeon of France, serving kings Henry II, Francis II, Charles IX, and Henry III. While basic prostheses existed long before him, Paré revolutionized them, rendering them far more functional and paving the way for modern prosthetics. He created both upper and lower prostheses, including an above-knee prosthesis with knee lock control and adjustable harness, as well as a prosthetic hand operated by springs used in battle by a French army captain [4]. Paré was also among the first to observe and document the beneficial effects of maggots on wound beds [4]. The larvae of the common green bottle fly specifically are highly effective at disrupting biofilm and digesting devitalized tissue while sparing healthy tissue. This therapy continued to develop over time and is today approved in many countries around the world, in the form of sterile, laboratory-grown larvae. These are kept in what would best be described as tea bags, such that they can still consume unhealthy tissue, yet they are easily removed from the patient when their work is done.

Napoleon Bonaparte's surgeon, Dominique Jean Larrey (1766–1842), also changed history in the course of war. He took notice of the extraordinarily high fatality rate of gangrene among their soldiers and strongly advocated for the early amputation of an unsalvageable limb [4]. His advice saved many lives, both then as well as into the future. Many doctors during the American Civil War adopted his strategy, and their amputees had an overall 75% survival rate despite unsanitary conditions and limited resources [25]. Larrey also aided in the development of the ambulance. Determining that it was easier to perform an amputation while a patient was still in shock from the injury, with higher endorphins and adrenaline, he outfitted a carriage with spring suspensions for smoother driving, so that he could pick up the wounded during battle and treat them there. First used during the Battle of Metz in 1793, this "ambulance volantes," or "flying hospital," was so successful that he was then tasked with organizing all fourteen armies of the French Republic [4].

Early Modern Great Britain

The concept of hyperbaric oxygen therapy, used widely today for a variety of types of wounds as well as other conditions, had its origins in Great Britain. In 1664, an English physician named Nathaniel Henshaw noted the health benefits of fresh air and published a treatise that described a "domicilium," or "air chamber." This chamber would utilize a pair of organ bellows to "condense," or increase the pressure of the air inside. He believed that breathing in condensed air would improve health and could treat a number of diseases, particularly of the pulmonary and digestive systems. [11]. Based on the mechanical impracticalities of the treatise's description, Clarke [5] asserts that Henshaw's machine was never built and remained only a concept for the next 170 years. The credit for the first construction and use of a hyperbaric chamber goes to the French. In the 1830s, Victor Junod, Émile Tabarié, and Charles Pravaz built a "compressed air bath," or "bain d'aire comprimé," using a spherical 1.4 m copper chamber and a steam-powered pump. They tested this treatment on patients with pulmonary conditions and found such success that the hyperbaric chamber found popularity across the United States and Europe [11]. In the context of wound healing today, hyperbaric oxygen has been demonstrated to improve neovascularization, stimulate fibroblast proliferation and migration, and enhance collagen cross-linking. Through these diverse mechanisms, hyperbaric oxygen significantly improves healing outcomes for those with vascular ulcers, including diabetic foot ulcers, delayed radiation injury, compromised skin grafts and flaps, and refractory osteomyelitis [19].

Scurvy, with symptoms of weakness, bleeding gums, slow-healing wounds, and organ failure, was one of the leading causes of morbidity and mortality of British sailors until the eighteenth century. In 1747, James Lind conducted one of the first recorded clinical trials—an experiment to test various treatments for this ailment. He took 12 patients, divided them into pairs, and gave each set a different treatment. One pair each drank a half pint of seawater daily. Another pair took a preparation of herbs and spices recommended by a hospital surgeon. Other treatments included a quart of cider daily, a diluted solution of sulphuric acid three times daily, and two spoonfuls of vinegar three times daily. Finally, as James Woodall (1569–1643) previously suggested, fresh fruit as a potential treatment, one pair was given two oranges and one lemon each daily. Only six days into treatment, those who received the citrus fruits recovered so well that one returned to duty and the other was appointed as a nurse to help care for the other patients [16]. This knowledge was ultimately incorporated into routine operations in the British Royal Navy. Lemon juice or lime juice became mandatory on their ships, earning British sailors the nickname "Limeys." The treatment of scurvy gave them an advantage over other nations, including the French during the Napoleonic wars. His research demonstrated the critical importance of vitamin C in wound healing as well as wound prevention.

John Hunter (Fig. 1.6), the personal physician for King Charles III, also made great advancements in wound care research, particularly in neovascularization. He

Fig. 1.6 "Portrait of John Hunter" (Smithsonian Institution, bequest of George Campbell Cooper, 1788, CC0)

speculated that blood supply was required for tissue growth. He took a deer, ligated the carotid artery on one side of its neck, and found that the antler growth on that side became stunted in comparison to the other side. In time, the antler growth became normal again, and after sacrificing the deer, he discovered that a network of new vessels had formed around his ligature, thus restoring blood flow. Hunter named this phenomenon angiogenesis and used this lesson in his clinical practice. A notable example was his patient with a large popliteal aneurysm at risk of rupture. To avoid amputation (which also carried a high mortality risk), Hunter instead chose to ligate his femoral artery. As they had hoped, the patient developed collateral circulation and was able to keep his leg [6]. Angiogenesis is now recognized as one of the primary forces of the proliferation stage of wound healing.

Twentieth Century

As the spread of information became globalized, more recent years created an explosion of discovery in the field of wound care. One of the most impactful moments in medicine was the day Alexander Fleming, a Scottish physician-scientist, accidentally left a petri dish of staphylococci by the window before leaving on vacation. When he returned, he found that the bacterial growth left a wide margin in certain areas near mold colonies that had grown from spores that floated in. He isolated this "mold juice," and determined that it came from a species of the *Penicillium* genus. Found to be an effective treatment against many illnesses caused by gram-positive bacteria, it was mass-produced with the help of Howard Florey and Ernst Chain. For their work, they won the 1945 Nobel Prize in medicine and changed the lives of millions [32].

While the discovery of antibiotics came about by accident, compression therapy was developed very much on purpose. Conrad Jobst, a German immigrant to the United States, was a brilliant engineer who patented forty different machines, from hydraulic presses to compression pumps. He personally suffered from severe venous insufficiency from a young age and developed leg ulcers and venous stasis dermatitis as a result of it. He worked with physicians for years, but the regular elastic stockings, the best treatment of the time, failed to improve his condition. While swimming, he noticed that graduated compression, with higher pressure around the foot and ankle and lower pressure higher up on the leg, provided substantial relief. Working with physicians from Henry Ford Hospital and physiologists from the Wright Patterson Aerospace Laboratory, he developed the first graduated compression system stockings. He used himself as a testing model and healed his own ulcers in a matter of weeks. He patented this invention in 1951, which required each pair to be custom-measured by a physician for each individual patient [20].

While negative pressure wound therapy is a recent development, suction in various forms has been used for millennia. In India, a mainstay of treatment of venomous snake bites was to attempt to suck out the venom from the wound with their mouths [21]. In Rome, the practice of wound sucking was used not just for venom but also for war injuries and other types of wounds. It was noted that they saw such success with this practice that they considered these wound sucking individuals to be of utmost importance to the army. In 1890, Gustav Bier, a German surgeon, developed a type of cup system with attached suction bulbs, allowing negative pressure to suck away wound secretions [22]. In 1986, Dr. Nail Bagautdinov of the Soviet Union described a wound suction system with foam dressings to help with drainage and closure of infected wounds [9].

Despite the long history of suction being used in medicine, none of these devices worked like the modern NPWT devices due to the lack of a proper wound interface. In the 1990s, the Vacuum Assisted Closure System (V.A.C. ® KCI, San Antonio, TX), was invented and patented by Dr. Argenta and Dr. Morykwas at Wake Forest University in North Carolina. The system involves an open-cell hydrophobic polyurethane foam dressing with 400–600-micron pore diameter, which collapses by a factor of ten in volume under suction of −125 mmHg relative to atmospheric pressure. The foam is placed over the wound bed, adhered and sealed with an adhesive drape, and attached via tubing to the vacuum device. A series of papers from Argenta and Morykwas, as well as Orgill, were the first to extensively study the properties of the foam and found that the foam-pressure system promoted wound healing through mechanobiology. In a concept called microdeformation, the pressure differential and mismatch in elasticity between the tissues and the sponge change the shape of the cells in the wound bed, altering microtubules and microfilaments and promoting cell proliferation through regulation of gene pathways. The system yields benefit through macrodeformation as well; the compressibility of the foam effectively shrinks the wound, promoting granulation and contracture of the wound edges through a cascade of cellular signaling effects. Furthermore, it improves edema in the tissues, thereby improving circulation, which, along with removal of exudate, significantly improves bacterial clearance [2, 26]. Published in Annals of Plastic Surgery, their articles became some of the most cited papers in medicine. This breakthrough led to an explosion of negative pressure wound therapy use and research over the last 30 years. The V.A.C.® was also incorporated into the United States military starting in 2004 for the treatment of combat wounds and other casualties [7].

After the early flaps and grafts of ancient India and the porcine xenografts developed in the early 1800s, progressive need for skin replacement due to war injuries and burns promoted the advancement of skin substitutes in the early twentieth century. In 1886, Dr. Karl Thiersch presented his method of allografting to the German Surgical Association. It involved taking large sheets of epidermis with a thin layer of dermis from a donor to graft [27]. While a significant advancement, the allografts faced the problem of frequent rejection and the inability to replace larger areas of skin loss. Peter Medawar and Macfarlane Burnet made strides in combating rejection through their research. Using animal models, they demonstrated that the rejection of skin grafts was related to an immune response by the host, and more likely to occur with unrelated individuals and with repeated donation occurrences from the same individual. They developed the concept of acquired immune tolerance, noting that if foreign skin is introduced during a critical period early in development, rejection is less likely. For their findings, which helped build the foundation of transplant medicine, Medawar and Burnet won a Nobel Prize in 1960 [29].

Twentieth-century advances in critical care and the adoption of early excision and grafting in burn care created a larger need for skin coverage. Numerous burn centers developed skin banks for cadaver allografts. Observing that the allografts lasted much longer in burn patients and other immunocompromised individuals, Charles Cuono and others set out to further the understanding of skin rejection. This ultimately led to the discovery of the importance of Langerhans cells in rejection. Their research was the foundation for decellularized dermal matrices, such as AlloDerm™ [17].

In large body surface burns, the epidermis is also limited for donor sites. It became clear that the solution was a bilaminate skin that included both the epidermis and the dermis. Cuono, Langdon, and McGuire [8] treated a patient with a 55% body surface area burn with cadaver skin and abraded the surface of the allografts, which effectively removed the antigen-presenting cells concentrated in the epidermis. They then resurfaced the allograft with autogenous keratinocyte cultures, and the patient healed with complete take of the graft epidermis and dermis. This was one of the first examples of a bilaminate solution for skin loss.

Even with rejection concerns aside, split-thickness skin grafting carries limitations. Without dermis, they tend to form contractions over the healing process and lead to loss of function. In the search for a solution to this issue, cell culturing was explored. In 1895, Mangolt invented epithelial cell seeding; however, until the mid-1970s, the ability to proliferate cell cultures was very limited. In 1975, Rheinwald and Green [28] at the Massachusetts Institute of Technology discovered that normal human epidermal keratinocytes were able to grow rapidly, stratify, and be serially propagable when cultured with 3 T3 fibroblast feeder cells. Using mouse fibroblasts on a collagen matrix with added epithelial cells, Bell et al. [3] grew living tissue in vitro. When the tissues were replanted into wounds on the mice, they grew as full-thickness skin. Gallico et al. [13] harvested cells from two patients with extensive full-thickness burns and grew sheets of epithelium over two weeks using the keratinocyte cell expansion technique. The sheets were placed back onto each respective patient's wounds with successful healing. Unfortunately, these epithelial grafts created in vivo were fragile and not an ideal solution for coverage.

Searching for ways to help burn wounds heal faster, Burke and Yannas at Harvard Medical School and the Massachusetts Institute of Technology took a different approach. They retro-engineered all the physical and biological properties of skin (appropriate bending and rigidity, ability to act as a bacterial barrier, nonantigenic, etc.), as well as desirable logistical properties (low cost, long shelf life, easy storage), and applied these principles to create an artificial skin [35]. They ultimately created this dermal substitute out of a bioengineered collagen and chondroitin-6-sulfate matrix with a silicone-based outer layer, which became known as Integra ® Dermal Regeneration Template, the first bioengineered regenerative medicine product to be commercially available [27]. While they noted that their healing process was slower than traditional skin grafting, they found that this new skin behaved more similarly to original skin, with higher functional and aesthetic scores at long-term follow-up [15]. Their product set the model in the development of synthetic skin substitutes as well as bioartificial organs [36].

Conclusion

A major advantage we carry today is that, unlike many of our ancestors, we now have access to a wealth of medical texts written before us from all over the world, now translated into our own languages. There is much to be learned from our past, and it is critical to do so to propel ourselves into a richer tomorrow. Some concepts, like the winning combination of lint, honey, and grease, have changed surprisingly little. Others, such as negative pressure wound therapy, have similarities to ancient medical techniques, but unique combinations with modern wound interfaces create a synergistic effect. Other aspects of our wound care history, from chants and rituals to the simple healing touch, are so intertwined with the human experience that they transcend all tests of time. With the global spread of information, we now have unprecedented ability to collaborate and build on each other's research. Using effective communication and respect for each other and our past, we can improve our methods and advance wound care into the future.

References

1. Ahmad W, Aquil Z, Alam SS (2020) Historical background of wound care. Hamdan Med J 13(4):189–195
2. Argenta LC, Morykwas MJ (1997) Vacuum-assisted closure: a new method for wound control and treatment: clinical experience. Ann Plast Surg 38(6):563–577
3. Bell E, Ehrlich HP, Buttle DJ, Nakatsuji T (1981) Living tissue formed in vitro and accepted as skin-equivalent tissue of full thickness. Science 211(4486):1052–1054
4. Broughton, G., Burris, D. (2010). War and medicine: a brief history of the military's contribution to wound care through world war I
5. Clarke R (2024) Nathanial Henshaw: not history's pioneering hyperbaric practitioner. Undersea Hyperb Med 51(2)
6. Cohen IK (2007) Lessons from the history of wound healing. Clin Dermatol 25(1):3–8
7. Couch KS, Stojadinovic A (2011) Negative-pressure wound therapy in the military: lessons learned. Plast Reconstr Surg 127:117S–130S
8. Cuono C, Langdon R, McGuire J (1986) Use of cultured epidermal autografts and dermal allografts as skin replacement after burn injury. Lancet 327(8490):1123–1124
9. Daar DA, Wirth GA, Evans GR, Carmean M, Gordon IL (2017) The Bagautdinov dressing method: negative pressure wound therapy in a patient with an allergy to acrylate adhesive. Int Wound J 14(1):198–202
10. Dave T, Habte A, Vora V, Sheikh MQ, Sanker V, Gopal SV (2024) Sushruta: The Father of Indian Surgical History. Plast Reconstr Surg 12(4):e5715. https://doi.org/10.1097/GOX.0000000000005715
11. Featherstone PJ, Ball CM (2021) The therapeutic use of air under hyperbaric pressure. Anaesth Intensive Care 49(3):159–161
12. Forrest RD (1982) Early history of wound treatment. J R Soc Med 75(3):198–205
13. Gallico GG, O'Connor NE, Banks-Sclegel S (1981) Grafting of burns with cultured epithelium prepared from autologous epithelial cells. Lancet 1:75–78
14. George Broughton II, Janis JE, Attinger CE (2006) A brief history of wound care. Plast Reconstr Surg 117(7S):6S–11S
15. Giovannini UM, Teot L (2020) Long-term follow-up comparison of two different bi-layer dermal substitutes in tissue regeneration: clinical outcomes and histological findings. Int Wound J 17(5):1545–1547. https://doi.org/10.1111/iwj.13381
16. Hughes RE (1975) James Lind and the cure of scurvy: an experimental approach. Med Hist 19(4):342–351
17. Jansen LA, De Caigny P, Guay NA, Lineaweaver WC, Shokrollahi K (2013) The evidence base for the acellular dermal matrix AlloDerm: a systematic review. Ann Plast Surg 70(5):587–594
18. Kong Y (2010) Huangdi Neijing: a synopsis with commentaries. Chinese University Press, Hong Kong
19. Lam G, Fontaine R, Ross FL, Chiu ES (2017) Hyperbaric oxygen therapy: exploring the clinical evidence. Adv Skin Wound Care 30(4):181–190
20. Loss LJ, Lurie F, Kasper G, Brunicardi FC, Sferra JJ (2023) Conrad Jobst: from illness to innovation. Am Surg 89(2):316–318
21. Majno G (1975) The healing hand: man and wound in the ancient world. Harvard University Press
22. Miller C (2013) The history of negative pressure wound therapy (NPWT): from "lip service" to the modern vacuum system. J Am Coll Clin Wound Spec 4(3):61–62. https://doi.org/10.1016/j.jccw.2013.11.002
23. New International Version Bible. (2011). Zondervan. (Original work published 1978)
24. Ninan N, Forget A, Shastri VP, Voelcker NH, Blencowe A (2016) Antibacterial and anti-inflammatory pH-responsive tannic acid-carboxylated agarose composite hydrogels for wound healing. ACS Appl Mater Interfaces 8(42):28511–28521

25. Prechtel-Kluskens C (2014) Did your Union Civil War ancestor have an artificial limb? NGS Mag 40(1):41–45
26. Orgill DP, Bayer LR (2011) Update on negative-pressure wound therapy. Plast Reconstr Surg 127:105S–115S
27. Oualla-Bachiri W, Fernández-González A, Quiñones-Vico MI, Arias-Santiago S (2020) From grafts to human bioengineered vascularized skin substitutes. Int J Mol Sci 21(21):8197
28. Rheinwald JG, Green H (1975) Serial cultivation of strains of human epidermal keratinocytes: the formation of keratinizing colonies from single cells. Cell 6(3):331–343. https://doi.org/10.1016/s0092-8674(75)80001-8. PMID: 1052771
29. Ribatti D (2015) Peter Brian Medawar and the discovery of acquired immunological tolerance. Immunol Lett 167(2):63–66
30. Sandoval C, Rios G, Sepulveda N, Salvo J, Souza-Mello V, Farias J (2022) Effectiveness of copper nanoparticles in wound healing process using in vivo and in vitro studies: a systematic review. Pharmaceutics 14(9):1838
31. Shah JB (2011) The history of wound care. J Am Coll Certif Wound Spec 3(3):65–66
32. Tan SY, Tatsumura Y (2015) Alexander Fleming (1881–1955): discoverer of penicillin. Singapore Med J 56(7):366–367. https://doi.org/10.11622/smedj.2015105
33. Trinkaus E, Villotte S (2017) External auditory exostoses and hearing loss in the Shanidar 1 Neandertal. PLoS One 12(10):e0186684
34. Vogt PM, Andree C, Breuing K, Liu PY, Slama J, Helo G, Eriksson E (1995) Dry, moist, and wet skin wound repair. Ann Plast Surg 34(5):493–500
35. Yannas IV, Burke JF (1980) Design of an artificial skin. I. Basic design principles. J Biomed Mater Res 14(1):65–81
36. Yannas IV (2014) Tissue and organ regeneration in adults: extension of the paradigm to several organs. Springer
37. Yapijakis C (2009) Hippocrates of Kos, the father of clinical medicine, and Asclepiades of Bithynia, the father of molecular medicine. In Vivo 23(4):507–514
38. Walburn J, Vedhara K, Hankins M, Rixon L, Weinman J (2009) Psychological stress and wound healing in humans: a systematic review and meta-analysis. J Psychosom Res 67(3):253–271

Chapter 2
Anatomy, Histology and Embryonic Development of the Skin

Marika Berardini, Luisa Gesualdi, Francesco De Francesco, and Giulia Ricci

Abstract The human skin, covering the human body, has multiple and fundamental functions. Besides its role as a mechanical barrier, it protects the human body against chemical, physical (such as UV rays), and biological aggression. Moreover, it actively participates in immune response, sensitive reception, vitamin D metabolism, body temperature, and electrolyte regulation. This complex and crucial organ has been extensively studied due to its role in health and disease, and this chapter aims to summarize the anatomical and histological features of the human skin together with a description of the embryonic development of its parts.

Keywords Epidermis · Dermis · Hypodermis · Skin glands · Hair follicle

---

Authors Marika Berardini and Luisa Gesualdi contributed equally as first authors in the chapter. Authors Francesco De Francesco and Giulia Ricci contributed equally as senior authors in the chapter.

M. Berardini · L. Gesualdi
Department of Anatomy, Histology, Forensic-Medicine and Orthopedics, Section of Histology and Embryology, "Sapienza" University of Rome, Rome, Italy
e-mail: marika.berardini@uniroma1.it; luisa.gesualdi@uniroma1.it

F. De Francesco
Department of Reconstructive Surgery and Hand Surgery, Azienda Ospedaliera Universitaria delle Marche, Ancona, Italy
e-mail: francesco.defrancesco@ospedaliriuniti.marche.it

G. Ricci (✉)
Department of Experimental Medicine, University of Campania "Luigi Vanvitelli", Naples, Italy
e-mail: giulia.ricci@unicampania.it

F. De Francesco, K. Turksen (eds.), *Management and Strategies for Wound Healing*, Stem Cell Biology and Regenerative Medicine 79,
https://doi.org/10.1007/978-3-032-12442-5_2

Abbreviations

AC	Adenylyl cyclase
ACTH	Adrenocorticotropic hormone
APUD	Amine precursor uptake and decarboxylation
c-AMP	Cyclic adenosine monophosphate
ECM	Extracellular matrix
ERS	Epithelial root sheath
FCTS	Fibrous connective tissue sheath
IL-1, IL-4, IL-6, IL-13	Interleukin-1, 4, 6, 13
K-1, K-5, K-10, K-14	Keratin-1, 5, 10, 14
L-DOPA	L-3,4- DiOxyPhenylAlanine
MC1R	Melanocortin 1-receptor
MHC-I,II	Major histocompatibility complex-I,II
MMPs	Matrix metalloproteinases
SLRPs	Small leucine-rich proteoglycans
TGF-β	Transforming growth factor-beta
TIMPs	Tissue inhibitors of metalloproteinases
TNF	Tumor necrosis factor
UV	Ultraviolet
UVB	Ultraviolet radiation B
α-MSH	Alpha-melanocyte stimulating hormone

The integument covers all the external surface of the human body: its overall area measures approximately 1.5 to 2 m^2 in the adult. It is the second largest organ of the human body being the adsorptive surface area of the small intestine the first one for its epithelial extension. However, it should be mentioned that considering folds and invaginations, such as hair follicles and sweat ducts, the total surface area of the integument system increases up to 25 m^2 [1–3]. Considering all skin appendages, the integument's weight represents 7% of the total human body.

At the histological level, the skin consists of three layers: (1) a stratified squamous keratinized surface epithelium, called *epidermis*, (2) an inner compact and fibrous connective layer, called *dermis*, and (3) a loose subcutaneous connective tissue, called *hypodermis*, that attaches skin to the deeper structures of the body (Fig. 2.1).

Considering that the integumentary system covers the surface of the body, it serves mainly as a protective barrier against the following: (1) injuries, (2) pathogens, (3) dehydration, and (4) ultraviolet radiation. Moreover, this complex system has a role in the regulation of body temperature, ion excretion, vitamin D synthesis, and sensory reception.

Indeed, skin thickness varies regionally, from 0.5 to 4 mm. Thick skin on the palms of the hands and soles of the feet is glabrous, whereas thin skin, which covers

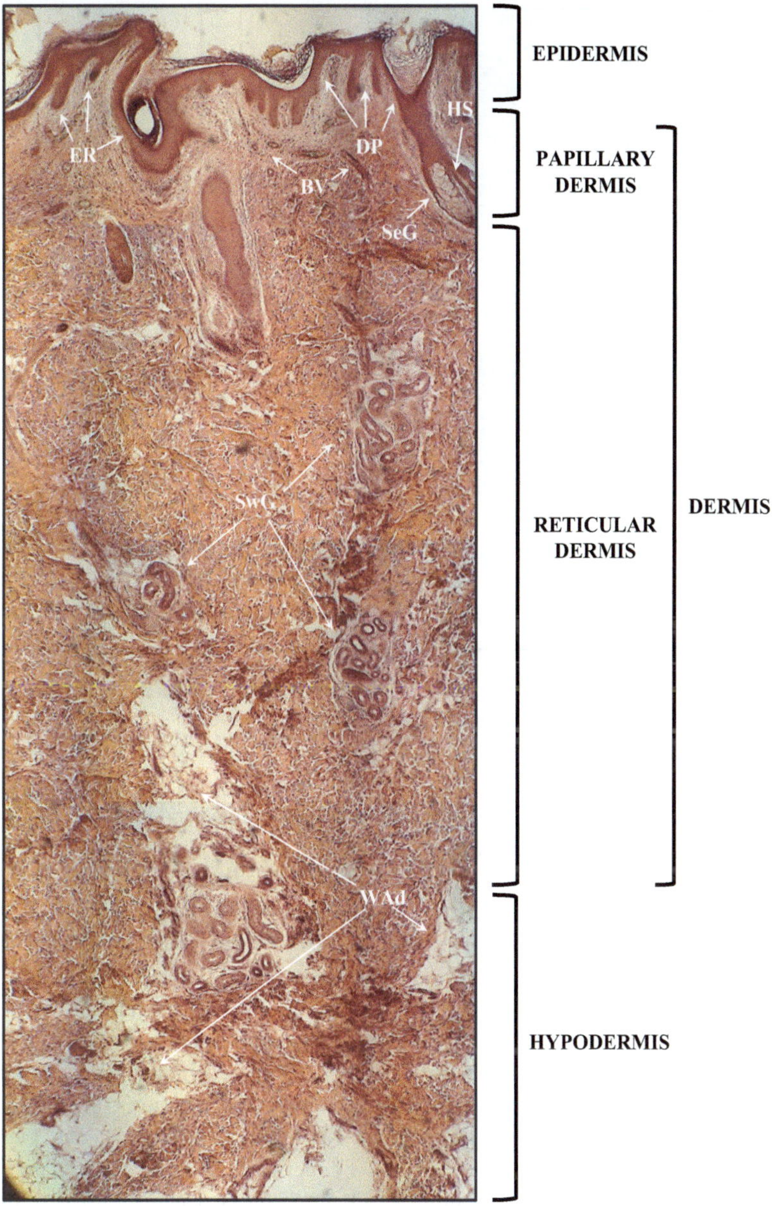

Fig. 2.1 Reconstruction of the integumentary system with low magnification images of a histological hematoxylin- and eosin-stained section of the human skin (original microphotograp by Prof. Giulia Ricci). ER: Epidermal Ridges; BV: Blood Vessel; DP: Dermal Papilla; HS: Hair Shaft; SeG: Sebaceous Gland; SwG: Sweat Gland; WAd: White Adipose tissue

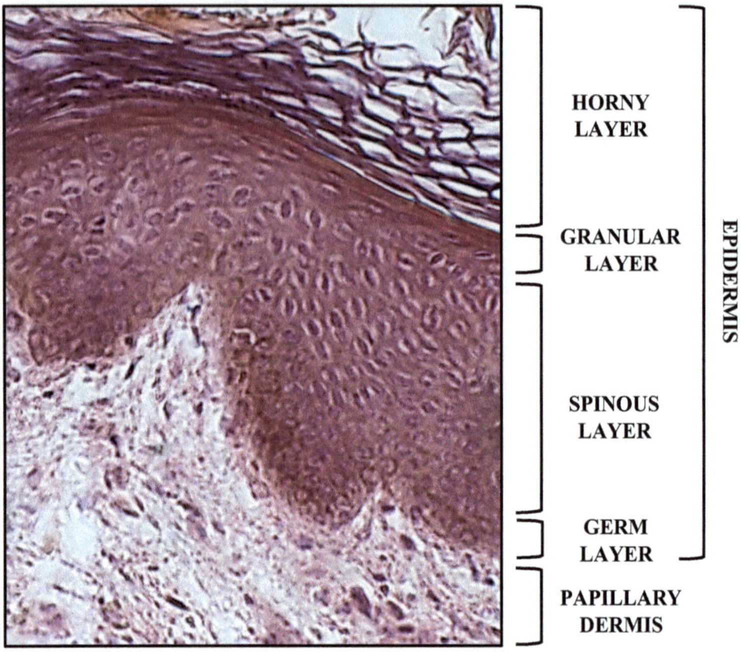

Fig. 2.2 Haematoxylin- and eosin-stained histological section of the human epidermis illustrating the epidermal layers (original microphotograph by Prof. Giulia Ricci)

most of the remaining body surface, contains hair follicles and its epidermis is thinner, having fewer cellular layers.

The dermo-epidermal junction is not straight: this border appears highly corrugated and features many extensions of epidermis downward toward the dermis, called *epidermal ridges*, that correspond, at the other side, to upward projections of the dermis that are called *dermal papillae* (Figs. 2.1, 2.2 and 2.3). The *basement membrane* at the dermo-epidermal junction usually requires special light or electron microscopy techniques to be visible. This specialized supporting zone of extracellular matrix consists of (1) lamina lucida and (2) lamina densa together constituting the basal lamina. This structure at the dermo–epidermal junction consists of extracellular matrix proteins such as collagen IV, laminins, nidogens, and perlecan. Laminin scaffolds connect the collagen IV network to nidogens and perlecan. Specifically, laminin 332 and laminin 511 are expressed here, with laminin 332 being crucial for anchoring properties (hemidesmosome formation) by interacting with $\alpha3\beta1$ and $\alpha6\beta4$ integrins on basal keratinocytes [4]. A deeper reticular lamina, made mainly of type I collagen fibres, merges the basal lamina with the underlying connective tissue.

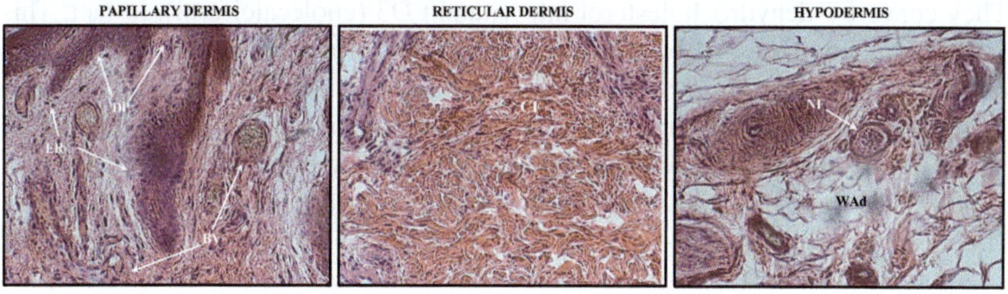

Fig. 2.3 Haematoxylin- and eosin-stained histological sections of the human skin illustrating the different parts of the dermis and the hypodermis (original microphotograph by Prof. Giulia Ricci). ER: Epidermal Ridges; BV: Blood Vessel; DP: Dermal Papilla; CF: Collagen Fibers; NF: Nerve Fiber; WAd: White Adipose tissue

The Epidermis

The epidermis represents the surface epithelial layer of the skin and consists of four distinct layers. The layer closer to the dermis is called *basal layer (stratum germinativum or germ layer)* and consists of a single layer of cuboidal to columnar epithelial cells, mainly keratinoblasts, resting on a basement membrane. These cells have oval nuclei that often show mitotic figures; they continuously undergo cell division to replace cells that move outward through the epidermis. Keratinoblasts of the basal layer are firmly attached to the underlying basement membrane by hemidesmosomes. Hemidesmosomes have only one intracytoplasmic attachment plaque to which keratin-based tonofilaments attach and take contact with integrin receptors anchoring to the basal lamina. The subsequent layer, *the spinous layer (stratum spinosum)*, comprises several layers of polyhedral cells, which gradually become more flattened as they approach the surface. Processes of adjacent cells are attached by desmosomes. The next layer, *the granular layer (stratum granulosum)*, consists of three to five layers of flattened cells aligned to the epidermal surface. The outermost layer, *the horny layer (stratum corneum)*, is made of dead, anucleate cornified cells (Fig. 2.2). In thick skin is observable an additional, translucent and lightly eosinophilic layer, absent in thin skin, *called stratum lucidum,* that is made of few layers of tightly packed squamous dead cells that lack organelles and nuclei. It serves an essential function by offering an additional layer of protection against friction and moisture loss, thereby enhancing skin defense mechanisms [5].

The epidermis is made of several cell lineages. *Keratinocytes* are the predominant cells of the epidermis. They originate from the basal layer and differentiate along all the other layers during the keratinization process (see below). Keratinocytes also contribute to ultraviolet B (UVB) rays' absorption and to Vitamin D activation.

They convert 7-dehydrocholesterol into vitamin D3 (cholecalciferol), in fact. This vitamin is then processed by the liver and kidneys to produce its active form, calcitriol, which regulates calcium and phosphate levels in the human body. The normal time for turnover of keratinocytes from the basal layer to the stratum corneum varies from 20 to 75 days, depending on epidermis thickness; however, transit times may be even more rapid in some diseases, such as psoriasis, in which transit time is about 8 days.

Melanocytes are responsible for the synthesis of the skin pigment, that is, the melanin. These cells derive from neural crest cells and are located near keratinocytes of the basal layer. UVB light stimulates melanin synthesis and secretion, thanks to the activation of the melanocyte enzyme tyrosinase, which converts the amino acid tyrosine. Tyrosine conversion occurs in specific granules called melanosomes. Melanosomes are transferred to the neighbouring epidermal cells through cytocrine secretion, where keratinocytes phagocytose the tips of melanocyte processes.

Merkel cells are mechanoreceptors cells placed in the epidermis basal layer where they interact with free nerve endings in the skin. These cells serve are sensitive cells to light touch thanks to their cytoskeleton and their binding to adjoining keratinocytes through desmosomes.

Langerhans cells are dendritic cells responsible for the skin's cellular immune defence being involved in antigen presentation. These cells are the only cells of the epidermis that do not derive from ectodermal germ cells being of mesenchymal origin: they derive from mononuclear phagocytic system bone marrow stem cells, in fact. Langerhans cells are found in the stratum spinosum of the epidermis, express Major Histocompatibility Complex MHC I and MHC II molecules.

Keratinocytes and the Keratinization Process

Over 95% of epidermis cells belong to the *keratinocyte* lineage, and the process of their differentiation is called keratinization. This name is due to the name of the main class of tonofilament protein of these cells: the keratins. Keratin pairs form intermediate filaments by combining an acidic keratin with a basic keratin. The tonofilaments of keratinoblasts of the basal layer are made of keratins 5 and 14 (K5, K14). As differentiation occurs, in the spinous layer, keratins 5 and 14 are replaced by keratins 1 and 10 (K1, K10), and the keratinocytes change morphology, acquiring several membrane extrusions. This type of keratin forms thicker bundles within the cells [6]. Moreover, keratinocytes in the upper part of the spinous layer start to produce electron-dense keratohyalin granules containing the protein filaggrin, which can cross-link keratin-based to nofilaments. The keratinocytes of the spinous layer are connected by desmosomes. These complex intercellular junctions mediate and enhance cell adhesion by anchoring keratin filaments to keratinocyte plasma membranes. By linking tonofilament bundles of adjacent cells, desmosomes provide the epidermis with structural continuity and mechanical strength.

Keratinocytes of the spinous layer present a second group of granules called lamellar bodies. These inclusions contain a mixture of glycosphingolipids, phospholipids, and ceramides assembled within Golgi-derived bodies. The contents of lamellar bodies are then secreted through exocytosis into the extracellular spaces between the stratum granulosum and corneum, most likely forming an intercellular barrier to water loss and contributing to skin impermeability.

Within the granular layer (which consists of three or four rows of cells), keratinocytes flatten because of their cytoskeletal modifications and accumulate keratohyalin granules with histidine- and cysteine-rich proteins that are tonofilament-associated proteins. These proteins are pro-filaggrin, filaggrin, involucrin, epiligrin, and trichohyalin. The keratohyaline granules help aggregate keratin filaments into a homogenous keratin matrix, converting granular cells into cornified cells in the horny layer. The formed keratin matrix is firmly bound to desmosomes and stabilizes these adherens junctions' attachment plaques (corneo-desmosomes), being integrated into a part of the cornified cell envelopes [7]. A secondary kind of junctions is located in the stratum granulosum, the epidermal tight junctions, which appear to play an important role in epidermis integrity, and impermeability, as well as important structures for epidermis defencing barrier function [8].

Cornified cells that reach the epidermis surface are gradually detached when corneo-desmosomes disintegrate. This process is guaranteed by the activation of degrading proteases. A complex interplay of serine proteases (kallikreins) and cysteine proteases (cathepsins) excreted through the vesiculo-tubular system of lamellar granules is orchestrated, resulting in the progressive digestion of the corneo-desmosomes in fact [8]. The proteinase activity of kallikrein-related serine peptidase is triggered by lowered pH near the surface.

The most superficial cells are continuously shed in a process known as desquamation and are continuously replaced by the differentiation of cells of the basal layer. The processes of skin shedding and desquamation vary slightly by body region. Hairless skin comprises more layers, with the addition of the stratum lucidum. Thus, keratinocytes in body regions with hairless skin go through more layers before reaching the surface [9].

Histology and Function of Melanocytes

As already mentioned, *melanocytes* are melanin-producing cells whose activity determines the pigmentation of skin and hair. Melanin has a protective role against the damaging effects of UV radiation on DNA. Since melanocyte number declines significantly in skin and hair with aging, the protection against UV rays is impaired in older human beings. Each melanocyte is surrounded by approximately 36 keratinocytes to which it establishes contact via dendritic processes. Melanin is synthesized in membrane-bound organelles known as melanosomes in which is included tyrosinase enzyme. This enzyme catalyses the gradual oxidation of L-tyrosine to L-3,4-dioxyphenylalanine (L-DOPA) and sequentially the transformation of L-DOPA to melanin pigment. Melanin conversion is a gradual process that is mirrored by melanosome differentiation stages: four myelinization stages are observable according to pigment content. Immature pre-melanosomes belong to stages I and II, whereas melanized melanosomes belong to stages III and IV. During their differentiation, melanosomes redistribute along melanocyte microtubules to the ends of their dendritic processes. Keratinocytes then phagocytose the melanocyte dendritic tips through the cytocrine process: in this way, melanosomes are transferred into keratinocyte cytoplasm and form a supranuclear protective cap [10]. Two major forms of melanin are found in humans: eumelanin, which is brown to black, and pheomelanin, which is yellow to red. Both molecules are derived from tyrosine.

Melanin production is stimulated both by physical and chemical stimuli [11]. The main physical stimuli to melanin production are UV rays that trigger the increased expression of Melanocortin Receptor 1 (MC1R) in melanocytes through a cyclic Adenosine Monophosphate (cAMP)-dependent activation pathway. The MC1R is a member of G-protein-coupled receptors that is activated by the Melanocyte Stimulating Hormone (α-MSH) and Adreno-CorticoTropic Hormone (ACTH). These hormones are secreted by the pituitary gland, even if α-MSH is also secreted by keratinocytes. The binding of α-MSH to MC1R activates the adenylyl cyclase (AC)-dependent signalling pathway that enhance the expression of melanogenesis enzyme genes [12].

Histology and Function of Merkel Cells

Merkel cells are placed along the dermo-epidermal junction and account for less than 5% of the total cell population in the epidermis. These postmitotic cells, which connect with somatosensory afferent nerve fibres, play a crucial role in the perception of light touch. The Merkel cell in apposition with the afferent sensory nerve endings is also referred to as *Merkel's corpuscle,* in fact. By using electron microscopy, the Merkel cells' surface reveals microvilli-like projections by which they interact with the surrounding keratinocytes. Several dense core vesicles of about 50–110 nm are concentrated near the juncture with the nerve ending. These vesicles are considered fundamental for the mechano-transduction mechanism. The Merkel cell cytoskeletal modification, due to the skin touch, triggers the exocytosis of these vesicles that contain a sort of neurotransmitter, which in turn causes membrane depolarization in the nerve endings via ionotropic receptors/ligand-gated ion channels.

Another key feature of these cells is the presence of electron-dense granules in the cytoplasm located away from the Golgi areas and concentrated near nerve terminations. Endocrine function is attributed to these cells since these granules show tight similarities with the dense-core granules of the amine precursor uptake and decarboxylation (APUD) system. It has been hypothesized that Merkel cells may discharge endocrine constituents via autonomic neural regulation [13]. Moreover, this endo-paracrine function of Merkel cells together with the physical relationship of these cells with the hair follicle gave rise to the hypothesis of a role of Merkel cells in the induction and regeneration of hair follicle stem cells.

Histology and Function of Langerhans Cells

Langerhans cells are dendritic cells that colonize the epidermis after their migration from bone marrow. These cells monitor and capture invading antigens, then enter the dermis and migrate to the paracortical region of skin-draining lymph nodes, where they induce an immune response via antigen presentation to T lymphocytes. Langerhans cells are found in the superficial layers of the stratum spinosum and in the stratum granulosum of the epidermis. They form an intercommunicating network with keratinocytes thanks to their dendritic processes emanating from their round-shaped cell body. Their cytoplasm contains unique cytoplasmic inclusions known as Birbeck granules, which look like tennis rackets and are best resolved by electron microscopy. These consist of multilayered membranes that contain the transmembrane protein langerin. These inclusions are thought to be cell membrane folding generated during antigen phagocytosis and processing [14, 15].

The Dermis

The dermis is a vascularized connective tissue beneath the epidermal germinal layer. The epidermis-dermis edge is not straight, and finger-like projections of the dermis toward the epidermis form the *dermal papillae*. The dermis provides mechanical and resilient support to the integumentary system. It is made of two parts: (1) the *papillary dermis*, made of highly vascular loose connective tissue directly in touch with the epidermis, and (2) the *reticular dermis*, the deeper layer made of dense connective tissue that forms the main mass of the dermis (Figs. 2.1, 2.2 and 2.3) [16].

The *fibrous extracellular matrix* of the dermis is mainly composed of fibril-forming collagens I, III, and V and elastic fibres. Collagen-based fibres guarantee tensile strength (Fig. 2.3), whereas elastic fibres give flexibility and the ability to recover form after mechanical stretch. Indeed, the elastic fibres system comprises elastic, elaunin, and oxytalan fibres [17, 18], being oxytalan fibres made of microfibrils without associated elastin, elaunin fibres made of microfibrils associated with small amounts of elastin, and elastic fibres made of elastin peripherally associated with small numbers of fibrillin-based microfibrils. In the dermis, elaunin fibres are horizontally arranged and found near the junction of the papillary and reticular dermis, whereas oxytalan fibres are perpendicular elastic fibres found mainly in the papillary dermis [4, 19]. *Langer lines*, or cleavage lines, correspond to the alignment of collagen and elastic fibres in the reticular dermis.

Ground substance of the dermis is featured by proteoglycans and glycoprotein expression. *Proteoglycans* are space-filling molecules, responsible for the hydration and ion homeostasis regulation. Moreover, these molecules sequester in the dermis growth factors, chemokines, and signalling molecules [20–22]. The dermal proteoglycans are mainly the following: versican, decorin, biglycan, lumican, keratocan, fibromodulin, and the Small Leucine-rich Repeat Proteoglycans (SLRPs). Perlecan and the syndecans, have also been found localized at hair follicles and are also associated with the dermis growth phase [19, 23].

Even *glycoproteins* of the dermis ground substance have multiple functions: (1) to stabilize collagen fibres during deposition, (2) to add elasticity or enhance rigidity to the dermis, and (3) to provide docking sites by acting as ligands for cell surface receptors. Fibronectin is the dominant glycoprotein in the interstitial matrix of the dermis and is mainly produced by fibroblasts and keratinocytes [21, 24]. Laminin, instead, is the principal glycoprotein component of the epidermis-dermis basement membrane (see above) [25].

The dermal extracellular matrix is produced and secreted by *dermal fibroblasts,* which are the principal cells of the dermis, even if their expression profiles vary in the different regions of the dermis [26, 27], and six transcriptionally distinct fibroblast subsets have been found [28]. Fibroblasts also synthesize a great variety of enzymes responsible for matrix remodelling and degradation: Matrix Metalloproteases (MMPs), Tissue Inhibitors of Metalloproteases (TIMPs), serine proteases (such as plasmins), granzymes, elastases, and cathepsins [21, 29].

Besides fibroblasts, *dermal white adipocytes* are recognizable in small groups in the reticular dermis. These cells are revealed to have specific and pleiotropic roles: they provide body insulation and energy storage as well as other white adipocyte depots, however, several studies brought out their role in hair follicle regeneration, defence against infection, and wound healing [30–35].

Other cell types are found within the dermis, with immune functions such as, *macrophages*, and *mast cells*. Notably, histiocytes are tissue macrophages resident within the connective tissue with homeostatic functions [36]. Even if these cells are known to switch from an inflammatory-immune phenotype (M1) to and anti-inflammatory homeostatic phenotype (M2) and vice versa [37], dermal resident macrophages exhibit mainly anti-inflammatory properties [36, 38].

On the contrary, mast cells are inflammatory cells located mainly in the perivascular regions of the dermis. They secrete vasoactive and proinflammatory mediators modulating vascular permeability, inflammatory reactions, and extracellular matrix remodelling [39, 40].

In addition, the dermis houses even other *immune cells* such as dendritic cells, lymphocytes, and eosinophils [41]. Generally, immune cells within the dermis secrete various cytokines and growth factors (including interleukin 1, 4, 6, 13 (IL-1, 4, 6,13); transforming growth factor-beta (TGF-β); tumour necrosis factor (TNF)) that can either promote extracellular matrix (ECM) synthesis or initiate its degradation [29, 42].

The dermis is a highly vascularized tissue, being endothelial cells and pericytes fundamental and cells of the dermal connective tissue with pleiotropic roles [43–45]. *Blood vessels* have the role to supply nutrients to all the integumentary system, with specific regard to the epidermis' avascular epithelium. A subepidermal network of capillaries is present immediately under the epidermis in the papillary dermis, in fact. Capillaries also surround the matrix of hair follicles and are closely associated with sweat and sebaceous glands (see below). Arteriovenous anastomoses, particularly in the dermis of fingers, lips, and toes, directly connect arterioles and venules without a capillary network. These vascular shunts, known as glomera bodies, involve coiled arteriole ends surrounded by modified smooth muscle cells acting as sphincters. They help regulate peripheral temperature by diverting blood from the superficial to the deep plexus under autonomic control. Skin lymphatic vessels accompany venules in deep and superficial vascular plexuses.

The dermis contains a network of *sensory nerves* and sympathetic innervation to sweat glands, vascular smooth muscles, and arrector pili (hair erector) muscles. Myelinated nerve fibres provide innervation to the epidermis and encapsulate sensory receptors in the dermis, such as *Meissner and Pacinian corpuscles*. Meissner corpuscles are mechanoreceptors located in dermal papillae, which mediate touch sense (as well as Merkel cells do) and receive a myelinated nerve fibre that loses its myelin sheath as it ends within it. These structures are numerous in palms and soles are featured by their elongated pinecone-like shape and consisting of a central axon, differentiated Schwann cells forming the lamellar cells, and an incomplete fibrous capsule [46]. Otherwise, Pacinian corpuscles are larger encapsulated receptors in deeper regions of dermis and subcutaneous tissue, which detect deep pressure and vibrations, consisting of multiple layers of loosely arranged concentric lamellae. A single myelinated nerve fibre supplies each corpuscle and loses its myelin sheath as it enters the receptor [47].

Finally, the dermis houses skin appendages of epithelial origin, such as hair follicles and glands, which will be described in the following paragraphs.

Dermis-Associated Skin Appendages

Sweat Glands

Sweat glands are simple, coiled tubular glands with a secretory adenomere and narrower excretory ducts. These structures are classified by three main types: eccrine, apocrine, and apoeccrine glands. Featured by cholinergic innervation, *eccrine sweat glands* secrete water and electrolytes in response to thermal and emotional stimuli, having a thermoregulatory role, by evaporative heat loss, and aiding ion excretion. They are present throughout the body but are absent at the skin of the external genitalia. The tightly convoluted secretory part of a gland deep in the dermis consists of

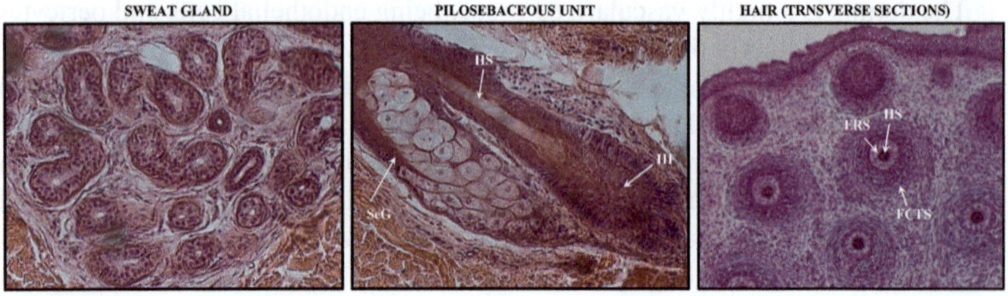

Fig. 2.4 Haematoxylin- and eosin-stained histological sections of the human skin illustrating the main skin appendages (original microphotograph by Prof. Giulia Ricci). HS: Hair Shaft; HF: Hair Follicle; SeG: Sebaceous Gland; ERS: Epithelial Root Sheath; FCTS: Fibrous Connective Tissue Sheath

secretory cells and smaller, intensely eosinophilic myoepithelial cells. Myoepithelial cells are mainly contractile and help expel sweat into the lumen of the adenomere. The duct has a smaller diameter than does the secretory part and lacks myoepithelial cells (Fig. 2.4) [48, 49].

Apocrine sweat glands are located in the axilla, nipples, face, scalp, external genitalia, and the perineum regions. These glands differ from eccrine sweat glands because they are larger, less coiled, and anastomosed, forming a tubular network. In addition, they open into hair follicles instead of onto the epidermal surface as eccrine glands do. They present more numerous myoepithelial cells with respect to eccrine glands, and these cells share a basement membrane with secretory epithelium. The height of secretory cells varies according to their state of secretion. Apocrine sweat glands produce a viscous, lipid-rich secretion, enriched with proteins, sugars, and ammonia, and their functions are mainly related to the production of pheromones. They equally respond to adrenergic and cholinergic stimuli, but their activity is also regulated by sex steroids [50].

Apoeccrine glands, as the name suggests, share properties with both eccrine and apocrine glands. They are intermediate in size and mainly located in the axilla. This type of gland produces abundant saltwater secretions similar to eccrine sweat. Their distal duct connects the adenomere secretion directly to skin surface. Unlikely, their secretion has a significant role in thermoregulation, since evaporation is inefficient in the axilla, and it is commonly accepted that their activity is mainly related to the acquisition of skin scent as apocrine sweat glands. The innervation of the apocrine gland is still poorly studied, but in vitro models suggest the apocrine gland is more sensitive to cholinergic than adrenergic stimuli [51, 52].

Sebaceous Glands

Sebaceous glands are typically found alongside hair follicles and are placed between one hair follicle and its arrector pili muscle within the dermis (as shown below) (Fig. 2.4). Indeed, most sebaceous glands release their secretions into the upper part of the hair follicle near the hair shaft, and therefore these glands are considered part of a unitary structure called the pilosebaceous unit.

Sebaceous glands are holocrine glands in which part of the secretory product, known as sebum, is made of lipid-rich decomposed cells. Sebum is a complex mixture of lipids, including glycerides, free fatty acids, and cholesterol, synthesized in abundant smooth endoplasmic reticulum of sebaceous epithelial secretory cells as they start to differentiate. It aggregates as lipid droplets in the well-developed Golgi complex in sebaceous terminal differentiated cells. The molecular properties of this secretion made it difficult to stain, and the sebaceous cells appear pale in stained histological sections (Fig. 2.4). Each sebaceous gland contains a peripheral layer of cuboidal cells (similar to epidermal basal cells) with spherical nuclei placed on a thin basement membrane. These mitotically active stem cells differentiate into the larger sebum-producing cells that are gradually moved to the gland's centre. The larger cells are polyhedral with pyknotic nuclei: they accumulate lipid in the cytoplasm, and gradually disintegrate, being the cell debris part of the secretory product. Indeed, holocrine secretion involves the breakdown of the entire sebaceous cell, being lysosomal enzymes responsible for this autolysis. Cell breakdown occurs as the final step in the differentiation and enlargement process. Sebaceous cells move to the centre of the adenomere propelled by continuous proliferation of the basal cell layer. Sebaceous glands do not have myoepithelial cells, but they are connected to the arrector pili (hair erector) muscle. When this smooth muscle contracts, it compresses the gland, pushing sebum into the follicle neck. The renewal rate of sebaceous gland lobules is 21–25 days. A thin connective tissue capsule surrounds each gland [53].

Sebaceous glands are under hormonal control and enlarge during puberty, when they produce a substantial amount of sebum, which may lead to the development of acne in adolescents.

In recent years, sebaceous glands have gained recognition as dynamic endocrine organs: they respond to hormonal (e.g., androgens) and neuroendocrine (e.g., corticotropin-releasing hormone) stimuli, and modulate immune responses via antimicrobial peptides and cytokines-regulating cutaneous inflammation. The various functions of these structures have prompted several scientists to refer to them as "the brain of the skin," due to their involvement in both physiological and pathological skin conditions [54].

Hair Follicles

The pilosebaceous unit consists of the hair, hair follicle, an associated arrector pili muscle, and a sebaceous gland. An apocrine sweat gland may be associated with a hair follicle. In this unit, *hair follicles* are responsible for the production of hair (Fig. 2.4). The hair follicle in the long axis has three segments: the permanent upper infundibulum and middle isthmus, and the deepest inferior segment that produces hair [55]. The *infundibulum* is the upper part of the follicle, starting at the skin surface and ending at the sebaceous duct opening. The *isthmus* lies between the sebaceous duct opening and the bulge, which is marked by the insertion of the arrector pili muscle and contains epidermal stem cells in the outer root sheath. The *inferior segment* extends from the bulge to the base of the follicle. This segment includes the bulb, which contains the follicular matrix (germinal matrix) surrounding the sides and top of the dermal papilla. The dermal papilla contains capillaries and interacts with the matrix, which has the highest mitotic rate of any organ. The matrix is where matrix keratinocytes proliferate to form the hair shaft. Mixed among the matrix cells, melanocytes provide the hair shaft colour.

In transverse section, the hair shaft consists of concentric layers: an inner core, the *medulla*, consisting of two or three layers of keratinized cells, and the *cortex,* which in coloured hair contains flattened keratinized cells with pigment granules between cells. Moving outward, the shaft *cuticle* is found. It is made of a thin, but highly protective, colourless layer of keratinized cells that encompasses the vulnerable cortex and medulla. The cells surrounding the hair shaft are the epithelial root sheath (ERS), which originates from the epidermis, and the outer fibrous connective tissue sheath (FCTS) derived from the dermis (Fig. 2.4) [56, 57].

Hair growth occurs in cycles, with the histologic appearance of follicles varying according to growth phase. The active growth period, the anagen stage, lasts about three years. During a three-week period of regression, called the catagen phase, hair growth ceases, and the follicle undergoes involution. A resting period of about 12 weeks, called the telogen phase, is featured by the absence of the lower part of the follicle. In this phase, the recruitment of a novel germinal matrix takes place. This cycle ensures that entirely new hair shafts continue to be produced.

Hair follicles house nerve endings, making them sensitive to touch and vibrations, enabling individuals to detect external stimuli and respond accordingly. Hair also serves as a physical barrier, protecting the skin from environmental damage, UV radiation, and mechanical injury [58].

Hypodermis

The hypodermis is placed between the reticular layer of the dermis and panniculus carnosus in vertebrates [59]. Hypodermis is mainly made of white adipose connective tissue whose thickness varies in different parts across the human body depending on reproductive hormone concentration and genetic signature, but its thickness decreases with age [60]. Several scientific data brought out the hypothesis that hypodermis is more than a fatty depot: besides the role of energy homeostasis, body temperature regulation, mechanical force lubrication, and tissue connection, this tissue has been revealed to be an important stem cell niche involved in wound healing, as well as a hormone and adipokine source [26, 61]. Moreover, the hypodermis is populated by local and blood-derived immune-cells that have an important role in tissue defence and inflammation.

Embryonic Development of the Skin

The skin develops mainly from two sources: (a) the epidermis develops from surface epi-ectoderm (Fig. 2.5) and (b) the dermis develops from the underlying mesoderm (Fig. 2.6). Moreover, the dermis of the ventral and lateral part of the body takes origin from the somatic lateral mesoderm, whereas the dorsal part originates from paraxial mesoderm somite-derived dermatomes. The mesoderm is also responsible for the formation of epidermal Langerhans cells. In addition to the above sources, it should be mentioned that the melanoblasts of epidermis, as well as part of the head and neck dermis connective tissue, develop from the neural crest cells of neuroectodermal origin [62]. Notably, the skin appendages derive from ectoderm specialization (placodes) that invaginate into the mesoderm and subsequently differentiate. In the development of these structures, a close epithelial–mesenchyme interplay occurs.

Embryonic Development of the Epidermis

After gastrulation at the third week of embryonic development, the surface epi-ectoderm consists of a single layer of cells. During the second month, epi-ectoderm proliferating cells form a second layer called periderm, constituted by flattened cells. The deeper layer of cuboidal cells (or basal layer) maintains staminal properties. The further proliferation of the basal epi-ectodermal layer cells allows the formation of a third intermediate layer. The formation of the four layers is acquired at the fourth month of embryonic development. The periderm disappears when the stratum corneum is formed. Periderm and desquamating superficial layers of epidermis get mixed up with secretions of sebaceous glands to form a whitish sticky substance called vernix caseosa, which covers the skin of the foetus. The vernix caseosa prevents the skin from maceration by amniotic fluid and protects integumentary system integrity (Fig. 2.5) [63].

The main cells of the early epidermis are keratinoblasts. The melanoblasts, deriving from neural crest cells, migrate in the basal layer as early as the eighth week of embryonic development, and this migration appears to be concluded by the third month. Langerhans cells appear in human foetal skin at the sixth week of gestation but demonstrate functional activity at around 12th–14th weeks of gestation. Merkel cells are cells associated with sensory nerve fibres along the dermal-epidermal junction and appear in human foetal skin at 18th week of gestation. The origin of Merkel cells is debated: some studies report they arise within the ectodermal cells of the epidermis, whereas other authors hypothesized that these cells migrate from neural crest cells [64, 65].

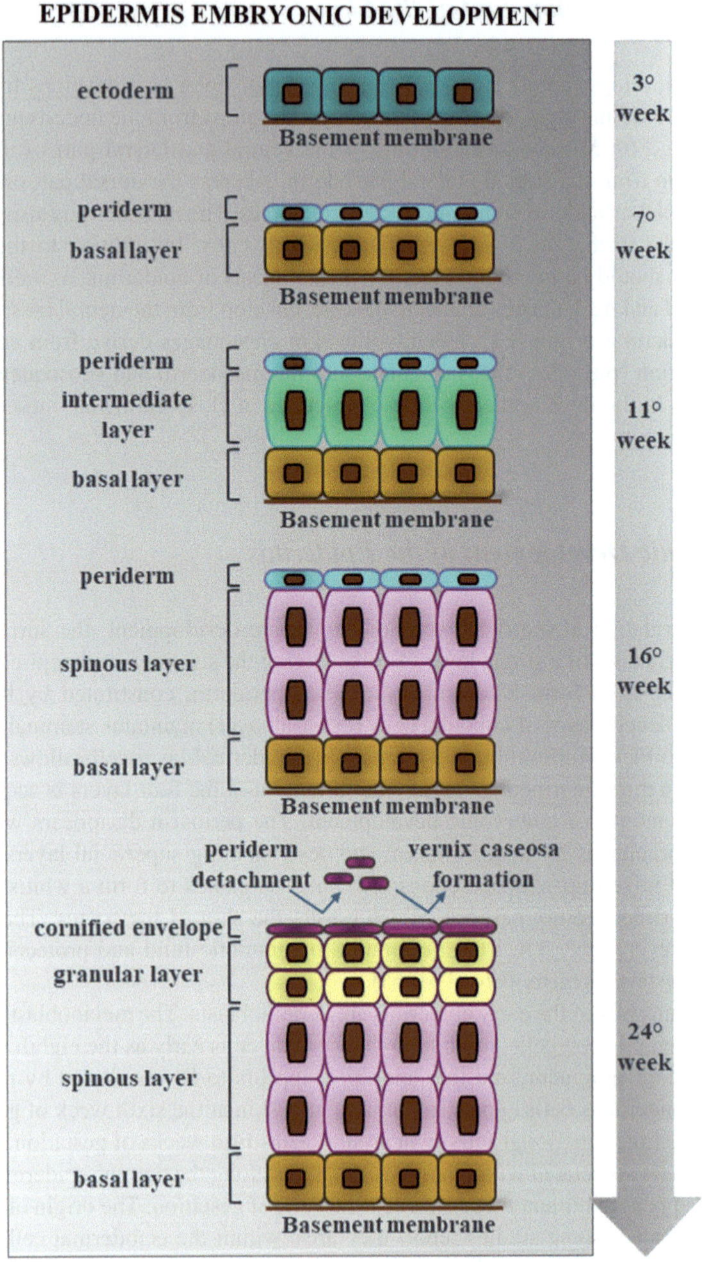

Fig. 2.5 Schematic representation of the stages of epidermis embryonic development (drawn by Luisa Gesualdi, PhD)

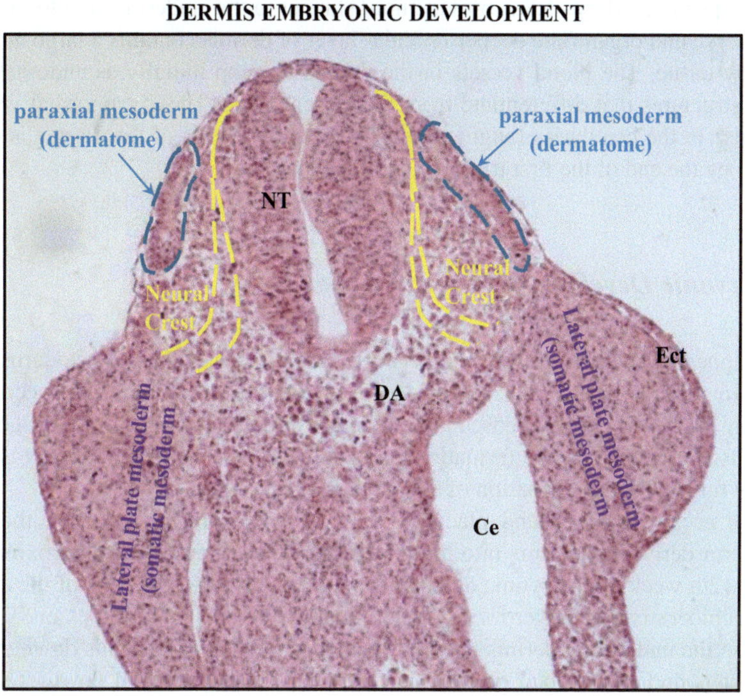

Fig. 2.6 Transverse histological section of a mammalian embryo (original microphotograph by Prof. Giulia Ricci). The image illustrates, with different colours, the embryonic structures that give rise to the dermis. The abbreviations indicate other parts of the embryo body, such as NT: Neural Tube; DA: Dorsal Aorta; Ce: Celom

Embryonic Development of the Dermis

As already mentioned, the dermis develops from the mesenchyme underneath the surface ectoderm. In particular, this mesenchyme is derived from three sources: (a) paraxial mesoderm (the dermatome) responsible for the development of the dermis of the dorsal part of the body, (b) somatopleuric layer of lateral plate mesoderm responsible for the development of the dermis of the lateral and ventral part of the body, and (c) neural crest cells (the ecto-mesenchyme) responsible for the head and neck connective tissue development (Fig. 2.6).

The mesenchymal cells differentiate into fibroblasts that form connective tissue fibres (collagen and elastic fibres) and amorphous ground substance of the connective tissue. The irregular ridges at the dermo-epidermal boundary, called dermal papillae, form during the third and fourth months; these structures project into the epidermis and interdigitate with the epidermal ridges. Most of the dermal papillae of the papillary layer of dermis contain a small loop of capillary plexus and a sensory receptor (sensory nerve end organ). Neural crest cells, responsible for specialized sensory nerve endings, migrate into the dermis during dermal development.

The loops of capillary plexus provide nourishment to the epidermis and to the sensory nerve end organ. The deeper reticular layer of dermis contains a large amount of fatty tissue. The blood vessels in the dermis develop initially as endothelium-lined structures that differentiate from the mesenchyme. These primordial vessels give rise to the capillaries (angiogenesis). The angiogenesis of the dermis is completed by the end of the first trimester of pregnancy [66].

Embryonic Development of Skin Appendages

Skin appendages develop in the embryo as invaginations of the epi-ectoderm after the formation of specific placodes. Placodes are epi-ectoderm thickening accompanied by the underlying condensed mesenchyme. This mesenchyme is, in most cases, responsible for the placode formation, and epithelial-mesenchyme interplay is necessary for the correct formation of skin appendages [67].

The eccrine sweat gland placodes develop from a downgrowth of the epi-ectoderm-derived epidermis into the underlying developing dermis approximately at the 12th week of embryonic development. In this process, the cells of the developing epidermis germ layer form a solid mass of epithelial cells that extend downward in the underlying dermis to form the bud of a sweat gland. This downgrowth elongates, and its terminal part becomes coiled. Later, this solid downgrowth is canalized to form a lumen. The terminal coiled part of the downgrowth forms the secretory part of the sweat gland, while the proximal straight part forms the duct of the sweat gland. The site of the downgrowth from the surface epithelium forms pore of the duct of the sweat gland. Eccrine sweat glands begin functioning shortly after birth [68, 69].

Apocrine sweat glands form in a similar way in association with the hair development in the whole body. Shortly before birth, they regress except at the level of axilla, nipples, face, scalp, external genitalia, and the perineum. Notably, they begin to function at puberty. Since apocrine sweat glands develop from the same epidermal buds that form the hair follicles, they open in the hair follicle lumen instead of opening on the skin surface.

Hair follicles start to develop from the hair placode at the third month of gestational age at the level of the eyelids, eyebrows, and upper lip. By the fifth month of development, the hair buds are present in all regions of the body and will continue to increase in number throughout gestation. Under the action of the mesenchyme, the ectodermal cells of the hair bud (placode) proliferate and deepen to form the axis of the hair. Subsequently, the basal part expands, forming the epithelial part of the *bulb*, and the dermis invaginates into it, forming the *dermal papilla*. The layer of ectodermal cells in contact with the dermal papilla, the follicular matrix (or germinal matrix), continues to proliferate, giving rise to the cells that will undergo a special keratinization process, forming the hair shaft (Fig. 2.7). The latter stretches from the base to the surface through the follicular canal. The eruption of hair on the body surface has a cephalo-caudal pattern. At about seven months, the body is

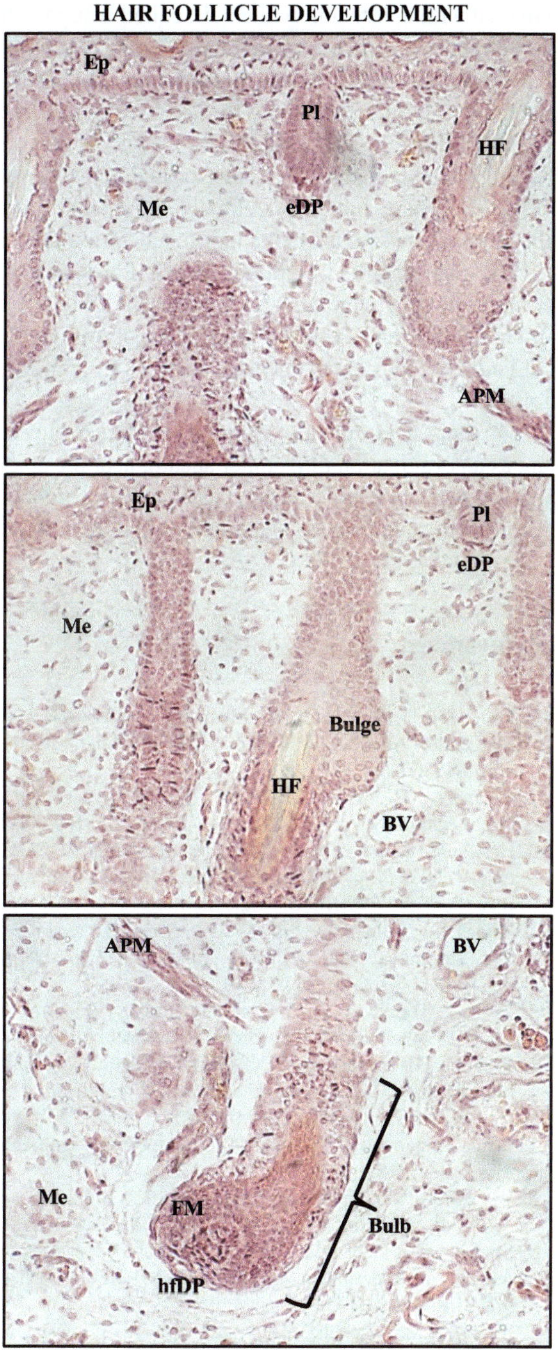

Fig. 2.7 Haematoxylin- and eosin-stained histological sections of the human developing skin illustrating different stages of hair follicle formation (original microphotograph by Prof. Giulia Ricci). HS: Hair Shaft; BV: Blood Vessel; Me: Mesenchyme; Ep: Epidermis; Pl: Placode; eDP: early Dermal Papilla; hfDp: hair follicle Dermal Papilla; APM: Arrector Pili Muscle; FM: Follicular Matrix

covered with a very thin white hair, called lanugo, which is replaced shortly before birth by thicker and pigmented hairs due to the infiltration of melanocytes into the hair bulb. The melanoblasts migrate into the hair bulbs and differentiate into the melanocytes, in fact. Melanin formed by these melanocytes is transferred to proliferating cells of the germinal matrix [55].

The hair follicle stem cells are located in a *bulge* just below the sebaceous gland (Fig. 2.7). These cells are cyclically stimulated to form a new germ matrix during the telogen phase to make the hair regeneration process possible.

The small bundle of smooth muscle fibres of *arrector pili muscle* differentiate from the mesenchyme surrounding the hair follicle and attach the dermal root sheath of hair follicle and papillary layer of the dermis (Fig. 2.7).

The sebaceous gland develops as a bud from the epithelial root sheath of the hair follicle.

Most sebaceous glands originate four weeks after the beginning of hair lengthening as an evagination of its outer sheath, near the epidermal surface. The bud grows into the surrounding dermis and divides into a number of capillary loop branches to form primordia of several adenomeres and their associated ducts [67].

References

1. Lopez-Ojeda W, Pandey A, Alhajj M, Oakley AM (2025) Anatomy, skin (integument). StatPearls. Accessed: 13 Jun 2025. [Online]. Available: https://pubmed.ncbi.nlm.nih.gov/28723009/
2. Helander HF, Fändriks L (2014) Surface area of the digestive tract-revisited. Scand J Gastroenterol 49(6):681–689. https://doi.org/10.3109/00365521.2014.898326
3. Gallegos-Alcalá P, Jiménez M, Cervantes-García D, Salinas E (2021) The keratinocyte as a crucial cell in the predisposition, onset, progression, therapy and study of the atopic dermatitis. Int J Mol Sci 22(19):10661. https://doi.org/10.3390/IJMS221910661
4. Krieg T, Aumailley M (2011) The extracellular matrix of the dermis: flexible structures with dynamic functions. Exp Dermatol 20(8):689–695. https://doi.org/10.1111/J.1600-0625.2011.01313.X
5. Dodia P, Arida MA (2023) Dermatopathology epidermis histology. StatPearls. Accessed: 13 Jun 2025. [Online]. Available: https://www.ncbi.nlm.nih.gov/books/NBK592423/
6. Bragulla HH, Homberger DG (2009) Structure and functions of keratin proteins in simple, stratified, keratinized and cornified epithelia. J Anat 214(4):516–559. https://doi.org/10.1111/J.1469-7580.2009.01066.X
7. Ishida-Yamamoto A, Igawa S (2015) The biology and regulation of corneodesmosomes. Cell Tissue Res 360(3):477–482. https://doi.org/10.1007/S00441-014-2037-Z
8. Haftek M (2015) Epidermal barrier disorders and corneodesmosome defects. Cell Tissue Res 360(3):483–490. https://doi.org/10.1007/S00441-014-2019-1
9. Simpson CL, Patel DM, Green KJ (2011) Deconstructing the skin: cytoarchitectural determinants of epidermal morphogenesis. Nat Rev Mol Cell Biol 12(9):565–580. https://doi.org/10.1038/nrm3175
10. Cichorek M, Wachulska M, Stasiewicz A, Tymińska A (2013) Skin melanocytes: biology and development. Postepy Dermatol Alergol 30(1):30–41. https://doi.org/10.5114/PDIA.2013.33376

11. D'Mello SAN, Finlay GJ, Baguley BC, Askarian-Amiri ME (2016) Signaling pathways in melanogenesis. Int J Mol Sci 17(7):1144. https://doi.org/10.3390/IJMS17071144
12. Dall'Olmo L, Papa N, Surdo NC, Marigo I, Mocellin S (2023) Alpha-melanocyte stimulating hormone (α-MSH): biology, clinical relevance and implication in melanoma. J Transl Med 21(1):1–9. https://doi.org/10.1186/S12967-023-04405-Y/FIGURES/1
13. Abraham J, Mathew S (2019) Merkel cells: a collective review of current concepts. Int J Appl Basic Med Res 9(1):9. https://doi.org/10.4103/IJABMR.IJABMR_34_18
14. Merad M, Ginhoux F, Collin M (2008) Origin, homeostasis and function of Langerhans cells and other langerin-expressing dendritic cells. Nat Rev Immunol 8(12):935–947. https://doi.org/10.1038/NRI2455
15. Neagu M et al (2022) Langerhans cells—revising their role in skin pathologies. J Pers Med 12(12)(12):2072. https://doi.org/10.3390/JPM12122072
16. Brown TM, Krishnamurthy K (2022) Histology, dermis. StatPearls. Accessed: 13 Jun 2025. [Online]. Available: https://www.ncbi.nlm.nih.gov/books/NBK535346/
17. Kielty CM, Sherratt MJ, Shuttleworth CA (2002) Elastic fibres. J Cell Sci 115(14):2817–2828. https://doi.org/10.1242/JCS.115.14.2817
18. Kielty CM, Sherratt MJ, Marson A, Baldock C (2005) Fibrillin microfibrils. Adv Protein Chem 70:405–436. https://doi.org/10.1016/S0065-3233(05)70012-7
19. Huang J et al (2022) Dermal extracellular matrix molecules in skin development, homeostasis, wound regeneration and diseases. Semin Cell Dev Biol 128:137–144. https://doi.org/10.1016/J.SEMCDB.2022.02.027
20. Hynes RO (1979) The extracellular matrix: not just pretty fibrils. Science 326(5957): 1216–1219., Nov. 2009. https://doi.org/10.1126/SCIENCE.1176009
21. Bonnans C, Chou J, Werb Z (2014) Remodelling the extracellular matrix in development and disease. Nat Rev Mol Cell Biol 15(12):786–801. https://doi.org/10.1038/NRM3904
22. Dyer BA et al (2020) Characterization and clinical validation of patient-specific three-dimensional printed tissue-equivalent bolus for radiotherapy of head and neck malignancies involving skin. Phys Med 77:138–145. https://doi.org/10.1016/j.cjmp.2020.08.010
23. Smith MM, Melrose J (2015) Proteoglycans in normal and healing skin. Adv Wound Care (New Rochelle) 4(3):152–173. https://doi.org/10.1089/WOUND.2013.0464
24. Uitto J (1989) Connective tissue biochemistry of the aging dermis: age-associated alterations in collagen and elastin. Clin Geriatr Med 5(1):127–148. https://doi.org/10.1016/S0749-0690(18)30700-6
25. Pfisterer K, Shaw LE, Symmank D, Weninger W (2021) The extracellular matrix in skin inflammation and infection. Front Cell Dev Biol 9:682414. https://doi.org/10.3389/FCELL.2021.682414/REFERENCE
26. Haydont V et al (2020) Fibroblasts from the human skin dermo-hypodermal junction are distinct from dermal papillary and reticular fibroblasts and from mesenchymal stem cells and exhibit a specific molecular profile related to extracellular matrix organization and modeling. Cells 9(2):368. https://doi.org/10.3390/CELLS9020368
27. Mellody KT, Horn J, Watson REB (2018) Heterogenity of dermal fibroblasts: skin equivalent models that aim to recreate the cutaneous microenvironment. Br J Dermatol 179(2):248–250. https://doi.org/10.1111/BJD.16773
28. Vorstandlechner V et al (2020) Deciphering the functional heterogeneity of skin fibroblasts using single- cell RNA sequencing. FASEB J 34(3):3677–3692. https://doi.org/10.1096/FJ.201902001RR
29. Bhattacharjee T, Datta SS (2019) Bacterial hopping and trapping in porous media. Nat Commun 10(1):1–9. https://doi.org/10.1038/s41467-019-10115-1
30. Kruglikov IL, Scherer PE (2016) Dermal adipocytes: from irrelevance to metabolic targets? Trends Endocrinol Metab 27(1):1–10. https://doi.org/10.1016/j.tem.2015.11.002
31. Driskell RR, Jahoda CAB, Chuong CM, Watt FM, Horsley V (2014) Defining dermal adipose tissue. Exp Dermatol 23(9):629–631. https://doi.org/10.1111/EXD.12450

32. Chen SX, Zhang LJ, Gallo RL (2019) Dermal white adipose tissue: a newly recognized layer of skin innate defense. J Invest Dermatol 139(5):1002–1009. https://doi.org/10.1016/J.JID.2018.12.031
33. Alexander CM et al (2015) Dermal white adipose tissue: a new component of the thermogenic response. J Lipid Res 56(11):2061–2069. https://doi.org/10.1194/jlr.R062893
34. Kasza I et al (2014) Syndecan-1 is required to maintain intradermal fat and prevent cold stress. PLoS Genet 10(8):e1004514. https://doi.org/10.1371/JOURNAL.PGEN.1004514
35. Zhang L j et al (2019) Age-related loss of innate immune antimicrobial function of dermal fat is mediated by transforming growth factor Beta. Immunity 50(1):121–136.e5. https://doi.org/10.1016/j.immuni.2018.11.003
36. Lee SH, Sacks DL (2024) Resilience of dermis resident macrophages to inflammatory challenges. Exp Mol Med 56(10):2105–2112. https://doi.org/10.1038/S12276-024-01313-z
37. Mills CD, Kincaid K, Alt JM, Heilman MJ, Hill AM (2000) M-1/M-2 macrophages and the Th1/Th2 paradigm. J Immunol 164(12):6166–6173. https://doi.org/10.4049/JIMMUNOL.164.12.6166
38. Lee SH et al (2020) M2-like, dermal macrophages are maintained via IL-4/CCL24-mediated cooperative interaction with eosinophils in cutaneous leishmaniasis. Sci Immunol 5(46):4415. https://doi.org/10.1126/SCIIMMUNOL.AAZ4415
39. Wilgus TA, Wulff BC (2014) The importance of mast cells in dermal scarring. Adv Wound Care (New Rochelle) 3(4):356–365. https://doi.org/10.1089/WOUND.2013.0457
40. Wang X, Zhang P, Tang Y, Chen Y, Zhou E, Gao K (2024) Mast cells: a double-edged sword in inflammation and fibrosis. Front Cell Dev Biol 12:1466491. https://doi.org/10.3389/FCELL.2024.1466491/BIBTEX
41. Nguyen AV, Soulika AM (2019) The dynamics of the skin's immune system. Int J Mol Sci 20(8):1811. https://doi.org/10.3390/IJMS20081811
42. Liguori M, Solinas G, Germano G, Mantovani A, Allavena P (2011) Tumor-associated macrophages as incessant builders and destroyers of the cancer stroma. Cancers (Basel) 3(4):3740–3761. https://doi.org/10.3390/CANCERS3043740
43. Geevarghese A, Herman IM (2014) Pericyte-endothelial crosstalk: implications and opportunities for advanced cellular therapies. Transl Res 163(4):296–306. https://doi.org/10.1016/j.trsl.2014.01.011
44. Morikawa S, Iribar H, Gutiérrez-Rivera A, Ezaki T, Izeta A (2019) Pericytes in cutaneous wound healing. Adv Exp Med Biol 1147:1–63. https://doi.org/10.1007/978-3-030-16908-4_1
45. Paquet-Fifield S et al (2009) A role for pericytes as microenvironmental regulators of human skin tissue regeneration. J Clin Invest 119(9):2795–2806. https://doi.org/10.1172/JCI38535
46. Vega JA, García-Suárez O, Montaño JA, Pardo B, Cobo JM (2009) The Meissner and Pacinian sensory corpuscles revisited new data from the last decade. Microsc Res Tech 72(4):299–309. https://doi.org/10.1002/JEMT.20651
47. Ziolkowski LH et al (2025) The inner core enables transient touch detection in the Pacinian corpuscle. Sci Adv 11(9). https://doi.org/10.1126/sciadv.adt4837
48. Baker LB (2019) Physiology of sweat gland function: the roles of sweating and sweat composition in human health. Temperature 6(3):211–259. https://doi.org/10.1080/23328940.2019.1632145
49. Wilke K, Martin A, Terstegen L, Biel SS (2009) Neurobiology of skin appendages: eccrine, apocrine, and apoeccrine sweat glands. In: Neuroimmunology of the Skin: Basic Science to Clinical Practice, pp 167–175. https://doi.org/10.1007/978-3-540-35989-0_15
50. Beier K, Ginez I, Schaller H (2005) Localization of steroid hormone receptors in the apocrine sweat glands of the human axilla. Histochem Cell Biol 123(1):61–65. https://doi.org/10.1007/S00418-004-0736-3
51. Sato K, Leidal R, Sato F (1987) Morphology and development of an apoeccrine sweat gland in human axillae. Am J Physiol Regul Integr Comp Physiol 252(1):21. https://doi.org/10.1152/AJPREGU.1987.252.1.R166

52. Sato K, Sato F (1987) Sweat secretion by human axillary apoeccrine sweat gland in vitro. Am J Physiol Regul Integr Comp Physiol 252(1):21. https://doi.org/10.1152/AJPREGU.1987.252.1.R181
53. Hoover E, Aslam S, Krishnamurthy K (2022) Physiology, sebaceous glands. StatPearls. Accessed: 13 Jun 2025. [Online]. Available: https://www.ncbi.nlm.nih.gov/books/NBK499819/
54. Simbulan-Rosenthal M, Mosca S, Ottaviani M, Briganti S, Di Nardo A, Flori E (2025) The Sebaceous Gland: a key player in the balance between homeostasis and inflammatory skin diseases. Cells 14(10):747. https://doi.org/10.3390/CELLS14100747
55. Schneider MR, Schmidt-Ullrich R, Paus R (2009) The Hair Follicle as a Dynamic Miniorgan. Curr Biol 19(3). https://doi.org/10.1016/j.cub.2008.12.005
56. Glover JD et al (2017) Hierarchical patterning modes orchestrate hair follicle morphogenesis. PLoS Biol 15(7). https://doi.org/10.1371/JOURNAL.PBIO.2002117
57. Hamada K, Randall VA (2006) Inhibitory autocrine factors produced by the mesenchyme-derived hair follicle dermal papilla may be a key to male pattern baldness. Br J Dermatol 154(4):609–618. https://doi.org/10.1111/J.1365-2133.2006.07144.X
58. Panteleyev AA (2018) Functional anatomy of the hair follicle: the secondary hair germ. Exp Dermatol 27(7):701–720. https://doi.org/10.1111/EXD.13666
59. Naldaiz-Gastesi N, Bahri OA, López de Munain A, McCullagh KJA, Izeta A (2018) The panniculus carnosus muscle: an evolutionary enigma at the intersection of distinct research fields. J Anat 233(3):275–288. https://doi.org/10.1111/joa.12840
60. Guerrero-Juarez CF, Plikus MV (2018) Emerging nonmetabolic functions of skin fat. Nat Rev Endocrinol 14(3):163–173. https://doi.org/10.1038/NRENDO.2017.162
61. Liu M, Lu F, Feng J (2024) Aging and homeostasis of the hypodermis in the age-related deterioration of skin function. Cell Death Dis 15(6). https://doi.org/10.1038/S41419-024-06818-Z
62. Hu MS et al (2018) Embryonic skin development and repair. Organogenesis 14(1):46–63. https://doi.org/10.1080/15476278.2017.1421882
63. Dermitzakis I et al (2024) Skin development and disease: a molecular perspective. Curr Issues Mol Biol 46(8):8239–8267. https://doi.org/10.3390/CIMB46080487
64. Szeder V, Grim M, Halata Z, Sieber-Blum M (2003) Neural crest origin of mammalian Merkel cells. Dev Biol 253(2):258–263. https://doi.org/10.1016/S0012-1606(02)00015-5
65. Van Keymeulen A et al (2009) Epidermal progenitors give rise to Merkel cells during embryonic development and adult homeostasis. J Cell Biol 187(1):91–100. https://doi.org/10.1083/JCB.200907080
66. Lee LW, Holbrook KA (2019) Embryogenesis of the skin. In: Harper's textbook of pediatric dermatology, pp 1–35. https://doi.org/10.1002/9781119142812.CH1
67. Liu S, Zhang H, Duan E (2013) Epidermal development in mammals: key regulators, signals from beneath, and stem cells. Int J Mol Sci 14(6):10869–10895. https://doi.org/10.3390/IJMS140610869
68. Cui CY, Schlessinger D (2015) Eccrine sweat gland development and sweat secretion. Exp Dermatol 24(9):644–650. https://doi.org/10.1111/exd.12773
69. Lu C, Fuchs E (2014) Sweat gland progenitors in development, homeostasis, and wound repair. Cold Spring Harb Perspect Med 4(2). https://doi.org/10.1101/CSHPERSPECT.A015222

Chapter 3
Pathophysiology of Wound Healing

Ali Golchin, Khan Sharun, and Aynaz Mihanfar

Abstract Wound healing is a complex cascade of interactions among different cells, growth factors, and extracellular matrix components, and currently some of its aspects are still not fully understood. While the physiopathology of wound healing outlines the main phases of tissue regeneration, it also highlights the influence of many internal and external variables on cellular and tissue microenvironments. The body's protective mechanisms are most beneficial when they occur during the process of wound healing. The wound healing process is made up of four important phases, including hemostasis, inflammation, proliferation, and remodeling, which eventually repair or regenerate the injured tissue. Although the wounds are created in different organs and tissues, cutaneous wound healing is introduced as a fundamental framework for describing the wound-healing process. Nonetheless, several types of wounds exist within the cutaneous tissue, which create distinctions in this fundamental framework. In this chapter, we want to review the physiopathology of wound healing and the factors that influence this process. Furthermore, our attention will be directed toward the pathophysiology of specific skin wound types.

A. Golchin (✉)
Iranian Tissue Engineering and Regenerative Medicine Society (ITERMS), Tehran, Iran

Regenerative Medicine Group (REMED), Universal Scientific Education and Research Network (USERN), Tehran, Iran
e-mail: agolchin.vet10@yahoo.com

K. Sharun
Graduate Institute of Medicine, Yuan Ze University, Taoyuan, Taiwan

Regenerative Medicine Group, Department of Health Science and Technology, Aalborg University, Gistrup, Denmark
e-mail: sharunk@hst.aau.dk

A. Mihanfar
Iranian Tissue Engineering and Regenerative Medicine Society (ITERMS), Tehran, Iran

Department of Clinical Biochemistry, School of Medicine, Urmia University of Medical Sciences, Urmia, Iran
e-mail: ainazmihanfar2013@gmail.com

F. De Francesco, K. Turksen (eds.), *Management and Strategies for Wound Healing*, Stem Cell Biology and Regenerative Medicine 79,
https://doi.org/10.1007/978-3-032-12442-5_3

Keywords Wound Healing · Hemostasis · Inflammation · Cell Proliferation · Extracellular Matrix

Abbreviations

ECM	Extracellular Matrix
EGF	Epidermal Growth Factor
EPC	Endothelial Progenitor Cell
FGF	Fibroblast Growth Factor
HIF-1α	Hypoxia-Inducible Factor 1-alpha
IL-1	Interleukin-1
IL-6	Interleukin-6
MMPs	Matrix Metalloproteinases
MSC	Mesenchymal Stem Cell
NF-κB	Nuclear Factor Kappa B
NO	Nitric Oxide
PDGF	Platelet-Derived Growth Factor
TGF-β	Transforming Growth Factor Beta
TNF-α	Tumor Necrosis Factor Alpha
VEGF	Vascular Endothelial Growth Factor

Introduction: The Delicate Balance of Normal Wound Healing

The human body possesses innate defenses and mechanisms essential for health and survival. The wound healing process is a primary mechanism for monitoring and preserving human body health. A wound refers to an injury, with or without disruption of structural integrity, in which the physiological function of tissues and organs is reduced or lost [1]. The process of wound healing is complex, comprising several phases and the interaction of diverse cells and biofactors to repair injured tissue. Wounds can occur in different organs, such as the skin, liver, heart, lungs, and brain. Hence, the cascade of events that occur after a myocardial infarction closely resembles that which occurs following a spinal cord injury, a cutaneous trauma, or a hepatic injury, despite the distinct nature of the injuries and the varied target organs [2]. Typically, the innate mechanism of wound healing restores the injured area without any visible signs or external intervention. However, under certain circumstances, such as chronic wounds or serious injuries, healthcare intervention may be required [3, 4]. Consequently, it is essential to acknowledge that wounds occur in all sizes and types. Based on the different options, like causing factors, size, and healing process, wounds can be characterized into multiple types. Wound healing consists of four interrelated phases: homeostasis, inflammation, proliferation, and remodeling [3]. However, various wound types necessitate tailored care and therapy to facilitate optimal healing and minimize infection. Therefore, knowledge of the physiopathology of wound healing is important, and it can help healthcare providers determine the most effective treatment plan. An understanding of the pathophysiology of wound healing elucidates the functions of various cells and biological substances involved, potentially facilitating the creation of innovative therapeutic strategies for enhanced wound management. For instance, it can demonstrate the function of mesenchymal stem cells in promoting the identified process of tissue regeneration and wound healing, and then highlight the method of MSCs or derived exosome delivery systems as potential treatments for chronic wounds [5, 6]. In this chapter, we want to review the physiopathology of wound healing and the factors that influence this process.

The Pathophysiology of Wound Healing

Wound classification may rely on different criteria, including etiology, depth, healing duration, and contamination; however, irrespective of type, healing occurs through a meticulously regulated, dynamic sequence of immunological and biological processes [7]. These events occur as an overlapping cascade of meticulously regulated mechanisms, each characterized by specific cellular and molecular activities within the wound bed. Although the interactions among cells (e.g., platelets, neutrophils, macrophages, fibroblasts, keratinocytes) and biomolecules (e.g., cytokines, growth factors, extracellular matrix proteins, proteases) evolve continuously [3, 7], they can be broadly divided into four interconnected phases: Hemostasis, Inflammation, Proliferation, Remodeling (Fig. 3.1). However, the rate and intensity of injury are the paramount factors that delineate the timetable of healing phases.

Hemostasis During hemostasis, exposed collagen at the injury site activates both the intrinsic (contact activation) and extrinsic (tissue factor) coagulation pathways. Activated endothelial cells and platelets collaborate to form a fibrin-dense clot, while a precisely regulated interaction between pro-coagulant factors and the fibrinolytic system maintains vascular integrity and reduces hemorrhage [3, 7].

Inflammation The inflammatory phase commences immediately post-injury and entails a coordinated influx of immune cells—initially neutrophils, followed by monocytes that differentiate into macrophages, and then lymphocytes—into the wound bed [3]. It can be subdivided into:

Early inflammation (0–24 h, up to 72 h):

- Polymorphonuclear neutrophils (PMNs) arrive first to phagocytose debris and bacteria, generate reactive oxygen species, and release proteases that help clear the wound of pathogens.

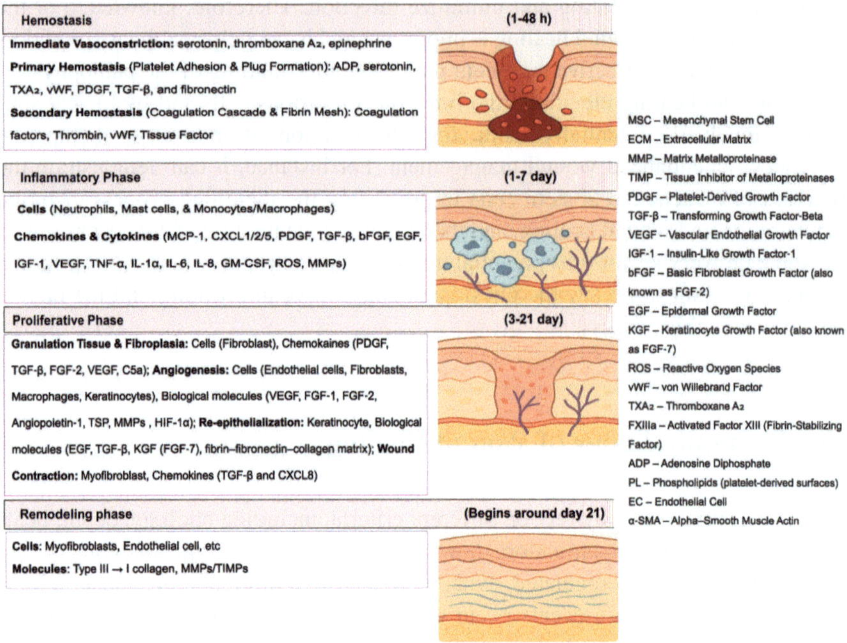

Fig. 3.1 Wound healing phases and summary of main key factors during phases and stages

Late inflammation (24–72 h onward, up to 7 days):

- Monocyte-derived macrophages dominate. These cells continue phagocytosis, secrete a cocktail of cytokines (e.g., TNF-α, IL-1β, IL-6, IL-10) and growth factors (e.g., PDGF, TGF-β, VEGF), and regulate the transition to tissue repair by:
  - Stimulating fibroblast proliferation and collagen synthesis
  - Recruiting endothelial cells to promote angiogenesis
  - Activating keratinocytes for re-epithelialization
- Lymphocytes (T and B cells) arrive later to help modulate inflammation and support remodeling.

However, the shift from neutrophil- to macrophage-predominant infiltrate indicates the termination of the inflammation phase and signals the onset of the proliferative phase [8, 9]. Local stem cells, particularly MSCs, initiate activation throughout wound healing from the inflammatory phase to the remodeling phase, while their primary function occurs in the proliferation phase, represented by cytokine release and stemness potential [2].

Proliferation The proliferation phase begins when hemostasis is achieved and the early immune response has eliminated debris and pathogens [10]. This phase, commencing approximately 3 days post-injury and enduring for about 10 to 21 days, reconstructs the wound bed by four synchronized processes:

- Fibroplasia and ECM Deposition
  Fibroblast migration and activation: Fibroblasts from the wound edges and adjacent stroma migrate into the defect and commence the synthesis of a new extracellular matrix [11]. However, there is some evidence of differentiation of stem cells and progenitor cells into fibroblasts in this phase [11]. Stem cells, particularly mesenchymal stem cells, can reach the injury site from adjacent sites and peripheral blood. Fibroblasts initially deposit a temporary scaffold abundant in type III collagen, fibronectin, and proteoglycans, progressively substituting the fibrin–fibronectin clot [8, 10].
- Angiogenesis and Granulation Tissue Formation
  Endothelial sprouting: Driven by VEGF, FGF-2, and PDGF, new capillaries develop from existing vessels, creating a highly vascularized granulation tissue that delivers oxygen and nutrients.
- Re-epithelialization
  Keratinocyte proliferation and migration: Keratinocytes near the wound periphery detach, proliferate under the influence of EGF and KGF, and migrate over the granulation tissue to restore a complete epidermal barrier [12]. This phenomenon adheres to a natural process whereby stem cells from various regions of the dermis facilitate the keratinocytes' proliferation process [8].
- Wound Contraction
  Myofibroblast differentiation: A subset of fibroblasts acquires α-SMA and contractile properties, generating mechanical forces that draw wound margins together and reduce defect size [12].
  At the final stage of this phase, the wound is populated with granulation tissue, the epidermis is regenerated, and the mechanical stability of the repair is enhanced, preparing for prolonged maturation and remodeling.

Remodeling The remodeling phase begins as granulation tissue production concludes, usually around day 21 post-injury, and may last for up to 12 months or more [7]. During this period:

- Collagen maturation and reorganization**:** Type III collagen is gradually supplanted by more robust, extensively cross-linked type I collagen. Fibrils thicken, align with tension lines, and expand in diameter, so as to augment the tensile strength of the tissue.
- ECM turnover: Hyaluronic acid and fibronectin from the provisional matrix undergo degradation, whereas matrix metalloproteinases (MMP-1, −2, −9, etc.), regulated by tissue inhibitors of metalloproteinases (TIMPs), modify the extracellular matrix to enhance fiber architecture.
- Cellular pruning: Myofibroblasts, surplus endothelial cells, and other extra reparative cells undergo apoptosis upon the completion of their functions, leading to the regression of unnecessary capillaries, which diminishes scar vascularity and erythema.
- Biochemical regulation: A coordinated network of growth factors (e.g., TGF-β, PDGF), cytokines, integrins, chemokines, keratins, and other extracellular macromolecules precisely regulates matrix deposition, degradation, and cell–matrix interactions to shape the mature scar.

Overall, the remodeling phase converts the temporary wound bed into a mechanically resilient scar by disassembling initial cell–cell and cell–matrix interactions, optimizing collagen structure, and modulating the equilibrium between synthesis and degradation [7, 9, 12]. A comprehensive understanding of these molecular pathways is essential for formulating medicines that reduce fibrosis and enhance tissue regeneration.

The Pathophysiology of Specific Wound Types

This section provides a comprehensive overview of the distinct mechanisms that govern the development and progression of four major wound types. Diabetic foot ulcers emerge from a triad of neuropathy, vasculopathy, and immunopathy, each contributing to impaired healing and increased infection risk. Pressure ulcers result from sustained ischemia, reperfusion injury, and mechanical tissue deformation, particularly in immobilized patients. Venous leg ulcers are driven by chronic inflammation and microangiopathy associated with venous hypertension and valvular dysfunction. In contrast, arterial ulcers primarily stem from severe ischemia due to peripheral arterial disease, where diminished perfusion severely compromises tissue viability. Understanding these unique pathophysiological pathways is essential for the development of targeted and effective therapeutic strategies.

Diabetic Foot Ulcers: A Confluence of Neuropathy, Vasculopathy, and Immunopathy

Diabetic foot ulcers (DFUs) arise from a complex interplay of neuropathy, vasculopathy, and immunopathy driven by chronic hyperglycemia (Fig. 3.2). Prolonged elevated blood glucose initiates metabolic cascades that damage peripheral nerves through multiple pathways. The polyol pathway converts excess glucose to sorbitol via aldose reductase, depleting intracellular antioxidants like glutathione and inducing oxidative stress, which impairs nerve conduction and promotes axonal degeneration [13]. Neuropathy is the most common and debilitating complication of diabetes and results in pain, decreased motility, and amputation. Diabetic neuropathy encompasses a variety of forms whose impact ranges from discomfort to death. Hyperglycemia induces oxidative stress in diabetic neurons and results in activation of multiple biochemical pathways. These activated pathways are a major source of damage and are potential therapeutic targets in diabetic neuropathy. Though therapies are available to alleviate the symptoms of diabetic neuropathy, few options are available to eliminate the root causes. The immense physical, psychological, and economic cost of diabetic neuropathy underscores the need for causally targeted therapies. Concurrent activation of protein kinase C (PKC) in hyperglycemic

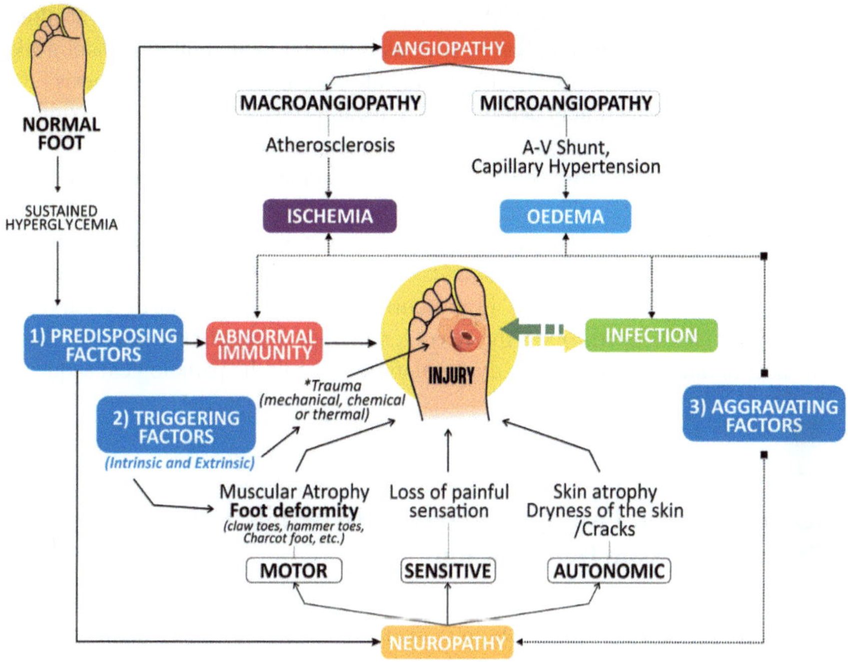

Fig. 3.2 Key risk factors for diabetic foot ulcer (DFU) development include angiopathy and neuropathy, which, along with intrinsic (e.g., foot deformity) and extrinsic (e.g., trauma) triggers, compromise skin integrity. Aggravating factors like ischemia, immune dysfunction, and neuropathy contribute to infection. (Reproduced from Ramirez-Acuña et al. 2019 under the Creative Commons Attribution (CC BY) license [21])

conditions exacerbates vascular dysfunction by reducing nitric oxide (NO) bioavailability, leading to endothelial injury and impaired vasodilation [14, 15]. Advanced glycation end products (AGEs), formed via non-enzymatic glycation of proteins, bind to receptors (RAGE) on endothelial and Schwann cells, triggering pro-inflammatory cytokine release and further oxidative damage [16]. These processes collectively cause sensory-motor neuropathy, characterized by loss of protective sensation, and autonomic neuropathy, which reduces sweat gland function, creating dry, fissured skin prone to ulceration [13].

(16)Vasculopathy in DFUs is marked by accelerated atherosclerosis, particularly in below-the-knee arteries, driven by hyperglycemia-induced endothelial dysfunction, platelet aggregation, and chronic inflammation [17]. Reduced blood flow from arterial stenosis and microvascular rarefaction creates tissue ischemia, impairing wound healing and increasing infection risk [18]. Immunopathic contributions involve dysregulated neutrophil extracellular trap (NET) formation, which perpetuates inflammation and delays repair by damaging surrounding tissues [19]. Macrophage polarization defects, specifically, persistent M1 dominance over anti-inflammatory M2 phenotypes-sustain a pro-inflammatory milieu, while impaired phagocytosis and excessive protease secretion degrade extracellular matrix components critical for healing [20]. Hyperglycemia also alters small extracellular vesicle (sEV) cargo in macrophages, such as miR-503, which inhibits insulin-like growth factor 1 receptor (IGF1R) in endothelial cells, hampering angiogenesis [20]. Mitochondrial dysfunction in neurons and Schwann cells, exacerbated by oxidative stress, further compromises energy production and axonal transport, while PARP activation due to DNA damage depletes NAD+, worsening cellular repair mechanisms [16].

The convergence of these pathways creates a vicious cycle: neuropathy masks trauma, vasculopathy limits nutrient delivery, and immunopathy impedes infection control. This triad explains DFUs' high recurrence rates (65% within 3–5 years) and amputation risks, underscoring the need for therapies targeting oxidative stress, AGE accumulation, and immune dysregulation to break the pathological sequence [18, 20].

Pressure Ulcers: Pathophysiology of Ischemia, Reperfusion Injury, and Tissue Deformation

Pressure ulcers develop through an intricate pathophysiology involving three interconnected mechanisms: tissue deformation, ischemia, and reperfusion injury, which collectively drive cellular damage and tissue necrosis. Tissue deformation, the primary initiating factor, occurs when mechanical loading compresses soft tissues between bony prominences and external surfaces, inducing cellular distortion that directly disrupts membrane integrity, cytoskeletal organization, and osmotic balance [22]. This mechanical strain generates localized stress exceeding tissue tolerance thresholds, particularly in metabolically active muscle cells, which are more susceptible to deformation-induced damage than skin due to lower stress tolerance [23, 24]. Deep pressure sores (DPS) are associated with inadequate soft tissue perfusion and excessive tissue deformation over critical time durations, as well as with ischemia-reperfusion cycles and deficiency of the lymphatic system. Muscle tissue shows the lowest tolerance to pressure injuries, compared with more superficial tissues. In this communication, we present new histopathology data for muscle tissue of albino (Sprague–Dawley) rats exposed to pressures for 15 or 30 min. These data are superimposed with an extensive literature review of all previous histopathology reported for albino rat skeletal muscles subjected to pressure. The pooled data enabled a new mathematical characterization of the pressure-time threshold for cell death in striated muscle of rats, in the form of a sigmoid pressure-time relation, which extends the previous pressure-time relation to the shorter exposure periods. We found that for pressure exposures shorter than 1 h, the magnitude of pressure is the important factor for causing cell death, and the exposure time has little or no effect: even relatively short exposures (15 min – 1 h) to pressures greater than 32 kPa (240 mmHg) cause cell death in rat muscle tissue. For exposures of 2 h or over, again, the magnitude of pressure is the important factor for causing cell death:

pressures greater than 9 kPa (67 mmHg) applied for over 2 h consistently cause muscle cell death. For the intermediate exposures (between 1 and 2 h), the magnitude of cell-death-causing pressure strongly depends on the time of exposure, i.e., critical pressure levels drop from 32 to 9 kPa. The present sigmoidal pressure-time cell death threshold is useful for the design of studies in albino rat models of DPS, and may also be helpful in numerical simulations of DPS development, where there is often a need to extrapolate from tissue pressures to biological damage [24]. A rat model was used to distinguish between the different factors that contribute to muscle tissue damage related to deep pressure ulcers that develop after compressive loading. The separate and combined effects of ischemia and deformation were studied. Loading was applied to the hindlimb of rats for 2 h. Muscle tissue was examined using MR imaging (MRI) and histology. An MR-compatible loading device allowed simultaneous loading and measurement of tissue status. Two separate loading protocols incorporated uniaxial loading, resulting in tissue compression and ischemic loading. Uniaxial loading was applied to the tibialis anterior by means of an indenter, and ischemic loading was accomplished with an inflatable tourniquet. Deformation of the muscle tissue during uniaxial loading was measured using MR tagging. Compression of the tissues for 2 h led to increased T2 values, which were correlated to necrotic regions in the tibialis anterior. Perfusion measurements, by means of contrast-enhanced MRI, indicated a large ischemic region during indentation. Pure ischemic loading for 2 h led to reversible tissue changes. From the MR-tagging experiments, local strain fields were calculated. A 4.5-mm deformation, corresponding to a surface pressure of 150 kPa, resulted in maximum shear strain up to 1.0. There was a good correlation between the location of damage and the location of high shear strain. It was concluded that the large deformations, in conjunction with ischemia, provided the main trigger for irreversible muscle damage [23]. Role of ischemia and deformation in the onset of compression-induced deep tissue injury: MRI-based studies in a rat model [21, 22]. Prolonged compression causes irreversible cell membrane rupture, organelle dysfunction, and necrotic death, even in the absence of ischemia, initiating a "damage cascade" that propagates injury [22]. Concurrently, sustained pressure exceeding arterial capillary pressure (approximately 32 mmHg) impedes blood flow, triggering ischemic injury as tissues are deprived of oxygen and nutrients [25]. Hypoxia forces cells to switch to anaerobic metabolism, leading to ATP depletion, lactic acid accumulation, and intracellular calcium overload, which disrupts ionic homeostasis, activates proteolytic enzymes, and induces cytotoxic edema [26].

Ischemic damage is exacerbated by endothelial dysfunction, which promotes microvascular leakage and inflammation, while accumulated metabolites like histamine and bradykinin further increase vascular permeability, amplifying interstitial edema and mechanical stress on surrounding tissues [27]. Upon pressure relief, reperfusion paradoxically worsens injury through reactive oxygen species (ROS) overproduction, as oxygen re-enters compromised tissues and interacts with damaged mitochondrial electron transport chains, generating superoxide radicals that overwhelm endogenous antioxidants [28]. These ROS oxidize lipids, proteins, and DNA, while activating NF-κB pathways that upregulate pro-inflammatory cytokines (e.g., TNF-α, IL-1β), recruiting neutrophils that release myeloperoxidase and proteases, further degrading extracellular matrix components [29]. Reperfusion also induces endoplasmic reticulum (ER) stress due to calcium dysregulation and protein misfolding, activating the unfolded protein response (UPR) and caspase-mediated apoptosis [30]. Cyclic pressure application and relief create repeated ischemia-reperfusion cycles, each exacerbating oxidative stress, inflammatory signaling, and apoptotic cell death [27]. Additionally, interstitial edema from inflammation increases tissue stiffness, reducing lymphatic drainage and amplifying mechanical deformation during subsequent pressure cycles [23]. The synergistic interplay of these mechanisms-direct mechanical damage, metabolic starvation, and oxidative-inflammatory cascades-culminates in a self-perpetuating cycle of tissue destruction.

Deep tissue injury often precedes visible skin changes, as muscle necrosis progresses toward the skin surface, masked by initially intact epidermis [24]. Critically, the duration and magnitude of pressure, rather than continuous application alone, determine injury severity, with cyclic loading causing more damage due to cumulative reperfusion injury [22]. Advanced imaging and biomechanical models confirm that tissue deformation redistributes stress unevenly, concentrating damage near bony interfaces, while ischemia-reperfusion injury spreads oxidative damage radially, expanding the injury zone [25, 28]. This pathophysiology underscores that pressure ulcers are not merely "ischemic wounds" but result from multifactorial interactions between physical forces, cellular vulnerability, and biochemical pathways, necessitating prevention strategies that mitigate both mechanical strain and reperfusion-associated oxidative stress [26, 29].

Venous Leg Ulcers: Chronic Inflammation and Microangiopathy as Key Pathophysiological Drivers

Venous leg ulcers (VLUs) arise from a complex interplay of hemodynamic and inflammatory mechanisms initiated by venous hypertension, which develops due to venous reflux, obstruction, or calf muscle pump failure [31]. Chronic venous hypertension increases ambulatory venous pressure, triggering microangiopathy characterized by capillary dilation, reduced capillary density, and impaired oxygen diffusion [32]. These microcirculatory changes are exacerbated by altered shear stress on endothelial cells, which disrupts nitric oxide production and upregulates adhesion molecules like ICAM-1, VCAM-1, and E-selectin, promoting leukocyte adhesion and transmigration into perivascular tissues [33]. Persistent venous stasis leads to extravasation of fibrinogen and other macromolecules, forming pericapillary fibrin cuffs that entrap growth factors like TGF-β and VEGF, impairing tissue repair [34].

Concurrently, chronic inflammation is perpetuated by M1 macrophage polarization, which drives the release of pro-inflammatory cytokines (IL-1α, IFN-γ, TNF-α) and matrix metalloproteinases (MMPs), particularly MMP-1, MMP-8, and MMP-9, causing excessive proteolytic degradation of the extracellular matrix [35]. This proteolytic imbalance is compounded by reduced tissue inhibitors of metalloproteinases (TIMPs), leading to collagen fragmentation and failure of granulation tissue maturation [36]. Leukocyte trapping in the microcirculation further amplifies inflammation, as activated neutrophils release reactive oxygen species and elastase, damaging endothelial cells and perpetuating tissue hypoxia [37]. Genetic polymorphisms in genes regulating iron metabolism (HFE), coagulation (Factor XIII), and matrix remodeling (MMP12) exacerbate susceptibility to ulcer formation and delayed healing [31, 38].

Venous hypertension also induces lymphatic dysfunction, reducing clearance of pro-inflammatory mediators and contributing to dermal fibrosis and lipodermatosclerosis [39]. Iron overload from erythrocyte extravasation sustains oxidative stress via Fenton reactions, while dysregulated fibroblast-to-myofibroblast differentiation increases tissue tension, further impairing microvascular perfusion [35]. Collectively, these pathophysiological cascades create a self-perpetuating cycle of inflammation, protease activation, and tissue breakdown, rendering VLUs resistant to healing without targeted compression therapy or interventions addressing underlying venous insufficiency [40].

Arterial Ulcers: The Dominant Role of Ischemic Pathophysiology

Arterial ulcers, characterized by tissue necrosis due to inadequate arterial perfusion, arise primarily from ischemic pathophysiology driven by compromised blood flow to the extremities, most commonly secondary to peripheral arterial disease (PAD) stemming from atherosclerosis [41]. The pathophysiological cascade begins with atherosclerotic plaque accumulation within arterial walls, mediated by the retention of apolipoprotein B-containing lipoproteins in the subendothelial space, which progressively narrows the vascular lumen and reduces distal blood flow [41, 42]. Hemodynamically significant stenosis occurs when arterial diameter decreases by 50%, equivalent to a 75% reduction in cross-sectional area, creating flow-limiting ischemia that disproportionately affects the lower extremities due to their reliance on collateral circulation, a hallmark of chronic limb-threatening ischemia (CLTI) [43]. Chronic hypoperfusion deprives tissues of oxygen and nutrients, triggering cellular metabolic failure characterized by impaired ATP production, ionic pump dysfunction, and intracellular acidosis, as observed in hypoxic wound environments [40, 44]. Resultant tissue hypoxia activates inflammatory pathways, with increased expression of adhesion molecules promoting leukocyte recruitment and platelet aggregation, further obstructing microvascular flow through capillary plugging and endothelial damage [45].

Concurrently, ischemia-reperfusion injury exacerbates oxidative stress through reactive oxygen species (ROS) generation, causing lipid peroxidation and mitochondrial dysfunction, which amplify cellular damage [46]. These disturbances impair all phases of wound healing: hypoxia disrupts fibroblast proliferation and collagen synthesis, while diminished angiogenesis and elevated matrix metalloproteinases (MMPs) destabilize the extracellular matrix, as seen in proteomic analyses of chronic wounds [42, 47]. Transcutaneous oxygen pressure ($TcPO_2$) levels below 30 mm Hg create an environment where oxygen-dependent processes, such as proline hydroxylation in collagen formation, become physiologically unattainable, halting repair mechanisms [47]. The ischemic milieu promotes necrosis over apoptosis, leading to uncontrolled cell death and MMP-driven degradation of wound beds, perpetuated by chronic inflammation [42, 47].

Clinically, this manifests as painful ulcers with "punched-out" morphology, pale bases lacking granulation tissue, and minimal bleeding-features consistent with CLTI diagnostic criteria [48]. Locations at terminal arterial watershed zones (toes, lateral malleoli) reflect inadequate collateral perfusion, as defined in vascular anatomic staging systems [48]. Persistent ischemia creates a self-perpetuating cycle: tissue necrosis increases metabolic demand while microthrombosis and edema further compromise circulation, establishing a microenvironment resistant to healing without revascularization [47]. This pathophysiology underscores the necessity of interventions targeting vascular insufficiency, as underscored by global guidelines emphasizing hemodynamic restoration over topical care alone [48].

Conclusion and Summary

Wound healing is a complex process involving different cells, growth factors, and extracellular matrix components. Generally, it involves four phases: hemostasis, inflammation, proliferation, and remodeling, which repair or regenerate injured tissue. The physiopathology of wound healing outlines the main phases of tissue regeneration, and it is influenced by internal and external variables. However, the cutaneous wound healing framework is fundamental, and several types of wounds exist within this tissue. Advancements in molecular and cellular understanding of the pathophysiology of wound healing have led to major improvements in therapeutic strategies and wound care. Understanding the molecular and cellular processes involved in wound healing is crucial for developing effective treatment modalities. By dissecting the intricate mechanisms of tissue repair, researchers can target specific pathways to enhance healing outcomes for various types of skin wounds. This chapter discussed the pathophysiology of wound healing and the distinct forms of wounds.

References

1. Wounds and injuries. MeSH. NCBI [Internet]. [cited 2025 Jun 17]. Available from: https://www.ncbi.nlm.nih.gov/mesh?Db=mesh&Cmd=DetailsSearch&Term=%22Wounds+and+Injuries%22%5BMeSH+Terms%5D
2. Gurtner GC, Werner S, Barrandon Y, Longaker MT (2008) Wound repair and regeneration. Nature 453(7193):314–321
3. Golchin A, Shaikhnia F, Heidari F, Mahdi D, Hemmati Y, Tayebi L (2024) Cell-derived materials for wound healing. In: Maia FR, Oliveira JM, Reis RL (eds) Handbook of the extracellular matrix [Internet]. Springer International Publishing, Cham, pp 717–738 [cited 2025 Jun 17]. Available from: https://link.springer.com/10.1007/978-3-031-56363-8_36
4. Golchin A (2022) Stem cell technology and skin disorders: from stem cell biology to clinical applications. Stem Cell Rev Rep 18(6):1881–1882
5. Rahnama M, Heidari M, Poursalehi Z, Golchin A (2024) Global trends of exosomes application in clinical trials: a scoping review. Stem Cell Rev Rep 20(8):2165–2193
6. Golchin A, Rekabgardan M, Taheri RA, Nourani MR (2018) Promotion of cell-based therapy: special focus on the cooperation of mesenchymal stem cell therapy and gene therapy for clinical trial studies. Adv Exp Med Biol 1119:103–118
7. Golchin A, Hosseinzadeh S, Ardeshirylajimi A (2018) The exosomes released from different cell types and their effects in wound healing. J Cell Biochem 119(7):5043–5052
8. Schultz GS, Chin GA, Moldawer L, Diegelmann RF (2011) Principles of wound healing. In: Mechanisms of vascular disease: a reference book for vascular specialists [Internet]. University of Adelaide Press. [cited 2025 Jun 17]. Available from: https://www.ncbi.nlm.nih.gov/books/NBK534261/
9. Landén NX, Li D, Ståhle M (2016) Transition from inflammation to proliferation: a critical step during wound healing. Cell Mol Life Sci CMLS 73(20):3861–3885
10. Wallace H, Bhimji S, Toney-Butler TJ, Swearingen S (2019) Wound Healing Phases. In [cited 2025 Jun 17]. Available from: https://www.semanticscholar.org/paper/Wound-Healing-Phases-Wallace-Bhimji/5c283b16286e5fb4887990bef4cdc290dc1c8dd9
11. Rahnama M, Ghasemzadeh N, Ebrahimi Y, Golchin A (2024) A comprehensive evaluation of dermal fibroblast therapy in clinical trials for treating skin disorders and cosmetic applications: a scoping review. Stem Cell Res Ther 15(1):318
12. Wound healing: cellular mechanisms and pathological outcomes. Open Biol [Internet]. [cited 2025 Jun 17]. Available from: https://royalsocietypublishing.org/doi/10.1098/rsob.200223
13. Edwards JL, Vincent AM, Cheng HT, Feldman EL (2008) Diabetic neuropathy: mechanisms to management. Pharmacol Ther 120(1):1–34
14. Das Evcimen N, King GL (2007) The role of protein kinase C activation and the vascular complications of diabetes. Pharmacol Res 55(6):498–510
15. Kizub IV, Klymenko KI, Soloviev AI (2014) Protein kinase C in enhanced vascular tone in diabetes mellitus. Int J Cardiol 174(2):230–242
16. Brownlee M (2005) The pathobiology of diabetic complications: a unifying mechanism. Diabetes 54(6):1615–1625
17. Mota RI, Morgan SE, Bahnson EM (2020) Diabetic vasculopathy: macro and microvascular injury. Curr Pathobiol Rep 8(1):1–14
18. Salvatore T, Galiero R, Caturano A, Vetrano E, Loffredo G, Rinaldi L et al (2022) Coronary microvascular dysfunction in diabetes mellitus: Pathogenetic mechanisms and potential therapeutic options. Biomedicine 10(9):2274
19. Ibrahim I, Nuermaimaiti Y, Maimaituxun G, Luo X, Maimaituxun M, Akbar A et al (2024) Neutrophil extracellular traps (NETs) are associated with type 2 diabetes and diabetic foot ulcer related amputation: a prospective cohort study. Diabetes Ther Res Treat Educ Diabetes Relat Disord 15(6):1333–1348
20. Louiselle AE, Niemiec SM, Zgheib C, Liechty KW (2021) Macrophage polarization and diabetic wound healing. Transl Res J Lab Clin Med 236:109–116

21. Ramirez-Acuña JM, Cardenas-Cadena SA, Marquez-Salas PA, Garza-Veloz I, Perez-Favila A, Cid-Baez MA et al (2019) Diabetic foot ulcers: current advances in antimicrobial therapies and emerging treatments. Antibiot Basel Switz 8(4):193
22. Bhattacharya S, Mishra RK (2015) Pressure ulcers: current understanding and newer modalities of treatment. Indian J Plast Surg 48(1):4–16
23. Stekelenburg A, Strijkers GJ, Parusel H, Bader DL, Nicolay K, Oomens CW (2007) Role of ischemia and deformation in the onset of compression-induced deep tissue injury: MRI-based studies in a rat model. J Appl Physiol 102(5):2002–2011
24. Linder-Ganz E, Engelberg S, Scheinowitz M, Gefen A (2006) Pressure-time cell death threshold for albino rat skeletal muscles as related to pressure sore biomechanics. J Biomech 39(14):2725–2732
25. Bouten CV, Oomens CW, Baaijens FP, Bader DL (2003) The etiology of pressure ulcers: skin deep or muscle bound? Arch Phys Med Rehabil 84(4):616–619
26. Peirce SM, Skalak TC, Rodeheaver GT (2000) Ischemia-reperfusion injury in chronic pressure ulcer formation: a skin model in the rat. Wound Repair Regen 8(1):68–76
27. Jan YK, Lee B, Liao F, Foreman RD (2012) Local cooling reduces skin ischemia under surface pressure in rats: an assessment by wavelet analysis of laser Doppler blood flow oscillations. Physiol Meas 33(10):1733–1745
28. Kerrigan CL, Stotland MA (1993) Ischemia reperfusion injury: a review. Microsurgery 14(3):165–175
29. Yagi H, Tan J, Tuan RS (2013) Polyphenols suppress hydrogen peroxide-induced oxidative stress in human bone-marrow derived mesenchymal stem cells. J Cell Biochem 114(5):1163–1173
30. Inside the neutrophil phagosome: oxidants, myeloperoxidase, and bacterial killing. Blood. American Society of Hematology [Internet]. [cited 2025 Jun 18]. Available from: https://ashpublications.org/blood/article/92/9/3007/261317/Inside-the-Neutrophil-Phagosome-Oxidants
31. SIGN [Internet]. [cited 2025 Jun 18]. Management of chronic venous leg ulcers. Available from: https://testing36.scot.nhs.uk
32. Coleridge Smith PD, Thomas P, Scurr JH, Dormandy JA (1988) Causes of venous ulceration: a new hypothesis. Br Med J Clin Res Ed 296(6638):1726–1727
33. The "trap" hypothesis of venous ulceration. PubMed [Internet]. [cited 2025 Jun 18]. Available from: https://pubmed.ncbi.nlm.nih.gov/7682272/
34. Mulder G, Jones R, Cederholm-Williams S, Cherry G, Ryan T (1993) Fibrin cuff lysis in chronic venous ulcers treated with a hydrocolloid dressing. Int J Dermatol 32(4):304–306
35. Galiano RD, Tepper OM, Pelo CR, Bhatt KA, Callaghan M, Bastidas N et al (2004) Topical vascular endothelial growth factor accelerates diabetic wound healing through increased angiogenesis and by mobilizing and recruiting bone marrow-derived cells. Am J Pathol 164(6):1935–1947
36. Simplifying venous leg ulcer management: consensus recommendations. Wounds International [Internet]. [cited 2025 Jun 18]. Available from: https://woundsinternational.com/consensus-documents/simplifying-venous-leg-ulcer-management-consensus-recommendations/
37. Bradbury AW, Murie JA, Ruckley CV (1993) Role of the leucocyte in the pathogenesis of vascular disease. Br J Surg 80(12):1503–1512
38. Mustoe TA, O'Shaughnessy K, Kloeters O (2006) Chronic wound pathogenesis and current treatment strategies: a unifying hypothesis. Plast Reconstr Surg 117(7 Suppl):35S–41S
39. Vasudevan B (2014) Venous leg ulcers: pathophysiology and classification. Indian Dermatol Online J 5(3):366–370
40. Diagnosis and treatment of venous leg ulcers: S2k Guideline of the German Society of Phlebology and Lymphology (DGPL) e.V. Valesky. 2024. JDDG: Journal der Deutschen Dermatologischen Gesellschaft. Wiley Online Library [Internet]. [cited 2025 Jun 18]. Available from: https://onlinelibrary.wiley.com/doi/full/10.1111/ddg.15415

41. Linton MF, Yancey PG, Davies SS, Jerome WG, Linton EF, Song WL et al (2000) The role of lipids and lipoproteins in atherosclerosis. In: Feingold KR, Ahmed SF, Anawalt B, Blackman MR, Boyce A, Chrousos G, et al (eds) Endotext [Internet]. MDText.com, Inc., South Dartmouth [cited 2025 Jun 18]. Available from: http://www.ncbi.nlm.nih.gov/books/NBK343489/
42. Reyes-Soffer G, Ginsberg HN, Berglund L, Duell PB, Heffron SP, Kamstrup PR et al (2022) Lipoprotein(a): a genetically determined, causal, and prevalent risk factor for atherosclerotic cardiovascular disease: a scientific statement from the American Heart Association. Arterioscler Thromb Vasc Biol 42(1):e48–e60
43. Global vascular guidelines on the management of chronic limb-threatening ischemia. PubMed [Internet]. [cited 2025 Jun 18]. Available from: https://pubmed.ncbi.nlm.nih.gov/31159978/
44. Reducing ischemic kidney injury through application of a synchronization modulation electric field to maintain Na+/K+-ATPase functions. PubMed [Internet]. [cited 2025 Jun 18]. Available from: https://pubmed.ncbi.nlm.nih.gov/35263146/
45. Epidemiology of peripheral artery disease and polyvascular disease. PubMed [Internet]. [cited 2025 Jun 18]. Available from: https://pubmed.ncbi.nlm.nih.gov/34110907/
46. Redox dysregulation in the pathogenesis of chronic venous ulceration. PubMed [Internet]. [cited 2025 Jun 18]. Available from: https://pubmed.ncbi.nlm.nih.gov/31560951/
47. Molecular pathophysiology of chronic wounds: current state and future directions. PubMed [Internet]. [cited 2025 Jun 22]. Available from: https://pubmed.ncbi.nlm.nih.gov/36123031/
48. Conte MS, Bradbury AW, Kolh P, White JV, Dick F, Fitridge R et al (2019) Global vascular guidelines on the management of chronic limb-threatening ischemia. J Vasc Surg 69(6S):3S–125S.e40

[illegible]

Chapter 4
Clinical Assessment of Hard-to-Heal Wounds: Tools, Evidence, and Multidisciplinary Strategies

Francesco De Francesco and Michele Riccio

Abstract Wound assessment tool (BWAT) has been developed to standardize wound evaluation and monitor healing progress. Beyond clinical inspection, advanced diagnostic modalities have been increasingly incorporated into wound assessment protocols. Techniques such as tissue oxygen measurement (transcutaneous oxygen monitoring), near-infrared spectroscopy, and fluorescence imaging allow clinicians to better understand wound pathophysiology and tailor interventions accordingly.

Keywords Wound assessment · Clinical Care · Hard-to-heal wounds · Risk assessment · Wound management · Imaging · Evaluation · Multidisciplinary care · Diagnostic tools · Digital wound care

Abbreviabbtion

AI	Artificial Intelligence
BWAT	Bates-Jensen Wound Assessment Tool
CCVUQ-e	Charing Cross Venous Ulcer Questionnaire (English version)
CLSM	Confocal Laser Scanning Microscopy
CRP	C-Reactive Protein
DNA	Deoxyribonucleic Acid
EHR	Electronic Health Record
FISH	Fluorescence In Situ Hybridization
HPR	Healing Progression Rate
IL-6	Interleukin-6

F. De Francesco (✉) · M. Riccio
Department of Reconstructive Surgery and Hand Surgery, Azienda Ospedaliera Universitaria delle Marche, Ancona, Italy
e-mail: francesco.defrancesco@ospedaliriuniti.marche.it

F. De Francesco, K. Turksen (eds.), *Management and Strategies for Wound Healing*, Stem Cell Biology and Regenerative Medicine 79,
https://doi.org/10.1007/978-3-032-12442-5_4

MMP	Matrix Metalloproteinase
MRSA	Methicillin-Resistant *Staphylococcus aureus*
NERDS	Non-healing, Exudate, Red tissue, Debris, Smell
NGF	Nerve Growth Factor
NGS	Next-Generation Sequencing
NPUAP	National Pressure Ulcer Advisory Panel
NPWT	Negative Pressure Wound Therapy
PUSH	Pressure Ulcer Scale for Healing
QoL	Quality of Life
RESVECH	Resultados Esperados de la Valoración y Evolución de la Cicatrización de las Heridas Crónicas
RNA	Ribonucleic Acid
STONES	Size, Temperature, Os (bone), New areas, Exudate, Smell
TIME	Tissue, Inflammation/Infection, Moisture, Edge
TIMER	Tissue management, Infection control, Moisture balance, Edge of wound, Repair-regeneration
TNF-α	Tumor Necrosis Factor-alpha
WBS	Wound Bed Score
WES	Wound Evaluation Scale

Introduction

Hard-to-heal wounds, often referred to as chronic wounds, represent a significant and growing burden for healthcare systems worldwide [1]. These wounds include venous leg ulcers, pressure ulcers, diabetic foot ulcers, and non-healing surgical wounds. Characterized by their failure to progress through the normal phases of healing in an orderly and timely manner, they are often complicated by persistent inflammation, infection, and impaired tissue regeneration. The definition of a hard-to-heal wound typically encompasses any wound that has failed to heal within a predictable timeframe, despite appropriate and comprehensive care [2]. Chronic wounds are frequently associated with systemic comorbidities such as diabetes mellitus, peripheral arterial disease, obesity, and immunosuppression. These comorbidities not only predispose individuals to wound development but also complicate the healing process, often leading to prolonged disability and increased healthcare resource utilization [3].

The global prevalence of hard-to-heal wounds is on the rise, primarily due to an aging population and the increasing incidence of chronic diseases. In China, a recent multicenter study identified comorbidities such as paralysis and diabetes as key prognostic factors in non-traumatic wounds [4]. In the United States, approximately 6.5 million individuals are estimated to live with chronic wounds, generating direct healthcare costs exceeding $25 billion annually [5]. This trend is mirrored globally. In India, rural populations show a notable burden of chronic wounds driven by limited access to care [6], while in Australia, up to 450,000 individuals are affected, placing a $3 billion strain on the healthcare system [7]. Across Europe, prevalence studies from countries such as Germany [8], Austria [9], Italy [10], Spain [11], Portugal [12], UK [13], Malta [14], and Finland [15] confirm the substantial impact of chronic wounds in both institutional and community settings. Similar trends are evident in emerging regions; for instance, a recent multicenter analysis from South Africa emphasized both the prevalence and prolonged healing trajectories of hard-to-heal wounds in under-resourced environments [16]. Together, these data underscore the global nature of the challenge and highlight the urgent need for scalable, evidence-based wound care strategies. These figures are expected to rise as life expectancy increases and chronic disease prevalence escalates. Beyond direct healthcare costs, hard-to-heal wounds also impose significant indirect costs related to reduced quality of life, loss of productivity, and psychosocial distress among affected individuals. The burden is particularly pronounced in lower-income populations and countries with limited access to advanced wound care technologies.

Understanding the pathophysiology underlying hard-to-heal wounds is essential for developing effective management strategies. A key feature is the persistence of an inflammatory microenvironment within the wound bed. Chronic inflammation is often driven by the presence of microbial biofilms, which form structured microbial communities embedded within a self-produced extracellular matrix. These biofilms are notoriously resistant to antibiotics and host immune responses, thus perpetuating the inflammatory state and impeding healing [17]. Tissue hypoxia is another critical factor. Insufficient oxygen delivery to the wound site impairs cellular activities necessary for tissue repair, including fibroblast proliferation, collagen synthesis, and angiogenesis. Furthermore, systemic factors such as poor glycemic control in diabetic patients exacerbate oxidative stress and microvascular dysfunction, further delaying healing [18].

Effective management of hard-to-heal wounds hinges on comprehensive and accurate assessment. Wound assessment encompasses a detailed evaluation of wound etiology, dimensions, tissue types present, exudate amount and type, signs of infection, and the condition of the surrounding skin [19]. Several validated tools, such as the Pressure Ulcer Scale for Healing (PUSH) and the Bates-Jensen Wound Assessment Tool (BWAT), have been developed to standardize wound evaluation and monitor healing progress. Beyond clinical inspection, advanced diagnostic modalities have been increasingly incorporated into wound assessment protocols. Techniques such as tissue oxygen measurement (transcutaneous oxygen monitoring, TCOM), near-infrared spectroscopy (NIRS), and fluorescence imaging allow clinicians to better understand wound pathophysiology and tailor interventions accordingly. Recently, artificial intelligence (AI) has emerged as a promising adjunct in wound assessment. AI-driven platforms can analyze digital wound images to quantify wound dimensions, detect tissue types, and predict healing trajectories with high accuracy. These technologies hold potential to standardize wound evaluation, reduce interobserver variability, and ultimately improve patient outcomes [20].

This chapter aims to provide a comprehensive overview of the assessment of hard-to-heal wounds. It will discuss the underlying mechanisms that contribute to delayed healing, highlight the key clinical and technological approaches to wound evaluation, and emphasize the importance of a multidisciplinary framework in optimizing wound care outcomes. Throughout, evidence-based practices and recent advancements in wound assessment methodologies will be explored to equip clinicians and healthcare providers with the necessary tools for managing this challenging patient population.

Pathogenesis of Hard-to-Heal Wounds

Hard-to-heal wounds, often synonymous with chronic wounds, represent a significant clinical challenge due to their prolonged healing times and resistance to standard therapeutic interventions. Understanding the underlying pathophysiological mechanisms is crucial for developing effective assessment and management strategies [21].

A hallmark of hard-to-heal wounds is the persistence of inflammation beyond the initial phases of wound healing [22]. This chronic inflammatory state is often exacerbated by the presence of microbial biofilms—structured communities of microorganisms encased within a self-produced extracellular matrix. Biofilms adhere to the wound bed, providing a protective environment for pathogens against host immune responses and antimicrobial agents, thereby perpetuating inflammation and delaying healing. Adequate oxygenation is vital for cellular activities involved in wound repair, including collagen synthesis, fibroblast proliferation, and angiogenesis. In hard-to-heal wounds, tissue hypoxia is a common feature, resulting from factors such as vascular insufficiency or edema. Hypoxic conditions impair the function of immune cells and reduce the efficacy of antimicrobial agents, further complicating the healing process. Cellular senescence, characterized by the cessation of cell division and altered secretory profiles, contributes to the chronicity of wounds. Senescent cells secrete pro-inflammatory cytokines and matrix metalloproteinases (MMPs), enzymes that degrade extracellular matrix components. Elevated MMP activity disrupts the balance between matrix deposition and degradation, leading to a non-healing wound environment. Systemic conditions such as diabetes mellitus, peripheral arterial disease, and immunosuppression significantly influence wound healing [23, 24]. Hyperglycemia, for instance, impairs leukocyte function, reduces collagen synthesis, and promotes the formation of advanced glycation end-products, all of which hinder the repair process. Additionally, comorbidities can alter the inflammatory response and vascular supply, further complicating wound healing.

To systematically assess and manage hard-to-heal wounds, the TIME framework [25–27] was introduced, encompassing four critical components:

- *Tissue*: Assessment of non-viable or deficient tissue that may impede healing.
- *Infection/Inflammation*: Identification and management of infection and inflammation.
- *Moisture*: Maintenance of optimal moisture balance to promote healing.
- *Edge*: Evaluation of the wound edges to ensure progression towards closure.

This structured approach facilitates targeted interventions aimed at addressing specific barriers to healing. Recognizing the need to incorporate regenerative strategies into wound management, De Francesco and Ogawa proposed an extension of the TIME framework, introducing the TIMER model. The additional "R" stands for "Regeneration," emphasizing the importance of activating and supporting tissue regeneration processes to promote complete and lasting wound healing. The TIMER framework thus includes:

- *Tissue*
- *Infection/Inflammation*
- *Moisture*
- *Edge*
- *Regeneration*

This model underscores the integration of regenerative therapies, such as cellular therapies, growth factors, and bioactive biomaterials, into standard wound care practices [28]. Understanding the multifaceted pathogenesis of hard-to-heal wounds informs clinical decision-making and highlights the necessity for a comprehensive, multidisciplinary approach. Incorporating frameworks like TIME and TIMER into clinical practice enables healthcare professionals to systematically identify and address the various factors impeding wound healing, thereby improving patient outcomes.

Clinical Assessment Methodologies

Effective assessment of hard-to-heal wounds is pivotal for guiding treatment strategies, monitoring healing progress, and evaluating therapeutic outcomes. A comprehensive evaluation encompasses visual inspection, measurement techniques, standardized assessment scales, and advanced imaging modalities. This section delves into these methodologies, highlighting their clinical utility and supporting evidence.

Visual inspection remains the cornerstone of wound assessment. Clinicians evaluate parameters such as wound size, depth, color, exudate, tissue type, and the condition of the surrounding skin. The presence of necrotic tissue, granulation tissue, slough, or signs of infection provides insights into the wound's healing stage and potential complications. Accurate documentation of these observations is essential. Standardized descriptors and terminology enhance communication among healthcare providers and facilitate consistent monitoring over time. Tools like the Wound Infection Continuum and the Triangle of Wound Assessment offer structured frameworks for evaluating wound characteristics and guiding clinical decisions.

Wound Measurement Techniques (Fig. 4.1)

Quantifying wound dimensions is critical for assessing healing progress. Traditional methods include manual measurements using rulers to determine length and width, and calculating the area by multiplying these dimensions. However, this approach assumes a rectangular shape and may not accurately represent irregular wound contours. Planimetry, involving tracing the wound perimeter onto a transparent sheet and calculating the area, offers improved accuracy. Digital planimetry systems have further enhanced precision and reproducibility. Studies have demonstrated the reliability of digital planimetry in measuring wound areas, emphasizing its superiority over manual methods [29, 30]. Three-dimensional (3D) imaging technologies have emerged as advanced tools for wound measurement. Devices like the Eykona and Silhouette cameras capture 3D images, enabling accurate assessment of wound area and volume. Research indicates that 3D imaging provides reliable measurements, with high correlation to traditional methods [31]. However, factors such as cost and technical complexity may limit widespread adoption.

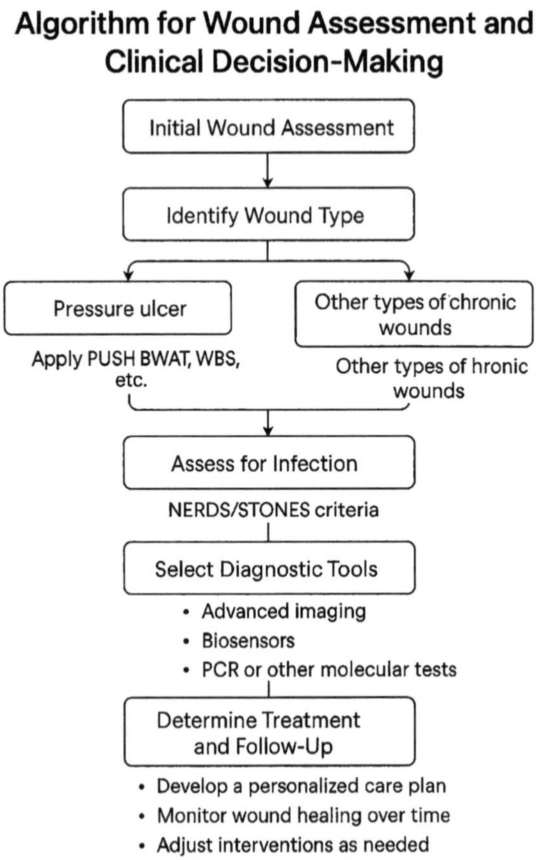

Fig. 4.1 Algorithm for wound assessment and clinical decision-making. This flowchart outlines a structured approach to wound assessment and clinical management. It begins with the initial evaluation of the wound, followed by classification (e.g., pressure ulcers or other chronic wounds), application of validated assessment tools (PUSH, BWAT, WBS), infection screening using NERDS/STONES criteria, and selection of advanced diagnostic technologies such as imaging, biosensors, or molecular assays. The process culminates in the formulation of an individualized treatment plan with ongoing monitoring and therapeutic adjustments

Standardized Assessment Scales

Standardized scales facilitate objective evaluation of wound healing and risk assessment. Several validated tools are utilized in clinical practice:

Pressure Ulcer Scale for Healing (PUSH): Developed by the National Pressure Ulcer Advisory Panel, the PUSH tool assesses pressure ulcer healing by evaluating wound size, exudate amount, and tissue type. Each parameter is scored, and the total score reflects the healing status. The PUSH tool has demonstrated reliability and sensitivity in monitoring pressure ulcer healing [32].

Bates-Jensen Wound Assessment Tool (BWAT): The BWAT evaluates various wound characteristics, including size, depth, edges, undermining, necrotic tissue type and amount, exudate type and amount, skin color surrounding the wound, peripheral tissue edema and induration, granulation tissue, and epithelialization. Each item is scored, with higher scores indicating more severe wounds. The BWAT has been validated for assessing different wound types and is sensitive to changes over time [33].

Wound Evaluation Scale (WES): The WES assesses wound healing based on six clinical variables: wound edge, exudate, surrounding skin color, peripheral edema, granulation tissue, and epithelialization. Each variable is scored, and the total score indicates the healing status. The WES has been utilized in clinical studies to evaluate wound healing outcomes [34].

Healing Progression Rate (HPR) Tool: The HPR tool, a modified version of the PUSH tool, was developed to enhance objectivity in assessing wound healing. It incorporates additional parameters and has demonstrated improved sensitivity in detecting changes in wound status compared to the PUSH tool [35].

Braden Scale: The Braden Scale is a widely utilized tool for predicting the risk of developing pressure injuries. It assesses six domains: sensory perception, moisture, activity, mobility, nutrition, and friction/shear. Each domain is scored, and lower total scores indicate higher risk. The Braden Scale has demonstrated high sensitivity and specificity in various clinical settings, making it a cornerstone of pressure injury prevention protocols [36].

Norton Scale: The Norton Scale is one of the earliest tools developed for pressure injury risk assessment. It evaluates five factors: physical condition, mental state, activity, mobility, and incontinence. Each category is scored from 1 to 4, and a lower cumulative score indicates a higher risk of developing pressure injuries. Although somewhat less detailed than the Braden Scale, the Norton Scale remains in use in certain healthcare settings, particularly in Europe [37].

Waterlow Scale: The Waterlow Scale is another commonly employed risk assessment tool, particularly in the United Kingdom. It incorporates a broader range of factors, including age, sex, body mass index, continence, mobility, appetite, and special risk categories such as neurological deficit and medication use. A higher score indicates a greater risk of pressure injury development. Despite its widespread use, concerns have been raised regarding its predictive accuracy, suggesting that it should be combined with clinical judgment [38].

NPUAP/EPUAP Pressure Injury Classification: In addition to risk assessment scales, classification systems like the National Pressure Ulcer Advisory Panel (NPUAP) and the European Pressure Ulcer Advisory Panel (EPUAP) guidelines are critical for grading the severity of pressure injuries. The NPUAP classification categorizes pressure injuries into four stages based on the depth of tissue damage:

- *Stage 1*: Non-blanchable erythema of intact skin.
- *Stage 2*: Partial-thickness skin loss with exposed dermis.
- *Stage 3*: Full-thickness skin loss.
- *Stage 4*: Full-thickness skin and tissue loss, often involving muscle, tendon, or bone.

Accurate staging is essential for selecting appropriate interventions and for monitoring healing progression [39].

Wound Bed Score (WBS): The Wound Bed Score (WBS), developed by Falanga and colleagues [40], is a validated tool designed to assess the wound bed's condition and predict healing outcomes in chronic wounds. The WBS evaluates eight parameters: healing edges, presence of black eschar, granulation tissue depth, exudate amount, edema, periwound dermatitis, periwound callus or fibrosis, and wound bed color (pale or pink). Each parameter is scored, with the total score ranging from 0 to 16, where higher scores indicate a healthier wound bed and a better prognosis for healing. In a study involving 177 patients with venous ulcers, the WBS demonstrated a significant correlation with complete wound closure. Wounds that achieved full closure had higher WBS scores compared to those that did not heal. Specifically, each unit increase in the WBS was associated with a 22.8% increase in the odds of healing, underscoring the score's predictive validity. The WBS's comprehensive assessment of both the wound bed and periwound skin makes it a valuable tool for clinicians. It facilitates standardized documentation, guides treatment decisions, and allows for monitoring of wound healing over time. The WBS has been utilized in various clinical settings and research studies, highlighting its versatility and applicability in wound care management.

NERDS and STONES: In addition to formal assessment tools that measure wound healing progression or risk of pressure injury development, clinicians must also evaluate for signs of infection—a critical determinant of healing outcomes in chronic wounds [41, 42].

To aid in this process, the NERDS and STONES framework has been developed as a validated, observation-based clinical tool to help distinguish between local and deep/systemic wound infections. The NERDS acronym identifies five signs that are suggestive of superficial colonization or local infection. When three or more signs are present, the wound is likely infected at a superficial level, and topical antimicrobials are indicated: N—Non-healing wound; E—Exudate (increased amount); R—Red, friable granulation tissue; D—Debris (yellow slough or biofilm); S—Smell (foul odor). The STONES acronym is used when a wound presents with signs of deeper infection, requiring systemic antibiotic therapy. The presence of three or more signs suggests that the infection may involve underlying tissues or the bloodstream: S—Size increasing; T—Temperature (warmth of wound/periwound); O–Os (bone exposure or probe to bone); N—New areas of breakdown; E—Exudate (purulent or changing); S—Smell (persistent or worsening odor). This framework is particularly helpful in community and outpatient wound care settings where access to advanced diagnostics may be limited. It supports early recognition and appropriate antimicrobial selection, especially when used in conjunction with wound cultures and clinical judgment (Table 4.1).

Table 4.1 Wound assessment scales

Scale name	Purpose	Key parameters assessed	Context of use	Reference
Pressure Ulcer Scale for Healing (PUSH)	Healing assessment	Wound size, exudate amount, tissue type	Monitoring healing of pressure ulcers	[32]
Bates-Jensen Wound Assessment Tool (BWAT)	Healing assessment	Size, depth, edges, undermining, tissue type, exudate, periwound skin color, edema, granulation tissue, epithelialization	Comprehensive wound assessment across wound types	[33]
Wound Evaluation Scale (WES)	Healing assessment	Wound edge, exudate, surrounding skin color, peripheral edema, granulation tissue, epithelialization	Clinical trials, healing outcomes evaluation	[34]
Healing Progression Rate (HPR) Tool	Healing monitoring	Modified parameters from PUSH with enhanced sensitivity	Monitoring subtle changes in wound healing progression	[35]
Braden Scale	Risk assessment	Sensory perception, moisture, activity, mobility, nutrition, friction/shear	Predicting pressure injury risk in clinical settings	[36]
Norton Scale	Risk assessment	Physical condition, mental state, activity, mobility, incontinence	Early identification of patients at risk for pressure injuries	[37]
Waterlow Scale	Risk assessment	Age, sex, BMI, continence, mobility, appetite, neurological status, medication use	Risk prediction for pressure injuries (especially UK hospitals)	[38]
NPUAP/EPUAP Pressure Injury Classification	Injury severity staging	Depth and extent of tissue damage: 4 stages	Classifying severity of pressure ulcers	[39]
Wound Bed Score (WBS)	Healing potential assessment	Wound edges, black eschar, granulation depth, exudate amount, edema, periwound dermatitis, callus/fibrosis, wound bed color	Prediction of healing outcomes in chronic wounds	[40]
NERDS/STONES	Identify infection (local/ systemic)	NERDS: local signs; STONES: systemic/deep signs	Any setting for infection decision-making	

The diverse array of wound assessment and risk evaluation tools reflects the complexity of chronic wound management. While each scale offers specific advantages, their optimal use depends largely on the clinical context, patient population, and intended outcomes.

Healing Assessment Tools such as the Pressure Ulcer Scale for Healing (PUSH), Bates-Jensen Wound Assessment Tool (BWAT), Wound Evaluation Scale (WES), Healing Progression Rate (HPR) Tool, and Wound Bed Score (WBS) are designed to monitor wound healing over time.

- The *PUSH tool* is ideal for routine monitoring of pressure ulcers in clinical settings where rapid and simple evaluation is needed.
- The *BWAT* provides a more comprehensive and detailed wound description, making it particularly useful in specialized wound care centers and research studies where a thorough assessment of wound status is required.
- The *WES* is commonly used in clinical trials and outcome studies to quantitatively assess wound healing progression, although its limited set of variables may restrict its sensitivity in complex wounds.
- The *HPR tool*, as a modification of the PUSH tool, is advantageous in settings where detecting small but clinically significant changes in healing is important.
- The *WBS* is uniquely positioned to predict healing potential, particularly in chronic venous ulcers, by assessing the wound bed and periwound tissue in a structured manner.

Risk Assessment Scales, including the *Braden*, *Norton*, and *Waterlow* scales, focus primarily on the prevention of pressure injuries by identifying at-risk patients.

- The *Braden Scale* remains the gold standard in many healthcare settings worldwide due to its high predictive validity and ease of use.
- The *Norton Scale*, although simpler, is still employed in long-term care facilities and in some European countries where rapid risk assessment is prioritized.
- The *Waterlow Scale* offers a broader risk profile, accounting for multiple comorbidities and patient characteristics, but requires careful interpretation alongside clinical judgment due to variability in its predictive accuracy.

Classification Systems like the *NPUAP/EPUAP Pressure Injury Staging* provide a standardized framework for categorizing the severity of pressure ulcers. Accurate staging is critical for selecting appropriate treatment strategies and for legal and documentation purposes in clinical practice.

In addition to these tools, the *NERDS and STONES clinical framework* has been developed to assist in the *identification and classification of wound infections.* NERDS identifies superficial infection based on non-healing, exudate, red granulation, debris, and smell—indicating the need for topical antimicrobial therapy when three or more signs are present. STONES, on the other hand, signals deep or systemic infection based on increasing size, warmth, bone involvement, new areas of breakdown, exudate changes, and persistent odor—prompting systemic antibiotic therapy when three or more criteria are met. This framework is particularly valuable in both inpatient and outpatient settings where timely recognition of infection is essential for effective wound management. A growing trend in wound care assessment involves *integrating structured documentation systems* and combining *objective and subjective assessment methods*. For instance, protocols that evaluate *wound history, comorbidities, and risk factors* alongside standardized clinical tools offer a more comprehensive picture of healing dynamics in patients with multimorbidity [43]. The importance of tailoring assessment strategies to the *presence of comorbidities* is increasingly recognized in the literature.

Clinical Contexts for Optimal Use

- *Acute Care Hospitals*: Braden Scale for pressure injury prevention; PUSH or BWAT for monitoring existing pressure ulcers; NERDS/STONES for early infection detection.
- *Wound Care Clinics*: BWAT or WBS for detailed wound assessment; HPR tool for tracking subtle improvements during treatment; NERDS/STONES for infection-related decision-making.
- *Home Health and Community Care*: PUSH tool for simple and rapid monitoring by home healthcare nurses; NERDS/STONES to guide topical vs. systemic antimicrobial use.
- *Research Settings*: WES and BWAT are frequently used for evaluating interventions in clinical trials.
- *Telemedicine and Remote Monitoring*: Tools like PUSH and digital adaptations of BWAT are increasingly employed in telehealth platforms for standardized remote wound assessment; the NERDS/STONES framework can be applied using remote visual inspection and patient-reported symptoms.

The advent of telemedicine and digital wound care platforms has transformed the application of wound assessment tools.

Many scales, including *PUSH* and *BWAT*, have been adapted into *electronic health records (EHRs)* and *mobile applications*, allowing for standardized data collection and improved communication across multidisciplinary teams. Similarly, the NERDS/STONES framework can be effectively used in remote care settings to support clinical decision-making when access to in-person diagnostics is limited (Fig. 4.2).

Digital photography and *3D wound imaging technologies* facilitate remote evaluation of wound size, depth, and tissue characteristics, supporting the objective scoring provided by traditional tools. Moreover, *artificial intelligence (AI)* is beginning to automate wound assessment, using image recognition algorithms to calculate parameters similar to those assessed by the PUSH, BWAT, and WBS tools, potentially increasing accuracy and reducing human error.

In telehealth programs, particularly those managing chronic wounds remotely, the use of standardized scales ensures consistency in documentation, enhances patient engagement by visualizing healing progress, and supports reimbursement

Clinical Context vs. Best-Fit Assessment Tools

CLINICAL CONTEXT	RECOMMENDED TOOLS	PURPOSE
Acute Care / Hospital Inpatient Units	Braden Scale, PUSH Tool	Risk prevention, healing monitoring
Specialized Wound Care Clinics	BWAT, WBS, HPR Tool	Detalled wound assessment and tracking
Long-Term Care / Nursing Homes	Norton Scale, Braden Scale	Rapid risk identification
Clinical Research / Trials	PUSH Tool, WES	Simple wound monitoring
Telemedicine / Remote Monitorinig	WES, BWAT	Quantitative outcome measurement
Rehabilitation Units / Post-Op Wards	PUSH Tool (digital), BWAT (app-based), 3D Imaging	Standardized remote tracking
Venous / Diabetic Ulcer Programs	Wound Bed Score (WBS), BWAT	Prognosis and healing potential

Fig. 4.2 Clinical context vs. best-fit assessment tools. This table summarizes the alignment between specific clinical settings and the most appropriate wound assessment tools. It highlights the practical application of validated scales—such as the PUSH Tool, BWAT, and WBS—based on the context of care (e.g., acute care, telemedicine, long-term care). The goal is to guide clinicians in selecting the most effective instruments for risk prevention, wound monitoring, and outcome measurement

processes by providing objective evidence of clinical outcomes. Overall, the strategic application of these validated tools, combined with emerging technologies, promises to enhance the quality, efficiency, and accessibility of wound care in modern clinical practice.

Wound Photography and 3D Imaging

Photographic documentation is integral to wound assessment, providing visual records for monitoring progression and facilitating communication among healthcare providers. Standardized photography protocols, including consistent lighting, distance, and angles, are essential for accurate comparisons over time. Digital photography has demonstrated high reliability in wound area measurement, with reported accuracy ranging from $R^2 = 0.86$ *to* 0.99, depending on technique standardization [44]. Advancements in imaging technology have introduced 3D wound imaging, offering detailed visualization of wound topography. 3D imaging systems capture depth and volume information, enabling precise measurements and enhancing the assessment of complex wounds. Studies have validated the reliability of 3D imaging in wound assessment, highlighting its potential in clinical practice [30]. Furthermore, integration of 3D imaging with software applications allows for automated analysis and tracking of wound healing. These technologies support objective evaluation and may improve the accuracy of assessments, particularly in research settings. Noninvasive biophysical methods such as *Laser Doppler Flowmetry and high-frequency ultrasonography* offer additional insights into local blood flow and tissue structure. These tools allow clinicians to assess skin perfusion and tissue echogenicity, supporting early identification of ischemia or fibrosis in chronic wounds [45].

Clinical Assessment Methodologies Biomarkes and Advanced Diagnostics in Hard-to-Heal Wounds (Fig. 4.3)

Hard-to-heal wounds present a multifactorial clinical challenge, often characterized by persistent inflammation, microbial colonization, and impaired tissue regeneration. The advent of biomarker analysis and advanced diagnostic tools has significantly enhanced our ability to understand the wound microenvironment, stratify risk, and guide personalized interventions. This section explores key developments in inflammatory biomarkers, emerging molecular diagnostics, and biofilm detection technologies, with emphasis on their relevance to clinical practice [46].

Inflammatory biomarkers offer clinicians a window into the local and systemic status of wound healing. However, a distinction must be made between biomarkers currently employed in everyday clinical practice and those still emerging or under investigation.

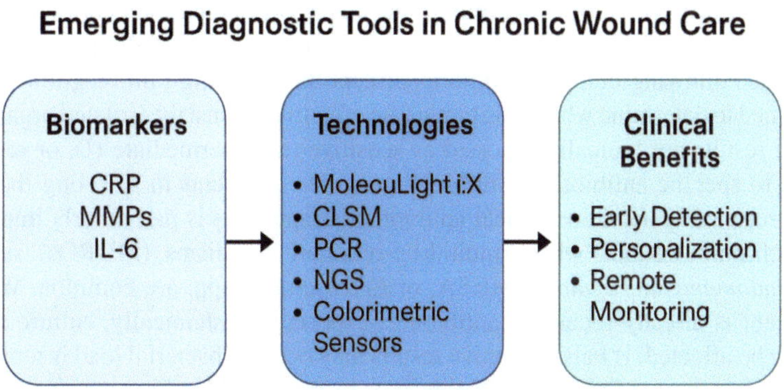

Fig. 4.3 Emerging diagnostic tools in chronic wound care. This flowchart illustrates the integration of emerging diagnostic strategies into chronic wound management. It categorizes current innovations into three domains: biomarkers (e.g., CRP, MMPs, IL-6), advanced diagnostic technologies (e.g., CLSM, MolecuLight i:X, PCR, NGS), and associated clinical benefits, such as early infection detection, personalized treatment planning, and remote monitoring capabilities

Biomarkers Used in Routine Clinical Practice

- *C-Reactive Protein (CRP):* CRP is widely utilized as a systemic marker of inflammation. Elevated CRP levels are associated with wound infection, sepsis, or delayed healing. It is commonly measured in blood tests and serves as a general indicator of inflammatory burden in chronic wound patients [47].
- *Bacterial Swabs for Culture Testing:* Microbiological diagnosis plays a critical role in managing chronic wounds, particularly when infection is suspected or confirmed. Among the available diagnostic tools, wound swab cultures remain the most widely used in clinical practice for identifying microbial colonization and guiding antimicrobial therapy [48]. Several techniques exist for obtaining wound swabs, and the method of collection significantly affects the quality and interpretability of the culture results:

    1. *Levine Technique (recommended)*:
        - This method [49, 50, 51] involves cleaning the wound with sterile saline, then pressing the swab firmly against a 1 cm^2 area of granulating tissue to absorb fluid from deeper layers, while rotating the swab. This technique minimizes surface contamination and increases the likelihood of isolating clinically relevant organisms.
    2. *Zig-Zag or Z-Stroke Technique (less specific)*:
        - In this method [50, 51], the swab is rolled over the wound bed in a zig-zag fashion, covering a large area. It's easier to perform but more prone to contamination from skin flora or non-pathogenic colonizers.
    3. *Curettage or Tissue Biopsy (gold standard, less common)*:
        - When a deeper infection is suspected, tissue biopsy or curettage may be used. These provide a more accurate representation of pathogenic organisms but are invasive and typically reserved for complex cases [52, 53].

    Prior to swab collection, necrotic tissue and debris should be debrided, and the wound gently rinsed with sterile saline to reduce the risk of sampling contaminants. Following culture, antibiotic susceptibility testing (antibiogram) is performed to determine which antibiotics are effective against the isolated organisms. The results are typically reported as sensitive (S), intermediate (I), or resistant (R) to specific antibiotics. Antibiograms guide clinicians in selecting the most appropriate systemic or topical antibiotic therapy. This is particularly important in chronic wounds where multidrug-resistant organisms (MDROs) such as *Pseudomonas aeruginosa*, MRSA, or *Enterococcus* spp. are common. When a patient is already receiving antibiotics, especially systemically, culture results may be affected: i) False-negative results may occur if bacterial load is temporarily reduced by antibiotics; ii) Delayed or incomplete identification of anaerobic or slow-growing pathogens is more likely. In these cases, ideally, the swab should be obtained before the next scheduled dose or before antibiotic initiation, whenever possible. If not feasible, clinicians must interpret culture results in the context of the clinical picture, and may consider serial cultures, wound scoring tools (e.g., NERDS/STONES), and clinical signs of infection. In selected cases, a

repeat culture after stopping antibiotics (48–72 hours) may be warranted, especially in non-healing wounds with persistent signs of infection.
- *Matrix Metalloproteinases (MMPs):* Particularly MMP-9, these enzymes are involved in extracellular matrix degradation. In chronic wounds, overexpression of MMPs can impair tissue regeneration. Some point-of-care test kits and diagnostic dressings now include MMP detection to assess wound chronicity risk [54].

Emerging Biomarkers and Technologies

- *Interleukin-6 (IL-6) and Tumor Necrosis Factor-Alpha (TNF-α):* These cytokines are strongly upregulated in chronic wounds and serve as markers of persistent inflammation. Although promising, their use remains primarily in research due to assay complexity and cost [55].
- *Calprotectin and Procalcitonin:* Both biomarkers are being studied as indicators of infection and inflammation. While used in other medical contexts, their role in wound diagnostics is still under validation [56].
- *Metabolic Indicators (pH and Lactate):* A declining pH and elevated lactate in wound exudate are emerging as markers of infection and delayed healing. Innovative dressings equipped with microelectrodes or colorimetric sensors are now able to monitor these markers continuously [57, 58].

Emerging Molecular Techniques

The integration of molecular biology into wound diagnostics has provided deeper insight into microbial dynamics, host response, and healing potential.

- *Polymerase Chain Reaction (PCR):* PCR enhances detection sensitivity for pathogens, including bacteria that may not grow in standard cultures. It is increasingly used in specialized settings for rapid identification of infective organisms and resistance genes.
- *Next-Generation Sequencing (NGS):* NGS allows comprehensive mapping of the wound microbiome, identifying microbial species and abundance. This technique can differentiate between pathogenic and commensal organisms, enabling precise antimicrobial stewardship [59].
- *Fluorescence* In Situ *Hybridization (FISH):* FISH uses DNA/RNA probes to locate bacterial cells within tissue samples or biofilms. It provides high-resolution visualization, helpful in severe or non-responsive infections [60].
- *Biosensors and Smart Dressings:* These devices detect local biochemical changes such as enzyme activity, temperature shifts, or pH changes. Some experimental biosensors can wirelessly transmit data to clinicians, enabling real-time monitoring and early intervention [61].

While many of these techniques are still in early-phase adoption, they represent a major step toward precision wound diagnostics.

Biofilm Diagnostics

Biofilms, which are structured communities of microorganisms embedded in an extracellular matrix, are present in over 60% of chronic wounds and significantly contribute to poor healing outcomes. Their detection is therefore vital in guiding appropriate wound management.

- *Confocal Laser Scanning Microscopy (CLSM):* CLSM remains the gold standard for in vitro biofilm visualization, offering detailed structural information. However, it is limited by cost and accessibility, making it more suitable for research than routine use [62].
- *Autofluorescence Imaging (*e.g., *MolecuLight i:X):* These handheld imaging tools can detect bacterial presence based on fluorescence signals without requiring contrast agents. Clinically, they are used to identify bacterial hotspots before debridement or sampling [63].
- *Colorimetric Sensors:* Simple, innovative sensors—such as CO_2-sensitive indicators—can detect microbial metabolic activity through a color change. These are being incorporated into advanced wound dressings as early infection alerts [64].
- *Molecular Probes:* Specific DNA, RNA, or peptide-based probes can bind to biofilm structures, enabling real-time detection and quantification via imaging or fluorescence. While still in clinical trials, these probes hold promise for highly selective biofilm diagnostics [65].

The integration of biomarker analysis and advanced diagnostics into wound care offers clear benefits: (i) Earlier infection detection; (ii) More accurate treatment targeting; (iii) Improved healing prediction and patient outcomes. Routine clinical use is currently limited to CRP testing, swab cultures, and some MMP-based diagnostics, while molecular and sensor-based methods are gradually entering specialized settings. The future lies in combining conventional clinical observation (e.g., via PUSH, BWAT) with biochemical and imaging diagnostics, powered by telemedicine and smart wound care platforms. This approach will facilitate more personalized, proactive, and cost-effective management of hard-to-heal wounds.

Multidisciplinary Approach to Wound Care

Effective management of chronic and complex wounds necessitates a multidisciplinary approach, integrating the expertise of various healthcare professionals to address the biological, vascular, biomechanical, and psychosocial dimensions of wound healing. This collaborative strategy not only enhances clinical outcomes but also improves patient adherence, reduces healing times, and minimizes complications [66–71]. The main healthcare professionals who play a key role in the management of hard-to-healing wounds are: (i) Wound Care Nurses: Often serving as care coordinators, nurses perform initial and ongoing wound assessments, deliver hands-on care, apply advanced dressings, monitor for infection, and provide essential education to patients and families. Their role is critical in ensuring continuity and consistency of care; (ii) Internists / General Physicians: Internists help identify and manage systemic factors affecting wound healing, including diabetes, malnutrition, anemia, autoimmune conditions, and chronic infections. Their holistic medical oversight is essential for optimizing the patient's general condition; (iii) Vascular Surgeons: Wounds of arterial or mixed etiology often require evaluation by vascular surgeons. Their role includes non-invasive vascular studies and, when indicated, surgical or endovascular revascularization procedures to restore perfusion and support wound healing; (iv) Plastic Surgeons: Plastic and reconstructive surgeons are involved in wound closure strategies, such as skin grafts, flaps, and negative pressure-assisted wound therapy (NPWT). Their input is often needed for complex or surgically challenging wounds; (v) Dermatologists: Dermatologists contribute expertise in diagnosing inflammatory and neoplastic skin conditions that may mimic or complicate chronic wounds. They help tailor topical therapies and address co-existing dermatoses; (vi) Podiatrists: In diabetic foot ulcer care, podiatrists manage offloading strategies, perform conservative sharp debridement, and implement preventive foot care. They are essential in minimizing the risk of amputation; (vii) Physiotherapists: Physical therapists play a crucial role in mobilization, offloading, prevention of contractures, and enhancement of circulation through therapeutic movement. Their contribution can reduce pressure-related complications and support functional recovery.

A structured interdisciplinary wound care team ensures that all aspects of the wound and the patient are addressed: (i) Improved Communication: Team meetings and shared records foster dialogue between disciplines, aligning goals and enhancing decision-making; (ii) Comprehensive Management: Multidimensional care, including medical, surgical, nutritional, and rehabilitative support, leads to more effective and patient-centered outcomes; (iii) Evidence-Based Benefit: Studies show that interdisciplinary wound care teams reduce healing time, recurrence rates, hospitalization duration, and healthcare costs [72–75].

Implementing and sustaining a multidisciplinary wound care team is a best practice model that has proven to significantly elevate the standard and effectiveness of chronic wound management.

Conclusions and Future Perspectives

Recent advances in technology are reshaping the landscape of wound care, offering transformative possibilities in the assessment, monitoring, and treatment of chronic wounds. Among the most promising innovations are smart wound dressings, which integrate biosensors capable of measuring critical parameters such as moisture, pH, and temperature directly at the wound site. These dressings enable the real-time collection of data, allowing clinicians to detect early signs of infection or imbalance and to tailor interventions more precisely and promptly to the patient's evolving needs.

In parallel, advanced imaging modalities are enhancing diagnostic accuracy and clinical decision-making. Techniques such as thermal imaging and three-dimensional wound mapping offer a detailed visualization of wound dimensions, depth, and tissue viability, enabling clinicians to objectively track healing trajectories and evaluate the effectiveness of therapeutic interventions over time.

Moreover, the integration of artificial intelligence (AI) into wound assessment protocols is gaining significant traction. AI-powered algorithms are now being employed to analyze wound images, automate the classification of wound types and stages, and even predict healing outcomes based on patient-specific clinical and demographic data. These tools not only increase diagnostic consistency but also support personalized treatment planning, helping clinicians to allocate resources more efficiently and to anticipate complications before they arise.

In the realm of treatment, regenerative medicine is opening new frontiers. Growth factor therapies, involving the application of bioengineered molecules that stimulate cell proliferation and tissue repair, have shown encouraging results in accelerating the healing of chronic wounds. Stem cell-based interventions are also being explored, with the potential to regenerate damaged tissue and modulate the inflammatory environment of non-healing wounds. Additionally, bioengineered skin substitutes—whether synthetic or derived from biological sources—are being developed to restore the structural and functional integrity of the skin in cases of extensive tissue loss.

Collectively, these emerging technologies and therapeutic strategies, when integrated into a multidisciplinary model of care, hold the promise of significantly improving the effectiveness of wound management. By combining cutting-edge science with collaborative clinical expertise, the future of wound care is poised to deliver better outcomes, enhanced patient quality of life, and more sustainable healthcare systems.

References

1. Martinengo L, Olsson M, Järbrink K, Bajpai R, Soljak M, Upton Z, Schmidtchen A, Car J (2018) Prevalence of chronic wounds in the general population: systematic review and meta-analysis of observational studies. Ann Epidemiol 28(10):731–738. https://doi.org/10.1016/j.annepidem.2018.10.005
2. Atkin L, Bućko Z, Conde Montero E, Cutting K, Moffatt C, Probst A, Romanelli M, Schultz GS, Tettelbach W (2019) Implementing TIMERS: the race against hard-to-heal wounds. J Wound Care 28. https://doi.org/10.12968/jowc.2019.28.Sup3a.S1
3. Jain M, Bhogar K, Baral P, Gaind R (2024) Evaluation of risk factors associated with hard-to-heal wound infection in a tertiary care hospital. J Wound Care 33(3):180–188. https://doi.org/10.12968/jowc.2024.33.3.180

4. Zhou JQ, Huang LF, Lu YC et al (2022) Prevalence and prognosis of hard-to-heal wounds with comorbidities in China. J Wound Care 31(Sup10):S7–S15. https://doi.org/10.12968/jowc.2022.31.Sup10.S7
5. Nussbaum SR, Carter MJ, Fife CE et al (2023) Human wound and its burden: updated 2022 compendium of estimates. Adv Wound Care 12(3):123–132. https://doi.org/10.1089/wound.2023.0150
6. Sharma R, Singh S, Kumar A et al (2024) Epidemiological profile of chronic wounds in an Indian population: a community-based cross-sectional observational study. Cureus 16(9):e68684. https://doi.org/10.7759/cureus.68684
7. Chandan K, Prentice J et al (2019) Chronic wounds in Australia: a systematic review of key epidemiological and clinical parameters. Int Wound J 16(1):84–94. https://doi.org/10.1111/iwj.12996
8. Raeder K, Jachan DE, Müller-Werdan U, Lahmann NA (2020) Prevalence and risk factors of chronic wounds in nursing homes in Germany: A Cross-Sectional Study. Int Wound J 17(5):1128–1134. https://doi.org/10.1111/iwj.13486. Epub 2020 Aug 19. PMID: 32815303; PMCID: PMC7949346
9. Morasek K, Himmelsbach R, Schneider C, Ritschl V, Wieser S, Pötscher A, Stamm T (2024) Prevalence and incidence of chronic wounds in Austria - a population-based real-world data analysis. Eur J Public Health 34(Supplement_3). https://doi.org/10.1093/eurpub/ckae144.1498
10. Monari P, Gualdi G, Rovaris S, Venturuzzo A, Rovati C, Pinton PC (2022) Etiology of chronic leg ulcers in a dermatologic wound clinic: a retrospective observational study. Wound Manag Prev 68(8):12–15. PMID: 35926144
11. Díaz-Herrera MÁ, Martínez-Riera JR, Verdú-Soriano J, Capillas-Pérez RM, Pont-García C, Tenllado-Pérez S, Cunillera-Puértolas O, Berenguer-Pérez M, Gea-Caballero V (2021) Multicentre study of chronic wounds point prevalence in primary health care in the southern metropolitan area of Barcelona. J Clin Med 10(4):797. https://doi.org/10.3390/jcm10040797. PMID: 33669397; PMCID: PMC7920417
12. Furtado KAX, Infante P, Sobral A, Gaspar P, Eliseu G, Lopes M (2020) Prevalence of acute and chronic wounds - with emphasis on pressure ulcers - in integrated continuing care units in Alentejo, Portugal. Int Wound J 17(4):1002–1010. https://doi.org/10.1111/iwj.13364. Epub 2020 Apr 13. PMID: 32285622; PMCID: PMC7949279
13. Guest JF, Fuller GW, Vowden P (2020) Cohort study evaluating the burden of wounds to the UK's National Health Service in 2017/2018: update from 2012/2013. BMJ Open 10(12):e045253. https://doi.org/10.1136/bmjopen-2020-045253. PMID: 33371051; PMCID: PMC7757484
14. Gabriele A, Holloway S, Ward CS (2024) Wounds in a correctional facility in Malta: a point prevalence study. J Wound Care 33(12):934–940. https://doi.org/10.12968/jowc.2023.0220
15. Ahmajärvi KM, Isoherranen KM, Mäkelä A, Venermo M (2019) A change in the prevalence and the etiological factors of chronic wounds in Helsinki metropolitan area during 2008–2016. Int Wound J 16(2):522–526. https://doi.org/10.1111/iwj.13077. Epub 2019 Jan 22. PMID: 30672095; PMCID: PMC7948615
16. Idensohn PJ, Gilbert K, Boodhoo K, Howard K, van de Vyver M (2025) The prevalence, aetiology and healing trajectories of hard-to-heal wounds in South Africa. Int Wound J 22(3):e70155. https://doi.org/10.1111/iwj.70155
17. Goswami AG, Basu S, Banerjee T, Shukla VK (2023) Biofilm and wound healing: from bench to bedside. Eur J Med Res 28(1):1–10. https://doi.org/10.1186/s40001-023-01121-7
18. Hong WX, Hu MS, Esquivel M, Liang GY, Rennert RC, McArdle A, Paik KJ, Duscher D, Gurtner GC, Lorenz HP, Longaker MT (2014) The role of hypoxia-inducible factor in wound healing. Adv Wound Care (New Rochelle) 3(5):390–399. https://doi.org/10.1089/wound.2013.0520
19. Nagle SM, Stevens KA, Wilbraham SC (2025) Wound Assessment. [Updated 2023 Jun 26]. In: StatPearls [Internet]. Treasure Island: StatPearls Publishing. Available from: https://www.ncbi.nlm.nih.gov/books/NBK482198/

20. Rippon MG, Fleming L, Chen T et al (2024) Artificial intelligence in wound care: diagnosis, assessment and treatment of hard-to-heal wounds: a narrative review. J Wound Care 33(4):229–242. https://doi.org/10.12968/jowc.2024.33.4.229
21. Frykberg RG, Banks J (2015) Challenges in the treatment of chronic wounds. Adv Wound Care (New Rochelle) 4(9):560–582. https://doi.org/10.1089/wound.2015.0635. PMID: 26339534; PMCID: PMC4528992
22. Wallace HA, Basehore BM, Zito PM (2025) Wound healing phases. [Updated 2023 Jun 12]. In: StatPearls [Internet]. Treasure Island: StatPearls Publishing. Available from: https://www.ncbi.nlm.nih.gov/books/NBK470443/
23. Guo S, Dipietro LA (2010) Factors affecting wound healing. J Dent Res 89(3):219–229. https://doi.org/10.1177/0022034509359125. Epub 2010 Feb 5. PMID: 20139336; PMCID: PMC2903966
24. Thomas Hess C (2011) Checklist for factors affecting wound healing. Adv Skin Wound Care 24(4):192. https://doi.org/10.1097/01.ASW.0000396300.04173.ec
25. Schultz GS, Barillo DJ, Mozingo DW, Chin GA, Wound Bed Advisory Board Members (2004) Wound bed preparation and a brief history of TIME. Int Wound J 1(1):19–32. https://doi.org/10.1111/j.1742-481x.2004.00008.x. PMID: 16722894; PMCID: PMC7951422
26. Fletcher J (2007) Wound assessment and the TIME framework. Br J Nurs 16(8):462–464, 446. https://doi.org/10.12968/bjon.2007.16.8.23415. PMID: 17551428
27. Leaper DJ, Schultz G, Carville K, Fletcher J, Swanson T, Drake R (2012) Extending the TIME concept: what have we learned in the past 10 years? Int Wound J 9 Suppl 2(Suppl 2):1–19. https://doi.org/10.1111/j.1742-481X.2012.01097.x. PMID: 23145905; PMCID: PMC7950760
28. De Francesco F, Ogawa R (2024) From time to Timer in wound healing through the regeneration. Adv Exp Med Biol 1470:1–18. https://doi.org/10.1007/5584_2024_815
29. Williams KJ, Sounderajah V, Dharmarajah B, Thapar A, Davies AH (2017) Simulated wound assessment using digital planimetry versus 3D cameras. Ann Vasc Surg 40:277–284. https://doi.org/10.1016/j.avsg.2016.10.029
30. Savage JM, Jeffery SL (2013) Use of 3D photography in complex-wound assessment. J Wound Care 22(3):156–160. https://doi.org/10.12968/jowc.2013.22.3.156
31. Gee Kee EL, Kimble RM, Stockton KA (2015) 3D photography is a reliable burn wound area assessment tool compared to digital planimetry in very young children. Burns 41(6):1286–1290. https://doi.org/10.1016/j.burns.2015.01.020
32. Günes UY (2009) A prospective study evaluating the Pressure Ulcer Scale for Healing (PUSH Tool) to assess stage II, stage III, and stage IV pressure ulcers. Ostomy Wound Manage 55(5):48–52. PMID: 19471048
33. Gould LJ, Serena TE, Sinha S (2021) Development of the BWAT-CUA scale to assess wounds in patients with calciphylaxis. Diagnostics 11(4):730. https://doi.org/10.3390/diagnostics11040730
34. Quinn JV, Wells GA (1998) An assessment of clinical wound evaluation scales. Acad Emerg Med 5(6):583–586. https://doi.org/10.1111/j.1553-2712.1998.tb02465.x
35. Young DL, Estocado N, Feng D, Black J (2017) The development and preliminary validity testing of the healing progression rate tool. Ostomy Wound Manage 63(8):32–44. PMID: 28933702
36. Bergstrom N, Braden BJ, Laguzza A, Holman V (1987) The Braden scale for predicting pressure sore risk. Nurs Res 36(4):205–210. PMID: 3299278
37. Defloor T, De Bacquer D, Grypdonck MH (2005) The effect of various combinations of turning and pressure reducing devices on the incidence of pressure ulcers. Int J Nurs Stud 42(1):37–46. https://doi.org/10.1016/j.ijnurstu.2004.05.001
38. Waterlow J (1985) Pressure sores: a risk assessment card. Nurs Times 81(48):49–55. PMID: 3851917
39. National Pressure Ulcer Advisory Panel (2016) European pressure ulcer advisory panel, and Pan Pacific pressure injury alliance. Prevention and Treatment of Pressure Ulcers: Clinical Practice Guideline. Cambridge Media

40. Falanga V, Saap LJ, Ozonoff A (2006) Wound bed score and its correlation with healing of chronic wounds. Dermatol Ther 19(6):383–390. https://doi.org/10.1111/j.1529-8019.2006.00054.x
41. Sibbald RG, Woo K, Ayello EA (2006) Increased bacterial burden and infection: the story of NERDS and STONES. Adv Skin Wound Care 19(8):447–461. wounds-uk.com+5woundsasia.com+5PubMed+5
42. Woo KY, Alam T, Marin J (2009) A cross-sectional validation study of using NERDS and STONEES to assess bacterial burden. Ostomy Wound Manage 55(8):40–48. snfwoundcare.com+4clwk.ca+4HMP Global Learning Network+4
43. Bolton L, McNees P, van Rijswijk L, de Leon J, Lyder C, Kobza L, Edman K,Scheurich A, Shannon R, Toth M (2004) Wound outcomes study group. Wound-healing outcomes using standardizedassessment and care in clinical practice. J Wound Ostomy Continence Nurs 31(2):65–71. https://doi.org/10.1097/00152192-200403000-00005. PMID: 15209428
44. Miot HA, Fernandes LP, Jorge EN, Pivotto DR, Nogueira CR, Mazeto GM (2009) Comparative evaluation of oculometric variables in Graves' ophthalmopathy. Clinics (Sao Paulo) 64(9):885–889. https://doi.org/10.1590/S1807-59322009000900009. PMID: 19759882; PMCID: PMC2745144
45. Timar-Banu O, Beauregard H, Tousignant J, Lassonde M, Harris P, Viau G, Vachon L, Levy E, Abribat T (2001) Development of noninvasive and quantitative methodologies for the assessment of chronic ulcers and scars in humans. Wound Repair Regen 9(2):123–132. https://doi.org/10.1046/j.1524-475x.2001.00123.x. PMID: 11350650
46. Weigelt MA, Lev-Tov HA, Tomic-Canic M, Lee WD, Williams R, Strasfeld D, Kirsner RS, Herman IM (2022) Advanced wound diagnostics: toward transforming wound care into precision medicine. Adv Wound Care (New Rochelle) 11(6):330–359. https://doi.org/10.1089/wound.2020.1319. Epub 2021 Jul 21. PMID: 34128387; PMCID: PMC8982127
47. Wolcott RD, Rhoads DD (2010) A study of biofilm-based wound management in subjects with critical limb ischemia. J Wound Care 19(7):285–292. https://doi.org/10.12968/jowc.2010.19.7.77785
48. Schultz GS, Sibbald RG, Falanga V et al (2003) Wound bed preparation: a systematic approach to wound management. Wound Repair Regen 11(Suppl 1):S1–S28. https://doi.org/10.1046/j.1524-475X.11.s2.1.x
49. Levine N, Robert B, Lindberg R, Mason A, Basil A, Pruitt B, Colonel MC (1976) The quantitative swab culture and smear: a quick, simple method for determining the number of viable aerobic bacteria on open wounds. J Trauma 16(2):89–94
50. Angel DE, Lloyd P, Carville K, Santamaria N (2011) The clinical efficacy of two semi-quantitative wound-swabbing techniques in identifying the causative organism(s) in infected cutaneous wounds. Int Wound J 8(2):176–185. https://doi.org/10.1111/j.1742-481X.2010.00765.x. Epub 2011 Feb 9. PMID: 21303456; PMCID: PMC7950681
51. Gardner S, Frantz R, Saltzman C, Hillis S, Park H, Scherubel M (2006) Diagnostic validity of three swab techniques for identifying chronic wound infection. Wound Repair Regen 14:548–557
52. Smith ME, Robinowitz N, Chaulk P, Johnson K. Comparison of chronic wound culture techniques: swab versus curetted tissue for microbial recovery Br J Community Nurs 2014;Suppl(90):S22-S26. doi: https://doi.org/10.12968/bjcn.2014.19.Sup9.S22. PMID: 25192558; PMCID: PMC4267254
53. Huang Y, Cao Y, Zou M, Luo X, Jiang Y, Xue Y, Gao F (2016) A comparison of tissue versus swab culturing of infected diabetic foot wounds. Int J Endocrinol 2016:8198714. https://doi.org/10.1155/2016/8198714. Epub 2016 Mar 30. PMID: 27123004; PMCID: PMC4829715
54. Towler MA et al (2024) Comparative analysis of electrochemical and optical sensors for chronic wound biomarkers. Biosens Bioelectron 200:114067. https://doi.org/10.1016/j.bios.2022.114067
55. Mahmoud NN, Hamad K, Al Shibitini A, Juma S, Sharifi S, Gould L, Mahmoudi M. Investigating inflammatory markers in wound healing: understanding implications and

identifying artifacts. ACS Pharmacol Transl Sci 2024;7(1):18–27. https://doi.org/10.1021/acsptsci.3c00336. PMID: 38230290; PMCID: PMC10789122

56. Ingram JR, Cawley S, Coulman E, Gregory C, Thomas-Jones E, Pickles T, Cannings-John R, Francis NA, Harding K, Hood K, Piguet V (2018) Levels of wound calprotectin and other inflammatory biomarkers aid in deciding which patients with a diabetic foot ulcer need antibiotic therapy (INDUCE study). Diabet Med 35(2):255–261. https://doi.org/10.1111/dme.13431. Epub 2017 Aug 15. PMID: 28734103; PMCID: PMC5811820
57. Darvishi M et al (2021) Advances in the sensing and treatment of wound biofilms. Angew Chem Int Ed Engl 60(9):4242–4251. https://doi.org/10.1002/anie.202112218
58. Haller HL, Sander F, Popp D, Rapp M, Hartmann B, Demircan M, Nischwitz SP, Kamolz LP (2021) Oxygen, pH, lactate, and metabolism-how old knowledge and new insights might be combined for new wound treatment. Medicina (Kaunas) 57(11):1190. https://doi.org/10.3390/medicina57111190. PMID: 34833408; PMCID: PMC8617754
59. Wang Z, Feng C, Chang G, Liu H, Zhang W (2025) Enhancing early diagnosis and monitoring of wound infections caused by multiple bacteria in tissues through digital PCR integration with cutaneous infection biomarkers. BMC Infect Dis 25(1):372. https://doi.org/10.1186/s12879-025-10761-5. PMID: 40098100; PMCID: PMC11917136
60. Barbosa A, Miranda S, Azevedo NF, Cerqueira L, Azevedo AS (2023) Imaging biofilms using fluorescence *in situ* hybridization: seeing is believing. Front Cell Infect Microbiol 13:1195803. https://doi.org/10.3389/fcimb.2023.1195803. PMID: 37284501; PMCID: PMC10239779
61. Vo DK, Trinh KTL (2025) Advances in wearable biosensors for wound healing and infection monitoring. Biosensors (Basel) 15(3):139. https://doi.org/10.3390/bios15030139. PMID: 40136936; PMCID: PMC11940385
62. Lawrence JR, Neu TR (1999) Confocal laser scanning microscopy for analysis of microbial biofilms. Methods Enzymol 310:131–144. https://doi.org/10.1016/s0076-6879(99)10011-9. PMID: 10547787
63. Le L, Baer M, Briggs P, Bullock N, Cole W, DiMarco D, Hamil R, Harrell K, Kasper M, Li W, Patel K, Sabo M, Thibodeaux K, Serena TE (2021) Diagnostic accuracy of point-of-care fluorescence imaging for the detection of bacterial burden in wounds: results from the 350-patient fluorescence imaging assessment and guidance trial. Adv Wound Care (New Rochelle) 10(3):123–136. https://doi.org/10.1089/wound.2020.1272. Epub 2020 Sep 25. PMID: 32870774; PMCID: PMC7876364
64. Yusufu D, Magee E, Gilmore B, Mills A (2022) Non-invasive, 3D printed, colourimetric, early wound-infection indicator. Chem Commun (Camb) 58(3):439–442. https://doi.org/10.1039/d1cc06147j. PMID: 34901973
65. Zhao M, Li B, Zhang H, Zhang F (2020) Activatable fluorescence sensors for *in vivo* bio-detection in the second near-infrared window. Chem Sci 12(10):3448–3459. https://doi.org/10.1039/d0sc04789a. PMID: 34163618; PMCID: PMC8179418
66. Geese F, Schmitt KU (2023) Interprofessional collaboration in complex patient care transition: a qualitative multi-perspective analysis. Healthcare (Basel) 11(3):359. https://doi.org/10.3390/healthcare11030359. PMID: 36766934; PMCID: PMC9914692
67. Ennis WJ (2012) Wound care specialization: the current status and future plans to move wound care into the medical community. Adv Wound Care (New Rochelle) 1(5):184–188. https://doi.org/10.1089/wound.2011.0346. PMID: 24527303; PMCID: PMC3839023
68. Long CD, Granick MS (1998) A multidisciplinary approach to wound care in the hospitalized patient. Clin Plast Surg 25(3):425–431. PMID: 9696902
69. Flores AM, Mell MW, Dalman RL, Chandra V (2019) Benefit of multidisciplinary wound care center on the volume and outcomes of a vascular surgery practice. J Vasc Surg 70(5):1612–1619. https://doi.org/10.1016/j.jvs.2019.01.087. Epub 2019 May 29. PMID: 31153696
70. Rivolo M, Staines K (2021) Cost effectiveness of a specialist wound care service. J Wound Care 30(9):685–692. https://doi.org/10.12968/jowc.2021.30.9.685. PMID: 34554836

71. Vu T, Harris A, Duncan G, Sussman G (2007) Cost-effectiveness of multidisciplinary wound care in nursing homes: a pseudo-randomized pragmatic cluster trial. Fam Pract 24(4):372–379. https://doi.org/10.1093/fampra/cmm024. Epub 2007 Jun 29. PMID: 17602174
72. Gu YH, Wang X, Sun SS (2022) Benefits of multidisciplinary collaborative care team-based nursing services in treating pressure injury wounds in cerebral infarction patients. World J Clin Cases 10(1):43–50. https://doi.org/10.12998/wjcc.v10.i1.43. PMID: 35071504; PMCID: PMC8727259
73. Lindholm C, Searle R (2016) Wound management for the 21st century: combining effectiveness and efficiency. Int Wound J 13 Suppl 2(Suppl 2):5–15. https://doi.org/10.1111/iwj.12623. PMID: 27460943; PMCID: PMC7949725
74. Edwards H, Finlayson K, Courtney M, Graves N, Gibb M, Parker C (2013) Health service pathways for patients with chronic leg ulcers: identifying effective pathways for facilitation of evidence based wound care. BMC Health Serv Res 13:86. https://doi.org/10.1186/1472-6963-13-86. PMID: 23510505; PMCID: PMC3599619
75. Moore Z, Butcher G, Corbett LQ, McGuiness W, Snyder RJ, van Acker K, Weller C (2014) AAWC, AWMA, EWMA position paper: managing wounds as a team. J Wound Care 23(5 Suppl):S1–S38

Chapter 5
Cellular and Molecular Mechanisms in Wound Healing

Md Sariful Islam Howlader and Hiranmoy Das

Abstract Wound healing is a highly dynamic and regulated biological process that is essential for restoring tissue integrity and function following an injury. It progresses through four overlapping phases: hemostasis, inflammation, proliferation, and remodeling, all coordinated by complex cellular and molecular interactions. Key cellular players such as keratinocytes, fibroblasts, endothelial cells, and immune cells communicate through cytokines, chemokines, and growth factors, including platelet-derived growth factor (PDGF), vascular endothelial growth factor (VEGF), and transforming growth factor-beta (TGF-β), to facilitate tissue repair. While acute wounds follow a predictable healing trajectory, chronic wounds often stall due to ongoing inflammation, impaired angiogenesis, and extracellular matrix (ECM) dysfunction, frequently exacerbated by diabetes, vascular diseases, and infections. This chapter explores the cellular and molecular mechanisms of wound healing, focusing on key signaling pathways, immune regulation, and therapeutic interventions. Emerging strategies, including stem cell therapy, bioengineered scaffolds, and gene editing, offer promising solutions for accelerating wound repair. Advances in nanotechnology and smart biomaterials enhance drug delivery and wound monitoring, while artificial intelligence (AI) enables personalized treatment approaches. By integrating regenerative medicine, bioengineering, and precision therapeutics, innovative wound healing strategies can revolutionize chronic wound management, ultimately improving clinical outcomes and patient quality of life.

Keywords Inflammation · Proliferation · Remodeling · Wound repair · Cytokines · Immune cells

---

M. S. I. Howlader · H. Das (✉)
Department of Pharmaceutical Sciences, Jerry H. Hodge School of Pharmacy, Texas Tech University Health Sciences Center, Amarillo, TX, USA
e-mail: hiranmoy.das@ttuhsc.edu

F. De Francesco, K. Turksen (eds.), *Management and Strategies for Wound Healing*, Stem Cell Biology and Regenerative Medicine 79,
https://doi.org/10.1007/978-3-032-12442-5_5

Introduction

Wound healing is a highly regulated biological process for restoring tissue's structural and functional integrity following injury and damage. It involves a sequence of overlapping and dynamic phases, including hemostasis, inflammation, proliferation, and remodeling, where each phase is characterized by specific cellular activities and molecular signaling pathways. These mechanisms coordinate the repair of tissue damage while maintaining homeostasis [1]. While acute wounds typically follow a predictable healing trajectory, chronic wounds are marked by dysregulated processes that hinder repair, posing significant challenges in clinical management.

Acute wounds progress through the phases of healing in an orderly manner. Hemostasis initiates repair by forming a fibrin clot and releasing growth factors, followed by the inflammatory phase, which clears pathogens and debris. The proliferative phase restores tissue structure through angiogenesis, fibroblast activation, and re-epithelialization. Finally, the remodeling phase strengthens and reorganizes the extracellular matrix (ECM), leading to scar formation [2, 3]. However, chronic wounds fail to complete these phases efficiently due to prolonged inflammation, impaired angiogenesis, or dysfunction of fibroblasts. Diabetes, vascular insufficiency, and infection exacerbate this pathology, preventing wound closure and tissue regeneration [4].

Cellular and molecular interactions underpin wound healing, with key players including keratinocytes, fibroblasts, [5] endothelial cells, and immune cells. These cells communicate via signaling molecules such as cytokines, chemokines, and growth factors, including vascular endothelial growth factor (VEGF), platelet-derived growth factor (PDGF), and transforming growth factor-beta (TGF-β) [5]. Dysregulation of these signals disrupts the balance between inflammation and tissue repair, a hallmark of chronic wound states [6].

Emerging therapeutic strategies aim to address the challenges of chronic wound healing. Advances in gene therapy, stem cell applications, and bioengineered delivery systems offer promising avenues to restore healing pathways. For example, stem cells provide regenerative potential through differentiation, paracrine signaling, and immunomodulation, making them viable options for treating non-healing wounds [7]. Furthermore, bioengineered scaffolds and controlled-release systems enhance the stability and efficacy of growth factor and cytokine therapies, addressing issues related to delivery and degradation in the wound microenvironment [7].

This chapter comprehensively overviews the cellular and molecular mechanisms that regulate acute and chronic wound healing. Exploring key signaling pathways, cellular functions, and therapeutic innovations highlights the intricacies of tissue repair and the potential of emerging technologies to improve outcomes in wound management.

Molecular Mechanisms

Wound healing is a complex and highly regulated process that restores tissue integrity and function following injury. It involves a tightly coordinated sequence of cellular and molecular events, ensuring that the damaged tissue is repaired effectively. The repair process is guided by intricate signaling networks involving growth factors, cytokines, chemokines, adhesion molecules, and extracellular matrix (ECM) components [8]. These molecular players facilitate crucial cellular activities such as cell migration, proliferation, ECM remodeling, and immune system modulation, all of which are essential for tissue restoration. Any disruptions or imbalances in these mechanisms can result in chronic wounds, excessive fibrosis, or impaired tissue regeneration, leading to poor clinical outcomes [9].

Wound healing occurs through four interconnected and overlapping phases: hemostasis, inflammation, proliferation, and remodeling (maturation) [2]. Each phase is driven by distinct molecular pathways that coordinate the activities of various cell types, including platelets, neutrophils, macrophages, fibroblasts, keratinocytes, and endothelial cells [10]. The transition from one phase to the next is dependent on the precise balance of pro-inflammatory and anti-inflammatory signals, growth factor release, and ECM remodeling [11]. Failure to regulate these interactions properly can contribute to pathological conditions such as chronic non-healing wounds or hypertrophic scarring [12].

This chapter explores the molecular mechanisms that govern wound healing, with a focus on the roles of growth factors, cytokines, immune regulation, and ECM components in each phase of tissue repair. Understanding these mechanisms is crucial for developing therapeutic interventions aimed at improving wound healing, particularly in cases of diabetic ulcers, pressure sores, and other chronic wounds that fail to resolve through natural processes [13] (Fig. 5.1).

Hemostasis

Hemostasis is the immediate response to vascular injury, to prevent further blood loss and initiate tissue repair. Upon injury, vasoconstriction occurs, reducing blood flow to the wound site. Platelets play a pivotal role during this phase by adhering to exposed collagen in the damaged vasculature, forming a temporary plug. Upon activation, platelets release a plethora of soluble mediators, including PDGF, TGF-β, TGF-α, basic fibroblast growth factor (bFGF), insulin-like growth factor-1 (IGF-1), and VEGF. These growth factors recruit and activate other cells, such as neutrophils and macrophages, to initiate downstream phases of healing [2].

Clot formation follows platelet activation and involves the coagulation cascade, culminating in the conversion of fibrinogen to fibrin. The resultant fibrin clot prevents further bleeding and acts as a provisional ECM scaffold, providing structural

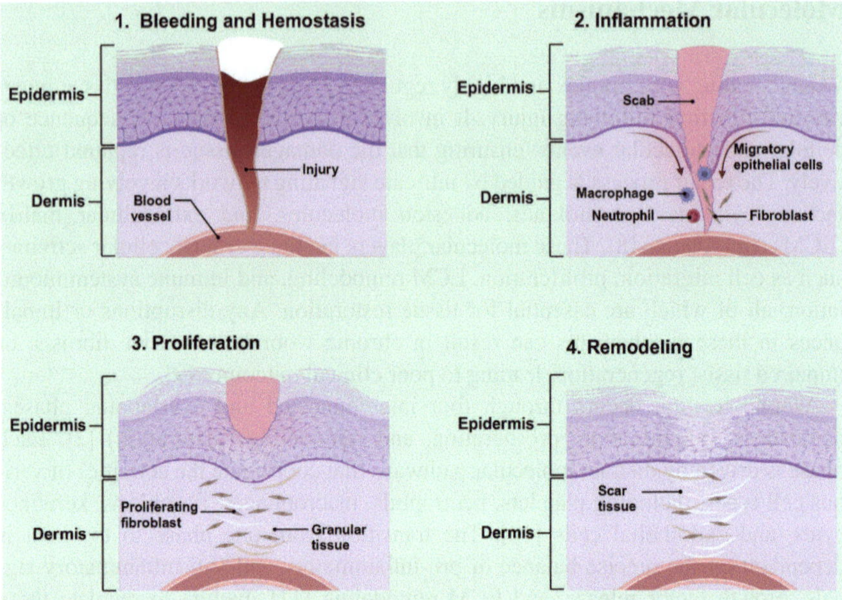

Fig. 5.1 Wound healing stages: (1) Bleeding *and Hemost*asis: This stage involves the formation of a fibrin clot to halt bleeding and create a temporary matrix for cell migration. Platelets release growth factors to recruit immune cells to the site of injury. (2) *Inflammation*: Immune cells are recruited to clear pathogens and cellular debris. Pro-inflammatory cytokines are released to trigger the activation of signaling pathways. (3) *Prolif*eration: Fibroblasts, endothelial cells, and keratinocytes proliferate to form new tissue. Induce angiogenesis to supply oxygen and nutrients; fibroblasts deposit extracellular matrix proteins to rebuild the wound site. (4) *Remodeling*: Extracellular matrix is utilized, and excess collagen is removed to strengthen the newly formed tissue. This phase restores tissue integrity and functionality, completing the healing process

support for migrating cells during the subsequent inflammatory and proliferative phases [1].

Inflammation

The inflammatory phase is the body's immediate response to injury, crucial for infection control and the clearance of cellular debris. This phase is initiated by mast cell degranulation, which releases histamine and other inflammatory mediators that induce vasodilation and increase vascular permeability, allowing immune cells to migrate to the wound site. These early events result in the characteristic signs of inflammation, including redness, heat, swelling, and pain [14].

Neutrophils are the first immune cells to infiltrate the wound, arriving in response to chemotactic signals such as PDGF, TGF-β, and complement component C5a. These cells engage in phagocytosis of pathogens, secrete proteases and ROS to break down damaged extracellular matrix (ECM), and play a key role in clearing microbial contamination [14].

As neutrophils complete their function, monocytes are recruited 24–48 hours post-injury, differentiating into macrophages upon arrival at the wound site, driven by monocyte chemoattractant protein-1 (MCP-1) [14]. Macrophages play a dual role in wound healing by transitioning between two phenotypes:

M1 Macrophages (pro-inflammatory)—Predominant in the early inflammatory phase, these macrophages secrete tumor necrosis factor-alpha (TNF-α), interleukin-1β (IL-1β), and interleukin-6 (IL-6), which amplify inflammation, enhance immune responses, and clear necrotic tissue through phagocytosis [14].

M2 Macrophages (Anti-inflammatory/Reparative)—As healing progresses, macrophages shift towards an M2 phenotype, characterized by the secretion of anti-inflammatory cytokines, including interleukin-10 (IL-10) and TGF-β, which resolve inflammation and promote tissue repair. Additionally, M2 macrophages release VEGF, PDGF, and TGF-β, stimulating angiogenesis, fibroblast activation, and ECM deposition, thus transitioning the wound toward the proliferative phase [14].

A delicate balance between pro-inflammatory (TNF-α, IL-1, IL-6) and anti-inflammatory (IL-10, TGF-β) signals determines whether inflammation gets effectively resolved or persists, leading to chronic wounds [4]. Efficient macrophage polarization is essential for seamless wound healing, as an extended M1 phase can prolong inflammation, while an early M2 shift may lead to excessive fibrosis.

This integrated macrophage discussion provides a structured and cohesive narrative, eliminating redundancy while clearly outlining their sequential and phenotypic roles in wound healing.

Proliferation

The proliferative phase focuses on rebuilding tissue through granulation tissue formation, angiogenesis, and re-epithelialization. Fibroblasts play a crucial role in this phase, migrating to the wound site in response to growth factors such as PDGF and TGF-β. These cells proliferate and synthesize ECM components, including collagen, glycosaminoglycans, and fibronectin, which provide structural integrity for new tissue formation [2].

Angiogenesis, an essential component of the proliferative phase, is primarily regulated by VEGF and bFGF, which drive endothelial cell migration and new capillary formation to ensure oxygen and nutrient supply for cellular repair. Additionally, keratinocytes migrate across the wound bed, stimulated by epidermal growth factor (EGF) and keratinocyte growth factor (KGF), contributing to the restoration of the epithelial barrier [11].

Initially, type III collagen is deposited to form granulation tissue, which is later remodeled by fibroblasts and matrix metalloproteinases (MMPs) to enhance tissue integrity [1].

Remodeling

The remodeling phase, also known as the maturation phase, involves ECM reorganization and strengthening of the repaired tissue. Myofibroblasts play a critical role in wound contraction, reducing the wound size by pulling the edges together. During this phase, type III collagen is replaced by type I collagen, which has greater tensile strength. Cross-linking of collagen fibrils, mediated by enzymes like lysyl oxidase, further enhances the durability of the repaired tissue [3].

Cell density decreases as fibroblasts, macrophages, and endothelial cells undergo apoptosis and excess capillaries regress. Proteolytic balance, maintained by MMPs and their inhibitors (tissue inhibitors of metalloproteinases, TIMPs), ensures appropriate ECM remodeling without excessive degradation or scarring [13]. Chronic wounds, however, often fail to progress through this phase, resulting in persistent open wounds.

Regulators of Inflammation

A network of growth factors, cytokines, chemokines, and proteases tightly regulates wound healing [13]:

- *Growth Factors*: PDGF, VEGF, FGF, EGF, and IGF regulate cell migration, proliferation, and ECM synthesis.
- *Cytokines*: IL-1, IL-6, TNF-α, and TGF-β mediate inflammation, cell recruitment, and tissue remodeling.
- *Chemokines*: Molecules such as monocyte chemoattractant protein-1 (MCP-1) and C-X-C motif chemokine ligand 8 (CXCL8) guide leukocyte trafficking and immune response.
- *Proteases and Inhibitors*: MMPs facilitate ECM degradation for cell migration and remodeling, while TIMPs prevent excessive ECM breakdown.

The molecular mechanisms of wound healing involve a complex interplay of cellular and biochemical events that proceed through hemostasis, inflammation, proliferation, and remodeling. While acute wounds typically follow a structured progression, chronic wounds often exhibit dysregulation in these processes, highlighting the importance of understanding the molecular regulators involved. Advances in therapeutic interventions targeting these mechanisms hold great promise for improving wound care and patient outcomes.

Cell Types in Wound Healing

Wound healing is a critical biological process that restores tissue structure and functionality following injury. This dynamic process is divided into four overlapping phases: hemostasis, inflammation, proliferation, and remodeling. Each phase involves distinct cellular and molecular mechanisms working in concert to ensure efficient tissue repair. The cells involved in wound healing, include neutrophils, monocytes, macrophages, T cells, B cells, fibroblasts, endothelial cells, and stem cells. These cells communicate through a network of growth factors, cytokines, and chemokines, collectively orchestrating a complex yet highly regulated healing process. While acute wounds follow a predictable healing trajectory, chronic wounds often exhibit dysregulated inflammation and impaired ECM remodeling, posing significant therapeutic challenges. Understanding the roles and interactions of key cellular players provides critical insights into the development of targeted therapies for chronic wound management [1, 13] (Fig. 5.2).

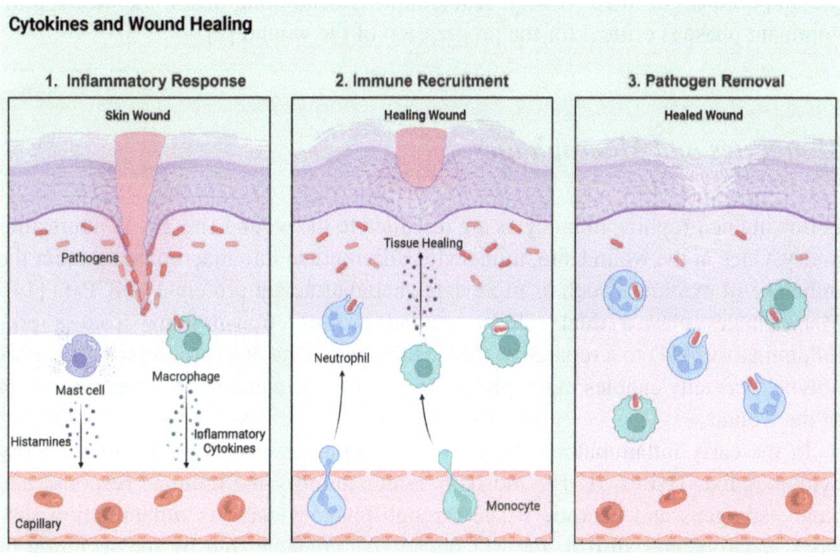

Fig. 5.2 Cytokines regulate wound healing: (1) Inflammatory *response:* pro-inflammatory cytokines are released by activated platelets and resident immune cells that trigger vascular permeability and the recruitment of neutrophils to the wound site. (2) *Immune recruitment:* chemokines recruit additional immune cells, including monocytes and macrophages. Macrophages release pro-inflammatory cytokines for pathogen clearance and anti-inflammatory cytokines to move the wound toward the proliferative phase. (3) *Pathogen removal:* cytokines stimulate the production of reactive oxygen species (ROS) and antimicrobial peptides to eliminate pathogens

Neutrophils

Neutrophils are the earliest immune cells recruited to the wound site, typically arriving within hours of tissue injury. Their recruitment is mediated by chemokines such as CXCL8 (IL-8) and complement components like C5a, which direct neutrophil migration to the wound via a chemotactic gradient [15]. Neutrophils play a pivotal role in the early inflammatory phase by performing several essential functions. First, they act as primary defenders against microbial invasion through phagocytosis, wherein neutrophils engulf and neutralize pathogens using ROS and antimicrobial peptides. Additionally, neutrophils release proteolytic enzymes such as MMPs and elastase, which degrade necrotic tissue and ECM components to prepare the wound bed for subsequent phases of healing. Moreover, neutrophils amplify the inflammatory response by releasing pro-inflammatory cytokines like IL-1β and TNF-α, which recruit and activate other immune cells, including monocytes and macrophages [2]. However, prolonged neutrophilic activity can cause collateral damage to healthy tissue due to excessive ROS and protease release, underscoring the importance of their timely resolution. Transitioning from the neutrophil-dominant phase is critical for the progression of the wound [2].

Monocytes and Macrophages

Following neutrophils, monocytes are recruited to the wound site 24–48 hours after injury. Once at the wound site, monocytes differentiate into macrophages under the influence of cytokines such as monocyte chemoattractant protein-1 (MCP-1) [14]. Macrophages play a dual role in wound healing, transitioning from a pro-inflammatory (M1) to a reparative (M2) phenotype as healing progresses. This phenotypic plasticity enables macrophages to adapt to the changing microenvironment of the wound.

In the early inflammatory phase, M1 macrophages produce pro-inflammatory cytokines like TNF-α, IL-1β, and IL-6, which amplify the immune response and clear pathogens and necrotic tissue through phagocytosis. As inflammation subsides, macrophages shift to the M2 phenotype, characterized by the secretion of anti-inflammatory cytokines such as interleukin-10 (IL-10) and transforming growth factor-beta (TGF-β). These cytokines play a pivotal role in dampening inflammation and promoting tissue repair. Additionally, macrophages secrete growth factors, including VEGF, PDGF, and TGF-β, which stimulate angiogenesis, fibroblast activation, and ECM deposition. This functional versatility makes macrophages indispensable in orchestrating the transition from the inflammation phase to the tissue repair and regeneration phase [6].

T Cells

T cells play a critical role in modulating the immune response during wound healing, with effector T cells and regulatory T cells (Tregs) performing distinct yet complementary functions. Effector T cells, including CD4+ and CD8+ subsets, secrete pro-inflammatory cytokines such as interferon-gamma (IFN-γ) and TNF-α, which enhance the recruitment and activation of macrophages and neutrophils to clear pathogens and cellular debris [6]. This activity is particularly important in the early stages of healing to control infection.

Conversely, Tregs contribute to immune resolution and tissue repair by producing anti-inflammatory cytokines, including IL-10 and TGF-β. These cytokines suppress excessive immune activation, preventing prolonged inflammation and tissue damage. T cells also interact with other cellular players, such as fibroblasts and macrophages, to regulate the ECM remodeling and angiogenesis process. By balancing pro-inflammatory and anti-inflammatory responses, T cells ensure the progression of wound healing while minimizing the risk of chronic inflammation [1].

B Cells

Although the role of B cells in wound healing is less well characterized than other immune cells, emerging evidence highlights their contribution to immune regulation and tissue repair. B cells produce antibodies that neutralize pathogens, reducing the risk of infection and facilitating debris clearance. Additionally, B cells secrete pro-inflammatory cytokines such as IL-6 and TNF-α, which influence the activity of immune cells, fibroblasts, and keratinocytes in the wound environment [16]. While the specific mechanisms by which B cells contribute to wound healing remain an active area of research, their role in maintaining immune homeostasis and supporting tissue repair warrants further exploration.

Fibroblasts

Fibroblasts are central to the proliferative phase of wound healing, primarily through their role in ECM production and tissue remodeling. Activated fibroblasts migrate into the wound bed in response to growth factors such as PDGF and TGF-β, where they secrete type III collagen, fibronectin, and glycosaminoglycans to form granulation tissue. This provisional matrix provides structural integrity to the regenerating tissue and supports the migration of endothelial cells and keratinocytes.

Fibroblasts also differentiate into myofibroblasts under the influence of TGF-β. Myofibroblasts produce contractile proteins such as alpha-smooth muscle actin (α-SMA), which facilitate wound contraction by pulling the wound edges together.

In addition to their structural role, fibroblasts release cytokines and growth factors, including VEGF and fibroblast growth factor (FGF), which promote angiogenesis and keratinocyte proliferation [17]. Dysregulated fibroblast activity, such as excessive collagen deposition or MMP overexpression, can lead to scarring or developing of chronic wounds.

Endothelial Cells

Endothelial cells are key players in angiogenesis, the process of new blood vessel formation that supplies oxygen and nutrients to the regenerating tissue. Angiogenesis is initiated by hypoxia-inducible factor-1 alpha (HIF-1α), which upregulates the expression of VEGF and FGF. These growth factors stimulate endothelial cell proliferation, migration, and differentiation, leading to the formation of capillary loops within the granulation tissue [18].

Newly formed blood vessels also increase vascular permeability, facilitating the delivery of immune cells, nutrients, and growth factors to the wound site. Impaired angiogenesis, commonly observed in chronic wounds, exacerbates hypoxia and delays healing. Enhancing endothelial cell function through targeted therapies can significantly improve wound outcomes [19].

Stem Cells

Stem cells, particularly mesenchymal stem cells (MSC), are promising for regenerative wound healing. MSC can differentiate into multiple cell types, including fibroblasts, keratinocytes, and endothelial cells, replenishing cellular populations necessary for repair. In addition to their differentiation potential, MSC secrete paracrine factors such as VEGF, PDGF, and IL-10, which modulate inflammation, stimulate angiogenesis, and enhance ECM production.

One of the most remarkable features of MSC is their immunomodulatory capacity. MSC suppress excessive T cell activation and promote macrophage polarization toward a reparative M2 phenotype, creating a pro-healing microenvironment [7]. Stem cell-based therapies, including MSC transplantation and bioengineered scaffolds, hold significant potential for addressing the challenges of chronic wound healing.

The coordinated actions of neutrophils, monocytes, macrophages, T cells, B cells, fibroblasts, endothelial cells, and stem cells ensure efficient tissue repair during wound healing. These cells communicate through molecular signals, including growth factors, cytokines, and chemokines, to regulate inflammation, angiogenesis, and ECM remodeling. Dysregulation of these processes can lead to chronic wounds, emphasizing the need for targeted interventions. Advances in stem cell therapy, bioengineered scaffolds, and growth factor delivery systems offer promising solutions for improving wound care outcomes. Continued research into the cellular and molecular mechanisms of wound healing will pave the way for innovative therapeutic strategies that enhance tissue repair and regeneration.

Signaling Pathways

Wound healing involves an intricate network of signaling pathways that regulate essential cellular functions, including migration, proliferation, differentiation, and ECM remodeling. These pathways mediate responses to growth factors, cytokines, and environmental stimuli, ensuring a coordinated repair process. Dysregulation of these pathways can contribute to chronic wounds or excessive scarring, underscoring their therapeutic relevance. Critical signaling pathways, including Akt/PI3K, SMAD, p38 MAPK, JAK/STAT, Wnt/β-catenin, and Notch, are discussed below, emphasizing their roles in wound healing and potential therapeutic implications.

Akt/PI3K

The Ak strain transforming (Akt), also known as protein kinase B and phosphoinositide 3-kinase (PI3K) signaling pathway is a signal transduction pathway that promotes survival and growth in response to extracellular signals and is a critical regulator of cell survival, proliferation, and migration, which are fundamental to wound healing. Activated downstream of PI3K by growth factors such as VEGF, PDGF, and EGF, Akt modulates diverse cellular processes.

In the context of wound healing, Akt promotes keratinocyte and fibroblast survival by phosphorylating and inhibiting pro-apoptotic proteins such as Bcl2-associated agonists of cell death (Bad) and Caspase-9. This anti-apoptotic function ensures the maintenance of a viable cell population at the wound site [7]. Akt also drives fibroblast and endothelial cell proliferation, which is essential for granulation tissue formation. In addition to cellular proliferation, Akt plays a pivotal role in angiogenesis by mediating VEGF signaling. VEGF-induced Akt activation enhances endothelial cell migration and tube formation. This process is critical for the formation of new blood vessels. Akt also stimulates endothelial nitric oxide synthase (eNOS), increasing nitric oxide (NO) production, which enhances vascular permeability and blood vessel stabilization [7].

Therapeutically, modulating the Akt/PI3K pathway has shown promise in chronic wound management. Gene therapy approaches and small molecules targeting Akt activation are being explored to restore angiogenesis and epithelialization in non-healing wounds.

SMAD

The suppressor of mother against decapentaplegic (SMAD) signaling pathway is a key mediator of TGF-β activity, regulating inflammation, ECM synthesis, and fibroblast function. Upon TGF-β ligand binding to its receptor, receptor-regulated SMADs (R-SMADs) such as SMAD2 and SMAD3 are phosphorylated and form complexes with SMAD4, translocating to the nucleus to regulate gene expression.

During wound healing, the SMAD pathway drives fibroblast differentiation into myofibroblasts, which secrete collagen and fibronectin, contributing to wound contraction and ECM deposition [20]. SMAD signaling also regulates MMPs and tissue inhibitors of metalloproteinases (TIMPs), ensuring a balance between ECM synthesis and degradation. Furthermore, SMAD activity modulates the inflammatory phase by suppressing excessive cytokine production and promoting macrophage polarization toward an anti-inflammatory phenotype, facilitating the transition to the proliferative phase.

However, dysregulation of TGF-β/SMAD signaling can result in pathological outcomes. Overactivation may lead to fibrosis and hypertrophic scarring, while inadequate signaling in chronic wounds impairs ECM synthesis and delays healing [16]. Therapies targeting SMAD signaling are under investigation to address these challenges, aiming to fine-tune the pathway for improved wound healing outcomes.

p38/MAPK

The p38 mitogen-activated protein kinase (p38 MAPK) pathway is activated in response to environmental stressors, growth factors, and cytokines, playing a pivotal role in inflammation, cell migration, and differentiation during wound healing. This pathway is particularly critical in the inflammatory and early proliferative phases.

p38 MAPK regulates the production of pro-inflammatory cytokines such as TNF-α and IL-1β, essential for pathogen clearance and debris removal [15]. Additionally, p38 MAPK enhances keratinocyte migration by reorganizing the cytoskeleton and upregulating adhesion molecules such as integrins, facilitating re-epithelialization [21]. Angiogenesis is another key function mediated by p38 MAPK, which promotes VEGF-induced endothelial cell migration and vascular remodeling, ensuring adequate blood supply to the wound bed.

Modulating p38 MAPK activity is a promising therapeutic strategy. In chronic wounds, inhibiting excessive inflammatory signaling while preserving pro-angiogenic and proliferative functions can accelerate healing. Small-molecule inhibitors targeting p38 MAPK are currently being developed for this purpose.

JAK/STAT

The Janus kinase-signal transducer and activator of transcription (JAK/STAT) signaling pathway is a central mediator of cytokine and growth factor signaling, influencing immune regulation and cell proliferation during wound repair. Activation of this pathway occurs when cytokines such as interferons and interleukins bind to their receptors, leading to the activation of JAK and subsequent phosphorylation of STAT proteins, which translocate to the nucleus to regulate target gene expression.

In wound healing, the JAK/STAT pathway modulates immune responses by regulating macrophage polarization and T-cell activation, creating an immune environment conducive to repair. STAT3 is a critical player, enhancing fibroblast proliferation and ECM synthesis, which are essential for granulation tissue formation [22]. Impaired JAK/STAT signaling has been linked to reduced immune responses and delayed fibroblast activation in chronic wounds, highlighting its therapeutic potential. Drugs targeting the JAK/STAT pathway may offer new avenues for managing immune dysregulation in non-healing wounds.

Wnt/β-Catenin

The Wnt/β-catenin pathway is essential for regulating cell proliferation, differentiation, and migration during wound healing, particularly in the epithelialization and remodeling phases. Wnt ligands activate this pathway by binding to Frizzled receptors, leading to the stabilization and nuclear translocation of β-catenin, which regulates the transcription of target genes.

In keratinocytes, Wnt signaling promotes proliferation and migration, facilitating re-epithelialization of the wound bed. In fibroblasts, Wnt/β-catenin activation stimulates proliferation and myofibroblast differentiation, contributing to wound contraction and ECM deposition [22]. However, the downregulation of Wnt signaling in chronic wounds impairs these processes, resulting in delayed healing. Therapies aimed at enhancing Wnt protein are being investigated to restore proper epithelial and fibroblast function in non-healing wounds.

Notch

Notch signaling regulates cell–cell communication, influencing cellular differentiation, and angiogenesis during wound repair. Activation occurs when Notch ligands bind to their receptors, triggering proteolytic cleavage and release of the Notch intracellular domain (NICD), which translocates to the nucleus to regulate gene expression.

Notch signaling plays a dual role in angiogenesis and keratinocyte differentiation. It interacts with VEGF signaling to regulate endothelial cell proliferation and vascular remodeling, ensuring proper blood vessel formation [13]. In keratinocytes, Notch signaling promotes differentiation, facilitating the formation of a functional epidermis. Dysregulation of this pathway can lead to impaired vascularization and defective epithelialization, making it a target for therapeutic intervention.

Integration of Pathways

These signaling pathways work in concert to ensure a coordinated wound-healing response. Crosstalk between pathways, such as Akt/PI3K and VEGF for angiogenesis or SMAD and p38 MAPK for ECM remodeling, highlights the complexity of their interactions. Dysregulation in one pathway often disrupts the entire repair process, as seen in chronic wounds.

Therapeutic strategies targeting these pathways include gene therapy, small-molecule inhibitors, and bioengineered delivery systems. For instance, gene therapy delivering VEGF or Akt can enhance angiogenesis, while p38 MAPK inhibitors can reduce excessive inflammation. Advanced biomaterials such as hydrogels and nanoparticles are being developed to deliver pathway-specific growth factors or inhibitors locally, offering sustained therapeutic effects.

Acute and Chronic Wound Healing

Wound healing is an essential biological process that restores the structural and functional integrity of the tissues following injury. The mechanisms underlying wound repair differ significantly between acute and chronic wounds, with acute wounds typically progressing through an orderly sequence of hemostasis, inflammation, proliferation, and remodeling. In contrast, chronic wounds exhibit dysregulated healing, often stalled at any stage of the healing process due to dysregulation of any of the intrinsic or extrinsic factors. This chapter delves into the cellular, molecular, and physiological differences between acute and chronic wound healing, drawing on key studies to highlight their distinct features.

Healing Process

The most fundamental distinction between acute and chronic wound healing lies in the duration and efficiency of the repair process. Acute wounds generally heal within days to weeks, depending on the severity of the injury, and progress predictably through the phases of wound healing. These wounds benefit from well-regulated cellular responses and molecular signaling cascades, ensuring timely repair and tissue restoration [6].

In contrast, chronic wounds fail to heal within three months and are characterized by an inability to transit smoothly through the healing phases. Factors such as diabetes, vascular insufficiency, and prolonged inflammation often contribute to this impaired process. As a result, chronic wounds persist as open lesions, posing a significant challenge to patient care and quality of life [6].

Inflammatory Response

Inflammation is a critical phase in wound healing, serving to clear pathogens and debris while preparing the wound bed for tissue repair. In acute wounds, the inflammatory response is transient and tightly controlled. Neutrophils are the first immune cells to arrive at the wound site, followed by monocytes that differentiate into macrophages. These macrophages play a pivotal role in resolving inflammation and promoting the transition to the proliferative phase [6].

Chronic wounds, however, exhibit prolonged and excessive inflammation. Neutrophils and macrophages are continuously recruited to the wound site, releasing pro-inflammatory cytokines, including TNF-α, IL-1, and IL-6. This persistent inflammatory environment disrupts the healing process, damaging the surrounding tissue and perpetuating the wound's chronic state [23].

Re-epithelialization

Re-epithelialization is a crucial step in restoring the skin's barrier function. In acute wounds, keratinocytes migrate efficiently across the wound bed, aided by MMPs that remodel the ECM. Growth factors such as EGF and hepatocyte growth factor (HGF) stimulate keratinocyte proliferation and migration, ensuring timely wound closure [6].

In chronic wounds, keratinocyte migration is impaired. These cells often enter a hyperproliferative state without successfully covering the wound. This dysfunction is partly attributed to reduced responsiveness to growth factors and downregulation of critical specific receptors. Additionally, cellular senescence may contribute to the inability of keratinocytes to re-epithelialize chronic wounds effectively [24].

Fibroblast Function and ECM Formation

Fibroblasts play a pivotal role in granulation tissue formation by synthesizing ECM components such as collagen and fibronectin. In acute wounds, fibroblasts proliferate actively and differentiate into myofibroblasts, which aid in wound contraction and ECM remodeling. This well-coordinated activity ensures the formation of a temporary wound matrix that supports cell migration and tissue repair [6].

Chronic wounds, however, exhibit dysfunctional fibroblast activity. Fibroblasts in these wounds often display reduced proliferation and enter a state of senescence, diminishing their capacity to produce ECM components. Elevated levels of MMPs in chronic wounds further degrade the ECM, preventing effective granulation tissue formation and undermining the structural integrity of the wound bed [6].

Angiogenesis and Oxygenation

Angiogenesis is essential for wound healing as it facilitates oxygen and nutrient delivery, particularly in hypoxic environments. Under normal wound-healing conditions, VEGF and FGF drive endothelial cell proliferation, forming capillary networks that sustain cellular activity during tissue repair [6]. However, in chronic wounds, angiogenesis is often impaired, leading to inadequate oxygenation, persistent inflammation, and delayed healing [25].

Microbial Infection and Biofilm Formation

Acute wounds typically resist infections due to an effective immune response. However, it was shown that chronic wounds are frequently colonized by bacteria, which can form biofilms. These biofilms shield pathogens from immune clearance and antimicrobial treatments, prolonging inflammation and further complicating the healing process. The presence of biofilms in chronic wounds is a major factor in their delayed healing and resistance to standard therapies [26].

Molecular Signals and Growth Factors

Growth factors play a vital role in coordinating cellular activities during wound healing. In acute wounds, growth factors such as PDGF, EGF, and HGF are upregulated to promote cell migration, proliferation, and ECM remodeling. These molecular signals are tightly regulated, ensuring timely progression through the healing phases [6].

Chronic wounds, by contrast, exhibit dysregulated growth factor signaling. Insensitivity to growth factors caused by receptor downregulation or excessive degradation by MMPs, which play a critically role in impairing the normal regenerative response. This lack of effective signaling contributes to the stalled healing observed in chronic wounds [16].

Scarring and Tissue Remodeling

The final stage of wound healing involves ECM remodeling and scar tissue formation. In acute wounds, fibroblasts and myofibroblasts reorganize collagen fibers, forming scar tissue that, while not fully functional, provides structural integrity to the repaired tissue [6].

Chronic wounds often fail to progress to effective remodeling. Instead of forming functional tissue, these wounds may exhibit abnormal collagen deposition or fibrosis, particularly if inflammation persists for extended periods. This failure to complete the remodeling phase underscores the challenges of managing chronic wounds [6].

Growth Factors in Wound Healing

Wound healing is a complex biological process involving a highly regulated interplay of cells, cytokines, and growth factors. Growth factors, which are signaling molecules that regulate cellular activities, play a pivotal role in coordinating the stages of wound repair, including cell proliferation, migration, differentiation, and matrix remodeling. We have explored and listed the key growth factors involved in wound healing, their mechanisms of action, and their potential applications in clinical settings in detail below.

Platelet-Derived Growth Factor

PDGF is one of the earliest growth factors released at the injury site, primarily by activated platelets. PDGF is critical in stimulating fibroblast proliferation and collagen synthesis, which are essential for granulation tissue formation. It also recruits other cells, such as smooth muscle cells and macrophages, to the wound site, enhancing the repair process [27]. The effectiveness of PDGF in chronic wound healing has made it a focus of therapeutic interventions. Clinical studies have demonstrated its potential to enhance wound closure rates in diabetic ulcers and other chronic wounds. However, challenges such as localized delivery and sustained activity remain areas of active research.

Vascular Endothelial Growth Factor

VEGF is a key angiogenic factor that stimulates endothelial cell migration and capillary formation, ensuring adequate tissue oxygenation. VEGF is upregulated in response to hypoxia and increases endothelial permeability to facilitate wound healing [18]. Impaired VEGF signaling in chronic wounds leads to insufficient angiogenesis and prolonged tissue damage [19]. Targeted VEGF therapies, including exogenous VEGF administration and gene therapy, are promising strategies to enhance vascularization and improve chronic wound healing outcomes.

Fibroblast Growth Factor

FGF is a family of growth factors involved in various cellular processes, including proliferation, migration, and differentiation. In the context of wound healing, FGF plays a crucial role in activating fibroblasts and endothelial cells, thereby supporting granulation tissue formation and angiogenesis [28]. FGF also facilitates keratinocyte migration, enhancing re-epithelialization. Its ability to stimulate multiple cell types makes it a versatile growth factor for skin repair. Clinical applications of FGF in wound healing have shown promise, particularly when used in bioengineered scaffolds or hydrogels to ensure controlled and sustained release.

Epidermal Growth Factor

EGF is a key player in promoting keratinocyte migration and proliferation, which are essential for re-epithelialization. EGF binds to its receptor (EGFR) on keratinocytes, triggering intracellular signaling pathways that regulate cytoskeletal rearrangement and cell migration. This enables keratinocytes to cover the wound bed efficiently, restoring the skin's barrier function [3]. The therapeutic potential of EGF has been demonstrated in managing chronic wounds, such as diabetic foot ulcers, where it accelerates healing. However, challenges related to its stability and delivery limit its widespread clinical use, prompting the development of advanced delivery systems such as nanoparticles and topical gels.

Insulin-Like Growth Factor

Insulin-like growth factor (IGF) is involved in promoting collagen synthesis and fibroblast proliferation, which are crucial for wound contraction and matrix remodeling. IGF also enhances keratinocyte migration and endothelial cell activity, contributing to re-epithelialization and angiogenesis [29]. Chronic wounds often exhibit reduced levels of IGF, which impairs these processes and delays healing. Restoring IGF levels through exogenous supplementation has shown the potential to accelerate wound repair. However, the development of effective delivery systems is essential to maximize its therapeutic benefits while minimizing potential side effects.

Clinical Applications

The application of growth factors in wound healing has opened new avenues for managing chronic wounds. Topical formulations, injectable delivery systems, and bioengineered scaffolds incorporating growth factors have been explored to enhance wound repair. For instance, recombinant PDGF has been approved for treating diabetic foot ulcers, demonstrating its clinical utility.

Despite their promise, growth factor therapies face significant challenges. High production costs and stability issues limit their accessibility and effectiveness. Growth factors are often prone to rapid degradation in the wound environment, requiring innovative delivery systems like hydrogels and nanoparticles for sustained release [5]. Moreover, excessive or inappropriate use of growth factors can lead to adverse effects, such as hypertrophic scarring or prolonged persistent inflammation.

Future research on growth factor therapies aims to overcome these challenges through advancements in biotechnology and bioengineering. The development of gene therapies and stem cell-based approaches that enable endogenous production of growth factors holds significant promise [22]. Furthermore, personalized medicine approaches that tailor growth factor therapies to individual patient profiles could optimize outcomes and reduce risks.

Growth factors such as PDGF, VEGF, FGF, EGF, and IGF play indispensable roles in wound healing, orchestrating cellular activities that drive tissue repair. While significant progress has been made in understanding their mechanisms and clinical applications, challenges related to delivery, stability, and cost must be addressed to realize their full potential. Continued innovation in growth factor-based therapies is expected to transform wound care, particularly for chronic and hard-to-heal wounds.

Cytokines and Challenges

Cytokines are small, soluble proteins that serve as critical mediators in the wound healing process. These signaling molecules regulate inflammation, cellular recruitment, proliferation, and ECM remodeling, thereby orchestrating tissue repair. Cytokines are broadly categorized into pro-inflammatory and anti-inflammatory types, each playing distinct roles in ensuring timely and effective wound healing.

However, an imbalance in cytokine levels or dysregulation of their signaling pathways can lead to delayed or impaired healing, as observed in chronic wounds. We examine the roles of key cytokines in wound repair and discuss the challenges associated with using cytokine and growth factor therapies for wound management in detail below.

Pro-inflammatory Cytokines

Pro-inflammatory cytokines such as IL-1, IL-6, and TNF-α are vital for initiating the inflammatory phase of wound healing. These cytokines are primarily secreted by activated immune cells in response to tissue injury. IL-1 and TNF-α stimulate the expression of adhesion molecules on endothelial cells, facilitating the recruitment of leukocytes to the wound site. IL-6 promotes the acute-phase response, enhancing the production of acute-phase proteins and stimulating immune cells for a continuing activation state [30].

While pro-inflammatory cytokines are essential for pathogen clearance and debris removal, excessive or prolonged inflammation can hinder the transition to the proliferative phase. Chronic wounds often exhibit elevated levels of IL-1, IL-6, and TNF-α, which perpetuate tissue damage and create a hostile environment for cellular repair processes [31]. Thus, a tightly regulated pro-inflammatory response is critical for successful wound healing.

Anti-Inflammatory Cytokines

Anti-inflammatory cytokines, including IL-10 and TGF-β, are instrumental in resolving inflammation and promoting the transition to the proliferative phase of wound healing. IL-10 suppresses the production of pro-inflammatory cytokines, thereby dampening excessive inflammation and minimizing tissue damage. It also enhances macrophage-mediated clearance of apoptotic cells, contributing to the resolution of inflammation [32].

TGF-β plays a dual role in the wound healing process. During the inflammatory phase, it aids in immune regulation by suppressing pro-inflammatory cytokine activity. As the healing process progresses, TGF-β performs various functions, such as stimulating the recruitment of fibroblasts, helping in proliferation, and mediating differentiation into myofibroblasts, facilitating collagen synthesis and ECM deposition. These activities are crucial for granulation tissue formation and wound contraction. Dysregulation of TGF-β signaling, however, can lead to fibrotic scarring or impaired healing, underscoring the need for balanced cytokine activity.

Chemokines

Chemokines are a subset of cytokines that regulate the migration of immune cells to the wound site. CXCL8 (also known as interleukin-8) is a prominent chemokine involved in recruiting neutrophils to the wound bed, enhancing the innate immune response. Chemokines such as monocyte chemoattractant protein-1 (MCP-1) play a similar role in attracting monocytes and macrophages to the site of injury [15]. These cells are essential for clearing pathogens, apoptotic cells, and necrotic tissue.

Dysregulated chemokine signaling can contribute to chronic wound pathophysiology. For example, excessive neutrophil recruitment mediated by CXCL8 can result in elevated levels of reactive oxygen species (ROS) and proteases, causing further tissue damage. Conversely, inadequate chemokine activity can impair immune cell migration, increasing the risk of infection and delayed healing [4].

Limitations

Despite the critical roles of cytokines and growth factors in wound healing, their therapeutic applications face significant challenges, particularly in the context of chronic wounds. Those challenges are elaborated below.

Delivery and Stability

One of the primary limitations of cytokine and growth factor therapies is their rapid degradation in the wound environment. Proteolytic enzymes present in chronic wounds often degrade exogenously applied cytokines and growth factors, making it difficult to maintain effective concentrations at the site of injury [33]. Researchers are exploring controlled-release systems, such as hydrogels, nanoparticles, and encapsulated delivery vehicles, which can protect these molecules from degradation and provide sustained release over time.

Side Effects

High concentrations of growth factors or cytokines can lead to unintended side effects, including excessive inflammation, hypertrophic scarring, or fibrosis. For instance, overexpression of TGF-β can result in fibrotic tissue formation, compromising the functional and aesthetic outcomes of wound repair [11]. Thus, precise dosing and targeted delivery are essential to minimize adverse effects while maximizing therapeutic benefits.

Regulatory and Economic Challenges

The production of recombinant growth factors and cytokines is costly, limiting their accessibility for widespread clinical use. Additionally, stringent regulatory requirements for biologics complicate their approval process, delaying the translation of promising therapies from the laboratory to clinical practice [13]. Innovations in bioengineering, such as the development of biosimilar products and gene therapy approaches, have the potential to reduce costs and streamline regulatory approvals.

Current Therapies

Wound healing is a multifaceted biological process involving diverse therapeutic interventions aimed at restoring tissue integrity, managing infections, and promoting functional recovery. Traditional wound care remains a foundational approach, employing various strategies to create an optimal environment for natural healing. Hydrocolloid dressings, for instance, are widely used to maintain a moist wound environment conducive to autolytic debridement and epithelialization, especially in shallow wounds [34]. Similarly, hydrogels, composed of water-based polymers, are effective in hydrating dry or necrotic wounds, thereby facilitating granulation tissue formation. For wounds with moderate to heavy exudate, foam dressings provide excellent absorption and protection against maceration [35]. Silver-based dressings stand out for their antimicrobial properties, which are particularly valuable in preventing and managing wound infections. Topical antimicrobial agents, such as silver sulfadiazine, iodine-based products, and honey, are commonly employed to control bacterial load. Honey, specifically, has demonstrated not only antimicrobial effects but also anti-inflammatory and wound-healing properties [26].

Additionally, debridement remains critical for removing necrotic tissue and biofilms, creating a cleaner wound bed. Techniques include surgical, enzymatic, autolytic, and mechanical debridement, each tailored to wound characteristics [3]. There are several major therapeutic approaches, which are elaborated below.

Growth Factor Therapies

Growth factors are signaling proteins that enhance cellular activities such as migration, proliferation, and ECM remodeling, making them critical in wound healing. Recombinant PDGF, specifically becaplermin, has been FDA-approved for treating diabetic foot ulcers due to its ability to stimulate fibroblast proliferation and collagen synthesis, thereby accelerating granulation tissue formation. VEGF is essential for angiogenesis, promoting endothelial cell proliferation and migration, although challenges such as its short half-life and delivery issues have limited its clinical adoption [18]. EGF facilitates keratinocyte migration and proliferation, with clinical trials demonstrating its efficacy in re-epithelialization and managing chronic wounds, including diabetic ulcers. FGF also plays a significant role by activating fibroblasts, supporting angiogenesis, and promoting keratinocyte migration. Advanced bioengineered systems, such as hydrogels and nanoparticles, have been developed to address stability and sustained release issues associated with FGF [28].

Challenge and Limitations

Despite their therapeutic potential, growth factor-based therapies face several limitations that hinder their widespread clinical adoption, which are listed below.

(a) *Instability and Short Half-Life*
Growth factors such as VEGF, PDGF, and FGF have a short biological half-life, leading to rapid degradation and clearance from the wound site, thereby limiting their effectiveness [8]. VEGF, for example, is quickly degraded by proteases in the extracellular environment, necessitating frequent administration to sustain therapeutic effects [36]. This instability reduces the bioavailability of growth factors at the wound site and contributes to inconsistent treatment outcomes.

(b) *Ineffective Delivery for Bioavailability*
Traditional delivery methods, such as topical application and injection, result in poor retention and diffusion of growth factors in the wound bed. These challenges arise due to rapid clearance by proteases or limited penetration through the ECM [11]. Moreover, systemic administration often leads to off-target effects and potential toxicity, further complicating their clinical use [5].

(c) *Uncontrolled Angiogenesis Enhancing Fibrosis Risk*
While VEGF and FGF are potent angiogenic factors, excessive or uncontrolled angiogenesis can lead to vascular hyperplasia, which may result in abnormal vessel formation and increased risk of chronic inflammation [37]. Additionally, growth factors such as TGF-β can contribute to fibrotic tissue formation, leading to hypertrophic scars or keloids, which impair functional tissue regeneration [37, 38].

(d) *Higher Cost*
Recombinant growth factors are expensive to produce, and their large-scale manufacturing presents challenges related to purity, stability, and batch-to-batch variability [39]. These factors contribute to high treatment costs, making them less accessible for widespread clinical use, especially in low-resource settings.

(e) *Regulatory and Safety Concerns*
Given their potent biological activity, growth factors must undergo rigorous regulatory approval to ensure safety, efficacy, and consistency. There is also a theoretical risk of tumorigenicity, as excessive stimulation of cell proliferation may contribute to uncontrolled tissue growth or malignancies [40].

(f) *Future Directions for Overcoming Limitations*
To enhance the clinical viability of growth factor therapies, advanced delivery systems such as hydrogels, nanoparticles, and controlled-release scaffolds have been developed to improve stability and prolong bioavailability [7].

Gene therapy approaches that modulate endogenous growth factor expression also hold promise for effective wound healing while reducing the need for repeated administration [7]. Further research into personalized growth factor formulations based on individual patient profiles may help optimize therapeutic outcomes while minimizing adverse effects.

Bioengineered Skin Substitutes

Bioengineered skin substitutes address tissue loss by providing structural support and stimulating cellular healing processes. Acellular matrices, such as porcine-derived collagen scaffolds like Integra®, offer a framework for cell migration and ECM deposition, effectively aiding in the repair of partial- and full-thickness wounds [11]. Cellular skin substitutes, such as Apligraf® and Dermagraft®, deliver living cells like keratinocytes and fibroblasts, which secrete growth factors and cytokines to enhance healing. These products are particularly effective in chronic wounds by promoting cellular activity and reducing healing time.

Challenges and Limitations

Despite their promising potential, bioengineered skin substitutes face several challenges and limitations that impact their clinical efficacy, accessibility, and long-term success in wound healing.

(a) *Cost and Accessibility*
Bioengineered skin substitutes, particularly cellular-based products like Apligraf® and Dermagraft®, are expensive, with high costs associated with manufacturing, storage, and clinical application [41]. The production of viable, functional skin substitutes requires advanced cell culture facilities, stringent regulatory compliance, and specialized handling, making these therapies less accessible in resource-limited settings [42].

(b) *Limited Long-Term Survival and Integration*
While acellular matrices provide a scaffold for tissue regeneration, they rely heavily on host cell infiltration and remodeling for success. Cellular-based substitutes introduce keratinocytes and fibroblasts, yet their survival post-transplantation remains a challenge. Issues such as immune rejection, infection, and insufficient vascularization can lead to graft failure or suboptimal integration with the native tissue [43]. Additionally, some engineered constructs degrade too quickly before complete tissue regeneration, limiting their therapeutic impact [44].

(c) *Risk of Immune Rejection and Increasing Infections*
The immunogenicity of bioengineered skin substitutes varies depending on their source. Allogeneic or xenogeneic materials can trigger immune responses, leading to delayed healing or graft rejection [10]. Even decellularized scaffolds may contain residual antigens that provoke an immune response. Furthermore,

these skin substitutes create a moist wound environment that, if not properly managed, increases the risk of bacterial infections and biofilm formation [45].

(d) *Limited Vascularization and Oxygen Supply*
Successful wound healing relies on rapid vascularization to deliver oxygen and nutrients to the regenerating tissue. However, many bioengineered skin substitutes lack functional vasculature, which delays their integration and reduces their long-term effectiveness [46]. The absence of blood vessels in early-stage grafts can lead to hypoxia, delayed healing, or necrosis, particularly in large or deep wounds [47].

(e) *Challenges in Customization for Patient-Specific Needs*
Current commercially available skin substitutes offer generalized solutions rather than personalized approaches tailored to individual patient needs. Factors such as patient age, comorbidities (e.g., diabetes, vascular disease), and wound type can influence the success of treatment [48]. Advancements in 3D bioprinting and patient-derived stem cell therapies offer promise in overcoming this limitation, but they remain in early-stage research and are not yet widely available.

(f) *Future Directions for Overcoming Limitations*
To enhance clinical outcomes, next-generation bioengineered skin substitutes are incorporating vascularized constructs, immune-evasive biomaterials, and bioactive molecules that enhance cell survival, integration, and healing efficiency [6]. The use of 3D bioprinting, nanotechnology, and personalized regenerative medicine is being explored to create patient-specific solutions that improve long-term efficacy and reduce immune complications [49].

Antimicrobial and Anti-biofilm Therapies

Chronic wounds often harbor bacterial biofilms, which resist immune clearance and impede healing. Advanced antimicrobial strategies include antimicrobial peptides (AMPs), which disrupt bacterial membranes, and biofilm-disrupting agents, such as dispersin B and DNase, which degrade the ECM of biofilms to enhance antibiotic efficacy [34]. These approaches are particularly valuable for overcoming the persistent challenges associated with biofilm-dominated wound environments.

Challenges and Limitations
Despite significant advancements, antimicrobial and anti-biofilm therapies face several challenges and limitations that hinder their clinical effectiveness in chronic wound management.

(a) *Antibiotic Resistance and Limited Efficacy*
One of the major challenges in antimicrobial therapies is the emergence of multidrug-resistant (MDR) bacteria, including *Staphylococcus aureus*, *Pseudomonas aeruginosa*, and *Acinetobacter baumannii*. These pathogens exhibit intrinsic and acquired resistance to conventional antibiotics, reducing

the efficacy of standard treatments [50]. Additionally, biofilm-associated bacteria are up to 1000 times more resistant to antibiotics than planktonic bacteria, making eradication extremely difficult [51].

(b) *Biofilm Heterogeneity and Persistence*
Biofilms consist of heterogeneous bacterial populations embedded in an extracellular polymeric substance (EPS), which provides physical protection against immune responses and antimicrobial agents [52]. The presence of persister cells, which enter a dormant state, further contributes to biofilm resilience and recurrence of infection even after antibiotic treatment [53]. Targeting these resistant subpopulations remains a significant challenge.

(c) *Limited Penetration of Antimicrobial Agents*
The dense EPS matrix in biofilms restricts the diffusion of antibiotics and antimicrobial peptides (AMPs), reducing their bioavailability and efficacy [54]. This leads to suboptimal drug concentrations within the biofilm, allowing bacteria to survive and re-establish infections. Innovative delivery approaches, such as nanoformulations and enzyme-based biofilm disruptors, are being explored to enhance penetration [55].

(d) *Cytotoxicity and Biocompatibility Issues*
Many antimicrobial agents, including AMPs, silver nanoparticles, and quorum-sensing inhibitors, have demonstrated cytotoxic effects on host cells at higher concentrations [56]. The challenge is to balance antimicrobial potency while ensuring biocompatibility to avoid delayed wound healing and tissue damage [34]. Further research is needed to optimize dosing strategies and minimize toxicity.

(e) *Regulatory and Clinical Translation Barriers*
Despite promising preclinical studies, many novel antimicrobial and biofilm-disrupting agents face regulatory hurdles due to limited clinical trial data, safety concerns, and manufacturing complexities [57]. The lack of standardized testing models for biofilm-targeting therapies also complicates regulatory approval and commercial adoption [57].

(f) *Risk of Microbial Adaptation and Development of Resistance*
Bacteria rapidly adapt to antimicrobial pressure, leading to the evolution of biofilm-specific resistance mechanisms, including efflux pumps, enzymatic degradation, and genetic adaptations [58]. Prolonged use of antimicrobial agents may select for resistant biofilm-forming strains, necessitating combination therapies or novel treatment strategies to prevent resistance development [59].

(g) *Future Directions for Overcoming Limitations*
To enhance the efficacy of antimicrobial and anti-biofilm therapies, emerging strategies include:

- Nanotechnology-based drug delivery systems to improve antibiotic penetration.
- Enzyme-based therapies (e.g., dispersin B, DNase) to degrade biofilm matrices and enhance drug efficacy.

- Combination therapies integrating antibiotics with quorum-sensing inhibitors or AMPs.
- Personalized antimicrobial approaches using genomic profiling of wound microbiota to tailor treatments.

Emerging Therapies

Emerging therapeutic strategies address the limitations of current treatments through innovative approaches such as gene therapy, nanoparticle-based delivery systems, and smart biomaterials. Gene therapy allows the delivery of genes encoding growth factors or cytokines directly to the wound site, ensuring sustained production of bioactive molecules. Nanoparticles enhance therapeutic precision by enabling controlled delivery of growth factors, antimicrobials, and stem cells, improving efficacy while minimizing off-target effects [35]. Smart biomaterials represent a new frontier, designed to release therapeutic agents in response to wound-specific stimuli like hypoxia or pH changes, making personalized wound care more feasible.

Despite significant advancements, challenges persist in wound healing therapies, including the high costs of advanced treatments, limited accessibility, and the complexity of chronic wound microenvironments. Future directions include integrating therapies, such as combining stem cells with growth factors, and developing personalized approaches tailored to patient-specific factors to improve outcomes. The scalability and affordability of advanced treatments through bioengineering and regenerative medicine will be critical for transforming wound care into a more accessible and patient-centered field. Continued innovation in molecular biology and therapeutic technologies is essential to overcoming these challenges, paving the way for improved quality of life for patients with chronic wounds.

Challenges and Limitations
Despite their promising potential, emerging therapies, such as gene therapy, nanoparticle-based drug delivery, and smart biomaterials, face several challenges that hinder their widespread clinical adoption. These limitations include high costs, regulatory hurdles, safety concerns, and complexities in chronic wound environments.

(a) *Cost and Accessibility Constraints*
One of the most significant barriers to implementing advanced wound healing therapies is their high cost. Gene therapy, stem cell-based treatments, and bioengineered skin substitutes require sophisticated manufacturing processes, stringent quality control, and specialized expertise, making them financially burdensome for healthcare systems and patients [60]. Additionally, limited production scalability further restricts access to these innovative therapies in resource-limited settings [61].

(b) *Safety and Immune Response Challenges*
Gene therapy and nanoparticle-based drug delivery pose potential safety risks, including off-target effects, immune reactions, and unintended genetic modifications. Viral vectors used in gene therapy (e.g., adenoviral and lentiviral vectors) can trigger inflammatory and immunogenic responses, leading to potential toxicity or reduced treatment efficacy [62]. Similarly, nanoparticles have raised concerns regarding long-term biocompatibility, as certain nanoparticle formulations accumulate in tissues, potentially leading to cytotoxicity and inflammatory reactions.

(c) *Regulatory and Ethical Barriers*
Gene-editing technologies such as CRISPR/Cas9, though highly precise, face stringent regulatory scrutiny due to ethical considerations and potential off-target mutations [63]. The clinical translation of gene therapy for wound healing remains challenging due to a lack of standardized protocols and uncertainties in long-term safety [45]. Similarly, stem cell-based therapies face ethical concerns regarding embryonic stem cell sourcing and inconsistencies in cell differentiation and functionality [64].

(d) *Complexity of Chronic Wound Microenvironments*
Chronic wounds exhibit highly heterogeneous and dynamic microenvironments, making it difficult for emerging therapies to achieve consistent efficacy. Factors such as persistent inflammation, impaired angiogenesis, and bacterial biofilms create hostile conditions that limit the effectiveness of growth factors, stem cells, and biomaterials [45]. Additionally, variations in patient-specific factors (e.g., diabetes, vascular disease) necessitate personalized treatment approaches, which are still in the early developmental stages [65].

(e) *Stability and Controlled Release of Therapeutic Agents*
Smart biomaterials and nanoparticle-based drug delivery systems aim to improve therapeutic precision by ensuring controlled and localized drug release. However, challenges remain in achieving long-term stability and bioavailability of bioactive molecules (e.g., growth factors, cytokines) within the wound bed [35]. Degradation rates, release kinetics, and biomaterial compatibility must be optimized to ensure sustained therapeutic benefits [6].

(f) *Integration into Standard Clinical Practice*
Adopting emerging therapies into routine clinical practice is challenging due to the lack of robust clinical trial data, physician training, and standardized treatment protocols. Unlike conventional treatments such as antibiotics, debridement, and dressings, advanced regenerative therapies require specialized expertise, which can limit their widespread implementation in general healthcare settings [66].

Future Directions for Overcoming Limitations

To enhance clinical translation and accessibility of emerging wound healing therapies, future strategies should focus on:

- Developing cost-effective bioengineering approaches to improve affordability and scalability.

- Advancing non-viral gene delivery techniques to reduce immune reactions.
- Enhancing regulatory frameworks to streamline approvals for innovative therapies.
- Improving personalization of treatments using AI-driven diagnostics and precision medicine.
- Conducting large-scale clinical trials to validate the long-term safety and efficacy of these novel approaches.

As research in regenerative medicine and bioengineering advances, emerging therapies hold great promise in revolutionizing chronic wound care and improving patient outcomes. However, overcoming these challenges and limitations is critical to translating cutting-edge innovations into effective, widely accessible treatments.

Stem Cell Therapy

Stem cell therapy represents a groundbreaking advancement in wound healing by leveraging the regenerative potential of stem cells. MSC are particularly effective due to their ability to secrete growth factors, such as VEGF and PDGF, modulate immune responses, and differentiate into fibroblasts, endothelial cells, and keratinocytes. These cells promote tissue repair by suppressing inflammation, polarizing macrophages toward a reparative phenotype, and inhibiting T-cell activation [7]. Delivery methods for stem cell therapy include scaffolds, hydrogels, and injectable formulations, with innovative approaches like gene-modified MSC and 3D bioprinting significantly enhancing therapeutic outcomes.

Wound healing is a complex and dynamic process involving an orchestrated interaction of cells, cytokines, and ECM components. While acute wounds generally follow a predictable path of hemostasis, inflammation, proliferation, and remodeling, chronic wounds often remain stagnant due to disruptions in these mechanisms. Stem cell therapy has emerged as a revolutionary approach to enhancing wound repair, offering the ability to regenerate damaged tissue, modulate inflammation, and promote angiogenesis. We will discuss in detail the potential of stem cells as therapeutic agents, highlighting bioengineering approaches, gene therapy applications, stem cell types, mechanisms of action, and future directions in wound management in the following sections.

Bioengineering Approaches

Bioengineering plays a pivotal role in optimizing stem cell delivery for wound healing. Hydrogels and nanoparticles are at the forefront of this innovation, offering controlled release of stem cells and bioactive molecules. Hydrogels serve as scaffolds that maintain stem cell viability and support their migration and proliferation at the wound site. These biomaterials can be engineered to release growth factors, cytokines, or other therapeutic agents in response to specific wound conditions such as hypoxia or inflammation [22, 67–72] Nanoparticles further enhance precision by delivering stem cells directly to the wound site and improving therapeutic efficacy.

Additionally, advancements in 3D bioprinting have enabled the creation of bio-engineered scaffolds integrated with stem cells. This technology allows for developing tissue constructs tailored to the wound's dimensions, promoting the precise placement of stem cells and biomaterials for optimal healing outcomes. These bio-engineering approaches represent a significant leap toward more effective and targeted stem cell therapies.

Gene Therapy for Enhancing Stem Cell Functionality

Gene therapy is another promising avenue for enhancing the regenerative capabilities of stem cells. By enabling stem cells to produce growth factors, cytokines, or other bioactive molecules, gene therapy addresses challenges such as insufficient therapeutic molecule availability at the wound site. Viral vectors, including adenoviral and lentiviral systems, have been widely used to deliver therapeutic genes to stem cells, ensuring sustained production of growth factors like VEGF, FGF, and EGF [18]. However, non-viral vectors such as lipid nanoparticles and plasmids are gaining traction due to their lower risk of immune reactions and off-target effects.

Gene-modified stem cells offer an innovative solution by secreting bioactive molecules directly at the wound site. This reduces the need for repeated applications and enhances healing efficiency. Such strategies hold immense potential for addressing the effective healing of chronic wounds, where traditional therapeutic approaches often fail.

Sources and Delivery

The therapeutic potential of stem cells lies in their ability to differentiate into specific cell types, secrete bioactive molecules, and modulate immune responses.

Stem cells hold immense promise in regenerative medicine due to their ability to differentiate into specific cell types, secrete bioactive molecules, and modulate immune responses. Their therapeutic applications in wound healing extend across various tissue sources, facilitating tissue repair, angiogenesis, and inflammation reduction. This chapter highlights the potential of MSC from various sources in promoting wound repair and regeneration. Several types of stem cells have been utilized in wound healing (Fig. 5.3):

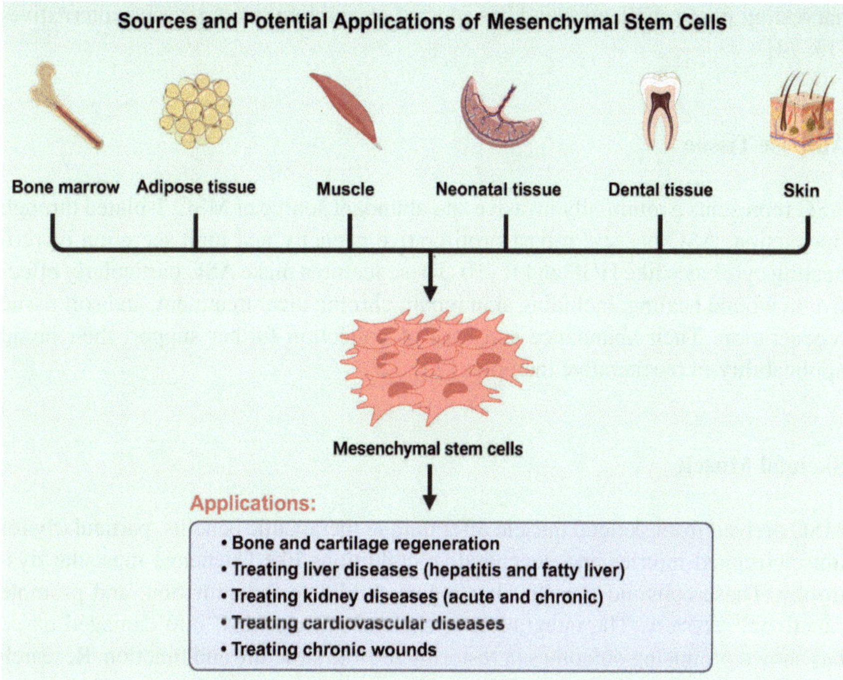

Fig. 5.3 Sources and potential application of mesenchymal stem cells. *Bone marrow*: Bone marrow is one of the most studied and widely used sources of MSC. These cells exhibit multipotent capabilities, enabling differentiation into osteoblasts, chondrocytes, and adipocytes. *Adipose tissue*: Adipose tissue is a rich and easily accessible source of MSC, known as adipose-derived stem cells. *Muscle:* MSC derived from skeletal muscle exhibit unique regenerative properties, especially in treating muscle-related injuries and degenerative diseases such as Duchenne muscular dystrophy. These cells contribute to muscle fiber repair and modulate local inflammation, enhancing functional recovery. *Neonatal tissues:* The umbilical cord, placental tissue, and amniotic fluid provide a noninvasive and abundant source of neonatal MSC to enhance proliferative and immunomodulatory activities. *Dental stem cells:* Dental tissues, including dental pulp, periodontal ligament, and exfoliated deciduous teeth, are rich sources of MSC. *Skins:* Skin-derived MSC are found in the dermal layer and exhibit strong regenerative potential for skin repair and wound healing by re-epithelialization, angiogenesis, and collagen synthesis for faster recovery from burns, ulcers, and other skin injuries

Bone Marrow

BM-MSC are among the most studied and extensively applied sources of MSC. These cells exhibit multipotent differentiation capabilities, enabling them to transform into osteoblasts, chondrocytes, and adipocytes. BM-MSC plays a significant role in treating fractures, cartilage regeneration, and musculoskeletal disorders. Their secretion of growth factors, such as VEGF and TGF-β, enhances tissue repair through angiogenesis and immune modulation. However, the invasive nature of BM harvesting poses challenges and has spurred research into less invasive alternatives [73, 74].

Adipose Tissue

ASC represents a minimally invasive and abundant source of MSC. Isolated through liposuction, ASC possess robust proliferative capacity and high secretion of pro-healing cytokines like HGF and IL-10. These features make ASC particularly effective in wound healing, including skin repair, chronic ulcer treatment, and soft tissue regeneration. Their abundance and ease of collection further support their broad applicability in regenerative therapies [75].

Skeletal Muscle

MSC derived from skeletal muscle offer unique therapeutic benefits, particularly for muscle-related injuries and degenerative conditions like Duchenne muscular dystrophy. These cells aid muscle fiber repair, modulate inflammation, and promote functional recovery. The integration of skeletal muscle MSC into damaged tissue has shown promising outcomes in restoring muscle structure and function. Research continues to explore their potential in broader regenerative applications [21].

Neonatal Tissues

The umbilical cord, placental tissue, and amniotic fluid are rich and ethically viable sources of MSC. These neonatal-derived MSC exhibit superior proliferative and immunomodulatory capacities compared to their adult counterparts, making them suitable for treating autoimmune diseases, inflammatory conditions, and tissue injuries. Furthermore, their low risk of immune rejection enhances their utility in allogeneic transplantation for cardiovascular, neurological, and dermatological applications [21].

Dental Stem Cells

Dental tissues, such as dental pulp, periodontal ligament, and exfoliated deciduous teeth, provide MSC with neurogenic and immunomodulatory properties. Dental pulp-derived stem cells (DPSC) secrete neurotrophic factors and promote angiogenesis, facilitating neural repair and craniofacial tissue engineering. The non-invasive collection methods, especially from exfoliated deciduous teeth, further strengthen the appeal of dental stem cells in regenerative medicine [76].

Skin

Skin-derived MSC, located in the dermal layer, are instrumental in wound healing and skin repair. These cells contribute to re-epithelialization, collagen synthesis, and angiogenesis, accelerating recovery from burns, ulcers, and other injuries. Their ease of access and potential for personalized applications have positioned them as vital players in dermatological regenerative therapies [77].

MSC from these sources demonstrate extensive applications in regenerative medicine. Their immunomodulatory properties allow suppression of excessive inflammation, while their secretion of growth factors promotes tissue repair and angiogenesis. MSC are widely applied in orthopedic, dermatological, neurological, and cardiovascular treatments, addressing conditions ranging from cartilage damage to neuron repair and cardiac regeneration. Their adaptability underscores their importance in innovative, personalized therapeutic approaches [74].

Mechanisms of Action

Various delivery methods have been employed to optimize stem cell application for better outcomes. The topical application involves incorporating stem cells into biocompatible scaffolds or hydrogels, which are applied directly to the wound bed. Injection techniques deliver stem cells into the wound margins or systemic circulation, particularly for hard-to-reach or chronic wounds. Encapsulation in biodegradable scaffolds, such as collagen or fibrin, ensures sustained release of stem cells while providing structural support. Spray delivery systems, ideal for treating large wound surfaces, enable even distribution of stem cells and are particularly beneficial for burns and ulcers [33].

(i) Mesenchymal stem cells: Known for their immunomodulatory properties, they release anti-inflammatory cytokines, promote angiogenesis, and differentiate into fibroblasts and endothelial cells essential for wound repair [7].
(ii) Epidermal stem cells: Contribute to re-epithelialization, restoring the skin's barrier function.
(iii) Adipose-derived stem Cells: Improve angiogenesis, ECM remodeling, and granulation tissue formation.
(iv) Bone marrow stem cells: Release bioactive molecules that stimulate fibroblasts, keratinocytes, and endothelial cells, accelerating wound closure.
(v) Induced pluripotent stem cells (iPSC): Provide a patient-specific source of stem cells reprogrammed into keratinocytes or fibroblasts for personalized therapies.

Stem cells contribute to wound healing through several mechanisms. They secrete anti-inflammatory cytokines like IL-10 and TGF-β, creating a pro-healing environment by reducing chronic inflammation [15]. Their angiogenic potential is mediated by releasing factors like VEGF, which stimulate the formation of new blood vessels, ensuring adequate oxygen and nutrient supply to the wound bed.

Paracrine signaling is another critical mechanism, where stem cells secrete growth factors such as EGF, FGF, and IGF. These factors enhance keratinocyte migration, fibroblast proliferation, and ECM remodeling. Additionally, stem cells differentiate into cell types that are essential for wound healing, such as keratinocytes, fibroblasts, or endothelial cells, directly contributing to tissue repair and regeneration.

Challenges and Considerations

Despite their enormous potential, stem cell therapies face several challenges. Immune rejection remains a significant concern, particularly with allogeneic transplantation. Strategies to evade immune responses or source autologous stem cells are being explored to address this issue [78]. Tumorigenicity, especially with iPSCs, requires stringent quality control measures to ensure safety. Furthermore, the high cost and complexity of stem cell therapies limit their accessibility. Advances in scalable production and delivery systems are essential to make these therapies more affordable.

Standardization and regulation also pose challenges, as developing consistent protocols and gaining regulatory approvals for stem cell products is a time-intensive process. Addressing these barriers is critical to translating stem cell therapies from research to clinical practice.

The future of stem cell-based wound therapies lies in personalization and innovation. Patient-specific stem cells, such as iPSC, offer the potential to tailor treatments to individual needs. Smart biomaterials capable of dynamically responding to wound conditions can further enhance therapeutic delivery. Combination therapies integrating stem cells with growth factors, gene therapy, or small molecules hold promise for synergistically accelerating wound healing.

Large-scale clinical trials are essential to validate the safety and efficacy of these emerging therapies. As research progresses, stem cell-based approaches are poised to transform wound management, offering hope for improved outcomes and enhanced patient care.

Stem cell delivery represents a transformative approach to wound healing, particularly for chronic and hard-to-heal wounds. By leveraging advances in bioengineering, gene therapy, and regenerative medicine, stem cell therapies offer unparalleled potential for restoring tissue integrity and function. Continued research and innovation in this field will undoubtedly shape the future of wound management and regenerative medicine.

Future Directions

The therapeutic landscape for wound healing is advancing rapidly, driven by innovations in molecular biology, bioengineering, and regenerative medicine. Chronic wounds, characterized by impaired healing processes, necessitate novel solutions that go beyond conventional treatments to enhance outcomes and improve quality of life. Advanced delivery systems represent a significant area of progress, offering responsive biomaterials like hydrogels and nanoparticles capable of delivering therapeutic agents in a controlled manner. Nanotechnology enables precision medicine through encapsulated agents, such as growth factors and stem cells, with smart nanoparticles designed to respond to ROS in chronic wounds [78]. The integration of 3D bioprinting further enhances these systems, allowing for the creation of custom scaffolds loaded with cells and growth factors tailored to the wound environment.

Gene and RNA therapies offer another promising frontier, enabling direct modulation of molecular pathways critical to wound repair. Gene therapy can upregulate essential factors like VEGF to improve angiogenesis, while RNA-based therapeutics, such as small interfering RNA (siRNA) and messenger RNA (mRNA), provide tools to target excessive inflammation or fibrosis [18]. Advances in stem cell therapy are also transforming wound care, with engineered stem cells genetically modified to enhance their regenerative potential. Additionally, exosome-based therapies provide a cell-free alternative to reduce the risks of tumorigenicity and immune rejection [7]. Combining stem cells with biomaterials, such as hydrogels embedded with growth factors, creates synergistic effects that support sustained therapeutic activity.

The integration of artificial intelligence (AI) and big data analytics is poised to revolutionize personalized wound care. AI-powered predictive models can analyze patient-specific data to tailor therapies, while machine learning tools aid in identifying novel therapeutic agents and optimizing formulations [34]. Wearable biosensors equipped with AI capabilities enable real-time monitoring of wound healing, providing clinicians with actionable insights for timely interventions. Personalized medicine approaches further enhance this landscape by tailoring treatments to individual genetic profiles and biomarkers, while 3D bioprinting facilitates the creation of scaffolds customized to specific wound geometries [6].

Bioengineered skin substitutes and artificial dermal layers are transformative innovations for complex wounds. Next-generation substitutes incorporate living cells, growth factors, and cytokines to mimic the structure and function of native skin more effectively. Emerging organoid technologies derived from patient-specific stem cells offer a personalized approach to skin reconstruction [79]. Chronic wounds often harbor biofilms that resist conventional treatments, and future therapies will focus on biofilm disruptors, targeted antimicrobials like bacteriophage therapy, and immune modulation to enhance clearance and promote tissue repair [34].

The interplay between the immune system and the microbiome presents additional therapeutic opportunities. Immune-modulating therapies, such as macrophage polarization from pro-inflammatory M1 to reparative M2 phenotypes, can resolve chronic inflammation and enhance healing. Microbiome engineering with probiotics or prebiotics is being explored to restore a healthy microbial balance critical for wound repair [26]. To ensure widespread access to these advanced therapies, efforts are underway to develop cost-effective solutions, including biosimilars of growth factors and automated production of bioengineered products. Telemedicine integration offers further potential to deliver these innovations to underserved regions, enhancing accessibility and equity in wound care [3, 29].

The future of wound healing therapies lies in the convergence of advancements across bioengineering, regenerative medicine, and precision healthcare. By leveraging innovations in delivery systems, gene editing, stem cell technology, and AI, the next generation of treatments aims to overcome the limitations of current approaches. Continued investment in research, interdisciplinary collaboration, and equitable access will drive the development of transformative solutions, ultimately revolutionizing wound management and improving patient outcomes in the decades to come.

Conclusion

Wound healing is a dynamic and multifaceted biological process essential for maintaining tissue integrity and function. While acute wounds typically progress through an orderly sequence of hemostasis, inflammation, proliferation, and remodeling, chronic wounds pose significant clinical challenges due to their dysregulated healing mechanisms. The key cellular and molecular regulators of wound repair, including keratinocytes, fibroblasts, endothelial cells, and immune cells, coordinate their activities through growth factors, cytokines, and chemokines. However, imbalances in these signaling pathways contribute to the persistence of chronic wounds.

Emerging therapeutic strategies offer promising avenues for addressing these challenges. Advances in bioengineering have enabled the development of controlled delivery systems such as hydrogels, nanoparticles, and scaffolds that stabilize growth factors and cytokines while enhancing their efficacy. Gene therapy and stem cell-based approaches have introduced innovative ways to restore regenerative potential, particularly in chronic and hard-to-heal wounds. Stem cells, with their capabilities for differentiation, paracrine secretion, and immune modulation, have revolutionized the therapeutic landscape, while bioengineered scaffolds and gene-modified cells offer the potential for personalized and sustained healing.

Despite these advancements, challenges such as immune rejection, cost, regulatory hurdles, and the complexity of chronic wound microenvironments remain. Future directions in wound healing research should focus on integrating gene therapy, stem cell technologies, and bioengineered systems to create personalized and targeted treatments. Additionally, further exploration of combination therapies and the application of advanced biomaterials will be critical to optimizing outcomes.

The integration of cellular and molecular insights with cutting-edge technologies heralds a new era in wound management. Continued research and innovation will drive the development of more effective and accessible therapies, ultimately improving the quality of life for patients suffering from chronic wounds. By harnessing the potential of regenerative medicine and precision healthcare, the field is poised to transform wound care into a more targeted and successful endeavor.

References

1. Martin P (1997) Wound healing--aiming for perfect skin regeneration. Science 276(5309):75–81
2. Gurtner GC, Werner S, Barrandon Y, Longaker MT (2008) Wound repair and regeneration. Nature 453(7193):314–321
3. Schultz GS, Wysocki A (2009) Interactions between extracellular matrix and growth factors in wound healing. Wound Repair Regen 17(2):153–162
4. Werner S, Grose R (2003) Regulation of wound healing by growth factors and cytokines. Physiol Rev 83(3):835–870
5. Eming SA, Martin P, Tomic-Canic M (2014) Wound repair and regeneration: mechanisms, signaling, and translation. Sci Transl Med 6(265):265sr6
6. Martin P, Nunan R (2015) Cellular and molecular mechanisms of repair in acute and chronic wound healing. Br J Dermatol 173(2):370–378
7. Wu Y, Chen L, Scott PG, Tredget EE (2007) Mesenchymal stem cells enhance wound healing through differentiation and angiogenesis. Stem Cells 25(10):2648–2659
8. Barrientos S, Stojadinovic O, Golinko MS, Brem H, Tomic-Canic M (2008) Growth factors and cytokines in wound healing. Wound Repair Regen 16(5):585–601
9. Eming SA, Brachvogel B, Odorisio T, Koch M (2007) Regulation of angiogenesis: wound healing as a model. Prog Histochem Cytochem 42(3):115–170
10. Koh TJ, DiPietro LA (2011) Inflammation and wound healing: the role of the macrophage. Expert Rev Mol Med 13:e23
11. Singer AJ, Clark RA (1999) Cutaneous wound healing. N Engl J Med 341(10):738–746
12. Brem H, Tomic-Canic M (2007) Cellular and molecular basis of wound healing in diabetes. J Clin Invest 117(5):1219–1222
13. Falanga V (2005) Wound healing and its impairment in the diabetic foot. Lancet 366(9498):1736–1743
14. Eming SA, Wynn TA, Martin P (2017) Inflammation and metabolism in tissue repair and regeneration. Science 356(6342):1026–1030
15. Eming SA, Krieg T, Davidson JM (2007) Inflammation in wound repair: molecular and cellular mechanisms. J Invest Dermatol 127(3):514–525
16. Pastar I, Stojadinovic O, Krzyzanowska A, Barrientos S, Stuelten C, Zimmerman K et al (2010) Attenuation of the transforming growth factor β-signaling pathway in chronic venous ulcers. Mol Med 16:92–101
17. Tomasek JJ, Gabbiani G, Hinz B, Chaponnier C, Brown RA (2002) Myofibroblasts and mechano-regulation of connective tissue remodelling. Nat Rev Mol Cell Biol 3(5):349–363
18. Ferrara N (1999) Molecular and biological properties of vascular endothelial growth factor. J Mol Med 77:527–543
19. Nissen NN, Polverini P, Koch AE, Volin MV, Gamelli RL, DiPietro LA (1998) Vascular endothelial growth factor mediates angiogenic activity during the proliferative phase of wound healing. Am J Pathol 152(6):1445
20. Shi Y, Massagué J (2003) Mechanisms of TGF-β signaling from cell membrane to the nucleus. Cell 113(6):685–700
21. Wang HS, Hung SC, Peng ST, Huang CC, Wei HM, Guo YJ et al (2004) Mesenchymal stem cells in the Wharton's jelly of the human umbilical cord. Stem Cells 22(7):1330–1337

22. Greene CJ, Anderson S, Barthels D, Howlader MSI, Kanji S, Sarkar J et al (2022) DPSC products accelerate wound healing in diabetic mice through induction of SMAD molecules. Cells 11(15):2409
23. Loots MA, Lamme EN, Zeegelaar J, Mekkes JR, Bos JD, Middelkoop E (1998) Differences in cellular infiltrate and extracellular matrix of chronic diabetic and venous ulcers versus acute wounds. J Invest Dermatol 111(5):850–857
24. Stojadinovic O, Pastar I, Vukelic S, Mahoney MG, Brennan D, Krzyzanowska A et al (2008) Deregulation of keratinocyte differentiation and activation: a hallmark of venous ulcers. J Cell Mol Med 12(6b):2675–2690
25. Herrick S, Sloan P, McGurk M, Freak L, McCollum C, Ferguson M (1992) Sequential changes in histologic pattern and extracellular matrix deposition during the healing of chronic venous ulcers. Am J Pathol 141(5):1085
26. Grice EA, Segre JA (2012) Interaction of the microbiome with the innate immune response in chronic wounds. In: Current topics in innate immunity II, pp 55–68
27. Heldin C-H, Westermark B (1999) Mechanism of action and in vivo role of platelet-derived growth factor. Physiol Rev 79:1283
28. Ornitz DM, Itoh N (2001) Fibroblast growth factors. Genome Biol 2:1–12
29. Schultz GS, Ladwig G, Wysocki A (2005) Extracellular matrix: review of its roles in acute and chronic wounds. World Wide Wounds 2005:1–18
30. Ashcroft GS, Horan MA, Ferguson M (1998) Aging alters the inflammatory and endothelial cell adhesion molecule profiles during human cutaneous wound healing. Lab Investig 78(1):47–58
31. Diegelmann RF, Evans MC (2004) Wound healing: an overview of acute, fibrotic and delayed healing. Front Biosci 9(1):283–289
32. Grotendorst GR (1997) Connective tissue growth factor: a mediator of TGF-β action on fibroblasts. Cytokine Growth Factor Rev 8(3):171–179
33. George Broughton I, Janis JE, Attinger CE (2006) Wound healing: an overview. Plast Reconstr Surg 117(7S):1e-S–32e-S
34. Bjarnsholt T (2013) The role of bacterial biofilms in chronic infections. APMIS 121:1–58
35. Boateng JS, Matthews KH, Stevens HN, Eccleston GM (2008) Wound healing dressings and drug delivery systems: a review. J Pharm Sci 97(8):2892–2923
36. Bao P, Kodra A, Tomic-Canic M, Golinko MS, Ehrlich HP, Brem H (2009) The role of vascular endothelial growth factor in wound healing. J Surg Res 153(2):347–358
37. Johnson KE, Wilgus TA (2014) Vascular endothelial growth factor and angiogenesis in the regulation of cutaneous wound repair. Adv Wound Care 3(10):647–661
38. Rippa AL, Kalabusheva EP, Vorotelyak EA (2019) Regeneration of dermis: scarring and cells involved. Cells 8(6):607
39. Li J, Chen J, Kirsner R (2007) Pathophysiology of acute wound healing. Clin Dermatol 25(1):9–18
40. Ferrara N, Gerber H-P, LeCouter J (2003) The biology of VEGF and its receptors. Nat Med 9(6):669–676
41. Witherel CE (2019) Biomaterial-mediated control over macrophage behavior for tissue regeneration. Drexel University
42. Murphy PS, Evans GR (2012) Advances in wound healing: a review of current wound healing products. Plast Surg Int 2012(1):190436
43. Clark RA (2001) Fibrin and wound healing. Ann N Y Acad Sci 936:355–367
44. Ojeh N, Pastar I, Tomic-Canic M, Stojadinovic O (2015) Stem cells in skin regeneration, wound healing, and their clinical applications. Int J Mol Sci 16(10):25476–25501
45. Frykberg RG, Banks J (2015) Challenges in the treatment of chronic wounds. Adv Wound Care 4(9):560–582
46. Las Heras K, Igartua M, Santos-Vizcaino E, Hernandez RM (2020) Chronic wounds: current status, available strategies and emerging therapeutic solutions. J Control Release 328:532–550

47. Opalenik SR, Davidson JM (2005) Fibroblast differentiation of bone marrow-derived cells during wound repair. FASEB J 19(11):1561–1563
48. Huang C, Holfeld J, Schaden W, Orgill D, Ogawa R (2013) Mechanotherapy: revisiting physical therapy and recruiting mechanobiology for a new era in medicine. Trends Mol Med 19(9):555–564
49. Lee V, Singh G, Trasatti JP, Bjornsson C, Xu X, Tran TN et al (2014) Design and fabrication of human skin by three-dimensional bioprinting. Tissue Eng Part C Methods 20(6):473–484
50. Czaplewski L, Bax R, Clokie M, Dawson M, Fairhead H, Fischetti VA et al (2016) Alternatives to antibiotics—a pipeline portfolio review. Lancet Infect Dis 16(2):239–251
51. Hall-Stoodley L, Costerton JW, Stoodley P (2004) Bacterial biofilms: from the natural environment to infectious diseases. Nat Rev Microbiol 2(2):95–108
52. Høiby N, Bjarnsholt T, Givskov M, Molin S, Ciofu O (2010) Antibiotic resistance of bacterial biofilms. Int J Antimicrob Agents 35(4):322–332
53. Lewis K (2007) Persister cells, dormancy and infectious disease. Nat Rev Microbiol 5(1):48–56
54. Sharma D, Misba L, Khan AU (2019) Antibiotics versus biofilm: an emerging battleground in microbial communities. Antimicrob Resist Infect Control 8:1–10
55. Koo H, Allan RN, Howlin RP, Stoodley P, Hall-Stoodley L (2017) Targeting microbial biofilms: current and prospective therapeutic strategies. Nat Rev Microbiol 15(12):740–755
56. Salwiczek M, Qu Y, Gardiner J, Strugnell RA, Lithgow T, McLean KM et al (2014) Emerging rules for effective antimicrobial coatings. Trends Biotechnol 32(2):82–90
57. Ghosh C, Sarkar P, Issa R, Haldar J (2019) Alternatives to conventional antibiotics in the era of antimicrobial resistance. Trends Microbiol 27(4):323–338
58. Singh R, Ray P, Das A, Sharma M (2010) Penetration of antibiotics through Staphylococcus aureus and Staphylococcus epidermidis biofilms. J Antimicrob Chemother 65(9):1955–1958
59. Yan J, Bassler BL (2019) Surviving as a community: antibiotic tolerance and persistence in bacterial biofilms. Cell Host Microbe 26(1):15–21
60. Veves A, Falanga V, Armstrong DG, Sabolinski ML, Study ADFU (2001) Graftskin, a human skin equivalent, is effective in the management of noninfected neuropathic diabetic foot ulcers: a prospective randomized multicenter clinical trial. Diabetes Care 24(2):290–295
61. Duscher D, Barrera J, Wong VW, Maan ZN, Whittam AJ, Januszyk M et al (2016) Stem cells in wound healing: the future of regenerative medicine? A mini-review. Gerontology 62(2):216–225
62. Ginn SL, Alexander IE, Edelstein ML, Abedi MR, Wixon J (2013) Gene therapy clinical trials worldwide to 2012–an update. J Gene Med 15(2):65–77
63. Doudna JA, Charpentier E (2014) The new frontier of genome engineering with CRISPR-Cas9. Science 346(6213):1258096
64. Volarevic V, Markovic BS, Gazdic M, Volarevic A, Jovicic N, Arsenijevic N et al (2018) Ethical and safety issues of stem cell-based therapy. Int J Med Sci 15(1):36
65. Nunan R, Harding KG, Martin P (2014) Clinical challenges of chronic wounds: searching for an optimal animal model to recapitulate their complexity. Dis Model Mech 7(11):1205–1213
66. Metcalfe AD, Ferguson MW (2007) Tissue engineering of replacement skin: the crossroads of biomaterials, wound healing, embryonic development, stem cells and regeneration. J R Soc Interface 4(14):413–437
67. Howlader MSI, Prateeksha P, Hansda S, Naidu P, Das M, Barthels D et al (2024) Secretory products of DPSC mitigate inflammatory effects in microglial cells by targeting MAPK pathway. Biomed Pharmacother 170:115971
68. Sarkar J, Das M, Howlader MSI, Prateeksha P, Barthels D, Das H (2022) Epigallocatechin-3-gallate inhibits osteoclastic differentiation by modulating mitophagy and mitochondrial functions. Cell Death Dis 13(10):908
69. Prateeksha P, Howlader MSI, Hansda S, Naidu P, Das M, Abo-Aziza F et al (2023) Secretome of dental pulp-derived stem cells reduces inflammation and proliferation of glioblastoma cells by deactivating Mapk-Akt pathway. Dis Res 3(2):74

70. Laha D, Sarkar J, Maity J, Pramanik A, Howlader MSI, Barthels D et al (2022) Polyphenolic compounds inhibit osteoclast differentiation while reducing autophagy through limiting ROS and the mitochondrial membrane potential. Biomolecules 12(9):1220
71. Barthels D, Prateeksha P, Nozohouri S, Villalba H, Zhang Y, Sharma S et al (2023) Dental pulp-derived stem cells preserve astrocyte health during induced gliosis by modulating mitochondrial activity and functions. Cell Mol Neurobiol 43(5):2105–2127
72. Bandy R, Shahi S, Quagraine N, Shahbazi Nia S, Howlader MSI, Srivenugopal K et al (2023) Mechanistic aspects of biphenyl urea-based analogues in triple-negative breast cancer cell lines. ACS Pharmacol Transl Sci 7(1):120–136
73. Pittenger MF, Mackay AM, Beck SC, Jaiswal RK, Douglas R, Mosca JD et al (1999) Multilineage potential of adult human mesenchymal stem cells. Science 284(5411):143–147
74. Caplan AI (1991) Mesenchymal stem cells. J Orthop Res 9(5):641–650
75. Zuk PA, Zhu M, Mizuno H, Huang J, Futrell JW, Katz AJ et al (2001) Multilineage cells from human adipose tissue: implications for cell-based therapies. Tissue Eng 7(2):211–228
76. Gronthos S, Mankani M, Brahim J, Robey PG, Shi S (2000) Postnatal human dental pulp stem cells (DPSCs) in vitro and in vivo. Proc Natl Acad Sci 97(25):13625–13630
77. Toma JG, Akhavan M, Fernandes KJ, Barnabé-Heider F, Sadikot A, Kaplan DR et al (2001) Isolation of multipotent adult stem cells from the dermis of mammalian skin. Nat Cell Biol 3(9):778–784
78. Mahla RS (2016) Stem cells applications in regenerative medicine and disease therapeutics. Int J Cell Biol 2016(1):6940283
79. Clark RA, Ghosh K, Tonnesen MG (2007) Tissue engineering for cutaneous wounds. J Invest Dermatol 127(5):1018–1029

Chapter 6
Mitochondrial Orchestration of Wound Healing: From Cellular Energetics to Therapeutic Innovation

Meghnad G. Joshi, Mrunal N. Damle, and Rakesh Kumar Sharma

Abstract In the event of injury, tissue repair necessitates the complex, multi-step process of wound healing. Recent discoveries highlight the critical coordinating role of mitochondria, transcending conventional ATP production. This review examines mitochondrial dynamics, quality control, and signalling in the context of wound healing, highlighting their impact at every stage, including cellular migration, immunological responses, tissue remodelling, and stem cell activation. Cell-type-specific and subcellular mitochondrial adaptations, the processes responsible for the insufficient healing of wounds, and the supplementary effects of environmental variables on mitochondrial functions are highlighted. The study discusses cutting edge therapeutic options, such as antioxidants, metabolic modulators, transplantation, biogenesis enhancers, and new methods, including photobiomodulation and delivery systems based on nanotechnology that improve mitochondrial function. Experimental methodologies and translational viewpoints provide a framework for newer mitochondrial-centric therapeutics, addressing clinical obstacles, biomarker advancement, and regulatory issues. Understanding how mitochondria control themselves could lead to a new therapeutic approach that speeds up and improves the wound healing.

Keywords Mitochondrial dynamics · Wound healing · Cellular energetics · Therapeutic innovation · Metabolic reprogramming

M. G. Joshi (✉) · M. N. Damle
Department of Stem Cells and Regenerative Medicine, D.Y. Patil Education Society (Deemed University), D. Y. Patil Vidyanagar, Kolhapur, Maharashtra, India

Stem Plus Biotech Pvt. Ltd, Sangli, Maharashtra, India

R. K. Sharma
Department of Obstetrics and Gynecology, D.Y. Patil Education Society Deemed University, D.Y. Patil Medical College, Kolhapur, Maharashtra, India

F. De Francesco, K. Turksen (eds.), *Management and Strategies for Wound Healing*, Stem Cell Biology and Regenerative Medicine 79,
https://doi.org/10.1007/978-3-032-12442-5_6

Abbreviations

AMPK	AMP-activated protein kinase
ATF4	Activating Transcription Factor 4
ATF5	Activating Transcription Factor 5
ATP	Adenosine triphosphate
CHOP-C/EBP	Homologous Protein
CRISPR	Clustered Regularly Interspaced Short Palindromic Repeats
DAMPs	Mitochondrial danger signals
DFUs	Diabetic foot ulcers
ECM	Extracellular matrix
EGCG	Epigallocatechin Gallate
ER	Endoplasmic reticulum
HIF-1α	hypoxia-inducible factor-1α
HMGB1	High Mobility Group Box 1
IMM	Inner mitochondrial membrane
IPC	Ischemic preconditioning
iPSC	Induced pluripotent stem cell
MAM	Mitochondria associated membrane
MCU	Mitochondrial calcium uniporter
MDPs	Mitochondrial-derived peptides
MFN2	Mitofusin 2
MIRO1	Mitochondrial Rho GTPase 1
MSCs	Mesenchymal stem cells
NCLX	Sodium-calcium-lithium exchanger
NFAT	Nuclear factor of activated T-cells
NAD^+	Nicotinamide adenine dinucleotide
Nrf2	Nuclear factor erythroid 2-related factor 2
OMM	Outer mitochondrial membrane
OPA1	Optic Atrophy 1
OXPHOS	Glycolysis and oxidative phosphorylation
PEMFs	Pulsed electromagnetic fields
PGC-1α	Peroxisome proliferator-activated receptor γ coactivator 1-alpha
ROS	Reactive oxygen species
sgRNA	Single guide RNA
SMAD	Suppressor of Mothers against Decapentaplegic
ssODN	Single-stranded oligodeoxyribonucleotide
TGF-β1	Transforming Growth Factor beta 1
TNTs	Tunnelling nanotubes
TPP	Triphenylphosphonium
VDACs	Voltage-dependent anion channels
VLUs	Venous leg ulcers

Introduction

Wound healing represents one of the most remarkable biological processes in which the tissue integrity is restored through a highly coordinated cascade of cellular and molecular events. This intricate process unfolds through four distinct yet overlapping phases: haemostasis, inflammation, proliferation, and remodelling (Fig. 6.1), each characterized by specific cellular activities and metabolic demands [37, 55, 171].

The haemostasis phase, which is clearly marked by the rapid vasoconstriction and the platelet accumulation to obtain primary haemostasis, starts as soon as the tissue injury occurs (Fig. 6.2). During this stage, a fibrin clot forms, which stops the blood loss and acts as a temporary scaffold for the processes of tissue repair and cellular infiltration that follow [99].

Following haemostasis, there is an inflammatory phase, and chemotaxis of white blood cells and coagulation cells both contribute to the inflammatory process by releasing a variety of mediators and cytokines, such as platelet-derived growth factor and other substances that encourage the breakdown of collagen and the transformation of fibroblasts [189].

Granulation tissue, which is composed of inflammatory cells, fibroblasts, and neovasculature embedded in a matrix of fibronectin, collagen, and other extracellular matrix components, is formed during the proliferative phase, which typically happens 3–5 days after injury [104]. This stage coincides with the inflammatory phase that came before it and marks a significant change from a mostly catabolic to an anabolic condition in the wound environment (Fig. 6.2). There is growing

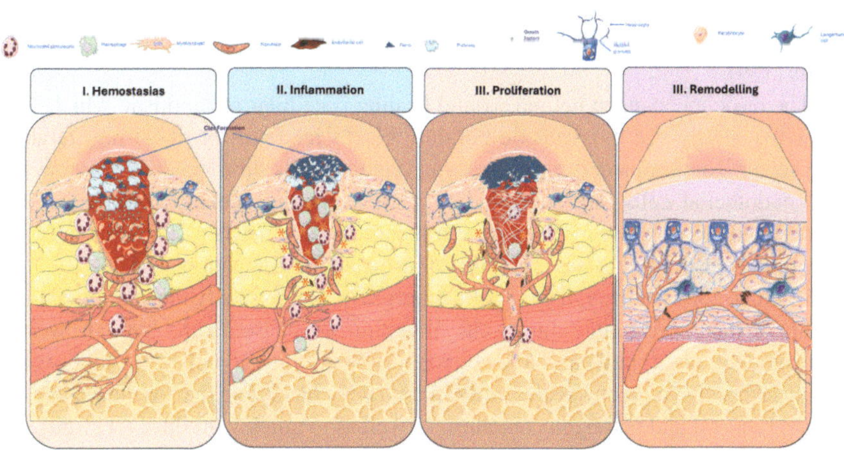

Fig. 6.1 Stages of wound healing. Schematic overview of the four key phases of wound healing: (I) Haemostasis (clot formation), (II) Inflammation (immune cell recruitment and cytokine release), (III) Proliferation (tissue regrowth and angiogenesis), and (IV) Remodelling (collagen reorganization and scar maturation). Key cell types and processes involved in each stage are highlighted

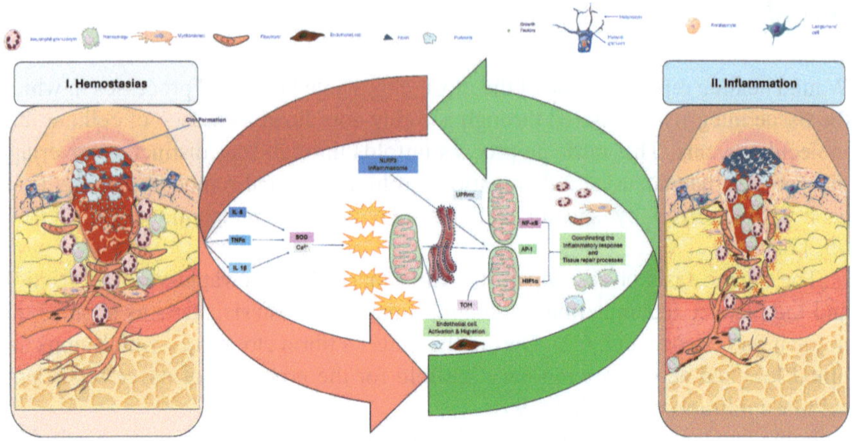

Fig. 6.2 Role of mitochondria in haemostasis and inflammation. Mitochondrial signalling pathways (e.g., mtROS, HIF1α, UPRmt) and molecular players (e.g., TNF-α, IL-6, NF-κB) coordinate the inflammatory response during wound healing. The diagram highlights interactions between platelets, neutrophils, macrophages, and endothelial cells, as well as activation of the *NLRP3 inflammasome* and growth factor release (e.g., fibrin, myofibroblasts)

evidence linking wound healing diseases to a disrupted transition from inflammation to proliferation, which has been identified as a critical phase in the healing process [42].

To recover mechanical strength and functionality, the newly produced tissue goes through structural reorganization during the last remodelling phase (Fig. 6.3), which also includes tissue maturation and the scar formation. Through a careful balance of the collagen synthesis and degradation, the extracellular matrix is continuously remodelled during this phase, which can last anywhere from months to years [139].

The basic mechanism behind wound healing is cellular metabolism, which supplies the energy and biosynthetic building blocks required for tissue regeneration and repair. Numerous cell types, such as immune cells, fibroblasts, keratinocytes, and endothelial cells, are subjected to high metabolic demands throughout the wound healing process; in order for these cells to perform their functions in tissue repair, they each need to undergo particular metabolic adaptations [12].

Immune cells go through metabolic reprogramming to satisfy the energy requirements of their active state during the inflammatory phase. In the hypoxic wound environment, neutrophils and M1 macrophages switch to glycolytic metabolism to produce reactive oxygen species and inflammatory mediators that are necessary for debris removal and pathogen clearance, as well as to quickly produce ATP [50]. On the other hand, a metabolic shift toward oxidative phosphorylation is necessary for the proliferative phase transition, especially in the M2 macrophages that support wound closure and tissue healing.

Fibroblasts serve as the primary functional cells in the proliferative phase of wound healing. A significant amount of energy is required for collagen production,

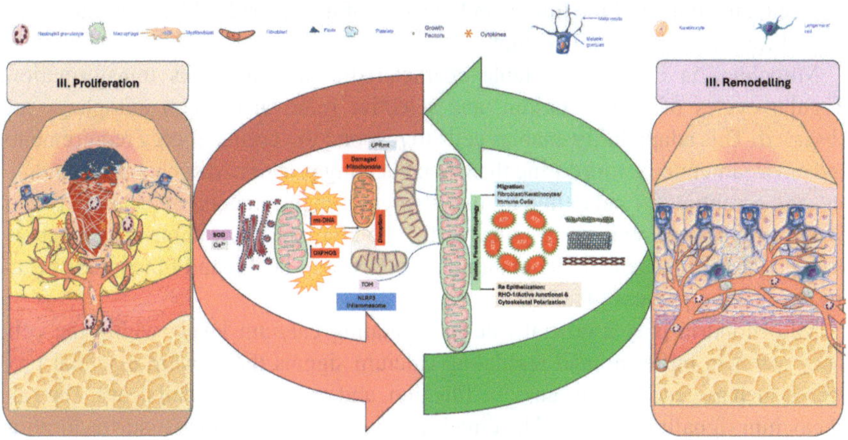

Fig. 6.3 Role of mitochondria in proliferation and remodelling. Mitochondrial dynamics regulate the *proliferation* phase (e.g., keratinocyte migration, angiogenesis) and *remodelling* phase (e.g., collagen deposition, immune polarization). Key components include metabolic transducers (e.g., *Higmetine*) and signalling pathways (e.g., *Pyrrole*) that drive functional polarization of immune cells and tissue repair

extracellular matrix formation, and cellular proliferation. Fibroblasts must concurrently regulate glucose utilization, amino acid synthesis, and lipid metabolism to fulfil these requirements and maintain cellular equilibrium [56]. Similarly, keratinocytes involved in the formation of new skin layers must align their energy production with the requirements of cellular movement and growth. Endothelial cells, crucial for angiogenesis, demonstrate metabolic flexibility. Depending on their activation state and signals from the environment, they can switch between glycolysis and oxidative metabolism. During the construction of new blood vessels, these cells undergo metabolic changes that promote growth, movement, and the production of vascular structures, all while maintaining the integrity of the vessels [32].

Mitochondria are frequently referred to as the "powerhouses" of the cell because of their involvement in ATP production via oxidative phosphorylation; however, their functions extend well beyond energy generation. Mitochondria are very important for many important functions, such as controlling metabolism, responding to stress, and deciding the fate of cells [38]. Recent mitochondrial research has shown that they are essential for cell communication, calcium balance, controlling apoptosis, and inflammatory responses, all of which are important for healing wounds. Mitochondria are integral to the production of reactive oxygen species (ROS), which serve dual functions as both beneficial and detrimental agents in the process of wound repair. Excessive reactive oxygen species (ROS) can harm tissues and slow down healing. On the other hand, controlled ROS generation works as a signalling system that affects gene expression, protein function, and whether a cell lives or dies [146]. Hence, for wound healing to progress effectively, it is very important to carefully control when and where mitochondrial ROS is generated. Consequently, precise management of the timing and location of mitochondrial ROS production is crucial for effective wound healing.

Mitochondria exhibit a notable characteristic of continuous transformation. Regular cycles of fusion, fission, and selective removal through mitophagy are observed. This adaptability enables cells to respond to varying energy requirements and environmental factors during the wound healing process. Mitochondrial dynamics are essential for adapting to the changing requirements of healing tissue. Mitophagy, the mechanism responsible for the elimination of damaged mitochondria, plays a crucial role in maintaining mitochondrial health and functionality during the healing process [41, 71].

Mitochondria play a crucial role in regulating calcium levels within the cell. They assist in buffering and regulating calcium signals that govern essential processes such as gene activity, enzyme function, and programmed cell death. The role of calcium signalling in wound healing is crucial because it guides tissue restoration and regulates cellular reaction to injury [120]. Another crucial role mitochondria perform is to regulate cell death via the intrinsic apoptotic pathway. The elimination of damaged or unwanted cells in a regulated manner is crucial for effective wound healing and tissue remodelling. Recent research highlights the significance of mitochondrial quality control mechanisms, such as the mitochondrial unfolded protein response (UPRmt), in aiding cells to sustain stability and function during the stress associated with wound repair [155].

The recognition of diverse mitochondrial functions has yielded significant insights into wound healing and has inspired novel treatment strategies. Mitochondrial transplantation has been demonstrated to enhance healing by reducing wound size, expediting granulation tissue formation, and promoting the growth of new skin, providing the first evidence of this therapy's efficacy in human wounds [102]. The expanding domain of mitochondrial medicine presents a viable approach for addressing both acute and chronic wounds by directly enhancing cellular energy, which is essential for tissue repair and regeneration.

Mitochondrial Biology and Its Relevance to Wound Healing

Structure and Function of Mitochondria

Mitochondria are small organelles within cells characterized by a double membrane structure, primarily recognized for their role in energy production. Additionally, they assist with various essential functions that maintain cellular viability and proper operations. Their structure is organized into an outer membrane (OMM), an inner membrane (IMM), and a space between the membranes known as cristae, which contains a central component [38, 66]. This provides variable functions to different components into separate areas, allowing the mitochondria to run several chemical processes at once.

The outer membrane acts as a selective barrier that only allows the passage of molecules up to 5 kDa through voltage-dependent anion channels (VDACs), although the inner membrane is impermeable to most ions and molecules, creating the electrochemical gradient important for ATP synthesis. The cristae are invaginations of the inner membrane; they increase the surface area for respiratory complex assembly and ATP production, with their morphology responding to cellular energy demands and metabolic conditions.

In the wound healing phenomenon, mitochondrial structure undergoes significant adaptations to meet the variable energy and biosynthetic demands of different healing phases with its functions. During the inflammatory phase, immune cells are required to produce rapid ATP generation to support their activated state, resulting in an increase of mitochondrial biogenesis and cristae density. Mitochondria in the proliferative phase show even more functional and structural adaptations due to the high energy need for tissue synthesis, including fibroblasts and keratinocytes, collagen production, cell division, and migration processes [12].

The mitochondrial respiratory chain, composed of five multi-subunit enzyme complexes (Complexes I–V), represents the core machinery for oxidative phosphorylation and ATP production. Recent studies have proved that mitochondrial respiration controls neoangiogenesis during wound healing and tumour growth, highlighting the critical role in wound healing of respiratory complex function in endothelial cell metabolism and vessel formation. Dysfunction of any component of this respiratory machinery can significantly impair wound healing by compromising the cellular energy production and changing redox homeostasis.

Mitochondrial Dynamics (Fusion, Fission, Mitophagy)

Mitochondrial dynamics include the continuous cycles of fusion, fission, and selective degradation (mitophagy), which permit mitochondria to adapt their morphology and circulation in response to cellular demands. To maintain the quality and functionality of mitochondria throughout the cell, they participate in repeated cycles of fusion and fission that mix the lipids and contents of a population of mitochondria. Mitochondrial fusion involves three essential GTPases: Mitofusin 1 and 2 (Mfn1/2) facilitate outer membrane fusion, while Optic Atrophy 1 (OPA1) is responsible for inner membrane fusion. This process facilitates the integration of mitochondrial components, such as proteins, lipids, and mitochondrial DNA, thereby sustaining mitochondrial function and averting the accumulation of defective organelles. High-energy output conditions can be optimized for metabolic requirements through the stimulation of mitochondrial fusion, facilitating component complementarity, and enhancing mitochondrial function. Studies demonstrated that the balance between fragmented and interconnected mitochondrial morphologies is crucial for sustaining mitochondrial functions and for modulating the network according to cellular requirements [51, 138]. Disruption of the fusion–fission balance has also been linked to impaired tissue regeneration and delayed wound healing, further underscoring the importance of mitochondrial dynamics in maintaining cellular homeostasis during repair processes [71].

Mitochondrial fission, primarily mediated by dynamin-related protein 1 (Drp1) and its adaptor proteins, facilitates the segregation of damaged mitochondrial components, thereby enabling their removal via mitophagy. Mitochondria may lose functionality under stress conditions, making mitochondrial fission crucial for the removal of defective components through mitophagy. However, the regulation of this process must be carefully balanced, as excessive mitochondrial fission can induce apoptosis, whereas moderate mitochondrial fusion can protect cells from oxidative stress [154].

Mitochondrial dynamics are essential in several stages of tissue restoration during wound healing. Mitochondrial dynamics are crucial for wound healing since they adjust to cellular requirements and external signals. In the inflammatory phase, immune cells rapidly undergo mitochondrial fission to enhance their activation and migration to areas of injury. The proliferative phase necessitates synchronized fusion events to meet the substantial energy requirements of tissue synthesis and cellular multiplication. Dysregulation of mitochondrial dynamics has been associated with chronic wound aetiology, wherein an altered balance of fusion and fission contributes to ongoing inflammation and delayed healing [62].

One important quality control mechanism that eliminates damaged or malfunctioning mitochondria from the cellular population is mitophagy, or the selective autophagy of mitochondria. During wound healing, the cells are subjected to high levels of oxidative stress and metabolic demands that may compromise mitochondrial integrity [41]. Important components in burn and other wound healing are mitochondrial dysfunction and mitophagy are known to be vital elements of the tissue recovery process. Maintaining a robust mitochondrial population that can meet the energy and biosynthetic needs of wound healing is ensured by effective mitophagy.

Reactive Oxygen Species (ROS) Signalling in Wound Repair

Mitochondria serve as producers and targets of reactive oxygen species, playing dual roles in wound healing through both beneficial signalling functions and potentially harmful oxidative damage. mtROS signalling is essential for wound healing at both single-cell and multi-cell levels, demonstrating the complex and nuanced role of mitochondrial ROS in tissue repair processes [185].

At normal levels, the reactive oxygen species (ROS) produced by mitochondria act as important signals that help control different parts of the wound healing process. A rise in mitochondrial ROS can stabilize a protein called HIF1α, which then shifts macrophages into the M1 type. M1 macrophages facilitate angiogenesis and initiate inflammation, both of which are critical early processes in the healing response. This signalling facilitates the transition from the inflammatory phase to tissue repair and angiogenesis [60].

Mitochondrial ROS generation is tightly controlled by the respiratory chain, the cell's antioxidant defences, and its overall energy level. In the process of wound healing, the regulated production of ROS serves multiple beneficial functions. They assist in combating microbes, initiate signals that provoke inflammation and promote the formation of new blood vessels and affect gene activity by activating specific redox-sensitive transcription factors. Mitochondrial ROS activate significant signalling pathways such as NF-κB, AP-1, and HIF-1α.

These pathways regulate the inflammatory response and facilitate tissue repair. M1 macrophages facilitate angiogenesis and initiate inflammation, both of which are critical early processes in the healing response. This signalling facilitates the transition from the inflammatory phase to tissue repair and angiogenesis [60]. However, when ROS production is too high or lasts too long, it can specifically hinder the healing process. Sustained mitochondrial ROS overproduction has been shown to impair keratinocyte migration and perpetuate inflammatory signaling in chronic wound environments [86]. Excessive ROS can damage important cell components, maintaining prolonged inflammation and hampering the wound healing process. It is important to maintain the balance between helpful ROS signalling and harmful oxidative stress during wound healing. The body relies on antioxidant systems to keep this balance in check. These include enzymatic antioxidants like superoxide dismutase, catalase, and glutathione peroxidase, as well as non-enzymatic ones such as glutathione, vitamin C, and vitamin E [45]. Most recent evidence indicates that increased mtROS can induce the relocalization of essential glycolytic enzymes to the mitochondrial surface, implying a redox-sensitive mechanism that spatially reconfigures cellular metabolism under stress. This ROS-dependent metabolic reorganization may enhance substrate channelling between glycolysis and oxidative phosphorylation, a process likely significant in the high-energy requirements of wound healing and tissue regeneration [44].

Recent studies have shown that both the timing and location of mitochondrial ROS production matter a lot during wound healing. In the early inflammatory phase, a quick burst of ROS helps kill microbes and signals immune cells to come to the injury site. Later on, ROS levels need to be carefully reduced to allow the wound to move into the proliferative and remodelling phases. If this timing is off or the regulation is poor, it can lead to chronic wounds and poor healing.

Calcium Homeostasis and Apoptosis Regulation

Mitochondria are important for keeping calcium levels in cells balanced and controlling apoptosis, both of which are important for healing wounds. The mitochondrial calcium handling system consists of the mitochondrial calcium uniporter (MCU), which takes in calcium, and the sodium-calcium-lithium exchanger (NCLX), which allows calcium out. Mitochondria serve as versatile calcium buffers, modulating cytosolic calcium concentrations and enabling essential cellular signalling pathways [172]. Calcium signalling is crucial in various aspects of wound healing, including the control of cell motility, proliferation, differentiation, and apoptosis. Mitochondria take in calcium, which changes the production of ATP and ROS and turns on certain calcium-sensitive enzymes that help repair tissue. For effective healing, the timing and location of mitochondrial calcium handling must be precise to ensure the coordination of complex cellular responses [164].

Mitochondria are crucial for controlling the intrinsic apoptotic pathway, which is necessary for proper wound healing. This pathway helps get rid of damaged cells, lowers inflammation, and guides the right way for tissues to remodel. The mitochondrial permeability transition pore (mPTP) is an important part of this process. It is a protein complex that goes through both mitochondrial membranes. When the mPTP opens, the mitochondria swell, cytochrome c is released, and the caspase cascade is turned on. This all leads to programmed cell death, or apoptosis [10].

During the healing process, apoptosis must be carefully controlled. Too much cell death can slow down healing, while inadequate cell death can also cause difficulties. During the initial inflammatory phase, the controlled apoptosis of immune cells, including neutrophils, is crucial for reducing inflammation and facilitating the healing process. Apoptosis allows for the elimination of surplus fibroblasts and myofibroblasts during the remodelling phase. This stops excessive scar tissue from forming and helps restore the normal structure of the tissue [10]. Disturbance of mitochondrial control over calcium and apoptosis may lead to compromised wound healing. Too much apoptosis can damage tissue and slow down healing, while a lack of apoptosis can cause long-lasting inflammation or too much scarring.

Improved knowledge of how mitochondria control these functions may therefore aid in the development of novel therapies for chronic or difficult-to-heal wounds.

Mitochondrial Quality Control Mechanisms (UPRmt, Protein Import/Export)

The maintenance of mitochondrial quality is crucial for ensuring that mitochondria remain functional and effective, particularly during the process of wound healing, when cellular energy and support demands are high. The mitochondrial unfolded protein response (UPRmt) is a crucial signalling pathway that enables the cell to manage stress occurring within the mitochondria. Upon the misfolding or damage of proteins within the mitochondrial matrix, the UPRmt is activated, initiating signals directed towards the nucleus. This results in an enhanced synthesis of protective molecules such as chaperones and proteases, which assist in refolding or eliminating defective proteins, thereby restoring equilibrium [64].

Mitochondria rely on their own proteostasis systems to maintain protein quality, and when these systems are overwhelmed, the UPRmt helps restore order. In mammals, this pathway involves important transcription factors such as Activating Transcription Factor 4 (ATF4), Activating Transcription Factor 5 (ATF5), and C/EBP Homologous Protein (CHOP). These factors help turn on genes that support mitochondrial repair and stress adaptation. This mitochondrial-to-nucleus communication, also called retrograde signalling, has also been studied in cancer cells, where mitochondrial stress from misfolded proteins or high ROS levels activates the UPRmt. A similar response likely plays a role in wound healing, where cells face significant stress and energy demands that challenge mitochondrial stability [64].

Another important part of mitochondrial quality control is protein import. Since most mitochondrial proteins are made in the nucleus, they must be delivered into mitochondria correctly. This is handled by the translocase of the outer membrane (TOM) and the translocase of the inner membrane (TIM) complexes, which move over a thousand nuclear-encoded proteins into the mitochondria. If this import process does not work properly, mitochondrial function and cell metabolism can suffer—something especially problematic during wound healing [5].

Mitochondria also need to get rid of damaged or unneeded proteins, and they do this through protein export pathways. Some of the exported proteins even act as signals to help the cell respond to stress. Recent research has identified several of these export systems, including the release of mitochondrial-derived peptides, which may help regulate broader cellular stress responses.

During wound healing, both the UPRmt and the protein import/export systems must adapt to meet higher energy demands, support new mitochondrial growth, and handle increased cellular stress. When these quality control mechanisms fail, mitochondria can become dysfunctional, slowing metabolism and delaying healing. Studying these pathways gives us a better understanding of what goes wrong in chronic wounds—and how we might fix it with targeted treatments.

Mitochondrial-Nuclear Communication and Retrograde Signalling

Mitochondria and the nucleus constantly communicate with each other so that cells can deal with stress and changing energy needs. This is especially important when healing a wound. This complex communication system keeps mitochondria working effectively and makes sure the cell can adjust to different metabolic conditions. Mitochondrial proteostasis, or protein balance, is a key part of this system. Retrograde signalling, in which mitochondria communicate to the nucleus to turn on genes that make mitochondrial chaperones and proteases, keeps it in surveillance [173].

This two-way interaction allows mitochondria to edit the expression of nuclear genes in addition to responding to signals communicated by the nucleus. When there is stress or a challenge with generating energy or checking the quality of proteins, mitochondria send signals back to the nucleus. Changes in mitochondrial ATP levels, ROS, calcium levels, or the balance of protein folding may initiate these signals. This often leads to the activation of large signalling networks inside the cell. This includes changes in important metabolites, signalling by ROS, calcium release, and the movement of molecules that come from mitochondria.

Key metabolites, including acetyl-CoA, α-ketoglutarate, succinate, and the NAD^+/NADH ratio, play a crucial role in these signalling pathways. These molecules not only provide energy for cellular processes but also play a crucial role in regulating gene expression through epigenetic modifications. In the process of wound healing, these signals play a crucial role in the regulation of genes that are essential for energy production, inflammation, tissue construction, and repair mechanisms. The citrate–malate cycle generates acetyl-CoA, which plays a crucial role in histone acetylation and lipid synthesis, both essential processes for healing [108, 145].

Mitochondrial ROS production is a prominent trigger of retrograde signalling. Reactive oxygen species not only indicate that there is an issue within the mitochondria, but they also initiate the activation of transcription factors, signalling proteins, and epigenetic modifications that influence nuclear gene expression, aiding the cell in its adaptation process. Calcium serves as an essential signalling molecule. The release of calcium by mitochondria can trigger certain transcription factors that translocate to the nucleus, thereby activating genes that respond to calcium signals. The NFAT (nuclear factor of activated T-cells) family and CaMKs (calcium/calmodulin-dependent protein kinases) represent two significant components in the realm of calcium-dependent mitochondrial-nuclear communication [140].

Recent studies have identified mitochondrial-derived peptides (MDPs) as a new class of signalling molecules. These small peptides are generated in the mitochondria and then sent to other parts of the cell, where they are crucial in regulating gene expression and reacting to stress. During the process of wound healing, MDPs may be very important in controlling cellular metabolism and the processes that initiate repair.

In general, these retrograde signalling pathways help cells' gene expression and metabolic processes stay in synchronized with the current state of their mitochondria. When this communication collapses, cells are unable to react to stress as well, which could slow down healing and make tissue repair less effective. A better understanding of these pathways could lead to the development of new drugs for chronic wounds and better ways to help tissue rejuvenation.

Mitochondrial Heterogeneity in Wound Healing

Cell-Type Specific Mitochondrial Adaptations in Wound Healing

Wound healing requires significant amounts of energy, and mitochondria's ability to adapt is very important for different types of cells to meet these needs as they shift. Mitochondria exhibit considerable functional versatility across diverse cell types involved in wound healing, underscoring their unique roles in energy production, redox signalling, and biosynthetic pathways. Every cellular component, such as macrophages, fibroblasts, keratinocytes, and endothelial cells, demonstrates unique mitochondrial metabolic plasticity that corresponds to its changing function during different phases of healing [62, 185]. Fibroblasts are very important for installing out the ECM and speeding up both glycolysis and OXPHOS during the proliferative phase. Their mitochondria change in ways that make it easier for collagen to be produced, along with for cells to become myofibroblasts. These changes involve more biogenesis and fission. Endothelial cells similarly deploy mitochondrial fragmentation and localized ATP hotspots to fuel angiogenic sprouting in the hypoxic granulation tissue [92]. The specificity of this cell type is highlighted by unique mitochondrial proteomes, stress responses, and antioxidant defences, which all demonstrate adaptation to different microenvironments and functional needs during the process of wound repair [62, 185]. Mitochondrial dynamics in keratinocytes undergo changes during re-epithelialization, where enhanced fission facilitates migration through ROS-mediated activation of pro-migratory kinases such as FAK and Src [14, 62, 177]. Fibroblasts, in contrast, rely on mitochondrial OXPHOS to sustain collagen synthesis, with impaired ATP production linked to fibrosis in chronic wounds [106]. Macrophages display perhaps the most striking metabolic flexibility M1-polarized cells favour glycolysis and accumulate fragmented mitochondria producing inflammatory ROS, while M2 macrophages adopt elongated mitochondria for oxidative metabolism to support tissue repair [177]. Keratinocytes, integral to re-epithelialization, must rapidly migrate and proliferate a feat demanding dynamic mitochondrial remodelling. During the progression wound, keratinocyte mitochondria increase fusion and local ATP production, which powers cytoskeletal rearrangement and directional migration. Importantly, mtROS produced here acts as a second messenger in the modulation of actin and cell–cell adhesion dynamics, expediting closure of the wound edge [14, 62, 177]. Cacace proposed that mitochondrial respiration and the OXPHOS complex content in PBMCs are compromised by inadequate glycaemic management as type 2 diabetes progresses. T2D patients with inadequate glycaemic control are more susceptible to vascular complications because of these alterations, which take place concurrently with increased neutrophil endothelial contacts and adhesion molecule levels [14]. Recent research involving *mito-Keima* mice has shown that endothelial cells in healing wounds keep a reserve of mitochondria ready for mitophagy, which helps support the formation of new blood vessels when there is low oxygen [92]. However, these adaptations often disturb in diabetes people. For instance, diabetic neutrophils exhibit excessive mitochondrial ROS that perpetuates inflammation, a phenotype reversed by mitochondrially targeted SOD2 [20, 196].

Altogether, these findings underscore not just mitochondria's role as cell powerhouses, but as metabolic sentinels finally tuned to the spatial and temporal dynamics of cutaneous repair.

Tissue-Resident Versus Recruited Cell Mitochondrial Differences

There is an array of significant differences between tissue-resident and recruited cells during wound healing. Tissue-resident immune cells differ from their circulating leukocyte counterparts in that they are extremely heterogenous and plastic with distinct mitochondrial capabilities. These resident cells can adapt to their tissue-specific environment due to specialized metabolic profiles. Tissue-resident macrophages and fibroblasts possess greater coupling efficiency and superior oxidative mitochondria, adapted to the unique redox and metabolic potential, dignified for rapid response to injury [90, 94]. After the tissue injury, circulating monocytes are activated and directed to the wound bed. Differentiation between wound-associated macrophages and the fibroblasts often involves a noticeable reprogramming of their mitochondrial networks. Interestingly, these recruited cells may have elevated mitochondrial fission and glycolytic rates at first, a phenotype reliable with immediate pro-inflammatory, antimicrobial, or migratory demands. On the other hand, resident cells are generally influenced by on more balanced mitochondrial morphology and substrate flexibility, properties that promote reparative and anti-inflammatory processes.

Tissue-resident macrophages can quickly detect minor cellular damage and isolate the injury, representing rapid mitochondrial activation patterns distinct from recruited neutrophils and other circulating immune cells. This immediate response capability reflects their unique mitochondrial positioning and the metabolic readiness [25].

Qin et al. highlight that macrophages are regulated by mitochondria and its polarization and function—key factors of immune responses during inflammation and tissue repair. In pro-inflammatory (M1) macrophages, mitochondrial oxidative phosphorylation (OXPHOS) is suppressed, resulting in the dependence of cells on glycolysis for energy.

This metabolic shift is carefully allied with increased production of mitochondrial reactive oxygen species (mtROS) and also accumulates metabolites like succinate, resulting in promoting inflammatory cytokine production and microbial defence. On the other hand, OXPHOS inhibition also stops M1 macrophages from effectively changing into M2 macrophage phenotypes [135].

While anti-inflammatory macrophages M2, which are essential and important for tissue repair and resolution, retain intact TCA cycle and powerful mitochondrial OXPHOS. This supports their anti-inflammatory functions, effective energy production, and tissue remodelling properties. This suggests that treatments targeted at re-establishing appropriate mitochondrial activity can reduce inflammation and facilitate efficient wound healing by positioning mitochondrial metabolism as a signal and a therapeutic target in inflammatory diseases [81, 112].

Intriguingly, neutrophils rely mainly on glycolysis for energy; their mitochondria have a distinct membrane potential ($\Delta\psi$m) that plays an important role beyond classical ATP production. The stable membrane potential supports essential functions like the control of apoptosis, chemotaxis, and niche adaptation, particularly in hypoxic conditions, despite the mitochondrial respiratory contribution to total ATP synthesis and oxygen utilization in neutrophils being essentially minimal (less than 15%). In vivo, the occurrence and variation of membrane potential ($\Delta\Psi$m) stimulate neutrophil migration speed, calcium signalling, and the controlled production of mitochondrial reactive oxygen species, resulting in antimicrobial defence and immune regulation. Thus, mitochondrial membrane potential in neutrophils is not simply an epiphenomenon but a critical controller of their viability, function, and adaptability in immune responses [25].

The division of labour is characterized by tissue-resident cells that tolerate tissue homeostasis and repair, while recruited cells deliver urgent inflammatory support before the adaptation of mitochondria to the wound site. This interplay, shaped by the intrinsic mitochondrial programming and the evolving wound niche, represents a critical axis for effective and balanced tissue repair.

Subcellular Mitochondrial Positioning and Function

The correct arrangement of bioenergetic and signalling components is necessary for the healing of wounds. The placement of mitochondria in cells is closely related to the specific roles they play in healing wounds. Mitochondria frequently reposition within cells to regions of elevated ATP demand, such as the leading edge of migrating keratinocytes or the tip of contractile myofibroblasts [90]. The spatial reorganization is very important for making certain that cellular repair mechanisms work together. Targeted mitochondrial redistribution toward high-energy-demand regions such as the leading edge of migrating keratinocytes is essential for efficient wound closure and cytoskeletal remodeling [91]. The placement of subcellular mitochondria is dynamically regulated during the processes of wound healing. Xu et al. [185] claim that tissue injuries induce a sudden elevation in cytosolic Ca^{2+}, resulting in mitochondrial fragmentation and a subsequent reliance on DRP-1 or MIRO-1. The broken-up mitochondria then make more mtROS and help the actin ring close.

Small and fragmented mitochondria are directed to wound sites in cells that are healing, showing that mitochondria are actively moving around within cells to meet local energy and signalling requirements. The correct arrangement of bioenergetic and signalling components is necessary for the healing of wounds. The placement of mitochondria in cells is closely related to the specific roles they play in healing wounds. Mitochondria frequently reposition within cells to regions of elevated ATP demand, such as the leading edge of migrating keratinocytes or the tip of contractile myofibroblasts [90]. The spatial reorganization is very important for making certain that cellular repair mechanisms work together. The placement of subcellular mitochondria is dynamically regulated during the processes of wound healing. Xu et al. [185] claim that tissue injuries induce a sudden elevation in cytosolic Ca^{2+}, resulting in mitochondrial fragmentation and a subsequent reliance on DRP-1 or MIRO-1. The broken-up mitochondria then make more mtROS and help the actin ring closure.

Small and fragmented mitochondria are directed to wound sites in cells that are healing, showing that mitochondria are actively moving around within cells to meet local energy and signalling requirements [94]. This positioning allows for localized ATP production and reactive oxygen species generation necessary for wound repair processes. The endoplasmic reticulum-mitochondria association also plays a regulatory role in wound healing. ER-mitochondria association negatively affects wound healing by regulating NLRP3 activation, indicating that subcellular organelle interactions significantly influence inflammatory responses during tissue repair [83]. Mitochondrial Bioenergetics of Functional Wound Closure is dependent on macrophage–keratinocyte exosomal crosstalk, demonstrating that efficient wound healing relies on a critical crosstalk between macrophages and keratinocytes at the wound site, specifically mediated by exosomes [152]. During early wound reepithelialization, macrophage-derived exosomes enriched with the mitochondrial outer membrane protein TOMM70 are transferred to leading-edge keratinocytes. TOMM70 plays a vital role in maintaining mitochondrial protein import and mitochondrial quality control. In wounded, hypoxic skin, endogenous TOMM70 becomes depleted in keratinocytes; however, exosomal delivery of TOMM70 from macrophages restores mitochondrial function, supports glycolytic ATP production, and powers keratinocyte migration, which is essential for wound closure [152].

Compartmentalization is not limited to cytoplasmic domains; mitochondrial-nuclear and mitochondrial-ER contacts also dynamically change, coordinating metabolic adaptation and stress responses during wound healing [62, 83]. These organellar contact sites serve as platforms for calcium signalling, lipid metabolism, and retrograde signalling pathways that are essential for cellular adaptation to injury conditions [142].

Advanced imaging and single-cell technologies continue to reveal the complex spatial organization of mitochondria within healing tissues. Spatiotemporal single-cell analysis of human skin wound healing deepens understanding of tissue repair mechanisms, providing new insights into how mitochondrial positioning coordinates with cellular function during the healing process [89]. Recent research by Landry et al. [77] indicates that disrupted ER–mitochondria coupling is a common manifestation in neurodegenerative diseases, significantly affecting cellular energy metabolism and calcium homeostasis. They discovered that the depletion of VAPB induces the activation of the integrated stress response and a significant disruption of ER–ER–mitochondria contacts, resulting in mitochondrial dysfunction.

When the ER and mitochondria are not connected, the mitochondria take up less calcium, which makes oxidative phosphorylation and ATP production less effective. The findings indicate a convergent mechanism that connects stress signalling to metabolic disruption through the physical uncoupling of ER–mitochondria interfaces [77].

The subcellular choreography reflected in mitochondrial shape, movement, and networking epitomizes the remarkable adaptability required for effective wound repair, merging bioenergetics, signalling, and structural plasticity in time and space [94, 185]. The importance of correct mitochondrial organization and function has been brought to light by recent research [100]. This study highlights the fact that inadequate wound healing might set the foundation for chronic illnesses.

Mitochondria in Different Phases of Wound Healing

Inflammatory Phase

Mitochondrial Regulation of Immune Cell Function (Macrophages, Neutrophils)

The inflammatory phase represents the initial response of the body following tissue damage. At this phase of wound healing, immune cells such as macrophages and neutrophils are rapidly mobilized to the affected area to assist in debris clearance and infection prevention. The immune functions of these cells are well-documented; however, their activity is meticulously controlled by mitochondria, which are historically viewed solely as energy producers but are now acknowledged as crucial regulators of immune function. Recent research conducted by Hunt and colleagues [117, 118] has demonstrated that mitochondria play a role beyond mere energy production; they are intricately involved in regulating immune cell behaviour, functioning as signalling centres that facilitate the activation and direction of immune responses during the healing process. A 2024 study by O'Siorain et al. [117] found that mitochondrial metabolism shifts depending on the time of day, driven by the body's internal circadian clock. These daily fluctuations directly affect how immune cells like macrophages activate inflammasomes such as NLRP3, which are essential for launching inflammatory responses. This means that even the timing of an injury could influence the strength and nature of the body's immune response, thanks to mitochondria (Fig. 6.2).

Macrophages change from the pro-inflammatory "M1" type to the anti-inflammatory "M2" type as wound healing advances to support tissue repair. This transformation is not arbitrary; rather, it depends heavily on the ability of the cell to reorganize its metabolism and mitochondrial function. Marques et al. [98] suggest that mitochondria have emerged as crucial signal-processing platforms, connecting metabolism and immune function in essential ways for the healing process.

ROS and NLRP3 Inflammasome Activation

The NLRP3 inflammasome is like a central hub that connects mitochondrial stress to inflammation when wounds are healing. It is important to make sure that the body's immune system is strong enough to keep it safe without getting in the way of healing.

A recent study by D'Achille et al. [31] found that the endoplasmic reticulum (ER) and mitochondria work together to activate NLRP3. When this link between the ER and mitochondria breaks down, it helps to lower inflammation and speed up healing. This discovery opens the door to potential treatments that might reduce inflammation and speed up healing.

There are several ways that mitochondria change how NLRP3 works. Paik et al. [118] claim that they achieve this by producing ROS, releasing mitochondrial DNA, and modifying cellular metabolism. The inflammasome helps cytokines like IL-1 derive out of cells, which creates a controlled inflammatory environment that is important for healing.

In the context of wound healing, NLRP3 also helps immune cells (macrophages) and skin cells (fibroblasts) communicate with each other. Zhao et al. [195] demonstrated that the lack of NLRP3 leads to delayed wound healing, marked by dimin ished collagen synthesis and compromised skin regeneration. This shows how important it is to maintain inflammasome activity in check for proper tissue repair.

Li et al. [80] proposed that excessive amounts of a beneficial factor could be detrimental. The group demonstrated that increased NLRP3 activation obstructs the healing process in older mice. Importantly, they found that NLRP3 inhibitors helped wounds heal faster. It reveals that healing depends on finding the right level of inflammasome activity: enough to fight infection and speed up repair, but not so much that it causes damage or delays.

Proliferative Phase

Role in Fibroblast Activation and Collagen Synthesis

During the proliferative stage of wound healing, the body begins to regenerate damaged tissue. In this phase, fibroblasts proliferate, synthesize collagen, and deposit the extracellular matrix, which are essential processes in the formation of new tissue. Mitochondria are essential for powering demanding cellular activities, serving not only as energy sources but also as signalling hubs that coordinate cellular behaviour and environmental responses.

Collagen production requires significant energy and amino acids. Fibroblasts must alter their metabolism to meet demands, with mitochondria adapting to facilitate the increased need for protein synthesis (Fig. 6.3). Talbott et al. [162] demonstrated that fibroblasts exhibit heterogeneity; distinct subtypes possess unique metabolic profiles aligned with their functions in the healing process. Some entities concentrate on ECM production, whereas others engage in signalling or tissue remodelling.

The interaction between macrophages and fibroblasts is essential for effective healing. Mitochondrial signals significantly influence this interaction, particularly via inflammasome pathways. Zhao et al. [195] demonstrated that the NLRP3 inflammasome is integral to this process. When functioning optimally, it facilitates macrophages in transmitting appropriate signals to fibroblasts, thereby prompting their activation and subsequent collagen production at the precise moment. This precise synchronization prevents excessive fibroblast responses, which could result in scarring or fibrosis.

Mitochondrial Support for Keratinocyte Migration and Re-epithelialization

Re-epithelialization is a key part of the proliferative phase of wound healing. It is the process where keratinocytes (the main cells in the skin's outer layer) move in, multiply, and rebuild the damaged surface. For this to happen smoothly, keratinocytes need both energy and tightly regulated signalling, much of which is supplied by mitochondria.

To migrate across the wound bed, keratinocytes need to constantly reorganize their internal structure, adjust their membranes, and form new attachments, all of which are energy-intensive. Mitochondria help by moving to where energy demand is highest within the cell, delivering localized ATP. They also generate reactive oxygen species (ROS), which, in controlled amounts, act as signalling cues to promote keratinocyte movement and growth.

Keratinocytes work with other types of cells to do their tasks. Their ability to heal is affected by how they interact with other cells at the site of the wound. Sharma et al. [152] recently showed that exosomes help keratinocytes make energy by allowing macrophages and keratinocytes to talk to each other. This kind of crosstalk allows skin cells to get the nutrients they need to heal wounds quickly.

Keratinocytes not only make a physical barrier, but they also help the immune system collaborate during the healing process. Piipponen et al. [126] state that these cells are integral to immune signalling and demonstrate dynamic responses to environmental stimuli. Their metabolic flexibility, made possible by adaptable mitochondrial function, allows them to change roles as demanded. This helps with tissue repair and communication, and coordination with other cell types at the wound site.

Angiogenesis and Endothelial Cell Metabolism

Angiogenesis, the genesis of new blood vessels, is an energy-intensive process during wound healing. The process requires both a quick increase in the number of endothelial cells and a carefully controlled change in the way they consume energy. Mitochondria are very important in this process because they give endothelial cells energy and control the signals that are required for generating new blood vessels.

Schiffmann et al. [147] conducted research that showed mitochondrial respiration plays a direct and essential role in angiogenesis during wound healing and tumour growth. The research demonstrated that the transition of blood vessels from a quiescent to an active, proliferative state is meticulously governed by mitochondrial function, specifically complex III of the electron transport chain, which is crucial for endothelial cell proliferation and sprouting.

During angiogenesis, endothelial cells undergo substantial metabolic reprogramming to keep growing quickly and maintain their functions. They need to find a good balance between glycolysis and oxidative phosphorylation. This metabolic adaptability enables the formation of new blood vessels to meet the increased oxygen and nutrient demands of regenerating tissue.

Recent research underscores the capacity of stem cell-derived endothelial cells to promote angiogenesis. Induced pluripotent stem cell (iPSC)-derived endothelial cells have shown the capacity to expedite wound closure by enhancing the formation of blood vessels. Their effectiveness is largely contingent upon optimal mitochondrial function and accelerated metabolic activity. When that are employed with biomaterial scaffolds, these cells are a powerful way to improve blood flow and tissue repair in wound healing [27].

Remodelling Phase

Mitochondrial Influence on Myofibroblast Function and Scar Formation

The remodelling phase, which can last anywhere from 3 weeks following an injury to several months or even a year, is the last stage of wound healing. The wound site experiences significant restructuring during this period. A mature extracellular matrix gradually replaces temporary tissue, and apoptosis eliminates unnecessary cells. During this stage, mitochondria remain essential because they not only provide the energy required for these functions but also have an impact on the choices cells make regarding removal, repair, or survival.

The myofibroblast is a key player in this stage. Through the process of the production and organization of ECM components, these specialized cells are in charge of reshaping the tissue and drawing the edges of the wound together. Fibroblasts must transform into myofibroblasts in order to accomplish this, a process that requires drastic metabolic adjustments and is supported by highly active mitochondria. These cells' energy systems must produce a lot of contractile proteins, such as alpha-smooth muscle actin, and matrix-remodelling enzymes (Fig. 6.3).

It is interesting to note that mitochondria not only promote this transformation but also aid in its regulation. During the remodelling phase, fibroblast behaviour is still influenced by the NLRP3 inflammasome, an immune response component that is strongly linked to mitochondrial activity. According to research by Zhao et al. [195], NLRP3 regulation is essential for efficient and natural wound closure. Excessive scarring or tissue lacking the strength required for long-term healing can result from a disturbance in this balance.

ECM Remodelling and Resolution of Inflammation

The resolution of inflammation and the remodelling of the extracellular matrix (ECM) are interrelated processes that depend on precisely coordinated cellular activities. Mitochondria are central to cellular function, recognized for their role in energy production, as well as their importance in regulating apoptosis, generating signalling molecules, and regulating metabolic processes [16, 49]. Recent studies indicate that dysfunctional mitochondria in macrophages can extend inflammation and hinder tissue repair. A primary mechanism involves the disruption of efferocytosis, the process responsible for the clearance of dead cells [200].

As tissues move into the remodelling phase of wound healing, inflammation must taper off but not at the expense of ongoing tissue repair. This balance requires mitochondria to work in a finely tuned way; they must support energy production for matrix-producing cells, while also helping clear out unneeded inflammatory cells through apoptosis. However, damaged mitochondria can actually trigger inflammation themselves, which can delay healing [180]. In fact, leaking mitochondrial DNA has been identified as a driver of chronic inflammation in aging and age-related diseases, underscoring the need to keep mitochondrial integrity intact during tissue remodelling [39].

Communication between different types of cells like fibroblasts, immune cells, and endothelial cells is key during this phase, and mitochondria help guide this cross-talk. For example, mitochondrial signalling helps coordinate how fibroblasts respond to inflammatory cues, reshape the ECM, and avoid excessive scarring. Recent findings suggest that the stiffness of the ECM itself can influence fibroblast behaviour making them more active and increasing their secretion of important cytokines involved in inflammation and repair [165]. Fibroblasts do not just build the ECM they constantly adjust it in response to the environment, including inflammation. Their goal is to rebuild tissue, exclude lingering immune cells, and minimize fibrosis [67]. In fact, ECM composition and remodelling actively shape each stage of wound healing, from inflammation to tissue maturation [33].

Interestingly, the role of mitochondria extends beyond wound healing. In chronic neuroinflammatory diseases, they act as hubs that control immune responses, influence cell survival, and regulate inflammasome activation [133]. These discoveries have opened the door to new therapeutic strategies. For example, using mitochondrial-targeted antioxidants to reduce reactive oxygen species (ROS) has been shown to improve healing and reduce mortality in experimental models [15].

Metabolic Reprogramming During Each Phase (Glycolysis Versus OXPHOS)

One of the most important ways cells change to meet the needs of each stage of wound healing is by reprogramming their metabolism. Cells change their ability to make energy during the healing process, switching between glycolysis and OXPHOS. Mitochondria are very important for managing these changes.

During the initial inflammatory phase, many cells increase glycolysis, which is a faster way to make energy, to meet the immediate needs of immune activation and cell growth. This phenomenon, known as the Warburg effect, is acknowledged in cancer biology and signifies a more extensive reaction to cellular stress and activation. Macrophages have different energy profiles depending on the activity they do: M1 (pro-inflammatory) macrophages mostly use glycolysis, while M2 (pro-repair) macrophages rely more on oxidative metabolism.

Energy needs go up considerably during the proliferative phase of wound healing. Cells are actively dividing, making proteins, and building new tissue. They need both quick energy and important raw materials for biosynthesis. This phase typically incorporates both glycolysis and oxidative phosphorylation. Fine-tuned metabolic switching between glycolysis and oxidative phosphorylation ensures optimal ATP production and biosynthetic support during active tissue regeneration [160]. Mitochondria not only make ATP, but they also make important metabolic intermediates. Wang et al. [174] deliver extensive insights into the modifications of cellular metabolism during the wound healing process. The study demonstrated that metabolic reprogramming, which enables cells to adjust their energy consumption in response to stress or signals, is crucial for the healing process. There is a change in cellular activity during the remodelling phase of wound healing. The main goal now is to improve the tissue, generate and organize the extracellular matrix, and maintain the structure. During this stage, energy production must be steady and effective, and oxidative metabolism must become progressively more important. For effective wound healing while maintaining tissue function, it is important to coordinate metabolic changes across different types of cells.

Mitochondrial Role in Stem Cell Activation and Differentiation

Stem cells are very important for healing wounds because they may differentiate into different types of cells that are required for tissue repair and regeneration. Mitochondrial function is significantly associated with stem cell behaviour, influencing their activation state and differentiation capacity [192]. Recent research has demonstrated that mitochondrial dynamics, which are important for tissue healing, control regeneration throughout the body by regulating stem cell pluripotency and mitonuclear balance [121]. The research has elucidated intricate mechanisms by which stem cells promote wound healing. Mesenchymal stem cells (MSCs) are important therapeutic targets because they may differentiate into different types of cells and send signals to other cells in a very effective way. The effectiveness of wound healing by MSCs is intricately linked to their mitochondrial functionality and metabolic potential.

Platelets boost the ability of mesenchymal stem cells to heal wounds by directly transferring mitochondria and changing their metabolism. Research has demonstrated that platelets can deliver functional mitochondria to mesenchymal stem cells (MSCs), thereby augmenting their bioenergetic capacity and therapeutic potential. This mitochondrial transfer mechanism represents an innovative strategy for enhancing stem cell efficacy in wound healing scenarios [79].

The metabolic condition of stem cells influences their differentiation capacity and therapeutic effectiveness. Diabetes and smoking, common comorbidities in individuals with chronic wounds, significantly impair mitochondrial function in patient-derived mesenchymal stem cells (MSCs). Research has demonstrated that smoking status affects mitochondrial function and synthetic abilities in mesenchymal stem cells derived from diabetic patients with arterial insufficiency, highlighting the importance of patient-specific factors in evaluating the therapeutic potential of stem cells [103].

Research conducted in 2025 demonstrated that stem cells can acquire mitochondria through tunnelling nanotubes, gap junctional channels, and extracellular vesicles, especially in response to stressors such as inflammation, hypoxia, and irradiation. These modified stem cells indicate a greater ability to divide and grow [88]. A recent study showed that the EZH2-H3K27me3-PPARγ pathway improves mitochondrial function and delivery from adipose-derived stem cell spheres. This gives us new information about how epigenetics controls mitochondrial dynamics in stem cells [24].

Endothelial progenitor cells represent a crucial stem cell population in wound healing, particularly in promoting angiogenesis. For these types of cells to become mature endothelial cells and help new blood vessels form, their mitochondria must work properly [92]. Research demonstrated that endothelial cells derived from iPSCs promote angiogenesis and speed up wound healing. Their effectiveness is closely related to the capacity of mitochondria and metabolic flexibility [27]. Recent discoveries underscore the significance of endothelial cell angiogenesis in both physiological and pathological contexts, such as wound healing, with mitochondria playing a vital role in endothelial functionality and vessel development [92]. Mesenchymal stem cells can sense mitochondria released from damaged cells as danger signals, which turns on their ability to respond. This exemplifies the complex intercellular mitochondrial communication networks that govern tissue repair [95].

Cross-Talk Between Mitochondria and Other Organelles (ER, Peroxisomes)

The interaction between mitochondria and other organelles is crucial for cellular function in the context of wound healing. These organelles do not function in isolation; rather, they continuously interact through physical connections, shared metabolic processes, and the exchange of signalling molecules. These partnerships facilitate the coordination of essential processes such as calcium signalling, lipid metabolism, oxidative stress responses, and quality control mechanisms, all of which are crucial for effective tissue repair [35].

ER–Mitochondria Interactions

The relationship between the endoplasmic reticulum (ER) and mitochondria is significant during the process of wound healing. A recent study by Licini et al. [83] demonstrated that ER-mitochondria contact sites are essential for the proper activation of the NLRP3 inflammasome, a critical element in inflammatory signalling. Reducing these contact points decreases inflammation and enhances wound healing, indicating that ER–mitochondria communication may serve as a potential target for future therapies. Additional studies have further highlighted that alterations in ER–mitochondria communication significantly influence redox balance and inflammatory signaling during tissue repair [84].

Research by Shiiba et al. [153] demonstrates that mitochondria can transfer between cells, such as from fibroblasts to keratinocytes, via structures known as tunnelling nanotubes (TNTs). This exchange compensates for energy deficits in damaged tissues and is regulated by endoplasmic reticulum stress pathways. The disruption of ER-mitochondria communication hinders the proper formation of nanotubes, leading to delayed wound closure, a notable concern especially in diabetic wounds. Endoni et al. [43] provided insights into the interplay between ER stress and mitochondrial dysfunction in chronic, non-healing wounds. Single-cell RNA sequencing revealed a significant reduction of key MAM (mitochondria-associated membrane) proteins, such as MFN2, in non-healing ulcers, exacerbating oxidative stress. Restoration of ER–mitochondria interactions through pharmacological enhancement of MFN2 activity resulted in improved healing in mouse models, indicating potential therapeutic applications.

Mitochondria-associated membranes serve as critical hubs for cellular functions, facilitating calcium management, lipid synthesis, and metabolic coordination. During wound healing, these interactions must be precisely regulated to accommodate the changing energy and repair requirements at each phase. The regulation of endoplasmic reticulum-mitochondria connections is essential for cellular stress management, efficient protein folding, and metabolic flexibility, which are critical for effective tissue recovery.

Peroxisome–Mitochondria Coordination

Peroxisomes collaborate with mitochondria to regulate metabolism and oxidative stress in the context of wound healing. Both organelles participate in the degradation of fatty acids, the detoxification of harmful by-products, and the regulation of ROS. Coordination is particularly vital during the inflammatory phase, characterized by a significant increase in oxidative stress within cells. Peroxisomes and mitochondria collaboratively address the changing metabolic requirements of healing tissue. Peroxisomes are essential for the synthesis of plasmalogens and other lipids critical for membrane integrity and cellular signalling, whereas mitochondria function as the primary energy generators of the cell. They help cells stay functional and effective even when they are under a lot of stress by sharing these functions.

During the initial phases of wound healing, ROS concentrations increase as an outcome of the immune response. Peroxisomes regulate oxidative stress by degrading hydrogen peroxide via enzymes such as catalase and peroxiredoxins. This promotes the mitochondria's intrinsic antioxidant mechanisms, supporting the protection of proteins, DNA, and membranes [149]. Peroxisomes simultaneously degrade long-chain fatty acids into smaller molecules suitable for mitochondrial energy utilization, establishing a metabolic relay between both organelles [35].

Peroxisomes are responsible for synthesizing specialized lipids such as plasmalogens, which facilitate membrane flexibility, vesicle transport, and redox signalling. Lipids play a crucial role in stabilizing mitochondrial membranes and supporting the structural integrity of respiratory complexes, thereby enhancing energy production during periods of elevated tissue energy demands [3, 115]. Mitochondrial stress can induce the expansion of peroxisomes, establishing a protective mechanism that contributes to cellular homeostasis [17, 161].

Recent studies indicate that peroxisomes and mitochondria can establish physical connections via membrane contact sites, analogous to the interactions between mitochondria and the endoplasmic reticulum. The connections facilitate the direct exchange of metabolites and redox signals, enabling cells to modify their metabolic activity in response to alterations in the wound environment [18, 26]. This partnership is essential for managing oxidative stress in inflammation and facilitating adequate tissue repair.

The collaboration between peroxisomes and mitochondria facilitates essential functions in key cell types involved in wound healing, such as keratinocytes, fibroblasts, and immune cells, enabling them to manage stress, generate energy efficiently, and perform repair processes. Targeting this inter-organelle network may provide novel approaches to enhance healing, particularly in chronic wounds such as diabetic ulcers or age-related slow-healing skin injuries.

Autophagy and Organellar Quality Control

The different organelles inside a cell do not work in isolation, especially during wound healing, when stress is high and the need for tight quality control is greatest. Mitochondria, the endoplasmic reticulum (ER), and peroxisomes link up through dynamic contact sites called mitochondria-associated membranes (MAMs). These junctions act as busy hubs where calcium is handed off, lipids are exchanged, and metabolic signals are relayed. Just as importantly, they serve as launchpads for selective autophagy. When a mitochondrion is damaged, key autophagy proteins are recruited right at the MAM to start its orderly removal (mitophagy). If MAM integrity falters under wound-healing stress, mitophagy slows, ER stress rises, and overall recovery is delayed.

Peroxisomes and mitochondria establish a comparable relationship. Metabolites are exchanged, the detoxification of reactive oxygen species (ROS) is collaboratively managed, and the degradation of defective organelles is coordinated. Peroxisomal autophagy (pexophagy) is intricately linked with mitophagy and reticulophagy (ER-selective autophagy), forming a comprehensive cellular mechanism that eliminates damaged peroxisomes, mitochondria, and endoplasmic reticulum to maintain overall cellular health.

Autophagy is equally important at the tissue surface. Keratinocyte autophagy initiates the activation of both keratinocytes and fibroblasts, enhancing their functionality and collaboration in the process of wound closure [132]. When collaborative quality-control pathways fail, defective organelles accumulate, reactive oxygen species levels increase, metabolism becomes dysregulated, and healing is impeded. Coordinated regulation of selective autophagy pathways, including mitophagy and pexophagy, is therefore critical to preserve organelle integrity during tissue repair [150].

Shared regulators, including DRP1 and the mitofusins, synchronize organelle fission, fusion, and turnover, thereby directly connecting selective autophagy to overall cellular homeostasis. A recent study indicated that enhancing autophagy pathways, stabilizing contact sites such as mitochondria-associated membranes (MAMs), or targeting shared quality-control proteins may provide novel therapeutic approaches for chronic wounds and fibrotic conditions. Advancements in single-cell and other "omics" technologies enable researchers to monitor autophagic flux and organelle crosstalk in real time [17, 76, 176]. These findings facilitate the development of regenerative therapies aimed at optimizing quality-control mechanisms and, consequently, enhancing the speed of human wound healing.

Mitochondrial Dysfunction and Impaired Wound Healing

Chronic Wounds: Diabetic Ulcers, Venous Leg Ulcers, and Mitochondrial Defects

In chronic wounds, particularly diabetic foot ulcers and venous leg ulcers, a significant barrier to healing is the ongoing dysfunction of mitochondria. Cells residing in the skin, along with immune cells that infiltrate the wound, demonstrate evidence of mitochondrial stress and damage. Mitochondria frequently exhibit fragmentation or become trapped in irregular fission/fusion cycles, resulting in decreased ATP production efficiency and an increased tendency to release injurious ROS [15, 184, 194]. Clinical and experimental studies further demonstrate that metabolic dysregulation in diabetic conditions exacerbates mitochondrial biogenesis defects and respiratory inefficiency in chronic wounds [111]. With low energy and high oxidative stress, the cells struggle to carry out essential repair tasks like keratinocyte migration, proper immune response, new blood vessel growth, and matrix rebuilding.

In diabetic wounds, the high-glucose environment creates additional stress. Mitochondrial DNA becomes damaged, membrane potential drops, and mitophagy, the process that removes dysfunctional mitochondria, breaks down. These failures combine to weaken the wound's ability to regenerate [62, 113]. As a result, inflammation lingers. Immune cells stay locked in a damaging pro-inflammatory state, and fibroblasts lose their ability to produce and remodel collagen effectively. On top of this, excessive mitochondrial ROS and the release of mitochondrial danger signals (DAMPs) keep the inflammatory cycle going, blocking the wound from progressing into the next healing stages.

In DFUs, high blood sugar levels drive mitochondrial fragmentation, often by overactivating the Drp1 protein, which interferes with keratinocyte movement and wound closure [2, 199]. Meanwhile, fibroblasts from VLUs show a marked drop in mitochondrial membrane potential likely due to reduced activity of PGC-1α, a key regulator of mitochondrial biogenesis. Encouragingly, some of this dysfunction can be reversed with NAD^+ supplementation, suggesting possible therapeutic avenues [193].

Age-related Decline in Mitochondrial Function and Delayed Healing

With advancing age, the ability of the skin to heal decreases, largely attributed to a decline in mitochondrial function. Research indicates that aging skin exhibits approximately 40% reduced mitochondrial respiratory capacity, primarily attributed to the accumulation of mtDNA mutations and compromised mitophagy [131,

159]. Age-associated mitochondrial decline is also characterized by increased oxidative damage and diminished adaptive stress responses, contributing to delayed cutaneous repair [129]. A significant concern is that aged fibroblasts exhibit a poor response to hypoxic conditions during the healing process. The activation of the SIRT3-LKB1-AMPK pathway presents a challenge, as it is essential for transitioning metabolism towards glycolysis, a vital initial phase in the healing process [134, 188].

This mitochondrial decline affects all the major skin cells: fibroblasts, keratinocytes, and endothelial cells. As a result, there is less ATP available, reduced cell growth, poor blood vessel formation, and slower collagen production [62, 69]. These changes explain why aged skin does not bounce back the way young skin does after an injury [68].

There is promising evidence from mitochondrial transfer studies in older mice. When young, healthy mitochondria were introduced early in the wound healing process before the cells became fully senescent, they helped restore collagen production [40]. Nonetheless, the age of the cells is crucial. Once damage occurs to an elderly organism, reversal becomes significantly more challenging. Another area of debate involves circulating mitochondrial DNA. Some researchers believe that high levels of mtDNA in the bloodstream can delay healing, acting like distress signals [69, 125], while others see these changes as side effects, not causes of poor healing.

Wounds that have aged tend to exhibit decreased responses to stress. There is decreased activity in key pathways, including HIF-1α and AMPK, which typically aid in the development of new blood vessels and controlling the flow of energy. Research indicates that enhancing mitophagy or modifying metabolism through interventions such as metformin may facilitate improved wound healing in aged models. This indicates that focusing on mitochondrial health may emerge as a crucial approach in addressing slow-healing wounds in the elderly population [167, 184].

Oxidative Stress and Persistent Inflammation

Reactive oxygen species (ROS) have a dual role in wound healing, exhibiting both beneficial and detrimental effects, contingent upon their concentration and duration of presence. In healthy wounds, short bursts of mitochondrial ROS actually help by sending signals that promote defence and tissue repair. But when ROS levels stay high and aren't properly controlled, like in chronic wounds, they become harmful. This ongoing oxidative stress damages proteins, fats, and DNA, and keeps inflammation going instead of letting the wound heal.

In real terms, mitochondrial dysfunction actively contributes to the development of chronic wounds like diabetic ulcers. In diabetes, neutrophils generate too much superoxide due to overactive NOX2 and malfunctioning mitochondria. This sets off a vicious cycle where oxidative stress feeds more inflammation [21, 141]. One study even found that excess mitochondrial ROS oxidizes a protein called PTEN in keratinocytes, which then overactivates Akt and disrupts normal cell movement, making it harder for skin cells to close the wound [63, 109].

More importantly, damaged mitochondria release mtDNA into the surrounding tissue when they are not adequately removed by mitophagy. This triggers immune pathways like cGAS-STING, leading to increased production of IFN-γ, which harms local stem cells and stalls repair [21, 141]. Exciting new research using spatial transcriptomics has even shown that these "ROS hotspots" cluster around senescent fibroblasts at the edges of non-healing wounds, suggesting that targeting specific areas with antioxidant therapy might help break the cycle [170].

This ongoing oxidative stress environment tends to keep immune cells like macrophages in a pro-inflammatory M1 state and draws in more neutrophils. Together, they release proteases that break down tissue even further. This kind of sustained damage is seen not just in diabetic wounds but also in the aged skin of older adults, where high ROS levels block cell migration and prevent the wound from moving into the healing phase [63, 109].

Environmental Factors Influencing Mitochondrial Function in Wound Healing

Oxygen Tension and Mitochondrial Adaptation

The levels of oxygen present in a wound play a crucial role in the healing process, primarily because oxygen plays vital roles in mitochondrial activity (Fig. 6.4). Immediately following an injury, there is damage to blood vessels, resulting in a decrease in oxygen supply, which causes the tissue to enter a hypoxic state. In response to stress, cells activate survival strategies that are mediated by mitochondria and essential regulators such as hypoxia-inducible factor-1α (HIF-1α). In conditions of low oxygen, HIF-1α becomes stabilized and activates genes that facilitate tissue survival and recovery. This process promotes the growth of blood vessels, alters metabolic pathways, and enhances cell survival [19, 150].

However, mechanical forces also influence mitochondria, in addition to oxygen levels. The skin adjacent to joints is perpetually subjected to stretching, whereas pressure ulcers develop in areas where tissue is crushed. Cells can sense these physical forces through structures like integrins, the cytoskeleton, and ion channels. These cues travel inside the cell and affect how mitochondria behave, how they move, change shape (through fusion and fission), and supply energy where it is needed most [59, 101].

The wound environment constantly shifts, with oxygen gradients changing from one area to another. Healing depends on enough oxygen reaching the tissue, because oxygen is needed for cells to grow, for new blood vessels to form, and for the body to rebuild proteins that restore skin structure [143]. To adapt to low oxygen, HIF-1α helps cells conserve oxygen by slowing down mitochondrial respiration. It does this by activating a gene called PDK1, which reroutes metabolism from oxygen-based energy production to glycolysis, a method that does not need oxygen [123].

In chronic wounds, hypoxia frequently persists for an extended duration. Prolonged low oxygen levels can diminish HIF-1α activity and its capacity to initiate healing responses. This weakens the formation of new blood vessels and slows down tissue repair [105]. Essentially, cells get stuck in a state where they can't fully respond to the lack of oxygen, and healing grinds to a halt.

Interestingly, HIF-1α not only helps cells adjust to low oxygen it also protects them from oxidative stress by targeting mitochondria directly. This shows how closely connected oxygen sensing and mitochondrial function really are in wound healing. Consequently, tailored oxygen therapy, administering oxygen to the hypoxic regions of a wound, may serve as an effective treatment. It leverages the body's inherent hypoxic response to initiate healing in challenging wounds [150].

Mechanical Stress and Mitochondrial Mechanosensing

The pressures exerted on a wound, varying from 2 to 20 kPa, are not merely passive background pressure. They are increasingly recognized for their active function in regulating cellular behaviour and facilitating wound healing. This process, called mechanotransduction, is how cells convert physical forces like stretching, compression, or shear into biochemical signals. It turns out that skin cells like fibroblasts and keratinocytes are highly responsive to these kinds of mechanical cues [179].

Recent studies indicate that mechanical forces directly affect essential biological processes related to skin repair and regeneration [75]. Mitochondria, traditionally regarded solely as the cell's power plants, are increasingly recognized as significant contributors to cellular responses to mechanical changes [62].

In fact, many types of cells involved in wound healing, like fibroblasts, epithelial cells, endothelial cells, and immune cells like macrophages, can sense the stiffness and tension in their environment. They relay this mechanical information all the way to the nucleus, where it can change which genes are turned on or off [36]. This involves complex coordination between the cytoskeleton, integrins, and mechanosensitive ion channels.

An interesting finding is that macrophages may sense the rigidity of the extracellular matrix and respond to growth hormones only through cytoskeletal alterations, without the necessity of integrin receptors [105]. This demonstrates the precision with which cells sense their surroundings (Fig. 6.4).

Mechanical forces even influence mitochondrial positioning within cells. In cells that are moving or contracting, like those helping to close a wound, mitochondria are actively relocated just beneath the cell membrane, right where energy is most needed. This redistribution, guided by proteins like MIRO1, ensures that enough ATP and calcium are available at focal adhesions and actin-rich protrusions to support movement and tissue contraction.

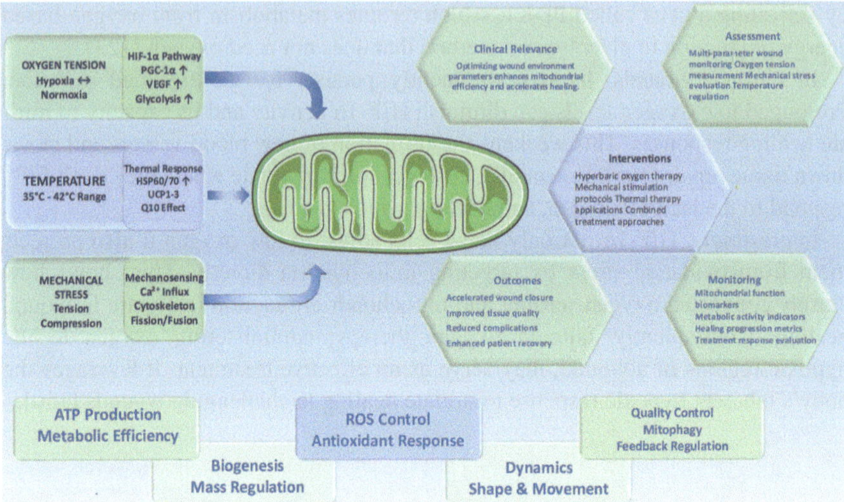

Fig. 6.4 Environmental factors influencing mitochondrial function in wound healing. Schematic representation of how *oxygen tension* (hypoxia/normoxia), *temperature* (35–42 °C), and *mechanical stress* (tension/compression) modulate mitochondrial pathways to impact wound healing. Key mechanisms include: *hypoxia-driven responses*: HIF-1α activation, PGC-1α upregulation, and VEGF promotion. *Thermal effects*: HSP60/70 induction and UCP1–3 activity. And *mechanotransduction*: Ca^{2+} influx and cytoskeletal reorganization. Clinical interventions (e.g., hyperbaric oxygen, thermal therapy) and monitoring strategies (mitochondrial biomarkers, metabolic activity) are shown to optimize healing outcomes

Nonetheless, excessive or insufficient mechanical stress could negatively influence the healing process. It could lead to enduring injuries or considerable scarring. Knowledge of the mechanisms by which cells detect and respond to mechanical forces may enhance the creation of advanced treatments that refine the wound environment to promote effective healing (Fig. 6.4).

Temperature Effects on Mitochondrial Function in Wound Healing

Temperature is a factor that is frequently underestimated in the context of wound healing; however, it exerts a significant influence, particularly through its effects on mitochondrial function. Extreme temperatures, whether high or low, can impact mitochondrial energy production, influence the generation of ROS, and affect the ability for tissue regeneration. During localised increases in tissue temperature, as observed in burns or areas of active inflammation, mitochondria increase their oxygen consumption and ATP production to meet the elevated energy demands. Nonetheless, a trade-off is present with rising temperatures; the efficiency of mitochondria declines. The process involves an increased consumption of oxygen for every ATP molecule generated, while a portion of energy is lost due to proton leakage across the mitochondrial membrane. The observed inefficiency leads to an elevation in ROS, which can have either advantageous or adverse effects based on the circumstances (Fig. 6.4).

Nonetheless, increased concentrations of ROS, especially under conditions of heat stress, may result in tissue injury and intensify inflammatory responses. Conversely, excessive cooling of the wound may also present significant issues. Mild hypothermia reduces mitochondrial metabolism, a significant concern for older adults or individuals with impaired blood flow, as their mitochondrial function is already compromised. Reduced temperatures can impair vital immune responses, decrease collagen synthesis by fibroblasts, and hinder the formation of new blood vessels, all of which are essential for effective wound healing [166].

Mitochondria can adjust to moderate temperature changes; however, extreme conditions, such as those encountered in deep burns or frostbite, may exceed their adaptive capacity, resulting in cell death and suboptimal healing outcomes. Careful management of wound temperature may significantly impact wound care outcomes. Recent studies explore the optimization of the thermal environment in the wound bed to enhance mitochondrial function and improve conditions for tissue repair [61, 177].

Therapeutic Strategies Targeting Mitochondria for Enhanced Wound Healing

Mitochondrial Antioxidants (MitoQ, SkQ1)

Targeting mitochondria with antioxidants is becoming an interesting approach to enhance wound healing. One such compound, MitoQ, is a powerful antioxidant known to support mitochondrial health and regulate redox balance. Although research directly linking MitoQ to wound healing is still developing, early studies suggest it may help protect the skin's barrier function by reducing mitochondrial DNA damage. This protective effect appears to work through the Nrf2/ARE signalling pathway, which is involved in cellular defence against oxidative stress [136].

Another mitochondria-targeted antioxidant, SkQ1, has shown even more direct promise in improving skin repair. SkQ1 speeds up wound closure, enhances the formation of new epithelial tissue, supports granulation and blood vessel growth, and helps regulate inflammation. Its main benefit lies in controlling excess mitochondrial reactive oxygen species (mtROS), a common problem in wounds that can stall both the early inflammatory response and later tissue regeneration if left unchecked (Fig. 6.5).

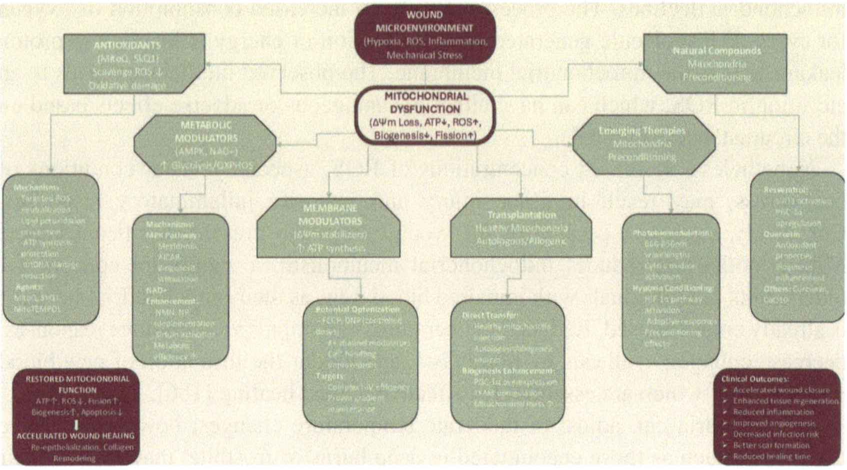

Fig. 6.5 Therapeutic strategies targeting mitochondria for enhanced wound healing. Overview of mitochondria-directed therapies, including: *Antioxidants* (e.g., MitoQ, SK01) to scavenge ROS. *Metabolic modulators* (e.g., AMPK activators, NAD^+ boosters) to improve glycolysis/OxPHOS balance. *Mitochondrial transplantation* (autologous/allogenic) and *preconditioning* (e.g., SIRT1/PGC-1α activation). Mechanistic targets (fission/fusion regulation, lipid peroxidation prevention) and clinical outcomes (reduced inflammation, improved angiogenesis) are highlighted. The figure emphasizes combinatorial approaches to restore mitochondrial function (ATP↑, ROS↓) and accelerate tissue repair

In animal models of diabetes and aging, where wound healing is often delayed, SkQ1 helps normalize the healing process. It supports everything from calming excessive inflammation to promoting the transformation of fibroblasts into myofibroblasts and guiding the formation of new extracellular matrix, key steps for proper tissue repair [29, 34, 156].

Metabolic Modulators (AMPK Activators, NAD+ Boosters)

Metabolic dysregulation is a hallmark of impaired wound healing. Boosters of nicotinamide adenine dinucleotide (NAD^+) and activators of AMP-activated protein kinase (AMPK) both enhance cellular metabolism crucial for wound repair. AMP-activated protein kinase (AMPK) activators (metformin, AICAR) and NAD^+ boosters (nicotinamide riboside, NMN) enhance mitochondrial efficiency by promoting ATP production and mitigating oxidative stress (Fig. 6.5). NAD^+ boosters (nicotinamide riboside) combined with metabolic interventions like exercise upregulate SIRT1 and PGC-1α, promoting mitochondrial biogenesis, reducing proinflammatory factors, and supporting tissue regeneration. AMPK activation supports muscle satellite cell renewal and differentiation, indirectly facilitating mitochondrial function and improving wound healing outcomes [57, 160]. Concrete evidence shows improved tendon repair and skeletal muscle regeneration via these pathways, though more research is required on direct cutaneous wound models. Recent studies indicate that AMPK activation accelerates diabetic wound closure by improving keratinocyte proliferation and collagen deposition [97].

Mitochondrial Transplantation and Biogenesis Enhancers (PGC-1α)

Mitochondrial biogenesis replaces damaged mitochondria and restores cellular energy balance. Mitochondrial transplantation has emerged as a groundbreaking therapeutic approach resulting in accelerated wound healing by reducing wound size, increasing granulation tissue, and hastening epithelialization [163]. This is the first clinical demonstration of the therapeutic efficacy of mitochondrial transplantation in wound healing.

PGC-1α is a central regulator that triggers biogenesis by activating NRF1 and mitochondrial transcription factor A (TFAM), leading to improved tissue repair and reduced oxidative stress [191]. PGC-1α activity also modulates the inflammatory microenvironment critical in healing, and its upregulation following injury reduces neuronal and tissue damage, which holds promise for future clinical interventions in wound healing [1].

Emerging Therapies: Photobiomodulation (Low-Level Laser Therapy) and Hypoxia Conditioning

Photobiomodulation (PBM), using low-level lasers or LED light, is effective in accelerating superficial wound healing, reducing inflammation, and modulating mitochondrial function (Fig. 6.5). Properly dosed PBM stimulates cellular proliferation, migration, and protein synthesis, but inappropriate parameters may impair healing [78]. Prado et al. reported that blue light at a low energy density (<20 J/cm^2) stimulates different cell types and proteins involved in healing [129].

Hypoxia conditioning and tissue preconditioning (using controlled, repeated cycles of low oxygen) can adapt tissues for increased resistance to injury, partly by upregulating mitochondrial pathways and cytoprotective proteins such as heme oxygenase-1 (HO-1) [73].

Mitochondrial Membrane Potential Modulators

Therapeutic strategies targeting mitochondrial membrane potential aim to restore optimal mitochondrial function. Agents that influence mitochondrial membrane potential, such as sodium ionophores (monensin A), can expedite wound healing (Fig. 6.5). Bai et al. reveal that Na^+ is a determinant factor in skin wound healing. The Na^+ ionophores-triggered dual supplementation of Na^+ and H^+ within the mitochondrial matrix compels ROS generation and energy metabolism, ultimately favouring productive wound healing programs [7]. These compounds enhance mitochondrial energy production and controlled ROS generation, which are critical for cell proliferation, angiogenesis, and collagen deposition during wound closure, especially in chronic and diabetic wounds.

Preconditioning Strategies (Ischemic, Pharmacological)

Preconditioning approaches prepare tissues for improved wound healing responses through controlled mitochondrial stress (Fig. 6.5). Preconditioning strategies such as ischemic preconditioning (IPC) and remote IPC confer resilience to tissues by activating protective genes involved in metabolism, stress response, and angiogenesis [30]. Pharmacological preconditioning, using agents such as curcumin, additionally assists in the induction of cytoprotective proteins. Both modalities demonstrate beneficial effects on tissue perfusion, neovascularization, and enhanced healing; however, the translational outcomes in clinical cutaneous wounds continue to develop [73].

Natural Compounds Targeting Mitochondria (Resveratrol, Quercetin)

Natural compounds that influence mitochondrial function present promising therapeutic opportunities; however, further research is required to explore the roles of resveratrol and quercetin in the context of wound healing (Fig. 6.5). Targeting may encompass the enhancement of glycolysis to optimize haemostasis and inflammatory response, the regulation of ROS dynamics, and the maintenance of appropriate mitochondrial dynamics, especially in aged populations [62]. Natural polyphenols and phytochemicals have attracted interest due to their capacity to influence mitochondrial function, diminish oxidative stress, and promote tissue regeneration in both chronic and acute wounds. These compounds operate via various mechanisms, such as promoting mitochondrial biogenesis, enhancing antioxidant defence, and modulating anti-inflammatory signalling, positioning them as potential adjuvants in wound care [23].

Resveratrol, an effective activator of SIRT1, promotes mitochondrial biogenesis through the PGC-1α/NRF2 pathway, leading to increased ATP production and decreased ROS levels in diabetic wounds [151]. Topical resveratrol formulations promote accelerated wound closure in diabetic rodent models through the enhancement of angiogenesis and collagen synthesis [122]. Quercetin is a flavonoid that binds to free iron, thereby inhibiting mitochondrial damage caused by the Fenton reaction [9]. Additionally, it stimulates AMPK, enhancing mitochondrial metabolism and autophagy in aged wounds [13]. The combination of quercetin with nanoparticle delivery systems enhances its bioavailability and facilitates deeper penetration in wound applications [8].

Curcumin helps to restore mitochondrial membrane potential ($\Delta\Psi m$) by upregulating mitochondrial fusion proteins (MFN2, OPA1) [190]. It disrupts bacterial biofilms in chronic wounds while protecting host cell mitochondria [65]. A recent report of Wu et al. demonstrated therapeutic benefits of curcumin for psoriasis. Curcumin treatment for 10 days reduced epidermal hyperplasia significantly, decreased keratinocyte proliferation, and restored immune homeostasis in lesional skin [182]. Berberine is known to activate TFAM (mitochondrial transcription factor A), promoting mtDNA replication and respiration [46]. It improves diabetic wound healing by reducing mtROS and enhancing fibroblast migration [198]. Berberine-loaded nanofibers accelerated diabetic wound healing by anti-inflammatory and anti-oxidative activities in a diabetic mouse model. Additionally, in the rabbit ear model, Berberine inhibited the scar hyperplasia through the TGF-β1/Smad3 signalling pathway [93]. Recent report of Gu et al. on antibacterial hydrogels with dual-colorimetric sensing capabilities for advanced wound management. The hydrogel demonstrated potent antibacterial activity against Gram-positive and Gram-negative bacteria, Berberine specifically inhibited bacterial FabH via hydrogen bonding and hydrophobic interactions [54].

Epigallocatechin Gallate (EGCG) helps to stabilizes HIF-1α under hypoxia, supporting angiogenesis in ischemic wounds [107]. It modulates Drp1-mediated fission, preventing excessive mitochondrial fragmentation in stressed keratinocytes [197]. Ni et al. designed a dressing incorporating vancomycin and EGCG in a nanofibre membrane for synergistic antimicrobial efficiency. This technology exhibited controlled release of vancomycin and EGCG at wound sites along with their synergistic antibacterial effects, and promptly healed the wounds [114]. Most recent study of An et al. showed microcapsule-loading aligned ECM-like scaffold with microfluidics enhanced wound healing through the promotion of γδ T cells, which act on keratinocytes and fibroblasts via the AREG-EGFR pathway [4].

Experimental Models and Analytical Approaches

In Vitro Models to Study Mitochondrial Function in Wound Healing

Many researchers are currently investigating mitochondrial behaviour during tissue repair, employing laboratory models that replicate the essential cellular and molecular processes involved in wound healing. In vitro systems provide a controlled environment for the comprehensive investigation of specific processes. Interestingly, new findings have shown for the first time that pulsed electromagnetic fields (PEMFs) can reprogram fibroblast metabolism, shifting their energy production from mitochondrial respiration to glycolysis. This shift is also linked to increased vesicle transport, a process essential for wound healing in lab settings [6].

The predominant in vitro models employed include primary cultures of skin cells, including keratinocytes, fibroblasts, and endothelial cells. Under controlled conditions, these cells may be subjected to injury to investigate the mechanisms of healing responses. Scratch assays continue to be the foremost method for evaluating cellular migration and the effectiveness of wound closure. Nonetheless, contemporary methodologies are increasingly incorporating sophisticated instruments, including micropatterned surfaces and microfluidic "wound-on-a-chip" systems, which more accurately emulate the three-dimensional architecture of real skin [53, 130, 186]. The development of organotypic skin models represents a significant advancement in the field. The models integrate various cell types organized to closely replicate actual human skin [48]. Through the application of mitochondria-targeted antioxidants or the silencing of specific mitochondrial genes, scientists can precisely determine the impact of mitochondria on tissue repair using pharmaceuticals or gene-editing tools in such systems.

These models hold significant importance for investigating the role of mitochondrial dysfunction in impaired healing, a critical area of focus when examining chronic wounds in diabetic or elderly individuals.

Animal Models of Wound Healing and Mitochondrial Assessment

Animal models are still very important for studying how mitochondria help wounds heal on a larger scale. Full-thickness skin wounds in mice and rats are among the most commonly utilized models, offering an extensive view of the different phases of healing.

Recently, several original studies have investigated the process of wound healing in humans by creating small wounds in healthy subjects and collecting tissue samples from the wound margins at different time points. By analysing gene expression in these samples, researchers gain a deeper insight into the mechanisms that facilitate tissue repair in the body [199]. Animal models continue to be the preferred method for examining intricate mechanisms, such as gene manipulation and comprehensive evaluation of mitochondrial function.

Specialized models, including diabetic mice (such as db/db or ob/ob strains, or those treated with streptozotocin), are particularly valuable for examining the role of compromised mitochondrial function in relation to impaired healing. These models closely resemble the metabolic difficulties encountered by individuals with diabetes. The examination of wound healing in aged animals elucidates the mechanisms by which aging mitochondria may hinder tissue repair processes [7].

There are many diabetic wound models, but no one acknowledges for certain, which one best represents human diabetic wounds for testing new treatments. The concept of mitochondrial transfer or transplantation is gaining traction in clinical research. These methods try to restore cells in damaged tissues to work again, especially in wounds that are hard to heal, while simultaneously attempting to avoid immune reactions or rejection of donor mitochondria [87, 199].

Imaging and Assays for Mitochondrial Activity

In the last 10 years, there have been significant advancements in the tools and methods employed to investigate mitochondrial function in the context of wound healing. Both local ROS production (mitochondrial electron-leak, NOX, xanthine oxidase) and compromised antioxidant defences are reflected in oxidative stress in injured tissues. Accurate measurement, which distinguishes species (e.g., superoxide vs. $H2O_2$), cellular compartments (mitochondrial vs. cytosolic), and functional consequences (ATP production, membrane potential), requires combining spatially resolved imaging with quantitative biochemical or respirometric assays. Mitochondria serve as the powerhouses of the cell, playing a crucial role in energy production and cell signalling. Consequently, any dysfunction in these organelles is associated with a variety of diseases. Understanding their behaviour during wound healing is essential [34].

Membrane Potential Checks

Fluorescent dyes such as TMRE and JC-1 are commonly employed to assess mitochondrial health. This allows for the assessment of mitochondrial membrane functionality in real-time. In the case of TMRE, healthy mitochondria exhibiting a robust membrane potential are represented as yellow-white, whereas those with diminished potential are indicated in red or dark [72]. This provides insight into the activity levels of mitochondria during various stages of wound repair.

Measuring Oxygen Use

High-resolution respirometry, which evaluates variables like maximal respiratory capacity, ATP-linked respiration, and basal respiration, can be used to measure mitochondrial oxygen consumption. Measurements of oxygen consumption during cellular processes, such as wound healing, offer vital information about mitochondrial efficiency, coupling status, and metabolic adaptability. Both oxidative phosphorylation and glycolytic contributions to cellular energetics can be assessed through real-time oxygen consumption analysis, which is frequently carried out using devices like the Seahorse XF Analyzer or Oroboros O2k [168]. A Seahorse XF analyser quantifies cellular oxygen consumption, serving as a critical indicator of mitochondrial activity. This enables the assessment of cellular energy production, efficiency, and stress levels, all of which are critical factors in the wound healing process [28].

Tracking Oxidative Stress (ROS)

Mitochondria are constantly changing in shape and function, which impacts their activity. One key aspect we track is the production of ROS, which can be harmful if not balanced. To detect ROS, use of tools like MitoSOX Red (for superoxide), H2DCFDA (for general ROS), and other specific sensors [187]. Both small-molecule and genetically encoded probes are used to image ROS; each has unique advantages and disadvantages. With the drawback of pH sensitivity that necessitates suitable controls, the HyPer family offers ratiometric, reversible, real-time $H2O_2$ detection that is targetable to mitochondria or the cytosol. roGFP-Orp1 and roGFP-GRX are helpful for mapping cellular redox potential because they allow for long-term live imaging of glutathione redox status or H_2O_2 levels. While DHE detects intracellular superoxide, it should be used in conjunction with HPLC or LC-MS to verify specificity. MitoSOX is a popular small-molecule probe for mitochondrial superoxide, but it needs to be carefully validated to prevent artifacts from non-specific oxidation and dependence on mitochondrial membrane potential. For $H2O_2$ imaging in tissues, redox-sensitive dyes such as PF1 and Peroxy Orange/Green work well, but they require rigorous adherence to specificity standards. Best practices for all techniques include oxidant and ROS-scavenger controls, pH calibration, and adjustment for the effects of membrane potential, mitochondrial content, and probe loading [111].

Advanced Imaging

With confocal and super-resolution microscopes, we can now watch mitochondria in action. Techniques like FRAP (Fluorescence Recovery After Photobleaching) help us study how connected and mobile mitochondria are. Live-cell imaging with special fluorescent markers shows how their shape changes as the wound heals.

Interestingly, short, controlled openings of the mitochondrial permeability transition pore (mPTP) seem to help maintain mitochondrial health. However, if this process goes wrong, it can lead to damage due to oxidative stress [117, 118].

By combining all these methods from simple cell cultures to complex animal models, we are getting a much clearer picture of how mitochondria influence wound healing. These insights are also paving the way for new treatment strategies.

Clinical Implications and Translational Perspectives of Mitochondrial Therapies in Wound Healing

Current Evidence from Clinical Studies

Mitochondrial therapies have recently gained a lot of attention, especially in the field of wound healing. Research now shows that mitochondria from platelets can boost the ability of mesenchymal stem cells (MSCs) to form new blood vessels, helping repair tissues and heal wounds, especially in long-term muscle injuries [163]. This is a shift from traditional treatments toward more targeted, mechanism-based strategies [96, 128].

Several clinical trials are exploring how mitochondrial-based treatments can help in conditions that share similar healing pathways. For example, KH176 is currently being tested in Phase 3 trials (NCT02544217), and elamipretide, a small peptide that protects mitochondrial structure and reduces oxidative stress, has already gone through Phase 1/2 trials [70]. These studies lay the groundwork for applying these therapies to wound healing. The most promising results so far come from studies on direct mitochondrial transplantation. These have shown that mitochondrial behaviour, how they respond to changing needs and stress, plays a key role in healing, suggesting that targeting these processes might improve patient outcomes [62, 66].

The translation of these promising preclinical findings to human studies is limited, primarily relying on case reports and small-scale pilot studies for current evidence.

Challenges in Translating Mitochondrial Therapies to Clinics

The transition of mitochondrial therapies from laboratory research to clinical application encounters several critical challenges that require systematic resolution. The main challenge resides in the intricate nature of mitochondrial biology, wherein dysfunction can present through various pathways and mechanisms. Mitochondrial dysfunction may result from genetic defects impacting essential mitochondrial components or from impaired pathways that disrupt mitochondrial homeostasis, complicating the development of universal therapeutic strategies [66].

Manufacturing and quality control constitute a significant obstacle. Mitochondrial therapies differ from conventional pharmaceuticals in that they typically involve living organelles or intricate molecular systems, necessitating specialized handling, storage, and delivery techniques. The heterogeneity of mitochondrial populations within tissues introduces additional complexity, as therapeutic responses may differ markedly among patients and even across various tissue types within the same individual [22].

The regulatory pathways for mitochondrial-based therapies are inadequately defined, especially concerning methods that involve organelle transplantation or genetic modifications. The distinctive characteristics of these interventions frequently occupy a space that straddles established regulatory classifications, resulting in ambiguity for researchers and clinicians alike. The absence of standardized outcome measures and biomarkers related to mitochondrial function in wound healing impedes the design and interpretation of clinical trials [22, 66].

Personalized Approaches in Patients with Mitochondrial Defects

Individual genetic backgrounds, metabolic profiles, and disease phenotypes must all be considered when developing therapeutic methods for mitochondrial illnesses due to their unique characteristics. Mitochondrial diseases affect approximately 4300 individuals, manifesting at different ages and involving various organ systems. The pathways through which mitochondrial abnormalities trigger a variety of human diseases remain incompletely elucidated [96]. This complexity underscores the importance of precision medicine approaches in the advancement of mitochondrial therapies. Customized therapeutic approaches frequently begin with the analysis of genetic information. A considerable number of individuals with mitochondrial disease have certain mutations, and finding them might help medical practitioners decide on the best therapy. Metabolic profiling represents a crucial phase in the research process. Various defects result in different patterns within an individual's metabolism. In addition to diagnosing the root cause, testing these can assist in determining the most effective treatments.

The development of patient-specific stem cell models, such as induced pluripotent stem cells (iPSCs), represents a highly promising approach. These cells serve as a valuable resource for evaluating treatments in laboratory settings prior to their application in clinical patients [58].

Biomarkers for Mitochondrial Dysfunction in Wound Healing

Establishing and confirming biomarkers for mitochondrial dysfunction in wound healing events is important for the progress of therapeutic uses. Mitophagy is crucial for cellular stability, as it selectively removes damaged mitochondria. Evidence from diverse disease models suggests that pharmaceutical therapies targeting mitophagy may constitute a promising therapeutic option [128]. This emphasizes the importance of developing biomarkers to assess mitochondrial quality assurance mechanisms [128].

Modern biomarker techniques focus on a few key areas. Circulating biomarkers, including mitochondrial DNA fragments, certain metabolites, and proteins produced from injured mitochondria, offer non-invasive methods for assessing mitochondrial function [52]. Systemic markers provide information on the health of mitochondria and may be linked to how efficiently wounds heal. However, the specificity of these findings for wound-related mitochondrial dysfunction requires further validation.

Tissue-based biomarkers offer a more direct assessment of mitochondrial function inside the wound environment. Clinically, tissue-based molecular markers linked to the Wnt signalling system, like nuclear β-catenin and c-myc, have been connected to poor healing. The Wnt pathway is very important for skin development and maintenance [85, 178].

Functional markers that measure the efficiency of energy production, oxidative stress levels, and mitochondrial respiratory capacity provide direct assessments of how efficiently mitochondria function. To assess mitochondrial function, sophisticated techniques such as real-time metabolic flux analysis and high-resolution respirometry can be applied to circulating cells or lesion tissue samples. The challenge lies in creating standardized methods that may be used in many different clinical contexts while yet keeping the integrity of the analysis.

Cost-Effectiveness Considerations of Mitochondrial Therapies

The economic valuation of mitochondrial therapies involves distinct challenges originating from their novel mechanisms of action, intricate manufacturing processes, and the chronic nature of the conditions they target. Conventional pharmacoeconomic models may fail to fully represent the value proposition of these interventions, especially considering their capacity for disease modification as opposed to merely addressing symptoms.

The manufacturing costs associated with mitochondrial therapies generally exceed those of conventional pharmaceuticals, attributable to the complexities involved in handling cellular components and advanced targeting systems. Autologous methods utilizing a patient's own mitochondria necessitate personalized processing and quality assurance, thereby substantially raising costs on a per-patient basis. Allogeneic methods may provide cost efficiencies; however, they also present increased complexity concerning immunocompatibility and standardization.

Nonetheless, the possibility of substantial long-term cost savings warrants consideration. Chronic wounds impose a significant economic burden on healthcare systems, with individual cases frequently necessitating prolonged treatment durations spanning months or years. Chronic wounds, resulting from factors such as diabetes and burns, pose significant challenges and financial burdens. Mitochondrial therapies that enhance healing or minimize chronic conditions may yield cost savings sufficient to warrant elevated initial treatment expenses [22].

The value-based healthcare framework offers a more suitable perspective for assessing mitochondrial therapies. This approach considers direct medical costs as well as enhancements in quality of life, functional outcomes, and productivity. Successful mitochondrial therapy for wound healing applications may reduce hospitalization durations, lower infection rates, and enhance long-term functional outcomes, thereby contributing to overall healthcare value.

Regulatory Challenges for Mitochondrial-Based Treatments

The regulatory framework for mitochondrial-based therapeutics poses complex issues that require careful management to ensure safety and efficacy while fostering innovation. The current regulatory frameworks were mostly created for traditional small-molecule medicines and might not be able to fully handle the unique features of mitochondrial therapy [66].

However, classification continues to be a big obstacle to overcome. Mitochondrial medicines can be grouped into several regulatory groups based on how they work and what they are made of. Purified mitochondria intended for transplantation may be classified as cellular therapy products, while tiny compounds targeting mitochondria may be regarded as conventional medications. Combination products that have both cellular and device parts make regulatory pathways more complicated [66].

Future Directions and Emerging Trends in Mitochondrial Research for Wound Healing

Nanotechnology and Mitochondria-Targeted Delivery Systems

Recent advancements in nanotechnology are transforming drug delivery to mitochondria, an organelle safeguarded by dual membranes and located deep within the cell. Researchers are currently employing nanoparticles decorated with specific targeting compounds such as triphenylphosphonium (TPP), peptides, or aptamers that overcome this challenge. These techniques enhance drug delivery precision, augmenting stability and bioavailability while mitigating adverse effects and toxicity, a prevalent concern with conventional treatments [11].

While numerous techniques are being investigated for neurological and metabolic issues, they are equally pertinent to wound healing and tissue restoration.

Nanocarriers can simultaneously deliver many therapeutic agents, such as drugs, proteins, or genes, enabling a multifaceted approach to addressing mitochondrial issues [74]. Nonetheless, challenges remain, including the release of medications from carriers within the cell, assuring long-term safety, and determining methods for scaling up production. Innovations in computer modelling and rapid screening techniques are anticipated to facilitate the transition of these technologies into clinical application [82].

Gene Editing and Mitochondrial Engineering (e.g., CRISPR-mtDNA)

Gene editing allows for novel approaches for the treatment of mitochondrial dysfunction. Tools like CRISPR/Cas are being adapted to target mitochondrial DNA (mtDNA), which could help correct harmful mutations or boost mitochondrial performance [157, 181].

One of the biggest challenges has been getting these tools, especially the guide RNAs and Cas enzymes, into the mitochondria. But recent efforts using viral vectors, peptide-based carriers, and nanotech-based methods are starting to show promise [47, 124]. While most of this research is still in early stages, the possibilities range from treating inherited mitochondrial diseases to improving healing in aged or damaged tissues. There's also growing interest in engineering mitochondria by introducing synthetic or healthy donor mitochondria to help cells regain their energy-producing capacity [74].

Unanswered Questions and Research Gaps

Despite all the progress, there are still many unanswered questions:

- *How exactly do therapeutic materials enter mitochondria?*
  The molecular mechanisms by which sgRNAs, ssODNs, or big proteins traverse the mitochondrial double membrane (e.g., through PNPASE or stem loop-mediated import) remain inadequately characterized and require stringent experimental validation.
- *How efficient is mitochondrial gene editing?*
  Delivering the CRISPR system into mitochondria remains difficult, and unwanted off-target edits are a real concern.
- *Is it safe long-term?*
  A thorough investigation is needed to determine the biocompatibility of new nanocarriers, any possible immunological reactions, and the development of off-target substances in non-mitochondrial compartments.
- *How do they perform in large animals or humans?*

The majority of techniques do not possess safety evidence for primates or humans. Advanced models must address off-target effects, oncogenicity, and germline transmission hazards.
- *How accurately can we target diseased mitochondria?*
  Delivering therapy only to the damaged mitochondria within complex tissues is still a big technical challenge.
- *How well do combination therapies work?*
  Using gene editing, drugs, and metabolic treatments together is still in early stages.
- *What are the regulatory and clinical hurdles?*
  Before any of these approaches can be used widely, they must pass rigorous testing, clinical trials, and meet regulatory standards.

Innovative approaches for mitochondria, such as programmable lipid nanoparticles, targeted base editors, and CRISPR platforms, present significant opportunities for the precise treatment of mitochondrial disorders. Successful translation relies on addressing critical knowledge gaps in delivery methods, import mechanisms, editing effectiveness and specificity, long-term heteroplasmy management, and safety, particularly in large animal or human contexts.

Conclusion

Mitochondria are not only the cell's source of energy they are also very important for healing wounds at all four stages: haemostasis, inflammation, proliferation, and remodelling. They play an important part in the healing process by giving the body energy, controlling immunological responses, and sending signals between cells. When bleeding stops, mitochondria release ATP and ROS from platelets to help form clots and start sending signals for healing. Mitochondria help govern how immune cells like macrophages act during inflammation. Depending on how they make energy, they can either help calm inflammation or keep it going. In the following stage of proliferation, mitochondrial energy helps with crucial repair processes like the growth of fibroblasts, the movement of keratinocytes, and the formation of new blood vessels. Mitochondria can even be sent from stem cells to damaged cells, which helps save them and get them functioning again. Finally, during tissue remodelling, mitochondria help reconstruct the ECM, direct the production of myofibroblasts, and keep ROS levels low to control scarring or fibrosis.

Mitochondrial dysfunction can disrupt this tightly regulated sequence. Chronic hyperglycaemia, aging, hypoxia, ischemia-reperfusion injury, persistent inflammation, oxidative stress, and the accumulation of mitochondrial DNA (mtDNA) mutations are factors that impair mitochondrial dynamics, including fission and fusion, as well as mitophagy and biogenesis. Disruptions result in diminished ATP synthesis, increased reactive oxygen species (ROS), activation of the inflammasome, and immune dysregulation, all of which contribute to chronic, non-healing wounds, especially in diabetic ulcers, pressure injuries, and irradiated tissues.

A variety of therapeutic strategies aimed at mitochondria are currently being explored to address these issues. Mitochondria-targeted antioxidants (e.g., MitoQ, SkQ1), agents that promote mitochondrial biogenesis (e.g., resveratrol, PGC-1α activators), and pharmacological inducers of mitophagy (e.g., urolithin A) are included in this category. Innovative approaches, including mitochondrial transplantation, MSC-based therapies, and the transfer of healthy mitochondria through tunnelling nanotubes or extracellular vesicles, have shown potential for tissue regeneration. Furthermore, photobiomodulation and electroceutical devices influence mitochondrial function and redox balance to facilitate tissue repair.

These advances offer significant opportunities from both clinical and translational perspectives. Early-stage clinical trials have been conducted by several substances, especially for radiation-induced wounds and diabetic foot ulcers. Key challenges persist, including the standardization of mitochondrial function assays in clinical settings, patient stratification implementing mitochondrial biomarkers, the development of scalable delivery systems, and the comprehension of tissue-specific mitochondrial responses.

Looking ahead, the future of wound healing may lie in personalized, mitochondria-focused medicine. New tools like mitochondria-targeting nanoparticles, CRISPR-based gene editing, and synthetic biology are opening doors for more precise treatments. When combined with cutting-edge techniques like 3D-printed scaffolds, lab-grown skin models (organoids), and AI-powered data analysis, these innovations could change how we approach wound care entirely.

As our knowledge advances, mitochondria will progressively be regarded not merely as solitary powerhouses but as coordinators of cellular resilience and regeneration. Leveraging these potentialities necessitates collaborative endeavours across various fields, including mitochondrial biology, regenerative medicine, biomaterials science, and computational biology. By combining basic mitochondrial biology with translational science and new biotechnologies, we get closer to creating mitochondria-centred regeneration techniques that potentially change the way patients with complicated wounds are treated.

Acknowledgments Dr. Meghnad G. Joshi acknowledges the research funding support from the D.Y. Patil Education Society (Deemed to be University) (DYPES/DU/R&D/3104). The schematic art pieces used to produce the figures of this article were provided by Servier Medical Art. Servier Medical Art by Servier is licensed under a Creative Commons Attribution 3.0 Unported License.

Declarations Use of AI and AI-Assisted Technology

During the preparation of this work, the author(s) used Quillbot to improve the English language and flow of the manuscript. After using this tool, the author(s) reviewed and edited the content as needed and take(s) full responsibility for the content of the published article.

Conflicts of Interest All authors have read the journal's authorship agreement and conflict of interest policy. The authors have no potential conflicts of interest to declare.

References

1. Abu Shelbayeh O, Arroum T, Morris S, Busch KB (2023) PGC-1α is a master regulator of mitochondrial lifecycle and ROS stress response. Antioxidants (Basel) 12(5):1075
2. Aghayants S, Zhu J, Yu J, Tao R, Li S, Zhou S, Zhou Y, Zhu Z (2024) The emerging modulators of non-coding RNAs in diabetic wound healing. Front Endocrinol 15:1465975
3. Almsherqi ZA (2021) Potential role of plasmalogens in the modulation of biomembrane morphology. Front Cell Dev Biol 9:673917. https://doi.org/10.3389/fcell.2021.673917
4. An T, Xu Z, Yang Y, Li X, Chen S, Yang X, Man Y, Hu C, Qu Y (2026) EGCG-releasing Nanofibrous scaffold enhances wound healing via γδ T cells modulation. Biomaterials 325:123543
5. Araiso Y, Imai K, Endo T (2022) Role of the TOM complex in protein import into mitochondria: structural views. Annu Rev Biochem 91(1):679–703
6. Azadeh SS, Neghab HK (2025) The role of mitochondria in chronic wound healing (Mitotherapy): signaling and therapeutic implications. Arch Biotechnol Biomed 9(1):001–009
7. Bai L, Wu L, Zhang C, Liu Z, Ma L, Ni J, He D, Zhu M, Peng S, Liu X, Yu H, Lei Y, Luo Y, Zhang Y, Wang X, Wei G, Li Y (2024) Replenishment of mitochondrial Na^+ and H^+ by ionophores potentiates cutaneous wound healing in diabetes. Mater Today Bio 26:101056
8. Baiomy RFE (2024) Quercetin nanoparticles as a therapeutic approach: pharmacological actions and potential applications in therapy. BioTechnologia (Pozn) 105(4):377–393
9. Bardestani A, Ebrahimpour S, Esmaeili A, Esmaeili A (2021) Quercetin attenuates neurotoxicity induced by iron oxide nanoparticles. J Nanobiotechnol. 19(1):327
10. Bernardi P, Gerle C, Halestrap AP, Jonas EA, Karch J, Mnatsakanyan N, Pavlov E, Sheu SS, Soukas AA (2023) Identity, structure, and function of the mitochondrial permeability transition pore: controversies, consensus, recent advances, and future directions. Cell Death Differ 30(8):1869–1885
11. Buck AC, Maarman GJ, Dube A, Bardien S (2025) Mitochondria targeted nanoparticles for the treatment of mitochondrial dysfunction-associated brain disorders. Front Bioeng Biotechnol 13:1563701
12. Buck DW, Dumanian GA, Mustoe TA (2021) The role of cellular metabolism in wound healing. Plast Reconstr Surg 147(1):15e–24e
13. Buyse C, Mignion L, Joudiou N, Melloul S, Driesschaert B, Gallez B (2024) Sensitive simultaneous measurements of oxygenation and extracellular pH by EPR using a stable monophosphonated trityl radical and lithium phthalocyanine. Free Radic Biol Med 213:11–18
14. Cacace J, Luna-Marco C, Hermo-Argibay A, Pesantes-Somogyi C, Hernández-López AO, Pelechá-Salvador M, Bañuls C, Apostolova N, Miguel-Rodríguez L, Morillas C, Rocha M, Rovira-Llopis M, Víctor VM (2025) Poor glycaemic control in type 2 diabetes compromises leukocyte oxygen consumption rate, OXPHOS complex content and neutrophil-endothelial interactions. Redox Biol 81:103516
15. Cai S, Zhao M, Zhou B, Yoshii A, Bugg D, Villet O, Sahu A, Olson GS, Davis J, Tian R (2023) Mitochondrial dysfunction in macrophages promotes inflammation and suppresses repair after myocardial infarction. J Clin Invest 133(4):e159498
16. Cardoso AR, Chausse B, da Cunha FM et al (2012) Mitochondrial compartmentalization of redox processes. Free Radic Biol Med 52(11–12):2201–2208
17. Carmichael RE (2025) Peroxisome dynamics and inter-organelle interactions in neuronal health and disease. Front Mol Neurosci 18:1603632
18. Casas-Martinez JC, Samali A, McDonagh B (2024) Redox regulation of UPR signalling and mitochondrial ER contact sites. Cell Mol Life Sci 81(1):250
19. Castilla DM, Liu ZJ, Velazquez OC (2012) Oxygen: implications for wound healing. Adv Wound Care (New Rochelle). 1(6):225–230
20. Caturano A, D'Angelo M, Mormone A, Russo V, Mollica MP, Salvatore T, Galiero R, Rinaldi L, Vetrano E, Marfella R, Monda M, Giordano A, Sasso FC (2023) Oxidative stress in type

2 diabetes: impacts from pathogenesis to lifestyle modifications. Curr Issues Mol Biol 45(8):6651–6666
21. Cavaillon JM, Singer M (eds) (2018) Inflammation, 4 volume set: from molecular and cellular mechanisms to the clinic, vol 1. Wiley
22. Chaudhari LR, Damle MN, Sharma RK, Joshi MG (2024) Chapter 13 - Limitations of mitochondrial transplantation. In: Ostojic SM, Kubat GB, Ulger O, Gunaydin S (eds) Translational and applied bioenergetics, mitochondrial transplantation and transfer. Academic Press, pp 281–311
23. Chen Q, Ruan D, Shi J, Du D, Bian C (2023) The multifaceted roles of natural products in mitochondrial dysfunction. Front Pharmacol 14:1093038. https://doi.org/10.3389/fphar.2023.1093038
24. Chen S, Wei Q, Tang B, Lin SC, Lee JS, Vu TD, Vu HT, Ramasamy TS, Lin SP, Wu CC (2025) Enhanced mitochondrial function and delivery from adipose-derived stem cell spheres via the EZH2-H3K27me3-PPARγ pathway for advanced therapy. Stem Cell Res Ther 16:85
25. Chen S, Saeed AF, Liu Q, Jiang Q, Xu H, Xiao GG, Rao L, Duo Y (2023) Macrophages in immunoregulation and therapeutics. Signal Transduct Target Ther 8:207
26. Chen B, Lyssiotis CA, Shah YM (2025) Mitochondria-organelle crosstalk in establishing compartmentalized metabolic homeostasis. Mol Cell 85(8):1487–1508
27. Clayton ZE, Tan RP, Miravet MM, Lennartsson K, Cooke JP, Bursill CA, Wise SG, Patel S (2018) Induced pluripotent stem cell-derived endothelial cells promote angiogenesis and accelerate wound closure in a murine excisional wound healing model. Biosci Rep 38(4):BSR20180563
28. Connolly NMC, Theurey P, Adam-Vizi V et al (2018) Guidelines on experimental methods to assess mitochondrial dysfunction in cellular models of neurodegenerative diseases. Cell Death Differ 25:542–572
29. Cortés-Rojo C, Vargas-Vargas MA (2024) Don't give up on mitochondria as a target for the treatment of diabetes and its complications. World J Diabetes 15(10):2015
30. Cremers NA, Wever KE, Wong RJ, van Rheden RE, Vermeij EA, van Dam GM, Carels CE, Lundvig DM, Wagener FA (2017) Effects of remote ischemic preconditioning on heme oxygenase-1 expression and cutaneous wound repair. Int J Mol Sci 18(2):438
31. D'Achille G, Perrone M, Spadoni T, Graciotti L, Bigossi G, Provinciali M, Offidani A, Mattioli-Belmonte M, Cirioni O, Pinton P, Simonetti O, Marchi S (2024) ER-mitochondria association negatively affects wound healing by regulating NLRP3 activation. Cell Death Dis 15:407
32. De Bock K, Georgiadou M, Schoors S, Kuchnio A, Wong BW, Cantelmo AR et al (2013) Role of PFKFB3-driven glycolysis in vessel sprouting. Cell 154(3):651–663
33. de Castro Brás LE, Frangogiannis NG (2020) Extracellular matrix-derived peptides in tissue remodeling and fibrosis. Matrix Biol 91-92:176–187
34. Demyanenko IA, Popova EN, Zakharova VV, Ilyinskaya OP, Vasilieva TV, Romashchenko VP, Fedorov AV, Manskikh VN, Skulachev MV, Zinovkin RA, Pletjushkina OY, Skulachev VP, Chernyak BV (2015) Mitochondria-targeted antioxidant SkQ1 improves impaired dermal wound healing in old mice. Aging (Albany NY) 7(7):475–485
35. Di Cara F, Andreoletti P, Trompier D, Vejux A, Bülow MH, Sellin J, Lizard G, Cherkaoui-Malki M, Savary S (2019) Peroxisomes in immune response and inflammation. Int J Mol Sci 20(16):3877
36. Di X, Gao X, Peng L, Ai J, Jin X, Qi S, Li H, Wang K, Luo D (2023) Cellular mechanotransduction in health and diseases: from molecular mechanism to therapeutic targets. Sig Transduct Target Ther 8:282
37. Dias C, Borges J, Silva F (2024) The four stages of wound healing. Wound Evolution. Retrieved from: https://www.woundevolution.com/blog/the-four-stages-of-wound-healing/
38. Dieude M, Striegl H, Tyzack AJ, Wang A (2023) Multifaceted roles of mitochondria in wound healing and chronic wound pathogenesis. Front Cell Dev Biol 11:1252318
39. Ding W, Chen J, Zhao L, Wu S, Chen X, Chen H (2024) Mitochondrial DNA leakage triggers inflammation in age-related cardiovascular diseases. Front Cell Dev Biol 12:1287447

40. Dong Z, Liu X, Li S, Fu X (2025) Inhibition and rescue of hyperglycemia-induced cellular senescence by mitochondrial transfer from enucleated mesenchymal stem cell-derived microvesicles for chronic wound healing. Adv Sci 2025:e01612
41. Eldeeb MA, Thomas RA, Ragheb MA, Fallahi A, Fon EA (2022) Mitochondrial quality control in health and in Parkinson's disease. Physiol Rev 102(4):1721–1755
42. Eming SA, Wynn TA, Martin P (2017) Inflammation and metabolism in tissue repair and regeneration. Science 356(6342):1026–1030
43. Endoni BT, Koval OM, Allamargot C, Kortlever T, Qian L, Sadoski RJ, Juhr D, Grumbach IM (2025) MIRO1 is required for dynamic increases in mitochondria-ER contact sites and mitochondrial ATP during the cell cycle. Cells 14(7):482
44. Esparza-Moltó PB, Goswami AV, Bozkurt S, Münch C, Newman LE, Moyzis AG, Rojas GR, Guan D, Jones JR, Gage FH (2025) Shadel GS.ROS-dependent localization of glycolytic enzymes to mitochondria. Redox Biol:103812
45. Fadilah NI, Phang SJ, Kamaruzaman N, Salleh A, Zawani M, Sanyal A, Maarof M, Fauzi MB (2023) Antioxidant biomaterials in cutaneous wound healing and tissue regeneration: a critical review. Antioxidants 12(4):787
46. Fang X, Wu H, Wei J, Miao R, Zhang Y, Tian J (2022) Research progress on the pharmacological effects of berberine targeting mitochondria. Front Endocrinol (Lausanne) 13:982145
47. Faria R, Boisguérin P, Sousa Â, Costa D (2023) Delivery systems for mitochondrial gene therapy: a review. Pharmaceutics 15(2):572
48. Filippo MD, Karakaya T, Cheng F, Hennig P, Slaufova M, Boukamp P, Pascolo S, Maul JT, Kolm I, Levesque MP, Kündig T, Beer HD (2025) NLRP1 inflammasome activation in skin equivalents reveals mechanistic insights into the roles of keratinocytes in psoriasis. bioRxiv:2025.06.25.66150
49. Galluzzi L, Kepp O, Trojel-Hansen C, Kroemer G (2012) Mitochondrial control of cellular life, stress, and death. Circ Res 111(9):1198–1207
50. Galván-Peña S, O'Neill LA (2014) Metabolic reprogramming in macrophage polarization. Front Immunol 5:420
51. Gao S, Hu J (2021) Mitochondrial fusion: the machineries in and out. Trends Cell Biol 31(1):62–74
52. Gervasoni J, Primiano A, Cicchinelli M, Santucci L, Servidei S, Urbani A, Primiano G, Iavarone F (2024) Mitochondrial biomarkers in the omics era: a clinical-pathophysiological perspective. Int J Mol Sci 25(9):4855
53. Ghetu DM, Raymond K, Titorencu I, Simionescu M (2025) Innovative strategies: use of stromal cell-derived secretome for chronic wound therapy. Int J Mol Sci 26(12):5609
54. Gu L, Sun X, Pan J, Liu D, Huang L, Yu Y, Yu B, Cong H (2025) AIE-active PVA/berberine antibacterial hydrogel for wound healing, visual monitoring pH and dehydration. Biomaterials 323:123432
55. Guo S, DiPietro LA (2010) Factors affecting wound healing. J Dent Res 89(3):219–229
56. Hamann N, Härtel C, Köster S, Groth M, Celardo I (2018) Fibroblast metabolic regulation during wound healing. Cell Rep 25(11):2936–2943
57. Heidary MR, Samimi Z, Asgary S, Mohammadi P, Hozeifi S, Hoseinzadeh-Chahkandak F, Xu S, Farzaei MH (2022) Natural AMPK activators in cardiovascular disease prevention. Front Pharmacol 12:738420
58. Heiduschka S, Prigione A (2025) iPSC models of mitochondrial diseases. Neurobiol Dis 207(April):106822
59. Huang D, Chen S, Xiong D, Wang H, Zhu L, Wei Y, Li Y, Zou S (2023) Mitochondrial dynamics: working with the cytoskeleton and intracellular organelles to mediate mechanotransduction. Aging Dis 14(5):1511–1532
60. Huang X, Zheng L, Zhou Y, Hu S, Ning W, Li S, Lin Z, Huang S (2024) Controllable adaptive molybdate-oligosaccharide nanoparticles regulate M2 macrophage mitochondrial function and promote angiogenesis via PI3K/HIF-1α/VEGF pathway to accelerate diabetic wound healing. Adv Healthc Mater 13(3):2302256

61. Huang J, Fan C, Ma Y, Huang G (2024) Exploring thermal dynamics in wound healing: the impact of temperature and microenvironment. Clin Cosmet Investig Dermatol 17:1251–1258
62. Hunt M, Torres M, Bachar-Wikström E, Wikström JD (2023) Multifaceted roles of mitochondria in wound healing and chronic wound pathogenesis. Front Cell Dev Biol 11:1252318
63. Iheagwam FN, Joseph AJ, Adedoyin ED, Iheagwam OT, Ejoh SA (2025) Mitochondrial dysfunction in diabetes: shedding light on a widespread oversight. Pathophysiology 32(1):9
64. Inigo JR, Chandra D (2022) The mitochondrial unfolded protein response (UPRmt): shielding against toxicity to mitochondria in cancer. J Hematol Oncol 15(1):98
65. Javadi K, Ahmadi MH, Rajabnia M, Halaji M (2025) Effects of curcumin on biofilm production and associated gene in multidrug-resistant *Acinetobacter baumannii* isolated from hospitalized patients. Int J Mol Cell Med 14(1):567–575
66. Joshi MG, Damle MN, Sharma RK (2024) Chapter 12 - Application route of mitochondrial transplantation. In: Ostojic SM, Kubat GB, Ulger O, Gunaydin S (eds) Translational and applied bioenergetics, mitochondrial transplantation and transfer. Academic Press, pp 231–280
67. Karsdal M, Cox TR, Parker AL, Willumsen N, Sand JMB, Jenkins G, Hansen HH, Oldenburger A, Geillinger-Kaestle KE, Larsen AT, Black D, Genovese F, Eckersley A, Heinz A, Nyström A, Holm Nielsen S, Bennink L, Johannsson L, Bay-Jensen AC, Ricard-Blum S (2025) Advances in extracellular matrix-associated diagnostics and therapeutics. J Clin Med 14(6):1856
68. Kasperova BJ, Cinkajzlova A, Horvath L, Svoboda P, Haluzik M, Stemberkova Hubackova S (2025) Coming of age: could obesity-related metabolic complications be treated by targeting senescent cells? Front Cell Dev Biol 13:1622107
69. Khalid KA, Nawi AFM, Zulkifli N, Barkat MA, Hadi H (2022) Aging and wound healing of the skin: a review of clinical and pathophysiological hallmarks. Life (Basel) 12(12):2142
70. Koene S, Spaans E, Van Bortel L, Van Lancker G, Delafontaine B, Badilini F, Beyrath J, Smeitink J (2017) KH176 under development for rare mitochondrial disease: a first in man randomized controlled clinical trial in healthy male volunteers. Orphanet J Rare Dis 12(1):163
71. Kotiadis VN, Duchen MR, Osellame LD (2014) Mitochondrial quality control and communications with the nucleus are important in maintaining mitochondrial function and cell health. Biochim Biophys Acta 1840(4):1254–1265
72. Kowaltowski AJ, Abdulkader F (2024) How and when to measure mitochondrial inner membrane potentials. Biophys J 123(24):4150–4157
73. Krauss S, Rothenberger J, Mayer J, Sogorski A, Held M, Wahler T, Daigeler A, Kolbenschlag J (2018) Tissue conditioning - strategies to improve perfusion and reduce ischemia - reperfusion injury. Plast Aesthet Res 5:39
74. Kubat GB, Picone P, Tuncay E, Aryan L, Girgenti A, Palumbo L, Turkel I, Akat F, Singh KK, Nuzzo D (2025) Biotechnological approaches and therapeutic potential of mitochondria transfer and transplantation. Nat Commun 16:5709
75. Kuehlmann B, Bonham CA, Zucal I, Prantl L, Gurtner GC (2020) Mechanotransduction in wound healing and fibrosis. J Clin Med 9(5):1423
76. Kumar R, Islinger M, Worthy H, Carmichael R, Schrader M (2024) The peroxisome: an update on mysteries 3.0. Histochem Cell Biol 161(2):99–132
77. Landry C, Costanzo JP, Mitne-Neto M, Zatz M, Schaffer AE, Hatzoglou M, Muotri AR, Miranda HC (2025) Convergent activation of the integrated stress response and ER-mitochondria uncoupling in VAPB-associated ALS. EMBO Mol Med 17:2299
78. Lawrence J, Sorra K (2024) Photobiomodulation as medicine: low-level laser therapy (LLLT) for acute tissue injury or sport performance recovery. J Funct Morphol Kinesiol 9(4):181
79. Levoux J, Prola A, Lafuste P, Gervais M, Chevallier N, Koumaiha Z, Kefi K, Braud L, Schmitt A, Yacia A, Schirmann A, Hersant B, Sid-Ahmed M, Larbi SB, Komrskova K, Rohlena J, Relaix F, Neuzilm J, Rodriguez AM (2021) Platelets facilitate the wound-healing capability of mesenchymal stem cells by mitochondrial transfer and metabolic reprogramming. Cell Metab 33(2):283–299.e9

80. Li H, Wang Z, Zhou F, Zhang G, Feng X, Xiong Y, Wu Y (2023) Sustained activation of NLRP3 inflammasome contributes to delayed wound healing in aged mice. Int Immunopharmacol 116:109828
81. Li P, Fan Z, Huang Y, Luo L, Wu X (2025) Mitochondrial dynamics at the intersection of macrophage polarization and metabolism. Front Immunol 16:1520814
82. Li Y, Li X-M, Wei L-S, Ye J-F (2024) Advancements in mitochondrial-targeted nanotherapeutics: overcoming biological obstacles and optimizing drug delivery. Front Immunol 15:1451989
83. Licini C, Morroni G, Lucarini G, Vitto VAM, Orlando F, Missiroli S, D'Achille G, Perrone M, Spadoni T, Graciotti L, Bigossi G, Provinciali M, Offidani A, Mattioli-Belmonte M, Cirioni O, Pinton P, Simonetti O, Marchi S (2024) ER-mitochondria association negatively affects wound healing by regulating NLRP3 activation. Cell Death Dis 15(6):407
84. Licini C, Morroni G, Lucarini G, Maria Vitto VA, Orlando F, Missiroli S, Lismont C, Nordgren M, Van Veldhoven PP, Fransen M (2015) Redox interplay between mitochondria and peroxisomes. Front Cell Dev Biol 3:35
85. Lindley LE, Stojadinovic O, Pastar I, Tomic-Canic M (2016) Biology and biomarkers for wound healing. Plast Reconstr Surg 138(3 Suppl):18S–28S
86. Liu L, Zhao M, Jin X, Ney G, Yang KB, Cao J et al (2024) Mitochondrial unfolded protein response (UPRmt): what we know thus far. Front Cell Dev Biol 12:1405393
87. Liu Y, Xin L, Si W, Tan J, Zhu X, Chen X, Song J, Chen T, Zhai Q (2026) Nanoparticles with reprogramming of mitochondrial respiratory chain complex and epigenetic modifications functions for osteoporosis treatment. Biomaterials 324:123468
88. Liu Y, Dissanayaka WL, Yiu C (2025) Therapeutic implications of mitochondrial transfer on stem cell fate in regenerative medicine. J Transl Med 23:568
89. Liu Z, Bian X, Luo L, Björklund ÅK, Li L, Zhang L, Chen Y, Guo L, Gao J, Cao C, Wang J, He W, Xiao Y, Zhu L, Annusver K, Gopee NH, Basurto-Lozada D, Horsfall D, Bennett CL, Kasper M, Haniffa M, Sommar P, Li D, Landén NX (2025) Spatiotemporal single-cell roadmap of human skin wound healing. Cell Stem Cell 32(3):479–498.e8
90. López-Doménech G, Covill-Cooke C, Ivankovic D, Halff EF, Sheehan DF, Norkett R, Birsa N, Kittler JT (2018) Miro proteins coordinate microtubule- and actin-dependent mitochondrial transport and distribution. EMBO J 37(3):321–336
91. Luo Z, Li NN, Gu X, Zhou CX, Jiang ZZ, Luo XJ, Zhu HW, Zhu XY (2025) Interaction between mitochondrial oxidative stress and myocardial fibrosis in the context of diabetes. Front Endocrinol (Lausanne) 16:1596436
92. Luo Z, Yao J, Wang Z, Xu J (2023) Mitochondria in endothelial cells angiogenesis and function: current understanding and future perspectives. J Transl Med 21(1):441
93. Lv H, Wang S, He M, Zhu X, He P, Sun R, Jia Z, Wang Z, Zhao W, Zhong Z, Han Y (2025) Electrospun SF/PHBV nanofibers loaded with berberine as a bioactive wound dressing: accelerating diabetic wound healing and alleviating hypertrophic scar. Mater Des 251:113574
94. Madan S, Uttekar B, Chowdhary S, Rikhy R (2022) Mitochondria lead the way: mitochondrial dynamics and function in cellular movements in development and disease. Front Cell Dev Biol 9:781933. https://doi.org/10.3389/fcell.2021.781933
95. Mahrouf-Yorgov M, Augeul L, Da Silva CC, Jourdan M, Rigolet M, Manin S, Ferrera R, Ovize M, Henry A, Guguin A, Meningaud JP, Dubois-Randé JL, Motterlini R, Foresti R, Rodriguez AM (2017) Mesenchymal stem cells sense mitochondria released from damaged cells as danger signals to activate their rescue properties. Cell Death Differ 24(7):1224–1238
96. Main EN, Cruz TM, Bowlin GL (2023) Mitochondria as a therapeutic: a potential new frontier in driving the shift from tissue repair to regeneration. Regen Biomater 10:rbad070
97. Maleki MH, Siri M, Jafarabadi A, Rajabi M, Mazhari SA, Noori Z, Koohpeyma F, Dehghanian A, Esmaeili N, Aryanian Z, Dastghaib S (2024) Boosting wound healing in diabetic rats: the role of nicotinamide riboside and resveratrol in UPR modulation and pyroptosis inhibition. Int Immunopharmacol 132:112013

98. Marques E, Kramer R, Ryan DG (2024) Multifaceted mitochondria in innate immunity. npj Metab Health Dis 2:6
99. Martin P, Nunan R (2015) Cellular and molecular mechanisms of repair in acute and chronic wound healing. Br J Dermatol 173(2):370–378
100. Martin P, Pardo-Pastor C, Jenkins RG, Rosenblatt J (2024) Imperfect wound healing sets the stage for chronic diseases. Science 386:eadp2974
101. Martino F, Perestrelo AR, Vinarský V, Pagliari S, Forte G (2018) Cellular mechanotransduction: from tension to function. Front Physiol 9:824
102. McCully JD, Cowan DB, Emani SM, del Nido PJ (2024) Effects of mitochondrial transplantation on chronic pressure wound healing in a human patient. JACC Basic Transl Sci 9(3):328–330
103. McLaughlin D, Sasaki M, Hoffmann C, Brewster L, E Hekman K (2024) Smoking status impacts mitochondrial function and synthetic function in mesenchymal stem cells derived from diabetics with arterial insufficiency. Adv Wound Care (New Rochelle)
104. Alhajj M, Goyal A (2022) Physiology, granulation tissue. In: StatPearls [Internet]. StatPearls Publishing
105. Meizlish ML, Kimura Y, Pope SD, Matta R, Kim C, Philip NH, Meyaard L, Gonzale A, Medzhitov R (2024) Mechanosensing regulates tissue repair program in macrophages. Sci Adv 10:eadk6906
106. Miao X, Jiang P, Wang Z, Kong W, Feng L (2025) Mitochondrial transplantation: a novel therapeutic approach for treating diseases. MedComm 6(6):e70253
107. Mohammadi B, Esmaeilizade Z, Omrani MD (2022) The effect of co-treating human mesenchymal stem cells with epigallocatechin gallate and hypoxia-inducible factor-1 on the expression of RANKL/RANK/OPG signaling pathway, osteogenesis, and angiogenesis genes. Regen Eng Transl Med 8:117–124
108. Monzel AS, Enríquez JA, Picard M (2023) Multifaceted mitochondria: moving mitochondrial science beyond function and dysfunction. Nat Metab 5(4):546–562
109. Mu X, Wu X, He W, Liu Y, Wu F, Nie X (2022) Pyroptosis and inflammasomes in diabetic wound healing. Front Endocrinol (Lausanne) 13:950798
110. Murley A, Nunnari J (2016) The emerging network of mitochondria-organelle contacts. Mol Cell 61(5):648–653
111. Murphy MP, Bayir H, Belousov V et al (2022) Guidelines for measuring reactive oxygen species and oxidative damage in cells and in vivo. Nat Metab 4:651–662
112. Nassef MZ, Hanke JE, Hiller K (2021) Mitochondrial metabolism in macrophages. Am J Physiol Cell Physiol 321(6):C1070–C1081. https://doi.org/10.1152/ajpcell.00126.2021
113. Natarelli N, Gahoonia N, Aflatooni S, Bhatia S, Sivamani RK (2024) Dermatologic manifestations of mitochondrial dysfunction: a review of the literature. Int J Mol Sci 25(6):3303
114. Ni J, Chen Y, Zhang L, Wang R, Wu X, Khan NU, Xie F (2025) Epigallocatechin gallate and vancomycin loaded poly(vinyl)-pyrrolidone-gelatine nanofibers, conceivable curative approach for wound healing. Colloids Surf B: Biointerfaces 249:114506
115. Nielson JR, Rutter JP (2018) Lipid-mediated signals that regulate mitochondrial biology. J Biol Chem 293(20):7517–7521
116. Nunnari J, Suomalainen A (2012) Mitochondria: in sickness and in health. Cell 148(6):1145–1159
117. O'Siorain JR, Luo W, Smith BJ, Philp LK, Miotto PM, Day EA et al (2024) Time-of-day control of mitochondria regulates NLRP3 inflammasome activation in macrophages. FASEB J 38:e70235
118. Paik S, Kim JK, Shin HJ, Park EJ, Kim IS, Jo EK (2025) Updated insights into the molecular networks for NLRP3 inflammasome activation. Cell Mol Immunol 22:563–596
119. Palikaras K, Lionaki E, Tavernarakis N (2018) Mechanisms of mitophagy in cellular homeostasis, physiology and pathology. Nat Cell Biol 20(9):1013–1022

120. Palty R, Silverman WF, Hershfinkel M, Caporale T, Sensi SL, Parnis J, Sekler I (2010) NCLX is an essential component of mitochondrial Na+/Ca2+ exchange. Proc Natl Acad Sci 107(1):436–441
121. Pan X, Zhao Y, Li Y, Chen J, Zhang W, Yang L, Xiong YZ, Ying Y, Xu H, Zhang Y, Gao C, Sun Y, Li N, Chen L, Chen Z, Lei K (2024) Mitochondrial dynamics govern whole-body regeneration through stem cell pluripotency and mitonuclear balance. Nat Commun 15(1):10681
122. Pan Y, Xia M, Luo J, Lu S (2025) Resveratrol promotes wound healing by enhancing angiogenesis via inhibition of Ferroptosis. Food Sci Nutr 13(5):e70254
123. Papandreou I, Cairns RA, Fontana L, Lim AL, Denko NC (2006) HIF-1 mediates adaptation to hypoxia by actively downregulating mitochondrial oxygen consumption. Cell Metab 3(3):187–197
124. Phan HTL, Lee H, Kim K (2023) Trends and prospects in mitochondrial genome editing. Exp Mol Med 55(5):871–878
125. Picca A, Lezza AMS, Leeuwenburgh C, Pesce V, Calvani R, Bossola M, Manes-Gravina E, Landi F, Bernabei R, Marzetti E (2018) Circulating mitochondrial DNA at the crossroads of mitochondrial dysfunction and inflammation during aging and muscle wasting disorders. Rejuvenation Res 21(4):350–359
126. Piipponen M, Li D, Landén NX (2020) The immune functions of keratinocytes in skin wound healing. Int J Mol Sci 21(22):8790
127. Popov LD (2020) Mitochondrial biogenesis: an update. J Cell Mol Med 24(9):4892–4899
128. Prabhakaran HS, Hu D, He W, Luo G, Liou YC (2023) Mitochondrial dysfunction and mitophagy: crucial players in burn trauma and wound healing. Burns Trauma 11:tkad029
129. Prado TP, Zanchetta FC, Barbieri B, Aparecido C, Melo Lima MH, Araujo EP (2023) Photobiomodulation with blue light on wound healing: a scoping review. Life (Basel) 13(2):575
130. Pu Z, Zhang L, Yang H, Shao T, Wang D, Wang J, Yan Y, Si W (2025) Vitrification of 3D-MSCs encapsulated in GelMA hydrogel: improved cryosurvival, reduced cryoprotectant concentration, and enhanced wound healing. Int J Biol Macromol 296:139716
131. Qi X, Liu C, Si J, Yin B, Huang J, Wang X, Huang J, Sun H, Zhu C, Zhang (2024) A bioenergetically-active ploy (glycerol sebacate)-basedmultiblock hydrogel improved diabetic wound healing throughrevitalizing mitochondrial metabolism. Cell Prolif 57:e13613
132. Qiang L, Yang S, Cui YH, He YY (2020) Keratinocyte autophagy enables the activation of keratinocytes and fibroblastsand facilitates wound healing. Autophagy 17(9):2128–2143
133. Qin P, Sun Y, Li L (2024) Mitochondrial dysfunction in chronic neuroinflammatory diseases (Review). Int J Mol Med 53(5):47
134. Qin Z, Zhou S, Chang X, Liu R (2024) SIRT3 regulates mitochondrial function: a promising star target for cardiovascular disease therapy. Biomed Pharmacother 170:116004
135. Qin Q, Haba D, Takizawa C, Tomida S, Horinouchi A, Katagiri M et al (2025) Candidate biomarkers for hard-to-heal wounds revealed by single-cell RNA sequencing of wound fluid in murine wound models. Wound Repair Regen 33(3):e70038
136. Qsee HS, Shetty S, De S, Bharati S (2024) Mitochondria-targeted antioxidant, mitoQ protects hepatic tissue from N-nitrosodiethylamine-induced damage by modulating mitochondrial function and redox status. Adv Redox Res 12:100108
137. Quiros PM, Mottis A, Auwerx J (2016) Mitonuclear communication in homeostasis and stress. Nat Rev Mol Cell Biol 17(4):213–226
138. Ranieri M, Brajkovic S, Riboldi G, Ronchi D, Rizzo F, Bresolin N, Corti S, Comi GP (2013) Mitochondrial fusion proteins and human diseases. Neurol Res Int 2013(1):293893
139. Reinke JM, Sorg H (2012) Wound repair and regeneration. Eur Surg Res 49(1):35–43
140. Reisman EG, Hawley JA, Hoffman NJ (2024) Exercise-regulated mitochondrial and nuclear signalling networks in skeletal muscle. Sports Med 54(5):1097–1119
141. Ren J, Hua Y, Liao Z, Wen X, Wang Q (2025) Subcellular two-pronged targeting therapeutics: disrupting mitochondria-endoplasmic reticulum communication in inflammatory disorders. Available at SSRN 5358318

142. Rodrigues M, Nina Kosaric N, Clark A, Bonham CA, Gurtner GC (2019) Wound healing: a cellular perspective. Physiol Rev 99(1):665–706
143. Rodriguez PG, Felix FN, Woodley DT, Shim EK (2008) The role of oxygen in wound healing: a review of the literature. Dermatol Surg 34(9):1159–1169
144. Ruan L, Zhou C, Jin E, Kucharavy A, Zhang Y, Wen Z et al (2017) Cytosolic proteostasis through importing of misfolded proteins into mitochondria. Nature 543(7645):443–446
145. Russo M, Pileri F, Ghisletti S (2023) Novel insights into the role of acetyl-CoA producing enzymes in epigenetic regulation. Front Endocrinol 14:1272646
146. Schieber M, Chandel NS (2014) ROS function in redox signaling and oxidative stress. Curr Biol 24(10):R453–R462
147. Schiffmann LM, Werthenbach JP, Heintges-Kleinhofer F (2020) Mitochondrial respiration controls neoangiogenesis during wound healing and tumour growth. Nat Commun 11(1):3653
148. Schrader M, Costello J, Godinho LF, Islinger M (2020) Proliferation and fission of peroxisomes an update. Biochim Biophys Acta Mol Cell Res 1867(12):118769
149. Schrader M, Costello J, Godinho LF (2015) Peroxisome-mitochondria interplay and disease. J Inherit Metab Dis 38:681–702
150. Sen CK (2009) Wound healing essentials: let there be oxygen. Wound Repair Regen 17(1):1–18
151. Shaito A, Al-Mansoob M, Ahmad SMS, Haider MZ, Eid AH, Posadino AM, Pintus G, Giordo R (2023) Resveratrol-mediated regulation of mitochondria biogenesis-associated pathways in neurodegenerative diseases: molecular insights and potential therapeutic applications. Curr Neuropharmacol 21(5):1184–1201
152. Sharma A, Srivastava R, Gnyawali SC, Bhasme P, Anthony AJ, Xuan Y, Trinidad JC, Sen CK, Clemmer DE, Roy S, Ghatak S (2024) Mitochondrial bioenergetics of functional wound closure is dependent on macrophage–keratinocyte exosomal crosstalk. ACS Nano 18(44):30405–30420
153. Shiiba I, Ito N, Oshio H et al (2025) ER-mitochondria contacts mediate lipid radical transfer via RMDN3/PTPIP51 phosphorylation to reduce mitochondrial oxidative stress. Nat Commun 16:1508
154. Shirihai OS, Song M, Dorn GW (2015) How mitochondrial dynamism orchestrates mitophagy. Circ Res 116(11):1835–1849
155. Shpilka T, Haynes CM (2018) The mitochondrial UPR: mechanisms, physiological functions and implications in ageing. Nat Rev Mol Cell Biol 19(2):109–120
156. Skulachev VP, Vyssokikh MY, Chernyak BV, Averina OA, Andreev-Andrievskiy AA, Zinovkin RA, Lyamzaev KG, Marey MV, Egorov MV, Frolova OJ, Zorov DB, Skulachev MV, Sadovnichii VA (2023) Mitochondrion-targeted antioxidant SkQ1 prevents rapid animal death caused by highly diverse shocks. Sci Rep 13:4326
157. Soldatov VO, Kubekina MV, Skorkina MY, Belykh AE, Egorova TV, Korokin MV, Pokrovskiy MV, Deykin AV, Angelova PR (2022) Current advances in gene therapy of mitochondrial diseases. J Transl Med 20:562
158. Spinelli JB, Haigis MC (2018) The multifaceted contributions of mitochondria to cellular metabolism. Nat Cell Biol 20(7):745–754
159. Sprason C, Tucker T, Clancy D (2024) MtDNA deletions and aging. Front Aging 5:1359638
160. Su M, Qiu F, Li Y, Che T, Li N, Zhang S (2024) Mechanisms of the NAD+ salvage pathway in enhancing skeletal muscle function. Front Cell Dev Biol 12:1464815. https://doi.org/10.3389/fcell.2024.1464815
161. Sugiura A, Mattie S, Prudent J, McBride HM (2017) Newly born peroxisomes are a hybrid of mitochondrial and ER-derived pre-peroxisomes. Nature 542(7640):251–254
162. Talbott HE, Mascharak S, Griffin M, Wan DC, Longaker MT (2022) Wound healing, fibroblast heterogeneity, and fibrosis. Cell Stem Cell 29(8):1161–1180
163. Taner OF, Ulger O, Ersahin S, Baser NT, Genc O, Kubat GB (2024) Effects of mitochondrial transplantation on chronic pressure wound healing in a human patient. Cytotherapy 26(6):579–585

164. Terrell K, Choi S, Choi S (2023) Calcium's role and signaling in aging muscle, cellular senescence, and mineral interactions. Int J Mol Sci 24(23):17034
165. Tiskratok W, Chuinsiri N, Limraksasin P, Kyawsoewin M, Jitprasertwong P (2025) Extracellular matrix stiffness: mechanotransduction and mechanobiological response-driven strategies for biomedical applications targeting fibroblast inflammation. Polymers 17(6):822
166. Tissier R, Chenoune M, Pons S, Zini R, Darbera L, Lidouren F, Ghaleh B, Berdeaux A, Morin D (2013) Mild hypothermia reduces per-ischemic reactive oxygen species production and preserves mitochondrial respiratory complexes. Resuscitation 84(2):249–255
167. Toppila M, Ranta-Aho S, Kaarniranta K, Hytti M, Kauppinen A (2024) Metformin alleviates inflammation and induces mitophagy in human retinal pigment epithelium cells suffering from mitochondrial damage. Cell 13(17):1433
168. Torres-López M, Spiller PF, Gao L, García-Flores P, Murphy MP, Ortega-Sáenz P, López-Barneo J (2025) Acute oxygen sensing by arterial chemoreceptors with a mutant mitochondrial complex I ND6 subunit lacking reverse electron transport. FEBS Lett 599:1122–1134
169. Traa A, Machiela E, Rudich PD, Soo SK, Senchuk MM, Van Raamsdonk JM, Miller HA (2024) Overexpression of mitochondrial fission or mitochondrial fusion genes enhances resilience and extends longevity. Aging Cell 23(8):e14262
170. Valsami EA, Chu G, Guan M, Gilman J, Theocharidis G, Veves A (2025) The role of omics techniques in diabetic wound healing: recent insights into the application of single-cell RNA sequencing, bulk RNA sequencing, spatial transcriptomics, and proteomics. Adv Ther 42(7):3089–3110
171. Velnar T, Bailey T, Smrkolj V (2009) The wound healing process: an overview of the cellular and molecular mechanisms. J Int Med Res 37(5):1528–1542
172. Vishwanath AA (2023) The role of mitochondrial calcium homeostasis in presynaptic metabolism and neurotransmission: understanding how modulating mitochondrial metabolism affects neuronal physiology (Doctoral dissertation, Sorbonne Université)
173. Walker BR, Moraes CT (2022) Nuclear-mitochondrial interactions. Biomolecules 12(3):427
174. Wang Z, Zhao F, Xu C, Zhang Q, Ren H, Huang X, He C, Ma J, Wang Z (2024) Metabolic reprogramming in skin wound healing. Burns Trauma 12:tkad047
175. Wang Y, Hekimi S (2015) Mitochondrial dysfunction and longevity in animals: untangling the knot. Science 350(6265):1204–1207
176. Wei X, Manandhar L, Kim H, Chhetri A, Hwang J, Jang G, Park C, Park R (2025) Pexophagy and oxidative stress: focus on peroxisomal proteins and reactive oxygen species (ROS) signaling pathways. Antioxidants 14(2):126
177. Willenborg S, Sanin DE, Jais A, Ding X, Ulas T, Nüchel J, Eming SA (2021) Mitochondrial metabolism coordinates stage-specific repair processes in macrophages during wound healing. Cell Metab 33(12):2398–2414
178. Williams-Reid H, Johannesson A, Buis A (2025) Wound management, healing, and early prosthetic rehabilitation: part 3 - a scoping review of chemical biomarkers. Can Prosthet Orthot J 8(1):1
179. Wong VW, Longaker MT, Gurtner GC (2012) Soft tissue mechanotransduction in wound healing and fibrosis. Semin Cell Dev Biol 23(9):981–986
180. Wu K, Shieh J, Qin L, Guo JJ (2024) Mitochondrial mechanisms in the pathogenesis of chronic inflammatory musculoskeletal disorders. Cell Biosci 14:76
181. Wu SS, Li QC, Yin CQ, Xue W, Song CQ (2020) Advances in CRISPR/Cas-based gene therapy in human genetic diseases. Theranostics 10:4374–4382
182. Wu W, Liu Y, Peng S, Chen Y, Ye W, Wang W, Wang G, Lu C, Wu C, Pan X, Liu Y, Lu A, Peng T (2025) Mitochondria-targeted curcumin microneedles for psoriasis therapy via autophagy activation and epidermal hyperplasia suppression. Int J Pharm 682:126000
183. Wu S, Wang Y, Zou M, Yue Z, Li J, Zhang L et al (2024) Mitochondrial fusion–fission dynamics and its involvement in colorectal cancer. Mol Oncol 18(5):1133–1153
184. Xiong Y, Knoedler S, Alfertshofer M et al (2025) Mechanisms and therapeutic opportunities in metabolic aberrations of diabetic wounds: a narrative review. Cell Death Dis 16:341

185. Xu S, Li S, Bjorklund M, Xu S (2022) Mitochondrial fragmentation and ROS signaling in wound response and repair. Cell Regen 11(1):38. https://doi.org/10.1186/s13619-022-00141-8
186. Yang H, Chen Y, Rong Y (2025) Multifunctional hydrogel targeting senescence to accelerate diabetic wound healing through promoting angiogenesis. J Nanobiotechnol 23:177
187. Yin Y, Shen H (2022) Common methods in mitochondrial research (Review). Int J Mol Med 50:126
188. Yip JMX, Chiang GSH, Lee ICJ, Lehming-Teo R, Dai K, Dongol L, Wang LYT, Teo D, Seah GT, Lehming N (2025) Mitochondria and the repurposing of diabetes drugs for off-label health benefits. Int J Mol Sci 26(1):364
189. Young A, McNaught CE (2011) The physiology of wound healing. Surgery (Oxford) 29(10):475–479
190. Yu T, Dohl J, Wang L, Chen Y, Gasier HG, Deuster PA (2020) Curcumin ameliorates heat-induced injury through NADPH oxidase–dependent redox signaling and mitochondrial preservation in C2C12 myoblasts and mouse skeletal muscle. J Nutr 150(9):2257–2267
191. Yuan Y, Tian Y, Jiang H, Cai L-Y, Song J, Peng R, Zhang X-M (2023) Mechanism of PGC-1α-mediated mitochondrial biogenesis in cerebral ischemia–reperfusion injury. Front Mol Neurosci 16:1224964
192. Zhang H, Menzies KJ, Auwerx J (2018) The role of mitochondria in stem cell fate and aging. Development 145(8)
193. Zhang M, Wu J, Cai K, Liu Y, Lu B, Zhang J, Xu J, Gu C, ChenT (2024) From dysfunction to healing: advances in mitochondrial therapy for Osteoarthritis. J Transl Med 22:1013
194. Zhang X, Li H, Liu Y, Yu J, Zhang P, Yu P, Liu Y, Jia S, Ling L, Li P, Li L, Wang Y, Huang T, Jin G, Zhao Y, Ma G, Yuan Q, Zhu L, Zhang Z, Li H, Li W (2024) Acid-responsive CST@ NPs enhanced diabetic wound healing through rescuing mitochondrial dysfunction. Bioact Mater 44:269–282
195. Zhao J, Zhang S, Gong Z, Mao W, Bao W, Li Q, Bai Y, Gao F, Feng S (2025) NLRP3: a key regulator of skin wound healing and macrophage-fibroblast interactions in mice. Cell Commun Signal 23:55
196. Zhong J, Xu C, Gabbay-Benziv R, Lin X, Yang P (2016) Superoxide dismutase 2 overexpression alleviates maternal diabetes-induced neural tube defects, restores mitochondrial function and suppresses cellular stress in diabetic embryopathy. Free Radic Biol Med 96:234–244
197. Zhong M, Peng J, Xiang L, Yang X, Wang X, Zhu Y (2020) Epigallocatechin Gallate (EGCG) improves anti-angiogenic state, cell viability, and hypoxia-induced endothelial dysfunction by downregulating high mobility group box 1 (HMGB1) in preeclampsia. Med Sci Monit 26:e926924
198. Zhou R, Xiang C, Cao G, Xu H, Zhang Y, Yang H, Zhang J (2021) Berberine accelerated wound healing by restoring TrxR1/JNK in diabetes. Clin Sci Lond 135(4):613–627
199. Zong Y, Li H, Liao P, Pan Y, Zheng Y, Zhang C, Liu D, Zheng M, Gao J (2024) Mitochondrial dysfunction: mechanisms and advances in therapy. Signal Transduct Target Ther 9:124
200. Zuo W, Sun R, Ji Z, Ma G (2023) Macrophage-driven cardiac inflammation and healing: insights from homeostasis and myocardial infarction. Cell Mol Biol Lett 28(1):81

Chapter 7
Cutaneous Wound Epithelialization: Role of Cadherin- and Integrin-Mediated Adhesions

Katheryn E. Rothenberg and Martine Dunnwald

Abstract The skin is the largest organ in the body and protects internal organs from external insults. When the skin is wounded, the tissue responds rapidly to restore barrier function. One critical step of wound healing is epithelialization, in which epidermal cells called keratinocytes proliferate and migrate collectively to repair the lesion. To drive and coordinate movement, keratinocytes form adhesions to their neighboring cells and to the extracellular matrix (ECM). In this chapter, we summarize the current knowledge of two main classes of cell adhesion molecules in the context of wound epithelialization: cadherins, which mediate cell-cell adhesion, and integrins, which mediate cell–ECM adhesion. We draw from studies in diverse organisms, including *Drosophila*, *C elegans*, *Xenopus*, mice, and humans, to build a consensus understanding of how these adhesions drive coordinated epidermal migration. Cadherins mediate cell–cell adhesion and enable transmission of cytoskeletal forces between cells. Cadherins are initially downregulated at the wound edge to facilitate cell movement, then reinforced at focal sites to enable coordinated migration. Integrins mediate cell–ECM adhesion and are critical for generating traction forces that drive migration. During wound healing, integrin expression changes to interact with the provisional wound matrix, and integrins relocalize to contribute to cell-cell adhesion. Cadherin and integrin adhesions dynamically remodel in space and time and undergo complex crosstalk with each other and other adhesion structures. Understanding these molecular mechanisms by harnessing diverse model organisms and systems is necessary for developing novel therapeutic strategies to improve wound healing and prevent chronic wounds.

K. E. Rothenberg (✉)
Department of Biology, University of Iowa College of Liberal Arts and Sciences, Iowa City, IA, USA
e-mail: katheryn-rothenberg@uiowa.edu

M. Dunnwald
Department of Anatomy and Cell Biology, University of Iowa Carver College of Medicine, Iowa City, IA, USA

F. De Francesco, K. Turksen (eds.), *Management and Strategies for Wound Healing*, Stem Cell Biology and Regenerative Medicine 79,
https://doi.org/10.1007/978-3-032-12442-5_7

Keywords Skin · Epidermis · Wound epithelialization · Cell adhesion · Cadherin · Integrin

Abbreviations

ECM	Extracellular matrix
E-cadherin	Epithelial cadherin
EMT	Epithelial-to-mesenchymal transition
F-actin	Filamentous actin
N-cadherin	Neural cadherin
ROS	Reactive oxygen species

Introduction

The skin is the largest organ in the body, constituted of an epithelium (the epidermis) and a connective tissue (the dermis). The most important function of the skin is to protect internal organs from insults such as physical forces, pathogens, UV irradiation, and water loss [1]. Because a breach in this barrier is detrimental to the entire body, the skin responds quickly to wounding by initiating a series of overlapping yet orchestrated steps to heal the lesion (for a detailed recent review on wound healing, see Peña and Martin [2]). Following injury, a wave of calcium and hydrogen peroxide travels from the wound outward [3], promoting local release of growth factors. These growth factors initiate vasoconstriction and clotting to minimize blood loss as well as recruitment of inflammatory cells to eliminate pathogens and cellular debris. Extracellular matrix components from the plasma are deposited in the wound to temporarily repair the basement membrane. Keratinocytes, the epithelial cells of the epidermis, also contribute to this initial phase of wound healing by producing cytokines and growth factors. In the next stage of wound healing, keratinocytes proliferate and migrate collectively to repair the lesion, a stage often referred to as wound epithelialization. As the keratinocytes seal the wound, the fibroblasts in the underlying dermis remodel the extracellular matrix (ECM) proteins in an effort to recover the uninjured tissue structure.

While all stages of wound repair are important to regenerate functional skin, epithelialization of the wound is the key step in restoring the protective barrier. Failure to seal the epidermis leads to persistent infection, inflammation, and ultimately chronic wounds, a major healthcare burden [4]. The successful restoration of the epidermal barrier requires keratinocytes to move as a coordinated group. Importantly, this necessitates that cells form and maintain connections to each other and to the ECM. There are a variety of adhesive structures in epidermal cells: adherens junctions, tight junctions, and desmosomes connect neighboring cells [5],

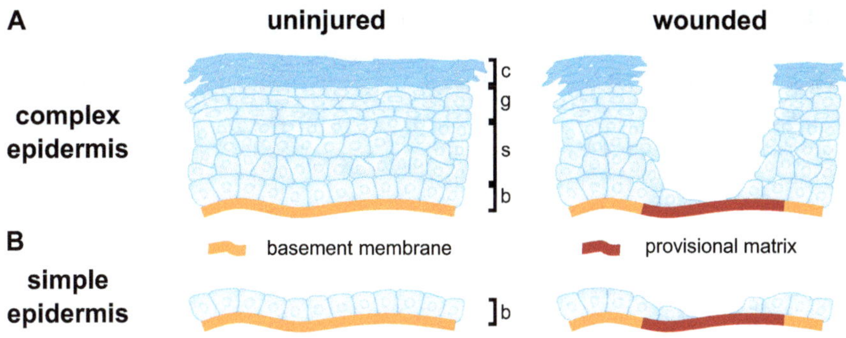

Fig. 7.1 Organization of complex and simple epidermis during homeostasis and epithelialization. (**A**) The complex epidermis is made up of multiple layers of keratinocytes. The basal cells (b) are in direct contact with the basement membrane. Above the basal layer is the spinous layer (s) made up of polyhedral keratinocytes. Next is the granular layer (g), which contains flattened keratinocytes. At the surface of the epidermis is the cornified layer (c) composed of dead keratinocytes that protect the living cells from abrasion, light, heat, and other insults. Epidermal wounds lead to the production of provisional matrix that helps initiate epithelialization—the process of keratinocyte migration to cover the wound bed. (**B**) Simple epidermis consists of a single layer of polyhedral epithelial cells in contact with the basement membrane, like basal cells (b). Wounds in simple epidermis have many features that are similar to complex epidermis, including epithelialization of the wound site from epidermal cells neighboring the lesion

while focal adhesions and hemidesmosomes anchor cells to the ECM [6]. These cellular adhesions are indispensable for keratinocytes to rebuild the injured epidermis. However, how adhesion structures dynamically remodel in response to injury and then return to their homeostatic function at the completion of epithelialization is still unclear.

Human epidermis and all adult vertebrate epidermis consist of four distinct stratified layers of keratinocytes that vary in gene expression, differentiation status, and function—from deep to superficial are the basal layer, spinous layer, granular layer, and cornified layer [7]. In this chapter, we will refer to this multilayered epidermis as "complex" (Fig. 7.1a), in opposition to monolayer or "simple" epidermis found in non-vertebrate organisms (Fig. 7.1b). Epithelialization in complex epidermis is only one stage of the healing process, and involves sealing of the tissue gap with keratinocytes derived from multiple layers before reconstruction of the stratified structure. In simple epidermis, the epithelization stage makes up most, if not all, of the wound healing process. Because of their ease of manipulation and imaging, simple epidermal models are ideally suited to the study of spatial and temporal dynamics of cellular adhesions, which is often challenging in complex multilayered epidermis.

In this chapter, we summarize the current knowledge of two main classes of cell adhesion molecules in the context of wound epithelialization: cadherins, which mediate cell–cell adhesion, and integrins, which mediate cell–ECM adhesion. We draw from studies in diverse organisms, including *Drosophila*, *C elegans*, *Xenopus*,

Table 7.1 Cell adhesion proteins and their gene names across species

Adhesion Protein	Human	Mouse	Xenopus	C elegans	Drosophila
E-cadherin	CDH1	Cdh1	cdh1	hmr-1	shg
N-cadherin	CDH2	Cdh2	cdh2	N/A	CadN
β-catenin	CTNNB1	Ctnnb1	ctnnb1	hmp-2	armadillo
α-catenin	CTNNA1 (αE)/ CTNNA2 (αN)	Ctnna1/ Ctnna2	ctnna1/ ctnna2	hmp-1	α-catenin
Desmoglein	DSG1/ DSG2/ DSG3/DSG4	Dsg1a/Dsg1b/ Dsg1c/Dsg2/ Dsg3/Dsg4	dsg1/dsg2	N/A	N/A
Desmocollin	DSC1/DSC2/ DSC3	Dsc1/Dsc2/ Dsc3	dsc3	N/A	N/A
Desmoplakin	DSP	Dsp	dsp	N/A	N/A
Plakoglobin	JUP	Jup	jup	N/A	N/A
FAT Atypical Cadherin	FAT1/FAT2/ FAT3/FAT4	Fat1/Fat2/ Fat3/Fat4	fat1/fat2/ fat3/fat4	cdh-3/ cdh-4	fat
Dachsous Atypical Cadherin	DCHS1/ DCHS2	Dchs1/ Dchs2	dchs1/ dchs2	cdh-1	ds
β-integrin	ITGB1/ ITGB4/ ITGB5/ ITGB6/ ITGB8	Itgb1/Itgb4/ Itgb5/Itgb6/ Itgb8	itgb/itgb4/ itgb5/ itgb6/ itgb8	pat-3	mys/βv
α-integrin	ITGA2/ ITGA3/ ITGA5/ ITGA6 ITGA9/ ITGAV	Itga2/Itga3/ Itga5/Itga6 Itga9/Itgav	itga2/ itga3/ itga5/ itga6/ itga9/itgav	ina-1/ pat-2	mew/if/ scab/ αPS4/ αPS5
Talin	TLN1/TLN2	Tln1/Tln2	tln1/tln2	tln-1	rhea
Plectin	PLEC	Plec	plec	vab-10a	N/A

mice, and humans, to build a consensus understanding of how these adhesions drive coordinated epidermal migration. Cadherins and integrins, and their associated proteins, are well-conserved across species (Table 7.1), enabling generalization of principles governing cell adhesion during epithelialization. A deeper understanding of these critical cellular structures at the molecular level could lead to new therapeutic targets for the treatment of chronic wounds.

Cadherin-Based Adhesions: Mechanically Integrating Neighboring Cells

Cadherins are calcium-dependent transmembrane proteins that interact with other cadherins in the same cell (*cis*) or in neighboring cells (*trans*) in a mutually cooperative manner [8]. *Cis* cadherin interactions contribute to clustering and assembly of an adhesion complex with mechanical heterogeneity, allowing for active remodeling [9]. *Trans* cadherin interactions permit cells to mechanically interface with their neighbors [10]. There are multiple subfamilies of cadherins, including classical cadherins such as epithelial (E)-cadherin and neural (N)-cadherin, desmosomal cadherins including desmogleins and desmocollins, and atypical cadherins like Fat and Dachsous that mediate cell adhesion and polarity [11–15]. In the epidermis, keratinocytes rely primarily on the classical E-cadherin and desmosomal cadherins to adhere to neighboring cells and maintain tissue integrity [5]. In the cytoplasm, these cadherins are connected to the cytoskeleton via structural adaptor proteins, which organize the cytoplasmic tails of the cadherins into an adhesion structure that becomes a hub for cytoskeletal engagement and biochemical signal transduction. E-cadherin links to the filamentous actin (F-actin) cytoskeleton via catenins in structures termed adherens junctions [16, 17], while desmosomal cadherins link to intermediate filaments via plakoglobin and desmoplakin in structures termed desmosomes [18, 19] (Fig. 7.2a). By integrating with the cytoskeleton, cadherin-based adhesions facilitate force transmission between cells. In complex epidermis, tight junctions, which are not cadherin-based, are localized within the granular layer, which is the outermost live cell layer to create a water-tight barrier. Adherens junctions and desmosomes localize between keratinocytes throughout the epidermis [19] (Fig. 7.2b). In simple epidermis, such as in *Drosophila* or *C. elegans*, all cell-cell junctions are organized along the apical-basal plane in a single cell, with species-dependent organization (Fig. 7.2c).

Loss of Cadherin Adhesion Initiates Epithelialization

Epithelialization of a wound requires epidermal cells to migrate into the tissue gap to restore the epithelial barrier. The conversion of the cells surrounding the wound from a relatively quiescent state to a migratory state is characteristic of a partial epithelial-to-mesenchymal transition (EMT): cells gain front-rear polarity, reorganize their cytoskeleton, and remodel adhesions to enable movement [20]. A hallmark of EMT is a transition away from E-cadherin expression [21], suggesting a reduction in cell–cell adhesion as cells become more mesenchymal-like. This is apparent in the more elongated cells at the wound edge, as well as in the presence of gaps between migrating cells and frequent neighbor exchanges [22, 23]. Downregulation of cell–cell adhesion is thought to permit cells to slide past each other more easily. However, epidermal cells must also maintain barrier function and

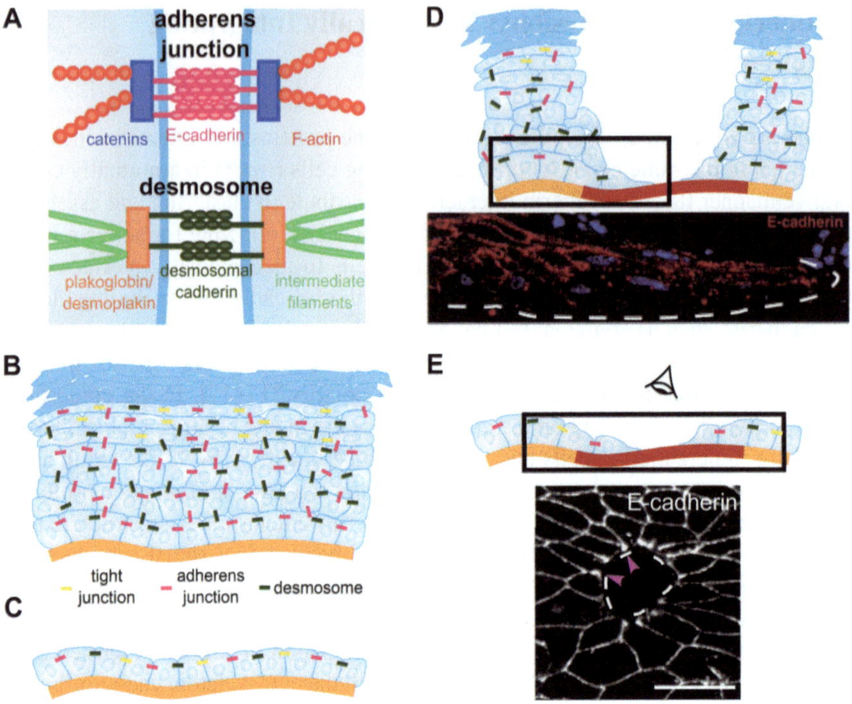

Fig. 7.2 Cell–cell adhesions in epidermal tissues. (**A**) Components of cadherin-mediated adhesions. Adherens junctions link E-cadherin in neighboring cells to F-actin via α- and β-catenin. Desmosomes link desmocollin and desmoglein in neighboring cells to intermediate filaments via plakoglobin and desmoplakin. (**B**) Localization of cell–cell adhesions within complex epidermis. Tight junctions are restricted to the granular layer, where they help form a water-tight barrier. Adherens junctions and desmosomes are found throughout the basal, spinous, and granular layers, connecting cells both horizontally and vertically. (**C**) Localization of cell–cell adhesions within simple epidermis. Tight junctions, adherens junctions, and desmosomes connect all neighboring cells within the monolayer of epidermal cells that make up the simple epidermis. (**D**) During epithelialization in complex epidermis, E-cadherin is downregulated in cells adjacent to the wound while desmosomes remain largely intact. Image of E-cadherin staining in migrating keratinocytes in wounded murine epidermis is reprinted under terms of the Creative Commons Attribution License (CC BY) [22]. Dashed white line denotes wound. (**E**) In simple epidermis undergoing epithelialization, E-cadherin is depleted from leading edge cells except in areas of lateral contacts along the wound edge. Fluorescence image depicts E-cadherin tagged with eGFP in healing *Drosophila* embryonic epidermis as viewed from above. Dashed white line denotes wound, magenta arrowheads indicate reinforced adherens junctions, scale bar indicates 10 μm

cannot fully lose connections with their neighbors. Epidermal cells, therefore, must optimize the use of their cadherin-based adhesions.

Immunostaining of wounds in mouse skin demonstrates a decrease in E-cadherin expression during the epithelialization stages of healing that persists 1–2 days after epithelial cells meet across the gap [22, 24] (Fig. 7.2d). While desmosomal cadherins are not downregulated during epithelialization [22], there is evidence that desmosomal cadherins switch to a less adhesive state upon wounding [25], with the overall effect of loosening the connections between neighboring cells. Correlated with these changes in adhesion, keratinocytes that lead the migration into the wound bed reduce actin stress fiber formation and exhibit gaps between neighbors [22]. These findings support the model that cells must reduce their cell–cell adhesion to enable movement, potentially by reducing contractility and providing more space to make cell shape changes. Testing this model requires a comprehensive understanding of the spatiotemporal dynamics of cell adhesion structures, which is challenging in stratified epidermis because of the complexity and depth of the tissue.

Due to their optical clarity and rapid healing, embryonic models are highly suitable for visualizing cadherin-based adhesion throughout the process of epithelialization. Some of the first embryonic wound healing experiments were performed in chick embryos. This work established that while actin and myosin accumulate along the edge of the healing wound, forming a supracellular cable that drives coordinated contractility, cadherins are depleted from the leading edge, localized only in small, sparse puncta [26]. This loss of cadherin localization to the leading edge of migrating epidermal cells has been observed in other simple wound healing models, such as in the *Drosophila* embryo [27, 28] and larva [29]. In fact, removal of cadherin from the wound edge via endocytosis appears to be a necessary prelude to the formation of a supracellular actomyosin cable [27, 28, 30]. Consequently, overexpression of E-cadherin, either genetically or through biochemical stimulation delays wound closure due to slower initiation of cell migration [27, 31].

Reduction in cadherin expression and localization at the leading edge of migrating epidermal cells may be triggered by early wound healing signals, including calcium influx and production of reactive oxygen species (ROS), both required for proper healing across species [32–34]. In fact, reducing calcium-dependent ROS signals causes defects in E-cadherin removal from the wound edge in *Drosophila* embryos [34], suggesting a conserved mechanism for cadherin downregulation and removal from the wound edge to trigger subsequent cytoskeletal reorganization and epidermal cell migration.

Focal Reinforcement of Cadherin Adhesions Drives Rapid Epithelialization

While transient downregulation of cadherin adhesions is necessary to initiate epithelialization, complete loss of cadherin-based adhesion is detrimental to completing wound repair in a timely manner and restoring epidermal homeostasis. Depletion of E-cadherin in murine epidermis leads to perinatal death due to loss of tissue integrity and barrier function, despite maintenance of desmosome adhesions [35]. Disrupting desmosomal adhesion components leads to slower epithelialization in *Xenopus* embryos [36] and adult mouse skin [37]. Similar defects are seen when disrupting cadherin function in the adult mouse cornea [38] and intestinal mucosa

[39], suggesting a conserved role for functional cadherin-based adhesion in the rate of epithelial healing.

How do we reconcile the need for downregulation of cadherin adhesion to initiate epithelialization with the need to maintain cell–cell adhesion for epithelialization to progress? The solution is for cells to spatially localize cadherins where they are needed most and to reinforce cadherin contacts at specific subcellular locations. E-cadherin is localized in "sparse puncta" in chick embryonic wounds [26], which also appear in *Drosophila* embryonic wounds at focal points between leading cells along the wound edge [40, 41] (Fig. 7.2e). These reinforced junctions appear to be sites of cytoskeletal assembly or anchorage [28], where forces are transmitted between neighboring cells via cadherin adhesions. Cytoskeletal force transmission between cells drives mechanosensitive signaling that promotes cell contractility and polarization. Additionally, supracellular actomyosin structures enable cell coordination at the wound edge. The lack of reinforcement of punctate adherens junctions leads to defects in wound closure, such as reduced actomyosin reinforcement and contractility, likely a consequence of interrupted force transmission between neighboring cells [41]. While desmosomal proteins have not been visualized in puncta, desmosomal cadherins may serve a similar role in maintaining mechanical integration of the epidermis via the intermediate filament cytoskeleton.

Epithelialization occurs through collective cell migration of a cohesive cell sheet, not by independent movement of epidermal cells. Thus, epidermal cells must coordinate their movements as a group via supracellular structures linked at cadherin-based adhesions. Mice with keratinocyte-specific disruption of the desmosomal protein Perp heal more slowly, despite individual keratinocytes being able to migrate more rapidly in vitro [37]. This difference between individual and collective cell migration suggests that the speed of migration is not as important to the epithelialization process as being able to collectively coordinate movement between neighboring cells. Epidermal cells rely on the ability to mechanically integrate with their neighbors to effectively migrate across a wound. The reliance on cadherin-based adhesion for short-range and long-range coordination is supported by data in vitro demonstrating that cells can transmit mechanical cues via cadherin adhesions over 7 mm [42]. Thus, in a large intact tissue such as the skin, cadherin-based adhesion must be maintained for coordinated migration and tissue-scale communication.

Cadherin Adhesions Have Seemingly Paradoxical Roles, But It Is All About Dynamics

It may seem counterintuitive that cadherins need to be both downregulated and reinforced to enable epithelialization. However, these two functions can be reconciled by considering their spatiotemporal dynamics. Initial damage signals at the wound site, like calcium and ROS, trigger downregulation of cadherins and their removal from the membrane. This adhesion remodeling is required for polarization and assembly of cytoskeletal material in epithelial cells at the leading edge, contributing to front-rear polarity of the surrounding cells. However, cadherin-based adhesions also serve as anchors for cytoskeletal structures and a means to transmit mechanical information between neighboring cells. Therefore, after initial removal from the membrane, cadherins are reinforced at focal sites along the wound edge, where they contribute to supracellular coordination of force generation and movement. This spatially biased cellular relocalization of E-cadherin during epithelialization occurs as cells transition into a more dynamic adhesive state. Overall, cadherins play complex roles in epithelialization not only through cellular expression changes but also by active remodeling to promote cell communication during epidermal tissue repair.

Integrin-Based Adhesion: How Cells Interact with the Extracellular Matrix

Cell adhesion to much of the extracellular matrix (ECM) is mediated by transmembrane heterodimeric integrin receptors. Each integrin receptor is made up of an alpha (α) subunit and a beta (β) subunit that together enable binding to ECM proteins on the exterior of the cell. There are at least 18 alpha subunits and 8 beta subunits in humans, and different heterodimer combinations enable cells to bind to many different ligands, including collagen, laminin, and fibronectin [43]. The cytoplasmic side of the integrin heterodimer connects to the cytoskeleton via adaptor proteins, much like cadherin-based adhesions. Integrins can be primed to bind a ligand by inside-out activation, where intracellular signaling causes a conformational change, increasing integrin binding affinity for ECM ligands [44]. In turn, binding to an extracellular ligand promotes integrin clustering and formation of an adhesion complex in the cytoplasm—termed outside-in signaling [44]. Epidermal cells constitutively express multiple integrin subunits that form heterodimers that primarily interact with different components of the basal lamina (collagen IV and laminin): α2β1, α3β1, α6β4, and αvβ5 [6]. Upon wounding, keratinocytes upregulate other integrin subunits that form heterodimers to bind provisional wound matrix: α5β1, α9β1, αvβ8, and αvβ6 [6]. Many of the integrin heterodimers are linked to the F-actin cytoskeleton via talin in structures commonly referred to as focal adhesions. While this term may not accurately describe all of the integrin-based structures found in the epidermis, we will use "focal adhesion" to generally refer to adhesion structures mediated by certain integrin heterodimers. The integrin heterodimer α6β4 links to intermediate filaments via plectin in structures called hemidesmosomes [45] (Fig. 7.3a).

Integrins are important for cell survival and the maintenance of the epidermis. In simple epithelium and single-layered epidermis of non-vertebrate organisms, integrins are typically localized to the basal and basolateral surfaces of cells, where they interact with basement membrane proteins (Fig. 7.3c). Loss of integrin-mediated adhesion often leads to cell death via anoikis [46]. In a more complex stratified

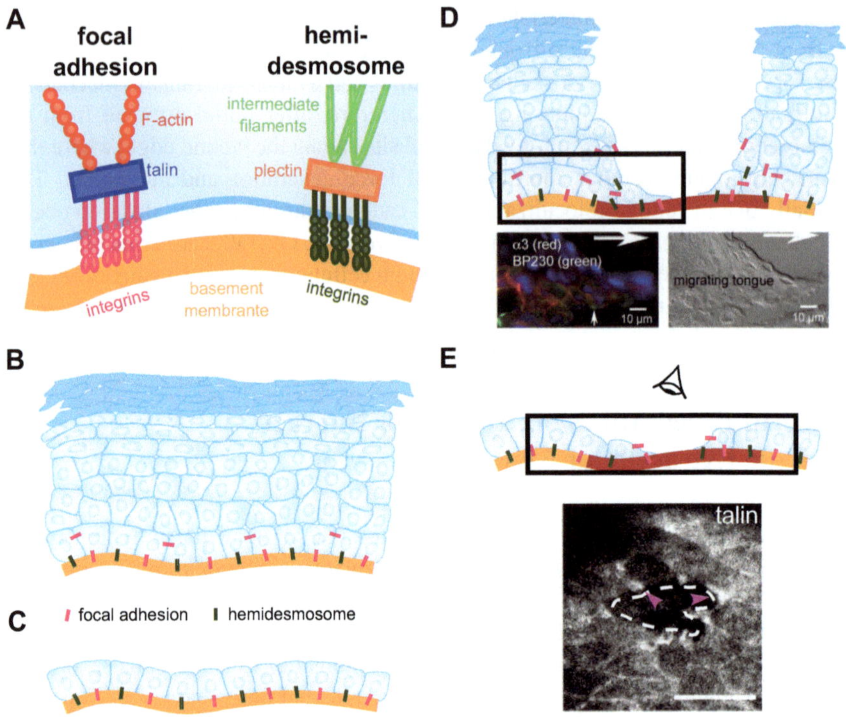

Fig. 7.3 Integrin-based adhesion in epidermal tissues. (**A**) Components of integrin-mediated adhesions. Focal adhesions link integrins bound to the ECM to F-actin via talin. Hemidesmosomes link integrins bound to the ECM to intermediate filaments via plectin. (**B**) Localization of integrin adhesions within complex epidermis. Integrin expression is restricted to the basal layer, where they adhere keratinocytes to the basement membrane and occasionally to neighboring cells. (**C**) Localization of integrin adhesions within simple epidermis. Integrins connect the basal surfaces of cells within the monolayer of epidermal cells to the basement membrane. (**D**) During epithelialization in complex epidermis, integrins are expressed in suprabasal layers of the epidermis at the wound edge and are not restricted to cell-basement membrane interaction sites. Image of integrin staining in migrating keratinocytes in wounded adult human skin is reprinted with permission from Sage Journals [55]. α3 integrin staining labels focal adhesion structures, here seen localized between neighboring cells and in suprabasal cells. BP230 staining labels hemidesmosomes, seen at focal contacts with the ECM (small white arrow). The large arrow indicates the direction of migration. (**E**) In simple epidermis undergoing epithelialization, integrins and associated proteins are not restricted to basal surfaces of cells and localize strongly to apical cell membranes. Fluorescence image depicts the integrin-binding protein talin tagged with eGFP in an apical view of healing *Drosophila* embryonic epidermis. Dashed white line denotes wound, magenta arrowheads indicate talin accumulation, scale bar indicates 10 µm

epidermis, integrin expression is restricted to the basal cell layer with limited expression in more superficial layers (Fig. 7.3b). Basal epidermal cells that lose contact with the basement membrane leave that compartment and proceed along the differentiation pathway, which shares some transcriptional similarities to the anoikis pathway [47]. Integrin-based adhesions are also instrumental in generating traction, especially during migration. Cells attach and exert force on the surrounding ECM, using it as a scaffold upon which to move [48]. The integrin-mediated coupling between the ECM and the cytoskeleton enables productive force generation.

Integrin-Based Adhesion Is Required for Epithelialization of Wounds

Knowing the essential roles played by integrins in cell polarity and migration, it is not surprising that integrin-based adhesions contribute to effective epithelialization during epidermal wound healing. In *Drosophila* embryonic epidermal wound healing, the α integrin subunit scab and the β integrin subunit myospheroid (*mys*) are essential for rapid wound closure [49, 50]. Scab is one of five α integrins and forms heterodimers with both myospheroid and the other β integrin subunit βv. *Scab* mutants have epidermal defects prior to wounding and then fail to close wounds in a timely manner [49]. RNAi-mediated knockdown of *mys* allows for the epidermis to form normally, but wound closure is significantly delayed with defects in actin organization [50]. Similar effects have been demonstrated in complex epidermis. In an epidermal-specific β1 integrin knockout mouse, which disrupts a majority of the integrin heterodimers found in the epidermis, skin wounds show significant delays in epithelialization and altered epidermal architecture once healed [51]. While the epidermal-specific knockout of the laminin-binding integrin α3β1 in mice leads to faster wound healing, the overall epidermis presents with inflammation and micro-blistering [52]. These results indicate that integrins enable epidermal migration to seal the tissue gap at a prescribed rate, resulting in healthy epidermal architecture.

Integrins in the epidermis appear to take on roles beyond basal ECM adhesion. During epithelialization of human skin wounds, α2, α3, α6, and β1 integrins, which are normally restricted to the basal layer of the epidermis, are expressed in suprabasal cells [53] (Fig. 7.3d). This induced expression may recruit cells from non-basal layers to navigate down to the wound bed and contribute to reforming the tissue. In *Drosophila* embryonic epidermal healing, the focal adhesion component talin (*rhea*) localizes along the apical leading edge of the epidermis, where it partially colocalizes with E-cadherin (*shg*) [50] (Fig. 7.3e). Similarly, cultured keratinocytes and human skin show α2β1 and α3β1 integrin localization primarily at cell-cell contacts [54, 55]. These findings are clear evidence that, in addition to a role as cell-ECM anchors, some integrins contribute to the adhesion between two neighboring cells. This is substantiated by the demonstration that β1-integrin blocking antibody applied to cultured keratinocytes leads to dissociation of cell–cell adhesions [54] and integrin adhesions at cell-cell contacts appear quite stable with little turnover in quiescent human skin [55]. Meanwhile, during wound healing, β4 integrins that are typically found in hemidesmosomes are localized to endosomes, cell-cell adhesions, filopodia, and fibrin-interacting adhesions [55]. These findings suggest that integrins fulfill multiple roles during epidermal migration and wound healing beyond their canonical basally restricted interaction with the ECM.

Integrins and ECM Are Dynamically Regulated Throughout the Entire Wound Healing Process

Compared to cadherin-based adhesions, little is known about the spatiotemporal dynamics of the integrin heterodimers during epithelialization. Many studies of integrins and ECM have focused on either the initial inflammatory response to wounding or the remodeling phase following epithelialization. In stratified epidermis, fibrin and fibronectin are deposited in early stages of wounding to facilitate clot formation and recruit immune cells. This deposition also forms the first matrix upon which keratinocytes migrate. In *Drosophila* embryonic epidermis, which predominantly associates with basement membrane ECM components, epidermal injury induces an increase in the deposition of collagen IV and laminin at the wound site [50]. Therefore, at the onset of epithelialization, the composition and organization of the ECM differ substantially from the uninjured epidermal tissue. Because of these ECM changes, leading migrating epidermal cells upregulate unique integrin subtypes to bind these proteins, such as $\alpha5\beta1$ and $\alpha v\beta6$ [6, 56]. Keratinocytes also secrete their own matrix to facilitate their migration on this temporary ECM, particularly laminin 5 [57]. Studies in *Xenopus* tail regeneration indicate that leading cells entering the wound bed have a unique transcriptional signature that includes ECM remodeling factors like matrix metalloproteinases, along with members of the TGFβ signaling pathway that may stimulate proliferation and migration of follower cells [58].

Integrins have also been well-studied for their role in returning the epithelium to a homeostatic state at the conclusion of epithelialization, which is important to prevent scarring and hyperproliferation. Once the provisional wound matrix is covered with laminin, keratinocytes become less migratory, ultimately returning to a more quiescent phenotype [57]. While necessary to initiate epidermal migration into the wound [59], fibronectin and other wound-specific ECM components can trigger a fibrotic phenotype. This response is dominated by dermal fibroblasts, which are recruited to the wound bed [60] and converted to a myofibroblast phenotype. Myofibroblasts increase secretion of cellular fibronectin and fibrillar collagen, creating a stiffer environment that induces scarring [61]. Although most myofibroblasts originate from dermal fibroblasts, there is potential for keratinocytes to undergo epithelial to mesenchymal transition and be converted into myofibroblasts [62]. In other tissues, this conversion occurs due to sustained changes in ECM composition, integrin expression, and downstream signaling. Thus, the plasticity of integrin expression in keratinocytes could be detrimental to the recovery of a quiescent phenotype, even while integrin versatility enables cells to adapt and migrate in a changing environment. Live monitoring of integrin localization throughout epithelialization, while monitoring cell fate in the healed tissue, could provide more insight into the impact that integrin dynamics play in successful healing.

Integrin-Based Adhesion—Maybe Less Is More

Wound healing requires a precise balance of integrin-based adhesion throughout the tissue: there must be enough integrin engagement to provide sufficient anchorage for migrating keratinocytes, yet sustained expression of wound-specific integrin heterodimers impairs recovery of healthy epithelialized skin. Interestingly, simple epidermis heals rapidly and without scarring—potentially due to the simple integrin-ECM network. In non-vertebrate organisms, epidermal cells have fewer integrin heterodimers (Table 7.1) and the ECM is primarily composed of basement membrane proteins. To promote quick, dynamic changes, cells in simple epidermis may repurpose the materials they have through changes in localization, making them easily reversible. Studying the role of integrin-based adhesion in simple epidermis can uncover the ways that integrin-based adhesion can positively contribute to wound healing, which can provide methods for abrogating the negative effects of integrin-ECM crosstalk.

Interplay Between Adhesion Structures During Wound Healing

In mature tissues, cell adhesions have distinct roles: they help cells adhere to each other or to the ECM, and help them link different cytoskeletal structures. In repairing tissues, however, cell adhesions exhibit cooperative functions with significant influence over each other. For example, desmoplakin-deficient keratinocytes have reduced adherens junction formation, and desmosomes are not anchored to keratin filaments [63]. In turn, E-cadherin interacting in *trans* with neighboring cells recruits desmoglein to the cell-cell interface through *cis* interactions, initiating desmosome assembly [64]. Eventually, as the cell-cell interface matures, E-cadherin leaves the desmosome and becomes restricted to adherens junctions [64]. Cadherin-based adhesions also interact with other cell-cell adhesive structures. For example, E-cadherin expression is required for the formation of claudin- and occludin-mediated tight junctions to create a water-tight epidermal barrier [35]. E-cadherin also interacts with the Eph–ephrin complexes. These complexes bridge neighboring cells and are typically associated with establishing tissue boundaries [65] and promoting axon guidance [66]. Ephrin-B1 and EphB2 are upregulated in cells at the wound leading edge, promoting the downregulation of E-cadherin necessary for cells to rapidly remodel cell adhesions during migration [22]. Organization of the cell-cell interface requires different structures working in concert, raising additional questions about how cells dynamically remodel all of their cell–cell adhesions during epithelialization and re-establish order at the conclusion of migration.

The interaction of cell–cell adhesions at cell boundaries is intuitive due to their spatial proximity; perhaps more surprising is the crosstalk between cadherin-based and integrin-based adhesions. Typically, these structures are thought to localize to different regions of the cell, making their physical interaction unlikely. However, in epidermal cells, integrins and cadherins share some spatial overlap—as seen with co-localization of integrins and cell-cell contacts in cultured keratinocytes and human skin [54, 55]. There is also evidence for these adhesions to compete for shared components. For example, in MDCK epithelial cells, adherens junctions and integrin adhesions compete for the linker protein vinculin [67, 68] and in mesenchymal stem cells, N-cadherin and integrin adhesions compete for actin binding [69]. One could imagine that migrating keratinocytes would prioritize reinforcement of integrin-based adhesions over cadherin-based adhesions at the onset of epithelialization for generating traction forces. Finally, both cadherin-based adhesions and integrin-based adhesions are linked to the cytoskeleton, which spans large distances across cells and can transmit mechanical information, providing additional mechanisms for crosstalk [70].

Because of this crosstalk between these integrin-based and cadherin-based adhesions, disrupting one type of adhesion affects the other. Many studies have reported the effects of disrupting integrin adhesions on the stability of cadherin adhesions. For instance, in *Drosophila* embryonic epidermis, disrupting integrin expression and engagement has deleterious effects on the E-cadherin reinforcement and F-actin assembly required to rapidly heal wounds [50]. In the extraembryonic epithelial tissue called the amnioserosa, knockout of the β-integrin *mys* results in altered distribution of E-cadherin and defects in cortical actomyosin networks [71]. Similar examples can be found in cells in vitro. In human HeLa cells, knockdown of integrin adhesion components paxillin and FAK disrupts the formation of N-cadherin adhesions due to altered Rac1 GTPase signaling [72]. In MDCK epithelial cells, Rac1 effectors acting downstream of integrin signaling are required for the recruitment of E-cadherin and the formation of strong cell-cell adhesion [73]. In turn, changes to cadherin-based adhesions can feed back on integrin adhesions. Examples include the promotion of ECM deposition when cadherin adhesions are under tension with concomitent recruitment, clustering, and activation of integrin complexes in both HeLa cells and *Xenopus* embryos [74, 75]. Changes to the extracellular matrix and integrin-based adhesion can thus directly feed back on cadherin-based adhesions and vice versa. Although many of these relationships have not been investigated during skin epithelialization, the concepts challenge our current understanding of cell adhesion. We must consider cell adhesions as interdependent structures that cannot be fully understood in isolation from each other.

Conclusions and Looking to the Future

Cell adhesions play a major role in epithelialization and are indispensable for proper skin wound repair. The general contributions of distinct adhesion structures are well-established: cadherin-based adhesions link neighboring cells, enabling coordination and maintenance of tissue integrity, whereas integrin-based adhesions link cells to the extracellular matrix, providing a means to generate movement. However, this model is somewhat simplistic for a few reasons. Cell adhesions are not static—they are constantly rearranging and remodeling even in homeostatic tissue. Therefore, it is critical to consider cell adhesions as dynamic structures both in space and time by measuring stability of structures and changes in localization. Another complexity is that cell adhesions function in concert with each other and with other subcellular structures. There is evidence of two-way crosstalk between cadherin-based and integrin-based adhesions, as well as between different types of cell-cell adhesions. We do not yet fully understand how these inter-adhesion communications occur, but they are essential to the epithelialization process. Finally, other transmembrane adhesion molecules outside of cadherins and integrins may prove to have crucial functions in epithelialization either cooperatively or independently.

Our ultimate goal is to understand the mechanisms of human skin epithelialization to provide better treatment in the clinic, but the study of this complex process directly in human skin is challenging. The use of different model organisms and in vitro systems provides complementary approaches that, when collated, will increase our knowledge of the molecular mechanisms driving epithelialization. In vitro cell culture models offer the best opportunity to acquire high spatiotemporal resolution images that uncover adhesion dynamics and physical protein interactions, yet they do not fully replicate native tissue composition and mechanics. In vivo studies in simple epithelia offer the advantage of an intact living tissue, robust genetic tools, and accessible imaging, while not capturing the full complexity of the human genome and proteome. In vivo studies in complex mammalian epidermis enable contexts closer to human tissue, but present challenges in imaging at high spatiotemporal resolution. Using diverse models across scales will further our understanding of the fundamental epithelialization processes and identify viable therapeutic targets that have effects on cell adhesions and their downstream signaling.

Acknowledgements We gratefully acknowledge Lindsey Rhea for the critical reading of the manuscript. Work in the Rothenberg Lab is supported by start-up funds from the University of Iowa and grant #25-6040 from the Roy J Carver Charitable Trust. Work in the Dunnwald Lab is supported by NSF IOS2414883. The images in Fig. 7.2e and 7.3e were acquired in the Rothenberg lab by Aiza Khan and Dakota Thompson, respectively.

References

1. Nemes Z, Steinert PM (1999) Bricks and mortar of the epidermal barrier. Exp Mol Med 31:5–19
2. Peña OA, Martin P (2024) Cellular and molecular mechanisms of skin wound healing. Nat Rev Mol Cell Biol 25:599–616
3. Niethammer P (2020) Early signals in wound healing. FASEB J 34:1–1
4. Sen CK et al (2009) Human skin wounds: a major and snowballing threat to public health and the economy. Wound Repair Regen 17:763–771

5. Sumigray KD, Lechler T (2015) Chapter Twelve - Cell adhesion in epidermal development and barrier formation. In: Yap AS (ed) Current topics in developmental biology, vol 112. Academic Press, pp 383–414
6. Watt FM (2002) Role of integrins in regulating epidermal adhesion, growth and differentiation. EMBO J 21:3919–3926
7. Akat E et al (2022) Comparison of vertebrate skin structure at class level: a review. Anat Rec 305:3543–3608
8. Thompson CJ, Vu VH, Leckband DE, Schwartz DK (2021) Cadherin cis and trans interactions are mutually cooperative. Proc Natl Acad Sci U S A 118:e2019845118
9. Huebner RJ et al (2021) Mechanical heterogeneity along single cell-cell junctions is driven by lateral clustering of cadherins during vertebrate axis elongation. elife 10:e65390
10. Harrison OJ et al (2011) The extracellular architecture of adherens junctions revealed by crystal structures of type I cadherins. Structure 19:244–256
11. Leckband DE, le Duc Q, Wang N, de Rooij J (2011) Mechanotransduction at cadherin-mediated adhesions. Curr Opin Cell Biol 23:523–530
12. Hoffman BD, Yap AS (2015) Towards a dynamic understanding of cadherin-based mechanobiology. Trends Cell Biol 25:803–814
13. Lecuit T, Yap AS (2015) E-cadherin junctions as active mechanical integrators in tissue dynamics. Nat Cell Biol 17:533–539
14. Zuidema A, Wang W, Sonnenberg A (2020) Crosstalk between cell adhesion complexes in regulation of mechanotransduction. Bioessays 42:e2000119
15. Troyanovsky SM (2022) Adherens junction: the ensemble of specialized cadherin clusters. Trends Cell Biol S0962-8924(22):00209–4. https://doi.org/10.1016/j.tcb.2022.08.007
16. Yamada S, Pokutta S, Drees F, Weis WI, Nelson WJ (2005) Deconstructing the cadherin-catenin-actin complex. Cell 123:889–901
17. Kwiatkowski AV et al (2010) In vitro and in vivo reconstitution of the cadherin–catenin–actin complex from Caenorhabditis elegans. Proc Natl Acad Sci 107:14591–14596
18. Al-Amoudi A et al (2011) The three-dimensional molecular structure of the desmosomal plaque. Proc Natl Acad Sci 108:6480–6485
19. Antiguas A, Paul BJ, Dunnwald M (2022) To stick or not to stick: adhesions in orofacial clefts. Biology 11:153
20. Haensel D, Dai X (2018) Epithelial-to-mesenchymal transition in cutaneous wound healing: where we are and where we are heading. Dev Dyn 247:473–480
21. Kalluri R, Weinberg RA (2009) The basics of epithelial-mesenchymal transition. J Clin Invest 119:1420–1428
22. Nunan R et al (2015) Ephrin-Bs drive junctional downregulation and actin stress fiber disassembly to enable wound re-epithelialization. Cell Rep 13:1380–1395
23. Tetley RJ, Staddon MF, Banerjee S, Mao Y (2018) Tissue fluidity promotes epithelial wound healing. bioRxiv
24. Kuwahara M, Hatoko M, Tada H, Tanaka A (2001) E-cadherin expression in wound healing of mouse skin. J Cutan Pathol 28:191–199
25. Garrod DR, Berika MY, Bardsley WF, Holmes D, Tabernero L (2005) Hyper-adhesion in desmosomes: its regulation in wound healing and possible relationship to cadherin crystal structure. J Cell Sci 118:5743–5754
26. Brock J, Midwinter K, Lewis J, Martin P (1996) Healing of incisional wounds in the embryonic chick wing bud: characterization of the actin purse-string and demonstration of a requirement for Rho activation. J Cell Biol 135:1097–1107
27. Hunter MV, Lee DM, Harris TJ, Fernandez-Gonzalez R (2015) Polarized E-cadherin endocytosis directs actomyosin remodeling during embryonic wound repair. J Cell Biol 210:801–816
28. Matsubayashi Y, Coulson-Gilmer C, Millard TH (2015) Endocytosis-dependent coordination of multiple actin regulators is required for wound healing. J Cell Biol 210:419–433
29. Tsai C-R, Galko MJ (2019) Casein kinase 1α decreases β-catenin levels at adherens junctions to facilitate wound closure in Drosophila larvae. Development 146:dev175133

30. Carvalho L, Jacinto A, Matova N (2014) The Toll/NF-kappaB signaling pathway is required for epidermal wound repair in Drosophila. Proc Natl Acad Sci UAS 111:E5373–E5382
31. Bablok M, Gellisch M, Brand-Saberi B, Morosan-Puopolo G (2022) Local glucocorticoid administration impairs embryonic wound healing. Biomedicine 10:3125
32. Love NR et al (2013) Amputation-induced reactive oxygen species are required for successful Xenopus tadpole tail regeneration. Nat Cell Biol 15:222–228
33. Xu S, Chisholm ADC (2014) Elegans epidermal wounding induces a mitochondrial ROS burst that promotes wound repair. Dev Cell 31:48–60
34. Hunter MV, Willoughby PM, Bruce AEE, Fernandez-Gonzalez R (2018) Oxidative stress orchestrates cell polarity to promote embryonic wound healing. Dev Cell 47:377–387 e4
35. Tunggal JA et al (2005) E-cadherin is essential for in vivo epidermal barrier function by regulating tight junctions. EMBO J 24:1146–1156
36. Kofron M, Heasman J, Lang SA, Wylie CC (2002) Plakoglobin is required for maintenance of the cortical actin skeleton in early Xenopus embryos and for cdc42-mediated wound healing. J Cell Biol 158:695–708
37. Beaudry VG et al (2010) Loss of the desmosomal component perp impairs wound healing in vivo. Dermatol Res Pract 2010:759731
38. Danjo Y, Gipson IK (1998) Actin 'purse string' filaments are anchored by E-cadherin-mediated adherens junctions at the leading edge of the epithelial wound, providing coordinated cell movement. J Cell Sci 111(Pt 22):3323–3332
39. Flemming S et al (2020) Desmocollin-2 promotes intestinal mucosal repair by controlling integrin-dependent cell adhesion and migration. Mol Biol Cell 31:407–418
40. Zulueta-Coarasa T, Tamada M, Lee EJ, Fernandez-Gonzalez R (2014) Automated multidimensional image analysis reveals a role for Abl in embryonic wound repair. Development 141:2901–2911
41. Rothenberg KE, Chen Y, McDonald JA, Fernandez-Gonzalez R (2023) Rap1 coordinates cell-cell adhesion and cytoskeletal reorganization to drive collective cell migration in vivo. Curr Biol 33:2587–2601.e5
42. Wu P et al (2024) Long range juxtacrine signalling through cadherin for collective cell orientation. Acta Biomater:S1742-7061(24)00627–5. https://doi.org/10.1016/j.actbio.2024.10.035
43. Takada Y, Ye X, Simon S (2007) The integrins. Genome Biol 8:215
44. Ginsberg MH, Partridge A, Shattil SJ (2005) Integrin regulation. Curr Opin Cell Biol 17:509–516
45. Sonnenberg A et al (1991) Integrin alpha 6/beta 4 complex is located in hemidesmosomes, suggesting a major role in epidermal cell-basement membrane adhesion. J Cell Biol 113:907–917
46. Frisch SM, Francis H (1994) Disruption of epithelial cell-matrix interactions induces apoptosis. J Cell Biol 124:619–626
47. Banno T, Blumenberg M (2014) Keratinocyte detachment-differentiation connection revisited, or anoikis-pityriasi nexus redux. PLoS One 9:e100279
48. De Pascalis C, Etienne-Manneville S (2017) Single and collective cell migration: the mechanics of adhesions. Mol Biol Cell 28:1833–1846
49. Campos I, Geiger JA, Santos AC, Carlos V, Jacinto A (2010) Genetic screen in Drosophila melanogaster uncovers a novel set of genes required for embryonic epithelial repair. Genetics 184:129
50. Ly M, Schimmer C, Hawkins R, Rothenberg K, Fernandez-Gonzalez R (2023) Integrin-based adhesions promote cell-cell junction and cytoskeletal remodelling to drive embryonic wound healing. J Cell Sci:jcs.261138. https://doi.org/10.1242/jcs.261138
51. Grose R et al (2002) A crucial role of beta 1 integrins for keratinocyte migration in vitro and during cutaneous wound repair. Development 129:2303–2315
52. Margadant C et al (2009) Integrin alpha3beta1 inhibits directional migration and wound re-epithelialization in the skin. J Cell Sci 122:278–288
53. Hertle MD, Kubler MD, Leigh IM, Watt FM (1992) Aberrant integrin expression during epidermal wound healing and in psoriatic epidermis. J Clin Invest 89:1892–1901

54. Larjava H et al (1990) Novel function for beta 1 integrins in keratinocyte cell-cell interactions. J Cell Biol 110:803–815
55. Underwood RA, Carter WG, Usui ML, Olerud JE (2009) Ultrastructural localization of integrin subunits beta4 and alpha3 within the migrating epithelial tongue of in vivo human wounds. J Histochem Cytochem Off J Histochem Soc 57:123–142
56. Larjava H, Salo T, Haapasalmi K, Kramer RH, Heino J (1993) Expression of integrins and basement membrane components by wound keratinocytes. J Clin Invest 92:1425–1435
57. Nguyen BP, Ryan MC, Gil SG, Carter WG (2000) Deposition of laminin 5 in epidermal wounds regulates integrin signaling and adhesion. Curr Opin Cell Biol 12:554–562
58. Sindelka R et al (2024) Characterization of regeneration initiating cells during Xenopus laevis tail regeneration. Genome Biol 25:251
59. Lenselink EA (2013) Role of fibronectin in normal wound healing. Int Wound J 12:313–316
60. Greiling D, Clark RAF (1997) Fibronectin provides a conduit for fibroblast transmigration from collagenous stroma into fibrin clot provisional matrix. J Cell Sci 110:861–870
61. To WS, Midwood KS (2011) Plasma and cellular fibronectin: distinct and independent functions during tissue repair. Fibrogenesis Tissue Repair 4:21
62. Darby IA, Zakuan N, Billet F, Desmoulière A (2015) The myofibroblast, a key cell in normal and pathological tissue repair. Cell Mol Life Sci CMLS 73:1145–1157
63. Vasioukhin V, Bowers E, Bauer C, Degenstein L, Fuchs E (2001) Desmoplakin is essential in epidermal sheet formation. Nat Cell Biol 3:1076–1085
64. Shafraz O et al (2018) E-cadherin binds to desmoglein to facilitate desmosome assembly. elife 7:e37629
65. Cayuso J, Xu Q, Wilkinson DG (2015) Mechanisms of boundary formation by Eph receptor and ephrin signaling. Dev Biol 401:122–131
66. Egea J, Klein R (2007) Bidirectional Eph–ephrin signaling during axon guidance. Trends Cell Biol 17:230–238
67. Bays JL et al (2014) Vinculin phosphorylation differentially regulates mechanotransduction at cell-cell and cell-matrix adhesions. J Cell Biol 205:251–263
68. Bejar-Padilla V et al (2022) α-Catenin dependent vinculin recruitment to adherens junctions is antagonistic to focal adhesions. Mol Biol Cell:mbc.E22-02-0071. https://doi.org/10.1091/mbc.E22-02-0071
69. Barcelona-Estaje E et al (2024) N-cadherin crosstalk with integrin weakens the molecular clutch in response to surface viscosity. Nat Commun 15:8824
70. Mui KL, Chen CS, Assoian RK (2016) The mechanical regulation of integrin-cadherin crosstalk organizes cells, signaling and forces. J Cell Sci 129:1093–1100
71. Goodwin K et al (2017) Cell–cell and cell–extracellular matrix adhesions cooperate to organize actomyosin networks and maintain force transmission during dorsal closure. Mol Biol Cell 28:1301–1310
72. Yano H et al (2004) Roles played by a subset of integrin signaling molecules in cadherin-based cell-cell adhesion. J Cell Biol 166:283–295
73. Toret CP, Collins C, Nelson WJ (2014) An Elmo-Dock complex locally controls Rho GTPases and actin remodeling during cadherin-mediated adhesion. J Cell Biol 207:577–587
74. Dzamba BJ, Jakab KR, Marsden M, Schwartz MA, DeSimone DW (2009) Cadherin adhesion, tissue tension, and noncanonical Wnt signaling regulate fibronectin matrix organization. Dev Cell 16:421–432
75. Hadjisavva R, Anastasiou O, Ioannou PS, Zheltkova M, Skourides PA (2022) Adherens junctions stimulate and spatially guide integrin activation and extracellular matrix deposition. Cell Rep 40:111091

Chapter 8
Cellular Activation of Surface G-Protein Coupled Receptors as Drug Targets for Skin Wound Healing

Apiwan Arinno and Pawin Pongkorpsakol

Abstract This chapter emphasizes the physiological and pharmacological roles of several G-protein-coupled receptors, including chemokine receptors, prostaglandin receptors, sphingosine-1-phosphate receptors, beta-adrenergic receptors, histamine receptors, purinergic receptors, frizzled receptors, zinc-sensing G-protein-coupled receptor 39 (GPR39), and sweet taste receptor T1R3 on skin wound healing. In addition, mechanisms underlying wound healing mediated by G-protein coupled receptors were also discussed based on accumulating lines of evidence at present. Furthermore, the future perspectives on these therapeutic targets for skin regeneration were additionally addressed.

Keywords Wound healing · G-protein coupled receptor · Cytokine receptor · Nutrient-sensing receptor · Proliferation · Migration

Abbreviations

AMPK	AMP-activated protein kinase
cAMP	Cyclic adenosine monophosphate
CXCL12	C-X-C motif chemokine ligand 12
CXCR4	CXC chemokine receptor type 4
DAG	Diacylglycerol
ECM	Extracellular matrix

A. Arinno
Department of Science, Technology and Innovation, Faculty of Science, Chulabhorn Royal Academy, Bangkok, Thailand

P. Pongkorpsakol (✉)
Princess Srisavangavadhana Faculty of Medicine, Chulabhorn Royal Academy, Bangkok, Thailand
e-mail: pawin.pon@cra.ac.th

F. De Francesco, K. Turksen (eds.), *Management and Strategies for Wound Healing*, Stem Cell Biology and Regenerative Medicine 79,
https://doi.org/10.1007/978-3-032-12442-5_8

EP	E-type prostanoid receptor
FZD	Frizzled receptor
GPCR	G-protein coupled receptor
GPR39	G-protein coupled receptor 39
IP3	Inositol 1,4,5-trisphosphate
PDGF	Platelet-derived growth factor
PGE_2	Prostaglandin E_2
PKA	Protein kinase A
PKC	Protein kinase C
PLCβ	Phospholipase Cβ
S1P	Sphingosine-1-phosphate
S1PR	S1P receptors
SIRT-1	Sirtuin-1
SMA	Smooth muscle actin
SphK1	Sphingosine kinase-1
TGF-β	Transforming growth factor-beta
VEGF	Vascular endothelial growth factor
β-AR	Beta-adrenergic receptor

Cellular Structure of Skin Cells

The skin is considered as the largest organ of the human body that serves as a crucial, physical barrier to compartmentalize and separate the human internal environment from the external milieu. Human skin is separately organized into three main layers, including the epidermis, dermis, and hypodermis. Each layer specifically contributes to establishing distinct structural and functional roles in protection, repair, and homeostasis. The outermost layer of skin is the epidermis, which can be further classified from superficial to deepest part of epidermis, into distinct five layers, including stratum corneum, stratum lucidum, stratum granulosum, stratum spinosum, and the stratum basale [1]. Keratinocytes are primarily found in epidermal layers of skin, with the main function of producing keratin to strengthen skin's physical barriers. There are three types of well-identified stem cell populations that have regenerative capacity to stabilize and repair skin surface, including epidermal stem cells, hair follicle stem cells, and melanocyte stem cells. Of note, epidermal stem cells support the stratified epidermal layers, whereas hair follicle stem cells are important for hair follicle growth. In addition, melanocyte stem cells are known to produce a number of melanocytes that are regulators of the appearance of skin and hair dark color. Skin stem cells inhabit specialized niches of diverse cell types that support tissue renewal in the steady state and repair after injury. They also detect and adapt to systemic health changes and external cues beyond the skin [2]. Indeed, the cellular structure of the skin layers was shown in Fig. 8.1.

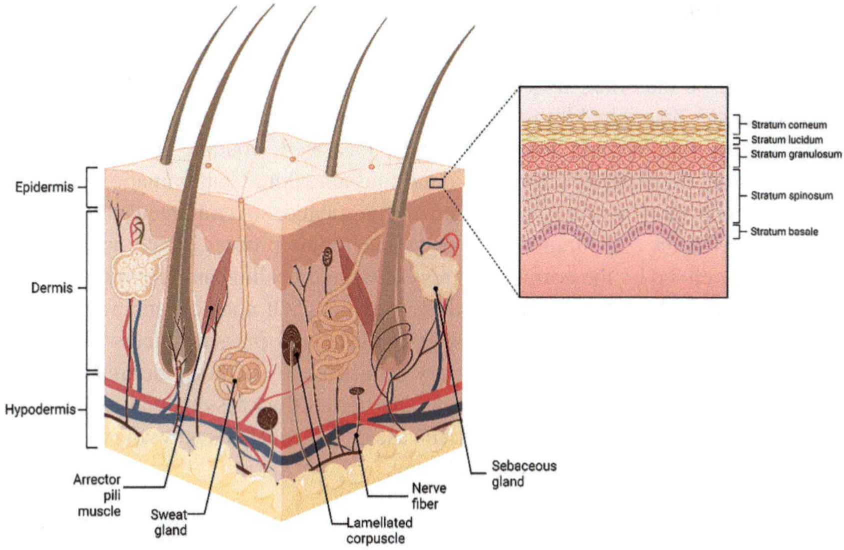

Fig. 8.1 The cellular structure of the skin layers is composed of epidermis, dermis, and hypodermis. There are also sweat glands and sebaceous glands in the dermis. In addition, epidermal layers are further subdivided into five layers, including stratum corneum, stratum lucidum, stratum granulosum, stratum spinosum, and the stratum basale

Introduction to Skin Wound Healing

Significances and Public Health Issues of Skin Wound

Skin wound affects the quality of life of people worldwide [3, 4]. Chronic wounds are defined as the incompetence of keratinocytes and their related cells to heal wounds, which may take a longer time than normal (approximately more than a month) to be completely closed [3]. Indeed, chronic wounds can be found in patients with diabetes mellitus and some infections, such as *Staphylococcus aureus* and *Pseudomonas aeruginosa* infections [3, 5–7]. In addition, sporadically prolonged wound healing can be affected by aging, nutrition deficiencies (i.e., deficiencies of protein, vitamin C, zinc, and iron), immunosuppression, autoimmune disease, and ischemia [5, 8–16]. The cost burden for patients with chronic wounds and comorbidity-associated poor quality of life is unpredictably substantial annually [4, 17]. Discovery and development of therapeutic and management strategies for improving wound repair are required.

Types of Wounds

Wounds can be classified by cause of injury, contamination level, duration, and depth of tissue damage. Almost all causes of wounds are mainly due to physical stress factors that include incised wounds, lacerated wounds, bruise, puncture wounds, burn, gunshot, penetrating wound and abrasion. Contamination levels of wounds can be classified into clean, clean-contaminated, contaminated, and infected wounds. Based on the duration of the wound repair process, acute and chronic wounds are defined by the normal progress of wound healing and prolonged wound repair that may take more than a month, respectively. Of note, chronic wounds are generally found in patients with diabetes mellitus and pressure ulcers (bedsore). Considering the depth of wounds, if injury occurs locally in the epidermal layers of the skin, it is called a superficial wound. The partial thickness wound is identified when damages are found in the dermis. In case of full full-thickness wound, injuries can destroy the structure of muscle and bone.

Mechanisms of Wound Healing

There are four major phases of wound healing that include hemostatic, inflammatory, proliferative, and remodeling phases [18, 19]. Hemostatic phase responds immediately after the wound occurs within a short period of time (few minutes), with the main goal of limiting rates of bleeding. Hemostatic phase is mediated by smooth muscle contraction, vasoconstriction, induced platelet aggregation, and fibrin-dependent clotting. In this phase, endothelium-derived von Willebrand factor mainly contributes to the enhancement of platelet adhesion and aggregation. In addition, platelet-derived growth factor (PDGF), vascular endothelial growth factor (VEGF), and transforming growth factor-beta (TGF-β) are also important for platelet activity, blood clotting, angiogenesis, and promoting immune cell recruitment to the wound site [20–24]. Thereafter, the inflammatory phase occurs within a day after tissue injury to at least three days. In this phase, immune cells (i.e., mast cells, neutrophils, and macrophages) are recruited to the lesional region to eliminate possible infections. Furthermore, these mentioned immune cells also secrete several types of cytokines and chemokines to promote positive feedback of immune cell penetration to the wound area and to activate crucial cells required for the proliferative phase [25]. During the proliferative phase of wound healing that occurs within 3 to 12 days after injury, there are two processes of keratinocytes that are migration and proliferation induced by either fibroblasts or several endogenous agonists [19]. In the last stage, the remodeling phase (duration of this phase up to 6 months), also known as maturation, is primarily mediated by collagen deposition and smooth muscle actin (SMA)-positive myofibroblast-dependent wound contraction [26–28]. Of note, the scar formation of the wound area can be promoted by TGF-β-dependent collagen deposition and a decrease in the extracellular matrix (ECM). During wound

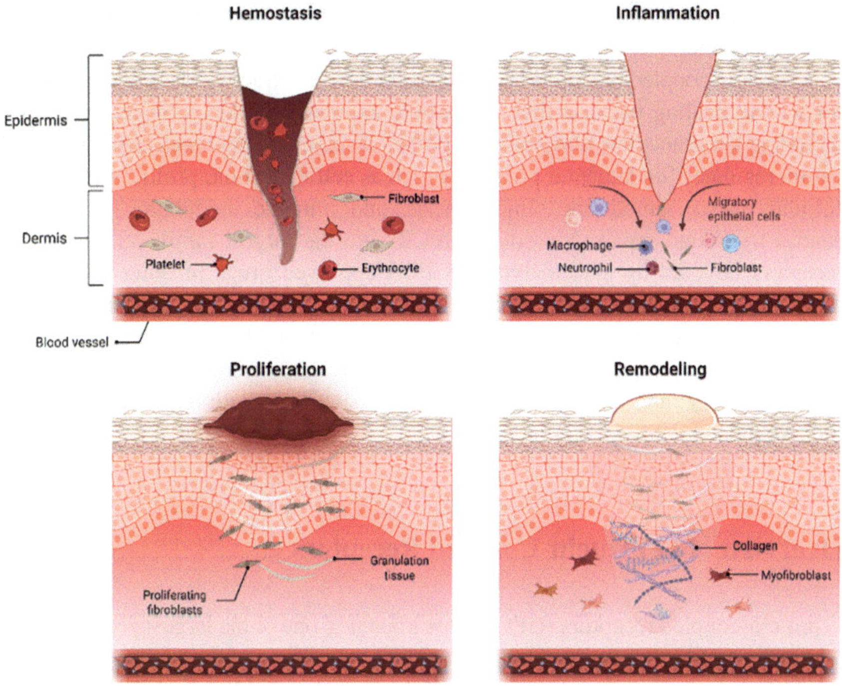

Fig. 8.2 Wound healing process includes homeostatic, inflammatory, proliferative, and remodeling phases

healing process, skin fibroblasts play important roles in many stages of epithelialization and tissue repair by generating the granulation tissue, secreting ECM, and stimulating keratinocyte migration and proliferation. The whole process of wound healing is shown in Fig. 8.2.

Current Gaps in Wound Healing Therapies and the Rationale for Targeting GPCRs

In spite of existing significant advances in wound care, current therapeutic strategies for skin wound healing remain circumscribed. Gold standard treatments such as dressings, debridement, negative pressure wound therapy, and growth factor or stem cell-based treatments often fail to achieve complete and timely healing, especially in chronic wounds such as diabetic ulcers, venous leg ulcers, and pressure sores. These conditions are further complicated by impaired angiogenesis, prolonged inflammation, microbial infection, and reduced cellular responsiveness in the wound microenvironment. Subsequently, there is a need for identifying novel therapeutic targets that can modulate multiple phases of wound repair in a coordinated and efficient manner.

G-protein coupled receptors (GPCRs) represent an attractive class of drug targets for wound healing because they are widely expressed on key skin-resident cells, including keratinocytes, fibroblasts, endothelial cells, and immune cells. Their activation regulates fundamental processes such as cell migration, proliferation, extracellular matrix remodeling, angiogenesis, and immune modulation, which are central to wound repair. Moreover, GPCRs are the largest family of membrane receptors targeted by several types of FDA-approved drugs in other therapeutic areas, offering a well-established pharmacological toolkit and opportunities for drug repurposing. These features collectively position GPCRs as promising candidates for the development of next-generation therapies aimed at accelerating skin wound healing and improving outcomes in patients with chronic or refractory wounds.

Overview of G Protein-Coupled Receptors

G protein-coupled receptors (GPCRs) are known as one of the largest protein receptor superfamilies that are generally located on the surface of the plasma membrane in many cell types to mediate extracellular signaling with crucial physiological impacts [29]. Normally, endogenous ligands of the GPCRs are hormones, neurotransmitters, odors, chemokines, and nutrients. GPCRs have been implicated as drug targets for several diseases [30]. Upon being agonized by the specific extracellular ligands, these receptors are induced to interact with either distinguished G proteins or β-arrestin, an increases in intracellular cAMP and calcium levels, or enhancement of phosphorylation of extracellular regulated protein kinases 1/2 (pERK1/2) [31]. Considering the structure of the GPCRs, they are known as the seven-transmembrane proteins. Based on the specific amino sequences and their three-dimensional structures, the GPCRs can be classified into Class A (Rhodopsin-like receptor), Class B1 (Secretin receptor family), Class B2 (Adhesion GPCRs), Class C (Metabotropic glutamate receptor), and Class F (Frizzled receptor) subfamilies [32–36].

Class A (Rhodopsin-like receptor) is the most common and abundant group, encompassing more than 80% of all identified GPCRs. Members of this group of GPCRs include adrenergic, histamine, chemokine, purinergic, and prostaglandin receptors, which regulate neurotransmission, immune responses, and cardiovascular functions. Class B1 (Secretin receptor family) receptors respond to peptide hormones such as glucagon, calcitonin, and corticotropin-releasing hormone, playing essential roles in metabolism and endocrine regulation. Class B2 (Adhesion GPCRs) feature unusually long extracellular domains that facilitate cell–cell and cell–matrix interactions, with latrophilins being representative members. Class C (Metabotropic glutamate receptor) receptors possess large N-terminal domains that bind ligands including glutamate, γ-aminobutyric acid (GABA), and taste molecules (e.g., T1Rs for sweet/umami). They are vital for neuronal signaling and sensory perception. Class F (Frizzled receptor) participates in developmental and oncogenic pathways via Wnt and Hedgehog signaling. Together, these classes highlight the remarkable diversity of GPCRs. Their structural features and functional specializations underpin critical aspects of physiology, while their druggability makes them central to therapeutic innovation.

Cellular Activation of GPRCs in Skin Cells

GPCRs are fundamental signaling proteins expressed across diverse skin cell populations, including keratinocytes, fibroblasts, melanocytes, and immune cells [37–40]. Their activation by extracellular ligands such as chemokines, lipids, and growth factors initiates a cascade of intracellular events mediated by heterotrimeric G proteins. These events converge on second messengers that orchestrate cellular responses essential for skin repair and wound healing. Upon ligand binding, GPCRs undergo conformational changes that facilitate GDP–GTP exchange on the Gα subunit, leading to dissociation of Gα and Gβγ subunits. This process triggers multiple downstream signaling axes. One of the most studied pathways is the activation of adenylyl cyclase (via $G\alpha_s$), which elevates intracellular cAMP levels, resulting in activation of protein kinase A (PKA)-mediated keratinocyte proliferation, migration, and re-epithelialization [41]. Conversely, GPCRs coupled to $G\alpha_i$ reduce cAMP, fine-tuning cellular responses to balance proliferation and differentiation during tissue regeneration. Another key pathway involves phospholipase Cβ (PLCβ) activation via $G\alpha_q$. PLCβ hydrolyzes phosphatidylinositol 4,5-bisphosphate into inositol 1,4,5-trisphosphate (IP_3) and diacylglycerol (DAG). IP_3 mobilizes intracellular Ca^{2+} stores, while DAG activates protein kinase C (PKC). These signals regulate cytoskeletal remodeling, directional migration, and secretion of growth factors, all critical for wound closure. Additionally, Ca^{2+} signaling modulates cell–cell junctions and enhances communication between keratinocytes and fibroblasts in the healing microenvironment [42]. The $G\alpha_{12/13}$–Rho GTPase axis represents another GPCR-driven pathway that influences actin cytoskeleton dynamics. This mechanism is vital for fibroblast contractility, collagen deposition, and overall wound matrix remodeling. Moreover, GPCR signaling through β-arrestin provides a parallel mechanism, functioning independently of G proteins to activate MAPK/ERK cascades, thereby promoting cell survival and angiogenesis. Importantly, lipid receptors such as sphingosine-1-phosphate (S1P) receptors and chemokine receptors (e.g., CXCR4) exemplify the relevance of GPCR signaling in skin repair. They integrate cAMP, Ca^{2+}, and Rho-mediated pathways to coordinate immune cell recruitment, vascular responses, and keratinocyte migration.

In conclusion, GPCR-mediated activation of second messengers such as cAMP, Ca^{2+}, DAG, and Rho GTPases forms the signaling backbone that drives wound healing in skin. These pathways collectively regulate proliferation, migration, angiogenesis, and matrix remodeling, underscoring the therapeutic potential of targeting GPCR signaling to enhance regenerative outcomes.

GPCR Subtypes as Drug Targets for Wound Healing

Chemokine Receptors

Chemokine is known as a class of cytokine family that play crucial roles in potentiating the migration (chemotaxis) of immune cells during inflammation. Most of the actions of chemokines are mediated by agonistic responses of the GPCRs on the target cells. The C-X-C motif chemokine ligand 12 (CXCL12), also known as the stromal cell-derived factor 1 (SDF1), is one of the chemokines that is an endogenous agonist of the CXC chemokine receptor type 4 (CXCR4) [43, 44]. Indeed, CXCR4 belongs to class A of the GPCR family [45]. Cryo-EM reconstruction reveals the molecular interaction between CXCR4 and its $G\alpha_i$ subunit [45]. The CXCR4 is highly expressed in immune, epithelial, and stem cells [46, 47]. In cancer cells, CXCR4 activation promotes proliferation, migration, and invasion of the tumor environment [48]. Several lines of evidence suggest that CXCL12/CXCR4 signaling promotes wound healing. Of particular importance, the thiadiazine derivative, UCUF-965, which serves as an allosteric activator of CXCR4, can promote angiogenesis and accelerate wound healing in diabetic mice by promoting migration and activities of fibroblasts via G-protein-dependent a rise of $[Ca^{2+}]$ activation of PI3K/Akt and MAPK signaling [49, 50]. These effects were fully abolished by the treatment of CXCR4 antagonist AMD3100 [50, 51]. Moreover, the formulated CXCR4-loaded hydrogel dressing may effectively recover wounds [50]. In addition, moist exposed burn ointment (MEBO), an oil-based herbal paste for the management of burn wounds, was shown to heal wounds by activating the signaling pathways necessary for angiogenesis via the SDF-1/CXCR4-dependent signaling in a diabetic rat model [52]. Nevertheless, it is still unclear about the local compartmentalization of the CXCR4 complex since it was shown to be mainly coupled with the $G\alpha_i$ subunit but also promote the crosstalk between other G-protein and growth factor receptors.

Prostaglandin Receptors

Prostaglandin E2 (PGE_2) is known as a key mediator for initiating immune-mediated inflammation. Nonsteroidal anti-inflammatory drugs (NSAIDs) such as the selective cyclooxygenase-2 (COX-2) inhibitors are effectively and specifically able to suppress PGE_2 production and inflammation. Considering the process of PGE_2 generation, cyclooxygenase (COX) enzymes catalyze arachidonic acid to prostaglandin G2 (PGG_2) and prostaglandin H2 (PGH_2), which are converted to PGE_2 by prostaglandin E synthases (PGES) [53, 54]. It is well accepted that prostaglandin E2 (PGE_2) plays pivotal roles in the regulation of diverse biological processes in the human body, at least in part, via stimulating the E-type prostanoid (EP) receptor family. Concerning the receptors for recognizing PGE_2, there are four main types of

PGE2 receptors, including EP1, EP2, EP3, and EP4. Of note, EP2 and EP4 receptors are mainly coupled with $G\alpha_{s,}$ whereas the EP1 receptor is coupled with $G\alpha_q$ [55]. On the other hand, the EP3 receptor is coupled with $G\alpha_i$ [55]. Upon being activated, PGE_2 and its EP receptor elicit its related intracellular signaling to result in regeneration of tissue-to-organ levels at the injured sites and promote activity of stem cells, immune responses, and angiogenesis [56–58]. Indeed, there is some evidence suggesting that chitosan-encapsulated hydrogel containing PGE_2 was shown to improve healing capacity in a murine cutaneous wound model by inducing M2 macrophage polarization [59]. In free-floating collagen gels surrounded by human adult dermal fibroblast, EP2 receptor agonist (butaprost) and forskolin (an activator of adenylate cyclase that promotes cAMP production) shown to significantly reduce contraction of the free-floating collagen gels, suggesting that EP2 receptor may be important for wound remodeling without scar development in fetal wound healing, but not for adult because the wound may not be fully healed [60]. Of note, similar to the EP4 receptor, the EP2 receptor is mainly coupled with $G\alpha_s$ and cAMP signaling [61]. In contrast, the EP1 receptor agonist (17-phenyl-trinor-prostaglandin E2) promoted contraction of the free-floating collagen gels by enhancing the activities of both human fetal and adult dermal fibroblasts [60]. Furthermore, the rate of wound healing was delayed in EP3 receptor knock-out mice with lower VEGF expression and angiogenesis compared to wild-type mice [62]. Altogether, different EP receptor agonists may provide distinguished responses for wound healing at different physiological stages.

Sphingosine-1-Phosphate Receptors

Sphingosine-1-phosphate (S1P) is a metabolite of the membrane sphingolipids [63]. There are five types of S1P receptors (S1PRs), including S1PR1 to S1PR5 [64]. In fact, S1PRs are considered a group of GPCRs and have been identified as therapeutic targets for various diseases, i.e., lymphoma, lung cancer, asthma, liver fibrosis, alcoholic liver disease, hepatocellular carcinoma, and autoimmune encephalitis [64]. Therefore, agonists and antagonists of these S1PRs may have clinical applications for several related diseases [65, 66]. Moreover, delayed wound healing in association with reduced angiogenesis, and recruitment of immune cells to the wound area in sphingosine kinase-1 (SphK1) knockout mice were observed [67]. Of note, SphK1 is an enzyme that synthesizes S1P from sphingosine. In a hyperglycemic condition, S1P was shown to effectively increase the rate of skin wound healing in diabetic db/db mice, but not in normal mice, by inducing angiogenesis [66, 68]. In response to ginsenoside Rb1 treatment, S1P can be produced by keratinocytes and plays an important role in inducing skin wound healing by promoting keratinocyte-like HaCaT cell migration process via S1PR/ERK/NF-κB/MMP-dependent mechanism [69]. Based on recent evidence, S1PR which is expressed in both endothelial cells and keratinocytes, contributes to angiogenesis and keratinocyte migration, leading to an increase in the rate of wound healing.

Beta-Adrenergic Receptors

Beta-adrenergic receptors (β-ARs) are prototypical members of the GPCR superfamily, characterized by seven transmembrane α-helices connected by extracellular and intracellular loops. The extracellular domains are involved in ligand recognition, whereas the cytoplasmic regions interact with heterotrimeric G proteins to transmit signals across the plasma membrane. Upon agonist binding, β-ARs primarily couple to $G\alpha_s$ proteins, stimulating adenylyl cyclase and elevating intracellular cyclic AMP (cAMP). This second messenger activates protein kinase A (PKA) and other effectors, thereby regulating diverse cellular processes, including metabolism, contractility, and gene transcription, underscoring their importance in physiological and therapeutic contexts. In response to wound appearance, keratinocyte migration occurs in parallel with increases in tyrosine hydroxylase- and phenylethanolamine-N-methyl transferase-dependent epinephrine synthesis [70]. Of note, epinephrine is well accepted as β-AR agonist. Consistently, the β2-AR is known to be highly abundant in keratinocytes. An antagonist of the β-AR was previously shown to induce ERK phosphorylation, increase wound healing and keratinocyte migration [70]. Conversely, isoproterenol (β-AR agonist) treatment was demonstrated to significantly reduce keratinocyte migration and the colocalization between ERK phosphorylation and vinculin (The scaffolding protein that is required and essential for actin cytoskeleton movement and cell-to-cell adhesion sites, forming and maintaining focal adhesions and adherens junctions) at the lamellipodial edge of keratinocytes in a phosphatase (PP) 2A-dependent mechanism since these effects were suppressed by okadaic acid (PP2A inhibitor) [71]. However, agonism of β2-adrenergic receptor by Salbutamol/Formoterol was able to decrease human dermal fibroblast differentiation, angiogenesis, and inflammation, leading to a significant reduction of scar area and skin pigment production [72]. In addition, timolol (β2-adrenergic receptor antagonist) was also reported to improve wound repair in combined burn and radiation [73]. Therefore, we can conclude that β2-adrenergic receptor antagonists might be useful for the treatment of wound healing. On the other hand, β2-adrenergic receptor agonists may delay wound healing rate but reduce scarring.

Histamine Receptors

Histamine receptors are a family of GPCRs that mediate the diverse physiological and pathological effects of histamine in human tissues. Structurally, each receptor subtype is composed of a seven-transmembrane domain architecture typical of GPCRs, with extracellular loops responsible for ligand recognition and intracellular domains that interact with heterotrimeric G proteins. Upon histamine binding, the receptor undergoes conformational changes that activate distinct signaling cascades depending on the receptor subtype. To date, four major histamine receptor subtypes have been characterized, including H_1, H_2, H_3, and H_4 receptors. The H_1 receptor primarily couples to $G\alpha_{q/11}$ proteins, stimulating phospholipase C, inositol trisphosphate (IP_3) production, and calcium mobilization, leading to smooth muscle contraction and pro-inflammatory responses [74–78]. The H_2 receptor couples to $G\alpha_s$ proteins, activating adenylyl cyclase and elevating intracellular cAMP, which regulates gastric acid secretion and cardiac contractility [77]. In contrast, the H_3 receptor is linked to $G\alpha_{i/o}$ proteins, resulting in inhibition of adenylyl cyclase and modulation of neurotransmitter release in the central nervous system [79]. The H_4 receptor, also Gi/o-coupled, is enriched in immune cells and influences chemotaxis and cytokine production [78, 80, 81]. Collectively, histamine receptors illustrate how a single biogenic amine, through structurally related but functionally distinct GPCRs, orchestrates a wide spectrum of biological activities and provides multiple therapeutic targets.

In response to cutaneous injury, histamine histamine-releasing process is triggered in mast cells. The H_1 and H_2 histamine receptors are highly expressed in keratinocytes. In tape stripping of hairless mice, cimetidine and famotidine that are known as H_2 receptor antagonists were shown to be capable of increasing skin barrier function [82]. Similar effects were found in hairless mice in response to treatment with H_1 receptor antagonists, including tripelennamine and diphenhydramine [82]. Furthermore, topical uses of H_2 receptor antagonists were able to attenuate epidermal hyperplasia in hairless mice [82]. Of particular interest, histamine initiated proinflammatory responses in keratinocytes to promote wound healing and suppressed apoptotic rates via an H_1-dependent mechanism [83]. Agonism of H_1 and H_4 histamine receptors was previously reported to promote angiogenesis after wound occurrence, whereas H1 receptor agonism was shown to mediate epidermal epithelial-to-mesenchymal transition (EMT), resulting in cell migration [84]. In contrast, agonism of the H2 histamine receptor was capable of inhibiting the EMT process of the epidermal layers [84]. Therefore, histamine receptors serve as drug targets for skin regeneration, and their agonists or antagonists can be used for the treatment of wounds, but need to be further addressed.

Purinergic Receptors

Purinergic receptors represent a diverse family of cell surface receptors that mediate cellular responses to extracellular nucleotides such as ATP, ADP, UTP, and adenosine [85]. They are broadly divided into two main classes, including P1 (A1, A2A, A2B, and A3) receptors, which are selective for adenosine, and P2 receptors, which recognize nucleotides. The P2 family is further subdivided into P2X and P2Y receptors [86–88]. P2X receptors are ligand-gated ion channels composed of trimeric subunits that permit rapid flux of cations, while P2Y receptors are G protein–coupled receptors (GPCRs) with seven transmembrane domains that initiate intracellular signaling cascades. Functionally, purinergic receptors are critical in processes such as neurotransmission, inflammation, immune regulation, vascular tone, and tissue repair. Activation of P1 receptors often modulates cAMP levels through $G\alpha_s$ or $G\alpha_i$ proteins, whereas P2Y receptors signal via $G\alpha_q$ or $G\alpha_i$ proteins to regulate phospholipase C, inositol trisphosphate, diacylglycerol, and downstream calcium mobilization [89–91]. P2X receptors, in contrast, directly gate ion flux to influence membrane potential and secondary messenger systems [92, 93]. Together, these receptors provide a finely tuned system for sensing extracellular nucleotides, translating them into diverse physiological outcomes. Their structural variety and signaling complexity make purinergic receptors central to both normal physiology and potential therapeutic targeting in neurological, cardiovascular, and immune disorders.

In rat epidermis, P2Y1 and P2Y2 receptors are highly expressed in proliferating keratinocytes at the basal layer, whereas expression of the P2X5 receptor was positively correlated with epidermal keratinocytes at the stages of differentiation and proliferation [94]. In addition, the P2X7 receptor was mainly found in the stratum corneum of keratinocytes [94]. The P2Y1 receptor was demonstrated to be significantly upregulated in keratinocytes during the recovery period of wounds [94]. These findings suggested the relationship between expression levels of various types of purinergic receptors along the stages of skin tissue injury and regeneration. Furthermore, A2A adenosine receptor agonist was reported to enhance skin wound healing in both diabetic and normal animals, which might be useful for the treatment of diabetic foot ulcers [95, 96]. Of particular interest, wound size was generally closed faster in WT in consistent with increased expression of collagen I, α-SMA, and fibronectin than P2Y2 knockout mice [97]. In response to wound occurrence in vitro, P2Y2 receptor transcript expression was predominantly upregulated, and treatment with either ATP or UTP strongly promoted fibroblast migration and proliferation in WT, but not in P2Y2 receptor siRNA-transfected cells [97]. Similarly, in the keratinocyte-like HaCaT cells, AR-C118925XX (P2Y2 receptor antagonist) treatment was found to be capable of delaying wound closure [98]. These studies indicated that agonism of purinergic receptors, at least in part, in fibroblasts and keratinocytes may play pivotal roles in inducing wound healing.

Frizzled Receptors

Frizzled receptors (FZDs) are a family of GPCRs characterized by a seven-transmembrane domain similar to other types of GPCRs with an additional extracellular cysteine-rich domain (CRD) that recognizes Wnt proteins [99]. The CRD furnishes high-affinity to lipid-modified Wnt ligands, while the intracellular C-terminal tail interacts with Dishevelled (DVL) and other signaling proteins. The principal activators of FZDs are Wnt glycoproteins [100]. Upon ligand binding, FZDs initiate distinct signaling cascades including the canonical Wnt/β-catenin pathway, where β-catenin accumulates and regulates gene transcription and non-canonical pathways of planar cell polarity and Wnt/Ca^{2+} signaling, which modulate cytoskeletal organization, migration, and calcium flux. Expression of FZD receptors is widespread but cell-type specific. Different isoforms (FZD1–FZD10 in mammals) are found in embryonic stem cells, neural tissue, bone, intestine, and immune cells, often with overlapping yet context-dependent roles. Functionally, FZDs govern critical processes such as embryonic axis formation, organ development, stem cell maintenance, and tissue repair [99]. Aberrant FZD signaling contributes to pathological states, including cancer progression, fibrosis, and degenerative diseases [99]. Their central role in Wnt biology has positioned FZDs as promising therapeutic targets in regenerative medicine and oncology.

In psoriatic wounds, transcript and protein expression of WNT5A, FZD2, and FZD5 were profoundly upregulated [101–103]. Indeed, several lines of evidence reported that Wnt signaling was able to stimulate the Raf1/MEK/ERK pathway to promote keratinocyte proliferation, resulting in wound healing in diabetic conditions [104, 105]. Therefore, based on its proliferative property, Wnt signaling and its FZD have been proposed as therapeutic targets for enhancing wound healing.

Zinc-Sensing G-Protein Coupled Receptor 39 (GPR39)

Zinc has long been considered a crucial micronutrient human body necessary for various physiological processes, including cell differentiation, development, growth, and survival [106]. Zinc is highly abundant in injured skin [107, 108]. Of particular interest, the delayed wound healing process has generally been found together with zinc deficiency [9, 109–111]. Zinc treatment predominantly promoted wound healing in mouse models and humans [110]. In addition to being a cofactor of enzymes, zinc action and its physiological roles can be mediated by a specific receptor. The G-protein-coupled receptor 39 (GPR39) has been identified as a molecular representative of zinc-sensing receptor [112–115]. The GPR39 is expressed mainly in hippocampal neurons, epithelial cells in the gastrointestinal tract, and in skin that is tissue-specifically compartmentalized with several subunits of the G-protein, including $G\alpha_s$, $G\alpha_q$, $G\beta/\gamma$, G12/13, and G-protein-independent β-arrestin pathways [112, 116–119]. It is well accepted that the activation of GPR39 attenuated the severity of depression and promoted intestinal tight junction assembly [118, 119]. The functional expression of GPR39 was found in the keratinocyte-like HaCaT cells [37]. An agonist GPR39, TC-G 1008, was shown to promote keratinocyte proliferation and induce ERK phosphorylation [37]. However, GPR39-dependent keratinocyte proliferation and ERK phosphorylation were abrogated by co-treatment with wortmannin (PI3K inhibitor) and U0126 (MKK inhibitor) [37]. Furthermore, FR180204 (ERK inhibitor) treatment interfered with the effect of TC-G 1008-induced cell proliferation [37]. Hence, based on these findings, pharmacological agonism of GPR39 potentiates keratinocyte proliferation via a PI3K/MKK/ERK-dependent fashion [37]. Recently, GPR39 agonist, TC-G 1008, was also found to induce wound healing in keratinocyte-like HaCaT cells, but this effect was functionally suppressed following enzymatic inhibitions of protein kinase A

(PKA), AMP-activated protein kinase (AMPK), sirtuin-1 (SIRT-1), and ERK [120]. Of particular importance, inhibition of keratinocyte proliferation by mitomycin C was not able to interrupt the wound healing property of the GPR39 agonist. In addition, TC-G 1008 was effective in preventing keratinocyte apoptosis following UVB exposure by mechanisms similar manners as its wound healing impact [120]. Indeed, pharmacological activation of GPR39 promotes wound healing and protects against UVB-induced cellular damage of keratinocytes. Although the GPR39 agonist was previously shown to promote keratinocyte proliferation, this effect does not contribute to enhancing skin wound healing. Therefore, GPR39 agonists may be useful for the further development of the cellular enhancer of skin wound healing and other related protective effects.

Sweet Taste Receptor T1R3

Sucralose is an artificial sucrose derivative sweetener being as a food additive without caloric load. Sucralose effectively decreases calorie intake, controls body weight, and exhibits euglycemic effects in obese patients with diabetes mellitus, at least in part, by sensitizing insulin signaling. The sweet taste receptor T1R3 is known as a receptor for sucralose expressed in various epithelia such as the taste buds and in the gastrointestinal tract [121, 122]. Upon agonization, the heterodimeric complex of T1R3 and T1R2 receptor mediates their intracellular signaling via the Gα-gustducin, resulting in increasing intracellular cAMP level and $[Ca^{2+}]_i$ [121, 123, 124]. Recently, sucralose treatment in keratinocyte-like HaCaT cells was shown to effectively promote keratinocyte proliferation, accelerate skin wound healing, and prevent UVB-induced keratinocyte apoptosis [125]. Its proliferative effect of sucralose mainly enhanced the rate of wound healing [125]. The effects of sucralose on proliferation, wound healing, and protective impact against UVB radiation were fully diminished by inhibition of protein kinase A (PKA), AMP-activated protein kinase (AMPK), sirtuin-1 (SIRT-1), and extracellular signal-regulated kinase (ERK). Sucralose treatment-mediated functional expression of the phosphorylated AMPK and its relative ratio to the total form of AMPK was entirely suppressed by pretreatment with a PKA inhibitor. Sucralose treatment-induced an increase of the ratio of phosphorylated form of ERK to total ERK was decreased by pretreatment of inhibitors of AMPK and SIRT-1. Consistently, the enzymatic activity of SIRT-1 enhanced by sucralose was diminished by pretreatment with PKA and AMPK inhibitors as well. Therefore, it can be concluded that sucralose can effectively regenerate keratinocytes via the PKA/AMPK/SIRT-1/ERK pathway in a proliferation-dependent manner [125]. Because of the mechanisms of action of sucralose and its GPCR activation-related characteristics, the hypothesis arises regarding the potential roles of the sweet taste receptor T1R3 on skin regenerative effects.

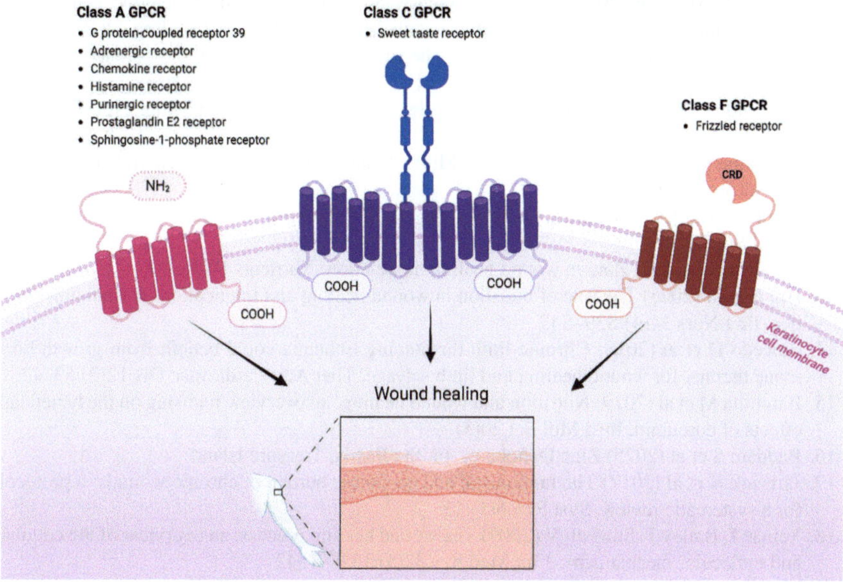

Fig. 8.3 G-protein coupled receptors as drug targets for skin wound healing

Future Perspective

Here, several surface GPCRs have been discussed in their roles as drug targets for wound healing (Fig. 8.3). Therefore, GPCR modulators may be useful for the treatment of skin wound healing. However, there are some concerns, including the delivery forms for the topical uses of GPCR modulators in wound treatment, large-scale production of the GPCR modulators, efficiency, and potential toxicity in patients, which need to be further investigated.

References

1. Yin H, Hu M, Li D (2023) Regulation of epidermal stratification and development by basal keratinocytes. J Cell Physiol 238(4):742–748
2. Hsu YC, Fuchs E (2022) Building and maintaining the skin. Cold Spring Harb Perspect Biol 14(7)
3. Sen CK (2023) Human wound and its burden: updated 2022 compendium of estimates. Adv Wound Care (New Rochelle) 12(12):657–670
4. Olsson M et al (2019) The humanistic and economic burden of chronic wounds: a systematic review. Wound Repair Regen 27(1):114–125
5. Uberoi A, McCready-Vangi A, Grice EA (2024) The wound microbiota: microbial mechanisms of impaired wound healing and infection. Nat Rev Microbiol 22(8):507–521
6. Hurlow J, Bowler PG (2022) Acute and chronic wound infections: microbiological, immunological, clinical and therapeutic distinctions. J Wound Care 31(5):436–445

7. Campbell AE et al (2023) Variable staphyloxanthin production by Staphylococcus aureus drives strain-dependent effects on diabetic wound-healing outcomes. Cell Rep 42(10):113281
8. Wright JA, Richards T, Srai SK (2014) The role of iron in the skin and cutaneous wound healing. Front Pharmacol 5:156
9. Stechmiller JK (2010) Understanding the role of nutrition and wound healing. Nutr Clin Pract 25(1):61–68
10. Razzaque MS, Wimalawansa SJ (2025) Minerals and human health: from deficiency to toxicity. Nutrients 17(3)
11. Moores, J (2013) Vitamin C: a wound healing perspective. Br J Community Nurs Suppl:S6, S8–11
12. Lin PH et al (2017) Zinc in wound healing modulation. Nutrients 10(1)
13. Hill B et al (2025) The role of nutrition in wound healing and implications for nursing practice. Br J Nurs 34(4):S39–S42
14. Caicedo D et al (2018) Chronic limb-threatening ischemia could benefit from growth hormone therapy for wound healing and limb salvage. Ther Adv Cardiovasc Dis 12(2):53–72
15. Barchitta M et al (2019) Nutrition and wound healing: an overview focusing on the beneficial effects of curcumin. Int J Mol Sci 20(5)
16. Baddam S et al (2025) Zinc Deficiency. In: StatPearls, Treasure Island
17. Jarbrink K et al (2017) The humanistic and economic burden of chronic wounds: a protocol for a systematic review. Syst Rev 6(1):15
18. Velnar T, Bailey T, Smrkolj V (2009) The wound healing process: an overview of the cellular and molecular mechanisms. J Int Med Res 37(5):1528–1542
19. Pena OA, Martin P (2024) Cellular and molecular mechanisms of skin wound healing. Nat Rev Mol Cell Biol 25(8):599–616
20. Tyavambiza C, Meyer M, Meyer S (2022) Cellular and molecular events of wound healing and the potential of silver based nanoformulations as wound healing agents. Bioengineering (Basel) 9(11)
21. Takeo M, Lee W, Ito M (2015) Wound healing and skin regeneration. Cold Spring Harb Perspect Med 5(1):a023267
22. Qing C (2017) The molecular biology in wound healing & non-healing wound. Chin J Traumatol 20(4):189–193
23. Etulain J (2018) Platelets in wound healing and regenerative medicine. Platelets 29(6):556–568
24. Canedo-Dorantes L, Canedo-Ayala M (2019) Skin acute wound healing: a comprehensive review. Int J Inflam 2019:3706315
25. Chen L et al (2018) Inflammatory responses and inflammation-associated diseases in organs. Oncotarget 9(6):7204–7218
26. Yoshida M et al (2012) TGF-beta-operated growth inhibition and translineage commitment into smooth muscle cells of periodontal ligament-derived endothelial progenitor cells through Smad- and p38 MAPK-dependent signals. Int J Biol Sci 8(7):1062–1074
27. Nguyen AV, Soulika AM (2019) The dynamics of the skin's immune system. Int J Mol Sci 20(8)
28. Li C et al (2013) Notch signal regulates corneal endothelial-to-mesenchymal transition. Am J Pathol 183(3):786–795
29. Insel PA et al (2019) GPCRomics: an approach to discover GPCR drug targets. Trends Pharmacol Sci 40(6):378–387
30. Sriram K, Insel PA (2018) G protein-coupled receptors as targets for approved drugs: how many targets and how many drugs? Mol Pharmacol 93(4):251–258
31. Wootten D et al (2018) Mechanisms of signalling and biased agonism in G protein-coupled receptors. Nat Rev Mol Cell Biol 19(10):638–653
32. Zhang M et al (2024) G protein-coupled receptors (GPCRs): advances in structures, mechanisms, and drug discovery. Signal Transduct Target Ther 9(1):88
33. Yang D et al (2021) G protein-coupled receptors: structure- and function-based drug discovery. Signal Transduct Target Ther 6(1):7

34. Schioth HB, Fredriksson R (2005) The GRAFS classification system of G-protein coupled receptors in comparative perspective. Gen Comp Endocrinol 142(1–2):94–101
35. Lawson AD (2015) G protein-coupled receptors - targets for fragment-based drug discovery. Curr Top Med Chem 15(24):2523–2527
36. Do HN et al (2022) Unique features of different classes of G-protein-coupled receptors revealed from sequence coevolutionary and structural analysis. Proteins 90(2):601–614
37. Satianrapapong W, Pongkorpsakol P, Muanprasat C (2020) A G-protein coupled receptor 39 agonist stimulates proliferation of keratinocytes via an ERK-dependent pathway. Biomed Pharmacother 127:110160
38. Pedro MP et al (2020) Activation of G-protein coupled receptor-Galphai signaling increases keratinocyte proliferation and reduces differentiation, leading to epidermal hyperplasia. J Invest Dermatol 140(6):1195–1203 e3
39. Pedro MP, Lund K, Iglesias-Bartolome R (2020) The landscape of GPCR signaling in the regulation of epidermal stem cell fate and skin homeostasis. Stem Cells 38:1520
40. Dwivedi NV et al (2023) GPCRs and fibroblast heterogeneity in fibroblast-associated diseases. FASEB J 37(8):e23101
41. Chun KS, Lao HC, Langenbach R (2010) The prostaglandin E2 receptor, EP2, stimulates keratinocyte proliferation in mouse skin by G protein-dependent and beta-arrestin1-dependent signaling pathways. J Biol Chem 285(51):39672–39681
42. Hoq MI et al (2011) Human catestatin enhances migration and proliferation of normal human epidermal keratinocytes. J Dermatol Sci 64(2):108–118
43. Zheng F et al (2017) The Axis of CXCR4/SDF-1 plays a role in colon cancer cell adhesion through regulation of the AKT and IGF1R signalling pathways. Anticancer Res 37(8):4361–4369
44. Teicher BA, Fricker SP (2010) CXCL12 (SDF-1)/CXCR4 pathway in cancer. Clin Cancer Res 16(11):2927–2931
45. Saotome K et al (2025) Structural insights into CXCR4 modulation and oligomerization. Nat Struct Mol Biol 32(2):315–325
46. Fui LW et al (2019) Understanding the multifaceted mechanisms of diabetic wound healing and therapeutic application of stem cells conditioned medium in the healing process. J Tissue Eng Regen Med 13(12):2218–2233
47. Cano Sanchez M et al (2018) Targeting oxidative stress and mitochondrial dysfunction in the treatment of impaired wound healing: a systematic review. Antioxidants (Basel) 7(8)
48. Chen CY et al (2024) Cancer-associated-fibroblast-mediated paracrine and autocrine SDF-1/CXCR4 signaling promotes stemness and aggressiveness of colorectal cancers. Cells 13(16)
49. Peddibhotla S et al (2023) Triazolothiadiazine derivative positively modulates CXCR4 signaling and improves diabetic wound healing. Biochem Pharmacol 216:115764
50. Wang L et al (2025) Mechanism underlying the involvement of CXCR4/CXCL12 in diabetic wound healing and prospects for responsive hydrogel-loaded CXCR4 formulations. Front Pharmacol 16:1561112
51. Sandhanam K et al (2024) Deciphering the interlinked CXCR4-mediated feedback loop among signaling pathways in diabetic wound healing. Curr Diabetes Rev
52. Zhang N et al (2025) Moist-exposed burn ointment promotes angiogenesis during full-thickness wound healing by activating the SDF-1/CXCR4 Axis in a hyperglycaemic rat model. Wound Repair Regen 33(4):e70070
53. Park JY, Pillinger MH, Abramson SB (2006) Prostaglandin E2 synthesis and secretion: the role of PGE2 synthases. Clin Immunol 119(3):229–240
54. Cheng H et al (2021) Role of prostaglandin E2 in tissue repair and regeneration. Theranostics 11(18):8836–8854
55. Gilman KE, Limesand KH (2021) The complex role of prostaglandin E(2)-EP receptor signaling in wound healing. Am J Physiol Regul Integr Comp Physiol 320(3):R287–R296
56. Palla AR et al (2021) Inhibition of prostaglandin-degrading enzyme 15-PGDH rejuvenates aged muscle mass and strength. Science 371(6528)

57. Galliot B et al (2017) Trends in tissue repair and regeneration. Development 144(3):357–364
58. Cao X et al (2020) IGF-1C hydrogel improves the therapeutic effects of MSCs on colitis in mice through PGE(2)-mediated M2 macrophage polarization. Theranostics 10(17):7697–7709
59. Zhang S et al (2018) Prostaglandin E(2) hydrogel improves cutaneous wound healing via M2 macrophages polarization. Theranostics 8(19):5348–5361
60. Parekh A et al (2007) Differential regulation of free-floating collagen gel contraction by human fetal and adult dermal fibroblasts in response to prostaglandin E2 mediated by an EP2/cAMP-dependent mechanism. Wound Repair Regen 15(3):390–398
61. Regan JW (2003) EP2 and EP4 prostanoid receptor signaling. Life Sci 74(2–3):143–153
62. Kamoshita E et al (2006) Recruitment of a prostaglandin E receptor subtype, EP3-expressing bone marrow cells is crucial in wound-induced angiogenesis. Am J Pathol 169(4):1458–1472
63. Obinata H, Hla T (2019) Sphingosine 1-phosphate and inflammation. Int Immunol 31(9):617–625
64. Chen H et al (2022) Sphingosine 1-phosphate receptor, a new therapeutic direction in different diseases. Biomed Pharmacother 153:113341
65. Chun J, Giovannoni G, Hunter SF (2021) Sphingosine 1-phosphate receptor modulator therapy for multiple sclerosis: differential downstream receptor signalling and clinical profile effects. Drugs 81(2):207–231
66. Cartier A, Hla T (2019) Sphingosine 1-phosphate: lipid signaling in pathology and therapy. Science 366(6463)
67. Aoki M et al (2019) Sphingosine-1-phosphate facilitates skin wound healing by increasing angiogenesis and inflammatory cell recruitment with less scar formation. Int J Mol Sci 20(14)
68. Kawanabe T et al (2007) Sphingosine 1-phosphate accelerates wound healing in diabetic mice. J Dermatol Sci 48(1):53–60
69. Shin KO et al (2018) Ginsenoside Rb1 enhances keratinocyte migration by a sphingosine-1-phosphate-dependent mechanism. J Med Food 21(11):1129–1136
70. Pullar CE, Rizzo A, Isseroff RR (2006) beta-Adrenergic receptor antagonists accelerate skin wound healing: evidence for a catecholamine synthesis network in the epidermis. J Biol Chem 281(30):21225–21235
71. Pullar CE et al (2006) Beta2-adrenergic receptor activation delays wound healing. FASEB J 20(1):76–86
72. Le Provost GS, Pullar CE (2015) beta2-adrenoceptor activation modulates skin wound healing processes to reduce scarring. J Invest Dermatol 135(1):279–288
73. Albrecht H et al (2018) The Beta 2 adrenergic receptor antagonist Timolol improves healing of combined burn and radiation wounds. Radiat Res 189(4):441–445
74. Smit MJ et al (1999) Molecular properties and signalling pathways of the histamine H1 receptor. Clin Exp Allergy 29(Suppl 3):19–28
75. Shiroishi M, Kobayashi T (2017) Structural analysis of the histamine H(1) receptor. Handb Exp Pharmacol 241:21–30
76. Shimamura T et al (2011) Structure of the human histamine H1 receptor complex with doxepin. Nature 475(7354):65–70
77. Morini G et al (1993) Coexistence of histamine H1 and H2 receptors coupled to distinct signal transduction pathways in isolated intestinal muscle cells. J Pharmacol Exp Ther 264(2):598–603
78. Beermann S et al (2015) Histamine H(1)- and H(4)-receptor signaling cooperatively regulate MAPK activation. Biochem Pharmacol 98(3):432–439
79. Nieto-Alamilla G et al (2016) The histamine H3 receptor: structure, pharmacology, and function. Mol Pharmacol 90(5):649–673
80. Zampeli E, Tiligada E (2009) The role of histamine H4 receptor in immune and inflammatory disorders. Br J Pharmacol 157(1):24–33
81. de Esch IJ et al (2005) The histamine H4 receptor as a new therapeutic target for inflammation. Trends Pharmacol Sci 26(9):462–469

82. Ashida Y, Denda M, Hirao T (2001) Histamine H1 and H2 receptor antagonists accelerate skin barrier repair and prevent epidermal hyperplasia induced by barrier disruption in a dry environment. J Invest Dermatol 116(2):261–265
83. Gutowska-Owsiak D et al (2014) Histamine enhances keratinocyte-mediated resolution of inflammation by promoting wound healing and response to infection. Clin Exp Dermatol 39(2):187–195
84. Yaron JR et al (2025) Histamine receptor agonism differentially induces immune and reparative healing responses in biomaterial-facilitated tissue repair. Biomaterials 315:122967
85. Idzko M, Ferrari D, Eltzschig HK (2014) Nucleotide signalling during inflammation. Nature 509(7500):310–317
86. Borea PA et al (2018) Pharmacology of adenosine receptors: the state of the art. Physiol Rev 98(3):1591–1625
87. Borah P et al (2019) P1 receptor agonists/antagonists in clinical trials - potential drug candidates of the future. Curr Pharm Des 25(26):2792–2807
88. Zemkova H (2023) Purinergic P2 receptors: structure and function 2.0. Int J Mol Sci 24(6)
89. von Kugelgen I (2021) Molecular pharmacology of P2Y receptor subtypes. Biochem Pharmacol 187:114361
90. Burnstock G (2006) Purinergic signalling. Br J Pharmacol 147 Suppl 1(Suppl 1):S172–S181
91. Burnstock G (2006) Purinergic signalling--an overview. In: Novartis Found Symp, vol 276, pp 26–48; discussion 48–57, 275–81
92. North RA (2002) Molecular physiology of P2X receptors. Physiol Rev 82(4):1013–1067
93. Khakh BS, North RA (2006) P2X receptors as cell-surface ATP sensors in health and disease. Nature 442(7102):527–532
94. Greig AV et al (2003) Purinergic receptor expression in the regeneration epidermis in a rat model of normal and delayed wound healing. Exp Dermatol 12(6):860–871
95. Valls MD, Cronstein BN, Montesinos MC (2009) Adenosine receptor agonists for promotion of dermal wound healing. Biochem Pharmacol 77(7):1117–1124
96. Montesinos MC, Desai-Merchant A, Cronstein BN (2015) Promotion of wound healing by an agonist of adenosine A2A receptor is dependent on tissue plasminogen activator. Inflammation 38(6):2036–2041
97. Jin H et al (2014) P2Y2 R activation by nucleotides promotes skin wound-healing process. Exp Dermatol 23(7):480–485
98. McEwan TB et al (2021) Autocrine regulation of wound healing by ATP release and P2Y(2) receptor activation. Life Sci 283:119850
99. Clevers H, Nusse R (2012) Wnt/beta-catenin signaling and disease. Cell 149(6):1192–1205
100. Voloshanenko O et al (2017) Mapping of Wnt-frizzled interactions by multiplex CRISPR targeting of receptor gene families. FASEB J 31(11):4832–4844
101. Zhang Y et al (2015) Wnt/beta-catenin and Wnt5a/ca pathways regulate proliferation and apoptosis of keratinocytes in psoriasis lesions. Cell Physiol Biochem 36(5):1890–1902
102. Reischl J et al (2007) Increased expression of Wnt5a in psoriatic plaques. J Invest Dermatol 127(1):163–169
103. Gudjonsson JE et al (2010) Evidence for altered Wnt signaling in psoriatic skin. J Invest Dermatol 130(7):1849–1859
104. Zuo C et al (2024) MiR-488-3p facilitates wound healing through CYP1B1-mediated Wnt/beta-catenin signaling pathway by targeting MeCP2. J Diabetes Investig 15(2):145–158
105. Pandey S et al (2025) Molecular insights into diabetic wound healing: focus on Wnt/beta-catenin and MAPK/ERK signaling pathways. Cytokine 191:156957
106. Ogawa Y et al (2018) Zinc and skin disorders. Nutrients 10(2)
107. Schwartz JR, Marsh RG, Draelos ZD (2005) Zinc and skin health: overview of physiology and pharmacology. Dermatologic Surg 31(7 Pt 2):837–847. discussion 847
108. Ogawa Y, Kawamura T, Shimada S (2016) Zinc and skin biology. Arch Biochem Biophys 611:113–119

109. Zorrilla P et al (2006) Low serum zinc level as a predictive factor of delayed wound healing in total hip replacement. Wound Repair Regen 14(2):119–122
110. Lansdown AB et al (2007) Zinc in wound healing: theoretical, experimental, and clinical aspects. Wound Repair Regen 15(1):2–16
111. Henzel JH, DeWeese MS, Lichti EL (1970) Zinc concentrations within healing wounds. Significance of postoperative zincuria on availability and requirements during tissue repair. Arch Surg 100(4):349–357
112. Pongkorpsakol P, Moonwiriyakit A, Muanprasat C (2017) Fatty acid and mineral receptors as drug targets for gastrointestinal disorders. Future Med Chem 9(3):315–334
113. McKee KK et al (1997) Cloning and characterization of two human G protein-coupled receptor genes (GPR38 and GPR39) related to the growth hormone secretagogue and neurotensin receptors. Genomics 46(3):426–434
114. Holst B et al (2007) GPR39 signaling is stimulated by zinc ions but not by obestatin. Endocrinology 148(1):13–20
115. Hershfinkel M (2018) The zinc sensing receptor, ZnR/GPR39, in health and disease. Int J Mol Sci 19(2)
116. Shimizu Y, Koyama R, Kawamoto T (2017) Rho kinase-dependent desensitization of GPR39; a unique mechanism of GPCR downregulation. Biochem Pharmacol 140:105–114
117. Santos-Zas I et al (2016) beta-Arrestin scaffolds and signaling elements essential for the obestatin/GPR39 system that determine the myogenic program in human myoblast cells. Cell Mol Life Sci 73(3):617–635
118. Pongkorpsakol P et al (2019) An agonist of a zinc-sensing receptor GPR39 enhances tight junction assembly in intestinal epithelial cells via an AMPK-dependent mechanism. Eur J Pharmacol 842:306–313
119. Mlyniec K et al (2015) GPR39 Zn(2+)-sensing receptor: a new target in antidepressant development? J Affect Disord 174:89–100
120. Chatkul P et al (2025) An agonist of zinc-sensing G-protein coupled receptor 39 accelerates skin wound healing and protects against UVB-induced keratinocyte damage. J Exp Pharmacol 17:571–585
121. Shi Q, Zhu X, Deng S (2021) Sweet taste receptor expression and its activation by sucralose to regulate glucose absorption in mouse duodenum. J Food Sci 86(2):540–545
122. Ahmad SY, Friel JK, Mackay DS (2020) Effect of sucralose and aspartame on glucose metabolism and gut hormones. Nutr Rev 78(9):725–746
123. Liauchonak I et al (2019) Non-nutritive sweeteners and their implications on the development of metabolic syndrome. Nutrients 11(3)
124. Laffitte A, Neiers F, Briand L (2014) Functional roles of the sweet taste receptor in oral and extraoral tissues. Curr Opin Clin Nutr Metab Care 17(4):379–385
125. Chatkul P et al (2025) Cellular mechanisms underlying the effects of sucralose on keratinocyte proliferation and its contributing roles in epidermal regeneration. J Funct Foods 131:106944

Chapter 9
Growth Factors and Cytokines: Unravelling Molecular Mechanisms for Therapeutic Innovations in Wound Care

Madhan Jeyaraman, Naveen Jeyaraman, Sathish Muthu, Shrideavi Murugan, and Sharun Khan

Abstract Cutaneous wound healing is a complex and dynamic process involving multiple cellular mechanisms and signaling pathways. The disruption of normal anatomy triggers a cascade of molecular events integrating various cell types, cytokines, growth factors, and extracellular matrix components. This intricate molecular interplay facilitates cell migration and proliferation of immune cells, endothelial cells, keratinocytes, and fibroblasts to the wound site, promoting tissue repair. Aberrant wound healing can result in chronic wounds or excessive scar formation, underscoring the need to understand the underlying biological mechanisms. Key processes such as ubiquitination, inflammation, immune response, cellular migration, proliferation, and extracellular matrix remodeling play pivotal roles. Growth factors like EGF and PDGF stimulate cellular functions essential for granulation tissue formation, angiogenesis, and re-epithelialization. Cytokines act as mediators, balancing pro- and anti-inflammatory responses to ensure proper tissue repair while

M. Jeyaraman (✉) · N. Jeyaraman
Department of Orthopaedics, ACS Medical College and Hospital, Dr MGR Educational and Research Institute, Chennai, Tamil Nadu, India

Orthopaedic Research Group, Coimbatore, Tamil Nadu, India

S. Muthu
Orthopaedic Research Group, Coimbatore, Tamil Nadu, India

Central Research Laboratory, Meenakshi Medical College Hospital and Research Institute, Meenakshi Academy of Higher Education and Research, Chennai, Tamil Nadu, India

S. Murugan
Department of Orthopaedics, Tirunelveli Medical College and Hospital, Tirunelveli, Tamil Nadu, India

S. Khan
Centre for Regenerative Nanomedicine, Northwestern University, Chicago, IL, USA

Graduate Institute of Medicine, Yuan Ze University, Taoyuan, Taiwan

F. De Francesco, K. Turksen (eds.), *Management and Strategies for Wound Healing*, Stem Cell Biology and Regenerative Medicine 79,
https://doi.org/10.1007/978-3-032-12442-5_9

preventing pathological consequences. Dysregulation in signaling pathways such as TGF-β, VEGF, and PI3K/AKT affects angiogenesis, cellular activity, and ECM remodeling, contributing to healing impairment. Innovative therapeutic strategies utilizing hydrogels, nanoparticles, and biomimetic systems show promise in optimizing the delivery and bioavailability of growth factors and cytokines. These approaches mimic natural wound healing dynamics, enhancing cellular responses and promoting tissue regeneration. Despite advancements, challenges such as co-morbidities, protein stability, and translational hurdles persist, highlighting the need for multidisciplinary efforts to bridge the gap between research and clinical application. Comprehensive insights into cellular behavior and molecular interactions offer potential solutions to achieve effective and personalized wound healing therapies.

Keywords Wound healing · Growth factors · Angiogenesis · Extracellular matrix

Abbreviations

AuNP	Gold nanoparticles
CO	Carbon monoxide
CuNP	Copper nanoparticles
ECM	Extracellular matrix
EGF	Epidermal growth factor
FGF	Fibroblast growth factor
G-CSF	Granulocyte colony-stimulating factor
GM-CSF	Granulocyte-macrophage colony-stimulating factor
H2S	Hydrogen sulfide
HB-EGF	Heparin-binding epidermal growth factor
HIF-1	Hypoxia-inducible factor-1
IKK	I-kappa-B kinase
IL	Interleukin
IP3	Inositol triphosphate
KGF	Keratinocyte growth factor
NF-κB	Nuclear factor kappa B
NLRP3	NOD-, LRR-, and pyrin domain-containing protein 3
NO	Nitric oxide
PDGF	Platelet-derived growth factor
RNAi	RNA interference
ROS	Reactive oxygen species
siRNA	Small interfering RNA
TGF-β	Transforming growth factor-beta
TNF-α	Tumor necrosis factor-alpha
VEGF	Vascular endothelial growth factor
ZnO NP	Zinc oxide nanoparticles

Introduction

Wounds remain a persistent clinical and public health challenge in daily life, with early and late consequences causing morbidity and mortality [1]. Due to an increase in the rate of occurrence and frequency in the elderly, they have a profound impact on the economy and society [1]. A wound is defined as a disruption or damage of normal anatomy or function, ranging from superficial breaks in the epithelium to deeper extension into the subcutaneous tissue and damage of muscles, tendons, vessels, nerves, and parenchymal organs, eventually involving the bone [2, 3]. Wound healing is a dynamic process that involves the integration of various cell types, soluble mediators (cytokines), and extracellular matrix [4]. This involves a complex interplay of vascular endothelial cells, fibroblasts, immune cells recruited to the site, associated extracellular matrix, and keratinocytes in this case [5]. In the event of aberrant wound healing, there is an occurrence of one of the two major outcomes: (a) becoming an ulcerative skin defect (chronic wound) or (b) excessive scar tissue formation (hypertrophic scar or keloid) [5]. An acute wound heals in the following sequence: immune cell migration to the site of the wound with locally invading pathogens, and this begins the healing process. This is then followed by upregulation of wound-associated genes by cut epidermal edges, causing cell migration. Following collective migration of cells, there occurs proliferation of local and blood-borne fibroblasts, which form the granulation tissue, providing structure and further signaling for depositing the extracellular matrix. Some of the fibroblasts differentiate to form myofibroblasts, which begin the process of wound contraction along with fibrin as shown in Fig. 9.1. The wound bed is now perfused with nutrients and oxygen via new blood vessels created by angiogenesis [5].

Some of the key molecular mechanisms involved in wound healing are

(a) *Ubiquitination:* This plays a critical role in epithelial wound healing by regulating key processes such as migration and proliferation. During the process of healing, the level of ubiquitin protein rises significantly, highlighting the importance of its post-translational modification. Amongst these, E3 ubiquitin ligase Nedd4 plays a key role in facilitating wound healing by degrading PTEN, a negative regulator of the Stat3 signaling pathway. This, in turn, upregulates the expression of the Stat3 signaling pathway, leading to cellular migration and proliferation, which are necessary for wound healing. Nedd4 is regulated by microRNAs from the mir30 and mir200 families, which are decreased significantly, leading to an uninterrupted rise in Nedd4 levels [6].

(b) *Inflammation and immune response:* They both play a crucial and dynamic role in the process of wound healing. They drive tissue repair in their normal state, but when dysregulated, they contribute to delayed healing and the formation of pathological scars. The release of DAMP (damage-associated molecular patterns) and PAMP (pathogen-associated molecular patterns) by activated resident immune cells following trauma recruits neutrophils and monocytes from the circulation. Neutrophils, through phagocytosis, formation of extracellular traps, and production of reactive oxygen species (ROS), clear microbes, and

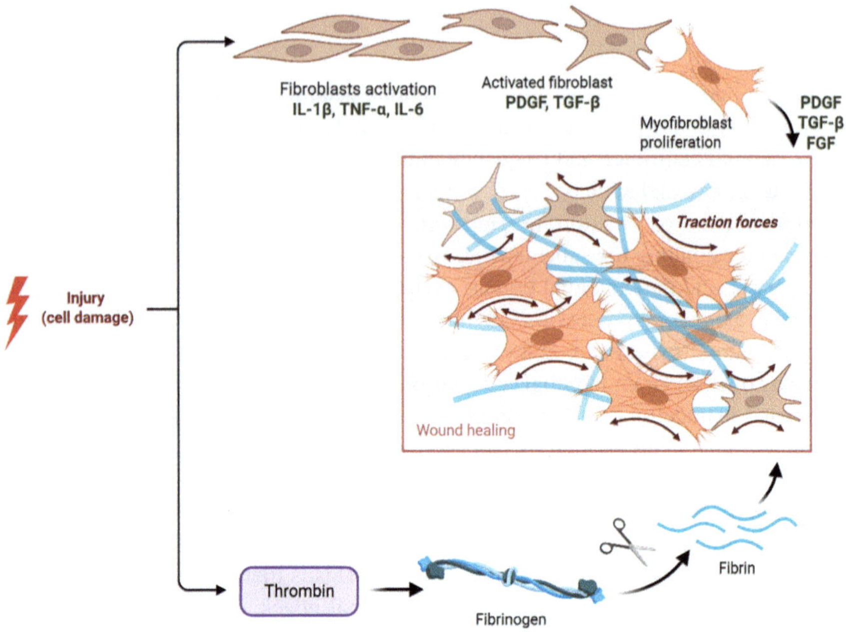

Fig. 9.1 Wound contraction by myofibroblasts with fibrin promoted by key cytokines PDGF, TGF-β, and FGF

debris at the wound site. Additionally, they release cytokines to propagate the inflammatory cascade.

Monocytes, after differentiating into macrophages, play a dual role in both initiating and resolving the inflammation. Initially establishing a pro-inflammatory cascade, they help clear apoptotic neutrophils and pathogens. In later stages, they potentiate an anti-inflammatory state, which promotes tissue repair by secreting factors like TGF-β and VEGF. These factors then support the proliferation of keratinocytes, activation of fibroblasts, and angiogenesis, leading to proliferative and remodeling stages of wound healing [7].

Mechanical cues from the environment sometimes amplify the inflammatory process, thus leading to increased immune cell recruitment and eventually promoting the formation of a scar [5]. In the event of a chronic wound, wounds occurring in the elderly, or those with diabetes, persistent and dysfunctional inflammatory processes occur. Neutrophils and macrophages do not phagocytose adequately, leading to a delay in progressing to the reparative stages. This eventually causes increased activity of protease, damage, and impaired keratinocyte migration. Hence, angiogenesis is disrupted and the ECM is poorly reconstructed [7]. The chronicity of pro-inflammatory factors preserves the wound in an unresolved inflammatory state, leading to treating these difficult wounds with novel immune-modulating therapies rather than conventional therapies [5, 7].

(c) *Cellular proliferation and migration:* These are key components of the wound healing process, especially during the proliferative stage. After the initial inflammatory response, keratinocytes, endothelial cells, fibroblasts, and other progenitor cells are activated, and these repopulate the wound site. Keratinocytes begin the process of re-epithelialization while fibroblasts initiate the deposition of ECM components like collagen, providing a base for tissue reconstruction [8].

Along with the formation of granulation tissue, new blood vessels form via angiogenesis, and this is supported by endothelial cell migration and proliferation. This neovascularization ensures adequate oxygen and nutrient supply to the regenerating tissues. Growth factors and matrix proteins, which are crucial for ECM remodeling and wound contraction, are produced by fibroblasts [9].

Additionally, modification of the macrophage phenotype at this stage determines whether regeneration or fibrosis of the wound site occurs. Macrophages, in addition to clearing cellular debris, also secrete factors supporting keratinocyte and fibroblastic activity [9].

(d) *Extracellular matrix remodeling:* This plays a vital role in wound healing by providing structural support and regulating key cellular activities. During the initial stages, they serve as a scaffold for the infiltrating immune cells and fibroblasts, which then clear the debris and initiate the regeneration process. As healing progresses, fibroblasts begin synthesizing new ECM components, particularly collagen, fibronectin, and proteoglycans. They then help form the granulation tissue, a temporary matrix facilitating re-epithelialization and angiogenesis. Key genes involved in ECM organization are COL6A3, COL4A1, THY1, and MMP11, which are upregulated post-injury. These genes play an important role in collagen production, cell adhesion, and matrix remodeling. Matrix metalloproteinases are of importance, as they enable the reshaping of tissue architecture whilst supporting cell migration. They also help in the degradation of damaged ECM products [9].

Besides structural processes, biochemical interactions of the ECM molecules with the cell surface receptors, such as integrins, activate the intracellular signaling pathways, which influence genetic expression and cellular behavior. This communication is crucial for the transition from the proliferative to the regenerative stage, where the temporary matrix is replaced by a more organized and resilient tissue [9].

(e) *Signaling pathways:* these are central to all stages of wound healing and are especially critical in the management of chronic wounds. Dysregulation of these pathways in chronic wounds leads to impaired cellular proliferation, angiogenesis, and incomplete tissue remodeling. Several pathways are identified for therapeutic targetings, such as PI3K/AKT, Wnt/β-catenin, TGF-β, Nrf2, Notch, and HIF-1 [10].

Key Growth Factors in Wound Healing

Epithelium, when wounded, repairs from both damaged wound edges and also sweat and hair gland appendages. Among the various genes upregulated post damage, the earliest upregulations include Ap1, Jun, Fos, and the Knox zinc transcription factors functioning as a part of machinery for transcriptional activation and causing subsequent upregulation of several hundred genes leading to a surge in cellular proliferation, which is accompanied by epidermal migration of keratinocytes at the junction between healthy granulation tissue and scab [11, 12]. The Polycomb family of epigenetic regulators causes the deposition of histone methylation marks, which silence the epidermal growth factor receptors, and these are removed soon after a wound is established, making these genes available for translation [13]. The epidermal growth factor plays an essential role in cellular regeneration and dermal wound healing by stimulating the proliferation and migration of keratinocytes. In addition to this, it also promotes fibroblast motility and the formation of granulation tissue. Having a high affinity for cell surface receptors and, in turn, inducing their dimerization, thus activating tyrosine kinase receptors in the cytoplasmic receptor domain, subsequently initiating a signal transduction that leads to DNA synthesis and cellular proliferation [14]. EGF is sourced from a variety of cells, including platelets. They stimulate a variety of processes such as angiogenesis, collagenase synthesis, fibroplasia, and fibronectin synthesis, starting right after the initial hemostasis platelet plug formation. Controlled release of EGF is mediated by expression of EGF receptor on the target cell [14]. Enhanced wound healing and reduced scar formation were noted in wounds treated with EGF in studies conducted in the 1970s [15, 16].

Among the first growth factors to be deposited at the wound site by the degranulation of platelets is the platelet-derived growth factor. They are synthesized by many cell types, such as endothelial progenitor cells, adipose-derived stem cells, macrophages, and platelets [17]. These contribute to angiogenesis and vasculogenesis under the influence of cytokines, growth factors, and hormones [17]. This exists in 3 isoforms (AA, BB, AB) composed of dimers built by polypeptide A and B chains [18]. They are provoked by the interaction between two specific cell receptor tyrosine kinases. Binding to ligands results in autophosphorylation and dimerization of the receptors. Stimulation of β receptors results in transduction of cell motility and formation of circular actin structure on the cell surface, whereas inhibition of chemotaxis of fibroblasts and smooth muscle cells is achieved by the α receptors [18]. They also stimulate multiple cellular processes such as chemotaxis, cell proliferation, differentiation, and survival, having an important role in the regulation of tissue pressure. This factor tends to be deficient in those with diabetes and chronic wounds, and exogenous administration of this factor accelerates wound closure and healing in both diabetic and non-diabetic wound models. During the inflammatory stage of healing, they promote the chemotaxis of neutrophils and macrophages, which produce ROS and remove bacterial contamination at the wound site. They may also stimulate the production of wound repair factors from macrophages [19].

Previous studies conducted to emphasize the role of PDGF in the process of angiogenesis, with full-thickness grafts dorsal wounds regenerating faster when treated with PDGF-B [20, 21]

Cytokine Dynamics in Wound Healing

Cytokines, both pro- and anti-inflammatory, play essential roles in the healing process, acting as mediators in the initial inflammatory stage. In the initial stages, there is an abundance of pro-inflammatory cytokines such as IL-6, IL-1β, and TNF-α, and various other chemokines, which play a crucial role by recruiting cells for the clearance of debris and promote the accumulation of growth factors [22]. This accumulation of a pro-inflammatory state is key to establishing an anti-infectious process. However, overproduction of the same cytokines may lead to deleterious consequences such as SIRS (systemic inflammatory response syndrome) [23]. As healing continues, there occurs a shift in the dynamic, which causes an increase in anti-inflammatory cytokines such as IL-4 and IL-10, promoting regeneration of the wound with processes like angiogenesis and wound remodeling [22]. These cytokines are activated by the presence of micro-organisms or their derived products such as membrane compounds, intracellular constituents that are released following their death, and toxins released in the vicinity of the wound [23]. The highest level of these cytokines is observed 12–24 h after a wound has occurred. Following completion of the early phase of wound repair, the mRNA of these cytokines returns to basal levels. They are expressed predominantly in polymorphonuclear leukocytes. At later stages of repair, there is an increase in IL-1α, IL-1β, and TNF-α in macrophages. TNF-α was also found in the edges of wounds, which contain hyperproliferative epithelium, and IL-1α in the keratinocytes of hair follicles [24]. Damage-associated molecular patterns and pathogen-associated molecular patterns activate Toll-like receptors (TLRs) and INF-δ on immune cells. This leads to further activation of various pathways such as NF-κB, IP3, and IKK, which drive the transcription process to release pro-inflammatory cytokines. The first mediators to be released include TNF-α, IL-1β, IL-6, and IL-8, which serve a multitude of purposes such as vasodilation, increased vascular permeability, enabling the migration of neutrophils and monocytes to the wound site. Macrophages amplify the inflammatory process by initiating the release of various pro-inflammatory cytokines. Nitric oxide and reactive oxygen species also enhance their tissue debriding function [25]. Accumulation of ROS, damage signals, and mitochondrial DNA then activates the NLRP3 inflammasome pathway in macrophages, which facilitates the proteolytic cleavage of pro-caspase 1 into active caspase 1, and this converts pro-inflammatory cytokines IL-1β and IL-18 into their active forms. These propagate the inflammatory signals, leading to a highly immunogenic state in the wound bed. Eventually, this has to be tapered to prevent progression to chronic inflammation and tissue damage [25].

Table 9.1 Role of cytokines in wound healing

Cytokine	Role in wound healing
IL-1	Initiates pro-inflammatory response and recruits immune cells. Promotes angiogenesis and granulation tissue formation
TNF-α	Drives inflammation, enhances vascular permeability, and recruits neutrophils and macrophages for debris clearance
IL-6	Supports immune cell activation, ECM remodeling, and angiogenesis during the inflammatory and proliferative phases
IL-10	Suppresses pro-inflammatory mediators and promotes tissue repair. Enhances fibroblastic activity and ECM synthesis
TGF-β	Resolves inflammation and facilitates fibroblast-to-myofibroblast differentiation for ECM synthesis and granulation
VEGF	Promotes angiogenesis, lymphangiogenesis, and vascular remodeling to restore oxygen and nutrient delivery
IL-4	Transition cytokine that fosters wound regeneration, angiogenesis, and remodeling phases

ECM extracellular matrix, *IL* interleukin, *TNF-α* tumor necrosis factor-alpha, *TGF-β* transforming growth factor-beta, *VEGF* vascular endothelial growth factor, *EGF* epidermal growth factor, *PDGF* platelet-derived growth factor

This is achieved by switching the macrophage phenotype from M1, which is pro-inflammatory, to a reparative M2 type. Resolution of pathogenic stimuli and rising anti-inflammatory cytokines like IL-10 and TGF-β1 help trigger this change. IL-10 suppresses the activity of NF-κB and downregulates the pro-inflammatory mediators, whereas TGF-β1 inhibits inflammation and promotes active tissue regeneration by stimulating fibroblasts, converting them into myofibroblasts. Myofibroblasts produce collagen and other ECM components, which are crucial in the formation of granulation tissue and re-epithelialization [25].

Anti-inflammatory cytokines also upregulate the levels of superoxide dismutase, catalase, and glutathione transferase, which neutralize the ROS and protect regenerating cells from damage. This also halts the NLRP3 inflammasome pathway, thus eventually stopping the pro-inflammatory cascade. As inflammation now resolves, indicated by the reduction of the C-reactive protein and fibrinogen levels, fibroblasts continue to lay down new ECM and promote angiogenesis to restore the skin barrier. As time passes, there occurs remodeling of the granulation tissue and replacement of weaker type 3 collagen by a stronger type 1 collagen [25]. Table 9.1 illustrates the role of each cytokine in the wound-healing process.

Mechanisms of Action

TGF-β and VEGF signaling pathways play a critical role in the tissue repair and regeneration process (Table 9.2). They are involved in various stages of the process, which include cell migration, proliferation, differentiation, and apoptosis, which are essential in maintaining tissue homeostasis and initiating repair after injury. TGF-β promotes the preservation of the ECM by potentiating the expression of profibrotic genes like type 1 collagen [26]. Its activation also leads to the accumulation of temporary collagen, essential for wound repair [27]. In addition to these functions, they also promote infiltration and proliferation of preexisting fibroblasts, generation of myofibroblasts, and inhibit collagenolysis [28].

Upon binding to receptors (TGFB1 and TGFB2), they trigger the formation of a complex, followed by activation of SMAD2 and SMAD3 proteins, which then form a complex with SMAD4. This complex is then transported to the nucleus to bind to transcription factors such as scleraxis. Fibrinogen and osteopontin are pro-inflammatory molecules that lead to upregulation of TGF-β via increased neutrophil count and IL-1β expression, respectively. Enhanced oxidative stress also led to its increased expression. TGF-β signaling is antagonized by transcription factor sharp-1, which binds to SMAD3. Additionally, microRNA 29 also downregulates the expression of pro-fibrotic genes [26]. TGF-β1 plays a predominant role in suppressing excess inflammation, which itself is an important aspect of tissue repair [27]. Deletion of the Tgfb1 gene in mice was found to cause rapid death after birth in mice as a result of excessive systemic inflammation [29].

Vascular endothelial growth factor (VEGF) signaling has a crucial role in tissue repair and wound healing by promoting cellular proliferation, angiogenesis, and lymphangiogenesis [30]. This is a heterodimeric heparin-binding protein. VEGF mRNA and protein tend to increase at the site of injury and remain the same for at least a week in wounds produced by surgical procedures. Keratinocytes produce VEGF at later stages of wound healing. Other sources of VEGF include platelets, neutrophils, and mast cells. Its transcription is upregulated by growth factors, pro-inflammatory cytokines, cellular stress, hormones, and placental growth factors. On the other hand, Fra-1, which is a member of the Fos transcription family, antagonizes the role of VEGF in the healing of skin flaps [31, 32].

VEGF during repair regulates vascular permeability, migration, and proliferation of pre-existing inflammatory cells and increases the population of inflammatory cells at the site of injury. It exerts its effect on VEGF-1 and VEGF-2 tyrosine kinase receptors, and also on a recently described neuropilin-1 receptor. These receptors are upregulated in dermal capillaries and macrophages in the event of skin repair. In comparison to VEGF-2, VEGF-1 acts as a molecule to which the ligand binds during angiogenesis [32]. Other pathways apart from these include PI3K/AKT, Wnt/β--catenin, Nrf2, Notch, HIF-1, gastrotransmitter-mediated, and lipid-based signaling.

The PI3K/AKT pathway plays a similar role as other pathways by promoting cell survival, migration, proliferation, and angiogenesis. This pathway is most active during the inflammatory and proliferative phases, thus enhancing keratinocyte and fibroblastic activity. Various natural compounds like kaempferol, naringin, and ginsenoside Rg1 activate this pathway, thereby leading to a reduction in inflammation and improved wound closure. One other factor that potentiates this pathway is exosomes containing microRNAs such as miR-21-3p, which causes enhanced fibroblast function and epithelial repair [10].

The Wnt/β-catenin signaling pathway is also involved in stem cell activation, cell proliferation, and angiogenesis. At the wound site, activation of Wnt leads to transcription of genes promoting fibroblast and keratinocyte functioning, leading to the formation of granulation tissue and re-epithelialization. Alterations occurring in this pathway have shown potential in shifting chronic wounds to regenerative healing [10]. One pathway that is of utmost importance in ensuring adequate wound healing in diabetic and chronic wounds is the Nrf2 pathway. Stimulating this pathway leads to the expression of antioxidant genes and heme oxygenase-1, which protects against ROS-induced damage. They also indirectly support keratinocyte migration and ECM remodeling. Activating this pathway in an oxidant stress-prone environment leads to enhanced healing outcomes [10].

The fate of various cellular decisions is in the hands of the Notch pathway, with key involvement of this pathway in keratinocyte differentiation, angiogenesis, and polarization of macrophages. This also enables the change of macrophage phenotype from Type 1 to Type 2 phenotype, progressing toward the resolution of inflammatory surroundings and promoting tissue regeneration [10]. Hypoxia-inducible factor -1 (HIF-1) gets activated in response to reduced oxygen levels at the wound site. This also promotes angiogenesis chiefly via downstream targets like VEGF. They are also involved in other components of wound healing, such as modulating factors for tissue oxygenation, tissue remodeling, and migration of fibroblasts. In wounds with poor vascularization, activation of HIF-1 has been shown to restore the healing capacity and also support regeneration [10].

Gastrotransmitters like nitric oxide (NO), carbon monoxide (CO), and hydrogen sulfide (H_2S) are small signaling molecules with big roles in modulating inflammation, vascular tone, and oxidative stress. Low concentrations of NO promote endothelial permeability and migration of inflammatory cells, while increased levels suppress excess inflammation. Hydrogels and nanoparticles incorporated with NO-releasing materials have shown promise in enhanced wound healing [33]. The role of lipid-based signaling pathways in wound healing is an emerging area of interest. Bioactive lipids like Prostaglandins, leukotrienes, lipoxins, and resolvins play a stage-specific role in hemostasis, inflammation, and proliferation. PG-E2 contributes towards the promotion of fibroblast proliferation and angiogenesis, whereas leukotrienes play a role in the recruitment of immune cells. Resolvins and Maresins- specialized pro-resolving mediators actively resolve inflammation and enable tissue regeneration. These lipid mediators are secreted by Extracellular vesicles, which act as carriers of signaling molecules and also influence the behavior of keratinocytes and matrix deposition [34].

Cytokines and growth factors have synergistic and antagonistic effects on each other during wound healing. These interactions ensure a coordinated progression through all the stages of wound healing. Any disruption in this may lead to aberrations in wound healing in the form of wound chronicity and fibrosis [35–37]. Early in wound healing, IL-1 released by keratinocytes and leukocytes has a positive feedback effect on HGF in fibroblasts, which enhances angiogenesis, granulation tissue formation, and re-epithelialization [37]. In return, platelet-derived TGF-β1 stimulates pro-inflammatory cytokines like IL-6 and IL-1, enabling immune cell recruitment and matrix deposition [35, 37]. This highlights a feedforward loop with cytokines and growth factors, each enabling the expression of the other. Similarly,

EGF and TGF-α, which stimulate keratinocyte migration and proliferation under the effect of cytokines IL-1 and TNF-α, also provide an environment that influences the local expression of IL-1 and IL-6, creating a local environment favorable for epithelial regeneration [35, 36].

In chronic wounds, there is a dysregulation of this interactive signaling between the two. Studies show that the degradation of EGFR by proteases and loss of TGF-β signaling in keratinocytes impair their responsiveness, despite the abundance of cytokine signaling. This eventually leads to a non-healing state [35, 37]. Upregulation of adhesion molecules by VEGF also requires cytokine signaling, and the absence of the signaling leads to impaired angiogenesis and matrix remodeling [37] (Table 9.2).

Table 9.2 Role of cytokine-based signaling pathways in wound healing

Mechanism	Role in tissue repair and wound healing
TGF-β signaling pathway	Promotes preservation of ECM by potentiating profibrotic gene expression (e.g., type 1 collagen). Enhances temporary collagen accumulation essential for repair. Stimulates fibroblast infiltration and myofibroblast generation and inhibits collagenolysis. Involved in suppressing excessive inflammation and maintaining tissue homeostasis. Transcriptional activation via SMAD2/3-SMAD4 complex regulates fibroblast proliferation and ECM production. Antagonized by factors like sharp-1 and microRNA 29, ensuring balance in fibrosis and inflammation
VEGF signaling pathway	Critical for angiogenesis, lymphangiogenesis, and vascular permeability regulation. Promotes endothelial cell migration and proliferation, supporting oxygen and nutrient delivery to regenerating tissue. Exerts effects via VEGF receptors (VEGF-1, VEGF-2, and neuropilin-1). Produced by keratinocytes, platelets, and immune cells in later wound stages. Upregulated by pro-inflammatory cytokines, cellular stress, and growth factors. Antagonized by Fra-1 transcription factor in specific scenarios (e.g., skin flap healing)
PI3K/AKT pathway	Active during inflammatory and proliferative phases, promoting cell survival, migration, and angiogenesis. Activated by microRNA-containing exosomes to enhance fibroblast function and epithelial repair
Wnt/β--catenin pathway	Regulates stem cell activation, fibroblast/keratinocyte proliferation, and granulation tissue formation. Potentially shifts chronic wounds toward regenerative healing
Nrf2 pathway	Induces antioxidant gene expression and protects against ROS-induced damage. Supports keratinocyte migration and ECM remodeling, ensuring enhanced healing outcomes
Notch pathway	Key in keratinocyte differentiation, macrophage phenotype switching (M1 to M2), and angiogenesis, promoting inflammation resolution and regeneration
HIF-1 pathway	Activates under hypoxic conditions to restore healing capacity in poorly vascularized wounds. Promotes VEGF-driven angiogenesis and fibroblast migration

ECM extracellular matrix, *EGF* epidermal growth factor, *PDGF* platelet-derived growth factor, *VEGF* vascular endothelial growth factor, *IL* interleukin, *TNF-α* tumor necrosis factor-alpha, *TGF-β* transforming growth factor-beta, *ROS* reactive oxygen species, *NF-κB* nuclear factor kappa B, *NLRP3* NOD-, LRR-, and pyrin domain-containing protein 3, *HIF-1* hypoxia-inducible factor-1, *IP3* inositol triphosphate, *IKK* I-kappa-B kinase, *NO* nitric oxide

Therapeutic Implications

Though the therapeutic use of growth factors and cytokines in wound healing is a very promising strategy, factors such as poor protein stability, short in vivo half-lives, and lack of sustained release at wound sites make its implementation difficult [38, 39]. To address these issues, novel topical and controlled release systems have been developed to enable localized, prolonged delivery of the biologic molecules mimicking the natural wound cascade.

Hydrogels are among the most studied carriers and mimic the ECM and provide a moist, protective environment for sustained GF delivery. EGF and FGF-loaded hydrogels have enhanced the process of angiogenesis, granulation tissue formation, and re-epithelialization [38]. Their high water content and porous structure allow them to mimic the natural ECM, thus facilitating cell migration, proliferation, and nutrient exchange. They also maintain adequate hydration at the wound site, which leads to accelerated epithelization, reduction in pain, and minimizes scab formation [40]. They also exhibit excellent biocompatibility and biodegradability. They possess properties like stretchability, adhesion, and self-healing, which make them suitable for use on movable areas like joints, where durability and flexibility are critical to prevent secondary injuries [41]. Advanced hydrogels also possess antimicrobial and antioxidant properties, which enhance their healing potential by preventing infection and reducing oxidative stress [41].

They can also be modified through custom physical and chemical cross-linking to control their structure, function, and degradation rate. This makes them an ideal option for treating acute and chronic wounds, burns, diabetic ulcers, and surgical incisions [40].

PDGF-BB delivered in carboxymethylcellulose hydrogel has shown superior effects in wound healing when compared to conventional topical application of the same [38]. Other systems include ECM-mimic materials, nanoparticles, and microspheres, which shield GFs from proteolytic degradation and ensure their gradual release. Natural polymers such as chitosan, fibrin, and hyaluronic-based hydrogels are being successfully used to deliver GF with prolonged bioavailability [39]. Nanoparticles offer targeted, controlled, and enhanced delivery of therapeutic agents while simultaneously protecting healing through their intrinsic properties. Their small size, accompanied by large surface area and customizable chemical characteristics, allows them to easily penetrate through tissue and interact with cellular and molecular pathways associated with healing [42]. Silver nanoparticles are known for their strong antimicrobial and anti-inflammatory effects. They also accelerate wound healing by promoting fibroblastic activity, keratinocyte migration, and myofibroblast differentiation while preventing the onset of infection [42].

Besides silver, copper (CuNP), gold (AuNP), and Zinc oxide (ZnO NP) have also had promising effects in chronic and diabetic wound models. While AuNP enhances re-epithelialization, CuNP mimics hypoxia to promote VEGF production, thus improving vascularity. Cerium oxide and silicon-based NPs offer antioxidant and pro-angiogenic effects, and these are of great value in oxidative stress-prone chronic wounds [43]. They also act as drug delivery systems for GFs, siRNA, and drugs. Their capability to encapsulate improves the stability of the drug and better release profiles, which improve the therapeutic outcome when compared to the free form of drugs [43].

Biomimetic and coacervate-based systems are yet another way to enhance wound healing through controlled, localized delivery of growth factors. Researchers have developed a cell-responsive delivery system by anchoring keratinocyte growth factor to fibrin matrices via peptides recognized by factor XIII and plasmin. This allows KGF to be released in response to cellular activity, particularly during the degeneration of the matrix system by plasmin, thus ensuring precise delivery of the drug as needed. In vitro, this led to a significant increase in epithelial cell proliferation and accelerated wound healing. In vivo, while using a humanized skin graft model in mice, fibrin-bound KGF led to faster and more complete re-epithelialization of wounds when compared to free KGF. This system mimicked the natural healing dynamic by enabling the KGF to persist in the wound environment and promoting its gradual release, thus supporting tissue regeneration and granulation tissue formation [44].

On the other hand, coacervate-based delivery systems, which utilize heparin-binding EGF-like growth factor, are formed by compounding heparin and a synthetic polycation, which protects it from degeneration and enables sustained release. In vitro, they enhance keratinocyte migration and maintain their proliferative potential. In a murine full-thickness wound model, they enabled accelerated wound closure, promoted early and complete re-epithelialization, and enhanced granulation tissue development and angiogenesis. When compared with free HB-EGF, they ensured optimal growth factor levels and bioactivity throughout the healing process [45].

Dual delivery systems have also been developed, and these show improved depth of protein penetration and duration of therapeutic action [38, 39]. Another strategy involves the engineering of GFs with ECM binding domains or fusion proteins, which improves their retention at the wound site while simultaneously protecting them from enzymatic breakdown [39].

Cytokines such as IL-10, TGF-β, and VEGF have been experimented with to provide a controlled release system at the wound site. IL-10, due to its short half-life, requires sustained delivery via biomaterials like hydrogels or ECM scaffolds [39, 46]. TGF-β has dual roles in wound healing. TGF-β1 is linked to fibrosis, and TGF-β3 is linked to scarring. TGF-β3 is delivered via injectable systems or ECM binding scaffolds, and these have shown promising results in modulating fibroblastic activity and wound remodeling without the use of excess ECM [39, 47]. Improved neovascularization of ischemic wounds is achieved by local delivery of VEGF through hydrogels and gene vectors [38, 39, 48].

Additionally, GM–CSF and G–CSF applied topically or subcutaneously enhance immune cell recruitment and epithelial cell repair in chronic wounds. However, the results have remained inconsistent due to poor retention at the wound site, which may be solved by effective delivery at the wound site [46, 48]. Cytokine-blocking strategies have also been under research, with the use of antibodies of RNAi against TNF-α, IL-6, and other pro-inflammatory mediators. Though these are more often used in autoimmune diseases, the use of chronic wounds may aid in the reduction of inflammation and promote healing, which is further improved when paired with GF therapy [46]. The key novel therapeutic systems and their strategies used in promoting wound healing are outlined in Table 9.3.

Table 9.3 Novel therapeutic systems and their strategies employed in wound healing

Therapeutic system	Mechanism	Benefits	Applications
Hydrogels	Mimic ECM provides a moist environment and supports cell migration and nutrient exchange. Modified with custom cross-linking for controlled release	Enhances angiogenesis, re-epithelialization, hydration, and pain reduction, and prevents scab formation	Acute/chronic wounds, burns, diabetic ulcers, and surgical incisions
PDGF-BB in hydrogels	Delivered via carboxymethylcellulose hydrogels for sustained release	Superior healing effects compared to conventional application	Chronic wounds, diabetic ulcers
Natural polymers (chitosan, fibrin, hyaluronic acid)	Deliver growth factors with prolonged bioavailability and shield against degradation	Enhanced stability and prolonged therapeutic action	Wound healing and tissue regeneration
Nanoparticles (e.g., silver, copper, gold)	Targeted, controlled delivery with intrinsic antimicrobial, anti-inflammatory, and angiogenic effects	Accelerate fibroblast/keratinocyte activity, prevent infection, and improve vascularity and granulation tissue	Chronic wounds, oxidative stress-prone conditions, and diabetic ulcers
Biomimetic delivery systems	Cell-responsive delivery anchoring growth factors to fibrin matrices	Gradual release supports epithelial proliferation, granulation tissue formation, and tissue regeneration	Chronic wounds, humanized skin graft models in vivo
Coacervate-based systems	Protect heparin-binding GFs using synthetic polycation, enabling sustained release	Promotes re-epithelialization, angiogenesis, and granulation tissue with prolonged GF bioactivity	Murine full-thickness wound models and chronic wounds
Dual delivery systems	Combine complementary delivery techniques for improved protein penetration and retention at wound sites	Enhanced depth and duration of therapeutic action	Complicated wounds require prolonged action
Cytokine delivery (IL-10, TGF-β, VEGF)	Hydrogels and ECM scaffolds provide controlled release; engineered forms with ECM-binding domains	Modulates inflammation, promotes fibroblastic activity, and neovascularization, and reduces fibrosis/scarring	Chronic wounds, ischemic wounds, diabetic ulcers
Cytokine blocking strategies	Antibodies/RNAi targeting TNF-α and IL-6 to reduce excessive inflammation	Promotes wound healing in chronic wounds when paired with growth factor therapy	Autoimmune-related chronic wounds, inflammatory wound conditions

ECM extracellular matrix, *PDGF-BB* platelet-derived growth factor-BB, *GF* growth factor, *KGF* keratinocyte growth factor, *HB-EGF* heparin-binding epidermal growth factor, *IL-10* interleukin-10, *TGF-β* transforming growth factor-beta, *VEGF* vascular endothelial growth factor, *RNAi* RNA interference, *TNF-α* tumor necrosis factor-alpha, *IL-6* interleukin-6

Several challenges are encountered in the translation of molecular insights into clinical applications of wound healing. These include:

1. The intricacy and complexity of the wound healing process, which includes various cellular and molecular compounds, make it difficult to precisely use therapeutic agents [5, 49, 50].
2. The presence of co-morbidities in patients complicates treatment. This also necessitates the use of diversified and personalized approaches to wound healing, making their development more challenging [51].
3. Current animal models do not fully replicate the complexity of human beings, making it difficult to translate the findings and results onto human beings [51].
4. Though various novel and innovative strategies have been developed, their implementation is hampered by challenges encountered in delivering the therapy or regulations preventing their implementation [5, 50].

Overcoming these hurdles requires a multidisciplinary approach, improved preclinical trials, and better regulations for the implementation of therapy and personalized therapy.

Conclusion

Cutaneous wound healing exemplifies a finely tuned interplay of cellular mechanisms, signaling pathways, and biomolecular interactions. Growth factors and cytokines orchestrate the repair process by regulating immune responses, cellular proliferation, ECM remodeling, and angiogenesis. Despite advancements in therapeutic strategies such as hydrogels, nanoparticles, and biomimetic systems, challenges, including comorbidities and translational hurdles, remain. Bridging research with clinical application demands a multidisciplinary approach, innovative delivery systems, and personalized therapies. Unlocking the intricacies of wound healing mechanisms holds the key to overcoming these obstacles and advancing effective, targeted treatments to enhance patient outcomes and revolutionize wound care practices globally.

References

1. Velnar T, Bailey T, Smrkolj V (2009) The wound healing process: an overview of the cellular and molecular mechanisms. J Int Med Res 37(5):1528–1542. https://doi.org/10.1177/147323000903700531
2. Robson MC, Steed DL, Franz MG (2001) Wound healing: biologic features and approaches to maximize healing trajectories. Curr Probl Surg 38(2):72–140. https://doi.org/10.1067/msg.2001.111167
3. Alonso JE, Lee J, Burgess AR, Browner BD (1996) The management of complex orthopedic injuries. Surg Clin North Am 76(4):879–903. https://doi.org/10.1016/s0039-6109(05)70486-2

4. Ganapathy N, Venkataraman SS, Daniel R, Aravind RJ, Kumarakrishnan VB (2012) Molecular biology of wound healing. J Pharm Bioallied Sci 4(Suppl 2):S334–S337. https://doi.org/10.4103/0975-7406.100294
5. Martin P, Nunan R (2015) Cellular and molecular mechanisms of repair in acute and chronic wound healing. Br J Dermatol 173(2):370–378. https://doi.org/10.1111/bjd.13954
6. Ling X, Xu W, Tang J, Cao Q, Luo G, Chen X et al (2024) The role of ubiquitination and the E3 ligase Nedd4 in regulating corneal epithelial wound healing. Invest Ophthalmol Vis Sci 65(6):29. https://doi.org/10.1167/iovs.65.6.29
7. Wilkinson HN, Hardman MJ (2020) Wound healing: cellular mechanisms and pathological outcomes. Open Biol 10(9):200223. https://doi.org/10.1098/rsob.200223
8. Jimi S, Saparov A, Takagi S (2021) Editorial: cellular and molecular mechanisms at the proliferation stage in wound healing: from scarring to tissue regeneration. Front Cell Dev Biol 9:659089. https://doi.org/10.3389/fcell.2021.659089
9. Wang M, Zhang J, Qiao C, Yan S, Wu G (2024) Comparative analysis of human and mouse transcriptomes during skin wound healing. Front Cell Dev Biol 12:1486493. https://doi.org/10.3389/fcell.2024.1486493
10. Bonnici L, Suleiman S, Schembri-Wismayer P, Cassar A (2023) Targeting signalling pathways in chronic wound healing. Int J Mol Sci 25(1):50. https://doi.org/10.3390/ijms25010050
11. Martin P, Nobes CD (1992) An early molecular component of the wound healing response in rat embryos--induction of c-fos protein in cells at the epidermal wound margin. Mech Dev 38(3):209–215. https://doi.org/10.1016/0925-4773(92)90054-n
12. Grose R, Harris BS, Cooper L, Topilko P, Martin P (2002) Immediate early genes krox-24 and krox-20 are rapidly up-regulated after wounding in the embryonic and adult mouse. Dev Dyn 223(3):371–378. https://doi.org/10.1002/dvdy.10064
13. Shaw T, Martin P (2009) Epigenetic reprogramming during wound healing: loss of polycomb-mediated silencing may enable upregulation of repair genes. EMBO Rep 10(8):881–886. https://doi.org/10.1038/embor.2009.102
14. Hardwicke J, Schmaljohann D, Boyce D, Thomas D (2008) Epidermal growth factor therapy and wound healing--past, present and future perspectives. Surg 6(3):172–177. https://doi.org/10.1016/s1479-666x(08)80114-x
15. Starkey RH, Cohen S, Orth DN (1975) Epidermal growth factor: identification of a new hormone in human urine. Science (New York, N.Y.) 189(4205):800–802. https://doi.org/10.1126/science.1172293
16. Franklin JD, Lynch JB (1979) Effects of topical applications of epidermal growth factor on wound healing. Experimental study on rabbit ears. Plast Reconstr Surg 64(6):766–770. https://doi.org/10.1097/00006534-197912000-00003
17. Wu L-W, Chen W-L, Huang S-M, Chan JY-H (2019) Platelet-derived growth factor-AA is a substantial factor in the ability of adipose-derived stem cells and endothelial progenitor cells to enhance wound healing. FASEB J 33(2):2388–2395. https://doi.org/10.1096/fj.201800658R
18. Rajkumar VS, Shiwen X, Bostrom M, Leoni P, Muddle J, Ivarsson M et al (2006) Platelet-derived growth factor-β receptor activation is essential for fibroblast and pericyte recruitment during cutaneous wound healing. Am J Pathol 169(6):2254–2265. https://doi.org/10.2353/ajpath.2006.060196
19. Kaltalioglu K, Coskun-Cevher S (2015) A bioactive molecule in a complex wound healing process: platelet-derived growth factor. Int J Dermatol 54(8):972–977. https://doi.org/10.1111/ijd.12731
20. Lee JA, Conejero JA, Mason JM, Parrett BM, Wear-Maggitti KD, Grant RT et al (2005) Lentiviral transfection with the PDGF-B gene improves diabetic wound healing. Plast Reconstr Surg 116(2):532–538. https://doi.org/10.1097/01.prs.0000172892.78964.49
21. Man L-X, Park JC, Terry MJ, Mason JM, Burrell WA, Liu F et al (2005) Lentiviral gene therapy with platelet-derived growth factor B sustains accelerated healing of diabetic wounds over time. Ann Plast Surg 55(1):81–86; discussion 86. https://doi.org/10.1097/01.sap.0000168211.76318.fa

22. Mahmoud NN, Hamad K, Al Shibitini A, Juma S, Sharifi S, Gould L et al (2024) Investigating inflammatory markers in wound healing: understanding implications and identifying artifacts. ACS Pharmacol Transl Sci 7(1):18–27. https://doi.org/10.1021/acsptsci.3c00336
23. Cavaillon J-M, Adib-Conquy M (2002) The pro-inflammatory cytokine cascade. In: Marshall JC, Cohen J (eds) Immune response in the critically ill. Springer, Berlin, Heidelberg, pp 37–66. https://doi.org/10.1007/978-3-642-57210-4_4
24. Hübner G, Brauchle M, Smola H, Madlener M, Fässler R, Werner S (1996) Differential regulation of pro-inflammatory cytokines during wound healing in normal and glucocorticoid-treated mice. Cytokine 8(7):548–556. https://doi.org/10.1006/cyto.1996.0074
25. Lopes FB, Sarandy MM, Novaes RD, Valacchi G, Gonçalves RV (2024) OxInflammatory responses in the wound healing process: a systematic review. Antioxidants (Basel, Switzerland) 13(7):823. https://doi.org/10.3390/antiox13070823
26. Ismaeel A, Kim J-S, Kirk JS, Smith RS, Bohannon WT, Koutakis P (2019) Role of transforming growth factor-β in skeletal muscle fibrosis: a review. Int J Mol Sci 20(10):2446. https://doi.org/10.3390/ijms20102446
27. Kim KK, Sheppard D, Chapman HA (2018) TGF-β1 signaling and tissue fibrosis. Cold Spring Harb Perspect Biol 10(4):a022293. https://doi.org/10.1101/cshperspect.a022293
28. Ren L-L, Li X-J, Duan T-T, Li Z-H, Yang J-Z, Zhang Y-M et al (2023) Transforming growth factor-β signaling: from tissue fibrosis to therapeutic opportunities. Chem Biol Interact 369:110289. https://doi.org/10.1016/j.cbi.2022.110289
29. Shull MM, Ormsby I, Kier AB, Pawlowskr S, Diebold RJ, Yin M et al (1992) Targeted disruption of the mouse transforming growth factor-β1 gene results in multifocal inflammatory disease. Nature 359(6397):693–699. https://doi.org/10.1038/359693a0
30. Hong Y-K, Lange-Asschenfeldt B, Velasco P, Hirakawa S, Kunstfeld R, Brown LF et al (2004) VEGF-A promotes tissue repair-associated lymphatic vessel formation via VEGFR-2 and the alpha1beta1 and alpha2beta1 integrins. FASEB J 18(10):1111–1113. https://doi.org/10.1096/fj.03-1179fje
31. Johnson KE, Wilgus TA (2014) Vascular endothelial growth factor and angiogenesis in the regulation of cutaneous wound repair. Adv Wound Care 3(10):647–661. https://doi.org/10.1089/wound.2013.0517
32. Eming SA, Krieg T (2006) Molecular mechanisms of VEGF-A action during tissue repair. J Investig Dermatol Symp Proc 11(1):79–86. https://doi.org/10.1038/sj.jidsymp.5650016
33. Krizanova O, Penesova A, Sokol J, Hokynkova A, Samadian A, Babula P (2022) Signaling pathways in cutaneous wound healing. Front Physiol 13:1030851. https://doi.org/10.3389/fphys.2022.1030851
34. Pils V, Terlecki-Zaniewicz L, Schosserer M, Grillari J, Lämmermann I (2021) The role of lipid-based signalling in wound healing and senescence. Mech Ageing Dev 198:111527. https://doi.org/10.1016/j.mad.2021.111527
35. Barrientos S, Stojadinovic O, Golinko MS, Brem H, Tomic-Canic M (2008) Growth factors and cytokines in wound healing. Wound Repair Regen 16(5):585–601. https://doi.org/10.1111/j.1524-475X.2008.00410.x
36. Werner S, Grose R (2003) Regulation of wound healing by growth factors and cytokines. Physiol Rev 83(3):835–870. https://doi.org/10.1152/physrev.2003.83.3.835
37. Behm B, Babilas P, Landthaler M, Schreml S (2012) Cytokines, chemokines and growth factors in wound healing. J Eur Acad Dermatol Venereol 26(7):812–820. https://doi.org/10.1111/j.1468-3083.2011.04415.x
38. Nurkesh A, Jaguparov A, Jimi S, Saparov A (2020) Recent advances in the controlled release of growth factors and cytokines for improving cutaneous wound healing. Front Cell Dev Biol 8:638. https://doi.org/10.3389/fcell.2020.00638
39. Briquez PS, Hubbell JA, Martino MM (2015) Extracellular matrix-inspired growth factor delivery systems for skin wound healing. Adv Wound Care 4(8):479–489. https://doi.org/10.1089/wound.2014.0603

40. Huang C, Dong L, Zhao B, Lu Y, Huang S, Yuan Z et al (2022) Anti-inflammatory hydrogel dressings and skin wound healing. Clin Transl Med 12(11):e1094. https://doi.org/10.1002/ctm2.1094
41. Yu P, Wei L, Yang Z, Liu X, Ma H, Zhao J et al (2024) Hydrogel wound dressings accelerating healing process of wounds in movable parts. Int J Mol Sci 25(12):6610. https://doi.org/10.3390/ijms25126610
42. Nqakala ZB, Sibuyi NRS, Fadaka AO, Meyer M, Onani MO, Madiehe AM (2021) Advances in nanotechnology towards development of silver nanoparticle-based wound-healing agents. Int J Mol Sci 22(20):11272. https://doi.org/10.3390/ijms222011272
43. Blanco-Fernandez B, Castaño O, Mateos-Timoneda MÁ, Engel E, Pérez-Amodio S (2021) Nanotechnology approaches in chronic wound healing. Adv Wound Care 10(5):234–256. https://doi.org/10.1089/wound.2019.1094
44. Johnson NR, Wang Y (2013) Controlled delivery of heparin-binding EGF-like growth factor yields fast and comprehensive wound healing. J Control Release 166(2):124–129. https://doi.org/10.1016/j.jconrel.2012.11.004
45. Geer DJ, Swartz DD, Andreadis ST (2005) Biomimetic delivery of keratinocyte growth factor upon cellular demand for accelerated wound healing in vitro and in vivo. Am J Pathol 167(6):1575–1586
46. Berry-Kilgour C, Cabral J, Wise L (2021) Advancements in the delivery of growth factors and cytokines for the treatment of cutaneous wound indications. Adv Wound Care 10(11):596–622. https://doi.org/10.1089/wound.2020.1183
47. Legrand JMD, Martino MM (2022) Growth factor and cytokine delivery systems for wound healing. Cold Spring Harb Perspect Biol 14(8):a041234. https://doi.org/10.1101/cshperspect.a041234
48. Yamakawa S, Hayashida K (2019) Advances in surgical applications of growth factors for wound healing. Burns Trauma 7:10. https://doi.org/10.1186/s41038-019-0148-1
49. Rodrigues M, Kosaric N, Bonham CA, Gurtner GC (2019) Wound healing: a cellular perspective. Physiol Rev 99(1):665–706. https://doi.org/10.1152/physrev.00067.2017
50. Eming SA, Martin P, Tomic-Canic M (2014) Wound repair and regeneration: mechanisms, signaling, and translation. Sci Transl Med 6(265):265sr6. https://doi.org/10.1126/scitranslmed.3009337
51. Pastar I, Wong LL, Egger AN, Tomic-Canic M (2018) Descriptive vs mechanistic scientific approach to study wound healing and its inhibition: is there a value of translational research involving human subjects? Exp Dermatol 27(5):551–562. https://doi.org/10.1111/exd.13663

Chapter 10
Inflammation, Fibrosis, and Innate Immune Response in Diabetic Foot Ulcer

Gurkaranjot Singh, Zuhair Chaudhry, Sankalp Nigam, Dominick Shoha, and Vikrant Rai

Abstract Diabetic foot ulcers (DFUs) are serious complications of diabetes, characterized by chronic wounds that do not heal due to peripheral neuropathy, ischemia, and prolonged hyperglycemia. These ulcers greatly diminish the quality of life and lead to high lower-limb amputation rates and healthcare costs. The pathogenesis of diabetic foot ulcers involves systemic metabolic imbalance with local immune abnormalities contributing to impaired healing. Chronic inflammation, reduced growth factors, impaired angiogenesis, and extracellular matrix (ECM) remodeling defects lead to nonhealing DFUs. Immune dysregulation by hyperglycemia increases inflammation, exacerbated by impaired macrophage polarization and neutrophil function. States of oxidative stress lead to decreased endothelial function, whereas ECM degradation due to imbalances in matrix metalloproteinases impedes the formation and structural integrity of granulation tissue. DFUs frequently get infected with treatment, which gets complicated by biofilm formation containing pathogenic bacteria. This chapter discusses the complex interplay of inflammation, innate immune responses, and fibrosis in DFUs, highlighting the need for new drugs to tackle this health problem. A comprehension of molecular mechanisms that underlie the pathogenesis of DFU will help in creating interventions that are specific to enhance results for patients and eliminate the socioeconomic burden associated with complications from diabetes. Proper knowledge and comprehension of these mechanisms are essential for the development of targeted therapeutics and effective treatment options.

G. Singh · Z. Chaudhry · S. Nigam · D. Shoha
College of Osteopathic Medicine of the Pacific, Western University of Health Sciences, Pomona, CA, USA

V. Rai (✉)
Department of Translational Research, Western University of Health Sciences, Pomona, CA, USA
e-mail: vrai@westernu.edu

F. De Francesco, K. Turksen (eds.), *Management and Strategies for Wound Healing*, Stem Cell Biology and Regenerative Medicine 79,
https://doi.org/10.1007/978-3-032-12442-5_10

Keywords Diabetic foot ulcer · Chronic inflammation · Angiogenesis · Extracellular matrix · Immune response · Fibrosis

Introduction

Diabetic foot ulcers (DFUs) constitute an extreme and debilitating complication of diabetes mellitus (DM). DFUs are described as chronic wounds or ulcers on the feet that originate from a combination of diabetic peripheral neuropathy, peripheral arterial disease, and prolonged hyperglycemia [13]. Simple wear and tear of the foot or minor, unnoticed injuries cause ulcers in compromised tissue [50]. Peripheral neuropathy associated with DFUs decreases the sensation in the lower extremity. This causes progressive worsening of the tissue since the patient is unaware of the added pressures and forces that are being applied to the diabetic foot [28]. Furthermore, peripheral artery disease due to damage to endothelial cells and structural changes in the presence of hyperglycemia creates sclerotic vessels in association with various factors, such as the location of the sclerotic vessels, the severity of dyslipidemia, obesity, smoking, and family history [28, 57]. Decreased arterial elasticity, luminal diameter, endothelial damage, endothelial cell dysfunction, and vascular sclerosis decrease the blood flow to the lower extremities. As a result, a restrictive environment is created, preventing oxygen and nutrients from reaching the tissue. This leads to blue/purple skin, atrophic skin, and lower surface skin temperature, all facilitating the formation of foot ulcers.

The wound that appears on the skin surface is the primary hallmark of advanced diabetes. It further places itself at the center of the very serious global health problem due to its chronicity, non-healing, and associated morbidity and mortality. Wagner's Classification for DFUs is based on the severity of the advanced-stage infection, gangrene, or osteomyelitis [64], and higher stages are associated with multiple complications. Non-healing, recurrence, and amputation are common complications of DFUs primarily due to poor angiogenesis, persistent inflammation, and altered extracellular matrix formation. Foot ulcers result in both high rates of lower-limb amputations and significantly lower quality of life for those affected [53].

The World Health Organization (WHO) reports that the prevalence of diabetes was more than 537 million adults globally in 2022 and is forecast to rise to 783 million by 2045 [42]. This population develops DFU between 19% and 34% during their lifetime, with about 20% experiencing DFU at some point in life [44]. DFU's prevalence varies across regions worldwide, ranging from 10% to 30% in African countries. In Europe, DFU ranges from 1% to 17%. However, in the US, DFU frequency increased from 8% in 2013 to 52% in 2018 [20]. DFUs pose an issue as they are non-healing; the wound can remain unhealed for many months or even years without appropriate management. The longer a wound persists, the greater the risk of infection and sepsis, as well as amputation.

DFUs plague healthcare systems worldwide. In just the United States, DFUs already impose an estimated $9–13 billion per year in direct healthcare costs, on top of the already astonishing $327 billion spent on diabetes management [58]. More inpatient admissions accompanied by care in specialized wound care centers, surgical interventions, and long-term rehabilitation care are some of the services available for DFU patients. An economic burden is added by the indirect costs due to lost productivity through disability and premature death. Countries with low- to middle-income often have underdeveloped healthcare infrastructures, increasing the burden. The low availability of early diagnostic tools, advanced wound care products, and surgical expertise leads to delayed patient treatment, with some cases resulting in worse outcomes. Families face an immense financial burden, with most of them paying directly for the healthcare costs placed on them. This is not even supposed to be effective, considering the delayed presentation. These flag DFUs around the world are not just a medical issue but also an outstanding social crisis that sooner needs concern from healthcare and public policymakers [79].

Pathogenesis of Nonhealing Diabetic Ulcers

Diabetic foot ulcers (DFUs) are an example of the complex interplay between systemic metabolic abnormalities and the local immune response within the wound microenvironment. Chronic hyperglycemia triggers adverse biochemical processes that impair wound healing through multiple mechanisms. Diabetic foot ulcers exhibit impaired wound healing because of the interaction between systemic metabolic dysfunction and abnormalities within the local microenvironment of the wound. The orderly process of normal wound healing involves three interrelated and overlapping phases, which are inflammation, proliferation, and remodeling. However, in diabetic individuals, this cascade of wound healing is impaired at every step, prolonging the inflammatory phase because of chronic hyperglycemia and its downstream effects [38, 55, 69]. A complex interaction between systemic metabolic disturbances and local immune defects due to chronic hyperglycemia impairs immune cell function, promotes advanced glycation end-product formation, causes endothelial and fibroblast functional defects, and disrupts the balance between proteases, leading to impaired wound healing in diabetic patients [54, 66]. This all together leads to the chronicity of inflammation, decreased angiogenesis, and altered ECM formation (Fig. 10.1), causing nonhealing, recurrence, and amputation.

Chronic Inflammation

Chronic inflammation is a major hallmark feature of impaired wound healing in diabetes. In normal wound healing, inflammation occurs in the acute phase, which resolves rather quickly once phenotypic change occurs with macrophages; it

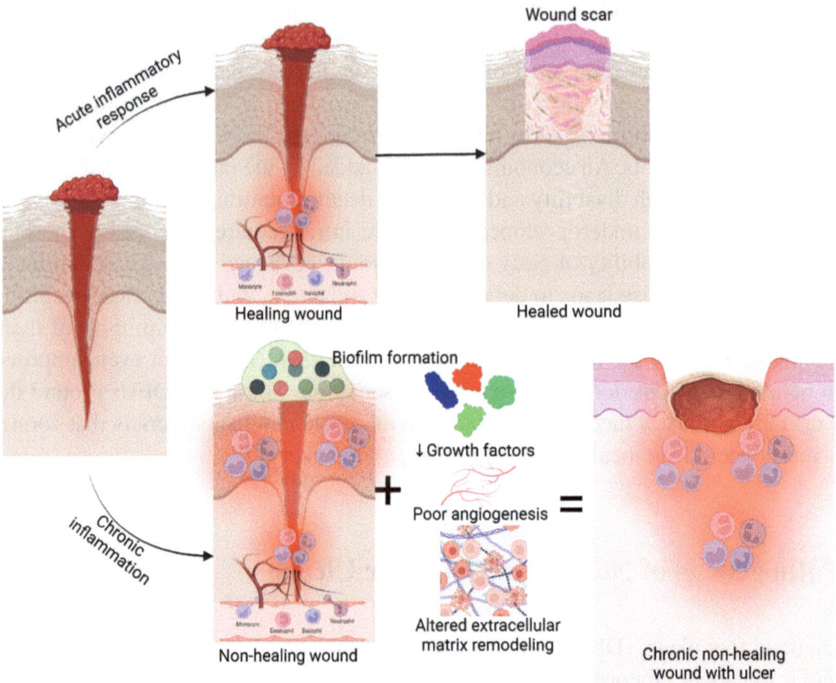

Fig. 10.1 Pathogenesis of healing and chronic nonhealing diabetic wound. Chronicity of inflammation, decreased angiogenesis, and altered ECM formation, along with altered fibroblasts, endothelial cells, and vascular smooth muscle cell function, result in chronic nonhealing wounds with ulcers

transitions from the pro-inflammatory (M1) phenotype into the anti-inflammatory (M2) phenotype to aid in promoting tissue repair. In diabetic wounds, however, sustained hyperglycemia disrupts immune cells by delaying or preventing this phenotypic transition. The inability of M1 to become an M2 macrophage is both a cause and consequence of the inflammation. When compared with the healthy counterpart, levels of pro-inflammatory cytokines such as IL-1β and TNFα are elevated in the diabetic wound condition, maintaining a high state of inflammation [7]. Hyperglycemia also impairs the adaptive immune response by causing T cells to differentiate toward a pro-inflammatory phenotype and reducing regulatory T cells (Tregs) essential for wound healing. This imbalance promotes a persistent inflammatory state that is detrimental to tissue repair [51]. Delayed healing is also facilitated by the altered function of neutrophils and innate immune responses. In diabetes, neutrophils exhibit impaired chemotaxis and phagocytosis, reducing the body's ability to eradicate pathogens and clear wound debris. This neutrophil dysfunction increases susceptibility to infection and delays the healing process [16]. Increased secretion of pro-inflammatory cytokines plays a major role in persistent inflammation.

Decreased Growth Factors

Vascular endothelial growth factor (VEGF), platelet-derived growth factor (PDGF), epidermal growth factor (EGF), and fibroblast growth factor (FGF) regulate the proliferation and differentiation of fibroblasts, keratinocytes, vascular smooth muscle cells, and endothelial cells regulate angiogenesis, fibroblast proliferation, and the development of the extracellular matrix during the process of wound healing [68, 69]. In diabetes, due to oxidative stress and advanced glycation end product formation, the availability of these growth factors is significantly reduced, and it hampers the cellular signaling pathways mandatory for regeneration and repair of tissues, resulting in nonhealing and chronicity of wounds.

Impaired Angiogenesis

In a normal, healthy wound, angiogenesis, regulated by various pro-and anti-angiogenic factors provides the supply line for oxygen and nutrients to the healing tissue [53, 55]. Angiogenesis plays a major role in wound healing in patients with DFUs because the formation of new blood vessels allows more blood supply into the region to deliver oxygen and nutrients necessary for proper healing. Specifically, studies have highlighted the downregulation of angiogenic growth factors VEGF and SDF-1α [61]. Nrf2, a transcription factor normally involved in the antioxidant response, is also an upstream factor involved in VEGF signaling [37]. When Nrf2 is knocked out, VEGF has reduced expression as well as HO-1, both of which lead to decreased angiogenesis. It was found that Nrf2 had a positive correlation with miRNA-27b, meaning that decreased expression of miRNA-27b, reduced Nrf2 expression and ultimately affected angiogenesis [61]. These impairments affect pro-angiogenic signaling and limit new vessel formation, which is vital in supplying nutrients and oxygen to the wound.

However, in wounds of diabetics, this vital event is impaired by the dysfunction of the endothelial cell in hyperglycemia, increased expression of anti-angiogenic factors, and decreased expression of pro-angiogenic factors. Endothelial cells exposed to high glucose concentration have low production of nitric oxide and are under oxidative stress. Such a condition makes these cells unresponsive to the pro-angiogenic effects of vascular endothelial growth factor (VEGF) and remain hypoxic for a long period due to poor angiogenesis. This results in delayed wound healing and increases the probability of necrosis or infection for the patient [45]. Endothelial dysfunction caused by hyperglycemia results in reduced bioavailability of nitric oxide, which hinders vasodilation and angiogenesis. This vascular disorder limits oxygen and nutrient delivery to the wound and impedes tissue regeneration [19]. Therefore, both molecular dysregulation and endothelial dysfunction increase the susceptibility to complications in diabetic wounds due to impaired angiogenesis.

Altered ECM Remodeling

During normal wound healing, the ECM provides structural support to cells, and at the same time acts as a reservoir for several growth factors and cytokines. ECM, largely composed of collagen, remodeling is a must for angiogenesis and proper wound healing; however is impaired in diabetic foot ulcers [25]. Decreased collagen production, an increase in MMPs, and disruption in the balance between matrix metalloproteinases (MMPs) and tissue inhibitors of matrix metalloproteinases (TIMPs) associated with chronic inflammation lead to the degradation of collagen, imbalances the matrix laydown and remodeling of the ECM in diabetic wounds due to altered collagen regulation [25]. The imbalance of MMP activity over TIMPs in diabetic wounds results in excessive degradation of ECM components. This degradation disrupts the structural integrity of the wound bed and impairs epithelial regeneration and the formation of granulation tissue [5, 19, 67]. Diabetic fibroblasts have reduced proliferation and synthetic activity, resulting in impaired extracellular matrix production and delayed wound contraction. These cellular dysfunctions impede the formation of granulation tissue, which is required for effective wound healing [19]. Further. Lower levels of fibronectin, which is involved in clotting with fibrin as well as overall wound healing by interacting with various cell types, such as lysyl oxidase (an enzyme that helps crosslink collagen fibers for ECM restoration) are decreased in diabetics. Fibroblasts help secrete proteins and collagen used in the ECM, which is essential for proper wound healing. In diabetic patients, however, fibroblasts can become senescent, rendering them ineffective and creating incomplete, dysfunctional ECM components [19]. Diabetes presents a wound environment that is non-healing due to chronic inflammation, protease imbalance, and fibroblast dysfunction.

Infectious Complications

Besides being chronically inflamed, DFUs are also a major site for infections as well as generalized tissue damage. DFUs are often found with the loss of sweat and oil glands in the lower extremity, which makes the affected foot keratinized and dry, making it more susceptible to fissures in the skin. This opening in the cutaneous tissue creates more entry spots for infection to enter the body [49]. Infections related to DFUs consist of both gram-positive (most commonly *S. aureus*) and gram-negative bacteria and enter the broken tissues on the foot. They then associate and wrap themselves with extracellular polymers to form biofilms in the ulcer, which have high resistance to antimicrobial agents [28]. It is also worth noting that patients with diabetes tend to have compromised immune systems. Therefore, if there is a budding infection that is accumulating in the diabetic foot, the weakened immune system is unable to fight it effectively due to reduced numbers of leukocytes. Additionally, the hyperglycemic state of diabetic patients provides a favorable environment for microbes to grow, making infections even more plausible for DFUs. Lastly, DFUs can lead to states that can be extremely difficult to treat due to possible pathogenic bacterial colonization in the affected extremity. This colonization leads to the formation of a biofilm consisting of a polymer-based matrix that embeds the bacterial cells and renders them resistant to host immune defenses as well as administered antibiotics. Overall, biofilm formation leads to states of chronic inflammation and infection through the promotion of wound inflammation and resistance, which allows for further progression of the DFU [1].

Both aerobic and anaerobic bacteria play a role in the infection and biofilm formation in DFUs. The microbiological spectrum is broad and complex. The most common isolate in all studies done so far is *Staphylococcus aureus*, which was found in between 17.7% and 34.34% of all isolates. These gram-positive cocci can be highly virulent and have a high prevalence of methicillin resistance. *Escherichia coli* (10.9–21.16%), *Pseudomonas aeruginosa* (10.5–18.89%), and Proteus species (especially *Proteus mirabilis*) were the most frequent gram-negative isolates after *S. aureus* [4]. An epidemiological survey conducted recently indicates a change in the distribution of pathogens, where the prevalence of the gram-negative bacteria (52.4–66.7%) is greater than that of gram-positive organisms (34.16–43.4%) in most regions, particularly hot climate areas and developing countries [17]. Due to explicit resistance mechanisms demonstrated by most gram-negative species, this has substantial implications for therapy. 60–70% of infections in DFU cases are polymicrobial, involving several species of bacteria, which act synergistically to promote pathogenicity or establish a biofilm [4]. The aerobic and anaerobic bacteria in the polymicrobial communities perpetuate the continued existence of the microbiota in DFU. Aerobic bacteria account for an increasing proportion of the total bacterial count in infected DFUs. High rates of multidrug-resistant organisms (MDROs) have been reported in DFUs, with reports of up to 92.9% resistance among isolates to more than one antibiotic [4]. This will also associate the bacterial profile with the severity of ulcers because deep and enduring wounds tend to reflect a more diversified bacterial profile, including Gram-negative bacteria. Therefore, the clinical practitioners involved in the management of infections in patients with DFU are challenged by this changing microbial scenario, which further emphasizes the importance of acquiring appropriate cultures before starting antimicrobial therapy.

In addition to biofilm formation, tissue necrosis, infection spreading to surrounding tissues or the bloodstream (sepsis), and the development of osteomyelitis (bone infection) are other complications of infected DFUs. The spread of infection to deeper tissues leads to cellulitis, tissue necrosis, and potentially bone infection (osteomyelitis). In more severe cases, the infection may spread to the bloodstream, causing sepsis, a life-threatening condition. Bone infection can lead to chronic pain, delayed healing, and potentially require surgical intervention or even amputation [48]. Type of diabetes, duration of diabetes, mean age, ulcer severity (depth), polymicrobial infections, socioeconomic status, mean age, smoking, along with high BMI, body, and visceral fats play a critical role in the infection of DFUs [31]. In addition to these factors, as mentioned above, peripheral arterial disease, neuropathy, poor glycemic control, and poor foot hygiene also play a critical role in foot infection. Redness, heat, swelling, pain around the ulcer, purulent drainage (pus) from the wound, and systemic symptoms like fever, chills, and general malaise are common clinical symptoms of infected DFUs [6].

Immune Response During Wound Healing

Before exploring inflammation in diabetic foot ulcers, it is important to briefly review the normal inflammatory response in acute wound healing. With acute wounds, the inflammatory response is characterized first by neutrophils rushing to the wound area, phagocytosing damaged extracellular matrix components, foreign substances, and necrotic cells [2]. Neutrophil numbers tend to peak around day 2 following the wound and decrease from then onwards. From days 2 to 7, a larger number of monocytes and macrophages are recruited to clear out remaining cellular debris, remove any pathogens, and clear neutrophils that contain digested microbes (Fig. 10.2). Following this, they begin to initiate the repair process of the wound through the secretion of cytokines, as well as factors such as VEGF, which help promote the blood supply needed for restoration to the area, as well as attracting fibroblasts to help rebuild the damaged tissue [2]. Resolution of inflammation is an active process that includes the generation of specialized pro-resolving mediators (SPMs) like resolvins, protectins, and maresins. These mediators enhance the removal of apoptotic neutrophils by macrophages (efferocytosis) and modulate the phenotype of macrophages from pro-inflammatory to pro-resolving (Fig. 10.2). In diabetic wounds, including diabetic foot ulcers (DFUs), defective SPM production disrupts these pathways, leading to persistent neutrophil accumulation, prolonged M1 macrophage dominance, and unchecked pro-inflammatory signaling (e.g., NF-κB activation, TNF-α/IL-1β release). This failure creates a self-perpetuating cycle of chronic inflammation, characterized by protease-rich microenvironments, oxidative stress, and impaired angiogenesis, which directly drives non-healing and therapeutic resistance [8, 41].

During wound healing, the first line of defense against pathogens is comprised of a network of cellular and molecular components, coordinating the early phases of wound repair [41]. Neutrophils, macrophages, and dendritic cells, key plates in this process, work together to eliminate microbial threats and promote tissue regeneration [41, 63]. Pathogen-associated molecular patterns (PAMPS) and damage-associated molecular patterns (DAMPs) activate pattern recognition receptors (PRRs) during injury [43]. These PRRs include Toll-Like, NOD, and C-type lectin receptors. As a result, proinflammatory cytokines are released, attracting neutrophils and macrophages to the wound site [34, 43]. Usually, within hours of an injury, neutrophils are among the first immune cells to reach the wound site and play an essential role in the removal of pathogens through phagocytosis, degranulation, and the creation of neutrophil extracellular traps (NETs) [41, 64]. All stages of wound healing, including proliferation, inflammation, and remodeling, depend on

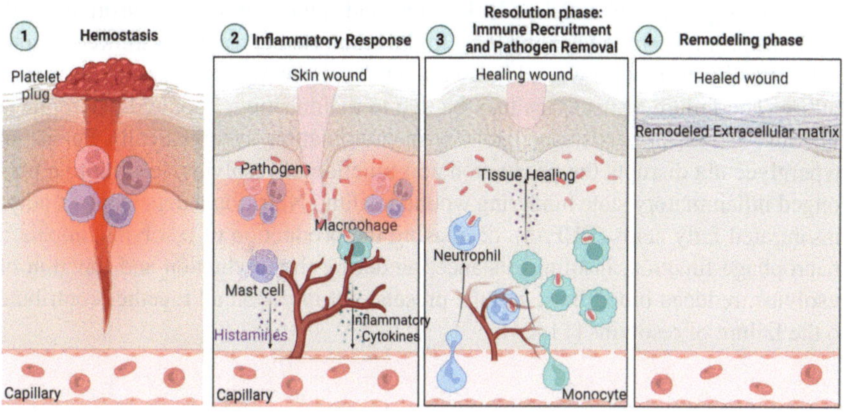

Fig. 10.2 Immune response in normal wound healing. The process includes four sequential phases: hemostasis, inflammation, resolution, and remodeling. Neutrophils and macrophages are recruited to the wound site, where they secrete cytokines to guide immune cell clearance, pathogen removal, and tissue repair. Fibroblasts and vascular smooth muscle cells contribute to extracellular matrix remodeling and wound closure

macrophages following neutrophils in arrival [34]. They contribute to the elimination of pathogens, removal of debris, and synthesis of growth factors that facilitate tissue repair [34]. During normal physiological wound healing, macrophages experience a phenotypic transition from pro-inflammatory (M1) to anti-inflammatory (M2) phenotypes, and this is essential for the transition from the inflammatory to the proliferative phase of wound healing by attenuating inflammation, ultimately driving tissue repair [8, 34, 41].

Dendritic cells (DCs) form a link between innate and adaptive immune responses. DCs break down and present antigens to T cells, activating the adaptive immune response [43, 70], and are involved in the activation of innate and adaptive immune systems in case of acute injury during wound healing. More recent studies have also pointed out the role played by innate lymphoid cells (ILCs) in wound healing. ILC2s, which are activated by IL-33, promote re-epithelialization and wound healing. Invariant NK T cells assist in wound healing by inducing the production of IFN-γ, the release of growth factors such as VEGF and TGF-β, greater collagen deposition, myofibroblast differentiation, and promotion of blood vessel formation [43]. The coordinated action of these innate immune elements is very important in successful wound healing. In the diabetic state, this balance is lost, and healing is impaired, with a heightened susceptibility to infection [70].

In a healing wound, innate immune responses and innate microbial killing and clearance may be enhanced by pro-resolving mediators called "resolvins". Resolvins are part of a group of specialized pro-resolving lipid mediators (SPM), including lipoxins (LX), resolvins (RV), protectins (PD), and maresins (MR). A decreased infiltration of neutrophils, transmigration of immune cells, adipose tissue macrophages, and platelet activation, increased phagocytic activity of macrophages and apoptosis of polymorphonuclear leukocytes, and improved insulin sensitivity with resolvins promote wound healing [66]. Initial results using resolvins showed beneficial results by promoting wound healing in DFUs. However, there are incidences of failure. The failure of resolvins may be due to malnutrition, impaired macrophage function, and insulin resistance, alone or in combination, synergistically. Persistent hyperglycemia disrupts the production and function of resolvins, leading to a prolonged inflammatory state hindering wound healing. Nutritional deficiency of polyunsaturated fatty acids (PUFAs), decreasing the production of resolvins, impaired macrophage function, insulin resistance, hindering the production and function of resolvins, reduced blood flow, and the presence of infection all together contribute to the failure of resolvins [11, 24].

Immune Response in Diabetic Wounds

In diabetic foot ulcers (DFUs), the immune response is profoundly dysregulated, leading to impaired neutrophil and macrophage function, heightened susceptibility to infection, and persistent biofilm formation [60, 63]. This immune dysfunction drives chronic inflammation and inhibits resolution, as feedback loops between immune cells, cytokines, and damaged tissue perpetuate the inflammatory state [8, 41]. Hyperglycemia further exacerbates these effects by promoting atherosclerosis and impairing nutrient and oxygen delivery to the wound [7]. It also disrupts protein synthesis, keratinocyte migration and proliferation, endothelial cell function, and cytoskeletal organization, all of which are necessary for re-epithelialization and effective wound closure [38]. Additionally, hyperglycemia induces excessive reactive oxygen species (ROS) production via the protein kinase C, polyol, and AGE pathways. While low ROS levels can support early wound healing, excessive ROS in diabetic patients causes peripheral nerve damage, vascular dysfunction, and oxidative stress, thereby hindering wound repair and contributing to non-healing ulcers [7].

Neutrophil dysfunction is a defining characteristic of DFUs. These innate immune cells typically play a critical role in early wound defense through chemotaxis, phagocytosis, and the formation of neutrophil extracellular traps (NETs). However, in diabetic conditions, neutrophils exhibit impaired chemotaxis, reduced phagocytic activity, and defective NET formation [79]. These impairments have been linked to reduced expression of the transcription factor FOXM1, which is essential for neutrophil recruitment and activation, and TREM1, a neutrophil-specific modulator of inflammation. Together, the TREM1/FOXM1 pathway is emerging as a critical regulator of NET formation and neutrophil function in diabetic wound healing [63]. As a result, neutrophils in DFUs demonstrate delayed migration to the wound site and an impaired capacity to clear pathogens and cellular debris. This leaves the wound susceptible to infection and contributes to sustained inflammation, which is a common and serious complication in DFU pathology [60]. The chronic presence of dysfunctional neutrophils amplifies inflammation and leads to overproduction of ROS, pro-inflammatory cytokines, and proteases, further damaging tissue and impairing repair mechanisms. The failure of neutrophils to transition appropriately also affects the recruitment and polarization of macrophages, which represent another key component of immune dysfunction in DFUs.

Macrophages in healthy wound healing follow a temporal transition from pro-inflammatory M1 to anti-inflammatory M2 phenotypes. In DFUs, this M1-to-M2 transition is often delayed or completely disrupted, prolonging the inflammatory phase and delaying entry into tissue repair [34]. DFU macrophages exhibit abnormal polarization profiles, impaired phagocytic function, and persistent production of inflammatory cytokines such as TNF-α, IL-1β, and IL-6 [8, 40, 41]. These cytokines establish a self-sustaining loop of inflammation that recruits more immune cells and inhibits wound progression. Furthermore, impaired efferocytosis—the process by which macrophages clear apoptotic neutrophils—contributes to a build-up of inflammatory debris and cellular stress, exacerbating wound chronicity.

The inflammatory landscape in DFUs is further intensified by the abnormal persistence of mast cells. While these cells are normally active in early wound healing, in DFUs, they persist abnormally and maintain a pro-inflammatory state [46]. Mast cells, in conjunction with elevated levels of TGF-β, CXCL1, and IL-36γ, promote increased infiltration of neutrophils and macrophages into the wound bed. Although these immune cells are essential for initiating wound healing, their excessive and prolonged activity in DFUs leads to further disruption of tissue repair processes.

Along with neutrophil and macrophage dysfunction, T-cell dysregulation also plays a significant role in perpetuating inflammation in DFUs. Helper T cells in diabetic wounds exhibit overexpression of the CXCR3 ligand, which promotes excessive T-cell activation and cytokine secretion [59]. In addition, effector T cells accumulate in DFUs and contribute to a heightened pro-inflammatory milieu. CCR4, a chemokine receptor involved in the migration of regulatory T cells (Tregs), is also upregulated. This paradoxically increases circulating Tregs while reducing their suppressive activity, leading to continued production of pro-inflammatory cytokines such as IL-1, IL-6, IL-12, and TNF-α [56]. These cytokines further stimulate IL-6 and IL-8 secretion and recruit additional immune cells to the wound site, perpetuating the inflammatory cascade and impairing the wound's ability to transition into proliferative and remodeling phases [54]. Emerging evidence also highlights the involvement of other immune cells. Innate lymphoid cells (ILCs), particularly ILC2s, may contribute to wound healing; however, their roles in DFUs are not yet clearly defined. Similarly, invariant natural killer (NK) T cells, which normally help regulate neutrophil-driven inflammation, may become dysregulated in diabetic conditions and exacerbate inflammatory responses [43].

Another hallmark of DFU pathophysiology is excessive production of matrix metalloproteinases (MMPs), particularly MMP-9, which is secreted by neutrophils. While MMPs are essential for ECM remodeling, in diabetic wounds, they are often overexpressed, leading to degradation of ECM components and growth factors required for tissue regeneration [46]. Chronic inflammation amplifies MMP expression through cytokine and ROS signaling, establishing a cycle of tissue destruction and inflammation [8, 40]. This persistent ECM breakdown prevents progression into later wound-healing stages. Additional molecular disruptions in DFUs include decreased expression of Mfge8, a glycoprotein involved in anti-inflammatory signaling and clearance of apoptotic cells. Reduced Mfge8 levels contribute to the dominance of pro-inflammatory pathways in the wound environment [46]. This inflammatory milieu also promotes cellular senescence in endothelial cells and fibroblasts. Senescent cells accumulate in chronic wounds and secrete pro-inflammatory cytokines, chemokines, and proteases—a collective referred to as the senescence-associated secretory phenotype (SASP). SASP factors further aggravate inflammation, inhibit the proliferation and migration of keratinocytes and fibroblasts, and block normal wound resolution [8].

In conclusion, the innate and adaptive immune systems play central roles in the chronic non-healing state of diabetic foot ulcers. Neutrophil and macrophage dysfunction, T-cell dysregulation, biofilm persistence, excessive MMP activity, and senescence-driven inflammation all contribute to impaired pathogen clearance, unresolved inflammation, and blocked transition to tissue repair. A deeper understanding of these interconnected immune pathways may inform the development of targeted therapies aimed at restoring immune balance and facilitating successful wound healing in diabetic patients.

Fibrosis in DFUs

Wound fibrosis or excessive scar tissue formation is an impediment to healing, leading to a chronic nonhealing wound and often amputation. Fibrosis is characterized by excessive accumulation of ECM proteins, mainly collagen, leading to scar formation. Fibrosis is a part of normal wound healing; however, its dysregulation in DFUs results in the overproduction of scar tissue, hindering wound healing. Fibrosis of the healing wound is a major concern because it results in impaired wound healing, decreased oxygenation of the healing tissue, increased risk of infection, and increased risk of amputation [18, 39, 47, 62, 84]. As discussed earlier, dysregulated inflammation, impaired ECM remodeling, impaired fibroblast function, and association with infection, ischemia, and impaired immune response contribute to nonhealing and may also contribute to wound fibrosis.

The process of wound healing is mediated by a delicate balance between inflammation, angiogenesis, tissue regeneration, and collagen repair. The intricate balance between extracellular matrix (ECM) synthesis and degradation is essential for proper wound healing and tissue repair. However, when there is an excessive amount of ECM deposition due to dysregulated wound and connective tissue repair response, it becomes fibrotic [81]. In normal wound healing, fibroblasts and myofibroblasts lead the way in fibrotic response. Once the body detects an injury, the first thing the body does is to start its coagulation cascade to prevent blood loss. Platelets become activated as they come into contact with collagen and von Willebrand factor once the endothelium is exposed. These activated platelets will release growth factors such as platelet-derived growth factor (PDGF) and transforming growth factor-β1

(TGF-β1)-β1, which stimulate fibroblasts [77]. Fibroblasts reside in tissues in a quiescent status and respond to physical demands by phenotypic switch to proliferative fibroblasts, synthesizing, degrading, and remodeling the ECM [14]. With TGF-B1 serving as a key mediator, fibroblasts will turn from their quiescent form into proto-myofibroblasts, a transition state, and then into their active form, myofibroblasts. Myofibroblasts acquire a network of contractile actin-myosin complexes rich in α-smooth muscle actin (α-SMA), which is the differentiating factor between proto-myofibroblasts [10, 33]. With the α-SMA, myofibroblasts can exert a contractile force to close the wound [52]. Myofibroblasts will produce ECM components that include collagen types I and III primarily, elastin, and proteoglycans such as fibronectin, laminin, and tenascin [12, 32]. Once myofibroblasts produce enough ECM for tissue stability, different mechanisms come into play to stop the production of matrix components. This phase is a controlled reduction in myofibroblast activity and ECM production, which is critical in preventing excessive scarring and fibrosis [55].

Cellular senescence is a major player when it comes to controlling myofibroblasts. Cellular senescence is the permanent end of the cell cycle [3]. As wound healing progresses, myofibroblasts are driven into senescence, converting them from ECM-producing cells to ECM-degrading cells [29]. Endothelial cells will also promote myofibroblast senescence by platelet-derived growth factor alpha polypeptide a (PDGF-AA) in the senescence-associated secretory phenotype (SASP) [3]. Another protein that induces cellular senescence is the matricellular protein CCN1/CYR61, which is expressed at wound sites. Through cell adhesion receptors, integrin α6β1, and heparan sulfate proteoglycans, CCN1 induces DNA damage response and p53 activation and activates the RAC1-NOX1 complex to induce reactive oxygen species (ROS) generation [30]. However, when these mechanisms go unchecked, they can lead to fibrosis. In diabetic foot ulcers, there is an imbalance in healing, which leads to fibrosis. This imbalance is led by TGF-β. As stated before, TGF-β is the initiator of fibroblasts into myofibroblasts. Dysregulation of TGF-β, leads to diabetic foot ulcers due to chronic inflammation in the wound [72]. Excessive TGF-β leads myofibroblasts to create more collagen and ECM, which leads to scar tissue formation, thus delaying wound healing [44].

In diabetic patients, chronic hyperglycemia leads to increased advanced glycation end products (AGEs). AGEs are the product of glycation of proteins, a reaction between reducing sugars and amino acid residues of protein, through the Maillard Reaction. Because of chronic hyperglycemia, proteins are more prone to these reactions, producing AGEs such as N-carboxymethyl-lysine (CML), N-carboxyethyl-lysine, and pentosidine [26]. AGEs contribute to fibrosis in DFUs through several mechanisms. AGEs directly crosslink ECM proteins, usually collagen, leading to increased tissue stiffness and reduced elasticity [82]. This impairs cell migration and tissue remodeling. Through their Receptor for Advanced Glycation End-products (RAGE), AGEs promote inflammation and oxidative stress through signaling pathways, including MAPK, p38, SAPK/JNK, ERK1/2, and JAK/STAT pathways, among which NF-κB is the most important. AGEs play a role in fibrosis, particularly in the context of diabetes. Prolongs hyperglycemia increases AGEs, TBG-β, and RAGEs, creating a pro-fibrotic environment in DFUs. Activation of the AGE-RAGE pathway triggers downstream signaling, including NF-κB, which promotes the expression of proinflammatory cytokines such as TNF-α, IL-1β, and IL-6, contributing to tissue damage. Blocking this pathway has been shown to improve healing in diabetic mice [22, 75].

The interplay between persistently high TGF-β levels and AGE accumulation creates a complex pro-fibrotic environment in DFUs. TGF-β not only stimulates ECM production but also induces the expression of RAGE, potentially amplifying the detrimental effects of AGEs. Conversely, AGEs can enhance TGF-β signaling, creating a feed-forward loop that perpetuates fibrosis. The persistent interaction of high TGF-β levels and AGE accumulation promotes a profibrotic environment in DFUs and complicates chronic wound healing. TGF-β not only stimulates ECM production but also regulates RAGE, thereby enhancing AGE-induced fibrotic signaling. Conversely, AGEs enhance TGF-β activity and create a self-sustaining fibrotic cycle. During wound remodeling, fibroblasts and endothelial cells contribute to ECM deposition and wound contraction, thereby shaping scar morphology [18]. Myofibroblasts expressing α-SMA produce collagen that switches from type III to type I stretch. However, in hypertrophic scars, excessive inflammation and mechanical stress promote pathological collagen deposition and lead to impaired scar tissue architecture. Normal wound healing requires timely apoptosis of myofibroblasts, but delaying this apoptosis under fibrotic conditions prolongs excessive collagen production. Antifibrotic therapies target these processes. Pirfenidone, a pyridine derivative, has been shown to inhibit the transition of fibroblasts to myofibroblasts. Matrix metalloproteinases (MMP-1, MMP-3) promote collagen degradation and are being investigated as therapeutic targets. Recent studies have highlighted the role of the ubiquitin-proteasome system in TGF-β regulation, with E3 ligases such as TTC3 regulating myofibroblast differentiation. In addition, plasminogen activator inhibitor-1 (PAI-1) links immune responses to fibrosis progression [18].

In conclusion, fibrosis and extracellular matrix dynamics play a critical role in the pathogenesis of diabetic foot ulcers. The dysregulation of normal wound healing processes, driven by persistently high TGF-β levels and AGE accumulation, leads to excessive collagen deposition, abnormal ECM remodeling, and impaired tissue function. Understanding the complex interplay between these factors is crucial for developing effective therapeutic strategies to improve wound healing outcomes in diabetic patients. By targeting key pathways involved in fibrosis and ECM dynamics, it may be possible to restore the balance between tissue repair and regeneration, ultimately reducing the burden of diabetic foot ulcers and improving patient quality of life.

Current and Emerging Therapeutic Approaches

The primary hurdle in the treatment of diabetic foot ulcers (DFUs) is achieving a balance between inflammation and tissue repair. In a typical wound, inflammation initiates a cascade of cell proliferation and remodeling, ultimately leading to healing. However, this process is disrupted in diabetic wounds, and these wounds may be stalled in a stage of inflammation, unable to transition into the proliferative and remodeling stages. In addition to chronic inflammation, diabetic wounds are prone to excessive extracellular matrix deposition and fibrosis, which further impairs tissue healing and regeneration [65]. As a result, targeting inflammation has become a key focus for therapeutic development in DFU.

SPMs such as resolvins, protectins, and maresins play critical roles in immune response modulation and tissue regeneration. SPMs help by promoting the clearance of apoptotic neutrophils, reducing the levels of reactive oxygen and nitrogen species, and inducing anti-inflammatory mediators, thereby reducing prolonged inflammation. As mentioned in the section on immune response, SPM production is often impaired in diabetic wounds, leading to failure in the resolution phase of inflammation in chronic conditions such as diabetes. Administration of exogenous SPMs in preclinical animal models has been shown to effectively resolve inflammation without an immunosuppressive effect in the lungs. While SPMs have not been studied directly in the context of DFUs, their potent effect on immune modulation in inflammatory conditions offers a promising therapeutic option [35].

Recent studies have also highlighted the potential of mesenchymal stem cell (MSC)-derived exosomes to modulate inflammation by promoting macrophage polarization toward a pro-resolving M2 phenotype [15, 71, 85]. These findings suggest a dual role for exosomes in both immune regulation and tissue repair, which will be discussed in more detail in the subsequent section on regenerative therapies. Given their established role in ECM remodeling and inflammation, MMP modulation has emerged as a promising therapeutic strategy. MMP-9, as discussed, displays an inflammatory effect along with TGF-β, and is overexpressed in DFUs. Conversely, MMP-8 has been identified as beneficial for the repair of wounds by significantly reducing inflammatory cytokines such as IL-6. In preclinical trials, selective inhibition of MMP-9 and promotion of MMP-8 have been shown to decrease inflammation while enhancing angiogenesis and re-epithelialization of diabetic wounds when administered alone, with an even greater effect with co-administration. Despite the efficacy in preclinical trials involving MMP inhibitors, the potential for off-target effects remains a major concern with treatment, and further investigation addressing safety in human subjects is required [21].

In addition to inflammatory modulation, targets for fibrosis prevention in DFU include components responsible for matrix remodeling. As mentioned, the differentiation of fibroblasts into active myofibroblasts by TGF-β is left unchecked in diabetic patients, contributing to fibrosis in DFU [44, 72]. Pharmacological inhibitors of fibrosis, such as Pirfenidone, act selectively as TGF-β inhibitors to combat their fibrotic effect in diabetic wounds. Phase 3 clinical trials have shown that topical administration of Pirfenidone, in addition to conventional treatment, significantly improves healing outcomes in chronic, non-infected DFUs [27].

A promising therapeutic to mitigate the excessive scarring in DFU involves the use of exosomes derived from human amniotic fluid stem cells (hAFSC-exo), which contain a range of microRNAs capable of suppressing TGF-β signaling. These exosomes inhibit the conversion of fibroblasts into active matrix-secreting myofibroblasts, preventing excessive ECM deposition and fibrotic scarring [80]. Exosomes are extracellular vesicles containing a multitude of non-coding RNA species such as miRNA, lncRNA, circRNA, proteins, and various complexes intended to aid in intercellular communication. Exosomes are capable of activating signaling cascades among cells involved in a variety of physiological processes, including wound healing, immune regulation, and fibrosis modulation [74, 78, 83]. While inhibitors of TGF-β signaling show promise in wound healing for DFUs, the timing of administration is critical. TGF-β is essential during the inflammatory phase to initiate repair, and inhibition should only occur after wound maturation. This presents a challenge as it would require careful monitoring to determine the wound's progression through inflammation, proliferation, and remodeling stages [21]. Mesenchymal stem cells (MSCs) have emerged as a potential alternative to surgical interventions, with benefits to both wound healing and fibrosis reduction. Notably, laboratory studies suggest that the therapeutic effects of MSCs may be largely mediated by the exosomes they secrete, prompting interest in exosome-based therapies [23, 74, 78, 83]. The use of (MSC)-derived exosomes and hAFSC-Exo has been discussed previously regarding their effects on M2 phenotypes and TGF-β signaling, respectively.

Exosomal ncRNAs, such as miRNA contained in exosomes, can regulate inflammation and may promote re-epithelialization and ECM remodeling. In animal models, ncRNA was shown to reduce inflammation by promoting the polarization of macrophages to M2 phenotypes, a key complication of the hyperglycemic state in diabetes that perpetuates chronic inflammation [23]. Additionally, tRNA-derived fragments (tRFs), also found in exosomes, are capable of regulating the inflammatory response via NF-KB inhibition.

Treatment of adipose-derived stem cells (ADSC) with hypoxic media shows greater levels of exosome secretion in vitro, allowing for the laboratory concentration of anti-inflammatory, fibroblast-forming exosomes for use in DFU therapy [23]. In mouse subjects exposed to hypoxia-induced adipose stem cell exosomes (HypADSCs-exo), inhibition of inflammation and acceleration of diabetic wound healing were achieved with greater blood perfusion to healing tissues [73, 76]. Treatment with exosomes may offer a promising cell-free therapy to mitigate fibrosis, scarring, and inflammation in diabetic feet. Although cell-free therapies are promising, the process of exosome isolation involves a heavy workload and accompanying costs due to the complexity of exosome-containing samples as well as the heterogeneity among exosomes themselves [36]. Various laboratory methods exist to purify exosomes from protein aggregates and other exosomes with overlapping biochemical properties, but scalability and standardization remain major obstacles. Large-scale isolation without altering bioactivity is difficult to achieve, which hinders clinical translation and the development of standardized protocols for therapeutic use [9].

Conclusion

The complex interplay between inflammation, immune response, and fibrosis in diabetic foot ulcers presents challenges for wound healing. Chronic inflammation seen in DFU leads to elevated proinflammatory cytokines and impaired immune cell function, resulting in persistent tissue damage. This prolonged inflammatory phase disrupts the normal wound-healing process, leading to reduced angiogenesis and altered extracellular matrix remodeling. This problem is further exacerbated by the dysregulated immune response in DFUs. Impaired neutrophil function and T-cell imbalance lead to increased susceptibility to infection and delayed healing. In addition, biofilms formed by colonizing bacteria create a protective environment that resists host immune defenses and antibiotic therapy. Fibrosis resulting from altered ECM remodeling is another barrier to wound healing. Imbalances between matrix metalloproteinases (MMPs) and their inhibitors lead to excessive degradation of ECM components and compromise of the structural integrity of the wound bed. This, coupled with reduced collagen production and impaired fibroblast function, impedes the formation of healthy granulation tissue required for wound closure. Given the complexity of these interactions, innovative therapeutic approaches are urgently needed to address multiple aspects of the impaired healing process in DFUs. Future research should focus on developing interventions that simultaneously modulate inflammation, enhance immune function, and promote appropriate remodeling of the extracellular matrix. This integrated strategy may be key to improving healing outcomes and reducing the significant burden of diabetic foot ulcers. Overall, when treating DFUs, it is important to get an early diagnosis for a better long-term prognosis and consider targeting multiple pathways simultaneously, which can lead to more optimal outcomes for the patient.

Acknowledgments None.

Supplementary Materials Not applicable.

Author Contributions Conceptualization, G.S. and V.R.; methodology, G.S., Z.C., S.N., and D.S.; writing—original draft preparation, G.S., Z.C., S.N., D.S, and V.R.; writing—review and editing, V.R.; All authors have read and agreed to the published version of the manuscript.

Institutional Review Board Statement Not applicable.

Informed Consent Statement Not applicable.

Data Availability Statement Not applicable.

Conflicts of Interest The authors declare no conflicts of interest.

Funding This research received no external funding.

References

1. Afonso AC, Oliveira D, Saavedra MJ, Borges A, Simoes M (2021) Biofilms in diabetic foot ulcers: impact, risk factors and control strategies. Int J Mol Sci 22
2. Aitcheson SM, Frentiu FD, Hurn SE, Edwards K, Murray RZ (2021) Skin wound healing: normal macrophage function and macrophage dysfunction in diabetic wounds. Molecules 26
3. Andrade AM, Sun M, Gasek NS, Hargis GR, Sharafieh R, Xu M (2022) Role of senescent cells in cutaneous wound healing. Biology (Basel) 11
4. Atlaw A, Kebede HB, Abdela AA, Woldeamanuel Y (2022) Bacterial isolates from diabetic foot ulcers and their antimicrobial resistance profile from selected hospitals in Addis Ababa, Ethiopia. Front Endocrinol (Lausanne) 13:987487
5. Ayuk SM, Abrahamse H, Houreld NN (2016) The role of matrix metalloproteinases in diabetic wound healing in relation to photobiomodulation. J Diabetes Res 2016:2897656
6. Boulton AJ, Armstrong DG, Hardman MJ, Malone M, Embil JM, Attinger CE, Lipsky BA, Aragón-Sánchez J, Li HK, Schultz G (2020) Diagnosis and management of diabetic foot infections. American Diabetes Association
7. Burgess JL, Wyant WA, Abdo Abujamra B, Kirsner RS, Jozic I (2021) Diabetic wound-healing science. Medicina (Kaunas) 57
8. Clayton SM, Shafikhani SH, Soulika AM (2024) Macrophage and neutrophil dysfunction in diabetic wounds. Adv Wound Care (New Rochelle) 13:463–484
9. Colao IL, Corteling R, Bracewell D, Wall I (2018) Manufacturing exosomes: a promising therapeutic platform. Trends Mol Med 24:242–256
10. D'Urso M, Kurniawan NA (2020) Mechanical and physical regulation of fibroblast-myofibroblast transition: from cellular mechanoresponse to tissue pathology. Front Bioeng Biotechnol 8:609653
11. Da Porto A, Miranda C, Brosolo G, Zanette G, Michelli A, Ros RD (2022) Nutritional supplementation on wound healing in diabetic foot: what is known and what is new? World J Diabetes 13:940–948
12. Darby IA, Laverdet B, Bonte F, Desmouliere A (2014) Fibroblasts and myofibroblasts in wound healing. Clin Cosmet Investig Dermatol 7:301–311
13. Deng H, Li B, Shen Q, Zhang C, Kuang L, Chen R, Wang S, Ma Z, Li G (2023) Mechanisms of diabetic foot ulceration: a review. J Diabetes 15:299–312
14. Do NT, Lee SY, Lee YS, Shin C, Kim D, Lee TG, Son JG, Kim SH (2023) Time-sequential fibroblast-to-myofibroblast transition in elastin-variable 3D hydrogel environments by collagen networks. Biomater Res 27:103
15. Dong Z, Fu Y, Cai Z, Dai H, He Y (2025) Recent advances in adipose-derived mesenchymal stem cell-derived exosomes for regulating macrophage polarization. Front Immunol 16:1525466
16. Dowey R, Iqbal A, Heller SR, Sabroe I, Prince LR (2021) A bittersweet response to infection in diabetes; targeting neutrophils to modify inflammation and improve host immunity. Front Immunol 12:678771
17. Du F, Ma J, Gong H, Bista R, Zha P, Ren Y, Gao Y, Chen D, Ran X, Wang C (2022) Microbial infection and antibiotic susceptibility of diabetic foot ulcer in China: literature review. Front Endocrinol (Lausanne) 13:881659
18. El Ayadi A, Jay JW, Prasai A (2020) Current approaches targeting the wound healing phases to attenuate fibrosis and scarring. Int J Mol Sci 21
19. Fang Y, Shen J, Yao M, Beagley KW, Hambly BD, Bao S (2010) Granulocyte-macrophage colony-stimulating factor enhances wound healing in diabetes via upregulation of proinflammatory cytokines. Br J Dermatol 162:478–486
20. Gamboa-Antinolo FM (2023) Diabetic foot ulcers: a growing global health emergency in the COVID-19 era. Intern Emerg Med 18:1259–1261

21. Gao M, Nguyen TT, Suckow MA, Wolter WR, Gooyit M, Mobashery S, Chang M (2015) Acceleration of diabetic wound healing using a novel protease-anti-protease combination therapy. Proc Natl Acad Sci USA 112:15226–15231
22. Gumede DB, Abrahamse H, Houreld NN (2024) Targeting Wnt/beta-catenin signaling and its interplay with TGF-beta and Notch signaling pathways for the treatment of chronic wounds. Cell Commun Signal 22:244
23. He X, Dong Z, Cao Y, Wang H, Liu S, Liao L, Jin Y, Yuan L, Li B (2019) MSC-derived exosome promotes M2 polarization and enhances cutaneous wound healing. Stem Cells Int 2019:7132708
24. Hellmann J, Tang Y, Spite M (2012) Proresolving lipid mediators and diabetic wound healing. Curr Opin Endocrinol Diabetes Obes 19:104–108
25. Huang Y, Kyriakides TR (2020) The role of extracellular matrix in the pathophysiology of diabetic wounds. Matrix Biol Plus 6-7:100037
26. Huijberts MS, Schaper NC, Schalkwijk CG (2008) Advanced glycation end products and diabetic foot disease. Diabetes Metab Res Rev 24(Suppl 1):S19–S24
27. Janka-Zires M, Almeda-Valdes P, Uribe-Wiechers AC, Juarez-Comboni SC, Lopez-Gutierrez J, Escobar-Jimenez JJ, Gomez-Perez FJ (2016) Topical administration of pirfenidone increases healing of chronic diabetic foot ulcers: a randomized crossover study. J Diabetes Res 2016:7340641
28. Jiang P, Li Q, Luo Y, Luo F, Che Q, Lu Z, Yang S, Yang Y, Chen X, Cai Y (2023) Current status and progress in research on dressing management for diabetic foot ulcer. Front Endocrinol (Lausanne) 14:1221705
29. Jun JI, Lau LF (2010) Cellular senescence controls fibrosis in wound healing. Aging (Albany NY) 2:627–631
30. Jun JI, Lau LF (2010) The matricellular protein CCN1 induces fibroblast senescence and restricts fibrosis in cutaneous wound healing. Nat Cell Biol 12:676–685
31. Khan MS, Azam M, Khan MN, Syed F, Ali SHB, Malik TA, Alnasser SMA, Ahmad A, Karimulla S, Qamar R (2023) Identification of contributing factors, microorganisms and antimicrobial resistance involved in the complication of diabetic foot ulcer treatment. Microb Pathog 184:106363
32. Klingberg F, Hinz B, White ES (2013) The myofibroblast matrix: implications for tissue repair and fibrosis. J Pathol 229:298–309
33. Knoedler S, Broichhausen S, Guo R, Dai R, Knoedler L, Kauke Navarro M, Diatta F, Pomahac B, Machens HG, Jiang D, Rinkevich Y (2023) Fibroblasts - the cellular choreographers of wound healing. Front Immunol 14:1233800
34. Krzyszczyk P, Schloss R, Palmer A, Berthiaume F (2018) The role of macrophages in acute and chronic wound healing and interventions to promote pro-wound healing phenotypes. Front Physiol 9:419
35. Li J, Xiao Y, Zhang Y, Li S, Zhao M, Xia T, Meng H (2023) Pulmonary delivery of specialized pro-resolving mediators-based nanotherapeutics attenuates pulmonary fibrosis in preclinical animal models. ACS Nano 17:15354–15370
36. Li P, Kaslan M, Lee SH, Yao J, Gao Z (2017) Progress in exosome isolation techniques. Theranostics 7:789–804
37. Li Z, Chen L, Yang S, Han J, Zheng Y, Chen Z, Shi X, Yang J (2024) Glucose and pH dual-responsive hydrogels with antibacterial, reactive oxygen species scavenging, and angiogenesis properties for promoting the healing of infected diabetic foot ulcers. Acta Biomater 190:205–218
38. Littig JPB, Moellmer R, Agrawal DK, Rai V (2023) Future applications of exosomes delivering resolvins and cytokines in facilitating diabetic foot ulcer healing. World J Diabetes 14:35–47
39. Liu Y, Liu Y, He W, Mu X, Wu X, Deng J, Nie X (2022) Fibroblasts: immunomodulatory factors in refractory diabetic wound healing. Front Immunol 13:918223

40. Lu YZ, Nayer B, Singh SK, Alshoubaki YK, Yuan E, Park AJ, Maruyama K, Akira S, Martino MM (2024) CGRP sensory neurons promote tissue healing via neutrophils and macrophages. Nature 628:604–611
41. Macleod AS, Mansbridge JN (2016) The innate immune system in acute and chronic wounds. Adv Wound Care (New Rochelle) 5:65–78
42. Magliano DJ, Boyko EJ (2022) IDF diabetes atlas
43. Mamun AA, Shao C, Geng P, Wang S, Xiao J (2024) Recent advances in molecular mechanisms of skin wound healing and its treatments. Front Immunol 15:1395479
44. Mcdermott K, Fang M, Boulton AJM, Selvin E, Hicks CW (2023) Etiology, epidemiology, and disparities in the burden of diabetic foot ulcers. Diabetes Care 46:209–221
45. Mieczkowski M, Mrozikiewicz-Rakowska B, Kowara M, Kleibert M, Czupryniak L (2022) The problem of wound healing in diabetes-from molecular pathways to the design of an animal model. Int J Mol Sci 23
46. Mohsin F, Javaid S, Tariq M, Mustafa M (2024) Molecular immunological mechanisms of impaired wound healing in diabetic foot ulcers (DFU), current therapeutic strategies and future directions. Int Immunopharmacol 139:112713
47. Monaghan MG, Borah R, Thomsen C, Browne S (2023) Thou shall not heal: overcoming the non-healing behaviour of diabetic foot ulcers by engineering the inflammatory microenvironment. Adv Drug Deliv Rev 203:115120
48. Murphy-Lavoie HM, Ramsey A, Nguyen M, Singh S (2023) Diabetic foot infections. In: StatPearls [Internet]. StatPearls Publishing
49. Noor S, Khan RU, Ahmad J (2017) Understanding diabetic foot infection and its management. Diabetes Metab Syndr 11:149–156
50. Packer CF, Ali SA, Manna B (2018) Diabetic ulcer
51. Pezhman L, Tahrani A, Chimen M (2021) Dysregulation of leukocyte trafficking in type 2 diabetes: mechanisms and potential therapeutic avenues. Front Cell Dev Biol 9:624184
52. Potekaev NN, Borzykh OB, Medvedev GV, Pushkin DV, Petrova MM, Petrov AV, Dmitrenko DV, Karpova EI, Demina OM, Shnayder NA (2021) The role of extracellular matrix in skin wound healing. J Clin Med 10
53. Rai V, Le H, Agrawal DK (2023) Novel mediators regulating angiogenesis in diabetic foot ulcer healing. Can J Physiol Pharmacol 101:488–501
54. Rai V, Moellmer R, Agrawal DK (2022) The role of CXCL8 in chronic nonhealing diabetic foot ulcers and phenotypic changes in fibroblasts: a molecular perspective. Mol Biol Rep 49:1565–1572
55. Rai V, Moellmer R, Agrawal DK (2023) Role of fibroblast plasticity and heterogeneity in modulating angiogenesis and healing in the diabetic foot ulcer. Mol Biol Rep 50:1913–1929
56. Raziyeva K, Kim Y, Zharkinbekov Z, Kassymbek K, Jimi S, Saparov A (2021) Immunology of acute and chronic wound healing. Biomolecules 11
57. Ren B, Li B, Pan T, Zhao E, Ju S, Li X, Li X, Zhu Y, Cai Y, Huang L, Fu W, Dong Z (2023) Risk factors for at-risk foot and peripheral artery disease among the population with diabetes: a multicommunity-based cross-sectional study. Diabetes Res Clin Pract 203:110869
58. Rice JB, Desai U, Cummings AK, Birnbaum HG, Skornicki M, Parsons NB (2014) Burden of diabetic foot ulcers for medicare and private insurers. Diabetes Care 37:651–658
59. Rong Y, Yang H, Xu H, Li S, Wang P, Wang Z, Zhang Y, Zhu W, Tang B, Zhu J, Hu Z (2022) Bioinformatic analysis reveals hub immune-related genes of diabetic foot ulcers. Front Surg 9:878965
60. Roy R, Zayas J, Singh SK, Delgado K, Wood SJ, Mohamed MF, Frausto DM, Albalawi YA, Price TP, Estupinian R, Giurini EF, Kuzel TM, Zloza A, Reiser J, Shafikhani SH (2022) Overriding impaired FPR chemotaxis signaling in diabetic neutrophil stimulates infection control in murine diabetic wound. elife 11
61. Sakshi S, Jayasuriya R, Sathish Kumar RC, Umapathy D, Gopinathan A, Balamurugan R, Ganesan K, Ramkumar KM (2023) MicroRNA-27b impairs Nrf2-mediated angiogenesis in the progression of diabetic foot ulcer. J Clin Med 12

62. Sawaya AP, Stone RC, Brooks SR, Pastar I, Jozic I, Hasneen K, O'Neill K, Mehdizadeh S, Head CR, Strbo N, Morasso MI, Tomic-Canic M (2020) Deregulated immune cell recruitment orchestrated by FOXM1 impairs human diabetic wound healing. Nat Commun 11:4678
63. Sawaya AP, Stone RC, Mehdizadeh S, Pastar I, Worrell S, Balukoff NC, Kaplan MJ, Tomic-Canic M, Morasso MI (2022) FOXM1 network in association with TREM1 suppression regulates NET formation in diabetic foot ulcers. EMBO Rep 23:e54558
64. Shah P, Inturi R, Anne D, Jadhav D, Viswambharan V, Khadilkar R, Dnyanmote A, Shahi S (2022) Wagner's classification as a tool for treating diabetic foot ulcers: our observations at a suburban teaching hospital. Cureus 14:e21501
65. Shen J, Zhao X, Zhong Y, Yang P, Gao P, Wu X, Wang X, An W (2022) Exosomal ncRNAs: the pivotal players in diabetic wound healing. Front Immunol 13:1005307
66. Shofler D, Rai V, Mansager S, Cramer K, Agrawal DK (2021) Impact of resolvin mediators in the immunopathology of diabetes and wound healing. Expert Rev Clin Immunol 17:681–690
67. Singh D, Rai V, Agrawal DK (2023) Regulation of collagen I and collagen III in tissue injury and regeneration. Cardiol Cardiovasc Med 7:5–16
68. Smith J, Rai V (2024) Novel factors regulating proliferation, migration, and differentiation of fibroblasts, keratinocytes, and vascular smooth muscle cells during wound healing. Biomedicine 12
69. Smith J, Rai V (2024) Platelet-rich plasma in diabetic foot ulcer healing: contemplating the facts. Int J Mol Sci 25
70. Song J, Hu L, Liu B, Jiang N, Huang H, Luo J, Wang L, Zeng J, Huang F, Huang M, Cai L, Tang L, Chen S, Chen Y, Wu A, Zheng S, Chen Q (2022) The emerging role of immune cells and targeted therapeutic strategies in diabetic wounds healing. J Inflamm Res 15:4119–4138
71. Teo KYW, Zhang S, Loh JT, Lai RC, Hey HWD, Lam KP, Lim SK, Toh WS (2023) Mesenchymal stromal cell exosomes mediate M2-like macrophage polarization through CD73/Ecto-5'-nucleotidase activity. Pharmaceutics 15
72. Téot L, Mustoe TA, Middelkoop E, Gauglitz GG (2020) Textbook on scar management: state of the art management and emerging technologies. Springer
73. Wang J, Wu H, Peng Y, Zhao Y, Qin Y, Zhang Y, Xiao Z (2021) Hypoxia adipose stem cell-derived exosomes promote high-quality healing of diabetic wound involves activation of PI3K/Akt pathways. J Nanobiotechnol 19:202
74. Wang M, Yu F, Li P, Wang K (2020) Emerging function and clinical significance of exosomal circRNAs in cancer. Mol Ther Nucleic Acids 21:367–383
75. Wang Q, Zhu G, Cao X, Dong J, Song F, Niu Y (2017) Blocking AGE-RAGE signaling improved functional disorders of macrophages in diabetic wound. J Diabetes Res 2017:1428537
76. Wang Z, Feng C, Liu H, Xia Y, Shan M, Hao Y (2025) Hypoxia-induced adipose derived stem cells-derived exosomes promote diabetic wound healing through circ-0001747/miR-199a-5p/HIF-1alpha axis. Arch Dermatol Res 317:456
77. Wynn TA, Ramalingam TR (2012) Mechanisms of fibrosis: therapeutic translation for fibrotic disease. Nat Med 18:1028–1040
78. Zeng QL, Liu DW (2021) Mesenchymal stem cell-derived exosomes: an emerging therapeutic strategy for normal and chronic wound healing. World J Clin Cases 9:6218–6233
79. Zhang P, Lu J, Jing Y, Tang S, Zhu D, Bi Y (2017) Global epidemiology of diabetic foot ulceration: a systematic review and meta-analysis (dagger). Ann Med 49:106–116
80. Zhang Y, Yan J, Liu Y, Chen Z, Li X, Tang L, Li J, Duan M, Zhang G (2021) Human amniotic fluid stem cell-derived exosomes as a novel cell-free therapy for cutaneous regeneration. Front Cell Dev Biol 9:685873
81. Zhao M, Wang L, Wang M, Zhou S, Lu Y, Cui H, Racanelli AC, Zhang L, Ye T, Ding B, Zhang B, Yang J, Yao Y (2022) Targeting fibrosis, mechanisms and cilinical trials. Signal Transduct Target Ther 7:206
82. Zhou M, Zhang Y, Shi L, Li L, Zhang D, Gong Z, Wu Q (2024) Activation and modulation of the AGEs-RAGE axis: implications for inflammatory pathologies and therapeutic interventions - a review. Pharmacol Res 206:107282

83. Zhou X, Xie F, Wang L, Zhang L, Zhang S, Fang M, Zhou F (2020) The function and clinical application of extracellular vesicles in innate immune regulation. Cell Mol Immunol 17:323–334
84. Zhu Y, Lu J, Wang S, Xu D, Wu M, Xian S, Zhang W, Tong X, Liu Y, Huang J, Jiang L, Guo X, Xie S, Gu M, Jin S, Ma Y, Huang R, Xiao S, Ji S (2023) Mapping intellectual structure and research hotspots in the field of fibroblast-associated DFUs: a bibliometric analysis. Front Endocrinol (Lausanne) 14:1109456
85. Zubair M, Abouelnazar FA, Iqbal MA, Pan J, Zheng X, Chen T, Shen W, Yin J, Yan Y, Liu P, Mao F, Chu Y (2025) Mesenchymal stem cell-derived exosomes as a plausible immunomodulatory therapeutic tool for inflammatory diseases. Front Cell Dev Biol 13:1563427

Chapter 11
Preclinical Models for Wound Healing and Repair Studies

Rochelle Bitolas and David A. Brown

Abstract Preclinical models are essential tools in wound healing research, enabling both mechanistic insights and the evaluation of therapeutic safety prior to human use. To effectively translate findings, these models aim to replicate the human wound environment as closely as possible while reducing confounding factors. Over time, researchers have developed a range of in vitro, ex vivo, and in vivo systems to capture the complex mechanisms of wound healing. These models have contributed to major clinical advances, improving our ability to predict, prevent, and treat wound-related complications. They also help understand why certain wounds heal better or worse depending on timing, wound environment, and underlying pathology. However, preclinical models have inherent limitations due to anatomic and physiological differences between humans and animals, which vary based on species type. These challenges can be mitigated by selecting an appropriate model type and weighing the advantages and limitations of each based on the specific injury type or phenotype desired. Ongoing work is focused on standardizing in vivo protocols for reproducibility and advancing in vitro models to better mimic humans and reduce animal use for in vivo models. This chapter explores the evolution of preclinical wound models and current applications in wound healing research.

Keywords Preclinical model · Animal model · Wound healing · Scar formation · Murine wound healing · Hypertrophic scar model · Regeneration · In vivo model · In vitro model

---

R. Bitolas (✉)
Duke University School of Medicine, Durham, NC, USA
e-mail: Rochelle.bitolas@duke.edu

D. A. Brown
Duke University School of Medicine, Durham, NC, USA

Duke Department of Surgery, Division of Plastic, Maxillofacial, and Oral Surgery, Duke University Medical Center, Durham, NC, USA

F. De Francesco, K. Turksen (eds.), *Management and Strategies for Wound Healing*, Stem Cell Biology and Regenerative Medicine 79,
https://doi.org/10.1007/978-3-032-12442-5_11

Abbreviations

AAV	Adeno-Associated Virus
AHA	Algan Hemostatic Agent
AWA	Animal Waelfare Act
CRISPR	Clustered Regularly Interspaced Short Palindromic Repeats
FDA	Food and Drug Administration
H&E	Hematoxylin and Eosin
HSEs	Human Skin Equivalents
IACUC	Institutional Animal Care and Use Committee
IND	Investigational New Drug
IRB	Institutional Review Board
NIH	National Institutes of Health
NPWT	Negative pressure wound therapy
NRC	National Research Council
NSAID	Nonsteroidal anti-inflammatory drugs
PDGF	Platelet derived growth factor
POSS	polyhedral oligomeric sesquioxide
scRNA-seq	Single-cell RNA sequencing
snRNA-seq	Single-nucleus RNA sequencing
STZ	Streptozocin
TGF-β3	Transforming Growth Factor Beta 3
TNF-α	Tumor Necrosis Factor alpha

Introduction

The landscape of wound healing is vast and widely uncharted, filled with intricate interactions and dynamic processes. Before embarking on the journey to deeply understand this process, it is essential to first have some familiarity with the terrain, much like needing a compass to navigate unknown territory. Research into wound healing is similar in that preclinical models act as pathways that give researchers a solid footing, enabling them to move forward with greater confidence. Each step through these models provides clarity, refining our maps and gaining further detail about the road ahead. As the landscape becomes better understood and safe paths are established, others are better able to navigate these landscapes with efficiency. Preclinical models are vital exploration tools and provide the foundation needed to deepen our understanding of wound healing to build bridges necessary to translate these insights from the bench into clinical practice.

Preclinical models serve as a critical research tool by simulating human physiology and pathology. Studying wound healing inherently involves tissue damage, making it unethical to subject humans to harm for research purposes [1]. Preclinical models are essential for advancing wound healing research and developing new treatments [2]. These models play an instrumental role in understanding the biological mechanisms behind wound healing. While translating findings to clinical applications remains challenging, preclinical research continues to enhance our knowledge of tissue repair and drive therapeutic innovations to improve wound healing outcomes.

Evolution of Preclinical Models

Understanding the progression of wound healing research highlights its progression toward replicating the process in a controlled, reproducible, and generalizable environment. Early studies identified key cell populations in wound healing, such as Abercrombie and Heaysman's work on contact inhibition of locomotion, which explained how fibroblasts at the wound edge migrate to close the wounds [3]. Significant advances in cell culture technology further enhanced in vitro wound healing research, particularly in 1975, when Rheinwald and Green established the serial culture of human epidermal keratinocytes [4]. These breakthroughs laid the foundation for the development of 2D wound models commonly known as "scratch assays", which have since become a widely used technique in wound healing research [5]. Scratch assays involve culturing keratinocytes or fibroblasts in a monolayer, where a wound is artificially created by scaping the cell layer using a pipette tip, scraper tool, or laser. The extent of wound closure is then assessed by observing cell migration into the wounded area. However, despite widespread use, this model has limitations in accurately replicating in vivo wound healing, as the damaged underlying material may hinder cell migration and does not fully capture the complexity of wound healing in a living system [2].

The development of 3D wound models, also known as Human Skin Equivalents (HSEs) began in 1979 with the creation of the fibroblast-populated collagen lattice by Bell el al [6]. This model consists of multiple layers, including fibroblasts embedded in a collagen type I hydrogel, where the fibroblasts were able to simulate wound contraction [6]. To ensure experimental consistency, standardized wounds can be created in this 3D model using specialized wounding devices. Following the collagen lattice model, Bell further advanced HSEs in 1981 with the creation of full-thickness skin equivalents grown in vitro [7]. In this model, dermal fibroblasts embedded in collagen were layered beneath an epidermal layer of human keratinocytes, providing a more accurate representation of dermal-epidermal interactions than traditional 2D modes [2].

In vivo models provide a more representative way to study the dynamics of human wound healing, offering deeper insights into the underlying cellular and molecular mechanisms [8]. These models allow scientists to observe a wide variety of cells and their behavior under physiological conditions [8]. Early studies of animal physiology can be traced back to ancient Greece and Rome, where second-century physicians like Galen performed live animal dissections to study inflammation and tissue responses to injury [9].

Significant advancements came in the nineteenth and twentieth centuries, particularly with German pathologist Julius Cohnheim's work in 1867. Cohnheim investigated inflammation at the cellular level using a transparent frog mesentery model with injectable dye, allowing real-time observation of white blood cell migration to the site of injury [10]. By mid-twentieth century, animal-based in vivo models played a crucial role in advancing wound healing research. One of the most notable studies was conducted by George D. Winter in 1962, who explored the impact of moisture in wound healing. Using domestic pigs, he created superficial wounds exposed to air or covered with occlusive dressing, ultimately demonstrating that moist environments promote faster healing [11]. Another landmark study by Thomas Hunt in 1967 examined the effects of oxygen tension and pH on wound healing. Using a rabbit in vivo model, he observed that oxygen tension increased over time, significantly influencing the healing process [12]. Modern in vivo preclinical models for wound healing research are broadly categorized into two groups: small animal models, which include mice, rats, and rabbits, and large animal models, which include pigs, sheep, and dogs [13]. Each of these models will be further discussed throughout this chapter.

The advancement of both in vitro models and ex vivo models has led to the development of a hybrid approach known as ex vivo models or humanized models. Ex vivo models integrate both in vitro and in vivo components by using human or animal tissues maintained in in vitro conditions [14]. This method allows researchers to preserve tissue architecture while controlling confounding factors more effectively than in vivo models. At the same time, ex vivo models offer greater physiological relevance compared to traditional in vitro models [15]. However, while ex vivo models provide a balance between in vitro and in vivo approaches, they still lack systemic effects that influence the full wounding process [15]. Early ex vivo studies using organ cultures to model wound healing demonstrated that wounded epithelium in ex vivo models closely mimicked in vivo wound healing processes during the first 48 h [16]. Since then, further refinements have been made to enhance the accuracy and applicability of ex vivo models in wound healing research. As seen in Fig. 11.1, each of these models has its own specific applications, along with distinct advantages and limitations.

Using preclinical wound models is essential for conducting wound healing research with appropriate controls that ensure generalizability and reproducibility [17]. Reproducibility is supported through the establishment of standardized wound healing models, which help maintain consistency in experimental results and ensure both internal and external validity. Table 11.1 highlights model techniques aimed at replicating specific pathologic conditions. Preclinical models allow researchers to compare results across different experiments, facilitating meta-analyses and systematic reviews that contribute to advancements in wound healing research [17].

Preclinical models also enable the controlled implementation of experimental variables. Standardized wound conditions can be created with consistent wound dimensions, such as depth, using standard surgical tools to ensure uniformity across replicates [20]. Additionally, preclinical models help regulate highly variable real-world factors such as age, gender, environmental influences, and genetics [2]. By

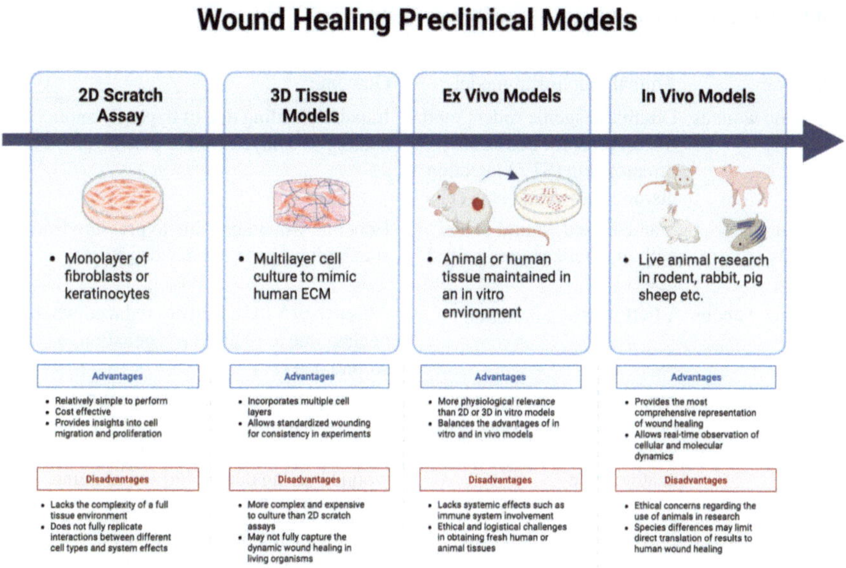

Fig. 11.1 Wound healing preclinical models. Various types of wound models are illustrated along with the advantages and disadvantages of each. (Created in BioRender. Bitolas, R. (2025) https://BioRender.com/dzsvw63)

selecting specific animal strains with desired genetic or phenotypic traits, researchers can control genetic factors, including the use of knockout models to investigate specific gene functions. For example, immunocompromised animals can be used to eliminate immune cell interference in wound healing studies [22].

Reducing variability in experimental conditions minimizes the number of replicates required, leading to lower costs and a smaller number of animals subjected to distress. The successful translation of preclinical findings to human wound healing depends on carefully simulating human wound conditions in experimental models. For example, a tool that has helped enhance clinical translatability includes the db/db mouse model, which mimics the pathological wound healing processes seen in diabetic patients [18].

The advantages of preclinical models in terms of reproducibility, generalizability, and controllability have led to several successful clinical applications in wound healing. Preclinical wound healing models have helped bridge the gap between experimental research and human treatments. One example is Becaplermin (Regranex), a topical gel containing platelet-derived growth factor (PDGF) that stimulates fibroblast and keratinocyte migration [27]. This therapeutic advancement was developed using genetically modified animal models and is now an FDA-approved treatment for chronic diabetic ulcers [28].

Additionally, Integra Dermal Regeneration Template, a bioengineered skin substitute approved by the FDA in 1996, is widely used in reconstructive surgeries and

Table 11.1 Common modifications and disease models

Human pathology	Animal preclinical model	Outcome
Diabetic wounds	Db/db transgenic rodent model (leptin receptor deficiency) Streptozocin (STZ) injection disrupts pancreatic beta cells	Impaired healing due to hyperglycemia leading to delayed wound closure [18, 19]
Pressure ulcer wounds	Standardized wound model with magnetized pressurized device	Ischemic wound bed due to pressure from standardized pressure device [20]
Ischemic venous ulcer	VEGF, HIF-1α mutation	Absence of VEGF leads to reduced wound healing due to reduced oxygenation to wound bed [21]
Immune compromised patients	Foxn1, IL2 deletion injection of immunosuppressant drugs such as prednisone, azathioprine	Athymic/ immunosuppressed animal, lacking mature T cells model of immunocompromised pathology. Delayed wound healing due to muffled immune response [22, 23]
Chronic infected wound	Porcine biofilm-infected wound model	Wound healing process complicated by inflammation from polymicrobial infection [24]
Hypertrophic scarring and keloids	TGF-B1 overexpressed mouse P53 knockout	Increased fibrosis in scar formation due to excessive fibroblast activity [25, 26]

Table summarizing modeling techniques used to replicate specific pathological wound healing conditions

burn treatments. This innovation was first tested in preclinical animal models before being successfully implemented in human clinical trials [29]. Another significant wound care product is Grafix, a biological wound dressing derived from placental tissue. Extensive preclinical testing was essential to confirm that Grafix would not cause toxicity or immune rejection in humans [30].

Vashe® a hypochlorous acid-based topical solution, is commonly used to prevent wound contamination. Preclinical studies, including biofilm models in rodents, helped establish its efficacy and safety [24].

Another major advancement in wound healing is negative pressure wound therapy (NPWT). Preclinical trials using porcine models of cutaneous wounds were conducted to measure improvements in blood flow, granulation tissue formation, and bacterial clearance, all of which contributed to its widespread adoption in clinical settings [31]. A more recent innovation in wound care is Natrox, a topical device that delivers oxygen directly to wounded tissue. The development of this device was made possible by studying its effects on animal models of hypoxic wounds [32]. Similar preclinical research on hypoxic wound healing in animals contributed to refining oxygen-based wound therapies [33].

Utility of Preclinical Models in Wound Healing Research

Preclinical models play a crucial role in tracking the stages of the wound healing process. Macroscopic techniques used for monitoring wound progression include wound tracing, caliper measurements, and photographic documentation at different stages [34]. Microscopic tools provide a more detailed classification of wound healing stages, such as histological analysis using H&E stain, Masson's trichrome stain, and collagen deposition staining [35].

At the molecular and cellular level, more advanced techniques provide further granularity in wound characterization. Immunohistochemistry and flow cytometry can identify specific cell types involved in the wound healing process [35]. Additionally, PCR and ELISA assays are commonly used to quantify cytokines and growth factors present at different healing stages [35]. More recently, single-nucleus RNA sequencing (snRNA-seq) and single-cell RNA sequencing (scRNA-seq) have allowed researchers to track gene expression dynamics at various stages of wound healing [36].

Preclinical modeling provides a unique opportunity to compare wound healing under different conditions. By using animal models, researchers can investigate why some tissues heal without scarring, while others develop hypertrophic scars [37]. These controlled models help limit confounding factors, enabling direct comparisons of different healing processes. Understanding these differences can contribute to the development of targeted therapies aimed at promoting scarless healing or improving wound healing in pathological conditions [37].

One key application of preclinical modeling is fetal wound healing research [38]. Unlike adult wound healing, which typically results in fibrotic scarring, fetal wounds can heal without forming scars at specific developmental time points [39]. Scarless fetal wound healing is characterized by lower inflammation, increased hyaluronic acid levels, elevated growth factors like TGF-β3, and a higher collagen III to collagen I ratio, leading to more organized tissue regeneration [38]. Fetal surgery models in animals are currently being used to further investigate the biological pathways underlying scarless healing, which could inform the development of future therapeutic interventions for humans [40].

Another example of minimal-scarring wound healing comes from mucosal tissues [41]. Unlike cutaneous wounds, mucosal tissues, such as oral, vaginal, and intestinal epithelium, heal with minimal scarring [41]. Factors contributing to this process include reduced inflammation, increased hyaluronic acid and laminin levels, and enhanced fibroblast proliferation and migration [42]. Preclinical models used to study scarless mucosal healing include rodent, pig, and rabbit models of oral mucosal wounds, intestinal resections, and ulcer-induced healing in the gut [43].

Preclinical Models as a Bridge to Clinical Trials

The FDA mandates that preclinical models serve as a prerequisite for human clinical trials to ensure that potential therapeutics are safe and effective before being tested in humans [2]. Preclinical models provide critical data on potential toxic effects and risks associated with a treatment [2]. These studies allow researchers to evaluate the efficacy of treatments by assessing their ability to enhance tissue repair in animal models while also identifying toxicities or adverse effects to ensure the safety of an intervention before advancing to human trials [44].

Using preclinical studies as a required benchmark before human clinical trials helps conserve valuable resources, including time and money, while preventing negative health consequences that could arise if an intervention's safety was not properly established [45]. Additionally, results from these experiments help determine the appropriate starting dose for human trials [45]. Figure 11.2 illustrates the steps necessary for developing and introducing a wound-healing intervention into clinical applications. This process involves multiple steps over several years, including preclinical testing through either in vitro or in vivo studies. These studies help identify promising therapeutic factors and evaluate potential toxic effects at various doses [46].

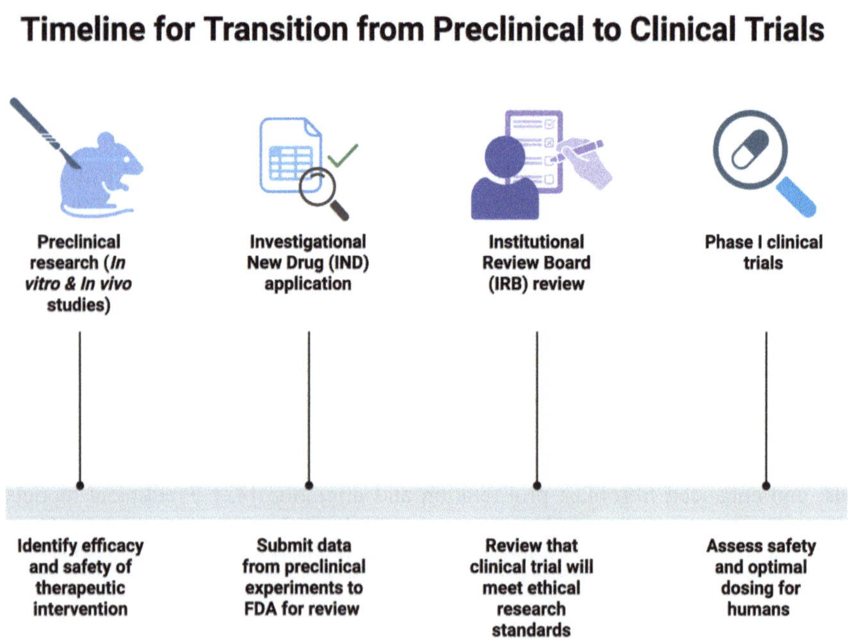

Fig. 11.2 Timeline for transition from preclinical to clinical trials. Timeline illustrating the role of animal studies as a critical precursor to human clinical trials in the drug approval process. (Created in BioRender. Bitolas, R. (2025) https://BioRender.com/nptrbmc)

The transition to human clinical trials requires regulatory approval through an Investigational New Drug (IND) application submitted to the FDA in the United States. This submission includes laboratory and animal data to determine whether sufficient evidence supports the safety of testing the intervention in humans [45]. In addition to FDA approval, an Institutional Review Board (IRB) must review and approve the study to ensure it meets ethical guidelines. Once these requirements have been met, Phase I clinical trials involving human participants can begin.

Limitations of Preclinical Models

Inherent limitations of preclinical models stem from the fact that animal models differ from humans in both anatomy and physiology. Table 11.2 discusses the key variations, including epidermal thickness. On average, human epidermis is thicker than mouse epidermis, while rabbit epidermis is slightly more similar to mice but remains thinner than human skin.[13] Among preclinical models, pig epidermis most closely resembles human epidermal thickness [50]. Immunological differences also exist between humans and animals, particularly in inflammatory responses. Humans sometimes exhibit lower levels of inflammatory pathways, which are less pronounced compared to rodent models. This discrepancy can affect the applicability of certain therapeutic interventions that may work in animals but not translate effectively to human physiology [35]. Additionally, murine wound models show distinct differences in chemokine production during the inflammatory stage of wound healing, further limiting direct translation to human treatments [51].

Table 11.2 Comparison of animal models for wound healing research

Animal model	Epidermal thickness (relative to human)	Wound healing mechanism	Hair follicle density	Panniculus carnosus
Human [47]	50–120 μm	Re epithelialization	Varied	Absent
Mice [47]	10–20 μm	Contraction	High (fur covered)	Present
Rats [47]	10–20 μm	Contraction	High (fur covered)	Present
Rabbits [47]	9–10 μm	Less dense than mice/rats	Absent	Present
Pigs [47]	15–80 μm	Re epithelialization	Sparse	Absent
Sheep [48]	25 μm	Contraction	High (wool covered)	Absent
Dogs [49]	10 μm	Mostly contraction (depending on region of body)	Depends on breed	Present

Table summarizing the anatomical and physiologic differences among animal species commonly used in wound healing research

Among preclinical models, pigs exhibit the most similar biological responses to wound healing and share the closest skin anatomy with humans, making findings from pig models more likely to be transferable to human applications [35]. Another major difference is in the re-epithelialization process. In humans, granulation tissue formation is slower, whereas in murine models, rapid wound contraction allows for quicker closure with minimal granulation tissue formation [13]. There are also differences in skin appendages across species. Humans have a much sparser distribution of hair compared to rodents and rabbits.[13]In terms of sweat glands, pigs have a distribution more similar to humans, whereas rabbits exhibit sweat gland physiology that more closely resembles rodents. This distinction is particularly important when studying burn injuries and thermoregulation in relation to wound healing [13].

Tissue oxygenation levels also vary across species, with rodents exhibiting lower oxygenation levels during healing due to their higher metabolic rates [52]. Their rapid metabolism also results in faster drug absorption, shorter half-life, and quicker excretion, making it challenging to directly translate effective dosages of wound healing therapies from rodents to humans [53]. Beyond biological differences, logistical challenges also impact preclinical wound healing research. The costs associated with animal care, including husbandry, feeding, veterinary services, and specialized equipment, significantly contribute to overall research expenses. Personnel requirements also vary depending on the size and type of animal model used [54]. While larger animals more closely resemble human physiology, they also require greater resources, making them more expensive to maintain.

Preclinical Model Insight into the Stages of Wound Healing

Each major stage of wound healing provides an opportunity to study and develop therapeutic interventions, as illustrated in Fig. 11.3. Hemostasis is the first immediate response to cutaneous injury and is crucial for preventing excessive blood loss [55]. This phase is essential for the subsequent steps of wound healing to occur. However, studying hemostasis in vitro is challenging because it relies on systemic feedback signals that are difficult to replicate in controlled environments [56]. Preclinical models are used to test therapeutic agents such as hemostatic dressings, as the effectiveness of this phase directly impacts the rate of wound healing and regenerative outcomes. Rodent models are commonly used to study clot formation and vasoconstriction because their clotting physiology, including complement and coagulation factors, is well-characterized and shares regulatory similarities with human systems [57]. These models typically involve tail vein bleeding, excisional wounds, and laser-induced thrombosis [56]. One study assessed the effect of the Algan Hemostatic Agent (AHA) on bleeding time in a rat tail hemorrhage model, where hemorrhage was induced by transecting the tail and applying either a saline-soaked sponge (control) or the AHA sponge. Results showed that the AHA sponge significantly reduced bleeding time [58].

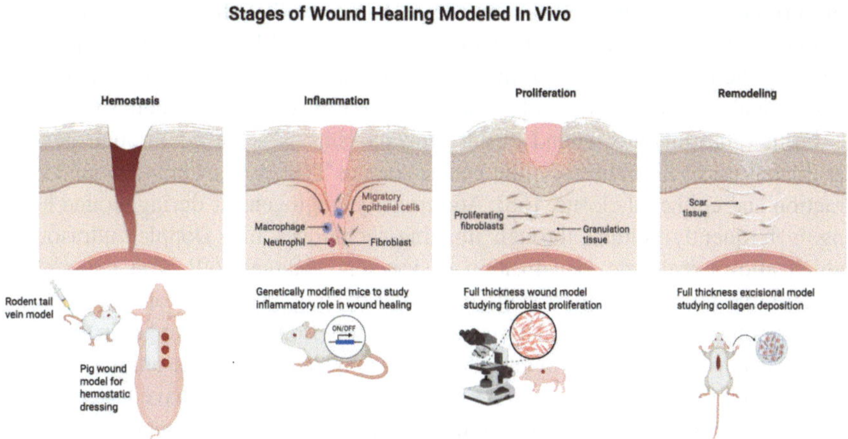

Fig. 11.3 Stages of wound healing modeled in vivo. Stages of wound healing and corresponding areas of animal research that align with each phase of the healing process. (Created in BioRender. Bitolas, R. (2025) https://BioRender.com/iulx5ti)

Pig models are also used to study hemostasis in wound healing due to their similarities to human skin in anatomy and physiology. For example, one study assessed the efficacy of the hemostatic agent polyhedral oligomeric sesquioxide (POSS) by comparing its effects on early wound repair in full-thickness wounds versus saline alone, demonstrating accelerated wound healing [59]. The inflammatory stage immediately follows hemostasis and plays a key role in clearing pathogens, removing necrotic tissue, and preparing for tissue regeneration. As with hemostasis, this phase is difficult to study in vitro due to the complexity of immune signaling pathways [60]. However, in vivo studies provide valuable insight into the dynamic processes occurring during inflammation, which cannot be fully captured in vitro models [61].

Preclinical models, particularly genetically modified mice, allow researchers to investigate the role of specific cytokines and how their deficiencies impact wound healing [62]. Additionally, cellular inflammatory responses have been studied by introducing medications such as glucocorticoids, immunosuppressants, and NSAIDs to evaluate how externally modulating the immune response affects the inflammatory phase and subsequent wound healing stages [63]. In some cases, antibodies are introduced to block specific cytokines without altering genetic expression. For example, infliximab, a TNF-α inhibitor, has been shown to improve wound healing outcomes [64]. Similarly, cellular interventions using clodronate liposome treatment allow for the selective depletion of macrophages, enabling researchers to study their role in inflammation and wound healing [65].

The proliferative phase is best studied in vivo due to the dynamic interactions required for extracellular matrix remodeling and vascularization [66]. Preclinical studies focus on mechanisms behind angiogenesis, granulation tissue formation, and fibroblast activity [66]. However, rodent models heal faster than humans and rely primarily on wound contraction rather than re-epithelialization, making them an imperfect representative of human wound healing [13]. Common methods for studying the proliferative phase include full-thickness wound models followed by histological analysis to assess fibroblast proliferation and collagen deposition [67].

Further insight into wound closure mechanisms is achieved through genetic modifications of fibroblast-specific genes, such as TGF-β, to examine wound contraction and epithelial closure [68]. Additionally, angiogenesis during wound healing is frequently studied through imaging techniques like Doppler ultrasound, which allows real-time visualization of vascular changes [69]. The remodeling phase is the final and longest stage of wound healing and is primarily studied by focusing on scar formation and collagen deposition. Long-term changes in extracellular matrix remodeling can be observed in vivo, provided that the animal survives long enough to allow for evaluation [70]. Rodent models, which have faster wound healing processes, enable efficient studies on remodeling by allowing researchers to track changes over a shorter time frame. Full-thickness excisional wound models in rodents are commonly used to study collagen transition from type III (early scar) to type I (mature scar). These changes are analyzed using histological staining methods such as Masson's trichrome and Picrosirius red, which help assess collagen fiber organization [71].

While rodent models offer efficiency in studying remodeling, they do not fully replicate the excessive fibroblast activity and collagen deposition observed in human hypertrophic scarring. Instead, porcine wound models are preferred for studying fibrotic scar formation, as pigs exhibit hypertrophic scarring during this phase. Conversely, myofibroblast activity, which plays a significant role in wound contraction, is more effectively studied in rodent models, as their wound healing process is predominantly characterized by contraction rather than granulation tissue formation [72].

Types of Preclinical Wound Modeling

Just as wounds emerge from many different types of injuries, modeling techniques offer a broad range of tools to replicate various types of tissue damage. Some of these models are shown in Fig. 11.4. Full-thickness wounds involve the epidermis, dermis, and subcutaneous tissue and are expected to progress through all stages of wound healing [13]. Animal models can replicate these wounds using circular or square-shaped excisions, linear incisions, or punch biopsies on the dorsum. In these models, one wound can serve as a control with no intervention. To prevent contraction from closing the wound too quickly and to better model human scarring, splinting can be used. This technique involves placing a silicone ring around the wound to maintain its size and delay closure [66]. Splinting is also useful for modeling chronic wounds, as it prolongs the healing process to better reflect the delayed wound healing seen in humans [73].

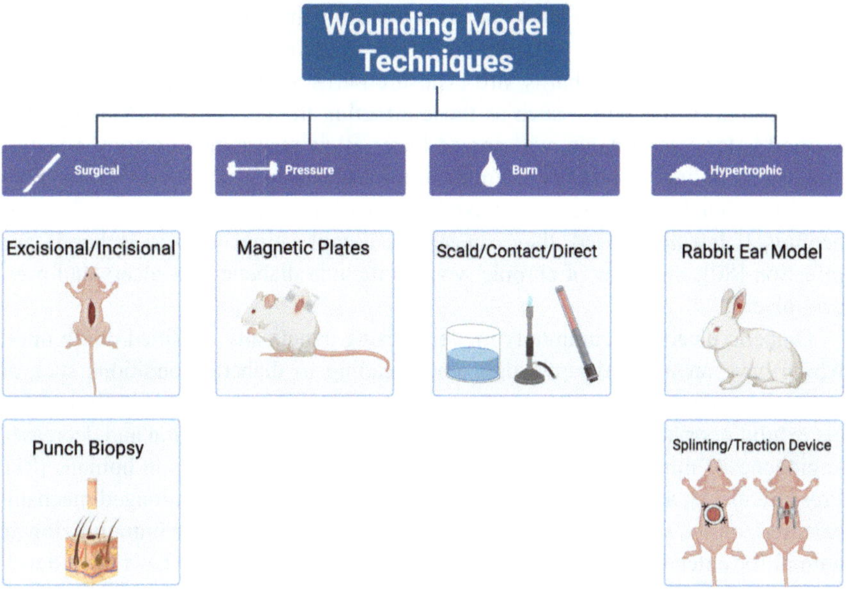

Fig. 11.4 Wound model techniques. Common wound modeling techniques used to replicate human skin injuries. (Created in BioRender. Bitolas, R. (2025) https://BioRender.com/w585wth)

Porcine models are frequently used for studying larger incisional wounds on the dorsum, as their skin anatomy and physiology closely resemble human skin. Unlike rodent models, splinting is not necessary in pigs, making them well-suited for preclinical therapy testing [8]. In rabbit models, full-thickness excisional wounds are often created on the back or ears [74]. Since the ears lack underlying muscle, wound contraction is minimal, eliminating the need for splinting. As a result, rabbit ear models are particularly useful for studying re-epithelialization processes rather than wound contraction [74].

Burn wounds can be classified into first-degree burns, which affect only the epidermis, second-degree or partial-thickness burns, which extend into the dermis, and full-thickness burns, which involve the entire epidermis and dermis and may reach subcutaneous tissue. Preclinical models are commonly used to study burn pathophysiology, grafting techniques, infection control, and fibrosis following burn injuries [75]. One widely used burn model is the scald burn model, which involves applying hot water to the skin at a controlled temperature for a set period. This technique allows researchers to create partial-thickness and full-thickness burns while maintaining control over burn depth, temperature, and exposure time [76]. Another method is the contact burn model, in which a heated metal surface is pressed against the skin for a predetermined duration. This model is particularly useful for studying localized burn injuries and site-specific healing processes. However, one limitation of this method is variability in burn depth between replicates [77].

The direct fire exposure model involves exposing the animal to an open flame for a set period, making it a valuable model for studying full-thickness burns [78]. Unlike scald and contact burns, fire exposure burns are not as localized but better replicate real-world burns, such as those affecting the human face. Although this method is less controllable, widespread burns allow researchers to study systemic effects, including respiratory complications associated with severe burns [79]. Chronic wounds result from a delayed progression through the stages of wound healing. If left unresolved, they can lead to complications such as ischemia and infection [80]. Subtypes of chronic wounds include diabetic foot ulcers and pressure ulcers.

Diabetic ulcers are commonly modeled using genetically modified db/db mice, which have an induced leptin deficiency leading to diabetic conditions such as hyperglycemia, insulin resistance, and obesity [81]. Wounds created in these models exhibit poor healing outcomes, including increased inflammation and decreased angiogenesis, mirroring the characteristics of diabetic foot ulcers in humans [81]. Pressure ulcers, another form of chronic wounds, occur due to prolonged mechanical stress on the skin, often resulting from limited mobility. Since immobilizing an animal for extended periods is not practical, alternative methods have been developed to replicate these conditions in a shorter time frame. One such approach is the magnetic compression model, in which two opposing magnets are placed on both sides of the skin to apply pressure, leading to ischemia and necrosis [82].

Chronic wounds are also highly susceptible to bacterial infections, which further delay healing. These infections can be replicated in preclinical models through inoculation with common wound pathogens, such as Pseudomonas aeruginosa *and* Staphylococcus aureus [18]. Samples can then be collected to confirm successful bacterial colonization through culture analysis. Hypertrophic wounds result from excessive fibroblast activity, prolonged inflammation, and excessive extracellular matrix deposition, leading to raised, stiffened tissue [83]. Hypertrophic wound models are valuable for studying tissue fibrosis and abnormal collagen organization [83].

In vivo models are essential for studying the systemic responses that contribute to hypertrophic scar formation. Applying mechanical tension to wounds, such as through splinting or implanted traction devices, can induce hypertrophic scarring in rodent models, where it would not naturally occur [26]. The rabbit ear hypertrophic scar model involves creating full-thickness excisional wounds on the ventral ear, which leads to the formation of hypertrophic scars [84]. In porcine models, deep dermal burns reliably produce hypertrophic scarring, making them a relevant model for studying human burn-induced fibrosis [85].

Animal Models in Pre-clinical Wound Healing Research

Small Animals

Mice

Mice are a commonly used rodent model for wound healing due to their cost-effectiveness, small size, and fast healing process, which typically lasts about 7 days, as shown in Table 11.3. This rapid healing also allows for shorter study durations [35]. Genetic modifications are possible in mice, as their widely sequenced genome enables transgenic manipulation, including gene knockouts and the insertion of human genes to mimic human diseases [86]. Several methods exist for genetic modification in mice. DNA microinjection introduces foreign genes into embryos, while embryonic stem cell modification involves inserting foreign DNA into embryonic stem cells to create transgenic lines [87]. Genes can also be

Table 11.3 Animal model characteristics

Animal model	Advantages	Disadvantages
Mice	Cost effective Small size allows high-throughput studies Rapid healing allows shorter study duration Genetic modification possible	Heal by contraction Naturally lack hypertrophic scars
Rats	Larger body size allows for multiple wounds per animal Slower healing than mice, more similar to humans	Heals with some degree of contractions Fewer genetic modification tools More expensive than mice
Rabbits	Healing process closer to humans Larger body size Rabbit ear model allows for hypertrophic scar studies	More expensive than rodents Limited genetic modification tools
Pigs	Most similar skin structure and healing process to humans Useful for testing biomaterials	Expensive to maintain Require specialized facilities Limited genetic modification options
Sheep	Longer epidermal turnover rate allows for extended observation Easier venous access compared to pigs	Greater wound contraction than pigs Limited genetic modifications than pigs
Dog	Skin structure closer to humans than rodents Useful for reconstructive surgery studies	High interbreed variability Ethical concerns
Zebrafish	Rapid wound healing Regenerative scarless healing Transparent skin allows real time imaging Genetic modifications possible	Cannot model hypertrophic scars or fibrotic conditions

Table summarizing advantages and limitations of different animal species used in wound healing research delivered through viral vectors such as Adeno-Associated Virus (AAV), which allows gene introduction into specific tissues [88]. Additionally, CRISPR techniques can be used to alter specific genetic sequences, enabling researchers to induce or track changes throughout the wound healing process [89].

Mice serve various applications in wound healing research. They are widely used to study signaling pathways involved in wound healing and fibrosis, particularly pathways regulated by TGF-β and IL-1. These models are also used to test potential antifibrotic treatments for safety and efficacy before advancing to studies in larger animals [90]. Furthermore, immunodeficient mouse strains can be developed to accept human stem cells or human skin grafts, allowing for the evaluation of cell-based therapies and regenerative biomaterials [91]. Despite these advantages, mice have limitations as a model for human wound healing. Unlike humans, mice primarily heal by contraction rather than epithelialization due to the presence of the panniculus carnosus, a skeletal muscle layer absent in humans. This contraction-driven healing prevents the development of hypertrophic scars unless specific interventions are introduced [2].

Rat

Rats are also commonly used as rodent models for wound healing. They are larger in size compared to mice, which allows for the creation of larger wounds and makes surgical manipulation easier [13]. Like mice, rats have a widely sequenced genome that enables transgenic manipulation, though the availability of transgenic rat lines is more limited than that of mice [54, 92]. A key advantage of using rats in wound healing research is that their wounds heal more slowly than those of mice, making their healing process more similar to human wound healing and allowing for a longer observation period [35]. While gene expression patterns in rats are similar to those in mice, the duration and concentration of gene expression can differ due to the prolonged wound healing process in rats [51].

Although rats form minimal scars similar to mice, they exhibit slightly less wound contraction. Despite being closer to human wound healing than mice, their healing process still involves contraction, which differs from human re-epithelialization. Compared to mice, rats have fewer options for genetic modifications, making it easier to deliver gene-targeting therapies in mice than in rats. Additionally, rats are more expensive than mice and require more resources for maintenance.

Rabbit

Rabbit wound models are widely used in wound healing and hypertrophic scar research [84]. Among smaller animal models such as rats and mice, rabbits more closely resemble human wound healing characteristics [93]. This is because their healing process is primarily driven by re-epithelialization rather than contraction.

Additionally, their larger body size allows for the accommodation of multiple wounds, and the tensile strength of rabbit skin better mimics human wound healing physiology compared to rodents. Unlike mice, rabbits lack a panniculus carnosus layer, which contributes to the reduced contraction seen in their healing process [94].

A unique advantage of rabbit wound models is the use of rabbit ears, which provide a well-vascularized area where wounds heal without contraction. This is due to the cartilaginous composition of the ear, which lacks underlying muscle, preventing the wound from contracting [95]. Because of these advantages, rabbits are a suitable small animal model for studying hypertrophic scars, fibrosis, and angiogenesis in wound healing research [13].

However, compared to other small animal models, rabbits present certain limitations. They are more expensive than rodents, require larger housing, and have fewer available genetic modification tools. While the rabbit ear wound healing model is particularly useful for studying angiogenesis, its application is somewhat limited by these factors.

Large Animals

Pig

Pig models offer the closest anatomical and physiological resemblance to human skin. The characteristics that make pig skin advantageous for modeling human wound healing include epidermal thickness and re-epithelialization rate [8]. Their sparse distribution of hair follicles more closely resembles human wound healing than the dense hair coverage seen in rodents [8]. The wound healing process in pigs is marked by a similar epidermal turnover rate, sparse hair distribution, the absence of a panniculus carnosus, and a comparable distribution of blood vessels in the underlying dermis [77]. Pigs exhibit re-epithelialization between days 13 and 18 post-injury [96]. Angiogenesis occurs within the first 4 days following partial-thickness wounds and can be identified using Doppler imaging. These anatomical and physiological characteristics make pigs highly suitable for various wound healing applications.

One major advantage of pig models is their similar immune cell response to humans, with several conserved chemokines between human and pig wound healing, making them particularly useful for studying the inflammatory phase of wound healing [97]. Histologically, hypertrophic scars in pigs exhibit an absence of rete pegs, which is also a characteristic of human hypertrophic scarring [98]. Their similarity in wound healing physiology makes them an excellent candidate for investigating hypertrophic scarring mechanisms and potential interventions. Additionally, pigs are well-suited for testing biomaterials, especially bioengineered skin substitutes [99].

Despite these advantages, using pigs as a model for wound healing comes with limitations. Pigs are not as cost-effective as small animal models and present logistical challenges in wound healing studies [77]. Their similarity to human anatomy and physiology also comes with the downside of requiring complex surgical procedures and anesthesia management [100]. Additionally, pigs offer fewer options for targeted genetic interventions compared to small animal models, making certain experimental modifications more difficult.

Sheep

Sheep are another large animal model that can be used for studying human wound healing. Like pigs, they offer several anatomical and physiological similarities to human wound healing, making them a more relevant model than small animals. However, unlike pigs, sheep have a longer epidermal turnover rate, with full epithelialization occurring between 4 and 6 weeks [101]. One factor influencing this slower healing process is the difference in hair follicles. While pigs have sparse hair follicles, sheep have wool follicles, which can complicate the wound healing process due to their structural differences from human hair follicles. Additionally, sheep exhibit greater wound contraction than pigs, likely due to their looser skin. Studies have shown that sheep wounds can undergo 78–81% contraction within 2 weeks post-injury [101]. Despite their similarities to human wound healing, sheep have limited flexibility for gene-targeted wound healing studies and are less cost-effective than small animal transgenic models [102]. Although sheep provide easy peripheral venous access and have sufficient dorsal skin surface to accommodate multiple lesions, they are still not as logistically convenient to work with as other large animal models [103].

Dog

The use of dogs as a preclinical model has historically been applied in reconstructive surgery advancements and biomaterial studies [104]. However, dogs are primarily classified as companion animals and are not widely used for wound healing research [105]. While they are less commonly utilized than pigs, dogs offer unique advantages for studying surgical wound healing, particularly in procedures involving flaps and grafts [106]. A major challenge in using dogs as a preclinical model is the high degree of interbreed variability, including differences in skin thickness, hair coverage, and skin tension [104]. Anatomically, the epidermal and dermal thickness of dogs is more similar to humans than rodents, although it is generally thinner than human skin [104]. One key limitation is that dogs possess a panniculus carnosus, which enables wound contraction, particularly on their dorsal surface [107]. However, wounds inflicted on the extremities of dogs heal with minimal contraction due to the absence of the panniculus carnosus in these areas [108]. In a preclinical context, dogs have been used for studying the implementation of skin flaps, which closely mimic human surgical wound healing [106]. While their use in wound healing research is limited, they remain valuable for specific applications related to reconstructive surgery and biomaterial testing.

Zebrafish

Zebrafish (*Danio rerio*) are of significant interest in regenerative medicine and wound healing research. Unlike adult mammals, zebrafish can regenerate skin, tissues, and organs, healing without scar formation [109]. While they share the least anatomical and physiological similarities to human wound healing among preclinical animal models, they provide valuable insights into regenerative wound healing mechanisms [110]. Their small size and rapid wound healing process, which typically lasts between 1 and 3 days, make zebrafish a convenient wound healing model. Re-epithelialization begins within just a few hours, allowing for the study of rapid tissue regeneration [109]. Zebrafish are widely used in scar-free wound healing research, offering key insights into cellular and molecular mechanisms of regeneration. Their transparent skin allows for live imaging and tracking of cellular migration under a microscope, facilitating real-time visualization of the healing process [111]. they are easily genetically modified, enabling researchers to manipulate genes involved in wound healing and regeneration [112].

Despite these advantages, zebrafish have several limitations as a preclinical wound healing model. They cannot effectively model a wide range of human diseases. For example, fibrosis-related skin conditions cannot able to be studied in this species because their tissue does not develop fibrosis. Additionally, zebrafish lack a keratinized stratum corneum, which is present in humans, making them unsuitable for testing repair therapies. Furthermore, they cannot replicate fibrotic pathologies such as hypertrophic scars [109].

Ethical Considerations in Animal Model Research

Regulations governing the use of animals in preclinical wound healing research are essential for maintaining the integrity and well-being of animals while ensuring a safe environment for all participants [17]. The main principles of ethical animal research were first proposed by microbiologists Rex Burch and William Russell and continue to be followed today to uphold both the welfare of research animals and the validity of scientific studies [113]. These principles are collectively known as the Three Rs—Replacement, Reduction, and Refinement—which serve as a framework for ethical animal use in research [114].

Replacement refers to the principle that animals should only be used when necessary, and their use should be avoided or substituted with alternative methods whenever possible. These alternatives include in vitro models, computational simulations, and the use of pre-existing datasets [115]. Reduction emphasizes minimizing the number of animals used in experiments while maintaining the statistical power needed for valid results [115]. Refinement focuses on optimizing experimental methods to ensure the humane treatment of animals, including pain reduction techniques, appropriate anesthesia and analgesia, humane euthanasia, and careful post-operative care planning [115].

These principles were formally incorporated into US federal law through the Animal Welfare Act (AWA) of 1966, which is enforced by the United States Department of Agriculture (USDA). This legislation established standards for the treatment of animals used in research and led to the creation of the Institutional Animal Care and Use Committee (IACUC), which oversees animal research practices to ensure compliance with ethical guidelines [116]. Institutional systems also uphold these principles, requiring that all research involving animals be reviewed and approved by IACUC or an equivalent oversight body. When submitting a research proposal that involves animal use, researchers must justify the necessity of using animals, specify humane endpoints, and describe pain management strategies. In the United States, the AWA and NIH guidelines govern the ethical treatment of research animals. Additionally, international guidelines, such as the National Research Council (NRC) Guide, establish specific standards for proper animal care in research settings [117].

Future Directions for Animal Models

Keeping in mind the use of preclinical models and the welfare of animals in advancing research, future advancements can provide alternatives that reduce the number of animals used in wound healing studies. There are many opportunities for improving preclinical animal models, particularly in enhancing the translational quality of data. One promising direction is the ability to establish transgenic models in animals beyond rodents, such as pigs, allowing researchers to study models that are genetically and physiologically more similar to humans [118, 119].

Additionally, improvements in 3D human skin models offer alternatives to reduce reliance on animal testing. Innovations such as integrating a microvascular system into a simulated vasculature within a collagen matrix have enhanced the physiological accuracy of in vitro models [119]. Furthermore, static human skin models have been expanded with microfluidic skin-on-a-chip devices, which culture skin under perfusion to regulate fluid flow, nutrient supply, and oxygen gradients, improving the relevance of these models for wound healing research [119].

Standardizing and increasing the reproducibility of preclinical wound healing studies is another critical area of advancement. The Wound Reporting in Animal and Human Preclinical Models guideline has been introduced to improve the consistency of preclinical wound healing studies [17]. This recently published guideline provides a checklist and reporting templates for commonly used preclinical models, ensuring more detailed documentation, better reproducibility of results, and improved consistency across studies. The implementation of such guidelines can facilitate cross-study comparisons and enhance the overall reliability of animal wound healing research [17]. Technological advancements are also paving the way for non-invasive monitoring devices that minimize pain and distress while providing access to more comprehensive data beyond traditional tissue collection and euthanasia. Emerging technologies such as smart bandages and hydrogel biosensors enable real-time, in vivo data collection in a minimally invasive manner, reducing the need for frequent biopsies or euthanasia [120].

Conclusion

Preclinical wound healing and repair models have been powerful tools in understanding the complex processes involved in wound repair. These models have contributed significantly to the study of the cellular and molecular mechanisms underlying wound healing at each stage, helping researchers better understand the systemic interactions essential to the healing process. They have also played a crucial role in the development and approval of new therapeutic agents, ensuring safety and efficacy before human trials.

Preclinical models, including both in vitro and in vivo methods, have evolved to mimic and study various types of wounds, such as full-thickness wounds, burn wounds, chronic wounds, and hypertrophic scars. Additionally, advancements such as genetic modifications and transgenic species have allowed researchers to replicate different human disease states within these models. Various species are used in preclinical wound healing experiments, each offering unique advantages and limitations. Efforts toward standardization of these models have improved the reproducibility and translational value of research, ensuring that findings are more applicable to human wound healing.

The field of preclinical wound healing research is continuously evolving to align with the 3Rs principles (Replacement, Reduction, and Refinement) and to develop models that require fewer live animals while improving research efficiency. Advances in physiologically accurate 3D in vitro wound healing models are helping to reduce reliance on animal testing. Furthermore, established guidelines for standardizing preclinical research are enhancing the reproducibility of wound healing studies, while non-invasive monitoring technologies are enabling less invasive research methods in live animal models.

References

1. Varkey B (2021) Principles of clinical ethics and their application to practice. Med Princ Pract 30(1):17–28. https://doi.org/10.1159/000509119
2. Sarian M, Zulkefli N, Zain S, Maniam S, Fakurazi S (2023) A review with updated perspectives on in vitro and in vivo wound healing models. Turk J Biol 47(4):236–246. https://doi.org/10.55730/1300-0152.2659

3. Abercrombie M, Heaysman JE (1953) Observations on the social behaviour of cells in tissue culture. I. Speed of movement of chick heart fibroblasts in relation to their mutual contacts. Exp Cell Res 5(1):111–131. https://doi.org/10.1016/0014-4827(53)90098-6
4. Rheinwald JG, Green H (1975) Serial cultivation of strains of human epidermal keratinocytes: the formation of keratinizing colonies from single cells. Cell 6(3):331–343. https://doi.org/10.1016/s0092-8674(75)80001-8
5. Rodriguez LG, Wu X, Guan JL (2005) Wound-healing assay. Methods Mol Biol 294:23–29. https://doi.org/10.1385/1-59259-860-9:023
6. Bell E, Ivarsson B, Merrill C (1979) Production of a tissue-like structure by contraction of collagen lattices by human fibroblasts of different proliferative potential in vitro. Proc Natl Acad Sci USA 76(3):1274–1278. https://doi.org/10.1073/pnas.76.3.1274
7. Bell E, Ehrlich HP, Buttle DJ, Nakatsuji T (1981) Living tissue formed in vitro and accepted as skin-equivalent tissue of full thickness. Science 211(4486):1052–1054. https://doi.org/10.1126/science.7008197
8. Sullivan TP, Eaglstein WH, Davis SC, Mertz P (2001) The pig as a model for human wound healing. Wound Repair Regen 9(2):66–76. https://doi.org/10.1046/j.1524-475x.2001.00066.x
9. Dunn PM (2003) Galen (ad 129–200) of Pergamun: anatomist and experimental physiologist. Arch Dis Child Fetal Neonatal Ed 88(5):F441–F443. https://doi.org/10.1136/fn.88.5.F441
10. Cavaillon JM (2021) Once upon a time, inflammation. J Venom Anim Toxins Incl Trop Dis 27:e20200147. https://doi.org/10.1590/1678-9199-JVATITD-2020-0147
11. Winter GD (1962) Formation of the scab and the rate of epithelization of superficial wounds in the skin of the young domestic pig. Nature 193(4812):293–294. https://doi.org/10.1038/193293a0
12. Hunt TK, Twomey P, Zederfeldt B, Dunphy JE (1967) Respiratory gas tensions and pH in healing wounds. Am J Surg 114(2):302–307. https://doi.org/10.1016/0002-9610(67)90388-1
13. Grada A, Mervis J, Falanga V (2018) Research techniques made simple: animal models of wound healing. J Invest Dermatol 138(10):2095–2105.e1. https://doi.org/10.1016/j.jid.2018.08.005
14. Neves LMG, Wilgus TA, Bayat A (2023) In vitro, ex vivo, and in vivo approaches for investigation of skin scarring: human and animal models. Adv Wound Care 12(2):97–116. https://doi.org/10.1089/wound.2021.0139
15. Hofmann E, Fink J, Pignet AL et al (2023) Human in vitro skin models for wound healing and wound healing disorders. Biomedicine 11(4):1056. https://doi.org/10.3390/biomedicines11041056
16. Gibbins JR (1969) Migration of stratified squamous epithelium after injury in vivo and in organ culture. Pathology 1(3):225. https://doi.org/10.3109/00313026909071300
17. Ojeh N, Vecin NM, Pastar I et al (2025) The wound reporting in animal and human preclinical studies (WRAHPS) guidelines. Wound Repair Regen 33(1):e13232. https://doi.org/10.1111/wrr.13232
18. Zhao G, Hochwalt PC, Usui ML et al (2010) Delayed wound healing in diabetic (db/db) mice with Pseudomonas aeruginosa biofilm challenge: a model for the study of chronic wounds. Wound Repair Regen 18(5):467–477. https://doi.org/10.1111/j.1524-475X.2010.00608.x
19. Furman BL (2015) Streptozotocin-induced diabetic models in mice and rats. Curr Protoc Pharmacol 70(1):5.47.1–5.47.20. https://doi.org/10.1002/0471141755.ph0547s70
20. Tuca AC, Bernardelli de Mattos I, Funk M et al (2024) A standardized porcine model for partial-thickness wound healing studies: design, characterization, model validation, and histological insights. Int J Mol Sci 25(14):7658. https://doi.org/10.3390/ijms25147658
21. Duscher D, Maan ZN, Whittam AJ et al (2015) Fibroblast-specific deletion of hypoxia inducible factor-1 critically impairs murine cutaneous neovascularization and wound healing. Plast Reconstr Surg 136(5):1004–1013. https://doi.org/10.1097/PRS.0000000000001699
22. Alshehabat M, Hananeh W, Ismail ZB, Rmilah SA, Abeeleh MA (2020) Wound healing in immunocompromised dogs: a comparison between the healing effects of moist

exposed burn ointment and honey. Vet World 13(12):2793–2797. https://doi.org/10.14202/vetworld.2020.2793-2797
23. Wei X, Lai Y, Li B et al (2017) CRISPR/Cas9-mediated deletion of Foxn1 in NOD/SCID/IL2rg−/− mice results in severe immunodeficiency. Sci Rep 7(1):7720. https://doi.org/10.1038/s41598-017-08337-8
24. Robson MC, Payne WG, Ko F et al (2007) Hypochlorous acid as a potential wound care agent: part II. Stabilized Hypochlorous acid: its role in decreasing tissue bacterial bioburden and overcoming the inhibition of infection on wound healing. J Burns Wounds 6:e6
25. Liu Y, Li Y, Li N et al (2016) TGF-β1 promotes scar fibroblasts proliferation and transdifferentiation via up-regulating MicroRNA-21. Sci Rep 6(1):32231. https://doi.org/10.1038/srep32231
26. Aarabi S, Bhatt KA, Shi Y et al (2007) Mechanical load initiates hypertrophic scar formation through decreased cellular apoptosis. FASEB J 21(12):3250–3261. https://doi.org/10.1096/fj.07-8218com
27. LeGrand EK (1998) Preclinical promise of becaplermin (rhPDGF-BB) in wound healing. Am J Surg 176(2A Suppl):48S–54S. https://doi.org/10.1016/s0002-9610(98)00177-9
28. Cohen MA, Eaglstein WH (2001) Recombinant human platelet-derived growth factor gel speeds healing of acute full-thickness punch biopsy wounds. J Am Acad Dermatol 45(6):857–862. https://doi.org/10.1067/mjd.2001.117721
29. Integra® Dermal Regeneration Template. Accessed June 7, 2025. https://products.integralife.com/integra-dermal-regeneration-template/product/wound-reconstruction-care-inpatient-acute-or-integra-dermal-regeneration-template
30. Silini AR, Cargnoni A, Magatti M, Pianta S, Parolini O (2015) The long path of human placenta, and its derivatives, in regenerative medicine. Front Bioeng Biotechnol 3:162. https://doi.org/10.3389/fbioe.2015.00162
31. Morykwas MJ, Argenta LC, Shelton-Brown EI, McGuirt W (1997) Vacuum assisted closure: a new method for wound control and treatment: animal studies and basic foundation. Ann Plast Surg 38(6):553–562. https://doi.org/10.1097/00000637-199706000-00001
32. Tang TY, Mak MYQ, Yap CJQ et al (2024) An observational clinical trial examining the effect of topical oxygen therapy (Natrox™) on the rates of healing of chronic DiAbetic foot ulcers (OTONAL trial). Int J Low Extrem Wounds 23(2):326–337. https://doi.org/10.1177/15347346211053694
33. Fries RB, Wallace WA, Roy S et al (2005) Dermal excisional wound healing in pigs following treatment with topically applied pure oxygen. Mutat Res 579(1–2):172–181. https://doi.org/10.1016/j.mrfmmm.2005.02.023
34. Huang CX, Siwan E, Fox SL, Longfield M, Twigg SM, Min D (2023) Comparison of digital and traditional skin wound closure assessment methods in mice. Lab Anim Res 39(1):25. https://doi.org/10.1186/s42826-023-00176-1
35. Masson-Meyers DS, Andrade TAM, Caetano GF et al (2020) Experimental models and methods for cutaneous wound healing assessment. Int J Exp Pathol 101(1–2):21–37. https://doi.org/10.1111/iep.12346
36. Merrick D, Sakers A, Irgebay Z et al (2019) Identification of a mesenchymal progenitor cell hierarchy in adipose tissue. Science 364(6438):eaav2501. https://doi.org/10.1126/science.aav2501
37. Gurtner GC, Chapman MA (2016) Regenerative medicine: charting a new course in wound healing. Adv Wound Care 5(7):314–328. https://doi.org/10.1089/wound.2015.0663
38. Larson BJ, Longaker MT, Lorenz HP (2010) Scarless fetal wound healing: a basic science review. Plast Reconstr Surg 126(4):1172. https://doi.org/10.1097/PRS.0b013e3181eae781
39. Longaker MT, Whitby DJ, Adzick NS et al (1990) Studies in fetal wound healing, VI. Second and early third trimester fetal wounds demonstrate rapid collagen deposition without scar formation. J Pediatr Surg 25(1):63–68; discussion 68–69. https://doi.org/10.1016/s0022-3468(05)80165-4

40. Gomes MLNP, Krijnen PAJ, Middelkoop E, Niessen HWM, Boekema BKHL (2025) Fetal skin wound healing: key extracellular matrix components and regulators in scarless healing. J Invest Dermatol 145(2):280–302. https://doi.org/10.1016/j.jid.2024.05.027
41. Pereira D, Sequeira I (2021) A scarless healing tale: comparing homeostasis and wound healing of oral mucosa with skin and Oesophagus. Front Cell Dev Biol 9. https://doi.org/10.3389/fcell.2021.682143
42. Chen L, Arbieva ZH, Guo S, Marucha PT, Mustoe TA, DiPietro LA (2010) Positional differences in the wound transcriptome of skin and oral mucosa. BMC Genomics 11:471. https://doi.org/10.1186/1471-2164-11-471
43. Toma AI, Fuller JM, Willett NJ, Goudy SL (2021) Oral wound healing models and emerging regenerative therapies. Transl Res 236:17–34. https://doi.org/10.1016/j.trsl.2021.06.003
44. Preclinical studies. Toolkit. Accessed June 7, 2025. https://toolkit.ncats.nih.gov/glossary/preclinical-studies
45. Commissioner O of the. Step 2: Preclinical Research. FDA. Published online April 18, 2019. Accessed June 7, 2025. https://www.fda.gov/patients/drug-development-process/step-2-preclinical-research
46. Freedman BR, Hwang C, Talbot S, Hibler B, Matoori S, Mooney DJ (2023) Breakthrough treatments for accelerated wound healing. Sci Adv 9(20):eade7007. https://doi.org/10.1126/sciadv.ade7007
47. Wei JCJ, Edwards GA, Martin DJ, Huang H, Crichton ML, Kendall MAF (2017) Allometric scaling of skin thickness, elasticity, viscoelasticity to mass for micro-medical device translation: from mice, rats, rabbits, pigs to humans. Sci Rep 7(1):15885. https://doi.org/10.1038/s41598-017-15830-7
48. Sinclair AN, Filan SJ (1989) Lipid ingestion from sheep epidermis by *Psoroptes ovis* (Acari: Psoroptidae). Vet Parasitol 31(2):149–164. https://doi.org/10.1016/0304-4017(89)90030-7
49. Lloyd DH, Garthwaite G (1982) Epidermal structure and surface topography of canine skin. Res Vet Sci 33(1):99–104. https://doi.org/10.1016/S0034-5288(18)32367-1
50. Ahmad N (2023) In vitro and in vivo characterization methods for evaluation of modern wound dressings. Pharmaceutics 15(1):42. https://doi.org/10.3390/pharmaceutics15010042
51. Vinaik R, Aijaz A, Jeschke MG (2022) Chapter 9 - Small animal models of thermal injury. In: Montrose DC (ed) Methods in cell biology, Experimental models of infection, inflammation and injury, vol 168. Academic Press, pp 161–189. https://doi.org/10.1016/bs.mcb.2021.12.014
52. Perlman RL (2016) Mouse models of human disease: an evolutionary perspective. Evol Med Public Health 2016(1):170–176. https://doi.org/10.1093/emph/eow014
53. Anisimov VN, Ukraintseva SV, Yashin AI (2005) Cancer in rodents: does it tell us about cancer in humans? Nat Rev Cancer 5(10):807–819. https://doi.org/10.1038/nrc1715
54. Ansell DM, Holden KA, Hardman MJ (2012) Animal models of wound repair: are they cutting it? Exp Dermatol 21(8):581–585. https://doi.org/10.1111/j.1600-0625.2012.01540.x
55. Bayer IS (2022) Advances in fibrin-based materials in wound repair: a review. Molecules 27(14):4504. https://doi.org/10.3390/molecules27144504
56. Mohammed BM, Monroe DM, Gailani D (2020) Mouse models of hemostasis. Platelets 31(4):417–422. https://doi.org/10.1080/09537104.2020.1719056
57. Murine models in the evaluation of heparan sulfate-based anticoagulants | Springer Nature Experiments. Accessed June 8, 2025. https://experiments.springernature.com/articles/10.1007/978-1-0716-1398-6_59
58. Gedar Totuk ÖM, Güzel ŞE, Ekici H et al (2020) Effects of Algan Hemostatic Agent on bleeding time in a rat tail hemorrhage model. Ulus Travma Acil Cerrahi Derg 26(6):853–858. https://doi.org/10.14744/tjtes.2020.50384
59. Gerling KA, Kersey AJ, Lauria AL et al (2023) Evaluation of novel hemostatic agents in a coagulopathic swine model of junctional hemorrhage. J Trauma Acute Care Surg 95(2S Suppl 1):S144–S151. https://doi.org/10.1097/TA.0000000000004071

60. Wang H, Yang L (2023) Applications of injectable hemostatic materials in wound healing: principles, strategies, performance requirements, and future perspectives. Theranostics 13(13):4615–4635. https://doi.org/10.7150/thno.86930
61. Vågesjö E, Öhnstedt E, Mortier A et al (2018) Accelerated wound healing in mice by on-site production and delivery of CXCL12 by transformed lactic acid bacteria. Proc Natl Acad Sci 115(8):1895–1900. https://doi.org/10.1073/pnas.1716580115
62. Impaired wound healing with defective expression of chemokines and recruitment of myeloid cells in TLR3-deficient mice | The Journal of Immunology | Oxford Academic. Accessed June 8, 2025. https://academic.oup.com/jimmunol/article-abstract/186/6/3710/7990109?redirectedFrom=fulltext
63. Karner L, Drechsler S, Metzger M et al (2020) Contamination of wounds with fecal bacteria in immuno-suppressed mice. Sci Rep 10(1):11494. https://doi.org/10.1038/s41598-020-68323-5
64. Streit M (2004) Dramatic improvement of wound closure in nonhealing chronic leg ulcers after topical application of an antitumor necrosis factor-alpha monoclonal antibody (infliximab)1. J Am Acad Dermatol 50(3):P169. https://doi.org/10.1016/j.jaad.2003.10.599
65. van Amerongen MJ, Harmsen MC, van Rooijen N, Petersen AH, van Luyn MJA (2007) Macrophage depletion impairs wound healing and increases left ventricular remodeling after myocardial injury in mice. Am J Pathol 170(3):818–829. https://doi.org/10.2353/ajpath.2007.060547
66. Pignet AL, Schellnegger M, Hecker A, Kamolz LP, Kotzbeck P (2024) Modeling wound chronicity in vivo: the translational challenge to capture the complexity of chronic wounds. J Invest Dermatol 144(7):1454–1470. https://doi.org/10.1016/j.jid.2023.11.024
67. Wound healing modulation through the local application of powder collagen-derived treatments in an excisional cutaneous murine model. Accessed June 8, 2025. https://www.mdpi.com/2227-9059/10/5/960
68. Martinez-Ferrer M, Afshar-Sherif AR, Uwamariya C, de Crombrugghe B, Davidson JM, Bhowmick NA (2010) Dermal transforming growth factor-β responsiveness mediates wound contraction and epithelial closure. Am J Pathol 176(1):98–107. https://doi.org/10.2353/ajpath.2010.090283
69. Nurrachman AS, Azhari A, Epsilawati L, Pramanik F (2023) Temporal pattern of micro-CT angiography vascular parameters and VEGF mRNA expression in fracture healing: a radiograph and molecular comparison. Eur J Dent 17:283 295. https://doi.org/10.1055/s-0042-1757466
70. Saeed S, Martins-Green M (2024) Assessing animal models to study impaired and chronic wounds. Int J Mol Sci 25(7):3837. https://doi.org/10.3390/ijms25073837
71. Sicari BM, Johnson SA, Siu BF et al (2012) The effect of source animal age upon the *in vivo* remodeling characteristics of an extracellular matrix scaffold. Biomaterials 33(22):5524–5533. https://doi.org/10.1016/j.biomaterials.2012.04.017
72. Brewer CM, Nelson BR, Wakenight P et al (2021) Adaptations in Hippo-Yap signaling and myofibroblast fate underlie scar-free ear appendage wound healing in spiny mice. Dev Cell 56(19):2722–2740.e6. https://doi.org/10.1016/j.devcel.2021.09.008
73. Galiano RD, Michaels J, Dobryansky M, Levine JP, Gurtner GC (2004) Quantitative and reproducible murine model of excisional wound healing. Wound Repair Regen 12(4):485–492. https://doi.org/10.1111/j.1067-1927.2004.12404.x
74. Fourcaudot A (2017) Silver sulfadiazine retards wound healing and increases hypertrophic scarring in a rabbit ear excisional wound model. J Burn Care Res. https://doi.org/10.1097/BCR.0000000000000406
75. Schaefer TJ, Szymanski KD (2025) Burn evaluation and management (Archived). In: StatPearls. StatPearls Publishing. Accessed June 8, 2025. http://www.ncbi.nlm.nih.gov/books/NBK430741/

76. Venter NG, Monte-Alto-Costa A, Marques RG (2015) A new model for the standardization of experimental burn wounds. Burns 41(3):542–547. https://doi.org/10.1016/j.burns.2014.08.002
77. Abdullahi A, Amini-Nik S, Jeschke MG (2014) Animal models in burn research. Cell Mol Life Sci 71(17):3241–3255. https://doi.org/10.1007/s00018-014-1612-5
78. Stieritz DD, Holder IA (1975) Experimental studies of the pathogenesis of infections due to Pseudomonas aeruginosa: description of a burned mouse model. J Infect Dis 131(6):688–691. https://doi.org/10.1093/infdis/131.6.688
79. Hao D, Nourbakhsh M (2021) Recent advances in experimental burn models. Biology 10(6):526. https://doi.org/10.3390/biology10060526
80. Frykberg RG, Banks J (2015) Challenges in the treatment of chronic wounds. Adv Wound Care 4(9):560–582. https://doi.org/10.1089/wound.2015.0635
81. Sanapalli BKR, Yele V, Singh MK, Thaggikuppe Krishnamurthy P, Karri VVSR (2021) Preclinical models of diabetic wound healing: a critical review. Biomed Pharmacother 142:111946. https://doi.org/10.1016/j.biopha.2021.111946
82. Stadler I, Zhang RY, Oskoui P, Whittaker MBS, Lanzafame RJ (2004) Development of a simple, noninvasive, clinically relevant model of pressure ulcers in the mouse. J Invest Surg. https://doi.org/10.1080/08941930490472046
83. Rössler S, Nischwitz SP, Luze H, Holzer-Geissler JCJ, Zrim R, Kamolz LP (2022) In vivo models for hypertrophic scars—a systematic review. Medicina 58(6):736. https://doi.org/10.3390/medicina58060736
84. Nabai L, Ghahary A (2017) Hypertrophic scarring in the rabbit ear: a practical model for studying dermal fibrosis. Methods Mol Biol 1627:81–89. https://doi.org/10.1007/978-1-4939-7113-8_6
85. Ning X, Yang K, Shi W, Xu C (2020) Comparison of hypertrophic scarring on a red Duroc pig and a Guangxi Mini Bama pig. Scars Burns Heal 6:2059513120930903. https://doi.org/10.1177/2059513120930903
86. Hu MS, Cheng J, Borrelli MR et al (2018) An improved humanized mouse model for excisional wound healing using double transgenic mice. Adv Wound Care 7(1):11–17. https://doi.org/10.1089/wound.2017.0772
87. DeMayo JL, Wang J, Liang D, Zhang R, DeMayo FJ (2012) Genetically engineered mice by pronuclear DNA microinjection. Curr Protoc Mouse Biol 2(3):245–262. https://doi.org/10.1002/9780470942390.mo110168
88. Huang Q, Chan KY, Wu J et al (2024) An AAV capsid reprogrammed to bind human transferrin receptor mediates brain-wide gene delivery. Science 384(6701):1220–1227. https://doi.org/10.1126/science.adm8386
89. Henn D, Zhao D, Sivaraj D et al (2023) Cas9-mediated knockout of Ndrg2 enhances the regenerative potential of dendritic cells for wound healing. Nat Commun 14(1):4729. https://doi.org/10.1038/s41467-023-40519-z
90. Vegesna V, McBride WH, Taylor JM, Withers HR (1995) The effect of interleukin-1 beta or transforming growth factor-beta on radiation-impaired murine skin wound healing. J Surg Res 59(6):699–704. https://doi.org/10.1006/jsre.1995.1226
91. Abbas DB, Griffin M, Fahy EJ et al (2024) Establishing a xenograft model with CD-1 nude mice to study human skin wound repair. Plast Reconstr Surg 153(1):121–128. https://doi.org/10.1097/PRS.0000000000010465
92. Pradhan BS, Majumdar SS (2016) An efficient method for generation of transgenic rats avoiding embryo manipulation. Mol Ther Nucleic Acids 5. https://doi.org/10.1038/mtna.2016.9
93. Sharun K, Banu SA, Mamachan M et al (2024) Development and characterization of contraction-suppressed full-thickness skin wound model in rabbits. Tissue Cell 90:102482. https://doi.org/10.1016/j.tice.2024.102482
94. Parnell LKS, Volk SW (2019) The evolution of animal models in wound healing research: 1993–2017. Adv Wound Care (New Rochelle) 8(12):692–702. https://doi.org/10.1089/wound.2019.1098

95. Li Q, Lyu C, Chen D et al (2023) Gallic acid treats hypertrophic scar in rabbit ears via the TGF-β/Smad and TRPC3 signaling pathways. Pharmaceuticals 16(11):1514. https://doi.org/10.3390/ph16111514
96. Kuo TY, Huang CC, Shieh SJ et al (2022) Skin wound healing assessment via an optimized wound array model in miniature pigs. Sci Rep 12(1):445. https://doi.org/10.1038/s41598-021-03855-y
97. Summerfield A, Meurens F, Ricklin ME (2015) The immunology of the porcine skin and its value as a model for human skin. Mol Immunol 66(1):14–21. https://doi.org/10.1016/j.molimm.2014.10.023
98. Mistry R, Veres M, Issa F (2022) A systematic review comparing animal and human scarring models. Front Surg 9. https://doi.org/10.3389/fsurg.2022.711094
99. Patil P, Martin JR, Sarett SM et al (2017) Porcine ischemic wound-healing model for preclinical testing of degradable biomaterials. Tissue Eng Part C Methods 23(11):754–762. https://doi.org/10.1089/ten.TEC.2017.0202
100. Elloso M, Hutter MF, Jeschke N et al (2025) Challenges of porcine wound models: a review. Int J Transl Med 5(1):4. https://doi.org/10.3390/ijtm5010004
101. Martinello T, Gomiero C, Perazzi A et al (2018) Allogeneic mesenchymal stem cells improve the wound healing process of sheep skin. BMC Vet Res 14(1):202. https://doi.org/10.1186/s12917-018-1527-8
102. Banstola A, Reynolds JNJ (2022) The sheep as a large animal model for the investigation and treatment of human disorders. Biology 11(9):1251. https://doi.org/10.3390/biology11091251
103. Elia R, Maruccia M, Di Summa PG et al (2024) Conventional versus regenerative methods for wound healing: a comparative experimental study on a sheep model. Medicina (Kaunas) 60(11):1836. https://doi.org/10.3390/medicina60111836
104. Volk SW, Bohling MW (2013) Comparative wound healing--are the small animal veterinarian's clinical patients an improved translational model for human wound healing research? Wound Repair Regen 21(3):372–381. https://doi.org/10.1111/wrr.12049
105. Kiani AK, Pheby D, Henehan G et al (2022) Ethical considerations regarding animal experimentation. J Prev Med Hyg 63(2S3):E255. https://doi.org/10.15167/2421-4248/jpmh2022.63.2S3.2768
106. Mathes DW, Noland M, Graves S, Schlenker R, Miwongtum T, Storb R (2010) A preclinical canine model for composite tissue transplantation. J Reconstr Microsurg 26(3):201–207. https://doi.org/10.1055/s-0030-1247717
107. The panniculus carnosus muscle: an evolutionary enigma at the intersection of distinct research fields - Naldaiz-Gastesi - 2018 - Journal of Anatomy - Wiley Online Library. Accessed June 8, 2025. https://onlinelibrary-wiley-com.proxy.lib.duke.edu/doi/10.1111/joa.12840
108. Wilmink JM, Nederbragt H, van Weeren PR, Stolk PW, Barneveld A (2001) Differences in wound contraction between horses and ponies: the in vitro contraction capacity of fibroblasts. Equine Vet J 33(5):499–505. https://doi.org/10.2746/042516401776254817
109. Richardson R, Slanchev K, Kraus C, Knyphausen P, Eming S, Hammerschmidt M (2013) Adult zebrafish as a model system for cutaneous wound-healing research. J Invest Dermatol 133(6):1655–1665. https://doi.org/10.1038/jid.2013.16
110. Santoso F, De Leon MP, Kao WC et al (2024) Appendage-resident epithelial cells expedite wound healing response in adult zebrafish. Curr Biol 34(16):3603–3615.e4. https://doi.org/10.1016/j.cub.2024.06.051
111. Lumaquin D, Johns E, Montal E et al (eds) (2021) An in vivo reporter for tracking lipid droplet dynamics in transparent zebrafish. elife 10:e64744. https://doi.org/10.7554/eLife.64744
112. Silva Brito R, Canedo A, Farias D, Rocha TL (2022) Transgenic zebrafish (*Danio rerio*) as an emerging model system in ecotoxicology and toxicology: historical review, recent advances, and trends. Sci Total Environ 848:157665. https://doi.org/10.1016/j.scitotenv.2022.157665
113. Hubrecht RC, Carter E (2019) The 3Rs and humane experimental technique: implementing change. Animals 9(10):754. https://doi.org/10.3390/ani9100754

114. The principles of humane experimental technique. Accessed June 8, 2025. https://norecopa.no/textbase/the-principles-of-humane-experimental-technique
115. The 3Rs | NC3Rs. Accessed June 8, 2025. https://nc3rs.org.uk/who-we-are/3rs
116. Animal Welfare Act | National Agricultural Library. Accessed June 8, 2025. https://www.nal.usda.gov/animal-health-and-welfare/animal-welfare-act
117. Carbone L (2011) Pain in laboratory animals: the ethical and regulatory imperatives. PLoS One 6(9):e21578. https://doi.org/10.1371/journal.pone.0021578
118. Boverhof DR, Chamberlain MP, Elcombe CR et al (2011) Transgenic animal models in toxicology: historical perspectives and future outlook. Toxicol Sci 121(2):207–233. https://doi.org/10.1093/toxsci/kfr075
119. Flynn K, Mahmoud NN, Sharifi S, Gould LJ, Mahmoudi M (2023) Chronic wound healing models. ACS Pharmacol Transl Sci 6(5):783–801. https://doi.org/10.1021/acsptsci.3c00030
120. Thermoelectric porous laser-induced graphene-based strain-temperature decoupling and self-powered sensing | Nature Communications. Accessed June 8, 2025. https://www.nature.com/articles/s41467-024-55790-x

Chapter 12
Preclinical Insights into Wound Healing: Models, Mechanisms, and Translational Strategies

Naveen Jeyaraman, Mainak Roy, Sathish Muthu, Sharun Khan, and Madhan Jeyaraman

Abstract Wound healing is a complex biological process involving multiple overlapping stages, such as hemostasis, inflammation, proliferation, and remodeling, each orchestrated through intricate cellular and molecular interactions. Preclinical models are indispensable for unraveling these processes and assessing potential therapies. Rodent models, due to their ease of genetic manipulation and low maintenance costs, remain the cornerstone for early-phase investigations, although their healing mechanism differs substantially from humans. In contrast, porcine models offer a closer anatomical and immunological resemblance to human skin, making them ideal for translational studies despite logistical and ethical challenges. Emerging technologies like 3D bioprinted skin, organoids, and microfluidic skin-on-a-chip platforms are revolutionizing preclinical research by providing physiologically relevant and ethically acceptable alternatives. Advanced imaging techniques such as Optical Coherence Tomography and multiphoton microscopy now allow non-invasive, real-time monitoring of healing, while artificial intelligence-driven analysis enhances the accuracy and reproducibility of data interpretation. Despite significant progress, issues related to reproducibility, standardization, and

N. Jeyaraman (✉) · M. Jeyaraman
Department of Orthopaedics, ACS Medical College and Hospital, Dr MGR Educational and Research Institute, Chennai, Tamil Nadu, India

Orthopaedic Research Group, Coimbatore, Tamil Nadu, India

M. Roy
Department of Orthopaedics, All India Institute of Medical Sciences, New Delhi, India

S. Muthu
Orthopaedic Research Group, Coimbatore, Tamil Nadu, India

Central Research Laboratory, Meenakshi Medical College Hospital and Research Institute, Meenakshi Academy of Higher Education and Research, Chennai, Tamil Nadu, India

S. Khan
Centre for Regenerative Nanomedicine, Northwestern University, Chicago, IL, USA

Graduate Institute of Medicine, Yuan Ze University, Taoyuan, Taiwan

F. De Francesco, K. Turksen (eds.), *Management and Strategies for Wound Healing*, Stem Cell Biology and Regenerative Medicine 79,
https://doi.org/10.1007/978-3-032-12442-5_12

the limited ability of animal models to fully emulate human wound pathology continue to pose challenges. Integration of cutting-edge biotechnologies with traditional models and a focus on ethical research practices are critical for bridging the translational gap and accelerating the development of effective wound healing therapies. This review synthesizes the current landscape of preclinical wound healing research, highlighting traditional models, innovative technologies, assessment techniques, and the ongoing efforts to enhance the translational relevance of findings for clinical application.

Keywords Wound healing · Extracellular matrix · Histology · Animal models

Abbreviations

ECM	Extracellular Matrix
IL	Interleukin
TNF-α	Tumor Necrosis Factor-Alpha
TGF-β	Transforming Growth Factor-Beta
VEGF	Vascular Endothelial Growth Factor
PDGF	Platelet-Derived Growth Factor
MMP	Matrix Metalloproteinase
TIMP	Tissue Inhibitor of Metalloproteinases
ELISA	Enzyme-Linked Immunosorbent Assay
qPCR	Quantitative Polymerase Chain Reaction
NF-κB	Nuclear Factor Kappa B
MAPK	Mitogen-Activated Protein Kinase
miR	microRNA
LC-MS/MS	Liquid Chromatography-Tandem Mass Spectrometry
iTRAQ	Isobaric Tags for Relative and Absolute Quantitation
TMT	Tandem Mass Tags
PSCs	Pluripotent Stem Cells
AI	Artificial Intelligence
ML	Machine Learning
CNNs	Convolutional Neural Networks
SHG	Second Harmonic Generation
3D	Three-Dimensional

Introduction

The restoration of skin integrity represents a complex and critical biological process. This process progresses through a temporally regulated sequence of biological events. First, the body stops the bleeding (hemostasis), then it deals with any initial swelling and potential infection (inflammation). Following this, new tissue starts to grow (proliferation), which is then refined and strengthened (remodeling), ultimately leading to the skin regaining its protective abilities. A whole team of different cells is involved—skin cells (keratinocytes), the structural support cells (fibroblasts), blood vessel lining cells (endothelial cells), and various immune system players. They all work together, guided by the scaffolding of the extracellular matrix and a host of signaling molecules, to get the skin back to its job of protecting us, providing structural integrity, and acting as an immune barrier [1, 2]. Recent progress in areas like tissue engineering (building new tissues), regenerative medicine (helping the body heal itself), and the development of new drugs has provided us with a toolkit to help wounds heal better and faster. However, before we can try these new approaches on people, they need to be thoroughly tested in lab-based models. These models are incredibly important for understanding how a treatment works, how well it might work, and whether it is safe. By creating controlled scenarios that mimic how human wounds behave, we can observe the healing process, spot any potential problems, and figure out the best way to administer the treatment [3, 4]. Often, researchers use a combination of lab-dish experiments (in vitro) and studies in living organisms (in vivo). The in vivo models are particularly useful because they can better replicate the complex and ever-changing nature of wound healing. Pigs and rodents are often the go-to choices for these models because they can be adapted to study different kinds of wounds, like pressure sores, burns, surgical incisions, and wounds that just won't heal. It's important to remember that no single research model is flawless; each comes with its own set of strengths and weaknesses. Ultimately, the precise aims of the investigation, the necessity for reliable and reproducible findings, and our ethical responsibilities all play a role in deciding which model is the most suitable [5, 6]. The landscape of how animals are investigated for wound healing before it reaches human testing has undergone a significant transformation lately. Thrilling innovations, like 3D-printed artificial skin, miniature skin-like tissues grown in the lab (organoids originating from human cells), and microfluidic "skin-on-a-chip" systems, are providing us with models that much more closely resemble real human skin and can offer more dependable predictions about how treatments will work. These developments are not only assisting us in lessening our dependence on animal studies but also in gaining a more authentic understanding of human skin biology. Simultaneously, advancements in data visualization and analysis are allowing us to monitor the healing process with remarkable accuracy, often in real-time, and without directly intervening with the wound.

Commonly Used Models

Rodent Models

When it comes to studying how wounds heal in the lab before moving to human trials, rodents—particularly mice and rats—are by far the most popular animal choice. This widespread preference isn't random; it stems from a mix of practicalities and scientific benefits. They do not cost a fortune to look after, are generally easy for researchers to handle, and reproduce quite quickly, and, importantly, there are a huge collection of genetic tools and modified animal lines (transgenic models) available for them. These advantages enable precise genetic manipulation for dissecting the roles of individual genes and signaling pathways in tissue regeneration. This level of genetic control makes rodent models incredibly valuable for digging deep into the molecular and cellular processes that drive wound repair [7]. Researchers commonly use several well-established rodent models to investigate different kinds of injuries, including wounds that go all the way through the skin (full-thickness excisional wounds), surgical-like cuts (incisional wounds), burn injuries, and persistent wounds like diabetic ulcers. Among these, the excisional wound model is particularly prevalent. In this setup, a circular piece of skin is surgically removed, usually from the animal's back. However, one challenge in using rodents is their natural tendency to heal wounds primarily through contraction—a process that does not mirror how human wounds heal. To better mimic human wound healing, researchers often place silicone splints or stents around the wound. These barriers reduce contraction and instead promote healing through keratinocyte migration and granulation tissue development, which more closely resembles the human process [8].

Rodents also offer a practical platform for testing novel therapeutics, including small molecules, growth factors, stem cell therapies, and advanced wound dressings. The availability of diabetic and immunocompromised rodent strains further enables the simulation of delayed healing and infection-prone wound environments. For instance, streptozotocin-induced diabetic mice serve as a robust model for investigating non-healing diabetic foot ulcers. Additionally, nude mice (athymic) allow for human cell transplantation studies without immune rejection [9].

Despite their versatility, rodent models do present limitations. Anatomically, rodent skin is thinner and more elastic than human skin and contains a subcutaneous panniculus carnosus muscle layer that enhances wound contraction—an element absent in humans. Immunologically, cytokine expression patterns and immune cell behavior in rodents differ from those observed in humans, potentially impacting translational relevance. Therefore, while rodents are indispensable for early-phase discovery and mechanistic studies, their findings must be interpreted cautiously when predicting clinical outcomes [10].

Porcine Models

Porcine models are increasingly recognized as the most physiologically relevant large-animal model for human skin wound healing due to the close anatomical, physiological, and immunological similarities between pig and human skin. Compared to rodents, porcine skin shares several critical characteristics with human skin, including similar epidermal thickness, dermal–epidermal junction structure, dermal collagen density, and the presence of hair follicles and sweat glands distributed in a comparable pattern. Furthermore, pigs exhibit minimal wound contraction, making them especially suitable for modeling human-like re-epithelialization and granulation tissue formation [11].

Porcine models are particularly valuable for studying various types of wounds, including partial- and full-thickness wounds, thermal burns, ischemic ulcers, and pressure sores. In experimental protocols, wounds are usually created on the dorsal or flank area of the pig and monitored over time using advanced imaging, histology, and biomechanical testing. The larger surface area of pigs allows researchers to test multiple wound conditions and treatments on a single animal, thereby reducing inter-animal variability [12].

Miniature pig breeds such as Yucatan or Göttingen pigs are often preferred due to their manageable size, docile nature, and slower growth rate. Their suitability for long-term studies and ease of handling make them practical alternatives to full-sized farm pigs, especially in facilities with space limitations. Furthermore, the pig's immune response—including cytokine release, macrophage polarization, and lymphocyte activation—closely mirrors that of humans, rendering these models highly translational for immunomodulatory therapies [13].

However, the advantages of porcine models come at a cost. Pigs are expensive to purchase and maintain, require specialized housing, and are subject to stricter ethical and veterinary oversight. Their use is often limited to later stages of preclinical research when therapeutic candidates have already demonstrated promise in smaller animal models. Nevertheless, for interventions aimed at clinical translation—especially those involving skin substitutes, biomaterials, or surgical techniques—porcine models remain indispensable [14].

Canine and Non-human Primate Models

Although ethically controversial and seldom used today, canine and non-human primate models were historically applied in wound-healing research due to their physiological similarity to humans. Current ethical standards, availability of alternative models, and high costs have significantly reduced their use in favor of more humane options like porcine or 3D humanized models [5].

Overall, these alternative animal models expand the toolset for researchers, offering specific advantages based on anatomical, physiological, or functional criteria. They serve as important bridges between rodent studies and larger translational animal models like pigs, particularly in areas requiring specialized endpoints or long-term evaluation [15].

A summary of the animal models described above, their advantages and limitations is presented in Table 12.1.

Table 12.1 Pre-clinical models for wound healing studies

Model	Key features	Advantages	Limitations	Applications
Rodent models	Widely used; genetic tools available (e.g., transgenic models); cost-effective; rapid reproduction	Enables precise gene manipulation; useful for studying molecular and cellular mechanisms in wound repair	Skin is thinner and more elastic, healing primarily through contraction rather than re-epithelialization	Testing novel therapeutics, diabetic wound models (e.g., streptozotocin-induced mice), and immunocompromised states
Porcine models	Anatomically, physiologically, and immunologically similar to human skin, with minimal wound contraction	Closely mimics human-like re-epithelialization and granulation tissue formation; larger surface area for testing	High cost and ethical considerations; specialized housing required	Preclinical studies for skin substitutes, biomaterials, and surgical techniques
Miniature pig breeds	Smaller size, docile nature, and slower growth rate	Manageable for long-term studies; translational relevance for immunomodulatory therapies	Requires careful maintenance; high expenses are involved	Studying partial/full-thickness wounds, thermal burns, ischemic ulcers, and pressure sores
Canine models	Historically used for physiological similarity to humans	Useful for specialized endpoints in earlier studies	Rarely used now due to ethical concerns and the availability of alternative models	Limited use; previously applied in wound-healing research
Non-human primate models	Physiological relevance to humans	High translational potential for advanced studies	Ethical concerns, high costs, and reduced usage	Rarely utilized; replaced by porcine or advanced humanized models

Techniques for Assessment

Histological Analysis

Histology provides a foundational method for evaluating wound healing at the tissue and cellular level. Microscopic examination of stained tissue specimens remains a cornerstone method for histopathological assessment. This technique lets researchers zoom in on key events like the regrowth of the skin's surface layer (re-epithelialization), the formation of the structural protein collagen, the development of new blood vessels, the arrival of immune cells, and any changes happening in the supporting network outside the cells (extracellular matrix) as time goes by. By observing these characteristics at different points in the healing process, scientists can collect both descriptive and measurable information to follow the progress and see how the wound responds to different treatments.

When it comes to staining techniques, Hematoxylin and Eosin (H&E) are still a go-to method for getting an overall picture of the tissue structure. It clearly shows when the surface layer of the skin has closed and where the inflammatory cells are located. On the other hand, Masson's Trichrome stain is often used when the focus is on seeing collagen. This makes it especially helpful for evaluating how much scar tissue is forming and its pattern. For a more detailed look at collagen maturity, Picrosirius Red staining—especially under polarized light—can distinguish between type I collagen (more mature) and type III collagen (less mature), which helps evaluate the quality of tissue repair [16].

Immunohistochemistry (IHC) adds another layer of precision by allowing researchers to detect specific proteins involved in healing. Markers like CD31 (PECAM-1) are used to highlight new blood vessel formation, Ki67 points to actively dividing cells, and alpha-smooth muscle actin (α-SMA) identifies activated myofibroblasts, which are key players in contraction and matrix restructuring. Today, with the advent of digital pathology and advanced image analysis tools, much of this assessment can be automated, improving consistency and reducing subjective interpretation [17].

Although histological analysis is time-consuming and requires sacrificing the tissue sample, its ability to reveal detailed cellular and structural changes makes it an irreplaceable tool in wound-healing studies. It bridges the gap between what can be seen on the surface and what is happening at the microscopic level.

Molecular and Genetic Analyses

Molecular and genetic analyses provide an in-depth look at the cellular and biochemical mechanisms that drive wound healing. These methods empower researchers to measure how active genes are and how much protein they're producing. They also allow scientists to delve into the inner workings of cellular communication networks (signaling pathways) and identify specific molecules that could be targets for new therapies. Getting this kind of detailed understanding is vital for figuring out how different treatments affect key processes like inflammation, the regrowth of tissue, and the development of scars [18, 19].

Gene Expression Profiling

When it comes to studying the activity of genes involved in wound healing, quantitative PCR (qPCR) is still a fundamental tool. It's frequently used to measure the amounts of important signaling molecules like IL-1β and TNF-α (cytokines), growth promoters such as VEGF and TGF-β1 (growth factors), enzymes that break down the tissue matrix (MMPs), and structural components like collagen types I and III. By following how the activity of these genes changes throughout the different stages of healing, researchers can gain a clearer picture of the interplay between inflammation (both promoting and resolving it) and the processes that lead to scarring and new tissue growth. For a broader perspective, techniques like microarrays and RNA sequencing offer a complete overview of gene activity, showing all the gene networks that are switched on or off during healing. These high-powered methods have revealed key differences between wounds that heal normally and those that become chronic, as well as distinctions between patients who tend to heal well and those who don't [20].

Protein Expression and Signalling Pathways

To see and measure specific proteins that play a role in wound healing, scientists often use techniques called Western blotting and enzyme-linked immunosorbent assays (ELISA). These methods can detect things like the structural components outside cells (extracellular matrix or ECM), such as fibronectin and laminin, inflammatory signals like IL-6 and MCP-1, and important communication molecules within cells like p38 MAPK, ERK, and NF-κB. By comparing the amounts of these proteins in treated wounds versus untreated ones, researchers can figure out if a treatment is working and if it is having any unexpected effects. Furthermore, more advanced techniques like multiplex immunoassays and Luminex platforms allow for the simultaneous measurement of many different signaling molecules (like cytokines or chemokines) in small amounts of tissue or fluid, which is useful when tracking changes over time in a study [21, 22].

Epigenetic and microRNA Analysis

New research is revealing that factors beyond just our genes themselves—things like how our DNA is packaged (DNA methylation and histone modification)—also play a big part in how wounds behave. For instance, wounds that just won't heal often show messed-up patterns of histone acetylation, which can prevent the genes needed for healing from being active. At the same time, tiny RNA molecules called microRNAs (like miR-21 and miR-146a) are key controllers of inflammation, the growth of new blood vessels, and the formation of scar tissue. Checking how much of these microRNAs are present in wound tissue can give us new clues about why some wounds heal and others do not, and it can even point us toward potential new treatments.

Overall, these kinds of detailed molecular and genetic analyses deepen our understanding of the fundamental biology of wound healing. They help us discover biological markers that can predict how a wound will heal and guide the development of more targeted and effective therapies. These tools essentially bridge our observations of surface-level wound changes to the intricate molecular processes occurring beneath it [23].

Biomarker Discovery and Proteomics

The discovery of reliable biomarkers is a pivotal goal in wound-healing research, as it enables objective monitoring of healing progression, prediction of clinical outcomes, and personalization of therapeutic interventions. Proteomic technologies, in particular, have emerged as powerful tools to identify and quantify thousands of proteins in wound tissues, fluids, and cells, shedding light on complex biological networks that govern tissue repair [24].

The Role of Biomarkers in Wound Healing

Biomarkers can be categorized into diagnostic, prognostic, and predictive types. Diagnostic biomarkers help distinguish between acute and chronic wounds, or between infected and non-infected wounds. Prognostic biomarkers indicate the likelihood of healing or scarring, while predictive biomarkers help assess the expected response to a given treatment. For instance, elevated matrix metalloproteinases (MMP-9) and reduced levels of tissue inhibitors of metalloproteinases (TIMPs) have been associated with chronic, non-healing wounds.

Levels of growth factors such as VEGF, EGF, and PDGF can serve as indicators of angiogenesis and epithelial proliferation, while inflammatory cytokines like TNF-α and IL-1β reflect immune status. These can be detected in wound exudates, plasma, or biopsies and quantified using ELISA, multiplex bead assays, or protein microarrays [25].

Proteomics for Unbiased Discovery

Unlike targeted approaches, proteomics allows for unbiased profiling of the wound proteome. Mass spectrometry (MS)-based techniques, including liquid chromatography-tandem MS (LC–MS/MS), can identify protein expression patterns associated with different wound types, healing stages, or treatment responses. This is especially useful in discovering novel mediators that might be overlooked in candidate-based studies. Label-free quantification and isobaric tags (iTRAQ, TMT) further enhance the power of proteomic comparisons, making it possible to identify up- or downregulated proteins in response to therapeutic agents such as hydrogels, stem cells, or bioactive peptides [26].

A summary of techniques used in the analysis of wound healing in preclinical models is presented in Table 12.2.

Table 12.2 Analysis techniques for wound healing in preclinical models

Technique	Key features	Advantages	Limitations	Applications
Histological analysis	Examines tissue and cellular changes via staining techniques (e.g., H&E, Masson's trichrome, Picrosirius red)	Provides detailed insight into re-epithelialization, collagen formation, angiogenesis, and ECM remodeling	Time-consuming; requires tissue sample sacrifice	Evaluating structural and cellular changes during wound healing
Molecular and genetic analyses	Measures gene and protein activity and examines signaling pathways driving wound healing	Reveals biochemical and cellular mechanisms; identifies therapeutic targets	Complex and resource-intensive techniques	Studying inflammatory processes, tissue regrowth, and scar formation
Gene expression profiling	Quantifies activity of key genes (e.g., cytokines, growth factors, MMPs, collagen) via qPCR, microarrays, RNA sequencing	Offers detailed temporal and spatial gene activity data; differentiates between normal and chronic wounds	Limited to specific gene panels; requires specialized equipment	Tracking stages of wound healing, evaluating genetic responses to treatment
Protein expression and signaling pathways	Detects proteins (e.g., fibronectin, laminin, IL-6, NF-κB) via Western blotting, ELISA, and multiplex assays	Enables comparative analysis of treated vs untreated wounds; simultaneous measurement of multiple proteins	May miss low-abundance proteins; costly and technically demanding	Assessing therapeutic efficacy and unintended effects; tracking molecular changes over time
Epigenetic and microRNA analysis	Investigates DNA methylation, histone modification, and microRNA regulation (e.g., miR-21, miR-146a)	Explains non-genetic factors affecting wound healing; identifies new therapeutic targets	Requires advanced tools and techniques; data interpretation is challenging	Studying chronic wound mechanisms, identifying potential interventions for impaired healing
Biomarker discovery and proteomics	Identifies and quantifies proteins via proteomic technologies to monitor healing progression	Enables the discovery of reliable biomarkers for objective clinical assessments	High complexity and cost; requires large datasets for meaningful conclusions	Predicting clinical outcomes; personalizing therapeutic strategies

H&E Hematoxylin and Eosin, *ECM* Extracellular Matrix, *IL* Interleukin, *NF-κB* Nuclear Factor Kappa B, *MMP* Matrix Metalloproteinases, *qPCR* Quantitative Polymerase Chain Reaction, *ELISA* Enzyme-Linked Immunosorbent Assay, *miR* microRNA

Application of Advanced Imaging

Optical Coherence Tomography (OCT)

Optical Coherence Tomography is a powerful, non-invasive imaging technique that allows real-time, high-resolution cross-sectional imaging of wound tissues, often referred to as "optical biopsies." Operating on the principles of low-coherence interferometry, OCT can generate images at micrometer-scale resolution up to a few millimeters in depth, making it ideal for assessing superficial tissue structures such as the epidermis, dermis, and wound margins. In the world of preclinical wound healing research, Optical Coherence Tomography (OCT) offers a way to keep tabs on important processes like the regrowth of the skin surface (re-epithelialization), the formation of new healing tissue (granulation tissue), the growth of new blood vessels (angiogenesis), and swelling (edema) over time—all without needing to sacrifice the animals. This is especially useful for studies that need to track changes at multiple time points to understand how a treatment is working. It also cuts down on the need for taking repeated tissue samples (biopsies), which helps keep the tissue structure intact and makes the experimental results more consistent. Nowadays, many advanced OCT systems also come with extra imaging abilities, like Doppler OCT to measure blood flow and polarization-sensitive OCT to assess the organization of collagen fibers. These sophisticated features give us a much more detailed picture of what's happening within the wound, and when combined with computer-based image analysis, they allow for measurements that are objective and can be reliably reproduced [27, 28].

Fluorescence Microscopy

In preclinical wound healing research, fluorescence microscopy is a go-to technique that lets scientists see and follow specific molecules, cells, or processes as they happen. By using fluorescently tagged antibodies, genetic markers that glow, or dyes that stain living cells, researchers can watch things like the movement of immune cells, the formation of new blood vessels, where stem cells are located, and even the presence of microbes, all at a cellular level. Intravital microscopy, a cool type of fluorescence imaging, takes this a step further by allowing us to observe live tissues in real-time in anesthetized animals. This has been especially helpful for studying dynamic events like how white blood cells interact with blood vessel linings, how fibroblasts move around, and how blood vessels reshape themselves. More advanced techniques like confocal and two-photon microscopy let us see deeper into tissues with less damage from the light and with sharper images, making them perfect for tracking changes in thicker tissue samples over time. Moreover, by integrating microscopy techniques with genetically engineered animals that feature fluorescent markers on particular cells—such as green-glowing macrophages or red-glowing endothelial cells—fluorescence microscopy becomes an exceptionally potent tool for examining the timing and mechanisms of healing in living organisms [29].

Multiphoton and Intravital Imaging

Recent advancements in microscopy have unlocked unprecedented ways to observe biological processes within living tissue. Among these innovations, multiphoton microscopy (MPM) and intravital imaging stand out as transformative tools in studying wound healing prior to human trials. These techniques enable real-time visualization of cellular and molecular activity with exceptional detail, even deep within living tissues. They provide insights into the healing dynamics that traditional tissue staining and surface-level imaging simply cannot achieve [30].

Multiphoton Microscopy (MPM)

Multiphoton microscopy (MPM) is a clever technique that uses near-infrared light to make fluorescent tags glow deep inside tissues, letting us see things up to a millimeter down with minimal damage to the cells. This allows researchers to visualize structural elements like collagen fibers using a process called second harmonic generation (SHG), as well as track what cells are doing using fluorescent labels. This approach is especially useful for studying skin, as its abundance of collagen makes it possible to observe how the deeper layers (the dermis) regenerate during the wound-healing process. One of the big pluses of MPM is that it lets us monitor the structure of collagen without having to cut out tissue samples over time. In wound healing, how the collagen fibers line up, how dense they are, and how they're organized are really important signs of how mature the new tissue is and how good the scar will be. SHG imaging can follow these changes over time without disturbing the wound. MPM also allows us to watch the dynamic movement of immune cells, the formation of new blood vessels, and the migration of skin cells (keratinocytes). By using genetically modified mice with fluorescent proteins—for example, macrophages that glow green (CX3CR1-GFP) or blood vessel cells that glow green (Tie2-GFP)—researchers can directly see how these cells move, interact, and behave within the healing wound [31].

Intravital Imaging

Intravital imaging is a collection of microscopy techniques, often using confocal or multiphoton microscopes, that let scientists watch what's happening inside live, anesthetized animals. When it comes to studying wound healing, this approach lets researchers follow the same wound over several days. This gives us a timeline of

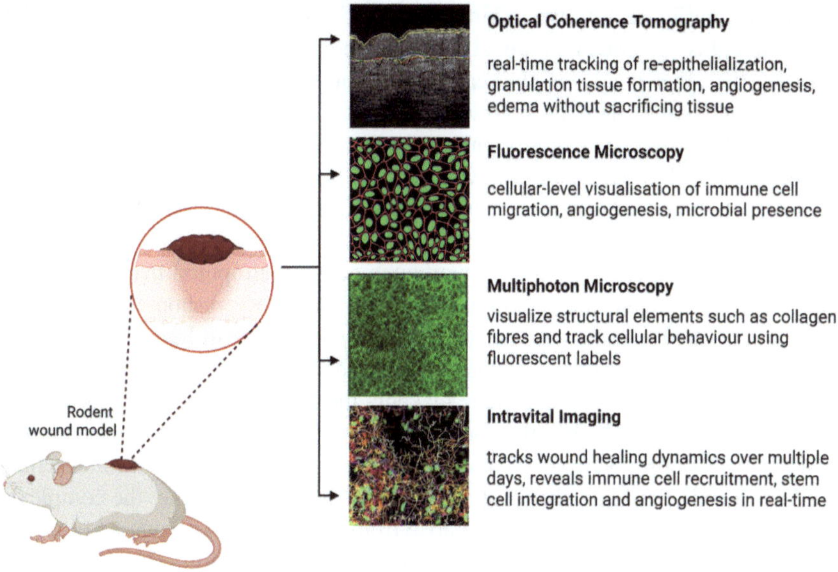

Fig. 12.1 Application of advanced imaging in pre-clinical wound healing studies

cellular events, like how immune cells arrive at the wound, how stem cells integrate into the tissue, and how new blood vessels form. This method has been incredibly important for understanding the step-by-step process of how immune cells are recruited and what they do. For example, it has shown us how quickly a type of immune cell called neutrophils rushes to the injury, how another type called macrophages changes their behavior from promoting inflammation to promoting healing, and how the leading cells of new blood vessels (endothelial tip cells) guide their growth. To make intravital imaging easier, researchers often implant special "windows" or dorsal skinfold chambers on mice. These devices hold the tissue still and provide a clear view while keeping the tissue in its natural living state [30] as shown in Fig. 12.1.

Challenges and Ethical Considerations

Limitations of Animal Models

Even though animal models have been crucial for helping us understand how wounds heal, it is important to recognize that they have some significant drawbacks. The main challenge is that no single animal model can perfectly mimic the intricate and multifaceted process of wound healing in humans. Variations in body structure, immune system function, and overall physiology can significantly influence how effectively findings from animal studies apply to human scenarios. For instance, rodents primarily heal by their skin contracting—they have a muscle layer called the panniculus carnosus that quickly pulls the edges of a wound together. Human skin, on the other hand, mainly heals by forming new tissue from the bottom up (granulation tissue) and the regrowth of the surface layer (re-epithelialization). Because of this, studies that rely only on rodent data might give us an overly optimistic view of how fast and how well wounds can heal in people [29]. When it comes to the immune response, there are also differences in the types of signaling molecules produced (cytokine expression), how certain immune cells (macrophages) behave, and how long other immune cells (neutrophils) live. Such differences can complicate the interpretation of inflammatory responses observed in animal models. Even pig models, which are more similar to humans in terms of anatomy, don't fully capture the wide range of factors that can influence healing in human populations, like age, genetic makeup, metabolic health, and other existing medical conditions [32]. It's also tough to accurately recreate chronic wounds, like diabetic ulcers or pressure sores, in animal models. Although diabetes can be chemically induced in rodents, the resulting wounds fail to fully capture the complexity of human diabetic foot ulcers, which arise from factors such as impaired blood circulation, nerve damage, infections, and metabolic dysfunction. These limitations emphasize the need for additional or alternative research models that are a closer match to human biology [33, 34].

Reproducibility and Standardization Issues

Reproducibility is a cornerstone of scientific integrity, yet it remains a persistent challenge in preclinical wound-healing studies. The complexity of biological systems, coupled with variability in experimental design, animal models, and measurement techniques, can lead to inconsistencies across studies—even when similar hypotheses are tested. Standardization is thus essential for ensuring that research findings are reliable, translatable, and meaningful across laboratories and institutions [35].

Variability in Wound Models

One major source of inconsistency lies in the creation of wound models. Variations in wound size, depth, location, and the tools used (e.g., biopsy punches, scalpel blades, lasers) can significantly alter healing dynamics. For example, a 6 mm excisional wound in the dorsal skin of a mouse may heal at a different rate than a similar wound placed closer to the limb due to differences in vascularity and tissue tension. Moreover, the use or absence of splinting techniques to prevent contraction in rodent models can drastically impact outcomes. Without splints, rodents primarily heal by contraction—a mechanism that doesn't mirror human wound healing. Studies failing to standardize this parameter may report faster but less clinically relevant healing outcomes [36].

Assessment Timing and Endpoints

Another critical factor is the timing of wound assessments. Healing progresses in overlapping stages, and data collected at one-time point may not reflect the full picture. Early studies might capture inflammation or granulation, while later evaluations may highlight remodeling and scarring. Inconsistencies in timing can therefore lead to conflicting conclusions. Furthermore, differences in outcome measures—whether macroscopic (wound size reduction), histological (re-epithelialization, collagen deposition), or molecular (cytokine levels)—can make it difficult to compare results across studies. A treatment showing improved collagen alignment but no change in wound closure might still be therapeutically valuable, but its efficacy could be overlooked if only surface healing is measured [37].

Innovations in Preclinical Studies

3D Bioprinted Models

3D bioprinting has revolutionized the field of tissue engineering by enabling the fabrication of highly organized, layered constructs that mimic the native architecture of human skin. This technology uses bioinks composed of hydrogels, cells (e.g., keratinocytes, fibroblasts, melanocytes), and growth factors to print full-thickness skin analogs with precision and reproducibility. These models can be designed to replicate key features of human skin, including epidermal and dermal compartments, basement membrane integrity, and vascular-like networks. The modularity of 3D bioprinting allows for the inclusion of patient-specific cells, making it an ideal platform for personalized medicine and disease modeling. Bioprinted skin models have numerous applications in wound-healing research. They can be used to screen pharmaceutical compounds, assess the efficacy of wound dressings and skin substitutes, and model infectious or inflammatory conditions. Additionally, integration with dynamic bioreactors enables the simulation of mechanical forces, fluid perfusion, and oxygen gradients, further enhancing physiological relevance. While still in the developmental phase, bioprinted constructs offer clear advantages over traditional cell cultures, including structural complexity, higher predictive value, and reduced reliance on animal testing. However, limitations persist, particularly the absence of immune and vascular systems, which are crucial for studying inflammation, infection, and regeneration [38].

Organoids

Skin organoids represent another cutting-edge innovation in preclinical modeling. Organoids are like miniature, self-assembling 3D versions of our tissues, and they can be grown from very versatile stem cells (pluripotent stem cells or PSCs) or more specialized adult progenitor cells [39]. These tiny tissue models can recreate key aspects of real skin, including its layered surface (stratified epithelium), the supportive cells underneath (dermal fibroblasts), and even components like hair follicles or pigment cells.

Compared to growing cells in flat, two-dimensional layers (traditional monolayer cultures), organoids provide a much more lifelike environment for studying how wounds heal. This is because they better support the complex ways cells interact with each other and with the surrounding structural proteins (cell-matrix interactions). These models are especially useful for looking at how skin develops, studying genetic skin disorders, and seeing how skin responds to different drugs. What's even more exciting is that researchers have been able to grow skin organoids together with immune cells and blood vessels. This allows for a more realistic simulation of the immune environment and the development of the blood supply that happens when a wound is healing [40, 41].

One really promising use of organoid technology is to create organoids from a specific patient's cells to model chronic wounds or abnormal scarring, like keloids. These personalized models could help us understand why healing goes wrong in some individuals and potentially lead to the development of treatments tailored to that specific person. Plus, organoids can be grown in large numbers and tested in automated systems, making them ideal for screening different substances that might help wound healing—whether they work by reducing scarring, promoting tissue regeneration, or fighting infection [42].

Microfluidic Skin-on-a-Chip Systems

With advancements in tissue engineering and regenerative medicine, researchers are turning to microfluidic technologies, commonly known as "organ-on-a-chip" platforms, to create miniature models of human skin that closely mimic its natural functions. These systems try to copy the natural structure, the different types of cells, and how real skin functions, offering a strong alternative to traditional animal testing and static lab cultures. "Skin-on-a-chip" specifically refers to a tiny engineered device that brings together human skin cells—like the surface cells (keratinocytes), the supportive cells (fibroblasts), and the blood vessel lining cells (endothelial cells)—within a 3D framework [43, 44]. Tiny channels are built in to mimic blood flow, the delivery of nutrients, and the removal of waste. These platforms usually have two or more sections separated by a thin, porous barrier, with one side acting like the junction between the outer and inner layers of skin, and the other simulating interactions with blood vessels or the immune system. Some advanced models even include mechanical actuators that simulate stretching or compression, effectively mimicking the natural movements of skin. This adds another layer of physiological relevance, particularly for evaluating wound dressings, skin adhesives, or mechanical stress responses during healing [45].

Skin-on-a-chip systems are highly suitable for modeling acute wounds, burns, or chronic ulcers, and offer real-time monitoring of healing-related events such as re-epithelialization, immune infiltration, and vascular responses. Since these systems are built using human-derived cells, they better replicate human-specific responses than animal models, enhancing translational accuracy. These chips can also be used to study infection dynamics, by introducing pathogens like *Staphylococcus aureus* or *Pseudomonas aeruginosa* into the system and observing host–pathogen interactions under controlled flow. In the context of chronic wounds, integration with diabetic or hypoxic conditions enables the exploration of impaired healing mechanisms. Furthermore, skin-on-a-chip devices support high-throughput drug screening, making them attractive platforms for testing antimicrobial agents, growth factors, and biomaterials with minimal reagent use and faster data generation [46].

Artificial Intelligence and Digital Imaging in Wound Assessment

As preclinical wound-healing research generates increasingly complex datasets—from histological images to time-series measurements—artificial intelligence (AI) and machine learning (ML) are becoming essential tools for enhancing analysis, accuracy, and predictive modeling. These new technologies are changing how researchers understand their findings, keep track of the healing process, and evaluate how well treatments are working, whether they're using animal models or sophisticated lab-based systems [47].

Automated Image Analysis and Wound Measurement

In the past, measuring the size of a wound was often done by hand, using tools like calipers or simply estimating by looking at photographs. However, these methods are quite subjective and can easily lead to mistakes because of human interpretation. AI-powered image processing tools now offer automated and objective quantification of wound parameters such as area, perimeter, granulation tissue coverage, and epithelialization rates. Deep learning algorithms, particularly convolutional neural networks (CNNs), can be trained on annotated image datasets to segment wound boundaries and track changes over time with high accuracy. These tools reduce variability, improve reproducibility, and significantly speed up data processing in longitudinal studies. Such methods are being integrated into mobile applications and digital pathology platforms, enabling researchers and clinicians to upload wound images and receive instant measurements and healing trajectory predictions—supporting both preclinical and translational studies [48].

Histological and Microscopic Data Interpretation

Artificial intelligence has significantly advanced histopathological image analysis, enhancing the precision and reproducibility of morphological assessments. High-resolution digital slide scanners coupled with ML models can analyze stained tissue sections for features such as:

- Inflammatory cell density
- Collagen deposition patterns
- Vascular density (e.g., CD31 staining)
- Granulation tissue volume
- Epidermal thickness

By training AI systems on validated datasets, these platforms can distinguish subtle pathological differences between experimental groups, improving the sensitivity of analysis. Furthermore, AI tools like cell segmentation algorithms are crucial in high-content fluorescence microscopy, where manual counting of nuclei, immune cells, or signal intensities would be laborious and inconsistent [49].

Predictive Modeling of Healing Outcomes

One of the most exciting frontiers is the use of AI to predict healing trajectories based on initial wound characteristics, biomarker profiles, and treatment types. By combining imaging, molecular data, and metadata (e.g., age, comorbidity, wound location), machine learning models can forecast whether a wound is likely to heal, form a scar, or become chronic. Such predictive tools can enhance decision-making and reduce reliance on long-term animal experiments.

In preclinical studies, this could allow researchers to identify promising therapeutics earlier, optimize study designs, and reduce resource use. In clinical settings, these algorithms may eventually help personalize wound care by tailoring treatments based on predicted responses [50].

A summary of the key innovations and their distinctive advantages is presented in Table 12.3.

Translational Relevance

Many promising interventions that demonstrate efficacy in animal or in vitro models fail to yield similar results in human trials. This disconnect underscores the importance of critically evaluating the translational relevance of preclinical models and refining strategies to bridge the preclinical-clinical gap.

There are remarkable examples of preclinical research directly informing successful clinical applications. For instance, the development of recombinant platelet-derived growth factor (PDGF) gel (becaplermin) for diabetic foot ulcers was

Table 12.3 Key innovations in pre-clinical research and their advantages

Innovation	Key features	Advantages	Applications
3D bioprinted models	Fabrication of full-thickness skin analogues using bioinks (hydrogels, cells, growth factors)	Mimics human skin structure with epidermal/dermal compartments; allows patient-specific modelling; reduces animal testing	Screening pharmaceutical compounds, testing wound dressings and skin substitutes, and disease modelling
Organoids	Miniature 3D versions of skin tissue grown from stem or progenitor cells	Supports lifelike cell-matrix interactions; enables simulation of immune and vascular environments	Studying genetic skin disorders, chronic wounds, scarring (e.g., keloids), and personalized medicine
Microfluidic skin-on-a-chip systems	Miniaturized platforms integrating human skin cells with 3D frameworks and fluidic channels	Simulates physiological conditions (blood flow, nutrient delivery); supports high-throughput drug screening	Acute wounds, burns, chronic ulcers, infection dynamics and impaired healing in diabetic wounds
AI and digital imaging	AI-powered tools for wound measurement, histological analysis, and predictive modelling	Enhances data accuracy, reproducibility, and analysis speed; predicts healing outcomes from molecular and imaging data	Automated wound assessment, histological analysis, personalizing wound care, and therapeutic optimization

3D Three-Dimensional, *AI* Artificial Intelligence

supported by strong preclinical data in rodent and porcine models showing enhanced granulation tissue and accelerated wound closure [51]. Similarly, negative pressure wound therapy (NPWT) transitioned from porcine trials into wide clinical use based on its demonstrated effects on wound perfusion, exudate removal, and tissue deformation [52].

Conversely, there are multiple instances where therapies showed promise in animal studies but failed during clinical evaluation. Many cell-based therapies, including mesenchymal stem cells (MSCs), have shown impressive regenerative potential in rodents but have yielded inconsistent or modest outcomes in humans. In some cases, the immune privilege of these cells in animal models does not hold in the human context, or the wound microenvironment in chronic human ulcers proves far more complex and resistant to intervention [53]. A key reason for translational failure is the difference in biology between animal models and human patients. Rodents, for example, have a fundamentally different healing mechanism that emphasizes contraction over re-epithelialization. Even pigs, which closely mimic human skin structure, are typically young and healthy animals, unlike many human patients with comorbidities such as diabetes, vascular disease, or advanced age. Moreover, human chronic wounds are often biofilm-infected, ischemic, and embedded in systemic inflammatory contexts that are difficult to replicate in preclinical models. These complexities influence immune responses, matrix turnover, and angiogenesis—factors that profoundly affect healing outcomes and may not be fully captured in simplified animal systems [54].

Conclusion

Preclinical models are essential tools in wound-healing research, offering insights into inflammation, angiogenesis, fibrosis, and tissue repair. While rodent and porcine models remain valuable for mechanistic and translational studies, their anatomical and immunological differences from humans highlight the need for more predictive platforms. Emerging technologies like 3D bioprinted skin and organoids provide human-relevant models with greater physiological accuracy. By integrating in vivo, in vitro, and ex vivo systems within ethical frameworks, researchers can advance personalized and effective therapies for both acute and chronic wounds.

References

1. Ucuzian AA, Gassman AA, East AT, Greisler HP (2010) Molecular mediators of angiogenesis. J Burn Care Res 31(1):158. https://doi.org/10.1097/BCR.0b013e3181c7ed82
2. Azari Z, Gorgani S, Hosseini SA, Wang AZ, Kim H-W, Kargozar S (2023) The role of immune cells in therapeutic angiogenesis: concepts in tissue engineering. Curr Opin Biomed Eng 28:100470. https://doi.org/10.1016/j.cobme.2023.100470
3. Parnell LKS, Volk SW (2019) The evolution of animal models in wound healing research: 1993–2017. Adv Wound Care 8(12):692–702. https://doi.org/10.1089/wound.2019.1098
4. Grada A, Mervis J, Falanga V (2018) Research techniques made simple: animal models of wound healing. J Invest Dermatol 138(10):2095–2105.e1. https://doi.org/10.1016/j.jid.2018.08.005
5. Kiani AK, Pheby D, Henehan G, Brown R, Sieving P, Sykora P et al (2022) Ethical considerations regarding animal experimentation. J Prev Med Hyg 63(2 Suppl 3):E255–E266. https://doi.org/10.15167/2421-4248/jpmh2022.63.2S3.2768
6. Naderi MM, Sarvari A, Milanifar A, Boroujeni SB, Akhondi MM (2012) Regulations and ethical considerations in animal experiments: international Laws and Islamic perspectives. Avicenna J Med Biotechnol 4(3):114–120
7. Hall B, Limaye A, Kulkarni AB (2009) Overview: generation of gene knockout mice. Curr Protoc Cell Biol/editorial board, Juan S. Bonifacino … [et al.], CHAPTER, Unit-19.1217. https://doi.org/10.1002/0471143030.cb1912s44
8. Raghavan JV, Dorai VK, Sagar SK, Sivaraman A, R KS, Jhunjhunwala S (2022) Immunomodulatory bandage for accelerated healing of diabetic wounds. ACS Bio Med Chem Au 2(4):409–418. https://doi.org/10.1021/acsbiomedchemau.1c00063
9. Kumar R, Kumar V, Mohan A, Gupta G, Kashyap V (2024) Translational research in the generation of therapeutic medicine for wound healing: a review. Discover Med 1(1):158. https://doi.org/10.1007/s44337-024-00142-3
10. Vandamme TF (2014) Use of rodents as models of human diseases. J Pharm Bioallied Sci 6(1):2–9. https://doi.org/10.4103/0975-7406.124301
11. Patil P, Martin JR, Sarett SM, Pollins AC, Cardwell NL, Davidson JM et al (2017) Porcine ischemic wound-healing model for preclinical testing of degradable biomaterials. Tissue Eng Part C Methods 23(11):754–762. https://doi.org/10.1089/ten.tec.2017.0202
12. Porcine Models of Cutaneous Wound Healing|ILAR Journal|Oxford Academic. https://academic.oup.com/ilarjournal/article/56/1/127/661626. Accessed 12 Apr 2025
13. Pig Breed – an overview|ScienceDirect Topics. https://www.sciencedirect.com/topics/immunology-and-microbiology/pig-breed. Accessed 12 Apr 2025

14. Elloso M, Hutter MF, Jeschke N, Rix G, Chen Y, Douglas A et al (2025) Challenges of porcine wound models: a review. Int J Transl Med 5(1):4. https://doi.org/10.3390/ijtm5010004
15. Mukherjee P, Roy S, Ghosh D, Nandi SK (2022) Role of animal models in biomedical research: a review. Lab Anim Res 38:18. https://doi.org/10.1186/s42826-022-00128-1
16. Santos TS, dos Santos IDD, Pereira-Filho RN, Gomes SVF, Lima-Verde IB, Marques MN et al (2021) Histological evidence of wound healing improvement in rats treated with oral administration of hydroalcoholic extract of Vitis labrusca. Curr Issues Mol Biol 43(1):335–352. https://doi.org/10.3390/cimb43010028
17. Balderas-Cordero D, Canales-Alvarez O, Sánchez-Sánchez R, Cabrera-Wrooman A, Canales-Martinez MM, Rodriguez-Monroy MA (2023) Anti-inflammatory and histological analysis of skin wound healing through topical application of Mexican Propolis. Int J Mol Sci 24(14):11831. https://doi.org/10.3390/ijms241411831
18. Tsai C-R, Wang Y, Galko MJ (2018) Crawling wounded: molecular genetic insights into wound healing from Drosophila larvae. Int J Dev Biol 62(6-7-8):479–489. https://doi.org/10.1387/ijdb.180085mg
19. Galko MJ, Krasnow MA (2004) Cellular and genetic analysis of wound healing in Drosophila larvae. PLoS Biol 2(8):e239. https://doi.org/10.1371/journal.pbio.0020239
20. Akhavan-Sigari R (2023) Genes involved in wound healing: a narrative review. Onkologia i Radioterapia 17(9)
21. Krizanova O, Penesova A, Sokol J, Hokynkova A, Samadian A, Babula P (2022) Signaling pathways in cutaneous wound healing. Front Physiol 13:1030851. https://doi.org/10.3389/fphys.2022.1030851
22. Bonnici L, Suleiman S, Schembri-Wismayer P, Cassar A (2024) Targeting signalling pathways in chronic wound healing. Int J Mol Sci 25(1):50. https://doi.org/10.3390/ijms25010050
23. Azevedo ML, Silveira RG, Nedel F, Lund RG (2023) MicroRNAs expressed during normal wound healing and their associated pathways: a systematic review and bioinformatics analysis. PLoS One 18(4):e0281913. https://doi.org/10.1371/journal.pone.0281913
24. Lindley LE, Stojadinovic O, Pastar I, Tomic-Canic M (2016) Biology and biomarkers for wound healing. Plast Reconstr Surg 138(3):18S–28S. https://doi.org/10.1097/PRS.0000000000002682
25. Shah J, Mohd Y, Omar E, Pai DR, Sood S (2012) Cellular events and biomarkers of wound healing. Indian J Plast Surg 45(2):220–228. https://doi.org/10.4103/0970-0358.101282
26. Baldan-Martin M, Martin-Rojas T, Corbacho-Alonso N, Lopez JA, Sastre-Oliva T, Gil-Dones F et al (2020) Comprehensive proteomic profiling of pressure ulcers in patients with spinal cord injury identifies a specific protein pattern of pathology. Adv Wound Care 9(5):277–294. https://doi.org/10.1089/wound.2019.0968
27. Glinos GD, Verne SH, Aldahan AS, Liang L, Nouri K, Elliot S et al (2017) Optical coherence tomography for assessment of epithelialization in a human ex vivo wound model. Wound Repair Regen 25(6):1017–1026. https://doi.org/10.1111/wrr.12600
28. Meyer JH, Larsen PP, Strack C, Harmening WM, Krohne TU, Holz FG et al (2019) Optical coherence tomography angiography (OCT-A) in an animal model of laser-induced choroidal neovascularization. Exp Eye Res 184:162–171. https://doi.org/10.1016/j.exer.2019.04.002
29. Grambow E, Sorg H, Sorg CGG, Strüder D (2021) Experimental models to study skin wound healing with a focus on angiogenesis. Med Sci 9(3):55. https://doi.org/10.3390/medsci9030055
30. Turk M, Biernaskie J, Mahoney DJ, Jenne CN (2022) Intravital microscopy techniques to image wound healing in mouse skin. Methods Mol Biol (Clifton, N.J.) 2440:165–180. https://doi.org/10.1007/978-1-0716-2051-9_10
31. Jones JD, Ramser HE, Woessner AE, Quinn KP (2018) In vivo multiphoton microscopy detects longitudinal metabolic changes associated with delayed skin wound healing. Commun Biol 1(1):1–8. https://doi.org/10.1038/s42003-018-0206-4
32. Barroca NCB, Della Santa G, Suchecki D, García-Cairasco N, Umeoka EH d L (2022) Challenges in the use of animal models and perspectives for a translational view of stress

and psychopathologies. Neurosci Biobehav Rev 140:104771. https://doi.org/10.1016/j.neubiorev.2022.104771
33. Landi M, Everitt J, Berridge B (2021) Bioethical, reproducibility, and translational challenges of animal models. ILAR J 62(1–2):60–65. https://doi.org/10.1093/ilar/ilaa027
34. McGonigle P, Ruggeri B (2014) Animal models of human disease: challenges in enabling translation. Biochem Pharmacol 87(1):162–171. https://doi.org/10.1016/j.bcp.2013.08.006
35. Sanapalli BKR, Yele V, Singh MK, Thaggikuppe Krishnamurthy P, Karri VVSR (2021) Preclinical models of diabetic wound healing: a critical review. Biomed Pharmacother 142:111946. https://doi.org/10.1016/j.biopha.2021.111946
36. Saeed S, Martins-Green M (2024) Assessing animal models to study impaired and chronic wounds. Int J Mol Sci 25(7):3837. https://doi.org/10.3390/ijms25073837
37. Schultz GS, Chin GA, Moldawer L, Diegelmann RF (2011) Principles of wound healing. In: Fitridge R, Thompson M (eds) Mechanisms of vascular disease: a reference book for vascular specialists. University of Adelaide Press, Adelaide
38. Tripathi S, Mandal SS, Bauri S, Maiti P (2022) 3D bioprinting and its innovative approach for biomedical applications. MedComm 4(1):e194. https://doi.org/10.1002/mco2.194
39. Wang Z, Zhao F, Lang H, Ren H, Zhang Q, Huang X et al (2025) Organoids in skin wound healing. Burns Trauma 13:tkae077. https://doi.org/10.1093/burnst/tkae077
40. Wang W, Liu P, Zhu W, Li T, Wang Y, Wang Y et al (2025) Skin organoid transplantation promotes tissue repair with scarless in frostbite. Protein Cell 16(4):240–259. https://doi.org/10.1093/procel/pwae055
41. Lee J, Koehler KR (2021) Skin organoids: a new human model for developmental and translational research. Exp Dermatol 30(4):613–620. https://doi.org/10.1111/exd.14292
42. Zhang T, Sheng S, Cai W, Yang H, Li J, Niu L et al (2024) 3-D bioprinted human-derived skin organoids accelerate full-thickness skin defects repair. Bioact Mater 42:257–269. https://doi.org/10.1016/j.bioactmat.2024.08.036
43. Cho SW, Malick H, Kim SJ, Grattoni A (2024) Advances in skin-on-a-chip technologies for dermatological disease modeling. J Invest Dermatol 144(8):1707–1715. https://doi.org/10.1016/j.jid.2024.01.031
44. Wufuer M, Lee G, Hur W, Jeon B, Kim BJ, Choi TH et al (2016) Skin-on-a-chip model simulating inflammation, edema and drug-based treatment. Sci Rep 6(1):37471. https://doi.org/10.1038/srep37471
45. Zoio P, Oliva A (2022) Skin-on-a-chip technology: microengineering physiologically relevant in vitro skin models. Pharmaceutics 14(3):682. https://doi.org/10.3390/pharmaceutics14030682
46. Shabestani MG, Ertl P, Rothbauer M (2021) Microfluidic and lab-on-a-chip systems for cutaneous wound healing studies. Pharmaceutics 13(6):793. https://doi.org/10.3390/pharmaceutics13060793
47. Anisuzzaman DM, Wang C, Rostami B, Gopalakrishnan S, Niezgoda J, Yu Z (2022) Image-based artificial intelligence in wound assessment: a systematic review. Adv Wound Care 11(12):687–709. https://doi.org/10.1089/wound.2021.0091
48. Chairat S, Chaichulee S, Dissaneewate T, Wangkulangkul P, Kongpanichakul L (2023) AI-assisted assessment of wound tissue with automatic color and measurement calibration on images taken with a smartphone. Healthcare 11(2):273. https://doi.org/10.3390/healthcare11020273
49. Komura D, Ochi M, Ishikawa S (2025) Machine learning methods for histopathological image analysis: updates in 2024. Comput Struct Biotechnol J 27:383–400. https://doi.org/10.1016/j.csbj.2024.12.033
50. Chen M-Y, Cao M-Q, Xu T-Y (2024) Progress in the application of artificial intelligence in skin wound assessment and prediction of healing time. Am J Transl Res 16(7):2765–2776. https://doi.org/10.62347/MYHE3488
51. Nagai MK, Embil JM (2002) Becaplermin: recombinant platelet derived growth factor, a new treatment for healing diabetic foot ulcers. Expert Opin Biol Ther 2(2):211–218. https://doi.org/10.1517/14712598.2.2.211

52. Yurista G, Novebri NA, Irianiwati, Seswandhana MR (2024) Impact of negative pressure wound therapy on inflammatory cell counts in porcine deep dermal burn wound models. Chin J Plast Reconstr Surg 6(4):166–171. https://doi.org/10.1016/j.cjprs.2024.09.002
53. Ineichen BV, Furrer E, Grüninger SL, Zürrer WE, Macleod MR (2024) Analysis of animal-to-human translation shows that only 5% of animal-tested therapeutic interventions obtain regulatory approval for human applications. PLoS Biol 22(6):e3002667. https://doi.org/10.1371/journal.pbio.3002667
54. Akhtar A (2015) The flaws and human harms of animal experimentation. Camb Q Healthc Ethics 24(4):407–419. https://doi.org/10.1017/S0963180115000079

Chapter 13
Extracellular Vesicles as Functional Mediators of Wound Healing

Heidi Yuan and Luisa A. DiPietro

Abstract The course of wound healing, during all of its phases, involves the intricate coordination of both canonical and non-canonical signaling mechanisms that enable distinct cell populations in the skin to guide each other towards repair. Among the non-canonical signaling modalities, extracellular vesicles (EVs), which are constitutively released by nearly all cells, have surfaced as important mediators of this intercellular communication. By trafficking a host of signaling molecules, including nucleic acids, bioactive lipids, and proteins, EVs can influence the phenotype of neighboring cells upon uptake. In the context of cutaneous repair, stem cell-derived EVs have demonstrated therapeutic potential for treating chronic and non-healing wounds. While this underscores the regenerative capacity of exogenously administered EVs as a cell-free treatment, it is equally important to recognize the contribution of endogenously produced EVs in a native wound environment to the healing process. In fact, these vesicle-bound parcels of information are released by epithelial cells, immune cells, vascular cells, and matrix-depositing cells, linking their function to the complex network of signaling and communication that activates upon tissue injury. This review will synthesize the existing evidence on the function of skin cell- and stem cell-derived EVs in wound healing and tissue repair, and provide an outlook on the fundamental and translational approaches to studying EVs in skin regeneration.

Keywords Skin regeneration · Exosome · Intercellular communication · Paracrine signaling · Tissue repair

---

H. Yuan · L. A. DiPietro (✉)
Center for Wound Healing and Tissue Regeneration, University of Illinois Chicago, Chicago, IL, USA
e-mail: ldipiet@uic.edu

F. De Francesco, K. Turksen (eds.), *Management and Strategies for Wound Healing*, Stem Cell Biology and Regenerative Medicine 79,
https://doi.org/10.1007/978-3-032-12442-5_13

Abbreviations

ADSC	Adipose-derived stem cell
AKT1	(RAC-alpha serine/threonine-protein kinase)
ALIX	ALG-2 interacting protein X
apoEVs	Apoptotic mesenchymal stem cell-derived extracellular vesicles
BMDCs	Bone marrow-derived dendritic cells
BMSC-EVs	Bone marrow-derived mesenchymal stem cells-derived extracellular vesicles
CD163	Cluster of Differentiation 163
CD206	Cluster of Differentiation 206
CD3	Cluster of Differentiation 3
CD4	Cluster of Differentiation 4
CD68	Cluster of Differentiation 68
CD8	Cluster of Differentiation 8
CXCL2	Chemokine (C-X-C motif) ligand 2
CXCL8	Chemokine (C-X-C motif) ligand 8
CXCR7	Chemokine (C-X-C motif) Receptor
D0	Day 0
D14	Day 14
D3	Day 3
D7	Day 7
DLL4	Delta-like ligand 4
DPC	Dermal papillary cells
EC	Endothelial cell
ECM	Extracellular matrix
EGF	Epidermal growth factor
ERK1/2	Extracellular Signal-Regulated Kinase 1/2
ESCRT	Endosomal sorting complexes required for transport
EV	Extracellular vesicle
FB	Fibroblast
FGF	Fibroblast growth factor
FHC	Fetal human cells
FN1	Fibronectin 1
GFP	Green fluorescent protein
GSK-3β	Glycogen synthase kinase-3 beta
GSTM2	Glutathione S-Transferase Mu 2
HATMSC1	Human Adipose Tissue Mesenchymal Stem Cells 1
HDAC8	Histone deacetylase 8
hTERT	Human telomerase reverse transcriptase
HUVEC	Human Umbilical Vein Endothelial Cells
IFNγ	Interferon-gamma
IL	Interleukin
ILV	Intraluminal vesicle

ISEV	International Society for Extracellular Vesicles
ITCH	Itchy E3 ubiquitin protein ligase
ITGBL1	Integrin subunit beta like 1
JAK	Janus Kinase
K1	Keratin 1
K10	Keratin 10
KC	Keratinocyte
LATS1	Large tumor suppressor kinase 1
LED	Light emitting diode
LPS	Lipopolysaccharide
LRG1	Leucine-rich α-2-glycoprotein 1
MAPK	Mitogen-activated protein kinases
MC	Macrophage
MCP-1	Monocyte chemoattractant protein-1
MHC	Major histocompatibility complexes
MITF	Microphthalmia-associated transcription factor
MMP9	Matrix Metalloproteinase 9
MSC	Mesenchymal stem cell
mTOR	Mammalian target of rapamycin
MVB	Multivesicular body
NF-κB	Nuclear factor kappa B
PDGF	Platelet-derived growth factor
PI3k	Phosphoinositide 3-kinases
pknox1	PBX/Knotted 1 Homeobox 1
PTEN	Phosphatase and TENsin homolog
PVA	Polyvinyl alcohol
ROS	Reactive oxygen species
SOX9	SRY-box transcription factor 9
SPRY1	Sprouty homolog 1
STAT	Signal Transducer and Activator of Transcription
TANK	TRAF family member-associated NF-κB activator
TBK	TANK-binding kinase 1
TGF-β	Transforming growth factor-beta
TGFβR1	Transforming Growth Factor Beta Receptor 1
THP-1	Tohoku Hospital Pediatrics-1
TIMP-1	Tissue inhibitor matrix metalloproteinase 1
TLR4	Toll-like receptor 4
TNF-α	Tumor Necrosis Factor Alpha
UVB	Ultraviolet B
VEGF	Vascular endothelial growth factor
YAP	Yes-associated protein 1
α-SMA	Alpha smooth muscle actin

Introduction

Wound healing is an essential and tightly coordinated biological process for restoring tissue integrity and function after injury. It is canonically comprised of four sequential yet overlapping phases that each involve specific cellular players and events: hemostasis, inflammation, proliferation, and remodeling [1, 2]. Immediately after tissue injury occurs, hemostasis begins, in which vasoconstriction and platelet aggregation take place. This leads to the formation of a fibrin clot, which can serve as a provisional matrix for the open wound and act as a reservoir for growth factors [3]. This phase is promptly succeeded by inflammation, characterized by the influx of neutrophils and macrophages that extravasate to the site of injury and clear pathogens and debris via phagocytosis. These inflammatory cells also secrete cytokines and chemokines to initiate pro-inflammatory signaling cascades among neighboring skin cells [4]. As the inflammation recedes, the subsequent proliferative phase begins. This phase involves fibroblast proliferation and deposition of extracellular matrix (ECM), while endothelial cells undergo extensive angiogenesis to re-vascularize and restore perfusion to the wound bed. At the same time, keratinocytes migrate to cover the wound in a process known as re-epithelialization [5]. A plethora of soluble mediators, including fibroblast growth factor (FGF), transforming growth factor-beta (TGF-β), and vascular endothelial growth factor (VEGF) contribute to these processes during this phase [5, 6]. Lastly, in the remodeling phase, the newly formed matrix matures by realignment and crosslinking of collagen. In particular, type III collagen is replaced by the more mechanically stable type I collagen, thus improving the structural integrity and tensile strength of the wounded skin [7].

In normal physiological conditions, the end result of the response to injury is rapid and functional wound repair. However, in conditions like diabetes, aging, or ischemia, wound healing is severely impaired, leading to chronic wounds that persist due to prolonged inflammation, compromised angiogenesis, altered epithelial cell function, and the unbalanced regulation of ECM deposition [8, 9]. Chronic wounds remain a clinical and economic burden, and are associated with increased morbidity, healthcare costs, and lower quality of life [10]. Thus, it is crucial to understand the underlying molecular mechanisms and mediators that govern wound repair in an effort to develop effective therapeutic interventions that can improve healing.

Recent efforts in wound healing research have identified mesenchymal stem cells (MSC) and their extracellular vesicles (EVs) as promising regenerative therapies because of their ability to modulate many key steps involved in wound healing. They have been shown to have immunomodulatory and pro-angiogenic effects, while enhancing the cellular survival of many key cell populations required for repair [11, 12]. A growing emphasis has also been placed on studying native skin cell-derived EVs and their role in facilitating intercellular communication between different cell populations [13, 14], which is integral to processes like tissue regeneration and wound healing. In light of this developing interest, this review will synthesize the existing evidence associated with native skin cell-derived EVs and their contribution to skin wound healing, and will provide a summary of the major functional mechanisms by which exogenous stem cell-derived EVs exert their therapeutic effects.

EV Biogenesis and Classification

Extracellular vesicles (EVs) are membrane-bound, non-replicating particles secreted by nearly all cell types [15]. They are implicated in intercellular communication through the transfer of bioactive cargo in the form of lipids, proteins, and non-coding RNAs (such as microRNAs). In addition, the contents of EVs reflect the physiological or pathological state of their parent cell, making them useful candidates for therapeutic applications and biomarkers in diagnostics [16]. Broadly, they can be classified into three categories based on their size, biogenesis, and molecular components: exosomes, microvesicles, and apoptotic bodies [17, 18]. A brief description of each class of EVs is provided below.

Exosomes

Small EVs that are below 200 nm in diameter and originate from the endosomal pathway are classified as exosomes. Exosomes are formed from the inward budding of the endosomal membrane, which form what are known as intraluminal vesicles (ILVs) that are encapsulated within larger, multivesicular bodies (MVBs). When these MVBs fuse with the outer plasma membrane, the ILVs are subsequently released into the extracellular space as exosomes [19]. Mechanisms of exosome secretion involve specific subunits of the endosomal sorting complexes required for transport (ESCRT) machinery, which is complemented by the cytoplasmic sorting and membrane budding activities of syntenin and the accessory protein ALG-2 interacting protein X (ALIX). Alternatively, exosomes can also be formed in an ESCRT-independent fashion, involving tetraspanins, such as CD9, CD63, and CD81, which interact with different transmembrane proteins and lipids (e.g., ceramide and sphingomyelin) to generate dynamic membrane microdomains that favor budding [19, 20].

Microvesicles

In contrast to exosomes, microvesicles, also deemed ectosomes or shedding vesicles, range from 100 to 1000 nm in diameter and are generated by the direct outward budding or fission of the plasma membrane. The shedding of microvesicles is dictated by structural rearrangements in the plasma membrane driven by calcium-dependent enzymatic machineries such as flippases and floppases (formally known as aminophospholipid translocases), scramblases, and calpain. These enzymes are responsible for exposing phosphatidylserine domains from the inner lipid bilayer to the outer cell surface, which bends the membrane and the underlying actin cytoskeleton, inducing vesicle release [19, 21]. Additional key regulators of this process include ARF6, RhoA, and myosin-1a, which orchestrate the detachment of budding vesicles from the membrane [22].

Apoptotic Bodies

During apoptosis, or programmed cell death, cells release apoptotic bodies, which are the largest class of EVs, ranging from 1 to 5 μm in size [23]. These distinct subcellular membrane-bound vesicles contain nuclear fragments, cellular organelles (e.g., mitochondria and endoplasmic reticulum), as well as other cytoplasmic remnants. The secretion of apoptotic bodies has been suggested to play a role in the clearance of apoptotic cells by phagocytes [24, 25].

EV Classification Challenges and ISEV Guidelines

Despite the classification above, size and surface markers often overlap between EV subtypes, making them difficult to distinguish unambiguously and isolate from one another. The International Society for Extracellular Vesicles (ISEV) recommends using the term "extracellular vesicles" as a general descriptor, along with specific parameters such as size (small vs. medium/large EVs), density, or biogenesis pathway, where possible [15, 17].

Endogenous (Native) Skin Cell-Derived EVs in Wound Healing

Many predominant cell types in the wound are known to secrete EVs that elicit functional changes among neighboring cells (Table 13.1). This section will detail these functional effects and explore their underlying mechanisms, focusing on EVs derived from four major cell types involved in wound healing—macrophages, keratinocytes, fibroblasts, and endothelial cells (Fig. 13.1).

Table 13.1 Functional effects of endogenous (native) skin-cell derived EVs on different cell types involved in wound healing by cell type

Donor cell	Donor cell state or modifications	Recipient	Mechanism	Effect	References
Macrophages		Endothelial cells		Promoted high glucose-treated HUVECs in assays of tube formation and scratch closure and proliferation, reduced inflammatory marker expression, and promoted diabetic wound healing in vivo	[26]
	M2 polarized mouse macrophages	Macrophages, fibroblasts, endothelial cells		Enhanced conversion of M1 to M2 macrophages in vitro and in vivo, promoted fibroblast migration and EC tube formation, and accelerated wound healing in vivo	[27]
	M2 macrophages	Fibroblasts	lncRNA-ASLNCS5088/miR-200c-3p/↑GLS and α-SMA	Fibroblast expression of ECM proteins (FN1, Col1, Col3, and α-SMA) increased in response to M2 macrophage EV treatment	[31]
	Unpolarized Raw264.7 murine macrophage cell line	Dermal papilla cells	Wnt3a/Wnt7b	EVs attached to dermal papillary membrane and were internalized, increased cell migration and proliferation, and promoted hair shaft elongation and growth in vivo	[30]
	Unpolarized Raw264.7 murine macrophage cell line	Endothelial cells	VEGF, Wnt3a, miR-130a	Increased endothelial cell proliferation, migration, and tube formation in vitro and in vivo	[29]
	M2 macrophages	Intestinal epithelial cells (FHC cells)	miR-590-3p/LATS1/YAP/β-catenin	Promoted proliferation and scratch wound closure of FHC cells	[34]
	M2 macrophages	Fibroblasts	LINC01605/↓miR-493-3p/↑AKT1	Promoted proliferation, migration, and invasion of fibroblasts	[32]
	M2 macrophage EVs loaded in paintable alginate hydrogels	HUVECs, macrophages		Promoted tube formation and migration of HUVECs, and accelerated wound closure	[28]
	M2 macrophages (THP-1)	Hypertrophic scar Fibroblasts	CXCL2/CXCR7/mTOR pathway	Promoted proliferation, migration, and expression of fibrotic proteins (Col1, Col3, α-SMA) and autophagy	[33]

(continued)

Table 13.1 (continued)

Donor cell	Donor cell state or modifications	Recipient	Mechanism	Effect	References
Fibroblasts	L929 fibroblasts	Fibroblasts & endothelial cells		Enhanced proliferation, migration, and scarless wound healing-related genes in fibroblasts, and increased migration and tube formation in endothelial cells (led to improved collagen formation and vascularity in mouse wounds)	[62]
		Fibroblasts & endothelial cells	miR-130a, miR-210, VEGF-D, CXCL8	Increased proliferation, migration, and expression of FGF, MMP2, and α-SMA, and increased proliferation and tube formation in endothelial cells, improved cell density, deposition and maturation of collagen in wounds in vivo	[63]
	High and normal glucose	Endothelial cells	GSK-3β/β-catenin	For normal glucose, EVs promoted migration and angiogenesis in vitro and in vivo, upregulating the GSK-3β/β-catenin pathway (opposite effect for high glucose)	[70]
		Fibroblasts	FGF-2	Increased proliferation and migration, activation of ERK/STAT pathways, and improved mouse wound healing	[64]
	High glucose	Keratinocytes	Lnc-URIDS/YTHDF2/MMP9	Elevated expression of MMP9 in primary rat keratinocytes, leading to collagen degradation	[72]
	GSTM2 overexpression	Keratinocytes & fibroblasts	GSTM2/NACA/ROS-ERK-ETS-Cyclin D pathway	Improved migration and proliferation in keratinocytes while promoting oxidative phosphorylation in aged fibroblasts	[75]
	Senescent	Keratinocytes	miR-23a-3p	Short-term treatment enhanced migration while long-term treatment inhibited differentiation	[73]
	Young and senescent	Keratinocytes		Young fibroblast-derived EVs promoted keratinocyte differentiation while senescent ones promoted weaker differentiation, increased IL-6 inflammatory markers	[74]

	BJ-5ta	Fibroblasts	MAPK/AP-1/MMP1	Protected BJ-5ta fibroblasts from UVB-mediated oxidative stress and cellular damage, promoting collagen synthesis and reversing cellular senescence	[65]
	Hypoxic	Macrophages		Enhanced macrophage sensitivity to M1 and M2 phenotypes under distinct microenvironmental cues, and rescued M2 macrophage polarization in diabetic wounds	[67]
	Autologous diabetic	Endothelial cells	AKT/β-catenin	Improved HUVEC migration, proliferation, and tube formation, and stimulated diabetic rat wound closure in vivo	[66]
		Dermal papillary and outer root sheath cells	Wnt3a/β-catenin/Axin2 and Lef1	Promoted proliferation and activation of Wnt/β-catenin signaling in DP cells, and promoted proliferation, migration, and differentiation of ORS cells while elongating hair shaft	[77]
	Pre-treated with FGF-2 and PDGF	Dermal papillary cells	Norrin/Frizzled4	Stimulated hair growth ex vivo, activated Norrin secretion in DPCs and consequently transferred Fzd4 receptors to follicular keratinocytes	[76]
	3D spheroid cultured	Fibroblasts	TIMP-1	Promoted scratch closure of fibroblasts and reduced wrinkle formation in UVB-induced skin photoaging in nude mice	[78]
	Normal and UVB-treated	Fibroblasts		Increased proliferation under both normal and UVB-treated conditions, reduced senescence, reduced UVB-induced intracellular ROS production, MMP1 expression, and promoted collagen production	[134]
	Dermal Fibroblasts from Chenghua pigs (thick skin)	Fibroblasts	miRNA-218, ITGBL1, TGFβ1-SMAD2/3	Increased proliferation, migration, collagen contraction and synthesis in vitro, and improved skin thickness and collagen I content in mouse skin in vivo	[80]
	TGF-β-stimulated healthy control fibroblasts or SSc fibroblasts	Macrophages		SSc fibroblast EVs stimulated M2 polarization (increased CD163 and CD206 expression), pro-inflammatory and fibrotic cytokine expression, and stimulated macrophage ability to activate SSc fibroblasts (increasing IL-6, Col1, Col3, and FN1 expression)	[68]
	SSc fibroblasts vs healthy fibroblasts	Keratinocytes	TBK/JAK/STAT	SSc fibroblast EVs upregulated type I IFN signature genes compared to healthy controls	[69]

(continued)

Table 13.1 (continued)

Donor cell	Donor cell state or modifications	Recipient	Mechanism	Effect	References
Keratinocytes	Black/Caucasian/Caucasian + UVB-stimulated	Melanocytes	miR-3196, miR-203a	Phototype-dependent upregulation of pigment formation by increased pigment production, tyrosinase activity, and MITF and Rab27a gene expression	[55]
		Melanocytes	H19, miR-675	Decreased MITF and pigment production	[57]
		Melanocytes	miR-330-5p	Decreased tyrosinase activity and pigment production	[59]
	UVB-irradiated	Melanocytes		Constitutive regulation of melanocyte activity, even in absence of UVB	[56]
	LED (585 nm)-irradiated	Melanocytes	miR-675	Significant upregulation of H19 and miR-675 following irradiation, and downregulation of melanin production, and TYR/MITF/TRP1 expression	[58]
	UVB-irradiated	Melanocytes	hsa-miR-644a, hsa-miR-365b-5p, and hsa-miR-29c-3p upregulated melanogenesis, while hsa-miR-18a-5p, hsa-miR-197-5p, and hsa-miR-4281 downregulated melanogenesis	Promoted melanogenesis and upregulated MITF, TRP1, TRP2, and TYR gene expression	[60]
	Murine keratinocyte cell line MPEK (+/− IFNγ) steady-state vs. inflammatory conditions	Dendritic cells (DCs, bone marrow-derived) and T cells	Antigen-specific information (ovalbumin)	Induced maturation of BMDCs (increased CD40 expression and production of IL-6, IL-10, and IL-12), no effect on T cell activation	[44]

	HaCaT	Fibroblasts		Upregulated MMP1, IL-6, IL-8, and TGF-β-associated genes, activated ERK1/2, JNK, Smad, and p38 signaling, stimulated fibroblast migration and induced fibroblast-mediated endothelial tube formation	[45]
	HaCaT	T cells	MHC I and II	Increased T cell proliferation via indirect contact (transwell), and following IFNγ stimulation of HaCaT and loading of *S. aureus* enterotoxin B, the EVs also increased CD4+ and CD8+ T cell proliferation	[43]
	miR-21-overexpressing HaCaT	Macrophages	miR-21	Promoted direct conversion of macrophages to fibroblast-like cells (activated Col1A1 promoter and FSP1 expression)	[37]
	Psoriatic cytokine-treated primary keratinocytes	Neutrophils	NF-κB and p38 MAPK pathway	Promoted NETosis and expression of pro-inflammatory factors in neutrophils	[40]
	Normal (vector), miR-21-overexpressing, and miR-21-inhibited	Fibroblasts and endothelial cells	miR-21/PTEN/RECK → MAPK/ERK signaling	Enhanced fibroblast migration and endothelial cell tube formation, upregulated expression of α-SMA and N-cadherin, and IL-6 and IL-8, and improved wound healing in diabetic rats in vivo	[47]
	GFP-labeled (via tissue nano-transfection) isolated from mouse skin wound edge	Macrophages	Surface N-glycans	Surface N-glycans on these EVs selectively enhanced macrophage engulfment and stimulated macrophage polarization from M1 to M2 phenotype	[38]
	HaCaT	HaCaT and fibroblasts	ERK1/2, p38, and MAPK	Enhanced keratinocyte and fibroblast migration in scratch wounds	[46]
	Aged vs young primary keratinocytes	Keratinocytes	miR-30a (differentially enhanced in aged keratinocytes)	EVs derived from aged keratinocytes significantly impaired recipient cell migration and proliferation and upregulated anti-proliferative p21 expression and differentiation markers K1, K10 (early differentiation) and loricrin (late differentiation)	[50]
	HaCaT cultured in high glucose	HUVECs	LINC01435/YY1/HDAC8	Impaired tube formation, angiogenesis, and migration	[49]

(continued)

Table 13.1 (continued)

Donor cell	Donor cell state or modifications	Recipient	Mechanism	Effect	References
	M5 (IL-17A, IL-22, TNF-α)-stimulated HaCaT (for psoriatic induction)	Macrophages	LRG1/TGFβR1	Increased IL-1β and TNF-α expression in macrophages and promoted M1 polarization/inflammatory phenotype associated with psoriatic dermatitis)	[41]
	Diabetic primary keratinocyte exosomes vs healthy controls	LPS & IFN-γ stimulated macrophages		Diabetic KC-EVs demonstrated impaired uptake by macrophages and persistent inflammatory marker expression (cytokines CXCL5, IL-18, VEGF, and TIM3) and pro-inflammatory phenotype (iNOS) compared to controls	[39]
	HaCaT cultured in high glucose	Fibroblasts	LINC01435	Impaired fibroblast proliferation, migration and collagen production via upregulation of autophagy	[48]
	HaCaT	Macrophages		Increased conversion of THP-1 macrophages to CD80+ and CD163+ macrophages, increased IL-10 and TGF-β expression	[42]
Endothelial cells	Hypoxic	Fibroblast	LOXL2	Increased fibroblast contraction in vitro	[83]
	AGE-treated	Fibroblast	miR-106b-5p/ERK1/2	Promoted fibroblast autophagy and collagen degradation	[84]
		Diabetic skin Fibroblast	YAP nuclear translocation and activation of PI3K/Akt/mTOR	Reduced senescence, enhanced proliferation and wound healing	[86]
	miR-106b overexpression	Fibroblasts & keratinocytes	miR-106b inhibition of JMJD3	Reduced adhesion and viability of both cell types	[135]
	Apoptotic	Fibroblasts & endothelial cells		No effect on fibroblasts, improved proliferation, migration, and tube formation in HUVECs	[136]
	Hypoxic	Endothelial cells and macrophages	lncHAR1B/BHLE23/KL4	Increased M2 macrophage polarization and tube formation among HUVECs, improved wound healing in diabetic mice	[85]
	LPS-induced (rat aortic endothelial cells)	Macrophages	TLR4/NF-κB	Increased viability and expression of M1 phenotype markers (iNOS and CD68)	[88]
	Quiescent	Monocytes	miR-10a inhibition of the NF-κB pathway	Decreased pro-inflammatory response and activation	[87]

Abbreviations: SSc **systemic sclerosis,** *ORS* **outer root sheath,** *DPCs* **dermal papillary cells**

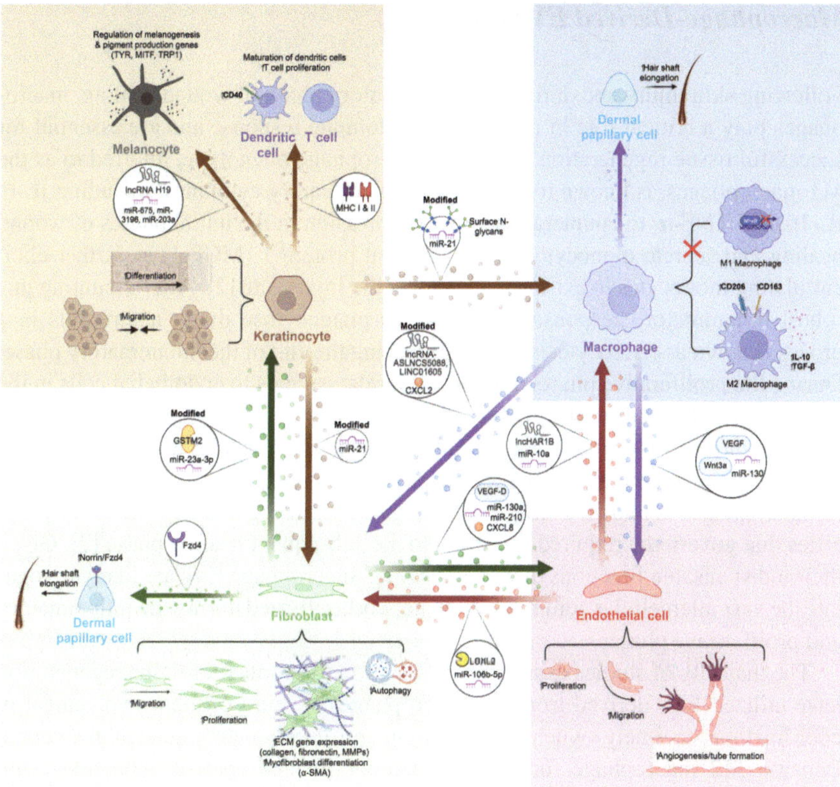

Fig. 13.1 Summary of predominant EV-mediated intercellular communication in wound healing. Four major cell types involved in skin wound healing are keratinocytes (KCs), macrophages (MCs), fibroblasts (FBs), and endothelial cells (ECs). These four cell types secrete extracellular vesicles (EVs) that influence healing-related functions of other neighboring cell types. Color of arrows correspond to origin cell type from which EVs are released. Callouts indicate evidence of important cargo that may contribute to the functional effects on recipient cell types. "Modified" cargo refers to components that were experimentally enriched in the EVs prior to addition to recipient cells. *Tan box:* KCs receive EVs from FBs, which promote KC differentiation and ultimately stratification. FB-EVs overexpressing antioxidant enzymes such as GSTM2 improve KC proliferation and migration, while senescent FB-EVs contain miR-23a-3p, which also transiently enhances KC migration. KC-EVs also play a role in regulating melanogenesis through transfer of lncRNA H19 and certain microRNAs, and can possess antigen-presenting properties via the expression of MHC class molecules. *Purple box:* KC-EVs have been found to contain superficial N-glycans that facilitate MC-specific uptake. FB-EVs and KC-EVs both play a role in polarizing MCs towards the M2 phenotype, demonstrated by increased CD206, CD163, and Arg1 expression. Consequently, IL-10 and TGF-β expression are upregulated. EC-EVs can inhibit activation of NF-κB in MCs through miRNA-10a, further promoting MCs towards the anti-inflammatory phenotype. MC-EVs also contribute to hair growth and elongation through dermal papillary cell (DPC) stimulation. *Green box:* FBs receive EV signals from MCs, KCs, and ECs. MC-EVs contain lncRNAs that promote FB proliferation, migration, and gene expression of key ECM regulators such as fibronectin, collagen I, and collagen III. Myofibroblast activation, identified through increased α-SMA expression, is also enhanced by MC-EVs. Chemokines, such as CXCL2, in MC-EVs mediate fibroblast autophagy. EC-EVs contain lysyl oxidase-like 2 (LOXL2), which

(continued)

Macrophage-Derived EVs (MC-EVs)

Following skin injury and during the inflammatory phase of wound healing, macrophages play a critical role in regulating the immune response and are essential for successful tissue regeneration. The early macrophage phenotype, referred to as the M1 macrophages, is known to release pro-inflammatory cytokines, including IL-6, IL-1β, and TNF-α, to counteract pathogenic invasion in the initial phases of wound healing and secrete monocyte chemoattractant protein-1 (MCP-1) to further elicit circulating monocytes to extravasate towards the injury site [2]. After mounting this robust inflammatory response, macrophages phagocytose dying neutrophils in a process known as efferocytosis, thereby marking the end of the inflammatory phase. During the proliferative phase, macrophages relay signals to endothelial cells in the form of growth factors like VEGF and PDGF to facilitate angiogenesis and re-vascularization of the wound bed. Simultaneously, macrophages interact with fibroblasts to enhance granulation tissue and extracellular matrix deposition. This subset of macrophages is known as M2 macrophages, possessing anti-inflammatory properties that govern the required transition to the subsequent healing phase [2]. Given their indispensable functions in wound healing, macrophage EVs are likely to regulate the vast intercellular communication network activated during the inflammatory and proliferative phases.

The majority of studies examining MC-EVs in the context of tissue regeneration have utilized EVs derived from M2 macrophages because of their involvement in coordinating the timely switch from a pro- to anti-inflammatory state in the wound. Because M2 macrophage functionality temporally and spatially coincides with angiogenesis in the wound, endothelial cells have often been studied as a recipient cell type. M2 MC-EVs have been shown to reverse impairments in endothelial cell function caused by hyperglycemic culture conditions, including proliferation, migration, tube formation, as well as inflammatory marker gene expression caused [26]. These MC-EVs also promoted diabetic wound healing in vivo [26]. Two other studies support this observation, demonstrating that MC-EVs obtained from M2 macrophages exhibit enhanced angiogenic potential among endothelial cells and possess pro-healing and regenerative properties in wound healing [27, 28]. However, the mechanism of this effect has not been well defined. One study that collected MC-EVs from unpolarized Raw264.7 cells, a murine macrophage cell line, for treatment of endothelial cells, also documented enhanced EC proliferation, migration, and tube formation response both in vitro and in vivo [29]. These unpolarized

Fig. 13.1 (continued) enhances FB contraction, and miR-106b-5p, which further promotes FB autophagy. DPCs are stimulated by FB-EVs to activate Wnt/β-catenin signaling. Additionally, FB-EVs upregulate secretion of Norrin in DPCs through direct transfer of its receptor Fzd4. *Pink box:* ECs, responsible for wound angiogenesis, receive pro-angiogenic cargo, such as VEGF, miRNA-130, and CXCL8, from MC-EVs, which can boost proliferation, migration, and tube formation. This is a simplified rendition based on current knowledge. See tables and text for complete details. *EV* extracellular vesicles, *KC* keratinocyte, *MC* macrophage, *FB* fibroblast, *EC* endothelial cell, *DPC* dermal papillary cell

MC-EVs were found to contain specific microRNAs (miR-210, miR-126, and miR-130a), VEGF, and Wnt3a that may contribute to the pro-angiogenic response [29]. Furthermore, an additional study from the same group demonstrated that these MC-EVs induced a consistent pro-survival phenotype in dermal papillary cells, accompanied by increased expression of VEGF and other growth factors [30]. Altogether, these findings point to a clear supportive role for MC-EVs in the re-vascularization of regenerated tissue.

Fibroblasts are another population of cells that both send and receive paracrine signals from macrophages during skin regeneration. A handful of studies have demonstrated that fibroblasts become activated in response to M2 MC-EVs [27, 31, 32]. Specifically, fibroblasts exhibited improved proliferation and migration following M2 MC-EV treatment [27, 32], while the expression of key ECM-related proteins, including fibronectin, collagens I and III, and α-SMA were markedly enhanced [31]. Long non-coding RNAs have been suggested to partially mediate this effect—in particular, lncRNA-ASLNCS5088 from the EVs can bind miR-200c-3p and subsequently upregulate glutaminase levels and promote ECM synthesis [31]. Alternatively, LINC01695 has been shown to target miR-493-3p and de-repress transcription of AKT1, enhancing fibroblast function [32]. More recently, Shi et al. found that M2 MC-EVs could specifically promote hypertrophic scar fibroblast function by mediating cellular autophagy through the CXCL2/CXCR7/mTOR signaling pathway [33]. When M2 MC-EVs were injected into scars in vivo, the collagen density and proportion of α-SMA-positive cells were significantly increased [33]. These studies collectively highlight a targeted yet diverse repertoire of non-coding RNA found within the cargo of M2 MC-EVs that de-repress transcriptional signatures implicated in fibroblast activation.

There is additional evidence to suggest that MC-EVs can influence the activity of epithelial cells. One study by Deng et al. found that M2 MC-EVs could directly promote the proliferation and scratch wound closure of fetal human cells (FHC), an intestinal epithelial cell line, through miR-590-3p-mediated repression of LATS1 [34]. This led to activation of YAP/β-catenin signaling [34], suggesting that intestinal mucosal repair may share similar transcriptomic patterns with cutaneous repair. While this specific communication and pathway has yet to be investigated in the context of skin regeneration, MC-EVs seem likely to serve as an important signaling mechanism to relay feedback to keratinocytes and further direct efforts at re-epithelialization.

Keratinocyte-Derived EVs (KC-EVs)

Keratinocytes play a critical role in re-epithelialization of the wound surface and maintaining skin immunity. Several lines of evidence suggest that keratinocyte-derived EVs (KC-EVs) can affect the functionality of multiple different cell types involved in skin wound healing.

As a primary barrier of defense against injury, keratinocytes within the epithelium are likely to communicate with various immune cells in the process of wound healing. One of these is macrophages, which have been proposed to exhibit considerable levels of plasticity following infiltration to the wound site [35, 36]. A study found that these cells of myeloid origin were capable of converting to fibroblast-like cells of the granulation tissue during wound healing, and that KC-EVs containing abundant levels of miR-21 positively regulate this phenotypic change [37]. Using a miR-21$^{fl/fl}$K14^{Cre} mouse model, keratinocyte-specific knockdown of miR-21 led to nearly a 50% reduction in the expression of this miRNA in total EVs collected from the wound. This knockdown not only abrogated the conversion of CD68-positive myeloid cells to FSP1-positive fibroblast-like cells, but also impaired healing and collagen maturation in the wound [37]. Although this does not rule out the possibility of other wound cell types contributing miR-21-rich EVs that support macrophage plasticity, it validates the importance of the keratinocyte-macrophage communication link required for optimal healing. In support of this, another study by the same research group found that macrophages selectively engulf KC-EVs due to specific N-glycans present on the EV surface [38]. These macrophages subsequently exhibit a phenotypic shift from a pro-inflammatory state to a pro-resolution state [38]. On the other hand, the authors demonstrated that KC-EVs isolated from diabetic wound fluid not only exhibited weaker uptake by wound macrophages, but also failed to promote their timely conversion to a pro-resolution state and instead adopted a persistent pro-inflammatory phenotype [39]. Taken together, these studies underscore a pivotal role for EVs in the keratinocyte-directed mechanisms of macrophage activity during wound healing.

Epithelial-immune crosstalk has also been studied in other contexts related to skin wound healing and cutaneous homeostasis. KC-EVs derived from human epidermal primary keratinocytes cultured under psoriatic-like inflammatory conditions were endocytosed by neutrophils and significantly induced NETosis followed by elevated expression of inflammatory cytokines IL-6, IL-8, and TNF-α. This effect was determined to be mediated by NF-κB and p38 MAPK signaling induced by psoriatic KC-EVs [40]. Another study utilized the M5 cytokine cocktail (IL-17A, IL-22, and TNF-α) to stimulate psoriatic inflammation in vitro among HaCaT cells, a spontaneously immortalized human skin keratinocyte cell line, to observe the effect of their EVs on macrophage function. These HaCaT-derived EVs produced a similar upregulation of inflammatory cytokine (IL-1β and TNF-α) gene expression, preceding a persistent M1 phenotype among THP-1 macrophages [41]. This was shown to be mediated in part by the Leucine-rich α-2-glycoprotein 1 (LRG1) enriched in the KC-EVs, which modulates TGFβR1 activity [41]. Yet another study examined the effect of native unstimulated HaCaT-derived EVs, in comparison to EVs from HeLa or cervical cancer cell lines, on macrophages. As compared to EVs from HeLa cells, HaCaT-derived EVs appear to increase anti-inflammatory IL-10 expression, while tumor cells upregulated pro-inflammatory IL-6 expression. TGF-β expression was increased in macrophages in response to EV treatment from all keratinocyte types. Overall, this suggests that KC-EVs can promote a moderately pro-inflammatory M1 phenotype, with slight differences in effect specific to the cell line used [42]. Together, these studies suggest that KC-EVs, both at baseline or primed with inflammatory stimuli, can relay signals to phagocytic cell types to assist the onset of an early innate immune response.

As non-professional antigen-presenting cells, keratinocytes can also secrete EVs containing the major histocompatibility complexes (MHC) I and II, and are therefore capable of eliciting an adaptive antigen-specific immune response in effector CD4+ and CD8+ T cells [43]. In one study, a murine keratinocyte cell line (MPEK) stimulated with IFNγ secreted EVs was shown to carry antigen-specific information that could be transferred to bone marrow-derived dendritic cells (BMDCs) and induce a mature phenotype [44]. Surprisingly, these EV-treated BMDCs failed to induce T cell proliferation and activation [44]. However, a subsequent study utilized IFNγ-stimulated and *Staphylococcus aureus* enterotoxin B-loaded HaCaT cells and investigated their function on memory T cells. These HaCaT cells, co-cultured with recipient T cells, were found to secrete EVs that promoted the proliferation of resting CD3+, CD4+, and CD8+ T cells [43]. This suggests that keratinocytes can directly stimulate helper T cell proliferation through their secreted EVs, without intermediate intervention or communication by dendritic cells.

HaCaT cells have often been used as a source of EVs to investigate crosstalk between fibroblasts, endothelial cells, and other keratinocytes. In accordance with the study examining how psoriatic KC-EVs affect neutrophils [40], two other studies have identified that HaCaT-derived EVs can enhance fibroblast migration likely through upregulation of ERK1/2, p38, and downstream MAPK signaling [45, 46]. Furthermore, HaCaT-derived EVs appear to consistently upregulate expression of cytokines IL-6 and IL-8, as well as pro-fibrotic genes including TGF-β and α-SMA in fibroblasts [45, 47]. This could potentially be explained by the presence of miR-21 within the EV cargo, which regulates these signaling pathways by targeting and repressing PTEN [47]. On the contrary, under hyperglycemic conditions, HaCaT-derived EVs have been shown to impede fibroblast proliferation, migration, and collagen production by triggering autophagy, possibly through the transfer of LINC01435 [48]. This same lncRNA from HaCaT-derived EVs was also found to impair angiogenesis and tube formation in vitro. LINC01435 was demonstrated to bind directly to the transcription factor YY1, subsequently upregulating histone deacetylase 8 (HDAC8) and modulating downstream pro-angiogenic gene expression [49]. In sum, KC-EVs have been implicated in mediating not only the immune response associated with wound healing, but also the formation of granulation tissue during the proliferative phase.

Recipient keratinocyte cells are also influenced by their own EVs, which can dictate their ability to migrate, proliferate, and differentiate. Nedachi et al. elucidated the influence of aging on epidermal barrier function by comparing EVs from primary keratinocytes derived from aged and young subjects. MiR-30a was found to be differentially upregulated in aged KC EVs, resulting in increased expression of anti-proliferative genes like p21 and differentiation markers K1 and K10. Aged KC-EVs also significantly impeded mouse skin wound healing in vivo [50]. As such, autocrine signaling via KC-EVs appears to be critical for directing keratinocyte differentiation, which controls the epidermal stratification essential for maintaining the integrity of the skin barrier [51].

The role of keratinocytes in skin pigment production has also been well documented in the literature. Although the contribution of pigment production and melanocyte activity to wound healing has not been well studied, evidence suggests that disruption of this keratinocyte–melanocyte communication, canonically referred to as the epidermal–melanin unit [52], can impair healing outcomes [53, 54]. EVs have been implicated in the crosstalk between keratinocytes and epidermal melanocytes. A seminal study conducted by Lo Cicero et al. found that KC-EVs promote melanin synthesis by increasing the expression and activity of multiple proteins involved in melanogenesis, including tyrosinase, microphthalmia-associated transcription factor (MITF), and Rab27a. Of note, the effects of KC-EVs were dependent on skin phototype—specifically, keratinocytes derived from black patients or UVB-treated keratinocytes from Caucasian patients were significantly more effective at producing a pigmented phenotype [55]. However, paracrine EV-mediated melanocyte stimulation from keratinocytes appears to be constitutively active, suggesting that even at a baseline, KC-EVs regulate pigmentation and melanocyte activity to maintain skin homeostasis [56]. At the same time, untreated native keratinocytes were found to contain high levels of miR-675 in their EVs, which had the opposite effect of repressing MITF expression and inhibiting pigmentation in recipient melanocytes [57]. MiR-675 in KC-EVs also appears to be upregulated in conjunction with H19 following LED (585 nm) exposure to keratinocytes [58]. In a similar vein, another study identified that KC-EVs expressed miR-330-5p that was capable of being transferred to melanocytes and effectively inhibited tyrosinase expression and pigmentation [59]. More recently, comprehensive miRNA sequencing of UVB-irradiated and non-UVB-irradiated KC-EVs demonstrated that miR-644a, miR-365b-5p, and miR-29c-3p upregulate melanogenesis, while miR-18a-5p, miR-197-5p, and miR-4281 downregulate it [60]. It is thus clear that the skin phototype and UVB exposure of the keratinocytes from which the EVs are derived determine the resulting pigment production of melanocytes. This communication mechanism is important not only for the regulation of skin pigmentation but potentially also for the maintenance of an intact skin barrier that is requisite for optimal healing.

Fibroblast-Derived EVs (FB-EVs)

Fibroblasts are the predominant cell type responsible for dermal matrix reconstitution, wound contraction, and collagen maturation [61] in wound healing. A multitude of studies have defined the role of fibroblast-derived EVs (FB-EVs) in crosstalk with other skin cells as well as in wound healing, photoaging, and hair follicle formation. A proportion of these studies have examined FB-EVs cultured from naïve primary fibroblasts or fibroblast cell lines [62–65]. One study found that FB-EVs isolated from L929 fibroblasts enhanced the proliferation, migration, and expression of scarless wound healing-related genes in recipient fibroblasts, while also promoting tube formation and migration among endothelial cells. This led to improved collagen formation and vascularity in mouse wounds [62]. The same group later showed that FB-EVs derived from primary human fibroblasts contain miR-130a, miR-210, and pro-angiogenic growth factors such as VEGF-D and CXCL8 that may explain this effect [63]. Additionally, autologous dermal FB-EVs from diabetic mice were shown to reverse the functional impairments in angiogenesis caused by high glucose treatment among HUVECs by regulating the AKT/β-catenin signaling pathway [66]. Another study found that FB-EVs contain high levels of FGF-2, preventing it from degradation. As a result, recipient fibroblasts experienced increased proliferation and migration, and mouse wounds treated with these FB-EVs containing FGF-2 demonstrated accelerated wound healing [64]. These studies underscore the contribution of native, endogenous FB-EVs to injury-induced fibroblast functions and their involvement in supporting parallel processes like blood vessel formation in wounds.

Additional studies have examined FB-EVs isolated from distinct culture conditions to mimic disease states associated with impaired healing. These include states such as hypoxia, fibrotic conditions like systemic sclerosis, hyperglycemia associated with diabetes, and cellular senescence as a model of aging skin. Hypoxic and fibrotic FB-EVs have been shown to influence macrophage polarization dynamics in a similar fashion. In particular, hypoxic FB-EVs enhance macrophage sensitivity to both M1 and M2 phenotypes under distinct environmental cues that correspond to the sequential phases of wound repair [67]. Hypoxic FB-EVs were also able to rescue M2 macrophage polarization in vivo in diabetic mouse wounds [67]. Likewise, EVs derived from primary fibroblasts isolated from systemic sclerosis patients orchestrated reciprocal functional modulation between macrophages and fibroblasts [68]. Bhandari et al. showed that when stimulated with TGF-β, fibroblasts from systemic sclerosis patients (but not healthy fibroblasts) produce EVs that increase CD163 and CD206 surface marker expression among macrophages (correlating with the M2 phenotype), upregulate macrophage pro-inflammatory and fibrotic cytokine expression, and further induce macrophage ability to activate contractile ECM-related genes (COL1A1, FN1, and α-SMA) [68]. Furthermore, keratinocytes have also been shown to take up systemic sclerosis FB-EVs, upregulating a strong type I IFN gene expression profile that is mediated by the TANK-binding kinase (TBK)/JAK/STAT signaling axis [69]. Taken together, hypoxic and fibrotic baseline states prime fibroblasts to generate EVs that can both enhance and exacerbate the immune response through intercellular communication with macrophages and distinct subsets of cells.

Hyperglycemic conditions also appear to generate FB-EVs that can impair recipient endothelial cell and keratinocyte activity. Endothelial cells treated with hyperglycemic FB-EVs demonstrate diminished migration and angiogenesis both in vitro and in vivo when compared to their non-hyperglycemic counterparts. This activity was shown to be dependent on the activation of the GSK-3β [70], a known negative regulator involved in the hormonal control of glucose homeostasis and Wnt/β-catenin signaling [71]. When treated with hyperglycemic FB-EVs, primary rat keratinocytes likewise showed impaired healing-related functions, including elevated expression of MMP9, leading to excessive collagen degradation, likely mediated by lnc-URIDS and YTH domain family protein-2 [72]. Thus, in contexts where glucose imbalances disrupt healing outcomes, FB-EVs modified to target metabolic regulatory pathways such as GSK-3β/β-catenin and or collagen synthesis may be an effective approach.

Cellular senescence also affects the function of FB-EVs. For instance, FB-EVs derived from senescent fibroblasts, either obtained from aged skin or replicative senescence, significantly impair keratinocyte differentiation capacity [73, 74]. When modulated to overexpress genes with antioxidant functions, such as GSTM2, however, FB-EVs can improve migration and proliferation in keratinocytes while also promoting enhanced oxidative phosphorylation in aged and senescent recipient fibroblasts [75]. Both senescence and hyperglycemia are cellular conditions involved in pathological healing. These studies provide evidence to suggest that targeting specific pathways through EV modifications could be a viable therapeutic approach.

FB-EVs have also been studied in skin functions relevant to wound repair, such as hair follicle growth and regeneration. FB-EVs containing growth factors (FGF-2 and PDGFA) were shown to provide a crucial communication axis between fibroblasts, dermal papillary cells (DPCs), and follicular keratinocytes by activating Norrin secretion in DPCs and transferring its associated receptor, Frizzled4, to the epidermal layer of the hair follicle [76]. Subsequently, it was found that FB-EVs stimulated the proliferation of dermal papillary cells by activating Wnt3a/β-catenin signaling, thereby stimulating downstream targets Axin1 and Lef1. The EVs also promoted proliferation, migration, and differentiation of the outer root sheath cells, thereby elongating the hair shaft and promoting follicle regeneration [77].

The ability of FB-EVs to improve skin photoaging, another function relevant to skin regeneration, has also been studied. Mechanisms involved in improving or reversing skin photoaging at the cellular level mimic the processes required for optimal tissue repair within a wound [78]. Interestingly, Hu et al. compared FB-EVs derived from fibroblasts cultured in a 3D spheroid form versus those isolated from a 2D monolayer. Those obtained from the spheroid culture method were superior in facilitating recipient fibroblast scratch closure, and contained higher levels of miRNAs related to pro-fibrotic processes and the tissue inhibitor matrix metalloproteinase 1 (TIMP-1). As a result, when injected into the skin of UVB-induced photoaged nude mice, these spheroid FB-EVs reduced wrinkle formation and significantly improved dermal thickness [78]. Another study extended these findings by demonstrating that FB-EVs mitigate UVB-induced functional impairments in fibroblasts, rescuing proliferation, stimulating collagen production, and minimizing senescence and intracellular ROS production [79]. Yet another study highlighted the ability of FB-EVs derived from Chenghua pigs, known for their dermal thickness, to stimulate fibroblast proliferation, migration, and contraction and synthesis of collagen in vitro, while enhancing skin thickness and collagen I content in vivo. These FB-EVs contained high levels of miR-218 and ITGBL1, which modulated TGFβ1-SMAD2/3 signaling in recipient fibroblasts [80].

Endothelial Cell-Derived EVs (EC-EVs)

A key aspect of healing wounds is angiogenesis, which re-vascularizes the site of injury [81]. During angiogenesis, endothelial cells release a substantial quantity of extracellular vesicles, which contain a cocktail of microRNAs, functional matrix metalloproteinases (e.g., MMPs and TIMPs), tissue factors, Notch ligands (e.g., delta-like ligand 4), and surface-level phosphatidylserine groups that can modulate signaling pathways within recipient endothelial cells [82]. These EV components likely contribute to the robust pro-angiogenic response that takes place in early wounds, while also providing a timely signal for the vascular pruning that reverts capillary density back to a normal state [82].

Endothelial cell-derived extracellular vesicles (EC-EVs), despite their essential role in angiogenesis and re-vascularization, are among the least studied in the context of wound healing. Only a handful of studies have investigated the role of EC-EVs in wound healing. Most of these have used EVs derived from endothelial cells with some modification or pre-treatment, and focused on fibroblasts as the recipient cell population. The resulting effects of EC-EVs on fibroblasts appear to depend heavily on the donor cell state. For example, endothelial cells cultured under hypoxic conditions resulted in increased fibroblast contraction and collagen cross-linking [83] while those pre-treated with advanced glycation end products enhanced fibroblast senescence and autophagy, and inhibited collagen synthesis [84]. Another study of hypoxic EC-EVs showed that recipient macrophage polarization towards the anti-inflammatory M2 phenotype was elevated, and HUVECs demonstrated enhanced angiogenesis, leading to improved healing in diabetic mice [85]. No existing studies have examined the role or effects of EVs derived from naïve or untreated endothelial cells obtained from skin microvasculature. Because EC-EVs can be found in circulation, a few studies have isolated them from blood samples and studied their effects on various tissues and cell types. EC-EVs isolated from the blood plasma of healthy donors were able to ameliorate functional deficits in primary skin fibroblasts from diabetic patients and improve wound healing in diabetic mice [86]. This pro-healing effect may be partially mediated by the anti-inflammatory properties of quiescent EC-EVs on immune cells, such as monocytes and macrophages, which have been examined in other contexts like vascular disease. For instance, circulating EC-EVs isolated from mouse plasma were found to suppress monocyte activation through miR-10a-mediated inhibition of NF-κB pro-inflammatory signaling [87]. In contrast, inflammatory EC-EVs from rat aortic endothelial cells preconditioned with lipopolysaccharide (LPS) shifted NR8383 macrophages towards the pro-inflammatory M1 phenotype [88], highlighting that the endothelial cell state and culture conditions are critical determinants of resulting EC-EV function. In sum, these studies point to a potential role for EC-EVs in modulating communication between inflammatory cells and matrix-depositing cells, likely in the name of orchestrating angiogenic cues released by various cell types in the wound.

Stem Cell-Derived EVs and Effects on Wound Healing

The past decade has witnessed an exponentially growing interest in the therapeutic potential of stem cell-derived EVs on tissue regeneration and wound healing. The majority of studies have pinpointed mesenchymal stem cells (MSCs) as harboring remarkable regenerative potential in tissue repair processes and cutaneous wound healing. MSCs are found not only in bone marrow, but also in adipose tissue, umbilical cord, amniotic fluid, dental pulp, placenta, and Wharton's jelly [89]. MSC-EVs, in comparison to MSCs themselves, demonstrate greater therapeutic potential by overcoming the concerns associated with cell-based therapy or transplant [90]. In particular, MSC-EVs exhibit reduced immunogenicity, retention of parent cell biological function and activity, and higher biocompatibility and scalability [90] compared to MSCs alone. Acknowledging this massive interest, a plethora of studies have already provided comprehensive reviews on the role of MSC-EVs in skin tissue regeneration [12, 14, 89–93]. Here, we will provide a brief summary of the major mechanisms by which MSC-EVs have been shown to facilitate tissue regeneration and wound healing (Table 13.2 and Fig. 13.2).

Immunomodulation and Inflammation Resolution

First, MSC-EVs possess immunomodulatory properties and can influence cellular players and pathways involved in inflammation resolution. More specifically, bone marrow-derived MSC-EV treatment of wounds was shown to elicit M2 macrophage polarization, likely through the transfer of miR-223, which targets pknox1 [94]. Another method by which MSC-EVs can modulate macrophage plasticity is via regulation of the TLR4/NF-κB signaling pathway [95, 96]. LPS-preconditioned MSCs were shown to produce EVs that induced M2 macrophage polarization more effectively than their untreated counterparts, and they contained upregulated levels of let-7b in their cargo that effectively suppresses the TLR4/NF-κB/STAT3/AKT pathway [96]. Umbilical cord MSC-EVs were found to contain high levels of miR-181c, which likewise was capable of suppressing TLR4 and downstream pathway targets within macrophages [95]. Collectively, these studies highlight that MSC-EVs can induce M2 macrophage polarization and downregulate known pro-inflammatory pathways to promote the resolution of wound inflammation.

Enhancement of Fibroblast and Keratinocyte Proliferation

Next, MSC-EVs can influence fibroblast activity, the synthesis of granulation tissue, and keratinocyte re-epithelialization during the proliferative phase of healing. One study found that Jagged1, a Notch signaling pathway ligand, contributes

Table 13.2 Major mechanisms of MSC-EVs in skin wound healing

Major mechanism	Molecular mediator (if applicable)	Function	References
Immunomodulation and inflammation resolution	miR-223	Promotes M2 macrophage polarization, reduces pro-inflammatory cytokines	[94]
	Let-7b	Suppresses TLR4/NF-κB pathway, drives M2 polarization	[96]
	miR-181c	Modulates macrophage phenotype via TLR4 targeting	[95]
Enhancement of fibroblast and keratinocyte proliferation	Jagged1	Activates Notch signaling in fibroblasts, increasing migration, proliferation, and ECM molecule gene expression	[97]
	lncRNA H19	Promotes fibroblast migration, proliferation via activation of the PI3K/Akt signaling pathway	[98–100]
	miR-21	Promotes keratinocyte migration and proliferation and intracellular expression of tissue proteases MMP9 and TIMP2	[101]
	miR-27b	Promotes fibroblast and keratinocyte migration through inhibition of E3 ubiquitin ligase ITCH	[102]
Stimulation of angiogenesis	miR-125a	Stimulates endothelial cell proliferation and angiogenesis	[103]
	miR-126, miR-210, miR-296, EGF, and FGF-2	Stimulates endothelial cell proliferation and angiogenesis	[104]
	miR-21	Promotes angiogenic signaling via PTEN/PI3K/Akt and SPRY	[105]
ECM remodeling and scar reduction		Regulating the ratios of TGF-β3 to TGF-β1, collagen III to collagen I, and MMP3 to TIMP1 in wound tissue, and reduces α-SMA expression	[106]
		Induces early wound collagen I and III secretion while inhibiting late wound collagen deposition	[107]
	miR-192-5p	IL-17RA inhibition and downregulation of SMAD2/3 phosphorylation in dermal fibroblasts, restraining excess collagen deposition	[108]
Oxidative stress reduction and mitochondrial protection		Restore pro-angiogenic functions within endothelial cells and reduce intracellular ROS production	[109]
	Functional mitochondria	Inhibition of NETosis-induced ferroptosis of endothelial cells	[110]

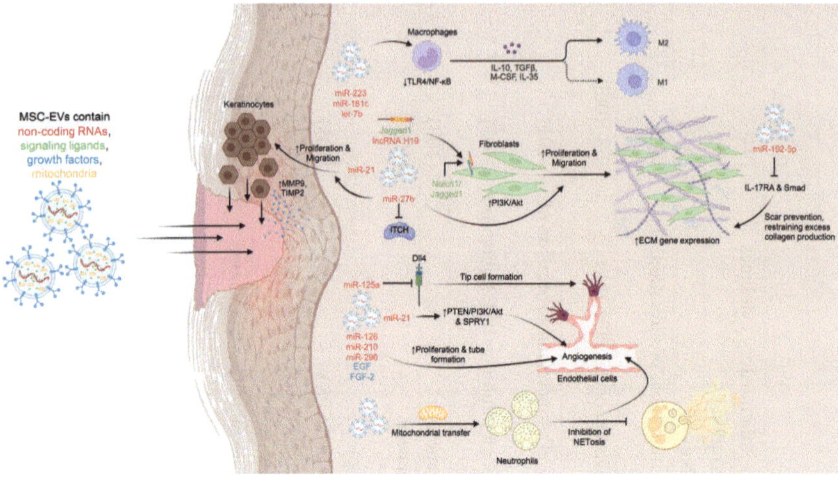

Fig. 13.2 Mechanisms of MSC-EV-mediated regulation of wound cell functions. MSC-EVs contain a panel of signaling molecules that modulate important processes involved in wound repair. MSC-EV activity, described here from top to bottom, includes the transfer of immunomodulatory and anti-inflammatory miRNAs such as miR-223, miR-181c, and let-7b, all of which promote M2 macrophage polarization during the initial inflammatory stage of healing. Next, MSC-EVs regulate re-epithelialization and dermal matrix reconstitution by enhancing fibroblast and keratinocyte activity and proliferation. This is achieved through Jagged1, which activates Notch signaling in fibroblasts, increasing their proliferation, migration, and ECM gene expression, and lncRNA H19, which activates canonical PI3K/Akt signaling. MiR-21 further enhances keratinocyte activity and the intracellular expression of tissue-degrading proteases MMP9 and TIMP2. Inhibition of the E3 ubiquitin ligase ITCH occurs via miR-27b transfer, ultimately amplifying this pro-proliferative effect. In the later remodeling stage of healing, other miRNAs, such as miR-192-5p work to downregulate fibrotic signaling through inhibition of IL-17RA and Smad, leading to minimal scar formation. Transfer of angiogenic cargo also implicates MSC-EVs in the regulation of endothelial cell function. MiR-125a inhibits Delta-like ligand 4 (Dll4), which promotes endothelial tip cell specification. Additional miRNAs such as miR-21, miR-126, miR-210, miR-296, as well as growth factors like EGF and FGF-2 augment cell proliferation pathways and increase tube formation capacity. Lastly, MSC-EVs contain functional mitochondria that can be transferred directly to neutrophils, inhibiting NETosis-induced ferroptosis of endothelial cells and restoring angiogenic activity. This is a simplified rendition based on current knowledge. See tables and text for complete details

significantly to the function of fetal dermal MSC-EVs, and is responsible for activating proliferation, migration, and gene expression of ECM molecules (collagen I and III, fibronectin, and elastin) in the wound [97]. Similarly, ADSC-EVs can promote fibroblast migration and proliferation in a dose-dependent manner, while increasing the gene and protein expression of collagens, MMP1, FGF-2, and TGF-β1 via activation of the PI3k/Akt signaling pathway [98, 99]. In addition, these same EVs were found to enhance fibroblast function through the transfer of lncRNA H19, which effectively binds miR-19b to de-repress transcriptional expression of SOX9 to promote wound healing [100]. Keratinocyte activity is likewise enhanced by adipose-derived stem cell (ADSC)-EVs, which also contain high levels of miR-21, effectively promoting HaCaT migration and proliferation and upregulating intracellular expression of MMP9 and TIMP2 [101]. Furthermore, umbilical cord MSC-EVs can stimulate both fibroblast and keratinocyte migration and proliferation via miR-27b, inhibiting the E3 ubiquitin ligase ITCH and enhancing cutaneous regeneration [102]. Altogether, MSC-EVs have the ability to improve the proliferation and migration of multiple key cell populations during matrix formation and re-epithelialization, setting the stage for the subsequent remodeling phase.

Stimulation of Angiogenesis

The influence of MSC-EVs on angiogenesis has been well-documented in the literature. For example, ADSC-EVs were shown to be enriched for miR-125a, which represses the anti-angiogenic delta-like ligand 4 (DLL4) to promote endothelial cell tube formation and tip cell differentiation [103]. Alternatively, human hTERT-immortalized adipose tissue-derived MSC line HATMSC1 produced EVs that stimulated dermal endothelial cell proliferation and tube formation in vitro. These EVs were enriched for pro-angiogenic miR-210, miR-126, and miR-296, as well as membrane-bound trophic factors such as EGF and FGF-2, suggesting that both the membranous and intra-vesicular components of the EVs may contribute to enhanced endothelial cell activity [104]. Additionally, umbilical cord blood MSC-EVs, like ADSC-EVs, contain high levels of miR-21, which mediates endothelial activity through modulating PTEN and sprouty homolog 1 (SPRY1) expression [105]. Thus, MSC-EVs deliver both miRNAs and functional growth factors to recipient endothelial cells, which work to re-vascularize the wound bed.

ECM Remodeling and Scar Reduction

Balancing cellular responses during the final remodeling stage of healing is essential for minimizing skin scar formation. MSC-EVs appear to play a role in modulating tissue architecture to reduce excessive ECM deposition by adjusting collagen synthesis and repressing cues that trigger fibrotic signaling. ADSC-EVs promote scarless cutaneous repair by regulating the ratios of TGF-β3 to TGF-β1, collagen III to collagen I, and MMP3 to TIMP1 [106]. At the same time, the expression of α-SMA in D14 and D21 wounds was also decreased following ADSC-EV treatment, suggesting that these EVs can restrain excessive myofibroblast activity to minimize skin contracture and scar formation [106]. Another study corroborated these findings by demonstrating that during early wound healing, ADSC-EVs promote collagen I and III secretion in vivo, while in the later phases, it appears to inhibit or preclude excess collagen deposition in wounds [107]. Moreover, ADSC-EVs were shown to contain miR-192-5p, which targets IL-17RA and downregulates SMAD2/3 phosphorylation in dermal fibroblasts, ultimately accelerating wound healing and reducing potential hypertrophic scar formation [108].

Oxidative Stress Reduction and Mitochondrial Protection

In addition to enhancing key cellular players at each stage of healing, MSC-EVs can also counteract excess oxidative stress and restore mitochondrial function. Hyperglycemia induces elevated oxidative stress damage to endothelial cells, but this effect can be reversed with the addition of umbilical cord MSC-EVs, which restore pro-angiogenic functions and reduce reactive oxygen species (ROS) production [109]. Accordingly, Lu et al. showed that umbilical cord MSC-EVs can inhibit neutrophil extracellular trap formation (NETosis)-induced ferroptosis of endothelial cells [110], a process elevated in diabetic wounds [111]. In their study, they found that MSC-EVs can even transfer functional mitochondria to neutrophils and thereby improve cellular response to ROS and oxidative stress [110].

Preclinical Efficacy of MSC-EVs on Diabetic and Non-diabetic Wound Healing

Preclinical rodent models of wound healing have been extensively utilized to investigate the therapeutic efficacy of MSC-EVs, native or modified, on regenerative outcomes. To succinctly review the large body of preclinical studies on this topic, we have summarized the key findings obtained from ten meta-analyses published to date (Table 13.3).

The majority of the studies evaluated have focused on diabetic rodent wound healing as a primary outcome. Bailey et al. identified that diabetic rat and mouse wounds that were treated with MSC-EVs demonstrated significantly improved wound closure. They also found that healing was further enhanced using MSC-EVs enriched with non-coding RNAs or microRNAs. Secondary outcomes pooled analysis showed that blood vessel density, re-epithelialization, and collagen deposition were all improved with a large effect following MSC-EV treatment [112]. Al-Masawa et al. analyzed 26 studies of both diabetic and non-diabetic wound models in response to MSC-EV treatment and found that overall wound closure, reduction of scar width and blood vessel density were significantly increased [113]. Qiao et al. analyzed 15 animal studies investigating MSC-EVs on diabetic wound healing. Consistently, they found a large effect for the ability of MSC-EVs obtained from various sources, including umbilical cord and vein, adipose tissue, bone marrow, menstrual blood, synovium, and gingiva, on wound repair outcomes. They also conducted a subgroup analysis that demonstrated an enhanced closure rate if the EVs were loaded into a novel dressing, such as a hydrogel or scaffold. Additional outcomes included neovascularization and scar width assessments, both of which favored EV treatment compared to controls, and exhibited a similar pattern of improvement when a subgroup analysis was conducted on dressing-administered EVs [114]. Another meta-analysis examined the primary outcome of wound closure at three different time points (D3, D7, and D14 ± 2 days at each time point) [115].

Table 13.3 Meta-analyses examining efficacy of stem-cell-derived EVs on preclinical rodent models of wound healing

Meta-analysis	EV type	Records included in quantitative analysis	ROB assessment	Animal models	Outcomes investigated
Soltani et al. [118]	ADSC-EVs	12	SYRCLE—unclear designation for the majority of attributes for most studies	Diabetic rat and mouse models	1. Diabetic wound healing 2. Angiogenesis 3. Collagen deposition
Bailey et al. [112]	MSC-EVs	10	SYRCLE—unclear designation for the majority of attributes for most studies	Diabetic rat and mouse models	1. Wound closure 2. Blood vessel density and number 3. Re-epithelialization 4. Collagen deposition 5. Scar width 6. Inflammation
Qiao et al. [114]	MSC-EVs, EVs from patient serum, macrophage cell lines, bovine milk, and human fibroblasts	21	Cochrane—unclear designation for the majority of attributes for most studies	Diabetic rat and mouse models	1. Wound closure 2. Blood vessel density and number 3. Re-epithelialization 4. Collagen deposition 5. Scar width 6. Inflammation
Yue et al. [116]	MSC-EVs and rat fibroblasts	19	SYRCLE—unclear designation for the majority of attributes for most studies	Type II diabetic rats and mice only	1. Wound closure 2. Blood vessel density and number 3. Re-epithelialization 4. Collagen deposition 5. Inflammation

(continued)

Table 13.3 (continued)

Meta-analysis	EV type	Records included in quantitative analysis	ROB assessment	Animal models	Outcomes investigated
Gunjan et al. [115]	MSC-EVs	18	(1) Selection of articles (Selection bias), (2) Selective reporting (reporting bias) (3) deviation from intended intervention (Performance bias), (4) Binding of outcome assessment (Detection bias) (5) Incomplete outcome data (Attrition bias) and (6) Baseline characteristics—most studies were deemed low risk	Diabetic rat and mouse models	1. Wound closure at D3 ± 2 2. Wound closure at D7 ± 2 3. Wound closure at D14 ± 2
Yuan et al. [117]	ADSC-EVs	13	SYRCLE—moderate ROB designation for the majority of attributes for most studies	Non-diabetic mice, rats, and rabbits	1. Wound closure 2. Vascularization 3. Fibrosis 4. Dermal thickness 5. Proliferation
Y. Zhu et al. [123]	MSC-EVs and Apoptotic EVs	31	SYRCLE—unclear designation for the majority of attributes for most studies	Diabetic and non-diabetic mice and rats	1. Wound closure 2. Collagen deposition 3. Blood vessel density 4. Scar width
Du et al. [120]	BMSC-EVs	11	SYRCLE—unclear designation for the majority of attributes for most studies	Diabetic mice and rats	1. Wound closure 2. Angiogenesis (blood vessel number) 3. Angiogenic markers (CD31, α-SMA)

Al-Masawa et al. [113]	MSC-EVs	26	SYRCLE—unclear designation for the majority of attributes for most studies	Diabetic and non-diabetic mice and rats	1. Wound closure 2. Scar formation 3. Blood vessel density
J. Wei et al. [119]	ADSC-EVs	19	STAIR—unclear designation for the majority of attributes for most studies	Diabetic and non-diabetic mice	1. Wound closure at D7 2. Wound closure at D14 3. Collagen deposition 4. Blood vessel density

Abbreviations. MSC-EVs mesenchymal stem cell-derived extracellular vesicles, *ADSC-EVs* adipose tissue derived stem cell-derived extracellular vesicles, *BMSC-EVs* bone marrow mesenchymal stem cell-derived extracellular vesicles, *ROB* risk of bias, *SYRCLE* SYstematic Review Centre for Laboratory animal Experimentation, *STAIR* Stroke Therapy Academic Industry Roundtable recommendations

With this approach, they authors show that across various animal models of diabetic wound healing, and regardless of EV source, the significant improvement in wound closure rate persists throughout the early and late phases of healing.

A meta-analysis by Yue et al. distinguishes between models of type I and type II diabetes, and focused on type II diabetic models. In their analysis, they found that MSC-EVs resulted in significantly improved diabetic wound healing, with relatively low heterogeneity, possibly due to their exclusion of all non-type II diabetic wound models. Of note, a further decrease in heterogeneity was observed when they stratified their analysis based on MSC origin, suggesting that overall heterogeneity may likely be due to EV source [116]. Likewise, all secondary outcomes, including angiogenesis, re-epithelialization, collagen deposition and inflammation were significantly improved following MSC-EV treatment.

Some meta-analyses analyzed the efficacy of specific cell- or tissue- derived EVs. Among these is adipose-derived-EVs, or ADSC-EVs. Yuan et al. focused on ADSC-EVs and their effects on non-diabetic, normal in vivo wound healing models and found that the therapeutic efficacy of ADSC-EVs was still greater compared to controls [117]. A subgroup analyses found that expression of fibrotic proteins, such as collagens I and III, as well as TGF-β, were increased for earlier time points of healing (D0–7) and decreased at later time points (D14 onwards) in response to ADSC treatment. This analysis highlights the efficacy of ADSC-EVs in fine-tuning the wound milieu in a time- and phase-dependent manner to support optimal healing outcomes. Another study aimed to examine the preclinical efficacy of adipose derived MSC-EVs, or ADSC-EVs, on diabetic wound healing [118]. Among the outcomes examined were wound closure, angiogenesis, and collagen deposition. Quantitative meta-analysis of 12 research studies recognized a large effect of ADSC-EVs on all three investigated outcomes—accelerated wound healing, improved neovascularization, and collagen deposition. Consistently, Wei et al. examined post-wounding D7 and D14 closure rates of both diabetic and non-diabetic wound models in response to ADSC-EV treatment, and determined that significant improvements in healing were observed, including blood vessel density and collagen deposition overall [119].

Bone marrow-derived EVs, or BMSC-EVs, are an alternative popular source of therapeutic EVs. Du et al. conducted a meta-analysis to determine the efficacy of specifically BMSC-EVs, on diabetic ulcer or wound repair. Consistently, BMSC-EVs produced a significant improvement in healing among diabetic wound models, especially when combined with novel materials or containing modified cargo [120].

EVs derived from apoptotic MSCs—deemed apoEVs—have also emerged as having potential regenerative properties, with greater potency and effect than small EVs alone [121, 122]. One meta-analysis attempted to establish a consensus on the whether these MSC-apoEVs were more effective than standard MSC-EVs in healing both normal and diabetic wounds [123]. Through sub-group analyses, the authors demonstrated that apoEVs exhibit greater healing potential in both diabetic and non-diabetic wounds, although the study numbers are limited. They also showed that the intradermal or subcutaneous injection method of administering the EV treatments was more effective compared to dressing applications or coverings over the wounds [123].

Although efforts have been made to standardize the systems and reporting methods of EV research [15, 17], adherence to guidelines and transparency in reporting remain continuous challenges in the field [113]. Consistent across all meta-analyses was the conclusion that there is an overall lack of clarity and transparency in reporting, which led to an overall "unclear" designation for risk of bias in these preclinical studies. This designation indicates that the approaches towards sample selection, allocation, blinding, and outcome reporting were not well defined or completely missing from most of the studies included in the quantitative analysis [124]. As clinical translation of EV-based therapies for wound repair emerges at the forefront of the agenda, more effort is needed to ensure that the fundamental studies conducted in this field are rigorous and transparent. With adequate attention towards standardizing EV-related methodologies, increasing clarity in reporting, and assessing levels of risk at each stage of research, EVs possess great potential for translation into clinical use.

Limitations, Challenges, and Perspectives on the Study of EVs in Wounds

Despite the advancements made in the last few decades in understanding how EVs function in the wound healing landscape, important questions remain unanswered and continue to pose challenges for a complete and thorough picture of wound-related EV dynamics. Many of the studies included in this review have utilized EVs isolated from tissue culture to investigate their effect on recipient cells. Because EVs are sensitive to cues conferred by their spatial context and tissue microenvironment, this methodological approach falls short of capturing many of the endogenous signals provided by the wound milieu in vivo that could dictate EV content and functionality. Park and colleagues have attempted to characterize the cell-specific EVs in wound fluid using a polyvinyl alcohol (PVA) sponge implant technique, identifying a large contribution of macrophage-secreted EVs in the interstitial fluid of murine wounds [125]. Their study was the first of its kind to develop a "lineage tracing" mouse model of EV secretion, in which macrophage-, keratinocyte-, and endothelial cell-derived EVs are GFP-labeled and can be isolated from the tissue [125]. To fully understand the relevance of cell-type-specific EV contribution to healing wounds, additional animal models in which cell-specific EV secretion can be knocked down or inhibited are required. However, developing a model of this caliber would involve taking into account the complexity of EV secretion pathways, which renders this task rather challenging. Although past studies have attempted to define the contribution of EVs to cellular functions and wound repair in vivo, most of these have utilized pharmacological inhibitors of EV secretion [31], such as GW4869, which would universally dampen EV release across all cell types.

Tracking EV release is only one side of the story—monitoring tissue and cellular uptake is another. Most studies to date have shown that cellular uptake of EVs occurs in tissue culture, but few have recapitulated this finding with the same cells in vivo [125]. Using a fused, multiplex fluorescent and bioluminescent EV membrane reporter, Lai et al. previously showed that tumor cells interact with resident non-tumor cells via the transfer of functional mRNA cargo [126]. Monitoring in vivo EV uptake at the tumor parenchyma border successfully captured trafficking of EV clusters into surrounding motile cells, likely infiltrating immune cells, though this cell type was not visualized in their study [126]. As the technologies for EV tracking continue to expand [127], more emphasis on the translation of these imaging methods for EV biodistribution studies in humans will be required.

An additional consideration, and especially important given the prevalent role of the ECM as both a regulator and product of wound healing, is the interaction between EVs and the matrix. It is well known that cellular production of EVs is a function of matrix stiffness [128]. The biogenesis of EVs by cells is also linked to the physicochemical properties of the ECM that can affect specific steps implicated in EV formation and release. For example, lipid rafts, which facilitate MVB formation, are more readily available when not utilized to counteract the mechanical tension imparted onto cells [129]. One class of lipid rafts, caveolae, is formed in response to actin-cytoskeleton dynamics [130], which can further serve as a barrier to EV release [128, 129]. Furthermore, the movement of EVs is well established to occur via Brownian motion, or passive diffusion [131]. Thus, EV transport through tissues and the interstitial space is further influenced by ECM pore size, cleavage and remodeling of the ECM, and selective binding of EVs to ECM receptors and proteins [132]. Accordingly, it has been shown that some EVs can bind and tether to growth factors like FGF-2, which can be sequestered in the ECM or extracellular space, in the absence of cells, thereby protecting them from protease-induced degradation [64]. Moreover, many studies in this field have examined EV-mediated interactions between skin cells that are physiologically separated by the epidermal basement membrane, such as keratinocyte-macrophage [38], fibroblast-dermal papillary cell [76], and even melanocyte-fibroblast [133] communication. While this membrane is disrupted in models of wound healing, evaluating the contribution of the resulting wound ECM to EV dynamics will be paramount for providing a complete understanding of the activity and function of these EVs.

Conclusion

This review summarizes the current findings on the functional effects of native skin cell and stem cell-derived EVs on cutaneous wound repair and regeneration. The current state of wound healing-related extracellular vesicle research, whether fundamental or translational in nature, advances the understanding of non-canonical intercellular communication networks in tissue repair. By understanding the properties and cargo of EVs derived from various cell types, we can gain better insight into the multifaceted cellular communication mechanisms that underpin both physiological and pathophysiological healing patterns. EVs comprise only one of the many paracrine signaling methods that take place between cells, and are likely to complement the plethora of canonical cell-cell communication mechanisms involving growth factors, cytokines, and chemokines. As not only messengers between cells, but also therapeutic delivery vehicles and diagnostic markers for disease, these nano-sized particles provide a novel lens for studying the wound and related regenerative processes. With sufficient efforts to overcome the existing challenges and questions in the field, EVs hold enormous potential for expanding our fundamental knowledge on how cells communicate and reimagining the clinical approaches for tissue regeneration.

Acknowledgments This research was funded by NIH R35-GM139603 (LAD) and F31-AR083830 (HY). Illustrations were created with Biorender.

Conflict of Interest The authors declare that they have no conflicts of interest.

Author Contributions HY and LAD contributed to the conception, drafting, and critical revision of this manuscript. All authors gave their final approval and agree to be accountable for all aspects of the work.

References

1. Gurtner GC, Werner S, Barrandon Y, Longaker MT (2008) Wound repair and regeneration. Nature 453:314–321. https://doi.org/10.1038/nature07039
2. Rodrigues M, Kosaric N, Bonham CA, Gurtner GC (2019) Wound healing: a cellular perspective. Physiol Rev 99:665–706. https://doi.org/10.1152/physrev.00067.2017
3. Eming SA, Krieg T, Davidson JM (2007) Inflammation in wound repair: molecular and cellular mechanisms. J Invest Dermatol 127:514–525. https://doi.org/10.1038/sj.jid.5700701
4. Martin P, Nunan R (2015) Cellular and molecular mechanisms of repair in acute and chronic wound healing. Br J Dermatol 173:370–378. https://doi.org/10.1111/bjd.13954
5. Barrientos S, Stojadinovic O, Golinko MS, Brem H, Tomic-Canic M (2008) Perspective article: growth factors and cytokines in wound healing. Wound Repair Regen 16:585–601. https://doi.org/10.1111/j.1524-475X.2008.00410.x
6. Tonnesen MG, Feng X, Clark RA (2000) Angiogenesis in wound healing. J Investig Dermatol Symp Proc 5:40–46. https://doi.org/10.1046/j.1087-0024.2000.00014.x
7. Xue M, Jackson CJ (2015) Extracellular matrix reorganization during wound healing and its impact on abnormal scarring. Adv Wound Care 4:119–136. https://doi.org/10.1089/wound.2013.0485
8. Falanga V (2005) Wound healing and its impairment in the diabetic foot. Lancet 366:1736–1743. https://doi.org/10.1016/S0140-6736(05)67700-8
9. Frykberg RG, Banks J (2015) Challenges in the treatment of chronic wounds. Adv Wound Care (New Rochelle) 4:560–582. https://doi.org/10.1089/wound.2015.0635
10. Sen CK, Gordillo GM, Roy S, Kirsner R, Lambert L, Hunt TK, Gottrup F, Gurtner GC, Longaker MT (2009) Human skin wounds: a major and snowballing threat to public health and the economy. Wound Repair Regen 17:763–771. https://doi.org/10.1111/j.1524-475X.2009.00543.x
11. Phinney DG, Pittenger MF (2017) Concise review: MSC-derived exosomes for cell-free therapy. Stem Cells 35:851–858. https://doi.org/10.1002/stem.2575

12. Nallakumarasamy A, Jeyaraman M, Maffulli N, Jeyaraman N, Suresh V, Ravichandran S, Gupta M, Potty AG, El-Amin SF, Khanna M et al (2022) Mesenchymal stromal cell-derived extracellular vesicles in wound healing. Life (Basel) 12:1733. https://doi.org/10.3390/life12111733
13. Wang M, Wu P, Huang J, Liu W, Qian H, Sun Y, Shi H (2022) Skin cell-derived extracellular vesicles: a promising therapeutic strategy for cutaneous injury. Burns Trauma 10:tkac037. https://doi.org/10.1093/burnst/tkac037
14. Narauskaitė D, Vydmantaitė G, Rusteikaitė J, Sampath R, Rudaitytė A, Stašytė G, Aparicio Calvente MI, Jekabsone A (2021) Extracellular vesicles in skin wound healing. Pharmaceuticals 14:811. https://doi.org/10.3390/ph14080811
15. Welsh JA, Goberdhan DCI, O'Driscoll L, Buzas EI, Blenkiron C, Bussolati B, Cai H, Di Vizio D, Driedonks TAP, Erdbrügger U et al (2024) Minimal information for studies of extracellular vesicles (MISEV2023): from basic to advanced approaches. J Extracell Vesicles 13:e12404. https://doi.org/10.1002/jev2.12404
16. Kalluri R, LeBleu VS (2020) The biology, function, and biomedical applications of exosomes. Science 367:eaau6977. https://doi.org/10.1126/science.aau6977
17. Théry C, Witwer KW, Aikawa E, Alcaraz MJ, Anderson JD, Andriantsitohaina R, Antoniou A, Arab T, Archer F, Atkin-Smith GK et al (2018) Minimal information for studies of extracellular vesicles 2018 (MISEV2018): a position statement of the International Society for Extracellular Vesicles and update of the MISEV2014 guidelines. J Extracell Vesicles 7:1535750. https://doi.org/10.1080/20013078.2018.1535750
18. Yáñez-Mó M, Siljander PR-M, Andreu Z, Bedina Zavec A, Borràs FE, Buzas EI, Buzas K, Casal E, Cappello F, Carvalho J et al (2015) Biological properties of extracellular vesicles and their physiological functions. J Extracell Vesicles 4:27066. https://doi.org/10.3402/jev.v4.27066
19. van Niel G, D'Angelo G, Raposo G (2018) Shedding light on the cell biology of extracellular vesicles. Nat Rev Mol Cell Biol 19:213–228. https://doi.org/10.1038/nrm.2017.125
20. Colombo M, Raposo G, Théry C (2014) Biogenesis, secretion, and intercellular interactions of exosomes and other extracellular vesicles. Annu Rev Cell Dev Biol 30:255–289. https://doi.org/10.1146/annurev-cellbio-101512-122326
21. Muralidharan-Chari V, Clancy JW, Sedgwick A, D'Souza-Schorey C (2010) Microvesicles: mediators of extracellular communication during cancer progression. J Cell Sci 123:1603–1611. https://doi.org/10.1242/jcs.064386
22. Tricarico C, Clancy J, D'Souza-Schorey C (2017) Biology and biogenesis of shed microvesicles. Small GTPases 8:220–232. https://doi.org/10.1080/21541248.2016.1215283
23. Tixeira R, Caruso S, Paone S, Baxter AA, Atkin-Smith GK, Hulett MD, Poon IKH (2017) Defining the morphologic features and products of cell disassembly during apoptosis. Apoptosis 22:475–477. https://doi.org/10.1007/s10495-017-1345-7
24. Jiang L, Paone S, Caruso S, Atkin-Smith GK, Phan TK, Hulett MD, Poon IKH (2017) Determining the contents and cell origins of apoptotic bodies by flow cytometry. Sci Rep 7:14444. https://doi.org/10.1038/s41598-017-14305-z
25. Orlando KA, Stone NL, Pittman RN (2006) Rho kinase regulates fragmentation and phagocytosis of apoptotic cells. Exp Cell Res 312:5–15. https://doi.org/10.1016/j.yexcr.2005.09.012
26. Li M, Wang T, Tian H, Wei G, Zhao L, Shi Y (2019) Macrophage-derived exosomes accelerate wound healing through their anti-inflammation effects in a diabetic rat model. Artif Cells Nanomed Biotechnol 47:3793–3803. https://doi.org/10.1080/21691401.2019.1669617
27. Kim H, Wang SY, Kwak G, Yang Y, Kwon IC, Kim SH (2019) Exosome-guided phenotypic switch of M1 to M2 macrophages for cutaneous wound healing. Adv Sci 6:1900513. https://doi.org/10.1002/advs.201900513
28. Li L, Wang Z, Wang K, Fu S, Li D, Wang M, Cao Y, Zhu H, Li Z, Weng L et al (2023) Paintable bioactive extracellular vesicle ink for wound healing. ACS Appl Mater Interfaces 15:25427–25436. https://doi.org/10.1021/acsami.3c03630

29. Gangadaran P, Rajendran RL, Oh JM, Hong CM, Jeong SY, Lee S-W, Lee J, Ahn B-C (2020) Extracellular vesicles derived from macrophage promote angiogenesis in vitro and accelerate new vasculature formation in vivo. Exp Cell Res 394:112146. https://doi.org/10.1016/j.yexcr.2020.112146
30. Rajendran RL, Gangadaran P, Seo CH, Kwack MH, Oh JM, Lee HW, Gopal A, Sung YK, Jeong SY, Lee S-W et al (2020) Macrophage-derived extracellular vesicle promotes hair growth. Cells 9:856. https://doi.org/10.3390/cells9040856
31. Chen J, Zhou R, Liang Y, Fu X, Wang D, Wang C (2019) Blockade of lncRNA-ASLNCS5088–enriched exosome generation in M2 macrophages by GW4869 dampens the effect of M2 macrophages on orchestrating fibroblast activation. FASEB J 33:12200–12212. https://doi.org/10.1096/fj.201901610
32. Zhu Z, Chen B, Peng L, Gao S, Guo J, Zhu X (2021) Blockade of LINC01605-enriched exosome generation in M2 -macrophages impairs M2 macrophage-induced proliferation, migration, and invasion of human dermal fibroblasts. Int J Immunopathol Pharmacol 35:20587384211016724. https://doi.org/10.1177/20587384211016724
33. Shi M, Zhang L, Bi F, Ma X (2024) Exosomes derived from M2 macrophages promote fibroblast autophagy to contribute to hypertrophic scar formation via CXCL2/CXCR7/mTOR pathway. Hum Exp Toxicol 43:09603271241303320. https://doi.org/10.1177/09603271241303320
34. Deng F, Yan J, Lu J, Luo M, Xia P, Liu S, Wang X, Zhi F, Liu D (2021) M2 macrophage-derived exosomal miR-590-3p attenuates DSS-induced mucosal damage and promotes epithelial repair via the LATS1/YAP/ β-catenin signalling axis. J Crohns Colitis 15:665–677. https://doi.org/10.1093/ecco-jcc/jjaa214
35. Mosser DM, Edwards JP (2008) Exploring the full spectrum of macrophage activation. Nat Rev Immunol 8:958–969. https://doi.org/10.1038/nri2448
36. Bailey AS, Willenbring H, Jiang S, Anderson DA, Schroeder DA, Wong MH, Grompe M, Fleming WH (2006) Myeloid lineage progenitors give rise to vascular endothelium. Proc Natl Acad Sci USA 103:13156–13161. https://doi.org/10.1073/pnas.0604203103
37. Sinha M, Sen CK, Singh K, Das A, Ghatak S, Rhea B, Blackstone B, Powell HM, Khanna S, Roy S (2018) Direct conversion of injury-site myeloid cells to fibroblast-like cells of granulation tissue. Nat Commun 9:936. https://doi.org/10.1038/s41467-018-03208-w
38. Zhou X, Brown BA, Siegel AP, El Masry MS, Zeng X, Song W, Das A, Khandelwal P, Clark A, Singh K et al (2020) Exosome-mediated crosstalk between keratinocytes and macrophages in cutaneous wound healing. ACS Nano 14:12732–12748. https://doi.org/10.1021/acsnano.0c03064
39. Guda PR, Sharma A, Anthony AJ, El Masry MS, Couse AD, Ghatak PD, Das A, Timsina L, Trinidad JC, Roy S et al (2023) Nanoscopic and functional characterization of keratinocyte-originating exosomes in the wound fluid of non-diabetic and diabetic chronic wound patients. Nano Today 52:101954. https://doi.org/10.1016/j.nantod.2023.101954
40. Jiang M, Fang H, Shao S, Dang E, Zhang J, Qiao P, Yang A, Wang G (2019) Keratinocyte exosomes activate neutrophils and enhance skin inflammation in psoriasis. FASEB J 33:13241–13253. https://doi.org/10.1096/fj.201900642R
41. Jiang W, Zhang T, Qiu Y, Liu Q, Chen X, Wang Q, Min X, Ouyang L, Jia S, Lu Q et al (2024) Keratinocyte-to-macrophage communication exacerbate psoriasiform dermatitis via LRG1-enriched extracellular vesicles. Theranostics 14:1049–1064. https://doi.org/10.7150/thno.89180
42. Acevedo-Sánchez V, Rodríguez-Hernández RM, Aguilar-Ruíz SR, Torres-Aguilar H, Pina-Canseco S, Chávez-Olmos P, Garrido E, Baltiérrez-Hoyos R, Romero-Tlalolini MA (2025) Keratinocyte-derived extracellular vesicles induce macrophage polarization toward an M1-like phenotype. Biochem Biophys Res Commun 758:151659. https://doi.org/10.1016/j.bbrc.2025.151659
43. Cai X-W, Zhu R, Ran L, Li Y-Q, Huang K, Peng J, He W, Zhou C-L, Wang R-P (2017) A novel non-contact communication between human keratinocytes and T cells: exosomes

derived from keratinocytes support superantigen-induced proliferation of resting T cells. Mol Med Rep 16:7032–7038. https://doi.org/10.3892/mmr.2017.7492

44. Kotzerke K, Mempel M, Aung T, Wulf GG, Urlaub H, Wenzel D, Schön MP, Braun A (2013) Immunostimulatory activity of murine keratinocyte-derived exosomes. Exp Dermatol 22:650–655. https://doi.org/10.1111/exd.12230
45. Huang P, Bi J, Owen GR, Chen W, Rokka A, Koivisto L, Heino J, Häkkinen L, Larjava H (2015) Keratinocyte microvesicles regulate the expression of multiple genes in dermal fibroblasts. J Invest Dermatol 135:3051–3059. https://doi.org/10.1038/jid.2015.320
46. Glady A, Vandebroek A, Yasui M (2021) Human keratinocyte-derived extracellular vesicles activate the MAPKinase pathway and promote cell migration and proliferation in vitro. Inflamm Regen 41:4. https://doi.org/10.1186/s41232-021-00154-x
47. Li Q, Zhao H, Chen W, Huang P, Bi J (2019) Human keratinocyte-derived microvesicle miRNA-21 promotes skin wound healing in diabetic rats through facilitating fibroblast function and angiogenesis. Int J Biochem Cell Biol 114:105570. https://doi.org/10.1016/j.biocel.2019.105570
48. Hong X, Cai L, Li L, Zheng D, Lin J, Liang Z, Fu W, Liang D, Zeng T, Sun K et al (2025) Keratinocyte-derived small extracellular vesicles delay diabetic wound healing by triggering fibroblasts autophagy. Arch Physiol Biochem 131:11–23. https://doi.org/10.1080/13813455.2024.2358020
49. Fu W, Liang D, Wu X, Chen H, Hong X, Wang J, Zhu T, Zeng T, Lin W, Chen S et al (2022) Long noncoding RNA LINC01435 impedes diabetic wound healing by facilitating YY1-mediated HDAC8 expression. iScience *25*:104006. https://doi.org/10.1016/j.isci.2022.104006
50. Nedachi T, Bonod C, Rorteau J, Chinoune W, Ishiuchi Y, Hughes S, Gillet B, Bechetoille N, Sigaudo-Roussel D, Lamartine J (2023) Chronological aging impacts abundance, function and microRNA content of extracellular vesicles produced by human epidermal keratinocytes. Aging 15:12702–12722. https://doi.org/10.18632/aging.205245
51. Gutowska-Owsiak D, Podobas EI, Eggeling C, Ogg GS, Bernardino de la Serna J (2020) Addressing differentiation in live human keratinocytes by assessment of membrane packing order. Front Cell Dev Biol 8:573230. https://doi.org/10.3389/fcell.2020.573230
52. Yamaguchi Y, Brenner M, Hearing VJ (2007) The regulation of skin pigmentation. J Biol Chem 282:27557–27561. https://doi.org/10.1074/jbc.R700026200
53. Gupta R, Priya A, Chowdhary M, Batra VV, Jyotsna, Nagarajan P, Gokhale RS, Singh A (2023) Pigmented skin exhibits accelerated wound healing compared to the nonpigmented skin in Guinea pig model. iScience 26. https://doi.org/10.1016/j.isci.2023.108159
54. Gupta A, Chauhan A, Priya A, Mantri B, Wadhokar M, Dalave K, Shah B, Gokhale RS, Batra VV, Singh A (2020) Lesional skin in vitiligo exhibits delayed in vivo reepithelialization compared to the nonlesional skin. Wound Repair Regen 28:307–314. https://doi.org/10.1111/wrr.12798
55. Lo Cicero A, Delevoye C, Gilles-Marsens F, Loew D, Dingli F, Guéré C, André N, Vié K, van Niel G, Raposo G (2015) Exosomes released by keratinocytes modulate melanocyte pigmentation. Nat Commun 6:7506. https://doi.org/10.1038/ncomms8506
56. Takano K, Hachiya A, Murase D, Tanabe H, Kasamatsu S, Takahashi Y, Moriwaki S, Hase T (2020) Quantitative changes in the secretion of exosomes from keratinocytes homeostatically regulate skin pigmentation in a paracrine manner. J Dermatol 47:265–276. https://doi.org/10.1111/1346-8138.15202
57. Kim N-H, Choi S-H, Kim C-H, Lee CH, Lee TR, Lee A-Y (2014) Reduced MiR-675 in exosome in H19 RNA-related melanogenesis via MITF as a direct target. J Invest Dermatol 134:1075–1082. https://doi.org/10.1038/jid.2013.478
58. Jin S, Chen L, Xu Z, Xing X, Zhang C, Xiang L (2020) 585 nm light-emitting diodes inhibit melanogenesis through upregulating H19/miR-675 axis in LEDs-irradiated keratinocytes by paracrine effect. J Dermatol Sci 98:102–108. https://doi.org/10.1016/j.jdermsci.2020.03.002

59. Liu Y, Xue L, Gao H, Chang L, Yu X, Zhu Z, He X, Geng J, Dong Y, Li H et al (2019) Exosomal miRNA derived from keratinocytes regulates pigmentation in melanocytes. J Dermatol Sci 93:159–167. https://doi.org/10.1016/j.jdermsci.2019.02.001
60. Yoon J-H, Jo C-S, Hwang J-S (2024) Comprehensive analysis of exosomal MicroRNAs derived from UVB-irradiated keratinocytes as potential melanogenesis regulators. Int J Mol Sci 25:3095. https://doi.org/10.3390/ijms25063095
61. Tomasek JJ, Gabbiani G, Hinz B, Chaponnier C, Brown RA (2002) Myofibroblasts and mechano-regulation of connective tissue remodelling. Nat Rev Mol Cell Biol 3:349–363. https://doi.org/10.1038/nrm809
62. Oh EJ, Gangadaran P, Rajendran RL, Kim HM, Oh JM, Choi KY, Chung HY, Ahn B-C (2021) Extracellular vesicles derived from fibroblasts promote wound healing by optimizing fibroblast and endothelial cellular functions. Stem Cells 39:266–279. https://doi.org/10.1002/stem.3310
63. Gangadaran P, Oh EJ, Rajendran RL, Kim HM, Oh JM, Kwak S, Hong CM, Choi KY, Chung HY, Ahn B-C (2022) Identification of angiogenic cargoes in human fibroblasts-derived extracellular vesicles and induction of wound healing. Pharmaceuticals (Basel) 15:702. https://doi.org/10.3390/ph15060702
64. Petit I, Levy A, Estrach S, Féral CC, Trentin AG, Dingli F, Loew D, Qu J, Zhou H, Théry C et al (2022) Fibroblast growth factor-2 bound to specific dermal fibroblast-derived extracellular vesicles is protected from degradation. Sci Rep 12:22131. https://doi.org/10.1038/s41598-022-26217-8
65. Park AY, Lee JO, Jang Y, Kim Y-J, Lee JM, Kim S-Y, Kim BJ, Yoo KH (2023) Exosomes derived from human dermal fibroblasts protect against UVB-induced skin photoaging. Int J Mol Med 52:1–15. https://doi.org/10.3892/ijmm.2023.5323
66. Han X, Wu P, Li L, Sahal HM, Ji C, Zhang J, Wang Y, Wang Q, Qian H, Shi H et al (2021) Exosomes derived from autologous dermal fibroblasts promote diabetic cutaneous wound healing through the Akt/β-catenin pathway. Cell Cycle 20:616–629. https://doi.org/10.1080/15384101.2021.1894813
67. Chen C, Yang J, Shang R, Tang Y, Cai X, Chen Y, Liu Z, Hu W, Zhang W, Zhang X et al (2025) Orchestration of macrophage polarization dynamics by fibroblast-secreted exosomes during skin wound healing. J Invest Dermatol 145:171–184.e6. https://doi.org/10.1016/j.jid.2024.05.007
68. Bhandari R, Yang H, Kosarek NN, Smith AE, Garlick JA, Hinchcliff M, Whitfield ML, Pioli PA (2023) Human dermal fibroblast-derived exosomes induce macrophage activation in systemic sclerosis. Rheumatology 62:SI114–SI124. https://doi.org/10.1093/rheumatology/keac453
69. Bryon J, Wasson CW, Koeppen K, Chandler F, Willis LF, Di Donato S, Klein E, Zeqiraj E, Ross RL, Del Galdo F (2025) Systemic sclerosis dermal fibroblast exosomes trigger type 1 interferon responses in keratinocytes via a TBK/JAK/STAT signaling axis. Arthritis Rheumatol 77:322–334. https://doi.org/10.1002/art.43029
70. Bian X, Li B, Tang H, Li Q, Hu W, Wei Q, Ma K, Yang Y, Li H, Fu X et al (2022) Extracellular vesicles derived from fibroblasts induced with or without high glucose exert opposite effects on wound healing and angiogenesis. Front Surg 9:1065172. https://doi.org/10.3389/fsurg.2022.1065172
71. Frame S, Cohen P, Biondi RM (2001) A common phosphate binding site explains the unique substrate specificity of GSK3 and its inactivation by phosphorylation. Mol Cell 7:1321–1327. https://doi.org/10.1016/s1097-2765(01)00253-2
72. Wu Y, Wu X, Wang J, Chen S, Chen H, Liu J, Zeng T, Hu M, Liang Y, Sun K et al (2023) Fibroblast-derived extracellular vesicle-packaged long noncoding RNA upregulated in diabetic skin enhances keratinocyte MMP-9 expression and delays diabetic wound healing. Lab Investig 103:100019. https://doi.org/10.1016/j.labinv.2022.100019
73. Terlecki-Zaniewicz L, Pils V, Bobbili MR, Lämmermann I, Perrotta I, Grillenberger T, Schwestka J, Weiß K, Pum D, Arcalis E et al (2019) Extracellular vesicles in human skin:

cross-talk from senescent fibroblasts to keratinocytes by miRNAs. J Invest Dermatol 139:2425–2436.e5. https://doi.org/10.1016/j.jid.2019.05.015
74. Choi E-J, Kil IS, Cho E-G (2020) Extracellular vesicles derived from senescent fibroblasts attenuate the dermal effect on keratinocyte differentiation. Int J Mol Sci 21:1022. https://doi.org/10.3390/ijms21031022
75. Wu H, Yao Z, Li H, Zhang L, Zhao Y, Li Y, Wu Y, Zhang Z, Xie J, Ding F et al (2024) Improving dermal fibroblast-to-epidermis communications and aging wound repair through extracellular vesicle-mediated delivery of Gstm2 mRNA. J Nanobiotechnol 22:307. https://doi.org/10.1186/s12951-024-02541-1
76. le Riche A, Aberdam E, Marchand L, Frank E, Jahoda C, Petit I, Bordes S, Closs B, Aberdam D (2019) Extracellular vesicles from activated dermal fibroblasts stimulate hair follicle growth through dermal papilla-secreted Norrin. Stem Cells 37:1166–1175. https://doi.org/10.1002/stem.3043
77. Rajendran RL, Gangadaran P, Kwack MH, Oh JM, Hong CM, Sung YK, Lee J, Ahn B-C (2021) Human fibroblast-derived extracellular vesicles promote hair growth in cultured human hair follicles. FEBS Lett 595:942–953. https://doi.org/10.1002/1873-3468.14050
78. Hu S, Li Z, Cores J, Huang K, Su T, Dinh P-U, Cheng K (2019) Needle-free injection of exosomes derived from human dermal fibroblast spheroids ameliorates skin photoaging. ACS Nano 13:11273–11282. https://doi.org/10.1021/acsnano.9b04384
79. Deng C-C, Hu Y-F, Zhu D-H, Cheng Q, Gu J-J, Feng Q-L, Zhang L-X, Xu Y-P, Wang D, Rong Z et al (2021) Single-cell RNA-seq reveals fibroblast heterogeneity and increased mesenchymal fibroblasts in human fibrotic skin diseases. Nat Commun 12:3709. https://doi.org/10.1038/s41467-021-24110-y
80. Zou Q, Zhang M, Yuan R, Wang Y, Gong Z, Shi R, Li Y, Fei K, Luo C, Xiong Y et al (2022) Small extracellular vesicles derived from dermal fibroblasts promote fibroblast activity and skin development through carrying miR-218 and ITGBL1. J Nanobiotechnol 20:296. https://doi.org/10.1186/s12951-022-01499-2
81. DiPietro LA (2016) Angiogenesis and wound repair: when enough is enough. J Leukoc Biol 100:979–984. https://doi.org/10.1189/jlb.4MR0316-102R
82. Todorova D, Simoncini S, Lacroix R, Sabatier F, Dignat-George F (2017) Extracellular vesicles in angiogenesis. Circ Res 120:1658–1673. https://doi.org/10.1161/CIRCRESAHA.117.309681
83. de Jong OG, van Balkom BWM, Gremmels H, Verhaar MC (2016) Exosomes from hypoxic endothelial cells have increased collagen crosslinking activity through up-regulation of lysyl oxidase-like 2. J Cell Mol Med 20:342–350. https://doi.org/10.1111/jcmm.12730
84. Zeng T, Wang X, Wang W, Feng Q, Lao G, Liang Y, Wang C, Zhou J, Chen Y, Liu J et al (2019) Endothelial cell-derived small extracellular vesicles suppress cutaneous wound healing through regulating fibroblasts autophagy. Clin Sci 133:CS20190008. https://doi.org/10.1042/CS20190008
85. Cheng P, Xie X, Hu L, Zhou W, Mi B, Xiong Y, Xue H, Zhang K, Zhang Y, Hu Y et al (2024) Hypoxia endothelial cells-derived exosomes facilitate diabetic wound healing through improving endothelial cell function and promoting M2 macrophages polarization. Bioact Mater 33:157–173. https://doi.org/10.1016/j.bioactmat.2023.10.020
86. Wei F, Wang A, Wang Q, Han W, Rong R, Wang L, Liu S, Zhang Y, Dong C, Li Y (2020) Plasma endothelial cells-derived extracellular vesicles promote wound healing in diabetes through YAP and the PI3K/Akt/mTOR pathway. Aging 12:12002–12018. https://doi.org/10.18632/aging.103366
87. Njock M-S, Cheng HS, Dang LT, Nazari-Jahantigh M, Lau AC, Boudreau E, Roufaiel M, Cybulsky MI, Schober A, Fish JE (2015) Endothelial cells suppress monocyte activation through secretion of extracellular vesicles containing antiinflammatory microRNAs. Blood 125:3202–3212. https://doi.org/10.1182/blood-2014-11-611046

88. He Z, Greven J, Shi Y, Qin K, Zhao Q, Zhang X, Buhl EM, Eschweiler J, Hildebrand F, Balmayor ER (2023) Extracellular vesicles derived from endothelial cells modulate macrophage phenotype in vitro. Eur J Med Res 28:506. https://doi.org/10.1186/s40001-023-01427-6
89. Nikfarjam S, Rezaie J, Zolbanin NM, Jafari R (2020) Mesenchymal stem cell derived-exosomes: a modern approach in translational medicine. J Transl Med 18:449. https://doi.org/10.1186/s12967-020-02622-3
90. Kou M, Huang L, Yang J, Chiang Z, Chen S, Liu J, Guo L, Zhang X, Zhou X, Xu X et al (2022) Mesenchymal stem cell-derived extracellular vesicles for immunomodulation and regeneration: a next generation therapeutic tool? Cell Death Dis 13:1–16. https://doi.org/10.1038/s41419-022-05034-x
91. Zhou C, Zhang B, Yang Y, Jiang Q, Li T, Gong J, Tang H, Zhang Q (2023) Stem cell-derived exosomes: emerging therapeutic opportunities for wound healing. Stem Cell Res Ther 14:107. https://doi.org/10.1186/s13287-023-03345-0
92. Hade MD, Suire CN, Suo Z (2021) Mesenchymal stem cell-derived exosomes: applications in regenerative medicine. Cells 10:1959. https://doi.org/10.3390/cells10081959
93. Ha DH, Kim H-K, Lee J, Kwon HH, Park G-H, Yang SH, Jung JY, Choi H, Lee JH, Sung S et al (2020) Mesenchymal stem/stromal cell-derived exosomes for immunomodulatory therapeutics and skin regeneration. Cells 9:1157. https://doi.org/10.3390/cells9051157
94. He X, Dong Z, Cao Y, Wang H, Liu S, Liao L, Jin Y, Yuan L, Li B (2019) MSC-derived exosome promotes M2 polarization and enhances cutaneous wound healing. Stem Cells Int 2019:7132708. https://doi.org/10.1155/2019/7132708
95. Li X, Liu L, Yang J, Yu Y, Chai J, Wang L, Ma L, Yin H (2016) Exosome derived from human umbilical cord mesenchymal stem cell mediates MiR-181c attenuating burn-induced excessive inflammation. EBioMedicine 8:72–82. https://doi.org/10.1016/j.ebiom.2016.04.030
96. Ti D, Hao H, Tong C, Liu J, Dong L, Zheng J, Zhao Y, Liu H, Fu X, Han W (2015) LPS-preconditioned mesenchymal stromal cells modify macrophage polarization for resolution of chronic inflammation via exosome-shuttled let-7b. J Transl Med 13:308. https://doi.org/10.1186/s12967-015-0642-6
97. Wang X, Jiao Y, Pan Y, Zhang L, Gong H, Qi Y, Wang M, Gong H, Shao M, Wang X et al (2019) Fetal dermal mesenchymal stem cell-derived exosomes accelerate cutaneous wound healing by activating notch signaling. Stem Cells Int 2019:2402916. https://doi.org/10.1155/2019/2402916
98. Zhang W, Bai X, Zhao B, Li Y, Zhang Y, Li Z, Wang X, Luo L, Han F, Zhang J et al (2018) Cell-free therapy based on adipose tissue stem cell-derived exosomes promotes wound healing via the PI3K/Akt signaling pathway. Exp Cell Res 370:333–342. https://doi.org/10.1016/j.yexcr.2018.06.035
99. Ren S, Chen J, Duscher D, Liu Y, Guo G, Kang Y, Xiong H, Zhan P, Wang Y, Wang C et al (2019) Microvesicles from human adipose stem cells promote wound healing by optimizing cellular functions via AKT and ERK signaling pathways. Stem Cell Res Ther 10:47. https://doi.org/10.1186/s13287-019-1152-x
100. Qian L, Pi L, Fang B-R, Meng X-X (2021) Adipose mesenchymal stem cell-derived exosomes accelerate skin wound healing via the lncRNA H19/miR-19b/SOX9 axis. Lab Investig 101:1254–1266. https://doi.org/10.1038/s41374-021-00611-8
101. Yang C, Luo L, Bai X, Shen K, Liu K, Wang J, Hu D (2020) Highly-expressed micoRNA-21 in adipose derived stem cell exosomes can enhance the migration and proliferation of the HaCaT cells by increasing the MMP-9 expression through the PI3K/AKT pathway. Arch Biochem Biophys 681:108259. https://doi.org/10.1016/j.abb.2020.108259
102. Cheng S, Xi Z, Chen G, Liu K, Ma R, Zhou C (2020) Extracellular vesicle-carried microRNA-27b derived from mesenchymal stem cells accelerates cutaneous wound healing via E3 ubiquitin ligase ITCH. J Cell Mol Med 24:11254–11271. https://doi.org/10.1111/jcmm.15692

103. Liang X, Zhang L, Wang S, Han Q, Zhao RC (2016) Exosomes secreted by mesenchymal stem cells promote endothelial cell angiogenesis by transferring miR-125a. J Cell Sci 129:2182–2189. https://doi.org/10.1242/jcs.170373
104. Krawczenko A, Bielawska-Pohl A, Paprocka M, Kraskiewicz H, Szyposzynska A, Wojdat E, Klimczak A (2020) Microvesicles from human immortalized cell lines of endothelial progenitor cells and mesenchymal stem/stromal cells of adipose tissue origin as carriers of bioactive factors facilitating angiogenesis. Stem Cells Int 2020:1289380. https://doi.org/10.1155/2020/1289380
105. Hu Y, Rao S-S, Wang Z-X, Cao J, Tan Y-J, Luo J, Li H-M, Zhang W-S, Chen C-Y, Xie H (2018) Exosomes from human umbilical cord blood accelerate cutaneous wound healing through miR-21-3p-mediated promotion of angiogenesis and fibroblast function. Theranostics 8:169–184. https://doi.org/10.7150/thno.21234
106. Wang L, Hu L, Zhou X, Xiong Z, Zhang C, Shehada HMA, Hu B, Song J, Chen L (2017) Exosomes secreted by human adipose mesenchymal stem cells promote scarless cutaneous repair by regulating extracellular matrix remodelling. Sci Rep 7:13321. https://doi.org/10.1038/s41598-017-12919-x
107. Hu L, Wang J, Zhou X, Xiong Z, Zhao J, Yu R, Huang F, Zhang H, Chen L (2016) Exosomes derived from human adipose mensenchymal stem cells accelerates cutaneous wound healing via optimizing the characteristics of fibroblasts. Sci Rep 6:32993. https://doi.org/10.1038/srep32993
108. Li Y, Zhang J, Shi J, Liu K, Wang X, Jia Y, He T, Shen K, Wang Y, Liu J et al (2021) Exosomes derived from human adipose mesenchymal stem cells attenuate hypertrophic scar fibrosis by miR-192-5p/IL-17RA/Smad axis. Stem Cell Res Ther 12:221. https://doi.org/10.1186/s13287-021-02290-0
109. Yan C, Xv Y, Lin Z, Endo Y, Xue H, Hu Y, Hu L, Chen L, Cao F, Zhou W et al (2022) Human umbilical cord mesenchymal stem cell-derived exosomes accelerate diabetic wound healing via ameliorating oxidative stress and promoting angiogenesis. Front Bioeng Biotechnol 10. https://doi.org/10.3389/fbioe.2022.829868
110. Lu W, Li X, Wang Z, Zhao C, Li Q, Zhang L, Yang S (2024) Mesenchymal stem cell-derived extracellular vesicles accelerate diabetic wound healing by inhibiting NET-induced ferroptosis of endothelial cells. Int J Biol Sci 20:3515–3529. https://doi.org/10.7150/ijbs.97150
111. Wong SL, Demers M, Martinod K, Gallant M, Wang Y, Goldfine AB, Kahn CR, Wagner DD (2015) Diabetes primes neutrophils to undergo NETosis, which impairs wound healing. Nat Med 21:815–819. https://doi.org/10.1038/nm.3887
112. Bailey AJM, Li H, Kirkham AM, Tieu A, Maganti HB, Shorr R, Fergusson DA, Lalu MM, Elomazzen H, Allan DS (2022) MSC-derived extracellular vesicles to heal diabetic wounds: a systematic review and meta-analysis of preclinical animal studies. Stem Cell Rev and Rep 18:968–979. https://doi.org/10.1007/s12015-021-10164-4
113. Al-Masawa ME, Alshawsh MA, Ng CY, Ng AMH, Foo JB, Vijakumaran U, Subramaniam R, Ghani NAA, Witwer KW, Law JX (2022) Efficacy and safety of small extracellular vesicle interventions in wound healing and skin regeneration: a systematic review and meta-analysis of animal studies. Theranostics 12:6455–6508. https://doi.org/10.7150/thno.73436
114. Qiao Z, Wang X, Zhao H, Deng Y, Zeng W, Yang K, Chen H, Yan Q, Li C, Wu J et al (2023) The effectiveness of cell-derived exosome therapy for diabetic wound: a systematic review and meta-analysis. Ageing Res Rev 85:101858. https://doi.org/10.1016/j.arr.2023.101858
115. Gunjan H, Pandey RP, Mukherjee R, Chang C-M (2024) Advanced meta-analysis on therapeutic strategies of mesenchymal derived exosome for diabetic chronic wound healing and tissue remodeling. Mol Cell Probes 77:101974. https://doi.org/10.1016/j.mcp.2024.101974
116. Yue G, Li Y, Liu Z, Yu S, Cao Y, Wang X (2024) Efficacy of MSC-derived small extracellular vesicles in treating type II diabetic cutaneous wounds: a systematic review and meta-analysis of animal models. Front Endocrinol 15. https://doi.org/10.3389/fendo.2024.1375632
117. Yuan T, Meijia L, Xinyao C, Xinyue C, Lijun H (2023) Exosome derived from human adipose-derived stem cell improve wound healing quality: a systematic review and meta-

analysis of preclinical animal studies. Int Wound J 20:2424–2439. https://doi.org/10.1111/iwj.14081
118. Soltani S, Zahedi A, Vergara AJS, Noli M, Soltysik FM, Pociot F, Yarani R (2024) Preclinical therapeutic efficacy of extracellular vesicles derived from adipose-derived mesenchymal stromal/stem cells in diabetic wounds: a systematic review and meta-analysis. Stem Cell Rev Rep 20:2016–2031. https://doi.org/10.1007/s12015-024-10753-z
119. Wei J, He T, Shen K, Xu Z, Han J, Yang X (2024) Adipose stem cell-derived exosomes in the treatment of wound healing in preclinical animal models: a meta-analysis. Burns Trauma 12:tkae025. https://doi.org/10.1093/burnst/tkae025
120. Du Y, Yan S, Sun Y, Han X, Shi H, Fan W, Liu G (2024) Extracellular vesicles secreted by bone marrow stem cells mediate angiogenesis for the treatment of diabetic ulcers: a systematic review and meta-analysis of preclinical studies. Heliyon 10. https://doi.org/10.1016/j.heliyon.2024.e25762
121. Xin L, Wei C, Tong X, Dai Y, Huang D, Chen J, Ma L, Zhang S (2022) In situ delivery of apoptotic bodies derived from mesenchymal stem cells via a hyaluronic acid hydrogel: a therapy for intrauterine adhesions. Bioact Mater 12:107–119. https://doi.org/10.1016/j.bioactmat.2021.10.025
122. Ma Q, Liang M, Wu Y, Ding N, Duan L, Yu T, Bai Y, Kang F, Dong S, Xu J et al (2019) Mature osteoclast-derived apoptotic bodies promote osteogenic differentiation via RANKL-mediated reverse signaling. J Biol Chem 294:11240–11247. https://doi.org/10.1074/jbc.RA119.007625
123. Zhu Y, Yang H, Xue Z, Tang H, Chen X, Liao Y (2025) Mesenchymal stem cells-derived small extracellular vesicles and apoptotic extracellular vesicles for wound healing and skin regeneration: a systematic review and meta-analysis of preclinical studies. J Transl Med 23:364. https://doi.org/10.1186/s12967-024-05744-0
124. Hooijmans CR, Rovers MM, de Vries RB, Leenaars M, Ritskes-Hoitinga M, Langendam MW (2014) SYRCLE's risk of bias tool for animal studies. BMC Med Res Methodol 14:43. https://doi.org/10.1186/1471-2288-14-43
125. Park DJ, Choi W, Sayeed S, Dorschner RA, Rainaldi J, Ho K, Kezios J, Nolan JP, Mali P, Costantini T et al (2024) Defining the activity of pro-reparative extracellular vesicles in wound healing based on miRNA payloads and cell type-specific lineage mapping. Mol Ther 32:3059–3079. https://doi.org/10.1016/j.ymthe.2024.02.019
126. Lai CP, Kim EY, Badr CE, Weissleder R, Mempel TR, Tannous BA, Breakefield XO (2015) Visualization and tracking of tumour extracellular vesicle delivery and RNA translation using multiplexed reporters. Nat Commun 6:7029. https://doi.org/10.1038/ncomms8029
127. Liu Q, Huang J, Xia J, Liang Y, Li G (2022) Tracking tools of extracellular vesicles for biomedical research. Front Bioeng Biotechnol 10:943712. https://doi.org/10.3389/fbioe.2022.943712
128. Lenzini S, Debnath K, Joshi JC, Wong SW, Srivastava K, Geng X, Cho IS, Song A, Bargi R, Lee JC et al (2021) Cell–matrix interactions regulate functional extracellular vesicle secretion from mesenchymal stromal cells. ACS Nano 15:17439–17452. https://doi.org/10.1021/acsnano.1c03231
129. Debnath K, Heras KL, Rivera A, Lenzini S, Shin J-W (2023) Extracellular vesicle-matrix interactions. Nat Rev Mater 8:390–402. https://doi.org/10.1038/s41578-023-00551-3
130. Sinha B, Köster D, Ruez R, Gonnord P, Bastiani M, Abankwa D, Stan RV, Butler-Browne G, Vedie B, Johannes L et al (2011) Cells respond to mechanical stress by rapid disassembly of caveolae. Cell 144:402–413. https://doi.org/10.1016/j.cell.2010.12.031
131. Dragovic RA, Gardiner C, Brooks AS, Tannetta DS, Ferguson DJP, Hole P, Carr B, Redman CWG, Harris AL, Dobson PJ et al (2011) Sizing and phenotyping of cellular vesicles using nanoparticle tracking analysis. Nanomedicine Nanotechnol Biol Med 7:780–788. https://doi.org/10.1016/j.nano.2011.04.003
132. Arif S, Moulin VJ (2023) Extracellular vesicles on the move: traversing the complex matrix of tissues. Eur J Cell Biol 102:151372. https://doi.org/10.1016/j.ejcb.2023.151372

133. Shen Z, Shao J, Sun J, Xu J (2022) Exosomes released by melanocytes modulate fibroblasts to promote keloid formation: a pilot study. J Zhejiang Univ Sci B 23:699–704. https://doi.org/10.1631/jzus.B2200036
134. Deng M, Yu TZ, Li D et al (2020) Human umbilical cord mesenchymal stem cell-derivedand dermal fibroblast-derived extracellular vesicles protect dermal fibroblasts from ultravioletradiation-induced photoaging in vitro. Photochem Photobiol Sci 19(3):406–14. https://doi.org/10.1039/c9pp00421a
135. Qi L, Lu Y, Wang Z, Zhang G (2021) MicroRNA-106b derived from endothelialcell–secreted extracellular vesicles prevents skin wound healing by inhibiting JMJD3 and RIPK3. J Cell Mol Med 25(10):4551–61. https://doi.org/10.1111/jcmm.16037
136. Liu J, Dong J, Pei X (2024) Apoptotic extracellular vesicles derived from human umbilicalvein endothelial cells promote skin repair by enhancing angiogenesis: From death to regeneration. Int J Nanomedicine 19:415–28. https://doi.org/10.2147/IJN.S441453

Chapter 14
Stem Cells and Extracellular Vesicles for the Treatment of Chronic Wounds and Peripheral Vascular Diseases

Barbara Zavan and Letizia Ferroni

Abstract Chronic wounds arise from a combination of factors, including impaired macrophage and fibroblast function, dysregulated cytokine signaling, and insufficient angiogenesis. Non-healing wounds, including diabetic foot ulcer, venous leg ulcer resulting from peripheral artery disease or critical limb ischemia, remain a significant and growing healthcare challenge, particularly in aging populations. Existing treatments often fail to address these multifactorial conditions. Although stem cell-based therapies show promising results, several challenges remain related to consistent cell sourcing, scalable expansion techniques, optimized delivery strategies, universally accepted protocols for cell characterization, and potency testing. Recent research has shifted toward exploring stem-cell-derived extracellular vesicles (EVs) as carriers of bioactive factors. EVs exhibit high tissue-targeting ability, biocompatibility, stability in circulation, and low immunogenicity. Their therapeutic functions include modulation of immune responses and inflammation, stimulation of angiogenesis, acceleration of cell proliferation, and regulation of collagen remodeling, making them suitable for regenerative medicine. Nevertheless, current preclinical studies show poor reproducibility in animal models, and limited clinical data exist regarding EV efficacy.

Overall, stem cell and EV-based therapies hold transformative potential for chronic wound care. Multidisciplinary collaboration will be necessary for these biological therapies to become integral components of personalized, effective, and accessible treatment strategies for patients with chronic, non-healing wounds.

Keywords Diabetic foot ulcer · Venous leg ulcer · Critical limb ischemia · Peripheral artery disease · Extracellular vesicle · Stem cell

---

B. Zavan (✉)
Translational Medicine Department, University of Ferrara, Ferrara, Italy
e-mail: barbara.zavan@unife.it

L. Ferroni (✉)
Maria Cecilia Hospital, GVM Care & Research, Ravenna, Italy
e-mail: lferroni@gvmnet.it

F. De Francesco, K. Turksen (eds.), *Management and Strategies for Wound Healing*, Stem Cell Biology and Regenerative Medicine 79,
https://doi.org/10.1007/978-3-032-12442-5_14

Abbreviation

ABI	Ankle-brachial index
AD-MSCs	Adipose-derived mesenchymal stem cells
AD-SVF	Adipose tissue derived stromal vascular fraction
bFGF	Basic fibroblast growth factor
BM-MSCs	Bone marrow-derived mesenchymal stem cells
circRNAs	Circular RNAs
CLI	Critical limb ischemia
CXCL5	C-X-C motif chemokine 5
DAMPs	Damage-associate molecular patterns
DFU	Diabetic foot ulcer
DPCs	Dermal papilla cells
DSCs	Dermal sheath cells
dUC	Differential ultracentrifugation
EGF	Epidermal growth factor
ESCRT	Endosomal sorting complex required for transport
EVs	Extracellular vesicles
FGF2	Fibroblast growth factor 2
GAGs	Glycosaminoglycans
GelMA	Gelatin methacryloyl
HPMC	Hydroxypropyl methylcellulose
ICAM1	Intercellular adhesion molecule 1
IDO	Indoleamine 2,3-dioxygenase
IFN-γ	Interferon gamma
IGF	Insulin-like growth factor
IL-1β	Interleukin 1 beta
IL-6	Interleukin 6
ILVs	Intraluminal vesicles
ISEV	International society for extracellular vesicles
KGF	Keratinocyte growth factor
lncRNAs	Long non-coding RNAs
MALAT1	Metastasis-associated lung adenocarcinoma transcript 1
MAPK	Mitogen-activated protein kinases
MHC	Major histocompatibility complex
miRNAs	MicroRNAs
MMPs	Metalloproteinases
MSC-EVs	Mesenchymal stem cell-derived EVs
MSCs	Mesenchymal stem cells
MVBs	Multivesicular bodies
NF-κB	Nuclear factor kappa-light-chain-enhancer of activated B cells
NGF	Nerve growth factor
NO	Nitric oxide
PAD	Peripheral artery disease

PAMPs	Pathogen-associated molecular patterns
PDGF	Platelet-derived growth factor
PDGFB	Platelet-derived growth factor beta
PGE2	Prostaglandin E2
PLMA	Poly-L-lysine methacryloyl
P-MSCs	Placenta-derived mesenchymal stem cells
PPARγ	Peroxisome proliferator-activated receptor gamma
SEC	Size-exclusion chromatography
STAT3	Signal transducer and activator of transcription 3
STZ	Streptozotocin
TBI	Toe brachial index
TcPO2	Transcutaneous oxygen pressure
TGF-β	Transforming growth factor beta
TIMPs	Tissue inhibitors of metalloproteinases
TLRs	Toll-like receptors
TNFα	Tumor necrosis factor alpha
UC-MSCs	Umbilical cord-derived mesenchymal stem cells
VCAM1	Vascular cell adhesion molecule 1
VEGF	Vascular endothelial growth factor

Introduction

Injury to tissue or organs results in the disruption of tissue continuity, thereby forming a wound. Wounds are broadly classified into two categories: open and closed. Open wounds are characterized by a break in the protective surface of the body, typically the skin or mucous membranes, exposing the underlying tissues to potential contamination by foreign materials. Common types of open wounds include punctures, incisions, abrasions, and lacerations. In contrast, closed wounds, such as hematomas and contusions, typically result from blunt force trauma. Although the skin remains intact, significant internal damage, including hemorrhage or injury to muscles, bones, or organs, may occur.

Wounds are further categorized based on the duration and pattern of healing into acute and chronic types. Acute wounds typically undergo a timely and orderly reparative process, leading to both anatomical and functional restoration, generally within 4 weeks. Conversely, chronic wounds fail to progress through the normal stages of healing and persist for extended periods, often exceeding 3 months.

All wounds initially present as acute, but may become chronic if the healing process is interrupted, particularly during the inflammatory phase, when the body mounts a protective response to injury. Several intrinsic and extrinsic factors can contribute to the chronicity of wounds, including advanced age, immobility, infection, and comorbid conditions such as diabetes mellitus, obesity, hyperlipidemia, and peripheral vascular disease. Examples of chronic wounds include those resulting from radiation injury, ischemia, infection, and various ulcerative conditions.

Among these, arterial, venous, pressure, and diabetic ulcers are the most prevalent types affecting the lower extremities.

Diagnosis of ulcerative wounds typically involves a thorough clinical examination, vascular assessment, palpation of pedal pulses, and measurement of the ankle-brachial index. Non-healing or atypical ulcers warrant biopsy to exclude malignancy or other underlying pathology. Effective management of chronic ulcers involves several key components: debridement of necrotic tissue, control of infection, cleansing of wound edges, and maintenance of an optimal moisture balance. Subsequent treatment should be tailored to the specific ulcer type. Arterial ulcers require prompt referral for vascular intervention. Venous ulcers are managed through compression therapy, limb elevation, and physical activity. Pressure ulcers necessitate offloading of the affected area, whereas diabetic foot ulcers are treated by both offloading and addressing any associated peripheral arterial disease [1].

Wound healing is a complex and highly regulated physiological process aimed at restoring tissue integrity and function. This process occurs in four overlapping yet distinct phases: hemostasis, inflammation, proliferation, and remodeling. The progression through these phases is dependent on a coordinated cascade of cellular signaling events, involving diverse cell types such as neutrophils, monocytes, macrophages, fibroblasts, keratinocytes, endothelial cells, and mesenchymal stem cells (MSCs). Within these broad stages, a series of events, including chemotaxis, phagocytosis, neocollagenesis, degradation, and remodeling of collagen, are coordinated through the involvement of multiple cells that release growth factors, cytokines, and chemokines. Neovascularization, reepithelialization, and the synthesis of proteoglycans and glycosaminoglycans (GAGs) are also critical events to wound healing [2].

Following injury, the clotting cascade is rapidly activated, forming a fibrin clot that ensures hemostasis and provides a scaffold for incoming inflammatory and repair cells. Platelets trapped in the clot also release growth factors and chemokines into the local wound environment. As a consequence of injury, the resident cells are exposed to damage-associated molecular patterns (DAMPs), molecules (e.g., intracellular proteins, DNA, and RNA) released by cells undergoing necrosis, and pathogen-associated molecular patterns (PAMPs), which are pathogen-specific molecules not found in the host (e.g., bacterial essential polysaccharides and polynucleotides). These molecules are recognized by pattern recognition receptors like toll-like receptors (TLRs), which are constitutively expressed on host cells. TLR activation triggers intracellular pathways, including nuclear factor kappa-light-chain-enhancer of activated B cells (NF-κB) and mitogen-activated protein kinases (MAPK) pathways, which lead to the expression of cytokines, chemokines, and antimicrobial peptides to initiate and perpetuate the inflammatory response [3].

During early inflammation (within 5 days post-injury), neutrophils are recruited to the wound site. Chemokines upregulate adhesion molecules, vascular cell adhesion molecule 1 (VCAM1) and intercellular adhesion molecule 1 (ICAM1), on endothelial cells, which drive neutrophil adhesion to the vessel wall and subsequent extravasation. Neutrophils eliminate pathogens and clear damaged tissue by releasing antimicrobial substances and proteases. In the late inflammatory phase, macrophages remove pathogens, dead neutrophils, and damaged tissue. They also secrete chemokines, cytokines, and growth factors to promote the transition of the wound into the proliferative phase. Precisely, tissue macrophages originate from the differentiation of blood monocytes. During the inflammatory phase, monocytes are rapidly recruited to the inflamed site and differentiate to tissue macrophages with pro-inflammatory (M1) or anti-inflammatory (M2) phenotype. M1 macrophages drive pathogen and debris clearance, and pro-inflammatory mediator secretion; whereas, M2 macrophages support tissue repair by producing extracellular matrix (ECM) components and anti-inflammatory cytokines like IL-10. Elevated M2 macrophage levels in the recovery site correlate with improved healing and reduced fibrosis and scarring, mainly due to the reduction of inflammatory mediators [4].

Chronic wounds result from sustained pro-inflammatory cytokines, such as interleukin (IL)-1β, IL-6, and tumor necrosis factor-α (TNFα). These cytokines enhance the activity of metalloproteinases (MMPs), causing the degradation of the ECM and growth factors, hindering cell migration and the proliferative phase. Furthermore, the persistent bacteria exacerbate the inflammation as they attract the influx of pro-inflammatory cells to the wound site. These events lead to a prolonged inflammatory phase and to a non-healing chronic wound [5].

In the proliferative phase, granulation tissue replaces the provisional matrix. This tissue, composed of fibroblasts, macrophages, granulocytes, blood vessels, and collagen, partially restores tissue structure and function. Fibroblasts are essential for the formation of the granulation tissue. They migrate to the wound in response to cytokines and growth factors, including transforming growth factor beta (TGF)-β, basic fibroblast growth factor (bFGF), and platelet-derived growth factor (PDGF), released by platelets and macrophages. In the wound matrix, fibroblasts proliferate and synthesize MMPs to remodel the initial matrix while depositing ECM components, including collagen, GAGs, proteoglycans, and fibronectin [6].

The re-epithelialization requires the migration and proliferation of keratinocytes from the wound-edge. The migration is triggered by the loss of contact inhibition and physical tension at cell adhesion structures, and allowed by the degradation of the ECM carried out by collagenase and elastase. Re-epithelialization is further driven by growth factors like keratinocyte growth factor (KGF), epidermal growth factor (EGF), insulin-like growth factor (IGF), nerve growth factor (NGF), and nitric oxide (NO) [7]. Actual wound healing also requires neovascularization; conversely, insufficient vascularization is a common sign of chronic non-healing wounds. Neovascularization occurs by vasculogenesis, the generation of new vessels from endothelial progenitor cells, and angiogenesis, the formation of new vessels from existing blood vessels. From the early to the late stage of the healing process, keratinocytes produce the most proangiogenic factor, the vascular endothelial growth factor (VEGF). VEGF is also expressed by activated fibroblasts, mast cells, and macrophages [8].

The remodeling phase overlaps with proliferation and can extend over months. In this final phase, the granulation tissue matures into scar tissue, increasing tensile strength. As remodeling progresses, the density of vessels, GAGs, proteoglycans, and cells declines, while collagen undergoes modifications to enhance tensile strength. Type III collagen, synthesized at high levels in granulation tissue, is replaced by type I collagen. Under normal conditions, proteolytic enzymes, mainly MMPs, reorganize collagen fibers while specific inhibitors tightly regulate the destructive actions of the proteolytic enzymes to prevent excessive degradation [9].

Stem Cell Therapies in Chronic Wound Management

Mesenchymal stem cells (MSCs) are multipotent stem cells with self-renewal capability, multidirectional differentiation potential, and well-recognized immunomodulatory and anti-inflammatory properties. According to the International Society for Cellular Therapy, the minimal criteria to define human MSC are plastic-adhesion when maintained in standard culture conditions, expression of surface markers CD73, CD90, and CD105, and lack of expression of CD11b or CD14, CD19 or CD79alpha, CD34, CD45, and HLA-DR [10]. Additionally, they must demonstrate in vitro differentiation into osteoblasts, adipocytes, and chondroblasts [11–14]. MSCs can be readily isolated from a variety of adult tissues, such as bone marrow, adipose tissue, umbilical cord, and dental pulp, and expanded in vitro [15, 16]. These properties make MSCs highly valuable in regenerative medicine, particularly in tissue repair and wound healing.

Endogenous MSCs play a pivotal role in wound healing by coordinating several signaling pathways essential for the resolution of the process. They regulate oxidative stress, immune cell infiltration, and inflammatory factor secretion in the wound microenvironment. This results in enhanced cell migration, granulation tissue formation, re-epithelialization, and angiogenesis, fostering a regenerative rather than fibrotic healing response [17]. Endogenous MSCs such as dermal papilla cells (DPCs) and dermal sheath cells (DSCs) are mobilized to the wound site following injury, primarily in response to platelet-derived growth factor B (PDGFB). These cells secrete bFGF and C-X-C motif chemokine 5 (CXCL5), inducing their conversion into myofibroblasts to support ECM synthesis and tissue repair [18]. However, in chronic wounds such as diabetic ulcers, the endogenous MSC population is often insufficient due to the lack of dermal tissue and tissue damage. Furthermore, MSCs residing close to chronic ulcers show reduced proliferation and migration capacities compared to those isolated from the dermis of healthy individuals [19].

The clinical application of exogenous MSCs in the treatment of chronic wounds has achieved promising results [20, 21]. Lacking major histocompatibility complex (MHC) class II molecules, MSCs have low immunogenicity, meaning they are less likely to be recognized and attacked by the immune system and therefore not rejected after transplantation. They migrate and home into the injury site via chemokine receptors, adhesion molecules, integrins, and growth factor receptors, where they exert immunomodulatory, anti-inflammatory, anti-apoptotic, and anti-fibrotic and pro-angiogenic effects [22]. The transplantation of allogenic or autologous bone marrow-derived MSCs (BM-MSCs) is not the only option; adipose-derived MSCs (AD-MSCs), placenta-derived MSCs (P-MSCs), and umbilical cord-derived MSCs (UC-MSCs) have comparable clinical potential. They are good alternatives because they share similar biological properties to BM-MSCs, they can be collected with minimally invasive procedures as they are abundantly available, and their use is free of ethical controversies [22]. Once at the wound site, MSCs can differentiate into endothelial cells, keratinocytes, and pericytes, thanks to their extensive differentiation capabilities [23]. MSCs also exert paracrine function, releasing bioactive molecules including growth factors, cytokines, chemokines, and extracellular vesicles (EVs) that modulate inflammation and stimulate tissue repair [24]. These factors promote angiogenesis, primarily through enhanced VEGF secretion, which supports oxygen and nutrient supply to the regenerating tissue [25]. VEGF, TGF-β, IL-6, and other cytokines released from newborn blood vessels regulate inflammation, reduce fibrosis, and stimulate fibroblast activity and collagen synthesis. Additionally, MSCs regulate ECM remodeling by secreting MMPs and tissue inhibitors of metalloproteinases (TIMPs), which balance ECM synthesis and degradation, ultimately minimizing scarring and fibrosis. MSCs also mitigate keloid formation by reducing fibroblast proliferation and ECM component synthesis, improving restoration and a more functionally and aesthetically favorable healing outcome [24].

MSCs modulate the activity of various immune cells both directly and via paracrine mechanisms. MSCs are recognized quickly and non-specifically by circulating monocytes and tissue-resident macrophages. When alive, dead, apoptotic, or fragmented MSCs are phagocytized by monocytes or macrophages, the phagocytic ability of them is significantly enhanced and they acquire anti-inflammatory M2 phenotypes. These M2 macrophages increase the secretion of anti-inflammatory factors, including IL-12 and tumor necrosis factor-α (TNF-α), inhibiting the inflammatory response and accelerating tissue repair. De facto, monocytes and macrophages are pivotal for extending the immunomodulatory effects of MSCs [26]. MSCs also modulate the proliferation and function of T cells, B cells, natural killer and dendritic cells. The immunomodulatory effects are mediated by cell–cell interactions and soluble factors such as prostaglandin E2 (PGE2), NO, IDO, TGF-β, and IFN-γ [27].

In clinical practice, MSCs are administered using various approaches tailored to the specific condition being treated. Common methods are intravenous infusion, localized injections directly into injured tissue, or via implantation into biomaterials that support integration and controlled release at them the injury site [28–31].

Numerous clinical trials (Table 14.1) have been performed to demonstrate the safety and efficacy of MSCs in the treatment of peripheral arterial disease (PAD), critical limb ischemia (CLI), diabetic foot ulcer (DFU), and venous leg ulcer [32–34].

PAD is a prevalent and progressive cardiovascular condition characterized by the narrowing or occlusion of peripheral arteries, most commonly affecting the lower extremities. The primary etiological factor is atherosclerosis, which leads to diminished perfusion of the limbs. Globally, PAD incidence rates increase with advancing age and the presence of established cardiovascular risk factors such as diabetes, hypertension, and smoking [35]. The clinical spectrum of PAD ranges from asymptomatic disease to intermittent claudication, rest pain, and CLI. CLI is the most

Table 14.1 Clinical trials exploring the safety and efficacy of stem cells in the treatment of chronic wounds (from 2015 to 2025)

NCT number	Study phase	Study type	Source of stem cells	Condition	Measured outcomes	Study start	Status of trial
NCT06843122	Phase 2	Interventional, Randomized, Double-Blind,	Allogeneic AD-MSCs in fibrin solution	DFU	Change in wound size and incidence of adverse events within 6 weeks.	2025	Recruiting
NCT06231771	Phase 1 Phase 2	Interventional, Randomized, Parallel Assignment, Placebo Controlled, Double Blind	Expanded allogeneic Wharton-Jelly derived MSCs	Non-Healing DFU	Complete healing/closure of the ulcer within 6-weeks.	2024	Recruiting
NCT06359912	Phase 1	Interventional, Randomized, Parallel Assignment	Allogeneic Endothelial Progenitor Cells (EPCs)	CLI	Incidence of adverse events within 1 year. Improvement in painless walking and intermittent claudication distance, blood flow at the affected limb, ulcer area, quality of life after 6 month.	2024	Recruiting
NCT05854641	Phase 4	Interventional, Single Group Assignment, Practice-Based, Feasibility Study	Expanded allogeneic MSCs	CLI with PAD	Change in ischemic rest pain, ulcer size, ABI, total walking distance, and angiogenesis. Incidence of several adverse events, abnormal laboratory test results, and major amputation-free survival within 24 months.	2024	Recruiting
NCT05962931	Phase 1 Phase 2	Interventional, Randomized, Parallel Assignment, Multicenter, Controlled	Allogeneic AD-MSCs	Venous Leg Ulcer	Remission of the ulcer and incidence of adverse events within 1 year.	2023	Recruiting
NCT06174597	Early Phase 1	Interventional, Randomized, Sequential Assignment	UC-MSCs	CLI	Change in hindlimb flow index, TcPO2, and ABI after 24 months.	2023	Recruiting

NCT05549609	Phase 1 Phase 2	Interventional, RandomisedRandomized, MulticentreMulticenter, Single-Blind	Expanded, allogeneic, integrin α10β1-selected AD-MSCs	Venous Leg Ulcer	Adverse events, local tolerability, wound area reduction (absolute and percentage) compared to baseline, pain perception after 4 months.	2022	Recruiting
NCT05595681	Phase 1	Interventional, Randomized, Parallel Assignment, Single Center	AD-MSCs	DFU	Time to complete ulcer healing, percentage wound area reduction, adverse events, and serious adverse events within 12 weeks.	2022	Recruiting
NCT06373809	Early Phase 1	Interventional, Non-Randomized, Parallel Assignment	P-MSCs	Chronic Ulcers and Neuropathy	Ulcer healing, hemodynamic and anatomical effects on foot arteries, and changes in the sensory perceptions of the foot within 12 months.	2021	Completed
NCT04466007	Phase 2	Interventional, Randomized, Placebo-Controlled, Double-Blind, Multicenter	Allogeneic AD-MSCs	CLI	Incidence of adverse events, neuropathic symptoms, revascularizations, and amputations within 1 year.	2021	Completed
NCT04497805	Phase 2	Interventional, Randomized, Parallel Assignment	Allogeneic MSCs in hydrogel sheet	DFU	Complete wound closure within 12 weeks.	2020	Recruiting
NCT05610865	Phase 1	Interventional, Randomized, Parallel Assignment	Autologous AD-SVF and expanded AD-MSCs	DFU	Complete wound closure within 3 months.	2020	Recruiting

(continued)

Table 14.1 (continued)

NCT number	Study phase	Study type	Source of stem cells	Condition	Measured outcomes	Study start	Status of trial
NCT04661644	Phase 1 Phase 2	Interventional, Non-Randomized, Sequential Assignment	Allogeneic AD-MSCs	CLI	Changes in ischemic pain, pain-free walking distance, TBI, ABI, and ulcer size, Incidence of abnormal laboratory test results, blood pressure, temperature, and physical condition at week 4, 12, and 24 compared to baseline	2020	Completed
NCT03754465	Phase 2	Interventional, Randomized, Comparator-Controlled, Double-Blind, Multicenter	Allogeneic AD-MSCs in a hydrogel sheet	DFU	Incidence of complete wound closure and changes in wound size within 12 weeks compared to baseline	2019	Completed
NCT03865394	Phase 1 Phase 2	Interventional, Non-Randomized, Parallel Assignment	Allogeneic AD-MSCs in fibrin gel	DFU	Changes in wound size and morphology, pro-angiogenic factor expression, wound-associated pain, quality of life, and incidence of adverse events within 8 weeks.	2019	Completed
NCT04104451	Phase 1	Interventional, Non-Randomized, Sequential Assignment	Expanded UC-MSCs	DFU	Incidence of serious adverse reactions within 4 months.	2019	Completed
NCT03370874	Phase 3	Interventional, Randomized, Parallel-Group, Placebo-Controlled, Double Blind, Multicenter	Allogeneic MSCs in a hydrogel sheet	DFU	Complete wound closure and change in wound size and depth compared to baseline within 12 weeks.	2018	Completed
NCT03257098	Phase 1 Phase 2	Interventional, Single Arm, Multicenter	ABCB5-positive MSCs	Chronic Venous Ulcer	Change in wound size and incidence of adverse event up to 12 months.	2017	Completed
NCT03267784	Phase 1 Phase 2	Interventional, Single Arm, Multicenter	Allogeneic ABCB5-positive MSCs	Neuropathic DFU	Change in wound surface area and incidence of adverse events up to 12 months.	2017	Completed

NCT02619877	Phase 2	Interventional, Randomized, Parallel-Group, Comparator-Controlled, Single-Blind, Multicenter	Allogeneic MSCs in a hydrogel sheet	DFU	Re-epithelialization during 8 weeks. Time to re-epithelialization and Adverse Events during 12 weeks.	2015	Completed
NCT02336646	Phase 1	Interventional, Randomized, Single Group Assignment, Double Blind	BM-MSCs cultured in hypoxia	CLI	Incidence of adverse events. Change in ischemic pain, TcPO2, ABI, and total walking distance after 6 weeks.	2015	Completed
NCT03455335	Phase 1	Interventional, Non-Randomized, Sequential Assignment	BM-MSCs	CLI	Incidence of serious adverse events. Change in TcPO2, ABI, ischemic rest pain, ulcer size, and quality of life within 12 months.	2015	Completed

ABI ankle-brachial index, *AD-MSCs* adipose-derived mesenchymal stem cells, *AD-SVF* adipose tissue derived stromal vascular fraction, *BM-MSCs* bone marrow-derived MSCs, *CLI* critical limb ischemia, *DFU* diabetic foot ulcer, *P-MSCs* placenta-derived MSCs, *UC-MSCs* umbilical cord-derived MSCs, *TBI* toe brachial index, *PAD* peripheral artery disease, *TcPO2* transcutaneous oxygen pressure severe manifestation of PAD and is associated with substantial morbidity, including a high risk of limb loss and mortality, thereby imposing a significant burden on patient quality of life and healthcare systems. Current treatment strategies for PAD include risk factor modification, pharmacological interventions, and revascularization procedures. Nevertheless, pharmacological treatments often provide only modest improvements in walking distance and quality of life. Revascularization procedures, such as angioplasty, stenting, or bypass surgery, may not be suitable for all patients, particularly those with diffuse disease. Moreover, a significant proportion of patients with CLI are not candidates for conventional revascularization or experience suboptimal outcomes despite intervention [36]. These limitations highlight the need for alternative therapeutic approaches to enhance blood flow to ischemic areas.

A phase I clinical study [37] investigated the safety and efficacy of the intramuscular administration of P-MSCs in CLI patients. The treatments were well tolerated, and after 6 months, patients showed a significant reduction in IL-1 and IFN-γ serum levels, suggesting reduced inflammation. The blood lymphocyte profile of treated participants did not significantly differ, suggesting that the injection of allogeneic cells did not cause T-cell proliferation. Additionally, there was a clinically significant improvement in pain relief, ulcer healing, and walking distance. However, angiography did not reveal significant new vessel formation [37]. Another study [38], a prospective, double-blind randomized placebo-controlled multicenter study investigated bone marrow-derived MSCs for angiogenesis and immunomodulation in CLI patients who were not suitable for revascularization. The MSC treatment group showed significant improvements in Ankle Brachial Index (ABI) and ankle pressure compared to the placebo group, with no major differences in serious adverse events between the two groups. Serious adverse events resulted in death, infected gangrene, and amputations that were related to disease progression but not to stem cell transplantation [38]. Moreover, in a phase I trial, intra-arterial administration of MSCs in CLI patients resulted in significant pain relief, increased ABI, and improved oxygen levels (TcPO2), with 86% of patients experiencing limb salvage and partial or complete ulcer healing [39].

DFU is a chronic, non-healing wound that results from a combination of macro- and micro-vascular complications, as well as peripheral neuropathy commonly observed in patients with diabetes. DFU serves as a major antecedent to lower-extremity amputations, preceding approximately 84% of diabetes-related lower-leg amputations, and are associated with substantial morbidity and mortality [40]. Current management strategies for DFU encompass sharp debridement, pressure off-loading, and patient education on diabetic foot care. Contemporary therapeutic approaches are grounded in these principles and include comprehensive local wound care with surgical debridement, the use of dressings that maintain a moist wound environment, mechanical off-loading, vascular evaluation, infection management, and strict glycemic control [41]. Nevertheless, despite ongoing advancements, DFU treatment remains a significant clinical challenge, largely due to the limited efficacy of available therapies.

In a phase I pilot study [42], patients with PAD and incurable DFU were treated with human UC-MSCs, in combination with conventional treatments. Topical and intravenous UC-MSC administrations were well tolerated, with only 16% of patients experiencing mild transient fever within 3 days, likely related to intravenous MSC infusion. Remarkably, more than 95% of the ulcer area showed significant closure within 1.5 months after treatment, and no patients required amputation due to DFU recurrence over a 3-year follow-up. The symptoms of CLI were alleviated and the duration of re-hospitalization for DFU was 2 years [42]. In another phase 1/2 trial [43], the safety and efficacy of allogeneic BM-MSC derivatives combined with wound dressings were evaluated in patients with grade 1 and 2 DFUs. Patients with chronic non-healing ulcers were randomized to receive MSC derivatives, MSCs, or conventional treatment with dressing. The wound-healing process was evaluated until the complete closure of the ulcers. Compared to conventional treatments, MSC derivatives or MSCs achieved greater wound closure, enhanced skin regeneration rate and greater ulcer-free survival, all without significant adverse events [43].

Leg ulcers of the lower extremities can arise from a variety of underlying etiologies, including chronic venous insufficiency, arterial disease, diabetes, pressure injuries, trauma, allergic reactions, and inflammatory conditions. Among these, chronic venous leg ulcers are particularly vulnerable to microbial colonization and subsequent infection, which may lead to significant complications such as delayed wound healing, ulcer expansion, severe pain, and systemic infection. Timely and accurate assessment, along with evidence-based management, is essential to optimize clinical outcomes in affected patients. Key principles in the management of chronic venous leg ulcers include identifying and addressing the underlying etiology, evaluating arterial perfusion, controlling systemic comorbidities and local infections, performing appropriate wound debridement, and minimizing mechanical pressure on the affected area. Standard wound care is typically initiated for a four-week period; however, approximately 25% of chronic venous leg ulcers do not demonstrate significant healing within this timeframe. Furthermore, even with the application of advanced therapeutic strategies, over 60% of ulcers fail to achieve complete healing. Ulcers that do not show at least a 50% reduction in area after 4 weeks of standard care should be re-evaluated, and advanced therapies may be considered in the absence of other underlying pathologies [44]. Despite the growing use of novel wound care technologies, robust evidence supporting their efficacy remains limited, with few interventions rigorously evaluated in randomized controlled trials. Notably, a randomized, controlled, double-blinded clinical trial conducted following full FDA approval for an investigational new drug evaluated the efficacy of BM-MSCs delivered in fibrin matrix, fibrin spray, or saline in patients with chronic wounds [21]. MSC treatment accelerated healing rates by about tenfold compared to saline or fibrin treatments, with no serious adverse events. Instead, a bioengineered bilayered living cellular construct skin substitute made of MSCs was also tested in a randomized controlled trial [45]. Compared to standard of care, the living cellular construct displayed a shift from the chronic non-healing ulcer to a healing milieu resembling that of an acute healing wound. Ulcers treated with the living cellular construct revealed modulation of inflammation, growth factor signaling, keratinocyte activation, and attenuation of Wnt/β-catenin signaling.

While MSC-based therapies show promising results for wound healing, several challenges remain. One major obstacle is the lack of universally accepted criteria for defining the MSC phenotype and functional properties in clinical applications. More trials are needed to demonstrate the potential therapeutic benefits of MSCs in vast cohorts of patients. Furthermore, significant variability in MSC populations, delivery protocols, and wound models across studies complicates the interpretation of results, making it difficult to assess the optimal cell origin, quantity, and delivery method. Additionally, there is limited evidence to suggest that MSCs directly differentiate into resident cutaneous cells during wound healing. However, accumulating evidence points to the role of MSCs in secreting bioactive factors through endocrine and paracrine signaling, which helps reduce inflammation and promote tissue repair. Based on these findings, recent research has shifted toward exploring the EVs as carriers of bioactive factors released by MSCs during wound healing.

Extracellular Vesicles in Chronic Wound Management

As previously mentioned, the therapeutic capacity of MSCs in wound healing and tissue regeneration is also associated with their paracrine activity, through the release of soluble factors and EVs that act by stimulating endogenous repair and regeneration mechanisms.

EVs are lipid bilayer-enclosed particles secreted by all cell types, including MSCs. According to the *International Society for Extracellular Vesicles* (ISEV), EVs are defined as membrane-bound structures with a diameter of less than a micrometer and incapable of self-replication [46]. Since EVs play essential roles in intercellular communication and are ubiquitously present in the extracellular space, they have attracted considerable attention for their diagnostic and therapeutic potential [47].

EVs are broadly categorized based on their size and biogenesis into small EVs (<200 nm) and large EVs (>200 nm). More specifically, they include three primary subtypes: apoptotic bodies, microvesicles, and exosomes. These subpopulations differ in origin, molecular composition, and functional roles [46]. Apoptotic bodies (>1000 nm), the largest subtype of EVs, are generated via plasma membrane blebbing during programmed cell death. These vesicles encapsulate cellular contents, including nuclear fragments and organelles, and can be internalized by recipient cells. They contribute to a range of biological processes, including tissue regeneration, embryonic development, immune modulation, and tumor progression [48]. Microvesicles (100–1000 nm in diameter), also known as ectosomes, originate through outward budding and fission of the plasma membrane at lipid-enriched microdomains. This process is tightly regulated by cytoskeletal remodeling, calcium influx, and enzymatic signaling pathways, and can be triggered by external stimuli such as hypoxia, oxidative stress, irradiation, or mechanical shear forces.

Microvesicles express distinct markers such as CD40 ligand, Annexin A1, and Annexin V [49]. Exosomes, typically ranging from 30 to 100 nm in diameter, are generated within the endosomal system. Their biogenesis begins with the inward budding of the limiting membrane of early endosomes, resulting in the formation of intraluminal vesicles (ILVs) within multivesicular bodies (MVBs). Upon maturation, MVBs may either fuse with lysosomes for degradation or with the plasma membrane to release ILVs as exosomes into the extracellular space. This process is orchestrated by the Endosomal Sorting Complex Required for Transport (ESCRT) machinery, in conjunction with accessory proteins such as Alix, TSG101, VPS4, and VTA-1. Tetraspanins, including CD63, CD9, and CD81, as well as proteins such as FLOT-1, are commonly used as exosomal markers due to their critical roles in vesicle formation and cargo selection [50].

Although EVs originate via distinct biogenetic pathways, all subtypes are capable of transporting a diverse array of bioactive molecules that directly reflect the metabolic and functional state of the parental cells. MSC-derived EVs (MSC-EVs) carry a variety of molecular cargoes, including metabolites, proteins (e.g., chemokines, growth factors), DNA, and various forms of RNA such as microRNAs (miRNAs), circular RNAs (circRNAs), and long non-coding RNAs (lncRNAs). These vesicles can be internalized by recipient cells, including fibroblasts, keratinocytes, immune cells, and endothelial cells, thereby promoting wound healing through mechanisms such as enhanced local circulation, stimulation of angiogenesis, attenuation of inflammatory responses, and recruitment of endogenous stem cells to the site of injury [51]. In addition, they facilitate re-epithelialization and enhance ECM remodeling by upregulating the synthesis of collagen and elastin. They also contribute to immune regulation by promoting macrophage polarization toward the anti-inflammatory M2 phenotype [52]. Furthermore, the therapeutic efficacy of MSC-EVs can be augmented through various strategies, including the targeted engineering of EV content, preconditioning of MSCs, or modification of surface receptors to enhance targeting and delivery efficiency [53].

Several studies attribute their regenerative efficacy primarily to their miRNA content. For example, UC-MSC-EVs are enriched in miR-21-5p, miR-125b-5p, and miR-150-5p, while EVs from decidua-derived MSCs are abundant in miR-23a-3p and miR-146a-5p. These miRNAs contribute to wound repair by various mechanisms: miR-21 and miR-150 suppress PTEN expression, enhancing tissue regeneration; miR-125 downregulates p53 to reduce hypoxia-induced apoptosis; and miR-23 and miR-146 inhibit TGF-β signaling, thereby preventing myofibroblast differentiation [54]. Additionally, AD-MSC-EVs overexpressing miR-146a-5p exhibit pro-angiogenic effects by targeting JAZF1 in endothelial cells [55], while bone BM-MSC-EVs enriched with miR-4645-5p promote diabetic wound healing via M2 macrophage polarization and inhibition of SREBP2 [56]. EVs engineered to carry miR-17-5p suppress cellular senescence and enhance endothelial and fibroblast proliferation and migration by targeting p21 and PTEN [57]. Likewise, AD-MSC-EVs loaded with miR-21-5p facilitate keratinocyte proliferation and migration through activation of the Wnt/β-catenin pathway and promote in vivo wound healing via enhanced re-epithelialization, collagen remodeling, angiogenesis, and vessel maturation [58].

MSC-EVs also deliver circRNAs, which function as miRNA sponges and regulate gene expression. For example, mmu_circ_0000250 in AD-MSC-EVs promotes diabetic wound healing through the mmu_circ_0000250/miR-128-3p/SIRT1 axis by sponging miR-128-3p and upregulating SIRT1 [59]. Similarly, mmu_circ_0001052 enhances angiogenesis in DFUs via the miR-106a-5p/FGF4/p38MAPK pathway. In endothelial cells, these EVs inhibit apoptosis and miR-106a-5p expression while promoting proliferation, migration, angiogenesis, and the expression of FGF4, VEGF, and phospho-p38 [60].

lncRNAs are transcripts exceeding 200 nucleotides in length that do not encode proteins but play critical roles in the regulation of gene expression and various cellular processes. These molecules exert their functions through interactions with DNA, RNA, and proteins, thereby influencing chromatin architecture, transcriptional activity, and post-transcriptional modifications. One such lncRNA, metastasis-associated lung adenocarcinoma transcript 1 (MALAT1), has been implicated in endothelial cell migration and angiogenesis across multiple disease contexts. Recent evidence demonstrates that AD-MSC-EVs enhance the proliferation and migration of human skin fibroblasts, in part by upregulating MALAT1 expression. Silencing MALAT1 in human ADSCs resulted in reduced fibroblast viability and migration, increased apoptosis, and impaired angiogenesis, primarily through upregulation of miR-378a. Overexpression of miR-378a was shown to suppress fibroblast phenotypes by downregulating fibroblast growth factor 2 (FGF2), suggesting that MALAT1-containing EVs promote wound healing via the MALAT1/miR-378a/FGF2 regulatory axis [61]. Another lncRNA, H19, delivered by AD-MSC-EVs, has also been shown to accelerate cutaneous wound healing. EVs enriched in H19 promoted M2 macrophage polarization, which subsequently enhanced fibroblast proliferation, migration, and endothelial cell-mediated angiogenesis. Depletion of H19 from these EVs abrogated these effects. Mechanistically, miR-130b-3p, identified as a direct target of H19, was found to regulate peroxisome proliferator-activated receptor gamma (PPARγ) and signal transducer and activator of transcription 3 (STAT3), key mediators involved in macrophage polarization and tissue regeneration. These findings indicate that AD-MSC-EVs facilitate wound healing through the H19/miR-130b-3p/PPARγ/STAT3 axis, highlighting the therapeutic relevance of EV-carried lncRNAs in modulating the wound microenvironment [62].

Due to their unique biological and physicochemical properties, MSC-EVs represent an ideal approach for cell-free therapeutic in regenerative medicine, particularly tissue repair and wound healing. MSC-EVs offer a valid and safer alternative to conventional stem cell-based therapies. While preserving the non-immunogenic characteristics of their parent cells, MSC-EVs lack the ability to self-replicate, thereby mitigating the risk of tumorigenesis associated with live cell transplantation [63]. Furthermore, the lipid bilayer encapsulating EVs serves as a protective barrier against proteolytic degradation, facilitating the stable and efficient delivery of their bioactive cargo to recipient cells. Beyond their biological advantages, MSC-EVs also present distinct technical benefits over cell-based therapies. Their production, handling, and storage are more amenable to standardization and scalability. Notably, the structural resilience of MSC-EVs allows them to maintain integrity through freeze-thaw cycles, enabling long-term storage without significant loss of functionality or bioactivity [64].

To date, clinical trials evaluating the safety and efficacy of EVs in wound treatment remain limited and are largely ongoing (Table 14.2). Nonetheless, emerging clinical and preclinical data increasingly support the therapeutic potential of MSC-EVs in promoting wound healing and tissue regeneration. A prospective, double-blind, randomized, split-face clinical trial [65] evaluated the efficacy and safety of MSC-EVs as an adjunct to fractional CO_2 laser therapy in the treatment of atrophic acne scars. Twenty-five patients underwent three treatment sessions, with one side of the face treated with an MSC-EV gel and the other side with a control gel. After 12 weeks, the EV-treated side showed significantly greater clinical improvement. In addition, erythema and post-treatment downtime were significantly reduced on the EV-treated side, indicating enhanced recovery and patient tolerance. These findings suggest a synergistic effect when combining MSC-EVs with laser resurfacing for scar management. Similarly, a case series exploring the use of MSC-EVs in routine clinical practice demonstrated rapid and effective wound healing following both aesthetic procedures and traumatic injuries [66]. Topical application of these EVs resulted in accelerated recovery, reduced pain, erythema, swelling, and downtime, with no adverse events reported. Notably, patients who had previously undergone comparable procedures without EV-therapy reported superior outcomes with EV use. These preliminary findings advocate for further controlled studies to validate the efficacy and broaden clinical applications of MSC-derived EVs in dermatologic and aesthetic medicine.

In a separate case report, MSC-EVs were successfully used to treat a chronic pressure ulcer in a patient unresponsive to 9 months of conventional care [67]. Following six subcutaneous EV injections over an 8-week period, complete wound closure was achieved, underscoring the regenerative potential of MSC-EVs in complex and refractory wound cases. Given the high recurrence rates and surgical limitations in pressure ulcer management, this cell-free therapeutic approach offers a promising alternative that warrants systematic clinical evaluation.

Collectively, these studies underscore the growing potential of EV-based therapies across a wide range of clinical applications, including chronic wound management. However, they also highlight the pressing need for larger, well-controlled trials to refine delivery strategies and establish standardized protocols for EV-based regenerative treatments.

Preclinical in vitro and in vivo studies have sought to enhance the delivery and administration of EVs. Despite these efforts, direct injection methods such as intravenous, intraperitoneal, subcutaneous, or intramuscular administration have consistently demonstrated rapid clearance from circulation, with EVs accumulating in the liver, spleen, lungs, and gastrointestinal tract as early as 2 h post-administration. Topical applications may result in even shorter EV half-lives due to environmental exposure and fluid-mediated elimination [68].

Table 14.2 Clinical trials exploring the safety and efficacy of extracellular vesicles in the treatment of chronic wounds

NCT number	Study phase	Study type	EV source	Condition	Study aim	Measured outcomes	Study start	Status
NCT06812637	Phase 1	Interventional, prospective, randomized, double-blind, placebo-controlled clinical trial	Wharton's Jelly-derived MSCs	DFU	Evaluation of safety and efficacy	Percent of wound closure compared to the baseline after 20 weeks	2024	Completed
NCT05475418	Pilot	Interventional, Single Group Assignment	AD-MSCs	Chronic wound	Evaluation of the safety and efficacy of EV mixed with sterile hydrogel and applied directly to the wound surface and covered with an inert protective dressing.	Percentage of wound healing at 4 weeks.	2022	Completed
NCT06825884	Phase 1	Interventional, Single Group Assignment	Not specified	DFU	Safety and efficacy evaluation of injection into DFU not-healed with standard of care.	Percentage of wound area reduction after 12 weeks	2024	Active
NCT04652531	N/A	Interventional, Non-Randomized	Autologous Serum-derived EV	Venous Trophic Lesions	Evaluation of the safety and efficacy of peri-wound injected EV in Venous Trophic Lesions Not Responsive to Conventional Treatments	Changes in the ulcer area and pain perception from baseline to 8 weeks	2020	Recruiting
NCT06319287	Phase 2	Interventional, Multi-center, Open Label, Randomized, Controlled Clinical Trial	Purified EV Product (lyophilized powder derived from apheresed platelets)	DFU	Evaluation of safety and efficacy	Percentage of wound area reduction, wound characteristics, pain perception, and quality of life after 12 weeks	2024	Recruiting

Given these challenges, extending the half-life of EVs at the target site is crucial to achieving therapeutic efficacy. In response, research has increasingly focused on the development of biomaterials designed to incorporate EVs and enable their sustained, localized release [69]. Among these biomaterials, hydrogels have emerged as particularly promising candidates. Their favorable properties include high porosity, biocompatibility, biodegradability, and a structure that mimics natural ECM. Furthermore, their tunable mechanical and physical characteristics make them highly versatile for therapeutic applications [70]. Hydrogels not only provide a barrier against microbial infection but also serve as multifunctional platforms that support tissue regeneration throughout their degradation process. Once applied, they integrate seamlessly with the wound bed, offering adhesion sites that support new tissue formation, while their degradation timeline can be synchronized with tissue growth [71]. The incorporation of EVs into hydrogels offers a protective matrix that protects EVs from premature degradation while facilitating targeted and localized delivery to the wound site. This approach not only enhances the stability and therapeutic efficacy of EVs but also significantly reduces the required dosage compared to systemic administration, which often suffers from low bioavailability and necessitates repeated dosing [72, 73]. Various composite hydrogels have been developed for the fabrication of advanced wound dressings that integrate EVs (Table 14.3).

Table 14.3 Animal model study exploring the efficacy of extracellular vesicles in the treatment of wounds

EV Source	Hydrogel	Animal model	Main results	References
AD-MSCs	Injectable adhesive thermosensitive pH-responsive multifunctional polysaccharide-based hydrogel	Full-thickness cutaneous wound in STZ-induced diabetic mouse	Enhanced cell proliferation, granulation tissue formation, collagen deposition, remodeling, and re-epithelialization. Fast healing with less scar tissue formation	[71]
AD-MSCs	Injectable biocompatible thermosensitive Pluronic F-127 hydrogel	Full-thickness cutaneous wound in a mouse model	Promoted skin wound healing by enhancing re-epithelialization and collagen synthesis, and reducing inflammation	[78]
AD-MSCs	Bioprintable methacrylated hyaluronic acid hydrogel	Pressure ulcer in a genetic mouse model of type 2 diabetes (db/db)	Improved wound epithelialization, angiogenesis, and innervation due to the controlled release of AD-MSC-EVs from hydrogel, enhancing vesicle efficacy	[79]

(continued)

Table 14.3 (continued)

EV Source	Hydrogel	Animal model	Main results	References
Metformin-treated AD-MSCs	Hydrogel made of decellularized adipose ECM and methacrylated hyaluronic acid	Full-thickness cutaneous wound in a mouse model	Transport of AD-MSC-derived mitochondria into the wound. Enhanced ATP production, reduced ROS levels, and promoted polarization of macrophages toward M2-phenotype	[80]
BM-MSCs	Multifunctional hydrogel loaded with magnesium-doped bioactive glass (MgBG)	Full-thickness skin defect in STZ-induced diabetic rats	Release of both EVs and MgBG into the wound Significant antibacterial and anti-inflammatory activity. Promoted cell migration, angiogenesis, and collagen deposition	[81]
UC-MSCs	Antibacterial GelMA/PLMA hydrogel	Full-thickness cutaneous wound in STZ-induced diabetic mouse	Promoted healing of an infected diabetic wound. Enhanced angiogenesis and anti-inflammatory effect	[82]
UC-MSCs	Thermosensitive Pluronic F-127 hydrogel	Full-thickness cutaneous wound in STZ-induced diabetic rat	Significantly accelerated wound closure rate, enhanced regeneration of granulation tissue, and upregulated expression of growth factors	[83]
UC-MSCs	Biodegradable, thermos- and pH-sensitive Carboxymethyl chitosan hydrogel	Full-thickness skin defect mouse model	Improved wound closure, re-epithelialization rates, and collagen deposition	[84]
UC-MSCs	Collagen I/ carboxymethyl chitosan hydrogel	Full-thickness skin defect mouse model	Superior wound healing efficiency with 15% faster wound closure compared to controls. Increased fibroblast proliferation, alleviated inflammation responses, and promoted angiogenesis	[85]
UC-MSCs	Thermosensitive Pluronic F-127/ HPMC/chitosan/ collagen I hydrogel	Full-thickness cutaneous wound in STZ-induced diabetic mouse	Enhanced collagen and vascular regeneration and decreased local inflammation	[86]

(continued)

Table 14.3 (continued)

EV Source	Hydrogel	Animal model	Main results	References
BM-MSCs or AD-MSCs or UC-MSCs	Biocompatible methacrylated collagen hydrogel	Full-thickness excisional wound in rat model	The UC-MSC-EVs exhibited superior therapeutic potential. EV-loaded hydrogel promoted tissue repair by enhancing vascular reconstruction, collagen deposition, and inflammation resolution	[87]
UC-MSCs	Polyvinyl alcohol-based hydrogel cross-linked with ROS-responsive linkers	Full-thickness cutaneous wound in STZ-induced diabetic rat	ROS absorption and EV release. Promoted wound healing through the down-regulation of pro-inflammatory cytokines and M1-macrophages, and enhanced angiogenesis, neurogenesis, and collagen synthesis	[88]

AD-MSCs adipose-derived mesenchymal stem cells, *BM-MSCs* bone marrow-derived MSCs, *HPMC* hydroxypropyl methylcellulose, *GelMA* gelatin methacryloyl, *PLMA* poly-L-lysine methacryloyl, *STZ* streptozotocin, *UC-MSCs* umbilical cord-derived MSCs

Conclusion and Future Directions

Chronic wounds arise from a combination of factors, including impaired macrophage and fibroblast function, dysregulated cytokine signaling, and insufficient angiogenesis. Existing treatments often fail to address the multifactorial nature of these conditions. Non-healing wounds remain a significant and growing healthcare challenge, particularly in aging populations, making the need for more effective treatments urgent.

Stem cell-based therapies offer a novel paradigm in the management of chronic wounds, particularly for patients who have exhausted conventional treatment options. Evidence from both preclinical investigations and clinical trials has demonstrated that MSCs play pivotal roles in promoting angiogenesis, tissue regeneration, and wound repair. These therapeutic effects are mediated through a combination of direct cellular incorporation into damaged tissues, paracrine signaling pathways, and immunomodulatory mechanisms. The synergistic action of these processes contributes to accelerated healing and improvements in patient quality of life.

The therapeutic potential of stem cell-based interventions for chronic wound management is substantial; however, their broader clinical integration remains hindered by persistent challenges. These include issues related to consistent cell sourcing, scalable expansion techniques, optimized delivery strategies, and the absence of universally accepted protocols for cell characterization and potency testing. Additionally, the absence of international, multicenter clinical registries hinders efforts to harmonize outcome measures and to establish comprehensive long-term safety and efficacy profiles.

Future research should focus on conducting large-scale, well-designed clinical trials to rigorously evaluate efficacy and safety, alongside efforts to define optimal cell origin, dosages, and administration regimens. In addition, strategies aimed at enhancing cell survival, engraftment, and functional integration within ischemic environments are critical. Stem cells exhibit considerable heterogeneity, and existing technologies are insufficient to standardize or monitor cell characteristics at scale. Differences in donor source, tissue origin, and host conditions can significantly affect proliferative capacity, differentiation potential, and gene expression profiles. Autologous transplantation is preferable to reduce immunogenic risk; however, the regenerative functionality of stem cells may be compromised in elderly or diabetic individuals. Advances in manufacturing, including the development of off-the-shelf allogeneic cell products with consistent quality and potency, have the potential to significantly improve the widespread clinical application.

Several technical barriers have been successfully addressed, notably the reliance on fetal bovine serum in clinical-grade stem cell expansion. The development of serum-free and chemically defined media, often supplemented with synthetic growth factors and nutrients, has provided a more consistent and clinically acceptable alternative. Concurrently, the incorporation of xenogeneic-free or human-derived components has further reduced the risks associated with animal-derived products [74]. Bioreactor systems are increasingly employed for large-scale stem cell culture under these optimized conditions, offering standardized, reproducible, and scalable platforms suitable for therapeutic applications. Despite these advancements, cell expansion technologies remain time-consuming and cost-intensive, presenting ongoing challenges for widespread clinical implementation.

Advancement in stem cell delivery and implantation will be achieved with the integration of innovative technologies. The development of bioengineered combination therapies that incorporate optimized cell populations within biomaterial scaffolds, augmented with growth factors, can improve retention and therapeutic efficacy. An array of chemical and natural compounds has been identified to enhance stem cell transplantation and tethering while preserving regenerative potential [75]. Advanced imaging modalities such as contrast-enhanced ultrasound and multiparametric MRI will support immediate monitoring of engraftment and therapeutic response, facilitating precision medicine applications. Concurrently, artificial intelligence-based predictive models may aid in identifying appropriate candidates for cell therapy and tailoring treatment strategies based on individual clinical phenotypes.

Beyond biological and technical challenges, economic and structural barriers significantly hinder the advancement and clinical implementation of stem cell therapies. The conventional pharmaceutical development model is designed primarily for small-molecule drugs, and it is not readily applicable to the complexities of cell-based products. Regulatory frameworks remain dynamic and evolving, with agencies such as the US Food and Drug Administration adopting tiered oversight approaches aimed at balancing innovation with patient safety. Furthermore, the field continues to struggle with limited venture capital investment, largely due to the high financial risk and extended timelines associated with return on investment. The substantial upfront costs of development, combined with intricate intellectual property considerations and regulatory uncertainty, further constrain commercial scalability. Addressing these multifaceted challenges is essential to narrowing the gap between promising experimental outcomes and their successful clinical translation.

In parallel, EVs derived from stem cells are gaining traction as potent agents for chronic wound healing due to their favorable biological characteristics. EVs exhibit high tissue-targeting ability, biocompatibility, stability in circulation, and low immunogenicity, making them suitable for regenerative medicine. Their therapeutic functions include modulation of immune responses and inflammation, stimulation of angiogenesis, acceleration of cell proliferation, and regulation of collagen remodeling. These effects can be enhanced by incorporating EVs into hydrogels or dressings. Nevertheless, current preclinical studies show poor reproducibility in animal models, and limited clinical data exist regarding EV efficacy. Quality control is essential to ensure the consistency, safety, and efficacy of EV preparations. According to the ISEV, quality control parameters should include vesicle number, size distribution, surface marker expression, and purity. Nevertheless, universally accepted standards for EV characterization and potency assessment are lacking.

EVs inherit most of their biological properties from parent cells, but their biogenesis and mechanisms of action remain insufficiently understood. A key obstacle to clinical translation is the lack of standardization and reproducibility in EV preparation, purification, and characterization. The main EV isolation techniques are ultracentrifugation, precipitation, ultrafiltration, immunoaffinity capture, and size-exclusion chromatography (SEC). Despite ongoing advances, novel methods such as microfluidics and tangential flow filtration are underutilized. Studies also demonstrate significant variation in isolation protocols, resulting in inconsistent product quality. Among existing methods, only differential ultracentrifugation (dUC) and SEC are considered clinically acceptable [76].

Effective clinical application of MSC-EVs demands stringent quality control, including standardization of cell source, culture conditions, and isolation methods. However, no universally accepted procedures exist for the identification, storage, or transport of EVs. Producing sufficient quantities of high-quality EVs remains a major challenge due to low yields and the complex interplay between MSC origin, proliferative capacity, and the surrounding microenvironment. Standard two-dimensional culture systems are insufficient for large-scale production and may compromise MSC quality over multiple passages. Three-dimensional culture techniques may overcome these limitations, increasing EV yield while preserving function.

The potency of EVs, a critical factor for clinical application, depends on their molecular content, which varies with cellular conditions and metabolic state. Despite the growing body of research, the specific roles and mechanisms of EV cargo remain unclear. Most existing studies analyze single EV components or target specific skin cells, overlooking the systemic, multifactorial nature of wound healing. A more integrative approach is required to understand how different EV constituents interact with the complex cellular environment of chronic wounds.

Moreover, EVs have potential as drug delivery systems due to their nanoscale size, intrinsic biocompatibility, and low immunogenicity. Their ability to evade the reticuloendothelial system makes them superior to synthetic nanocarriers. Engineering strategies aimed at enhancing therapeutic outcomes include modifying parent cells, loading EVs with drugs, and improving delivery mechanisms [77]. Additionally, the combination of EVs with biomaterials such as scaffolds and wound dressings provides a conducive environment for cell growth and tissue regeneration.

Despite promising results, clinical translation is hampered by unresolved questions, including optimal cell sources, dosing regimens, and frequency of administration. Mouse models, frequently used in preclinical research, fail to replicate the complexity of human wound healing, particularly in conditions like keloid formation. Lastly, combination therapies, such as the use of MSC-EVs with small molecule activators, may provide synergistic effects and accelerate healing. Thus, future efforts should focus on developing standardized protocols, scaling up production, and conducting rigorous clinical trials to validate the safety and efficacy of EV-based therapies in wound healing.

In summary, both stem cell-based and EV-based therapies hold transformative potential for the treatment of chronic wounds. Realizing this potential will require multidisciplinary collaboration to overcome current limitations in manufacturing, quality control, regulatory oversight, and mechanistic understanding. As research continues to advance, these biologic therapies may become integral components of personalized, effective, and accessible treatment strategies for patients with chronic, non-healing wounds.

References

1. Bowers S, Franco E (2020) Chronic wounds: evaluation and management. Am Fam Physician 101(3):159–166
2. Casado-Díaz A, Quesada-Gómez JM, Dorado G (2020) Extracellular vesicles derived from mesenchymal stem cells (MSC) in regenerative medicine: applications in skin wound healing. Front Bioeng Biotechnol 8:146
3. Landén NX, Li D, Ståhle M (2016) Transition from inflammation to proliferation: a critical step during wound healing. Cell Mol Life Sci 73(20):3861–3885
4. Dash SP, Gupta S, Sarangi PP (2024) Monocytes and macrophages: origin, homing, differentiation, and functionality during inflammation. Heliyon 10(8):e29686
5. Eming SA, Martin P, Tomic-Canic M (2014) Wound repair and regeneration: mechanisms, signaling, and translation. Sci Transl Med 6(265):265sr6
6. Wilkinson HN, Hardman MJ (2020) Wound healing: cellular mechanisms and pathological outcomes. Open Biol 10(9):200223
7. El Ghalbzouri A, Hensbergen P, Gibbs S, Kempenaar J, van der Schors R, Ponec M (2004) Fibroblasts facilitate re-epithelialization in wounded human skin equivalents. Lab Investig 84(1):102–112

8. Johnson KE, Wilgus TA (2014) Vascular endothelial growth factor and angiogenesis in the regulation of cutaneous wound repair. Adv Wound Care (New Rochelle) 3(10):647–661
9. Xue M, Jackson CJ (2015) Extracellular matrix reorganization during wound healing and its impact on abnormal scarring. Adv Wound Care (New Rochelle) 4(3):119–136
10. Dominici M, Le Blanc K, Mueller I, Slaper-Cortenbach I, Marini F, Krause D et al (2006) Minimal criteria for defining multipotent mesenchymal stromal cells. The International Society for Cellular Therapy position statement. Cytotherapy 8(4):315–317
11. Morganti C, Bonora M, Marchi S, Ferroni L, Gardin C, Wieckowski MR et al (2020) Citrate mediates crosstalk between mitochondria and the nucleus to promote human mesenchymal stem cell in vitro osteogenesis. Cells 9(4)
12. Lambertini E, Penolazzi L, Morganti C, Lisignoli G, Zini N, Angelozzi M et al (2015) Osteogenic differentiation of human MSCs: specific occupancy of the mitochondrial DNA by NFATc1 transcription factor. Int J Biochem Cell Biol 64:212–219
13. Ferroni L, Gardin C, Bellin G, Vindigni V, Mortellaro C, Zavan B (2019) Bovine pericardium membrane as new tool for mesenchymal stem cells commitment. J Tissue Eng Regen Med 13(10):1805–1814
14. Morganti C, Missiroli S, Lebiedzinska-Arciszewska M, Ferroni L, Morganti L, Perrone M et al (2019) Regulation of PKCβ levels and autophagy by PML is essential for high-glucose-dependent mesenchymal stem cell adipogenesis. Int J Obes 43(5):963–973
15. Ferroni L, De Francesco F, Pinton P, Gardin C, Zavan B (2022) Methods to isolate adipose tissue-derived stem cells. Methods Cell Biol 171:215–228
16. Bressan E, Ferroni L, Gardin C, Pinton P, Stellini E, Botticelli D et al (2012) Donor age-related biological properties of human dental pulp stem cells change in nanostructured scaffolds. PLoS One 7(11):e49146
17. Guillamat-Prats R (2021) The role of MSC in wound healing, scarring and regeneration. Cells 10(7)
18. Nedeau AE, Bauer RJ, Gallagher K, Chen H, Liu ZJ, Velazquez OC (2008) A CXCL5- and bFGF-dependent effect of PDGF-B-activated fibroblasts in promoting trafficking and differentiation of bone marrow-derived mesenchymal stem cells. Exp Cell Res 314(11–12):2176–2186
19. Ferroni L, Gardin C, Dalla Paola L, Campo G, Cimaglia P, Bellin G et al (2019) Characterization of dermal stem cells of diabetic patients. Cells 8(7)
20. Askø Andersen J, Rasmussen A, Frimodt-Møller M, Engberg S, Steeneveld E, Kirketerp-Møller K et al (2022) Novel topical allogeneic bone-marrow-derived mesenchymal stem cell treatment of hard-to-heal diabetic foot ulcers: a proof of concept study. Stem Cell Res Ther 13(1):280
21. Falanga V, Grada A, Otero-Vinas M, Lin X, Yufit T, Fiore D et al (2022) Autologous cultured bone marrow-derived mesenchymal stem cells in a fibrin spray to treat venous ulcers: a randomized controlled double-blind pilot study. Surg Technol Int 40:47–54
22. Li C, Zhao H, Cheng L, Wang B (2021) Allogeneic vs. autologous mesenchymal stem/stromal cells in their medication practice. Cell Biosci 11(1):187
23. Fu X, Fang L, Li X, Cheng B, Sheng Z (2006) Enhanced wound-healing quality with bone marrow mesenchymal stem cells autografting after skin injury. Wound Repair Regen 14(3):325–335
24. Wu S, Sun S, Fu W, Yang Z, Yao H, Zhang Z (2024) The role and prospects of mesenchymal stem cells in skin repair and regeneration. Biomedicine 12(4)
25. Du F, Liu M, Wang J, Hu L, Zeng D, Zhou S et al (2023) Metformin coordinates with mesenchymal cells to promote VEGF-mediated angiogenesis in diabetic wound healing through Akt/mTOR activation. Metabolism 140:155398
26. Abumaree MH, Al Jumah MA, Kalionis B, Jawdat D, Al Khaldi A, Abomaray FM et al (2013) Human placental mesenchymal stem cells (pMSCs) play a role as immune suppressive cells by shifting macrophage differentiation from inflammatory M1 to anti-inflammatory M2 macrophages. Stem Cell Rev Rep 9(5):620–641

27. Gardin C, Ferroni L, Bellin G, Rubini G, Barosio S, Zavan B (2018) Therapeutic potential of autologous adipose-derived stem cells for the treatment of liver disease. Int J Mol Sci 19(12)
28. Camponogara F, Zanotti F, Trentini M, Tiengo E, Zanolla I, Pishavar E et al (2022) Biomaterials for regenerative medicine in Italy: brief state of the art of the principal research centers. Int J Mol Sci 23(15)
29. Bressan E, Ferroni L, Gardin C, Sbricoli L, Gobbato L, Ludovichetti FS et al (2014) Graphene based scaffolds effects on stem cells commitment. J Transl Med 12:296
30. Casadei A, Epis R, Ferroni L, Tocco I, Gardin C, Bressan E et al (2012) Adipose tissue regeneration: a state of the art. J Biomed Biotechnol 2012:462543
31. Gardin C, Vindigni V, Bressan E, Ferroni L, Nalesso E, Puppa AD et al (2011) Hyaluronan and fibrin biomaterial as scaffolds for neuronal differentiation of adult stem cells derived from adipose tissue and skin. Int J Mol Sci 12(10):6749–6764
32. Kerstan A, Dieter K, Niebergall-Roth E, Klingele S, Jünger M, Hasslacher C et al (2022) Translational development of ABCB5. Stem Cell Res Ther 13(1):455
33. Moon KC, Suh HS, Kim KB, Han SK, Young KW, Lee JW et al (2019) Potential of allogeneic adipose-derived stem cell-hydrogel complex for treating diabetic foot ulcers. Diabetes 68(4):837–846
34. Dash NR, Dash SN, Routray P, Mohapatra S, Mohapatra PC (2009) Targeting nonhealing ulcers of lower extremity in human through autologous bone marrow-derived mesenchymal stem cells. Rejuvenation Res 12(5):359–366
35. Aday AW, Matsushita K (2021) Epidemiology of peripheral artery disease and polyvascular disease. Circ Res 128(12):1818–1832
36. Olin JW, Sealove BA (2010) Peripheral artery disease: current insight into the disease and its diagnosis and management. Mayo Clin Proc 85(7):678–692
37. Shirbaghaee Z, Heidari Keshel S, Rasouli M, Valizadeh M, Hashemi Nazari SS, Hassani M et al (2023) Report of a phase 1 clinical trial for safety assessment of human placental mesenchymal stem cells therapy in patients with critical limb ischemia (CLI). Stem Cell Res Ther 14(1):174
38. Gupta PK, Chullikana A, Parakh R, Desai S, Das A, Gottipamula S et al (2013) A double blind randomized placebo controlled phase I/II study assessing the safety and efficacy of allogeneic bone marrow derived mesenchymal stem cell in critical limb ischemia. J Transl Med 11:143
39. Das AK, Bin Abdullah BJ, Dhillon SS, Vijanari A, Anoop CH, Gupta PK (2013) Intra-arterial allogeneic mesenchymal stem cells for critical limb ischemia are safe and efficacious: report of a phase I study. World J Surg 37(4):915–922
40. Emanuelli T, Burgeiro A, Carvalho E (2016) Effects of insulin on the skin: possible healing benefits for diabetic foot ulcers. Arch Dermatol Res 308(10):677–694
41. Everett E, Mathioudakis N (2018) Update on management of diabetic foot ulcers. Ann N Y Acad Sci 1411(1):153–165
42. Zhang C, Huang L, Wang X, Zhou X, Zhang X, Li L et al (2022) Topical and intravenous administration of human umbilical cord mesenchymal stem cells in patients with diabetic foot ulcer and peripheral arterial disease: a phase I pilot study with a 3-year follow-up. Stem Cell Res Ther 13(1):451
43. Arango-Rodríguez ML, Solarte-David VA, Becerra-Bayona SM, Callegari E, Paez MD, Sossa CL et al (2022) Role of mesenchymal stromal cells derivatives in diabetic foot ulcers: a controlled randomized phase 1/2 clinical trial. Cytotherapy 24(10):1035–1048
44. Ren SY, Liu YS, Zhu GJ, Liu M, Shi SH, Ren XD et al (2020) Strategies and challenges in the treatment of chronic venous leg ulcers. World J Clin Cases 8(21):5070–5085
45. Stone RC, Stojadinovic O, Rosa AM, Ramirez HA, Badiavas E, Blumenberg M et al (2017) A bioengineered living cell construct activates an acute wound healing response in venous leg ulcers. Sci Transl Med 9(371)
46. Welsh JA, Goberdhan DCI, O'Driscoll L, Buzas EI, Blenkiron C, Bussolati B et al (2024) Minimal information for studies of extracellular vesicles (MISEV2023): from basic to advanced approaches. J Extracell Vesicles 13(2):e12404

47. Gardin C, Ferroni L, Chachques JC, Zavan B (2020) Could mesenchymal stem cell-derived exosomes be a therapeutic option for critically ill COVID-19 patients? J Clin Med 9(9)
48. Gardin C, Ferroni L, Leo S, Tremoli E, Zavan B (2022) Platelet-derived exosomes in atherosclerosis. Int J Mol Sci 23(20)
49. Leo S, Tremoli E, Ferroni L, Zavan B (2023) Role of epicardial adipose tissue secretome on cardiovascular diseases. Biomedicine 11(6)
50. Bellin G, Gardin C, Ferroni L, Chachques JC, Rogante M, Mitrecic D et al (2019) Exosome in cardiovascular diseases: a complex world full of hope. Cells 8(2)
51. Zhou C, Zhang B, Yang Y, Jiang Q, Li T, Gong J et al (2023) Stem cell-derived exosomes: emerging therapeutic opportunities for wound healing. Stem Cell Res Ther 14(1):107
52. He X, Dong Z, Cao Y, Wang H, Liu S, Liao L et al (2019) MSC-derived exosome promotes M2 polarization and enhances cutaneous wound healing. Stem Cells Int 2019:7132708
53. Qiu X, Liu J, Zheng C, Su Y, Bao L, Zhu B et al (2020) Exosomes released from educated mesenchymal stem cells accelerate cutaneous wound healing via promoting angiogenesis. Cell Prolif 53(8):e12830
54. Pan W, Wang W, Wang P, Chen D, Liu S, Zhang L et al (2024) Sustained delivery of extracellular vesicles using UiO-66-NH2 crosslinked hydrogel for accelerating chronic diabetic wound-healing. Mater Des 238:112688
55. Che D, Xiang X, Xie J, Chen Z, Bao Q, Cao D (2024) Exosomes derived from adipose stem cells enhance angiogenesis in diabetic wound via miR-146a-5p/JAZF1 axis. Stem Cell Rev Rep 20(4):1026–1039
56. Shi Y, Wang S, Wang K, Yang R, Liu D, Liao H et al (2024) Relieving macrophage dysfunction by inhibiting SREBP2 activity: a hypoxic mesenchymal stem cells-derived exosomes loaded multifunctional hydrogel for accelerated diabetic wound healing. Small 20(25):e2309276
57. Wei Q, Su J, Meng S, Wang Y, Ma K, Li B et al (2024) MiR-17-5p-engineered sEVs encapsulated in GelMA hydrogel facilitated diabetic wound healing by targeting PTEN and p21. Adv Sci (Weinh) 11(13):e2307761
58. Lv Q, Deng J, Chen Y, Wang Y, Liu B, Liu J (2020) Engineered human adipose stem-cell-derived exosomes loaded with miR-21-5p to promote diabetic cutaneous wound healing. Mol Pharm 17(5):1723–1733
59. Shi R, Jin Y, Hu W, Lian W, Cao C, Han S et al (2020) Exosomes derived from mmu_circ_0000250-modified adipose-derived mesenchymal stem cells promote wound healing in diabetic mice by inducing miR-128-3p/SIRT1-mediated autophagy. Am J Physiol Cell Physiol 318(5):C848–CC56
60. Liang ZH, Pan NF, Lin SS, Qiu ZY, Liang P, Wang J et al (2022) Exosomes from mmu_circ_0001052-modified adipose-derived stem cells promote angiogenesis of DFU via miR-106a-5p and FGF4/p38MAPK pathway. Stem Cell Res Ther 13(1):336
61. Pi L, Yang L, Fang BR, Meng XX, Qian L (2022) LncRNA MALAT1 from human adipose-derived stem cell exosomes accelerates wound healing via miR-378a/FGF2 axis. Regen Med 17(9):627–641
62. Li B, Qian L, Pi L, Meng X (2023) A therapeutic role of exosomal lncRNA H19 from adipose mesenchymal stem cells in cutaneous wound healing by triggering macrophage M2 polarization. Cytokine 165:156175
63. Zhang Y, Wu D, Zhou C, Bai M, Wan Y, Zheng Q et al (2024) Engineered extracellular vesicles for tissue repair and regeneration. Burns Trauma 12:tkae062
64. Bari E, Perteghella S, Di Silvestre D, Sorlini M, Catenacci L, Sorrenti M et al (2018) Pilot production of mesenchymal stem/stromal freeze-dried secretome for cell-free regenerative nanomedicine: a validated GMP-compliant process. Cells 7(11)
65. Kwon HH, Yang SH, Lee J, Park BC, Park KY, Jung JY et al (2020) Combination treatment with human adipose tissue stem cell-derived exosomes and fractional CO2 laser for acne scars: a 12-week prospective, double-blind, randomized, split-face study. Acta Derm Venereol 100(18):adv00310

66. Peredo M, Shivananjappa S (2024) Topical human mesenchymal stem cell-derived exosomes for acceleration of wound healing following tissue trauma and aesthetic procedures: a case series. J Drugs Dermatol 23(4):281–284
67. Messa GE, Tiongco RP, Lau FH (2022) Treatment of a recurrent ischial ulcer with injected exosomes. J Surg Case Rep 2022(6):rjac271
68. Kang M, Jordan V, Blenkiron C, Chamley LW (2021) Biodistribution of extracellular vesicles following administration into animals: a systematic review. J Extracell Vesicles 10(8):e12085
69. Chachques JC, Gardin C, Lila N, Ferroni L, Migonney V, Falentin-Daudre C et al (2021) Elastomeric cardiowrap scaffolds functionalized with mesenchymal stem cells-derived exosomes induce a positive modulation in the inflammatory and wound healing response of mesenchymal stem cell and macrophage. Biomedicine 9(7)
70. Lee JH (2018) Injectable hydrogels delivering therapeutic agents for disease treatment and tissue engineering. Biomater Res 22:27
71. Wang M, Wang C, Chen M, Xi Y, Cheng W, Mao C et al (2019) Efficient angiogenesis-based diabetic wound healing/skin reconstruction through bioactive antibacterial adhesive ultraviolet shielding nanodressing with exosome release. ACS Nano 13(9):10279–10293
72. Hernandez MJ, Gaetani R, Pieters VM, Ng NW, Chang AE, Martin TR et al (2018) Decellularized extracellular matrix hydrogels as a delivery platform for MicroRNA and extracellular vesicle therapeutics. Adv Ther (Weinh) 1(3)
73. Ferroni L, Gardin C, D'Amora U, Calzà L, Ronca A, Tremoli E et al (2022) Exosomes of mesenchymal stem cells delivered from methacrylated hyaluronic acid patch improve the regenerative properties of endothelial and dermal cells. Biomater Adv 139:213000
74. Gottipamula S, Muttigi MS, Kolkundkar U, Seetharam RN (2013) Serum-free media for the production of human mesenchymal stromal cells: a review. Cell Prolif 46(6):608–627
75. Kharbikar BN, Mohindra P, Desai TA (2022) Biomaterials to enhance stem cell transplantation. Cell Stem Cell 29(5):692–721
76. Akbar A, Malekian F, Baghban N, Kodam SP, Ullah M (2022) Methodologies to isolate and purify clinical grade extracellular vesicles for medical applications. Cells 11(2)
77. Cheng W, Xu C, Su Y, Shen Y, Yang Q, Zhao Y et al (2023) Engineered extracellular vesicles: a potential treatment for regeneration. iScience 26(11):108282
78. Zhou Y, Zhang XL, Lu ST, Zhang NY, Zhang HJ, Zhang J (2022) Human adipose-derived mesenchymal stem cells-derived exosomes encapsulated in pluronic F127 hydrogel promote wound healing and regeneration. Stem Cell Res Ther 13(1):407
79. Ferroni L, D'Amora U, Gardin C, Leo S, Dalla Paola L, Tremoli E et al (2023) Stem cell-derived small extracellular vesicles embedded into methacrylated hyaluronic acid wound dressings accelerate wound repair in a pressure model of diabetic ulcer. J Nanobiotechnol 21(1):469
80. Yao WD, Zhou JN, Tang C, Zhang JL, Chen ZY, Li Y et al (2024) Hydrogel microneedle patches loaded with stem cell mitochondria-enriched microvesicles boost the chronic wound healing. ACS Nano 18(39):26733–26750
81. Wang Z, Sun Z, Zhu S, Qin Z, Yin X, Ding Y et al (2025) A multifunctional hydrogel loaded with magnesium-doped bioactive glass-induced vesicle clusters enhances diabetic wound healing by promoting intracellular delivery of extracellular vesicles. Bioact Mater 50:30–46
82. Wang K, He Q, Yang M, Qiao Q, Chen J, Song J et al (2024) Glycoengineered extracellular vesicles released from antibacterial hydrogel facilitate diabetic wound healing by promoting angiogenesis. J Extracell Vesicles 13(11):e70013
83. Yang J, Chen Z, Pan D, Li H, Shen J (2020) Umbilical cord-derived mesenchymal stem cell-derived exosomes combined Pluronic F127 hydrogel promote chronic diabetic wound healing and complete skin regeneration. Int J Nanomedicine 15:5911–5926
84. Li Q, Gong S, Yao W, Yang Z, Wang R, Yu Z et al (2021) Exosome loaded genipin crosslinked hydrogel facilitates full thickness cutaneous wound healing in rat animal model. Drug Deliv 28(1):884–893

85. Wu Q, Guo Y, Li H, Zhang D, Wang S, Hou J et al (2024) Recombinant human collagen I/carboxymethyl chitosan hydrogel loaded with long-term released hUCMSCs derived exosomes promotes skin wound repair. Int J Biol Macromol 265(Pt 1):130843
86. He S, Li Z, Xie L, Lin R, Yan B, Li B et al (2025) Biomimetic gene delivery system coupled with extracellular vesicle-encapsulated AAV for improving diabetic wound through promoting vascularization and remodeling of inflammatory microenvironment. J Nanobiotechnol 23(1):242
87. Zeng X, Gan J, Huang D, Zhao Y, Sun L (2025) Recombinant human collagen hydrogels with different stem cell-derived exosomes encapsulation for wound treatment. J Nanobiotechnol 23(1):241
88. Zhang Y, Xu Y, Hu W, Ma X, Hu J, Ye Y et al (2025) An adhesive and self-healing ROS-scavenging hydrogel loading with hMSC-derived exosomes for diabetic wound healing. Int J Pharm 672:125315

Chapter 15
The Role of Bone Marrow-Derived Cells (BMDCs) During Wound Healing

Junko Okano and Hideto Kojima

Abstract Wounds can be caused by trauma, surgical procedures, or other medical conditions. Skin wounds compromise barrier function, increasing susceptibility to bacterial and viral infections that often delay healing and worsen the condition. Bone marrow-derived cells play important roles in wound healing. Platelets aid in hemostasis, while leukocytes and macrophages defend against cytotoxins and pathogens. Bone marrow-derived hematopoietic stem/progenitor cells that survive this hyperacute phase form tissue-specific vascular, neural, and lymphatic niches and conduct spatiotemporal remodeling according to the site of wound development. However, in severe injuries beyond biological wound healing abilities, systemic diseases, such as diabetes and arteriosclerotic diseases, or aging, acute healing is delayed, leading to chronic inflammation and fibrosis, hindering full recovery. In this chapter, we elucidate the nerves, blood vessels, and lymphatic pathways linking the bone marrow and skin as a communication axis, highlighting the cellular and molecular signals exchanged between them. It emphasizes the role of bone marrow-derived stem/progenitor cells in skin repair and provides insights into the mechanisms and countermeasures of delayed wound healing in chronic disease conditions.

Keywords Hematopoietic stem cells/progenitor cells (HSPCs) · Bone-marrow derived cells · Skin stem cells · Neovascularization · Lymphangiogenesis

---

J. Okano (✉)
Department of Plastic and Reconstructive Surgery, Shiga University of Medical Science, Otsu, Japan
e-mail: jokano@belle.shiga-med.ac.jp

H. Kojima (✉)
Department of Biocommunication Development, Shiga University of Medical Science, Otsu, Japan
e-mail: kojima@belle.shiga-med.ac.jp

F. De Francesco, K. Turksen (eds.), *Management and Strategies for Wound Healing*, Stem Cell Biology and Regenerative Medicine 79,
https://doi.org/10.1007/978-3-032-12442-5_15

Introduction

In vertebrates, mesodermal cells undergo hematopoiesis, vascularization, embryonic development, and organogenesis. Bone marrow-derived hematopoietic stem cells/progenitor cells (HSPCs) continuously migrate from embryogenesis to birth, supplying essential cells for vascularization and tissue development throughout the body. After birth, HSPCs remain in the bone marrow and continue to supply the progenitor cells necessary for blood and body tissues. Traditionally, tissue repair was believed to rely primarily on nearby resident stem cell (SCs); however, this remains poorly elucidated. Recent studies have clarified the cellular origins of the regeneration niche involved in wound healing. Particularly, HSPCs have been shown to migrate to wound sites, contribute to vascular niche reconstruction, and support tissue regeneration. At the wound site, blood vessels, nerves, and lymphatic vessels are simultaneously regenerated, and these systems form a step-by-step signaling loop with the bone marrow throughout the healing process. Additionally, fetal hematopoietic tissue cells, which are involved in embryonic development and organ formation, remain in the bone marrow even after birth and play a role in organ formation and the regeneration of injured tissue.

In this chapter, we focus on the origin of various SCs, especially HSPCs, which play a role in maintaining skin homeostasis, and discuss their role in wound healing. We also discuss recent advances in addressing delayed wound healing associated with diabetes, burns, trauma, and aging. As for the embryological origin and anatomical characteristics of bone marrow homeostasis, excellent review articles are recommended [1, 2, 49, 67].

Cell Players and Their Origins to Maintain Homeostasis in the Skin (Fig. 15.1)

The mammalian skin is composed of epidermis and dermis. Underneath the skin, the subcutaneous tissue, composed of adipocytes, is located, in which the network of blood vessels is abundant, and their vertical branches penetrate into the dermis. Keratinocytes, differentiating cells along the basal–apical axis, and fibroblasts are major populations in the epidermis and dermis, respectively. Conversely, there is a plethora of cell players derived from the bone marrow and skin SCs that maintain homeostasis. In this section, these cell populations, including their origins and the key molecules that determine their fate, will be discussed.

Skin SCs

Recent technological developments in science have led to the discovery of distinct populations of skin SC, such as interfollicular epidermis SCs (IFESCs), hair follicle SCs (HFSCs), sebaceous gland SCs, melanocyte SCs in the epidermis, neural crest

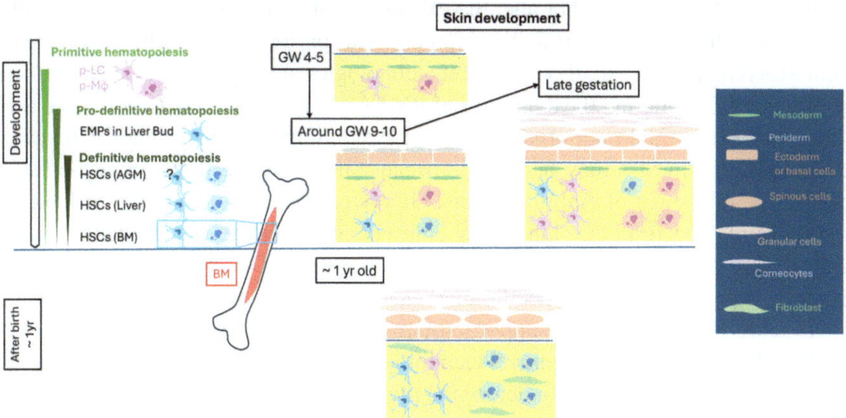

Fig. 15.1 Development of hematopoietic stem cells and their progeny (Langerhans cells (LC) and macrophages (MϕΦ)). During primitive hematopoiesis, the progeny of yolk-sac-derived primitive progenitors (p-LC or p-MΦ, colored in pink) migrate to the skin. As HSCs move into the liver, give rise through endothelial-to-hematopoietic transition (EHT), and migrate into the bone marrow (BM), the origin of LC or MΦ distributed in the skin alters, respectively. The progenies of LC or MΦ after the transient definitive hematopoiesis are colored in blue. EMP, erythroid-myeloid progenitors; HSCs, hematopoietic stem cells; GW, gestational weeks

SCs, mesenchymal SC-like dermal SCs, and dermal hematopoietic cells in the dermis [10]. Each SC name suggests the niche where they are located, because signaling from niche cells determines the fate of SCs.

Despite the existence of many local SCs, their origins remain unknown [10]. However, we speculate them from the first emergence of hematopoietic stem cells (HSCs) in the mammalian aorta-gonad-mesonephros (AGM) during embryogenesis [1]. The process of the first HSCs birth begins with the action on the arterial endothelium from niche cells, namely neural crest cells and mesenchymal stromal cells surrounding the endothelium of the AGM via catecholamines and endothelin. In response to niche cells, some subsets of the arterial endothelium begin to express *Notch1* and *Jagged1*. This cascade triggers the migration of yolk sac-derived megakaryocytes, which secrete TNF superfamily member 14. This, in turn, induces the expression of key transcription factors *Runx1* and *Gata2* in specific arterial endothelial cells, facilitating their transition into "pre-hematopoietic endothelial cells." The interaction between pre-hematopoietic endothelial cells and niche cells through *Shh* and *Wnt* signaling induces the endothelial-to-hematopoietic transition (EHT), marking the emergence of HSCs. Thus, HSCs originate from endothelial cells through the action of niche cells originating from ectodermal and/or mesodermal cells. Thereafter, during mid-embryogenesis, HSCs circulate to the liver, where they undergo expansion, maturation, and differentiation into both self-renewing HSCs and various progenitor cells (until approximately the second trimester in humans). Prenatally, expanded HSPCs migrate into all regions of the bone marrow and microscopically reside in their final niche, the arterioles [1, 49]. Therefore, the progeny of HSCs in the bone marrow, including HSCs per se are called bone marrow-derived cells (BMDCs).

Not only the behavior, but also the expression of genes in skin SCs is spatially and temporally dynamic during development. This partly explains why the origin of skin SCs remains unclear, though recent technologies are beginning to shed light on it. However, considering the dynamics of BMDCs, it is plausible to hypothesize that "the first" skin SCs do not originate from the skin during embryogenesis. Instead, the origin of skin SCs may be HSPCs because they migrate throughout the body during the mid-stage of embryogenesis. Otherwise, phenomena similar to EHT in the AGM may occur in the skin during embryogenesis. Notably, a substantial number of skin SCs, such as IFESCs, sebaceous gland SCs, melanocyte SCs, and HFSCs, express CD34 as a specific marker, which is also an important marker for HSPCs [1, 10]. This finding supports the hypothesis that skin SCs originate from HSPCs. Further studies using novel technologies and ideas to overcome the difficulties in this field are required to determine the origins of SCs.

Skin-Barrier Keepers

The skin is the largest organ in the body and functions as a barrier. It has two functions: one is the inside-out skin barrier that regulates proper evaporation, and the other is the outside-in skin barrier that protects against external stimuli, such as chemical compounds or pathogens. At the molecular level, the claudin family is known to play a crucial role in maintaining skin barrier homeostasis [14, 92]. Particularly, *Claudin-1* is essential for the inside-out skin barrier function, as its deletion results in fatal dehydration due to excessive transepidermal water loss. *Claudin-4* and *Claudin-7* are also important, being predominantly expressed in the suprabasal and basal layers of the epidermis, respectively [91, 92]. Furthermore, there is a plethora of immune cell populations to fight external invasions. These include macrophages, dermal dendritic cells (DCs), innate lymphoid cells (ILCs), γδ T cells in the dermis, Langerhans cells (LCs), and dendritic epidermal T cells (DETCs) in the epidermis in a steady state.

Macrophages (Fig. 15.1)

Macrophages, named from the Greek for "big eater," are phagocytes believed to be evolutionarily conserved from sponges to vertebrates [55]. Specifically, sponges, which have neither myeloid nor lymphoid lineage cells, have been demonstrated to have common clusters of genes in macrophages with those of vertebrates [55]. Macrophages developed independently from other HSC-derived lineages. This finding is supported by the independent origin of macrophages from HSCs. Between 4 and 5 gestational weeks (GW) in humans, primitive macrophages arising directly from yolk sac progenitors are observed in both the yolk sac and embryos. Therefore, these macrophages are the first residents in tissues [2]. Sequentially, a pro-definitive wave occurs in the hemogenic endothelial cells in the vessels of the yolk sac during the 3–5 GW. Erythroid-myeloid progenitors (EMP) develop from hemogenic endothelial cells through EHT. The expansion of EMPs in the yolk sac is followed by their movement to the liver via the circulation at 6 GW [96]. Enucleated erythrocytes, megakaryocytes, and monocytes differentiate from these EMPs. Additionally, monocytes reach individual organs and then differentiate into macrophages. In mice, c-Kit$^+$ CD41$^+$ CD16/32$^+$ CD93$^-$ EMPs can give rise to tissue-resident macrophages, while common EMPs are c-Kit$^+$, CD41$^+$, CD31$^+$, CD45$^+$, and CD61$^+$ [23, 24]. These new macrophages replace primitive macrophages and are tissue-resident macrophages generated by EMPs. However, after birth, fetal-derived macrophages are gradually replaced by adult monocyte- and bone marrow-derived macrophages, which migrate to the skin whenever homeostasis is disrupted by inflammation.

Dermal DCs

Dermal DCs are more capable of presenting antigens to T cells than macrophages. In the steady state, conventional DCs (cDCs), which originate from monocytes and dendritic cell progenitors (MDPs) generated by HSPCs in the bone marrow after birth, are present in the dermis [42]. In mice, cDCs are subdivided into two populations: cDC1 (Langerin$^+$) and cDC2 (CD11b$^+$), comprising approximately 10–20% and 70–80% of the cDC population, respectively [74]. cDC1 and cDC2 have distinct functions: cDC1 primarily activate CD8$^+$ T cells, while cDC2 mainly stimulate CD4$^+$ T cells.

ILCs

Innate ILCs are immune cells that maintain homeostasis in the skin, although they do not recognize antigens like DCs. ILCs are thought to be the counterparts of CD4$^+$ T cells with respect to the regulation of innate and immune cells via the secretion of cytokines and interleukins. ILCs are classified into three groups: ILC1 (Tbx21$^+$, producing IFN-γ), ILC2 (GATA-3$^+$, producing IL-5 and IL-13), and ILC3 (ROR γt$^+$, producing IL-17 and IL-22). All ILCs differentiate from HSPCs in the bone marrow during the perinatal period [19]. More precisely, common helper-like ILC progenitors, in which *Id2* is switched to common lymphoid progenitors (CLPs), give rise to ILCs in the skin [5].

LCs (Fig. 15.1)

Immune cells reside in both the epidermis and dermis, including Langerhans cells (LCs), named by Paul Langerhans in 1868. Initially mistaken for nerve cells due to their dendrites, LCs were later identified as a subset of dendritic cells capable of presenting antigens to naïve T cells [64].

LCs were once believed to originate from the bone marrow [34]. Currently, there is a broad consensus that LCs are derived from yolk sac-derived EMP before and after their transition into the liver as macrophages. However, unlike macrophages, which are gradually replaced by HSC-derived cells, LCs are self-renewing skin residents throughout life. Notably, when LCs are decreased in the skin such that a substantial number of LCs migrate into sentinel lymph nodes, BM-derived LCs, more precisely, Gr-1^{high} monocytes, have been shown to give rise to LCs. Such LCs are transient and *Id2*-negative, unlike embryo-derived LCs (eLCs), which are *Id2*-positive, but eLCs are deemed to be present in the skin in a steady state [81]. Although the morphological and functional characteristics of eLCs overlap with those of macrophages and DCs, recent hierarchical clustering has demonstrated that eLC characteristics are molecularly distinct from those of both DCs and macrophages [3, 46]. Collectively, LCs are a unique population that maintains skin homeostasis.

DETCs

The skin harbors immune cells for both innate and adaptive immunity. Among them, DETCs, a subset of γδ T cells, are well studied, though γδ T cells make up only 25% compared to the majority of αβ T cells. Notably, Gnetek et al. showed that DETCs progenitors are derived from yolk sac hematopoiesis during primitive and pro-definitive waves and mature in the thymus, whereas other lymphoid cells develop in the fetal liver and migrate to the thymus [18]. Mature DETCs commence to be observed in the epidermis around E16 in mice [22]. However, this is inconsistent with other studies in which γδ T cells share the same origin as T cells [4, 25]. Although we have described DETCs that are involved in homeostasis and wound healing in the skin, a substantial function of DETCs remains elusive. Importantly, the exact homolog to murine DETCs has not been discovered in humans, while Vδ 1 T cells are present in the epidermis and dermis in humans. Further studies are required to understand the biology of DETCs in the skin.

Vascular Vessels

To transport proteins, such as cytokines, growth factors, lipids, enzymes, and cells, a network of vascular and lymphatic capillary plexuses is distributed in the dermis. During heart development, vasculogenesis begins in endocardial progenitor cells, with angioblasts migrating to form the dorsal aorta and posterior cardinal vein [73].

Pericytes, which support capillaries, have diverse origins. Some arise from bone marrow-derived cells via TGF-β signaling, while others originate from mesenchymal or unknown progenitors [98]. Conversely, lymphatic endothelial progenitor cells, which are PROX1 and LYVE1 positive, are budding from the cardinal vein wall, and VEGF signaling, especially VEGF-C and -D, is involved in the differentiation and maturation of lymphatic endothelial cells (LEC), whereas VEGF-A affects vasculogenesis [73].

In summary, the skin comprises diverse cell populations of varied origins, all essential for maintaining homeostasis. While local cells can manage minor disruptions, severe wounds or systemic disorders lead to the recruitment of different cell populations—many bone marrow-derived—that migrate to the skin in a spatially and temporally regulated manner. In the next section, we discuss this phenomenon in detail.

Wound Healing (Fig. 15.2)

As the body's outermost organ, the skin forms a barrier between the internal and external environments and is constantly exposed to potential injury. Therefore, wound healing is indispensable for homeostasis. Wound healing occurs in four distinct phases: hemostasis, inflammation, proliferation, and remodeling, although the transition to each phase is continuous and frequently overlaps.

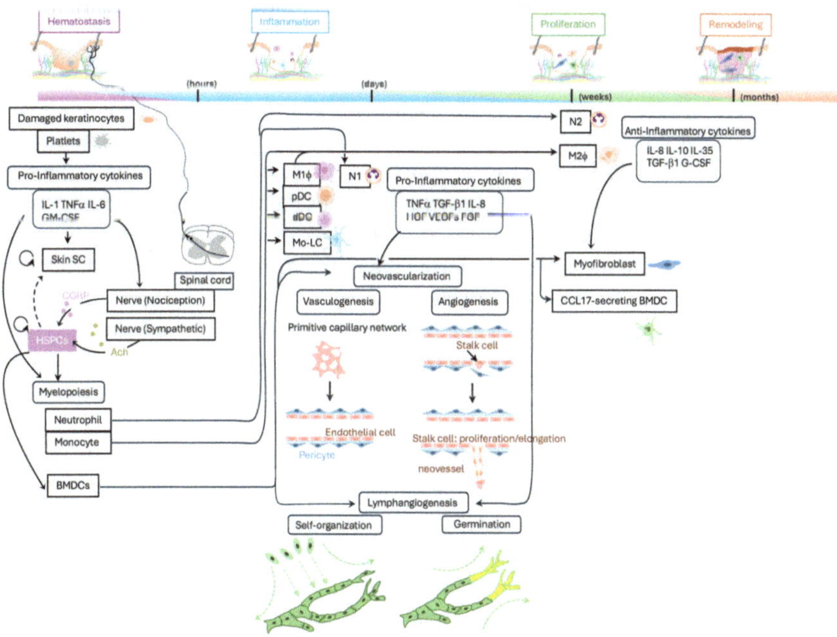

Fig. 15.2 Wound healing in the skin. The four phases occurring after the onset of a wound are shown along the horizontal axis. Underneath each phase, the cell players and their actions are indicated. pDC, plasmacytoid dendritic cells; dDC; dermal dendritic cells; Mo-LC, monocyte-derived Langerhans cells

Dynamic State in Hyperacute Phase

A skin wound can be regarded as a skin defect, indicating damaged keratinocytes and dermal tissue. The defect in the skin is immediately filled with dead keratinocytes, erythrocytes, and platelets, and is called a fibrin clot. The fibrin clot, later forming a scab, acts as a scaffold for immune cell accumulation essential for the next healing phase. Damaged keratinocytes, including platelets, secrete proinflammatory cytokines, including IL-1, IL-6, TNF-α, and GM-CSF, which allows immune cell migration to an injury site [31, 66]. Furthermore, these cytokines can reach HSPCs in the bone marrow to sense local injury. Perivascular niche cells sense thrombocytopenia from blood loss via IL-1, triggering HSPC priming toward the megakaryocyte lineage [45]. Recent studies have also shown that HSPCs directly differentiate into megakaryocytes, which are the source of platelets under stress conditions [16]. In addition to direct megakaryocyte differentiation, sensed HSPCs contribute to the generation of myeloid cells such as erythrocytes, neutrophils, monocytes, macrophages, and dendritic cells, which is known as emergency myelopoiesis. Thus, most immune cells that increase in number in a wound at this stage are progenies of BMDCs. Notably, specific subsets have been observed during tissue repair. Monocyte-derived dendritic cells (mo-DCs), characterized by DC-SIGN$^+$, MMR$^+$, and CD14$^+$ cells, were substantially increased during the inflammatory phase [80]. Mo-DCs produce TNF-α and iNOS to combat bacterial infections and are thus called Tip-DCs. Plasmacytoid DCs (pDCs) are a rare DC subset derived from HSPCs [38]. pDCs do not reside in the skin but in the blood or lymph nodes under a steady state. However, in response to the onset of a wound, pDCs produce a large amount of type-I IFN in the skin through the detection of viral components with intracellular TLR-7 and TLR-9, which causes acute inflammation necessary for a smooth transition to the next phase [21]. A subset of LCs is also specific to the inflammatory phase. This subset arises from Gr-1^{high} monocytes via HSPCs and is distributed throughout the epidermis during the inflammatory phase [81]. Monocyte-derived LCs (mo-LCs) are Id2-independent and short-term, whereas embryo-derived LCs (e-LCs) are Id-dependent and long-term. As eLCs migrate to lymph nodes, mo-LCs repopulate the epidermis. However, mo-LCs likely differ functionally from eLCs due to distinct distribution, morphology, and gene expression profiles.

Following a wound, HSPCs are influenced not only by circulating cytokines and chemokines but also by neurotransmitters. Calcitonin gene-related peptide (CGRP) from nociceptive nerves promotes HSPC mobilization from the bone marrow into circulation [15]. Conversely, neurotransmitters, including norepinephrine, neuropeptide Y, acetylcholine, and dopamine, regulate niche cells surrounding HSPCs [49]. Therefore, HSPCs can detect wound onset through multiple pathways. This transient (several days) activation of HSPCs and niche cells in the bone marrow is crucial for transitioning from hemostasis to inflammation, ensuring adequate myeloid cell delivery to the scab [94].

Skin SCs

It is plausible that skin SCs, as well as HSPCs, are activated in response to wound onset. Indeed, skin SC tracing analyses have revealed that IFESCs contribute to epithelialization rather than their progenitor cells, although their inducible factors have not yet been determined [50]. Furthermore, HFSCs with Keratin 15 in the bulge and secondary hair germ also migrate into the wound to support epithelialization, whereas this subset does not differentiate into keratinocytes under homeostasis [29, 39].

Altogether, heterogeneous cell populations derived from cells that maintain homeostasis and emerge specifically during an emergency contribute to the response to injury.

Neovascularization/Lymphangiogenesis

After injury, the immediate formation of neovascularization and lymphangiogenesis starts in a scab, a natural scaffold, as the transport of HSPCs, peptides, cytokines, and siRNA/lncRNA.

Neovascularization involves both vasculogenesis and angiogenesis, often used interchangeably but distinct in definition. Vasculogenesis involves BM-derived endothelial progenitor and mesenchymal cells forming new vessels. In adults, the exact source of EPCs remains unclear, with HSPCs and monocytes as potential candidates [47]. Angiogenesis is a process in which new vessels are formed by the sprouting or intussusception of existing vessels [93]. During angiogenesis, endothelial cells that contribute to budding are derived from local cells, whereas pericytes are derived from BMDCs [70]. Conversely, circulating BMDCs with the endothelial marker CD31 have been suggested to contribute to both vasculogenesis and angiogenesis. Other research groups have suggested that endothelial cells are derived from BMDCs [11, 36]. These controversies arise because neovascularization occurs in both the wound bed and edges during healing, making it difficult to distinguish between the two processes [48]. Additionally, as CD31 is positive with endothelial cells as well as pericytes, several critical markers overlap with different populations. Therefore, further research and technological developments are required. Despite differences in vasculogenesis and angiogenesis, the reconstruction of vascular networks is urgently necessary during wound healing. "Tortuous" microvessels start to contribute disproportionately to the onset of the wound [6]. Importantly, tortuous regions have been suggested to be hubs for sprouting neovessels. Notably, the hub may be determined by the pericytes [84]. As the wound matures, such vessels or the number of EPCs diminish, and functional vessels grow and distribute [82, 88]. Molecularly, not only VEGF signaling but also hypoxia-inducible factor 1-alpha (HIF-1α), upstream of VEGF, serves as a principal regulatory factor in promoting neovascularization in a wound [61]. Because damaged keratinocytes, macrophages, and fibroblasts express VEGF, including HIF-1α, those factors would stimulate BM-derived EPCs for their recruitment to a wound. Pericyte-endothelial interactions are also essential via gap junctions [35].

Compared to research on neovascularization during wound healing, little attention has been paid to the importance of lymphangiogenesis. Recently, lymphangiogenesis has been shown to be crucial for the drainage of serous fluid, peptides, and immune cells during wound healing because insufficient drainage due to the rupture of lymph vessels in a wound causes edema, which prolongs the inflammation phase [73]. Although VEGF-C, secreted by immune cells, such as macrophages, is a key molecule in embryogenesis, lymphangiogenesis in wounds remains unclear. There are two hypotheses to explain this.

1. Self-organization in which LECs migrate in a wound along the gradient axis of VEGF-C to form lymphatic vessels.
2. Germination, in which neo-lymph vessels bud from lymph vessels around a wound to extend into the wound.

It might be plausible that it depends on the condition of the wounds (i.e., depth, size, region, degree of infection, etc.) whether lymphangiogenesis could occur, while BM-derived LECs appear to be highly involved in the latter hypothesis [65]. Additionally, BM-derived endothelial progenitor cells have been suggested to be involved in corneal lymphangiogenesis [72]. Thus, regardless of which hypothesis is more accurate, BM-derived LECs seem to play a key role in lymphangiogenesis during wound healing.

Cell Players Peculiar to Wound Healing

While pro-inflammatory cells succeed in regulating infection caused by micropathogens in a wound, the proliferation phase starts in parallel with the inflammation phase. Recently, unlike immune cells that detect pathogen-associated molecular patterns (PAMPs) or damage-associated molecular patterns (DAMPs) to fight infection, other immune cell subsets have been identified that aid in tissue repair. Notably, the shift to the proliferation phase aligns with immune cell phenotype changes. For instance, neutrophil type 1 (N1) neutrophils in the inflammatory phase release IL-1β, IL-6, and TNF-α, clear debris and PAMPs/DAMPs, then undergo apoptosis or form neutrophil extracellular traps (NETs). Although N1 is characterized by $Ly6G^{+}CD206^{-}$, other neutrophils with $Ly6G^{+}CD206^{+}$ (neutrophil type 2; N2), which differentiate in the presence of anti-inflammatory factors such as IL-10, IL-35, TGF-β1, and G-CSF, are polarized and replaced in a wound in mice [76]. Furthermore, macrophage phenotype polarization from M1 to M2 occurs, consistent with the polarization from N1 to N2, which are called the N1-M1 and N2-M2 axes, respectively. The former has high phagocytic capacity, whereas the latter has the ability to regenerate tissue. Namely, N2 has an immunomodulatory effect by secreting anti-inflammatory factors as IL-8,

IL-10, and TGF-β1. The phenotypic switch from M1 to M2 involves a shift in arginine metabolism within macrophages. Arginine is metabolized into citrulline with NO in M1 and into ornithine in M2. Ornithine enables M2 macrophages to generate polyamines and proline, which are important for collagen synthesis [54]. Some subsets of M2 macrophages are derived from BMDC; Ly6C^{low} monocytes migrate out of the BM and patrol the luminal surface of the endothelium, differentiating into M2 macrophages [56, 62]. Conversely, a microenvironmental niche in local tissues allows macrophages to switch from M1 to M2. In this case, a substantial amount of M2 was derived from M1 [56]. Since lineage analyses of M2 in vivo are difficult because of the limitations of transgenic animals, the determination of the percentage of BMDC-derived M2 cells remains unknown, although the protocol for M2 differentiation from BMDCs in vitro has been established [98].

Myofibroblasts, which play an important role in the proliferation and remodeling phases, include multiple subsets, although their main function is summarized as contracting wounds by pulling and slacking the extracellular matrix (ECM), secreting growth factors such as PDGF and ECM, such as collagen, fibronectin, and laminin [79]. The importance of the macrophage-myofibroblast fibrotic niche, characterized by contacts enriched with cadherin-11 between these two populations, has been reported in lung tissues [44]. This could also be the case in the skin because recent technologies have unraveled this niche in the skin during wound healing [20]. Myofibroblasts have been known to be derived from multiple cell sources, whereas BMDCs are one of their origins. Transplantation of BMDCs labeled with green fluorescent protein (GFP) into recipient mice revealed that BMDCs were the source of myofibroblasts [53]. Particularly, a large population of myofibroblasts is thought to originate from HSPCs [58]. BM-derived myofibroblasts produce ECM more actively than those from fibroblasts [75]. Other sources of myofibroblasts include resident fibroblasts, epithelial cells, and endothelial cells through the epithelial-to-mesenchymal transition [7, 86].

Another novel BMDCs have been shown to contribute to tissue repair during remodeling [60]. This distinct subset, unlike macrophages and LCs, promotes keratinocyte proliferation via CCL17 secretion, suggesting an underexplored role of BMDCs in wound healing.

Impaired Wound Healing

“Normal wound healing” requires that the cell players and events described in the previous section act in concert sequentially. Conversely, any disruption can impair wound healing and lead to chronic wounds. This section briefly outlines such inhibitory factors and discusses common scenarios that cause healing failure.

Aberrance of the Initial Signals by Damaged Keratinocytes

Protracted secretion of pro-inflammatory cytokines as IL-1α, IL-6, IL-8, TNFα, and CCL2, which are produced by damaged keratinocytes, leads to chronic wounds [66]. These cytokines cause N1 and M1 buildup, blocking the shift to N2 and M2, thus prolonging inflammation. Excess pro-inflammatory cytokines also circulate, leading to HSPC hyperactivation. This type of activated myelopoiesis, called myeloid skewing, differs from normal wound healing. Myeloid skewing is defined as an unbalanced proportion of myeloid cell differentiation in HSCs under stress, leading to the disabled migration of proper immune cell populations. Consequently, chronic inflammation persists in the wound, and HSC function is compromised (referred to as HSC exhaustion).

Absent Contribution of Skin SCs in a Wound

During the proliferation phase, IFSC and HFSC recruitment around the wound is necessary for epithelialization [29, 39]. Particularly, the deletion of $Lgr6^{+}$ IFSCs and HFSCs has been shown to result in impaired wound healing [26].

A subset of BMDCs reported to generate skin SCs can migrate into the epidermis, although their presence among keratinocytes is rare and even disputed under steady-state conditions [28, 87]. In most cases, target cells are labeled with fluorescence by utilizing a promoter that is activated in the cells of interest to trace cell lineages to generate transgenic animals. However, the promoter is frequently inactivated, especially in highly differentiated cells such as keratinocytes. For instance, in bone marrow transplants using GFP-labeled donor mice, the ubiquitous promoter may become inactive when BMDCs reach the epidermis and begin switching their fates. Thus, research on the roles of BMDCs in skin SCs, particularly in IFESCs, is sporadic, whereas the disturbance of BMDCs migration into the epidermis can easily cause impaired wound healing [28].

Disruption of the BMDC System

As previously described, BMDCs play a central role in wound healing. The aberrant BMDCs system fails to cause myelopoiesis, resulting in failure of the inflammation phase in the wound. Moreover, the possibility that both neovascularization and lymphangiogenesis deteriorate in a wound indicates that the following phases, proliferation and remodeling, never occur. The pathogenesis of aberrant BMDCs depends on the individual disease, as described below.

Impairment of Neovascularization or Lymphangiogensis

Impaired vasculogenesis, angiogenesis, and lymphangiogenesis hinder wound healing. Reduced HIF-1α/VEGF signaling from dysfunctional or fewer macrophages leads to poor neovascularization, while abnormal ECs, pericytes, or LECs also disrupt vessel formation.

Notably, even if neovascularization and lymphangiogenesis are properly developed, a deep or extensive wound could become chronic because it is beyond the range of biological healing.

Impaired Transition from the Proliferation Phase to the Remodeling Phase

Inhibition of N1/N2 and/or M1/M2 polarization causes prolonged inflammation, leading to a chronic state. Alternatively, myofibroblast dysfunction, such as impaired cell contact or collagen synthesis, leads to wound stagnation and subsequent fibrosis.

Identifying the underlying causes of impaired wound healing, such as metabolic diseases, severe injuries, and aging, is essential for developing new treatments. The following sections focus on studies exploring these correlations.

Diabetes (Fig. 15.3)

Diabetic foot ulcers are a global issue, with an 11% mortality risk and an estimated economic burden of $5.2 billion worldwide in 2023 [63]. In fact, recent data indicate that diabetic foot ulcers account for over 40% of chronic wounds in hospitals [68]. Therefore, a discussion on the etiology of impaired wounds caused by diabetes is unavoidable.

Hyperglycemia, the primary disease process, can cause all the phenomena (section “Impaired wound healing”). Recently, methylglyoxal (MGO), a highly reactive dicarbonyl compound that forms advanced glycation end products (AGEs), has been implicated in diabetes [77]. Accordingly, MGO accumulation may be relevant in all phases of wound healing, leading to chronic wounds in diabetes. Although MGO initially forms in endothelial cells under high glucose conditions, it also accumulates in the skin and wounds of patients with diabetes [32]. This phenomenon could explain why skin dysfunction, characterized by an impaired skin barrier and aberrant keratinocyte differentiation, occurs prior to a wound in diabetic mice [59]. Notably, claudin-1, along with ZO-1, an adaptor protein that interacts with Claudin-1 in the cytoplasm, is significantly up-regulated under high-glucose conditions [59, 97]. AGEs produced by binding MGO to proteins, lipids, or DNA cause reactive oxidative stress, accompanied by IL-8 secretion, in addition to

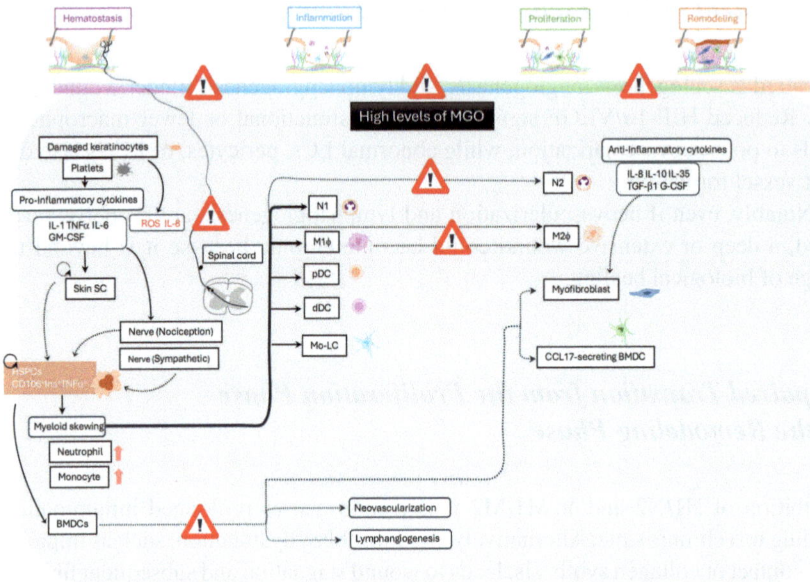

Fig. 15.3 Impaired wound healing in the skin of diabetes. The four phases occurring after the onset of a wound are shown along the horizontal axis, though the transition from one to the next phase is frequently impaired, as suggested by impaired icons. The concentration of methylglyoxal (MGO) is high in the skin of diabetes (colored in light green). Dotted lines indicate that the actions are inhibited

pro-inflammatory cytokines, by keratinocytes in wounds in diabetes. Reactive oxygen species (ROS) and IL-8 allow excessive and prolonged N1 and M1 infiltration, leading to failure of the transition from N1 or M1 to N2 or M2, respectively. Furthermore, it inhibits the migration of keratinocytes by upregulating matrix metalloproteinases [12]. Thus, it is plausible that MGO inhibits the activation of skin SCs during wound healing.

Conversely, hyperglycemia causes an abnormal subset of BMDCs, which is never detected in normal mice, CD106^{+}proinsulin^{+}TNF-α^{+} HSPCs [33]. This subset migrates out of the BM and fuses with local cells such as peripheral nerves and intestinal cells [13, 33, 40, 89]. CD106^{+} proinsulin^{+}TNF-α^{+} BMDCs accompany epigenetic changes, namely, altered expression pattern of histone deacetylases (HDACs), including upregulation of HDAC types 1, 3, 4, and 8 and Sirt-1, Sirt-6, and Sirt-7, including downregulation of HDAC types 2, 9, and 11 [37]. The novel finding that treatment with an HDAC inhibitor under blood sugar regulation led to complete remission of diabetes indicates the possibility of healing impaired wounds, although further studies are warranted for treatments addressed to diabetes complications.

Overall, there is impaired wound healing in patients with diabetes. Considering the severity of diabetes, the degree of the wound, and the patient's age, individual wounds include further diversity depending on their life stages, even in the same patients. Thus, it seems unlikely that there is a uniform method for an impaired wound in patients with diabetes; therefore, it is a fascinating strategy that treatment with Hdac inhibitors under proper blood sugar control results in the complete remission of diabetes per se.

Burn and Trauma (Fig. 15.4)

Although burn injuries have different causes, such as heat, cold, chemical compounds, or electricity, common pathophysiological characteristics are protein denaturation and coagulative necrosis. Burn severity depends on the depth and size of the injury. Burns are classified into three categories: first-degree burns (epidermis alone denatures); second-degree burns (epidermis and partial dermis denatures); and third-degree burns (epidermis and dermis denatures). Second-degree burns are classified as IIa and IIb depending on the depth of denaturation in the dermis. Specialists should determine which classification corresponds to the burns in patients. First-degree and IIa burns are usually healed through conservative treatments in 2 weeks unless infection by pathogens deepens the wound, while IIb and third-degree burns frequently require surgical intervention. This is because all components of the cells in the skin, including skin SCs, denature in IIb and third degree burn injuries, allowing pathogens to immediately reach, proliferate, and invade a wound before the accumulation of immune cells.

Currently, there are few publications on the effects of such severe burns on BMDCs. The protein denaturation and necrosis of large tissues trigger a persistent increase in the levels of catecholamine, cortisol, glucagon, dopamine, and pro-inflammatory factors such as IL-1, IL-6, TNFα, and ROS. Because their secretions continue for as long as a year after burn injury [30], there is a prolonged inflammatory phase in severe burns, contributing to impaired wound healing. Such stress conditions appear to be one of the pathogeneses of increased megakaryocytes in the BM of burn patients, consistent with the finding that HSPCs differentiate directly into megakaryocytes in response to stress [16, 57].

Besides, erythroid differentiation from HSPCs is inhibited (burn-induced anemia). Erythroid differentiation in the BM was significantly decreased in mice without debridement; more precisely, this explains the absence of bleeding after surgical intervention, after receiving 15% total burn surface area on the dorsal region [69]. One possible mechanism is that the deprived nociceptive sympathetic nerves supplying the skin due to burns lead to loss of CGRP signals to the central nervous system via the dorsal root ganglia, which in turn makes it impossible to transfer signals via sympathetic nerves supplying arterioles in the BM. Since CGRP promotes HSPC mobilization [69], disruption of this pathway caused by burns seems to impair the egress of HSPCs from the BM to the peripheral blood.

Subsequently, myeloid skewing occurs, accompanied by the arrest of myeloid maturation [83, 95]. One of the pathogeneses of aberrant myeloid skewing can be explained by the decreased expression of G-CSF receptors in BMDCs, a cytokine

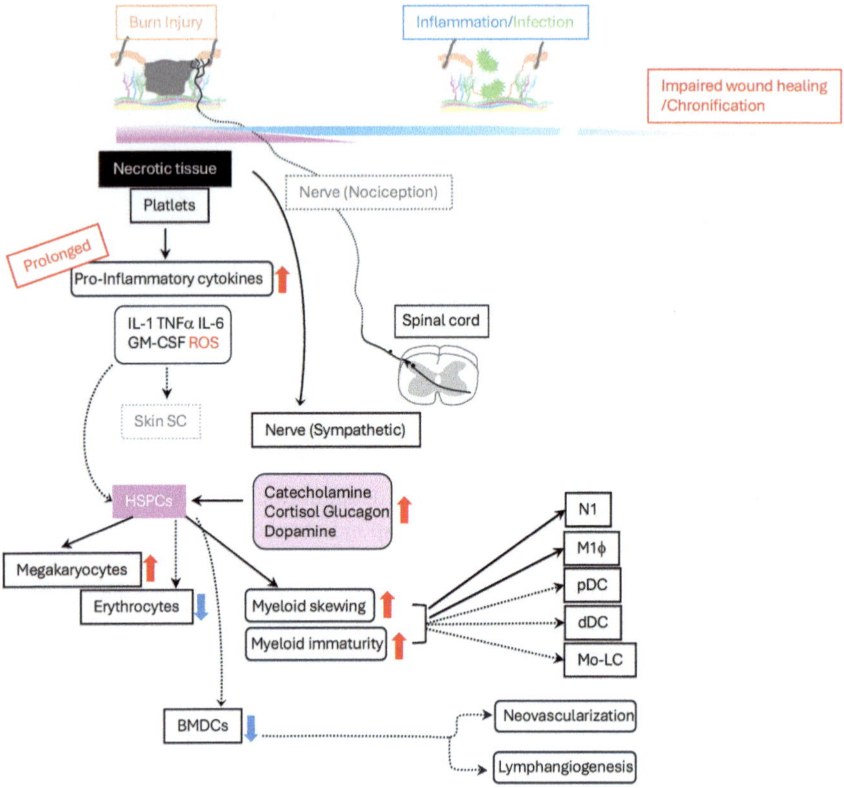

Fig. 15.4 Impaired wound healing in burnt skin. In severe burns, infection frequently causes a wound to prolonged inflammation, namely, chronification. After an injury, a wound is filled with eschar (colored in grey) instead of a fibrin clot. During the inflammation phase, bacteria (colored in green) are easily invaded in burnt skin. Dotted lines indicate that the actions are inhibited. Red and blue arrows are the actions or cell players that are activated or deactivated, respectively

that stimulates BMDCs toward the differentiation of granulocytes [94]. Furthermore, myeloid skewing caused by severe burns appears to respond to bacterial attacks, which invade substantial skin defects. However, immature granulocytes are unable to overcome infection, indicating a transition to impaired wound healing, that is, a chronic wound. Myeloid skewing prevents HSPCs from differentiating towards EPCs, leading to impaired neovascularization and lymphangiogenesis.

Chronic wounds in patients with burns allow proliferation of bacteria, followed by their invasion into vessels and bacteremia. As the function of BMDCs in burns deviates from that during normal wound healing, bacteremia can easily deteriorate into sepsis or even death. Collectively, BMDC regulation may shed light on treatments for severe burns.

Aging (Fig. 15.5)

It is well established that pro-inflammatory cytokines as TNF-α, IL-6, and IL-1 are upregulated in aged skin, a phenomenon referred to as "inflammaging" [52]. This chronic increase in pro-inflammatory cytokines attenuates the expression of proteins essential for maintaining skin integrity, including tight junction components and collagens. In particular, the downregulation of *Claudin-4* in response to inflammaging causes skin barrier dysfunction in aged skin [51]. Additionally, the age-related decline in collagen XVII governing impacts IFESC stemness [43]. More precisely, decreased collagen XVII in the aged epidermis–dermis junction deteriorated IFESC differentiation. Furthermore, *Pentraxin 3* (*Ptx3*) secreted by aged fibroblasts attenuates IFESCs [27]. Considering that the progenies of IFESCs, located around a wound, are activated by pro-inflammatory signals secreted by damaged keratinocytes and fibroblasts, these alterations in IFESCs may be associated with impaired wound healing in aged skin. Conversely, HFSCs retain undifferentiated fates in aged mice, whereas the number of HFSCs was lower than that in young mice [17]. Furthermore, aged HFSCs were not only reprogrammed but also regenerated in combination with a niche in the young dermis, whereas young HFSCs failed to regenerate HFs with a niche in the old dermis.

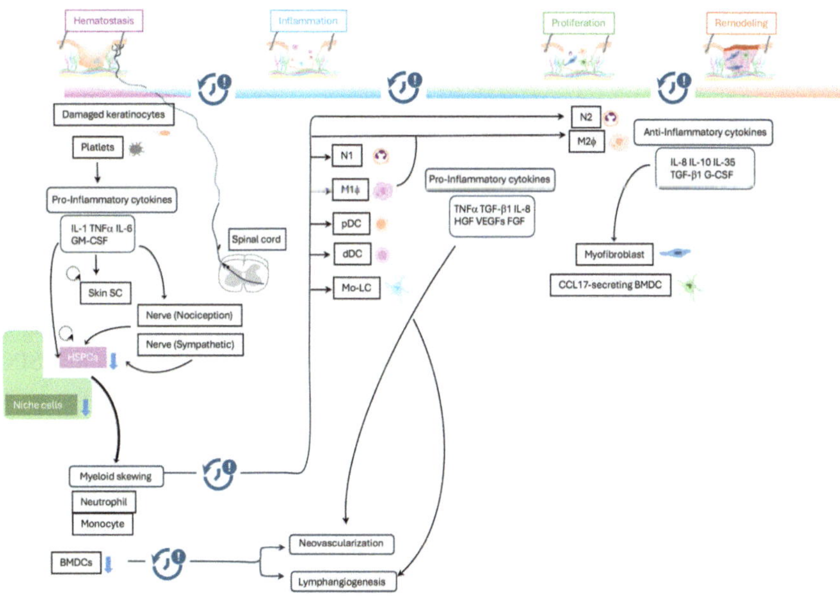

Fig. 15.5 Impaired wound healing in aged skin. The four phases occurring after the onset of a wound are shown along the horizontal axis, though the transition from one to the next phase is frequently delayed, as suggested by time-delayed icons. The functions and cell numbers in the niche surrounding HSPCs are significantly decreased, as suggested by blue arrows. Dotted lines indicate that the actions are inhibited

Aging leads to heterogeneous alterations in subsets of skin SCs, but the detriment of BMDCs associated with aging is a critical factor in impaired wound healing. HSPC homeostasis is maintained by crosstalk with niche cells, such as MSCs, Schwann cells, endothelial/lymphatic cells, megakaryocytes, and immune cells [49]. Aging alters the epigenetic programming of HSPCs per se through histone methylation/acetylation and DNA methylation [94]. For instance, the trimethylation of histone 3 lysine 4 is a critical epigenetic marker that is upregulated in aged HSPCs [85]. Furthermore, DNA methylation disturbs the balance between self-renewal and differentiation of aged HSPCs. Consequently, these alterations in the epigenome of HSPCs shift towards myeloid skewing [41, 85]. Furthermore, age-related alterations in niche cells surrounding HSPCs cause significant damage to HSPCs. Notably, degeneration of sympathetic nerves secondary to aging undergoes the decreased norepinephrine signal from the central nervous system, causing abnormal MSCs expansion and myeloid skewing in HSPCs, because MSCs and HSPCs have β3 adrenergic receptors (AR) and β_2AR, respectively. Aberrant MSCs expansion leads to a decline in osteoblasts but an increase in adipocytes. Age-related alterations in arterioles are characterized by a decrease in vascular tone and an increase in leakiness, suggesting EPC dysfunction, which has been poorly explored. Thus, these age-related alterations result in a decrease in the niche in the BM, leading to a decrease in HSPC pools. Because both aged EPC and MSCs secrete pro-inflammatory cytokines towards HSPCs [49], the regulation of niche cells appears crucial for the shift from impaired wounds to the normal wound healing process.

Altogether, the tendency of delayed wound healing in aged skin appears to be greatly influenced by the aging of the microenvironments surrounding skin SCs and HSPCs rather than that of skin SCs and HSPCs per se. These findings indicate that niche cells could be targets for novel treatments for impaired wound healing in aged skin.

Strategy for an Impaired Wound Healing

Impaired wound healing is not only attributed to local factors but also to HSPC dysfunction or the niche surrounding HSPCs in the BM. In the latter case, local treatments focused on the wound allowed it to transition into a chronic state.

First, the pathogenesis of impaired wound healing in patients with diabetes is discussed, in which the underlying cause is a systemic disease. In this case, wound chronification is a complication of a systemic disease, indicating that the treatment of diabetes per se is necessary and sufficient for wound healing. However, the treatment aim of many systemic diseases in a clinical setting is to control complications. In this respect, our finding that animal models of diabetes experience complete remission through BMDC treatment suggests the possibility that other systemic diseases are also “SC diseases.” Accordingly, normalization of the HSPC/BMDC system could be the key to reverting an impaired wound to a normal course.

Another example of impaired wound healing in burns has been discussed as representative of trauma. Indeed, BMDC function is severely affected by trauma but can be controlled if medical interventions are properly performed earlier. Instead, infection prevention and control appear to be critical for chronification. As antibiotic abuse has led to the growth of multiple drug-resistant bacteria worldwide, alternative strategies are required. The first is bacteriophage therapy (hereafter referred to as phage therapy). Bacteriophages are viruses that infect bacteria and archaea and are not harmful to mammals in most cases [71]. The narrow spectrum of individual bacteriophages that target bacteria significantly decreases the emergence of multiple drug-resistant bacteria, including adverse effects such as diarrhea, frequently caused by antibiotics owing to their broad spectrum. Considering that several case reports have been published in which bacteriophages overcame multiple drug-resistant bacteria in humans [8, 78], phage therapies could shed light on infection control in wounds.

The final examples of impaired wound healing include aging, neither systemic diseases nor trauma. Although aging is defined as a gradual decline in physical and mental capacity and is believed to be a natural course of life, it is noteworthy that the World Health Organization has included aging associated with a decline in intrinsic capacity as a disease code [90]. The concept of aging as a disease remains controversial; however, impaired wound healing in aged patients may be treated, including treatment of HSPCs/BMDCs and the niches surrounding them.

In summary, a stringent management of wounds is vital; however, the entire human body should be put into perspective. This is also evident from the fact that the skin is composed of heterogeneous subsets. Notably, substantial subsets are progenies of BMDCs. Considering that BMDCs circulate in blood and lymph vessels to reach their destinations, they may play a role in whole-body regeneration. Synchronization of the nervous, immune, and endocrine systems is warranted. Also, there may be some truth that those synchronized systems include BMDCs; the "BMDC system" could be regarded as the fourth system that is responsible for the maintenance of homeostasis.

References

1. Cain TL, Derecka M, McKinney-Freeman S (2025) The role of the haematopoietic stem cell niche in development and ageing. https://doi.org/10.1038/s41580-024-00770-8. PMID: 39256623
2. Canu G, Ruhrberg C (2021) First blood: the endothelial origins of hematopoietic progenitors. In: Angiogenesis, vol 24(2). Springer Science and Business Media B.V, pp 199–211. https://doi.org/10.1007/s10456-021-09783-9
3. Carpentier S, Vu Manh TP, Chelbi R, Henri S, Malissen B, Haniffa M, Ginhoux F, Dalod M (2016) Comparative genomics analysis of mononuclear phagocyte subsets confirms homology between lymphoid tissue-resident and dermal XCR1+ DCs in mouse and human and distinguishes them from Langerhans cells. J Immunol Methods 432:35–49. https://doi.org/10.1016/j.jim.2016.02.023

4. Castillo-González R, Cibrian D, Sánchez-Madrid F (2021) Dissecting the complexity of γδ T-cell subsets in skin homeostasis, inflammation, and malignancy. J Allergy Clin Immunol 147(6):2030–2042). Mosby Inc. https://doi.org/10.1016/j.jaci.2020.11.023
5. Cherrier DE, Serafini N, Di Santo JP (2018) Innate lymphoid cell development: a T cell perspective. Immunity 48(6):1091–1103). Cell Press. https://doi.org/10.1016/j.immuni.2018.05.010
6. Chong DC, Yu Z, Brighton HE, Bear JE, Bautch VL (2017) Tortuous microvessels contribute to wound healing via sprouting angiogenesis. Arterioscler Thromb Vasc Biol 37(10):1903–1912. https://doi.org/10.1161/ATVBAHA.117.309993
7. Darby IA, Laverdet B, Bonté F, Desmoulière A (2014) Fibroblasts and myofibroblasts in wound healing. Clin Cosmet Investig Dermatol 7:301–311). Dove Medical Press Ltd. https://doi.org/10.2147/CCID.S50046
8. Dedrick RM, Guerrero-Bustamante CA, Garlena RA, Russell DA, Ford K, Harris K, Gilmour KC, Soothill J, Jacobs-Sera D, Schooley RT, Hatfull GF, Spencer H (2019) Engineered bacteriophages for treatment of a patient with a disseminated drug-resistant mycobacterium abscessus. Nat Med 25(5):730–733. https://doi.org/10.1038/s41591-019-0437-z
9. Dermitzakis I, Kampitsi DD, Manthou ME, Evangelidis P, Vakirlis E, Meditskou S, Theotokis P (2024) Ontogeny of skin stem cells and molecular underpinnings. https://doi.org/10.3390/cimb46080481
10. Díaz-García D, Filipová A, Garza-Veloz I, Martinez-Fierro ML (2021) A beginner's introduction to skin stem cells and wound healing. Int J Mol Sci 22(20) MDPI. https://doi.org/10.3390/ijms222011030
11. Eming SA, Martin P, Tomic-Canic M (2014) Wound repair and regeneration: mechanisms, signaling, and translation. Sci Transl Med 6(265) American Association for the Advancement of Science. https://doi.org/10.1126/scitranslmed.3009337
12. Fang WC, Lan CCE (2023) The epidermal keratinocyte as a therapeutic target for management of diabetic wounds. Int J Mol Sci 24(5) Multidisciplinary Digital Publishing Institute (MDPI). https://doi.org/10.3390/ijms24054290
13. Fujimiya M, Kojima H, Ichinose M, Arai R, Kimura H, Kashiwagi A, Chan L (2007) Fusion of proinsulin-producing bone marrow-derived cells with hepatocytes in diabetes. Proc Natl Acad Sci USA 104(10):4030–4035. https://doi.org/10.1073/pnas.0700220104
14. Furuse M, Hata M, Furuse K, Yoshida Y, Haratake A, Sugitani Y, Noda T, Kubo A, Tsukita S (2002) Claudin-based tight junctions are crucial for the mammalian epidermal barrier: a lesson from claudin-1-deficient mice. J Cell Biol 156(6):1099–1111. https://doi.org/10.1083/jcb.200110122
15. Gao X, Zhang D, Xu C, Li H, Caron KM, Frenette PS (2021) Nociceptive nerves regulate haematopoietic stem cell mobilization. Nature 589(7843):591–596. https://doi.org/10.1038/s41586-020-03057-y
16. Garyn CM, Bover O, Murray JW, Ma J, Salas-Briceno K, Ross SR, Snoeck HW (2024) G2 arrest primes hematopoietic stem cells for megakaryopoiesis. Cell Rep 43(7). https://doi.org/10.1016/j.celrep.2024.114388
17. Ge Y, Miao Y, Gur-Cohen S, Gomez N, Yang H, Nikolova M, Polak L, Hu Y, Verma A, Elemento O, Krueger JG, Fuchs E, Christiano AM, Tyner AL, Wagers AJ, Yi R (2020) The aging skin microenvironment dictates stem cell behavior. Proc Natl Acad Sci USA. https://doi.org/10.1073/pnas.1901720117/-/DCSupplemental
18. Gentek R, Ghigo C, Hoeffel G, Jorquera A, Msallam R, Wienert S, Klauschen F, Ginhoux F, Bajénoff M (2018) Epidermal γδ T cells originate from yolk sac hematopoiesis and clonally self-renew in the adult. J Exp Med 215(12):2994–3005. https://doi.org/10.1084/jem.20181206
19. Ghaedi M, Takei F (2021) Innate lymphoid cell development. J Allergy Clin Immunol 147(5):1549–1560). Mosby Inc. https://doi.org/10.1016/j.jaci.2021.03.009
20. Gopee NH, Winheim E, Olabi B, Admane C, Foster AR, Huang N, Botting RA, Torabi F, Sumanaweera D, Le AP, Kim J, Verger L, Stephenson E, Adão D, Ganier C, Gim KY, Serdy SA, Deakin C, Goh I et al (2024) A prenatal skin atlas reveals immune regulation of human skin morphogenesis. Nature 635(8039):679–689. https://doi.org/10.1038/s41586-024-08002-x

21. Gregorio J, Meller S, Conrad C, Di Nardo A, Homey B, Lauerma A, Arai N, Gallo RL, DiGiovanni J, Gilliet M (2010) Plasmacytoid dendritic cells sense skin injury and promote wound healing through type i interferons. J Exp Med 207(13):2921–2930. https://doi.org/10.1084/jem.20101102
22. Havran WL, Allison JP (1990) Origin of Thy-1+ dendritic epidermal cells of adult mice from fetal thymic precursors. Nature 344(6261):68–70. https://doi.org/10.1038/344068a0
23. Hoeffel G, Chen J, Lavin Y, Low D, Almeida FF, See P, Beaudin AE, Lum J, Low I, Forsberg EC, Poidinger M, Zolezzi F, Larbi A, Ng LG, Chan JKY, Greter M, Becher B, Samokhvalov IM, Merad M, Ginhoux F (2015) C-Myb+ Erythro-myeloid progenitor-derived fetal monocytes give rise to adult tissue-resident macrophages. Immunity 42(4):665–678. https://doi.org/10.1016/j.immuni.2015.03.011
24. Hoeffel G, Ginhoux F (2018) Fetal monocytes and the origins of tissue-resident macrophages. Cell Immunol 330:5–15. https://doi.org/10.1016/j.cellimm.2018.01.001
25. Hu Y, Hu Q, Li Y, Lu L, Xiang Z, Yin Z, Kabelitz D, Wu Y (2023) γδ T cells: origin and fate, subsets, diseases and immunotherapy. Signal Transduct Target Ther 8(1) Springer Nature. https://doi.org/10.1038/s41392-023-01653-8
26. Huang S, Kuri P, Aubert Y, Brewster M, Li N, Farrelly O, Rice G, Bae H, Prouty S, Dentchev T, Luo W, Capell BC, Rompolas P (2021) Lgr6 marks epidermal stem cells with a nerve-dependent role in wound re-epithelialization. Cell Stem Cell 28(9):1582–1596.e6. https://doi.org/10.1016/j.stem.2021.05.007
27. Ichijo R, Maki K, Kabata M, Murata T, Nagasaka A, Ishihara S, Haga H, Honda T, Adachi T, Yamamoto T, Toyoshima F (2022) Vasculature atrophy causes a stiffened microenvironment that augments epidermal stem cell differentiation in aged skin. https://doi.org/10.1038/s43587-022-00244-6
28. Inokuma D, Abe R, Fujita Y, Sasaki M, Shibaki A, Nakamura H, McMillan JR, Shimizu T, Shimizu H (2006) CTACK/CCL27 accelerates skin regeneration via accumulation of bone marrow-derived keratinocytes. Stem Cells 24(12):2810–2816. https://doi.org/10.1634/stemcells.2006-0264
29. Ito M, Liu Y, Yang Z, Nguyen J, Liang F, Morris RJ, Cotsarelis G (2005) Stem cells in the hair follicle bulge contribute to wound repair but not to homeostasis of the epidermis. Nat Med 11(12):1351–1354. https://doi.org/10.1038/nm1328
30. Jeschke MG, Chinkes DL, Finnerty CC, Kulp G, Suman OE, Norbury WB, Branski LK, Gauglitz GG, Mlcak RP, Herndon DN (2008) Pathophysiologic response to severe burn injury. Ann Surg 248(3):387–400. https://doi.org/10.1097/SLA.0b013e3181856241
31. Jian Y, Li Y, Zhang Y, Tang M, Deng M, Liu C, Cheng M, Xiao S, Deng C, Wei Z (2024) Lymphangiogenesis: novel strategies to promote cutaneous wound healing. Burns Trauma 12). Oxford University Press. https://doi.org/10.1093/burnst/tkae040
32. Jiang M, Yakupu A, Guan, H., Dong J, Liu Y, Song F, Tang J, Tian M, Niu Y, Lu, S (2022) Pyridoxamine ameliorates methylglyoxal-induced macrophage dysfunction to facilitate tissue repair in diabetic wounds. https://doi.org/10.1111/iwj.13597
33. Katagi M, Terashima T, Ohashi N, Nakae Y, Yamada A, Nakagawa T, Miyazawa I, Maegawa H, Okano J, Suzuki Y, Fujino K, Eguchi Y, Kojima H (2021) Malfunctioning CD106-positive, short-term hematopoietic stem cells trigger diabetic neuropathy in mice by cell fusion. Commun Biol:1–14. https://doi.org/10.1038/s42003-021-02082-5
34. Katz SI, Tamaki K, Sachs DH (1979) Epidermal Langerhans cells are derived from cells originating in bone marrow [30]. Nature 282(5736):324–326. https://doi.org/10.1038/282324a0
35. Kikuchi-Taura A, Okinaka Y, Takeuchi Y, Ogawa Y, Maeda M, Kataoka Y, Yasui T, Kimura T, Gul S, Claussen C, Boltze J, Taguchi A (2020) Bone marrow mononuclear cells activate angiogenesis via gap junction-mediated cell-cell interaction. Stroke 51(4):1279–1289. https://doi.org/10.1161/STROKEAHA.119.028072
36. Kim H, Cho HJ, Kim SW, Liu B, Choi YJ, Lee J, Sohn YD, Lee MY, Houge MA, Yoon YS (2010) CD31+ cells represent highly angiogenic and vasculogenic cells in bone marrow: novel role of nonendothelial CD31+ cells in neovascularization and their therapeutic

effects on ischemic vascular disease. Circ Res 107(5):602–614. https://doi.org/10.1161/CIRCRESAHA.110.218396
37. Kojima H, Katagi M, Okano J, Nakae Y, Ohashi N, Fujino K, Miyazawa I, Nakagawa T (2023) Complete remission of diabetes with a transient HDAC inhibitor and insulin in streptozotocin mice. Commun Biol 6(1):637. https://doi.org/10.1038/s42003-023-05010-x
38. Laustsen A, Van Der Sluis RM, Gris-Oliver A, Sánchez Hernández S, Cemalovic E, Tang HQ, Henning Pedersen L, Uldbjerg N, Jakobsen MR, Bak RO (2021) Ascorbic acid supports ex vivo generation of plasmacytoid dendritic cells from circulating hematopoietic stem cells. elife 10:e65528–e65549. https://doi.org/10.7554/eLife
39. Levy V, Lindon C, Zheng Y, Harfe BD, Morgan BA (2007) Epidermal stem cells arise from the hair follicle after wounding. FASEB J 21(7):1358–1366. https://doi.org/10.1096/fj.06-6926com
40. Li Y, Kojima H, Fujino K, Matsumura K, Katagi M, Urabe H, Chan L, Eguchi Y, Zhao L, Kimura H (2011) Homing of the bone marrow-derived interstitial cells of Cajal is decreased in diabetic mouse intestine. J Gastroenterol Hepatol 26(6):1072–1078. https://doi.org/10.1111/j.1440-1746.2011.06670.x
41. Ling C, Rönn T (2019) Epigenetics in human obesity and type 2 diabetes. https://doi.org/10.1016/j.cmet.2019.03.009
42. Liu K, Nussenzweig MC (2010) Origin and development of dendritic cells. Immunol Rev 234(1):45–54. https://doi.org/10.1111/j.0105-2896.2009.00879.x
43. Liu Y, Ho C, Wen D, Sun J, Huang L, Gao Y, Li Q, Zhang Y (2022) Targeting the stem cell niche: role of collagen XVII in skin aging and wound repair. Theranostics 12(15):6446–6454. https://doi.org/10.7150/thno.78016
44. Lodyga M, Cambridge E, Karvonen HM, Pakshir P, Wu B, Boo S, Kiebalo M, Kaarteenaho R, Glogauer M, Kapoor M, Ask K, Hinz B (2019) Cadherin-11–mediated adhesion of macrophages to myofibroblasts establishes a profibrotic niche of active TGF. Sci Signal 12(564). https://doi.org/10.1126/scisignal.aao3469
45. Luis TC, Barkas N, Carrelha J, Giustacchini A, Mazzi S, Norfo R, Wu B, Aliouat A, Guerrero JA, Rodriguez-Meira A, Bouriez-Jones T, Macaulay IC, Jasztal M, Zhu G, Ni H, Robson MJ, Blakely RD, Mead AJ, Nerlov C et al (2023) Perivascular niche cells sense thrombocytopenia and activate hematopoietic stem cells in an IL-1 dependent manner. Nat Commun 14(1). https://doi.org/10.1038/s41467-023-41691-y
46. Mabbott NA, Kenneth Baillie J, Hume DA, Freeman TC (2010) Meta-analysis of lineage-specific gene expression signatures in mouse leukocyte populations. Immunobiology 215(9–10):724–736. https://doi.org/10.1016/j.imbio.2010.05.012
47. Marçola M, Rodrigues CE (2015) Endothelial progenitor cells in tumor angiogenesis: another brick in the wall. Stem Cells Int 2015). Hindawi Publishing Corporation. https://doi.org/10.1155/2015/832649
48. Markiewski MM, Daugherity E, Karbowniczek M, Reese B (2020) The role of complement in angiogenesis. Antibodies 9(4):1–14). MDPI. https://doi.org/10.3390/antib9040067
49. Maryanovich M (2024) Systemic and local regulation of hematopoietic homeostasis in health and disease. Nat Cardiovasc Res. https://doi.org/10.1038/s44161-024-00482-4
50. Mascré G, Dekoninck S, Drogat B, Youssef KK, Broheé S, Sotiropoulou P a, Simons BD, Blanpain C (2012) Distinct contribution of stem and progenitor cells to epidermal maintenance. Nature 489(7415):257–262. https://doi.org/10.1038/nature11393
51. Matsuda S, Miwa M, Tanabe M, Kobayashi M, Shu S, Yoshino Y, Tada N, Itoh A, Ikari A (2025) Sirtuin-2 is involved in the regulation of claudin-4 expression and paracellular barrier function in keratinocytes. J Cell Biochem 126(3). https://doi.org/10.1002/jcb.70027
52. Michaud M, Balardy L, Moulis G, Gaudin C, Peyrot C, Vellas B, Cesari M, Nourhashemi F (2013) Proinflammatory cytokines, aging, and age-related diseases. J Am Med Dir Assoc 14(12):877–882). Elsevier Inc. https://doi.org/10.1016/j.jamda.2013.05.009

53. Mori L, Bellini A, Stacey MA, Schmidt M, Mattoli S (2005) Fibrocytes contribute to the myofibroblast population in wounded skin and originate from the bone marrow. Exp Cell Res 304(1):81–90. https://doi.org/10.1016/j.yexcr.2004.11.011
54. Morris SM (2007) Arginine metabolism: boundaries of our knowledge. J Nutr 137:1–3
55. Nagahata Y, Masuda K, Nishimura Y, Ikawa T, Kawaoka S, Kitawaki T, Nannya Y, Ogawa S, Suga H, Satou Y, Takaori-Kondo A, Kawamoto H (2022) Tracing the evolutionary history of blood cells to the unicellular ancestor of animals. Blood 140:2611. http://amigo.geneontology
56. Nahrendorf M, Swirski FK, Aikawa E, Stangenberg L, Wurdinger T, Figueiredo JL, Libby P, Weissleder R, Pittet MJ (2007) The healing myocardium sequentially mobilizes two monocyte subsets with divergent and complementary functions. J Exp Med 204(12):3037–3047. https://doi.org/10.1084/jem.20070885
57. Noetzli LJ, French SL, Machlus KR (2019) New insights into the differentiation of megakaryocytes from hematopoietic progenitors. Arterioscler Thromb Vasc Biol 39(7):1288–1300. https://doi.org/10.1161/ATVBAHA.119.312129
58. Ogawa M, Larue AC, Drake CJ (2006) Hematopoietic origin of fibroblasts/myofibroblasts: its pathophysiologic implications. https://doi.org/10.1182/blood-2006-04
59. Okano J, Kojima H, Katagi M, Nakagawa T, Nakae Y, Terashima T, Kurakane T, Kubota M, Maegawa H, Udagawa J (2016) Hyperglycemia induces skin barrier dysfunctions with impairment of epidermal integrity in non-wounded skin of type 1 diabetic mice. PLoS One 11(11):e0166215. https://doi.org/10.1371/journal.pone.0166215
60. Okano J, Nakae Y, Nakagawa T, Katagi M, Terashima T, Nagakubo D, Nakayama T, Yoshie O, Suzuki Y, Kojima H (2021) A novel role for bone marrow-derived cells to recover damaged keratinocytes from radiation-induced injury. Sci Rep 11(1):5653. https://doi.org/10.1038/s41598-021-84818-1
61. Oladipupo S, Hu S, Kovalski J, Yao J, Santeford A, Sohn RE, Shohet R, Maslov K, Wang LV, Arbeit JM (2011) VEGF is essential for hypoxia-inducible factor-mediated neovascularization but dispensable for endothelial sprouting. Proc Natl Acad Sci USA 108(32):13264–13269. https://doi.org/10.1073/pnas.1101321108
62. Olingy CE, San Emeterio CL, Ogle ME, Krieger JR, Bruce AC, Pfau DD, Jordan BT, Peirce SM, Botchwey EA (2017) Non-classical monocytes are biased progenitors of wound healing macrophages during soft tissue injury. Sci Rep 7:447–463. https://doi.org/10.1038/s41598-017-00477-1
63. Olsson M, Järbrink K, Divakar U, Bajpai R, Upton Z, Schmidtchen A, Car J (2019) The humanistic and economic burden of chronic wounds: a systematic review. Wound Repair Regen 27(1):114–125). Blackwell Publishing Inc. https://doi.org/10.1111/wrr.12683
64. Otsuka M, Egawa G, Kabashima K (2018) Uncovering the mysteries of Langerhans cells, inflammatory dendritic epidermal cells, and monocyte-derived langerhans cell-like cells in the epidermis. Front Immunol 9(July):1768. https://doi.org/10.3389/fimmu.2018.01768
65. Park C, Lee JY, Yoon Y (2011) Role of bone marrow-derived lymphatic endothelial progenitor cells for lymphatic neovascularization. Trends Cardiovasc Med 21(5):135–140. https://doi.org/10.1016/j.tcm.2012.04.002.Role
66. Piipponen M, Li D, Landén NX (2020) The immune functions of keratinocytes in skin wound healing. Int J Mol Sci 21(22):1–26). MDPI AG. https://doi.org/10.3390/ijms21228790
67. Pinho S, Frenette PS (2019) Haematopoietic stem cell activity and interactions with the niche. Nat Rev Mol Cell Biol 20(5):303–320). Nature Publishing Group. https://doi.org/10.1038/s41580-019-0103-9
68. Popescu V, Cauni V, Petrutescu MS, Rustin MM, Bocai R, Turculet CR, Doran H, Patrascu T, Lazar AM, Cretoiu D, Varlas VN, Mastalier B (2023) Chronic wound management: from gauze to homologous cellular matrix. Biomedicines 11(9) Multidisciplinary Digital Publishing Institute (MDPI). https://doi.org/10.3390/biomedicines11092457
69. Posluszny JA, Muthumalaiappan K, Kini AR, Szilagyi A, He LK, Li Y, Gamelli RL, Shankar R (2011) Burn injury dampens erythroid cell production through reprioritizing

bone marrow hematopoietic response. J Trauma 71(5):1288–1296. https://doi.org/10.1097/TA.0b013e31822e2803
70. Rajantie I, Ilmonen M, Alminaite A, Ozerdem U, Alitalo K, Salven P (2004) Adult bone marrow-derived cells recruited during angiogenesis comprise precursors for periendothelial vascular mural cells. Blood 104(7):2084–2086. https://doi.org/10.1182/blood-2004-01-0336
71. Ranveer SA, Dasriya V, Ahmad MF, Dhillon HS, Samtiya M, Shama E, Anand T, Dhewa T, Chaudhary V, Chaudhary P, Behare P, Ram C, Puniya DV, Khedkar GD, Raposo A, Han H, Puniya AK (2024) Positive and negative aspects of bacteriophages and their immense role in the food chain. NPJ Sci Food 8(1) Nature Research. https://doi.org/10.1038/s41538-023-00245-8
72. Religa P, Cao R, Bjorndahl M, Zhou Z, Zhu Z, Cao Y (2005) Presence of bone marrow-derived circulating progenitor endothelial cells in the newly formed lymphatic vessels. https://doi.org/10.1182/blood-2005-01
73. Renò F, Sabbatini M (2023) Breaking a vicious circle: lymphangiogenesis as a new therapeutic target in wound healing. Biomedicine 11(3):1–13. https://doi.org/10.3390/biomedicines11030656
74. Rodrigues PF, Trsan T, Cvijetic G, Khantakova D, Panda SK, Liu Z, Ginhoux F, Cella M, Colonna M (2024) Progenitors of distinct lineages shape the diversity of mature type 2 conventional dendritic cells. Immunity 57(7):1567–1585.e5. https://doi.org/10.1016/j.immuni.2024.05.007
75. Saikia P, Crabb JS, Dibbin LL, Juszczak MJ, Willard B, Jang GF, Shiju TM, Crabb JW, Wilson SE (2020) Quantitative proteomic comparison of myofibroblasts derived from bone marrow and cornea. Sci Rep 10(1). https://doi.org/10.1038/s41598-020-73686-w
76. Sansores-España LD, Melgar-Rodríguez S, Vernal R, Carrillo-Ávila BA, Martínez-Aguilar VM, Díaz-Zúñiga J (2022) Neutrophil N1 and N2 subsets and their possible association with periodontitis: a scoping review. Int J Mol Sci 23(20) MDPI. https://doi.org/10.3390/ijms232012068
77. Schalkwijk CG, Stehouwer CDA (2020) Methylglyoxal, a highly reactive dicarbonyl compound, in diabetes, its vascular complications, and other age-related diseases. Physiol Rev 100(1):407–461. https://doi.org/10.1152/physrev.00001.2019
78. Schooley RT, Biswas B, Gill JJ, Hernandez-Morales A, Lancaster J, Lessor L, Barr JJ, Reed SL, Rohwer F, Benler S, Segall AM, Taplitz R, Smith DM, Kerr K, Kumaraswamy M, Nizet V, Lin L, McCauley MD, Strathdee SA et al (2017) Development and use of personalized bacteriophage-based therapeutic cocktails to treat a patient with a disseminated resistant Acinetobacter baumannii infection. Antimicrob Agents Chemother 61(10). https://doi.org/10.1128/AAC.00954-17
79. Schuster R, Younesi F, Ezzo M, Hinz B (2023) The role of Myofibroblasts in physiological and pathological tissue repair. Cold Spring Harb Perspect Biol 15(1). https://doi.org/10.1101/cshperspect.a041231
80. Serbina NV, Salazar-Mather TP, Biron CA, Kuziel WA, Pamer EG (2003) TNF/iNOS-producing dendritic cells mediate innate immune defense against bacterial infection. Immunity 19:59
81. Seré K, Baek J-H, Ober-Blöbaum J, Müller-Newen G, Tacke F, Yokota Y, Zenke M, Hieronymus T (2012) Two distinct types of Langerhans cells populate the skin during steady state and inflammation. Immunity 37(5):905–916. https://doi.org/10.1016/j.immuni.2012.07.019
82. Shi Z, Yao C, Shui Y, Li S, Yan H (2023) Research progress on the mechanism of angiogenesis in wound repair and regeneration. Front Physiol 14. https://doi.org/10.3389/fphys.2023.1284981
83. Shoup M, Weisenberger JM, Wang JL, Pyle JM, Gamelli RL, Shankar R (1998) Mechanisms of neutropenia involving myeloid maturation arrest in burn sepsis. Ann Surg 228(1) Lippincott-Raven Publishers
84. Stapor PC, Sweat RS, Dashti DC, Betancourt AM, Murfee WL (2014) Pericyte dynamics during angiogenesis: new insights from new identities. J Vasc Res 51(3):163–174). S. Karger AG. https://doi.org/10.1159/000362276
85. Sun D, Luo M, Jeong M, Rodriguez B, Xia Z, Hannah R, Wang H, Le T, Faull KF, Chen R, Gu H, Bock C, Meissner A, Göttgens B, Darlington GJ, Li W, Goodell MA (2014) Epigenomic profiling of young and aged HSCs reveals concerted changes during aging that reinforce self-renewal. Cell Stem Cell 14(5):673–688. https://doi.org/10.1016/j.stem.2014.03.002

86. Tai Y, Woods EL, Dally J, Kong D, Steadman R, Moseley R, Midgley AC (2021) Myofibroblasts: function, formation, and scope of molecular therapies for skin fibrosis. Biomolecules 11(8) MDPI. https://doi.org/10.3390/biom11081095
87. Tamai K, Yamazaki T, Chino T, Ishii M, Otsuru S, Kikuchi Y, Iinuma S, Saga K, Nimura K, Shimbo T, Umegaki N, Katayama I, Miyazaki J, Takeda J, McGrath J a, Uitto J, Kaneda Y (2011) PDGFRalpha-positive cells in bone marrow are mobilized by high mobility group box 1 (HMGB1) to regenerate injured epithelia. Proc Natl Acad Sci USA 108(16):6609–6614. https://doi.org/10.1073/pnas.1016753108
88. Tepper OM, Capla JM, Galiano RD, Ceradini DJ, Callaghan MJ, Kleinman ME, Gurtner GC (2005) Adult vasculogenesis occurs through in situ recruitment, proliferation, and tubulization of circulating bone marrow-derived cells. https://doi.org/10.1182/blood
89. Terashima T, Kojima H, Chan L (2012) Bone marrow expression of poly(ADP-ribose) polymerase underlies diabetic neuropathy via hematopoietic-neuronal cell fusion. FASEB J 26(1):295–308. https://doi.org/10.1096/fj.11-186262
90. The Lancet Healthy Longevity (2022) Is ageing a disease? Lancet Healthy Longev 3(7):e448). Elsevier Ltd. https://doi.org/10.1016/S2666-7568(22)00154-4
91. Turksen K (2013) All about claudins. Tissue Barriers 1(3):e26750. https://doi.org/10.4161/tisb.26750
92. Turksen K, Troy TC (2004) Barriers built on claudins. J Cell Sci 117(12):2435–2447. https://doi.org/10.1242/jcs.01235
93. Uccelli A, Wolff T, Valente P, Di Maggio N, Pellegrino M, Gürke L, Banfi A, Gianni-Barrera R (2019) Vascular endothelial growth factor biology for regenerative angiogenesis. Swiss Med Wkly (Vol. 149, p. w20011). NLM (Medline). https://doi.org/10.4414/smw.2019.20011
94. Urao N, Liu J, Takahashi K, Ganesh G (2022) Hematopoietic stem cells in wound healing response. Adv Wound Care 11(11):598–621. https://doi.org/10.1089/wound.2021.0065
95. Wallner SF, Warren GH (1985) The haematopoietic response to burning: an autopsy study. Burns 12(1):22–27. https://doi.org/10.1016/0305-4179(85)90179-2
96. Wu Y, Hirschi KK (2021) Tissue-resident macrophage development and function. Front Cell Dev Biol 8). Frontiers Media S.A. https://doi.org/10.3389/fcell.2020.617879
97. Yamada K, Matsushita K, Wang J, Kanekura T (2018) Topical glucose induces claudin-1 and filaggrin expression in a mouse model of atopic dermatitis and in keratinocyte culture, exerting anti-inflammatory effects by repairing skin barrier function. Acta Derm Venereol 98(1):19–25. https://doi.org/10.2340/00015555-2807
98. Yamazaki T, Nalbandian A, Uchida Y, Li W, Arnold TD, Kubota Y, Yamamoto S, Ema M, Mukouyama Y, Suke (2017) Tissue myeloid progenitors differentiate into pericytes through TGF-β signaling in developing skin vasculature. Cell Rep 18(12):2991–3004. https://doi.org/10.1016/j.celrep.2017.02.069
99. Yang F, Wang T, Du P, Fan H, Dong X, Guo H (2020) M2 bone marrow-derived macrophage-derived exosomes shuffle microRNA-21 to accelerate immune escape of glioma by modulating PEG3. Cancer Cell Int 20(1). https://doi.org/10.1186/s12935-020-1163-9

Chapter 16
The Role of Adipose-Derived Stem Cells in Accelerating and Refining Wound Healing

Francesco De Francesco

Abstract Adipose-derived stem cells (ASCs) represent a transformative tool in the evolving field of regenerative medicine, particularly in the context of wound healing. Their abundant availability, ease of harvest, and potent biological properties make them a compelling candidate for therapeutic applications in both acute and chronic wounds. This chapter provides a comprehensive overview of the biological characteristics of ASCs—including their origin in the stromal vascular fraction (SVF), surface phenotype, immunomodulatory secretome, and multipotent differentiation capacity. We examine the mechanisms through which ASCs modulate inflammation, stimulate angiogenesis, remodel the extracellular matrix, and prevent excessive fibrosis and scarring.

Both preclinical and clinical studies are explored in detail, demonstrating that ASCs enhance tissue regeneration, accelerate wound closure, and improve the quality of healed skin. From diabetic ulcers to radiation-induced injuries, evidence supports their efficacy and safety, even in allogeneic applications. The chapter also delves into the challenges surrounding standardization, regulatory frameworks, and long-term safety, while highlighting the ongoing clinical trials that aim to translate this technology into routine medical practice.

In synthesizing current scientific data and clinical insights, this chapter underscores the promise of ASCs as a next-generation cell-based therapy poised to redefine wound care across a range of indications.

Keywords Adipose-derived stem cells (ASCs) · Wound healing · Regenerative medicine · Stromal vascular fraction (SVF) · Angiogenesis · Immunomodulation · Extracellular matrix remodeling · Cell-based therapy

---

F. De Francesco (✉)
Department of Reconstructive Surgery and Hand Surgery, Azienda Ospedaliera Universitaria delle Marche, Ancona, Italy
e-mail: francesco.defrancesco@ospedaliriuniti.marche.it

F. De Francesco, K. Turksen (eds.), *Management and Strategies for Wound Healing*, Stem Cell Biology and Regenerative Medicine 79,
https://doi.org/10.1007/978-3-032-12442-5_16

Abbreviations

ADSCs	Adipose-derived stem cells
ALP	Alkaline phosphatase
ANGPT-1	Angiopoietin-1
bFGF	Basic fibroblast growth factor
CD	Cluster of differentiation (e.g., CD73, CD90, CD105)
ECM	Extracellular matrix
GFAP	Glial fibrillary acidic protein
HIF-1α	hypoxia-inducible factor-1 alpha
IDO	Indoleamine 2,3-dioxygenase
IL-10	Interleukin-10
MAP 2	Microtubule-associated protein 2
MMPs	Matrix metalloproteinases
MSCs	Mesenchymal stem cells
NO	Nitric oxide
PGE2	Prostaglandin E2
PPARγ	Peroxisome proliferator-activated receptor gamma
SDF-1	Stromal cell-derived factor 1
SVF	Stromal vascular fraction
TGF-β	Transforming growth factor-beta
TIMPs	Tissue inhibitors of metalloproteinases
VEGF	Vascular endothelial growth factor
vWF	von Willebrand Factor

Introduction

Regenerative medicine has emerged as a transformative frontier in contemporary healthcare, capitalizing on the body's inherent potential for self-repair through biological therapies that aim not only to replace damaged tissues but also to restore their native structure and function [1]. This paradigm shift moves beyond symptom management or mechanical replacement—hallmarks of traditional medicine—and instead focuses on rebuilding living tissue through the activation or transplantation of biological components. At the core of this discipline lies a multidisciplinary approach that synthesizes advances in molecular biology, biomaterials science, tissue engineering, and cellular therapy [2]. Molecular biology provides insight into gene expression, signaling pathways, and cell-to-cell interactions that govern tissue repair [3]. Tissue engineering contributes by designing scaffolds and biomimetic environments that support cell adhesion, migration, and differentiation [4]. Meanwhile, cell-based therapies introduce either autologous or allogeneic cells capable of repopulating or modulating damaged tissue environments [5]. These integrated strategies enable regenerative medicine to tackle challenges previously considered intractable, including chronic wounds, degenerative diseases, and extensive tissue loss following trauma or surgical resection. Moreover, the field is evolving in parallel with cutting-edge technologies such as gene editing (e.g., CRISPR/Cas9), 3D bioprinting, and bioreactor-based tissue cultivation, which have further expanded the therapeutic potential. These tools not only enhance the precision and personalization of regenerative interventions but also allow the recreation of complex tissue architectures, including vascularized or innervated constructs that are functionally integrated with host tissues [6]. In clinical practice, regenerative medicine holds the promise of reducing the need for lifelong pharmacotherapy, minimizing donor site morbidity, and avoiding complications associated with prosthetic or synthetic materials. Particularly in wound healing, regenerative strategies aim to accelerate tissue regeneration, reduce inflammation, and achieve superior aesthetic and functional outcomes—goals that are especially crucial in patients with impaired healing capacity [7]. Among these strategies, stem cell-based therapies have become central due to their multipotency, ability to secrete bioactive molecules, and capacity to modulate inflammatory and immune responses [8].

Stem cells can be broadly categorized into three major types: embryonic stem cells (ESCs), induced pluripotent stem cells (iPSCs), and adult (or somatic) stem cells [9]. Each type is defined by its developmental origin and differentiation potential. ESCs, derived from the inner cell mass of blastocysts, are pluripotent and can give rise to virtually all cell types in the body. However, their clinical use is hampered by ethical concerns, potential for teratoma formation, and immunogenicity when used in allogeneic settings. In contrast, iPSCs are adult somatic cells reprogrammed to a pluripotent state via the introduction of specific transcription factors. They offer similar differentiation capabilities to ESCs while circumventing ethical issues. Despite their promise, iPSCs still face significant translational barriers, including genomic instability, risk of tumorigenicity, and challenges in large-scale, standardized production. Adult stem cells, also known as multipotent stem cells, are more restricted in their differentiation potential but are considered safer for clinical applications. Among these, mesenchymal stem cells (MSCs) have gained considerable traction in regenerative medicine due to their unique combination of immunomodulatory, anti-inflammatory, and trophic properties. MSCs are capable of differentiating into osteoblasts, chondrocytes, adipocytes, and, under specific conditions, even neural or epithelial-like cells. MSCs can be isolated from several tissue sources, each with its own set of advantages and limitations: (i) Bone marrow-derived MSCs (BM-MSCs) are the most extensively studied but are limited by low yield and an invasive harvesting procedure. Additionally, the proliferative capacity of BM-MSCs declines with donor age [10]; (ii) Umbilical cord-derived MSCs (UC-MSCs), often harvested from Wharton's jelly, provide a non-invasive, ethically acceptable source and exhibit high proliferative rates and lower immunogenicity. However, their differentiation potential can be variable [11]; (iii) Adipose-derived stem cells (ASCs) are abundant and easily accessible via liposuction, yielding a significantly higher number of viable cells per volume compared to bone marrow. Their collection is less invasive and more acceptable to patients, which contributes to their growing popularity in clinical protocols [12].

However, in recent years, adipose-derived stem cells (ASCs) have emerged as a preferred MSC source due to several practical and biological benefits [13]. Adipose tissue is a rich and accessible reservoir of regenerative cells. A standard liposuction procedure can yield millions of stem cells, significantly surpassing the number obtainable from bone marrow aspiration. Moreover, the procedure is minimally invasive and typically results in fewer complications. The stromal vascular fraction (SVF) obtained from adipose tissue is a heterogeneous mixture containing not only ADSCs but also endothelial cells, pericytes, macrophages, and other progenitor cells—all of which contribute to tissue regeneration in a synergistic fashion [14]. Biologically, adipose-derived stem cells (ASCs) are a subpopulation of mesenchymal stem cells (MSCs) with a well-characterized surface immunophenotype. They typically express positive markers such as CD73 (ecto-5′-nucleotidase), CD90 (Thy-1), and CD105 (endoglin), all of which are consistent with the minimal criteria for MSCs as defined by the International Society for Cellular Therapy (ISCT) [15, 16]. These markers are involved in cellular adhesion, extracellular matrix interaction, and modulation of the immune response. Notably, ASCs do not express hematopoietic or endothelial lineage markers such as CD45, CD34 (after expansion), CD14, or HLA-DR, which helps distinguish them from blood-derived cells and immune subsets. Interestingly, CD34 expression, while typically absent in cultured MSCs, is present in freshly isolated stromal vascular fraction (SVF) and early passage ASCs [12, 17, 18]. Its loss over successive passages is attributed to in vitro adaptation and is often used as a dynamic marker of stemness and culture progression. This phenotypic plasticity reflects the responsiveness of ASCs to their microenvironment and culturing conditions. Functionally, ASCs possess high proliferative potential, making them suitable for large-scale expansion, which is necessary for clinical applications. They maintain a stable karyotype over several passages and exhibit clonogenicity, forming colony-forming unit–fibroblasts (CFU-F) when plated at low densities. Beyond expansion, their hallmark lies in their multilineage differentiation capacity: under defined induction protocols, ASCs can differentiate into adipocytes, chondrocytes, and osteoblasts, thereby fulfilling the classical mesodermal lineage commitment [19]. Moreover, several studies have demonstrated the transdifferentiation potential of ASCs toward non-mesodermal lineages, including neuronal, epithelial, and endothelial-like cells, especially when cultured in neurogenic or angiogenic conditions [20–22]. This broader plasticity is attributed to their rich transcriptional landscape and epigenetic adaptability, which allows them to respond to external biochemical cues. Such flexibility not only enhances their therapeutic utility across a range of tissue types but also raises interesting questions about their developmental origin and hierarchical position among adult stem cells. The biological features of ASCs, combined with their relative abundance, ease of harvest, and low immunogenic profile, support their wide application in regenerative medicine, wound healing, and tissue engineering platforms. Their paracrine activity, detailed further in the next sections, adds another dimension to their mechanism of action, enabling tissue regeneration not just through direct differentiation but also through modulation of the local wound microenvironment [23]. These paracrine signals exert anti-inflammatory, angiogenic, anti-apoptotic, and immunomodulatory actions, making ASCs highly effective in complex wound environments. In contrast to embryonic stem cells, ASCs are considered ethically acceptable and present a significantly lower risk of teratoma formation. Additionally, their low immunogenicity allows for both autologous and allogeneic applications with reduced risk of host rejection [24].

In the specific context of wound healing, ASCs demonstrate potent therapeutic promise. Wounds, particularly chronic ones, pose a significant clinical challenge due to impaired tissue repair mechanisms, persistent inflammation, reduced vascularization, and the risk of infection. The ability of ASCs to home to injured tissues, release growth factors like vascular endothelial growth factor (VEGF), TGF-β, and hepatocyte growth factor (HGF), and differentiate into cells of the dermal lineage makes them ideal candidates for facilitating repair and regeneration. Several preclinical studies in animal models have shown that ADSC transplantation accelerates wound closure, enhances collagen remodeling, and increases neovascularization in both acute and diabetic wounds [25–28].

The clinical translation of these findings has led to a growing number of human trials exploring the use of ASCs in complex wound settings, including radiation ulcers, chronic diabetic foot ulcers, and postsurgical wounds [29–31]. Initial results suggest improved healing times, reduced pain, and better cosmetic outcomes with minimal adverse effects. Moreover, exosome-based therapies derived from ASCs are emerging as a cell-free alternative that preserves many of the regenerative benefits without the complexity of live-cell delivery [32].

Beyond direct application, the integration of ASCs with advanced biomaterials such as hydrogels, nanofibers, and scaffolds is revolutionizing the delivery and retention of these cells at wound sites. These combinations aim to mimic the native extracellular matrix and provide mechanical support while enhancing cell viability and local signaling. Engineered constructs like ASC spheroids or genetically modified ASCs further extend their functional versatility and efficacy in nonhealing wounds [33].

In summary, the field of regenerative wound care is undergoing a paradigm shift, with ASCs positioned at the forefront due to their biological potency, accessibility, and clinical applicability. Their use aligns perfectly with the modern goals of wound management: to accelerate healing, minimize scarring, and improve aesthetic and functional outcomes with minimal invasiveness. In the following sections, we will explore in detail the mechanisms of action, clinical evidence, delivery methods, and emerging strategies involving ASCs in the treatment of both acute and chronic wounds.

Biological Properties of Adipose-Derived Stem Cells

Origin and Isolation: The Stromal Vascular Fraction

Adipose tissue is not only a reservoir of energy but also a rich, complex source of regenerative cells. At the heart of this potential lies the stromal vascular fraction (SVF)—a heterogeneous mix of cellular subtypes obtained through enzymatic digestion of adipose tissue [34]. Unlike expanded cultures of adipose-derived stem cells (ASCs), which offer a more homogeneous population, the SVF preserves the native microenvironment and cellular diversity of the tissue, thereby enabling synergistic interactions essential for regenerative processes.

The SVF contains a wide array of cell types, each contributing distinct functional roles [35]. Among these, ASCs typically comprise 15–30% of the total SVF cell population. These multipotent cells are identifiable by their positive expression of CD73, CD90, and CD105, and the absence of hematopoietic markers such as CD45 and CD14 in culture-expanded conditions. ASCs are largely responsible for the regenerative and immunomodulatory capacities attributed to the SVF and are capable of differentiating into various mesodermal and even non-mesodermal lineages. Another significant fraction is composed of endothelial progenitor cells (EPCs) and mature endothelial cells, which can account for 10–20% of the SVF [36]. These cells, marked by surface proteins such as CD31, CD34, and VEGFR-2, are pivotal for promoting angiogenesis, especially in ischemic or poorly vascularized wound environments. Pericytes, a population of perivascular cells characterized by CD146, NG2, and PDGFR-β expression, are present at approximately 3–10% [37]. These cells not only stabilize microvasculature but also function as mesenchymal progenitors. Their interaction with endothelial cells has been shown to modulate vascular morphogenesis and tissue repair, emphasizing their critical role in maintaining vascular homeostasis within regenerative contexts. In addition to these, the SVF contains preadipocytes and fibroblasts, which are involved in extracellular matrix deposition and remodeling—essential processes in both wound closure and scar modulation. The immune component of the SVF, including macrophages, T-regulatory cells, and other leukocytes, makes up roughly 10–20% of the population [38]. These immune cells play a crucial role in shaping the inflammatory milieu, polarizing macrophages toward pro-regenerative M2 phenotypes, and modulating local immune responses.

This diverse cellular ecosystem of the SVF enhances its functionality beyond what any single cell type can achieve. The interplay between ASCs, pericytes, endothelial cells, and immune cells creates a dynamic environment conducive to vascular regeneration, immune modulation, and tissue remodeling. Importantly, this complexity offers a significant advantage for clinical applications in regenerative medicine, particularly in scenarios such as wound healing, where multifaceted repair processes are necessary. The preservation of this heterogeneity during SVF isolation is therefore considered a key factor in its therapeutic efficacy, distinguishing it from culture-expanded ASCs and highlighting its relevance in translational applications.

Adipose tissue can be harvested from various anatomical sites, with subcutaneous fat from the abdomen, flank, and thigh being the most common. Subcutaneous (SC) fat yields significantly higher SVF cell counts than visceral fat, with abdominal SC adipose tissue offering the highest ADSC yield per gram [39]. Harvesting is typically done using manual or water-assisted liposuction (WAL) under sterile conditions with tumescent anesthesia, taking care to limit trauma that could affect cell viability. Notably, excess vacuum pressure negatively impacts ASC yield and function. The isolation of ASCs from lipoaspirate involves a multi-step process [40–42]: (i) Washing and Filtration: The lipoaspirate is washed multiple times with phosphate-buffered saline (PBS) or lactated Ringer's solution to remove blood and oil residues; (ii) Enzymatic Digestion: The washed tissue is incubated with type I collagenase (0.075–0.2%, depending on the protocol) at 37 °C for 30–60 min under gentle agitation. Collagenase type IA or NB4 is frequently used due to its broad substrate specificity [43]; (iii) Neutralization and Centrifugation: Enzyme activity is neutralized using serum-containing medium, followed by centrifugation at 300–500 × g for 5–10 min. This step yields a cell pellet (SVF) beneath the adipocyte layer; (iv) RBC Lysis and Filtration: Red blood cells may be lysed using ammonium chloride, and the final product is filtered through a 70–100 μm mesh to eliminate debris.

The final SVF pellet contains approximately 400,000–1,000,000 nucleated cells per gram of adipose tissue. Flow cytometry analysis reveals the following approximate composition: 15–30% ASCs, 10–20% endothelial progenitors, 5–10% pericytes, and various immune subsets [44, 45].

Phenotypic Characteristics and Immunophenotyping

Adipose-derived stem cells (ASCs) have gained prominence in regenerative medicine due to their high yield, multipotency, and immunomodulatory capacity. Among the hallmarks defining their identity is a distinctive immunophenotypic signature, which reflects both their mesenchymal origin and functional versatility. This signature is not static—it evolves depending on donor source, adipose depot, culture passage, and expansion conditions. Understanding these phenotypic patterns is essential for standardizing ASC-based therapies and interpreting functional data across studies.

ASCs are plastic-adherent, fibroblast-like in appearance, and clonogenic under standard culture conditions. These basic properties fulfill the first criterion of the International Society for Cellular Therapy (ISCT) for defining mesenchymal stem cells (MSCs) [15]. From this morphological baseline, ASCs express a panel of surface markers that enable further classification. The ISCT stipulates that MSCs must exhibit positive expression of CD73, CD90, and CD105 and lack markers such as CD45, CD14 (or CD11b), CD19, CD34 (in expanded cells), and HLA-DR. ASCs fulfill these criteria after a few passages in culture, thereby aligning with bone marrow-derived MSCs in phenotypic characterization [46]. Notably, freshly isolated ASCs from the stromal vascular fraction (SVF) often retain CD34 expression, a hematopoietic and endothelial progenitor marker. This marker is progressively downregulated during in vitro culture, coinciding with the emergence of a more classical MSC profile [47]. The loss of CD34 is not merely a phenotypic transition—it may also reflect shifts in adhesion, migration, and cytokine secretion capacity. Beyond the canonical MSC markers, ASCs express a broad array of accessory antigens with distinct biological implications: CD29 (β1-integrin): Integral for cell-extracellular matrix (ECM) interaction and signal transduction, CD44 (hyaluronan receptor): Facilitates matrix adherence, inflammation resolution, and migration, CD166 (ALCAM): A cell adhesion molecule with roles in intercellular interaction, neural development, and immune regulation. These markers contribute to the dynamic behavior of ASCs in tissue repair scenarios, particularly within complex inflammatory or fibrotic environments [48]. In addition to surface antigens, ASCs express several intracellular proteins that underscore their mesenchymal lineage. Vimentin, a type III intermediate filament, is one of the most prominent, supporting cytoskeletal integrity and plasticity under stress or during migration [49]. Vimentin expression is maintained throughout in vitro expansion, and its presence is often used as a hallmark of mesodermal progenitor identity.

Immunophenotypic expression is not static; it varies significantly with culture passage, oxygen tension, substrate stiffness, and serum concentration. For example, CD105 expression has been shown to increase during early passages but may plateau or decline in prolonged culture, affecting the chondrogenic and angiogenic potential of the cells [50]. Similarly, hypoxic conditions can upregulate CD90 and CD44, enhancing the immunosuppressive and wound-healing properties of ASCs. Another layer of complexity arises from the anatomical origin of the adipose tissue. Subcutaneous, visceral, and mammary fat depots yield ADSCs with subtly different marker intensities and functional profiles. Abdominal ASCs may show stronger CD105 expression, while breast-derived cells exhibit more robust CD90/CD44 signals [46]. These phenotypic distinctions correlate with differences in proliferation rate, angiogenic factor secretion, and immunosuppressive potential. Interestingly, obesity and donor metabolic status further modify the immunophenotypic landscape of ADSCs. Studies indicate that ASCs from obese individuals may express altered levels of CD44 and CD166, influencing both adipogenic capacity and cytokine release profiles [48].

The immunophenotypic signature of ADSCs is not merely taxonomic—it reflects their functional potential in both differentiation and immunomodulation. High CD105 expression supports chondrogenic lineage commitment, while strong CD44/CD90 profiles correlate with immune interaction and resolution of inflammation. Downregulation of CD34, while facilitating mesenchymal transition, may also reduce endothelial mimicry and paracrine angiogenesis, thus influencing wound-healing efficacy. In terms of immunoregulation, the surface profile governs interactions with T cells, macrophages, and natural killer (NK) cells. For instance, CD73 contributes to adenosine production, a key anti-inflammatory mediator in immune regulation. Meanwhile, CD90 and CD44 modulate macrophage polarization, steering immune cells toward regenerative M2 phenotypes.

Immunomodulatory and Paracrine Mechanisms

One of the hallmark properties that elevate adipose-derived stem cells (ASCs) in regenerative medicine is their robust immunomodulatory potential. Unlike many stem cell types that exert their influence through differentiation alone, ASCs engage with the immune system via cell-cell contact and the secretion of bioactive factors, collectively sculpting a local environment conducive to tissue repair, inflammation resolution, and fibrosis attenuation. ASCs modulate a broad array of immune cells, including T-lymphocytes, B-cells, natural killer (NK) cells, and dendritic cells, largely through the secretion of potent soluble factors such as interleukin-10 (IL-10), transforming growth factor-beta (TGF-β), prostaglandin E2 (PGE2), indoleamine 2,3-dioxygenase (IDO), and nitric oxide (NO). These molecules orchestrate immunosuppressive cascades that dampen pro-inflammatory cytokine production, promote T-regulatory cell induction, and impair dendritic cell maturation [51, 52]. In particular, PGE2 and IDO have been demonstrated to synergistically increase IL-10 production in MSCs, enhancing the anti-inflammatory microenvironment. IL-10 itself contributes to macrophage polarization and suppresses Th1 responses, while TGF-β plays a dual role in immunosuppression and extracellular matrix remodeling. The upregulation of these factors is dynamic and can be influenced by environmental cues, including injury, hypoxia, and systemic inflammatory signals [53].

Beyond lymphoid cell modulation, macrophage polarization is a critical function of ADSCs. These cells preferentially skew macrophages toward the M2 phenotype, which is characterized by anti-inflammatory cytokine profiles, enhanced tissue remodeling, and pro-angiogenic behavior. M2 macrophages secrete IL-10, arginase-1, and VEGF, contributing to the attenuation of chronic inflammation and the promotion of wound healing [54–56]. ASCs achieve these effects largely via their secretome—a cocktail of cytokines, growth factors, and extracellular vesicles. Key components include vascular endothelial growth factor (VEGF), hepatocyte growth factor (HGF), insulin-like growth factor-1 (IGF-1), angiopoietin-1 (ANGPT-1), and CXCL12 (SDF-1α). These factors collectively promote angiogenesis, fibroblast activation, and epithelial repair, thereby establishing a pro-regenerative niche that accelerates tissue recovery [57, 58]. Interestingly, the source of adipose tissue and donor-related factors, such as obesity and diabetes, substantially alter the immunosuppressive efficacy of ASCs. ASCs derived from obese or diabetic individuals often show reduced expression of IDO, TGF-β, and PGE2, and diminished capacity to induce M2 polarization. This phenotypic alteration compromises their ability to regulate inflammation and may necessitate either preconditioning protocols or selection of healthy donor sources for allogeneic applications [59].

Thus, ASCs are not passive bystanders in wound environments; rather, they are active immunological architects, capable of reprogramming immune responses and fostering tissue resilience. Their effectiveness, however, hinges upon both intrinsic factors (tissue source, donor health) and extrinsic modulation (culture and delivery conditions), which must be carefully considered in clinical protocols.

Differentiation Capacity: Mesodermal and Beyond

Adipose-derived stem cells (ASCs) are multipotent mesenchymal stem cells capable of differentiating into various lineages under specific induction conditions. Their mesodermal lineage commitment includes adipocytes, osteoblasts, and chondrocytes, while non-mesodermal transdifferentiation extends to neuronal and endothelial phenotypes, offering wide applicability in regenerative medicine.

Adipogenesis is classically induced by culturing ASCs in a medium containing 3-isobutyl-1-methylxanthine (IBMX), dexamethasone, insulin, and indomethacin. This cocktail activates peroxisome proliferator-activated receptor gamma (PPARγ), the master regulator of adipocyte differentiation, leading to the accumulation of intracellular lipid droplets and the expression of adipocyte-specific markers [60]. Osteogenesis is induced by supplementing media with ascorbate-2-phosphate, β-glycerophosphate, and dexamethasone. Under these conditions, ADSCs express alkaline phosphatase (ALP) and osteocalcin and form mineralized extracellular matrixes, visualized by Alizarin Red S staining [61]. Chondrogenesis is promoted in high-density micromass cultures using transforming growth factor-beta 1 or 3 (TGF-β1/3). This results in the synthesis of cartilaginous matrix proteins, including type II collagen and aggrecan, both hallmarks of chondrocyte differentiation [62].

ASCs have also shown capacity to transdifferentiate into non-mesodermal lineages, particularly neuronal and endothelial cells. In neurogenic differentiation, ASCs are cultured in media supplemented with brain-derived neurotrophic factor (BDNF), nerve growth factor (NGF), and retinoic acid. Within days, cells acquire neuron-like morphologies and begin expressing markers such as βIII-tubulin, microtubule-associated protein 2 (MAP 2), and glial fibrillary acidic protein (GFAP) [63]. For endothelial differentiation, exposure to vascular endothelial growth factor (VEGF) and fibroblast growth factor-2 (FGF-2) is key. Differentiated cells express CD31 (PECAM-1) and von Willebrand factor (vWF) and are capable of forming capillary-like structures on Matrigel substrates—a functional hallmark of angiogenesis [64].

The differentiation trajectory of ASCs is not determined solely by cytokine cocktails; extrinsic biophysical and biochemical factors exert substantial influence: (i) Biomaterial scaffolds, particularly those mimicking extracellular matrix components (e.g., collagen, fibrin, hyaluronic acid), enhance lineage-specific differentiation by modulating cell adhesion, morphology, and mechanotransduction pathways; (ii) Hypoxic preconditioning (e.g., 1–5% O_2) has been shown to upregulate VEGF and enhance chondrogenesis and angiogenic potential by stabilizing hypoxia-inducible factor-1 alpha (HIF-1α); (iii) Mechanical stimuli such as cyclic stretching and shear stress can direct ASCs toward smooth muscle or endothelial phenotypes, a feature leveraged in bioengineered vascular constructs.

Collectively, this vast differentiation landscape underscores the versatility of ASCs in tissue engineering and wound-healing applications, particularly when integrated with customized scaffolds, growth factor gradients, and bioreactor systems to emulate in vivo-like conditions.

Mechanisms of Action in Wound Healing

Modulation of Inflammation

The inflammatory phase of wound healing represents a critical window for tissue decontamination and immune cell recruitment, but excessive or prolonged inflammation is a hallmark of chronic wounds and fibrotic outcomes. Adipose-derived stem cells (ASCs) exhibit profound immunomodulatory potential, reshaping the inflammatory microenvironment through both paracrine signaling and cell-cell interactions, thereby transitioning wounds toward regeneration.

At the molecular level, ASCs secrete a rich profile of anti-inflammatory cytokines and immunomodulatory mediators, including interleukin-10 (IL-10), transforming growth factor-beta (TGF-β), prostaglandin E2 (PGE2), indoleamine 2,3-dioxygenase (IDO), and nitric oxide (NO). These soluble factors regulate local immune dynamics by suppressing pro-inflammatory responses and promoting tissue-resolving phenotypes. IL-10 plays a central role in deactivating M1 macrophages, downregulating antigen presentation and NF-κB signaling pathways, and stimulating Treg development, thereby inducing tolerance within the wound microenvironment [65]. TGF-β, beyond its fibrogenic role, contributes to T-cell polarization, inducing FoxP3+ regulatory T cells (Tregs) and inhibiting Th1 and Th17 responses. It also regulates myeloid cell maturation, maintaining dendritic cells (DCs) in an immature, tolerogenic state. IDO catalyzes tryptophan degradation, limiting T-cell proliferation and facilitating immune quiescence. It further mediates Treg recruitment and NK cell inhibition [66]. PGE2, through EP2/EP4 receptors, suppresses DC maturation and induces M2 macrophage polarization, enhancing resolution-phase signaling [67]. NO, a small molecule with multifaceted roles regulates vascular tone, leukocyte migration, and microbial control while maintaining immunosuppression when released at moderate levels.

Macrophages, which play a pivotal role in wound outcomes, undergo functional polarization influenced by ASCs. ASCs promote a shift from M1 (pro-inflammatory) to M2 (anti-inflammatory) phenotypes via secreted IL-10, TGF-β, and PGE2. M2 macrophages facilitate tissue remodeling and angiogenesis and express surface markers such as CD206 and arginase-1. This phenotypic transition has been observed in vivo, where wounds treated with ADSCs or ADSC-derived exosomes exhibit reduced iNOS+ M1 cells and increased CD163+ M2 macrophages, correlating with enhanced re-epithelialization and granulation tissue formation [68].

ASCs also modulate NK cell cytotoxicity through downregulation of IL-2-mediated JAK-STAT and AKT signaling, as demonstrated by increased expression of CIS and DUSP4, intracellular inhibitors of activation pathways. This suppression reduces IFN-γ release and cytotoxic granule production, thus preventing collateral tissue damage during early wound inflammation [69]. Moreover, ASCs exhibit a suppressive effect on both CD4+ helper and CD8+ cytotoxic T cells, impairing their proliferation, cytokine release, and cytolytic capacity. This occurs through both direct contact (e.g., PD-L1 expression) and secreted molecules such as IDO and

TGF-β. CD4+ T cells are shifted away from Th1/Th17 polarization toward Treg or Th2 phenotypes, which secrete anti-inflammatory cytokines like IL-4 and IL-10. CD8+ T cells experience reduced activation and lower expression of IFN-γ and granzyme B, diminishing tissue cytotoxicity [67].

The clinical implications of ASC immunomodulation are profound. In chronic wounds such as diabetic foot ulcers, venous leg ulcers, and radiation-induced injuries, the failure to resolve inflammation underlies persistent tissue breakdown. ADSC administration has demonstrated: (i) accelerated wound closure due to early inflammatory resolution, (ii) reduced scar formation linked to controlled fibroblast activation, and (iii) enhanced vascularization driven by M2 macrophage-derived VEGF and IL-10.

Studies have shown that allogeneic ASCs, despite MHC mismatch, can still be used effectively due to their immune-evasive properties, making them promising for off-the-shelf therapies. Preclinical evidence and early-phase human trials support the safety and efficacy of ASC therapies in modulating immune dysfunction and restoring wound-healing capacity [70–73].

Promotion of Angiogenesis

One of the most vital aspects of wound healing is the restoration of a functional microvascular network to ensure adequate oxygenation and nutrient delivery to regenerating tissues. ASCs significantly contribute to neovascularization through their potent paracrine activity, which includes a diverse angiogenic secretome and the activation of multiple molecular pathways that collectively stimulate endothelial proliferation, migration, and morphogenesis.

ASCs secrete a broad spectrum of angiogenic factors, the most prominent being:

- Vascular Endothelial Growth Factor (VEGF): This is the principal angiogenic cytokine, binding to VEGFR-2 on endothelial cells to stimulate proliferation, increase vascular permeability, and induce sprouting angiogenesis. VEGF expression by ASCs is upregulated under hypoxic conditions, mimicking the ischemic microenvironment of chronic wounds [74].
- Basic Fibroblast Growth Factor (bFGF): Acts synergistically with VEGF to promote endothelial cell proliferation and differentiation. It also enhances the deposition of extracellular matrix components that serve as scaffolds for new vessel formation [75].
- Angiopoietin-1 (ANGPT-1): Critical for vessel stabilization, ANGPT-1 binds to the Tie2 receptor on endothelial cells, reducing leakage and promoting maturation of the nascent vascular network [76].
- Stromal Cell-Derived Factor 1 (SDF-1): Functions as a chemoattractant for endothelial progenitor cells (EPCs), facilitating their homing to the wound bed. SDF-1 also promotes pericyte recruitment, which is essential for vessel maturation and integrity [77].

These molecules act not only individually but also synergistically to modulate angiogenic cascades, including ERK1/2, PI3K/AKT, and MAPK signaling, which are crucial for endothelial survival and migration. ADSCs also secrete extracellular vesicles (EVs), including exosomes loaded with pro-angiogenic microRNAs (e.g., miR-126, miR-132), which further enhance endothelial cell activity and tube formation in vitro.

In ischemic and pressure ulcer models, ASCs have shown upregulation of the peroxisome proliferator-activated receptor beta/delta (PPARβ/δ) pathway [78]. This nuclear receptor modulates genes involved in fatty acid metabolism and angiogenesis. When activated, PPARβ/δ enhances VEGF transcription and improves tissue perfusion, particularly in hypoxic environments [79]. Moreover, hypoxia-inducible factor-1 alpha (HIF-1α) is stabilized in low-oxygen environments, further boosting VEGF, ANGPT-1, and SDF-1 expression. This makes hypoxia-preconditioned ASCs or ASCs embedded in oxygen-releasing scaffolds even more effective in driving neovascularization [80].

Preclinical studies consistently demonstrate that ASCs enhance angiogenesis in ischemic limbs [81], diabetic wounds [82], and radiation ulcers [83]. In murine diabetic wound models, local ASC injection resulted in (i) faster revascularization, with increased capillary density in histological sections; (ii) elevated levels of CD31+ endothelial cells and α-SMA+ pericytes; and (iii) significant upregulation of VEGF and bFGF mRNA at the wound site [84]. Additionally, ASCs embedded in biomaterial scaffolds (e.g., collagen, fibrin, hydrogel matrixes) have shown improved angiogenic outcomes compared to cell injections alone, suggesting the importance of scaffold-assisted localization and controlled release of angiogenic signals [85].

Clinical applications of ASC-rich fat grafts and SVF preparations have also reported increased vascularity in radiation-induced wounds, chronic leg ulcers, and mastectomy scars, correlating with improved healing rates and better cosmetic outcomes [86–88].

Fibroblast Activation and ECM Remodeling

Adipose-derived stem cells (ASCs) exert a profound influence on wound repair not only through angiogenesis and immunomodulation but also by modulating the behavior of dermal fibroblasts and orchestrating a dynamic restructuring of the extracellular matrix (ECM). Fibroblasts are central players in wound healing, responsible for synthesizing collagen, fibronectin, elastin, and other ECM proteins that define the architectural integrity and mechanical properties of the skin [89]. ASCs enhance both fibroblast proliferation and directional migration, facilitating their recruitment into the wound bed and supporting re-epithelialization and matrix deposition [90].

At a molecular level, this pro-regenerative interaction is mediated by a rich paracrine secretome that includes basic fibroblast growth factor (bFGF), transforming growth factor-β (TGF-β), hepatocyte growth factor (HGF), and platelet-derived growth factor (PDGF)—all of which are known to activate fibroblast function [91]. Notably, ASC-conditioned medium has been demonstrated to significantly increase fibroblast proliferation, migration, and contractility, making them ideal cellular partners for dermal regeneration.

A particularly critical aspect of ECM control involves the balance between matrix metalloproteinases (MMPs) and their endogenous inhibitors (TIMPs). MMPs are zinc-dependent endopeptidases that degrade structural components of the ECM, enabling its turnover and remodeling. Specifically, MMP-1 (collagenase), MMP-2 (gelatinase A), and MMP-9 (gelatinase B) are implicated in remodeling of collagen types I and III, as well as basement membrane components. ASCs can regulate the expression of these enzymes in fibroblasts, stimulating appropriate degradation of old or damaged matrix components to allow for deposition of new tissue [92]. At the same time, ASCs modulate the expression of TIMP-1 and TIMP-2, preserving matrix stability and preventing excessive ECM breakdown or fibrotic scarring.

One of the earliest studies to demonstrate this was conducted by Kim et al. [90], where conditioned media from ASCs was shown to increase the expression of collagen type I, fibronectin, and elastin genes in normal human dermal fibroblasts, alongside increased MMP-2 and TIMP-1 mRNA expression, indicating a sophisticated remodeling effect. These findings were correlated with improved skin texture, elasticity, and dermal thickness in both in vitro and in vivo skin models. Further, studies using three-dimensional wound-healing models have revealed that ASCs promote dermal fibroblast contractility, a critical component of wound contraction, without driving them toward a myofibroblast phenotype, which is associated with fibrosis and scar formation [93–95]. This suggests a nuanced role of ASCs in regulating the fibroblast phenotype, favoring regenerative rather than fibrotic pathways. Additionally, emerging data suggests that extracellular vesicles (EVs) and exosomes secreted by ASCs contain microRNAs (miRNAs) such as miR-21, miR-146a, and miR-29, which directly modulate ECM gene expression and fibroblast signaling pathways [96, 97]. For instance, miR-29 has been shown to suppress collagen overproduction, whereas miR-21 promotes wound closure through activation of AKT and ERK signaling in fibroblasts.

In pathological settings such as chronic wounds, where fibroblasts often display a senescent or pro-inflammatory phenotype, ASCs have demonstrated the capacity to "rescue" fibroblast function. By restoring proliferative and migratory abilities and modulating oxidative stress and cytokine production, ASCs re-establish a regenerative niche conducive to efficient ECM turnover and dermal repair. Thus, ASCs serve not merely as supportive stromal cells but as bioactive orchestrators of matrix regeneration. Their secretome, immunomodulatory functions, and epigenetic signaling converge to influence the composition, organization, and remodeling of the ECM—paving the way for functional and cosmetically optimal skin repair.

Antifibrotic and Anti-Scar Effects

ASCs play a central role in counteracting dermal fibrosis, particularly in hypertrophic scars and keloids, through multiple interrelated mechanisms. These include the inhibition of myofibroblast differentiation, suppression of TGF-β1 signaling, matrix remodeling via matrix metalloproteinases (MMPs), and paracrine regulation of fibroblast activity.

TGF-β1 is a key cytokine in fibrotic diseases, promoting fibroblast-to-myofibroblast transition and overproduction of extracellular matrix (ECM). ASCs can attenuate the TGF-β1/Smad pathway, limiting α-SMA and collagen I expression. Studies demonstrate that exosomes from human ADSCs suppress ECM production in keloid fibroblasts by downregulating TGF-β1 [98]. ASCs influence ECM turnover by upregulating MMP-1, MMP-2, and MMP-14 and downregulating their inhibitors (TIMPs). This balance ensures ECM degradation of fibrotic tissue without triggering instability. Spiekman et al. [99] showed that ADSCs inhibit TGF-β1–induced myofibroblast conversion in both normal dermal and keloid fibroblasts, suggesting strong paracrine antifibrotic signaling. Through suppression of TGF-β1 and associated transcriptional regulators (e.g., Smad3, Notch1), ASCs prevent fibroblasts from acquiring a myofibroblastic phenotype—cells characterized by α-SMA expression and high contractility. This was confirmed in the work of Han et al. [100], where ASCs downregulated Notch-1 and α-SMA expression in fibrotic scar fibroblasts.

In vivo models support these in vitro findings. Borovikova et al. (2018) documented reduced fibrosis and collagen deposition following ASC administration in murine dermal injury models [101]. Similarly, Wu et al. [102] highlighted that blocking ASC-derived exosome pathways reactivates TGF-β1/Smad signaling in keloid fibroblasts, confirming the causal antifibrotic action of ASCs. A 2024 study by Wang et al. further confirmed that ASC-derived factors suppress fibrosis by lowering collagen gene expression and upregulating MMP activity in scar tissues [103]. A separate comprehensive review by Alfarafisa and colleagues [104] validated the safety and therapeutic potential of ASC-based products for treating pathological scarring in both preclinical and clinical contexts.

In summary, ASCs represent a potent antifibrotic tool by interfering with fibrogenic cytokine pathways, modulating fibroblast behavior, and facilitating ECM remodeling. These multifactorial effects position them as key candidates in future anti-scar and wound-healing therapies.

Preclinical and Clinical Evidence

Numerous preclinical studies have consistently demonstrated that ASC administration significantly improves re-epithelialization, angiogenesis, and collagen matrix remodeling in both normal and diabetic wound settings:

- An et al. (2020) reported that ASCs isolated from diabetic mice improved cutaneous wound healing in streptozotocin-induced diabetic models. Treated wounds exhibited enhanced angiogenesis, granulation tissue development, and increased VEGF expression compared to controls [105].
- In a porcine model—closer to human skin physiology—Irons et al. (2018) demonstrated that treatment with ADSCs, either undifferentiated or endothelial-differentiated, led to superior wound closure, collagen deposition, and vascularization, indicating translational potential [106].
- Kim et al. (2011) found that human ASCs significantly improved healing of ischemic wounds in diabetic nude mice, increasing capillary density and accelerating epithelialization [107].
- Jiang et al. (2013) evaluated ASC therapy embedded in collagen scaffolds in diabetic pigs and observed enhanced angiogenesis and granulation, as well as higher wound tensile strength [108].
- Navone et al. (2014) used a decellularized silk fibroin scaffold seeded with ASCs in diabetic mice and noted not only improved wound contraction and re-epithelialization but also gene expression changes supporting neovascularization [109].
- In another compelling study, Wang et al. (2023) demonstrated that extracellular vesicles derived from ASCs enhanced angiogenesis collagen synthesis and accelerated wound closure in diabetic mice, reinforcing the importance of the ASC secretome in regenerative effects [110].
- In a diabetic rat model, Almalki (2022) highlighted that AMSCs (a type of ASC) accelerated wound closure, enhanced re-epithelialization, and stimulated angiogenesis. The study also noted significant granulation tissue development and better collagen matrix organization [111].
- Sun et al. (2020) demonstrated that ASCs derived from diabetic mice still retained therapeutic efficacy. When administered in diabetic wound models, these cells significantly improved angiogenesis and accelerated healing despite originating from metabolically impaired donors [112].
- Ma et al. (2023) used a novel bilayer cell patch combining ASCs and epidermal stem cells to treat diabetic wounds, revealing synergistic improvement in re-epithelialization and neovascularization. This approach optimized scar size and collagen organization in vivo [113].
- In their 2022 study, Hsu et al. applied exosomes derived from diabetic ASCs to db/db mice wounds, showing improved wound contraction, collagen density, and angiogenic markers. This highlights the role of ASC secretome in healing, especially under diabetic conditions [114].
- Ahmadieh-Yazdi et al. (2024) reviewed how both ADSCs and their exosomes promote vascularization and re-epithelialization in diabetic rats, reinforcing the benefit of using both cells and derivatives for chronic wound care [115].
- Deng et al. (2024) found that ASC-conditioned medium from angiopoietin-1 gene-modified MSCs led to increased collagen types I and III expression in diabetic mice, suggesting enhanced tensile strength and quality of regenerated dermis [116].

- A porcine model by Chou et al. (2020) demonstrated that ASCs, when combined with carbon dioxide-decellularized dermal matrix, supported rapid re-epithelialization, organized collagen deposition, and robust vascular development, underlining the scalability of ASC therapies for large wound surfaces [117].

These findings collectively underscore that ASCs improve wound healing via angiogenesis, ECM reorganization, immunomodulation, and stem cell integration, making them a cornerstone of future regenerative strategies.

Clinical Trials and Case Studies

Recent clinical evidence robustly supports the therapeutic potential of ASCs, particularly in treating chronic, ischemic, and radiation-induced wounds. These therapies include the use of stromal vascular fraction (SVF), expanded ASCs, and fat grafts enriched with regenerative cells.

- A systematic clinical review by Laloze et al. confirmed the efficacy of ASC-based therapies across a range of wounds, including diabetic foot ulcers, chronic ulcers, and radiation-induced injuries, emphasizing enhanced healing rates and improved patient outcomes [118].
- In a randomized clinical trial, Tanios et al. (2021) found that autologous SVF injections in chronic ulcers led to statistically significant reductions in wound size, improved granulation, and rapid re-epithelialization compared to standard care [29].
- A case report by Huayllani et al. (2021) described the successful treatment of radiation-induced ulcers using ASC-enriched fat grafts. The patient showed rapid wound closure, improved skin pliability, and sustained epithelial integrity without adverse effects [119].
- Similarly, Wu et al. (2024) highlighted the use of ASCs in radiotherapy-damaged skin, reporting increased neovascularization, reduced scarring, and enhanced aesthetic outcomes in irradiated tissue beds [120].
- In diabetic foot ulcers, Álvaro-Afonso et al. (2020) demonstrated improved epithelial regeneration and scar thickness reduction using AMSCs (a subset of ASCs), confirming safety and reproducibility across early-phase clinical studies [30].
- Further, Milan et al. (2024) reviewed multiple studies highlighting ASC-enriched interventions that significantly improved scar quality, vascularity, and elasticity, especially in burns and radiation-damaged tissues [121].
- A systematic review by Stachura et al. (2021) focused on ASC and SVF applications in scar therapy and wound management. Their synthesis confirmed improved outcomes in clinical endpoints such as wound closure speed, skin elasticity, and pigmentation in patients receiving ADSC-based treatments [122].
- Sakai et al. (2021) presented real-world data from autologous ASC therapy in chronic ulcers. Their observational study revealed enhanced wound bed preparation, lower infection rates, and favorable aesthetic outcomes post-healing [123].

Safety Profile

Adipose-derived stem cells (ASCs) are widely recognized for their exceptional safety, both in autologous and allogeneic applications, a feature critical for their increasing translation into clinical practice.

- ASCs exhibit low immunogenicity due to their reduced expression of major histocompatibility complex class II (MHC-II) and co-stimulatory molecules. This renders them capable of bypassing host immune surveillance, making them ideal for allogeneic "off-the-shelf" therapies. As summarized by Ceccarelli et al., ASCs exert profound immunomodulatory effects, dampening immune responses through paracrine mechanisms without eliciting adverse immune activation [124].
- Additionally, Farhana et al. (2023) reported that T-regulatory cell maturation and immune orientation in allograft scenarios are maintained in ASC applications, indicating their potential for transplantation without immunosuppression [68].

One of the pivotal concerns with stem cell-based therapies is the potential for oncogenic transformation. However, comprehensive reviews have consistently shown no evidence of tumor formation or malignant transformation following ASC use:

- Pikuła and colleagues (2013) stated that, despite wide clinical use, no malignant transformation events have been recorded for ASCs, even when applied in high-risk oncologic or irradiated tissues [125].
- Shukla et al. (2015) emphasized the oncologic safety of fat grafting and ADSC therapy in irradiated fields, noting the absence of neoplastic recurrence over prolonged follow-up periods in radiation injury patients [126].

A broad review by Gir et al. (2012) summarized early clinical trials involving ADSCs and concluded a lack of serious adverse events, highlighting the overall tolerability and reproducibility of these therapies across multiple wound healing and reconstructive contexts [127]. More recently, Wang et al. (2025) reinforced that ASC-based interventions in regenerative medicine are not associated with immune rejection or tumorigenesis and that results are consistent across both early- and mid-phase clinical trials [103].

Before clinical adoption, the safety of ASCs was thoroughly tested in animal models, including rabbit and rodent models, where no toxicity, teratoma formation, or systemic complications were observed. These studies set the foundation for first-in-human trials. In conclusion, an expansive body of preclinical and clinical evidence supports the safety of ASCs, including their use in allogeneic settings, without observed immunogenic or tumorigenic complications. This reinforces their suitability for routine and investigational use in wound healing and beyond.

Challenges and Regulatory Considerations

One of the most pressing challenges in ASC therapy is the lack of standardization in terms of cell isolation, processing, and dosing. While adipose tissue offers an abundant source of stem cells, there is considerable variability in the isolation protocols—ranging from enzymatic digestion (e.g., collagenase) to mechanical separation techniques. This inconsistency results in heterogeneity in cell yield, viability, and functionality. As noted by Zocchi et al., this lack of procedural uniformity hinders reproducibility across clinical studies and affects therapeutic predictability [128]. Furthermore, defining a therapeutic dose remains elusive. Unlike pharmaceuticals, where dosing is well-defined, stem cell therapies must account for donor variability, cell passage number, and application context, making standardization difficult [129].

The regulatory framework for ASCs diverges significantly between the United States and Europe:

In the United States, the FDA classifies minimally manipulated stem cell products (including SVF) under HCT/Ps (Human Cells, Tissues, and Cellular and Tissue-Based Products). However, if manipulation exceeds minimal thresholds or if the cells are intended for non-homologous use (i.e., use not consistent with the tissue's original function), they fall under more stringent Investigational New Drug (IND) protocols. Turner (2015) discusses how the FDA has cracked down on clinics marketing unapproved stem cell interventions, citing safety, efficacy, and legal concerns [130].

In Europe, ASCs are considered Advanced Therapy Medicinal Products (ATMPs) by the European Medicines Agency (EMA). This requires manufacturers to comply with Good Manufacturing Practice (GMP), conduct extensive preclinical and clinical testing, and undergo central authorization procedures. Raposio et al. (2017) highlight the complex legal framework and the challenges it poses for clinical translation in the EU [131].

Global harmonization of these regulatory frameworks remains a long-term goal, as emphasized by Mendicino et al. (2015), who recommend aligning FDA and EMA pathways for better access and patient safety [132].

ASCs, being autologous or adult-derived, largely bypass the ethical controversies associated with embryonic stem cells. However, the direct-to-consumer marketing of unproven stem cell products remains a significant ethical challenge. Clinics often circumvent regulatory oversight, offering therapies without substantiated efficacy, placing patients at risk. Pean et al. (2019) raise concerns about misleading advertisements and lack of post-treatment surveillance [133]. Additionally, long-term safety data are still scarce. Although short-term trials report excellent tolerability, large-scale longitudinal studies are needed to rule out delayed immunogenicity, fibrosis, or tumorigenesis—especially in allogeneic or genetically manipulated ASC applications [134].

Conclusion

Adipose-derived stem cells (ASCs) are at the forefront of regenerative medicine, offering a clinically relevant, versatile, and scalable solution for managing acute and chronic wounds. Their multipotent differentiation capacity, coupled with potent immunomodulatory, angiogenic, and antifibrotic properties, makes them uniquely suited to enhance tissue regeneration, accelerate healing, and improve both functional and aesthetic outcomes. Unlike other stem cell sources, the abundance and accessibility of adipose tissue provide a minimally invasive and high-yield harvesting approach, facilitating real-world clinical translation.

Over the past two decades, an expanding body of preclinical and clinical evidence has validated the efficacy and safety of ASCs in treating diabetic foot ulcers, radiation-induced injuries, venous ulcers, and postsurgical wounds. Clinical trials and case series have demonstrated improved re-epithelialization, neovascularization, collagen remodeling, and reduced scar formation, even in patients with compromised healing capacity. Furthermore, advancements in cell delivery systems, including hydrogel scaffolds, exosome formulations, and oxygen-releasing biomaterials, have amplified the regenerative impact of ASCs and opened new avenues for targeted therapies.

However, the widespread adoption of ASC-based therapies into routine clinical practice requires standardized protocols for cell isolation and dosing, long-term safety data, and regulatory alignment across jurisdictions. Navigating the complex regulatory terrain set forth by agencies like the FDA and EMA—particularly regarding the classification of ASCs as Advanced Therapy Medicinal Products (ATMPs)—remains a pivotal step in enabling broader clinical use.

In conclusion, ASCs embody a next-generation therapeutic platform that bridges experimental innovation with clinical practicality. With continued investment in clinical trials, regulatory refinement, and technology development, they are poised to become an integral component of evidence-based wound care, transforming outcomes for patients across diverse clinical settings.

References

1. Mooney DJ (2015) Regenerative medicine: current therapies and future directions. Proc Natl Acad Sci USA 112(47):14452–14459. https://doi.org/10.1073/pnas.1508520112. PMID: 26598661; PMCID: PMC4664309
2. Langer R, Vacanti JP (1993) Tissue engineering. Science 260(5110):920–926. https://doi.org/10.1126/science.8493529. PMID: 8493529
3. Armingol E, Officer A, Harismendy O, Lewis NE (2021) Deciphering cell-cell interactions and communication from gene expression. Nat Rev Genet 22(2):71–88. https://doi.org/10.1038/s41576-020-00292-x. Epub 2020 Nov 9. PMID: 33168968; PMCID: PMC7649713
4. Kim TG, Shin H, Lim DW (2012) Biomimetic scaffolds for tissue engineering. Adv Funct Mater 22:2446–2468. https://doi.org/10.1002/adfm.201103083

5. Karantalis V, Schulman IH, Balkan W, Hare JM (2015) Allogeneic cell therapy: a new paradigm in therapeutics. Circ Res 116(1):12–15. https://doi.org/10.1161/CIRCRESAHA.114.305495. PMID: 25552688; PMCID: PMC4411634
6. Goldenberg D, McLaughlin C, Koduru SV, Ravnic DJ (2021) Regenerative engineering: current applications and future perspectives. Front Surg 8:731031. https://doi.org/10.3389/fsurg.2021.731031. PMID: 34805257; PMCID: PMC8595140
7. Lacina L, Kolář M, Pfeiferová L, Gál P, Smetana K Jr (2024) Wound healing: insights into autoimmunity, ageing, and cancer ecosystems through inflammation and IL-6 modulation. Front Immunol 15:1403570. https://doi.org/10.3389/fimmu.2024.1403570. PMID: 39676864; PMCID: PMC11638159
8. Hussen BM, Taheri M, Yashooa RK, Abdullah GH, Abdullah SR, Kheder RK, Mustafa SA (2024) Revolutionizing medicine: recent developments and future prospects in stem-cell therapy. Int J Surg 110(12):8002–8024. https://doi.org/10.1097/JS9.0000000000002109. PMID: 39497543; PMCID: PMC11634165
9. Trounson A, McDonald C (2015) Stem cell therapies in clinical trials: progress and challenges. Cell Stem Cell 17(1):11–22. https://doi.org/10.1016/j.stem.2015.06.007. PMID: 26140604
10. Kassem M, Abdallah BM (2008) Human bone-marrow-derived mesenchymal stem cells: biological characteristics and potential role in therapy of degenerative diseases. Cell Tissue Res 331(1):157–163. https://doi.org/10.1007/s00441-007-0509-0. Epub 2007 Sep 25. PMID: 17896115
11. Nagamura-Inoue T, He H (2014) Umbilical cord-derived mesenchymal stem cells: their advantages and potential clinical utility. World J Stem Cells. 6(2):195–202. https://doi.org/10.4252/wjsc.v6.i2.195. PMID: 24772246; PMCID: PMC3999777
12. De Francesco F, Ricci G, D'Andrea F, Nicoletti GF, Ferraro GA (2015) Human adipose stem cells: from bench to bedside. Tissue Eng Part B Rev 21(6):572–584. https://doi.org/10.1089/ten.TEB.2014.0608. Epub 2015 Jun 17. PMID: 25953464
13. Bertozzi N, Simonacci F, Gricco MP, Grignaffini E, Raposio E (2017) The biological and clinical basis for the use of adipose-derived stem cells in the field of wound healing. Ann Med Surg (Lond) 20:41–48. https://doi.org/10.1016/j.amsu.2017.06.058. PMID: 28702186; PMCID: PMC5491486
14. El-Badawy A, Ahmed SM, El-Badri N (2017) Adipose-derived stem cell-based therapies in regenerative medicine. In: El-Badri N (ed) Advances in stem cell therapy. Stem cell biology and regenerative medicine. Humana Press
15. Dominici M, Le Blanc K, Mueller I, Slaper-Cortenbach I, Marini F, Krause D, Deans R, Keating A, Prockop D, Horwitz E (2006) Minimal criteria for defining multipotent mesenchymal stromal cells. The International Society for Cellular Therapy position statement. Cytotherapy 8(4):315–317. https://doi.org/10.1080/14653240600855905. PMID: 16923606
16. Choudhery MS, Mahmood R, Harris DT, Ahmad FJ (2022) Minimum criteria for defining induced mesenchymal stem cells. Cell Biol Int 46(6):986–989. https://doi.org/10.1002/cbin.11790. Epub 2022 Mar 27. PMID: 35293653
17. De Francesco F, Tirino V, Desiderio V, Ferraro G, D'Andrea F, Giuliano M, Libondi G, Pirozzi G, De Rosa A, Papaccio G (2009) Human CD34/CD90 ASCs are capable of growing as sphere clusters, producing high levels of VEGF and forming capillaries. PLoS One 4(8):e6537. https://doi.org/10.1371/journal.pone.0006537. PMID: 19657392; PMCID: PMC2717331
18. Ferraro GA, De Francesco F, Nicoletti G, Paino F, Desiderio V, Tirino V, D'Andrea F (2013) Human adipose CD34+ CD90+ stem cells and collagen scaffold constructs grafted in vivo fabricate loose connective and adipose tissues. J Cell Biochem 114(5):1039–1049. https://doi.org/10.1002/jcb.24443. PMID: 23129214
19. Ong WK, Chakraborty S, Sugii S (2021) Adipose tissue: understanding the heterogeneity of stem cells for regenerative medicine. Biomolecules. 11(7):918. https://doi.org/10.3390/biom11070918. PMID: 34206204; PMCID: PMC8301750

20. Baer PC (2011) Adipose-derived stem cells and their potential to differentiate into the epithelial lineage. Stem Cells Dev 20(10):1805–1816. https://doi.org/10.1089/scd.2011.0086. Epub 2011 Jun 1. PMID: 21495915
21. Anghileri E, Marconi S, Pignatelli A, Cifelli P, Galié M, Sbarbati A, Krampera M, Belluzzi O, Bonetti B (2008) Neuronal differentiation potential of human adipose-derived mesenchymal stem cells. Stem Cells Dev 17(5):909–916. https://doi.org/10.1089/scd.2007.0197. PMID: 18564036
22. Policha A, Zhang P, Chang L, Lamb K, Tulenko T, DiMuzio P (2014) Endothelial differentiation of diabetic adipose-derived stem cells. J Surg Res 192(2):656–663. https://doi.org/10.1016/j.jss.2014.06.041. Epub 2014 Jul 5. PMID: 25091340
23. Qin Y, Ge G, Yang P, Wang L, Qiao Y, Pan G, Yang H, Bai J, Cui W, Geng D (2023) An update on adipose-derived stem cells for regenerative medicine: where challenge meets opportunity. Adv Sci (Weinh) 10(20):e2207334. https://doi.org/10.1002/advs.202207334. Epub 2023 May 10. PMID: 37162248; PMCID: PMC10369252
24. Mazini L, Rochette L, Amine M, Malka G (2019) Regenerative capacity of adipose derived stem cells (ADSCs), comparison with mesenchymal stem cells (MSCs). Int J Mol Sci 20(10):2523. https://doi.org/10.3390/ijms20102523. PMID: 31121953; PMCID: PMC6566837
25. Kuo YR, Wang CT, Cheng JT, Kao GS, Chiang YC, Wang CJ (2016) Adipose-derived stem cells accelerate diabetic wound healing through the induction of autocrine and paracrine effects. Cell Transplant 25(1):71–81. https://doi.org/10.3727/096368915X687921. Epub 2015 Apr 7. PMID: 25853951
26. Chang SY, Lee JH, Oh SC, Lee MY, Lim NK (2023) Human fibroblast growth factor-treated adipose-derived stem cells facilitate wound healing and revascularization in rats with Streptozotocin-induced diabetes mellitus. Cells 12(8):1146. https://doi.org/10.3390/cells12081146. PMID: 37190055; PMCID: PMC10136967
27. Hamada T, Matsubara H, Yoshida Y, Ugaji S, Nomura I, Tsuchiya H (2019) Autologous adipose-derived stem cell transplantation enhances healing of wound with exposed bone in a rat model. PLoS One 14(5):e0214106. https://doi.org/10.1371/journal.pone.0214106. PMID: 31083652; PMCID: PMC6513073
28. Zografou A, Papadopoulos O, Tsigris C, Kavantzas N, Michalopoulos E, Chatzistamatiou T, Papassavas A, Stavropoulou-Gioka C, Dontas I, Perrea D (2013) Autologous transplantation of adipose-derived stem cells enhances skin graft survival and wound healing in diabetic rats. Ann Plast Surg 71(2):225–232. https://doi.org/10.1097/SAP.0b013e31826af01a. PMID: 23636118
29. Tanios E, Ahmed TM, Shafik EA, Sherif MF, Sayed D, Gaber N, Hassan Y (2021) Efficacy of adipose-derived stromal vascular fraction cells in the management of chronic ulcers: a randomized clinical trial. Regen Med 16(11):975–988. https://doi.org/10.2217/rme-2020-0207. Epub 2021 Oct 1. PMID: 34596433
30. Álvaro-Afonso FJ, Sanz-Corbalán I, Lázaro-Martínez JL, Kakagia D, Papanas N (2020) Adipose-derived mesenchymal stem cells in the treatment of diabetic foot ulcers: a review of preclinical and clinical studies. Angiology 71(9):853–863. https://doi.org/10.1177/0003319720939467. Epub 2020 Jul 29. PMID: 32723090
31. Moon KC, Suh HS, Kim KB, Han SK, Young KW, Lee JW, Kim MH (2019) Potential of allogeneic adipose-derived stem cell-hydrogel complex for treating diabetic foot ulcers. Diabetes 68(4):837–846. https://doi.org/10.2337/db18-0699. Epub 2019 Jan 24. PMID: 30679183
32. Papadopoulos KS, Piperi C, Korkolopoulou P (2024) Clinical applications of adipose-derived stem cell (ADSC) exosomes in tissue regeneration. Int J Mol Sci 25(11):5916. https://doi.org/10.3390/ijms25115916. PMID: 38892103; PMCID: PMC11172884
33. Yuan X, Li L, Liu H, Luo J, Zhao Y, Pan C, Zhang X, Chen Y, Gou M (2022) Strategies for improving adipose-derived stem cells for tissue regeneration. Burns Trauma 10:tkac028. https://doi.org/10.1093/burnst/tkac028. PMID: 35992369; PMCID: PMC9382096

34. Gentile P, Calabrese C, De Angelis B, Pizzicannella J, Kothari A, Garcovich S (2019) Impact of the different preparation methods to obtain human adipose-derived stromal vascular fraction cells (AD-SVFs) and human adipose-derived mesenchymal stem cells (AD-MSCs): enzymatic digestion versus mechanical centrifugation. Int J Mol Sci 20(21):5471. https://doi.org/10.3390/ijms20215471. PMID: 31684107; PMCID: PMC6862236
35. Ramakrishnan VM, Boyd NL (2018) The adipose stromal vascular fraction as a complex cellular source for tissue engineering applications. Tissue Eng Part B Rev 24(4):289–299. https://doi.org/10.1089/ten.TEB.2017.0061. Epub 2017 Apr 13. PMID: 28316259; PMCID: PMC6080106
36. Silva KR, Côrtes I, Liechocki S, Carneiro JR, Souza AA, Borojevic R, Maya-Monteiro CM, Baptista LS (2017) Characterization of stromal vascular fraction and adipose stem cells from subcutaneous, preperitoneal and visceral morbidly obese human adipose tissue depots. PLoS One 12(3):e0174115. https://doi.org/10.1371/journal.pone.0174115. PMID: 28323901; PMCID: PMC5360317
37. Zhu S, Chen M, Ying Y, Wu Q, Huang Z, Ni W, Wang X, Xu H, Bennett S, Xiao J, Xu J (2022) Versatile subtypes of pericytes and their roles in spinal cord injury repair, bone development and repair. Bone Res 10(1):30. https://doi.org/10.1038/s41413-022-00203-2. PMID: 35296645; PMCID: PMC8927336
38. Frazier T, Lee S, Bowles A, Semon J, Bunnell B, Wu X, Gimble J (2018) Gender and age-related cell compositional differences in C57BL/6 murine adipose tissue stromal vascular fraction. Adipocytes 7(3):183–189. https://doi.org/10.1080/21623945.2018.1460009. Epub 2018 Jun 8. PMID: 29882687; PMCID: PMC6224190
39. Senesi L, De Francesco F, Farinelli L, Manzotti S, Gagliardi G, Papalia GF, Riccio M, Gigante A (2019) Mechanical and enzymatic procedures to isolate the stromal vascular fraction from adipose tissue: preliminary results. Front Cell Dev Biol 7:88. https://doi.org/10.3389/fcell.2019.00088. PMID: 31231649; PMCID: PMC6565890
40. Francis MP, Sachs PC, Elmore LW, Holt SE (2010) Isolating adipose-derived mesenchymal stem cells from lipoaspirate blood and saline fraction. Organogenesis 6(1):11–14. https://doi.org/10.4161/org.6.1.10019. PMID: 20592860; PMCID: PMC2861738
41. Zhu M, Heydarkhan-Hagvall S, Hedrick M, Benhaim P, Zuk P (2013) Manual isolation of adipose-derived stem cells from human lipoaspirates. J Vis Exp 79:e50585. https://doi.org/10.3791/50585. PMID: 24121366; PMCID: PMC3935774
42. Palumbo P, Lombardi F, Siragusa G, Cifone MG, Cinque B, Giuliani M (2018) Methods of isolation, characterization and expansion of human adipose-derived stem cells (ASCs): An overview. Int J Mol Sci 19(7):1897. https://doi.org/10.3390/ijms19071897. PMID: 29958391; PMCID: PMC6073397
43. Ferroni L, De Francesco F, Pinton P, Gardin C, Zavan B (2022) Methods to isolate adipose tissue-derived stem cells. In: Methods Cell Biol, vol 171, pp 215–228. https://doi.org/10.1016/bs.mcb.2022.04.011. Epub 2022 Jun 6. PMID: 35953202
44. Condé-Green A, Kotamarti VS, Sherman LS, Keith JD, Lee ES, Granick MS, Rameshwar P (2016) Shift toward mechanical isolation of adipose-derived stromal vascular fraction: review of upcoming techniques. Plast Reconstr Surg Glob Open 4(9):e1017. https://doi.org/10.1097/GOX.0000000000001017. PMID: 27757339; PMCID: PMC5055005
45. Doornaert M, Colle J, De Maere E, Declercq H, Blondeel P (2018) Autologous fat grafting: latest insights. Ann Med Surg (Lond) 37:47–53. https://doi.org/10.1016/j.amsu.2018.10.016. PMID: 30622707; PMCID: PMC6318549
46. Maslennikov S, Maksym G (2023) The most commoncly used cell surface markers for determining mesenchymal stromal cells in stromal vascular fraction and bone marrow autologous concentrate: a systematic review. J Biotech Research 14:85–94
47. Watson JE, Patel NA, Carter G, Moor A, Patel R, Ghansah T, Mathur A, Murr MM, Bickford P, Gould LJ, Cooper DR (2014) Comparison of markers and functional attributes of human adipose-derived stem cells and dedifferentiated adipocyte cells from subcutaneous fat

of an obese diabetic donor. Adv Wound Care (New Rochelle). 3(3):219–228. https://doi.org/10.1089/wound.2013.0452. PMID: 24669358; PMCID: PMC3955970
48. Pachón-Peña G, Serena C, Ejarque M, Petriz J, Duran X, Oliva-Olivera W, Simó R, Tinahones FJ, Fernández-Veledo S, Vendrell J (2016) Obesity determines the Immunophenotypic profile and functional characteristics of human mesenchymal stem cells from adipose tissue. Stem Cells Transl Med 5(4):464–475. https://doi.org/10.5966/sctm.2015-0161. Epub 2016 Mar 8. PMID: 26956208; PMCID: PMC4798735
49. Iser IC, Ceschini SM, Onzi GR, Bertoni AP, Lenz G, Wink MR (2016) Conditioned medium from adipose-derived stem cells (ADSCs) promotes epithelial-to-mesenchymal-like transition (EMT-like) in glioma cells in vitro. Mol Neurobiol 53(10):7184–7199. https://doi.org/10.1007/s12035-015-9585-4. Epub 2015 Dec 19. PMID: 26687184
50. Zuk PA, Zhu M, Mizuno H, Huang J, Futrell JW, Katz AJ, Benhaim P, Lorenz HP, Hedrick MH (2001) Multilineage cells from human adipose tissue: implications for cell-based therapies. Tissue Eng 7(2):211–228. https://doi.org/10.1089/107632701300062859. PMID: 11304456
51. Kuppa SS, Kim HK, Kang JY, Lee SC, Seon JK (2022) Role of mesenchymal stem cells and their paracrine mediators in macrophage polarization: An approach to reduce inflammation in osteoarthritis. Int J Mol Sci 23(21):13016. https://doi.org/10.3390/ijms232113016. PMID: 36361805; PMCID: PMC9658630
52. Cai H, Guo H (2023) Mesenchymal stem cells and their Exocytotic vesicles. Int J Mol Sci 24(3):2085. https://doi.org/10.3390/ijms24032085. PMID: 36768406; PMCID: PMC9916886
53. Kyurkchiev D, Bochev I, Ivanova-Todorova E, Mourdjeva M, Oreshkova T, Belemezova K, Kyurkchiev S (2014) Secretion of immunoregulatory cytokines by mesenchymal stem cells. World J Stem Cells 6(5):552–570. https://doi.org/10.4252/wjsc.v6.i5.552. PMID: 25426252; PMCID: PMC4178255
54. Dong Z, Fu Y, Cai Z, Dai H, He Y (2025) Recent advances in adipose-derived mesenchymal stem cell-derived exosomes for regulating macrophage polarization. Front Immunol 16:1525466. https://doi.org/10.3389/fimmu.2025.1525466. PMID: 39963133; PMCID: PMC11830602
55. Krzyszczyk P, Schloss R, Palmer A, Berthiaume F (2018) The role of macrophages in acute and chronic wound healing and interventions to promote pro-wound healing phenotypes. Front Physiol 9:419. https://doi.org/10.3389/fphys.2018.00419. PMID: 29765329; PMCID: PMC5938667
56. Parisi L, Gini E, Baci D, Tremolati M, Fanuli M, Bassani B, Farronato G, Bruno A, Mortara L (2018) Macrophage polarization in chronic inflammatory diseases: killers or builders? J Immunol Res 2018:8917804. https://doi.org/10.1155/2018/8917804. PMID: 29507865; PMCID: PMC5821995
57. Trzyna A, Banaś-Ząbczyk A (2021) Adipose-derived stem cells Secretome and its potential application in "stem cell-free therapy". Biomolecules 11(6):878. https://doi.org/10.3390/biom11060878. PMID: 34199330; PMCID: PMC8231996
58. Skovronova R, Scaccia E, Calcat-I-Cervera S, Bussolati B, O'Brien T, Bieback K (2023) Adipose stromal cells bioproducts as cell-free therapies: manufacturing and therapeutic dose determine in vitro functionality. J Transl Med 21(1):723. https://doi.org/10.1186/s12967-023-04602-9. PMID: 37840135; PMCID: PMC10577984
59. Mahmoud M, Abdel-Rasheed M (2023) Influence of type 2 diabetes and obesity on adipose mesenchymal stem/stromal cell immunoregulation. Cell Tissue Res. 394(1):33–53. https://doi.org/10.1007/s00441-023-03801-6. Epub 2023 Jul 18. PMID: 37462786; PMCID: PMC10558386
60. Zuk PA, Zhu M, Ashjian P, De Ugarte DA, Huang JI, Mizuno H, Alfonso ZC, Fraser JK, Benhaim P, Hedrick MH (2002) Human adipose tissue is a source of multipotent stem cells. Mol Biol Cell 13(12):4279–4295. https://doi.org/10.1091/mbc.e02-02-0105. PMID: 12475952; PMCID: PMC138633

61. Khazaei S, Keshavarz G, Bozorgi A, Nazari H, Khazaei M (2022) Adipose tissue-derived stem cells: a comparative review on isolation, culture, and differentiation methods. Cell Tissue Bank 23(1):1–16. https://doi.org/10.1007/s10561-021-09905-z. Epub 2021 Feb 22. Erratum in: Cell Tissue Bank. 2021 Dec;22(4):737. doi: 10.1007/s10561-021-09928-6. PMID: 33616792
62. Jang Y, Jung H, Nam Y, Rim YA, Kim J, Jeong SH, Ju JH (2016) Centrifugal gravity-induced BMP4 induces chondrogenic differentiation of adipose-derived stem cells via SOX9 upregulation. Stem Cell Res Ther 7(1):184. https://doi.org/10.1186/s13287-016-0445-6. PMID: 27931264; PMCID: PMC5144493
63. Huang SJ, Fu RH, Shyu WC, Liu SP, Jong GP, Chiu YW, Wu HS, Tsou YA, Cheng CW, Lin SZ (2013) Adipose-derived stem cells: isolation, characterization, and differentiation potential. Cell Transplant 22(4):701–709. https://doi.org/10.3727/096368912X655127. Epub 2012 Oct 12. PMID: 23068312
64. Mazzocchi AR, Man AJ, DesOrmeaux JPS, Gaborski TR (2014) Porous membranes promote endothelial differentiation of adipose-derived stem cells and perivascular interactions. Cell Mol Bioeng. https://doi.org/10.1007/s12195-014-0354-7
65. Rehak L, Giurato L, Meloni M, Panunzi A, Manti GM, Uccioli L (2022) The immune-centric revolution in the diabetic foot: monocytes and lymphocytes role in wound healing and tissue regeneration-a narrative review. J Clin Med 11(3):889. https://doi.org/10.3390/jcm11030889. PMID: 35160339; PMCID: PMC8836882
66. Li X, Hetjens L, Wolter N, Li H, Shi X, Pich A (2023) Charge-reversible and biodegradable chitosan-based microgels for lysozyme-triggered release of vancomycin. J Adv Res 43:87–96. https://doi.org/10.1016/j.jare.2022.02.014. Epub 2022 Feb 24. PMID: 36585117; PMCID: PMC9811367
67. Heydari P, Zargar Kharazi A, Asgary S, Parham S (2022) Comparing the wound healing effect of a controlled release wound dressing containing curcumin/ciprofloxacin and simvastatin/ciprofloxacin in a rat model: a preclinical study. J Biomed Mater Res A 110(2):341–352. https://doi.org/10.1002/jbm.a.37292. Epub 2021 Aug 11. PMID: 34378857
68. Farhana S, Kai YC, Kadir R, Sulaiman WAW, Nordin NA, Nasir NAM (2023) The fate of adipose tissue and adipose-derived stem cells in allograft. Cell Tissue Res 394(2):269–292. https://doi.org/10.1007/s00441-023-03827-w. Epub 2023 Aug 25. PMID: 37624425
69. Ko E, Yoon T, Lee Y, Kim J, Park YB (2023) ADSC secretome constrains NK cell activity by attenuating IL-2-mediated JAK-STAT and AKT signaling pathway via upregulation of CIS and DUSP4. Stem Cell Res Ther 14(1):329. https://doi.org/10.1186/s13287-023-03516-z. PMID: 37964351; PMCID: PMC10648656
70. Larocca RA, Moraes-Vieira PM, Bassi EJ, Semedo P, de Almeida DC, da Silva MB, Thornley T, Pacheco-Silva A, Câmara NO (2013) Adipose tissue-derived mesenchymal stem cells increase skin allograft survival and inhibit Th-17 immune response. PLoS One 8(10):e76396. https://doi.org/10.1371/journal.pone.0076396. PMID: 24124557; PMCID: PMC3790669
71. Kim I, Bang SI, Lee SK, Park SY, Kim M, Ha H (2014) Clinical implication of allogenic implantation of adipogenic differentiated adipose-derived stem cells. Stem Cells Transl Med. 3(11):1312–1321. https://doi.org/10.5966/sctm.2014-0109. Epub 2014 Oct 1. PMID: 25273542; PMCID: PMC4214848
72. Zhang J, Bai X, Zhao B, Wang Y, Su L, Chang P, Wang X, Han S, Gao J, Hu X, Hu D, Liu X (2016) Allogeneic adipose-derived stem cells promote survival of fat grafts in immunocompetent diabetic rats. Cell Tissue Res 364(2):357–367. https://doi.org/10.1007/s00441-015-2334-1. Epub 2015 Dec 14. PMID: 26662284
73. Avivar-Valderas A, Martín-Martín C, Ramírez C, Del Río B, Menta R, Mancheño-Corvo P, Ortiz-Virumbrales M, Herrero-Méndez Á, Panés J, García-Olmo D, Castaner JL, Palacios I, Lombardo E, Dalemans W, DelaRosa O (2019) Dissecting allo-sensitization after local administration of human allogeneic adipose mesenchymal stem cells in perianal fistulas of Crohn's disease patients. Front Immunol 10:1244. https://doi.org/10.3389/fimmu.2019.01244. PMID: 31258526; PMCID: PMC6587893

74. Bekhite MM, Finkensieper A, Rebhan J, Huse S, Schultze-Mosgau S, Figulla HR, Sauer H, Wartenberg M (2014) Hypoxia, leptin, and vascular endothelial growth factor stimulate vascular endothelial cell differentiation of human adipose tissue-derived stem cells. Stem Cells Dev 23(4):333–351. https://doi.org/10.1089/scd.2013.0268. Epub 2013 Dec 3. PMID: 24134622
75. Stavri GT, Zachary IC, Baskerville PA, Martin JF, Erusalimsky JD (1995) Basic fibroblast growth factor upregulates the expression of vascular endothelial growth factor in vascular smooth muscle cells. Synergistic interaction with hypoxia. Circulation 92(1):11–14. https://doi.org/10.1161/01.cir.92.1.11. PMID: 7788904
76. Lemieux C, Maliba R, Favier J, Théorêt JF, Merhi Y, Sirois MG (2005) Angiopoietins can directly activate endothelial cells and neutrophils to promote proinflammatory responses. Blood 105(4):1523–1530. https://doi.org/10.1182/blood-2004-09-3531. Epub 2004 Oct 21. PMID: 15498854
77. Li JH, Li Y, Huang D, Yao M (2021) Role of stromal cell-derived factor-1 in endothelial progenitor cell-mediated vascular repair and regeneration. Tissue Eng Regen Med. 18(5):747–758. https://doi.org/10.1007/s13770-021-00366-9. Epub 2021 Aug 27. PMID: 34449064; PMCID: PMC8440704
78. Kapoor A, Collino M, Castiglia S, Fantozzi R, Thiemermann C (2010) Activation of peroxisome proliferator-activated receptor-beta/delta attenuates myocardial ischemia/reperfusion injury in the rat. Shock 34(2):117–124. https://doi.org/10.1097/SHK.0b013e3181cd86d6. PMID: 19997057
79. Eming SA, Murray PJ, Pearce EJ (2021) Metabolic orchestration of the wound healing response. Cell Metab 33(9):1726–1743. https://doi.org/10.1016/j.cmet.2021.07.017. Epub 2021 Aug 11. PMID: 34384520
80. Rennert RC, Sorkin M, Januszyk M, Duscher D, Kosaraju R, Chung MT, Lennon J, Radiya-Dixit A, Raghvendra S, Maan ZN, Hu MS, Rajadas J, Rodrigues M, Gurtner GC (2014) Diabetes impairs the angiogenic potential of adipose-derived stem cells by selectively depleting cellular subpopulations. Stem Cell Res Ther 5(3):79. https://doi.org/10.1186/scrt468. PMID: 24943716; PMCID: PMC4097831
81. Yoshida S, Yoshimoto H, Hirano A, Akita S (2016) Wound healing and angiogenesis through combined use of a vascularized tissue flap and adipose-derived stem cells in a rat Hindlimb irradiated ischemia model. Plast Reconstr Surg 137(5):1486–1497. https://doi.org/10.1097/PRS.0000000000002062. PMID: 27119923
82. Xiao S, Liu Z, Yao Y, Wei ZR, Wang D, Deng C (2019) Diabetic human adipose-derived stem cells accelerate pressure ulcer healing by inducing angiogenesis and neurogenesis. Stem Cells Dev 28(5):319–328. https://doi.org/10.1089/scd.2018.0245. Epub 2019 Feb 4. PMID: 30608025
83. Huang SP, Huang CH, Shyu JF, Lee HS, Chen SG, Chan JY, Huang SM (2013) Promotion of wound healing using adipose-derived stem cells in radiation ulcer of a rat model. J Biomed Sci 20(1):51. https://doi.org/10.1186/1423-0127-20-51. PMID: 23876213; PMCID: PMC3724586
84. Li Y, Leng Y, Liu Y, Zhong J, Li J, Zhang S, Li Z, Yang K, Kong X, Lao W, Bi C, Zhai A (2024) Advanced multifunctional hydrogels for diabetic foot ulcer healing: active substances and biological functions. J Diabetes 16(4):e13537. https://doi.org/10.1111/1753-0407.13537. PMID: 38599855; PMCID: PMC11006623
85. Chung E, Rytlewski JA, Merchant AG, Dhada KS, Lewis EW, Suggs LJ (2015) Fibrin-based 3D matrices induce angiogenic behavior of adipose-derived stem cells. Acta Biomater. 17:78–88. https://doi.org/10.1016/j.actbio.2015.01.012. Epub 2015 Jan 16. PMID: 25600400; PMCID: PMC4380681
86. Berry CE, Abbas DB, Lintel HA, Churukian AA, Griffin M, Guo JL, Cotterell AC, Parker JBL, Downer MA, Longaker MT, Wan DC (2024) Adipose-derived stromal cell-based therapies for radiation-induced fibrosis. Adv Wound Care (New Rochelle). 13(5):235–252.

https://doi.org/10.1089/wound.2022.0103. Epub 2022 Dec 9. PMID: 36345216; PMCID: PMC11304913
87. Deptuła M, Brzezicka A, Skoniecka A, Zieliński J, Pikuła M (2021) Adipose-derived stromal cells for nonhealing wounds: emerging opportunities and challenges. Med Res Rev 41(4):2130–2171. https://doi.org/10.1002/med.21789. Epub 2021 Feb 1. PMID: 33522005; PMCID: PMC8247932
88. Spiekman M, van Dongen JA, Willemsen JC, Hoppe DL, van der Lei B, Harmsen MC (2017) The power of fat and its adipose-derived stromal cells: emerging concepts for fibrotic scar treatment. J Tissue Eng Regen Med 11(11):3220–3235. https://doi.org/10.1002/term.2213. Epub 2017 Feb 3. PMID: 28156060; PMCID: PMC5724515
89. Tracy LE, Minasian RA, Caterson EJ (2016) Extracellular matrix and dermal fibroblast function in the healing wound. Adv Wound Care (New Rochelle) 5(3):119–136. https://doi.org/10.1089/wound.2014.0561. PMID: 26989578; PMCID: PMC4779293
90. Kim WS, Park BS, Sung JH, Yang JM, Park SB, Kwak SJ, Park JS (2007) Wound healing effect of adipose-derived stem cells: a critical role of secretory factors on human dermal fibroblasts. J Dermatol Sci 48(1):15–24. https://doi.org/10.1016/j.jdermsci.2007.05.018. Epub 2007 Jul 23. PMID: 17643966
91. Qin X, He J, Wang X, Wang J, Yang R, Chen X (2023) The functions and clinical application potential of exosomes derived from mesenchymal stem cells on wound repair: a review of recent research advances. Front Immunol 14:1256687. https://doi.org/10.3389/fimmu.2023.1256687. PMID: 37691943; PMCID: PMC10486026
92. Cho JW, Kang MC, Lee KS (2010) TGF-β1-treated ADSCs-CM promotes expression of type I collagen and MMP-1, migration of human skin fibroblasts, and wound healing in vitro and in vivo. Int J Mol Med 26(6):901–906. https://doi.org/10.3892/ijmm_00000540. PMID: 21042785
93. Monsuur HN, van den Broek LJ, Koolwijk P, Niessen FB, Gibbs S (2018) Endothelial cells enhance adipose mesenchymal stromal cell-mediated matrix contraction via ALK receptors and reduced follistatin: potential role of endothelial cells in skin fibrosis. J Cell Physiol 233(10):6714–6722. https://doi.org/10.1002/jcp.26494. Epub 2018 May 9. PMID: 29345319; PMCID: PMC6056025
94. Imai Y, Mori N, Nihashi Y, Kumagai Y, Shibuya Y, Oshima J, Sasaki M, Sasaki K, Aihara Y, Sekido M, Kida YS (2022) Therapeutic potential of adipose stem cell-derived conditioned medium on scar contraction model. Biomedicine 10(10):2388. https://doi.org/10.3390/biomedicines10102388. PMID: 36289649; PMCID: PMC9598573
95. Yu J, Wang MY, Tai HC, Cheng NC (2018) Cell sheet composed of adipose-derived stem cells demonstrates enhanced skin wound healing with reduced scar formation. Acta Biomater 77:191–200. https://doi.org/10.1016/j.actbio.2018.07.022. Epub 2018 Jul 12. PMID: 30017923
96. Asgarpour K, Shojaei Z, Amiri F, Ai J, Mahjoubin-Tehran M, Ghasemi F, ArefNezhad R, Hamblin MR, Mirzaei H (2020) Exosomal microRNAs derived from mesenchymal stem cells: cell-to-cell messages. Cell Commun Signal 18(1):149. https://doi.org/10.1186/s12964-020-00650-6. PMID: 32917227; PMCID: PMC7488404
97. Giannasi C, Cadelano F, Della Morte E, Baserga C, Mazzucato C, Niada S, Baj A (2024) Unlocking the therapeutic potential of adipose-derived stem cell Secretome in Oral and maxillofacial medicine: a composition-based perspective. Biology (Basel) 13(12):1016. https://doi.org/10.3390/biology13121016. PMID: 39765683; PMCID: PMC11673083
98. Li J, Li Z, Wang S, Bi J, Huo R (2022) Exosomes from human adipose-derived mesenchymal stem cells inhibit production of extracellular matrix in keloid fibroblasts via downregulating transforming growth factor-β2 and Notch 1 expression. Bioengineered 13(4):8515–8525. https://doi.org/10.1080/21655979.2022.2051838. PMID: 35333672; PMCID: PMC9161879
99. Spiekman M, Przybyt E, Plantinga JA, Gibbs S, van der Lei B, Harmsen MC (2014) Adipose tissue-derived stromal cells inhibit TGF-β1-induced differentiation of human dermal

fibroblasts and keloid scar-derived fibroblasts in a paracrine fashion. Plast Reconstr Surg 134(4):699–712. https://doi.org/10.1097/PRS.0000000000000504. PMID: 25357030
100. Han B, Fan J, Liu L, Tian J, Gan C, Yang Z, Jiao H, Zhang T, Liu Z, Zhang H (2019) Adipose-derived mesenchymal stem cells treatments for fibroblasts of fibrotic scar via downregulating TGF-β1 and Notch-1 expression enhanced by photobiomodulation therapy. Lasers Med Sci 34(1):1–10. https://doi.org/10.1007/s10103-018-2567-9. Epub 2018 Oct 26. PMID: 30367294
101. Borovikova AA, Ziegler ME, Banyard DA, Wirth GA, Paydar KZ, Evans GRD, Widgerow AD (2018) Adipose-derived tissue in the treatment of dermal fibrosis: Antifibrotic effects of adipose-derived stem cells. Ann Plast Surg 80(3):297–307. https://doi.org/10.1097/SAP.0000000000001278. PMID: 29309331
102. Wu ZY, Zhang HJ, Zhou ZH, Li ZP, Liao SM, Wu ZY, Huang HH, Shi YC (2021) The effect of inhibiting exosomes derived from adipose-derived stem cells via the TGF-β1/Smad pathway on the fibrosis of keloid fibroblasts. Gland Surg 10(3):1046–1056. https://doi.org/10.21037/gs-21-4. PMID: 33842249; PMCID: PMC8033055
103. Wang L, Jiang X, Zhao F, Duan P, Li Z, Luo Y (2025) A review of adipose-derived mesenchymal stem cells' impacts and challenges: metabolic regulation, tumor modulation, immunomodulation, regenerative medicine and genetic engineering therapies. Front Endocrinol (Lausanne) 16:1606847. https://doi.org/10.3389/fendo.2025.1606847. PMID: 40510466; PMCID: PMC12158724
104. Alfarafisa NM, Chou Y, Santika R, Riestiano BE, Soedjana H, Syamsunarno MRAA (2025) Adipose-derived stem cell products and combination therapies for the treatment of pathological scars: a review of current preclinical and clinical studies. Clin Cosmet Investig Dermatol 18:1309–1337. https://doi.org/10.2147/CCID.S511067. PMID: 40452780; PMCID: PMC12126117
105. An R, Zhang Y, Qiao Y, Song L, Wang H, Dong X (2020) Adipose stem cells isolated from diabetic mice improve cutaneous wound healing in streptozotocin-induced diabetic mice. Stem Cell Res Ther 11(1):120. https://doi.org/10.1186/s13287-020-01621-x. PMID: 32183899; PMCID: PMC7079496
106. Irons RF, Cahill KW, Rattigan DA, Marcotte JH, Fromer MW, Chang S, Zhang P, Behling EM, Behling KC, Caputo FJ (2018) Acceleration of diabetic wound healing with adipose-derived stem cells, endothelial-differentiated stem cells, and topical conditioned medium therapy in a swine model. J Vasc Surg 68(6S):115S–125S. https://doi.org/10.1016/j.jvs.2018.01.065. Epub 2018 May 9. PMID: 29753580
107. Kim EK, Li G, Lee TJ, Hong JP (2011) The effect of human adipose-derived stem cells on healing of ischemic wounds in a diabetic nude mouse model. Plast Reconstr Surg 128(2):387–394. https://doi.org/10.1097/PRS.0b013e31821e6de2. PMID: 21788830
108. Jiang S, Zhuang Y, Cai M, Wang X, Lin K (2023) Decellularized extracellular matrix: a promising strategy for skin repair and regeneration. Eng Reg 4:357–374. https://doi.org/10.1016/j.engreg.2023.05.001
109. Navone SE, Pascucci L, Dossena M, Ferri A, Invernici G, Acerbi F, Cristini S, Bedini G, Tosetti V, Ceserani V, Bonomi A, Pessina A, Freddi G, Alessandrino A, Ceccarelli P, Campanella R, Marfia G, Alessandri G, Parati EA (2014) Decellularized silk fibroin scaffold primed with adipose mesenchymal stromal cells improves wound healing in diabetic mice. Stem Cell Res Ther 5(1):7. https://doi.org/10.1186/scrt396. PMID: 24423450; PMCID: PMC4055150
110. Wang JW, Zhu YZ, Ouyang JY, Nie JY, Wang ZH, Wu S, Yang JM, Yi YY (2023) Adipose-derived stem cell extracellular vesicles improve wound closure and angiogenesis in diabetic mice. Plast Reconstr Surg 151(2):331–342. https://doi.org/10.1097/PRS.0000000000009840. Epub 2022 Nov 8. PMID: 36696316
111. Almalki SG (2022) Adipose-derived mesenchymal stem cells and wound healing: potential clinical applications in wound repair. Saudi Med J 43(10):1075–1086. https://doi.org/10.15537/smj.2022.43.10.20220522. PMID: 36261194; PMCID: PMC9994497

112. Sun Y, Song L, Zhang Y, Wang H, Dong X (2020) Adipose stem cells from type 2 diabetic mice exhibit therapeutic potential in wound healing. Stem Cell Res Ther. 11(1):298. https://doi.org/10.1186/s13287-020-01817-1. Erratum in: Stem Cell Res Ther 2022;13(1):82. doi: 10.1186/s13287-022-02778-3. PMID: 32680569; PMCID: PMC7368682
113. Ma T, Zhao Y, Shen G, Chai B, Wang W, Li X, Zhang Z, Meng Q (2023) Novel bilayer cell patch combining epidermal stem cells and angiogenic adipose stem cells for diabetic wound healing. J Control Release 359:315–325. https://doi.org/10.1016/j.jconrel.2023.06.010. Epub 2023 Jun 14. PMID: 37315690
114. Hsu HH, Wang AYL, Loh CYY, Pai AA, Kao HK (2022) Therapeutic potential of exosomes derived from diabetic adipose stem cells in cutaneous wound healing of db/db mice. Pharmaceutics 14(6):1206. https://doi.org/10.3390/pharmaceutics14061206. PMID: 35745779; PMCID: PMC9227821
115. Ahmadieh-Yazdi A, Karimi M, Afkhami E, Hajizadeh-Tafti F, Kuchakzadeh F, Yang P, Sheykhhasan M (2024) Unveiling therapeutic potential: adipose tissue-derived mesenchymal stem cells and their exosomes in the management of diabetes mellitus, wound healing, and chronic ulcers. Biochem Pharmacol 226:116399. https://doi.org/10.1016/j.bcp.2024.116399. Epub 2024 Jun 27. PMID: 38944396
116. Deng Q, Pan S, Du F, Sang H, Cai Z, Xu X, Wei Q, Yu S, Zhang J, Li C (2024) The effect of conditioned medium from Angiopoietin-1 gene-modified mesenchymal stem cells on wound healing in a diabetic mouse model. Bioengineering (Basel) 11(12):1244. https://doi.org/10.3390/bioengineering11121244. PMID: 39768062; PMCID: PMC11673525
117. Chou PR, Lin YN, Wu SH, Lin SD, Srinivasan P, Hsieh DJ, Huang SH (2020) Supercritical carbon dioxide-decellularized porcine acellular dermal matrix combined with autologous adipose-derived stem cells: its role in accelerated diabetic wound healing. Int J Med Sci 17(3):354–367. https://doi.org/10.7150/ijms.41155. PMID: 32132871; PMCID: PMC7053351
118. Laloze J, Fiévet L, Desmoulière A (2021) Adipose-derived mesenchymal stromal cells in regenerative medicine: state of play, current clinical trials, and future prospects. Adv Wound Care (New Rochelle). 10(1):24–48. https://doi.org/10.1089/wound.2020.1175. Epub 2020 Jun 2. PMID: 32470315; PMCID: PMC7698876
119. Huayllani MT, Ruiz-Garcia H, Boczar D, Avila FR, Lu X, Rinker BD, Moran SL, Sarabia-Estrada R, Quiñones-Hinojosa A, Forte AJ (2021) Adipose-derived stem cells therapy for radiation-induced skin injury. Ann Plast Surg 87(6):639–649. https://doi.org/10.1097/SAP.0000000000003039. PMID: 34724441
120. Wu SH, Shirado T, Mashiko T, Feng J, Asahi R, Kanayama K, Mori M, Chi D, Sunaga A, Sarukawa S, Yoshimura K (2018) Therapeutic effects of human adipose-derived products on impaired wound healing in irradiated tissue. Plast Reconstr Surg 142(2):383–391. https://doi.org/10.1097/PRS.0000000000004609. PMID: 29787514
121. Milan M, Takla J, Hayek F et al (2024) Regenerative potential of adipose derived stem cells in wound healing and scar management: a review. Curr Derm Rep 13:351–360. https://doi.org/10.1007/s13671-024-00442-y
122. Stachura A, Paskal W, Pawlik W, Mazurek MJ, Jaworowski J (2021) The use of adipose-derived stem cells (ADSCs) and stromal vascular fraction (SVF) in skin scar treatment-a systematic review of clinical studies. J Clin Med 10(16):3637. https://doi.org/10.3390/jcm10163637. PMID: 34441935; PMCID: PMC8396936
123. Sakai Y, Fukunishi S, Takamura M, Kawaguchi K, Inoue O, Usui S, Takashima S, Seki A, Asai A, Tsuchimoto Y, Nasti A, Bich Ho TT, Imai Y, Yoshimura K, Murayama T, Yamashita T, Arai K, Yamashita T, Mizukoshi E, Honda M, Wada T, Harada K, Higuchi K, Kaneko S (2021) Clinical trial of autologous adipose tissue-derived regenerative (stem) cells therapy for exploration of its safety and efficacy. Regen Ther 18:97–101. https://doi.org/10.1016/j.reth.2021.04.003. PMID: 34095367; PMCID: PMC8165289
124. Ceccarelli S, Pontecorvi P, Anastasiadou E, Napoli C, Marchese C (2020) Immunomodulatory effect of adipose-derived stem cells: the cutting edge of clinical application. Front Cell

Dev Biol 8:236. https://doi.org/10.3389/fcell.2020.00236. PMID: 32363193; PMCID: PMC7180192
125. Pikuła M, Marek-Trzonkowska N, Wardowska A, Renkielska A, Trzonkowski P (2013) Adipose tissue-derived stem cells in clinical applications. Expert Opin Biol Ther 13(10):1357–1370. https://doi.org/10.1517/14712598.2013.823153. Epub 2013 Aug 7. PMID: 23919743
126. Shukla L, Morrison WA, Shayan R (2015) Adipose-derived stem cells in radiotherapy injury: a new frontier. Front Surg 2:1. https://doi.org/10.3389/fsurg.2015.00001. PMID: 25674565; PMCID: PMC4309196
127. Gir P, Oni G, Brown SA, Mojallal A, Rohrich RJ (2012) Human adipose stem cells: current clinical applications. Plast Reconstr Surg 129(6):1277–1290. https://doi.org/10.1097/PRS.0b013e31824ecae6. PMID: 22634645
128. Zocchi ML, Facchin F, Pagani A et al (2022) New perspectives in regenerative medicine and surgery: the bioactive composite therapies (BACTs). Eur J Plast Surg 45:1–25. https://doi.org/10.1007/s00238-021-01874-6
129. Mazini L, Ezzoubi M, Malka G (2021) Overview of current adipose-derived stem cell (ADSCs) processing involved in therapeutic advancements: flow chart and regulation updates before and after COVID-19. Stem Cell Res Ther 12(1):1. https://doi.org/10.1186/s13287-020-02006-w. PMID: 33397467; PMCID: PMC7781178
130. Turner LG (2015) Federal regulatory oversight of US clinics marketing adipose-derived autologous stem cell interventions: insights from 3 new FDA draft guidance documents. Mayo Clin Proc 90(5):567–571. https://doi.org/10.1016/j.mayocp.2015.02.003. PMID: 25939934
131. Raposio E, Ciliberti R (2017) Clinical use of adipose-derived stem cells: European legislative issues. Ann Med Surg (Lond) 24:61–64. https://doi.org/10.1016/j.amsu.2017.11.002. PMID: 29204274; PMCID: PMC5709339
132. Mendicino M, Weber D (2015; Chapter 6) Regulatory and intellectual property considerations for therapeutic human stem cell-based regenerative medicine product development: a US perspective. In: Stem Cell-Based Tissue Repair. Wiley. https://doi.org/10.1002/9781118846193.ch6
133. Pean CA, Kingery MT, Strauss E, Bosco JA, Halbrecht J (2019) Direct-to-consumer advertising of stem cell clinics: ethical considerations and recommendations for the health-care community. J Bone Joint Surg Am 101(19):e103. https://doi.org/10.2106/JBJS.19.00266. PMID: 31577688
134. Malhotra P, Saraswat SK, Khatri N, Verma R, Datt R, Pandey S (2024) Chapter 23—Stem cell-based products and the regulatory hurdle race. In: Verma YK, Satija NK, Raghav PK, Tyagi N, Kumar S (eds) Stem cells. Academic Press, pp 371–384. https://doi.org/10.1016/B978-0-323-95545-4.00023-2

Chapter 17
Wound Dressings: Principles and Modern Applications

Apirag Chuangsuwanich and Pitawan Rachata

Abstract Wound healing is an inherently complex biological process, even in healthy individuals. In patients with chronic wounds, especially those with comorbidities such as diabetes mellitus, cardiovascular disease, renal insufficiency or advanced age, this process is often significantly impaired. An appropriate wound dressing plays a crucial role in optimizing the wound environment, promoting healing and preventing further complications.

Keywords Wound dressing · Chronic wound · Advance wound dressing

Introduction

Wound healing is an inherently complex biological process, even in healthy individuals. In patients with chronic wounds, especially those with comorbidities such as diabetes mellitus, cardiovascular disease, renal insufficiency or advanced age, this process is often significantly impaired. An appropriate wound dressing plays a crucial role in optimizing the wound environment, promoting healing and preventing further complications.

Selecting the correct type of dressing tailored to the specific wound condition is a critical component of effective wound management. Modern wound dressings aim to control the local wound environment, maintain moisture balance and protect against infection, thereby supporting the body's natural healing mechanisms.

A. Chuangsuwanich (✉)
Division of Plastic Surgery, Department of Surgery, Faculty of Medicine Siriraj Hospital, Mahidol University, Bangkok, Thailand

P. Rachata
Division of Plastic Surgery, Department of Surgery, School of Medicine, Mae Fah Luang University, Chiang Rai, Thailand

F. De Francesco, K. Turksen (eds.), *Management and Strategies for Wound Healing*, Stem Cell Biology and Regenerative Medicine 79,
https://doi.org/10.1007/978-3-032-12442-5_17

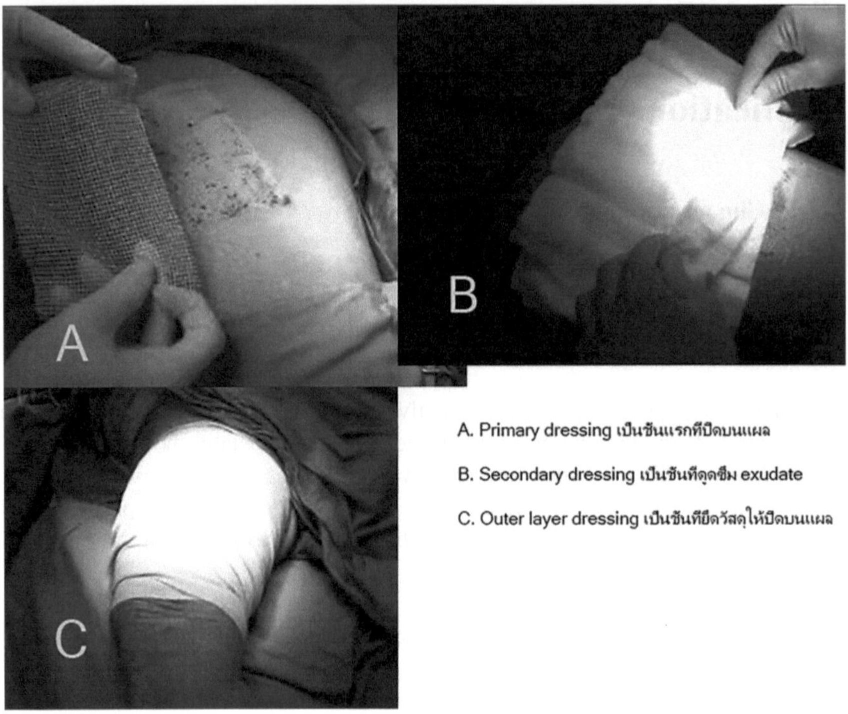

Fig. 17.1 Primary Dressing: Directly contacts the wound surface

Classification of Wound Dressings

Wound dressings can be categorized based on their function and physical characteristics (Fig. 17.1):

1. Primary Dressing: Directly contacts the wound surface.
2. Secondary Dressing: Applied over the primary dressing to provide additional protection, absorption or compression.
3. Occlusive Dressing: Impermeable to both water and gas, creates a sealed environment.
4. Semiocclusive Dressing: Permits the exchange of oxygen and water vapor while maintaining a moist wound environment.

Historical Perspective and Evolution

The practice of wound care has evolved substantially since World War I, when antiseptics such as iodine, Dakin's solution, carbolic acid (phenol) and mercurochrome became widely used to prevent infection. However, these agents often caused skin irritation.Early dressings consisted primarily of dry gauze. To mitigate tissue adherence, paraffin-impregnated gauze was introduced. A major paradigm shift occurred in the 1960s, when research demonstrated that moist wound environments promote faster and more effective healing compared to dry environments [1].

By the early 1980s, the principle of moist-environment wound healing became widely accepted. It was shown to enhance angiogenesis, granulation tissue formation and re-epithelialization [2].

Moisture Balance and Wound Healing

A dry wound environment leads to water loss at the wound surface and the formation of a scab—composed of coagulated blood and fibrinogen molecules—which impedes epithelial cell migration and slows healing. On the other hand, the condition of wound edema with excessive transudate arises from the accumulation of fluid at the wound site (plasma protein leak) along with the release of proteolytic enzymes and the presence of necrotic tissue and microbial toxins. These factors contribute to a chronic inflammatory state, impeding the healing process. To address these issues, various types of wound dressings have been developed to maintain an optimal moist environment—preventing both excessive dryness and excessive fluid retention. These dressings can be selected based on the quantity of wound exudate [3–5].

Transparent films and hydrocolloids were among the first dressings developed to enhance moisture retention [2]. Subsequently, more advanced wound dressing materials were developed in various forms such as alginates, hydrogels and foams. In addition to these dressing materials, various bioactive substances have been introduced to further support wound healing. These include growth factors and bioengineered dressings containing living cells or collagen matrices, which play a significant role in promoting tissue regeneration (Fig. 17.2).

Dressings designed to maintain an optimal hydration should exhibit low moisture vapor transmission rates, ideally below 35 $g/m^2/h$ [6].

Clinical Considerations in Dressing Selection

Prior to choosing a wound dressing, the clinician must establish clear treatment objectives and identify the wound's etiology [7]. This allows for a targeted approach in dressing selection based on factors such as:

1. Infection control
2. Debridement of necrotic tissue
3. Exudate management, each material has different exudate absorption properties (Table 17.1)

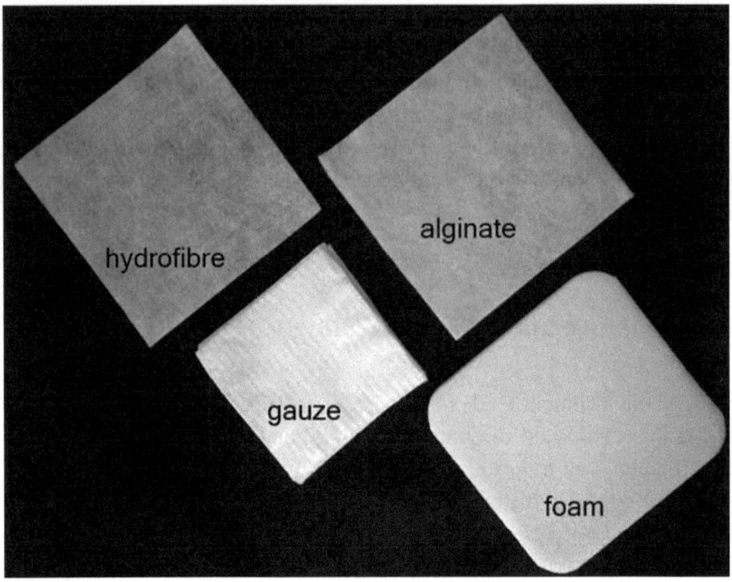

Fig. 17.2 Type of dressings

Table 17.1 Selection of wound dressings based on exudate level

Wound condition	Dry wound	Minimal exudate	Moderate exudate	Heavy exudate
Recommended Dressings	Non-adherent, Hydrogel, Film dressing	Hydrogel, Hydrocolloid	Hydrocolloid, Calcium alginate, Hydrofiber, Foam, Negative pressure wound therapy	Hydrofiber, Foam, Ostomy bag, Negative pressure wound therapy

4. Promotion of wound healing
5. Pain reduction during dressing changes
6. Odor control

Nevertheless, dressing selection should not rely solely on the material's prop erties. Key considerations include cost-effectiveness and clinical response. Most importantly, accurate wound assessment is essential to determine the wound conditions, thereby guiding the selection of the most appropriate dressing [3, 8] (Table 17.2).

As wound characteristics may change over time, dressing selection is a dynamic process that requires regular reassessment at appropriate intervals to ensure continued suitability and optimal healing outcomes.

Table 17.2 Recommended dressing selection based on wound conditions and characteristics

Wound condition	Treatment objective	Minimal exudate recommended dressing	Heavy exudate recommended dressing
Necrotic tissue	Moisture balance, Debridement	Wet-to-moist gauze, hydrogel, hydrocolloid	Alginate, Foam
Infected wound	Identify/Eliminate source, Infection control	Iodine/silver-impregnated dressing	Iodine/silver-impregnated dressing + Foam
Granulation/ Skin lost	Maintain moist environment, Infection prevention	Hydrocolloid	Alginate, Foam

Principles and Categories of Wound Dressings

A wide range of wound dressings is currently available; however, no single product possesses all the ideal properties for comprehensive wound management [9]. Selection should aim to enhance healing efficiency while minimizing patient discomfort. Practical factors such as ease of use and reduced frequency of dressing changes also influence clinical decision-making [10].

Wound dressings can be broadly categorized into two main types based on their interaction with the wound bed:

1. *Passive (Inert) Dressings*: These provide basic wound coverage and exudate absorption but do not directly interact with the wound environment to promote healing. They include materials such as gauze, cotton and nonadherent tulle. They primarily serve as physical barriers without influencing the biochemical healing process.
2. *Interactive Dressings*: These dressings actively support the wound healing process by maintaining moisture balance, modulating the wound microenvironment, controlling microbial load, promoting granulation tissue formation and enhancing re-epithelialization. Examples include hydrocolloids, hydrogels, foams, alginates and semipermeable films.

Passive(Inert) Dressings

Exudate-Absorbing Materials

This category of dressings is primarily designed to cover wounds and absorb exudate without actively promoting the healing process (Fig. 17.3). Typical materials include gauze, cotton and cotton-wrapped gauze pads. These may be used in conjunction with nonadherent mesh dressings—such as petrolatum-impregnated tulle—with or without topical antibiotics.

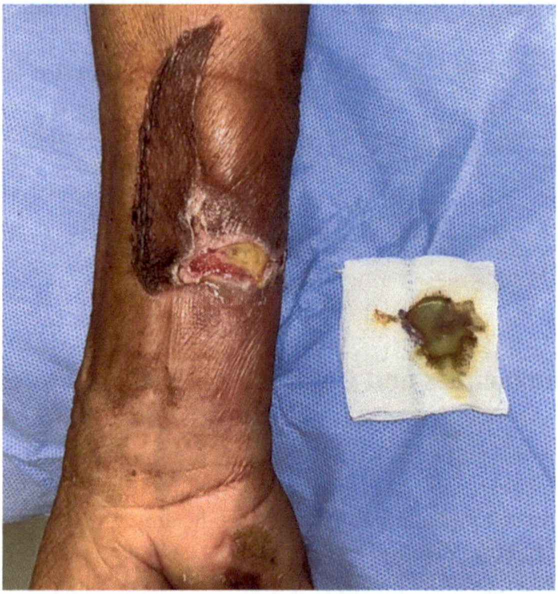

Fig. 17.3 Gauze absorbs exudate but does not retain it, potentially causing maceration of the surrounding skin

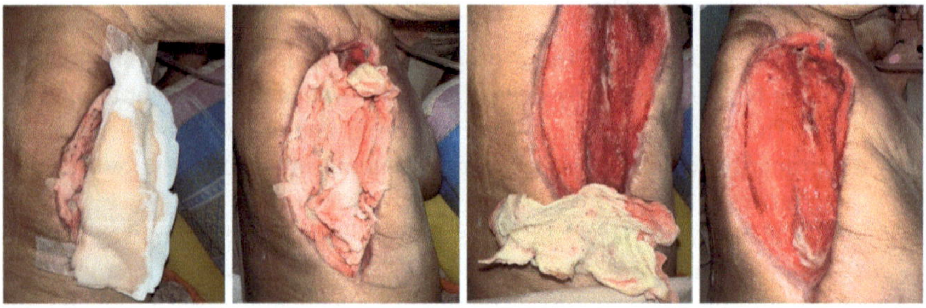

Fig. 17.4 Wet-to-dry dressing

In cases where there is a small amount of necrotic tissue, clinicians may employ the *wet-to-dry dressing* technique (Fig. 17.4). This involves applying saline-moistened gauze to the wound, which is allowed to dry. Upon removal, the adherent gauze mechanically debrides the wound by pulling away devitalized tissue—a method known as *mechanical debridement*.

Modern versions of these dressings are now commercially available with an integrated nonadherent film layer to reduce trauma during dressing changes—for example, **Melolin®**.

Nonadherent Dressings

Nonadherent dressings are primary dressings designed to minimize or prevent adherence to the wound surface (Fig. 17.5). They typically incorporate materials such as petrolatum or antibacterial ointments, which reduce tissue trauma during dressing changes. These dressings are generally not suitable for wounds with high levels of exudate. In such cases, a secondary dressing is required to manage exudate absorption.

Common indications for nonadherent dressings include donor and recipient sites of skin grafts, abrasions and partial-thickness wounds. Although these dressings are coated with substances to reduce adhesion, they do not completely eliminate the risk of sticking. Therefore, dressing removal must be performed gently to avoid patient discomfort and prevent disruption of newly formed epithelial tissue.

To address this issue, advanced nonadherent dressings have been developed using *silicone mesh technology*. These dressings feature a smooth, nonadherent surface that allows exudate to pass through to the secondary dressing layer without adhering to the wound bed, thereby further minimizing trauma.

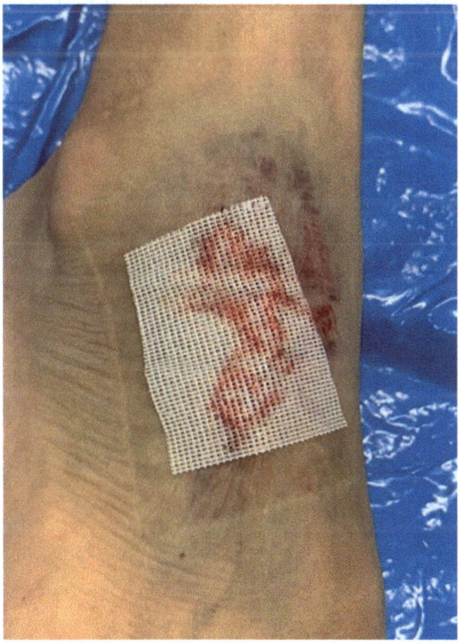

Fig. 17.5 Nonadherent dressings: These are materials designed not to stick to the wound bed, such as gauze impregnated with oil or coated with silicone. They are typically used as a contact layer or primary dressing to protect the wound and minimize trauma during dressing changes

Interactive Dressings

Semipermeable Film Dressings

Film dressings are thin, transparent polyurethane sheets that allow continuous visualization of the wound bed (Fig. 17.6). They are *semipermeable*, allowing the exchange of gases such as oxygen and water vapor, while providing an effective barrier against external contaminants including water and bacteria. These dressings adhere to the periwound skin via a pressure-sensitive adhesive.

Importantly, film dressings *do not contain absorptive components* and are therefore *only suitable for wounds with very scant exudate*. Their use is generally limited to dry or nearly dry wounds, or as a protective secondary dressing. Special caution is advised during removal, particularly in patients with fragile or atrophic skin, to prevent epidermal stripping [11].

Hydrogel

Hydrogels are wound dressings composed of 60–90% water, making them highly effective at hydrating the wound bed. They are particularly useful for *dry wounds with eschar*, as the high water content softens necrotic tissue, facilitating its removal through *autolytic debridement* (Fig. 17.7). This process selectively breaks down

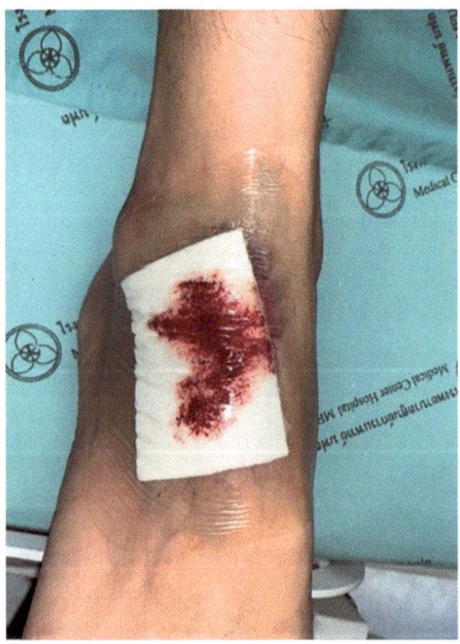

Fig. 17.6 Semipermeable film dressing: Transparent but nonabsorbent; unsuitable for wounds with moderate to heavy exudate

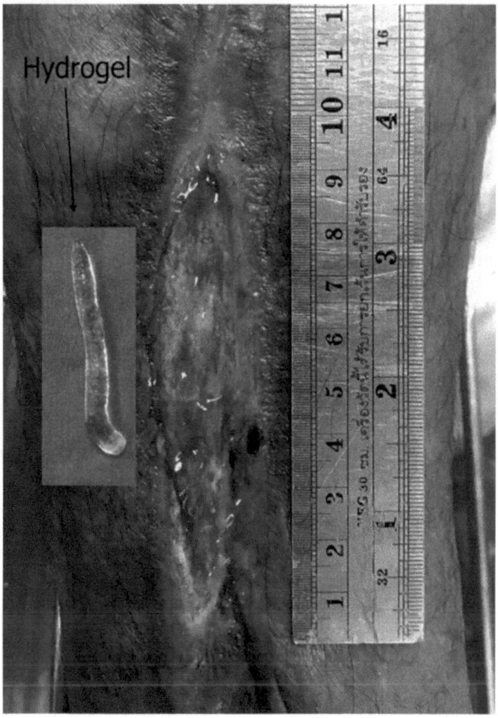

Fig. 17.7 Hydrogel dressing: A clear gel containing ~80% water; suitable for dry wound beds with minimal exudate

devitalized tissue without damaging viable granulation tissue or epithelial cells, thereby supporting *re-epithelialization* [12].

Hydrogels are available in two main forms:

- *Amorphous or semiliquid gels*, which can be applied directly to the wound.
- *Sheet or impregnated forms*, in which the gel is incorporated into a dressing matrix to reduce water evaporation.

A key contraindication is their use in combination with *larval therapy* [13]. This is due to the presence of *propylene glycol*, a preservative in many hydrogel formulations, which can be toxic to larvae used for biological debridement.

Hydrocolloid

Hydrocolloid dressings share similar functions with hydrogels but offer superior *exudate absorption*. Upon contact with wound fluid, the inner layer—typically composed of *sodium carboxymethylcellulose (CMC), gelatin* or similar hydrophilic agents—undergoes a gel-forming reaction that creates a moist interface over the wound bed (Fig. 17.8). The outer layer is usually made of a *polymeric film*, which may or may not be permeable to gases, depending on the product design.

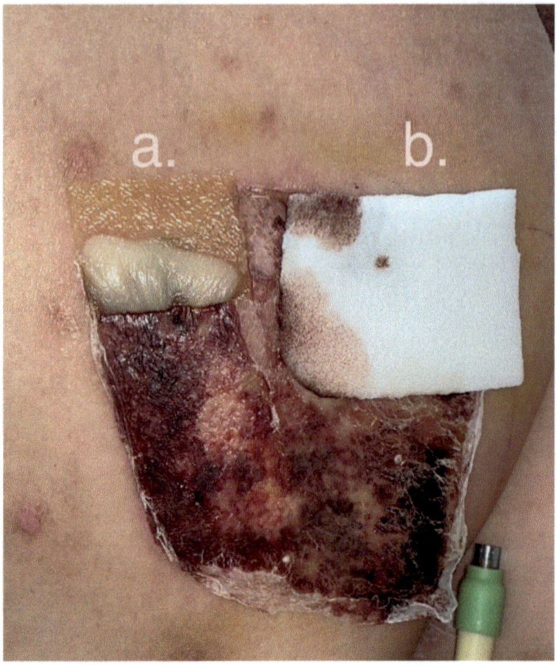

Fig. 17.8 Dressing comparison: (**a**) Sheet-form hydrocolloid—conforms well to wound surface and suitable for moderate exudate. (**b**) Foam dressing—also for moderate exudate, but offers superior absorption and cushioning

Hydrocolloids support *autolytic debridement* by maintaining a moist environment that facilitates the breakdown of necrotic tissue. However, they *should not be used on infected wounds* due to the occlusive nature of the dressing, which may exacerbate bacterial proliferation [14–16].

Prolonged use has been associated with hypergranulation tissue, which may necessitate discontinuation. Additionally, as exudate mixes with the hydrocolloid, it may form a thick, brown, malodorous gel. This appearance is not necessarily a sign of infection, although it may be mistaken as such by inexperienced clinicians.

Commercial hydrocolloid dressings are available in both *gel-based* and *sheet-form* variants, allowing flexibility based on wound size, depth and location.

Alginate

Alginate dressings are derived from seaweed and primarily composed of calcium salts. They have excellent fluid absorption capacity, making them ideal for *highly exudative wounds*. Their key mechanism involves an ion exchange reaction, in which calcium ions from the dressing are exchanged with sodium ions in wound exudate, forming a sodium alginate gel that conforms to the wound bed [17].

This gel formation helps maintain a moist environment and facilitates autolytic debridement. However, if used on wounds with *insufficient exudate*, the alginate fibers may not fully convert into gel and can leave residual fibers in the wound bed,

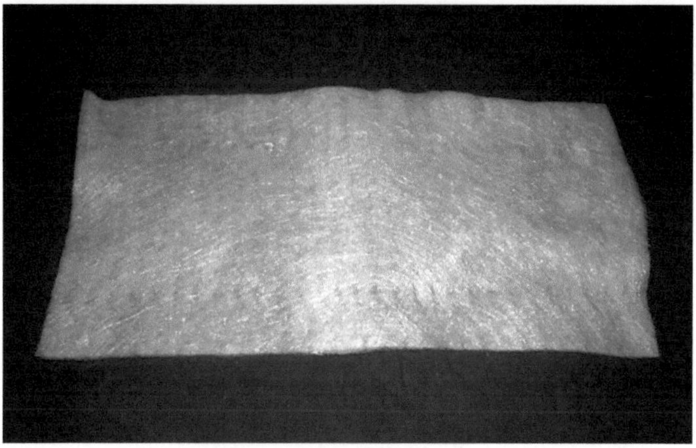

Fig. 17.9 Alginate dressing: Fiber or sheet form; ideal for wounds with moderate exudate

making removal difficult. Additionally, alginates have *hemostatic properties*, contributing to bleeding control in superficial wounds [19].

Caution should be exercised when using alginate dressings in deep or tunneled wounds, as residual fibers may be retained and potentially lead to chronic inflammation [18].

Alginate dressings serve as *primary dressings* and often require a *secondary dressing* such as a hydrocolloid or hydrofiber to secure and support the wound environment (Fig. 17.9).

Hydrofiber

Hydrofiber dressings are highly absorbent wound dressings composed primarily of *carboxymethylcellulose (CMC)*. They possess *greater fluid absorption capacity than alginates*, although both are considered highly absorbent [20, 21]. When in contact with wound exudate—especially in the presence of bacteria—hydrofibers transform into a soft, cohesive *gel*, creating an optimal moist wound environment and facilitating *autolytic debridement*.

Some formulations are *impregnated with silver* to enhance antimicrobial activity and control bioburden [22]. Hydrofiber dressings typically require a *secondary dressing* to secure them in place and manage additional exudate (Fig. 17.10)

Because they do not adhere to the wound bed, hydrofiber dressings are generally easy and painless to remove, improving patient comfort. They are available in various forms, including large flat sheets for extensive wounds and ribbon-shaped dressings for packing into deep or tunneled wounds.

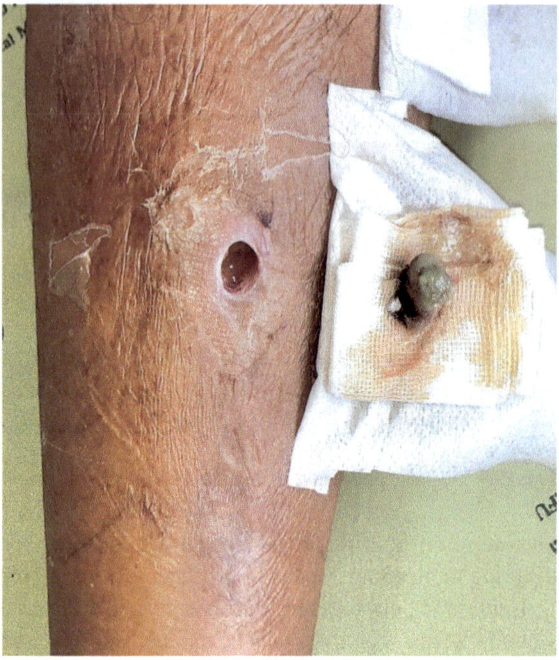

Fig. 17.10 Silver-impregnated hydrofiber dressing: conforms well to wound contours, suitable for cavity wounds with moderate to heavy exudate, and requires a secondary dressing

Foam

Foam dressings are typically made from *polyurethane* and are known for their *high absorbency*, making them effective in managing exudate and reducing the risk of maceration of the periwound skin [23, 24]. Depending on the level of exudate, some foam dressings can remain in place for up to 7 days, reducing the need for frequent dressing changes (Fig. 17.11).

These dressings are flexible and can *conform well to different anatomical contours*, making them suitable for various body locations. They can also be used to *pack deep wounds or cavities*, helping to reduce *dead space* and support moist wound healing.

Foam dressings are generally nonadhesive, requiring a secondary fixation such as medical tape or an adhesive border to secure them in place.

Maggot (Larval) Therapy

Maggot therapy, also known as *larval therapy*, involves the application of sterile fly larvae to a wound for the purpose of *debriding necrotic tissue*. This method has been used for over 1000 years, with historical records from various cultures. One of the earliest documented medical uses with observable clinical outcomes dates back

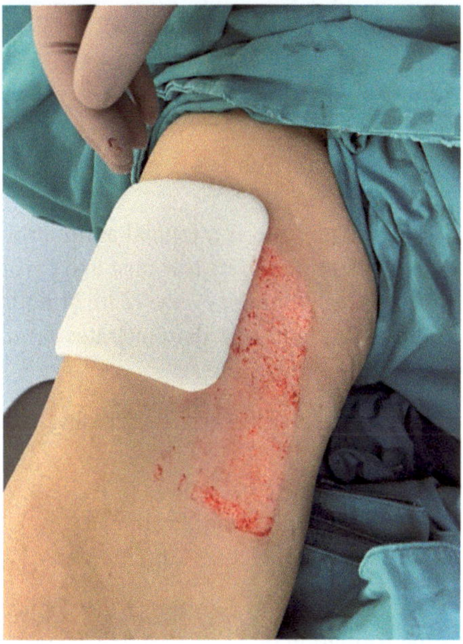

Fig. 17.11 Foam dressing has a sponge-like structure capable of absorbing and retaining exudate. Its thickness also helps reduce pressure over the wound area

to Ambroise Paré in 1557. Today, the species most commonly used is *Lucilia sericata* [25], which selectively debrides devitalized tissue without harming healthy tissue, and has been shown to *stimulate granulation tissue formation*—a response believed to be partially mediated by increased levels of *growth factors*, such as *hepatocyte growth factor* [26].

Before clinical application, larvae are *sterilized* to prevent transmission of infectious agents such as malaria, filariasis and trypanosomiasis. The larvae used are typically *2 mm in length* at the time of application. They are placed onto the wound and secured with an occlusive dressing, often a hydrocolloid, to contain them while allowing exudate absorption.

The *number of larvae* applied depends on the extent of necrosis, typically ranging from *200 to 600 larvae per session* [27]. Treatment usually lasts *about 3 days*, after which the larvae are removed. Studies have shown that the efficacy of debridement with maggot therapy is *comparable to conventional surgical debridement* [28, 29]. This method has re-emerged as an effective adjunct for managing chronic wounds, particularly in cases where surgical or enzymatic debridement is contraindicated or less effective.

Additional benefits of maggot therapy include [27]:

- Odor reduction in chronic wounds
- A nonsurgical alternative for patients who are not surgical candidates
- Cost savings, as therapy can often be performed on an outpatient basis
- Reduced antibiotic use, contributing to antimicrobial stewardship

Antimicrobial-Containing Dressings

Silver Compounds

Silver Sulfadiazine

Silver sulfadiazine is a topical antimicrobial agent commonly used in burn wound management due to its low cost and favorable safety profile. It is typically applied at a 1% concentration every 12 to 24 hours.

Silver sulfadiazine demonstrates *broad-spectrum antimicrobial activity*, effective against:

- *Gram-positive bacteria*, including *methicillin-resistant Staphylococcus aureus* (MRSA) [30]
- Gram-negative bacteria [31]
- *Fungi*, such as Candida albicans
- Occasionally, Herpes simplex virus

Mild *stinging or burning* at the application site may occur. A notable limitation is the formation of a *thin pseudoeschar* within 2–3 days, which may obscure wound assessment—particularly for inexperienced clinicians. If necessary, this layer can be gently removed to facilitate evaluation.

Systemic absorption of silver sulfadiazine is possible, but *toxicity is rare*. Infrequent cases of *transient leukopenia* have been reported, typically occurring 2–3 days after initiation [32, 33]. This condition is usually self-limiting and does not always require discontinuation. Leukocyte counts below 2000 cells/mm^3 are very rare.

In addition to its antimicrobial properties, silver sulfadiazine has been shown to:

- Accelerate re-epithelialization [34, 35]
- Promote neovascularization
- Suppress prolonged inflammation
- Enhance granulation tissue formation [36]
- Significantly reduce wound bacterial load [37]

Silver-Impregnated Dressings

These dressings are embedded or coated with silver in various forms, including *silver sulfadiazine, ion-releasing silver salts* or *nanocrystalline silver compounds* (Fig. 17.12). In nanocrystalline formulations, sterile water activates the controlled release of silver ions, enhancing the antimicrobial effectiveness (nanocrystalline silver releasing systems).

Silver dressings have demonstrated:

- Broad-spectrum antibacterial activity [38, 39]
- Reduction in exudate and frequency of dressing changes (typically changed every 3 days)

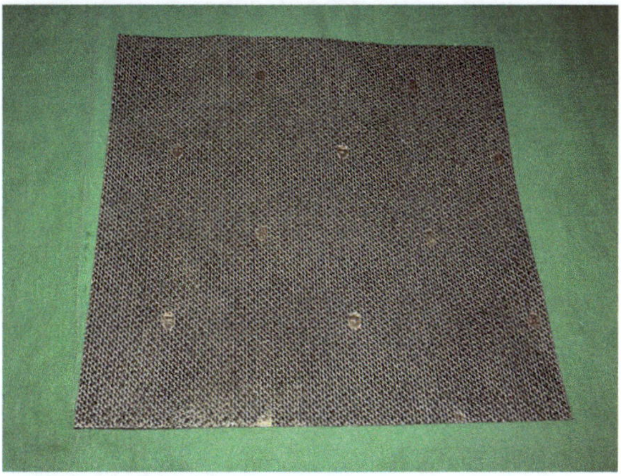

Fig. 17.12 Silver dressing provides broad-spectrum antimicrobial activity. It is available in various forms and can continuously release silver ions over several days

- Moisture retention, maintaining an optimal moist wound environment
- Enhanced re-epithelialization
- Shortened treatment time and improved clinical outcomes [40, 41]

These properties make silver-impregnated dressings particularly suitable for wounds with moderate to heavy exudate, wounds at risk of infection or those already colonized by bacteria.

Iodine-Based Compounds

Iodine, a member of the halogen group (like chlorine), is a broad-spectrum antiseptic effective against *bacteria, viruses, fungi* and *amoebae*. It is widely used in various wound care formulations.

The most commonly used preparation is *povidone-iodine* (e.g., Betadine® or Povidine Solution), which combines iodine with *anionic surface-active agents* to facilitate a *slow*, *sustained release* of iodine.

Effects on Wound Healing

The impact of povidone-iodine on wound healing remains *controversial*, with conflicting findings in the literature. Variability in study design—including iodine concentration, exposure time, wound type, and sample size—has led to mixed conclusions, some studies suggest delayed healing or inhibition of re-epithelialization, collagen synthesis and granulation tissue formation [42–44], others report no adverse impact on these processes. In fact, certain studies have observed enhanced revascularization, indicating a potentially beneficial effect on wound healing [45]

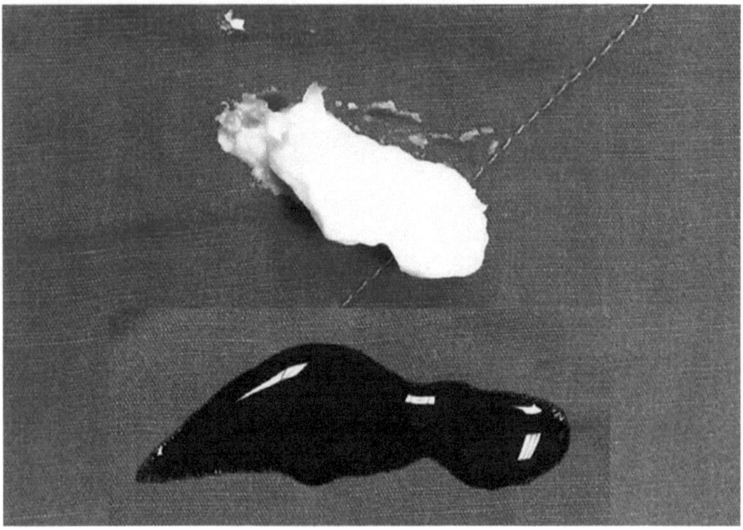

Fig. 17.13 Antibiotics, iodine, and silver can be formulated as topical creams for direct application to wounds

Another widely used iodine-based dressing is *cadexomer iodine* (Fig. 17.13). This formulation has demonstrated several benefits in *chronic wounds*, including:

- Reduction of biofilm formation [46]
- Decrease in bacterial load and infection rates [47–49]
- Promotion of autolytic debridement [50]
- Overall support of wound healing processes

Mechanism of Action

Iodine acts by *irreversibly binding to cellular proteins*, causing denaturation and cell death. Its efficacy is reduced in alkaline environments and in the presence of blood, pus or necrotic tissue in large quantities.

Honey

Honey is a natural product that has been used for wound care since ancient times due to its *broad-spectrum antimicrobial properties*. Its antibacterial effect is primarily attributed to its *high osmolarity* [51], which inhibits microbial growth by drawing water out of bacterial cells, and its *hydrogen peroxide content*, which contributes to oxidative antimicrobial activity [52, 53].

In modern wound care, honey is incorporated into various formulations, including gels for direct application and composite dressings such as honey-impregnated alginates or hydrocolloids.

Clinical outcomes associated with honey-based dressings have been mixed. Some studies report *accelerated wound healing*, while others show *no significant difference* compared to standard wound care methods [54–58].

Advanced Technologies in Wound Care

Negative Pressure Wound Therapy (NPWT)

Negative Pressure Wound Therapy (NPWT), also known as *Vacuum-Assisted Closure*, is a globally accepted innovation in plastic and reconstructive surgery (Fig. 17.14). It is especially useful for wounds that cannot be closed primarily. NPWT systems are available in both bedside-connected and portable formats.

Mechanisms of Action:

1. *Fluid removal*: Negative pressure removes *excess interstitial fluid*, which helps reduce edema [59]. Interstitial fluid is known to contain *matrix metalloproteinases* [60], which impair the wound healing process.
2. *Enhanced perfusion*: As edema is reduced, local blood flow improves, facilitating oxygen and nutrient delivery to the wound bed [59, 61, 62].

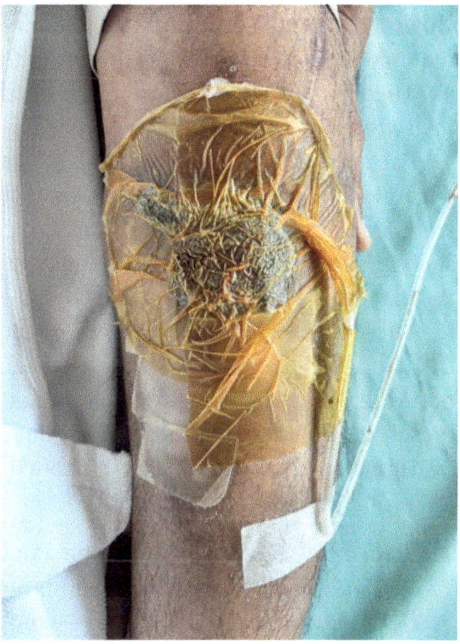

Fig. 17.14 Negative Pressure Wound Therapy (NPWT): a sealed system that accelerates healing by reducing edema, improving blood flow, lowering bacterial burden, and promoting granulation tissue

3. *Reduction of bacterial load and biofilm*: NPWT helps lower bacterial burden and disrupt biofilm formation, both of which contribute to delayed healing [63].
4. *Mechanical microstrain*: Negative pressure generates mechanical forces at the cellular level, promoting cellular migration and granulation tissue formation [64]. This results in increased granulation tissue growth.

Clinical Advantages:

- *Reduced dressing change frequency*: Dressings typically require replacement every *48–72 hours*. In clean, noninfected wounds, changes every 3 days are acceptable.
- *Pain reduction*: Less frequent dressing changes reduce patient discomfort and staff workload.

Key Considerations:

- NPWT should only be applied *after complete debridement* of necrotic tissue and *clearance of pus*. Silver dressing might be applied with NPWT to decrease or kill bacteria especially in chronic wound or compound fracture [65, 66].
- In wounds with *compromised vascularity*, NPWT may fail to stimulate granulation. If no clinical improvement is observed after *2 weeks*, alternative treatments should be considered.

Tissue-Engineered Skin Equivalents

In cases of significant skin loss, *autologous skin grafting* remains the gold standard. However, when insufficient donor skin is available or additional dermal support is required, *tissue-engineered skin substitutes* may be considered. These technologies, although promising, are currently limited to *advanced wound care centers* due to their *high cost and complex production requirements*.

One approach involves culturing the patient's own *keratinocytes or fibroblasts* to generate a skin-equivalent graft. Another involves using *decellularized allografts* from *human cadaveric skin*, processed to eliminate immunogenic cells while preserving structural proteins. These may be used to *enhance dermal regeneration* or *augment skin graft thickness*, although they remain costly.

Alternatively, *xenogenic materials*, such as *porcine or bovine collagen matrices* or cultured epithelial cells or mucosal cells, may be used to provide structural support (Fig. 17.15). A key advantage is that these allow the use of *thinner autografts*, preserving donor sites while maintaining sufficient mechanical strength.

Tissue-engineered skin equivalents can be categorized into three groups based on the skin layers they replicate:

1. *Epidermal substitutes*—containing only keratinocytes
2. *Dermal substitutes*—containing fibroblast-seeded scaffolds or dermal matrix
3. *Dermo-epidermal substitutes*—combining both epidermal and dermal components

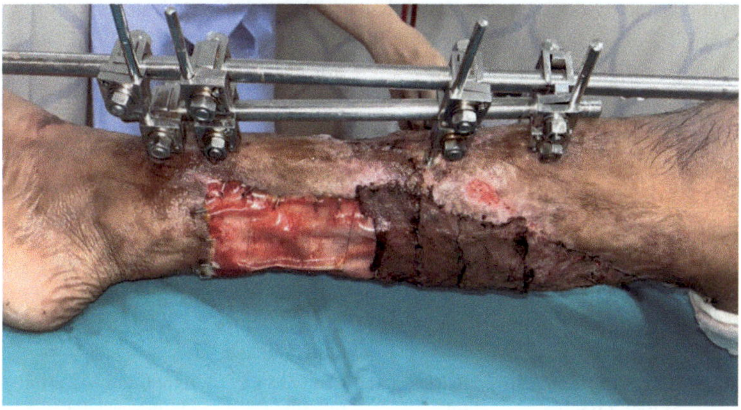

Fig. 17.15 Dermal substitute: A biomaterial derived from bovine collagen, used to replace or augment the dermal layer. It allows for the use of thinner split-thickness skin grafts, promoting faster donor site healing. At the recipient site, it restores dermal volume and elasticity, improving the quality of the reconstructed skin. Application usually requires a staged surgical approach

The characteristics and clinical applications of each type are summarized in Table 17.3 [67].

Growth Factors

Growth factors are biologically active polypeptides that play a pivotal role in regulating the wound healing cascade. They influence various cellular activities, including chemotaxis, cell proliferation, differentiation and angiogenesis, which are essential for tissue repair and regeneration [68, 69].

Several growth factors have been investigated for their therapeutic potential in chronic and complex wounds:

- *Platelet-Derived Growth Factor (PDGF)*: Stimulates fibroblast proliferation, granulation tissue formation and matrix deposition.
- *Epidermal Growth Factor*: Enhances keratinocyte and fibroblast migration, promoting re-epithelialization.
- *Vascular Endothelial Growth Factor*: Induces angiogenesis by promoting endothelial cell proliferation and new capillary formation.
- *Fibroblast Growth Factor*: Supports cell proliferation, angiogenesis and extracellular matrix remodeling.

While recombinant growth factors such as PDGF-BB (becaplermin) have received regulatory approval for use in diabetic foot ulcers, the overall clinical adoption remains limited due to high cost, the need for precise wound bed preparation and concerns over long-term safety. Current research continues to explore delivery systems (e.g., hydrogels, nanoparticles) and combination therapies to enhance their efficacy and bioavailability in wound environments [69, 70].

Table 17.3 Characteristics and properties of tissue-engineered skin equivalents

Type	Description	Uses	Advantages	Disadvantages	Example brands
Epidermal	Cultured allograft from patient's own skin	Full-thickness burns	Autograft capability	High cost, limited accessibility	Epicel
Dermal	Nylon mesh with porcine collagen and human fibroblasts attached to silicone membrane	Superficial Burns, ulcers	Transparent, permitting visualization; different sizes available; pores allow for fluid drainage	Possible adherence, infection risk	Biobrane
Dermal	Bilayered, with matrix of type I bovine collagen and chondroitin-6-sulfate (a glycosaminoglycan from shark cartilage) beneath a silicone epidermal sheet	Partial- and full-thickness chronic ulcers and surgical wounds	Ready to use, biomimetic	Expensive, requires staged procedures	Integra
Dermal	Porcine small intestine submucosa matrix	Chronic wounds (not third-degree burns)	Long shelf life, versatile	Not suitable for deep burns	OASIS
Dermal	Human fibroblast-seeded bioabsorbable polyglactin mesh scaffolds	Diabetic foot ulcers	Promotes complex wounds healing	Expensive, special storage (at −75 °C until use)	Dermagraft
Multilayer	Upper epidermal layer with differentiated keratinocytes; lower dermal layer with bovine type I collagen and human fibroblasts	Venous ulcers lasting for >4 weeks and diabetic foot ulcers lasting >3 weeks	Maintains moist environment, occlusive	Requires pre-order and a specific application is required	Apligraf

References

1. Weller C, Sussman G (2006) Wound dressings update. J Pharm Pract Res 36:318–324
2. Baranoski S, Ayello EA (2012) Wound dressings: an evolving art and science. Adv Skin Wound Care 25:87–92; quiz 92–4
3. Brett DW (2006) A review of moisture-control dressings in wound care. J Wound Ostomy Continence Nurs 33(6 Suppl):S3–S8
4. Lachenbruch C, VanGilder C (2012) Estimates of evaporation rates from wounds for various dressing/support surface combinations. Adv Skin Wound Care 25:29–36
5. Fulton JA, Blasiole KN, Cottingham T, Tornero M, Graves M, Smith LG et al (2012) Wound dressing absorption: a comparative study. Adv Skin Wound Care 25:315–320
6. Nemeth AJ, Eaglstein WH, Taylor JR, Peerson LJ, Falanga V (1991) Faster healing and less pain in skin biopsy sites treated with an occlusive dressing. Arch Dermatol 127:1679–1683

7. Ayello EA (2005) What does the wound say? Why determining etiology is essential for appropriate wound care. Adv Skin Wound Care 18:98–109; quiz 110–1
8. Cowan LJ, Stechmiller J (2009) Prevalence of wet-to-dry dressings in wound care. Adv Skin Wound Care 22:567–573
9. Paddle-Ledinek JE, Nasa Z, Cleland HJ (2006) Effect of different wound dressings on cell viability and proliferation. Plast Reconstr Surg 117(7 Suppl):110S–118S; discussion 119S–20S
10. Eichler R (2005) Homecare solutions: cost effective wound care with hydrogel dressings: 664. J Wound Ostomy Continence Nurs 32(3 Suppl 2):S21
11. Rakel BA, Bermel MA, Abbott LI, Baumler SK, Burger MR, Dawson CJ et al (1998) Split-thickness skin graft donor site care: a quantitative synthesis of the research. Appl Nurs Res 11:174–182
12. Reyzelman AM, Vartivarian M (2015) Evidence of intensive autolytic debridement with a self-adaptive wound dressing. Wounds 27:229–235
13. Thomas S, Andrews A (1999) The effect of hydrogel dressings on maggot development. J Wound Care 8:75–77
14. Heyneman A, Beele H, Vanderwee K, Defloor T (2008) A systematic review of the use of hydrocolloids in the treatment of pressure ulcers. J Clin Nurs 17:1164–1173
15. Gilchrist B, Reed C (1989) The bacteriology of chronic venous ulcers treated with occlusive hydrocolloid dressings. Br J Dermatol 121:337–344
16. Fletcher J (2003) The benefits of using hydrocolloids. Nurs Times 99:57
17. Yan S, Colin X, Coudray-Omnès C, Guido-Morin P, Kommala DR (2014) A budget impact analysis comparing a Hydrofiber® dressing to an alginate dressing in managing exuding venous leg ulcers in France. Int Wound J 11:304–313
18. Belmin J, Meaume S, Rabus MT, Bohbot S (2002) Sequential treatment with calcium alginate dressings and hydrocolloid dressings accelerates pressure ulcer healing in older subjects: a multicenter randomized trial of sequential versus nonsequential treatment with hydrocolloid dressings alone. J Am Geriatr Soc 50:269–274
19. Vaneau M, Chaby G, Guillot B, Martel P, Senet P, Téot L et al (2007) Consensus panel recommendations for chronic and acute wound dressings. Arch Dermatol 143:1291–1294
20. Armstrong SH, Brown DA, Hill E, Ruckley CV, Pucheault M, Hofman D (1995, Nov 21–24) A randomized, comparative clinical trial of a new hydrofibre dressing versus an alginate dressing in the treatment of exuding leg ulcers. Presented at: 5th European Conference on Advances in Wound Management Harrogate, UK
21. Armstrong SH, Ruckley CV (1997) Use of a fibrous dressing in exuding leg ulcers. J Wound Care 6:322–324
22. Leaper DJ, Durani P (2019) Silver in wound care: friend or foe? Plast Reconstr Surg Glob Open 7(8):e2373. https://doi.org/10.1097/GOX.0000000000002373
23. Banks V, Bale S, Harding K, Harding EF (1997) Evaluation of a new polyurethane foam dressing. J Wound Care 6:266–269
24. Dumville JC, Deshpande S, O'Meara S, Speak K (2011) Foam dressings for healing diabetic foot ulcers. In: Cochrane Database Syst Rev, vol 9, p CD009111
25. Chan DC, Fong DH, Leung JY, Patil NG, Leung GK (2007) Maggot debridement therapy in chronic wound care. Hong Kong Med J 13:382–386
26. Honda K, Okamoto K, Mochida Y, Ishioka K, Oka M, Maesato K et al (2011) A novel mechanism in maggot debridement therapy: protease in excretion/secretion promotes hepatocyte growth factor production. Am J Physiol Cell Physiol 301:C1423–C1430
27. Turkmen A, Graham K, McGrouther DA (2010) Therapeutic applications of the larvae for wound debridement. J Plast Reconstr Aesthet Surg 63:184–188
28. Sherman RA (2003) Maggot therapy for treating diabetic foot ulcers unresponsive to conventional therapy. Diabetes Care 26:446–451
29. Paul AG, Ahmad NW, Lee HL, Ariff AM, Saranum M, Naicker AS et al (2009) Maggot debridement therapy with *Lucilia cuprina*: a comparison with conventional debridement in diabetic foot ulcers. Int Wound J 6:39–46

30. Ito K, Saito A, Fujie T, Nishiwaki K, Miyazaki H, Kinoshita M et al (2015) Sustainable antimicrobial effect of silver sulfadiazine-loaded nanosheets on infection in a mouse model of partial-thickness burn injury. Acta Biomater 24:87–95
31. Hoffmann S (1984) Silver sulfadiazine: an antibacterial agent for topical use in burns. A review of the literature. Scand J Plast Reconstr Surg 18:119–126
32. Fuller FW, Engler PE (1988) Leukopenia in non-septic burn patients receiving topical 1% silver sulfadiazine cream therapy: a survey. J Burn Care Rehabil 9:606–609
33. Fuller FW (2009) The side effects of silver sulfadiazine. J Burn Care Res 30:464–470
34. Lari AR, Alaghehbandan R (2002) Silver sulfadiazine and fundermol in the topical treatment of burn wounds: an experimental comparative study in rats. Arch Iran Med 5:170–174
35. Srivastava P, Durgaprasad S (2008) Burn wound healing property of Cocos nucifera: an appraisal. Indian J Pharmacol 40:144–146
36. Han MC, Durmus AS, Karabulut E, Yaman I (2005) Effects of Turkish propolis and silver sulfadiazine on burn wound healing in rats. Rev Med Vet (Toulouse) 156:624
37. Bocchiotti G, Robotti E (1990) The use of a topical antibacterial agent (silver sulfadiazine) on soft-tissue wounds. Minerva Chir 45:677–681
38. Lee JH, Kwak JJ, Shin HB, Jung HW, Lee YK, Yeo ED et al (2013) Comparative efficacy of silver-containing dressing materials for treating MRSA-infected wounds in rats with streptozotocin-induced diabetes. Wounds 25:345–354
39. Lin YH, Hsu WS, Chung WY, Ko TH, Lin JH (2016) Silver-based wound dressings reduce bacterial burden and promote wound healing. Int Wound J 13:505–511
40. Khor WS, Thong HJ, Lew BL (2023) Randomized comparative study of silver hydrofiber versus silver sulfadiazine in partial-thickness burn wounds. J Cosmet Dermatol 22(1):148–155. https://doi.org/10.1111/jocd.15639
41. Mason SA, Jull AB, Vandal AC, Parag V, Walker N (2020) Meta-analysis of randomized controlled trials of silver sulfadiazine in the treatment of burn wounds. Burns 46(2):299–308. https://doi.org/10.1016/j.burns.2019.07.017
42. Faddis D, Daniel D, Boyer J (1977) Tissue toxicity of antiseptic solutions. A study of rabbit articular and periarticular tissues. J Trauma 17:895–897
43. Goldenheim PD (1993) An appraisal of povidone-iodine and wound healing. Postgrad Med J 69(Suppl 3):S97–S105
44. Bigliardi PL, Alsagoff SAL, El-Kafrawi HY, Pyon JK, Wa CTC, Villa MA (2017) Povidone iodine in wound healing: a review of current concepts and practices. Int J Surg 44:260–268. https://doi.org/10.1016/j.ijsu.2017.06.073
45. Mitani O, Nishikawa A, Kurokawa I, Gabazza EC, Ikeda M, Mizutani H (2016) Enhanced wound healing by topical application of ointment containing a low concentration of povidone-iodine. J Wound Care 25:521–529
46. Zhao Z, Xie X, Wang J, Tang Y, Zhang X (2023) Comparative effectiveness of silver- and iodine-containing dressings for wound healing: a systematic review and network meta-analysis. BMJ Open 14(8):e077902. https://doi.org/10.1136/bmjopen-2023-077902
47. Wolcott RD, Kennedy JP (2023) Medical mythology, misconceptions, and misinformation: does iodine impede wound healing? Wounds 35(3):E43–E45
48. Holloway GA, Johansen KH, Barnes RW, Pierce GE (1989) Multicenter trial of cadexomer iodine to treat venous stasis ulcer. West J Med 151:35–38
49. Nherera LM, Woodmansey E, Trueman P, Gibbons GW (2016) Estimating the clinical outcomes and cost differences between standard care with and without cadexomer iodine in the management of chronic venous leg ulcers using a Markov model. Ostomy Wound Manage 62:26–40
50. Williams RL (2009) Cadexomer iodine: an effective palliative dressing in chronic critical limb ischemia. Wounds 21:15–28
51. Chirife J, Herszage L, Joseph A, Kohn ES (1983) In vitro study of bacterial growth inhibition in concentrated sugar solutions: microbiological basis for the use of sugar in treating infected wounds. Antimicrob Agents Chemother 23:766–773

52. Molan PC (2001) Potential of honey in the treatment of wounds and burns. Am J Clin Dermatol 2:13–19
53. Bang LM, Buntting C, Molan P (2003) The effect of dilution on the rate of hydrogen peroxide production in honey and its implications for wound healing. J Altern Complement Med 9:267–273
54. Subrahmanyam M (1999) Early tangential excision and skin grafting of moderate burns is superior to honey dressing: a prospective randomised trial. Burns 25:729–731
55. Jull A, Walker N, Parag V, Molan P, Rodgers A (2008) Randomized clinical trial of honey-impregnated dressings for venous leg ulcers. Br J Surg 95:175–182
56. Wijesinghe M, Weatherall M, Perrin K, Beasley R (2009) Honey in the treatment of burns: a systematic review and meta-analysis of its efficacy. N Z Med J 122:47–60
57. Zbuchea A (2014) Up-to-date use of honey for burns treatment. Ann Burns Fire Disasters 27:22–30
58. Tirado DJ, Hudson NR, Maldonado CJ (2014) Efficacy of medical grade honey against multidrug-resistant organisms of operational significance: part I. J Trauma Acute Care Surg 77(Suppl 2):S204–S207
59. Ubbink DT, Westerbos SJ, Evans D, Land L, Vermeulen H (2008) Topical negative pressure for treating chronic wounds. In: Cochrane Database Syst Rev, vol 3, p CD001898
60. Glass GE, Murphy GF, Esmaeili A, Lai LM, Nanchahal J (2014) Systematic review of molecular mechanism of action of negative-pressure wound therapy. Br J Surg 101:1627–1636
61. Morykwas MJ, Argenta LC, Shelton-Brown EI, McGuirt W (1997) Vacuum-assisted closure: a new method for wound control and treatment: animal studies and basic foundation. Ann Plast Surg 38:553–562
62. Xia CY, Yu AX, Qi B, Zhou M, Li ZH, Wang WY (2014) Analysis of blood flow and local expression of angiogenesis-associated growth factors in infected wounds treated with negative pressure wound therapy. Mol Med Rep 9:1749–1754
63. Bradley BH, Cunningham M (2013) Biofilms in chronic wounds and the potential role of negative pressure wound therapy: an integrative review. J Wound Ostomy Continence Nurs 40:143–149
64. McNulty AK, Schmidt M, Feeley T, Kieswetter K (2007) Effects of negative pressure wound therapy on fibroblast viability, chemotactic signaling, and proliferation in a provisional wound (fibrin) matrix. Wound Repair Regen 15:838–846
65. Warriner R, Burrell R (2005) Infection and the chronic wound: a focus on silver. Adv Skin Wound Care 18(Suppl 1):2–12
66. Stinner DJ, Waterman SM, Masini BD, Wenke JC (2011) Silver dressings augment the ability of negative pressure wound therapy to reduce bacteria in a contaminated open fracture model. J Trauma 71(Suppl 1):S147–S150
67. Powers JG, Higham C, Broussard K, Phillips TJ (2016) Wound healing and treating wounds: chronic wound care and management. J Am Acad Dermatol 74:607–625; quiz 625–6
68. Safari M, Ghahari L, Zoroufchi MDBHZ (2014) Effects of epidermal growth factor, platelet derived growth factor and growth hormone on cultured rat keratinocytes cells in vitro. Pak J Biol Sci 17:931–936
69. Martí-Carvajal AJ, Gluud C, Nicola S, Simancas-Racines D, Reveiz L, Oliva P et al (2015) Growth factors for treating diabetic foot ulcers. Cochrane Database Syst Rev 10:CD008548
70. Niu Y, Li Q, Ding Y, Dong L, Wang C (2019) Engineered delivery strategies for enhanced control of growth factor activities in wound healing. Adv Drug Deliv Rev 146:190–208. https://doi.org/10.1016/j.addr.2018.06.002

Chapter 18
Current Biomaterials and Strategies in Tissue Engineering for Wound Healing

F. Re, E. Borsani, M. Bonetti, and F. Bonomini

Abstract The process of wound repair represents a challenge for clinical practitioners and researchers, as it represents the intrinsic capacity of the body to recover from injury. Skin tissue healing is divided into four distinct phases: hemostasis, inflammation, proliferation and remodeling. The transition between these phases is dependent on cell-to-cell interactions, which are mediated by both cytokines and chemokines. The healing process may be impaired by several factors, including wound size, infection, decreased vascularization or systemic illnesses. The conventional approach to the prompt healing of skin wounds employs the use of natural products, also called wound dressings, that possess anti-inflammatory, antioxidant and antimicrobial properties. In recent years, the field of regenerative medicine has become an outstanding area of research, aiming to develop artificial tissues and organ regeneration techniques to address significant challenges in human health. The combination of biocompatible scaffolds or structures that resemble the natural extracellular matrix together with stem cells, platelet derivatives, vascular stromal fraction and multiple biological factors has attracted extensive attention in the field of wound regenerative medicine. Indeed, scaffolds facilitate the delivery and application of stem cells for the purposes of tissue regeneration, as well as the upregulation of specific molecular signaling pathways that are pivotal for the repair of the skin.

F. Re (✉)
Department of Clinical and Experimental Sciences, University of Brescia, Bone Marrow Transplant Unit, ASST Spedali Civili, Brescia, Italy

Centro di Ricerca Emato-Oncologica AIL (CREA), ASST Spedali Civili, Brescia, Italy
e-mail: Federica.re@unibs.it

E. Borsani · M. Bonetti · F. Bonomini
Division of Anatomy and Physiopathology, Department of Clinical and Experimental Sciences, University of Brescia, Brescia, Italy
e-mail: elisa.borsani@unibs.it; matteo.bonetti@unibs.it; francesca.bonomini@unibs.it

F. De Francesco, K. Turksen (eds.), *Management and Strategies for Wound Healing*, Stem Cell Biology and Regenerative Medicine 79,
https://doi.org/10.1007/978-3-032-12442-5_18

Keywords Regenerative medicine · Skin substitutes · Scaffolds · biomaterials · Wound healing · stem cells · Mesenchymal stromal cells · Platelet derivatives

Abbreviations

ADSCs	MSCs from adipose tissue
bFGF	Basic fibroblast growth factor
BM-MSCs	MSCs from bone marrow
CEA	Cultured epidermal autograft
DAG	Diacylglycerol
ECM	Extracellular matrix
EGF	Endothelial growth factor
En-1	Engrailed-1
ESCRT	Endosomal sorting complex required for transport
EVs	Extracellular vesicles
FAK	Focal adhesion kinase
GDFE	Polydopamine-modified graphene oxide
GM-CSF	Granulocyte-macrophage colony-stimulating factor
GPA	Glycerol-preserved allografts
HIFs	Hypoxia inducible factors
HPL	Human platelet lysate
IGF-1	Insulin-like growth factor-1
iPSCs	Induced pluripotent stem cells
JAK-STAT3	Janus kinase/signal transducer and activator of transcription 3
KGF	Keratinocyte growth factor
MEK/ERK	Mitogen-activated protein kinase kinase/extracellular signal-regulated kinase
MMPs	Matrix metalloproteinases
mTORC	Mammalian target of rapamycin complex
PCL	Poly(caprolactone)
PDGF	Platelet-derived growth factor
PDMS	Polydimethylsiloxane
PEG	Polyethylene glycol
PEO	Polyethylene oxide
PGA	Polyglycolic acid
PI3K	phosphatidylinositol 3-kinase
PIP2	4,5-bisphosphate
PIP3	(3,4,5)-trisphosphate
PKCβ	Protein kinase C beta
PLA	Polylactic acid
PLGA	Poly(lactic-co-glycolic) acid
PRP	Platelet-rich plasma
PU	Polyurethane
PVP	Polyvinylpyrrolidone

SMA	Smooth muscle actin
SMAD	Small mother against decapentaplegic
SNARE	Soluble N-ethylmaleimide sensitive fusion protein (NSF) attachment receptor
SVF	Stromal vascular fraction
TGF-β	Transforming growth factor β
TGF-βR	TGF-β receptor
UC-MSCs	MSCs from umbilical cord
VEGF	Vascular endothelial growth factor
VEGFR	VEGF receptor
VLU	Healing venous leg ulcers
YAP/TAZ	Yes-associated protein/transcriptional coactivator with PDZ-binding motif

Mechanisms for Wound Healing

The skin plays a pivotal role as an organ that serves as the body's primary defense mechanism, safeguarding against external insults and regulating body temperature [1]. Furthermore, the skin performs a multitude of indispensable physiological functions, including the processing and metabolism of structural biomolecules such as lipids, proteins and glycans, as well as the production and secretion of hormones (vitamin D synthesis and peripheral conversion of prohormones) [2]. The skin also performs additional functions. It acts as a sensory organ due to the presence of receptors that enable the reception of various external stimuli, including tactile, thermal and painful sensations. The more superficial layer of the skin registers and transmits pressure (tactile) and painful and thermal stimuli, while the deeper layer transmits thermal and vibratory signals. The skin also serves as an excretory organ, allowing the excretion of salts, water and catabolites through sweating. Additionally, the skin has an absorption function, enabling the assimilation of active ingredients from the external environment into the cells and into the bloodstream. In addition, the skin performs a thermoregulatory function, which entails the regulation of body temperature through sweating and the presence of adipose tissue capable of regulating blood flow, increasing heat dispersion through vasodilation or decreasing it through vasoconstriction [3].

In normal condition, skin is a self-repairing organ that maintains homeostasis through a sophisticated and organized series of mechanisms. The regenerative property is crucial for the maintenance of the skin barrier during development, in homeostasis and for wound healing. This capacity is maintained by stem cells in the skin through the secretion of cytokines, chemotactic factors, extracellular matrix (ECM) remodelers, growth factors (GFs) and their inhibitors [4–6]. Wound healing is a complex process, comprising a series of cellular and molecular events. The process is divided into four overlapping phases: hemostasis, inflammation, proliferation and migration of stem cells to the wound site and, at last, re-epithelization (Fig. 18.1) [7]. Following the formation of a wound, damaged vessels are closed by blood clots

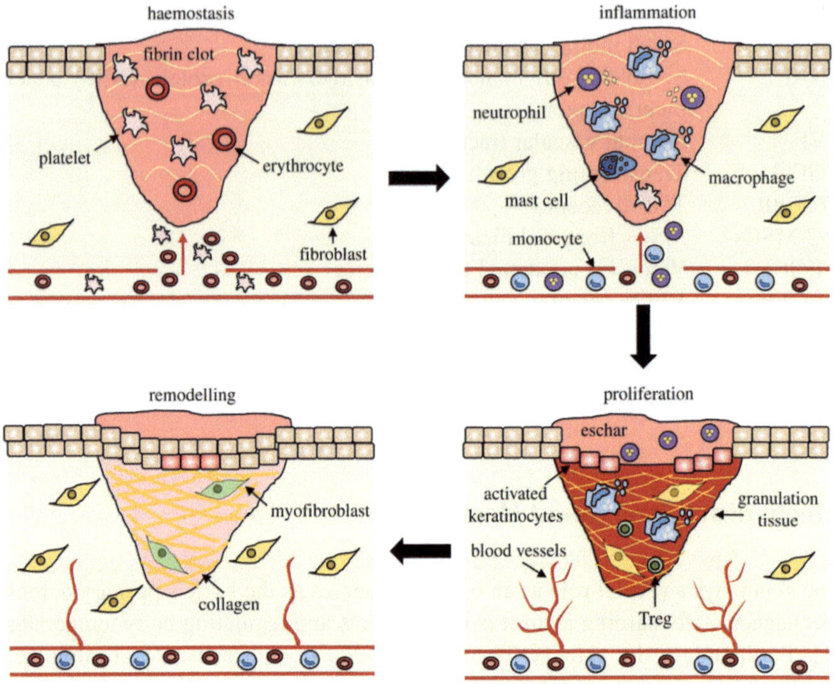

Fig. 18.1 Overview of the stages of wound repair and the major cellular components involved. The stages of wound healing involve an interaction among the ECM, platelets, fibroblast, keratinocytes, endothelial cells and macrophages. (Reproduced with the permission of Wilkinson HN et al. [3, 12])

(hemostasis phase), and macrophages and neutrophils are recruited at the wound site to eliminate dead cells and secrete growth factors necessary for cell proliferation (inflammation phase). The continued proliferation, migration and differentiation of cells, including keratinocytes, fibroblasts, stem cells and epithelial progenitors, facilitate wound repair, restoration of vascular networks and formation of granulation tissue. Fibroblasts are responsible for collagen production, which serves to restore the ECM. Meanwhile, matrix metalloproteinases facilitate the remodeling of the matrix itself, replacing collagen type 3 with collagen type 1, thereby strengthening the skin. The proliferation of keratinocytes is initiated predominantly at the wound edges, where a range of growth factors and cytokines, including epidermal growth factor (EGF), keratinocyte growth factor (KGF), vascular endothelial growth factor (VEGF) and transforming growth factor beta (TGF-β), are secreted to facilitate the healing process. During this process, cells that are no longer required are removed from the wound site, and the ECM is remodeled to restore the natural skin morphology. Minor wounds are readily resolved by the growth of cells within the wound bed, whereas severe skin wounds are more protracted in their healing process and are susceptible to complications such as prolonged inflammation and infection. Additionally, comorbidities, such as diabetes, have a deleterious impact on the healing cascade, thereby impairing tissue repair or even leading to wound chronicity. It is therefore pertinent to consider these factors when engineering the scaffolds [1, 8, 9].

The classification of a wound as acute or chronic is contingent upon the duration of the healing process. An acute wound is typically caused by a sudden skin injury from an accident or surgical procedure. It heals at a predictable and relatively short time frame (within 8–12 weeks). On the contrary, chronic wounds are typified by an imbalance between proinflammatory and anti-inflammatory stimuli which hinders the healing process. The development of chronic wounds may occur subsequent to various pathologies, including but not limited to: trauma, burns, prolonged pressure on the skin (pressure ulcers), improper function of the venous valves (venous ulcers) or metabolic disorders (diabetic foot ulcers) [10, 11].

In recent years, a variety of skin substitutes (autologous, allogenic and heterologous) of dermal, epidermal or dermo-epidermal origin have been employed in the context of skin regeneration. The current trend in this field is the utilization of scaffolds incorporating cell populations, such as stem cells, keratinocytes and fibroblasts (Fig. 18.2) [3, 9, 12, 13]. Stem cell seeding, also referred to as stem cell transplantation, entails the organization of stem cells into a scaffold and the subsequent provision of an adequate time frame for their proliferation. Once the

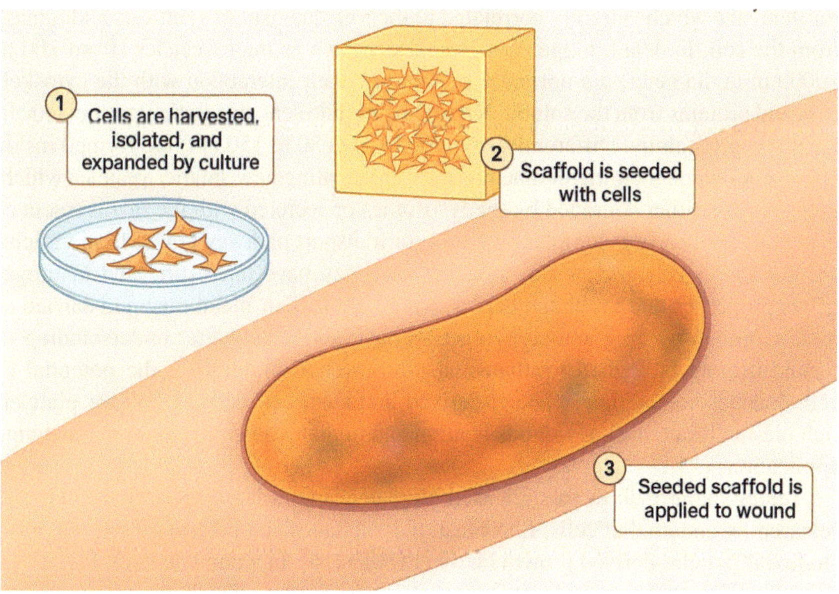

Fig. 18.2 The use of cells and scaffolds in tissue engineering for wound healing, (Reproduced with the permission of Downer M. et al. [12])

proliferative phase is complete, the seeded scaffold is positioned over the wound, where the accelerated proliferation of stem cells can facilitate the healing process [12].

Regulatory Signals in Wound Healing

The many cells involved in wound healing are recruited, activated and proliferated at the site of damage in a coordinated manner. The steps of this process are finely modulated by a variety of autocrine and paracrine signals secreted at wound site by surrounding tissues. When vessel integrity is compromised, platelets act as first responders by aggregating at the site of damage and starting the coagulation process to limit blood loss. This happens mainly as damaged endothelial cells activate platelets cytoskeletal reorganization and platelet-mediated conversion of fibrinogen to fibrin and of prothrombin to thrombin. Ultimately platelets aggregate and localize around the newly synthetized fibrin matrix at the site of damage. At the same time, platelets help the recruitment of immune and stem cells in the fibrin matrix and secrete a combination of growth factors, chemokines and cytokines together with lipids, mRNA and other regulatory RNAs.

These molecules are normally stored inside the cells and are secreted as paracrine signals or as cargo of extracellular vesicles (EVs) which may greatly vary in size, content, half-life and mechanism of release. EVs are commonly differentiated for their size which is tightly correlated to their mechanism of synthesis and release from the cell. In detail, bigger vesicles, also known as microvesicles (from 100 to 1000 nm in diameter) are normally secreted by their interaction with the cytoskeleton and proteins from the soluble N-ethylmaleimide sensitive fusion protein attachment receptor family. Differently, exosomes (from 30 to 150 nm) are formed inside endosomes when their membrane buds inward creating new smaller vesicles which, then, may be either degraded by the lysosomes or secreted with the involvement of the endosomal sorting complex required for transport pathway [14]. These vesicles are not exclusively secreted by platelets and may have interesting and heterogeneous properties due to the molecules expressed on their membrane and carried as their cargo. Harvesting and studying EVs have allowed a better understanding of wound-healing biological mechanisms. Moreover, it highlighted the potential of cells-loaded biomaterials, platelet-derived and stem cell-derived EVs or platelet-rich plasma lysate in skin wounds healing and in many other regenerative medicine applications [15].

Among the growth factors released by platelets (and later discovered to be also released by endothelial cells, fibroblast, macrophages and others), one of the most studied is platelet-derived growth factor (PDGF) [16]. In wound healing, PDGF is a ligand with chemotactic and mitogenic effects for dimeric PDGF receptors in fibroblasts, monocytes and macrophages. Therefore, PDGF positive impact in initiating and accelerating the wound healing process is mainly to be attributed to its effect on immune cells and stem cells recruitment. Similarly, during the proliferation and remodeling phases of wound healing, its beneficial effects are mainly due to fibroblast proliferation and neo-endothelium stabilization during angiogenesis. Among the other growth factors involved in wound healing, VEGF has received the most attention especially considering angiogenesis importance in vascular biology and cancer research. VEGF is considered the most important stimulus for angiogenesis; its production is extensive in wounded skin, and it is key to many steps of the wound healing process [17]. The majority of VEGF is secreted in a paracrine fashion by keratinocytes and myeloid cells in response to hypoxia and oxidative stress, which are abundant in skin wounds given the oxidants production and low oxygen levels induced by vessel damage. More specifically, VEGF expression is tightly induced by hypoxia inducible factors, transcriptional factors and central elements of hypoxia response [18]. Once secreted, free VEGF binds to VEGF receptor (VEGFR), a transmembrane protein with kinase activity capable of activating many proangiogenetic pathways. Among these, VEGFR can phosphorylate protein kinase C beta which then, through diacylglycerol (DAG) synthesis from phosphatidylinositol 4,5-bisphosphate (PIP2), activates the mitogen-activated protein kinase kinase/extracellular signal-regulated kinase (MEK/ERK) pathway that ultimately acts on many transcription factors to favor cell proliferation. Moreover, VEGF phosphorylates a phosphatidylinositol 3-kinase (PI3K) which then allows phosphatidylinositol [3–5]-trisphosphate (PIP3) synthesis from PIP2 which activates the Akt pathway that ultimately favors the mammalian target of rapamycin complex (mTORC) action, blocks autophagy and favors cell survival. Furthermore, VEGFR is able to strongly activate focal adhesion kinase (FAK) which favors cell migration during angiogenesis. In wound healing, angiogenesis also serves the purpose of increasing endothelial permeability and, therefore, improves proper recruitment of immune cells [19]. This initially favors infiltration of myeloid cells speeding-up angiogenesis, inflammation and the re-epithelialization process, while, later, it may help recruiting the immune cells that lower inflammation and restore skin tissue. Interestingly, VEGF levels regulation can strongly influence wound healing outcomes as downregulation leads to impaired or slower re-epithelization, while prolonged VEGF upregulation is often associated with high inflammation and some chronic wounds [20] (Fig. 18.3). An example of this can be found in a recent clinical trial where patients with nonhealing diabetic ulcers were treated using mesenchymal stem cells with the hope of re-establishing vascularization at wound site. These cells were previously exposed to hypoxic conditions to secrete VEGF and to express other endothelial cells markers, this improved angiogenesis and led to a significant improvement in wound surface area and probability of complete closure [21] [ID Clinical Trial: NCT03267784].

Initially, transforming growth factor beta (TGF-β) was considered to be a key element of the anti-inflammatory and re-epithelialization phase. This initial belief was motivated by a series of studies carried out in mice with compromised TGF-β signaling that resulted in systemic inflammation and a less efficient or delayed re-epithelialization of skin wounds [22]. Notably, in numerous animal models, such delay was easily reversed through locally delivered exogenous TGF-β, suggesting therapeutic potential for this growth factor [23]. Later, many human studies

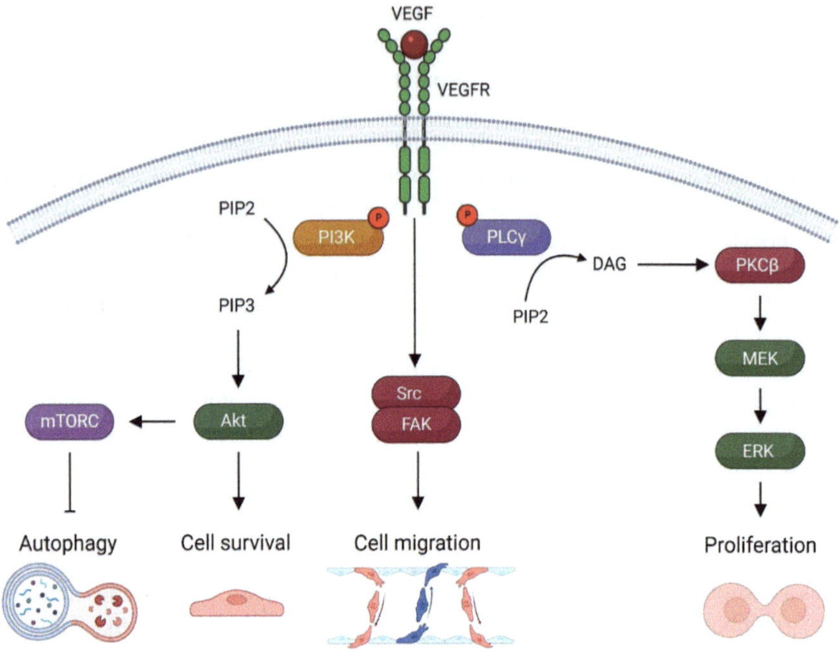

Fig. 18.3 Simplified representation of the VEGF signaling pathway in angiogenesis. VEGFR binds VEGF and activates by phosphorylation the Src/Fak pathway, the PI3K-mediated synthesis of PIP3 and the PLCg-mediated synthesis of DAG. These reactions activate the Akt/mTORC and MEK/ERK pathway respectively resulting in the blocking of autophagy and the activation of cell migration, survival and proliferation. Created in BioRender. (Borsani E. [2025] https://BioRender.com/yomqazx)

employing exogenous TGF-β challenged this conviction and proposed that TGF-β may act in a proinflammatory or anti-inflammatory manner depending based on the biological context [24]. Nowadays, in the context of wound healing, TGF-β is mostly regarded as a proinflammatory molecule which in later phases favors fibroblast differentiation into myofibroblast and therefore stimulate collagen synthesis, ECM remodeling both leading to scar tissue formation [25]. A recent clinical trial has gathered transcriptomic data from nonhealing tissue of venous leg ulcers (VLU) before and after successful treatment with a bilayered living cell construct (Apligraf). These data showed elevated expression levels for genes linked to fibrosis, inflammation and the TGF-β pathway in the nonhealing tissue; differently, successful treatment was paired with TGF-β pathway suppression [26] [ID Clinical Trial NCT01327937].

At wound site, free TGF-β binds to the TGF-β receptor (TGF-βR) which is directly involved in the activation of janus kinase/signal transducer and activator of transcription 3 (JAK-STAT3) and PI3K/Akt pathways. Moreover, TGF-βR can further influence fibroblast activity by employing different members of the small mother against decapentaplegic (SMAD) protein family as messengers and regulators. More specifically, SMAD2 and SMAD3, once phosphorylated by TGF-βR, can form the so-called SMAD complex and translocate to the nucleus to directly influence gene expression. As re-epithelization progresses through ECM remodeling, there is an increased need for the downregulation of TGF-β signaling. This is carried out mainly by SMAD6 and SMAD7 which can limit the SMAD complex localization in the nucleus or by activating Smurf, a protein that can favor TGF-βR degradation by ubiquitination. Dysregulation in the TGF-β pathway may cause the formation of hypertrophic scar tissue or, differently, a prolonged inflammatory state leading to delayed re-epithelialization or chronic wounds (Fig. 18.4).

Unfortunately, in adults, the wound healing process rarely ends with the recovery of initial skin appearance and functionality, differently, in most cases, it culminates with the remodeling of the ECM and the formation of fibrotic scar tissue at the site of damage. The resulting scar shows reduced strength and flexibility when compared to unwounded skin and lacks hair follicles or glands severely affecting natural skin functions. At the moment, many progresses have been made toward a better understanding of the mechanisms regulating the process of scar tissue development, however, treatment reliably achieving total skin recovery has yet to be developed. As briefly mentioned before, fibroblasts are responsible for most ECM remodeling,

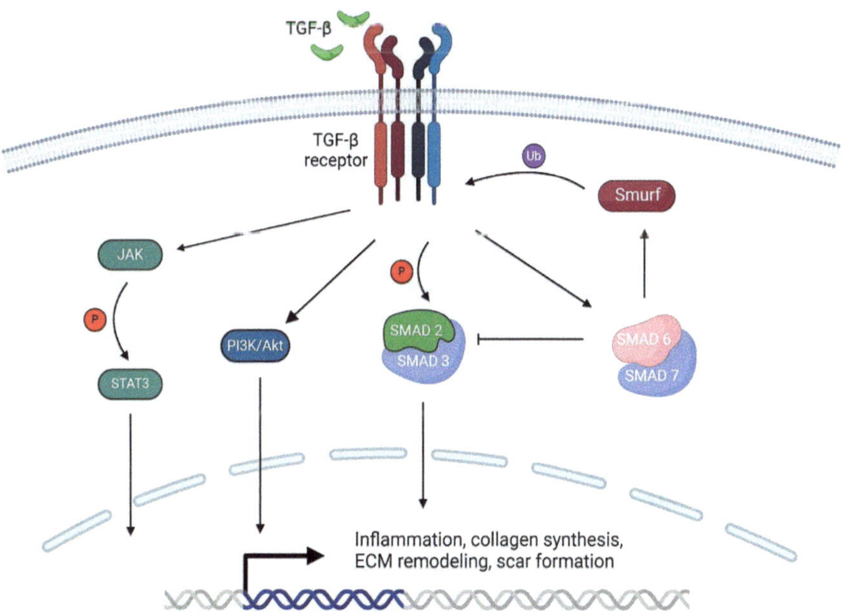

Fig. 18.4 Simplified representation of the TGF-β signaling pathway in fibroblast differentiation and wound healing. TGF-β R binds TGF-β and activates by phosphorylation the SMAD2-SMAD3 complex, the JAK-STAT3 and PI3K-Akt pathway. The SMAD complex activity is then regulated by the SMAD6-SMAD7 complex which can both directly inhibit the SMAD2-SMAD3 pathway and favor TGF-β receptor degradation by the Smurf ubiquitin ligase activity. Created in BioRender. (Borsani E [2025] https://BioRender.com/yomqazx)

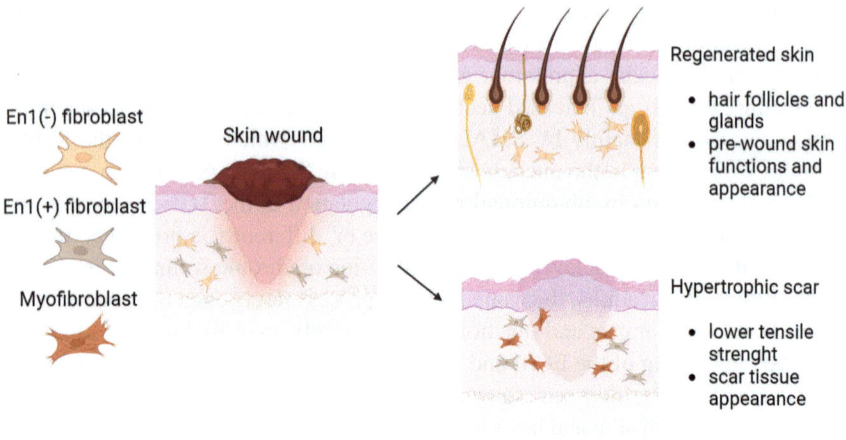

Fig. 18.5 En-1 impact on wound healing outcomes. En-1 downregulation in fibroblast favors complete wound healing, while En-1 positive fibroblast may differentiate in myofibroblast and lead to scar tissue formation and less desirable wound healing outcomes. Created in BioRender. (Borsani E [2025] https://BioRender.com/yomqazx)

collagen deposition and tissue closing at the wound site. In these cells, the upregulation of Engrailed-1 (En-1) has been described as a crucial step for the development of scar tissue as opposed to normal skin recovery [27]. Further research has linked En-1 upregulation to mechanical force-mediated integrin activation which then leads to nuclear translocation of the Yes-associated protein/transcriptional coactivator with PDZ-binding motif (YAP/TAZ) transcription factor through the activity of Rho and Rock kinases. These results may pave the way for new experimental approaches for total skin recovery involving wound dressing with "soft" biomaterials which do not trigger the integrin response [28], or a pharmacological inhibition of the YAP/TAZ transcription factor. Despite this being an interesting line of research with promising clinical implications, many questions regarding this approach and correct timing for YAP/TAZ activity modulation are yet to be answered [29] (Fig. 18.5).

Skin Substitutes Used in Clinical Practice

The field of clinical dermatology currently offers a diverse range of dermal substitutes. Skin substitutes are frequently benchmarked against split-thickness skin autografts, which represent the gold standard for wound healing. In a previous study, Nathoo et al. provided a list of qualities that comprise the "ideal skin substitute." These include resistance to infection, lack of antigenicity, cost efficiency, wide availability, durability, stability and versatility to cover a wide range of wound characteristics [28]. The necessity for enhanced wound healing and the constraints on

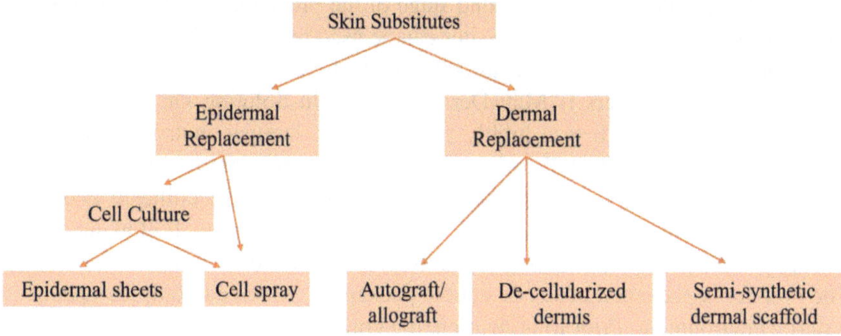

Fig. 18.6 Categories of skin substitutes. (Adapted from Nyame et al. (2015) [1, 30])

the availability of skin graft donor sites have given rise to the diverse and rapidly expanding field of skin substitutes (Fig. 18.6) [1].

Among the various commercial epidermal substitutes (autografts), cultured epidermal autograft (CEA) represents a well-established and standardized method of artificial skin coverage. This is created from the patients' own cells. CEA has been employed for the treatment of extensive burn wounds for over two decades. The various forms of CEA that are currently available for medical use: Epicel® (Genzyme Tissue Repair Corp., Cambridge, MA), Cellspray® (Avita Medical, Woburn, MA), Myskin® (Regenerys, Cambridge, UK) and Epidex® (EuroDerm Biotech & Aesthetics) [31].

Xenografts are dermal substitutes (cellular allografts) derived from animals, most commonly porcine skin. In the case of porcine-derived xenografts, pig skin is treated to remove the dermal layer and prepare it for grafting. These grafts are primarily indicated for clean partial-thickness burns for temporary coverage and for full-thickness wounds resulting from chronic ulcers and traumatic injuries [28]. The most prevalent synthetic skin substitute utilized in burn patients is Integra® (Integra Life Sciences Corp., Plainsboro, NJ, USA), an artificial skin substitute. Integra® consists of a bilaminar structure comprising shark-derived chondroitin-6-sulfate and bovine flexor tendon type 1 collagen, which have been cross-linked with glutaraldehyde. Additionally, Integra® is coated on one side with a silicone membrane (polysiloxane film), which mimics the epidermis and regulates wound moisture loss. Following the application of the collagen layer to a wound, it integrates with the surrounding tissue, resulting in the formation of new skin. This process typically takes between 3 and 6 weeks. At this juncture, the silicone layer can be detached, thereby creating space for the placement of a thin skin graft. Integra® is employed in cases of severe burns where skin tissue auto-transplantation is not a viable option [32]. Integra® is costly when compared to cadaveric allograft skin obtained from skin banks, and it is postulated that there is a high probability of initial failure. Conversely, it provides superior elasticity and aesthetic outcomes compared to a split-skin transplant, with diminished donor-site morbidity due to accelerated healing and minimal scarring. Furthermore, it eliminates the typical infection risk associated with allografts. The second stage of the procedure can be carried out at any convenient time, and the donor sites can be used repeatedly [33]. An alternative graft that has gained considerable attraction in the field of dermatology for the treatment of burns and ulcers is MatriDerm® (MedSkin Solution Dr. Suwelack AG, Billerbeck, Germany). It is a highly porous single laminar matrix composed of native (non-cross-linked) collagen and elastin, the two principal ECM. Elastin is produced by hydrolysis of bovine nuchal ligament, while the collagen is derived from bovine dermis and contains types 1, 2 and 5 collagen types. Following freeze-drying, the media is subjected to gamma irradiation and stored at room temperature. The product is packaged in sterile double packaging, which can only be opened in a sterile environment. Prior to utilization, the graft must be rehydrated in saline and positioned on the water's surface, rather than immersed, in order to prevent the formation of trapped air pockets, which could impede diffusion and compromise transplantation. The matrix is prepared for grafting when the surface undergoes a change from whitish to translucent. MatriDerm® has been demonstrated to control the formation of scars and function as a scaffold for tissue regeneration. Furthermore, it exhibits excellent hemostatic properties, thereby reducing the risk of subtransplant hematoma formation between the matrix and the wound bed. It also facilitates orderly healing and acts as a provisional adhesive barrier, exhibiting limited mechanical stability when wet but conforming effectively to the wound bed. MatriDerm® serves as a substitute for acellular tissue, wherein native, structurally unaltered collagen functions as a component of the ECM, facilitating the fusion of blood vessels and cells. Moreover, MatriDerm®'s limited thickness allows for nutrients diffusion through the graft which facilitates rapid vascularization. In contrast to other synthetic dermal matrices, MatriDerm® has been shown to integrate well, enabling one-step skin grafting in both animal models and humans and to provide an adequate structure to restore the skin and modulate the formation of scar tissue at the skin level [34–36]. Other dermal substitutes include Biobrane® (UDL Laboratories, Inc., Rockford, IL), a bilaminar nylon mesh filled with type 1 porcine collagen and covered with a thin lamina of silicone, which is used as a temporary skin substitute. Biobrane® has been employed to cover partial-thickness burns and skin graft donor sites and has also been used for wounds associated with congenital diseases, such as epidermolysis bullosa and hidradenitis suppurativa [37, 38]. Another such substitute is OASIS wound matrix (Smith & Nephew, Inc., Hull, UK), which is composed of an acellular ECM scaffold derived from porcine jejunum submucosa. OASIS is indicated for a wide variety of wounds, including partial and full-thickness wounds, pressure ulcers, venous ulcers, chronic vascular ulcers, diabetic ulcers, trauma, draining wounds and surgical wounds. Evidence suggests that this treatment has both clinical and histological benefits for wound healing. Furthermore, it has been demonstrated to be an effective method for managing Stage IV pressure ulcers [39]. Nevelia® (SYMATESE Biomateriaux, ZI Les Troques, Chaponost, France) represents a further example of a dermal substitute. This dermal substitute is composed of a three-dimensional porous matrix of stabilized bovine type 1 collagen, covered on top by a semipermeable silicone membrane that acts as a pseudoepidermis [40].

Other dermal substitutes, available as temporary human skin substitutes, comprise the cadaveric skin allograft and include cryopreserved allografts such as AlloDerm® (Life Cell, Branchburg, NJ) [32] and glycerol-preserved allografts [41], the most commonly used in clinical practice.

Artificial dermo-epidermal skin substitutes are the most advanced among all types of skin graft. These composite skin substitutes comprise both epidermal and dermal components, thereby emulating the histological structure of normal skin.

Most of these products are based on allogeneic skin cells being incorporated into a dermal scaffold [42]. The composite scaffolds include Apligraft® (Organogenesis, Inc., Canton, MA) and StrataGraft® (Stratatech, a Mallinckrodt Company). Apligraft® is a composite allograft comprising a layer of bovine collagen gel with neonatal fibroblasts for the dermis and an epidermal layer of neonatal keratinocytes [43, 44]. Apligraft® is an US Food and Drug Administration (FDA)-approved therapy and living cell construct demonstrating efficacy in healing venous leg ulcers (VLU), thanks to its antifibrotic and remodeling effects [26] [ID Clinical Trial: NCT01327937]. StrataGraft® is an allogenic living epidermal substitute that employs neonatal keratinocytes to generate a biologically functional, fully stratified epidermis. This provides a barrier that mimics the intact epidermis [45]. Among the cell-based skin substitutes, TransCyte®, produced by Advanced Tissue Sciences in La Jolla, California, is composed of human neonatal fibroblasts cultured on nylon mesh and a thin silicone layer of Biobrane® (Dow B. Hickam, Inc., Sugarland, TX) for a period of 3 to 6 weeks [46]. Dermagraft® (Advanced BioHealing Inc., La Jolla, CA) is a cellular dermal substitute comprising fibroblasts and keratinocytes derived from neonatal foreskins [41]. Hyalomatrix® (Fidia Advanced Biopolymers, Padua, Italy) is composed of a scaffold of nonwoven pads of hyaluronic acid benzyl ester, layered on top by a semipermeable membrane of silicone. Hyalograft® 3D (Fidia Advanced Biopolymers, Abano Terme, Italy) is an autologous dermal substitute comprised of autologous fibroblasts and covered with a silicone membrane.

Cells and Platelet Derivatives in Regenerative Medicine for Wound Healing

Cell-based skin substitutes are produced using fibroblasts (dermal grafts), keratinocytes (epidermal grafts) or both fibroblasts and keratinocytes (dermo-epidermal grafts). Fibroblasts and keratinocytes can be obtained by direct isolation from skin biopsy, by differentiation of stem cells into mature skin cells or by primary culture of skin cells purchased from a cell culture bank. Coculture systems of fibroblasts and keratinocytes have advantages over monoculture because fibroblasts promote keratinocyte growth and differentiation, while keratinocytes stimulate proliferation of the fibroblasts obtained from the skin cell culture bank. The main disadvantage of employing isolated primary cells is their low proliferative potential, which makes it difficult to obtain enough cells to produce cellular skin grafts [11, 47, 48]. A phase

1 clinical trial tested a two-layered allogeneic keratinocyte and fibroblast skin substitute in a small cohort of patients and showed wound healing and safety results in diabetic foot ulcers. The skin substitutes were applied twice a week for 6 months. In another phase I/II clinical trial, keratinocyte sheet transplantation was successfully used as an off-the-shelf product for the treatment of second-degree burns for 6 months [30, 49] [ID Clinical Trials: IRCT20080728001031N33; IRCT20080728001031N31].

Adult mesenchymal stromal cells (MSCs) are used as alternative sources of skin cells for the generation of living skin constructs. MSCs in skin tissues, coupled with their integral role in normal wound healing, secretory potential and ease of isolation, suggest that MSCs may be beneficial for chronic wound healing. It has been demonstrated that MSCs from bone marrow (BM-MSCs) have the ability to migrate to the injured tissue and differentiate into the required cell type, including fibroblasts. Indeed, up to 20% of fibroblasts at the wound site could be originally derived from BM-MSCs. MSCs target the damaged skin, inhibiting inflammation and increasing the proliferation and differentiation capacity of fibroblast, epidermal and endothelial cells. MSCs have antimicrobial activity, which is crucial for eliminating infections from wounds and are able to migrate in vivo in response to chemotactic signals that regulate inflammation, repair damaged tissues and promote regeneration. Overall, the use of these cells can improve wound healing and reduce scar formation [50]. Both MSCs differentiation and paracrine signaling are involved in tissue healing; differentiation facilitates the regeneration of damaged tissues, while paracrine signaling controls cellular responses to inflammation at the local level [51]. MSCs secrete a plethora of mediators for tissue repair, including growth factors, cytokines and chemokines such as VEGF, PDGF, basic fibroblast growth factor (bFGF), Granulocyte-macrophage colony-stimulating factor, EGF, KGF, insulin-like growth factor-1 (IGF-1) and TGF-β [50, 52]. These mitogens regulate the survival, proliferation, migration and gene expression of keratinocytes, dermal fibroblasts and endothelial cells [53]. In detail, EGF plays a role in the migration of keratinocytes, the proliferation of fibroblasts and the differentiation of these cells into granulation tissue, along with the improvement of wound healing. In the initial stages of the healing process, bFGF and aFGF, which are present in the wound fluid, facilitate the formation of new blood vessels, suggesting a potential paracrine control of epithelial cell development. VEGF is the principal agent responsible for the induction of skin regeneration, thanks to its capacity to promote endothelial cell migration and proliferation, basement membrane breakdown, increased vascular permeability and the development of new blood vessels. The overexpression of IGF-1 accelerates the process of wound healing in epidermic conditions, although the reduction of its expression has been linked to a delay in the healing of cutaneous wounds. PDGF is implicated in all phases of the healing process and is responsible for the differentiation of fibroblasts into myofibroblasts, which enables general tissue contraction. TGF-β is expressed at high levels by numerous cell types and has the capacity to act in both autocrine and paracrine forms. Autocrine activity involves the stimulation of its own synthesis within target cells, including keratinocytes, monocytes, macrophages and fibroblasts. Paracrine activity, on the other hand, induces the surrounding cells to synthesis and release other growth factors that are involved in skin healing and regeneration [4, 5, 10, 52, 54].

MSCs are useful in all phases of the wound healing process, both used alone or together with a biocompatible scaffold and have a multitude of clinical applications, including wound healing, burn treatment and skin rejuvenation [55]. From 2015, several studies have been conducted examining the potential of MSCs as a cell therapy in treating a range of wounds, including chronic varicose leg ulcers, traumatic heel lesions and diabetic foot ulcers [56]. However, the most promising results were obtained when cell-based skin substitutes have been developed for wounds, burns and ulcers [56, 57]. Despite preclinical and clinical research, only few commercial cell-based skin substitutes are commercially available. This is primarily due to their costs and toxicity issues associated with material [58]. For this reason, the use of MSCs in the treatment of cutaneous diseases and injuries is a promising therapeutic strategy, even when used alone. However, in the context of wounds and ulcers, the number of studies employing MSCs alone or cell-based skin substitutes is equivalent because of the small size of injuries, in comparison to severe burns. Moreover, allogeneic MSCs were mostly used instead of autologous cells thanks to their advantage on immunomodulatory capacity. Nevertheless, there is not a standardized number of cells, doses or routes of administration in clinical research of skin wound healing [59].

An analysis of the tissue sources revealed that MSCs for wound healing were preferentially isolated from adipose tissue, bone marrow or umbilical cord blood [60]. Allogenic MSCs from adipose tissue (ADSCs) are the cell population most investigated, and engraftment of cell-based skin substitutes with these cells was the preferred strategy [61]. This is due to the high availability and easy accessibility of adipose tissue and the less invasive tissue harvesting procedure compared to bone marrow. Many preclinical studies have also evaluated the use of secretome derived from MSCs, both from bone marrow and adipose tissue, alone or combined with scaffolds, with successful results in wound healing [62–64]. The secretome of allogenic BM-MSCs was used to enhance skin regeneration in patients with grade 2 diabetic foot ulcers. The same results as conventional cell therapy were obtained in terms of percentage of wound closure [65] [ID Clinical Trial: NCT124174455522].

Umbilical cord MSCs (UC-MSCs) were applied in clinical trials for the treatment of diabetic foot ulcers and skin burns due to their immunoregulatory, growth factors secreting, vasoactive and anti-inflammatory properties. UC-MSCs enhanced anti-infectious ability of the human body and promote angiogenesis and ulcer healing. Fetal fibroblasts were also seeded onto amniotic membranes for the treatment of burn patients, resulting in safe and tolerable treatment, with reduced pain intensity [66–68] [ID Clinical Trials: MR-32-21-015759; IRCT201302218177N6]. In fact, a new lyophilized human amniotic membrane device, AMNIODERM+®, has been shown to heal pressure ulcers, venous ulcers, diabetic ulcers, chronic vascular ulcers, surgical wounds and trauma wounds with a significant reduction in ulceration area after 12 weeks [69] [ID Clinical Trials: Nr. CIV-SK-22-10-041146].

In addition to ADSCs, stromal vascular fraction (SVF) has been employed in the field of wound healing. SVF is obtained by either digestion or mechanical extraction of adipose tissue, followed by several cycles of centrifugation. The SVF is a composite of stem cells (10–30% of which are ADSCs, pericytes and hematopoietic stem cells), mature cells (fibroblasts, muscle cells, endothelial cells and blood cells) as well as various cytokines and growth factors. SVF reveals strong immunomodulatory properties and the ability to promote skin regeneration, which is comparable to or even superior to the activity recorded for ADSCs [70].

Several randomized controlled trials have been conducted in clinical settings, providing substantial evidence in support of the efficacy of human fat products and affirming their high level of safety in the context of clinical wound management. The findings of these studies have demonstrated that the use of human fat products has been associated with accelerated healing rates, reduced healing times and an increased probability of achieving complete healing. The wounds that were treated with human fat products showed a reduction in lymphocyte infiltration and inflammation levels, which in turn resulted in increased collagen deposition and enhanced vascularization.

Specifically, adipose tissue transplantation was effective in healing ulcers in patients with systemic sclerosis, an autoimmune disease characterized by fibrotic changes in the affected organs and tissues [ID Clinical Trial: NCT03406988]. Local injection of microfragmented adipose tissue (Lipogems®) resulted in a safe, valid therapeutic option overcoming several criticisms of stem cell therapy for the treatment of minor diabetic foot amputations [ID Clinical Trial: NCT03276312]. Autologous fat grafting was shown to be a minimally invasive procedure with a beneficial effect on burn wound healing. This technique resulted in improved healing time with a lower incidence of scarring or contracture compared with conventional therapies [ID Clinical Trial: NCT03791710]. SVF was combined with fractional CO_2 laser to improve the treatment of burn scars, resulting in a treatment option for better management of these lesions [ID Clinical Trial: IRCT20210515051307N1]. SVF has also been encapsulated in fibrin-collagen hydrogel for diabetic wound healing. In fact, the endothelial progenitor cells contained in SVF generate intrinsic microvasculature in dermal and epidermal skin substitutes. The clinical data showed for the first time that permanent full-thickness skin grafts with SVF accelerate the wound healing process compared to commercially available dermal-epidermal skin grafts [ID Clinical Trial: IRCT2015110224834N1] [71–75].

Nevertheless, the divergent outcomes observed in clinical trials with regard to efficacy may be ascribed to the capacity of SVF to preserve a more extensive mesenchymal milieu, comprising cells that exhibit a plethora of reparative functions and maintain the presence of numerous cytokines [60].

In addition, a Phase I/IIa feasibility study used peripheral blood mononuclear cells to effectively and noninvasively regenerate tissue and vasculature in dermal ulcers, facilitating their epithelialization. This approach had several advantages, including the need for only 200 mL of peripheral blood with a minimally invasive procedure capable of restoring the vasculogenic function of the tissue, thereby reducing healing time [76] [ID Clinical Trial: UMIN000016665].

Regenerative procedures, incorporating also the utilization of platelet-rich plasma (PRP) or human platelet lysate (HPL) in conjunction with MSCs, have been demonstrated to be a successful treatment modality for chronic skin wounds and diabetic foot ulcers [77–79]. Platelet derivatives have a wide range of applications in various clinical fields, due to their function as a tissue sealant and delivery system for mitogenic and chemotactic growth factors. PDGF has been approved by the F DA for the treatment of diabetic foot ulcers [80]. PRP contains α-granules, platelets secrete several growth factors (GFs) after being activated, such as TGF-β, PDGF and VEGF [79]. GFs and other proteins contained in PRP, including adhesion molecules and chemokines, interact with the local environment to promote cell differentiation and proliferation. These processes are responsible for re-epithelization and angiogenesis via MSCs recruitment and ECM synthesis.

PRP has been used in several clinical scenario including deep dermal burns, diabetic chronic cutaneous ulcers, burn scars treated with fractional CO_2 laser, acute traumatic wounds, ulcers of multifactorial etiologies and ulcer healing in leprosy [81–88] [ID Clinical Trials: ISRCTN14946762; CTRI/2021/09/036230; NCT02352480; ISRCTN14946762; ChiCTR-TRC-09000325; IRCT2017030716148N4; UMIN000015689; ISRCTN14933421; IRCT20220808055641N1].

Platelet gel derived from umbilical cord was used to treat sclerotic skin ulcers in pediatric patients with chronic graft versus host disease following hematopoietic stem cell transplantation. The PRP was easily isolated from the umbilical cord, resulted safe and offered regenerative potential [66] [ID Clinical Trials: IRCT20190101042197N1].

Recently Devesvar et al. (2025) suggested the use of PRP as a more effective treatment option for chronic ulcers. This study emphasizes the role of PRP as a successful treatment option in healing efficiency, patient comfort and cost-effectiveness, considering this medication as a strategy for promoting better granulation tissue formation at both initial and endpoint assessments but also supporting sustained healing, higher healing rates and fewer complications [89]. Despite this, as revised by Manole et al. 2024, it must also be considered that the effectiveness of PRP use can be affected by several factors, both biological, such as the age of the patient or skin types, and nonbiological, such as the protocols used for its preparation and application. So, the field of PRP therapy is evolving, with ongoing research aimed at better understanding its mechanisms and optimizing its use in different medical conditions [90].

HPL is the growth factor-rich product released by platelets by freeze-thawing preparation from platelet concentrates [91]. HPL can be stored at low temperatures for extended periods, standardized for clinical applications and use in different clinical settings, which makes it a superior option to other platelet concentrates, such as PRP [92]. HPL have been demonstrated to be also a reservoir of EVs, including the release of platelets' exosome during collection and storage processes, or the generation of EVs during the preparation of the lysate [8]. EVs as a key effector of the regenerative activity of HPL have recently emerged, leading to mounting interest in evaluating EVs as a stand-alone clinical adjunct in various fields of regenerative medicine [93].

Even though a considerable number of studies have been conducted on the use of platelet-derived products in the treatment of wounds, there are still limitations, including low usage and limited external action time.

Biomaterials in Wound Healing

The scaffold, mimicking the ECM, serves as a temporary support structure to facilitate cell adhesion and thus promote skin regeneration. An ideal scaffold should be sufficiently flexible to conform to the wound's shape, offer a large surface area to volume ratio and sustain a moist environment that supports cell adhesion and promotes re-epithelialization [94]. For many years, wound care primarily used biomaterials as covering for the wound, therefore, the materials only played a passive role in supporting the healing process. In recent years, the advent of functional biomaterials has greatly expanded their potential in regenerative medicine [95], especially considering that the requirements for wound dressing vary significantly depending on the type of wound. However, certain common characteristics can be identified (Table 18.1). Most of these characteristics are largely impacted by the inherent properties of the materials used in their production. Finely tuning biomaterials composition has allowed them to better express their intrinsic proangiogenic, anti-inflammatory, antimicrobial and bioadhesive properties, with the effect to guide host cell behavior and largely influence the cellular microenvironment. Furthermore, their mechanical properties, such as viscoelasticity and stiffness, are usually customizable and play a crucial role in influencing cellular phenotype [95]. Moreover, maintaining an appropriate level of hydration is crucial for effective wound healing [96]. For dry wounds, dressings with a low vapor transmission rate are needed, while exuding wounds require dressings with higher permeability. Thus, optimal dressing must strike a balance between absorbing and releasing fluids [97]. However, completely blocking fluid evaporation should be avoided, as excessive fluid buildup beneath the dressing can lead to maceration and, eventually, infection [97]. Therefore, the chemo-physical properties of the dressings, such as porosity and morphology, should be tailored to the specific type of wound and the level of exudate it produces.

Biomaterials employed in scaffold preparation are commonly classified as natural, synthetic or hybrid types (Table 18.1).

Natural biomaterials are substances derived from biological sources, including animal, microbial [98] and vegetables, that can be applied in medical, engineering or environmental fields. Among these, fibrin-based scaffolds, commonly referred to as "sealants," mimic the final phases of the blood coagulation cascade by forming a stable, physiological fibrin clot that supports wound repair [99, 100]. These biologically derived materials, which originate from living organisms or biological processes, are able to support the restoration of damaged tissues and biological

Table 18.1 Relevant biomaterial properties to consider for the design of customized scaffolds in wound healing. (Created in BioRender. Borsani E [2025] https://BioRender.com/pj98tpz)

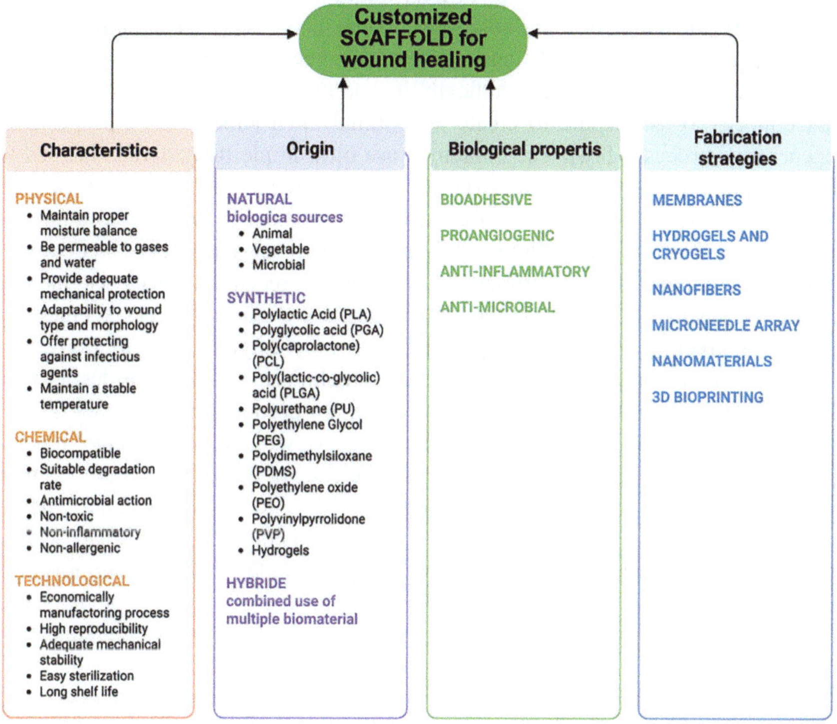

functions. Collagen, a principal component of the ECM, is one of the most extensively used alongside gelatin and silk [101]. Additional examples of animal-derived natural biomaterials include eggshell membranes, fibrin, hyaluronic acid or keratin [94, 102] which can release growth factors, known as important molecules.

Synthetic biomaterials are composed of nonbiological substances that are specifically engineered to interact with biological systems for medical or clinical applications. These materials are designed to mimic the structural and functional properties of native tissues, offering advantages such as tunable degradation kinetics, superior mechanical properties and high versatility for customization according to specific regenerative or therapeutic requirements. Notable synthetic biomaterials used in wound healing include polylactic acid (PLA) [103] and polyglycolic acid (PGA) [104], poly(caprolactone) [9], poly(lactic-co-glycolic) acid [105], polyurethane (PU) [106], polyethylene glycol [107], polydimethylsiloxane [108], polyethylene oxide [109] and polyvinylpyrrolidone [110]. Interestingly, PLA has demonstrated the ability to accelerate the skin regeneration process without inducing an inflammatory response in experimental mice [111], whereas PGA does not promote the expression of alpha smooth muscle actin production, considered as a marker associated with scar formation [112]. Moreover, another class of synthetic biomaterials is hydrogels, hydrophilic polymers able to absorb a great amount of water. Their ability to guarantee a moist wound environment creates a barrier to pathogen infiltration and promotes cellular migration. All these characteristics are considered critical in optimizing conditions for cutaneous repair [113]. Furthermore, synthetic polymers such as PU and silicone have been engineered to support gaseous exchange, exhibit mechanical flexibility and moisture-retentive characteristics that collectively contribute to create an optimal environment for a proper skin regenerative process [114]. The combined use of multiple biomaterials has led to the development of hybrid biomaterials [115], in which distinct properties of each component synergically enhance biological performance. For instance, a composite of cellulose, chitosan and cellulose poly(methyl methacrylate) fibers had demonstrated inherent antibacterial properties. Additionally, recent advances also include the development of biocompatible gelatin-based hybrid hydrogels synthesized through a green cross-linking approach using poly(ethylene glycol) diglycidyl ether, with the incorporation of carrageenan and chitosan polysaccharides into the polymeric matrix [116]. Bioadhesive properties are particularly important in providing a robust wound healing environment because wounds often secrete mucus, so moisture and slipperiness can hinder the adherence of dressings. To address this, various bioadhesive formulations have been developed to also enhance drug delivery at the wound site [117], among these, chitosan and poly(acrylic acid) are widely recognized for their strong mucoadhesive properties [118, 119]. Recent advancements include the development of polydopamine-coated gelatin–alginate hydrogel as a binding platform for bioactive molecules used for various tissue engineering applications [120].

Angiogenic potential is another essential property, especially in the context of chronic wounds, where impaired vascularization delays healing. Certain biomaterials show an interesting approach by releasing ions or molecules capable of stimulating vascularization in a poorly healing or chronic wound environment [121, 122]. For example, silica-based nanocomposite hydrogels exhibited both proangiogenic and antimicrobial properties, highlighting their relevance in chronic wound management [123]. Recently Ding et al. introduced a PU-based foam functionalized with different cationic groups (poly[ethylenimine], poly[L-lysine], and chitosan), which demonstrated enhanced antimicrobial efficacy [124]. Most biomaterials possess more than one bio-functionality. For example, polydopamine-modified graphene oxide (GDFE) hybrid hydrogels were fabricated to show proangiogenic, anti-inflammatory and antimicrobial properties [107]. The developed GDFE composite hydrogels reduced inflammation and stimulated M2 polarization of macrophages in diabetic wounds [125]. Therefore, various biomaterial platforms based on anti-inflammatory responses have been researched to provide a practical approach for inflammatory wounds to promote healing in chronic models.

The mechanical robustness of biomaterials, expressed as high stretchability, toughness and fatigue resistance, is an important parameter for developing wound patches particularly for chronic wounds. Mechanical performance is often improved by strategically combining two or three materials. For example, chitosan hydrogels are sometimes enforced with silver nanoparticles to significantly enhance their mechanical properties and structural stability [126]. A wide range of advanced wound dressing strategies have emerged in recent years, utilizing various biomaterials and fabrication techniques (Table 18.1) and employing both natural and synthetic polymers, as well as their composites, membranes or films are commonly employed as wound dressing. These are typically composed of thin, transparent adhesive sheets made from single or composite biopolymers, designed as a protective barrier but still allowing oxygen permeability.

Hydrogels and cryogels have recently gained significant attention as advanced wound dressing material because of their superior moisture retention capacity, biodegradability, biocompatibility and the ability to deliver therapeutic agents [77, 127–129]. Despite these advantages, certain limitations hinder their widespread application, including high production costs, lengthy synthesis procedures and the use of potentially hazardous cross-linking agents.

Nanomaterials are also promising for wound healing applications due to their small size, high surface-area-to-volume ratio and tunable physical and chemical properties [96]. Their ability to encapsulate a variety of molecules makes them excellent candidates as drug delivery vehicles for therapeutics [130]. In addition, many of these nanomaterials are stable, biodegradable, tunable and versatile. Among these nanofibers-based scaffolds represent a particularly promising class of biomaterials due to their nanoscale architecture, which finely resembles ECM and thus have increased ability to facilitate cell adhesion, growth and migration [131].

Microneedle arrays offer another innovative and minimally invasive approach to wound therapy. These devices penetrate the outermost skin layer (stratum corneum), allowing for localized and sustained delivery of therapeutics and controlled depths [132]. They are often fabricated from materials such as silicon, metals or biodegradable polymers.

Finally, 3D bioprinting represents a cutting-edge technology that enables the precise fabrication of complex, biologically relevant structures by integrating living cells, biomolecules and biomaterials in a layer-by-layer manner [133]. Although bioprinted hydrogels often suffer from low mechanical strength, necessitating frequent dressing changes and potentially reducing patient compliance, this technique offers distinct advantages over conventional fabrication methods. These include high reproducibility, design flexibility and fast construction tailored to wound morphology. For example, collagen has been widely used for generating bioprinted wound healing scaffolds [134, 135].

Bioengineering: Cell-Scaffold Pairing Interaction

A novel perspective in innovative skin regeneration strategies emphasizes not only the use of appropriate scaffolds but also their integration with cellular therapies. In this combined approach, the resulting construct is called a "bioengineered scaffold." This synergy between cells and scaffolds is pivotal in enhancing regenerative performance and improving clinical outcomes. Therefore, a comprehensive understanding of cells-scaffold interactions is essential for the optimization of skin regeneration strategies [136].

Scaffold production plays an important role in the material choosing; in fact, the material needs fundamental properties. First, the essential characteristic is biocompatibility, which is crucial for a correct biological response in the host organism, indeed bioactive materials usually interface with biological surroundings and influence the activities of cells [94]. Nevertheless, other different factors concur in a successful process of wound healing which must be considered in creating new materials: surface properties, mechanical properties, biodegradability and ECM deposition [137].

Bioengineered scaffolds can be classified as "cellular artificial skin grafts" when the scaffold is preseeded or has cells incorporated within its matrix [31, 138, 139]. Artificial skin substitutes can be classified as epidermal, dermal or dermo-epidermal, considering their structural composition [11, 31, 139, 140]. These materials have the potential to address fundamental limitations associated with donor skin rejection sometimes observed in allogeneic or xenogeneic grafts, thereby expanding their clinical applicability [139]. It is important to recognize that the implantation of bioengineered skin constructs initiates a sequence of host immune responses, firstly directed at the cellular components, and subsequently at the biomaterial constituents [139]. Although biomaterials may induce inflammatory reactions, their contribution to transplant rejection is generally minimal [141, 142]. Nevertheless, there is a paucity of studies specifically addressing host immune responses to biological scaffold materials, with existing literature sometimes reporting inflammatory outcomes. The removal or inactivation of immunogenic cellular elements (e.g., cells and cell-associated antigens) from the matrix scaffold can reduce early xenogeneic rejection and help prevent chronic immune reactions postimplantation. Additionally, biocompatibility may be improved by preseeding scaffolds with fibroblasts or other cell types to promote natural remodeling prior to clinical application [99, 143]. The emergence of induced pluripotent stem cells (iPSCs) has opened promising opportunities for stem cell banking, presenting a viable approach to overcome graft rejection and facilitating the development of autologous skin grafts. However, despite these advantages, the clinical translation of iPSC-based therapies remains limited due to the early stage of the technology and significant concerns regarding tumorigenic risk [144, 145]. Nevertheless, ongoing progress, particularly in the design of advanced biomaterials capable of integrating with stem cell-derived components, underscores the future potential for more personalized and effective skin substitute strategies [13, 140]. Today, bioprinting represents the most advanced strategies in skin tissue engineering, which enables the precise spatial deposition of cell-laden hydrogels to replicate the complex architecture of native skin. This technique allows for the organized placement of various cell types, such as keratinocytes, and fibroblasts, in configurations that closely mimic the structure and function of natural skin [146]. A major limitation of current artificial skin grafts is their capacity to regenerate only the epidermal, dermal, and hypodermal layers, without fully restoring skin appendages or neural networks. Most available skin substitutes are unable to reconstruct critical components such as hair follicles, sweat and sebaceous glands and peripheral nerves. The regeneration of these structures, essential for thermoregulation, sensory function and overall quality of life, remains one of the greatest challenges in skin tissue engineering and regenerative medicine. As a result, numerous studies have focused on the use of various stem cell types and differentiation-inducing factors to promote the regeneration of appendages and neural elements [147–151]. Although the biofabrication of skin tissue with functional appendages has not yet been broadly translated into clinical practice, advances in 3D bioprinting offer promising prospects. Owing to its inherent advantages, such as automation, precision, reproducibility, scalability and the customization of complex architectures, 3D bioprinting is anticipated to play a pivotal role in the future clinical application of fully functional skin constructs [152].

Numerous cellular artificial skin substitutes that restore epidermal and dermal structures are currently commercially available, typically in selected clinical cases and with a focus on the treatment of chronic wounds and burn injuries. However, the reconstruction of the subcutaneous layer, the hypodermis, remains largely unachieved, despite the existence of some experimental protocols addressing this component [153]. Dermagraft® is a cellular dermal substitute indicated for the treatment of burns and chronic wounds, including diabetic foot ulcers. It consists of human neonatal fibroblasts seeded onto a bioabsorbable polyglactin mesh scaffold [154]. BioSeed® represents a cellular epidermal skin substitute developed for chronic leg ulcer treatment, composed of autologous keratinocytes cultured within a fibrin sealant matrix [155]. Laserskin® is a thin, transparent epidermal graft intended for burn wounds and chronic full-thickness ulcers. It is fabricated from a benzyl esterified hyaluronan derivative, on which autologous keratinocytes are cultured [156].

These are some examples, but other are EpiDex® [157], Apligraf® [44], MySkin™, MyDerm™ [158], Hyalograft 3D® [159] and JACE® [160].

Although the industry may successfully bring these products to market, integration into the standard of care will depend on the availability of robust clinical evidence and demonstrated health economic benefits. Only through the fulfillment of these criteria can emerging technologies progress beyond the research phase and achieve meaningful clinical impact for patients [161].

Conclusions

The search for new treatments for wound healing is a challenging research domain. There have been promising outcomes with respect to wounds, burns and ulcers, with the development of numerous advanced tissue-engineered skin substitutes. Until now, most of these therapies have been based on acellular biomaterials. However, a novel perspective emphasizing the integration of cellular therapies with appropriate scaffolds is now emerging. This combined approach has resulted in a new type of construct known as a "bioengineered scaffold."

This chapter provides a comprehensive overview of the main techniques currently employed in wound regenerative medicine, including wound dressing, skin substitutes, scaffolds, platelet derivatives alongside stem cells, mesenchymal stromal cells, fibroblasts and keratinocytes, drawing on the findings from clinical trials. The employment of epidermal, dermal or a combination of both epidermal and dermal substitutes is also discussed. The coculturing of fibroblast and keratinocytes was employed for the purpose of facilitating in vitro skin tissue regeneration. Nevertheless, the regeneration of critical components, which are essential for thermoregulation, sensory function and overall quality of life, remains one of the most significant challenges in the fields of skin tissue engineering and regenerative medicine.

The development and modification of biomaterials is an ongoing process, with the aim of creating scaffolds that are more effective in promoting cell growth and tissue regeneration. Both natural and synthetic biomaterials can be deliberately fabricated with the intention of treating wounds, with consideration given to their respective physical and chemical properties. In this context, advances in 3D bioprinting offer a range of promising prospects. There is an extensive range of commercially available artificial skin substitutes that restore epidermal and dermal structures. However, at present, these are typically utilized in selected clinical cases with a focus on the treatment of chronic wounds and burn injuries.

The incorporation of stem cells (i.e., cells from skin biopsies and stromal vascular fraction, iPSCs) into organized biomaterials has been proven to enhance their capacity to repair impaired skin tissue and to promote key wound-healing factors such as epithelialization, granulation tissue formation, vascularization and angiogenesis. Despite the encouraging results observed in nonclinical studies, the utilization of customized biomaterials for stem cell-based interventions in clinical settings remains limited.

References

1. Nyame TT, Chiang HA, Leavitt T, Ozambela M, Orgill DP (2015) Tissue-engineered skin substitutes. Plast Reconstr Surg 136(6):1379–1388
2. Elias PM, Friend DS (1975) The permeability barrier in mammalian epidermis. J Cell Biol 65(1):180–191
3. Wilkinson HN, Hardman MJ (2020) Wound healing: cellular mechanisms and pathological outcomes. Open Biol 10(9):200223
4. Blanpain C, Fuchs E (2009) Epidermal homeostasis: a balancing act of stem cells in the skin. Nat Rev Mol Cell Biol 10(3):207–217
5. Blanpain C, Horsley V, Fuchs E (2007) Epithelial stem cells: turning over new leaves. Cell 128(3):445–458
6. Cappuzzello C, Doni A, Dander E, Pasqualini F, Nebuloni M, Bottazzi B et al (2016) Mesenchymal stromal cell-derived PTX3 promotes Wound healing via fibrin remodeling. J Invest Dermatol 136(1):293–300
7. Rittié L (2016) Cellular mechanisms of skin repair in humans and other mammals. J Cell Commun Signal 10(2):103–120

8. Mogoşanu GD, Grumezescu AM (2014) Natural and synthetic polymers for wounds and burns dressing. Int J Pharm 463(2):127–136
9. Chaudhari AA, Vig K, Baganizi DR, Sahu R, Dixit S, Dennis V et al (2016) Future prospects for scaffolding methods and biomaterials in skin tissue engineering: a review. Int J Mol Sci 17(12)
10. Hocking AM, Gibran NS (2010) Mesenchymal stem cells: paracrine signaling and differentiation during cutaneous wound repair. Exp Cell Res 316(14):2213–2219
11. Przekora A (2020) A concise review on tissue engineered artificial skin grafts for chronic wound treatment: can we reconstruct functional skin tissue in vitro? Cells. 9(7)
12. Downer M, Berry CE, Parker JB, Kameni L, Griffin M (2023) Current biomaterials for wound healing. Bioengineering (Basel) 10(12)
13. Oliveira C, Sousa D, Teixeira JA, Ferreira-Santos P, Botelho CM (2023) Polymeric biomaterials for wound healing. Front Bioeng Biotechnol 11:1136077
14. Doyle LM, Wang MZ (2019) Overview of extracellular vesicles, their origin, composition, purpose, and methods for exosome isolation and analysis. Cells 8(7)
15. Antich-Rosselló M, Forteza-Genestra MA, Monjo M, Ramis JM (2021) Platelet-derived extracellular vesicles for regenerative medicine. Int J Mol Sci 22(16)
16. Hosgood G (1993) Wound healing—the role of platelet-derived growth factor and transforming growth factor beta. Vet Surg 22(6):490–495
17. Johnson KE, Wilgus TA (2014) Vascular endothelial growth factor and angiogenesis in the regulation of cutaneous wound repair. Adv Wound Care (New Rochelle) 3(10):647–661
18. Stockmann C, Kirmse S, Helfrich I, Weidemann A, Takeda N, Doedens A et al (2011) A wound size-dependent effect of myeloid cell-derived vascular endothelial growth factor on wound healing. J Invest Dermatol 131(3):797–801
19. Elson DA, Ryan HE, Snow JW, Johnson R, Arbeit JM (2000) Coordinate up-regulation of hypoxia inducible factor (HIF)-1alpha and HIF-1 target genes during multi-stage epidermal carcinogenesis and wound healing. Cancer Res 60(21):6189–6195
20. Collins PD, Connolly DT, Williams TJ (1993) Characterization of the increase in vascular permeability induced by vascular permeability factor in vivo. Br J Pharmacol 109(1):195–199
21. Kerstan A, Dieter K, Niebergall-Roth E, Klingele S, Jünger M, Hasslacher C et al (2022) Translational development of ABCB5. Stem Cell Res Ther 13(1):455
22. Shull MM, Ormsby I, Kier AB, Pawlowski S, Diebold RJ, Yin M et al (1992) Targeted disruption of the mouse transforming growth factor-beta 1 gene results in multifocal inflammatory disease. Nature 359(6397):693–699
23. Beck LS, Deguzman L, Lee WP, Xu Y, McFatridge LA, Amento EP (1991) TGF-beta 1 accelerates wound healing: reversal of steroid-impaired healing in rats and rabbits. Growth Factors 5(4):295–304
24. Bennett SP, Griffiths GD, Schor AM, Leese GP, Schor SL (2003) Growth factors in the treatment of diabetic foot ulcers. Br J Surg 90(2):133–146
25. Gilbert RWD, Vickaryous MK, Viloria-Petit AM (2016) Signalling by transforming growth factor beta isoforms in wound healing and tissue regeneration. J Dev Biol 4(2)
26. Stone RC, Stojadinovic O, Sawaya AP, Glinos GD, Lindley LE, Pastar I et al (2020) A bioengineered living cell construct activates metallothionein/zinc/MMP8 and inhibits TGFβ to stimulate remodeling of fibrotic venous leg ulcers. Wound Repair Regen 28(2):164–176
27. Mascharak S, desJardins-Park HE, Davitt MF, Griffin M, Borrelli MR, Moore AL et al (2021) Preventing. Science 372(6540)
28. Kurotsu S, Sadahiro T, Fujita R, Tani H, Yamakawa H, Tamura F et al (2020) Soft matrix promotes cardiac reprogramming via inhibition of YAP/TAZ and suppression of fibroblast signatures. Stem Cell Reports 15(3):612–628
29. Jiang D, Rinkevich Y (2021) Converting fibroblastic fates leads to wound healing without scar. Signal Transduct Target Ther 6(1):332
30. Farzanbakhsh S, Shahrbaf MA, Madani H, Dahmardei M, Sadri B, Vosough M (2024) A single-center, open-labeled, randomized, 6-month, parallel-group study to assess the safety

and efficacy of allogeneic cultured keratinocyte sheet transplantation for deep second-degree burn wounds: rationale and design of phase I/II clinical trial. Trials 25(1):226
31. Dai C, Shih S, Khachemoune A (2020) Skin substitutes for acute and chronic wound healing: an updated review. J Dermatolog Treat 31(6):639–648
32. Nathoo R, Howe N, Cohen G (2014) Skin substitutes: an overview of the key players in wound management. J Clin Aesthet Dermatol 7(10):44–48
33. Jones I, Currie L, Martin R (2002) A guide to biological skin substitutes. Br J Plast Surg 55(3):185–193
34. Mantelakis A, McKean AR, Biju NE, Shet AS, Geh JLC (2024) Evolutions in the use of Integra® and MatriDerm® in excision and reconstruction of cutaneous malignancies. J Plast Reconstr Aesthet Surg 93:24–27
35. Min JH, Yun IS, Lew DH, Roh TS, Lee WJ (2014) The use of matriderm and autologous skin graft in the treatment of full thickness skin defects. Arch Plast Surg 41(4):330–336
36. Shah R, Rodrigues R, Phillips V, Khatib M (2024) The use of artificial dermal substitutes for repair of the donor site following harvesting of a radial forearm free flap: a systematic review. J Plast Reconstr Aesthet Surg 88:501–516
37. Melkun ET, Few JW (2005) The use of biosynthetic skin substitute (Biobrane) for axillary reconstruction after surgical excision for hidradenitis suppurativa. Plast Reconstr Surg 115(5):1385–1388
38. Shakespeare P, Shakespeare V (2002) Survey: use of skin substitute materials in UK burn treatment centres. Burns 28(4):295–297
39. Yeh DD, Nazarian RM, Demetri L, Mesar T, Dijkink S, Larentzakis A et al (2017) Histopathological assessment of OASIS ultra on critical-sized wound healing: a pilot study. J Cutan Pathol 44(6):523–529
40. Nicoletti G, Tresoldi MM, Malovini A, Visaggio M, Faga A, Scevola S (2018) Versatile use of dermal substitutes: a retrospective survey of 127 consecutive cases. Indian J Plast Surg 51(1):46–53
41. Halim AS, Khoo TL, Mohd Yussof SJ (2010) Biologic and synthetic skin substitutes: an overview. Indian J Plast Surg. 43(Suppl):S23–S28
42. Shevchenko RV, James SL, James SE (2010) A review of tissue-engineered skin bioconstructs available for skin reconstruction. J R Soc Interface 7(43):229–258
43. Límová M (2010) Active wound coverings: bioengineered skin and dermal substitutes. Surg Clin North Am 90(6):1237–1255
44. Trent JF, Kirsner RS (1998) Tissue engineered skin: Apligraf, a bi-layered living skin equivalent. Int J Clin Pract 52(6):408–413
45. Schurr MJ, Foster KN, Centanni JM, Comer AR, Wicks A, Gibson AL et al (2009) Phase I/II clinical evaluation of StrataGraft: a consistent, pathogen-free human skin substitute. J Trauma 66(3):866–873. discussion 73–74
46. Noordenbos J, Doré C, Hansbrough JF (1999) Safety and efficacy of TransCyte for the treatment of partial-thickness burns. J Burn Care Rehabil 20(4):275–281
47. Germain L, Larouche D, Nedelec B, Perreault I, Duranceau L, Bortoluzzi P et al (2018) Autologous bilayered self-assembled skin substitutes (SASSs) as permanent grafts: a case series of 14 severely burned patients indicating clinical effectiveness. Eur Cell Mater 36:128–141
48. Meuli M, Hartmann-Fritsch F, Hüging M, Marino D, Saglini M, Hynes S et al (2019) A cultured autologous dermo-epidermal skin substitute for full-thickness skin defects: a phase I, open, prospective clinical trial in children. Plast Reconstr Surg 144(1):188–198
49. Farzanbakhsh S, Amini MR, Madani H, Sadri B, Hassani SN, Fallah N et al (2024) Safety evaluation of bi-layered allogenic keratinocyte and fibroblast skin substitute for diabetic foot ulcers-SAFESKIN-DFU: a phase 1 clinical trial. Diabetes Obes Metab 26(11):5078–5086
50. Maxson S, Lopez EA, Yoo D, Danilkovitch-Miagkova A, Leroux MA (2012) Concise review: role of mesenchymal stem cells in wound repair. Stem Cells Transl Med 1(2):142–149

51. Gnecchi M, Zhang Z, Ni A, Dzau VJ (2008) Paracrine mechanisms in adult stem cell signaling and therapy. Circ Res 103(11):1204–1219
52. Yamakawa S, Hayashida K (2019) Advances in surgical applications of growth factors for wound healing. Burns Trauma 7:10
53. Smith AN, Willis E, Chan VT, Muffley LA, Isik FF, Gibran NS et al (2010) Mesenchymal stem cells induce dermal fibroblast responses to injury. Exp Cell Res 316(1):48–54
54. Rajkumar VS, Shiwen X, Bostrom M, Leoni P, Muddle J, Ivarsson M et al (2006) Platelet-derived growth factor-beta receptor activation is essential for fibroblast and pericyte recruitment during cutaneous wound healing. Am J Pathol 169(6):2254–2265
55. Jo H, Brito S, Kwak BM, Park S, Lee MG, Bin BH (2021) Applications of mesenchymal stem cells in skin regeneration and rejuvenation. Int J Mol Sci 22(5)
56. Sierra-Sánchez Á, Montero-Vilchez T, Quiñones-Vico MI, Sanchez-Diaz M, Arias-Santiago S (2021) Current advanced therapies based on human mesenchymal stem cells for skin diseases. Front Cell Dev Biol 9:643125
57. Hashemi SS, Mohammadi AA, Kabiri H, Hashempoor MR, Mahmoodi M, Amini M et al (2019) The healing effect of Wharton's jelly stem cells seeded on biological scaffold in chronic skin ulcers: a randomized clinical trial. J Cosmet Dermatol 18(6):1961–1967
58. Brockmann I, Ehrenpfordt J, Sturmheit T, Brandenburger M, Kruse C, Zille M et al (2018) Skin-derived stem cells for wound treatment using cultured epidermal autografts: clinical applications and challenges. Stem Cells Int 2018:4623615
59. Renesme L, Pierro M, Cobey KD, Mital R, Nangle K, Shorr R et al (2022) Definition and characteristics of mesenchymal stromal cells in preclinical and clinical studies: a scoping review. Stem Cells Transl Med 11(1):44–54
60. Xiao Y, Nie M, Xu W, Zhang J, Lei S, Wu D (2024) The efficiency of human fat products in wound healing: a systematic review and meta-analysis. Int Wound J 21(9):e70016
61. Diehl R, Ferrara F, Müller C, Dreyer AY, McLeod DD, Fricke S et al (2017) Immunosuppression for in vivo research: state-of-the-art protocols and experimental approaches. Cell Mol Immunol 14(2):146–179
62. Shabbir A, Cox A, Rodriguez-Menocal L, Salgado M, Van Badiavas E (2015) Mesenchymal stem cell exosomes induce proliferation and migration of normal and chronic wound fibroblasts, and enhance angiogenesis in vitro. Stem Cells Dev 24(14):1635–1647
63. Aryan A, Bayat M, Bonakdar S, Taheri S, Haghparast N, Bagheri M et al (2018) Human bone marrow mesenchymal stem cell conditioned medium promotes wound healing in deep second-degree burns in male rats. Cells Tissues Organs 206(6):317–329
64. Dalirfardouei R, Jamialahmadi K, Jafarian AH, Mahdipour E (2019) Promising effects of exosomes isolated from menstrual blood-derived mesenchymal stem cell on wound-healing process in diabetic mouse model. J Tissue Eng Regen Med 13(4):555–568
65. Arango-Rodríguez ML, Solarte-David VA, Becerra-Bayona SM, Callegari E, Paez MD, Sossa CL et al (2022) Role of mesenchymal stromal cells derivatives in diabetic foot ulcers: a controlled randomized phase 1/2 clinical trial. Cytotherapy 24(10):1035–1048
66. Mohseni R, Mahdavi Sharif P, Khosravi A, Taheri AR, Behfar M, Zarrabi M et al (2024) The application of umbilical cord blood-derived platelet gel for skin ulcers associated with chronic graft-versus-host disease in pediatrics: a randomized trial. Transplant Cell Ther 30(7):694.e1–694e10
67. Zhang J, Zhao B, Wei W, Wang D, Wang H, Zhang A et al (2023) Prospective, randomized, and controlled study of a human umbilical cord mesenchymal stem cell injection for treating diabetic foot ulcers. J Vis Exp 193
68. Momeni M, Fallah N, Bajouri A, Bagheri T, Orouji Z, Pahlevanpour P et al (2019) A randomized, double-blind, phase I clinical trial of fetal cell-based skin substitutes on healing of donor sites in burn patients. Burns 45(4):914–922
69. Schmiedova I, Slama P, Dembickaja A, Kozova B, Hyneckova V, Gogolkova S et al (2024) Clinical evaluation of AMNIODERM+. BioTech (Basel) 13(3)

70. Atalay S, Coruh A, Deniz K (2014) Stromal vascular fraction improves deep partial thickness burn wound healing. Burns 40(7):1375–1383
71. Abouzaid AM, El Mokadem ME, Aboubakr AK, Kassem MA, Al Shora AK, Solaiman A (2022) Effect of autologous fat transfer in acute burn wound management: a randomized controlled study. Burns 48(6):1368–1385
72. Nilforoushzadeh MA, Sisakht MM, Amirkhani MA, Seifalian AM, Banafshe HR, Verdi J et al (2020) Engineered skin graft with stromal vascular fraction cells encapsulated in fibrin-collagen hydrogel: a clinical study for diabetic wound healing. J Tissue Eng Regen Med 14(3):424–440
73. Roohaninasab M, Khodadad F, Sadeghzadeh-Bazargan A, Atefi N, Zare S, Jafarzadeh A et al (2023) Efficacy of fractional CO. Stem Cell Res Ther 14(1):269
74. Del Papa N, Di Luca G, Andracco R, Zaccara E, Maglione W, Pignataro F et al (2019) Regional grafting of autologous adipose tissue is effective in inducing prompt healing of indolent digital ulcers in patients with systemic sclerosis: results of a monocentric randomized controlled study. Arthritis Res Ther 21(1):7
75. Lonardi R, Leone N, Gennai S, Trevisi Borsari G, Covic T, Silingardi R (2019) Autologous micro-fragmented adipose tissue for the treatment of diabetic foot minor amputations: a randomized controlled single-center clinical trial (MiFrAADiF). Stem Cell Res Ther 10(1):223
76. Tanaka R, Fujimura S, Kado M, Fukuta T, Arita K, Hirano-Ito R et al (2022) Phase I/IIa feasibility trial of autologous quality- and quantity-cultured peripheral blood mononuclear cell therapy for non-healing extremity ulcers. Stem Cells Transl Med 11(2):146–158
77. He M, Chen T, Lv Y, Song P, Deng B, Guo X et al (2022) The role of allogeneic platelet-rich plasma in patients with diabetic foot ulcer: current perspectives and future challenges. Front Bioeng Biotechnol 10:993436
78. Mastrogiacomo M, Nardini M, Collina MC, Di Campli C, Filaci G, Cancedda R et al (2022) Innovative cell and platelet rich plasma therapies for diabetic foot ulcer treatment: the allogeneic approach. Front Bioeng Biotechnol 10:869408
79. Akbarzadeh S, McKenzie MB, Rahman MM, Cleland H (2021) Allogeneic platelet-rich plasma: is it safe and effective for Wound repair? Eur Surg Res 62(1):1–9
80. Harrison-Balestra C, Eaglstein WH, Falabela AF, Kirsner RS (2002) Recombinant human platelet-derived growth factor for refractory nondiabetic ulcers: a retrospective series. Dermatol Surg 28(8):755–759. discussion 9–60
81. Marck RE, Gardien KL, Stekelenburg CM, Vehmeijer M, Baas D, Tuinebreijer WE et al (2016) The application of platelet-rich plasma in the treatment of deep dermal burns: a randomized, double-blind, intra-patient controlled study. Wound Repair Regen 24(4):712–720
82. Gude W, Hagan D, Abood F, Clausen P (2019) Aurix gel is an effective intervention for chronic diabetic foot ulcers: a pragmatic randomized controlled trial. Adv Skin Wound Care 32(9):416–426
83. Li L, Chen D, Wang C, Yuan N, Wang Y, He L et al (2015) Autologous platelet-rich gel for treatment of diabetic chronic refractory cutaneous ulcers: a prospective, randomized clinical trial. Wound Repair Regen 23(4):495–505
84. Vaheb M, Karrabi M, Khajeh M, Asadi A, Shahrestanaki E, Sahebkar M (2021) Evaluation of the effect of platelet-rich fibrin on wound healing at split-thickness skin graft donor sites: a randomized, placebo-controlled, triple-blind study. Int J Low Extrem Wounds 20(1):29–36
85. Hersant B, SidAhmed-Mezi M, Bosc R, Meningaud JP (2017) Autologous platelet-rich plasma/thrombin gel combined with split-thickness skin graft to manage postinfectious skin defects: a randomized controlled study. Adv Skin Wound Care 30(11):502–508
86. Morimoto N, Kakudo N, Matsui M, Ogura T, Hara T, Suzuki K et al (2015) Exploratory clinical trial of combination wound therapy with a gelatin sheet and platelet-rich plasma in patients with chronic skin ulcers: study protocol. BMJ Open 5(5):e007733
87. Napit IB, Shrestha D, Choudhury S, Gkini E, Ilozumba O, Gill P et al (2024) A randomised trial of autologous blood products, leukocyte and platelet-rich fibrin (L-PRF), to promote ulcer healing in LEprosy: the TABLE trial. PLoS Negl Trop Dis 18(5):e0012088

88. Roohaninasab M, Jafarzadeh A, Sadeghzadeh-Bazargan A, Zare S, Nouri M, Nilforoushzadeh MA et al (2024) Evaluation of the efficacy, safety and satisfaction rates of platelet-rich plasma, non-cross-linked hyaluronic acid and the combination of platelet-rich plasma and non-cross-linked hyaluronic acid in patients with burn scars treated with fractional CO. Int Wound J 21(10):e70065
89. Devesvar MV, Shetty SK, Ramesh KS (2025) Comparative study between platelet-rich plasma, platelet-derived growth factor gel, and normal saline dressing in chronic ulcer healing. Ann Afr Med 24:406
90. Manole CG, Voiculescu VM, Soare C, Ceafalan LC, Gherghiceanu M, Hinescu ME (2024) Skin Telocytes could fundament the cellular mechanisms of wound healing in platelet-rich plasma administration. Cells 13(16)
91. Oeller M, Laner-Plamberger S, Krisch L, Rohde E, Strunk D, Schallmoser K (2021) Human platelet lysate for good manufacturing practice-compliant cell production. Int J Mol Sci 22(10)
92. Re F, Sartore L, Moulisova V, Cantini M, Almici C, Bianchetti A et al (2019) 3D gelatin-chitosan hybrid hydrogels combined with human platelet lysate highly support human mesenchymal stem cell proliferation and osteogenic differentiation. J Tissue Eng. 10:2041731419845852
93. Johnson J, Law SQK, Shojaee M, Hall AS, Bhuiyan S, Lim MBL et al (2023) First-in-human clinical trial of allogeneic, platelet-derived extracellular vesicles as a potential therapeutic for delayed wound healing. J Extracell Vesicles 12(7):e12332
94. Negut I, Dorcioman G, Grumezescu V (2020) Scaffolds for wound healing applications. Polymers (Basel). 12(9)
95. Ali Zahid A, Chakraborty A, Shamiya Y, Ravi SP, Paul A (2022) Leveraging the advancements in functional biomaterials and scaffold fabrication technologies for chronic wound healing applications. Mater Horiz 9(7):1850–1865
96. Bhattacharya D, Ghosh B, Mukhopadhyay M (2019) Development of nanotechnology for advancement and application in wound healing: a review. IET Nanobiotechnol 13(8):778–785
97. Ovington LG (2007) Advances in wound dressings. Clin Dermatol 25(1):33–38
98. Zarepour A, Gok B, Budama-Kilinc Y, Khosravi A, Iravani S, Zarrabi A (2024) Bacterial nanocelluloses as sustainable biomaterials for advanced wound healing and dressings. J Mater Chem B 12(48):12489–12507
99. Wong C, Inman E, Spaethe R, Helgerson S (2003) Fibrin-based biomaterials to deliver human growth factors. Thromb Haemost 89(3):573–582
100. Bayer IS (2022) Advances in fibrin-based materials in wound repair: a review. Molecules 27:27(14)
101. Chouhan D, Mandal BB (2020) Silk biomaterials in wound healing and skin regeneration therapeutics: from bench to bedside. Acta Biomater 103:24–51
102. Yang H, Song L, Zou Y, Sun D, Wang L, Yu Z et al (2021) Role of hyaluronic acids and potential as regenerative biomaterials in wound healing. ACS Appl Bio Mater 4(1):311–324
103. Rubio-Elizalde I, Bernáldez-Sarabia J, Moreno-Ulloa A, Vilanova C, Juárez P, Licea-Navarro A et al (2019) Scaffolds based on alginate-PEG methyl ether methacrylate-Moringa oleifera-Aloe vera for wound healing applications. Carbohydr Polym 206:455–467
104. Zhang H, Zhou L, Zhang W (2014) Control of scaffold degradation in tissue engineering: a review. Tissue Eng Part B Rev 20(5):492–502
105. Lee CH, Hung KC, Hsieh MJ, Chang SH, Juang JH, Hsieh IC et al (2020) Core-shell insulin-loaded nanofibrous scaffolds for repairing diabetic wounds. Nanomedicine 24:102123
106. Lee SM, Park IK, Kim YS, Kim HJ, Moon H, Mueller S et al (2016) Physical, morphological, and wound healing properties of a polyurethane foam-film dressing. Biomater Res 20:15
107. Xie J, Huang Y, Hu X, Wu X, Luo X, Wei P et al (2025) A constant filgotinib delivery adhesive platform based on polyethylene glycol (PEG) hydrogel for accelerating Wound healing via restoring macrophage mitochondrial homeostasis. Small 21(14):e2408791

108. Uysal K, Firat IS, Creutz T, Aydin IC, Artmann GM, Teusch N et al (2022) A novel in vitro wound healing assay using free-standing, ultra-thin PDMS membranes. Membranes (Basel) 13(1)
109. Aljohani MM, Abu-Rayyan A, Elsayed NH, Alatawi FA, Al-Anazi M, Mustafa SK et al (2023) One-pot microwave synthesis of chitosan-stabilized silver nanoparticles entrapped polyethylene oxide nanofibers, with their intrinsic antibacterial and antioxidant potency for wound healing. Int J Biol Macromol 235:123704
110. Li W, Wang J, Cheng Z, Yang G, Zhao C, Gao F et al (2022) Sandwich structure Aloin-PVP/Aloin-PVP-PLA/PLA as a wound dressing to accelerate wound healing. RSC Adv 12(42):27300–27308
111. Aoki S, Kinoshita M, Miyazaki H, Saito A, Fujie T, Iwaya K et al (2013) Application of poly-L-lactic acid nanosheet as a material for wound dressing. Plast Reconstr Surg 131(2):236–240
112. Kibe T, Koga T, Nishihara K, Fuchigami T, Yoshimura T, Taguchi T et al (2017) Examination of the early wound healing process under different wound dressing conditions. Oral Surg Oral Med Oral Pathol Oral Radiol 123(3):310–319
113. Yuan N, Shao K, Huang S, Chen C (2023) Chitosan, alginate, hyaluronic acid and other novel multifunctional hydrogel dressings for wound healing: a review. Int J Biol Macromol 240:124321
114. Ijaola AO, Akamo DO, Damiri F, Akisin CJ, Bamidele EA, Ajiboye EG et al (2022) Polymeric biomaterials for wound healing applications: a comprehensive review. J Biomater Sci Polym Ed 33(15):1998–2050
115. Adeli H, Khorasani MT, Parvazinia M (2019) Wound dressing based on electrospun PVA/chitosan/starch nanofibrous mats: fabrication, antibacterial and cytocompatibility evaluation and in vitro healing assay. Int J Biol Macromol 122:238–254
116. Pasini C, Re F, Trenta F, Russo D, Sartore L (2024) Gelatin-based scaffolds with carrageenan and chitosan for soft tissue regeneration. Gels 10(7)
117. Zhu J, Zhou H, Gerhard EM, Zhang S, Parra Rodríguez FI, Pan T et al (2023) Smart bioadhesives for wound healing and closure. Bioact Mater 19:360–375
118. Chopra S, Mahdi S, Kaur J, Iqbal Z, Talegaonkar S, Ahmad FJ (2006) Advances and potential applications of chitosan derivatives as mucoadhesive biomaterials in modern drug delivery. J Pharm Pharmacol 58(8):1021–1032
119. Laffleur F, Leder N, Barthelmes J (2016) In vitro evaluation of thio-poly acrylic acid for intraoral delivery. Drug Deliv 23(6):2065–2073
120. Pacelli S, Chakravarti AR, Modaresi S, Subham S, Burkey K, Kurlbaum C et al (2021) Investigation of human adipose-derived stem-cell behavior using a cell-instructive polydopamine-coated gelatin-alginate hydrogel. J Biomed Mater Res A 109(12):2597–2610
121. Xiang P, Jiang M, Chen X, Chen L, Cheng Y, Luo X et al (2024) Targeting Grancalcin accelerates Wound healing by improving angiogenesis in diabetes. Adv Sci (Weinh) 11(14):e2305856
122. Fuchs B, Mert S, Hofmann D, Kuhlmann C, Birt A, Wiggenhauser PS et al (2025) Bioactivated scaffolds promote angiogenesis and lymphangiogenesis for dermal regeneration in vivo. J Tissue Eng 16:20417314251317542
123. Li Y, Xu T, Tu Z, Dai W, Xue Y, Tang C et al (2020) Bioactive antibacterial silica-based nanocomposites hydrogel scaffolds with high angiogenesis for promoting diabetic wound healing and skin repair. Theranostics 10(11):4929–4943
124. Ding Y, Sun Z, Shi R, Cui H, Liu Y, Mao H et al (2019) Integrated endotoxin adsorption and antibacterial properties of cationic polyurethane foams for Wound healing. ACS Appl Mater Interfaces 11(3):2860–2869
125. Li Y, Xu T, Tu Z, Dai W, Xue Y, Tang C et al (2022) Erratum: bioactive antibacterial silica-based nanocomposites hydrogel scaffolds with high angiogenesis for promoting diabetic wound healing and skin repair: erratum. Theranostics 12(10):4599–4600
126. Mohammed AM, Hassan KT, Hassan OM (2023) Assessment of antimicrobial activity of chitosan/silver nanoparticles hydrogel and cryogel microspheres. Int J Biol Macromol 233:123580

127. Jones LO, Williams L, Boam T, Kalmet M, Oguike C, Hatton FL (2021) Cryogels: recent applications in 3D-bioprinting, injectable cryogels, drug delivery, and wound healing. Beilstein J Org Chem 17:2553–2569
128. Qi L, Zhang C, Wang B, Yin J, Yan S (2022) Progress in hydrogels for skin Wound repair. Macromol Biosci 22(7):e2100475
129. Wang F, Sun J, Shi H, Zhou J, Ma X, Song X et al (2023) Multifunctionalized alginate/polydopamine cryogel for hemostasis, antibacteria and promotion of wound healing. Int J Biol Macromol 224:1373–1381
130. Abdel Azim S, Whiting C, Friedman AJ (2024) Applications of nitric oxide-releasing nanomaterials in dermatology: skin infections and wound healing. Nitric Oxide 146:10–18
131. Lu B, Wang T, Li Z, Dai F, Lv L, Tang F et al (2016) Healing of skin wounds with a chitosan-gelatin sponge loaded with tannins and platelet-rich plasma. Int J Biol Macromol 82:884–891
132. Nguyen TD, Nguyen TH, Vo VT, Nguyen TQ (2024) Panoramic review on polymeric microneedle arrays for clinical applications. Biomed Microdevices 26(4):41
133. Deepa C, Bhatt A (2024) Skin substitutes: from conventional to 3D bioprinting. J Artif Organs 28:154
134. Wosicka-Frąckowiak H, Poniedziałek K, Woźny S, Kuprianowicz M, Nyga M, Jadach B et al (2024) Collagen and its derivatives serving biomedical purposes: a review. Polymers (Basel) 16(18)
135. Lim YS, Ok YJ, Hwang SY, Kwak JY, Yoon S (2019) Marine collagen as a promising biomaterial for biomedical applications. Mar Drugs 17(8)
136. Sukmana BI, Margiana R, Almajidi YQ, Almalki SG, Hjazi A, Shahab S et al (2023) Supporting wound healing by mesenchymal stem cells (MSCs) therapy in combination with scaffold, hydrogel, and matrix; State of the art. Pathol Res Pract 248:154575
137. Wei Q, An Y, Zhao X, Li M, Zhang J (2024) Three-dimensional bioprinting of tissue-engineered skin: biomaterials, fabrication techniques, challenging difficulties, and future directions: a review. Int J Biol Macromol 266(Pt 1):131281
138. Vig K, Chaudhari A, Tripathi S, Dixit S, Sahu R, Pillai S et al (2017) Advances in skin regeneration using tissue engineering. Int J Mol Sci 18(4)
139. Dixit S, Baganizi DR, Sahu R, Dosunmu E, Chaudhari A, Vig K et al (2017) Immunological challenges associated with artificial skin grafts: available solutions and stem cells in future design of synthetic skin. J Biol Eng 11:49
140. Kaur A, Midha S, Giri S, Mohanty S (2019) Functional skin grafts: where biomaterials meet stem cells. Stem Cells Int 2019:1286054
141. Archana D, Singh BK, Dutta J, Dutta PK (2015) Chitosan-PVP-nano silver oxide wound dressing: in vitro and in vivo evaluation. Int J Biol Macromol 73:49–57
142. Khalid A, Khan R, Ul-Islam M, Khan T, Wahid F (2017) Bacterial cellulose-zinc oxide nanocomposites as a novel dressing system for burn wounds. Carbohydr Polym 164:214–221
143. Debels H, Hamdi M, Abberton K, Morrison W (2015) Dermal matrices and bioengineered skin substitutes: a critical review of current options. Plast Reconstr Surg Glob Open 3(1):e284
144. Lee AS, Tang C, Rao MS, Weissman IL, Wu JC (2013) Tumorigenicity as a clinical hurdle for pluripotent stem cell therapies. Nat Med 19(8):998–1004
145. Hsieh MW, Wang WT, Lin CY, Kuo YR, Lee SS, Hou MF et al (2022) Stem cell-based therapeutic strategies in diabetic wound healing. Biomedicine 10(9)
146. Nycz M, Litwinska-Inglot I, Partyka M, Snopek-Mista C, Waniczek D (2025) Advancing burn care: the role of 3D printing in advanced dressing development. Dermatologic Surg 51:e22
147. Abaci HE, Coffman A, Doucet Y, Chen J, Jacków J, Wang E et al (2018) Tissue engineering of human hair follicles using a biomimetic developmental approach. Nat Commun 9(1):5301
148. Xu J, Su Y, Yang J, Ma H, Fang H, Zhu J et al (2025) A 3D bioprinted gelatin/quaternized chitosan/decellularized extracellular matrix based hybrid bionic scaffold with multifunctionality for infected full-thickness skin wound healing. Int J Biol Macromol 309(Pt 1):142816

149. Li H, Chen L, Zeng S, Li X, Zhang X, Lin C et al (2015) Matrigel basement membrane matrix induces eccrine sweat gland cells to reconstitute sweat gland-like structures in nude mice. Exp Cell Res 332(1):67–77
150. Wang A, Tang Z, Park IH, Zhu Y, Patel S, Daley GQ et al (2011) Induced pluripotent stem cells for neural tissue engineering. Biomaterials 32(22):5023–5032
151. Blais M, Grenier M, Berthod F (2009) Improvement of nerve regeneration in tissue-engineered skin enriched with schwann cells. J Invest Dermatol 129(12):2895–2900
152. Hosseini M, Koehler KR, Shafiee A (2022) Biofabrication of human skin with its appendages. Adv Healthc Mater 11(22):e2201626
153. Keck M, Gugerell A, Kober J (2025) Engineering a multilayered skin substitute with keratinocytes, fibroblasts, adipose derived stem cells and adipocytes. Methods Mol Biol 2922:135–142
154. Hart CE, Loewen-Rodriguez A, Lessem J (2012) Dermagraft: use in the treatment of chronic wounds. Adv Wound Care (New Rochelle). 1(3):138–141
155. Vanscheidt W, Ukat A, Horak V, Brüning H, Hunyadi J, Pavlicek R et al (2007) Treatment of recalcitrant venous leg ulcers with autologous keratinocytes in fibrin sealant: a multinational randomized controlled clinical trial. Wound Repair Regen 15(3):308–315
156. Lam PK, Chan ES, To EW, Lau CH, Yen SC, King WW (1999) Development and evaluation of a new composite Laserskin graft. J Trauma 47(5):918–922
157. Ortega-Zilic N, Hunziker T, Läuchli S, Mayer DO, Huber C, Baumann Conzett K et al (2010) EpiDex® Swiss field trial 2004–2008. Dermatology 221(4):365–372
158. Mohamed Haflah NH, Ng MH, Mohd Yunus MH, Naicker AS, Htwe O, Fahmi M et al (2017) The use of engineered bilayered skin (MyDerm). Int J Low Extrem Wounds 16(3):212–216
159. Binder H, Hoffman L, Zak L, Tiefenboeck T, Aldrian S, Albrecht C (2021) Clinical evaluation after matrix-associated autologous chondrocyte transplantation: a comparison of four different graft types. Bone Joint Res 10(7):370–379
160. Matsumura H, Matsushima A, Ueyama M, Kumagai N (2016) Application of the cultured epidermal autograft "JACE(®)" for treatment of severe burns: results of a 6-year multicenter surveillance in Japan. Burns 42(4):769–776
161. Freedman BR, Hwang C, Talbot S, Hibler B, Matoori S, Mooney DJ (2023) Breakthrough treatments for accelerated wound healing. Sci Adv 9(20):eade7007

Chapter 19
Biomaterials as Skin Substitutes

Eylül Odabaş-Kansız, Seyithan Kansız, Ayşe Eser Elçin, and Yaşar Murat Elçin

Abstract Skin substitutes have emerged as a critical component in the management of complex wounds, burns, and chronic ulcers. These biomaterials are designed to replicate the structural, functional, and physiological properties of the native skin and promote effective wound healing and tissue regeneration. Skin substitutes are generally divided into two main categories based on their source: biological and polymeric skin substitutes. Biological substitutes, sourced from human or animal tissues, closely resemble the natural extracellular matrix and are often favored for their biocompatibility. On the other hand, polymer-based substitutes offer advantages such as tunable properties, and ease of fabrication. This chapter provides a comprehensive overview of both types of skin substitutes, highlighting their clinical applications, strengths, and limitations. Furthermore, current research trends and innovations in the field are explored. The chapter also delves into the fabrication methods used in developing skin substitutes, discussing the advantages and challenges associated with each technique to give a holistic view of the field.

Keywords Regenerative medicine · Tissue engineering · Decellularization · 3D bioprinting · Stem cells · Autograft · Allograft · Xenograft · Natural polymers · Synthetic polymers · Electrospinning · Nanopatterning · Micropatterning · 4D bioprinting · Smart skin substitutes

E. Odabaş-Kansız · S. Kansız · A. E. Elçin
Tissue Engineering, Biomaterials and Nanobiotechnology Laboratory, Faculty of Science and Stem Cell Institute, Ankara University, Ankara, Türkiye

Y. M. Elçin (✉)
Tissue Engineering, Biomaterials and Nanobiotechnology Laboratory, Faculty of Science and Stem Cell Institute, Ankara University, Ankara, Türkiye

Biovalda Health Technologies, Inc., Ankara, Türkiye

F. De Francesco, K. Turksen (eds.), *Management and Strategies for Wound Healing*, Stem Cell Biology and Regenerative Medicine 79,
https://doi.org/10.1007/978-3-032-12442-5_19

Abbreviations

AgNPs	Silver nanoparticles
bFGF	Basic fibroblast growth factor
BLM	Bilayer membrane
CAD	Computer-aided design
CCL-2	Chemokine (C-C motif) ligand 2
CMC	Carboxymethyl cellulose
CNT	Carbon nanotube
CQDs	Carbon quantum dots
CS	Chitosan
CuNPs	Copper nanoparticles
CUR	Curcumin
DFU	Diabetic foot ulcer
dhACM	Dehydrated human amnion/chorion membrane
DHM	Dihydromyricetin
DPA	D-Panthenol
E. coli	Escherichia coli
ECM	Extracellular matrix
FGF	Fibroblast growth factor
FPA	Fibrinopeptide A
FPB	Fibrinopeptide B
GA	Glycolic acid
GAGs	Glycosaminoglycans
Gel-MA	Gelatin methacryloyl
GFs	Growth factors
GPIb-IX-V	Glycoprotein Ib-IX-V complex
GPVI	Glycoprotein VI
HA	Hyaluronic acid
$HA-NH_2$	Amine-modified hyaluronic acid
hASCs	Human adipose-derived stem cells
HCPU-NHS	Hyperbranched citrate-based poly (urethane–N-hydroxysuccinimide)
IL	Interleukin
IL-10	Interleukin 10
IL-6	Interleukin 6
kDa	Kilodalton
LA	Lactic acid
LMF	Low-molecular-weight fucoidan
MMPs	Matrix metalloproteinases
MPa	Megapascal
NF-κB	Nuclear factor kappa B
n-HA	Nano-hydroxyapatite
NPs	Nanoparticles
OD	Oxidized dextran

OLS	Oxidized lignosulfonate
PAA	Poly (acrylic acid)
PCL	Poly (caprolactone)
PDGF	Platelet-derived growth factor
PDMS	Poly (dimethylsiloxane)
PEG	Poly (ethylene glycol)
PEUU	Poly (ester-urethane) urea
PGA	Poly (glycolic acid)
PHMB	Poly (hexamethylene biguanide)
PL	Platelet lysate
PLA	Poly (lactic acid)
PLGA	Poly (lactic-co-glycolic acid)
PVA	Poly (vinyl alcohol)
PVP	Poly (vinylpyrrolidone)
QLG	Quercus infectoria gall extract
RGD	Arginine-glycine-aspartic acid
ROS	Reactive oxygen species
S. aureus	Staphylococcus aureus
SAL	Sterility assurance level
SBS	Solution blow spinning
SIS	Small intestinal submucosa
SSE	Semi-solid extrusion
TE	Tissue engineering
TGF-β	Transforming growth factor beta
TNF-α	Tumor necrosis factor alpha
UA	Uric acid
UBM	Urinary bladder matrix
VEGF	Vascular endothelial growth factor
VEGFR2	Vascular endothelial growth factor receptor 2
VIIa	Activated coagulation factor VII
X	Coagulation factor X
XG	Xanthan gum
3D	Three-dimensional
4D	Four-dimensional

Introduction

The development of advanced skin substitutes has emerged as a critical strategy in wound management, particularly for addressing the complex challenges associated with large, full-thickness skin defects [1]. Many of these substitutes have found their way into clinical practice, aiming to accelerate the closure of wounds and minimize extensive scarring. They are commonly applied in the treatment of conditions such as chronic leg ulcers, diabetic foot ulcers, full-thickness burns, and surgical or acute open wounds. Essentially, an ideal skin substitute is expected to not only mimic the native biological structure and composition but also create a favorable microenvironment to guide cellular infiltration and facilitate proper regeneration [2]. Designing skin substitutes that can successfully integrate into host tissue, modulate the immune response, and promote long-term functional recovery is an ongoing challenge. Advances in biomaterial science and biofabrication technologies have paved the way for the development of next-generation substitutes with improved biocompatibility, tunable degradation rates, and the ability to deliver bioactive molecules or cells. These innovations aim to accelerate healing in acute and chronic wound scenarios, reduce scarring, and restore the mechanical and aesthetic properties of skin sections.

History of Skin Substitutes

The history of skin grafting can be traced back to ancient India, with the earliest documented procedures dating to around 2500 BC [3]. In particular, the *Sushruta Samhita*, a seminal text in Ayurvedic medicine written around 600 BC, describes sophisticated reconstructive techniques involving full-thickness skin grafts [4, 5]. These grafts were primarily taken from the forehead or cheek and used to reconstruct noses that had been cut off as punishment for crime. This approach, often referred to as the "Ancient Indian Method," laid the foundation for reconstructive surgery centuries ahead of its time. Despite these early developments, the field made little progress for over two millennia. It was not until the nineteenth century that modern skin grafting techniques emerged. In 1804, the first documented use of xenografts (skin taken from another species) occurred. A major turning point came in 1869, when the Swiss-born French surgeon Jacques-Louis Reverdin presented his innovative "epidermal grafting" technique to the Société Impériale de Chirurgie in Paris. Reverdin introduced the concept of punch grafting, in which small pieces of epidermis and superficial dermis were transplanted into chronic wounds to promote epithelialization, marking the beginning of scientific approaches to skin regeneration. In the 1870s, allotransplantation came to the fore, largely thanks to the work of the German surgeon Carl Thiersch. At the Fifteenth Congress of the German Surgical Society in 1886, Thiersch described a method of harvesting thin layers of skin for grafting, including both the epidermis and part of the dermis. Although similar techniques had been used earlier by the French surgeon Ollier, Thiersch's contributions received greater recognition—even though Ollier's earlier work had not been properly acknowledged [6]. Yannas and Burke introduced the first artificial skin substitute designed as a regeneration template in 1981 [7]. This innovation consisted of a bilayered structure containing a temporary silicone layer simulating the epidermis and a porous fibrillar dermis made of collagen and chondroitin 6-sulfate, which was developed primarily to treat severe acute burn injuries. During the 1990s, significant efforts were made to develop a variety of tissue engineering (TE) products using different scaffold systems, including both biological and synthetic options [8]. During this period, several skin substitute products emerged, gained widespread use, and established a strong presence in their respective markets.

Today, skin substitutes are used effectively in the treatment of a variety of wound types, including severe burns, diabetic foot ulcers, skin cancer lesions, residual wounds, venous stasis ulcers, vasculopathic ulcers, radiation dermatitis, peristomal pyoderma gangrenosum, as well as surgical and traumatic wounds. In addition to their clinical applications, skin substitutes remain an active area of research with ongoing efforts to enhance their functionality, integrate more advanced biomaterials, and improve patient outcomes.

Structure of the Skin

The skin stands as the body's largest organ, playing a vital role in maintaining homeostasis and acting as a protective shield against external factors, helping to prevent infections and dehydration [9]. It consists of three main layers: the epidermis, the dermis, and the hypodermis (or subcutaneous tissue), each contributing distinct structural and functional roles. The epidermis, the skin's outermost layer, primarily consists of keratinocytes, which make up approximately the majority of its cellular structure [10]. This layer serves as a crucial barrier, protecting the body against pathogens, ultraviolet radiation, mechanical stress, and excessive water or ion loss. Additional cells, such as melanocytes, Langerhans cells, and Merkel cells, are distributed within specific sub-layers of the epidermis, supporting pigmentation, immune defense, and sensory functions [11, 12]. The dermis, mesenchymal layer that makes up about 90% of the skin's total weight, is located beneath the epidermis. The dermis is mainly composed of extracellular matrix (ECM) proteins, vascular endothelial cells, and fibroblasts, along with accessory structures, including sweat glands, sebaceous glands, hair follicles, blood vessels, and nerve endings [13]. Fibroblasts play a key role in synthesizing collagen and elastin, providing the skin with mechanical strength and elasticity. The dermis consists of two sub-layers: the papillary layer, which contains microvascular networks and nerve endings crucial for nourishment and sensation, and the reticular layer, a denser region rich in collagen fibers. The hypodermis (subcutaneous tissue), the innermost skin layer, is composed of loose connective tissue, adipocytes, blood vessels, and nerves [14]. It serves as a cushion against mechanical impact and a reservoir for energy storage. Rich in glycosaminoglycans (GAGs) and proteoglycans, the hypodermis also facilitates water retention within the tissue, contributing to skin hydration and overall structural integrity. Together, these layers form a dynamic organ essential for protection, sensation, and homeostasis.

Wound-Healing Mechanism

Wound healing is a highly intricate and tightly regulated process, integrating multiple cellular, molecular, and humoral mechanisms to restore the skin's structural integrity and barrier function. The process initiates immediately after the formation of a lesion and can continue for an extended period, potentially lasting several years [9]. Biologically, wound healing is a dynamic, multi-step regenerative process regulated by a variety of growth factors (GFs), cytokines, cells, essential metal ions (e.g., Mg^{2+}, Ca^{2+}, Zn^{2+}), and reactive oxygen species (ROS) [15, 16]. Also, the sequential synchronization of multiple cell types is essential for effective skin healing. This highly coordinated wound-healing process consists of multiple stages, including hemostasis, inflammation, proliferation, and remodeling/maturation.

Hemostasis Phase

The hemostasis phase, essential in wound healing, involves primary and secondary pathways that collectively prevent blood loss and initiate the repair process. Primary hemostasis is initiated by the exposure of subendothelial proteins such as collagen, von Willebrand factor, and fibronectin at the vascular injury site. This triggers platelet adhesion, activation, and aggregation, primarily mediated by surface receptors like GPIb-IX-V and GPVI, leading to the formation of a temporary platelet plug [17]. Secondary hemostasis involves the activation of the coagulation cascade, culminating in the conversion of soluble fibrinogen into insoluble fibrin strands. These fibrin strands strengthen the platelet plug, forming a stable thrombus. The extrinsic pathway is initiated when tissue factor binds to factor VIIa, subsequently activating factor X. Activated factor X then converts prothrombin into thrombin. Thrombin plays a central role by cleaving fibrinogen into fibrin and amplifying coagulation through positive feedback mechanisms, facilitating robust clot formation [18].

Inflammation Phase

The primary objective of this stage in the wound-healing process is to prevent infection. Neutrophils rapidly infiltrate the wound site within the first hour following the injury [19]. Their migration is facilitated by chemical signaling pathways, including the complement cascade, interleukin (IL) activation, and transforming growth factor-β (TGF-β) signaling, which guide neutrophils to the wound site through a process called chemotaxis [20]. They play a crucial role in eliminating bacteria and cellular debris through three primary mechanisms. They can directly engulf and digest foreign particles via phagocytosis. Additionally, they release cytotoxic substances such as lactoferrin, proteases, neutrophil elastase, and cathepsins to break down bacteria and necrotic tissue. Moreover, the production of oxygen-free radicals during neutrophil activity contributes to bacterial destruction and, when combined with chlorine, further aids in sterilizing the wound. After completing their function, neutrophils are either removed from the wound surface through apoptosis or are engulfed by macrophages. Macrophages arrive at the wound site following the neutrophils. They are recruited in response to chemical signals released by platelets and damaged cells. Macrophages play a vital role in regulating inflammation, promoting angiogenesis, and facilitating the development of granulation tissue by releasing GFs [21]. Lymphocytes enter the wound area to contribute to wound regulation by supporting ECM formation and collagen remodeling [22]. The inflammatory phase continues for as long as necessary to ensure complete clearance of bacteria and debris from the wound. However, prolonged inflammation can have detrimental effects, including excessive tissue damage, delayed proliferation, and the potential transition to a chronic wound state. In such cases anti-inflammatory agents help to suppress immune activity and facilitate the transition to the next phase of wound healing.

Proliferation Phase

The proliferation phase of wound healing involves the coordinated activation of keratinocytes, fibroblasts, macrophages, and endothelial cells, collectively facilitating wound closure, ECM synthesis, and the development of new blood vessels essential for tissue regeneration [23]. Following the injury, keratinocytes respond to changes in mechanical stress, electrical gradients, and various signaling molecules, such as hydrogen peroxide, and GFs [24]. These stimuli initiate their activation, triggering a partial epithelial-to-mesenchymal transition, which allows them to adopt a motile and invasive behavior. This change in polarity facilitates lateral migration across the wound bed, a process known as re-epithelialization [23]. The keratinocytes at the leading edge create a new epithelial layer. They navigate through debris and necrotic tissue using their integrin receptors to interact with the temporary matrix. As keratinocytes move, they release matrix metalloproteinases (MMPs), primarily MMP-1 and MMP-9, to aid migration by degrading the ECM. Additionally, they deposit new ECM proteins to reconstruct the basement membrane [25]. Plasmin and other proteases help dissolve the fibrin-rich preliminary matrix, enhancing their migration. When keratinocytes from opposing wound edges meet, they terminate migration, establish a thin epithelial layer, and rebuild the basement membrane before undergoing terminal differentiation to regenerate the epidermis. Fibroblasts are key players in replacing the provisional matrix with granulation tissue. These cells, recruited by signals from platelets, endothelial cells, and macrophages, differentiate into either ECM-producing fibroblasts or myofibroblasts that drive wound contraction. Their functional roles vary; while some fibroblasts aid in re-epithelialization, others take a role in ECM deposition. The formation of granulation tissue involves the degradation of the fibrin-rich matrix and its replacement with fibronectin, immature collagen, and proteoglycans. This matrix supports the migration of wound cells and the development of mature ECM and new blood vessels.

Angiogenesis is critical during this phase to support the metabolically active tissue. It is driven by hypoxia-induced factors, such as vascular endothelial growth factor (VEGF), fibroblast growth factor (FGF), and platelet-derived growth factor (PDGF), which stimulate microvascular endothelial cells to proliferate and sprout new vessels [26]. Macrophages are the key players that enhance angiogenesis by releasing MMPs and chemotactic signals promoting endothelial migration. They also play a role in vascular remodeling by fusing vessel tips, phagocytosing unnecessary vessels, and suppressing excessive vascular growth [23].

Remodeling/Maturation Phase

The remodeling phase, representing the final stage of wound healing, occurs concurrently with granulation tissue formation [27]. This phase may extend for a year or more, ensuring proper tissue maturation and functional recovery. Myofibroblasts play an essential role in wound contraction and ECM remodeling by acquiring a contractile phenotype, characterized by actin filaments, and producing key ECM components [28]. Their differentiation from fibroblasts is driven by TGF-β1 within the wound microenvironment and further regulated by ECM rigidity. This rigidity is often attributed to molecular remodeling processes, such as the transition from collagen III to collagen I, which enhances the tensile strength necessary for effective wound stabilization and closure. During the remodeling phase, macrophages and fibroblasts progressively undergo apoptosis, and angiogenesis comes to a halt. Ultimately, the remodeling phase concludes with the formation of mechanically strong, sparsely vascularized tissue.

Biological Skin Substitutes

Biological skin substitutes are typically composed of biologically derived materials that mimic the natural ECM of the skin. They provide the structural support necessary for new tissue growth and serve as a scaffold for cellular infiltration, angiogenesis, and dermal regeneration. These substitutes are primarily made from collagen, fibronectin, and GAGs, and may include components derived from human or animal tissues such as GFs and other metabolic molecules [2]. They are classified as autografts, allografts, and xenografts according to their source.

Autografts

In the management of both acute and chronic wounds, skin grafting remains a cornerstone technique, with autografts being one of the most widely used methods due to their biocompatibility and lack of immune rejection. Autografts obtained from the patient's healthy skin are typically used in two forms: full-thickness autografts, which include the entire epidermis and dermis, and split-thickness autografts, consisting of the epidermis and a superficial portion of the dermis [29]. As the tissue is derived from the patient's own body, autografts completely eliminate the risk of immune rejection and disease transmission [30, 31]. They offer excellent graft integration and high survival rates, supporting rapid vascularization and tissue remodeling. Additionally, autografts restore the skin's structural and functional integrity more effectively than alternative graft types, with lower infection risk and no need for immunosuppressive treatment. These features make autografts highly reliable for promoting efficient and safe wound healing. Despite their advantages, autografts are associated with several limitations. The most significant drawback is donor site morbidity, which can result in pain, scarring, infection, or delayed healing at the harvest site. In patients with extensive wounds or limited healthy skin, the availability of donor tissue becomes a major challenge. Additionally, the procedure may involve longer operative times, increased surgical complexity, and the potential for aesthetic or functional compromise at both the donor and recipient sites.

Acellular Skin Substitutes

Acellular skin substitutes are biomaterials designed to support skin regeneration in cases such as burns, chronic wounds, and reconstructive surgeries. They originate from donated cadaver or animal tissues, where all living cells have been removed through a process called decellularization [2, 32–34]. This process eliminates cellular components that could trigger an immune response while preserving the ECM structure mainly composed of collagen, elastin, proteoglycans, and GAGs, which play a crucial role in supporting tissue regeneration [35, 36]. Upon application, acellular matrices act as temporary scaffolds that integrate with the host tissue. Once implanted, these matrices integrate with host cells, guided by the GFs and mechanical cues retained within their structure, which direct cellular behavior and tissue regeneration [37]. They support fibroblast migration and vascular ingrowth, leading to the formation of neodermis. Over time, the matrix is gradually degraded and replaced by native tissue. Importantly, acellular matrices reduce scar formation and improve skin elasticity. Acellular skin substitutes are typically derived from human or animal sources [36]. The decellularization process ensures biocompatibility, making these matrices suitable for a range of clinical applications, including burn treatment and reconstructive surgery (Table 19.1).

Table 19.1 Classification of acellular skin substitutes based on origin and composition

Product name	Manufacturer	Product overview by manufacturer	Manufacturer claims
Acellular skin substitutes derived from donated human dermis			
Allopatch®	MTF biologics, Edison, NJ, USA	This human allograft skin is minimally processed to remove epidermal and dermal cells while preserving the ECM. It is aseptically prepared without terminal sterilization, ensuring structural integrity for tissue repair and reinforcement.	Designed for immediate use, it requires no refrigeration or pre-hydration, reducing procedural delays. Its natural biomechanical properties are maintained, and it has a three-year shelf life at ambient temperature.
DermaSpan™	Zimmer Biomet, Warsaw, IN, USA	Processed allograft dermal matrix designed for biocompatibility and biomechanical strength, supporting tissue repair in various surgical applications.	Sterile, ready-to-use, and does not require refrigeration, providing ease of storage and application.
AlloPatch® Pliable	MTF biologics, NJ, USA	Aseptically processed human reticular dermal tissue designed to cover acute and chronic wounds.	Processed from deep-cut tissue with the epidermis removed, maintaining an open architecture that facilitates cell infiltration and vascularization. Provides a structural framework for host tissue remodeling. Three-year shelf life at ambient temperature.
DermACELL®	LifeNet health®, Virginia Beach, VA, USA	Utilizes Matracell® technology to remove ≥97% of donor DNA while maintaining structural integrity, promoting cellular infiltration and revascularization.	Ready-to-use, room temperature storage, and terminally sterilized (SAL 10^{-6}) for safety and infection control.

(continued)

Table 19.1 (continued)

Product name	Manufacturer	Product overview by manufacturer	Manufacturer claims
Alloskin™ AC Acellular Dermal Matrix	AlloSource, Centennial, CO, USA	A meshed, decellularized human dermis graft designed to retain the natural components and structure of the dermal matrix, supporting tissue integration and wound healing.	Provides an optimal environment for cellular repopulation and revascularization. Designed for a single application, with a pliable structure that conforms to wound contours. Meshed (1:1) to facilitate fluid drainage, suitable for suturing or stapling. Ready-to-use and shelf-stable at room temperature. Does not require cryopreservation.
AlloSkin™ RT	AlloSource, Centennial, CO, USA	A sterile, meshed human dermal graft designed for acute and chronic wound treatment, offering mechanical protection and biological support for healing.	Electron beam irradiated, ready-to-use, shelf-stable graft with room temperature storage. Pliable yet durable for suturing or stapling, with 1:1 meshing to promote exudate drainage and wound adherence.
FlowerDerm™	Flower Orthopedics, Horsham, PA, USA	Decellularized meshed dermal graft designed to retain natural biological components and support revascularization and cellular repopulation.	Ready-to-use, flexible, and durable for wound adherence, suturing, and stapling. Mesh design promotes fluid drainage. Stored at room temperature with a 2-year shelf life.
Coll-e-derm™	Parametrics Medical, Leander, TX, USA	A decellularized human dermal allograft processed to retain biological properties essential for wound healing.	Supports neovascularization and cell proliferation with preserved collagen and elastin. Provides a stable scaffold for tissue remodeling and accelerated healing. Terminally sterile and lyophilized for easy storage.

(continued)

Table 19.1 (continued)

Product name	Manufacturer	Product overview by manufacturer	Manufacturer claims
Dermapure®	Tissue Regenix Group, San Antonio, TX, USA	Processed with dCELL® technology to remove up to 99% of DNA while preserving the native dermal structure, creating a biologically supportive scaffold.	Facilitates cell migration, vascularization, and epithelial regeneration, fully integrating into the wound bed. Stored at ambient temperature.
GammaGraft™	Promethean LifeSciences, Inc., PA, USA	Irradiated human skin allograft designed for temporary coverage of burns and chronic wounds, with room temperature storage.	Sterilized with proprietary technology for safety and ease of use. Forms a keratin layer that preserves moisture, eliminating the need for additional coverings.
GraftJacket™ RTM	Wright Medical Group N.V., Memphis, TN, USA	Provides a biocompatible scaffold for cell repopulation and revascularization, supporting tissue regeneration.	Freeze-dried using a proprietary method to preserve vascular channels and prevent ice crystal formation. Stored at room temperature with a 2-year shelf life.
hMatrix® ADM	Bacterin International, Inc., Belgrade, MT, USA	Acellular dermal matrix is processed to retain collagen and elastin, supporting neovascularization and cell proliferation for wound healing.	Offers strong suture retention, flexibility for precise placement, reduced inflammatory response, and a 5-year shelf life. Provided frozen with a sterility assurance level (SAL) of 10^{-6}.
InteguPly®	Aziyo Biologics, Silver Spring, MD, USA	Acellular dermal matrix designed to preserve the ECM structure while ensuring sterility (SAL 10^{-6}), used for chronic wound treatment.	Supports integumental tissue repair and chronic wound closure. Promotes tissue remodeling with a strong safety profile, having over 20,000 applications without reported adverse reactions.
Matrix HD® Allograft	RTI Surgical, Alachua, FL, USA	Acellular human dermis allograft processed with the Tutoplast™ tissue sterilization method, preserving its three-dimensional fiber structure and mechanical properties.	Supports tissue regeneration with a preserved matrix structure. Terminally sterilized (SAL 10^{-6}) with validated viral inactivation. Room temperature storage with a five-year shelf life and easy rehydration.

(continued)

Table 19.1 (continued)

Product name	Manufacturer	Product overview by manufacturer	Manufacturer claims
Human placental membrane-derived acellular skin substitutes			
AmnioFix® Amnion/Chorion Membrane Allograft	MiMedx Group, Marietta, GA, USA	Dehydrated human amnion/chorion membrane (dHACM) allograft containing ECM proteins, GFs, and cytokines to support tissue healing and regeneration.	Reduces scar tissue formation, modulates inflammation, and serves as a protective barrier membrane. Terminally sterilized for patient safety with a five-year shelf life at ambient temperature.
BioDFactor® Viable Tissue Matrix	Integra LifeSciences, Originally BioD, LLC, Princeton, NJ, USA	Cryopreserved flowable allograft derived from morselized amniotic tissue and amniotic fluid components.	Supports tissue repair with collagen, elastin, fibronectin, and GFs. Offers anti-adhesive and antimicrobial properties. Stored at −65 °C or colder, with a two-year shelf life.
Biovance® Amniotic Membrane Allograft	Alliqua Biomedical, Langhorne, PA, USA	Decellularized, dehydrated human amniotic membrane preserving ECM and basement membrane for cellular attachment and GF storage.	Provides a protective scaffold for tissue repair and chronic wound healing. Easy application with a five-year shelf life. Supports natural tissue restoration.
EpiCord®	MiMedx, Marietta, GA, USA	Dehydrated umbilical cord allograft processed for easy application and ambient storage.	Creates a protective environment for healing and provides a connective tissue matrix for tissue repair. Five-year shelf life.
Epifix®	MiMedx, Marietta, GA, USA	Dehydrated human amnion/chorion membrane allograft preserving ECM proteins, GFs, and cytokines. Used for chronic wound treatment.	Functions as a barrier membrane, reducing scar formation and inflammation while promoting healing. Terminally sterilized for safety. Five-year shelf life at ambient conditions.
PalinGen® Membrane & Hydromembrane	Amnio Technology LLC, Phoenix, AZ, USA	Placental-derived allograft processed to retain ECM proteins, GFs, and cytokines essential for wound healing.	Resorbable and sterile, designed for various wound sizes. Free of chorion, rich in hyaluronic acid, and stored at room temperature.
AmnioBand® Allograft Placental Matrix	MTF Biologics, Edison, NJ, USA	Retains the structural integrity of the ECM, serving as a wound covering.	Preserves GFs for wound healing, maintains natural tissue structure, and conforms to wound anatomy. Ready-to-use, with a three-year shelf life at ambient temperature.

(continued)

Table 19.1 (continued)

Product name	Manufacturer	Product overview by manufacturer	Manufacturer claims
Neox® Wound Allografts	Amniox Medical, Inc., Miami, FL, USA	Human umbilical cord and amniotic membrane allograft designed to preserve natural biological properties. Used as a wound covering for dermal ulcers and defects.	Provides regenerative healing properties in a 1 mm thick allograft for improved handling and fixation. Preserved using Cryotek® technology. Two-year shelf life, stored in a refrigerator or freezer.
NuShield®	Organogenesis, Inc., Canton, MA, USA	Dehydrated placental allograft designed to support healing across various wound sizes and types.	Preserved with BioLoc™ technology to maintain native membrane structure and bioactive components. Optimized for strength and flexibility. Five-year shelf life.
Xwrap® Amniotic Membrane-Derived Allograft	Applied Biologics™, Scottsdale, AZ, USA	A chorion-free, non-crosslinked amniotic membrane designed as a natural scaffold to support cellular migration, attachment, and proliferation.	Preserves amniotic membrane structure with essential collagen and proteins. Supports wound healing while reducing pain, fibrosis, and bacterial growth.
Integra® BioFix® Amniotic Membrane Allograft	Integra LifeSciences, Princeton, NJ, USA	Sterile, decellularized amniotic membrane allograft for wound coverage.	Preserves tissue structure with HydraTek® technology. Provides a collagen-rich scaffold for regeneration. Omnidirectional placement for effective coverage. Stored at ambient temperature.
Acellular/skin substitutes derived from animal tissue			
Integra® Matrix Wound Dressing; Originally Avagen Wound Dressing	Integra LifeSciences, Princeton, NJ, USA	A biodegradable matrix composed of crosslinked bovine tendon collagen and GAG, providing a scaffold for cellular infiltration and capillary growth.	Supports a healing environment with a single-layer collagen matrix. Stored at room temperature with a 24-month shelf life.
Kerecis™ Omega3 Wound (originally Merigen wound dressing)	Kerecis, Arlington, VA, USA	A sterile fish-derived dermal matrix composed of fish collagen, available in intact or meshed sheet form.	Provides a cost-effective skin substitute with improved clinical performance and reduced disease transmission risk. Stored at ≤25 °C.

(continued)

Table 19.1 (continued)

Product name	Manufacturer	Product overview by manufacturer	Manufacturer claims
PriMatrix® Dermal Repair Scaffold	Integra LifeSciences, Princeton, NJ, USA	A fetal bovine-derived acellular dermal matrix designed to support cellular repopulation and revascularization for wound healing. Contains type III collagen, crucial for tissue development and repair.	Processed to preserve natural dermal collagen fibers while eliminating contaminants and artificial crosslinks. Rapidly absorbs blood, binding cells and GFs to promote healing. Stored at room temperature.
Geistlich DermaGide™	Geistlich Pharma North America Inc., Princeton, NJ, USA	A porous, resorbable, 3D matrix derived from porcine tissue, designed for wound management, including diabetic ulcers, venous ulcers, and surgical wounds.	Supports angiogenesis and wound closure by inactivating MMPs while promoting fibroblast, keratinocyte, and endothelial cell migration and proliferation.
Oasis® Wound Matrix	Smith & Nephew, Inc., Fort Worth, TX, USA	A naturally derived ECM scaffold composed of porcine small intestinal submucosa (SIS), designed to support the body's wound healing processes.	Facilitates cellular migration and vascular growth. Stored at room temperature with a two-year shelf life.
Endoform™ Dermal Template	Hollister Wound Care, Libertyville, IL, USA	A terminally sterilized ovine collagen ECM designed as a dermal template, offering a culturally acceptable alternative to other animal-derived sources.	Reduces MMP activity, supports advanced wound care accessibility, and enables cost-efficient weekly applications. Stored between 15 °C and 40 °C in a clean, dry environment.
PuraPly® Antimicrobial (PuraPly® AM) Wound Matrix (formally called FortaDerm)	Organogenesis, Inc., Canton, MA, USA	A dry collagen sheet infused with 0.1% polyhexamethylene biguanide (PHMB) designed for wound management. Packaged sterile in single-use pouches.	Combines a purified native collagen matrix with PHMB to help prevent biofilm formation while supporting wound healing across various wound types. Stored at room temperature in a clean, dry location.
Cytal® Wound Matrix	Acell, Inc., Columbia, MD, USA	A wound care device made from porcine urinary bladder matrix (UBM) with an intact epithelial basement membrane, suitable for acute and chronic wounds.	Supports a healing environment by promoting functional tissue remodeling. Stored in a clean, dry environment between 15 °C–35 °C in its original, unopened packaging.

(continued)

Table 19.1 (continued)

Product name	Manufacturer	Product overview by manufacturer	Manufacturer claims
Acellular/skin substitutes composed of natural and synthetic materials			
Integra® Bilayer Matrix Wound Dressing	Integra LifeSciences Corp., Princeton, NJ, USA	Bovine collagen-GAG matrix with a silicone layer for inpatient wound care.	Supports cellular infiltration and capillary growth while regulating moisture loss. Room temperature storage, 24-month shelf life.
Integra® Dermal Regeneration Template and Integra Omnigraft Regeneration Template	Integra LifeSciences Corp., Princeton, NJ, USA	Two-layer structure with a silicone outer layer for wound closure and a bovine collagen-GAG inner matrix to support tissue regeneration.	Provides a bacterial barrier, controls fluid loss, and promotes cellular growth. A biodegradable matrix integrates with autologous dermal tissue. Room temperature storage, 24-month shelf life.
Acellular/skin substitutes composed of synthetic materials			
Hyalomatrix® Tissue Reconstruction Matrix	Anika therapeutics, Bedford, MA, USA	Bilayered wound dressing with a hyaluronic acid scaffold for cell growth and a silicone layer for moisture control.	Protects wounds, minimizes contamination, and allows easy monitoring. Room temperature storage.
Restrata™	Acera surgical, Inc., St. Louis, MO, USA	Restrata is a synthetic electrospun wound dressing that mimics the ECM, promoting cell infiltration and vascularization for healing.	Its synthetic construction offers advantages such as resistance to enzymatic degradation and no special storage requirements.

Allografts

Allografts refer to tissues or organs that are transplanted from one person (the donor) to another (the recipient) of the same species. In the context of skin substitutes, human allografts typically involve tissues harvested from deceased organ donors, which are then used in wound treatment. The origin of the tissue, whether dermal or placental, significantly influences the structural composition, biological activity, and clinical performance of the resulting matrix. Dermis-derived matrices are rich in structural proteins such as collagen and elastin, providing excellent mechanical strength and a scaffold that closely mimics the native skin environment. This makes them particularly suitable for applications requiring robust support, such as soft tissue reinforcement and burn reconstruction [38]. In contrast, placenta-derived matrices are thinner and naturally rich in GFs, cytokines, and anti-inflammatory molecules, making them highly effective in promoting wound healing, angiogenesis, and scar modulation [39]. A variety of acellular allograft skin substitutes derived from both dermal and placental tissues are widely used in clinical practice, each offering distinct advantages depending on the wound type and treatment goal (Table 19.1). However, the scarcity of donor numbers and increasing demand stand as major problems. Also, a potential immune response originating from incomplete decellularization or the presence of residual donor DNA/proteins may elicit an immune response in some recipients [40].

Xenografts

Xenografts, also known as heterografts, are biological tissue grafts obtained from a different species and used in human patients for wound healing and reconstructive purposes [41]. They are primarily applied as temporary wound coverings in the management of extensive burns, chronic wounds, and surgical defects. Their effectiveness in clinical applications is largely influenced by factors such as tissue source, decellularization techniques, and post-processing methods, which determine their success or failure in skin wound repair and other medical treatments. They are sourced from porcine (pig) or bovine (cow). For instance, clinically available products such as Oasis® and PriMatrix® are derived from porcine and bovine skin, respectively. Their key advantages include high availability and cost-effectiveness [42]. When properly processed, particularly through decellularization, xenografts can act as biocompatible scaffolds that allow initial cell infiltration. However, their major limitations involve the risk of immune rejection due to species differences and potential transmission of zoonotic diseases. Moreover, their integration into human tissue is generally less effective compared to autografts or allografts, which restricts their use to temporary applications rather than long-term tissue repair.

Polymer-Based Substitutes

Polymers are highly versatile materials with unique properties that make them suitable for biomedical applications. Their structural flexibility allows for easy chemical modifications, enabling the formation of three-dimensional scaffolds and the customization of surface functionalities to enhance biological interactions [43]. Moreover, polymers can be engineered to incorporate and release therapeutic agents in a controlled manner, enhancing tissue regeneration and accelerating wound healing. Polymers used in skin substitutes can be broadly classified according to their origin as either natural or synthetic. Natural polymers, derived from biological sources, are favored for their good biocompatibility, biodegradability, and intrinsic bioactivity, which often promote cell adhesion and tissue regeneration. Common examples include collagen, gelatin, chitosan, alginate, hyaluronic acid, and silk fibroin. They closely mimic the ECM and support physiological healing processes. In contrast, synthetic polymers such as poly(lactic acid) (PLA), poly(glycolic acid) (PGA), polycaprolactone (PCL), and polyethylene glycol (PEG) are engineered to provide consistent and tunable mechanical, chemical, and degradation properties [44]. Although they may lack inherent bioactivity, synthetic polymers offer greater control over scaffold architecture and are often combined with natural polymers to optimize performance in skin regeneration applications.

Natural Polymers

Natural polymers are commonly used materials in the development of skin substitutes due to their intrinsic biocompatibility, biodegradability, and ability to support cellular functions (Table 19.2). They are derived from biological sources such as microbial, plant, and animal biomass. These polymers partially mimic the native ECM, making them particularly suitable as a dermal substitute material (Table 19.2). Polymers such as collagen, gelatin, chitosan, alginate, hyaluronic acid, and silk fibroin are widely studied for their capacity to promote cell adhesion, proliferation, and migration [45]. In addition to their favorable biological interactions, many natural polymers possess hemostatic, antimicrobial, and anti-inflammatory properties that further enhance their therapeutic potential. However, their mechanical strength and stability can sometimes be limited, which has led to their frequent use in combination with synthetic polymers or crosslinking agents to improve performance. Overall, natural polymers offer a biologically inspired foundation for the design of effective and functional skin substitutes [46].

Collagen

Collagen, the most prevalent structural protein in the body, plays a vital role in wound healing by providing mechanical support and tensile strength to regenerating tissue [47]. During the repair process, fibroblasts and other cells synthesize collagen, which undergoes modifications to form complex structural arrangements. Initially, collagen III is produced to offer temporary support in the early stages of healing. As tissue matures, it is gradually replaced by collagen I, which dominates in fully healed skin. Throughout granulation tissue formation, collagen fibers are deposited in a disorganized manner. However, the lysyl oxidase enzyme facilitates covalent crosslinking, which reinforces and aligns the fibers, restoring tensile strength. This remodeling process continues for several months after wound closure, allowing the repaired tissue to regain approximately 80–85% of its original strength, providing the healing process remains uninterrupted. Different types of collagen contribute to skin structure and function. While fibrillar collagens such as types I, III, and V provide structural integrity, fibril-associated collagens including types XII, XIV, XVI, and VI support collagen network organization. Additionally, non-fibrillar collagens IV and XVIII, found in the basement membrane, contribute to skin stability and cellular interactions [48].

Beyond its structural role, collagen is also applied as a surface coating in scaffolds and matrices to enhance moisture retention and promote cell attachment. It contains arginine-glycine-aspartic acid (RGD) sequences, which interact with

Table 19.2 Different natural polymer-based biomaterials utilized in wound healing

Biomaterial type	Wound type	Application method	Results	Refs.
Gelatin methacrylate-based hydrogel	Chronic/acute wounds	Incorporation of 2D nanosheets with antimicrobial properties for enhanced wound-healing applications	Provides superior mechanical strength, antioxidant activity, and antimicrobial effects, supporting faster tissue regeneration and protection against infections in wound care	[76]
Alginate-based hydrogel	Postoperative melanoma wounds	CuS@Hydrogel for NIR-II photothermal therapy and wound healing	Enhances cell migration, angiogenesis, and skin regeneration while maintaining a moist wound environment	[77]
PVA/sodium alginate/glycine nanofibers	Topical and infected wounds	Electrospun alginate-based nanofibers with ZnO NPs for antibacterial wound-healing applications	Antimicrobial, moisture retention, epithelialization, high porosity, and mechanical strength	[78]
Hydrogel-based wound dressings	Chronic wounds, surgical wounds, and burn injuries	Alginate and hyaluronic acid-based hydrogels were designed to provide a moist wound environment, supporting wound debridement and healing	Promotes fibroblast proliferation, angiogenesis, and tissue regeneration, while maintaining hydration and reducing inflammation for improved wound closure	[79]
Composite hydrogel based on platelet lysate and hyaluronic acid, crosslinked with genipin	Soft tissue regeneration	Scaffolds were prepared by mixing PL with HA-NH_2 and crosslinking with genipin. hASCs were seeded onto the surface and cultured under standard conditions	The scaffold showed good mechanical properties, biocompatibility, and hemocompatibility, supporting cell adhesion and proliferation	[80]
Alginate hydrogel containing losartan	Full-thickness skin wounds	Alginate hydrogel was crosslinked with $CaCl_2$ and loaded with different concentrations of losartan for slow drug release and exudate absorption. A rat model was used to evaluate wound closure	The alginate hydrogel facilitated controlled drug release, maintained wound moisture, and improved tissue remodeling. The formulation containing 0.1 mg/mL losartan showed over 80% wound closure in two weeks, with reduced TGF-β1 and VEGF expression, leading to minimized fibrosis and enhanced regeneration	[81]

(continued)

Table 19.2 (continued)

Biomaterial type	Wound type	Application method	Results	Refs.
Gelatin-based composite	General wound healing, including full-thickness skin wounds	Gelatin was crosslinked with oxidized lignosulfonate (OLS) to enhance its stability and antibacterial properties. The gelatin/OLS films were tested in vitro and in vivo on mouse models to assess their effectiveness in wound healing	The gelatin/OLS film provided antibacterial activity against *E. coli* and *S. aureus*, improved mechanical stability, and promoted wound healing through enhanced re-epithelialization, collagen formation, inflammation reduction, and increased blood vessel density. The material supported fibroblast adhesion and migration, accelerating wound closure	[82]
Natural polysaccharide and protein-based hydrogels	Bacteria-infected wounds	Hydrogels composed of natural polysaccharides (e.g., sodium alginate, hyaluronic acid, chitosan) and proteins (e.g., gelatin, elastin) were modified with antibacterial agents and used for treating infected wounds	These hydrogels exhibited high biocompatibility, antibacterial activity, and moisture retention, effectively reducing bacterial infection, promoting granulation tissue formation, and accelerating wound closure. Some formulations included phototherapy-responsive hydrogels to further enhance antibacterial effects and wound healing outcomes	[83]
Alginate, curdlan, and agarose-based bilayer fibro-porous dressing	Full-thickness diabetic wounds	The bilayer dressing was fabricated using electrospinning and freeze-drying techniques, with a lower lyophilized porous layer (alginate, agarose, and polyvinyl alcohol) and a top electrospun layer (agarose, curdlan, and polyvinyl alcohol). Ciprofloxacin was incorporated into both layers for antibacterial properties	The bilayer dressing promoted moisture retention, antibacterial activity, and hemostasis, effectively inhibiting *S. aureus* and *E. coli*. It enhanced angiogenesis, cell proliferation, and fibroblast biocompatibility (>90%), accelerating wound closure (~95–100%) while regulating inflammation (↑IL-10, ↓IL-6) for improved healing	[84]

(continued)

Table 19.2 (continued)

Biomaterial type	Wound type	Application method	Results	Refs.
Gelatin/K--carrageenan methacryloyl hybrid hydrogel	Chronic wounds, deep and irregular wound defects	A dual crosslinked hydrogel combining methacrylated κ-carrageenan and gelatin via ionic and UV-induced crosslinking was developed. The injectable hydrogel provided self-healing properties and mimicked the ECM for wound-healing applications	The hydrogel exhibited high biocompatibility, self-healing, and shear-thinning properties, making it ideal for minimally invasive wound treatment. It provided a moist wound environment, promoted cell viability, facilitated fibroblast migration, and enhanced angiogenesis. In vitro scratch assays confirmed accelerated wound closure, indicating its effectiveness in skin regeneration	[85]
Composite cryogel scaffold composed of platelet lysate and oxidized dextran	TE applications	Scaffolds were prepared by combining platelet lysate with oxidized dextran, followed by cryogelation. hASCs were seeded onto the scaffold surface, and in vivo assessment was performed via subcutaneous implantation	The cryogels exhibited macroporosity (10–200 μm), elasticity, hemocompatibility, and degradability over 90–240 days depending on OD content. They supported cell adhesion, proliferation, chondrogenic differentiation, and matrix formation	[64]
Collagen/xanthan gum-based hydrogel	Chronic wounds	Semi interpenetrating polymer network hydrogel with polyurethane, forming an amorphous, fibrillar-granular structure. The hydrogel exhibited rapid gelation (15 min), high swelling (3100%), and effective drug encapsulation and controlled release	The hydrogel demonstrated excellent hemocompatibility, enhanced cell proliferation, and modulated inflammation by reducing TNF-α and CCL-2 while increasing TGF-β production in monocytes. It inhibited *E. coli* growth (up to 100%) and provided controlled release of therapeutic agents, making it a promising material for chronic wound healing	[86]

(continued)

Table 19.2 (continued)

Biomaterial type	Wound type	Application method	Results	Refs.
Chitosan (CS)-based porous scaffold with carbon nanotubes (CNT) and nano-hydroxyapatite (n-HA)	Full-thickness skin wounds	CS, CS/CNT, and CS/CNT/HA scaffolds were developed using freeze-drying techniques. The scaffolds were tested for biocompatibility, antibacterial activity, and photothermal properties, then applied to a full-thickness skin wound model in mice	The CS-based scaffolds showed high swelling capacity, good mechanical properties, and excellent biocompatibility. The addition of CNTs provided photothermal responsiveness, promoting wound healing through localized hyperthermia. CS/CNT/HA scaffolds exhibited superior skin regeneration due to the synergistic effect of CNTs' photothermal activity and n-HA's bioactivity	[87]
Chitosan-dextran-glycerol hydrogel	Acute and chronic wounds	The hydrogel was developed by incorporating chitosan, dextran, and glycerol with iron oxide nanoparticles, either pre-synthesized or formed in situ, to enhance wound healing and antimicrobial properties	The hydrogel exhibited strong biocompatibility and antibacterial activity, promoting tissue regeneration while preventing infections. The presence of chitosan contributed to hemostasis and regenerative effects, while dextran enhanced bioadhesion. Iron oxide nanoparticles provided additional antimicrobial properties and improved the hydrogel's structure for better skin regeneration	[88]

cellular integrins, fostering fibroblast and keratinocyte adhesion and migration in in vitro studies. In preclinical burn models, collagen-coated scaffolds have demonstrated accelerated wound re-epithelialization and enhanced healing, outperforming conventional dressings. These findings emphasize collagen's potential in advanced wound management and tissue regeneration applications [49].

Gelatin

Gelatin, a natural biopolymer derived from the partial hydrolysis of collagen, has long been utilized in the development of skin substitutes due to its excellent biocompatibility, biodegradability, and low immunogenicity [50]. Its structure closely mimics the ECM. As a skin substitute, gelatin offers several advantages. It can be easily processed into various forms, making it versatile for different wound-healing applications. Also, its ability to absorb exudates, maintain a moist wound environment, and allow gas exchange makes it particularly effective for promoting wound healing. Gelatin can also be loaded with bioactive agents like GFs, antimicrobials, or stem cells, further enhancing their therapeutic potential. Despite these benefits, gelatin alone may have limited mechanical strength and rapid degradation in physiological conditions, which necessitates modifications to improve its performance for long-term applications. To overcome these drawbacks, various strategies have been employed, such as chemical crosslinking using agents like genipin, glutaraldehyde, or carbodiimide, which enhance the mechanical integrity and slow down degradation. One widely adopted strategy is the development of gelatin methacryloyl (Gel-MA), a chemically modified derivative that introduces methacryloyl groups onto the gelatin backbone, enabling photo-crosslinking under mild conditions. Gel-MA retains the bioactive sequences of gelatin, such as RGD motifs, while providing enhanced mechanical properties and tunable degradation profiles [51].

Elastin

Elastin is an insoluble biopolymer generated through the crosslinking of its precursor, tropoelastin. Although it constitutes only 2–4% of the dry weight of the human skin dermis, elastin plays a crucial role in providing structural integrity, mechanical resilience, and facilitating cell signaling [52]. Elastin interacts directly with various cell types and plays a key role in regulating cellular activities such as migration, proliferation, and differentiation, as well as influencing the coagulation cascade. It has been shown to activate multiple biological pathways, including fibroblast migration and proliferation, smooth muscle cell expansion, keratinocyte migration, calcium transport, ECM turnover, and cell survival. Additionally, elastic fibers within the dermis contribute to matrix organization, cytokine production, and the regulation of cell phenotypes, further emphasizing their role in tissue homeostasis and wound healing [53].

Fibrin

Fibrin is a glycoprotein essential for blood coagulation, primarily synthesized by hepatocytes and formed through fibrinogen aggregation in the presence of thrombin. With a molecular weight of ~340 kDa, fibrin is composed of three interwoven polypeptide chains—Aα, Bβ, and γ—connected by disulfide bonds. Structurally, it features two expanded globular domains (D-domains) at both ends and a central E-domain, all linked by three α-helical peptide chains [54]. The coagulation process begins when external or internal stimuli activate coagulation factors, leading to the conversion of thromboplastin into thrombin in the presence of calcium ions. Thrombin binds to the E-domain of fibrinogen, cleaving fibrinopeptide A (FPA) from the N-terminus of the Aα chain and fibrinopeptide B (FPB) from the N-terminus of the Bβ chain, resulting in the formation of fibrin monomers. These monomers spontaneously assemble into insoluble fibrin, stabilized through non-covalent interactions between the D and E domains, while α-α and α-γ chain interactions facilitate further aggregation [55].

Keratin

Keratin is the predominant protein found in the cells of the epidermis, and a protective layer of keratin signifies the outermost surface of the skin. This group of proteins, known as keratins, consists of a family of filament-forming proteins that are rich in cysteine [56]. They can be divided into two primary categories: soft (epithelial) keratins and hard (trichocytic) keratins, based on the tissue types in which they are expressed. Soft keratins are primarily located in the epidermis, oral mucosa, and various other epithelial linings. These proteins form intricate networks of intermediate filaments, serving as scaffolds that help epithelial cells endure mechanical stress while maintaining their structural integrity. Additionally, soft keratins play vital roles in cell physiology, influencing processes like adhesion, migration, and differentiation. Their diverse functions are often mediated through interactions with numerous signaling proteins, which are enhanced by complex post-translational modifications. In contrast, hard keratins, found in epidermal appendages such as hair and feathers, are further classified into α-keratins, predominant in materials like sheep's wool, and β-keratins, characteristic of avian feathers and reptilian tissues. These hard keratins exhibit a higher sulfur content due to the increased proportion of cysteine residues in their amino acid sequences, making them structurally distinct from their soft counterparts [57].

Cellulose

Cellulose-derived biomaterials offer substantial potential in promoting skin wound regeneration. Cellulose is a structurally complex polysaccharide consisting of repeating glucose units, found in the cell walls of all plants and various algae [58]. The superior absorbency of cellulose-based biomaterials aids in managing wound exudate, minimizing fluid buildup, and lowering the risk of infection. By maintaining a moist healing environment, these materials enable tissue repair. Its modified form, carboxymethyl cellulose (CMC), can form a gel-like protective barrier upon contact with wound exudate, promoting faster healing while reducing discomfort. In addition to their barrier function, cellulose-based materials can be fabricated into three-dimensional structures that serve as scaffolds for granulation tissue formation, supporting cellular infiltration and tissue regeneration. These biomaterials also contribute to wound healing by modulating the inflammatory response and maintaining a moist, physiologically favorable environment.

Chitosan

Chitosan, a biodegradable, non-toxic, and biocompatible polysaccharide, is produced through the deacetylation of chitin, a structural component found in the exoskeletons of crustaceans and certain fungi. Its molecular structure consists of randomly distributed β-(1–4)-linked D-glucosamine and N-acetyl-D-glucosamine units, contributing to its versatility in biomedical applications [59]. This polymer exhibits several advantageous properties, including hemostasis, mucoadhesion, hydrophilicity, and non-toxicity. Chitosan's structural similarity to the ECM enhances cell adhesion and proliferation, establishing it as a valuable material in TE. Its derivatives are widely utilized in treating deep, open, and severe burn wounds due to their strong adhesion, hemostatic effects, and antimicrobial properties. Additionally, chitosan demonstrates excellent film-forming ability, along with antifungal and antibacterial activity, while offering superior mechanical strength and barrier functionality. These beneficial characteristics are particularly effective when key physicochemical parameters fall within optimal ranges, including a tensile strength of 22–61 MPa, a molecular weight between 90–880 kDa, and a degree of deacetylation between 65% and 95% [60]. Chitosan holds significant potential in wound healing by supporting new tissue formation and offering antimicrobial protection; however, further research is necessary to fully explore its therapeutic applications.

Carrageenan

Carrageenan, a naturally occurring polysaccharide extracted from red seaweed, has gained attention for its potential role in wound healing due to its distinctive properties. It consists of β-(1 → 3)-linked D-galactose and α-(1 → 4)-linked 3,6-anhydrogalactose [61]. These glycosidic bonds, along with varying degrees of sulfation, determine the gelation properties and the interaction of carrageenan with different ions, such as potassium and calcium. Upon application, it forms a gel-like protective matrix that serves as a barrier against external contaminants while maintaining a moist environment conducive to tissue repair. This matrix supports keratinocyte and fibroblast migration and proliferation, essential for effective wound healing. Beyond its structural role, carrageenan exhibits anti-inflammatory effects by reducing localized inflammation at the wound site. Additionally, it can regulate the release of GFs and cytokines, thereby promoting tissue regeneration and angiogenesis. These combined mechanisms contribute to accelerated wound closure, minimized scarring, and a more structured healing process. Carrageenan exists in different isomeric forms, including kappa-, iota-, and lambda-carrageenan, each possessing unique characteristics that define their suitability for various biomedical applications [58].

Dextran

Dextran is a polysaccharide that branches via α-(1 → 2), α-(1 → 3), and α-(1 → 4)-glycosidic bonds [62]. Clinical-grade dextran, which typically has a molecular weight between 40,000 and 70,000, exhibits high thermal stability, allowing it to withstand sterilization processes [63]. It has attracted interest in the development of skin substitutes due to its biocompatibility and biodegradability. Dextran can be used to create moisture-retaining environments that allow it to effectively absorb wound exudates, reduce discomfort, and promote faster wound healing. Its gel-forming ability with other natural polymers allows dextran-based materials to serve as scaffolds that promote cellular infiltration and tissue regeneration [64]. Despite its beneficial properties, dextran alone lacks sufficient bioactivity to actively promote wound healing. Recent studies have focused on adding bioactive agents to dextran-based scaffolds to enhance their regenerative potential.

Heparin

Heparin is a highly sulfated GAG, composed of repeating disaccharide units of glucosamine and uronic acid linked by β-(1 → 4) and α-(1 → 4)-glycosidic bonds [65]. Known for its anticoagulant properties, heparin has been extensively investigated for its role in wound healing, where it contributes to endothelial cell repair, regulates capillary circulation, and influences the overall healing process by inactivating various proteases [66]. Additionally, heparin plays a key role in dermal fibroblast proliferation, collagen synthesis, and the regulation of GFs such as TGF-β1 and bFGF, which are essential for tissue regeneration. Beyond its anticoagulant property, heparin affects local blood flow, protein-cell receptor interactions, and modulation of inflammatory responses. Moreover, it can inhibit pain-related neuropeptides from activating immune cells and regulate complement system activation.

The sulfate group of heparin may also interact with integrin adhesion molecules, affecting nuclear factor kappa B (NF-κB) signaling, which plays a role in inflammation and cellular response [67]. Despite its short biological half-life, advancements in nanotechnology have aimed to extend its duration of action and minimize adverse effects. Heparin has demonstrated safe and effective use as a complementary treatment for wound management in patients with DFU and critical limb conditions, offering potential benefits in improving tissue repair and microcirculation.

Alginate

Alginate is a water-soluble linear polysaccharide composed of β-D-mannuronic acid (M) and α-L-guluronic acid (G) residues, linked through 1–4-glycosidic bonds. Its structure consists of homogeneous poly-G and poly-M blocks as well as heterogeneous MG regions, each exhibiting distinct conformational properties [68]. It exhibits remarkable gelling properties, primarily attributed to the abundance of repeating G residues and metal ion-mediated interchain interactions such as the Ca^{2+} mediated egg-box model [69]. Furthermore, alginate-based hydrogels are biodegradable, gradually dissolving in the body as crosslinking ionic agents are released and exchanged with monovalent cations present in bodily fluids, leading to the breakdown of the hydrogel network. With its high absorbency and hemostatic properties, alginate is highly effective in managing exuding wounds and debridement of sloughed tissue. Despite its biocompatibility and non-immunogenicity, the lack of cell adhesion motifs limits its biological activity. However, modifying alginate chains with cell adhesion tripeptides, such as RGD (Arg-Gly-Asp), can promote cell adhesion [68].

Hyaluronic Acid

Hyaluronic acid (HA), also known as hyaluronan, is a naturally occurring polysaccharide composed of N-acetyl-D-glucosamine and D-glucuronic acid disaccharide units linked by β-1,4 and β-1,3 glycosidic bonds [70]. Its molecular weight covers a wide range from 10 kDa to several thousand kDa, which significantly influences its biological properties and applications. As a member of the GAG family, HA shares structural similarities with chondroitin sulfate, dermatan sulfate, and heparan sulfate. However, unlike these counterparts, it does not form covalent bonds with proteins. The main difference of hyaluronan from other major GAGs is the absence of sulfate groups, which contributes to its unique biochemical behavior. HA takes part in the wound-healing process. It contributes to tissue hydration, stabilization of the ECM, and maintenance of osmotic balance. HA facilitates cell movement and tissue regeneration by absorbing exudate. Despite its many benefits, HA also presents certain limitations in wound-healing applications. One major drawback is its short biological half-life, as it is rapidly degraded by enzymes such as hyaluronidases and by oxidative stress in the wound environment [71]. This instability often necessitates frequent reapplication or chemical modification to prolong its activity. Additionally, while high-molecular-weight HA exhibits anti-inflammatory properties, its low-molecular-weight fragments can have pro-inflammatory effects, potentially aggravating the wound site. Native HA also lacks sufficient mechanical strength, making it less suitable for use in deeper or more complex wounds unless combined with other structural materials. Moreover, the production of high-purity or medical-grade HA can be costly, and its incorporation into stable, effective formulations may require complex processing steps.

Xanthan Gum

Xanthan gum (XG) is a bacterial polysaccharide known for its exceptional water-absorbing capacity and structural adaptability. Composed of a β-1,4-D-glucose backbone with trisaccharide side chains, it features numerous hydroxyl and carboxyl groups, which contribute to its high hydrophilicity [72]. This property allows

XG-based hydrogels to retain large amounts of fluid, making them particularly beneficial for managing wound exudates and maintaining an optimal healing environment. A distinctive feature of XG is its five-fold helical conformation formed through non-covalent interactions between its functional groups [73]. This unique structure enables thermal and ionic responsiveness, allowing XG to undergo structural transitions based on environmental conditions. Beyond its absorbent and conformational properties, XG stands out due to its high viscosity even at low concentrations, along with remarkable thermal stability. Its biocompatibility, non-toxicity, and strong water-retention properties create a favorable environment for tissue regeneration. Various fabrication techniques have been employed to enhance XG's performance in wound healing, demonstrating its potential as a customizable and effective biomaterial. However, xanthan gum lacks intrinsic biological activity and may require functionalization or blending with more bioactive materials to improve efficacy. Additionally, its relatively weak mechanical strength and potential for microbial contamination necessitate careful formulation and preservation.

Fucoidan

Fucoidan is a structurally complex polysaccharide predominantly present in the cell walls of brown algae, composed mainly of sulfated fucose, with trace amounts of galactose, mannose, and xylose [74]. Its structure allows it to interact with GFs, similar to GAGs like heparin and chondroitin sulfate, protecting key regulators such as TGF-β1 from degradation and enhancing their bioavailability. Additionally, fucoidan modulates fibroblast activity, influencing collagen synthesis, angiogenesis, and tissue remodeling, which are crucial for wound repair. Studies have demonstrated that low-molecular-weight fucoidan (LMF) can regulate inflammatory responses, reduce neutrophil adhesion, and lower pro-inflammatory cytokine levels, leading to faster wound closure and improved tissue regeneration. Furthermore, its role in stimulating the proliferation of fibroblasts and epidermal cells has been linked to the upregulation of VEGFR2, TGF-β1, and MMP 9, all of which contribute to angiogenesis and ECM remodeling. The anticoagulant properties of fucoidan also influence microcirculation, facilitating nutrient and oxygen delivery to healing tissues [75].

Synthetic Polymers

Synthetic polymer-based skin substitutes are designed to imitate the properties of natural skin for use in wound healing and tissue regeneration. Unlike natural polymers, synthetic polymers, such as poly (ethylene glycol) (PEG), poly (urethane), poly (lactic acid) (PLA), and poly (glycolic acid) (PGA), can be tailored to exhibit specific mechanical, thermal, and chemical properties that are essential for supporting cell growth, reducing infection, and promoting healing in damaged skin. One of the primary advantages of synthetic polymer-based skin substitutes is their versatility [89]. These materials can be engineered to have varying degrees of porosity, flexibility, and biodegradability, making them suitable for different types of wounds, including chronic, acute, and burn wounds (Table 19.3). They can be functionalized with bioactive molecules such as GFs, antimicrobial agents, or heparin, which further enhance their effectiveness as skin substitutes. Despite their advantages, the use of synthetic polymers in skin substitutes poses potential risks in terms of their biocompatibility and immune response.

Poly (vinyl alcohol)

Poly (vinyl alcohol) (PVA) is a water-soluble synthetic polymer that has gained significant interest in the field of skin substitutes due to its unique properties, including biocompatibility, biodegradability, and the ability to be easily processed into a variety of forms. These characteristics make PVA an attractive material for use in skin substitutes. The hydrophilic nature of PVA allows it to absorb water, which helps to maintain a moist environment conducive to wound healing [90]. It can be processed into hydrogels or films and gradually breaks down over time, allowing for its replacement by the regenerating host tissue. However, it also has limitations such as mechanical weaknesses (e.g., brittleness and insufficient elasticity), limited cellular infiltration, and unpredictable degradation rates. Additionally, PVA lacks inherent bioactivity and may require additional bioactive factors to enhance tissue regeneration. To address these shortcomings, composite formulations incorporating natural or synthetic polymers have been employed to optimize PVA-based hydrogels, enhancing their physicochemical stability, mechanical stability, and bioactivity [91].

Poly (caprolactone)

Poly (caprolactone) (PCL) is a biodegradable synthetic polymer classified as an aliphatic polyester, consisting of repeating hexanoate units [92]. Its biodegradability and mechanical versatility make it a widely studied material in scaffold fabrication as a dermal substitute. PCL is recognized with its biocompatibility, mechanical strength, and slow degradation rate, which allows it to provide long-term support as tissues regenerate [93]. This slow degradation is advantageous in cases where prolonged scaffold support is required for tissue growth and repair. Moreover, it can be processed into various forms such as fibers, films, and scaffolds, using methods like electrospinning, 3D printing, and casting, making it highly versatile in dermal substitute applications. However, PCL also has limitations. Its hydrophobic nature can limit its ability to support cellular infiltration and moisture retention in wound-healing applications. Also, PCL lacks the bioactive signals necessary for optimal cell adhesion, migration, and proliferation, requiring surface modifications or the incorporation of bioactive agents to improve its interaction with cells. Despite these

Table 19.3 Various synthetic polymer-based biomaterials used in wound healing

Biomaterial type	Wound type	Application method	Results	Refs.
PVA-based nanofibers	General wound healing in a rat model	Electrospun nanofibers consisting of chitosan modified with graphene oxide and carbon quantum dot-doped TiO_2 were fabricated and applied as wound dressings	The nanofibrous mat exhibited hydrophilicity, antibacterial properties against gram-negative and gram-positive bacteria, good mechanical properties, and significant biocompatibility, leading to enhanced wound-healing efficiency in a rat model	[106]
PVA/PCL electrospun nanofibers	MRSA-infected and non-infected wounds	Electrospun nanofibers were fabricated using PCL as the supportive layer and PVA blended with *Quercus infectoria* (QLG) extract and biosynthesized CuNPs as the bioactive layer. The composite dressing was tested in vivo for wound-healing efficacy	The PCL/PVA/QLG/CuNPs nanofiber dressing exhibited strong antibacterial activity against MRSA, accelerated wound closure (77.6% in non-infected, 73.8% in infected wounds by day 5), and promoted angiogenesis and complete skin regeneration by day 10–15. It also maintained moisture, absorbed exudates, and supported gas exchange, making it a promising wound dressing	[107]
PVP/poly (acrylic acid) (PAA) hydrogel containing ZnO nanoparticles (NPs)	Excisional wound model in rats	The hydrogel, synthesized via free radical polymerization, was tested in a rat excisional wound model to assess wound contraction, antibacterial activity, and histological improvements	The hydrogel accelerated wound contraction, enhanced collagen deposition, and exhibited antibacterial activity, with ZnO NPs providing antimicrobial effects and PVP ensuring moisture retention	[108]
PVP/CS nanofiber membrane with dihydromyricetin (DHM)	Diabetic wounds	Electrospun nanofiber membranes incorporating PVP, CS, and DHM were fabricated and applied to diabetic wounds in mice to evaluate wound-healing efficacy	The PVP/CS/DHM nanofibers enhanced wound healing through antibacterial, antioxidant, and anti-inflammatory effects, promoting angiogenesis and autophagy. The dressing achieved 88.63% wound closure in 18 days, showing strong potential for diabetic wound treatment	[109]
PEG-CS-based hydrogel	Burn wounds	The hydrogel, composed of xanthan gum, gelatin, PEG, chitosan, and boric acid, was tested on rat burn wounds for wound contraction, moisture retention, and biodegradability over 28 days	The hydrogel showed high water content, controlled biodegradability, and optimal swelling, promoting wound closure, cell proliferation, and inflammation reduction, making it promising for burn wound treatment	[110]

PLGA/gelatin/$CaCO_3$/SiO_2/ Quercetin nanofiber patch	Diabetic wounds	The nanofiber patch, produced via electrospinning, integrates PLGA, gelatin, $CaCO_3$/SiO_2, and quercetin. It exhibits antibacterial properties, enhances fibroblast proliferation, and accelerates diabetic wound healing	The patch provides antioxidant and anti-inflammatory effects, promotes angiogenesis, enhances cellular adhesion, and retains moisture, leading to accelerated tissue regeneration in diabetic wounds	[111]
PGA-based bioabsorbable sheet	Oral mucosal defects	The PGA sheet, combined with fibrin glue, was used to cover oral mucosal defects post-surgery, supporting secondary healing and reducing the need for grafts or reconstruction	The bioabsorbable PGA sheet enhances epithelialization, minimizes scar contracture, and prevents infection while gradually degrading, ensuring effective wound coverage	[112]
PLA/beta-chitin whiskers/ silver NPs (AgNPs) bilayer membrane	Skin wounds	The bilayer membrane, produced via SBS and electrospinning, integrates beta-chitin whiskers for hydrophilicity and AgNPs for antibacterial activity. It was tested in vitro for biocompatibility and antimicrobial properties	The membrane balances moisture, absorbs exudates, and prevents dehydration while offering antibacterial protection, creating an optimal healing environment	[113]
Poly(caprolactone) (PCL)/ lignin-based 3D-printed dressing with curcumin (CUR) and D-Panthenol (DPA)	Skin wounds	The 3D-printed wound dressing, produced via SSE technology, integrates PCL, lignin, curcumin, and D-panthenol. It was tested for antimicrobial, antioxidant, and wound-healing properties, showing improved wound closure, reduced inflammation, and enhanced epithelialization in vivo	The dressing ensures sustained bioactive release, offering antimicrobial and antioxidant benefits. Curcumin and lignin combat bacteria and oxidative stress, while D-panthenol promotes fibroblast proliferation and moisture retention, optimizing wound healing	[114]
Citrate-based poly (urethane) (HCPU-NHS) adhesive	Skin incisions and soft tissue wounds	The adhesive, designed as a suture alternative, ensures strong wet and dry tissue adhesion. Its hydrophobic PCL layer enhances bonding, and UV crosslinking stabilizes the structure. In vivo tests showed superior wound closure	The poly (urethane) adhesive provides bioadhesion, biodegradability, and hemostasis. Citrate byproducts promote angiogenesis, while its liquid-tight sealing reduces infection risk and supports tissue regeneration	[115]

challenges, ongoing research to modify PCL's properties, including blending it with other materials or introducing bioactive components, holds promise for overcoming these limitations and enhancing its effectiveness in regenerative medicine [94].

Poly (lactic acid)

Poly (lactic acid), recognized for its high biocompatibility, controlled degradability, and favorable mechanical properties, has been FDA-approved for biomedical applications and widely utilized in the development of wound materials [95]. This polymer exhibits several key characteristics, including cost-effectiveness, non-toxicity, hydrophobicity, structural stability, excellent biocompatibility, and mechanical integrity, making it highly suitable for medical applications. As an aliphatic polyester derived from naturally occurring lactic acid, PLA undergoes enzymatic or hydrolytic degradation under physiological conditions, producing byproducts that are readily metabolized and absorbed by the body [96].

Poly (glycolic acid)

Poly (glycolic acid) (PGA) is a biodegradable polymer belonging to the family of linear aliphatic polyesters. It is known for its high biocompatibility and superior hydrophilicity compared to PLA [97]. PGA has been extensively studied for its potential in medical and TE applications. It offers advantages like high tensile strength, rapid biodegradability, and excellent mechanical properties, making it ideal for temporary scaffolds in wound healing. Despite its biocompatibility and biomedical potential, the clinical application of PGA is hindered by its rapid hydrolytic degradation, which leads to the accumulation of glycolic acid and carbon dioxide, ultimately reducing the local pH and increasing the risk of cellular and tissue necrosis. This acidic microenvironment has been associated with inflammatory responses, further limiting its widespread use in regenerative medicine. However, studies have highlighted PGA's role in neo-epithelium formation and inflammation modulation, making it a promising candidate for wound-healing applications. It has been employed in preventing delayed perforation, fistula closure in both open and endoscopic surgeries, and as a protective covering for wounds following partial glossectomy [98].

Poly (lactic-co-glycolic) Acid

Poly (lactic-co-glycolic acid) (PLGA) is a biodegradable synthetic polymer synthesized from the random polymerization of lactic acid (LA) and glycolic acid (GA), combining the advantages of both PLA and PGA. In moist wound environments, the ester bonds in PLGA hydrolyze into smaller fragments, which are metabolized and excreted by the body. Its degradation rate can be precisely controlled by adjusting the LA-to-GA ratio, with a higher GA content accelerating degradation due to its greater hydrophilicity compared to LA. This customizable degradation profile, along with its tunable physicochemical properties, makes PLGA a widely used scaffold material in wound-healing applications [99]. Additionally, its lactic acid byproduct induces mild acidosis at the healing site, which has been reported to reduce proteolytic activity, enhance oxygen availability, stimulate fibroblast proliferation, and support tissue regeneration. However, the application of PLGA as a standalone wound-healing scaffold is limited due to its hydrophobic surface and lack of specific cell-recognition sites, which hinder cell adhesion and proliferation. Blending PLGA with hydrophilic natural polymers has been a general approach to overcome these limitations [100]. A promising approach involves use of PLGA as an outer protective layer to shield against bacterial invasion, while hydrogels or other scaffold matrices provide a supportive inner environment that enhances wound healing. For instance, a double-layer membrane (BLM) scaffold consisting of an outer PLGA membrane and a lower alginate hydrogel layer was fabricated via 3D printing to imitate the epidermis and dermis of the skin. This BLM scaffold significantly improved skin regeneration, as demonstrated by increased neovascularization and enhanced collagen I/III deposition, highlighting its potential in advanced wound healing strategies [99].

Poly (ethylene glycol)

Poly (ethylene glycol) is a biocompatible, hydrophilic polymer that has gained attention as a potential dermal substitute [101]. Due to its favorable properties, such as non-toxicity, water retention, and ability to be easily modified, PEG is an ideal candidate for creating hydrogels and scaffolds used in wound healing and skin regeneration. PEG-based hydrogels can provide a moist environment conducive to tissue growth, helping to accelerate wound healing and improve recovery outcomes. One of the main advantages of PEG is its ability to be tailored to specific applications through changes in molecular weight and crosslinking density. This allows for control over its mechanical properties, degradation rate, and swelling behavior. Furthermore, PEG can be modified with bioactive agents such as GFs, peptides, or ECM components to enhance cellular adhesion, migration, and tissue integration, thereby addressing one of its main limitations: its intrinsic lack of bioactivity [102]. However, PEG's inert nature can be a disadvantage in certain dermal substitute applications, as it does not naturally promote robust cell adhesion or integration with the surrounding tissue. To overcome this, PEG scaffolds often require surface modifications or the addition of bioactive molecules to improve tissue interaction.

Poly (urethane)

Poly (urethane)s are a class of versatile polymers formed by the reaction of diisocyanates with polyols, resulting in a wide range of materials with varying properties [103]. They are widely used polymers in scaffold fabrication due to their high elasticity. They can be designed in both degradable and non-degradable forms, depending on their intended biomedical application. For instance, poly(ester-urethane) urea (PEUU) is a degradable variant that is particularly suitable for skin TE. Conversely, non-degradable polyurethane is primarily utilized for wound dressing applications, where nanofibrous scaffolds can be loaded with bioactive agents to enhance the healing process. These scaffolds can be functionalized with angiogenic, antioxidant, anti-inflammatory, or antimicrobial compounds to promote tissue regeneration and infection control [104]. Polyurethane polymers are increasingly being recognized for their potential as wound dressings. These polymers have a relatively simple chemical structure, which can be easily modified to introduce specific functional groups. Additionally, they are not commonly linked to issues such as disease transmission or immunogenicity. Polyurethane polymers, with their tunable soft and hard segments and modifiable chain extensions, are widely used in biological applications. By changing the ratio of hard and soft segments, the physicochemical properties of polyurethane can be tailored to suit specific needs [105].

Fabrication Techniques for the Preparation of Skin Substitutes

The fabrication technique employed in the production of skin substitutes plays a pivotal role in determining their final architecture and functional performance, as it directly influences critical parameters such as pore size, overall porosity, interconnectivity, mechanical strength, and elasticity. These properties dictate the scaffold's ability to support cell adhesion, proliferation, and migration, as well as its capacity to facilitate nutrient and oxygen diffusion, waste removal, and vascularization. Particularly, pore size affects cellular infiltration and growth. The ideal pore size for the regeneration of adult mammalian skin typically ranges from 20 to 125 μm [116]. Mechanical strength is also critical, as skin is continuously subjected to stretching, compression, and shear forces. Therefore, developed skin substitute material must possess sufficient tensile strength and elasticity to maintain structural integrity without tearing or collapsing. Also, an ideal skin substitute should be strong enough to protect the wound while being flexible enough to accommodate skin movement, particularly in joint areas. The degradation rate must be engineered to match the pace of tissue regeneration, ensuring that the scaffold provides temporary support without leaving behind harmful residues. Consequently, the method used to fabricate skin substitutes is not only a technical choice but also a strategic decision that directly influences their biological performance and clinical effectiveness [117].

Freeze-Drying (Lyophilization)

Freeze-drying is a commonly used technique for fabricating porous scaffolds from natural polymers such as collagen, gelatin, chitosan, and hyaluronic acid. The process involves freezing the polymer solution followed by sublimation of ice under vacuum, resulting in a highly porous structure. The pore size and orientation can be somewhat tuned by adjusting the freezing rate and polymer concentration, which are critical for cell migration, nutrient diffusion, and neovascularization. However, scaffolds produced via freeze-drying often lack mechanical strength and uniformity, making them unsuitable for load-bearing skin applications without post-processing such as crosslinking or reinforcement. Additionally, control over pore interconnectivity is limited, and batch-to-batch reproducibility can be challenging [117].

Electrospinning

Electrospinning is a process used to produce very fine fibers, often at the nanometer scale, by applying a high-voltage electric field to a liquid polymer solution or melt [118]. The electric field draws the liquid into a thin jet, which solidifies into fibers as it travels through the air and is collected on a conductive surface (Fig. 19.1). It enables the fabrication of nanofibrous scaffolds that closely resemble the ECM of the skin. Suitable materials include both synthetic (e.g., PCL, PLGA) and natural polymers (e.g., silk fibroin, gelatin, collagen), or their composites. These nanofibers promote cell adhesion, proliferation, and wound coverage due to their high surface area-to-volume ratio. The process also allows for surface modification and drug/GF incorporation. Nevertheless, electrospun scaffolds typically suffer from limited pore interconnectivity and depth, which restricts cellular infiltration and angiogenesis. Moreover, their mechanical properties are often insufficient without post-treatment, and large-scale production remains a challenge [2].

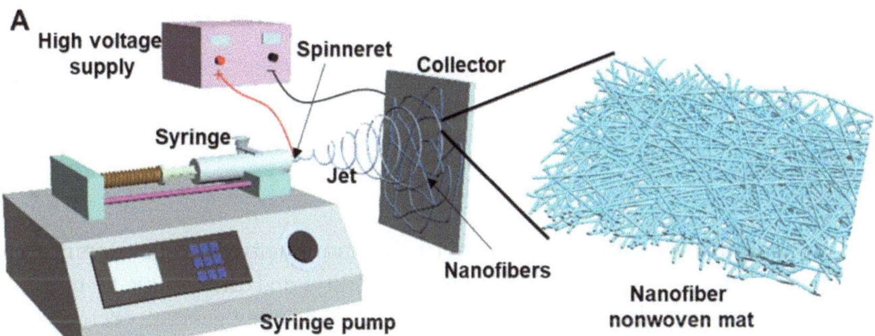

Fig. 19.1 Schematic illustration of electrospinning process [118] [CC BY 4.0]

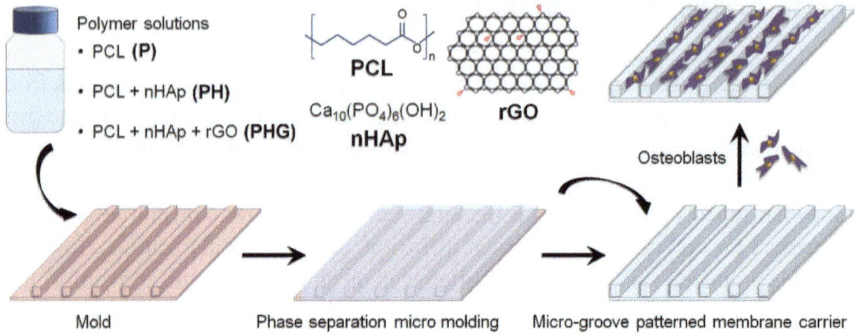

Fig. 19.2 Schematic illustration of micro-patterned fabrication technique for the composite film composed of PCL, nHAp, and rGO [122] [CC BY 4.0]

Nano/Micropatterning

Micropatterning is an emerging fabrication strategy in the development of advanced skin substitutes, offering precise control over the surface architecture [117]. This technique involves creating microscale patterns such as ridges, grooves, or grids on the scaffold surface to replicate the structural and functional complexity of native skin (Fig. 19.2). There are a variety of methods available to create these patterns, one of the most widely used being soft lithography. In this technique, a hydrogel is micromolded using a stamp typically made from poly(dimethylsiloxane) (PDMS) [119]. These micropatterns provide physical and biochemical cues that more closely resemble the native skin environment, ultimately enhancing the performance and regenerative potential of engineered skin substitutes. By mimicking the topographical cues, micro-patterned surfaces guide cellular organization and have been shown to improve cell-material interactions, stimulating more ECM production [120, 121]. These features contribute to better tissue integration and faster wound healing. Micropatterning technique can be used with natural or synthetic polymers such as collagen, gelatin, or PDMS to fabricate biocompatible and functional scaffolds that closely mimic native skin morphology and biomechanics. However, the micropatterning technique is restricted to creating features on a 2D surface or within a limited depth range.

3D Printing

3D printing has evolved into a powerful fabrication method for developing skin substitutes, offering high precision, customizability, and the ability to replicate complex skin architectures [123]. In this context, 3D printing generally refers to the layer-by-layer deposition of biomaterials without living cells to create scaffolds that mimic the ECM of native skin. These scaffolds can then support cell attachment, proliferation, and tissue regeneration when seeded with appropriate cell types post-fabrication. The key advantage of 3D printing in skin substitutes lies in its design flexibility and reproducibility, allowing for the creation of porous structures with controlled geometry, pore size, and interconnectivity parameters crucial for nutrient diffusion and tissue integration. Synthetic polymers such as PCL, PLA, and PLGA and natural polymers like Gel-MA, alginate, and chitosan composites are commonly used. Additionally, it allows surface modifications or incorporation of bioactive molecules to enhance the biological performance of the printed constructs. 3D printing techniques are classified according to the method of material deposition, and they are broadly categorized into extrusion-based, light-assisted (e.g., stereolithography or digital light processing), fused filament (e.g., fused filament fabrication), and inkjet-based printing techniques [124]. Despite its advantages in precision and structural control and variety in the techniques, uneven distribution and poor cell viability may occur due to post-cell seeding onto the scaffold after fabrication in the 3D printing applications.

3D Bioprinting

3D bioprinting represents an advanced technique capable of depositing cells, biomaterials, and signaling molecules in precise 3D architectures [125]. This method enables patient-specific designs by utilizing biocompatible materials that can support cell attachment and tissue formation. It allows for spatial control over cell distribution and the inclusion of vascular-like channels, offering promising outcomes in terms of integration and functionality (Fig. 19.3). The development of suitable bioink is a crucial aspect of this process, requiring careful consideration of printability, degradation kinetics, and mechanical stability. It is particularly advantageous for creating multilayered skin substitutes that mimic the stratified structure of the native skin [126]. Hydrogels such as Gel-MA, alginate, fibrin, and decellularized ECM-based bioinks are commonly used due to their printability and cell compatibility [127]. Similarly, 3D bioprinting techniques are categorized according to their printing mechanism. It offers several key advantages, including precise replication of the complex, multilayered structure of the native skin, customizability, and scalability of the process. 3D bioprinting allows for the design of patient-specific grafts based on imaging data, enabling personalized treatment options for wounds, burns, or other skin defects. Additionally, the technique supports automated and reproducible fabrication, reducing batch-to-batch variability and enhancing the standardization of tissue constructs for clinical applications.

However, the 3D bioprinting technology is still developing, and the complete replication of full-thickness skin with all its layers has yet to be achieved. The mechanical stability of 3D bioprinted constructs often remains low due to the inherent low mechanical strength of the hydrogels and requires reinforcement with synthetic polymers or composite strategies [129]. The process might be time-consuming, technically complex and necessitates expensive equipment and skilled personnel.

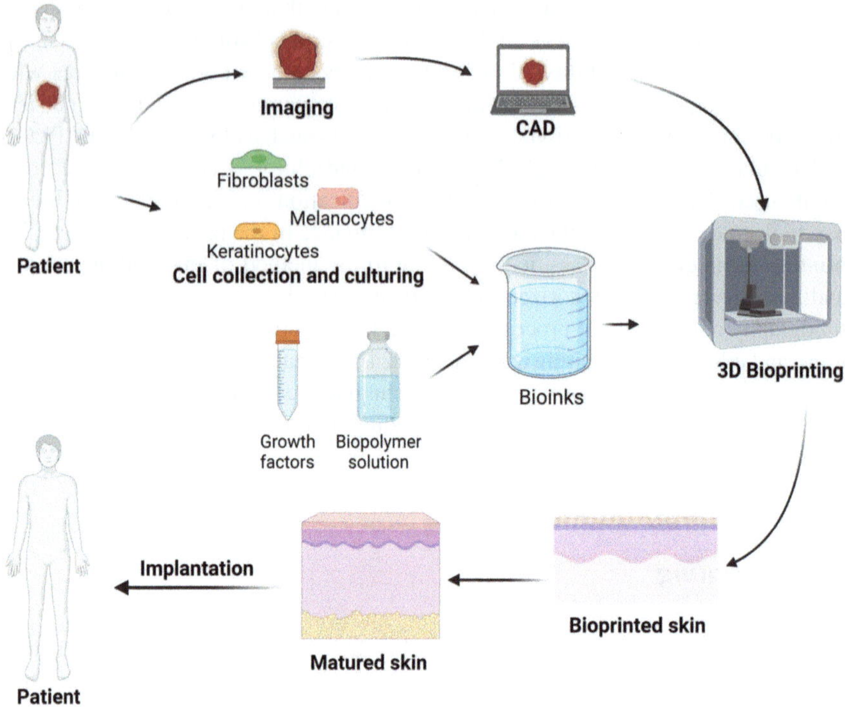

Fig. 19.3 Schematic illustration of the skin 3D bioprinting process [128] [CC BY 4.0]

Future Perspectives

Advancements in biotechnology, immunology, and TE have enabled the development of innovative and highly sophisticated skin substitutes for clinical applications. Innovative smart skin substitutes have emerged as a dual-function solution in chronic wound care, offering real-time monitoring while simultaneously enhancing the healing process. This approach integrates biomarker sensing technologies to monitor key wound parameters such as pH, temperature, uric acid (UA), ROS, and oxygen levels, enabling timely intervention for delayed healing and infections [130]. Among these, pH monitoring is particularly relevant, as chronic and infected wounds typically exhibit alkaline conditions (pH 7–9), promoting bacterial proliferation. Similarly, oxygen availability is essential for tissue regeneration, with hypoxic wounds ($\leq$20 mmHg) linked to impaired recovery [131]. Temperature fluctuations and UA levels also serve as indicators of infection and wound chronicity. To enable real-time monitoring of the wound pathophysiology, smart dressings incorporate electrochemical and optical sensors. Electrochemical systems, such as polyaniline-based pH sensors, provide wireless, battery-free monitoring, while temperature-sensitive dressings utilize thermoresistive materials for flexible, cost-effective sensing. Optical approaches, including colorimetric and fluorometric indicators, use pH-responsive dyes, natural pigments (e.g., anthocyanins, curcumin), and carbon quantum dots (CQDs), allowing for non-invasive digital wound analysis [132]. Also, smart dressings with therapeutic capabilities offer a dynamic approach to wound care by enabling the controlled delivery of therapeutic agents in response to internal or external stimuli. Internally triggered systems, particularly hydrogel-based platforms, respond to pathological changes in the wound environment, such as elevated ROS, pH, or temperature, by releasing bioactive compounds like antibiotics or GFs. Microneedle patches have further expanded this concept, providing responsive and localized drug delivery. On the other hand, externally triggered dressings offer clinicians precise control over treatment timing and location, using electrical, thermal, or optical signals. These include electro-responsive fibers and photothermal systems activated by near-infrared light, which can effectively eliminate infection while preserving tissue integrity. Together, these smart therapeutic dressings represent a significant advancement in personalized wound management, adapting treatment to the evolving conditions of the wound site [133].

4D bioprinting offers a dynamic new direction for future skin substitute strategies. 4D bioprinted systems are engineered to undergo dynamic transformations not only in shape but also in their physicochemical and biological properties in response to specific stimuli [134]. These smart constructs are made possible through the development of responsive bioinks with shape memory, self-healing capacity, and environmental sensitivity. By utilizing CAD-guided printing processes and materials responsive to triggers such as pH, temperature, light, and biomolecules, 4D bioprinting enables real-time adaptability and precise in situ control, which may significantly improve clinical outcomes. In the context of skin repair, ideal bioprinted grafts are expected to have a wide range of functions, from infection resistance and immune regulation to tolerance to wound exudates and mechanical stress, while also being cost-effective, durable, and user-friendly. The evolution of smart bioinks with biocompatibility, structural strength, and responsiveness is key to realizing the full therapeutic potential of 4D bioprinting in skin regeneration [135].

References

1. Ulrich MMW, Vlig M, Boekema BKHL (2016) 9—biomaterials for dermal substitutes. In: Wound Healing Biomaterials, pp 227–252. https://doi.org/10.1016/B978-1-78242-455-0.00009-4
2. Zhang M, Xing J, Zhong Y, Zhang T, Liu X, Xing D (2024) Advanced function, design and application of skin substitutes for skin regeneration. Mater Today Bio 24:100918. https://doi.org/10.1016/j.mtbio.2023.100918
3. Oualla-Bachiri W, Fernández-González A, Quiñones-Vico MI, Arias-Santiago S (2020) From grafts to human bioengineered vascularized skin substitutes. Int J Mol Sci 21:8197. https://doi.org/10.3390/ijms21218197
4. Imahara SD, Klein MB (2009) 5 - Skin grafts. In: Biomaterials for treating skin loss, pp 58–79. https://doi.org/10.1533/9781845695545.1.58

5. Ferreyra PR, Bolgiani A (2024) Fundamentals of skin grafts. In: Plastic and reconstructive surgery fundamentals: a case-based and comprehensive review, pp 47–55. https://doi.org/10.1007/978-3-031-61894-9_5
6. Davis JS (1941) The story of plastic surgery. Ann Surg 113(5):641–656. https://doi.org/10.1097/00000658-194105000-00001
7. Burke JF, Yannas IV, Quinby WC Jr, Bondoc CC, Jung WK (1981) Successful use of a physiologically acceptable artificial skin in the treatment of extensive burn injury. Ann Surg 194. https://doi.org/10.1097/00000658-198110000-00005
8. Hickerson WL, Compton C, Fletchall S, Smith LR (1994) Cultured epidermal autografts and allodermis combination for permanent burn wound coverage. Burns 20:S52–SS6. https://doi.org/10.1016/0305-4179(94)90091-4
9. Mamun AA, Shao C, Geng P, Wang S, Xiao J (2024) Recent advances in molecular mechanisms of skin wound healing and its treatments. Front Immunol 15:2024. https://doi.org/10.3389/fimmu.2024.1395479
10. Song S, Li F, Zhao B, Zhou M, Wang X (2025) Ultraviolet light causes skin cell senescence: from mechanism to prevention principle. Adv Biol 9:2400090. https://doi.org/10.1002/adbi.202400090
11. Bataille A, Le Gall C, Misery L, Talagas M (2022) Merkel cells are multimodal sensory cells: a review of study methods. Cells. 11
12. Brand A, Hovav A-H, Clausen BE (2023) Langerhans cells in the skin and oral mucosa: brothers in arms? Eur J Immunol 53:2149499. https://doi.org/10.1002/eji.202149499
13. Ye H, Rinkevich Y (2023) Fascia layer—a novel target for the application of biomaterials in skin wound healing. Int J Mol Sci 24:2936. https://doi.org/10.3390/ijms24032936
14. Liu M, Lu F, Feng J (2024) Aging and homeostasis of the hypodermis in the age-related deterioration of skin function. Cell Death Dis 15:443. https://doi.org/10.1038/s41419-024-06818-z
15. Linju MC, Rekha MR (2023) Role of inorganic ions in wound healing: an insight into the various approaches for localized delivery. Ther Delivery 14:649–667. https://doi.org/10.4155/tde-2023-0036
16. Wang G, Yang F, Zhou W, Xiao N, Luo M, Tang Z (2023) The initiation of oxidative stress and therapeutic strategies in wound healing. Biomed Pharmacother 157:114004. https://doi.org/10.1016/j.biopha.2022.114004
17. Cañedo-Dorantes L, Cañedo-Ayala M (2019) Skin acute wound healing: A comprehensive review. Int J Inflammation 2019:3706315. https://doi.org/10.1155/2019/3706315
18. Solarte David VA, Güiza-Argüello VR, Arango-Rodríguez ML, Sossa CL, Becerra-Bayona SM (2022) Decellularized tissues for wound healing: towards closing the gap between scaffold design and effective extracellular matrix remodeling. Front Bioeng Biotechnol 10:2022. https://doi.org/10.3389/fbioe.2022.821852
19. Las Heras K, Igartua M, Santos-Vizcaino E, Hernandez RM (2020) Chronic wounds: current status, available strategies and emerging therapeutic solutions. J Control Release 328:532–550. https://doi.org/10.1016/j.jconrel.2020.09.039
20. Wilgus TA, Roy S, McDaniel JC (2013) Neutrophils and wound repair: positive actions and negative reactions. Adv Wound Care 2:379–388. https://doi.org/10.1089/wound.2012.0383
21. Guan F, Wang R, Yi Z, Luo P, Liu W, Xie Y et al (2025) Tissue macrophages: origin, heterogenity, biological functions, diseases and therapeutic targets. Signal Transduct Target Ther 10:93. https://doi.org/10.1038/s41392-025-02124-y
22. Su L, Zheng J, Wang Y, Zhang W, Hu D (2019) Emerging progress on the mechanism and technology in wound repair. Biomed Pharmacother 117:109191. https://doi.org/10.1016/j.biopha.2019.109191
23. Wilkinson HN, Hardman MJ (2020) Wound healing: cellular mechanisms and pathological outcomes. Open Biol 10:200223. https://doi.org/10.1098/rsob.200223
24. Shaw TJ, Martin P (2016) Wound repair: a showcase for cell plasticity and migration. Curr Opin Cell Biol 42:29–37. https://doi.org/10.1016/j.ceb.2016.04.001

25. Rousselle P, Braye F, Dayan G (2019) Re-epithelialization of adult skin wounds: cellular mechanisms and therapeutic strategies. Adv Drug Deliv Rev 146:344–365. https://doi.org/10.1016/j.addr.2018.06.019
26. Raina N, Rani R, Gupta M (2021) Chapter 8—angiogenesis: aspects in wound healing. In: Chatterjee S (ed) Endothelial signaling Vasc Dysfunct dis. Academic Press, pp 77–90
27. Diller RB, Tabor AJ (2022) The Role of the Extracellular Matrix (ECM) in wound healing: a review. Biomimetics. 7
28. Monika P, Waiker PV, Chandraprabha MN, Rangarajan A, Murthy KNC (2021) Myofibroblast progeny in wound biology and wound healing studies. Wound Repair Regen 29:531–547. https://doi.org/10.1111/wrr.12937
29. Vecin NM, Kirsner RS. (2023) Skin substitutes as treatment for chronic wounds: current and future directions. Front Biomed; Volume 10 2023. https://doi.org/10.3389/fmed.2023.1154567, Front Med
30. Şeker Ş, Elçin AE, Elçin YM (2025) Current trends in the design and fabrication of PRP-based scaffolds for tissue engineering and regenerative medicine. Biomed Mater 20:022001. https://doi.org/10.1088/1748-605X/ada83f
31. Lalegül-Ülker Ö, Şeker Ş, Elçin AE, Elçin YM (2024) Encapsulation of MSCs in PRP-derived fibrin microbeads. In: Turksen K (ed) Stem cells and lineage commitment: methods and protocols. Springer US, New York, pp 85–93
32. Sierra-Sánchez Á, Kim KH, Blasco-Morente G, Arias-Santiago S (2021) Cellular human tissue-engineered skin substitutes investigated for deep and difficult to heal injuries. npj Regener Med 6:35. https://doi.org/10.1038/s41536-021-00144-0
33. Koç-Demir A, Elçin AE, Elçin YM (2024) Magnetic biocomposite scaffold based on decellularized tendon ECM and MNP-deposited halloysite nanotubes: physicochemical, thermal, rheological, mechanical and in vitro biological evaluations. Biomed Mater 19:035027. https://doi.org/10.1088/1748-605X/ad38ab
34. Parmaksiz M (2022) Decellularized tendon-based heparinized nanocomposite scaffolds for prospective regenerative applications: chemical, physical, thermal, mechanical and in vitro biological evaluations. J Mech Behav Biomed Mater 134:105387. https://doi.org/10.1016/j.jmbbm.2022.105387
35. Ercan H, Elçin AE, Elçin YM (2022) Assessment of dopamine-conjugated decellularized bovine tendon extracellular matrix as a bioadhesive. Mater Today Commun 31:103634. https://doi.org/10.1016/j.mtcomm.2022.103634
36. Parmaksiz M, Elcin AE, Elcin YM (2017) Decellularization of bovine small intestinal submucosa and its use for the healing of a critical-sized full-thickness skin defect, alone and in combination with stem cells, in a small rodent model. J Tissue Eng Regen Med 11:1754–1765. https://doi.org/10.1002/term.2071
37. Parmaksiz M, Elçin AE, Elçin YM (2020) Decellularized cell culture ECMs act as cell differentiation inducers. Stem Cell Rev Rep 16:569–584. https://doi.org/10.1007/s12015-020-09963-y
38. Xing H, Lee H, Luo L, Kyriakides TR (2020) Extracellular matrix-derived biomaterials in engineering cell function. Biotechnol Adv 42:107421. https://doi.org/10.1016/j.biotechadv.2019.107421
39. Roy A, Mantay M, Brannan C, Griffiths S (2022) Placental tissues as biomaterials in regenerative medicine. Biomed Res Int 2022:6751456. https://doi.org/10.1155/2022/6751456
40. Schlottmann F, Bucan V, Vogt PM, Krezdorn N (2021) A short history of skin grafting in burns: from the gold standard of autologous skin grafting to the possibilities of allogeneic skin grafting with immunomodulatory approaches. Medicina (B Aires) 57
41. Parmaksiz M, Dogan A, Odabas S, Elçin AE, Elçin YM (2016) Clinical applications of decellularized extracellular matrices for tissue engineering and regenerative medicine. Biomed Mater 11:022003. https://doi.org/10.1088/1748-6041/11/2/022003

42. Esmaeili A, Soleimani M, Rouhani M, Noorkhajavi G, Aghaei-Zarch SM, Hasannejad-Asl B et al (2024) Xenograft-based skin substitutes: A critical review. J Drug Delivery Sci Technol 95:105613. https://doi.org/10.1016/j.jddst.2024.105613
43. Sheikholeslam M, Wright MEE, Jeschke MG, Amini-Nik S (2018) Biomaterials for skin substitutes. Adv Healthc Mater 7:1700897. https://doi.org/10.1002/adhm.201700897
44. Chen J, Fan Y, Dong G, Zhou H, Du R, Tang X et al (2023) Designing biomimetic scaffolds for skin tissue engineering. Biomater Sci 11:3051–3076. https://doi.org/10.1039/D3BM00046J
45. Balavigneswaran CK, Selvaraj S, Vasudha TK, Iniyan S, Muthuvijayan V (2023) Tissue engineered skin substitutes: A comprehensive review of basic design, fabrication using 3D printing, recent advances and challenges. Biomater Adv 153:213570. https://doi.org/10.1016/j.bioadv.2023.213570
46. Negut I, Dorcioman G, Grumezescu V (2020) Scaffolds for wound healing applications. Polymers. 12
47. Ventura RD, Padalhin AR, Park CM, Lee BT (2019) Enhanced decellularization technique of porcine dermal ECM for tissue engineering applications. Mater Sci Eng C 104:109841. https://doi.org/10.1016/j.msec.2019.109841
48. Gould LJ (2015) Topical collagen-based biomaterials for chronic wounds: rationale and clinical application. Adv Wound Care 5:19–31. https://doi.org/10.1089/wound.2014.0595
49. Mathew-Steiner SS, Roy S, Sen CK (2021) Collagen in wound healing. Bioengineering 8
50. Naomi R, Bahari H, Ridzuan PM, Othman F (2021) Natural-based biomaterial for skin wound healing (gelatin vs. collagen): Expert review. Polymers. 13
51. Jin T, Fu Z, Zhou L, Chen L, Wang J, Wang L et al (2024) GelMA loaded with platelet lysate promotes skin regeneration and angiogenesis in pressure ulcers by activating STAT3. Sci Rep 14:18345. https://doi.org/10.1038/s41598-024-67304-2
52. Rnjak-Kovacina J, Weiss AS (2013) The role of elastin in wound healing and dermal substitute design. In: Kamolz L-P, Lumenta DB (eds) Dermal replacements in general, burn, and plastic surgery: tissue engineering in clinical practice. Springer, Vienna, pp 57–66
53. Gardeazabal L, Izeta A (2024) Elastin and collagen fibres in cutaneous wound healing. Exp Dermatol 33:e15052. https://doi.org/10.1111/exd.15052
54. Bayer IS (2022) Advances in Fibrin-based materials in wound repair: a review. Molecules. 27
55. Li S, Dan X, Chen H, Li T, Liu B, Ju Y et al (2024) Developing fibrin-based biomaterials/scaffolds in tissue engineering. Bioact Mater 40:597–623. https://doi.org/10.1016/j.bioactmat.2024.08.006
56. Ramey-Ward AN, Walthall HP, Smith S, Barrows TH (2024) Human keratin matrices promote wound healing by modulating skin cell expression of cytokines and growth factors. Wound Rep and Reg 32:257–267. https://doi.org/10.1111/wrr.13137
57. Konop M, Rybka M, Drapała A (2021) Keratin Biomaterials in skin wound healing, an old player in modern medicine: a mini review. Pharmaceutics. 13
58. Zhang H, Lin X, Cao X, Wang Y, Wang J, Zhao Y (2024) Developing natural polymers for skin wound healing. Bioact Mater 33:355–376. https://doi.org/10.1016/j.bioactmat.2023.11.012
59. ter Horst B, Moiemen NS, Grover LM (2019) 6—natural polymers: biomaterials for skin scaffolds. In: García-Gareta E (ed) Biomater Skin Repair Regener. Woodhead Publishing, pp 151–192
60. El-Saadony MT, Saad AM, Alkafaas SS, Dladla M, Ghosh S, Elkafas SS et al (2025) Chitosan, derivatives, and its nanoparticles: preparation, physicochemical properties, biological activities, and biomedical applications—A comprehensive review. Int J Biol Macromol:142832. https://doi.org/10.1016/j.ijbiomac.2025.142832
61. Amruth P, Akshay P, Rosemol Jacob M, Joy JM, Mathew S (2024) Developmental prospects of carrageenan-based wound dressing films: unveiling techno-functional properties and freeze-drying technology for the development of absorbent films—a review. Int J Biol Macromol 276:133668. https://doi.org/10.1016/j.ijbiomac.2024.133668

62. Luanda A, Badalamoole V (2023) Past, present and future of biomedical applications of dextran-based hydrogels: A review. Int J Biol Macromol 228:794–807. https://doi.org/10.1016/j.ijbiomac.2022.12.129
63. Karpagam B, Rajesh J, Rajagopal G (2024) Use of dextran in pharmaceutical and cosmetic compositions. In: Biopolymers in Pharmaceutical and Food Applications, pp 451–469
64. Şeker Ş, Elçin AE, Elçin YM (2020) Macroporous elastic cryogels based on platelet lysate and oxidized dextran as tissue engineering scaffold: in vitro and in vivo evaluations. Mater Sci Eng C 110:110703. https://doi.org/10.1016/j.msec.2020.110703
65. Oduah EI, Linhardt RJ, Sharfstein ST (2016) Heparin: Past, present, and future. Pharmaceuticals. 9
66. Su N, Xu T, Li X, Zheng H, Wu B, Zhang S et al (2022) Heparin and related substances for treating diabetic foot ulcers: A systematic review and meta-analysis. Front Endocrinol 13:2022. https://doi.org/10.3389/fendo.2022.749368
67. AlKhoury H, Hautmann A, Erdmann F, Zhou G, Stojanović S, Najman S, Groth T (2020) Study on the potential mechanism of anti-inflammatory activity of covalently immobilized hyaluronan and heparin. J Biomed Mater Res, Part A 108:1099–1111. https://doi.org/10.1002/jbm.a.36885
68. Zarei N, Hassanzadeh-Tabrizi SA (2023) Alginate/hyaluronic acid-based systems as a new generation of wound dressings: A review. Int J Biol Macromol 253:127249. https://doi.org/10.1016/j.ijbiomac.2023.127249
69. Cao L, Lu W, Mata A, Nishinari K, Fang Y (2020) Egg-box model-based gelation of alginate and pectin: A review. Carbohydr Polym 242:116389. https://doi.org/10.1016/j.carbpol.2020.116389
70. Antoszewska M, Sokolewicz EM, Barańska-Rybak W (2024) Wide use of hyaluronic acid in the process of wound healing a rapid review. Sci Pharm. 92
71. Petit N, Chang Y-yJ, Lobianco FA, Hodgkinson T, Browne S (2025) Hyaluronic acid as a versatile building block for the development of biofunctional hydrogels: in vitro models and preclinical innovations. Mater Today Bio 31:101596. https://doi.org/10.1016/j.mtbio.2025.101596
72. Tang S, Gong Z, Wang Z, Gao X, Zhang X (2022) Multifunctional hydrogels for wound dressings using xanthan gum and polyacrylamide. Int J Biol Macromol 217:944–955. https://doi.org/10.1016/j.ijbiomac.2022.07.181
73. Bueno VB, Petri DFS (2014) Xanthan hydrogel films: molecular conformation, charge density and protein carriers. Carbohydr Polym 101:897–904. https://doi.org/10.1016/j.carbpol.2013.10.039
74. Jeong J-W, Park D-J, Kim S-C, Kang HW, Lee B, Kim H-W et al (2025) Wound healing effect of fucoidan-loaded gelatin/oxidized carboxymethyl cellulose hydrogel. Int J Biol Macromol 286:138254. https://doi.org/10.1016/j.ijbiomac.2024.138254
75. Ghahtan N, Niloofar D, Mujib U, Arezoo K, Hassan H, Iraj N, Baghban N (2025) From seaweed to healing: the potential of fucoidan in wound therapy. Nat Prod Res 39:1345–1358. https://doi.org/10.1080/14786419.2024.2358387
76. Zaki-Germi S, Afshar D, Akbari A, Nikfarjam N (2025) Multifunctional bionanocomposite hydrogels based on gelatin methacrylate and polyphenolic 2D nanoparticles decorated with antimicrobial Bis(imidazolium)-based ionic liquids. Compos Sci Technol 262:111059. https://doi.org/10.1016/j.compscitech.2025.111059
77. Yin X, Fan T, Zheng N, Yang J, Ji T, Yan L et al (2025) Alginate-based hydrogel nanocomposites with CuS nanoparticles incorporation for NIR-II photothermal therapy of melanoma and wound healing. J Appl Polym Sci 142:e56783. https://doi.org/10.1002/app.56783
78. Kenawy E-R, Kamoun EA, Shehata A, El-Moslamy SH, Abdel-Rahman AAH (2025) Biosynthesized ZnO NPs loaded-electrospun PVA/sodium alginate/glycine nanofibers: synthesis, spinning optimization and antimicrobial activity evaluation. Sci Rep 15:1882. https://doi.org/10.1038/s41598-024-81939-1

79. Naghib SM, Hosseini SN, Beigi A (2024) 3D 4D printing of chitosan-based scaffolds for wound dressing applications. Carbohydr Polym Technol Appl 8:100594. https://doi.org/10.1016/j.carpta.2024.100594
80. Şeker Ş, Elçin AE, Elçin YM (2019) Autologous protein-based scaffold composed of platelet lysate and aminated hyaluronic acid. J Mater Sci Mater Med 30:127. https://doi.org/10.1007/s10856-019-6334-7
81. Zamani S, Salehi M, Abbaszadeh-Goudarzi G, Cheraghali D, Ehterami A, Esmaili S, Rezaei kolarijani N (2025) Evaluation effect of alginate hydrogel containing losartan on wound healing and gene expression. J Biomater Appl 39:762–788. https://doi.org/10.1177/08853282241292144
82. Najafloo R, Milan PB, Karimi A, Bagher Z, Kalmer RR, Ghasemian M, Faridi-Majidi R (2024) Crosslinking gelatin with robust inherent antibacterial natural polymer for wound healing. Int J Biol Macromol 280:136144. https://doi.org/10.1016/j.ijbiomac.2024.136144
83. Xu H, Che Y, Zhou R, Wang L, Huang J, Kong W et al (2024) Research progress of natural polysaccharide-based and natural protein-based hydrogels for bacteria-infected wound healing. Chem Eng J 496:153803. https://doi.org/10.1016/j.cej.2024.153803
84. Latiyan S, Kumar TSS, Doble M (2024) Functionally multifaceted alginate/curdlan/agarose-based bilayer fibro-porous dressings for addressing full-thickness diabetic wounds. Biomater Adv 157:213757. https://doi.org/10.1016/j.bioadv.2023.213757
85. Yilmaz-Aykut D, Torkay G, Kasgoz A, Shin SR, Bal-Ozturk A, Deligoz H (2023) Injectable and self-healing dual crosslinked gelatin/kappa-carrageenan methacryloyl hybrid hydrogels via host-guest supramolecular interaction for wound healing. J Biomed Mater Res, Part A 111:1921–1937. https://doi.org/10.1002/jbm.b.35295
86. Gutierrez-Reyes JE, Caldera-Villalobos M, Claudio-Rizo JA, Cabrera-Munguía DA, Becerra-Rodriguez JJ, Soriano-Corral F, Herrera-Guerrero A (2023) Smart collagen/xanthan gum-based hydrogels with antibacterial effect, drug release capacity and excellent performance in vitro bioactivity for wound healing application. Biomed Mater 18:035011. https://doi.org/10.1088/1748-605X/acc99c
87. Wang J, Fu S, Li H, Wu Y (2023) A CS-based composite scaffold with excellent photothermal effect and its application in full-thickness skin wound healing. Regener Biomater 10:rbad028. https://doi.org/10.1093/rb/rbad028
88. Chircov C, Bejenaru IT, Nicoară AI, Bîrcă AC, Oprea OC, Tihăuan B (2022) Chitosan-dextran-glycerol hydrogels loaded with iron oxide nanoparticles for wound dressing applications. Pharmaceutics. 14
89. Reddy MS, Ponnamma D, Choudhary R, Sadasivuni KK (2021) A Comparative review of natural and synthetic biopolymer composite scaffolds. Polymers. 13
90. Kamoun EA, Loutfy SA, Hussein Y, Kenawy E-RS (2021) Recent advances in PVA-polysaccharide based hydrogels and electrospun nanofibers in biomedical applications: A review. Int J Biol Macromol 187:755–768. https://doi.org/10.1016/j.ijbiomac.2021.08.002
91. Jin SG (2022) Production and application of biomaterials based on polyvinyl alcohol (PVA) as wound dressing. Chem Asian J 17:e202200595. https://doi.org/10.1002/asia.202200595
92. Raina N, Pahwa R, Khosla JK, Gupta PN, Gupta M (2022) Polycaprolactone-based materials in wound healing applications. Polym Bull 79:7041–7063. https://doi.org/10.1007/s00289-021-03865-w
93. Khor HL, Ng KW, Schantz JT, Phan T-T, Lim TC, Teoh SH, Hutmacher DW (2002) Poly(ε-caprolactone) films as a potential substrate for tissue engineering an epidermal equivalent. Mater Sci Eng C 20:71–75. https://doi.org/10.1016/S0928-4931(02)00015-2
94. Shen G, Gao B, Guo J, Xu W, Chen G, Huang S et al (2025) Dynamic culturing of large cell-loaded PCL/gelatin methacryloyl scaffolds for bone critical size defect repair. Int J Biol Macromol 298:139906. https://doi.org/10.1016/j.ijbiomac.2025.139906
95. Ehyaeirad N, Babolanimogadam N, Dadkhah M, Rezaie Shirmard L (2024) Polylactic acid films incorporated with nanochitosan, nanocellulose, and Ajwain essential oil: synthesis,

characterizations, with in-vitro and in-vivo antimicrobial properties for infected wound healing. Carbohydr Polym Technol Appl 7:100425. https://doi.org/10.1016/j.carpta.2024.100425
96. Castañeda-Rodríguez S, González-Torres M, Ribas-Aparicio RM, Del Prado-Audelo ML, Leyva-Gómez G, Gürer ES, Sharifi-Rad J (2023) Recent advances in modified poly (lactic acid) as tissue engineering materials. J Biol Eng 17:21. https://doi.org/10.1186/s13036-023-00338-8
97. Kohli N, Sharma V, Brown SJ, García-Gareta E (2019) 5—synthetic polymers for skin biomaterials. In: García-Gareta E (ed) Biomater skin repair regener. Woodhead Publishing, pp 125–149
98. Zhang J, Yang S, Yang X, Xi Z, Zhao L, Cen L et al (2018) Novel fabricating process for porous polyglycolic acid scaffolds by melt-foaming using supercritical carbon dioxide. ACS Biomater Sci Eng 4:694–706. https://doi.org/10.1021/acsbiomaterials.7b00692
99. Xu R, Fang Y, Zhang Z, Cao Y, Yan Y, Gan L et al (2023) Recent advances in biodegradable and biocompatible synthetic polymers used in skin wound healing. Materials 16
100. Ajmal G, Bonde GV, Mittal P, Pandey VK, Yadav N, Mishra B (2023) PLGA/gelatin-based electrospun nanofiber scaffold encapsulating antibacterial and antioxidant molecules for accelerated tissue regeneration. Mater Today Commun 35:105633. https://doi.org/10.1016/j.mtcomm.2023.105633
101. Bento C, Katz M, Santos MMM, Afonso CAM (2024) Striving for uniformity: A review on advances and challenges to achieve uniform polyethylene glycol. Org Process Res Dev 28:860–890. https://doi.org/10.1021/acs.oprd.3c00428
102. Mir M, Ali MN, Barakullah A, Gulzar A, Arshad M, Fatima S, Asad M (2018) Synthetic polymeric biomaterials for wound healing: a review. Prog Biomater 7:1–21. https://doi.org/10.1007/s40204-018-0083-4
103. Liu W, Li S, Wang B, Peng P, Gao C (2022) Physiologically responsive polyurethanes for tissue repair and regeneration. Adv Nano Biomed Res 2:2200061. https://doi.org/10.1002/anbr.202200061
104. Liang W, Ni N, Huang Y, Lin C (2023) An advanced review: polyurethane-related dressings for skin wound repair. Polymers. 15
105. Xu C, Hong Y (2022) Rational design of biodegradable thermoplastic polyurethanes for tissue repair. Bioact Mater 15:250–271. https://doi.org/10.1016/j.bioactmat.2021.11.029
106. Norouzi F, Pourmadadi M, Yazdian F, Khoshmaram K, Mohammadnejad J, Sanati MH et al (2022) PVA-based nanofibers containing chitosan modified with graphene oxide and carbon quantum dot-doped TiO_2 enhance wound healing in a rat model. J Funct Biomater. 13
107. Fahimirad S, Parastu S, Ali G, Abtahi H (2023) Wound healing performance of PVA/PCL based electrospun nanofiber incorporated green synthetized CuNPs and Quercus infectoria extracts. J Biomater Sci Polym Ed 34:277–301. https://doi.org/10.1080/09205063.2022.2116209
108. Shahrousvand M, Mirmasoudi SS, Pourmohammadi-Bejarpasi Z, Feizkhah A, Mobayen M, Hedayati M et al (2023) Polyacrylic acid/polyvinylpyrrolidone hydrogel wound dressing containing zinc oxide nanoparticles promote wound healing in a rat model of excision injury. Heliyon 9:e19230. https://doi.org/10.1016/j.heliyon.2023.e19230
109. Liu X, Wang S, Ding C, Zhao Y, Zhang S, Sun S et al (2024) Polyvinylpyrrolidone/chitosan-loaded dihydromyricetin-based nanofiber membrane promotes diabetic wound healing by anti-inflammatory and regulating autophagy-associated protein expression. Int J Biol Macromol 259:129160. https://doi.org/10.1016/j.ijbiomac.2023.129160
110. Hasan MM, Uddin MF, Zabin N, Shakil MS, Alam M, Achal FJ et al (2021) Fabrication and characterization of chitosan-polyethylene glycol (Ch-peg) based hydrogels and evaluation of their potency in rat skin wound model. Int J Biomater 2021:4877344. https://doi.org/10.1155/2021/4877344
111. Kumar AS, Prema D, Rao RG, Prakash J, Balashanmugam P, Devasena T, Venkatasubbu GD (2024) Fabrication of poly (lactic-co-glycolic acid)/gelatin electro spun nanofiber patch con-

taining CaCO3/SiO2 nanocomposite and quercetin for accelerated diabetic wound healing. Int J Biol Macromol 254:128060. https://doi.org/10.1016/j.ijbiomac.2023.128060
112. Udayakumar SIV, Kwon D, Kwon T-G, Paeng J-Y (2023) Secondary healing property using Neoveil®, a polyglycolic acid bioabsorbable sheet on the oral mucosal defects. J Stomatol Oral Maxillofac Surg 124:101259. https://doi.org/10.1016/j.jormas.2022.08.002
113. Grizzo A, dos Santos DM, da Costa VPV, Lopes RG, Inada NM, Correa DS, Campana-Filho SP (2023) Multifunctional bilayer membranes composed of poly(lactic acid), beta-chitin whiskers and silver nanoparticles for wound dressing applications. Int J Biol Macromol 251:126314. https://doi.org/10.1016/j.ijbiomac.2023.126314
114. Domínguez-Robles J, Cuartas-Gómez E, Dynes S, Utomo E, Anjani QK, Detamornrat U et al (2023) Poly(caprolactone)/lignin-based 3D-printed dressings loaded with a novel combination of bioactive agents for wound-healing applications. Sustain Mater Technol 35:e00581. https://doi.org/10.1016/j.susmat.2023.e00581
115. Li Y, Liu J, Lian C, Yang H, Zhang M, Wang Y, Dai H (2024) Bioactive citrate-based polyurethane tissue adhesive for fast sealing and promoted wound healing. Regener Biomater 11:rbad101. https://doi.org/10.1093/rb/rbad101
116. Zhang Z, Feng Y, Wang L, Liu D, Qin C, Shi Y (2022) A review of preparation methods of porous skin tissue engineering scaffolds. Mater Today Commun 32:104109. https://doi.org/10.1016/j.mtcomm.2022.104109
117. Nicholas MN, Jeschke MG, Amini-Nik S (2016) Methodologies in creating skin substitutes. Cell Mol Life Sci 73:3453–3472. https://doi.org/10.1007/s00018-016-2252-8
118. Li T, Sun M, Wu S (2022) State-of-the-art review of electrospun gelatin-based nanofiber dressings for wound healing applications. Nanomaterials 12
119. Verhulsel M, Vignes M, Descroix S, Malaquin L, Vignjevic DM, Viovy J-L (2014) A review of microfabrication and hydrogel engineering for micro-organs on chips. Biomater 35:1816–1832. https://doi.org/10.1016/j.biomaterials.2013.11.021
120. Ermis M, Antmen E, Hasirci V (2018) Micro and nanofabrication methods to control cell-substrate interactions and cell behavior: a review from the tissue engineering perspective. Bioact Mater 3:355–369. https://doi.org/10.1016/j.bioactmat.2018.05.005
121. Sezer S, Tüzün-Antepli B, Parmaksiz M, Bayramli-Öner B, Elçin AE, Elçin YM (2023) Development of a micro-patterned membrane consisting of a PCL/keratin/PEGDE ternary blend using PSµM for potential biotechnological applications. J Polym Res 30:320. https://doi.org/10.1007/s10965-023-03671-0
122. Tüzün-Antepli B, Şeker Ş, Elçin AE, Khang G, Elçin YM (2022) Evaluation of human osteoblasts on NIPS micro-patterned PCL carriers containing nanohydroxyapatite and reduced graphene oxide using PSµM. Molecules 27:7091. https://doi.org/10.3390/molecules27207091
123. Radmanesh S, Shabangiz S, Koupaei N, Hassanzadeh-Tabrizi SA (2022) 3D printed bio polymeric materials as a new perspective for wound dressing and skin tissue engineering applications: a review. J Polym Res 29:50. https://doi.org/10.1007/s10965-022-02899-6
124. Ngo TD, Kashani A, Imbalzano G, Nguyen KTQ, Hui D (2018) Additive manufacturing (3D printing): A review of materials, methods, applications and challenges. Compos Part B 143:172–196. https://doi.org/10.1016/j.compositesb.2018.02.012
125. Vurat MT, Ergun C, Elçin AE, Elçin YM (2020) 3D bioprinting of tissue models with customized bioinks. In: Chun HJ, Reis RL, Motta A, Khang G (eds) Bioinspired biomaterials: advances in tissue engineering and regenerative medicine. Springer, Singapore, pp 67–84
126. Kang H-W, Lee SJ, Ko IK, Kengla C, Yoo JJ, Atala A (2016) A 3D bioprinting system to produce human-scale tissue constructs with structural integrity. Nat Biotechnol 34:312–319. https://doi.org/10.1038/nbt.3413
127. Vurat MT, Şeker Ş, Lalegül-Ülker Ö, Parmaksiz M, Elçin AE, Elçin YM (2022) Development of a multicellular 3D-bioprinted microtissue model of human periodontal ligament-alveolar bone biointerface: towards a pre-clinical model of periodontal diseases and personalized periodontal tissue engineering. Genes Dis 9:1008–1023. https://doi.org/10.1016/j.gendis.2020.11.011

128. Cai R, Gimenez-Camino N, Xiao M, Bi S, DiVito KA (2023) Technological advances in three-dimensional skin tissue engineering. REVIEWS ON ADVANCED MATERIALS SCIENCE 62. https://doi.org/10.1515/rams-2022-0289
129. Cui X, Li J, Hartanto Y, Durham M, Tang J, Zhang H et al (2020) Advances in extrusion 3d bioprinting: A focus on multicomponent hydrogel-based bioinks. Adv Healthc Mater 9:1901648. https://doi.org/10.1002/adhm.201901648
130. Kuo S-H, Shen C-J, Shen C-F, Cheng C-M (2020) Role of pH value in clinically relevant diagnosis. Diagnostics. 10
131. Youssef K, Ullah A, Rezai P, Hasan A, Amirfazli A (2023) Recent advances in biosensors for real time monitoring of pH, temperature, and oxygen in chronic wounds. Mater Today Bio 22:100764. https://doi.org/10.1016/j.mtbio.2023.100764
132. Brown MS, Ashley B, Koh A (2018) Wearable technology for chronic wound monitoring: current dressings, advancements, and future prospects. Front Bioeng Biotechnol 6. https://doi.org/10.3389/fbioe.2018.00047
133. Horta-Velázquez A, Mota-Morales JD, Morales-Narváez E (2024) Next-generation of smart dressings: integrating multiplexed sensors and theranostic functions. Int J Biol Macromol 254:127737. https://doi.org/10.1016/j.ijbiomac.2023.127737
134. Shinde S, Mane R, Vardikar A, Dhumal A, Rajput A (2023) 4D printing: from emergence to innovation over 3D printing. Eur Polym J 197:112356. https://doi.org/10.1016/j.eurpolymj.2023.112356
135. Damiati LA, Alsudir SA, Mohammed RY, Majrashi MA, Albrahim SH, algethami A et al (2025) 4D printing in skin tissue engineering: a revolutionary approach to enhance wound healing and combat infections. Bioprinting 45:e00386. https://doi.org/10.1016/j.bprint.2025.e00386

Chapter 20
Innovative Wound Dressing

Apirag Chuangsuwanich and Pitawan Rachata

Abstract Traditional wound care strategies, including debridement, moist wound environments, and infection control, remain foundational (Schultz et al. Wound Repair Regen 11(Suppl 1):S1–S28, 2003). The causes of wounds such as arterial or venous in-sufficiency, off- loading or in-proper wound bed preparation should be corrected first. If they may not suffice for challenging wounds adjunctive or innovative modalities could apply. Recent advances in biomaterials, cellular therapy, bioengineering, and biophysical modalities have paved the way for innovative treatments that actively modulate the wound environment to accelerate healing. This chapter explores these novel approaches, evaluates their mechanisms, and examines their clinical relevance.

Keywords Hemostasis · Inflammation · Proliferation · Matuaration

Introduction

Wound healing is a complex physiological process essential to restoring skin integrity following injury. While acute wounds typically heal through a well-orchestrated cascade involving hemostasis, inflammation, proliferation, and remodeling (Fig. 20.1), chronic wounds often deviate from this trajectory due to underlying pathologies, such as diabetes, vascular disease, and infection. These nonhealing wounds represent a significant burden on healthcare systems worldwide and negatively impact patient quality of life [1].

A. Chuangsuwanich (✉)
Division of Plastic Surgery, Department of Surgery, Faculty of Medicine Siriraj Hospital
Mahidol University, Bangkok, Thailand

Mae Fah Luang Medical Center Hospital, Mae Fah Luang University, Chiang Rai, Thailand

P. Rachata
Mae Fah Luang Medical Center Hospital, Mae Fah Luang University, Chiang Rai, Thailand

F. De Francesco, K. Turksen (eds.), *Management and Strategies for Wound Healing*,
Stem Cell Biology and Regenerative Medicine 79,
https://doi.org/10.1007/978-3-032-12442-5_20

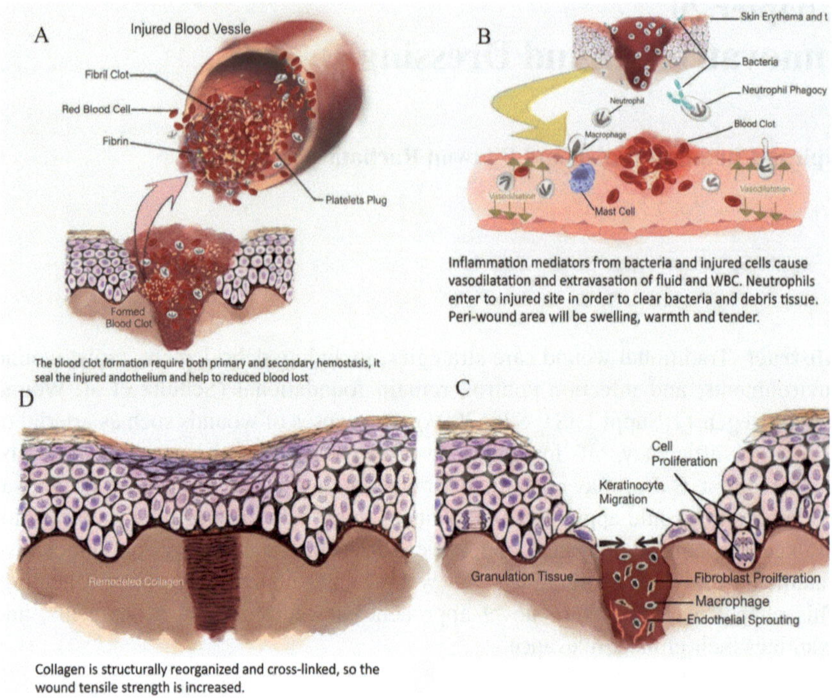

Fig. 20.1 Described four stages of wound healing (**a**) Hemostasis, (**b**) Inflammation, (**c**) Proliferation, (**d**) Matuaration

Traditional wound care strategies, including debridement, moist wound environments, and infection control, remain foundational [2]. The causes of wounds, such as arterial or venous in-sufficiency, off-loading, or in-proper wound bed preparation, should be corrected first. If they may not suffice for challenging wounds, adjunctive or innovative modalities could apply. Recent advances in biomaterials, cellular therapy, bioengineering, and biophysical modalities have paved the way for innovative treatments that actively modulate the wound environment to accelerate healing. This chapter explores these novel approaches, evaluates their mechanisms, and examines their clinical relevance.

Cell-Based Therapies

Stem Cell Therapy

Stem cells, particularly mesenchymal stem cells (MSCs) [3, 4], have shown immense promise in regenerative medicine. MSCs represent a cornerstone in regenerative medicine due to their multipotent differentiation ability and their potent

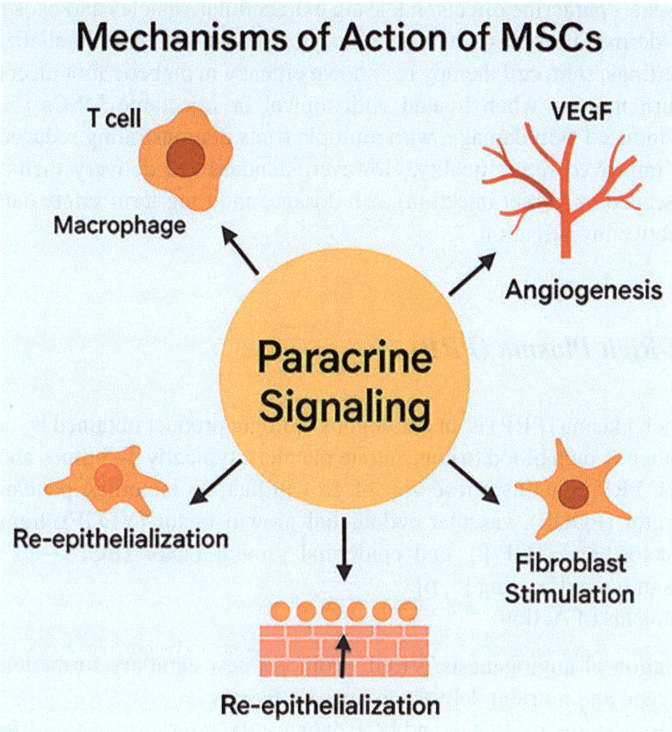

Fig. 20.2 Mesenchymal stem cell paracrine function

paracrine signaling functions (Fig. 20.2). They can be harvested from several sources, including bone mesenchymal marrow, adipose tissue, umbilical cord blood, and dental pulp, with each source offering unique advantages. For example, adipose-derived MSCs are more abundant and can be harvested with lower patient morbidity via liposuction compared to bone marrow extraction.

Mechanistically, MSCs facilitate wound repair through:

- Immunomodulation: suppressing excessive inflammatory responses by interacting with T cells and macrophages.
- Angiogenesis induction: secretion of pro-angiogenic factors such as VEGF, HGF, and ANG-1.
- Fibroblast stimulation: promoting dermal matrix remodeling and granulation tissue formation.
- Re-epithelialization: enhancing keratinocyte migration and proliferation.

Moreover, recent preclinical studies highlight the role of extracellular vesicles (EVs) and exosomes derived from MSCs as bioactive carriers, offering cell-free alternatives with similar regenerative capabilities but fewer regulatory and immunogenic hurdles.

MSCs exert paracrine effects, releasing extracellular vesicles and exosomes that stimulate dermal fibroblasts and keratinocytes, enhancing re-epithelialization. In clinical settings, stem cell therapy has shown efficacy in diabetic foot ulcers (DFUs) [5, 6], burn injuries when treated with topical or injectable MSCs [5, 6], and radiation-induced skin damage, with multiple trials demonstrating reduced healing time and improved tissue quality. However, standardized delivery methods (e.g., scaffold-seeded vs. direct injection), cell dosage, and long-term safety data remain areas of active investigation.

Platelet-Rich Plasma (PRP)

Platelet-rich plasma (PRP) is an autologous biologic product obtained by centrifuging the patient's own blood to concentrate platelets, typically 3–5 times above baseline levels. PRP contains a reservoir of growth factors, including platelet-derived growth factor (PDGF), vascular endothelial growth factor (VEGF), transforming growth factor-beta (TGF-β), and epidermal growth factor (EGF)—all essential mediators in wound healing [5, 6].

Mechanisms of Action:

- Stimulation of angiogenesis: VEGF promotes new capillary formation, critical for oxygen and nutrient delivery to hypoxic tissues.
- Cellular recruitment: PDGF and EGF enhance the migration and proliferation of fibroblasts and keratinocytes.
- Matrix remodeling: TGF-β modulates extracellular matrix (ECM) synthesis and collagen deposition.
- Antimicrobial effects: Platelets release proteins like thrombocidins that have intrinsic bactericidal activity, supporting infection control.

Clinical Applications:

PRP has been applied topically or injected into chronic wounds, including:

- Diabetic foot ulcers (DFUs): multiple RCTs demonstrate significantly improved wound closure rates and reduced recurrence.
- Pressure ulcers: shown to reduce healing time when combined with standard wound care.
- Surgical wound dehiscence: particularly useful in orthopedic and spinal surgeries.

Moreover, PRP preparation is relatively cost-effective and minimally invasive, making it appealing for outpatient and resource-limited settings (Figs. 20.3 and 20.4). However, variability in preparation techniques (e.g., leukocyte-rich vs. leukocyte-poor PRP), dosing frequency, and activation protocols (e.g., calcium chloride vs. thrombin) leads to inconsistent outcomes. Therefore, efforts are underway to standardize PRP protocols for wound care applications [7, 8].

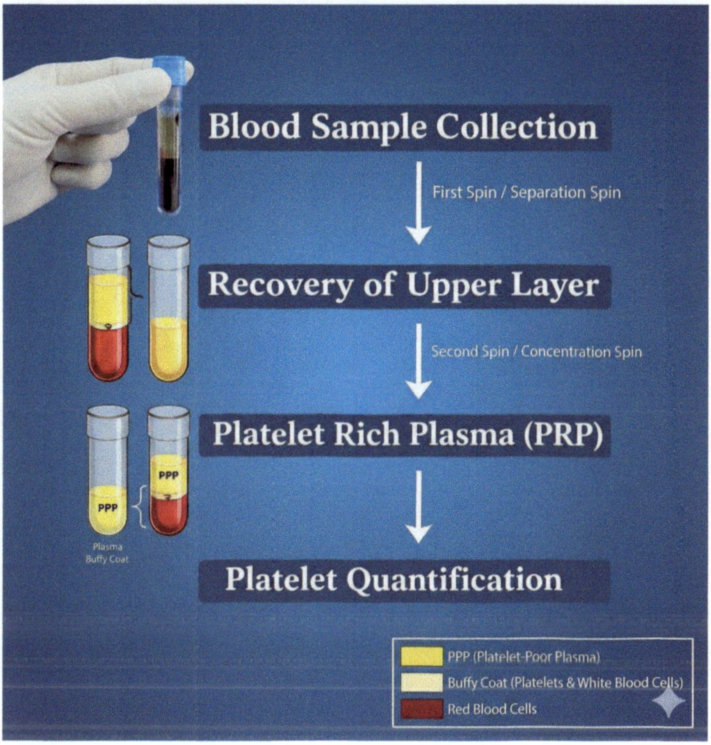

Fig. 20.3 Platelet rich plasma preparation

Fibroblast and Keratinocyte Therapy

This therapy involves the application of cultured skin cells, either autologous or allogenic, to stimulate the wound healing process. Fibroblasts, found in the dermis, are responsible for collagen production and extracellular matrix deposition, while keratinocytes are essential for epithelialization and barrier formation. Commercially available products such as Apligraf® and Dermagraft® [9, 10].

Mechanism of Action:

- Growth factor secretion: Both cell types release cytokines such as IL-6, FGF, and GM-CSF that enhance granulation tissue formation and epithelial coverage.
- Bio-scaffold interaction: These cells are often seeded onto biocompatible scaffolds or matrixes that integrate into the wound bed, supporting tissue regeneration.
- Inflammation modulation: Reduce pro-inflammatory cytokine levels and promote the resolution phase of healing.

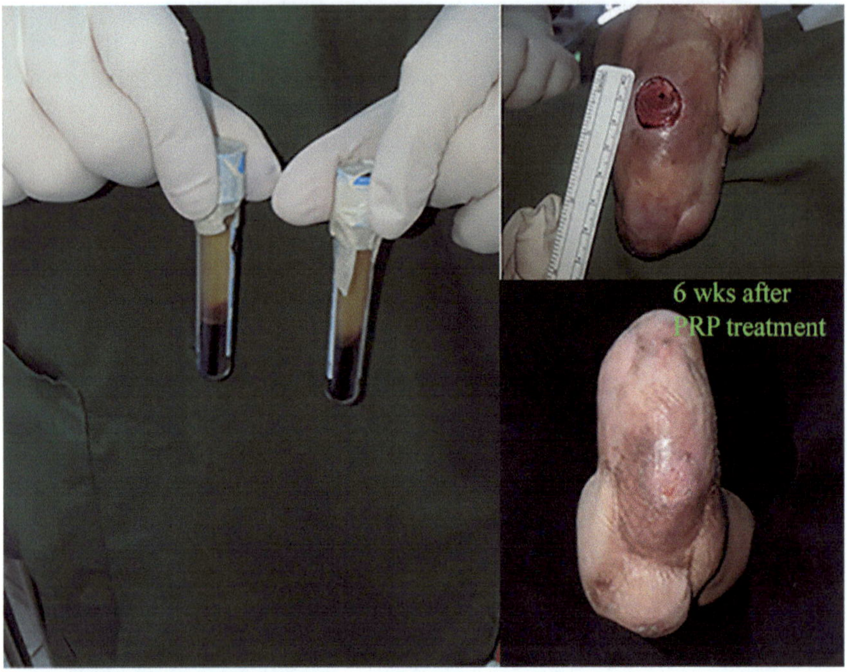

Fig. 20.4 Diabetic wounds treated with PRP

Commercial Applications:

- Apligraf®: A bilayered living skin equivalent combining neonatal fibroblasts in a bovine collagen matrix (dermal layer) with keratinocytes (epidermal layer). FDA-approved for venous leg ulcers and diabetic foot ulcers.
- Dermagraft®: A cryopreserved human fibroblast-derived dermal substitute placed into full-thickness chronic wounds (Fig. 20.5).

While effective, these products are high-cost and often limited to use in specialty centers. Additionally, immune rejection is minimal due to the absence of Langerhans cells and vascular components, though repeated applications may raise immunogenic concerns.

The comparison of cell-based therapy is demonstrated in Table 20.1.

Advanced Wound Dressings

Modern wound dressings have evolved beyond passive coverings into bioactive and responsive platforms that interact with the wound microenvironment. They aim to provide moisture balance, infection control, and biochemical stimulation to accelerate healing while minimizing trauma during dressing changes.

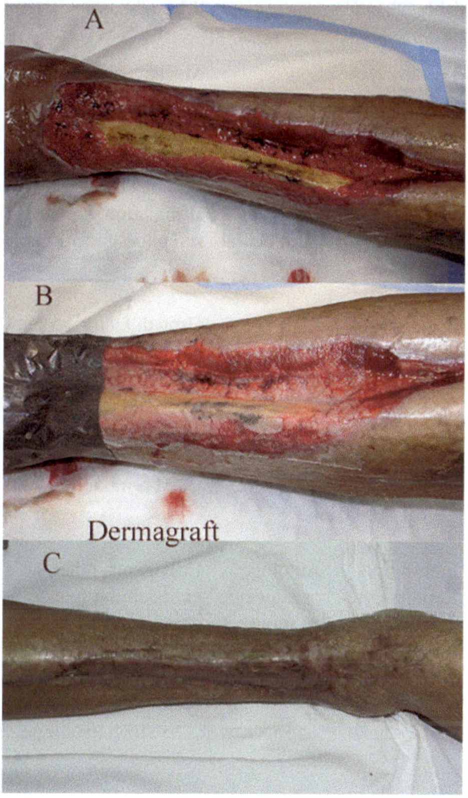

Fig. 20.5 Dermagraft on open wound

Bioactive Dressings

Modern wound dressings are not merely passive coverings. Bioactive dressings incorporate antimicrobial agents (e.g., silver, iodine), growth factors, or natural compounds (e.g., honey, alginate). They create a moist wound environment while simultaneously reducing microbial load and promoting tissue regeneration. New innovative hydrogel-based dressings incorporate with many particles such as exosome, drugs, or nanomaterials which can promote wound healing by antibacterial, promote tissue growth, or hemostasis [11, 12].

Key Types and Mechanisms:

- Antimicrobial dressings: Impregnated with silver, iodine, PHMB (polyhexamethylene biguanide), or honey—agents that target microbial biofilms and reduce wound bioburden without systemic antibiotic use.
- Growth factor-loaded dressings: Deliver molecules such as EGF or PDGF to directly stimulate cell proliferation and angiogenesis.

Table 20.1 Comparison of cell-based therapies

Therapy type	Source	Mechanism of action	Clinical applications	Advantages	Limitations
Mesenchymal Stem Cells (MSCs)	Bone marrow, adipose tissue, umbilical cord, dental pulp	Paracrine signaling, angiogenesis, immunomodulation, fibroblast stimulation, re-epithelialization	Diabetic foot ulcers, burns, radiation-induced wounds	Multipotent differentiation, strong regenerative capacity, paracrine effects	Regulatory hurdles, variable delivery methods, cost, need for cell culture
Platelet-Rich Plasma (PRP)	Autologous blood	Growth factor delivery (PDGF, VEGF, TGF-β, EGF) to enhance angiogenesis, proliferation, and ECM remodeling	Diabetic foot ulcers, pressure ulcers, surgical wound dehiscence	Cost-effective, minimally invasive, autologous (low risk of rejection)	Variability in preparation protocols, inconsistent outcomes, multiple sessions needed
Fibroblast/ Keratinocyte Therapy	Autologous or allogenic cultured skin cells	Collagen synthesis, extracellular matrix deposition, epithelialization, and growth factor secretion	Venous leg ulcers, diabetic foot ulcers, full-thickness chronic wounds	Directly provides essential skin cells for wound coverage and matrix formation	High cost, limited availability, potential for immune reaction in repeated use

- Natural compound-infused: Include alginate (from seaweed), chitosan (from crustaceans), and honey-based dressings that promote autolytic debridement, hemostasis, and tissue regeneration.

Hydrogel-Based Innovations:
Emerging hydrogel platforms can encapsulate exosomes, nanomaterials, or antibiotics, offering:

- Prolonged release profiles
- Responsive drug delivery (e.g., temperature-sensitive gels)
- Moisture retention to prevent desiccation and eschar formation

These dressings are particularly beneficial for diabetic foot ulcers, burns, and postoperative wounds. Studies have shown improved re-epithelialization rates, reduced pain scores, and shorter healing durations with bioactive dressings compared to standard care.

Smart Dressings

Smart dressings are engineered to respond to environmental stimuli such as pH, temperature, or moisture. For example, a pH-sensitive dressing can indicate infection onset, while thermo-responsive hydrogels release drugs in response to inflammation-induced heat [13, 14]. These dressings offer real-time monitoring and controlled drug delivery.

Functional Capabilities: (Fig. 20.6).

- pH-responsive polymers: Detect changes in wound pH—an early marker of infection—and may trigger a visual color change or release antimicrobial agents accordingly.
- Temperature-sensitive hydrogels: Release therapeutic agents when local inflammation elevates tissue temperature.
- Moisture-sensing matrices: Adjust water vapor permeability or signal the need for dressing change to avoid maceration.
- Sensor-integrated examples: Dual-sensor dressings that monitor both temperature and moisture have been shown to reduce dressing changes and improve clinician response times.
- Some prototypes include Bluetooth-enabled sensors that transmit wound data to mobile devices, allowing remote monitoring and care adjustments.

These systems are particularly useful in chronic, high-risk wounds (e.g., pressure injuries in immobile patients), where infection surveillance and moisture control are critical.

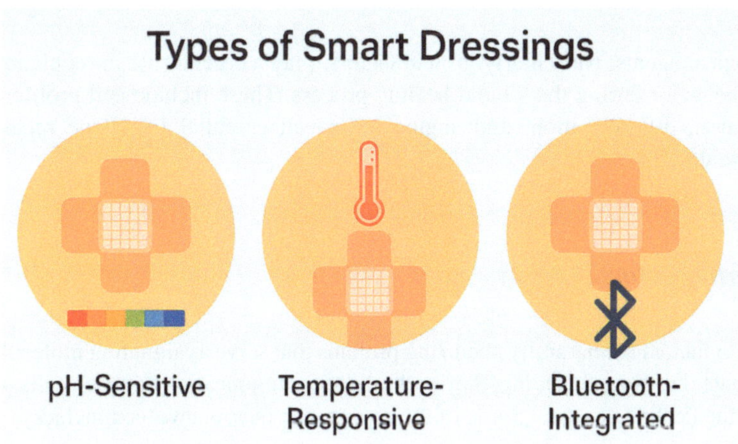

Fig. 20.6 Examples of sensor sensitive dressing

Nanofiber Materials

Extracellular matrix is a component for cell growth and a platform for new tissue laydown. Electrospun nanofibers mimic the extracellular matrix, providing a scaffold for cell migration and proliferation. They can be loaded with bioactive molecules or antibiotics [8, 15]. Nanotechnology also allows for increased surface area and better control over drug release kinetics, making them ideal for chronic wounds.

Properties and advantages:

- High surface area-to-volume ratio: allows dense loading of drugs, growth factors, or peptides.
- Porous microstructure: facilitates oxygen diffusion and exudate absorption.
- Biodegradability: some nanofibers degrade over time, eliminating the need for dressing removal.

Functional additives:

- Antimicrobials (e.g., ciprofloxacin, silver nanoparticles) for infection control
- Anti-inflammatory agents (e.g., curcumin or NSAIDs) for modulating local immune response
- Angiogenic peptides or siRNAs for stimulating vascular regeneration

Recent studies show nanofiber dressings accelerate healing in radiation-induced wounds, venous ulcers, and post-traumatic injuries, especially when integrated with controlled-release nanotechnology or gene therapy agents.

Biological Agents and Growth Factors

Biological agents, particularly growth factors, play a crucial role in regulating cellular behavior during the wound healing process. These include cell proliferation, migration, differentiation, and angiogenesis—all essential for tissue repair and regeneration.

Growth Factor Therapy

Growth factors are naturally occurring proteins that serve as signaling molecules. In the context of wound healing, they orchestrate communication between cells to promote the healing cascade. Some of the key growth factors involved include:

- PDGF (Platelet-Derived Growth Factor): promotes the recruitment and proliferation of fibroblasts and smooth muscle cells. It also stimulates extracellular matrix (ECM) production and angiogenesis.

- VEGF (Vascular Endothelial Growth Factor): a key regulator of angiogenesis. It stimulates the formation of new blood vessels from existing vasculature, essential for oxygen and nutrient delivery to healing tissues.
- EGF (Epidermal Growth Factor): enhances keratinocyte proliferation and migration, crucial for re-epithelialization of the wound surface [16, 17].

Clinical Application:

- One example of growth factor therapy in clinical use is Becaplermin gel®, a recombinant human PDGF (rhPDGF) formulation. It is FDA-approved for the treatment of diabetic foot ulcers, which are notoriously slow to heal due to poor vascularization and chronic inflammation. Becaplermin gel® [18], a recombinant PDGF gel, is approved for diabetic foot ulcers.

Challenges of Growth Factor Therapy:
Despite promising results, growth factor therapy faces several limitations:

- Short Half-Life: Many growth factors are rapidly degraded by enzymes present in the wound environment, especially in chronic wounds where proteolytic activity is elevated.
- Instability: These proteins are sensitive to temperature and pH, which can compromise their activity.
- Cost and Delivery Efficiency: Producing recombinant proteins is expensive, and delivering them effectively to the target site without systemic side effects is challenging.

Emerging Solutions:
To overcome these barriers, researchers are developing sustained-release delivery systems that protect the growth factors and release them gradually over time. These include:

- Liposomes: Tiny vesicles that encapsulate the growth factor and release it in a controlled manner.
- Hydrogels: Polymer-based structures that can hold and slowly release therapeutic agents while maintaining a moist environment for healing.
- Microspheres or nanoparticles: Biodegradable particles that can be engineered for specific release kinetics and targeted delivery. [19, 20].

Gene-Activated Matrices

Mechanism:

- Instead of applying the growth factors directly, plasmid DNA encoding these factors is incorporated into a scaffold (such as a collagen or synthetic polymer matrix).

- When placed in the wound, host cells take up the DNA and begin producing the growth factors in situ.
- This allows for sustained, localized production, minimizing systemic exposure and potential side effects.

Advantages:

- Prolonged therapeutic effect: The gene remains active in the local cells for an extended period, providing continuous stimulation for healing.
- Targeted action: Therapy is confined to the wound site, reducing risk of off-target effects.
- Reduced protein instability: Since the proteins are produced directly by cells at the wound site, issues with storage and degradation are reduced.

Applications and Research:

- Early-stage studies have shown promising results, particularly in ischemic wounds where blood supply is compromised.
- Benefits include increased angiogenesis (formation of new blood vessels) and collagen synthesis, both critical for effective wound closure and tissue strength. Preliminary studies show enhanced angiogenesis and collagen synthesis in ischemic wounds [21, 22].

Biophysical Modalities

Biophysical modalities leverage physical energy forms—such as pressure, electrical current, light, plasma, or oxygen—to influence the wound microenvironment. These interventions are particularly valuable for chronic, recalcitrant, or complex wounds where standard care fails. They are often used in combination with biological or topical therapies to enhance outcomes.

Negative Pressure Wound Therapy (NPWT)

NPWT, also known as vacuum-assisted closure (VAC), is a widely adopted technique in both inpatient and outpatient settings. It involves applying subatmospheric pressure to the wound through a sealed dressing connected to a vacuum pump. This technology had been developed more than 20 years but the application still expands its indication.

Mechanisms of action: [23–25]

- Mechanical microdeformation of the wound bed promotes cellular proliferation and granulation tissue formation.
- Fluid evacuation reduces edema and interstitial pressure, improving perfusion.

- Infection control by removing exudate and reducing bacterial load.
- Edge approximation accelerates contraction of large open wounds.

Clinical Applications:
NPWT has been validated in a variety of settings:

- Diabetic foot ulcers—significantly improved healing rates and reduced amputation risk.
- Surgical incisions—NPWT is widely used for complex wounds and post-surgical sites. It is proved that it reduced the complications in high-risk wounds [26, 27].
- Traumatic wounds, burns, pressure injuries, and skin graft fixation—supports tissue integration and reduces seroma formation (Fig. 20.7).

Innovations include:

- Instillation NPWT (NPWTi-d): combines negative pressure with intermittent delivery of antiseptic or antimicrobial fluids.
- Portable NPWT devices: enable home use, improving compliance and reducing hospitalization.

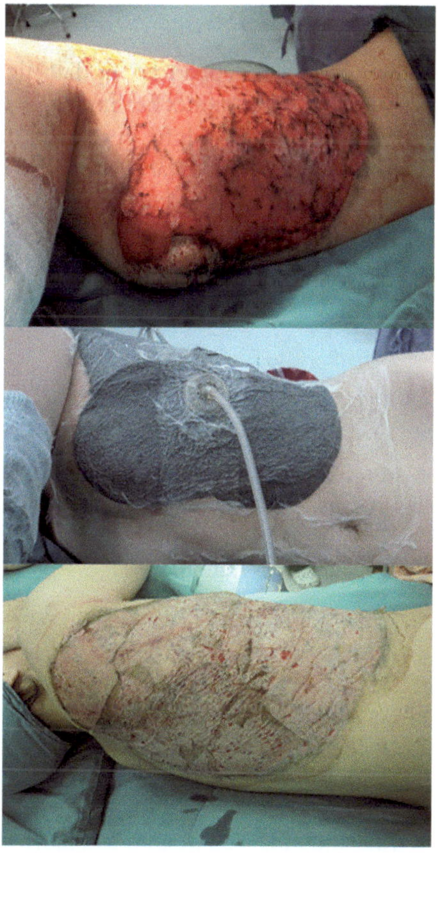

Fig. 20.7 Negative pressure wound therapy on open wound with skin graft

Electrical Stimulation (ES)

Electrical stimulation (ES) mimics the endogenous electrical fields that arise naturally during wound healing. These fields play a crucial role in guiding cell migration, modulating inflammation, and enhancing protein synthesis.
Mechanisms:
Electrical fields influence cell migration, protein synthesis, and angiogenesis [28, 29].

- Galvanotaxis: directs fibroblasts, keratinocytes, and macrophages toward the wound center.
- Angiogenesis induction via increased VEGF expression.
- Enhanced collagen synthesis and ECM remodeling.

Modalities used:

- Direct current (DC)
- Pulsed current (PC)
- High-voltage pulsed current (HVPC)

Evidence and Use:

- Meta-analyses support ES in diabetic ulcers, pressure ulcers, and venous leg ulcers, showing shorter healing times and increased closure rates [28–30].
- ES is particularly helpful in patients with spinal cord injury or limited mobility, where intrinsic electrical cues may be impaired.

Challenges include the need for specialized equipment, trained personnel, and standardized treatment protocols. However, the modality is non-invasive, low-risk, and compatible with most wound types.

Low-Level Laser Therapy (LLLT)

Also referred to as photobiomodulation, LLLT uses specific wavelengths of non-thermal, low-intensity light (typically in the red or near-infrared spectrum) to stimulate biological processes.

Mechanism of Action:

- Photons are absorbed by cytochrome c oxidase in mitochondria, enhancing ATP production.
- Increases cellular metabolism, fibroblast proliferation, and collagen synthesis.
- Anti-inflammatory effects reduce oxidative stress and support transition to the proliferative phase [31].

Clinical uses:

- Radiation-induced wounds and mucositis in oncology

- Neuropathic diabetic ulcers
- Burns and donor site wounds [32, 33].

Evidence shows significant reductions in pain, inflammation, and healing time. Treatment is painless and non-invasive, though effectiveness depends on wavelength, dosage, and application frequency.

Low-Temperature Atmospheric Pressure Plasma (LTAPP)

Plasma is the fourth state of matter, composed of ionized gases containing electrons, ions, reactive oxygen species, reactive nitrogen species, UV photons, and electromagnetic fields (Fig. 20.8). LTAPP is generated at room temperature and pressure, making it suitable for biological applications (Fig. 20.9).

Mechanisms:

- Antimicrobial effects: disrupts bacterial membranes, biofilms, and multidrug-resistant organisms.
- Angiogenesis promotion: stimulates VEGF and endothelial cell activity.
- Fibroblast activation and keratinocyte proliferation.
- Biofilm eradication: especially relevant in chronic wounds where biofilms impair healing [34, 35]. Ultraviolet light and other free particles are its components. LTAPP can reduce biofilm, stimulate angiogenesis, and kill pathogens, both gram positive and negative bacteria

Clinical Evidence:

- RCTs in pressure ulcers, infected wounds, and burns have shown improved healing rates and infection resolution (Fig. 20.10).
- Studies from Thailand and Europe have demonstrated biofilm clearance, pain reduction, and reduced need for antibiotics [36–38].

LTAPP is an emerging, non-contact modality that shows promise in resistant or colonized wounds, though access and cost remain limitations in many settings.

Topical Oxygen Therapy

Oxygen plays a central role in wound healing by supporting:

- ATP production
- Collagen synthesis
- Bacterial clearance
- Angiogenesis

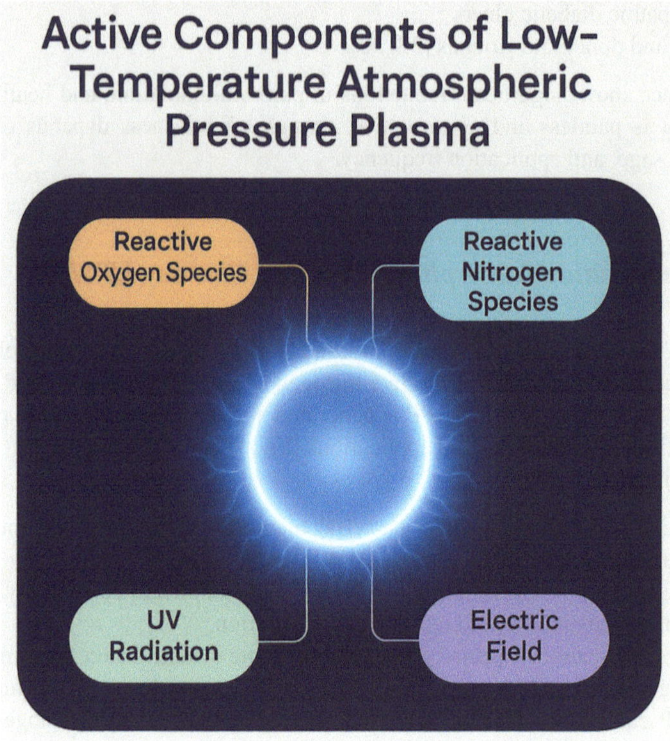

Fig. 20.8 Active components of low temperature atmospheric pressure plasma

While systemic oxygen (e.g., HBOT) has known applications, topical oxygen therapy delivers oxygen directly to the wound bed, bypassing vascular deficiencies [39, 40].

Delivery methods:

- Continuous diffusion oxygen therapy: low-flow oxygen through a tube into an occlusive dressing.
- Pressurized chambers or oxygen-permeable membranes.

Evidence and Applications:

- Effective in diabetic foot ulcers, particularly where peripheral arterial disease limits systemic oxygen delivery.
- Studies demonstrate increased granulation, epithelialization, and bacterial clearance.

TOT is noninvasive, portable, and generally well-tolerated. However, standardization, device variability, and reimbursement issues limit its broader implementation.

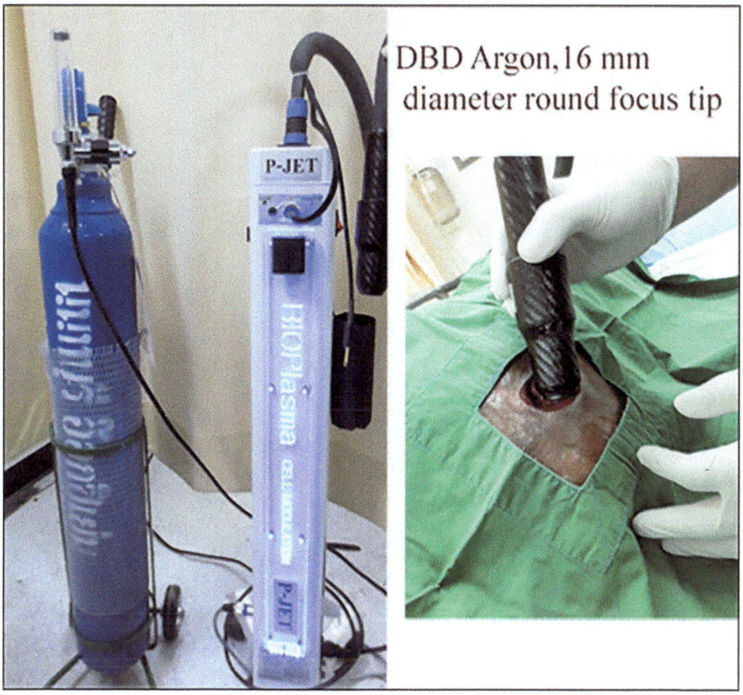

Fig. 20.9 Dielectric barrier discharge argon plasma device

3D Bioprinting and Tissue Engineering

3D bioprinting and tissue engineering represent cutting-edge technologies in regenerative medicine, offering customized solutions for skin regeneration and complex wound reconstruction. These approaches aim to create functional tissue substitutes that can replicate both the structure and function of native skin, enhancing repair in cases where traditional healing is impaired or impossible.

Principles of 3D Bioprinting in Wound Healing

3D bioprinting involves layer-by-layer deposition of bioinks—combinations of cells, biomaterials, and signaling molecules—to fabricate precise three-dimensional constructs that mimic the skin's architecture.

Components of Bioprinted Skin:

- Dermal layer: Typically constructed using fibroblasts embedded in natural polymers like collagen, gelatin, or hyaluronic acid.

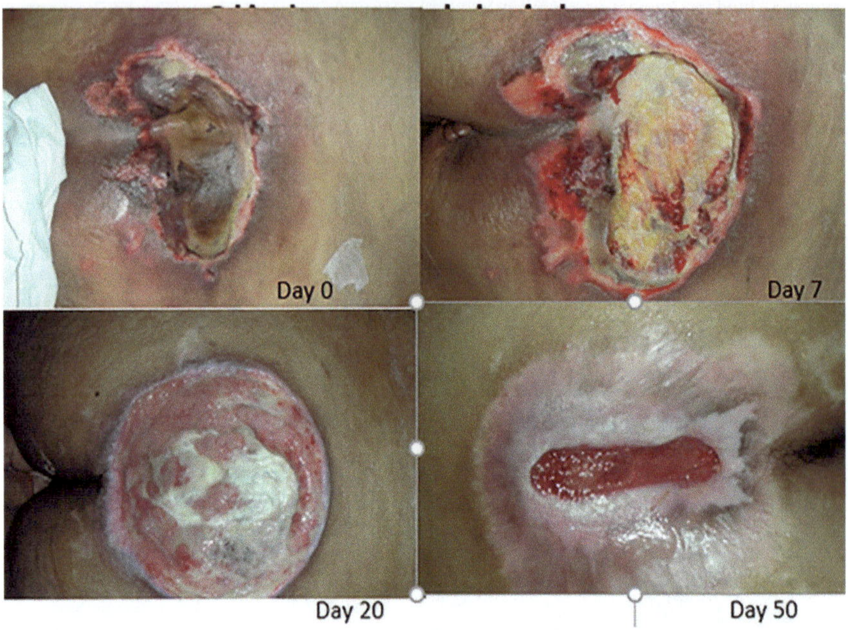

Fig. 20.10 A sacral pressure injury ulcer treated with LTAPP

- Epidermal layer: Composed of keratinocytes to promote re-epithelialization and barrier formation.
- Vascularization strategies: Include endothelial cell inclusion or angiogenic growth factor release (e.g., VEGF) to ensure blood supply and graft survival.

Bioprinters can be programmed to deposit cells in spatial patterns that replicate native skin anatomy — including sweat glands, vasculature, and hair follicle templates — although full functionality is still under development.

Clinical Applications of Bioprinted Skin Equivalents

Several commercially available skin substitutes based on tissue engineering principles have demonstrated success in clinical practice, even if not yet fully bioprinted:

- Integra® Dermal Regeneration Template: A bilayer matrix of bovine collagen and chondroitin-6-sulfate with a silicone outer layer. Commonly used in burns, trauma, and oncologic resections to promote neodermis formation.
- Matriderm®: A single-layer dermal substitute composed of a collagen-elastin matrix, supporting graft integration, angiogenesis, and fibroblast migration. Particularly beneficial for deep burns and full-thickness wounds, including pediatric facial injuries.

- Apligraf® and Dermagraft®: Skin equivalents combining allogenic keratinocytes and fibroblasts in 3D matrices; useful for diabetic foot ulcers and venous leg ulcers [41, 42].

These engineered products support:

- Rapid epithelial coverage
- Reduced scarring and contraction
- Enhanced cosmetic and functional recovery

Personalized Bioprinting: The Future of Skin Grafting

Recent advances now allow for customized tissue constructs using the patient's own cells (e.g., autologous fibroblasts or stem cells), significantly reducing the risk of immunologic rejection.

Innovations Include:

- In situ bioprinting: Portable bioprinters capable of directly depositing cell-laden bioinks onto the wound bed—already tested in military trauma and burn units.
- Multicellular constructs: Incorporating immune cells, vascular elements, or sweat gland progenitors for functional restoration.
- Integration with imaging tools: High-resolution scanning to map wound topography for precise bioink deposition.

Current Challenges:

- Ensuring vascular integration to maintain viability of large constructs
- Regulatory pathways for patient-specific bioprinted grafts
- High production costs, sterility assurance, and manufacturing scalability

Nevertheless, 3D bioprinting holds great promise for burn victims, oncologic resection wounds, chronic ulcers, and even cosmetic reconstruction in the near future.

Regenerative Medicine Approaches

Regenerative medicine focuses on restoring tissue function using the body's own healing mechanisms, either through direct cell replacement or by modulating the microenvironment. Recently, cell-free approaches, particularly those involving extracellular vesicles (EVs) such as exosomes, have gained attention due to their lower immunogenicity, ease of storage, and reduced regulatory complexity compared to live cell therapies.

Extracellular Vesicles and Exosomes

Exosomes are nanosized vesicles (30–150 nm) secreted by various cell types, especially mesenchymal stem cells (MSCs). They carry a rich cargo of proteins, lipids, mRNAs, and microRNAs that reflect the therapeutic potential of their parent cells.

Mechanisms of Action:

They offer a cell-free alternative to stem cell therapy with lower risk profiles [43, 44].

- Immunomodulation: Suppress excessive inflammation by altering macrophage phenotypes from M1 (pro-inflammatory) to M2 (pro-healing).
- Angiogenesis: Deliver pro-angiogenic miRNAs (e.g., miR-126, miR-210) and VEGF, stimulating new vessel formation.
- Fibroblast activation and migration: Enhance ECM synthesis and granulation tissue development.
- Keratinocyte proliferation: Promote re-epithelialization and barrier restoration.

Delivery Strategies:

- Hydrogel scaffolds and collagen matrices embedded with exosomes for sustained, localized release
- Spray formulations or topical gels applied directly to the wound
- Injectable exosome solutions for deep wound beds or tunneling ulcers

Clinical Progress:

- Preclinical studies in diabetic wounds, burns, and ischemic injuries show accelerated closure, reduced inflammation, and enhanced tissue remodeling.
- Human trials are ongoing, with early data suggesting improved healing without tumorigenic risk or immune rejection.

Exosomes are emerging as “cell-free stem cell therapy”—offering many of the regenerative benefits of MSCs, but with greater scalability, safety, and storage potential.

Cell-Free Therapies

In addition to exosomes, researchers are developing synthetic molecules and biologically inspired agents that replicate the paracrine effects of stem cells without using living cells.

Examples of Cell-Free Agents:

- Synthetic peptides: Mimic domains of growth factors or ECM proteins (e.g., laminin-derived peptides to enhance keratinocyte attachment).
- Recombinant growth factor mimetics: Engineered analogs that resist proteolysis and have longer half-lives in the wound environment.

- Small molecules: Compounds that activate signaling pathways such as Wnt/β--catenin, PI3K/Akt, or HIF-1α involved in cell migration, angiogenesis, and proliferation.

Advantages of Cell-Free Therapies:

- Easier manufacturing and quality control than cell-based products
- Longer shelf life and room-temperature stability
- Lower immunogenicity and no risk of live cell transformation or tumor formation
- Potential for mass production and cost-effective scale-up

Although most cell-free therapies remain in the preclinical or early clinical stages, they hold promise for standardized, off-the-shelf wound care products, especially in settings where cold chain logistics or stem cell isolation is impractical [45, 46].

Antimicrobial and Anti-biofilm Strategies

Infection is a leading cause of chronic wound persistence and delayed healing, particularly in diabetic ulcers, burns, and pressure injuries. Compounding this problem, microbial biofilms—structured communities of bacteria embedded in a self-produced matrix—are highly resistant to antibiotics and host defenses. Emerging antimicrobial strategies now go beyond conventional antibiotics, targeting resistant pathogens and biofilms with innovative and often biological interventions.

Antimicrobial Peptides (AMPs)

Antibiotic resistance is a global health problem. AMPs have emerged as a candidate against resistant pathogens. AMPs are part of the innate immune response and have broad-spectrum antimicrobial activity. Synthetic AMPs are being developed as topical agents that disrupt bacterial membranes and prevent biofilm formation [47, 48].

Bacteriophage Therapy

Bacteriophages (phages) are viruses that specifically infect and kill bacteria. They offer a targeted approach to managing bacterial infections, particularly in wounds colonized by MDR pathogens.

Benefits Over Antibiotics:

- Host specificity: Phages target only the infecting bacteria without harming human cells or beneficial microbiota.
- Biofilm penetration: Some phages produce depolymerase enzymes that degrade biofilm matrices.
- Self-replication: Phages multiply at the site of infection, increasing their own efficacy.
- Low resistance risk: Bacteria can evolve resistance, but phage cocktails or engineered phages can counter-adapt.

Applications in Wound Care:

- Topical phage gels or sprays have been successfully used in burn wounds, ulcers, and trauma-related infections.
- Phage therapy has shown benefit in treating infections caused by MRSA, *Acinetobacter baumannii*, and *Pseudomonas aeruginosa*.
- Early-phase clinical trials demonstrate favorable safety profiles, with minimal adverse effects [49, 50].

Challenges include regulatory approval, phage banking and matching, and ensuring purity and consistency in preparation.

Anti-Biofilm Hydrogels

Biofilms are structured microbial communities that attach to wound surfaces and secrete a protective extracellular polymeric substance (EPS). This matrix acts as a barrier to antibiotics and immune cells, significantly reducing healing potential.

Novel Hydrogel Approaches:

- Chelating agents (e.g., EDTA): disrupt EPS structure and enhance antibiotic penetration.
- Surfactants (e.g., poloxamers): break surface tension to remove biofilms mechanically.
- Enzyme-loaded hydrogels: contain DNase, dispersin B, or proteases that degrade the EPS matrix.
- Dual-function hydrogels: combine antibiotics with anti-biofilm agents, enhancing synergy.

Some hydrogels also incorporate silver nanoparticles, AMPs, or photothermal materials for enhanced microbial killing without systemic toxicity.

Clinical Relevance:

- Particularly effective in chronic venous ulcers, diabetic wounds, and postoperative infections with biofilm formation.
- Enhance the action of antibiotics and can restore sensitivity in previously resistant bacterial colonies.

Future designs may include smart hydrogels that respond to microbial load or pH changes to initiate release [51, 52].

Natural and Integrative Therapies

Natural therapies have long played a role in wound healing across cultures and continue to gain scientific validation. With growing concerns about antibiotic resistance, cost, and biocompatibility, natural compounds and traditional remedies are being increasingly incorporated into evidence-based integrative wound care.

Medical-Grade Honey

Honey has been used in wound healing for over 2,000 years, notably in ancient Egypt, Ayurveda, and traditional Chinese medicine (TCM). Medical-grade honey, particularly Manuka honey from New Zealand, is now widely used in modern wound care due to its unique antimicrobial, anti-inflammatory, and pro-healing properties. Manuka honey, in particular, has been extensively studied. It creates a hyperosmolar environment, promotes autolysis, and accelerates tissue repair [53].

Mechanisms of Action:

- Broad-spectrum antimicrobial activity: due to its low pH, osmotic effect, and presence of hydrogen peroxide and methylglyoxal (MGO).
- Autolytic debridement: promotes enzymatic breakdown of necrotic tissue, reducing the need for surgical debridement.
- Moist wound environment: maintains optimal hydration while controlling exudate.
- Stimulation of granulation and epithelialization: enhances fibroblast activity and angiogenesis.

Clinical Use:

- Available as impregnated dressings, ointments, or gels, medical-grade honey is effective for:
  - Diabetic foot ulcers
  - Burns and donor sites
  - Radiation wounds
  - Pressure injuries

Randomized trials and systematic reviews have shown honey improves healing rates, shortens time to closure, and reduces odor and pain in chronic wounds. It is generally safe and well-tolerated, though transient stinging may occur in sensitive individuals [54, 55].

Herbal and Traditional Medicines

Herbal preparations such as *Centella asiatica*, *Aloe vera*, and curcumin have shown wound healing potential. Traditional Chinese Medicine (TCM) and Ayurvedic compounds are being validated through modern pharmacological studies [56–58].

Examples of Evidence-Based Herbal Therapies:

- *Centella asiatica* (Gotu kola):
  - Rich in asiaticoside and madecassoside, which promote collagen synthesis and angiogenesis.
  - Used topically or orally in venous ulcers and post-operative wounds.
- *Aloe vera*:
  - Contains aloins, polysaccharides, and enzymes that reduce inflammation and support tissue repair.
  - Beneficial in burns, abrasions, and radiation-induced skin injuries.
- Curcumin (from turmeric):
  - A potent anti-inflammatory and antioxidant compound.
  - Reduces wound bioburden and stimulates fibroblast migration and collagen deposition.
  - Often incorporated into hydrogels, nanofibers, or topical creams.

Traditional Medical Systems:

- Traditional Chinese Medicine (TCM):
  - Utilizes herbs like *Angelica sinensis*, *Salvia miltiorrhiza*, and *Panax ginseng*.
  - Shown to improve wound angiogenesis and immune regulation.
- Ayurveda:
  - Includes wound-care formulations with Neem, Tulsi, Triphala, and Guggul.
  - Often used in infected or necrotic wounds with excellent antimicrobial properties.

Modern Integration:

- Herbal extracts are now being combined with biopolymer scaffolds, smart hydrogels, and nanocarriers to enhance stability and bioavailability.
- Regulatory-approved products with botanical actives are becoming more common in Asia and Europe, though standardization and quality control remain a challenge.

Future Directions and Challenges

The future of wound care is poised to become increasingly personalized, data-driven, and technologically advanced. One promising avenue is the integration of genomic and proteomic profiling, which may enable clinicians to stratify patients based on their biological healing potential, inflammatory response, and risk of chronicity. Coupled with machine learning algorithms, this personalized profiling could help predict wound healing trajectories, select optimal therapies, and reduce trial-and-error treatment.

Smart wound dressings represent another frontier. These next-generation materials are being designed to monitor real-time biomarkers—such as pH, oxygen levels, temperature, and specific inflammatory cytokines—directly from the wound environment. Such "intelligent" bandages can not only detect early signs of infection or delayed healing but also respond dynamically by releasing antibiotics, growth factors, or moisture-balancing agents when needed. Some systems are being developed with wireless transmission capabilities, enabling remote wound monitoring and reducing the need for frequent hospital visits, especially in elderly or immobile patients.

Bioengineered dressings and regenerative scaffolds are also advancing rapidly. These include nanofiber matrices, hydrogel composites, and 3D-bioprinted constructs that mimic the extracellular matrix and can deliver embedded stem cells, exosomes, or therapeutic peptides to modulate inflammation, angiogenesis, and re-epithelialization. As these materials evolve, the line between wound dressing and therapeutic platform becomes increasingly blurred.

Yet, with great innovation come significant challenges. Regulatory approval pathways for advanced therapies—such as gene-based products, autologous cell therapies, and AI-assisted devices—are complex, time-consuming, and not yet standardized globally. Additionally, cost remains a major barrier. Many cutting-edge products are expensive to manufacture and may not be reimbursed under current healthcare frameworks, making them inaccessible to patients in low-resource settings.

Long-term safety and efficacy data are also limited, particularly for novel biotechnologies. For example, while stem cell-infused dressings have shown promise in pilot studies, robust evidence from multicenter, randomized trials is still needed. Furthermore, the successful implementation of these innovations will require not only interdisciplinary collaboration between researchers, engineers, and clinicians, but also extensive training programs to ensure clinical personnel can safely and effectively use these new tools.

A key future consideration is sustainability and global equity. As the burden of chronic wounds grows—especially in aging and diabetic populations worldwide—we must ask how innovation can be harnessed without widening healthcare disparities. Solutions must be scalable, affordable, and adaptable to various care settings, including community clinics and home care. Eco-friendly wound care materials that reduce medical waste, reusable sensor systems, and digital platforms for remote consultation are examples of sustainable directions that align with global health priorities.

Ultimately, advancing wound care should not only focus on technological novelty but also on ensuring that innovation benefits all who suffer from wounds—regardless of geography, income, or background.

Conclusion

Innovations in wound healing are transforming a traditionally passive field into a dynamic, regenerative science. From stem cells to smart dressings, these approaches offer new hope for patients with chronic or complex wounds. Continued research, validation, and equitable access will determine the extent to which these breakthroughs improve clinical outcomes worldwide.

Key Points Summary

- Chronic wounds present a significant burden on healthcare systems and require innovative, multidisciplinary treatment approaches.
- Cell-based therapies such as MSCs, PRP, and cultured fibroblasts/keratinocytes have demonstrated efficacy in promoting wound healing via paracrine signaling, angiogenesis, and tissue regeneration.
- Advanced wound dressings, including smart and bioactive dressings, actively interact with the wound microenvironment to accelerate healing and reduce infection.
- Biophysical modalities like NPWT, electrical stimulation, low-level laser therapy, LTAPP, and topical oxygen therapy offer non-invasive adjuncts for enhancing wound repair.
- 3D bioprinting and tissue engineering are transforming wound care by enabling personalized skin substitutes with enhanced integration and function.
- Cell-free regenerative strategies, especially those involving EVs and exosomes, show promise as scalable, low-risk alternatives to stem cell therapy.
- Emerging antimicrobial strategies target biofilms and resistant pathogens using phages, AMPs, and specialized hydrogels.
- Natural and integrative therapies, including medical-grade honey and botanicals, are increasingly supported by scientific evidence and provide cost-effective wound care options.
- Future wound care will be personalized, tech-enabled, and sustainability-focused, though challenges in cost, regulation, and global access remain.

References

1. Frykberg RG, Banks J (2015) Challenges in the treatment of chronic wounds. Adv Wound Care 4(9):560–582
2. Schultz GS, Sibbald RG, Falanga V, Ayello EA, Dowsett C, Harding K et al (2003) Wound bed preparation: a systematic approach to wound management. Wound Repair Regen 11(Suppl 1):S1–S28
3. Duscher D, Barrera J, Wong VW, Maan ZN, Whittam AJ, Januszyk M et al (2016) Stem cells in wound healing: the future of regenerative medicine? a mini-review. Gerontology 62(2):216–225
4. Kosaraju R, Rennert RC, Maan ZN, Duscher D, Barrera J, Whittam AJ et al (2016) Adipose-derived stem cell-seeded hydrogels increase endogenous progenitor cell recruitment and neo-vascularization in wounds. Tissue Eng Part A 22(3–4):295–305
5. Etulain J (2018) Platelets in wound healing and regenerative medicine. Platelets 29(6):556–568
6. Oneto P, Etulain J (2021) PRP in wound healing applications. Platelets 32(2):189–199
7. Altalbawy FMA, Mukhlif BAM, Hussen A, Mohammed JS, S RJ, Singh A et al (2025) Regenerative potential of PRP-based scaffolds in chronic wound healing: mechanisms, advances, and therapeutic insights. Regen Therap 30:278–298
8. Meznerics FA, Fehérvári P, Dembrovszky F, Kovács KD, Kemény LV, Csupor D et al (2022) Platelet-rich plasma in chronic wound management: a systematic review and meta-analysis of randomized clinical trials. J Clin Med 11(24)
9. Allie DE, Hebert CJ, Lirtzman MD, Wyatt CH, Keller VA, Souther SM et al (2004) Novel treatment strategy for leg and sternal wound complications after coronary artery bypass graft surgery: bioengineered Apligraf. Ann Thorac Surg 78(2):673–678. discussion 8
10. Hart CE, Loewen-Rodriguez A, Lessem J (2012) Dermagraft: use in the treatment of chronic wounds. Adv Wound Care 1(3):138–141
11. Brumberg V, Astrelina T, Malivanova T, Samoilov A (2021) Modern wound dressings: hydrogel dressings. Biomedicines 9:9
12. Shang S, Zhuang K, Chen J, Zhang M, Jiang S, Li W (2024) A bioactive composite hydrogel dressing that promotes healing of both acute and chronic diabetic skin wounds. Bioact Mater 34:298–310
13. Chen Y, Wang X, Tao S, Wang Q, Ma PQ, Li ZB et al (2023) Research advances in smart responsive-hydrogel dressings with potential clinical diabetic wound healing properties. Mil Med Res 10(1):37
14. Mirani B, Hadisi Z, Pagan E, Dabiri SMH, van Rijt A, Almutairi L et al (2023) Smart dual-sensor wound dressing for monitoring cutaneous wounds. Adv Healthc Mater 12(18):e2203233
15. Zhang X, Meng Y, Gong B, Wang T, Lu Y, Zhang L et al (2022) Electrospun nanofibers for manipulating soft tissue regeneration. J Mater Chem B 10(37):7281–7308
16. Rozman P, Bolta Z (2007) Use of platelet growth factors in treating wounds and soft-tissue injuries. Acta dermatovenerologica Alpina, Pannonica, et Adriatica 16(4):156–165
17. Yeganeh PM, Tahmasebi S, Esmaeilzadeh A (2022) Cellular and biological factors involved in healing wounds and burns and treatment options in tissue engineering. Regen Med 17(6):401–418
18. Waycaster CR, Gilligan AM, Motley TA (2016) Cost-effectiveness of becaplermin gel on diabetic foot ulcer healing changes in wound surface area. J Am Podiatr Med Assoc 106(4):273–282
19. Creaney L, Hamilton B (2008) Growth factor delivery methods in the management of sports injuries: the state of play. Br J Sports Med 42(5):314–320
20. Laiva AL, O'Brien FJ, Keogh MB (2018) Innovations in gene and growth factor delivery systems for diabetic wound healing. J Tissue Eng Regen Med 12(1):e296–e312
21. D'Mello S, Elangovan S, Salem AK (2015) FGF2 gene activated matrices promote proliferation of bone marrow stromal cells. Arch Oral Biol 60(12):1742–1749

22. Doukas J, Blease K, Craig D, Ma C, Chandler LA, Sosnowski BA et al (2002) Delivery of FGF genes to wound repair cells enhances arteriogenesis and myogenesis in skeletal muscle. Mol Therap 5(5 Pt 1):517–527
23. Li W, Zheng J (2024) Negative pressure wound therapy for chronic wounds. Ann Plast Surg 93(2S Suppl 1):S19–s26
24. Pappalardo V, Frattini F, Ardita V, Rausei S (2019) Negative pressure therapy (NPWT) for management of surgical wounds: effects on wound healing and analysis of devices evolution. Surg Technol Int 34:56–67
25. Teot L, Ohura N (2021) Challenges and Management in Wound Care. Plast Reconstr Surg 147(1s-1):9s–15s
26. Lalezari S, Lee CJ, Borovikova AA, Banyard DA, Paydar KZ, Wirth GA et al (2017) Deconstructing negative pressure wound therapy. Int Wound J 14(4):649–657
27. Webster J, Liu Z, Norman G, Dumville JC, Chiverton L, Scuffham P et al (2019) Negative pressure wound therapy for surgical wounds healing by primary closure. Cochrane Database Syst Rev 3(3):Cd009261
28. Borges D, Pires R, Ferreira J, Dias-Neto M (2023) The effect of wound electrical stimulation in venous leg ulcer healing-a systematic review. J Vasc Surg Venous Lymphat Disord 11(5):1070–9.e1
29. Gardner SE, Frantz RA, Schmidt FL (1999) Effect of electrical stimulation on chronic wound healing: a meta-analysis. Wound Repair Regen 7(6):495–503
30. Lala D, Spaulding SJ, Burke SM, Houghton PE (2016) Electrical stimulation therapy for the treatment of pressure ulcers in individuals with spinal cord injury: a systematic review and meta-analysis. Int Wound J 13(6):1214–1226
31. Zecha JA, Raber-Durlacher JE, Nair RG, Epstein JB, Sonis ST, Elad S et al (2016) Low level laser therapy/photobiomodulation in the management of side effects of chemoradiation therapy in head and neck cancer: part 1: mechanisms of action, dosimetric, and safety considerations. Support Care Cancer 24(6):2781–2792
32. Zecha JA, Raber-Durlacher JE, Nair RG, Epstein JB, Elad S, Hamblin MR et al (2016) Low-level laser therapy/photobiomodulation in the management of side effects of chemoradiation therapy in head and neck cancer: part 2: proposed applications and treatment protocols. Support Care Cancer 24(6):2793–2805
33. Zein R, Selting W, Benedicenti S (2017) Effect of low-level laser therapy on bone regeneration during osseointegration and bone graft. Photomed Laser Surg 35(12):649–658
34. Duarte S, Panariello BHD (2020) Comprehensive biomedical applications of low temperature plasmas. Arch Biochem Biophys 693:108560
35. Cui HS, Cho YS, Joo SY, Mun CH, Seo CH, Kim JB (2019) Wound healing potential of low temperature plasma in human primary epidermal keratinocytes. Tissue Eng Regener Med 16(6):585–593
36. Chuangsuwanich A, Assadamongkol T, Boonyawan D (2016) The healing effect of low-temperature atmospheric-pressure plasma in pressure ulcer: a randomized controlled trial. Int J Low Extrem Wounds 15(4):313–319
37. Davis CC, Dias Panariello F, Panariello B (2025) Exploring the efficacy of low-temperature plasmas on oral biofilms: a scoping review. Med Sci (Basel) 13(2)
38. Tonaree W, Taweepraditpol S, Kongkunnavat N, Poungjantaradej N, Kotistienkul B, Yongsuvimol M et al (2025) The efficacy of low-temperature atmospheric-pressure plasma (LTAPP) in the multicenter treatment of pressure ulcers: a randomized controlled trial. Int J Low Extrem Wounds:15347346251323940
39. Sun XK, Li R, Yang XL, Yuan L (2022) Efficacy and safety of topical oxygen therapy for diabetic foot ulcers: an updated systematic review and meta-analysis. Int Wound J 19(8):2200–2209
40. Vulakh GM, Hingorani AP, Ascher E, Marks N (2023) Adjunctive topical oxygen therapy for wound healing in patients with peripheral arterial disease. Vascular 31(4):737–740
41. Hones KM, Hones J, Satteson ES, Chim H (2023) Treatment of complex extremity wounds with MatriDerm: first clinical experience in the US. J Wound Care 32(3):167–171

42. Jackson SR, Roman S (2019) Matriderm and split skin grafting for full-thickness pediatric facial burns. J Burn Care Res 40(2):251–254
43. Song Y, You Y, Xu X, Lu J, Huang X, Zhang J et al (2023) Adipose-derived mesenchymal stem cell-derived exosomes biopotentiated extracellular matrix hydrogels accelerate diabetic wound healing and skin regeneration. Adv Sci (Weinheim, Baden-Wurttemberg). 10(30):e2304023
44. Zhou C, Zhang B, Yang Y, Jiang Q, Li T, Gong J et al (2023) Stem cell-derived exosomes: emerging therapeutic opportunities for wound healing. Stem Cell Res Ther 14(1):107
45. Fani N, Moradi M, Zavari R, Parvizpour F, Soltani A, Arabpour Z et al (2024) Current advances in wound healing and regenerative medicine. Curr Stem Cell Res Ther 19(3):277–291
46. Malinda KM, Kleinman HK (1996) The laminins. Int J Biochem Cell Biol 28(9):957–959
47. Bucataru C, Ciobanasu C (2024) Antimicrobial peptides: opportunities and challenges in overcoming resistance. Microbiol Res 286:127822
48. Fry DE (2018) Antimicrobial Peptides. Surg Infect 19(8):804–811
49. Nikolich MP, Filippov AA (2020) Bacteriophage therapy: developments and directions. Antibiotics (Basel) 9(3)
50. Plumet L, Ahmad-Mansour N, Dunyach-Remy C, Kissa K, Sotto A, Lavigne JP et al (2022) Bacteriophage therapy for staphylococcus aureus infections: a review of animal models, treatments, and clinical trials. Front Cell Infect Microbiol 12:907314
51. Ammons MC, Ward LS, James GA (2011) Anti-biofilm efficacy of a lactoferrin/xylitol wound hydrogel used in combination with silver wound dressings. Int Wound J 8(3):268–273
52. Fathil MAM, Katas H (2023) Antibacterial, anti-biofilm and pro-migratory effects of double layered hydrogels packaged with Lactoferrin-DsiRNA-silver nanoparticles for chronic wound therapy. Pharmaceutics 15(3)
53. Mandal MD, Mandal S (2011) Honey: its medicinal property and antibacterial activity. Asian Pac J Trop Biomed 1(2):154–160
54. Holubová A, Chlupáčová L, Cetlová L, Cremers NAJ, Pokorná A (2021) Medical-grade honey as an alternative treatment for antibiotics in non-healing wounds-a prospective case series. Antibiotics (Basel) 10(8)
55. McArdle C, Coyle S, Santos D (2023) The impact of wound pH on the antibacterial properties of medical grade honey when applied to bacterial isolates present in common foot and ankle wounds. An in vitro study. J Foot Ankle Res 16(1):66
56. Chaudhary R (2023) The role of medicinal plants in the diabetic wound healing process. Curr Diabetes Rev 19(4):e240222201410
57. Hoveidamanesh S, Irilouzadian R, Bagheri T, Saboury M, Fahimi Tafreshi S, Ghadimi T et al (2024) A review on traditional medicine used for burn treatment. J Burn Care Res 45(6):1598–1606
58. Jangra N, Singla A, Puri V, Dheer D, Chopra H, Malik T et al (2025) Herbal bioactive-loaded biopolymeric formulations for wound healing applications. RSC Adv 15(16):12402–12442

Chapter 21
Innovative Treatments for Accelerated Wound Healing

David S. Salven and David A. Brown

Abstract The wound healing landscape is entering a new age, as our ability to characterize novel pathways and therapeutic targets continues to expand rapidly with the rise of single-cell transcriptomics technologies [1]. Our therapeutic toolbox has grown to encompass not only small molecule-based drugs but also innovative gene delivery and protein engineering techniques to target newly characterized pathways more precisely [2]. The realm of accelerated wound healing, a field that began with the experimental use of growth factors in the 1980s, is now a showcase of complex advancements in biomedical sciences designed to alter the biology of failed wound healing. Although growth factor-based therapies remain in use, emerging technologies will likely forever change the multi-billion-dollar industry dedicated to wound care [3]. Small molecule inhibitors can now target precise protein–protein interactions in wound healing pathways, and tunable biomaterials allow for the release of growth factors, metals, cellular products, and gene therapy vectors into the local wound environment. This chapter details the latest data surrounding the creative approaches currently in use and under investigation.

Keywords Accelerated wound healing · Growth factor therapy · Biomaterials · Gene therapy · Regenerative medicine

D. S. Salven · D. A. Brown (✉)
Division of Plastic, Maxillofacial, and Oral Surgery, Duke University School of Medicine, Durham, NC, USA
e-mail: david.a.brown@duke.edu

F. De Francesco, K. Turksen (eds.), *Management and Strategies for Wound Healing*, Stem Cell Biology and Regenerative Medicine 79,
https://doi.org/10.1007/978-3-032-12442-5_21

Introduction

The wound healing landscape is entering a new age, as our ability to characterize novel pathways and therapeutic targets continues to expand rapidly with the rise of single-cell transcriptomics technologies [1]. Our therapeutic toolbox has grown to encompass not only small molecule-based drugs but also innovative gene delivery and protein engineering techniques to target newly characterized pathways more precisely [2]. The realm of accelerated wound healing, a field that began with the experimental use of growth factors in the 1980s, is now a showcase of complex advancements in biomedical sciences designed to alter the biology of failed wound healing. Although growth factor-based therapies remain in use, emerging technologies will likely forever change the multi-billion-dollar industry dedicated to wound care [3]. Small molecule inhibitors can now target precise protein–protein interactions in wound healing pathways, and tunable biomaterials allow for the release of growth factors, metals, cellular products, and gene therapy vectors into the local wound environment. This chapter details the latest data surrounding the creative approaches currently in use and under investigation.

Wound healing is an intricate process that occurs in four key stages: hemostasis, inflammation, proliferation, and remodeling. When tissues are injured, hemostasis is achieved via constriction of blood vessels, aggregation of platelets, and formation of a fibrin clot [4]. Growth factors then serve as chemoattractants alongside various cytokines to draw immune cells to the wounded area [5]. This marks the beginning of the inflammation stage, which peaks 1–3 days after wound creation. Specifically, neutrophils arrive first, serving to phagocytose bacteria and other foreign material. Then, macrophages previously differentiated from monocytes arrive, continuing to degrade pathogenic material and dead wound tissue. These cells aid in the progression of wound healing by secreting other growth factors and cytokines. The proliferative phase follows, marked by fibroblast migration and extracellular matrix (ECM) protein deposition in the wound bed, forming granulation tissue over 3–12 days. The process of angiogenesis begins with the assistance of growth factors, and keratinocytes begin to proliferate across the healing wound. This marks the start of the re-epithelialization process. Myofibroblasts also begin to contract at this time, further approximating the wound edges. Finally, remodeling occurs, a step marked by matrix metalloproteinases (MMPs) and tissue inhibitors of metalloproteinases, which regulate the process of elastin and collagen organization [4]. Specifically, a portion of the collagen III in granulation tissue is replaced by more robust collagen I [6].

Within this process lie multiple opportunities for disease to interfere with wound healing. For example, 78–90% of chronic wounds are thought to contain biofilms [7]. Infection is believed to prolong the inflammatory phase, where neutrophil presence increases the level of MMPs that restructure the ECM while limiting the ability of antibiotics to penetrate infected tissue [4]. Increased protease activity resulting from inflammatory activity also leads to the degradation of required growth factors for wound healing [4]. Other common insults are ischemia, hypoxia, and failure of angiogenesis. Some degree of wound hypoxia is essential for stimulating vascular endothelial growth factor (VEGF) production necessary for angiogenesis [5, 6]. However, dysregulated VEGF levels associated with diabetic ulcers are thought to contribute to dysangiogenesis [5]. Additionally, oxidative stress, hyperglycemia, and inflammatory states have been implicated in cellular senescence, possibly contributing to poor wound healing. However, this connection has yet to be fully characterized [8].

Non-healing chronic wounds are those that fail to close properly after four to eight weeks [4, 9]. These are plagued by a time- and resource-intensive healing process, often not leading to complete wound closure. This negatively impacts patient populations worldwide, with approximately 1–2% of any developed country's population thought to be affected [10]. Although wounds are caused by a wide variety of procedures and conditions, diabetes is one of the most common illnesses leading to chronic wounds, as diabetic foot ulcers develop annually in 9.1 million to 26.1 million patients worldwide [11]. In this context, only 30–40% of wounds are said to be fully healed at 3 months. Once they heal, one-year recurrence rates are high, with approximately 42% recurring after one year and 65% recurring after five years. Not only does this lead to suffering on behalf of patients subject to complex wound care protocols, but it also increases financial strain on healthcare systems across the globe. Depending on the method used for calculation, the Medicare program in the United States spends between $28.1 billion and $96.8 billion on wounds each year [10]. In the case of diabetic foot ulcers, mortality can be high, with a 2023 meta-analysis of 124,376 patients estimating the five-year mortality rate to reach approximately 50% [12] (Fig. 21.1).

Growth Factor-Based Approaches to Accelerated Wound Healing

Several growth factors are known to play vital roles in wound healing, although few are used to augment healing in clinical practice. Investigation into whether growth factors could be used therapeutically began with a 1987 animal study involving wounded rats and transforming growth factor-β (TGF-β). The strength of healed incisions was increased in a dose-dependent manner with only one TGF-β application. At 5 days post-application, wound strength was 220% higher than control, and an approximate 3-day acceleration of wound healing was also noted [14]. Unfortunately, human clinical trial results did not demonstrate the same degree of promise [15].

Platelet-derived growth factor (PDGF) is a dimeric molecule of A- and B-chains bound by disulfide bonds. The A- and B-chains bind to α- and β-receptors, respectively, two tyrosine kinase receptors that have been shown to alter cell growth, actin remodeling, chemotaxis, calcium ion mobilization, communication via gap junctions, and induction of apoptosis [16]. Becaplermin, a PDGF-BB homodimer, is the

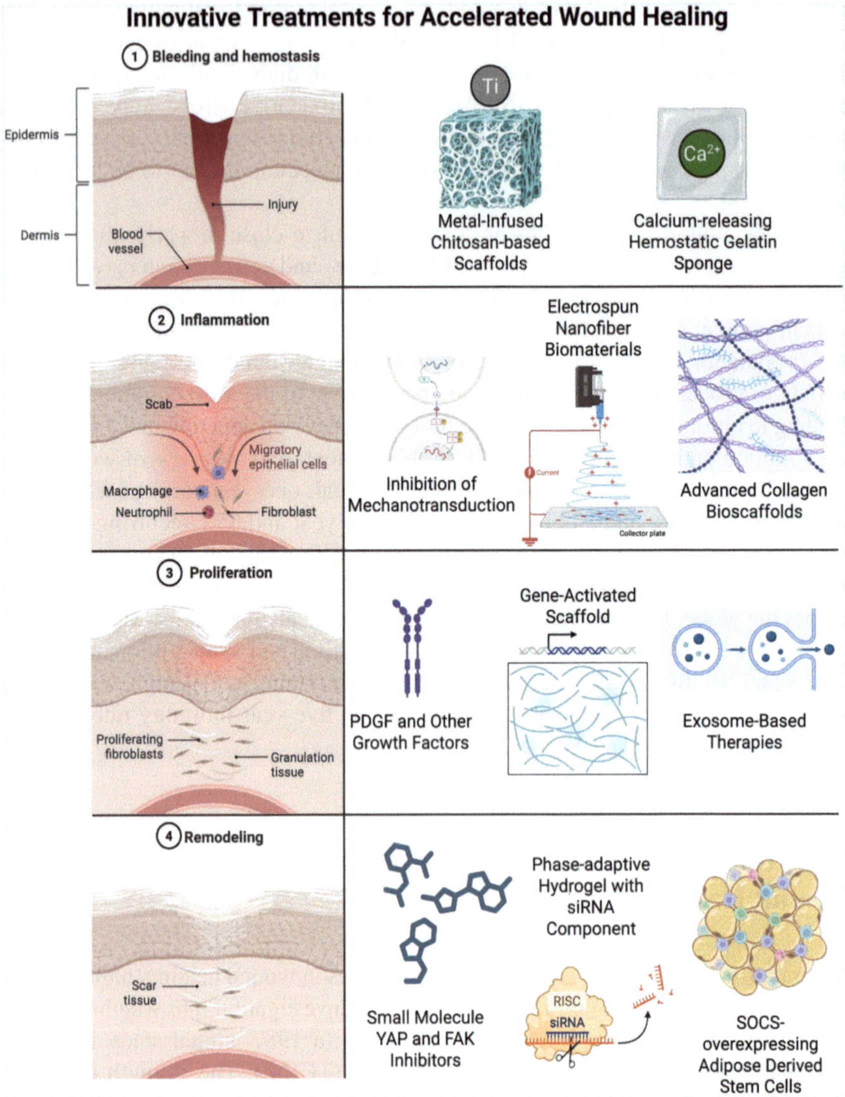

Fig. 21.1 The current pipeline of innovative treatments that accelerate wound healing spans all four stages of the healing process [13]

first and only growth factor approved by the United States Food and Drug Administration (FDA) for wound healing purposes and is supplied as a 100 μg/g gel [17, 18].

As early as 1992, a small PDGF-BB trial decreased ulcer depth and volume with minimal toxicity [19]. A 1995 landmark trial of 118 patients by Steed et al. demonstrated the efficacy of topical becaplermin gel in diabetic foot ulcers [20]. Subsequently, a combined analysis of 922 patients enrolled in randomized controlled trials showed improvements in complete wound healing for diabetic ulcers less than 10 cm^2 when compared to placebo (50% vs. 36%, $p < 0.007$), in addition to quickened healing time (14 weeks vs. 20 weeks, $p = 0.01$) [21]. In 2005, a retrospective cohort study by Margolis et al. of 24,898 patients with diabetic neuropathic foot ulcers identified 2394 who received PDGF. PDGF administration was associated with improved healing (RR = 1.32, 95% CI = 1.22–1.38) and decreased amputation risk (RR = 0.65, 95% CI = 0.54–0.78) [22]. Adequate wound debridement was a vital adjunct to PDGF administration in this patient population [23]. In the context of pressure ulcers, one small, randomized trial suggested that preoperative PDGF administration may improve the outcome of surgical treatment [24].

Despite positive data, the FDA placed a black box warning on the gel in 2008 [21–25]. At the time, there were concerns that patients with current malignancies who had also received three tubes of becaplermin had a death rate increased by fivefold. This warning was subsequently removed in 2018 after a matched cohort study failed to identify a significantly increased cancer mortality risk (HR = 1.2, 95% CI = 0.7–1.9) [26].

Epidermal growth factor (EGF) is another factor of interest in wound healing. EGF is thought to promote the migration of keratinocytes through extracellular matrix proteins, having been shown to increase the expression of the α-2 integrin [27]. A 1991 prospective, open-label crossover trial illustrated the use of an EGF-Silvadene combination to treat wounds that remained unhealed after conventional therapy and Silvadene alone [28]. In 1992, an ophthalmic solution containing mouse EGF effectively reduced the healing time of corneal epithelial defects [29]. In 2003, a Hong Kong-based trial involving 61 patients demonstrated a 95% ulcer healing rate by 3 months at a concentration of 0.04%, compared to 42% in the control arm [30]. In 2018, a larger trial that included 167 patients with chronic diabetic foot ulcers showed a larger proportion of patients with fully healed wounds (73.2%) compared to placebo (50.6%, $p = 0.001$). Significantly, administration of recombinant human EGF was associated with accelerated ulcer healing by approximately 28 days (HR = 2.13, $p < 0.001$) [31]. Although it continues to be used outside the US, it is not currently approved by the FDA. Despite positive trial data surrounding EGF and other growth factors, the scientific community has called for more rigorous clinical studies with larger sample sizes and consistency in measured endpoints [17, 18] (Table 21.1).

Small-Molecule Inhibitors in Accelerated Wound Healing

HIF-1α as a Small Molecule Target for Improved Wound Healing

The hypoxia-inducible factor-1 alpha (HIF-1α) pathway has gained attention among wound healing experts as a potential target for intervention with small molecule-based drugs. Although many signaling factors regulate the pathway, hypoxia

Table 21.1 Growth factor-based therapies for accelerated wound healing

Growth factor	Formulation/ delivery	Key pre-clinical evidence	Key clinical evidence	Regulatory status & notes	Main advantages	Main limitations
TGF-β	Single topical application (early animal studies)	Rat-incision model: One dose ↑ wound strength by 220% at day 5 and cut ~3 days off healing time	Human trials failed to replicate dramatic benefit	No regulatory approvals	Potent pro-healing signal	Scar risk, variable human efficacy
PDGF-BB (becaplermin)	0.01% topical gel (Regranex®)	–	Steed 1995 RCT ($n = 118$): faster closure of diabetic foot ulcers Meta-analysis ($n = 922$): 50% vs 36% complete healing, 6-week faster closure Large cohort ($n = 24{,}898$) ↓ amputation risk (RR 0.65)	FDA-approved (only U.S. growth-factor product); 2008 boxed warning removed in 2018	Robust human data; amputation-risk reduction	Daily application; cost; benefit mainly in ulcers <10 cm^2
EGF	Spray or cream (0.04%–0.05%), often with silver sulfadiazine	–	61-pt study (2003) 95% healed by 12 wks. vs 42% control Phase III RCT (2018, n = 167) 73% vs 51% healing; ~28-day faster closure	Widely used outside US; not FDA-approved	Accelerates re-epithelialization; good safety	Regulatory gap in US; heterogeneity in trial designs

generally stabilizes HIF-1α and leads to decreased ubiquitination and proteasomal degradation by the von Hippel-Lindau protein (pVHL) [32]. Consequently, the pVHL-dependent pathway has become of particular interest. Within the context of diabetic foot ulcers, hyperglycemia is thought to inhibit HIF-1α, and stabilization of HIF-1α has been proposed as a method of improving wound closure in these patients [33, 34].

Metal complexes have also been developed into lifesaving drugs. This is due in part to the groundwork laid by Barnett Rosenberg, Loretta Van Camp, and Thomas Krigas in 1965, where various group VIII*b* transition metals, including platinum in the form of cisplatin, were found to inhibit the division of *E. coli* [35]. Today, platinum-based small-molecule chemotherapeutics cisplatin, carboplatin, and oxaliplatin are used worldwide to treat cancer [36]. Together, these ideas have been incorporated into developing metal complex-containing small molecule-based drugs to improve wound healing.

A recent example is a cyclometalated iridium(III) complex used to stabilize HIF-1α and promote wound healing. In vitro studies suggested that the complex selectively inhibits interaction between HIF-1α and pVHL while also serving to upregulate HIF-1α, VEGF, glucose transporter type I (GLUT1), and erythropoietin (EPO) expression. The authors proceeded to investigate the effect of the drug on wild-type and diabetic mice, finding that the compound led to accelerated wound healing, augmented skin thickness, and collagen deposition, as well as improved skin perfusion in both control and diabetic mice. Additionally, protein expression in wounded skin mirrored that of in vitro experiments, with increased levels of HIF-1α, VEGF, GLUT1, and EPO. Of note, HIF-1α signaling was impaired significantly in diabetic mice. Levels of mRNA corresponding to various genes involved in cell motility, recruitment, and vascular growth were found to be increased. Together, these results offered promising insight into wound healing mediated by HIF-1α and pVHL while simultaneously demonstrating the utility of small molecules in accelerating the wound healing process [37].

Aside from pVHL, the SET7/9 lysine methyltransferase is vital to regulating HIF-1α activity, as SET7/9-mediated methylation of HIF-1α leads to its degradation. Aberrant SET7/9 activity is thought to be implicated in vascular changes associated with diabetes and other wound healing pathologies. It has, therefore, become a desired target of small molecule inhibitors, one of which includes an iridium(III) complex with acetonitrile ligands. In a model using human umbilical vein endothelial cells (HUVECs), the compound was found to increase the abundance of HIF-1α. An in vitro scratch assay was performed in HUVECs, demonstrating that the small molecule also promoted scratch closure despite some cytotoxic effects. The wound-closure effect was enhanced in the presence of homocysteine, which is thought to be elevated in diabetic physiology. A zebrafish-based in vivo model demonstrated the compound's ability to partially rescue angiogenesis inhibited by the VEGFR tyrosine kinase inhibitor II. Finally, in vivo studies in diabetic mice showed accelerated wound healing, more robust skin quality, improved angiogenesis, and upregulation of HIF-1α, VEGF, EPO, and GLUT1—another promising illustration of a small molecule's ability to hasten wound healing physiology [33].

Inhibition of Mechanotransduction to Accelerate Wound Healing

Mechanotransduction is the process by which physical forces applied to cells or tissues are translated into biochemical responses. Cellular responses occur as signals are transduced through numerous pathways, often involving a combination of focal adhesion kinase (FAK), Rho/Rho-associated kinase-1 (ROCK-1) signaling, and the Hippo pathway [38]. While FAK's function was initially understood to be related to cell motility, its role in mechanotransduction quickly became apparent [39]. Wong et al. performed a microarray analysis of mouse wounds that had undergone controlled tension, where FAK was a vital factor. A tamoxifen-inducible FAK knockout line of mice was created, with the knockout specific to dermal fibroblasts. When a hypertrophic scarring model was applied, the scars were found to have decreased area and cellularity compared to control mice. Levels of Tgf-β1 in experimental mice were lower, as were levels of the inflammatory chemokine MCP-1 and evidence of its associated cell trafficking activity. When a small molecule inhibitor of FAK (PF573228) was introduced to a culture of mechanically strained human fibroblasts, spread morphology, focal adhesion formation, motility, contraction, and α-smooth muscle actin induction were diminished. Additionally, ERK was found to be mechanoresponsive and reliant on FAK signaling, as inhibition led to reduced MCP-1. Finally, PF573228 was administered to wounded mice, significantly decreasing the scar area. Improvements were also noted in histological characteristics of the skin, including reduced epithelial and dermal proliferation [40].

The same group later expanded this model to test a different FAK inhibitor in the red Duroc pig, where the skin architecture and physiology of scarring are more similar to that of humans [39, 41]. Partial-thickness wounds were created in the animal, and the VS-6062 small molecule FAK inhibitor was administered via a pullulan-collagen hydrogel. Wounds treated with the FAK inhibitor closed approximately 10 days sooner than control wounds treated with empty carrier gels. Once healed, the skin was found to be similarly elastic to normal skin, with regeneration of hair follicles and other skin appendages. Furthermore, algorithmic analysis of collagen organization in the treatment group showed a "basket weave-like" array of collagen fibers similar to normal skin, while controls were found to contain elongated parallel fibers typical of scar tissue. A subsequent single-cell RNA sequencing (scRNAseq) analysis of human fibroblasts confirmed that mechanical strain led cells toward heterogeneous profibrotic clusters with enriched collagen, myofibroblast development, and MAPK-ERK-YAP signaling. Inhibition of FAK was associated with an antifibrotic state, with heightened expression of matrix-metalloproteinases, stanniocalcin-1, and other pro-regenerative factors [41].

While inhibition of FAK has been a focus of many studies, yes-associated protein (YAP) has also proven to be a viable small molecule target. As part of the Hippo pathway, YAP and transcriptional coactivator with PDZ-binding motif (TAZ) are responsible for transducing signals stemming from mechanical force on the extracellular matrix (ECM) [39]. Here, the authors used verteporfin, a YAP inhibitor FDA-approved for treating age-related macular degeneration and chronic central serous chorioretinopathy [42, 43]. Full-thickness hexagonal wounds were made on the dorsal aspect of red Duroc pigs before closure took place under high tension. With the administration of verteporfin, the macroscopic appearance of the resulting scar was more favorable than that of control wounds. Furthermore, histology revealed regeneration of skin appendages and a "basket weave-like" collagen pattern mirroring unwounded skin. Physical testing also revealed increased tensile strength and decreased Young's modulus. A subsequent scRNAseq analysis identified five different fibroblast clusters, each with either a pro-regenerative or profibrotic gene expression signature. The pro-regenerative cluster expressed intercellular adhesion molecule 1 (ICAM1), moesin (MSN), fatty acid binding protein 5 (FABP5), and interleukin-33 (IL-33). In contrast, the pro-fibrotic signature was characterized by marker genes, including the oncostatin M receptor (OSMR), α-smooth muscle actin 2 (ACTA2), sprouty homolog 2 (SPRY2), and insulin-like growth factor binding protein 2 (IGFBP2). IL-33 was also found to be critical for healing. When IL-33 was neutralized with an antibody, wounds treated with verteporfin in a human neonatal foreskin xenograft model experienced scarring and delayed closure [43] (Table 21.2).

Advanced Biomaterials as Wound Healing Scaffolds

Growth factor and small molecule administration have shown promise in creating a molecular stimulus for accelerated wound closure. However, research groups globally have begun to investigate bioactive materials and intelligent wound dressings to improve the substrate for healing or provide additional context about the biochemical state of the wound for clinicians to act upon. Several biomaterials have recently become popular, ranging from naturally derived polymers to fully synthetic biocompatible scaffolds. These may be combined with dressings containing chemical indicators or electronic components that assess the wound's state of closure.

Chitosan-Based Bioscaffolds

Among the most popular emerging materials is chitosan, which is derived from chitin—an abundant polysaccharide naturally found as a structural element of invertebrate cytoskeletons and particular cell walls. Chitosanases and chitinases are used in hydrolyzing chitin into chitosan, which has been used extensively in recent tissue

Table 21.2 Small molecule therapies for accelerated wound healing

Pathway/target	Small molecule inhibitor (class)	Direct molecular action	Experimental models	Key wound-healing findings
HIF-1α/pVHL	Cyclometalated iridium(III) complex	Blocks HIF-1α-pVHL interaction → stabilizes HIF-1α, ↑ VEGF, GLUT1, EPO	HUVECs (in vitro); wild-type & streptozotocin-diabetic mice (in vivo)	Faster closure; thicker epidermis; greater collagen deposition; improved skin perfusion
HIF-1α/SET7/9	Iridium(III)–acetonitrile complex	Inhibits SET7/9-mediated HIF-1α methylation and ↑ HIF-1α	HUVEC scratch assay; zebrafish angiogenesis model; diabetic mice	Scratch closure despite some cytotoxicity; rescued angiogenesis; accelerated healing with superior skin quality; ↑ HIF-1α, VEGF, EPO, GLUT1
Mechanotransduction/ FAK	PF573228 (pyrimidine-based FAK inhibitor)	Inhibits FAK to ↓ focal adhesions, α-SMA, ERK/ MCP-1	Mechanically strained human fibroblasts; hypertrophic-scar mouse model	Reduced cell spreading & contraction; smaller, less cellular scars; lower TGF-β1 & MCP-1; improved histology
Mechanotransduction/ FAK	VS-6062 delivered in pullulan–collagen hydrogel	Small molecule FAK inhibitor	Partial-thickness red Duroc pig wounds	~10 days faster closure; regenerated hair follicles & appendages; "basket-weave" collagen; antifibrotic scRNAseq signature
Hippo/YAP-TAZ	Verteporfin (benzoporphyrin derivative)	Inhibits YAP nuclear activity; IL-33 dependent	High-tension full-thickness red Duroc pig wounds; human neonatal foreskin xenograft ± anti-IL-33	Superior macroscopic scars; appendage & basket-weave collagen regeneration; ↑ tensile strength, ↓ young's modulus; IL-33 neutralization negated benefits

engineering applications. Potential advantages include intrinsic antimicrobial and angiogenic properties. Due to the polymer's positive charge, it binds negatively charged red blood cells, promoting hemostasis. Its structural components are the basis for essential chemical reactions that allow its use in biological settings. A detailed 2024 review by Haider et al. describes the chemical and physical methods used to modify and produce chitosan bioscaffolds, in addition to the numerous modifications that benefit wound healing [44].

Several protocols for chitosan scaffold synthesis have been documented, including lyophilization and 3D printing. Electrospinning, solvent casting, and particulate extraction are additional methods used in manufacturing. Electrospinning involves the application of a high-voltage electric field while casting the scaffold to facilitate the formation of ultrafine fibers. This has been said to increase the surface-to-volume ratio [45]. Solvent casting and particulate extraction involve casting the polymer with porogenic compounds, which are subsequently removed via leaching with a solvent or water, enabling fine control over pore size. Various modifications have been made to chitosan to mimic the extracellular matrix. For example, collagen, gelatin, fibrinogen, polylactic acid, polyethylene glycol, polyglycolic acid, and polycaprolactone have frequently been incorporated into these scaffolds. Bioceramics have also been included in chitosan-based wound healing scaffolds, improving mechanics, increasing bioactivity, and creating the potential for osteoinductivity. In addition to enhancing scaffold strength, the infusion of metal ions has an antibacterial effect. It also enables electrical conductivity, allowing the scaffold to play a role in newer therapies that may require electrical stimulation. Acetylation of chitosan provides a more hydrophobic, less antigenic scaffold that is mechanically sound with improved potential for cell adhesion and proliferation. Grafting is also used frequently, describing a chemical modification method involving exposure of the scaffold surface to low-temperature plasma, ultraviolet radiation, or other chemical processes to alter the scaffold's functionality. Lastly, laser irradiation has been used to modify scaffold morphology and cell adhesion [44].

A 2025 study illustrates the use of a chemically modified chitosan scaffold infused with metal ions in the context of cutaneous wound healing. The scaffold constructed was multilayered, with a base layer composed of polyacrylonitrile nanofibers synthesized via electrospinning. The base layer was allowed to react with di-ethylene-tri-amine, resulting in a nanofiber scaffold functionalized with amine groups. In this setting, amine functionalization was intended to provide a depot of positive charges, which would adhere more strongly to negatively charged cell membranes. This functionalized product was then electrosprayed with a titanium dioxide suspension in dimethyl-formamide, intended to confer an antibacterial effect onto the healing wound. Finally, a chitosan solution was electrospun onto the titanium-containing polyacrylonitrile scaffold. Scaffolds then underwent standard tests to analyze morphology, porosity, surface characteristics, and hydrophilicity. An in vitro analysis demonstrated that the scaffold promoted NIH3T3-E1 cell adhesion and proliferation. Additionally, titanium dioxide demonstrated activity against *Klebsiella pneumoniae*, *Shigella flexenri*, *Staphylococcus aureus*, and

Micrococcus luteus. While this study provides encouraging in vitro results, the scaffold has yet to be tested in vivo [46].

Other chitosan-based scaffolds have exhibited encouraging properties in vivo. For example, a 2023 study detailed the development and testing of a nanofibrous mat containing chitosan, polyethylene oxide, and copper ferrite in a mouse wound model. Unlike the previous study's layered approach to bioscaffold production, the authors aimed to combine all three components in solution before electrospinning. Once a mat of the desired thickness was produced, its chemical and physical properties were characterized via scanning electron microscopy, energy dispersive X-ray analysis, and Fourier transform infrared spectroscopy. After an in vitro cell monolayer scratch assay suggested expedited fibroblast migration in the presence of the scaffold, an in vivo analysis in wounded mice was conducted. The authors noted how wounds exposed to the scaffold exhibited more robust skin healing, with enhanced collagen synthesis and hair follicle development at the 14-day endpoint [47].

Several groups have focused their efforts on chitosan as a biologically derived base upon which other materials, such as non-biological polymers like polyethylene oxide, polyacrylonitrile, or metal ions, might be added to improve strength, biocompatibility, or wound closure rates. However, materials of cellular origin have also shown promise in functionalizing chitosan. One example is the loading of bioscaffolds with exosomes, which are extracellular vesicles released from cells when multivesicular bodies within the cell fuse with the plasma membrane [48]. A 2020 study set out to determine whether exosomes from isolated human endometrial stem cells would improve cutaneous wound healing when incorporated into a chitosan-glycerol hydrogel. An in vitro scratch assay revealed that treating human skin fibroblasts with exosomes improved cellular migration, as evidenced by complete scratch closure in exosome-treated groups compared to only 62% closure in untreated controls. When fibroblasts were grown with an exosome containing chitosan hydrogel, fibroblast proliferation was greater than both untreated controls and plain chitosan-glycerol hydrogels without exosomes. Finally, an in vivo wound healing assay was performed, comparing sterile gauze to chitosan hydrogels with and without exosomes. The exosome-containing hydrogel group reached a mean wound closure rate of 83.6 ± 4.9% at 14 days. Exosome-free chitosan-glycerol hydrogels were associated with 65.4 ± 4.6% closure at 14 days, whereas the gauze control led to only 51.5 ± 3.7% closure. Histological studies demonstrated identifiable dermal-epidermal junctions, abundant collagen, developing hair follicles, and more substantial evidence of angiogenesis [49].

A similar approach was taken in 2025, where a hydrogel was engineered from gellan gum and infused with "small extracellular vesicles" derived from adipose mesenchymal stem cells. As gellan gum is not reported to possess strong cell adhesion properties, it was functionalized with ethylenediamine, where additional amino groups would contribute to stability and elasticity. Degradation studies revealed a loss of less than 20% of the gel's initial mass at 21 days, suggesting appropriate stability for the wound healing timeline. Although the system was not tested in vivo, the authors mimicked a pro-inflammatory environment with TNF-α and IL-1β.

Hydrogels loaded with vesicles demonstrated rescue of cell viability to 84.21 ± 1.15%, compared to 63.67 ± 1.45% without extracellular vesicles ($p = 0.0016$). Furthermore, 48h gene expression studies showed increased expression of cell-cycle promotors Ki67 and CCND1, in addition to decreased cell cycle inhibitor CDKN1B expression. Expression of PTEN, a gene involved in cytoskeletal remodeling, was also increased, as were immunomodulatory cytokines TGF-β and IL-10. These observations suggest potential value in adding exosomes and other forms of stem cell-derived extracellular vesicles to hydrogels and bioscaffolds [50].

Collagen-Based Biomaterials

Collagen accounts for approximately one-third of the total protein in the human body and is a key component of the extracellular matrix. The combination of polypeptide chains into triple helices creates types of collagen specialized to various bodily tissues, with well-known types I-III forming fibrillar collagen abundant in skin, bone, cartilage, tendon, and blood vessels, and other types forming networks found in basement membranes [51]. Thus far, approximately 29 types of collagen have been identified [52]. Collagen has numerous biomedical applications, with uses ranging from wound dressings and hemostatic sponges to cell culture hydrogels and drug delivery depots [52]. The chemotactic effects of collagen types I-III on human dermal fibroblasts were established long ago. Since then, the tissue engineering community has become interested in collagen's utility in treating osteochondral and vascular diseases, in addition to the injury of the skin, cornea, and nerves [53, 54].

One of the most well-known uses of collagen-based products in wound healing lies in plastic and reconstructive surgery. Dermal substitutes serve as an implantable extracellular matrix that promotes tissue formation and are often used in the surgical treatment of traumatic injuries, burns, chronic wounds, and defects secondary to oncologic disease or resection where a split-thickness skin graft is required. Numerous substitutes exist today, each with varying collagen sources and additional additives. Collagen sources are often human, bovine, or porcine, with extra components, including elastin, glycosaminoglycans, and silicone [55–58]. Although current data suggests that the use of dermal substitutes delays initial re-epithelialization (7.30%, $p = 0.02$), their use has demonstrated improvement in specific long-term outcomes [59]. For example, a 2024 systematic review noted subjective scar quality in burn patients to be higher at 6 months (−1.95, $p < 0.01$) [59]. A 2025 randomized controlled trial comparing a collagen/elastin dermal matrix with split-thickness skin grafts noted more favorable skin extension ($p = 0.034$) and elasticity ($p = 0.036$) scores in patients receiving dermal matrix. However, quantification of scar quality showed no difference at 6- and 12-month endpoints [60].

While inclusion within a dermal substitute represents an established use of collagen in wound healing applications, numerous laboratories continue to combine collagen with innovative biomaterials to further its role in wound care. Recently, asiaticoside and epsilon-poly-L-lysine were mixed with collagen in a bilayer artificial skin scaffold. Asiaticoside is a compound extracted from *Centella asiatica* leaves that is thought to increase type I collagen synthesis by activating the TβRI kinase-independent Smad pathway in human dermal fibroblasts [61]. In contrast, it has been shown to inhibit type I and III collagen expression while increasing Smad7 expression upon exposure to keloid-derived fibroblasts [62]. The other component of this scaffold, epsilon-poly-L-lysine, is a cationic peptide with antimicrobial activity used as a food preservative [63]. Combined in a collagen scaffold, these compounds exhibited antimicrobial activity against *Staphylococcus aureus* and *Pseudomonas aeruginosa* in both an in vitro assay and ex vivo infected porcine skin model. Although the scaffold was found to accelerate in vivo wound closure within the first 14 days of healing, a statistically significant difference was no longer present as time progressed. However, the scaffold was found to modulate the TLR4, MAPK, and NF-kB signaling pathways, creating an anti-inflammatory state that may benefit wound healing [64].

In contrast to relying on biomaterials and other chemical adjuncts to alter gene expression within the wound healing environment, an alternative approach is to manufacture bioscaffolds that deliver genes directly to healing tissue. The stromal cell-derived factor 1α (SDF-1α) gene has recently been of particular interest. Also known as CXCL12, this gene has been implicated in wound revascularization and epithelial cell regrowth [65]. Specifically, when wounded areas enter a hypoxic state, hypoxia-inducible factor-1 (HIF-1) increases expression of SDF-1, allowing progenitor cells to migrate to the area [66]. Additionally, matrix metalloproteinases (MMP-9) from macrophages at diabetic wound sites are thought to inactivate CXCL12, limiting the ability of endothelial progenitor cells to aid in revascularization [67]. In a therapeutic context, delivery of SDF-1α via a lentiviral vector was shown to decrease the surface area of diabetic wounds, promote the formation of granulation tissue, and, eventually, full epithelialization [68]. In light of concerns surrounding the ability of lentiviral vectors to cause insertional mutagenesis or the formation of a replicating virus, it is understandable why research groups are investigating non-viral gene delivery methods.

For example, one group aimed to deliver SDF-1α into human Schwann cells using a collagen-chondroitin sulfate "gene-activated scaffold" [69]. When injured, Schwann cells typically dedifferentiate and activate pathways that facilitate repair [70]. The expression of the dedifferentiating transcription factor c-Jun increases when Schwann cells are injured, which helps suppress the expression of myelin [70]. After synthesizing, sterilizing, and crosslinking the freeze-dried scaffolds, plasmid DNA-polyethyleneimine (PEI) polyplexes were loaded via a soak-loading technique. SDF-1α, VEGF-A, and c-Jun expression increased when Schwann cells were cultured in 3D with the gene-containing scaffold. Notably, secreted VEGF collected from the 3D microenvironment was confirmed to be bioactive using a Matrigel assay that measured endothelial branching within the matrix. Furthermore, the scaffold encouraged remodeling of the basement matrix, defined by the degradation of fibronectin and collagen IV before synthesizing a laminin matrix [69]. A study subsequently performed by the same group investigated the effect of a similar SDF-1α gene-activated collagen scaffold on the function of a human endothelial cell and adipose-derived stem cell coculture. They reported enhanced cooperation between these cell types and increased expression of vasculogenic genes SDF-1α, CXCR4, and eNOS. Furthermore, they noted the organization of Schwann cells into Bünger band-like structures, signifying a propensity toward neurogenesis [71].

Although angiogenesis and neovascularization are vital components of the wound healing process, the less frequently discussed lymphatic system also plays an important role, helping regulate fluid balance, inflammation, and immune cell movement within tissues [72]. To this end, a bioactivated collagen scaffold seeded with photosynthetic, oxygen-producing cyanobacteria (*Synechococcus* sp. *PCC 7002*) was recently designed to promote angiogenesis and lymphangiogenesis in the setting of cutaneous wound healing. Specifically, bacteria-free scaffolds were compared to scaffolds containing wild-type bacteria and a third group with genetically modified bacteria expressing hyaluronic acid. After the scaffolds were successfully colonized, they were implanted into full-thickness mouse skin defects and exposed to light for 1 week. With light exposure, cyanobacteria enhanced vascularization and lymphatic tissue formation. This was assessed by quantifying lymphatic and endothelial cell markers LYVE01, VEGFR3, and CD31. Despite the success of this novel wound-healing intervention, the authors note several limitations, including a short study duration, the dependence of the bacteria on light, and the potential for immune-mediated reactions [73].

Gelatin-Based Biomaterials

Gelatin is a denatured form of collagen often derived from bovine and porcine tissue via acid or base extraction. Although this process creates an element of unpredictability in terms of fragment size and biochemical properties, gelatin demonstrates several key advantages in a biomedical context [74]. Specifically, its low immunogenicity and high capability of resorption improve its biocompatibility. Gelatin's structure also lends itself well to chemical modification, and its rheological properties have been tuned via the degree of substitution, polymer concentration, UV exposure, initiator concentration, and storage time [75]. Together, these properties enhance its utility in hydrogels or bioscaffolds designed to hasten wound healing or control drug delivery.

Like chitosan scaffolds, metal ions are frequently incorporated into gelatin-based biomaterials, and calcium has been heavily implicated in various aspects of wound healing [76, 77]. For example, gelatin matrices have been combined with carbonate apatite to create hemostatic sponges. In one 2025 study, engineers took advantage of β-type carbonate ion substitutions more like that found in the apatite of human bone. This also provided optimal calcium ion supply to the surrounding tissue. After synthesis, sponge variants underwent analysis to determine their physicochemical properties, such as swelling, degradability, and crosslinking.

Sponge variants were found to be under a 2% threshold for hemolysis and caused no skin irritation in rabbit dermal patch tests. The sponge variants were then applied to full-thickness wounds in rats, where collagen density and granulation scores were compared to commercially available porcine sponge and gauze. The carbonate apatite sponge was found to be efficacious in enhancing wound healing, although the authors call for additional research into the mechanism behind this effect [77].

Selenium nanoparticles were embedded into an electrospun polycaprolactone/gelatin nanofiber scaffold, along with vitamin E, to capitalize on synergistic antioxidant properties. Various in vitro assays were performed to determine the gel's release characteristics and cellular compatibility with 3T3 fibroblasts. Selenium and vitamin E were released at physiologic temperature and pH at 57% and 43% after 4 days, respectively. Reactive oxygen species were scavenged both in vitro and in vivo. When full-thickness wounds were created in Wistar rats, the selenium and vitamin E-containing scaffold contributed to complete re-epithelialization and greater collagen deposition [78] (Table 21.3).

Gene Therapy Methods for the Acceleration of Wound Healing

Overview of Common Gene Delivery Methods

Gene therapy has emerged as a way to accelerate wound repair, altering the intrinsic behavior of cells comprising the local wound area. This can be accomplished by adding a therapeutic gene, silencing a maladaptive gene expression pattern, or editing the existing genome [79]. However, several challenges have been discussed, likely slowing the implementation of gene therapy products in a clinical setting. These include cost, manufacturing complexity, variability between individual patient responses, and safety concerns in viral vector use [80].

Viruses are a standard method of delivering genetic material to cells, with the use of adenoviruses, adeno-associated viruses (AAV), and lentiviruses documented in hundreds of clinical trials. Adenoviruses are double-stranded DNA viruses with an icosahedral capsid engineered to encompass larger (10.5 kb) or multiple transgenes [81]. Advantages of adenoviruses include their high transduction efficiency, episomal residence of transduced DNA, a well-characterized production method, and diverse array of targetable tissues [82]. However, limitations include immunogenicity and transient length of gene expression [68, 70]. Recombinant adeno-associated viruses have been studied in treating various muscular dystrophies and hematologic conditions. These viruses are nonintegrating, allow for episomal gene expression, and have a favorable safety profile. However, they are limited by immunogenicity and a 5 kb vector capacity [81]. There are currently seven FDA-approved AAV-based therapies [83]. Unlike adenoviruses and AAVs, lentiviruses efficiently integrate their genome into that of the target tissue, leading to potential lifelong

Table 21.3 Bioactive scaffolds

Scaffold/dressing type	Core material(s) & key additives	Fabrication method	Intended biofunction(s)	Representative findings
Chitosan family	Chitosan ± collagen/gelatin/metal ions; acetylated or graft-modified	Lyophilization, electrospinning, 3D printing, solvent-casting/particulate extraction	Intrinsic hemostasis, antimicrobial & proangiogenic properties	Review [44] highlights tunable porosity, conductivity, exosome loading
Multilayer PAN-TiO_2-chitosan nanofiber	Amine-functionalized polyacrylonitrile base → TiO_2 spray → chitosan top	Sequential electrospinning/electrospraying	Stronger cell adhesion (+charge), antibacterial (TiO_2)	Promoted NIH3T3 adhesion; active vs *K. pneumoniae*, *S. aureus*; in vitro stage only
Chitosan/PEO/$CuFe_2O_4$ nanofiber mat	Chitosan + polyethylene oxide with $CuFe_2O_4$ nanoparticles	Electrospinning	Antibacterial, proangiogenic, O_2-releasing	In mice, ↑ collagen & hair-follicle formation at day 14
Exosome-loaded chitosan–glycerol hydrogel	Human endometrial stem cell exosomes	In situ gelling injectable	Boost fibroblast migration & proliferation	14-day wound closure 83.6% vs 65.4% (blank hydrogel)
EDA-functionalized gellan-gum hydrogel + MSC sEVs	Gellan gum + ethylenediamine; adipose MSC small EVs	Chemical modification + hydrogel casting	Elastic, stable matrix; immunomodulatory	Under inflammatory media, rescued cell viability to 84%
Collagen dermal substitutes	Human/porcine/bovine collagen ± elastin/GAG/silicone	Freeze-dry/cross link sheets	Template for neodermis; reduces contracture	Systematic review: better scar quality @ 6 mo; RCT: ↑ elasticity
Collagen-Asiaticoside/εPLL bilayer	Collagen scaffold with asiaticoside (pro-collagen) + ε-poly-L-lysine (antimicrobial)	Layer-by-layer casting	Anti-infective, anti-inflammatory	Early-phase accel. healing; modulates TLR4/MAPK/NF-κB
Gelatin-carbonate-apatite sponge	Type-B gelatin + β-carbonate apatite (Ca^{2+} source)	Foaming/lyophilization, dehydro-crosslink	Hemostatic, Ca-mediated signaling	In rats, ↑ granulation & collagen density
PCL/gelatin nanofiber + Se-NP + vit E	Polycaprolactone/gelatin electrospun with selenium NP, vitamin E	Electrospinning	ROS-scavenging antioxidant	Full re-epithelialization, denser collagen in Wistar rats
Photosynthetic collagen scaffold	Collagen seeded with Synechococcus (± HA gene)	Cell-laden scaffold culture	O_2 generation; angiogenesis & lymphangiogenesis	Light-exposed mice: ↑ CD31, LYVE-1, VEGFR-3 vascular markers

expression of that transgene. However, insertional mutagenesis remains a concern [84]. Non-viral gene therapy systems have also been investigated. Although transfection efficiency may be lower or transient, non-viral methods may be more suitable for the regeneration of specific tissues, as genes can be delivered to local areas via lipid-based vectors, inorganic vectors, or bioscaffolds [85].

Current Wound Healing Applications

As discussed previously, topical becaplermin (PDGF) remains the only growth factor-based treatment on the market in the United States to aid in healing diabetic ulcers [17, 18]. Hoping to avoid the challenges of frequent application of topical PDGF, one group utilized a lentiviral vector to induce expression of PDGF in the full-thickness wounds created on C57BIKsJ-*db/db* diabetic mice. Two third-generation lentiviral backbones were made—one containing the human PDGF gene and another containing GFP as a control. The effect of each lentivirus was compared to that of an empty control 21 days after injection into the base and edges of the wound. Both the empty and GFP-containing vectors induced minimal PDGF expression, and collagen within the healing wound visualized via picrosirius red staining was organized into thin, disorganized fibers. Vascular density quantified by CD31 immunohistochemistry was significantly higher in mice exposed to the PDGF-containing vector compared to empty and GFP-containing controls, suggesting an angiogenic effect [86].

Although lentiviruses remain an ideal vector for achieving sustained transgene expression, there remains a risk of insertional oncogenesis [84]. To circumvent this potential adverse effect, groups have turned their attention toward other means of genetically modifying wounded tissue. Hydrogels are an attractive option due to their superb biocompatibility and ability to fine-tune design to fit diverse applications [87]. A 2025 study used these properties to engineer an injectable, self-healing, and self-regulating hydrogel that sequentially released key materials. First, low-molecular ε-polylysine (εPL) and εPL-modified oxygen-deficient nanoceria were released for the antibacterial and antioxidant properties necessary to treat bacterial biofilm infection on chronic wounds. Next, a pro-regenerative growth factor (basic FGF) was released, along with a siRNA targeting c-Jun, which together promote cellular proliferation and limit the formation of fibrotic tissue. In vitro studies confirmed antioxidant activity and revealed a strong ability to inhibit biofilm growth with minimal hemolysis, impact on cell viability, or induction of inflammation in murine organs. In an MRSA-infected chronic full-thickness wound model in mice, skins exposed to the hydrogel exhibited mature hair follicles and glands. They were also found to possess greater elasticity, more even pigmentation, and evidence of neural regeneration. The gel accelerated wound closure and decreased scar elevation in a hypertrophic scar model using a rabbit ear. Finally, transcriptomic analysis confirmed successful siRNA knockdown of profibrotic genes involved in scar formation [88] (Table 21.4).

Table 21.4 Gene delivery strategies

Strategy/vector	Genetic cargo & goal	Delivery vehicle/context	Key pre-clinical outcome(s)	Safety/ limitations
Adenoviral vectors	Up to 10.5 kb therapeutic genes (episomal)	Direct viral transduction	High efficiency; transient expression	Strong innate immunity; short duration
AAV vectors	≤ 5 kb genes; episomal	Systemic or local viral injection	7 FDA-approved therapies; good safety	Limited payload; pre-existing antibodies
Lentiviral vectors	Integrating PDGF-B, SDF-1α, etc.	Local injection into wound edges	PDGF-B LV ↑ vascular density & closure in db/db mice; SDF-1α LV rescued diabetic angiogenesis	Insertional mutagenesis risk
Gene-activated collagen scaffold	Plasmid SDF-1α/PEI polyplex	Soak-loaded freeze-dried collagen	↑ Schwann-cell pro-regenerative phenotype; ↑ endothelial branching	Non-viral; lower efficiency
Phase-adaptive hydrogel (self-healing)	Sequential bFGF + siRNA–c-Jun	Injectable ε-poly-L-lysine/ nanoceria hybrid hydrogel	Ultrafast closure; hair follicles, neural regeneration; ↓ scars in rabbit ear	Complex to manufacture; early-stage
Electroporated ADMSC	SOCS gene cassette to dampen inflammation	Cells suspended in hydrogel, injected into pressure ulcer	↓ COX-2, iNOS; transcriptomics show accelerated transition to remodeling	Cell therapy logistics; patient variability
Non-viral PEI/ lipid systems	Naked plasmid, mRNA, siRNA	Lipoplex, polyplex, inorganic NPs	Tissue-specific targeting; scalable	Typically transient; lower transfection

Rather than introducing a virus or siRNA to alter gene expression in wound tissue directly, stem cells can be modified ex vivo to produce a wound-friendly gene expression profile when reintroduced to the damaged tissue. Recently, one group electroporated adipose-derived mesenchymal stem cells (ADMSC) with a plasmid designed to induce the expression of suppressors of cytokine signaling (SOCS). In a murine model of pressure ulcers, modified ADMSCs suspended in a small volume of hydrogel were injected into the ulcerated area. After exposure to modified ADSMCs, Western blot analysis revealed decreased pro-inflammatory COX-2, pERK, p-p38, and iNOS. Subsequent RNA sequencing supported the idea that genetic upregulation resulting from SOCS-modified ADMSCs promoted transition to a later stage of wound healing [89]. Together, these ideas highlight the feasibility of manufacturing effective genetically modified cell-based therapies without a requirement for viral vectors (Fig. 21.2).

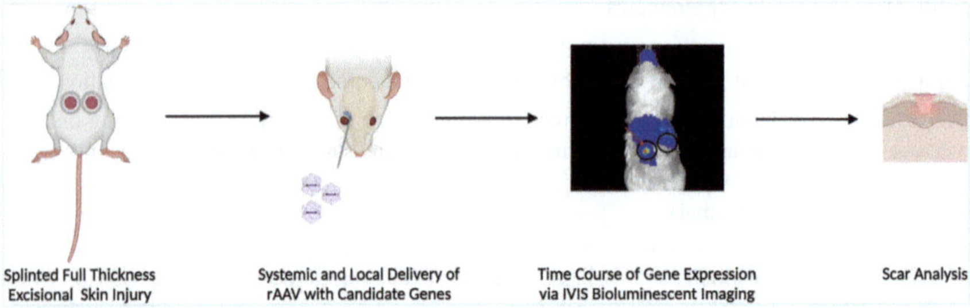

Fig. 21.2 Enhancer-based rAAV gene therapy in an adult mouse full-thickness splinted excision injury model

Summary and Future Directions

Wound healing is a complex, four-stage process involving hemostasis, inflammation, proliferation, and remodeling. Numerous cell types, including platelets, immune cells, fibroblasts, keratinocytes, and myofibroblasts, must all work in synchrony to advance healing to the next appropriate stage. Unfortunately, pathological processes frequently dysregulate the healing cascade, leading to the development of a chronic, non-healing wound that requires intensive treatment with high rates of recurrence. Growth factor-based therapies like becaplermin established that administration of exogenous growth factors can accelerate wound healing and improve outcomes. However, heterogeneous trial data has limited its widespread use.

As our understanding of the molecular mechanisms underlying the wound healing process has grown alongside biomedical engineering and gene therapy methods, promising regenerative medicine technologies have been developed. Advanced biocompatible materials, including chitosan-, collagen-, and gelatin-based bioscaffolds, provide maximum flexibility to engineers looking to incorporate synthetic polymers and metal ions to enhance regeneration. These scaffolds may also serve as a delivery device for gene therapies, which work to alter the cell's intrinsic machinery. Viral vectors offer a reliable method of gene delivery, and several AAV-based therapies have been FDA-approved. However, non-viral methods of controlling gene expression are currently under development, intending to limit the risk of immunogenicity, insertional oncogenesis, or other off-target effects.

Together, these advances reflect the scientific community's more nuanced understanding of wound healing. Currently, in vitro studies are plentiful, and several in vivo studies using rodent and porcine models have demonstrated strong preliminary data supporting their efficacy. However, transitioning these advances into the clinical space will require standardized, quantitative methods of evaluating wound healing and a careful cost-benefit analysis that accounts for both statistical and clinical significance.

References

1. Aldridge S, Teichmann SA (2020) Single cell transcriptomics comes of age. Nat Commun 11:4307
2. Gurevich EV, Gurevich VV (2014) Therapeutic potential of small molecules and engineered proteins. In: Gurevich VV (ed) Arrestins—pharmacology and therapeutic potential. Springer, Berlin, Heidelberg, pp 1–12. https://doi.org/10.1007/978-3-642-41199-1_1
3. Sen CK (2019) Human wounds and its burden: an updated compendium of estimates. Adv Wound Care 8:39–48
4. Fernandez-Guarino M, Naharro-Rodriguez J, Bacci S (2024) Aberrances of the wound healing process: a review. Cosmet 11:209
5. Wernick B, Nahirniak P, Stawicki SP (2025) Impaired wound healing. In: StatPearls. StatPearls Publishing, Treasure Island (FL)
6. Rodrigues M, Kosaric N, Bonham CA, Gurtner GC (2018) Wound healing: a cellular perspective. Physiol Rev. https://doi.org/10.1152/physrev.00067.2017
7. Davis SC et al (2025) Protection with a collagen wound matrix containing polyhexamethylene biguanide supports innate wound healing in biofilm-infected porcine wounds. Wound Repair Regen 33:e70025
8. Yang H, Zhang X, Xue B (2024) New insights into the role of cellular senescence and chronic wounds. Front Endocrinol 15
9. Sen CK (2021) Human wound and its burden: updated 2020 compendium of estimates. Adv Wound Care. https://doi.org/10.1089/wound.2021.0026
10. Nussbaum SR et al (2018) An economic evaluation of the impact, cost, and Medicare policy implications of chronic nonhealing wounds. Value Health 21:27–32
11. Armstrong DG, Boulton AJM, Bus SA (2017) Diabetic foot ulcers and their recurrence. N Engl J Med 376:2367–2375
12. Chen L, Sun S, Gao Y, Ran X (2023) Global mortality of diabetic foot ulcer: a systematic review and meta-analysis of observational studies. Diabetes Obes Metab 25:36–45
13. BioRender. https://app.biorender.com/illustrations/canvas-beta/68346e15a0e34bbc0def4f7b
14. Mustoe TA et al (1987) Accelerated healing of incisional wounds in rats induced by transforming growth factor-β. Science 237:1333–1336
15. Barrientos S, Stojadinovic O, Golinko MS, Brem H, Tomic-Canic M (2008) Perspective arti cle: growth factors and cytokines in wound healing. Wound Repair Regen 16:585–601
16. Heldin C-H, Westermark B (1999) Mechanism of action and in vivo role of platelet-derived growth factor. Physiol Rev 79:1283–1316
17. Veith AP, Henderson K, Spencer A, Sligar AD, Baker AB (2019) Therapeutic strategies for enhancing angiogenesis in wound healing. Adv Drug Deliv Rev 146:97–125
18. Yamakawa S, Hayashida K (2019) Advances in surgical applications of growth factors for wound healing. Burns Trauma 7:10
19. Robson MC et al (1992) Recombinant human platelet-derived growth factor-BB for the treatment of chronic pressure ulcers. Ann Plast Surg 29:193–201
20. Steed DL, Study Group, the D. U (1995) Clinical evaluation of recombinant human platelet—derived growth factor for the treatment of lower extremity diabetic ulcers. J Vasc Surg 21:71–81
21. Steed DL (2006) Clinical evaluation of recombinant human platelet-derived growth factor for the treatment of lower extremity ulcers. Plast Reconstr Surg 117:143S
22. Margolis DJ, Bartus C, Hoffstad O, Malay S, Berlin JA (2005) Effectiveness of recombinant human platelet-derived growth factor for the treatment of diabetic neuropathic foot ulcers. Wound Repair Regen 13:531–536
23. Steed DL, Donohoe D, Webster MW, Lindsley L (1996) Effect of extensive debridement and treatment on the healing of diabet ic foot ulcers. Diabetic Ulcer Study Group. J Am Coll Surg 183(1):61–64

24. Kallianinen LK, Hirshberg J, Marchant B, Rees RS (2000) Role of platelet-derived growth factor as an adjunct to surgery in the management of pressure ulcers. Plast Reconstr Surg 106:1243
25. 20181205 FDA approves removal of boxed warning from REGRANEX gel. https://www.smith-nephew.com/en/news/2018/12/05/20181205-fda-approves-removal-of-boxed-warning-from-regranex-gel
26. Ziyadeh N, Fife D, Walker AM, Wilkinson GS, Seeger JD (2011) A matched cohort study of the risk of cancer in users of Becaplermin. Adv Skin Wound Care 24:31
27. Chen JD et al (1993) Epidermal growth factor (EGF) promotes human keratinocyte locomotion on collagen by increasing the alpha 2 integrin subunit. Exp Cell Res 209:216–223
28. Brown GL, Curtsinger L, Jurkiewicz MJ, Nahai F, Schultz G (1991) Stimulation of healing of chronic wounds by epidermal growth factor. Plast Reconstr Surg 88:189
29. Pastor JC, Calonge M (1992) Epidermal growth factor and corneal wound healing: a multicenter study. Cornea 11:311
30. Tsang MW et al (2003) Human epidermal growth factor enhances healing of diabetic foot ulcers. Diabetes Care 26:1856–1861
31. Park KH et al (2018) Topical epidermal growth factor spray for the treatment of chronic diabetic foot ulcers: a phase III multicenter, double-blind, randomized, placebo-controlled trial. Diabetes Res Clin Pract 142:335–344
32. Masoud GN, Li W (2015) HIF-1α pathway: role, regulation and intervention for cancer therapy. Acta Pharm Sin B 5:378–389
33. Li G et al (2022) Homocysteine-targeting compounds as a new treatment strategy for diabetic wounds via inhibition of the histone methyltransferase SET7/9. Exp Mol Med 54:988–998
34. Gabryelska A, Karuga FF, Szmyd B, Białasiewicz P (2020) HIF-1α as a mediator of insulin resistance, T2DM, and its complications: potential links with obstructive sleep apnea. Front Physiol 11
35. Rosenberg B, Van Camp L, Krigas T (1965) Inhibition of cell division in Escherichia coli by electrolysis products from a platinum electrode. Nature 205:698–699
36. Muhammad N, Guo Z (2014) Metal-based anticancer chemotherapeutic agents. Curr Opin Chem Biol 19:144–153
37. Li G et al (2021) A small molecule HIF-1α stabilizer that accelerates diabetic wound healing. Nat Commun 12:3363
38. Martino F, Perestrelo AR, Vinarský V, Pagliari S, Forte G (2018) Cellular Mechanotransduction: from tension to function. Front Physiol 9
39. Berry CE et al (2023) The effects of mechanical force on fibroblast behavior in cutaneous injury. Front Surg 10
40. Wong VW et al (2012) Focal adhesion kinase links mechanical force to skin fibrosis via inflammatory signaling. Nat Med 18:148–152
41. Chen K et al (2021) Disrupting biological sensors of force promotes tissue regeneration in large organisms. Nat Commun 12:5256
42. Zaman M et al (2025) Safety and efficacy of half-dose and half-fluence photodynamic therapy in chronic central serous chorioretinopathy: a systematic review and meta-analysis. Am J Ophthalmol 271:233–242
43. Mascharak S et al (2025) Inhibiting mechanotransduction prevents scarring and yields regeneration in a large animal model. Sci Transl Med. https://doi.org/10.1126/scitranslmed.adt6387
44. Haider A et al (2024) Chitosan as a tool for tissue engineering and rehabilitation: recent developments and future perspectives—a review. Int J Biol Macromol 278:134172
45. Frontiers | Recent Advances in Bioengineered Scaffolds for Cutaneous Wound Healing. https://www.frontiersin.org/journals/bioengineering-and-biotechnology/articles/10.3389/fbioe.2022.841583/full
46. Khan R et al (2025) Preparation of amine-functionalized polyacrylonitrile-TiO2-chitosan multilayer nanofibers as a potential wound dressing: characterization and investigation of in vitro cell viability, proliferation and antibacterial study. Int J Biol Macromol 305:141006

47. Sharifi E, Jamaledin R, Familsattarian F, et al (2023) Bioactive chitosan/poly(ethyleneoxide)/CuFe2O4 nanofibers for potential wound healing. Environ Res 239(Pt 2):117448. https://doi.org/10.1016/j.envres.2023.117448
48. Edgar JR (2016) Q&A: what are exosomes, exactly? BMC Biol 14:46
49. Nooshabadi VT et al (2020) Impact of exosome-loaded chitosan hydrogel in wound repair and layered dermal reconstitution in mice animal model. J Biomed Mater Res A 108:2138–2149
50. Amine-Functionalized Gellan Gum-Based Hydrogel Loaded with Adipose Stem Cell-Derived Small Extracellular Vesicles: An In Vitro Proof of Concept for Enhancing Diabetic Foot Ulcer Healing. https://www.mdpi.com/2310-2861/11/2/119
51. Shoulders MD, Raines RT (2009) Collagen structure and stability. Annu Rev Biochem 78:929–958
52. Chattopadhyay S, Raines RT (2014) Collagen-based biomaterials for wound healing. Biopolymers 101:821–833
53. Postlethwaite AE, Seyer JM, Kang AH (1978) Chemotactic attraction of human fibroblasts to type I, II, and III collagens and collagen-derived peptides. Proc Natl Acad Sci USA 75:871–875
54. Parenteau-Bareil R, Gauvin R, Berthod F (2010) Collagen-based biomaterials for tissue engineering applications. Materials 3:1863–1887
55. Chang DK, Louis MR, Gimenez A, Reece EM (2019) The basics of integra dermal regeneration template and its expanding clinical applications. Semin Plast Surg 33:185–189
56. Dickson K et al (2023) A histological and clinical study of MatriDerm® use in burn reconstruction. J Burn Care Res 44:1100–1109
57. Glynou SP et al (2024) A comparison of acellular dermal matrices (ADM) efficacy and complication profile in women undergoing implant-based breast reconstruction: a systematic review and network meta-analysis. BMC Cancer 24:1598
58. Khan N, Wickman M, Schultz I (2024) A long-term evaluation of acellular dermal matrix for immediate implant-based breast reconstruction following risk-reducing mastectomy. Plast Reconstr Surg Glob Open 12:e5951
59. van den Bosch AS et al (2024) Outcomes of dermal substitutes in burns and burn scar reconstruction: a systematic review and meta-analysis. Wound Repair Regen 32:960–978
60. Gardien KLM et al (2023) Short- and long-term outcomes of an acellular dermal substitute versus standard of care in burns and reconstructions: a phase I/II Intrapatient randomized controlled trial. Adv Skin Wound Care 36:540
61. Lee J et al (2006) Asiaticoside induces human collagen I synthesis through TGFβ receptor I kinase (TβRI kinase)-independent Smad signaling. Planta Med 72:324–328
62. Tang B et al (2011) Asiaticoside suppresses collagen expression and TGF-β/Smad signaling through inducing Smad7 and inhibiting TGF-βRI and TGF-βRII in keloid fibroblasts. Arch Dermatol Res 303:563–572
63. Hagh Ranjbar H, Hosseini-Abari A, Ghasemi SM, Hafezi Birgani Z (2023) Antibacterial activity of epsilon-poly-l-lysine produced by Stenotrophomonas maltophilia HS4 and Paenibacillus polymyxa HS5, alone and in combination with bacteriophages. Microbiology 169:001363
64. A collagen-AS/εPLL bilayered artificial substitute regulates anti-inflammation and infection for initial inflamed wound healing—Biomaterials Science (RSC Publishing). https://pubs.rsc.org/en/content/articlelanding/2021/bm/d1bm01071a#tab1
65. Pereira RVS et al (2024) Protection of stromal cell-derived factor-1 SDF-1/CXCL12 against proteases yields improved skin wound healing. Front Immunol 15
66. Ceradini DJ et al (2004) Progenitor cell trafficking is regulated by hypoxic gradients through HIF-1 induction of SDF-1. Nat Med 10:858–864
67. Feng G, Hao D, Chai J (2014) Processing of CXCL12 impedes the recruitment of endothelial progenitor cells in diabetic wound healing. FEBS J 281:5054–5062
68. Badillo AT, Chung S, Zhang L, Zoltick P, Liechty KW (2007) Lentiviral gene transfer of SDF-1α to wounds improves diabetic wound healing. J Surg Res 143:35–42

69. Laiva AL, O'Brien FJ, Keogh MB (2021) SDF-1α gene-activated collagen scaffold drives functional differentiation of human Schwann cells for wound healing applications. Biotechnol Bioeng 118:725–736
70. Jessen KR, Mirsky R (2016) The repair Schwann cell and its function in regenerating nerves. J Physiol 594:3521–3531
71. Laiva AL, O'Brien FJ, Keogh MB (2021) SDF-1α gene-activated collagen scaffold enhances provasculogenic response in a coculture of human endothelial cells with human adipose-derived stromal cells. J Mater Sci Mater Med 32:26
72. Jian Y et al (2024) Lymphangiogenesis: novel strategies to promote cutaneous wound healing. Burns Trauma 12:tkae040
73. Fuchs B et al (2025) Bioactivated scaffolds promote angiogenesis and lymphangiogenesis for dermal regeneration in vivo. J Tissue Eng. https://doi.org/10.1177/20417314251317542
74. Olsen D et al (2003) Recombinant collagen and gelatin for drug delivery. Adv Drug Deliv Rev 55:1547–1567
75. Van Den Bulcke AI et al (2000) Structural and rheological properties of methacrylamide modified gelatin hydrogels. Biomacromolecules 1:31–38
76. Lansdown ABG (2002) Calcium: a potential central regulator in wound healing in the skin. Wound Repair Regen 10:271–285
77. Ana ID et al (2025) Newly developed calcium-enriched hemostatic sponge enhanced wound healing and regeneration. J Biomed Mater Res B Appl Biomater 113:e35568
78. Doostmohammadi M et al (2021) Bioactive anti-oxidative polycaprolactone/gelatin electrospun nanofibers containing selenium nanoparticles/vitamin E for wound dressing applications. J Biomater Appl 36:193–209
79. Shaabani E et al (2022) Gene therapy to enhance angiogenesis in chronic wounds. Mol Ther Nucleic Acids 29:871–899
80. Kohn DB, Chen YY, Spencer MJ (2023) Successes and challenges in clinical gene therapy. Gene Ther 30:738–746
81. Bulcha JT, Wang Y, Ma H, Tai PWL, Gao G (2021) Viral vector platforms within the gene therapy landscape. Signal Transduct Target Ther 6:1–24
82. Lee CS et al (2017) Adenovirus-mediated gene delivery: potential applications for gene and cell-based therapies in the new era of personalized medicine. Genes Dis 4:43–63
83. Byrne BJ et al (2025) Current clinical applications of AAV-mediated gene therapy. Mol Ther. https://doi.org/10.1016/j.ymthe.2025.04.045
84. Marcucci KT et al (2018) Retroviral and Lentiviral safety analysis of gene-modified T cell products and infused HIV and oncology patients. Mol Ther 26:269–279
85. Wu P et al (2018) Non-viral gene delivery systems for tissue repair and regeneration. J Transl Med 16:29
86. Lee JA et al (2005) Lentiviral transfection with the PDGF-B gene improves diabetic wound H ealing. Plast Reconstr Surg 116:532–538
87. Wu X et al (2025) DNA-based hydrogels for bone regeneration: a promising tool for bone organoids. Mater Today Bio 31:101502
88. Zhang F et al (2025) A dynamically phase-adaptive regulating hydrogel promotes ultrafast anti-fibrotic wound healing. Nat Commun 16:3738
89. Eom Y et al (2023) Therapeutic effects and underlying mechanism of SOCS-com gene-transfected ADMSCs in pressure ulcer mouse models. Cells 12:1840

Chapter 22
Regenerative Therapy for Burns

Ahmed T. El-Serafi, Raghad Alghazali, Iman M. Talaat, Folke Sjöberg, Moustafa Elmasry, Maha Saber-Ayad, and Ibrahim El-Serafi

Abstract Burn injuries represent a significant clinical challenge due to the skin's complex structure and essential barrier function. This chapter explores the evolving landscape of regenerative therapies aimed at enhancing skin repair and restoring function after burn trauma. The chapter focuses on the therapeutic potential of stem cell-based interventions in vitro and in vivo models. Various stem cell sources are evaluated for their regenerative capacity and immunomodulatory effects. The chapter also reviews current clinical trials applying these technologies in burn patients, focusing on safety profiles, efficacy outcomes, and translational chal-

A. T. El-Serafi (✉)
Department of Medical Laboratory Sciences, College of Health Sciences, University of Sharjah, Sharjah, United Arab Emirates

Department of Hand Surgery, Plastic Surgery and Burns, Linköping University Hospital, Linköping, Sweden

Department of Biomedical and Clinical Sciences, Linköping University, Linköping, Sweden
e-mail: ahmed.elserafy@liu.se

R. Alghazali
Department of Hand Surgery, Plastic Surgery and Burns, Linköping University Hospital, Linköping, Sweden

Department of Biomedical and Clinical Sciences, Linköping University, Linköping, Sweden

I. M. Talaat · M. Saber-Ayad
Clinical Sciences Department, College of Medicine, University of Sharjah, Sharjah, United Arab Emirates

F. Sjöberg
Department of Biomedical and Clinical Sciences, Linköping University, Linköping, Sweden

M. Elmasry
Department of Hand Surgery, Plastic Surgery and Burns, Linköping University Hospital, Linköping, Sweden

I. El-Serafi
Basic Medical Sciences Department, College of Medicine, Ajman University, Ajman, United Arab Emirates

F. De Francesco, K. Turksen (eds.), *Management and Strategies for Wound Healing*, Stem Cell Biology and Regenerative Medicine 79,
https://doi.org/10.1007/978-3-032-12442-5_22

lenges. Advances in biomaterials, including scaffolds, hydrogels, and bioengineered skin substitutes, are discussed in relation to their role in enhancing stem cell delivery and integration. Finally, the chapter addresses the regulatory landscape governing the clinical use of advanced therapy medicinal products (ATMPs) and highlights emerging directions in the field. Collectively, these insights offer a thorough analysis of current strategies and future directions in regenerative medicine for burn care.

Keywords Skin regeneration · Burns · Stem cells · Exosomes · Biomaterials · Advanced therapy

Introduction

The global incidence of burn injuries is estimated to be around nine million cases annually, resulting in over 100,000 deaths and economic losses exceeding USD 112 billion [1]. Burn wounds present a formidable challenge in the field of medical treatment due to the complex interplay of physiological and logistical factors that may hinder surgical intervention. The overall health status of the patient can severely limit treatment options, and in cases of extensive burns, the scarcity of available donor sites for autologous skin grafting poses additional difficulties [2]. To address these challenges, a range of medical and surgical strategies have been developed to enhance patient outcomes. Among these is the use of biomaterials designed as temporary wound coverings, which protect the injury site while fostering an optimal environment for healing [3]. Additionally, the advancement of dermal skin grafts has emerged as a promising technique, enabling more effective reconstruction of skin structural integrity [4]. Furthermore, cell-based therapies have gained traction as a significant area of research, offering alternative solutions for challenging cases that resist conventional treatment methods [5].

Despite recent progress, the continuous development of innovative strategies for managing burn wounds remains crucial, particularly in light of improved survival rates observed in recent years. One of the most promising avenues in this pursuit is stem cell therapy, a revolutionary regenerative approach that aims to accelerate wound healing and restore both the functional and aesthetic qualities of the injured skin. This strategy holds the potential to significantly improve the quality of life for burn survivors. Ultimately, sustained commitment to research and clinical translation of these novel therapies is essential to ensure that patients receive the most effective and compassionate care possible.

Current Approaches for Burn Skin Wound Management

Advances in regenerative medicine have introduced innovative treatment modalities for severe burns, encompassing both cell-based and cell-free therapies. These approaches address a critical clinical need by offering potential solutions for extensive and refractory burn wounds. Since the 1970s, the management of burn wounds, particularly those prone to prolonged healing, has evolved significantly, reflecting advancements in medical practices and technologies [6, 7]. One landmark advancement was the introduction of burn excision followed by grafting, a technique that has been demonstrated to reduce mortality and improve clinical outcomes in patients with severe burns. Initially, this approach involved the total excision of burned tissue, followed by immediate coverage of the affected areas using wide-meshed skin grafts, providing a foundational strategy for managing extensive burn wounds.

Successful grafting requires prior treatment and control of localized wound infections, as microbial colonization can compromise graft take and overall healing [8]. Despite its widespread use, skin grafting poses notable limitations, including donor site morbidity and the scarcity of available healthy skin in patients with extensive injuries [9, 10]. Moreover, the final aesthetic and functional outcomes remain suboptimal in many cases, often deemed unsatisfactory by both patients and clinicians [6, 7, 11]. To address these challenges, alternative techniques, including staged excision, temporary wound coverage using allografts and/or xenografts, and the use of dermal grafts, have been developed [4]. Among emerging techniques, enzymatic debridement has garnered attention as a less invasive alternative; however, the current body of scientific evidence regarding its efficacy is inconsistent and warrants further validation [12].

In parallel, advancements in the field of regenerative medicine have paved the way for the introduction of new therapeutic modalities specifically designed for patients suffering from severe injuries, including autologous and allogenic keratinocyte single-cell suspensions. These cells may be either cultured or freshly isolated and can be sprayed directly onto the wound site, often in combination with wide-meshed skin grafts or as a standalone intervention [13]. Cultured epidermis and compound skin analogues have been approved as life-saving therapies in select contexts, although their application in extensive or complex burns remains a subject of ongoing research [14]. One particularly promising direction involves the integration of cell therapies with biomaterials to create biocompatible, multilayered skin constructs. These advanced skin substitutes represent a significant leap forward in the management of major burn injuries and are currently progressing toward clinical application [3, 15].

Stem Cells and Skin Regeneration Under Physiological Conditions

Stem cells play a fundamental role in maintaining skin homeostasis and facilitating regeneration. Exposure to chemical and physical environmental stressors can compromise skin integrity, impairing its function as the body's first line of defense. In response to such insults, locally residing stem cells are primarily responsible for initiating skin repair, while systemically circulating stem cells are recruited to aid in tissue regeneration. These stem cells possess the capacity to differentiate into multiple skin cell types, thereby restoring the epidermal barrier and promoting the re-establishment of healthy skin.

A deeper understanding of the mechanisms that govern stem cell activation, migration, and differentiation in response to injury is critical for the development of effective therapeutic strategies. Such insights can inform the design of targeted regenerative interventions aimed at improving burn wound healing and optimizing skin repair under both physiological and pathological conditions.

Resident Stem Cells in the Skin

Skin-resident stem cells are categorized based on their location and are essential for the continuous regeneration of the epidermis, hair follicles, and sebaceous glands, thereby maintaining the structural and functional integrity of the skin. Upon tissue injury, these cells differentiate, initiating the repair process. Key resident stem cells include Epidermal Stem Cells (EpSCs), Hair Follicle Stem Cells (HFSCs), Sebaceous Gland Stem Cells (SGSCs), and Interfollicular Stem Cells (IFSCs). EpSCs are found in the basal layer of the epidermis and are essential for preserving the skin barrier and facilitating wound repair. They exhibit self-renewal capacity and can differentiate into keratinocytes and other skin-associated cells [16]. HFSCs reside in the hair follicle bulge and contribute to hair growth and cycling. HFSCs are multipotent and capable of differentiating into various epidermal and follicular cell types [17]. SGSCs share a common origin with HFSCs and are characterized by the expression of the Lrig1+ marker. SGSC can differentiate into sebocytes, playing a crucial role in sebaceous gland regeneration [18]. IFSCs are multipotent cells found in the interfollicular epidermis and play a crucial role in epidermal homeostasis and skin regeneration. Following injury, IFSCs migrate to the wound sites, contributing to early stages of re-epithelialization before other stem cell populations are recruited [19]. Additionally, adipose-derived stem cells (ADSCs), a subclass of mesenchymal stem cells (MSCs) located in the subcutaneous tissue, play an important role in wound repair due to their proximity to injury sites. ADSC-derived exosomes have been shown to contribute to extracellular matrix (ECM) remodeling by activating signaling pathways such as EPK/MAPK and inducing matrix metalloproteinase 3 (MMP-3) expression [20, 21].

Circulating Stem Cells in Skin Repair

Circulating stem cells, particularly mesenchymal stem cells (MSCs) derived from adult tissues such as bone marrow and adipose tissue, are central to systemic wound healing responses. Upon injury, these cells are mobilized by chemoattractant signals and exhibit homing behavior, allowing them to migrate to wound sites. Once localized, MSCs modulate the immune environment, release trophic factors, stimulate angiogenesis, and enhance epithelial regeneration [22, 23]. These properties position MSCs as promising agents for therapeutic applications in burn wound management.

Summary of Stem Cell Mechanisms in Burn Healing

Stem cells can contribute to burn wound healing through a range of mechanisms (Fig. 22.1):

- *Differentiation:* MSCs transform into key skin cell types, including keratinocytes and fibroblasts.
- *Paracrine Effects:* They secrete growth factors and cytokines that promote angiogenesis, modulate inflammation, and enhance cell migration.
- *Immunomodulation:* MSCs help regulate immune responses, preventing excessive inflammation and minimizing scar formation.

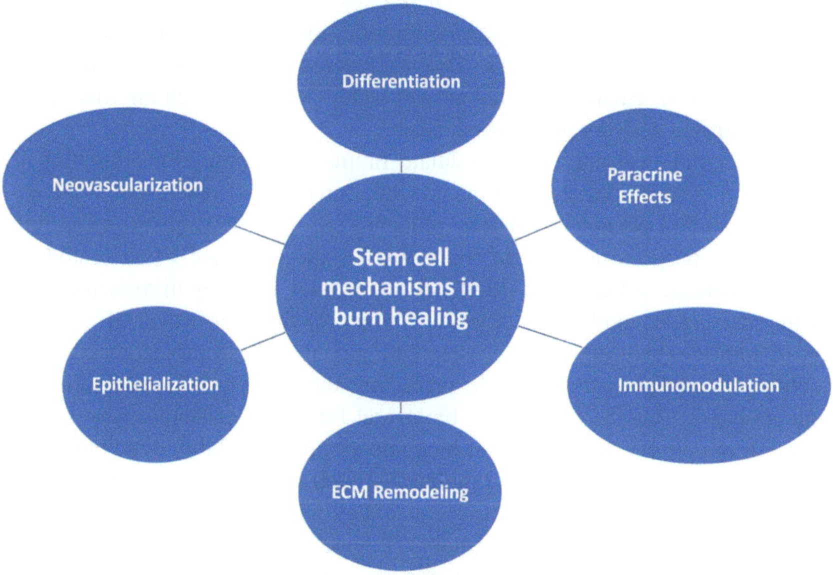

Fig. 22.1 Summary of the stem cell mechanisms in burn healing

- *ECM Remodeling:* Stem cells regulate collagen deposition and degradation, preventing fibrosis and promoting optimal wound healing.
- *Neovascularization:* MSCs stimulate the formation of new blood vessels, enhancing oxygen and nutrient delivery to the wound site.
- *Epithelialization:* They stimulate the proliferation and migration of epithelial cells, leading to rapid wound closure [24, 25].

Conditioned Media and Cell-Free Therapy

Stem cell-derived exosomes are nano-sized extracellular vesicles that facilitate intercellular communication by transferring bioactive molecules, such as proteins, lipids, and nucleic acids. These exosomes are essential for paracrine signaling, enabling stem cells to modulate both local and systemic target cells. A key aspect of their regenerative role is the delivery of growth factors and cytokines that stimulate cellular proliferation, differentiation, angiogenesis, and immune modulation. Additionally, exosomes exhibit anti-apoptotic properties, enhancing the survival of recipient cells under stress by transmitting protective molecular signals. They also contribute to ECM remodeling by regulating fibroblast activity, which reduces fibrosis and supports the restoration of normal tissue architecture. Recent studies have highlighted their role in neuroregeneration, further expanding the therapeutic potential of exosome-based therapies [26, 27].

Given their regenerative, anti-inflammatory, and immunomodulatory properties, MSC-derived exosomes have emerged as a promising cell-free therapeutic approach for burn wound healing (Fig. 22.2). These exosomes enhance wound healing by promoting fibroblast proliferation and migration, thereby accelerating wound closure and stimulating angiogenesis. This angiogenic effect is associated with the upregulation of vascular endothelial growth factor (VEGF), which improves perfusion and nutrient delivery to damaged tissue.

MSC-derived exosomes also modulate immune responses by promoting the polarization of macrophages from the pro-inflammatory M1 phenotype to the anti-inflammatory M2 phenotype, thereby reducing excessive inflammation and minimizing secondary tissue damage. Furthermore, they regulate ECM remodeling through balanced collagen deposition and degradation, reducing fibrosis and hypertrophic scarring via modulation of transforming growth factor-beta (TGF-β) signaling. In addition, exosomes also possess antioxidant and cytoprotective properties that mitigate oxidative stress, thereby protecting keratinocytes and endothelial cells from apoptosis. Beyond their innate biological functions, exosomes may serve as natural drug delivery vehicles. Their biocompatibility and ability to be engineered for targeted delivery of therapeutic molecules, such as anti-inflammatory agents or growth factors, make them attractive candidates for incorporation into advanced wound dressings or hydrogels for sustained release [28–30].

Conditioned media (CM), which is the secretome collected from cell cultures, contains a complex mixture of cytokines, chemokines, and growth factors that

Fig. 22.2 Role of stem cell-derived exosomes in burn management

regulate cellular signaling and promote tissue repair [31]. The composition of MSC-conditioned media (MSC-CM) is influenced by several factors, including the tissue source (e.g., bone marrow, adipose tissue, umbilical cord blood), the passage number of the cells, and the duration of media collection [32–34]. MSC-CM has been shown to promote the differentiation of local stem cells into keratinocytes, enhance angiogenesis, and modulate inflammation, thereby contributing to wound healing and skin regeneration. In vitro studies have demonstrated the therapeutic potential of MSC-CM in skin conditions, reporting significant wound closure rates following treatment.

Innovative Approaches in Burn Regenerative Therapy

Over the past decade, regenerative therapy for burn injuries has advanced beyond traditional autologous skin grafting toward sophisticated bioengineered strategies that combine stem cells with functional biomaterials. These approaches aim to stimulate endogenous repair mechanisms, particularly in severe or extensive burns where donor skin is limited. Technologies such as three-dimensional (3D) bioprinting, hydrogel-based delivery systems, multilayered scaffolds, and in vitro organotypic models have shown considerable promise in preclinical settings for enhancing skin regeneration and reducing scarring.

Three-Dimensional Bioprinting of Skin Constructs

3D bioprinting enables the fabrication of skin-like constructs through layer-by-layer deposition of bioinks, integrating biomaterials with living cells. Commonly used bioinks include polycaprolactone (PCL), polyethylene glycol (PEG), collagen, and chitosan, often embedded with MSCs to generate structurally organized skin-like constructs. These bioinks provide both structural support and biological cues for cell migration, differentiation, and angiogenesis.

Notably, chitosan-PEG-based hydrogels bioprinted with MSCs onto PCL/poly(L-lactide-co-ε-caprolactone) (PLCL) scaffolds have shown superior outcomes in burn models. These constructs improved wound closure, enhanced tissue integration, and facilitated paracrine signaling by MSCs, which accelerated epithelialization and reduced scar formation [35, 36].

Hydrogel-Based Biomaterial for Stem Cell Delivery in Cutaneous Repair

Hydrogels emulate the extracellular matrix, providing a hydrated, biocompatible microenvironment optimal for delivering stem cells and growth factors to wound sites. Stimuli-responsive hydrogels, such as those responsive to temperature or pH changes, are gaining popularity for their controlled release capabilities. For example, gelatin-chitosan hydrogels crosslinked with PEG have demonstrated the ability to preserve the viability of ADSCs and promote their differentiation into keratinocyte-like cells, thereby facilitating the re-epithelialization and restoration of the epidermal barrier [37, 38].

Scaffolds as a Base for Three-Dimensional Bioconstruction

Scaffolds provide essential mechanical support and biochemical environment required for skin tissue remodeling. Advanced bilayer scaffolds, often fabricated via electrospinning, are designed to mimic both the dermal and epidermal layers of human skin. Materials like silk fibroin, PCL, and polyurethane promote cell adhesion, infiltration, vascularization, and organized collagen deposition, all of which are crucial for scarless healing. Chitosan-modified scaffolds offer enhanced antibacterial properties and cell attachment, as well as improved mechanical integrity, supporting in vivo translation and reducing infection risks [39, 40].

Organotypic and 3D In Vitro Models

To bridge the translational gap between preclinical research and clinical application, 3D in vitro burn models serve as essential platforms for evaluating the efficacy and integration of engineered constructs. These organotypic constructs typically consist of fibroblasts, keratinocytes, and collagen matrices, recreating the structural and functional characteristics of native human skin. Organotypic models allow for the assessment of stem cell behavior, inflammatory responses, and re-epithelialization in a controlled setting without the ethical and logistical limitations of animal studies. These systems can also be beneficial for screening bioengineered therapies before clinical trials [38].

Although many of these bioengineered strategies remain at the preclinical or early translational stage, they offer tremendous potential for transforming burn treatment. Personalized skin constructs using a patient's cells and embedded in functional scaffolds could revolutionize treatment, especially for pediatric patients or individuals with extensive full-thickness burns. Despite challenges in manufacturing scalability, regulatory approval, and clinical standardization, advances in stem cell biology and material science continue to drive the field closer to clinical implementation.

Stem Cell Applications in Burns

Experimental Studies on Stem Cell Applications in Burns

In Vitro Studies

MSCs have emerged as promising candidates for enhancing skin regeneration, particularly in the context of burn injuries and chronic wound healing. Their multipotency, immunomodulatory capacity, and ability to secrete regenerative factors make them especially suited for integration with tissue engineering platforms. Recent in vitro studies have demonstrated that MSCs exhibit enhanced regenerative performance when cultured within three-dimensional (3D) skin equivalents, which more accurately mimic the native dermal and epidermal architecture compared to traditional two-dimensional (2D) monolayers.

For instance, Al-Shaibani et al. employed a 3D human skin model comprising dermal fibroblasts and keratinocytes to evaluate the effect of MSC-conditioned medium. Their findings indicated accelerated wound closure and enhanced epidermal cell differentiation within the matrix, suggesting potent paracrine activity of MSCs in a physiologically relevant setting [41]. Similarly, Li et al. reported that human ADSCs cultured on a collagen sponge scaffold under 3D conditions display superior epithelial differentiation compared to those maintained in 2D cultures, underscoring the pivotal role of spatial architecture in directing stem cell fate and improving the development of skin substitutes [42]. Furthermore, MSCs are increasingly considered as viable alternatives to autologous keratinocytes, which are limited by donor site availability and slow expansion rates, particularly in extensive burns [13]. Advanced 3D wound models have further integrated MSCs into bilayer protein-based scaffolds, such as silk fibroin, to facilitate scarless tissue regeneration. Gholipourmalekabadi et al. demonstrated that such construct successfully restored epidermal barrier function and promoted organized tissue regeneration in third-degree burn simulations [43].

Importantly, MSC secretome—comprising cytokines, growth factors, and extracellular vesicles—has been shown to exhibit enhanced therapeutic potency when cells are cultured in 3D systems. Gangadaran et al. showed that the secretome derived from 3D-cultured MSCs led to significantly improved wound healing in a murine burn model compared to 2D-derived secretions, likely due to the enriched bioactivity of the 3D-conditioned medium [44].

In a complementary approach, Schneider et al. developed an in vitro 3D epidermal burn model by applying a thermal rod heated to 83 °C for 7 min, inducing a wound covering 25% of the model's surface. This setup enabled precise metabolic and molecular analysis of keratinocyte responses to thermal stress and facilitated the assessment of therapeutic interventions over time. However, as this model replicates only the epidermal layer, its application remains limited to assessing first-degree burns [45].

Collectively, these studies underscore the importance of organotypic 3D skin models in elucidating the mechanistic underpinnings of MSC differentiation, paracrine signaling, and wound repair. They also represent a critical step in advancing translational tissue-engineered skin substitutes, offering new avenues for clinical applications in burn medicine.

Animal Models for Burn Research

Animal models are essential for advancing our understanding of the pathophysiology and therapeutic interventions related to thermal skin injuries. While in vitro systems offer controlled environments to study specific cellular responses, they often fail to replicate the complex physiological interactions seen in vivo. Therefore, animal models, particularly murine and porcine, are indispensable for investigating wound healing dynamics, immune responses, and the efficacy of regenerative treatments.

A major challenge in translational burn research has been the lack of a suitable animal model that accurately mimics the structural, immunological, and reparative characteristics of human skin wounds [46]. Various species (mice, rats, rabbits, and pigs) have been employed, each with unique advantages and limitations. Selecting the appropriate model requires careful consideration of translational relevance, including anatomical and physiological congruence with human skin, immune function, and the cost and ethical burden of animal use (Table 22.1) [47]. Rodents, particularly mice and rats, are commonly used due to their availability, ease of genetic manipulation, and cost-effectiveness. However, their skin structure differs significantly from that of humans. Mice possess a much thinner dermis and epidermis, high hair follicle density, and a subcutaneous muscle layer (panniculus carnosus) that enables rapid wound contraction, which is an uncommon healing mechanism in humans [48]. This contraction-based healing contrasts with human skin repair, which predominantly occurs through re-epithelialization and granulation tissue formation [49, 50]. Additionally, mice rarely develop hypertrophic or keloid scars, further limiting their applicability for modeling fibrotic wound outcomes. Mouse skin is also enriched in progenitor cells due to dense hair follicles, leading to enhanced keratinocyte activity not typically observed in humans [51, 52]. Furthermore, significant interspecies differences exist in immune responses. For example, mice lack the ability to produce human-like defensins, key antimicrobial peptides involved in maintaining barrier integrity and preventing infection during wound healing [53–55]. Similarly, essential chemokines such as CXCL8 (IL-8), CXCL7, and CXCL11, which regulate inflammation, angiogenesis, and epithelialization in humans, are absent in murine systems [53–56].

Rat models share several limitations with mice but are widely used in burn studies due to their larger body size, which facilitates surgical manipulation and sampling. A systematic review of preclinical burn studies between 2010 and 2020 found that 15 studies used rat models and 6 used mice, typically involving cohorts of approximately 30 animals per study [57]. The regenerative processes in rats also depend on wound contraction mediated by the panniculus carnosus, which accelerates closure but does not replicate human-like healing pathways [48, 58].

Among large animal models, pigs are, histologically and physiologically, the most similar to humans, with comparable epidermal thickness, dermal structure, and healing mechanisms. They are capable of developing hypertrophic scars and possess similar immune profiles, making them the most robust model for translational burn research (Table 22.1). Despite higher costs and ethical considerations, porcine models offer a critical bridge between rodent studies and clinical application [59].

The translational utility of animal models has been supported by meta-analyses of preclinical studies using stem cell therapies. A recent meta-analysis encompassing 400 burn wound animals, including 211 treated with stem cells, revealed that stem cell therapy significantly enhanced wound healing rates, angiogenesis, and vascular endothelial growth factor (VEGF) expression, while simultaneously reducing the expression of pro-inflammatory cytokines IL-1 and TNF-α [60]. Despite variation in stem cell sources and delivery protocols, mesenchymal stem cells (MSCs) consistently demonstrated beneficial effects across multiple studies [57].

The anti-inflammatory potential of stem cells was evident in most of the included studies, with 10 out of 12 studies reporting a reduction in inflammation, while two studies yielded inconclusive results. Neovascularization was enhanced in 12 of the 13 studies that investigated this parameter, indicating a strong angiogenic effect of stem cell therapy. Granulation tissue formation improved in 13 of 14 studies, suggesting accelerated tissue remodeling processes. Similarly, 11 out of 12 studies reported improved re-epithelialization of the wound bed, underscoring the

Table 22.1 Comparison of animal models in burn research

Factor	Humans	Rodents (mice/rats)	Pigs	Non-human primates (NHPs)
Epidermal & dermal thickness	Thick epidermis, well-defined dermis	Thin epidermis and dermis, less defined layers [87]	Similar epidermal and dermal thicknesses to humans [88]	Similar skin structure, but ethical concerns limit usage [89]
Wound healing mechanism	Re-epithelialization, granulation tissue, scar formation	Healing mainly by contraction due to the panniculus carnosus muscle [90]	Healing by re-epithelialization, granulation, and fibrosis, similar to humans [88]	Similar healing process, but limited availability and high ethical concerns [91]
Immune & inflammatory response	Balanced neutrophil-macrophage response, cytokine cascade activation	More neutrophil-dominant response, different cytokine profiles [91]	Similar inflammatory response and cytokine production in humans [88]	Closely mimics human immune response but is limited by ethical and logistical constraints [92]
Burn depth & injury progression	Zones of coagulation, stasis, and hyperemia	Rapid burn healing, minimal depth progression [87]	Develops classic burn wound zones, similar to humans [89]	Mimics human burn progression, but ethical limitations restrict use [93]
Skin architecture	Firmly attached	Loose	Firmly attached	Firmly attached
Hair cycle	Up to several years	3 weeks for mice	Uncertain or N/A	Uncertain or N/A
L-gluconolactone	Absent	Present in rats, relevant in wound healing, as vitamin C plays a vital role in collagen synthesis [94]	Uncertain or N/A	Uncertain or N/A

(continued)

Table 22.1 (continued)

Factor	Humans	Rodents (mice/rats)	Pigs	Non-human primates (NHPs)
Scar formation & fibrosis	Prone to hypertrophic scars and fibrosis	Minimal scarring due to contraction-based healing [90]	Develops hypertrophic scars similar to humans [88]	Scar formation is comparable to humans, but there is limited research due to ethical concerns [91]
Pain perception & ethical concerns	High pain perception requires intensive analgesia management	High metabolic rate, rapid systemic response, requires analgesia [87]	Similar pain response to humans, but requires careful ethical consideration [88]	High ethical concerns require specialized care and justification for use [89]

regenerative capabilities of MSCs. However, the impact on scar remodeling and fibrosis was more variable; five of eight studies documented a positive effect on tissue remodeling, while two were inconclusive and one reported no observable difference. Wound healing progression was most frequently assessed through non-invasive parameters, such as wound closure rate, while four studies incorporated histological scoring systems to evaluate tissue maturation and structural restoration [57].

However, interpretation of these results requires caution due to biases in study design. Issues such as lack of randomization, inadequate blinding, and selective reporting were prevalent across many preclinical studies. Furthermore, the inherent biological differences across species, especially regarding immune and reparative mechanisms, complicate data extrapolation to humans. The heterogeneity in cell preparation, application methods, and outcome measures also presents significant challenges for meta-analysis and reproducibility.

In conclusion, animal models remain indispensable tools in burn research but must be selected and interpreted with a thorough understanding of their species-specific limitations. Future studies should aim to standardize protocols, enhance methodological rigor, and prioritize models with high translational fidelity to optimize the path from bench to bedside.

Clinical Studies on Stem Cell Applications in Burns

The diversity in cell preparation and delivery methods noticed in preclinical studies extends into clinical trials, which can be classified by the mode of stem cell application.

Stem Cells Applied in Suspension

Administering stem cells in suspension remains a fundamental clinical approach. In a randomized clinical trial (NCT03686449), keratinocyte suspensions enriched with adipose stem cells (ADSCs) are compared with standard grafting, with the primary outcome being time to 95% wound closure. Similarly, another trial (NCT01443689) uses a combination of allogenic cord blood mononuclear and human umbilical cord mesenchymal stem cells (hUC-MSCs) to treat patients with moderate to severe full-thickness burns, aiming to assess long-term skin quality 6 months post-intervention. Additionally, a dose-escalation study (NCT02104713) is evaluating the safety of administering allogenic MSCs in patients with second-degree burns.

Stem Cells in Combination with Biomaterials

The use of biomaterials in conjunction with stem cells enhances therapeutic efficacy by offering structural support and improving cell viability. The biomaterials could be natural or synthetic in origin [3]. Natural biomaterials include decellularized amniotic membrane seeded with keratinocytes and amniotic epithelial stem cells, currently under evaluation in a trial (NCT05652816) comparing it to conventional split-thickness skin grafts in extensive burns. Another approach employs progenitor biological bandages, containing fibroblast progenitors embedded in collagen sheets, currently undergoing phase I/II trials for second-degree burns and donor sites (NCT05339490). Furthermore, umbilical cord-derived MSCs are being tested as biological additives to various skin grafts in an ongoing study (NCT06122532) to assess healing rates within 3–4 weeks.

Synthetic biomaterials include a long list of materials that can be custom-made to fit the wound's shape and size. A multi-phase trial series (NCT02394873, NCT03183622, NCT02619851, NCT03183648) has explored the application of ADSC-laden hydrogel sheets in deep second-degree burns, focusing on parameters such as re-epithelialization, scar formation, and safety. Another trial (NCT03113747) evaluates ADSC hydrogels as adjuncts to meshed autografts to target inflammation and epithelialization. Integra®, a dermal substitute, is being investigated in another study (NCT05344521) with and without autologous cells to assess its effect on healing outcomes and scar quality. Additionally, a trial using medical-grade collagen loaded with hUC-MSCs (NCT02672280) has targeted a range of skin defects, including burn wounds.

Stem Cell-Free Trials

To replicate stem cell benefits without live cell implantation, stem cell-free therapies are under investigation. A phase I trial (NCT05078385) assesses the safety of extracellular vesicles derived from bone marrow MSCs in burn wound healing.

Another study (NCT04235296) investigates the regenerative potential of MSC-conditioned media as a non-cellular alternative to conventional epidermal growth factors.

Stem Cells for Burn Scar Management

Several registered clinical trials focus on post-burn scar management. Trials such as NCT05004779, NCT04557514, and NCT03352297 assess interventions, including non-ablative laser paired with stem cell-conditioned media, as well as stromal vascular fraction (SVF)-rich lipoaspirate and nanofat injections, both of which are rich in ADSCs. These studies reported improvements in scar pigmentation, pliability, and overall appearance. A recruiting trial (NCT05610878) investigates the efficacy of allogenic ADSCs combined with autologous fat grafts for treating asymmetric facial scars. Furthermore, SVF application in combination with CO_2 laser therapy (IRCT20210515051307N1) has shown significant improvements in scar characteristics. Nanofat grafting, an advanced SVF isolation technique, was used as an adjunct to autologous fat grafting (NCT03791710), yielding promising outcomes in hospital stay duration, grafting requirements, and scar quality.

Collectively, these clinical trials (Table 22.2) reflect a paradigm shift toward regenerative and cell-free therapies in burn treatment, with an emphasis on stem cell-driven wound repair and immune modulation. Once results from these studies become fully available, they may pave the way for evidence-based integration of stem cell therapies into routine clinical practice. Moreover, these studies establish a foundation for developing future trials that are optimized for specific patient populations and local healthcare systems.

Key Clinical and Regulatory Hurdles in Bringing Regenerative Therapies for Burns to Market

Cell-based therapy for burns represents a transformative approach to addressing the significant clinical challenges posed by severe burns. However, the development and commercialization of these therapies encounter considerable clinical and regulatory obstacles. These hurdles range from the biological limitations of skin regeneration to the stringent regulatory requirements for therapeutic approval (Table 22.3). This section examines these challenges in detail, supported by insights from recent research.

Table 22.2 Summary of the clinical trials involving stem cell application in burns

Study title	number of recruits	Method of delivery	Active component	Phase	Status	Registration number
Effect of autologous fat grafting on acute burn wound healing	100	Topical	Nanofat	III	Completed	NCT03791710
Regenerative effects of human stem cell media after laser therapy in hypertrophic scar	30	Topical	Stem cell-conditioned media	N/A	Unknown	NCT05004779
Comparative study between fat injection and platelet-rich plasma in post-burn facial scar	60	Injection	Fat	N/A	Unknown	NCT04557514
Nanofat in post-burn scars on the face	48	Injection	Nanofat	N/A	Completed	NCT03352297
Efficacy of preconditioned adipose-derived stem cells in fat grafting	24	Injection	Adipose-derived stem cells	I	Recruiting	NCT05610878
A study to evaluate the safety of ALLO-ASC-DFU in the subjects with deep second-degree burn wound	5	Hydrogel sheet	Adipose-derived stem cells	I	Completed	NCT02394873
A follow-up study to evaluate the safety of ALLO-ASC-DFU in ALLO-ASC-BI-101 clinical trial	5	Hydrogel sheet	Adipose-derived stem cells (follow-up for study NCT02394873)	N/A	Completed	NCT03183622
A clinical trial to evaluate the safety and efficacy of ALLO-ASC-DFU for subjects with second-degree burn injury subjects	20	Hydrogel sheet	Adipose-derived stem cells	2	Completed	NCT02619851

A follow-up study to evaluate the safety of ALLO-ASC-DFU in ALLO-ASC-BI-201 clinical trial	14	Hydrogel sheet	Adipose-derived stem cells (follow-up for study NCT02619851)	N/A	Completed	NCT03183648
Umbilical cord mesenchymal stem cells for the repair of large area burn wounds	20	Topical	Umbilical cord mesenchymal stem cells	N/A	Not yet recruiting	NCT06122532
Biological skin graft with keratinocyte-stem cell co-culture for burn patients	20	Decellularized amnion bilayer	Autologous keratinocyte co-cultured with amniotic epithelial stem cells	I	Unknown	NCT05652816
Use of progenitor biological bandages in burn care	76	Horse collagen scaffold	Human skin progenitor cells	I/II	Not yet recruiting	NCT05339490
Allogeneic ADSCs and platelet-poor plasma fibrin hydrogel to treat the patients with burn wounds	20	Platelet-poor plasma fibrin hydrogel	Allogeneic cultured adipose-derived stem cells	I/II	Unknown	NCT03113747
Autologous keratinocyte suspension versus adipose-derived stem cell-keratinocyte suspension for post-burn raw area	33	Topical	Adipose-derived stem cell-keratinocyte suspension	N/A	Unknown	NCT03686449
A phase I, single-blind, randomized study of the safety and efficacy of cellularized Integra® using autologous burn-derived stem cells	29	Dermal bovine matrix	Autologous stem cells	I	Not yet recruiting	NCT05344521

(continued)

Table 22.2 (continued)

Study title	number of recruits	Method of delivery	Active component	Phase	Status	Registration number
Safety of extracellular vesicles for burn wounds	1	Topical	Extracellular vesicles from allogenic bone marrow-derived stem cells	I	Completed	NCT05078385
Mesenchymal stem cell-conditioned medium-derived pleiotropic factor in treating residual burn wound	30	Topical	Mesenchymal stem cell-conditioned medium	I	Unknown	NCT04235296
Evaluation of the effectiveness and satisfaction of fractional carbon dioxide laser and stromal vascular fraction in comparison with laser alone in the treatment of burn scars	10	Injection	Stromal vascular fraction	II/III	Recruitment completed	IRCT20210515051307N1
Allogenic stem cell therapy in patients with acute burn	20	Topical	Human cord blood mononuclear cells and human umbilical cord mesenchymal stem cells	I/II	Unknown	NCT01443689
Stem cell therapy to improve burn wound healing	15	Topical	Allogenic mesenchymal stem cells	I	Completed	NCT02104713
Safety and exploratory efficacy study of collagen membrane with mesenchymal stem cells in the treatment of skin defects	30	Collagen membrane	Umbilical cord-derived mesenchymal stem cells	I/II	Unknown	NCT02672280

Table 22.3 Comparison of current regenerative therapies for burns

Therapy type	Key features	Clinical status and challenges
Cultured skin substitutes	Bioengineered skin substitutes (e.g., Apligraf) composed of collagen and skin cells. Provide a ready-made epidermis/dermis	Approved for specific indications, such as diabetic foot ulcers, but limited by high production costs and immune rejection risks [9, 73, 95]
Spray-on skin	Autologous skin cell suspensions applied directly to the wound site, promoting rapid re-epithelialization	Offers improved cosmetic outcomes. Approved in Europe, Australia, and Canada; FDA-approved in 2018 (ReCell) [96]
Dermal regeneration templates	Bilayered scaffolds that promote dermal regeneration	Established for burn and complex wound management, with high engraftment rates and reduced scarring [61, 62]
Bioactive molecules	Biologically active compounds (e.g., growth factors) that regulate cellular processes to enhance skin regeneration	Emerging as a promising strategy but still in early stages of clinical translation [69]
3D bioprinting	Layer-by-layer deposition of cells and biomaterials to create functional skin tissue	In preclinical stages, with potential for large-scale production and personalized therapies [9, 75]
MSC-based therapies	Mesenchymal stem cells derived from bone marrow, adipose tissue, or umbilical cord, offering immunomodulatory and pro-angiogenic effects	Promising preclinical and early clinical data for burn wound healing but requires further clinical trials to establish safety and efficacy [97]

Practical Challenges

The limitations in current burn management strategies have driven the development of alternative strategies, such as biosynthetic scaffolds and skin substitutes, which aim to replicate the structure and function of natural skin [61, 62]. Yet, these innovations often fall short in recreating the full complexity of native skin, including its epidermal appendages like hair follicles and sweat glands, which are crucial for thermoregulation and sebum production [63]. The absence of these structures in regenerated skin can compromise physiological function and impede complete recovery [63].

Biomaterials remain indispensable in tissue engineering because they provide structural support and a conducive environment for wound healing. However, their efficacy depends on factors such as biocompatibility, mechanical strength, degradation kinetics, and the ability to deliver bioactive molecules in a controlled manner. Despite significant progress, challenges persist in replicating the biochemical and biomechanical properties of the native ECM [64]. For example, while natural polymers like collagen and chitosan are widely used due to their biodegradability and compatibility, their mechanical properties and degradation rates require further optimization to ensure effective wound healing [65].

Emerging techniques such as 3D-bioprinted skin constructs offer promising avenues to address these shortcomings but continue to face significant barriers to clinical translation, including inadequate vascularization and limited long-term stability [3, 66]. Scar formation is another major concern, as current treatment modalities often fail to prevent disfiguring scars, necessitating novel strategies for anti-scar healing [67]. Innovative approaches, such as artificial dermis engineered using silver-sulfur (Ag-S) bonds and recombinant proteins, have shown potential by reducing inflammation and promoting ECM remodeling. However, issues with their long-term stability and effectiveness in providing broad-spectrum antibacterial effects remain challenging [68].

Technological challenges further complicate the clinical translation of these therapies into clinical practice. For instance, bioprinting and spray-on skin technologies offer exciting potential for creating customized skin constructs. However, these methods still require enhanced reproducibility, precision, and extensive clinical testing before being widely adopted [9, 69]. Similarly, platelet-rich plasma therapy has emerged as a regenerative tool in burn care, yet its clinical potential is hindered by unresolved issues surrounding its preparation methods, standardization protocols, and mechanisms of action [70].

Cell-based therapies, including keratinocytes, fibroblasts, and MSCs, have shown promise in accelerating burn wound healing. However, several scientific and technical concerns must be resolved before these therapies become standard practice, including scalability, cell transport, and inter-patient variability [71, 72].

Regulatory Hurdles

In addition to clinical challenges, the translation of regenerative therapies for burns into clinical practice is significantly impeded by a complex regulatory landscape, with patient safety as the central priority. The approval process for new therapies, especially in burn management, requires adherence to strict guidelines to ensure patient safety and efficacy. Regulatory agencies, such as the U.S. Food and Drug Administration (FDA) and the European Medicines Agency (EMA), enforce rigorous guidelines to ensure the safety, efficacy, and quality of advanced therapies. Compliance with Good Manufacturing Practices (GMP) is mandatory, requiring validated manufacturing protocols, comprehensive biological characterization of cellular components, microbiological testing, and environmental quality control to ensure batch-to-batch consistency. Addressing scalability, cost-effectiveness, and reimbursement strategies is also crucial for widespread clinical adoption [73, 74]. Moreover, demonstrating the safety and therapeutic efficacy of skin substitutes and cell-based products through rigorous clinical trials is a prerequisite for commercialization [75].

Obtaining regulatory approval for novel biomaterials and cell therapies is a lengthy and costly process involving multiple stages of testing and evaluation (Fig. 22.3). Using stem cells in regenerative therapies adds further regulatory

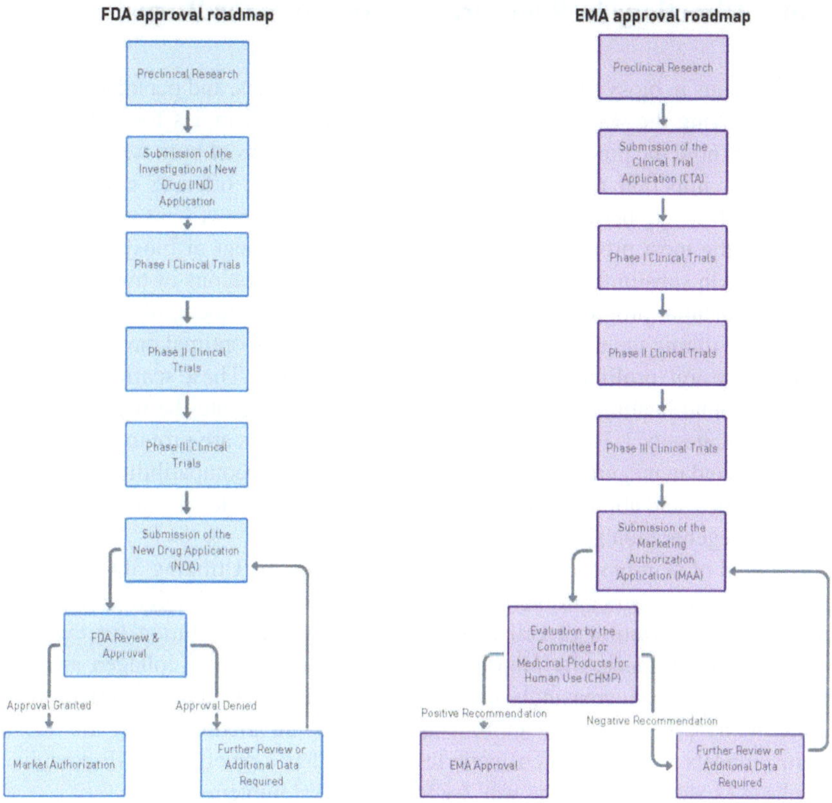

Fig. 22.3 Overview of drug approval pathways by the FDA and EMA. Both processes begin with preclinical research, during which safety and efficacy are initially assessed through laboratory and animal studies. *In the FDA pathway*, an Investigational New Drug application is submitted, enabling the start of a series of clinical trials. Upon successful completion, sponsors submit a New Drug Application, followed by an FDA review to determine whether the drug receives market authorization or if further data are required. *In the EMA pathway*, a Clinical Trial Application is submitted, followed by submission of a Marketing Authorization Application, which is reviewed by the Committee for Medicinal Products for Human Use (CHMP)

complexity. While autologous stem cell treatments offer personalized treatment potential, they are subjected to stringent regulations related to safety, efficacy, scalability, and cost-effectiveness. Similarly, the use of biologically active substances derived from allogenic sources, such as cord blood or umbilical cord, requires careful evaluation to mitigate immunological risks and ensure therapeutic efficacy [76].

In conclusion, while the clinical and regulatory hurdles for bringing regenerative therapies for burns to market are substantial, ongoing advancements in tissue engineering, biomaterials, and regulatory frameworks pave the way for more effective and accessible treatments. Overcoming these challenges will require continued research, enhanced collaboration between scientists and regulators, and a commitment to advancing the field of regenerative medicine.

Future Directions in Regenerative Therapies for Burns

Advancements in biosynthetic scaffolds, stem cell therapies, and personalized medicine are paving the way for the future of regenerative therapies for burns. These innovations hold the potential to address current limitations of burn treatment, such as scarring, infection risk, and delayed wound healing, offering new hope for patients with severe burns.

One of the most promising directions is the development of biosynthetic scaffolds and skin substitutes, designed to overcome the limitations of traditional skin grafts. Substantial progress has been made in engineering biocompatible materials that closely mimic the natural skin matrix and provide an optimal environment for cellular adhesion, proliferation, and differentiation [9, 67]. These scaffolds not only enhance wound healing but also reduce scarring and facilitate the regeneration of skin appendages [67]. For example, biosynthetic hydrogels are proving effective in chronic wound management owing to their remarkable biocompatibility, antimicrobial properties, and ability to maintain appropriate moisture levels, thereby enhancing tissue regeneration [77].

The field of skin bioprinting is also advancing rapidly. Utilizing 3D bioprinting techniques, researchers can fabricate large-scale skin grafts with defined thickness and spatial organization of cell types, offering a scalable solution for extensive burns [9]. For instance, 3D bioprinted scaffolds incorporating collagen and synthetic polymers are increasingly used to support cell adhesion and proliferation in severe burns [78]. Although current applications are predominantly limited to in vitro models for drug and cosmetic testing, the development of GMP-compliant bioinks is expected to pave the way for future clinical implementation.

Nanoengineered scaffolds and smart dressings represent another frontier in burn care. Nanofibers, which mimic the native extracellular matrix, have been shown to promote cell attachment and proliferation, resulting in faster healing and minimal scarring [79, 80]. These nanofibers can be embedded with bioactive agents like growth factors and antibiotics, further enhancing their therapeutic efficacy [81]. Smart materials, such as stimuli-responsive hydrogels and electroactive polymers, offer dynamic responses to environmental cues, such as pH, temperature, or mechanical stress, enabling real-time modulation of the wound microenvironment [82–84].

Induced pluripotent stem cells (iPSCs) are another promising approach. iPSCs-derived keratinocytes (iPSCs-KC) have shown a significant role in treating deep second-degree burns in murine models, reducing wound area, improving wound re-epithelialization, and lowering pro-inflammatory factors [85]. Similarly, exosomes derived from hiPSC mesenchymal stem cells (hiPSC-MSC-Exos) enhance the migration of human dermal fibroblasts and human umbilical vein endothelial cells (HUVECs), facilitating vascularization and accelerating re-epithelialization by reducing scar width and promoting collagen maturity [86].

Collectively, these advancements have the potential to shorten healing times, reduce complications, and improve both functional and aesthetic outcomes for burn patients. Integrating stem cell therapy and biosynthetic materials rapidly transforms burn treatment into a more effective, personalized, and technology-driven approach.

Conclusion

Regenerative therapies have made significant strides in burn care, evolving from traditional grafting techniques to biologically advanced interventions. Stem cell-based approaches, particularly those using MSCs, ADSCs, and iPSCs, have demonstrated strong potential in promoting wound healing, modulating inflammation, and reducing scarring. Their efficacy is further enhanced by innovative delivery systems such as hydrogels, nanofibers, and bioprinted scaffolds. Clinical trials reflect growing interest in both cell-based and cell-free therapies, although many remain in early phases. Critical regulatory and technological barriers, including issues of scalability, standardization, and long-term safety, must still be addressed. Future research should focus on refining these approaches and promoting international collaboration to enable their safe and effective integration into clinical practice. Continued investment in this field will be essential for transforming regenerative therapies from promising research into standard clinical practice.

References

1. Gerstl JVE, Ehsan AN, Lassaren P, Yearley A, Raykar NP, Anderson GA, Smith TR, Sabapathy SR, Ranganathan K (2024) The global macroeconomic burden of burn injuries. Plast Reconstr Surg 153(3):743–752. https://doi.org/10.1097/PRS.0000000000010595
2. Sjoberg F, Elmasry M, Abdelrahman I, Nyberg G, A TE, Ursing E, Steinvall I (2024) The impact and validity of the Berlin criteria on burn-induced ARDS: examining mortality rates, and inhalation injury influences. A single center observational cohort study. Burns 50(6):1528–1535. https://doi.org/10.1016/j.burns.2024.05.005
3. Nugud A, Alghfeli L, Elmasry M, El-Serafi I, El-Serafi AT (2022) Biomaterials as a vital frontier for stem cell-based tissue regeneration. Front Cell Dev Biol 10:713934. https://doi.org/10.3389/fcell.2022.713934
4. Dogan S, Elmasry M, El-Serafi A, Sjoberg F, Vuola J, Kankuri E, Grigoriadi MP, Valtonen J, Abdelrahman I, Steinvall I, Karlsson M, Olofsson P, Lindford A (2022) A prospective dual-Centre intra-individual controlled study for the treatment of burns comparing dermis graft with split-thickness skin auto-graft. Sci Rep 12(1):21666. https://doi.org/10.1038/s41598-022-25346-4
5. El-Serafi AT, El-Serafi IT, Elmasry M, Steinvall I, Sjoberg F (2017) Skin regeneration in three dimensions, current status, challenges and opportunities. Differentiation 96:26–29. https://doi.org/10.1016/j.diff.2017.06.002
6. Engrav LH, Heimbach DM, Reus JL, Harnar TJ, Marvin JA (1983) Early excision and grafting vs. nonoperative treatment of burns of indeterminant depth: a randomized prospective study. J Trauma 23(11):1001–1004. https://doi.org/10.1097/00005373-198311000-00007

7. Herndon DN, Barrow RE, Rutan RL, Rutan TC, Desai MH, Abston S (1989) A comparison of conservative versus early excision. Therapies in severely burned patients. Ann Surg 209(5):547–552; discussion 552–553. https://doi.org/10.1097/00000658-198905000-00006
8. Negut I, Grumezescu V, Grumezescu AM (2018) Treatment strategies for infected wounds. Molecules 23:9. https://doi.org/10.3390/molecules23092392
9. Rivers J (2013) Advances in skin regeneration. J Cutan Med Surg 17(2):75–76. https://doi.org/10.1177/120347541301700201
10. Schlottmann F, Lorbeer L (2024) Update burn surgery: overview of current multidisciplinary treatment concepts. Innov Surg Sci 9(4):181–190. https://doi.org/10.1515/iss-2024-0020
11. Xiao-Wu W, Herndon DN, Spies M, Sanford AP, Wolf SE (2002) Effects of delayed wound excision and grafting in severely burned children. Arch Surg 137(9):1049–1054. https://doi.org/10.1001/archsurg.137.9.1049
12. Siegwart LC, Bocker AH, Diehm YF, Kotsougiani-Fischer D, Erdmann S, Ziegler B, Kneser U, Hirche C, Fischer S (2021) Enzymatic debridement for burn wound care: interrater reliability and impact of experience in post-intervention therapy decision. J Burn Care Res 42(5):953–961. https://doi.org/10.1093/jbcr/iraa218
13. Karlsson M, Steinvall I, Olofsson P, Thorfinn J, Sjoberg F, Astrand L, Fayiz S, Khalaf A, Divyasree P, El-Serafi AT, Elmasry M (2020) Sprayed cultured autologous keratinocytes in the treatment of severe burns: a retrospective matched cohort study. Ann Burns Fire Disasters 33(2):134–142
14. Horch RE, Schmidt VJ (2012) Burn reconstruction: skin substitutes and tissue engineering. In: Kamolz L-P, Jeschke MG, Horch RE, Küntscher M, Brychta P (eds) Handbook of burns: reconstruction and rehabilitation, vol 2. Springer Vienna, Vienna. https://doi.org/10.1007/978-3-7091-0315-9_13
15. Larson KW, Austin CL, Thompson SJ (2020) Treatment of a full-thickness burn injury with NovoSorb biodegradable temporizing matrix and RECELL autologous skin cell suspension: A case series. J Burn Care Res 41(1):215–219. https://doi.org/10.1093/jbcr/irz179
16. Diaz-Garcia D, Filipova A, Garza-Veloz I, Martinez-Fierro ML (2021) A beginner's introduction to skin stem cells and wound healing. Int J Mol Sci 22(20). https://doi.org/10.3390/ijms222011030
17. Wang B, Liu XM, Liu ZN, Wang Y, Han X, Lian AB, Mu Y, Jin MH, Liu JY (2020) Human hair follicle-derived mesenchymal stem cells: isolation, expansion, and differentiation. World J Stem Cells 12(6):462–470. https://doi.org/10.4252/wjsc.v12.i6.462
18. Niemann C, Horsley V (2012) Development and homeostasis of the sebaceous gland. Semin Cell Dev Biol 23(8):928–936. https://doi.org/10.1016/j.semcdb.2012.08.010
19. Vagnozzi AN, Reiter JF, Wong SY (2015) Hair follicle and interfollicular epidermal stem cells make varying contributions to wound regeneration. Cell Cycle 14(21):3408–3417. https://doi.org/10.1080/15384101.2015.1090062
20. Li P, Guo X (2018) A review: therapeutic potential of adipose-derived stem cells in cutaneous wound healing and regeneration. Stem Cell Res Ther 9(1):302. https://doi.org/10.1186/s13287-018-1044-5
21. Shahin H, Belcastro L, Das J, Perdiki Grigoriadi M, Saager RB, Steinvall I, Sjoberg F, Olofsson P, Elmasry M, El-Serafi AT (2024) MicroRNA-155 mediates multiple gene regulations pertinent to the role of human adipose-derived mesenchymal stem cells in skin regeneration. Front Bioeng Biotechnol 12:1328504. https://doi.org/10.3389/fbioe.2024.1328504
22. Strong AL, Neumeister MW, Levi B (2017) Stem cells and tissue engineering: regeneration of the skin and its contents. Clin Plast Surg 44(3):635–650. https://doi.org/10.1016/j.cps.2017.02.020
23. Nourian Dehkordi A, Mirahmadi Babaheydari F, Chehelgerdi M, Raeisi Dehkordi S (2019) Skin tissue engineering: wound healing based on stem-cell-based therapeutic strategies. Stem Cell Res Ther 10(1):111. https://doi.org/10.1186/s13287-019-1212-2

24. Li Y, Xia WD, Van der Merwe L, Dai WT, Lin C (2020) Efficacy of stem cell therapy for burn wounds: a systematic review and meta-analysis of preclinical studies. Stem Cell Res Ther 11(1):322. https://doi.org/10.1186/s13287-020-01839-9
25. Wang M, Xu X, Lei X, Tan J, Xie H (2021) Mesenchymal stem cell-based therapy for burn wound healing. Burns Trauma 9:tkab002. https://doi.org/10.1093/burnst/tkab002
26. Kalluri R, LeBleu VS (2020) The biology, function, and biomedical applications of exosomes. Science 367(6478). https://doi.org/10.1126/science.aau6977
27. Kumar MA, Baba SK, Sadida HQ, Marzooqi SA, Jerobin J, Altemani FH, Algehainy N, Alanazi MA, Abou-Samra AB, Kumar R, Al-Shabeeb Akil AS, Macha MA, Mir R, Bhat AA (2024) Extracellular vesicles as tools and targets in therapy for diseases. Signal Transduct Target Ther 9(1):27. https://doi.org/10.1038/s41392-024-01735-1
28. Fakouri A, Razavi ZS, Mohammed AT, Hussein AHA, Afkhami H, Hooshiar MH (2024) Applications of mesenchymal stem cell-exosome components in wound infection healing: new insights. Burns Trauma 12:tkae021. https://doi.org/10.1093/burnst/tkae021
29. Wang Y, Ding H, Bai R, Li Q, Ren B, Lin P, Li C, Chen M, Xu X (2025) Exosomes from adipose-derived stem cells accelerate wound healing by increasing the release of IL-33 from macrophages. Stem Cell Res Ther 16(1):80. https://doi.org/10.1186/s13287-025-04203-x
30. Zhou C, Zhang B, Yang Y, Jiang Q, Li T, Gong J, Tang H, Zhang Q (2023) Stem cell-derived exosomes: emerging therapeutic opportunities for wound healing. Stem Cell Res Ther 14(1):107. https://doi.org/10.1186/s13287-023-03345-0
31. El-Serafi AT, El-Serafi I, Steinvall I, Sjoberg F, Elmasry M (2022) A systematic review of keratinocyte secretions: A regenerative perspective. Int J Mol Sci 23(14). https://doi.org/10.3390/ijms23147934
32. Shabbir A, Cox A, Rodriguez-Menocal L, Salgado M, Van Badiavas E (2015) Mesenchymal stem cell exosomes induce proliferation and migration of Normal and chronic wound fibroblasts, and enhance angiogenesis in vitro. Stem Cells Dev 24(14):1635–1647. https://doi.org/10.1089/scd.2014.0316
33. Im GB, Kim YH, Kim YJ, Kim SW, Jung E, Jeong GJ, Wang K, Kim J, Kim DI, Kim TH, Yi GR, Yu T, Bhang SH (2019) Enhancing the wound healing effect of conditioned medium collected from mesenchymal stem cells with high passage number using bioreducible nanoparticles. Int J Mol Sci 20(19). https://doi.org/10.3390/ijms20194835
34. Maguire G (2013) Stem cell therapy without the cells. Commun Integr Biol 6(6):e26631. https://doi.org/10.4161/cib.26631
35. Fang W, Yang M, Wang L, Li W, Liu M, Jin Y, Wang Y, Yang R, Wang Y, Zhang K, Fu Q (2023) Hydrogels for 3D bioprinting in tissue engineering and regenerative medicine: current progress and challenges. Int J Bioprint 9(5):759. https://doi.org/10.18063/ijb.759
36. Kaur G, Narayanan G, Garg D, Sachdev A, Matai I (2022) Biomaterials-based regenerative strategies for skin tissue wound healing. ACS Appl Bio Mater 5(5):2069–2106. https://doi.org/10.1021/acsabm.2c00035
37. Deptula M, Zawrzykraj M, Sawicka J, Banach-Kopec A, Tylingo R, Pikula M (2023) Application of 3D- printed hydrogels in wound healing and regenerative medicine. Biomed Pharmacother 167:115416. https://doi.org/10.1016/j.biopha.2023.115416
38. Lazaridou M, Bikiaris DN, Lamprou DA (2022) 3D bioprinted chitosan-based hydrogel scaffolds in tissue engineering and localised drug delivery. Pharmaceutics 14(9). https://doi.org/10.3390/pharmaceutics14091978
39. Hazrati R, Alizadeh E, Soltani S, Keyhanvar P, Davaran S (2024) Development of a composite hydrogel containing statistically optimized PDGF-loaded polymeric nanospheres for skin regeneration: in vitro evaluation and stem cell differentiation studies. ACS Omega 9(13):15114–15133. https://doi.org/10.1021/acsomega.3c09391
40. Beheshtizadeh N, Lotfibakhshaiesh N, Pazhouhnia Z, Hoseinpour M, Nafari M (2020) A review of 3D bio-printing for bone and skin tissue engineering: a commercial approach. J Mater Sci 55(9):3729–3749. https://doi.org/10.1007/s10853-019-04259-0

41. Al-Shaibani M, Wang XN, Tulah A, Crossland RE, Dickinson AM, Lovat PE (2024) role of mesenchymal stem cell conditioned medium on wound healing using a developed 3D skin model. J Wound Care 33(Sup8a):clxxxii–cxciii. https://doi.org/10.12968/jowc.2021.0397
42. Li M, Ma J, Gao Y, Dong M, Zheng Z, Li Y, Tan R, She Z, Yang L (2020) Epithelial differentiation of human adipose-derived stem cells (hASCs) undergoing three-dimensional (3D) cultivation with collagen sponge scaffold (CSS) via an indirect co-culture strategy. Stem Cell Res Ther 11(1):141. https://doi.org/10.1186/s13287-020-01645-3
43. Gholipourmalekabadi M, Seifalian AM, Urbanska AM, Omrani MD, Hardy JG, Madjd Z, Hashemi SM, Ghanbarian H, Brouki Milan P, Mozafari M, Reis RL, Kundu SC, Samadikuchaksaraei A (2018) 3D protein-based bilayer artificial skin for the guided scarless healing of third-degree burn wounds in vivo. Biomacromolecules 19(7):2409–2422. https://doi.org/10.1021/acs.biomac.7b01807
44. Gangadaran P, Oh EJ, Rajendran RL, Oh JM, Kim HM, Kwak S, Chung HY, Lee J, Ahn BC, Hong CM (2023) Three-dimensional culture conditioned bone marrow MSC secretome accelerates wound healing in a burn injury mouse model. Biochem Biophys Res Commun 673:87–95. https://doi.org/10.1016/j.bbrc.2023.05.088
45. Schneider V, Kruse D, de Mattos IB, Zöphel S, Tiltmann KK, Reigl A, Khan S, Funk M, Bodenschatz K, Groeber-Becker F (2021) A 3D in vitro model for burn wounds: monitoring of regeneration on the epidermal level. Biomedicine 9(9). https://doi.org/10.3390/biomedicines9091153
46. Abdullahi A, Amini-Nik S, Jeschke MG (2014) Animal models in burn research. Cell Mol Life Sci 71(17):3241–3255. https://doi.org/10.1007/s00018-014-1612-5
47. Liu Y, Zhu Y, Cui Y (2019) Challenges and opportunities towards fast-charging battery materials. Nat Energy 4(7):540–550. https://doi.org/10.1038/s41560-019-0405-3
48. Wong VW, Sorkin M, Glotzbach JP, Longaker MT, Gurtner GC (2011) Surgical approaches to create murine models of human wound healing. J Biomed Biotechnol 2011:969618. https://doi.org/10.1155/2011/969618
49. Gurtner GC, Werner S, Barrandon Y, Longaker MT (2008) Wound repair and regeneration. Nature 453(7193):314–321. https://doi.org/10.1038/nature07039
50. Li J, Chen J, Kirsner R (2007) Pathophysiology of acute wound healing. Clin Dermatol 25(1):9–18. https://doi.org/10.1016/j.clindermatol.2006.09.007
51. Caliari-Oliveira C, Yaochite JNU, Ramalho LNZ, Palma PVB, Carlos D, De Queiróz Cunha F, De Souza DA, Frade MAC, Covas DT, Malmegrim KCR, Oliveira MC, César Voltarelli J (2016) Xenogeneic mesenchymal stromal cells improve wound healing and modulate the immune response in an extensive burn model. Cell Transplant 25(2):201–215. https://doi.org/10.3727/096368915X688128
52. Hosni Ahmed H, Rashed LA, Mahfouz S, Elsayed Hussein R, Alkaffas M, Mostafa S, Abusree A (2017) Can mesenchymal stem cells pretreated with platelet-rich plasma modulate tissue remodeling in a rat with burned skin? Biochem Cell Biol 95(5):537–548. https://doi.org/10.1139/bcb-2016-0224
53. Mestas J, Hughes CC (2004) Of mice and not men: differences between mouse and human immunology. J Immunol 172(5):2731–2738. https://doi.org/10.4049/jimmunol.172.5.2731
54. Ganz T (2003) Defensins: antimicrobial peptides of innate immunity. Nat Rev Immunol 3(9):710–720. https://doi.org/10.1038/nri1180
55. Oppenheim JJ, Biragyn A, Kwak LW, Yang D (2003) Roles of antimicrobial peptides such as defensins in innate and adaptive immunity. Ann Rheum Dis 62(Suppl 2):ii17–ii21. https://doi.org/10.1136/ard.62.suppl_2.ii17
56. Werner S, Grose R (2003) Regulation of wound healing by growth factors and cytokines. Physiol Rev 83(3):835–870. https://doi.org/10.1152/physrev.2003.83.3.835
57. Henriksen JL, Sørensen NB, Fink T, Zachar V, Porsborg SR (2020) Systematic review of stem-cell-based therapy of burn wounds: lessons learned from animal and clinical studies. Cells 9(12):2545

58. Dorsett-Martin WA (2004) Rat models of skin wound healing: a review. Wound Repair Regen 12(6):591–599. https://doi.org/10.1111/j.1067-1927.2004.12601.x
59. Parnell LKS, Volk SW (2019) The evolution of animal models in wound healing research: 1993–2017. Adv Wound Care (New Rochelle) 8(12):692–702. https://doi.org/10.1089/wound.2019.1098
60. Qiao Y, Zhang Q, Peng Y, Qiao X, Yan J, Wang B, Zhu Z, Li Z, Zhang Y (2023) Effect of stem cell treatment on burn wounds: A systemic review and a meta-analysis. Int Wound J 20(1):8–17. https://doi.org/10.1111/iwj.13831
61. Muangman P, Engrav LH, Heimbach DM, Harunari N, Honari S, Gibran NS, Klein MB (2006) Complex wound management utilizing an artificial dermal matrix. Ann Plast Surg 57(2):199–202. https://doi.org/10.1097/01.sap.0000218636.61803.d6
62. Gupta S, Moiemen N, Fischer JP, Attinger C, Jeschke MG, Taupin P, Orgill DP (2024) Dermal regeneration template in the management and reconstruction of burn injuries and complex wounds: A review. Plast Reconstr Surg Glob Open 12(3):e5674. https://doi.org/10.1097/GOX.0000000000005674
63. Hebda PA (2009) Regeneration of epidermal appendages. Wound Repair Regen 17(3):293. https://doi.org/10.1111/j.1524-475X.2009.00495.x
64. Bhar B, Das E, Manikumar K, Mandal BB (2024) 3D bioprinted human skin model recapitulating native-like tissue maturation and immunocompetence as an advanced platform for skin sensitization assessment. Adv Healthc Mater 13(15):e2303312. https://doi.org/10.1002/adhm.202303312
65. Rajinikanth BS, Rajkumar DSR, K K, Vijayaragavan V (2024) Chitosan-based biomaterial in wound healing: a review. Cureus 16(2):e55193. https://doi.org/10.7759/cureus.55193
66. Gopakumar N, Ali AM, Oudda S, Singam A, Park S (2025) 3D-bioprinted skin tissues for improving wound healing: current status and perspective. Adv Exp Med Biol 1474:35–51. https://doi.org/10.1007/5584_2024_817
67. Xiong M, Yang X, Shi Z, Xiang J, Gao H, Ji S, Li Y, Pi W, Chen H, Zhang H, Wang M, Li Y, Hong Y, Liu D, Fu X, Dong Y, Sun X (2024) Programmable artificial skins accomplish antiscar healing with multiple appendage regeneration. Adv Mater 36(50):e2407322. https://doi.org/10.1002/adma.202407322
68. Wu Z, Li M, Liu Q, Sun J, Tao C, Jia B, Sun X, Zhang H, Liu K, Wang F (2025) Engineered multifunctional artificial dermis for infected burn wound healing. Adv Funct Mater 35(8):2415514. https://doi.org/10.1002/adfm.202415514
69. Chen D, Hou Q, Zhong L, Zhao Y, Li M, Fu X (2019) Bioactive molecules for skin repair and regeneration: progress and perspectives. Stem Cells Int 2019:6789823. https://doi.org/10.1155/2019/6789823
70. Hosseini MS, Nouri M, Zarrabi M, Fatemi MJ, Shpichka A, Timashev P, Hassan M, Vosough M (2023) Platelet-rich plasma in regenerative medicine: possible applications in management of burns and post-burn scars: a review. Cell J 25(5):281–290. https://doi.org/10.22074/cellj.2023.558213.1093
71. Domaszewska-Szostek AP, Krzyzanowska MO, Czarnecka AM, Siemionow M (2021) Local treatment of burns with cell-based therapies tested in clinical studies. J Clin Med 10(3). https://doi.org/10.3390/jcm10030396
72. Henriksen JL, Sorensen NB, Fink T, Zachar V, Porsborg SR (2020) systematic review of stem-cell-based therapy of burn wounds: lessons learned from animal and clinical studies. Cells 9(12). https://doi.org/10.3390/cells9122545
73. Laurent A, Lin P, Scaletta C, Hirt-Burri N, Michetti M, de Buys Roessingh AS, Raffoul W, She BR, Applegate LA (2020) Bringing safe and standardized cell therapies to industrialized processing for burns and wounds. Front Bioeng Biotechnol 8:581. https://doi.org/10.3389/fbioe.2020.00581
74. Lagerwall C, Shahin H, Abdallah S, Steinvall I, Elmasry M, Sjoberg F, El-Serafi AT (2021) Xeno-free workflow exhibits comparable efficiency and quality of keratinocytes isolated from human skin biopsies. Regen Ther 18:401–407. https://doi.org/10.1016/j.reth.2021.09.005

75. Chocarro-Wrona C, Lopez-Ruiz E, Peran M, Galvez-Martin P, Marchal JA (2019) Therapeutic strategies for skin regeneration based on biomedical substitutes. J Eur Acad Dermatol Venereol 33(3):484–496. https://doi.org/10.1111/jdv.15391
76. Hirai T, Yasuda S, Umezawa A, Sato Y (2023) Country-specific regulation and international standardization of cell-based therapeutic products derived from pluripotent stem cells. Stem Cell Reports 18(8):1573–1591. https://doi.org/10.1016/j.stemcr.2023.05.003
77. Alberts A, Bratu AG, Niculescu AG, Grumezescu AM (2025) New perspectives of hydrogels in chronic wound management. Molecules 30(3). https://doi.org/10.3390/molecules30030686
78. Antezana PE, Municoy S, Álvarez-Echazú MI, Santo-Orihuela PL, Catalano PN, Al-Tel TH, Kadumudi FB, Dolatshahi-Pirouz A, Orive G, Desimone MF (2022) the 3D bioprinted scaffolds for wound healing. Pharmaceutics 14(2). https://doi.org/10.3390/pharmaceutics14020464
79. Ziauddin T, Hussain A, Nazir U, Mahmood M, Hameed, Ramakrishna S, Abid S (2022) Nanoengineered therapeutic scaffolds for burn wound management. Curr Pharm Biotechnol 23(12):1417–1435. https://doi.org/10.2174/1389201023666220329162910
80. Zhang S, Yang W, Gong W, Lu Y, Yu DG, Liu P (2024) Recent progress of electrospun nanofibers as burning dressings. RSC Adv 14(20):14374–14391. https://doi.org/10.1039/d4ra01514b
81. Hermosilla J, Pastene-Navarrete E, Acevedo F (2021) Electrospun fibers loaded with natural bioactive compounds as a biomedical system for skin burn treatment. A review. Pharmaceutics 13(12). https://doi.org/10.3390/pharmaceutics13122054
82. Lei H, Zhu C, Fan D (2020) Optimization of human-like collagen composite polysaccharide hydrogel dressing preparation using response surface for burn repair. Carbohydr Polym 239:116249. https://doi.org/10.1016/j.carbpol.2020.116249
83. El-Husseiny HM, Mady EA, Hamabe L, Abugomaa A, Shimada K, Yoshida T, Tanaka T, Yokoi A, Elbadawy M, Tanaka R (2022) Smart/stimuli-responsive hydrogels: cutting-edge platforms for tissue engineering and other biomedical applications. Mater Today Bio 13:100186. https://doi.org/10.1016/j.mtbio.2021.100186
84. Nasra S, Pramanik S, Oza V, Kansara K, Kumar A (2024) Advancements in wound management: integrating nanotechnology and smart materials for enhanced therapeutic interventions. Discover Nano 19(1):159. https://doi.org/10.1186/s11671-024-04116-3
85. Wu LJ, Lin W, Liu JJ, Chen WX, He WJ, Shi Y, Liu X, Li K (2023) Transplantation of human induced pluripotent stem cell derived keratinocytes accelerates deep second-degree burn wound healing. World J Stem Cells 15(7):713–733. https://doi.org/10.4252/wjsc.v15.i7.713
86. Zhang J, Guan J, Niu X, Hu G, Guo S, Li Q, Xie Z, Zhang C, Wang Y (2015) Exosomes released from human induced pluripotent stem cells-derived MSCs facilitate cutaneous wound healing by promoting collagen synthesis and angiogenesis. J Transl Med 13:49. https://doi.org/10.1186/s12967-015-0417-0
87. Aijaz A, Vinaik R, Jeschke MG (2022) Chapter 10—large animal models of thermal injury. In: Montrose DC (ed) Methods in cell biology. Academic Press. https://doi.org/10.1016/bs.mcb.2021.12.015
88. Seaton M, Hocking A, Gibran NS (2015) Porcine models of cutaneous wound healing. ILAR J 56(1):127–138. https://doi.org/10.1093/ilar/ilv016
89. Żwierełło W, Piorun K, Skórka-Majewicz M, Maruszewska A, Antoniewski J, Gutowska I (2023) Burns: classification, pathophysiology, and treatment: a review. Int J Mol Sci 24(4). https://doi.org/10.3390/ijms24043749
90. Naldaiz-Gastesi N, Bahri OA, López de Munain A, McCullagh KJA, Izeta A (2018) The panniculus carnosus muscle: an evolutionary enigma at the intersection of distinct research fields. J Anat 233(3):275–288. https://doi.org/10.1111/joa.12840
91. Mulder PPG, Koenen HJPM, Vlig M, Joosten I, de Vries RBM, Boekema BKHL (2022) Burn-induced local and systemic immune response: systematic review and meta-analysis of animal studies. J Invest Dermatol 142(11):3093–3109.e15. https://doi.org/10.1016/j.jid.2022.05.004
92. D'Arpa P, Leung KP (2017) Toll-like receptor signaling in burn wound healing and scarring. Adv Wound Care (New Rochelle) 6(10):330–343. https://doi.org/10.1089/wound.2017.0733

93. Mulder PPG, Vlig M, Boekema B, Stoop MM, Pijpe A, van Zuijlen PPM, de Jong E, van Cranenbroek B, Joosten I, Koenen H, Ulrich MMW (2020) Persistent systemic inflammation in patients with severe burn injury is accompanied by influx of immature neutrophils and shifts in T cell subsets and cytokine profiles. Front Immunol 11:621222. https://doi.org/10.3389/fimmu.2020.621222
94. Peterkofsky B (1991) Ascorbate requirement for hydroxylation and secretion of procollagen: relationship to inhibition of collagen synthesis in scurvy. Am J Clin Nutr 54(6 Suppl):1135s–1140s. https://doi.org/10.1093/ajcn/54.6.1135s
95. Zaulyanov L, Kirsner RS (2007) A review of a bi-layered living cell treatment (Apligraf) in the treatment of venous leg ulcers and diabetic foot ulcers. Clin Interv Aging 2(1):93–98. https://doi.org/10.2147/ciia.2007.2.1.93
96. Holmes JH (2023) A brief history of RECELL(R) and its current indications. J Burn Care Res 44(Suppl_1):S48–S49. https://doi.org/10.1093/jbcr/irac121
97. Zhang Y, Hao H, Liu J, Fu X, Han W (2012) Repair and regeneration of skin injury by transplanting microparticles mixed with Wharton's jelly and MSCs from the human umbilical cord. Int J Low Extrem Wounds 11(4):264–270. https://doi.org/10.1177/1534734612463577

Chapter 23
Cultured Epithelial Autografts Based on a Fibrin Matrix System: Optimising Engraftment in Extensive Burns

Khong-Yik Chew, Alvin W. C. Chua, Keya Rani Roy, Seng-Teik Lee, Yann Barrandon, and Bien-Keem Tan

Abstract Since its inception, cultured epithelial autograft technology has become an integral part of the armamentarium in the treatment of extensive burns in Singapore. This report details the use of a fibrin matrix system for cell culture and highlights the technical aspects of successful clinical application.

Keywords Cell culture · Keratinocyte · Skin grafting · Meek micrografting

Abbreviations

CEA	Cultured epithelial autograft
CFE	Colony-forming efficiency
DEJ	Dermal-epidermal junction
DMEM	Dulbecco's Modified Eagle's Medium
hEKs	Human epidermal keratinocytes
pHEMA	Poly(2-hydroxyethylmethacrylate)
PPS	Plasma polymerised functional surface containing 20% carboxylic acid
SGH	Singapore General Hospital
TBSA	Total body surface area

K.-Y. Chew · A. W. C. Chua · K. R. Roy · S.-T. Lee · Y. Barrandon · B.-K. Tan (✉)
Department of Plastic, Reconstructive and Aesthetic Surgery, Singapore General Hospital, Musculoskeletal Sciences Academic Clinical Programmes, Duke-NUS Medical School, Singapore, Singapore

F. De Francesco, K. Turksen (eds.), *Management and Strategies for Wound Healing*, Stem Cell Biology and Regenerative Medicine 79,
https://doi.org/10.1007/978-3-032-12442-5_23

Introduction

In 1975, Rheinwald and Green first reported the successful serial cultivation of human epidermal keratinocytes (hEKs) in vitro and subsequently developed methods to expand these cells into epithelial sheets suitable for grafting [1, 2]. These autologous cultured sheets, known as cultured epithelial autografts (CEA), were later shown to be enduring, providing permanent coverage in extensive full-thickness burns [3, 4].

Cultured epithelial autograft technology was introduced from the Howard Green laboratory at the Harvard Medical School to the Singapore General Hospital (SGH) Burn Centre by S.T. Lee and Y. Barrandon in 1991 [5]. It was the first application of regenerative medicine using autologous human cultured stem cells in Southeast Asia and the second (to Japan) in Asia. This achievement highlighted the SGH Burn Centre's pioneering role in CEA clinical application. Since its inception, CEA technology has remained an integral part of the therapeutic arsenal for treating extensive burns and skin defects in Singapore and has contributed to numerous lives saved.

However, there have been reports of poor take and suboptimal clinical outcomes of CEA, even in experienced hands [6–10]. Poor outcomes could be attributed to unfamiliarity with technique, inadequate wound bed preparation, and wound colonisation, among other factors [11, 12]. The purpose of this report is to describe our experience in optimising CEA engraftment.

CEA Culture Technique

Our CEA culture technique is based on the Rheinwald and Green method [1], where human epidermal keratinocytes (hEKs) are grown in culture together with mouse fibroblast feeder cells [13]. This method effectively promotes the proliferative activity of hEKs and maintains their epidermal stem cell population. It is crucial for hEKs to proliferate without differentiating into mature cells, as differentiating would greatly reduce their regenerative potential.

The culture technique involves the following steps (Fig. 23.1):

1. Cell isolation: Human epidermal keratinocytes (hEKs) are isolated from partial- or full-thickness skin biopsies measuring 1 to 4 cm^2. The isolation process involves enzymatic treatment using a combination of dispase and trypsin/ethylenediaminetetraacetic acid (EDTA) to dissociate both tissue and cellular components.
2. Primary culture: The isolated cells are seeded into flask(s) pre-coated with lethally irradiated mouse feeder cells (3 T3-J2) for the primary cell culture of hEKs. The growth of hEKs is supported by a nutrient medium comprising Dulbecco's Modified Eagle's Medium (DMEM) and HAM's F12 in a 3:1 ratio, supplemented with foetal bovine serum, adenine, hydrocortisone, insulin, cholera toxin, triiodothyronine, and epidermal growth factor. Direct contact and co-

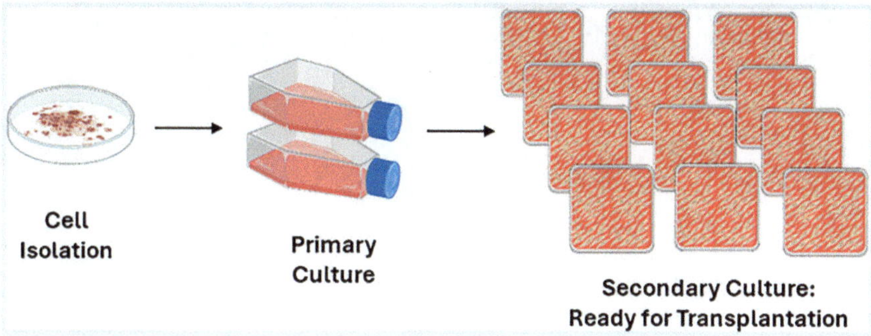

Fig. 23.1 An illustration of the CEA cultivation process. The cells are first isolated from skin biopsies. They are then expanded in primary cultures, before the final step of secondary culture to generate CEA sheets

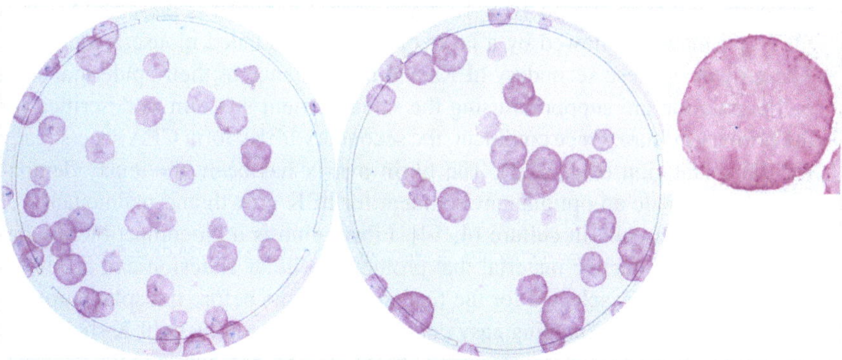

Fig. 23.2 Colony-forming efficiency of human keratinocytes derived from foreskin. The colony-forming efficiency assay is a simple way of evaluating the regenerative capacity of the keratinocytes and is the quality control for the CEA process in this specimen. Human keratinocytes (one hundred individual cells) were cultured for 12 days on 3 T3-J2 feeder cells in cFAD medium composed of a 3:1 mixture of DMEM 1× (Gibco) and Ham's F-12 medium (Gibco), 10% fetal bovine serum (SIGMA-ALDRICH), 1% penicillin and streptomycin (100X), hydrocortisone (0.4 μg/ml), cholera toxin (10–10 M), triiodothyronine (T3) (2×10^{-9} M), insulin (5 μg/ml, SIGMA-ALDRICH), and human epidermal growth factor (hEGF) 10 ng/ml. The cultured cells were fixed in 10% neutral buffered formalin and stained with 1% Rhodamine B. Inset, the perimeter of most of the colonies is nearly circular, and boundaries are darker, indicating smaller proliferating cells concentrated near the perimeter [22]

culture of hEKs with the mitotically inactivated mouse feeder cells are essential not only for supporting human keratinocyte growth, but also for maintaining the epidermal stem cell population within the hEK cultures. As the hEKs proliferate, they eventually displace the feeder cells, resulting in a confluent layer of keratinocytes with negligible mouse cell residues. The colony-forming efficiency assay is a simple way of evaluating the regenerative capacity of the keratinocytes and is the quality control for our CEA system (Fig. 23.2).

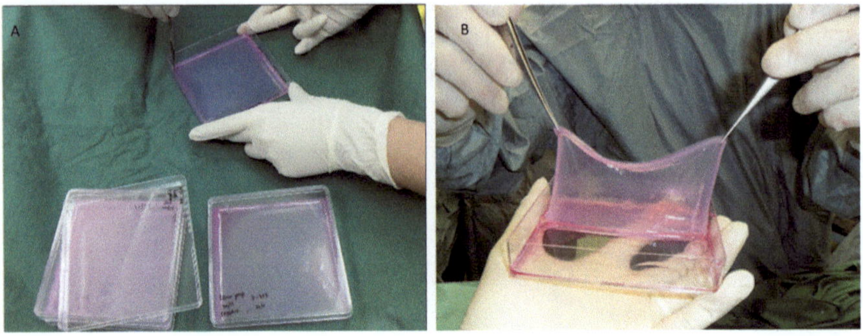

Fig. 23.3 (**a**) CEA is cultured on square culture dishes as shown. (**b**) The CEA is carried by a gelatinous fibrin mat, which is handled delicately during transfer

3. Secondary culture: On reaching confluence, hEKs from the primary culture are harvested and seeded onto 10 × 10 cm^2 dishes. Each dish is pre-coated with a fibrin gel matrix, followed by a layer of lethally irradiated mouse feeder cells. The growth of these secondary hEKs and maintenance of their epidermal stem cell population are supported using the same nutrient medium as described for the primary culture. Once confluent, the secondary hEKs form CEA sheets ready for transplantation (Fig. 23.3). The fibrin matrix has been previously demonstrated to provide an optimal environment for hEK growth and maintenance of epidermal stem cells in culture [4, 14]. Fibrin glue is a biocompatible, biodegradable, and bioactive material that promotes cellular adhesion and proliferation. This matrix is selected for the final culture stage before transplantation, as it serves as a carrier, enabling easy detachment of the confluent hEK sheet from the culture dish and a direct transfer onto the wound bed, while preserving cellular integrity (Fig. 23.4).

CEA Application Technique

Our indications for CEA are patients with >50% total body surface area (TBSA) burns with limited donor sites. Typically, patients with major burns undergo an immediate scrub down under general anesthesia within 24 hours of admission. The wounds are covered with biosynthetic dressing, Biobrane®, after a scrub down. At the same time, a 2 × 6 cm split-thickness skin biopsy of unburnt tissue is obtained aseptically and placed in culture media provided by the laboratory. This initial surgery also allows a thorough clinical assessment of the patient, not least the identification of deep areas that require excision and grafting.

Total excision is completed by post-burn day 7. All wounds are covered with cryopreserved sheet allografts from our skin bank [15] and inspected every other day to maximize engraftment of allografts. The sheet allografts are fenestrated but not meshed for easy removal, as they will ultimately be replaced with autografts. We

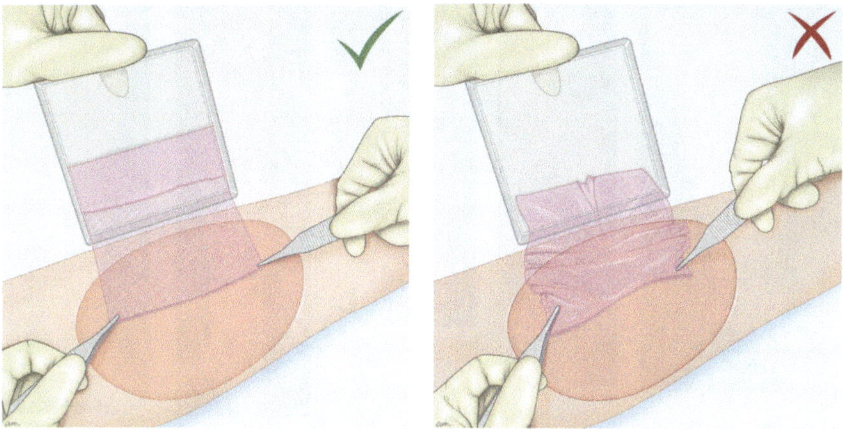

Fig. 23.4 Left: The correct transfer of CEA onto the wound bed. The CEA sheet is delicately grasped with two pairs of forceps, lifted from the culture dish, and pulled over to rest evenly on a well-prepared wound bed without tearing it. Right: Incorrect handling. The CEA sheet becomes folded, displacing the invisible sheet of keratinocytes on the fibrin mat

have observed that the meshed allograft, whilst haemostatic, does not seal the wound and is therefore inferior to sheet allografts. Furthermore, meshed allografts are highly adherent to the wound bed and, at the time of rejection, become exceedingly difficult to remove without causing bleeding. Hence, our preference for sheet allografts.

Our main autografting method for massive burns is the Meek micrografting method, which is carried out early in the second post-burn week. The Meek method can achieve an expansion of over 1:15, and we typically overlay the Meek micrografts with sheet allografts to secure them in place and seal the wound [16]. After 2 weeks, non-adherent allografts are replaced with new allografts. Adherent allografts are left in place but will be replaced after 3 weeks to prevent infection. With successful take of the micrografts, we have observed spreading epithelialisation from individual micrografts under the cover of the allografts. As epithelialisation occurs, the allografts naturally lift off and are trimmed away with a pair of iris scissors.

CEA is usually ready in 3 weeks. For any given patient, the first batch of CEA from the original skin biopsy consists of 8 to 20 sheets of CEA measuring 10 × 10 cm each, delivered on square petri dishes. Subsequent batches of CEA take 2 weeks to produce. We avoid applying CEA on a non-granulating wound bed, as the take is poor, owing to a deficiency of connective tissue elements. Topical negative pressure promotes granulation, and we routinely use this over fat and fascia to promote granulation tissue growth. Thus, 3 weeks after the initial injury is an optimal time interval for CEA application. Other wound bed conditions for optimal CEA take include the absence of bleeding, infection, and biofilm, as well as a wound devoid of silver nanoparticles [17, 18] and Meek spray adhesive, which is cytotoxic. In preparation for the CEA application, the wound is gently cleared of biofilm, which typically

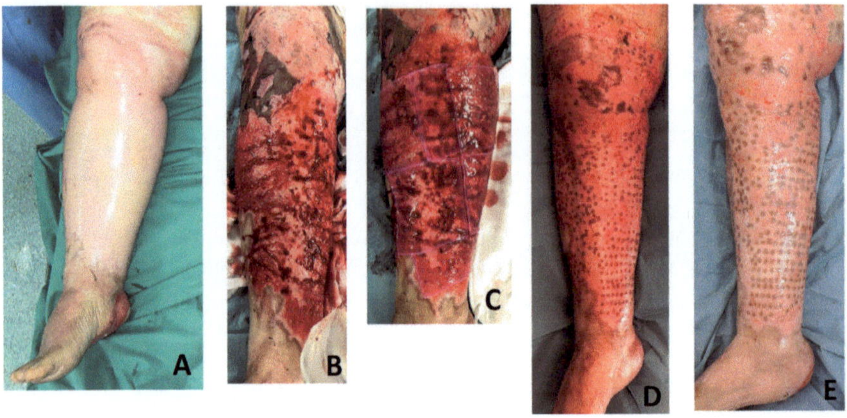

Fig. 23.5 This 70-year-old female sustained 56% TBSA burns. Her right leg was resurfaced with a combination of micrografts and CEA. (**a**) Appearance of her right leg on post-burn day 1 after scrub down. (**b**) Meek micrografts manually laid on her right leg on post-burn week 3. (**c**) The micrografts were overlaid with CEA. (**d**) 3 weeks post-CEA application, showing the wound dry and non-exudative, indicating epithelialisation. (**e**) At 6 weeks post-CEA, the leg is fully covered with a translucent sheen

resides between the expanding micrograft islands. A fine mist of slow-setting fibrin glue (Artiss®) is sprayed on the bed for haemostasis as well as to promote CEA adhesion.

CEA is delivered on gelatinous fibrin sheets. Transfer of the fibrin sheet onto the wound bed is accomplished using two pairs of blunt-tip forceps to first lift it from the culture dish. By gentle traction, it is then pulled over onto the wound. (Figs. 23.3 and 23.4). Care should be taken that the fibrin mat carrier does not tear, crinkle, or overturn during the transfer to avoid disrupting the invisible layer of keratinocytes resting on it. The CEA is covered with Urgotul lipo-colloid dressings. Our experience has shown that CEA placed on anterior wound beds has improved take rates, therefore CEA application on posterior body surfaces and joints is avoided [11].

CEA is used in combination with micrografts (Fig. 23.5), widely meshed autografts (Figs. 23.6 and 23.7), or both (Fig. 23.8) [19]. In the latter case, split-thickness autografts are meshed according to a 3:1 or 6:1 expansion ratio and overlaid with CEA sheets. When Meek micrografts are used, each micrograft is painstakingly applied onto the wound according to a 9:1 or greater expansion pattern before laying CEA on top. We do not use the Meek expansion method with the pre-folded nylon backing as it involves the application of an adhesive. The adhesive prevents early removal of the nylon gauze backing and is potentially toxic to CEA.

Dressing changes are performed beginning on the third or fifth postoperative day, depending on the level of wound exudation. For exudative wounds, inspection is done early and frequently to prevent loss of micrografts and CEA. By the postoperative second week, the CEA and micrografts would have taken, evidenced by a dry, healthy appearance of the wound bed (Figs. 23.5d and 23.8b). CEA which has taken is invisible, and the raw-looking areas may be mistaken for areas devoid of skin.

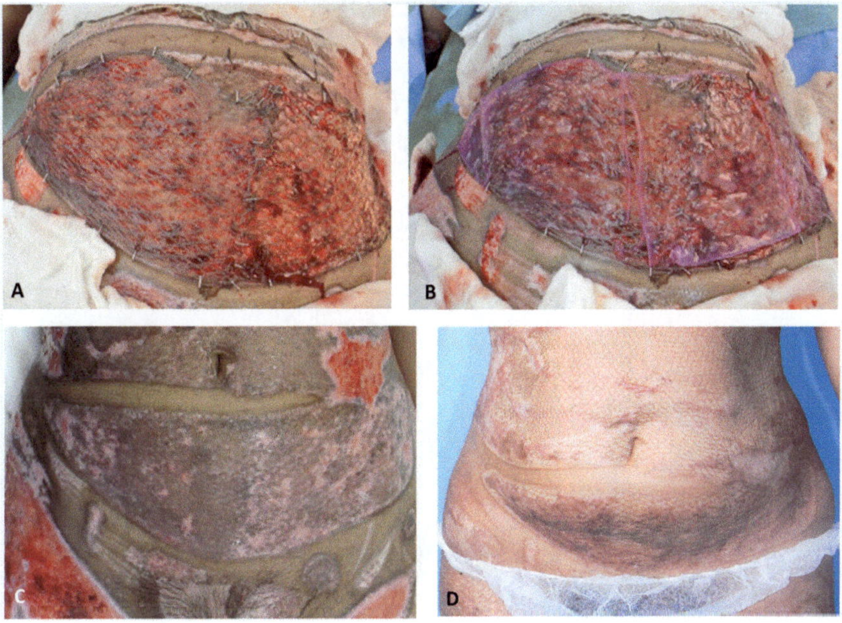

Fig. 23.6 This 19-year-old female sustained 40% TBSA burns. Her lower abdomen and thighs were covered with a combination of 3:1 mesh grafts and CEA. (**a**) Full-thickness abdominal burns exposing fat overlaid with 3:1 mesh autografts. (**b**) CEA was applied over the mesh grafts. (**c**) Complete epithelialisation was seen 1 month after CEA application. (**d**) Appearance of the abdomen 1 year postoperatively. There is healing with scar contracture

The same areas would eventually be repopulated with keratinocytes spreading radially from each micrograft as well as from the CEA. At this stage, wound handling is performed carefully using a non-touch technique. We routinely cover these areas with Urgotul dressings, followed by bulky dressings on top. By 3 to 4 weeks, well-taken CEA appears as a translucent sheen covering the entire wound with pigment islands (Figs. 23.5e and 23.8b). After 4 weeks, the grafted area, if fully re-epithelialized, is exposed.

Patients and Results

Between 2015 and 2023, 22 patients with extensive burns received CEA treatment following either micrografting or widely meshed autograft placement (Table 23.1). These patients presented with an average total body surface area (TBSA) burns of 66.2%, with 15 of them complicated by smoke inhalation injuries. The mortality rate was 31.8%. The CEA cultivation process, from the initial skin biopsy to CEA application, took an average of 21 days. Each patient received CEA sheets covering an average area of 1389 cm^2. The overall success rate of CEA integration on average was 70%. The burn survivors were hospitalised for an average of 78 days.

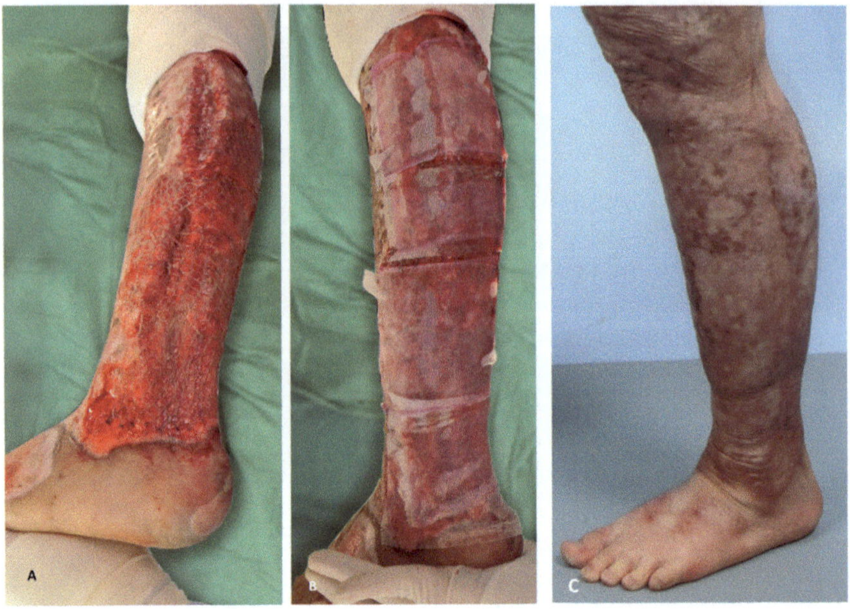

Fig. 23.7 This 19-year-old girl sustained 80% TBSA burns. The left leg had full-thickness burns and a fasciotomy wound. (**a**) The leg was prepped with one previous round of micrografts, and at the present surgery, loose pieces of 3:1 meshed autografts were applied. (**b**) The leg was then covered with CEA and dressed with Urgotul non-stick dressings. (**c**) Appearance of the same leg 1 year postoperatively

Discussion

Stem Cells and CEA

Adult stem cells—also known as tissue stem cells—are cells responsible for maintaining and renewing a target tissue or an organ for a lifetime. They are self-renewing and either multipotent, with the potential to generate several lineages (e.g. different blood cell lines), or unipotent, with the potential to generate a single lineage (e.g. epidermal cells). They reside in a specific microenvironment that contributes to maintaining stemness and stem cell fate. Skin contains both multipotent stem cells in the hair follicles [20, 21] and unipotent stem cells in the epidermis. Human epidermal stem cells, also called holoclones, are the first bona fide stem cells that can be massively multiplied in culture [1, 22] and autologously transplanted to provide permanent engraftment in large third-degree burn wounds [3]. Under appropriate culture conditions, the holoclone has tremendous growth potential [23], even to the extent of replacing the entire diseased epidermis in certain hereditary conditions. As a proof of concept, a single, genetically modified holoclone from recessive dystrophic epidermolysis bullosa was demonstrated to regenerate normal, functional epidermis in a mouse model [24].

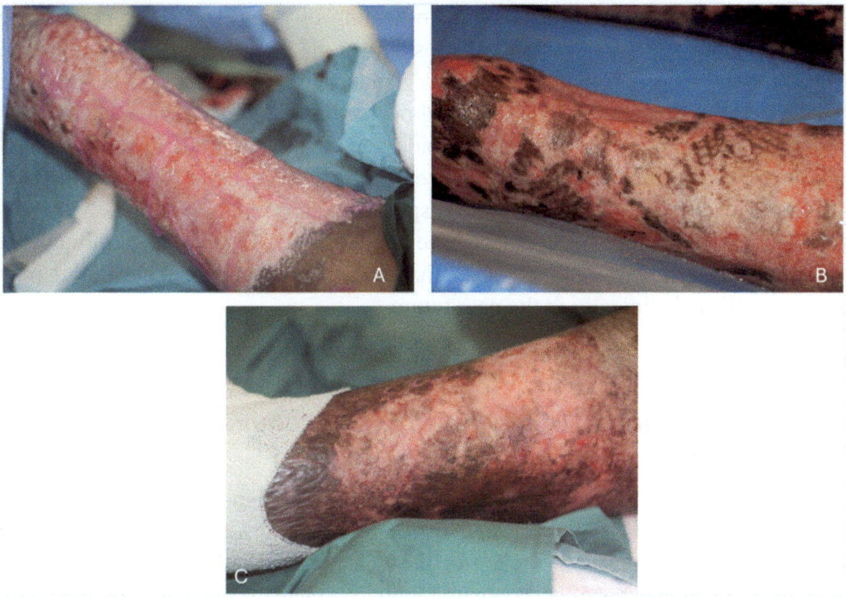

Fig. 23.8 This 34-year-old male sustained 71% TBSA burns. (**a**) The left thigh was initially covered with CEA, and 1 week later, a mixture of micrografts and widely meshed autografts was sparsely applied (not shown). (**b**) Appearance at 3 weeks. The thigh is dry and non-exudative, indicating epithelialisation. (**c**) Appearance at 5 weeks, showing complete healing and islands of pigmentation

Table 23.1 Use of CEA on burn patients between 2015 and 2023

	Total burn cases $n = 22$
Age	38.6 ± 17.6
Gender (Male:Female)	17:5
Type of burn	
Flame	22
Smoke inhalation (n, % of total)	15, 68.2%
% TBSA	66.2 ± 12.6
Baux score	119.9 ± 26.9
Mortality (n, % of total)	7, 31.8%
	Burn survivors $n = 15$
First CEA application day from biopsy collection (days)	20.9 ± 5.7
Area of CEA grafted per patient (cm^2)	1386.7 ± 956.8
Length of stay (days)	78.0 ± 37.4

Inadequate culture conditions or a poor wound bed can adversely impact stem cell behavior, resulting in rapid loss of stemness and consequently poor take [25]. CEA can either be used to promote wound healing or provide long-term permanent coverage. These two clinical applications rely on fundamentally different biological mechanisms [26]. CEA used to promote healing of a superficial wound acts by stimulating the migration of stem cells residing at the wound edges or in epidermal appendages (i.e., hair follicles, sebaceous glands, and sweat glands), whereas CEA used to treat a deep wound such as a third-degree burn relies on permanent engraftment and the self-renewing nature of epidermal stem cells. Thus far, transplanted autologous epidermal stem cells have been shown to regenerate the epidermis and its vital barrier function solely, whereas specialised appendages, such as hair follicles and sebaceous and sweat glands, have not. However, there are breakthroughs in animal studies [21, 27, 28] that indicate that epidermal appendage regeneration may be accomplished clinically in the near future.

Clinical Application Using CEA in Combination with Autografts

As reported in previous studies, the CEA take rate remains inconsistent, ranging from 0% to 100% [12]. A Japanese multicentre study showed that when CEA was combined with autografts, the take rate increased significantly [19]. Their average take for CEA alone was 52%, compared with 77% when CEA was combined with widely meshed autografts ($p < 0.0001$). Other investigators have also corroborated the efficacy of this hybrid technique [29]. Our experience mirrors that of the Japanese. We have observed that when CEA is combined with auto-micrografts or widely meshed autografts [30], epithelialisation is accelerated, with complete epithelialisation occurring after 3 to 4 weeks (Figs. 23.5, 23.6, 23.7, and 23.8). The case of Fig. 23.8 illustrates the delayed application of autografts after initial CEA application, owing to logistical constraints. Despite the delay, the CEA was supported, resulting in accelerated coverage of the thigh. In our micrografting technique [16], medium-thickness micrografts are preferred, as they disperse stem cells more efficiently. These micrografts are typically 9/1000th of an inch (0.23 mm) in thickness. We postulate that at this thickness, the basement layer is exposed and becomes a source of stem cells released to the wound. Thick micrografts take well but do not readily produce a halo-like ring of migrating keratinocytes in the early post-grafting phase.

Conversely, when CEA is exclusively used, the results are inferior (Fig. 23.9). The case of Fig. 23.9 had 70% TBSA burns and, due to a lack of donor skin, the abdomen was covered with CEA alone. One half of the abdomen had good take, and the other half poor, which attests to the inconsistency of CEA take rate in the absence of autologous elements.

At a molecular level, we postulate that the CEA receives autocrine and paracrine signaling from secreted growth factors and cytokines produced by various cells (keratinocytes, fibroblasts, endothelial cells, melanocytes, and lymphocytes) within the autografts [31, 32]. Furthermore, migrating keratinocytes are known to deposit components of the skin basement membrane [33]. Thus, at the cellular level, we postulate that migrating fibroblasts and keratinocytes from adjacent autografts

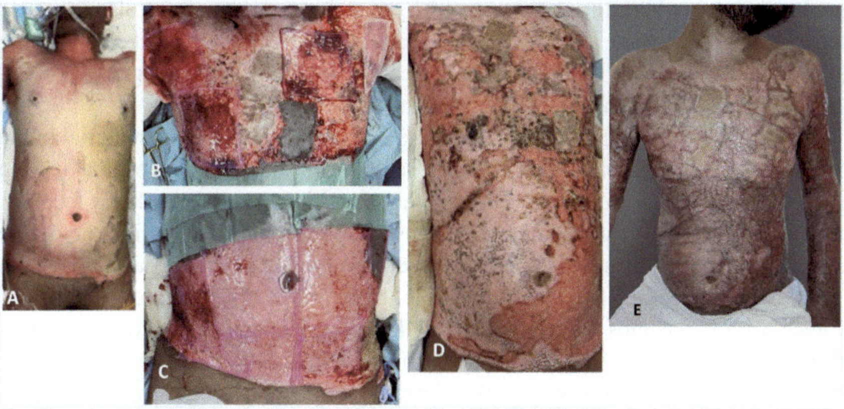

Fig. 23.9 This 29-year-old male sustained 70% TBSA burns. The chest was resurfaced with a combination of micrografts and CEA. The abdomen was resurfaced solely with CEA. (**a**) Appearance of his anterior torso on admission. (**b**) The chest, showing the presence of micrografts, which were covered with CEA. (**c**) The abdomen was solely covered with CEA owing to the shortage of autografts. (**d**) 1 month post-CEA. Notice the right lower abdomen has epithelialised, while the left remains raw, indicating poor CEA take there. (**e**) 4.5 months post-CEA showing full epithelialisation of the torso, albeit with more scarring on the left

contribute to the deposition of connective elements, effectively anchoring the CEA in place.

Other hybrid methods include CEA on allograft dermis [13, 34] or artificial dermis [35, 36]. However, in cases of massive burns, we avoid such techniques owing to the risk of infection. In our local humid climate, burn wounds become heavily colonised with bacteria, and allografts readily develop infection. Hence, our practice aims at completely removing all allografts and substituting them with automicrografts, once they have fulfilled their initial roles of wound seal and haemostasis. Allografts are highly immunogenic, and the inflammation they incite at the time of rejection impairs healing.

In terms of postoperative care, we use Urgotul to cover CEA. The lipocolloid coating in Urgotul is non-stick and lasts for a week. We avoid Tulle gras as the paraffin is absorbed within a day or two, which causes the gauze to adhere to the wound. Silver-containing dressings are prohibited because silver is cytotoxic to monolayer keratinocytes [17, 37]. Once CEA take is established, light dressings (e.g., Tulle gras or Urgotul), as opposed to occlusive ones, are preferred, since exposure to air promotes keratinisation.

Carriers for CEA

In the first two pioneering CEA transplantations performed by Dr. Howard Green's clinical collaborators [3, 38], confluent CEA sheets from the bases of culture vessels were harvested after dispase treatment. Following dispase treatment, CEA sheets rise to the top, where they are captured by matching sheets of petrolatum gauze through surface tension. The CEA is secured to the gauze with surgical clips. Owing to the delicate nature of CEA, which is typically 2 to 8 cells thick, the gauze backing is crucial for handling and application on the wound.

The transfer of enzymatically detached CEA sheets onto backing material in the traditional method involved numerous intricate, time-consuming steps, making the process laborious and complex. Owing to their fragility, there is a tendency for them to shrink, fold, fragment, and overturn. Thus, despite technical expertise, the risk of damaging CEA remains [14]. Additionally, enzymatic treatment to float CEA causes them to shrink, leading to a size reduction of up to 50% [39]. To address these challenges, various systems for improving cell delivery to wound beds have been developed over the years. Reported delivery methods include cell suspensions in different media, fibrin matrices, Biobrane®, Tegaderm™, synthetic hydrogels such as poly(2-hydroxyethylmethacrylate) (pHEMA), sterile medical-grade polymer coated with plasma-polymerised functional surface containing 20% carboxylic acid (PPS), Tegapore™, Myskin™, and chitin membrane [12].

Our preferred CEA carrier is the fibrin matrix. Fibrin is a natural protein that is a component of the blood clotting cascade. It is formed by the action of thrombin on fibrinogen, which causes it to polymerise. Polymerised fibrin is a meshwork of cross-linked fibrin, which forms a natural scaffold for cell attachment and growth.

Modified for transparency, visualisation, and charting of epithelial cell growth [4], the fibrin matrix has emerged as the most comprehensively evaluated substrate for optimal culture using the Rheinwald and Green method. The fibrin matrix is also a well-evaluated delivery system for human epithelial cells that has been shown to contain a robust population of holoclones [14]. Clinically, the fibrin mat is transferred en bloc (Figs. 23.3 and 23.4), obviating the need for additional enzymatic treatment of CEA sheets from the flask. After transfer, proteases secreted by the wound environment and cultured keratinocytes dissolve the fibrin, allowing the cells with the correct polarity to settle on the wound. These qualities make it ideal as a CEA carrier. This CEA-fibrin combination has demonstrated success in treating a variety of conditions, including major burns [4, 13, 14], complex full-thickness cutaneous defects [40], aplasia cutis congenita [36], junctional epidermolysis bullosa [41], and ocular injuries associated with limbal stem cell deficiency [42, 43].

Acknowledgements This work was supported by the Lee Seng Teik and Lee Hoo Leng Distinguished Professorship in Plastic Surgery and Regenerative Medicine, Duke-NUS Medical School, Singapore.

References

1. Rheinwald JG, Green H (1975) Serial cultivation of strains of human epidermal keratinocytes: the formation of keratinizing colonies from single cells. Cell 6(3):331–343
2. Green H, Kehinde O, Thomas J (1979) Growth of cultured human epidermal cells into multiple epithelia suitable for grafting. Proc Natl Acad Sci USA 76(11):5665–5668

3. Gallico GG 3rd, O'Connor NE, Compton CC, Kehinde O, Green H (1984) Permanent coverage of large burn wounds with autologous cultured human epithelium. N Engl J Med 311(7):448–451
4. Ronfard V, Rives JM, Neveux Y, Carsin H, Barrandon Y (2000) Long-term regeneration of human epidermis on third degree burns transplanted with autologous cultured epithelium grown on a fibrin matrix. Transplantation 70(11):1588–1598
5. Fei X, Seah CS, Lee ST (1991) Human keratinocyte cell culture for the burns patients--a preliminary report. Ann Acad Med Singapore 20(4):493–497
6. Munster AM, Weiner SH, Spence RJ (1990) Cultured epidermis for the coverage of massive burn wounds. A single center experience. Ann Surg 211(6):676–679; discussion 679–680
7. Herndon DN, Rutan RL (1992) Comparison of cultured epidermal autograft and massive excision with serial autografting plus homograft overlay. J Burn Care Rehabil 13(1):154–157
8. Clugston PA, Snelling CF, Macdonald IB, Maledy HL, Boyle JC, Germann E, Courtemanche AD, Wirtz P, Fitzpatrick DJ, Kester DA et al (1991) Cultured epithelial autografts: three years of clinical experience with eighteen patients. J Burn Care Rehabil 12(6):533–539
9. Barillo DJ, Nangle ME, Farrell K (1992) Preliminary experience with cultured epidermal autograft in a community hospital burn unit. J Burn Care Rehabil 13(1):158–165
10. Haith LR Jr, Patton ML, Goldman WT (1992) Cultured epidermal autograft and the treatment of the massive burn injury. J Burn Care Rehabil 13(1):142–146
11. Sood R, Roggy D, Zieger M, Balledux J, Chaudhari S, Koumanis DJ, Mir HS, Cohen A, Knipe C, Gabehart K et al (2010) Cultured epithelial autografts for coverage of large burn wounds in eighty-eight patients: the Indiana University experience. J Burn Care Res 31(4):559–568
12. Lo CH, Chong E, Akbarzadeh S, Brown WA, Cleland H (2019) A systematic review: current trends and take rates of cultured epithelial autografts in the treatment of patients with burn injuries. Wound Repair Regen 27(6):693–701
13. Chua AWC, Khoo YC, Truong TTH, Woo E, Tan BK, Chong SJ (2018) From skin allograft coverage to allograft-micrograft sandwich method: a retrospective review of severe burn patients who received conjunctive application of cultured epithelial autografts. Burns 44(5):1302–1307
14. Pellegrini G, Ranno R, Stracuzzi G, Bondanza S, Guerra L, Zambruno G, Micali G, De Luca M (1999) The control of epidermal stem cells (holoclones) in the treatment of massive full-thickness burns with autologous keratinocytes cultured on fibrin. Transplantation 68(6):868–879
15. Chua A, Song C, Chai A, Chan L, Tan KC (2004) The impact of skin banking and the use of its cadaveric skin allografts for severe burn victims in Singapore. Burns 30(7):696–700
16. Chong SJ, Choke A, Tan BK (2017) Technical tips to enhance micrografting results in burn surgery. Burns 43(5):983–986
17. Burd A, Kwok CH, Hung SC, Chan HS, Gu H, Lam WK, Huang L (2007) A comparative study of the cytotoxicity of silver-based dressings in monolayer cell, tissue explant, and animal models. Wound Repair Regen 15(1):94–104
18. Atiyeh BS, Costagliola M, Hayek SN, Dibo SA (2007) Effect of silver on burn wound infection control and healing: review of the literature. Burns 33(2):139–148
19. Matsumura H, Matsushima A, Ueyama M, Kumagai N (2016) Application of the cultured epidermal autograft "JACE(®)" for treatment of severe burns: results of a 6-year multicenter surveillance in Japan. Burns 42(4):769–776
20. Rochat A, Kobayashi K, Barrandon Y (1994) Location of stem cells of human hair follicles by clonal analysis. Cell 76(6):1063–1073
21. Oshima H, Rochat A, Kedzia C, Kobayashi K, Barrandon Y (2001) Morphogenesis and renewal of hair follicles from adult multipotent stem cells. Cell 104(2):233–245
22. Barrandon Y, Green H (1987) Three clonal types of keratinocyte with different capacities for multiplication. Proc Natl Acad Sci USA 84(8):2302–2306
23. Rochat A, Grasset N, Gorostidi F, Lathion S, Barrandon Y (2013) Chapter 65 - Regeneration of epidermis from adult human keratinocyte stem cells. In: Lanza R, Atala A (eds) Handbook of stem cells, 2nd edn. Academic Press, San Diego, pp 767–780

24. Droz-Georget Lathion S, Rochat A, Knott G, Recchia A, Martinet D, Ben Mohammed S, Grasset N, Zaffalon A, Besuchet Schmutz N, Savioz-Dayer E et al (2015) A single epidermal stem cell strategy for safe ex vivo gene therapy. EMBO Mol Med 7(4):380–393
25. Barrandon Y, Grasset N, Zaffalon A, Gorostidi F, Claudinot S, Droz-Georget SL, Nanba D, Rochat A (2012) Capturing epidermal stemness for regenerative medicine. Semin Cell Dev Biol 23(8):937–944
26. Gurtner GC, Werner S, Barrandon Y, Longaker MT (2008) Wound repair and regeneration. Nature 453(7193):314–321
27. Claudinot S, Nicolas M, Oshima H, Rochat A, Barrandon Y (2005) Long-term renewal of hair follicles from clonogenic multipotent stem cells. Proc Natl Acad Sci USA 102(41):14677–14682
28. Claudinot S, Sakabe JI, Oshima H, Gonneau C, Mitsiadis T, Littman D, Bonfanti P, Martens G, Nicolas M, Rochat A et al (2020) Tp63-expressing adult epithelial stem cells cross lineages boundaries revealing latent hairy skin competence. Nat Commun 11(1):5645
29. Braye F, Oddou L, Bertin-Maghit M, Belgacem S, Damour O, Spitalier P, Guillot M, Bouchard C, Gueugniaud PY, Goudeau M et al (2000) Widely meshed autograft associated with cultured autologous epithelium for the treatment of major burns in children: report of 12 cases. Eur J Pediatr Surg 10(1):35–40
30. Kreis RW, Mackie DP, Hermans RR, Vloemans AR (1994) Expansion techniques for skin grafts: comparison between mesh and Meek Island (sandwich-) grafts. Burns 20(Suppl 1):S39–S42
31. Werner S, Krieg T, Smola H (2007) Keratinocyte-fibroblast interactions in wound healing. J Invest Dermatol 127(5):998–1008
32. Peura M, Siltanen A, Saarinen I, Soots A, Bizik J, Vuola J, Harjula A, Kankuri E (2010) Paracrine factors from fibroblast aggregates in a fibrin-matrix carrier enhance keratinocyte viability and migration. J Biomed Mater Res A 95(2):658–664
33. Sudbeck BD, Pilcher BK, Welgus HG, Parks WC (1997) Induction and repression of collagenase-1 by keratinocytes is controlled by distinct components of different extracellular matrix compartments. J Biol Chem 272(35):22103–22110
34. Cuono C, Langdon R, McGuire J (1986) Use of cultured epidermal autografts and dermal allografts as skin replacement after burn injury. Lancet (London, England) 1(8490):1123–1124
35. Pandya AN, Woodward B, Parkhouse N (1998) The use of cultured autologous keratinocytes with integra in the resurfacing of acute burns. Plast Reconstr Surg 102(3):825–828; discussion 829–830
36. Ngeow A, Hui C, Chua AWC, Koh M, Ong YS, Poon WB (2022) Challenges in the management of extensive aplasia cutis congenita. BMJ Case Rep 15(2)
37. Chen L, Wu M, Jiang S, Zhang Y, Li R, Lu Y, Liu L, Wu G, Liu Y, Xie L et al (2019) Skin toxicity assessment of silver nanoparticles in a 3D epidermal model compared to 2D keratinocytes. Int J Nanomed 14:9707–9719
38. O'Connor N, Mulliken J, Banks-Schlegel S, Kehinde O, Green H (1981) Grafting of burns with cultured epithelium prepared from autologous epidermal cells. Lancet 317(8211):75–78
39. Chester DL, Balderson DS, Papini RP (2004) A review of keratinocyte delivery to the wound bed. J Burn Care Rehabil 25(3):266–275
40. Wong M, Chua A, Tan B-K (2013) Cultured epithelial autografts for the coverage of large wounds: minimizing skin graft donor sites in the sick patient. Eur J Plast Surg 36(6):371–376
41. Hirsch T, Rothoeft T, Teig N, Bauer JW, Pellegrini G, De Rosa L, Scaglione D, Reichelt J, Klausegger A, Kneisz D et al (2017) Regeneration of the entire human epidermis using transgenic stem cells. Nature 551(7680):327–332
42. Rama P, Bonini S, Lambiase A, Golisano O, Paterna P, De Luca M, Pellegrini G (2001) Autologous fibrin-cultured limbal stem cells permanently restore the corneal surface of patients with total limbal stem cell deficiency. Transplantation 72(9):1478–1485
43. Rama P, Matuska S, Paganoni G, Spinelli A, De Luca M, Pellegrini G (2010) Limbal stem-cell therapy and long-term corneal regeneration. N Engl J Med 363(2):147–155

Chapter 24
Healing Beyond the Surface: Regenerative Strategies in Complex Burn Treatment

Frederik Schlottmann, Frederik J. Kloss, Lisa Lorbeer, Moritz R. Milewski, and Peter M. Vogt

Abstract Severe burn injuries represent one of the most complex challenges in modern medicine, often requiring interdisciplinary management across intensive care, surgical reconstruction, infection control, and long-term rehabilitation. While early excision and grafting remain the cornerstone of acute surgical treatment, regenerative strategies are gaining increasing importance in improving both short- and long-term outcomes. This chapter explores the integration of regenerative medicine into the clinical management of complex burn injuries, with a particular focus on critically ill patients requiring intensive care. Regenerative medicine is not intended to replace surgical principles but to complement them—bridging the gap between life-saving intervention and tissue restoration. As research evolves, individualized regenerative concepts tailored to patient comorbidities, burn severity, and systemic response are likely to shape the future of burn care. This chapter provides a clinically grounded, forward-looking overview of regenerative strategies in complex burn treatment, aiming to support clinicians in evidence-based integration of innovative therapies into practice.

Keywords Burns · Burn Units · Intensive Care Units · Plastic Surgery Procedures · Rehabilitation · Skin Transplantation

Abbreviations

ABA	American Burn Association
ABSI	Abbreviated Burn Severity Index
ACLS	Advanced Cardiac Life Support

F. Schlottmann (✉) · F. J. Kloss · L. Lorbeer · M. R. Milewski · P. M. Vogt
Department of Plastic, Aesthetic, Hand and Reconstructive Surgery, Hannover Medical School, Hannover, Germany
e-mail: Schlottmann.Frederik@mh-hannover.de

F. De Francesco, K. Turksen (eds.), *Management and Strategies for Wound Healing*, Stem Cell Biology and Regenerative Medicine 79,
https://doi.org/10.1007/978-3-032-12442-5_24

ARDS	Acute respiratory distress syndrome
ATLS	Advanced Trauma Life Support
BTM	Biodegradable Temporizing Matrix
COVID	Coronavirus disease
CT	Computed tomography
DGV	German Society for Burn Medicine
ECMO	Extracorporeal membrane oxygenation
GBR	Global Burn Registry
ICU	Intensive care unit
PiCCO	Pulse Contour Cardiac Output
PRP	Platelet-rich plasma
RRT	Renal replacement therapy
TBSA	Total body surface area
VR	Virtual reality
WHO	World Health Organization
wIRA irradiation	Water-filtrated infrared-A irradiation

Introduction

With a total surface area of 1.5 to 2 m^2 and a weight of approximately 4 kilograms, the skin is one of the largest organs in the human body. It forms a protective barrier against physical and chemical environmental influences, prevents significant fluid loss and microbial contamination, and thus ensures the integrity of the human body. Additionally, it plays a vital role in thermoregulation [1, 2]. Burn trauma refers to tissue damage, most commonly affecting the skin, caused by external physical or chemical agents. For instance, thermal injury increases intradermal temperature, resulting in varying degrees of damage depending on the duration and intensity of heat exposure. As a rule of thumb, an intradermal temperature of 45 °C causes erythema, temperatures above 55 °C lead to blistering, and temperatures exceeding 60 °C result in tissue necrosis due to protein denaturation [3]. However, the full extent of the damage is not always immediately apparent after heat exposure, as residual heat can cause continued tissue injury, a phenomenon known as afterburn [3]. Clinically, several classifications of burn injuries are commonly used. One widely accepted example is Jackson's model, which divides a burn wound into three distinct zones: a central zone of necrosis, a surrounding zone of stasis, and an outer zone of hyperemia [3–5]. Burn depth is typically categorized into four degrees, which are summarized in Table 24.1, along with their respective clinical presentations and regenerative potential.

The wound healing process is essential for the survival of the human organism. Thanks to intensive research efforts over the past decades, the cellular and molecular mechanisms of wound healing are now fairly well understood [6, 7]. Wound healing is an evolutionarily highly conserved process that proceeds through roughly four phases, some of which overlap. These include first hemostasis, then the

Table 24.1 Overview of the four burn degrees with the corresponding clinical characteristics.

Degree	Affected skin layers	Clinical presentation	Regeneration / Healing
I	Epidermis	Erythema Edema Moderate persistent pain Hyperesthesia	Scar-free healing through desquamation of the skin within a maximum of 7 days
IIa	Epidermis Superficial dermis	Erythema Blister formation Severe pain Well-perfused wound bed	Re-epithelialization via preserved stem cells in skin appendages Scar-free healing, depending on the patient's age and wound care regimen within 10–14 days
IIb	Epidermis Superficial and profound Dermis	Blisters White-gray wound bed Non-perfused, necrotic tissue Minimal pain	Markedly delayed wound healing due to loss of stem cells in destroyed skin appendages Frequent development of hypertrophic scars Indication for surgical debridement and skin grafting
III	Epidermis Dermis Subdermal structures	Leather-like eschar formation Typically no blisters Complete destruction of all skin layers No pain No tactile sensation Loss of hair follicles	Healing of small areas through contraction and scar formation Indication for surgical debridement and skin grafting

inflammation phase with exudation and resorption, and the proliferation phase, that involves the growth of new blood vessels (angiogenesis), the production of collagen by fibroblasts, and the migration of epithelial cells to cover the wound as well as granulation tissue, a type of connective tissue that fills the wound bed during this phase. Finally, the remodeling phase involving reorganization and strengthening of collagen fibers completes the wound healing process—leading to wound contraction and reducing its size [7, 8]. The regeneration process of human skin by means of re-epithelialization is based on epidermal stem cells located in the stratum basale and can be triggered by various promoters [9]. Further studies have shown that stem cells located in the hair follicles are also involved in the regeneration process of the epithelium and are even essential for this process in some cases [10–12].

The wound healing process varies greatly between different species. Some amphibians, such as the axolotl (*Ambystoma mexicanum*), can completely regenerate full-thickness skin defects without scarring. Even skin pigmentation and secretory skin appendages can be regenerated [13, 14]. In comparison, only limited wound healing is possible in humans. The only exception is human fetuses, in which scar-free wound healing was first described by Rowlatt in 1979 [15]. Further studies have shown that, due to the naive fetal immune system, there is no comprehensive inflammatory response as observed in the wound area of adults [16, 17]. In adults, the wound healing process ends with the formation of scar tissue, which restores basic barrier functionality against environmental influences but is inferior to healthy skin [18, 19]. There is often also cosmetic impairment of the external appearance, which in cases of larger scars can lead to considerable psychosocial impairment for the individual [20]. Based on current scientific findings, intensive research is underway to influence or support the wound healing process [21]. This is likely to remain an important area of research for optimizing burn wound treatment in the upcoming years.

The history of wound management and skin grafting for burn injuries spans centuries [22]. In modern medicine, the treatment of burn injuries has evolved into a specialized field aimed at optimizing clinical outcomes through dedicated burn care units and interdisciplinary management strategies. Most epidemiological data on burns are currently collected at the national level. In 2018, the World Health Organization (WHO) launched the Global Burn Registry (GBR) [23]. Its purpose is to help improve the understanding of burn injury worldwide. As the register is still under construction, in January 2020, a total of 4307 cases were available that had been submitted by 30 burn units from 29 cities and 17 countries [23]. According to data from the German Society for Burn Medicine (DGV), approximately 4350 patients were treated in specialized burn intensive care units (ICUs) across Germany in 2016 due to severe burn injuries [24]. However, this figure likely underestimates the total number of burn injuries, as many cases do not require ICU-level care.

Most burn injuries occur in domestic and recreational settings, which account for up to 70% of burn injuries in adults and up to 95% in children [25]. A significant proportion of these incidents are preventable. Preventive strategies, including smoke detector mandates—introduced in Germany in 2016—have significantly reduced burn incidence in industrialized nations [26]. Additional prevention measures include standardized safety protocols in occupational environments and the use of flame-resistant protective clothing. These initiatives have also been shown to be cost-effective [26]. In recent years, social media has emerged as a potential platform for disseminating burn prevention information and raising public awareness [27].

The COVID-19 pandemic had a notable impact on both the healthcare system and burn care delivery. During the early stages of the pandemic, reductions in elective surgeries and procedural capacities were observed. Interestingly, compared to the early (2020) and late (2021) pandemic phases, more burn patients were managed on an outpatient basis in the pre-pandemic year of 2019 [28]. Despite this, an initial rise in inpatient burn interventions was recorded at the onset of the pandemic, followed by a gradual return to previous surgical patterns as elective procedures resumed [28].

Burn ICUs are equipped with advanced medical technologies and staffed by multidisciplinary teams, including burn surgeons, specialized nursing staff, physiotherapists, and mental health professionals [29]. Admission to these units is based on established criteria, including second-degree burns affecting ≥10% of the total body surface area (TBSA), third-degree burns, involvement of critical anatomical sites (e.g., face, hands, genitals), electrical and chemical burns, inhalation injuries, and complex comorbidities [25]. Additionally, patients requiring specialized psychological, psychiatric, or physical support may benefit from treatment in a

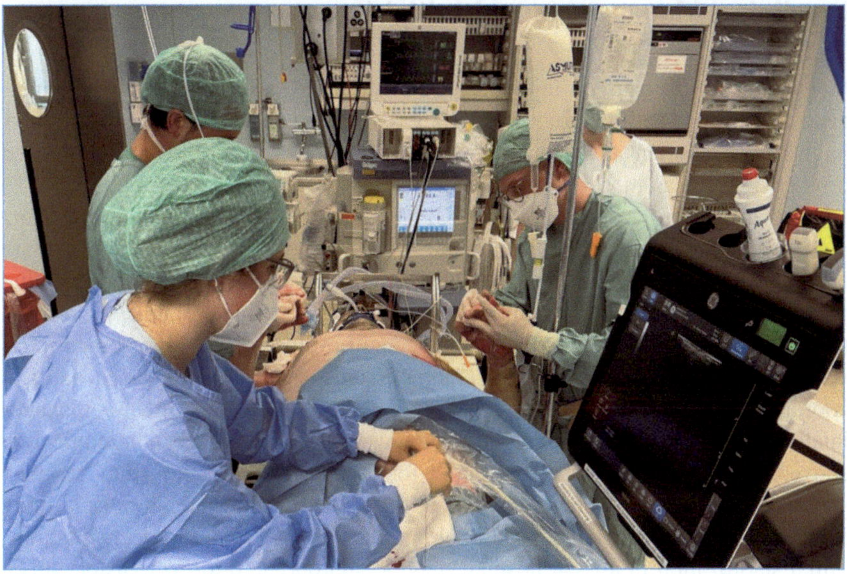

Fig. 24.1 Admission of a patient with severe burns to a burn intensive care unit by a multidisciplinary team

dedicated burn center. Admission can occur via referral from emergency departments, trauma units, or other healthcare institutions and should involve consultation with burn specialists to tailor the initial management strategy. Special attention is warranted for pediatric and geriatric patients due to their heightened vulnerability and specific care needs [30]. Pediatric burn centers are structured to provide child-centered care environments and tailored services (Fig. 24.1).

The costs of acute burn care vary internationally and are influenced by several factors, including patient age, mechanism and extent of injury, burn depth, and survival outcomes [31]. Higher treatment costs are associated with advanced age, inhalation injuries, electrical burns, and fatal outcomes. Average daily inpatient costs for burn care range between $3275.90 and $3677.35 in high-income countries [31].

According to 2022 data from the German burn registry, the leading causes of burn injuries were flame-related burns (including deflagrations, 44%), followed by scalds (21%), oil and fat burns (3%), blistering dermatological conditions (3%), explosions (2%), electrical burns, and chemical burns involving acids or alkalis (1% each) [25]. The comprehensive treatment of burn injuries follows a multidisciplinary approach aimed at early stabilization, effective pain management, wound care, and prevention of complications such as infections and organ dysfunction [29]. The overarching goal is to restore physical function, promote psychological recovery, and enhance long-term quality of life for burn survivors. The following sections will explore the specific components of this multimodal therapeutic strategy in detail.

Preclinical Management and Initial Care

Independent of the cause of injury, rapid and structured scene assessment by emergency personnel is critical for initiating appropriate management strategies. A thorough risk assessment that emphasizes responder safety and follows standardized protocols—such as those outlined in the Advanced Trauma Life Support (ATLS) and Advanced Cardiac Life Support (ACLS) guidelines—is essential for improving outcomes and reducing complications in burn victims [32]. Prehospital burn care encompasses the immediate interventions performed before hospital arrival, aiming to prevent secondary injury, stabilize the patient, and facilitate seamless transition to definitive care. Key components include swift evaluation, evidence-based interventions, and coordinated teamwork among first responders. In Germany, a national coordination center based in Hamburg—operated by the fire department—manages intensive care capacity for severely burned individuals [33]. Timely communication and precise documentation are essential for efficient coordination. Early contact with a specialized burn center allows for adequate preparation and multidisciplinary involvement, especially when additional injuries are suspected. In select situations, initial stabilization in a local hospital may be appropriate prior to transfer, which should occur only after the patient has been sufficiently stabilized.

Securing the airway is paramount; if airway compromise is suspected or confirmed, early intubation must be considered. Classic indications include respiratory failure, absence of protective airway reflexes, and imminent airway obstruction [32]. Established guidelines, such as those from the American Burn Association (ABA) and the Denver criteria, aid in identifying candidates for early airway management [34]. The Denver criteria extend ABA recommendations by including signs such as singed nasal or facial hair and suspected smoke inhalation. Importantly, indiscriminate intubation should be avoided; studies have shown that up to 40% of patients who were intubated without meeting formal criteria experienced associated complications [35].

Preventing hypothermia, the consequence of impaired thermoregulation, is another critical element, as it correlates with poor prognosis in burn patients [36]. Thermal management must begin at the scene, especially in cases involving combined thermal and mechanical trauma [37]. Active rewarming strategies are preferred during this phase, while any cooling measures should be discontinued to mitigate hypothermia risk [37]. The use of sterile, insulating materials such as warming blankets and dressings supports both thermoregulation and pain control [33].

Although no standardized prehospital analgesia protocol exists, intravenous administration is considered most effective. To avoid fluid overload, a conservative approach is recommended—adults should initially receive 1000 mL of warmed, balanced crystalloid solution within the first 2 hours. Established in-hospital resuscitation formulas, such as the Parkland-Baxter regimen, are not advised in the prehospital setting [38].

Prognostic Factors and Acute Phase Management in Severe Burn Injury

The prognosis of burn patients is predominantly determined by total body surface area (TBSA) affected, burn depth, concomitant injuries, pre-existing comorbidities, and the quality of initial and ongoing medical care. Burn ICUs are typically managed collaboratively by plastic surgeons and anesthesiologists [29]. Prognostic assessment at admission is supported by several scoring systems, with the Abbreviated Burn Severity Index (ABSI) being among the most validated [39]. ABSI incorporates sex, age, TBSA, inhalation injury, and third-degree burns to calculate a survival probability based on a 2–13 point scale. The Burn Mortality Prediction (BUMP) score was introduced in 2023 to improve mortality prediction based on data from the German burn registry and showed reliable results [40].

Arriving at the burn center, patients should be presented to the multidisciplinary team using standardized handover protocols analogous to polytrauma care. A detailed medical history should be obtained, including eyewitness accounts, when necessary, to understand the trauma mechanism. Regardless of the severity of burns, general trauma diagnostics should follow trauma room algorithms, including computed tomographic (CT) imaging if the mechanism is uncertain or complex [41] (Fig. 24.2).

Initial Management and Burn Assessment

After admission, the initial management includes airway stabilization, large-bore venous access, invasive hemodynamic monitoring, and early surgical evaluation of burn wounds. Burn depth assessment requires removal of vesicles and soot, shaving of singed hair, and complete antiseptic washing. To prevent hypothermia, the hydrotherapeutic care unit should be preheated, and procedures should be conducted efficiently [33].

Preclinical assessments of burn depth and extent must be critically reviewed by an experienced burn surgeon. If available, Laser Doppler Imaging (LDI) can assist in evaluating burns of indeterminate depth [42]. For documentation and forensic purposes, photographic records should be obtained [33]. Inhalation trauma can be confirmed via bronchoscopy, but a careful risk-benefit decision should be made when indicating bronchoscopy [43]. Early microbiological wound swabs are recommended to enable targeted antibiotic therapy during later phases of treatment, if indicated. However, a standardized empirical antibiotic regimen is not recommended [44].

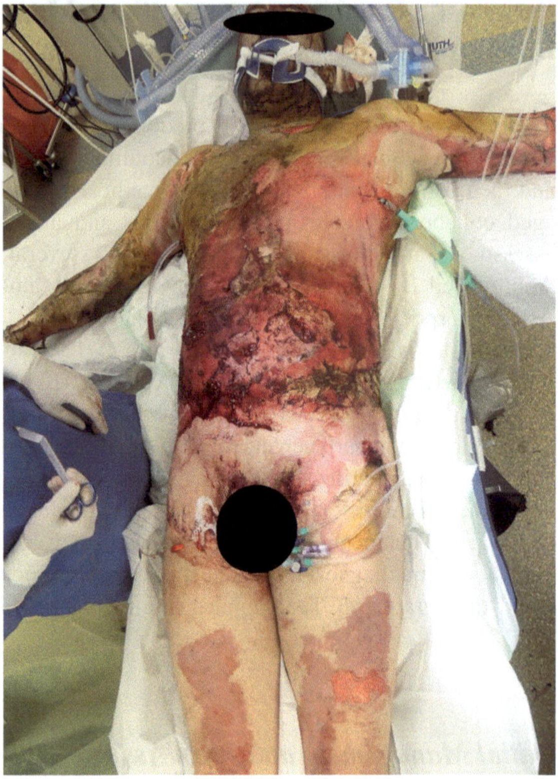

Fig. 24.2 Severe burn patient with deep, extensive burns and accompanying chest and open abdominal trauma

Burn Shock and Resuscitation

After initial stabilization, patients are admitted to the burn ICU. The subsequent acute phase is typically characterized by burn shock—a life-threatening systemic response to extensive burns, due to a massive release of cytokines and inflammatory response. Management includes balanced fluid resuscitation guided by hemodynamic parameters, vasopressor support, pain control, thermoregulation, oxygen therapy, and organ support. Over-resuscitation has emerged as a critical issue, contributing to increased morbidity and mortality [33, 45]. Crystalloid solutions are the cornerstone of fluid replacement, supplemented by the colloid solution human albumin as recommended by the American Burn Association (ABA) in large TBSA burns to reduce fluid volumes and improve urine output [46].

Effective fluid therapy in critically ill or hemodynamically unstable patients requires careful and continuous assessment to avoid the risks of both hypovolemia and fluid overload. Traditional static parameters such as central venous pressure have proven to be unreliable predictors of fluid responsiveness. Instead, dynamic monitoring methods provide more accurate and individualized guidance. One such method is Pulse Contour Cardiac Output (PiCCO) technology, which allows for minimally invasive, real-time monitoring of key hemodynamic parameters [47].

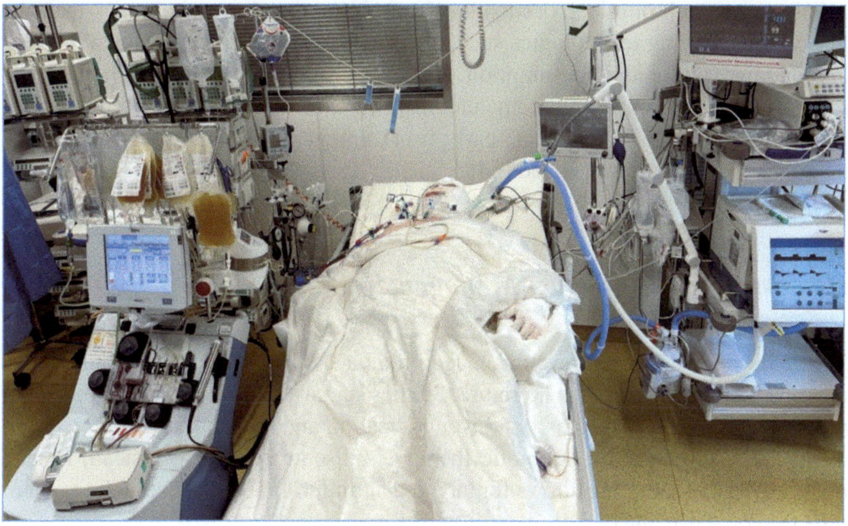

Fig. 24.3 Burn patient in an intensive care setting with mechanical ventilation and total plasma exchange therapy. Note the disposable blanket, the highly stacked infusion pump rack, and the mummy-like dressings that cover the patient, hinting at the extent of the injury

However, PiCCO should not be used in isolation. Echocardiography, particularly focused or point-of-care ultrasound, serves as an essential complementary tool. Parameters such as mean arterial pressure, lactate levels, capillary refill time, and urine output must be continuously interpreted in context with the hemodynamic data. A multimodal approach that integrates PiCCO-derived metrics, echocardiographic findings, and bedside clinical assessment provides the most comprehensive and precise framework for tailoring fluid therapy to the individual needs of the patient [47]. High-dose ascorbic acid may help reduce fluid requirements and tissue edema in patients with TBSA above 20%, although robust randomized controlled trials are lacking [48, 49]. If fluid resuscitation is insufficient to maintain organ perfusion, vasopressor therapy, particularly norepinephrine, is indicated [50]. Dobutamine may be considered in cases of low cardiac output with normotensive values [50]. Mechanical ventilation is not routinely indicated and should follow lung-protective strategies. Early tracheostomy should be considered if prolonged ventilation is anticipated [32] (Fig. 24.3).

Analgesia, Early Mobilization, and Adjunct Therapies

Pain management should employ a multimodal strategy, combining opioids with adjuvants and non-pharmacological interventions [51]. No single analgesic regimen has proven superior [51]. Early mobilization, ideally within the first 14 ICU days, has been shown to reduce delirium and mitigate muscle atrophy [52]. Interdisciplinary collaboration with physical, occupational, respiratory, and speech therapy teams is essential. Emerging technologies such as virtual reality (VR) have shown promise in reducing pain, opioid consumption, and anxiety in burn patients, warranting further research for routine implementation [53]. Given the impact of psychiatric comorbidities on ICU outcomes, early involvement of psychiatry and psychosomatic medicine is recommended [54].

Hemostasis and Infection Management

Severe burn patients may develop an acquired Factor XIII deficiency, although this remains underreported in the literature. Studies suggest that Factor XIII supplementation may improve hemostasis and wound healing and reduce transfusion requirements [55]. Prophylactic antibiotic therapy is discouraged [56]; instead, routine microbiological sampling allows for targeted antimicrobial treatment in the event of infection or sepsis [44].

Organ Support and Critical Care Considerations

Burn shock can progress to multi-organ failure. Neither the ABA nor German guidelines provide standardized recommendations for early renal replacement therapy (RRT) [46]. Although potential benefits exist, optimal RRT timing remains unclear, and further research is needed [57]. In select cases, plasma exchange is used to stabilize refractory burn shock by removing pro-inflammatory cytokines [58]. In cases of severe respiratory failure (e.g., ARDS), extracorporeal membrane oxygenation (ECMO) may be considered, though each case requires a careful risk-benefit analysis [59, 60].

Metabolic and Nutritional Considerations

After surviving the initial burn shock, patients may experience additional critical phases due to infections or organ complications [61]. Early enteral nutrition is recommended; if not feasible orally, feeding tubes or parenteral nutrition should be used [62]. Supplementation with trace elements and vitamins is essential [62]. In the post-shock phase, the hypermetabolic response to burns can be counteracted with pharmacological interventions. The synthetic anabolic steroid oxandrolone has demonstrated benefits in reducing mortality and hospital length of stay in patients with ABSI >10 points [63]. Beta-blocker therapy with propranolol is also widely accepted to modulate metabolic pathways and improve clinical outcomes following burn trauma [64].

Ethical Considerations

Family members should be included early in the care process to clarify legal guardianship and assess the patient's presumed wishes. In ethically complex situations, consultation with ethics committees and palliative care teams is advised [65].

Surgical Management of Burn Injuries

The comprehensive surgical management of burn injuries is multifaceted, focusing on facilitating wound healing, mitigating complications, and restoring both function and aesthetic outcomes. This process necessitates a collaborative, multidisciplinary approach involving plastic surgeons, anesthesiologists, specialized burn care nurses, and other allied healthcare professionals. Prior to any surgical intervention, a meticulous assessment of the burn's severity, depth, and TBSA is paramount. Initial patient stabilization, encompassing airway management, shock therapy, fluid resuscitation, and sepsis management, must precede surgical procedures.

Optimal Timing and Modalities of Surgical Debridement

Surgical debridement, the removal of devitalized or contaminated tissue, is crucial for promoting wound healing, reducing infection risk, and preparing the wound bed for subsequent interventions such as skin grafting. Debridement techniques include surgical approaches like tangential or epifascial excision, as well as non-surgical methods such as enzymatic or mechanical debridement. The advent of bromelain-based proteolytic enzyme debridement [e.g., NexoBrid™ (MediWound Ltd., Yavne, Israel)] in Europe in 2012 has significantly impacted the management of deep partial-thickness (2b degree burns) to full-thickness third-degree burns [66]. Systematic meta-analyses have demonstrated that NexoBrid™ offers a safe, rapid, and selective method for eschar removal, leading to reduced operative interventions, shorter hospital stays, and decreased incidences of wound-associated sepsis [66] (Fig. 24.4).

The optimal timing of surgical debridement in burn patients remains a subject of ongoing expert debate. A recent German register study revealed that early excision was performed in less than half of burn patients, with TBSA being the primary determinant of this decision [67]. Notably, while early excision did not reduce mortality, it was associated with a shorter inpatient treatment duration, and the time of admission served as a predictive factor for early versus delayed excision [67].

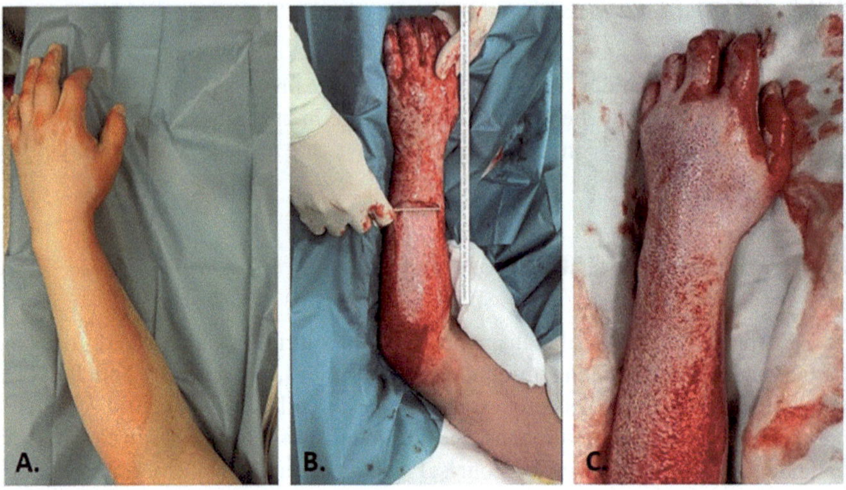

Fig. 24.4 Second- to third-degree burn of the left forearm and hand: before (**a**), during (**b**), and after enzymatic debridement (**c**) with NexoBrid™

Defect Coverage Strategies

Prior to debridement, an individualized and comprehensive defect coverage plan, including rescue options, is essential. Skin grafting techniques are the most prevalent surgical procedures for covering extensive burn wounds. Beyond these established methods, numerous research initiatives are exploring advanced approaches, including skin tissue engineering, 3D bioprinting, and immunoengineering of allogeneic skin grafts, aiming to introduce innovative therapeutic options in the near future [22].

Autologous Skin Grafts

Autologous skin grafting, a practice with over two centuries of history in European surgery, is the gold standard for burn wound coverage [22]. Continuous advancements in this field have solidified its role as a standardized therapeutic procedure, providing permanent wound coverage, promoting healing, and enhancing functional and cosmetic outcomes. However, for large burn areas, the limitations of autologous split-thickness skin grafting become apparent due to restricted donor site availability and increased donor site morbidity [68]. In specific anatomical regions, such as the face and hands, full-thickness skin grafts are utilized to achieve superior aesthetic results and reduce wound contracture [69] (Fig. 24.5).

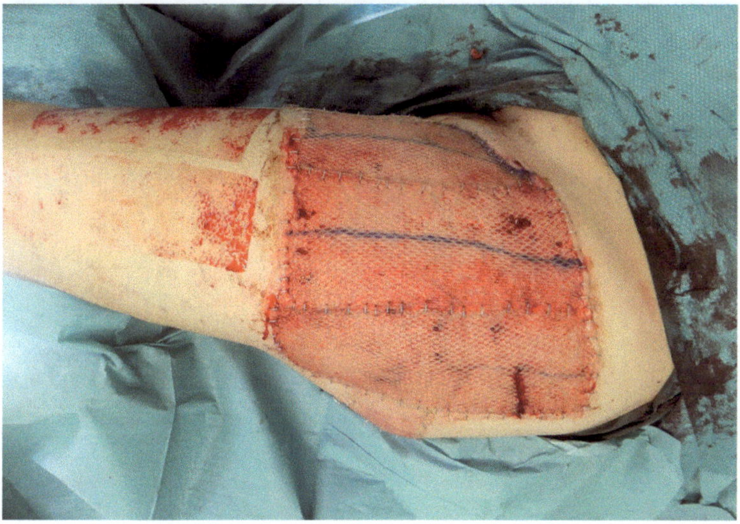

Fig. 24.5 Autologous split skin grafting after tangential excision of a burn in the left hip area. The split skin graft site can be seen on the left side of the image

Allogeneic Skin Grafts

Allogeneic skin grafts serve as a temporary wound covering for extensive burns when adequate autologous donor sites are unavailable. They effectively protect the wound from infection and prepare the wound bed for subsequent autologous grafting [70]. However, their application is challenged by the risk of immunological rejection, infection, and increased scarring [71]. Cryopreserved or glycerol-preserved skin allografts are commercially accessible globally through biobanks and skin tissue banks. Despite this, a nationwide supply network for these grafts is not yet established worldwide [72]. Specific national skin banks, such as the Hannover Skin Bank, are currently pioneering the production of vital allografts from post-bariatric surgery tissue donations, showing promising initial clinical outcomes [73]. While various preservation methods, including glycerolization, can reduce skin allograft immunogenicity, they also compromise cellular epidermal and dermal components [74]. Given that severely burned patients are typically immunocompromised in the acute phase, allografts often achieve temporary integration but are subsequently rejected upon immune system reconstitution [74] (Fig. 24.6).

Xenogeneic Skin Grafts

In addition to autologous grafting, xenogeneic skin grafts, such as porcine and fish skin, are employed as temporary coverings for large-area burns [71]. These grafts generally serve as a bridge until definitive autologous defect coverage can be achieved. Recent studies, including a randomized controlled Phase II trial, have

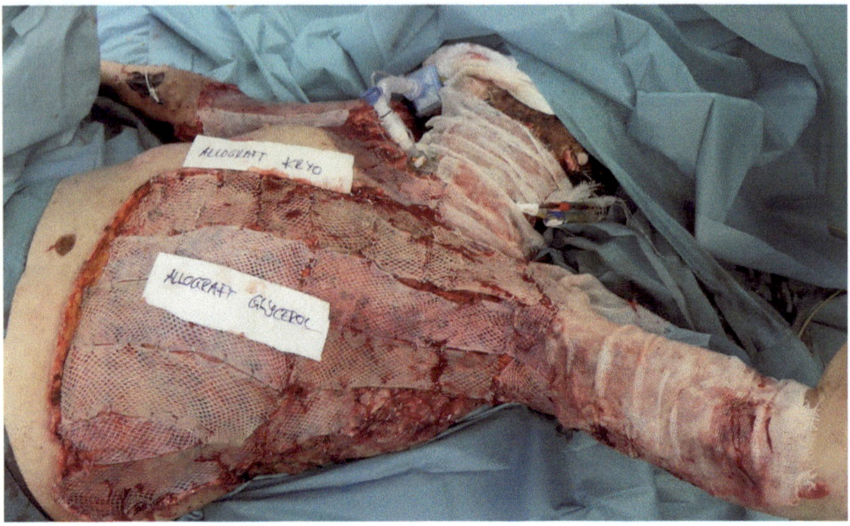

Fig. 24.6 Temporary defect coverage of epifascially debrided burns with commercially available glycerol-preserved allograft skin (Allograft Glycerol) and vital cryopreserved allograft skin (Allograft Cryo) from the Hannover Skin Bank

explored the use of Nile tilapia skin (*Oreochromis niloticus*) for complete re-epithelialization of burn wounds without the need for autologous split-thickness skin grafting [75]. While xenografts have shown promising clinical perspectives, further research is necessary to fully evaluate their potential in comparison to autologous skin grafting.

Biological and Synthetic Skin Substitutes

The market for biological and synthetic skin substitutes for burn wound treatment is extensive, with a plethora of products available globally [76]. These can be broadly categorized into epidermal and dermal substitutes. Among the established epidermal substitutes, SUPRATHEL® is recognized for its efficacy in treating superficial partial-thickness 2a-degree burns and scalds, demonstrating excellent wound healing, reduced costs, and improved mortality rates [77].

Dermal skin substitutes are generally classified as decellularized human or animal dermis constructs, artificial scaffolds made from various biomaterials, or entirely synthetic polymers [78]. A common limitation across many dermal substitutes is the mandatory requirement for subsequent application of an autologous skin graft to achieve complete dermo-epidermal replacement. NovoSorb® Biodegradable Temporizing Matrix (BTM), a fully synthetic dermal substitute comprising biodegradable polyurethane foam with a temporary non-degradable sealant, has been extensively studied for its safety and ability to achieve permanent wound closure in conjunction with autologous split-thickness skin grafts in a two-stage procedure

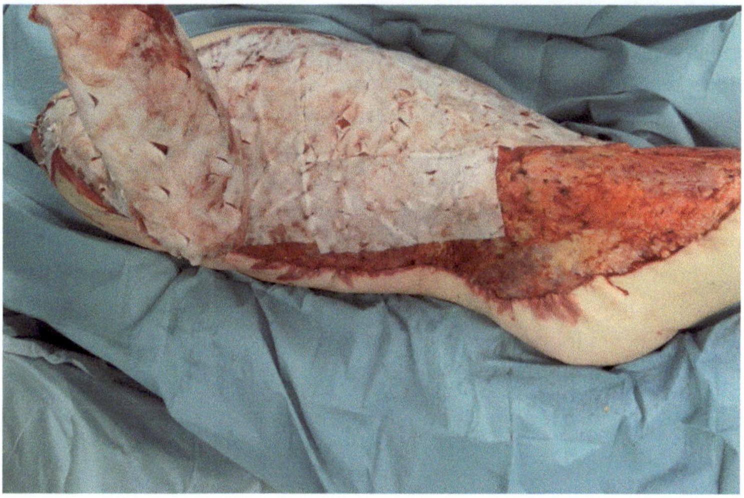

Fig. 24.7 Defect coverage of an epifascially debrided burn with BTM

[79, 80]. However, BTM has been associated with an integration and vascularization period of up to 3–4 weeks, leading to prolonged hospitalization and increased treatment costs [81] (Fig. 24.7).

MatriDerm®, a dermal, cell-free skin substitute composed of bovine collagen types I, III, V, and elastin, is commercially available for deep burn wounds. Extensive animal studies indicate its conversion into an autologous matrix within the recipient over several weeks [82]. MatriDerm® is available in various thicknesses, and its ability to facilitate single-stage defect coverage with simultaneous autologous split-thickness skin grafting is a significant clinical advantage [82]. Furthermore, improved scar quality and reduced wound contraction have been observed with its use [83].

Integra®, already developed in the 1980s as a collagen-glycosaminoglycan matrix, is a dermal skin substitute and consists of bovine collagen type I and shark chondroitin-6-sulfate, covered by an outer silicone membrane [84]. This matrix integrates and vascularizes within the recipient over approximately 3 weeks, after which the silicone layer is removed, and definitive autologous split-thickness skin grafting is performed [84]. Integra® has been linked to improved scar quality, objectively higher viscoelasticity, and shorter hospitalization times, though its prolonged integration time and less resistance to infection remain drawbacks [84].

Autologous Cultured Keratinocytes and Skin Grafts

Beyond traditional autologous skin grafts, autologous cultured keratinocytes are commercially available as spray suspensions (e.g., RECELL®) or cell sheets (e.g., Epicel®) from various manufacturers [85, 86]. These are clinically utilized in selected patient cases. Ongoing clinical trials for cultured autologous dermo-epidermal skin substitutes for severe burns show promising results [87]. However, the widespread adoption of cultured autografts is constrained by high production costs and cultivation times spanning several weeks.

Flap Surgery and Microsurgical Reconstruction

For deep burns or wounds involving critical functional structures like tendons, muscles, or bones, flap surgery is indispensable. While skin grafting and locoregional flaps are considered foundational, microsurgical free tissue transfer should be individually considered and tailored to the specific wound bed and surgical requirements [88].

Adjunctive Wound Treatment Concepts

The surgical management of burns also incorporates various topical therapies aimed at promoting skin regeneration and healing, reducing pain, and preventing infection. These include antimicrobial dressings, skin substitutes, growth factors, and specialized wound care products [78]. Given the extensive range of available wound dressings and topical agents, a detailed discussion of each is beyond the scope of this section. Water-filtered infrared-A (wIRA) irradiation can be clinically applied to enhance wound drying and re-epithelialization [89, 90]. In cases of burn wound colonization with multi-resistant pathogens, focused anti-infective therapy using specific bacteriophage strains has been successfully reported in isolated instances [90]. Initial findings on the use of atmospheric cold plasma and platelet-rich plasma (PRP) in burn wound treatment show promise [91–93]; however, further studies are required to conclusively assess their clinical value.

Rehabilitation and Post-Treatment Phase

Rehabilitation is a critical, ongoing process in the postoperative management of burn injuries, extending beyond the acute phase to optimize physical, psychological, and social recovery. Following discharge from intensive care, burn patients typically transfer to a general ward before moving to specialized rehabilitation or aftercare facilities. Germany, for instance, offers highly specialized clinics with standardized rehabilitation protocols [94]. This phase employs a multidisciplinary, multimodal approach—mirroring the acute phase—integrating physiotherapy, occupational therapy, and psychological care to support comprehensive recovery. Prerequisites for initiating rehabilitation include largely epithelialized defects, no immediate need for further surgical intervention, cardiorespiratory stability, and partial patient mobilization.

Different types of rehabilitation are tailored to specific recovery goals [95]:

- Primary rehabilitation commences immediately post-hospitalization, concentrating on scar and wound healing.
- Post-acute rehabilitation spans approximately 6–12 weeks, aiming to restore independence in daily living activities.
- Follow-up rehabilitation targets patients who have largely regained independence, focusing on improving function, fitness, and mental health over an anticipated 3 to 8 weeks. Inpatient treatment is indicated if outpatient therapy proves insufficient or if deficits persist.

Multimodal therapeutic strategies during rehabilitation encompass scar therapy, movement therapy, contracture treatment, psychological support for both patients and their families, pain management, treatment of comorbidities and secondary complications, and social reintegration.

Scar tissue often presents significant functional and aesthetic challenges. Consequently, scar management techniques such as massage, pressure garments, silicone sheets, laser therapy, or surgical scar revision are employed to optimize functional and cosmetic outcomes, thereby enhancing patient quality of life [96]. The disfiguring nature of visible scarring frequently leads to psychosocial impairments and social stigmatization [97]. Surgical scar revision may involve techniques like Z-plasty, autologous skin grafts, or percutaneous collagen induction and modulation through surgical needling [96]. Autologous fat grafting (lipofilling) is a minimally invasive treatment for painful, hypertrophic, or retracted scars, involving the injection of processed fat tissue into the hypodermis beneath the scar [98]. It can improve both scar tension and cosmetic appearance but often requires multiple sessions due to fat reabsorption rates of 10–70% [98]. Beyond scar therapy, lifelong skincare for grafted and donor sites, using moisturizing agents, is crucial [95]. Strict sun avoidance on affected areas is also imperative to prevent secondary malignancies and pigmentation disorders.

For patients experiencing limb loss or functional impairment, prosthetic devices and orthotic interventions may be prescribed to enhance mobility, functionality, and independence [99]. Bionic prostheses are also utilized in selected cases. Burn survivors frequently experience debilitating nociceptive, neuropathic, and mixed pain during and after rehabilitation. Effective pain management, adhering to WHO guidelines, is therefore paramount. Additive procedures, based on a multimodal treatment concept, are essential to mitigate the heightened risk of pain chronification.

Long-term follow-up care provides continuous support, monitoring, and interventions to address evolving needs, improve quality of life, and facilitate successful societal reintegration for burn survivors. Altered immune function in burn survivors can correlate with increased morbidity. A range of secondary pathologies, including elevated risks of cancer, cardiovascular disease, nervous system disorders, diabetes, musculoskeletal issues, gastrointestinal problems, and psychological disorders, are associated with long-term burn survival [100]. Patients with a history of ICU stays face an increased risk of both acute and chronic mental health disorders [101]. Post-traumatic stress disorder is frequently observed in patients with severe burns [102], and societal reintegration can exacerbate psychological stress [101]. Adults who sustained burns in childhood are at higher risk for both mental and physical illnesses. Continuous re-evaluation is indispensable during aftercare. Overall, burn survivors have a higher rate of suicidal ideation. While data on suicidality in survivors are limited, they often possess various risk factors linked to increased suicidality in the general population [103]. Even long after the acute phase and in advanced rehabilitation, occupational therapists are vital for supporting the recovery of social and professional participation, despite limited resources [104]. Many survivors only realize during rehabilitation that recovery is a progressive, lifelong journey, and a lack of understanding regarding wound care and rehabilitation progress often hinders their reintegration [100]. Broadly, environmental and personal factors significantly influence participation in employment and work performance.

Conclusion

The treatment of patients with severe burns remains an interdisciplinary challenge of great complexity. Major thermal trauma leads to systemic inflammatory responses and multiple organ failure and is highly susceptible to infection. A key challenge is to stabilize the patient early on and control fluid balance. At the same time, surgical treatment of large wounds requires a coordinated approach to debridement and skin coverage. Limited donor skin resources make the development of skin replacement procedures essential. Multidrug-resistant bacteria and nosocomial infections pose an increasing threat. Pain and intensive care must be individually tailored and continuously optimized. Future prospects lie in bioactive skin equivalents, improved topical antimicrobial therapy including bacteriophage applications, immunomodulatory therapies, and personalized medicine. Digitalization, for example through AI-supported decision-making algorithms, can also improve care. Holistic rehabilitation, including psychosocial aspects, remains central to the long-term quality of life of patients.

Acknowledgments The authors declare that they have no potential conflicts of interest.

References

1. Welsch U (2006) Haut. In: Welsch U (ed) Sobotta Lehrbuch Histologie: Zytologie, Histologie, Mikroskopische Anatomie, 2nd edn. Urban & Fischer Verlag/Elsevier GmbH, München & Jena, pp 549–568
2. Wolff H (2005) Struktur und Funktion der Haut. In: Braun-Falco O, Plewig G, Wolff HH, Burgdorf WHC, Landthaler M (eds) Dermatologie und Venerologie, 5th edn. Springer-Verlag, Berlin/Heidelberg, pp 3–5

3. Kolokythas P, Aust M (2011) Verbrennungsverletzungen. In: Vogt PM (ed) Praxis der Plastischen Chirurgie, 1st edn. Springer-Verlag, Berlin, Heidelberg, pp 650–666
4. Jackson DM (1970) In search of an acceptable burn classification. Br J Plast Surg 23:219–226
5. Jackson DM (1970) A historical review of the use of local physical signs in burns. Br J Plast Surg 23:211–218
6. Takeo M, Lee W, Ito M (2015) Wound healing and skin regeneration. Cold Spring Harb Perspect Med 5:a023267
7. Ellis S, Lin EJ, Tartar D (2018) Immunology of wound Healing. Curr Dermatol Rep 7:350–358
8. Yang S, Ma K, Geng Z, Sun X, Fu X (2015) Oriented cell division: new roles in guiding skin wound repair and regeneration. Biosci Rep 35
9. Mascré G, Dekoninck S, Drogat B, Youssef KK, Broheé S, Sotiropoulou PA et al (2012) Distinct contribution of stem and progenitor cells to epidermal maintenance. Nature 489:257–262
10. Levy V, Lindon C, Zheng Y, Harfe BD, Morgan BA (2007) Epidermal stem cells arise from the hair follicle after wounding. FASEB J 21:1358–1366
11. Taylor G, Lehrer MS, Jensen PJ, Sun TT, Lavker RM (2000) Involvement of follicular stem cells in forming not only the follicle but also the epidermis. Cell 102:451–461
12. Brownell I, Guevara E, Bai CB, Loomis CA, Joyner AL (2011) Nerve-derived sonic hedgehog defines a niche for hair follicle stem cells capable of becoming epidermal stem cells. Cell Stem Cell 8:552–565
13. Seifert AW, Monaghan JR, Voss SR, Maden M (2012) Skin regeneration in adult axolotls: a blueprint for scar-free healing in vertebrates. PLoS One 7:e32875
14. Menger B, Vogt PM, Allmeling C, Radtke C, Kuhbier JW, Reimers K (2011) AmbLOXe—an epidermal lipoxygenase of the Mexican axolotl in the context of amphibian regeneration and its impact on human wound closure in vitro. Ann Surg 253:410–418
15. Rowlatt U (1979) Intrauterine wound healing in a 20 week human fetus. Virchows Arch A Pathol Anat Histol 381:353–361
16. Satish L, Kathju S (2010) Cellular and molecular characteristics of scarless versus fibrotic wound healing. Dermatol Res Pract 2010:790234
17. Wulff BC, Parent AE, Meleski MA, DiPietro LA, Schrementi ME, Wilgus TA (2012) Mast cells contribute to scar formation during fetal wound Healing. J Invest Dermatol 132:458–465
18. Jones N (2010) Scar tissue. Curr Opin Otolaryngol Head Neck Surg 18:261–265
19. Corr DT, Hart DA (2013) Biomechanics of scar tissue and uninjured skin. Adv Wound Care (New Rochelle) 2:37–43
20. Königová R, Pondelícek I (1987) Psychological aspects of burns. Scand J Plast Reconstr Surg Hand Surg 21:311–314
21. Harding K (2015) Innovation and Wound Healing. J Wound Care 24:7–13
22. Schlottmann F, Bucan V, Vogt PM, Krezdorn N (2021) A short history of skin grafting in burns: from the gold standard of autologous skin grafting to the possibilities of allogeneic skin grafting with immunomodulatory approaches. Medicina (B Aires) 57:225
23. Gibson C, Bessey PQ, Gallagher JJ (2020) The global burn registry: a work in progress. J Burn Care Res 41:929–934
24. Deutsche Gesellschaft für Verbrennungsmedizin (2017) Jahresbericht 2017 Verbrennungsmedizin [Internet]. Available from: https://www.verbrennungsmedizin.de/verbrennungsregister/jahresberichte.php
25. Deutsche Gesellschaft für Verbrennungsmedizin (2022) Jahresbericht 2022 Verbrennungsmedizin [Internet]. Available from: https://verbrennungsmedizin.de/files/dgv_files/pdf/jahresbericht/Jahresbericht%202022%20gesamt.pdf
26. Haddix AC (2001) Cost effectiveness analysis of a smoke alarm giveaway program in Oklahoma City, Oklahoma. Inj Prev 7:276–281
27. Batra N, Colson CD, Alberto EC, Burd RS (2021) Using social Media for the Prevention of pediatric burn injuries: pilot design and usability study. JMIR Form Res 5:e23242

28. Yoon JS, Khoo KH, Akhavan AA, Lagziel T, Ha M, Cox CA et al (2022) Changes in burn surgery operative volume and metrics due to COVID-19. J Burn Care Res 43:1233–1240
29. Vogt PM, Mailänder P, Jostkleigrewe F, Reichert B, Adams HA (2007) Centers for severely burned patients in Germany--management structure and needs. Chirurg Suppl:411–413
30. Obed D, Knoedler S, Salim M, Gulbis N, Dastagir N, Dastagir K et al (2023) The modified 5-item frailty index as a predictor of complications in burn patients. JPRAS Open 36:62–71
31. Lam NN, Hung NT, Khanh NN (2023) Cost analysis and influencing factors amongst severe burn patients. Ann Burns Fire Disasters 36:183–188
32. Trupkovic T, Giessler G (2008) Das Verbrennungstrauma—Teil 1. Anaesthesist 57:898–907
33. Deutsche Gesellschaft für Verbrennungsmedizin (2018) S2k Leitlinie: Behandlung thermischer Verletzungen des Erwachsenen [Internet]. Available from: https://www.awmf.org/uploads/tx_szleitlinien/044-001l_S2k_Thermische__Verletzungen_Erwachsene_2018-12.pdf
34. Badulak JH, Schurr M, Sauaia A, Ivashchenko A, Peltz E (2018) Defining the criteria for intubation of the patient with thermal burns. Burns 44:531–538
35. Dingle LA, Wain RAJ, Bishop S, Soueid A, Sheikh Z (2021) Intubation in burns patients: a 5-year review of the Manchester regional burns centre experience. Burns 47:576–586
36. Ziegler B, Kenngott T, Fischer S, Hundeshagen G, Hartmann B, Horter J et al (2019) Early hypothermia as risk factor in severely burned patients: a retrospective outcome study. Burns 45:1895–1900
37. Singer AJ, Taira BR, Thode JHC, McCormack JE, Shapiro M, Aydin A et al (2010) The association between hypothermia, prehospital cooling, and mortality in burn victims. Acad Emerg Med 17:456–459
38. Ashworth HL (2001) Treatment before transfer: the patient with burns. Emerg Med J 18:349–351
39. Bartels P, Thamm OC, Elrod J, Fuchs P, Reinshagen K, German Burn Registry et al (2020) The ABSI is dead, long live the ABSI - reliable prediction of survival in burns with a modified Abbreviated Burn Severity Index. Burns 46:1272–1279
40. Bagheri M, Fuchs PC, Lefering R, Daniels M, Schulz A, the German Burn Registry et al (2023) The BUrn Mortality Prediction (BUMP) Score—an improved mortality prediction score based on data of the German burn registry. Burns 49:110–119
41. Ziegenthaler H, Neumann U, Fritzsche U, Sühnel B (2005) Polytraumatisierte Brandverletzte. Orthopade 34:906–916
42. Hop MJ, Stekelenburg CM, Hiddingh J, Kuipers HC, Middelkoop E, Nieuwenhuis MK et al (2016) Cost-effectiveness of laser Doppler imaging in burn care in The Netherlands. Plast Reconstr Surg 137:166e–176e
43. Ziegler B, Hundeshagen G, Uhlmann L, Will Marks P, Horter J, Kneser U et al (2019) Impact of diagnostic bronchoscopy in burned adults with suspected inhalation injury. Burns 45:1275–1282
44. Wellkamp L, Obed D, Enechukwu AOM, Bingoel AS, Dastagir K, Vogt PM (2023) Correlation between bacterial wound colonization and skin-graft loss in burn patients. J Burn Care Res 44:649–654
45. Peeters Y, Vandervelden S, Wise R, Malbrain MLNG (2015) An overview on fluid resuscitation and resuscitation endpoints in burns: past, present and future. Part 1 — historical background, resuscitation fluid and adjunctive treatment. Anestezjol Intens Ter 47:6–14
46. Cartotto R, Johnson LS, Savetamal A, Greenhalgh D, Kubasiak JC, Pham TN et al (2023) American burn association clinical practice guidelines on burn shock resuscitation. J Burn Care Res
47. Gong C, Zhang F, Li L, He F, Liu G-C, Zhu S-H et al (2017) The variation of hemodynamic parameters through PiCCO in the early stage after severe burns. J Burn Care Res 38:e966–e972

48. Malkoc A, Jong S, Fine K, Wong DT (2023) High-dose intravenous versus low-dose oral vitamin C in burn care: potential protective effects in the severely burned: a retrospective cohort study. Ann Med Surg 85:1523–1526
49. Cartotto R, Greenhalgh DG, Cancio C (2017) Burn state of the science. J Burn Care Res 38:e596–e604
50. Adibfar A, Camacho F, Rogers AD, Cartotto R (2021) The use of vasopressors during acute burn resuscitation. Burns 47:58–66
51. Girtler R, Gustorff B (2011) Schmerztherapie bei Verbrennungen. Anaesthesist 60:243–250
52. Cartotto R, Johnson L, Rood JM, Lorello D, Matherly A, Parry I et al (2023) Clinical practice guideline: early mobilization and rehabilitation of critically ill burn patients. J Burn Care Res 44:1–15
53. Werner E (2024) The use of virtual reality as a standard of care in verified burn centers. Burns 50:808–812
54. Obed D, Schroeter A, Gruber L, Salim M, Krezdorn N, Vogt PM (2023) Outcomes following burn injury in intensive care patients with major psychiatric disorders. Burns 49:830–837
55. Bucher F, Dastagir N, Obed D, Enechukwu A, Dieck T, Vogt PM et al (2023) Factor XIII: more than just a fibrin stabilizer for the burn patient? A matched-pair analysis. JPRAS Open 37:1–8
56. Ramos G, Cornistein W, Cerino GT, Nacif G (2017) Systemic antimicrobial prophylaxis in burn patients: systematic review. J Hosp Infect 97:105–114
57. Khandelwal A, Satariano M, Doshi K, Aggarwal P, Avasarala V, Sood A et al (2024) Management and outcomes of acute kidney injury due to burns: a literature review. J Burn Care Res 45:323–337
58. Mohan PB, Chittoria RK, Koliyath S, Pathan I, Thomas N, Kerakada N et al (2023) Therapeutic plasma exchange therapy in burns. J Cutan Aesthet Surg 16:259–261
59. Capoccia M, Maybauer DM, Brewer JM, Maybauer MO (2023) Extracorporeal life support in adult burn care: a systematic review. Int J Artif Organs 46:321–328
60. Bingoel AS, Schlottmann F, Plinke C, Dastagir K, Obed D, Enechukwu A et al (2024) Impact of different parameters on the outcome of vv-ECMO therapy in burn patients—a retrospective cohort study from a burn and high output ECMO center. Innov Surg Sci 10:73
61. Tejiram S, Shupp JW (2024) Sepsis in surgical patients: burn sepsis. Surgery 175:1259–1261
62. Martindale RG, McClave SA, Vanek VW, McCarthy M, Roberts P, Taylor B et al (2009) Guidelines for the provision and assessment of nutrition support therapy in the adult critically ill patient: Society of Critical Care Medicine and American Society for Parenteral and Enteral Nutrition: executive summary. Crit Care Med 37:1757–1761
63. Li H, Guo Y, Yang Z, Roy M, Guo Q (2016) The efficacy and safety of oxandrolone treatment for patients with severe burns: a systematic review and meta-analysis. Burns 42:717–727
64. Rehou S, de Brito Monteiro L, Auger C, Knuth CM, Abdullahi A, Stanojcic M et al (2023) Propranolol normalizes metabolomic signatures thereby improving outcomes after burn. Ann Surg 278:519–529
65. Orbay H, Corcos AC, Ziembicki JA, Egro FM (2024) Challenges in the management of large burns. Clin Plast Surg 51:319–327
66. Shoham Y, Gasteratos K, Singer AJ, Krieger Y, Silberstein E, Goverman J (2023) Bromelain-based enzymatic burn debridement: a systematic review of clinical studies on patient safety, efficacy and long-term outcomes. Int Wound J 20:4364–4383
67. Glaser J, Ziegler B, Hirche C, Tapking C, Haug V, Bliesener B et al (2021) The status quo of early burn wound excision: insights from the German burn registry. Burns 47:1259–1264
68. Rowan MP, Cancio LC, Elster EA, Burmeister DM, Rose LF, Natesan S et al (2015) Burn wound healing and treatment: review and advancements. Crit Care 19:243
69. Bogdanov SB, Gilevich IV, Melkonyan KI, Sotnichenko AS, Alekseenko SN, Porhanov VA (2021) Total full-thickness skin grafting for treating patients with extensive facial burn injury: a 10-year experience. Burns 47:1389–1398

70. Wang C, Zhang F, Lineaweaver WC (2020) Clinical applications of allograft skin in burn care. Ann Plast Surg 84:S158–S160
71. Cronin H, Goldstein G (2013) Biologic skin substitutes and their applications in dermatology. Dermatologic Surg 39:30–34
72. Heng WL, Wang QW, Sornarajah R, Tremblay J, Putri NM, Hamid SSA et al (2020) A review of skin banking guidelines and standards worldwide: towards the harmonization of guidelines for skin banking in therapeutic applications for the regions under the Asia Pacific Burn Association (APBA). Burns Trauma:8
73. Schlottmann F, Strauß S, Ziesing S, Reineke C, Ipaktchi R, Weyand B et al (2024) Organization of Hannover Skin Bank: sterile culture and procurement protocols for viable cryopreserved allogeneic skin grafts of living donors. Int Wound J 21:e14374
74. Horch R, Stark GB, Kopp J, Spilker G (1994) Cologne Burn Centre experiences with glycerol-preserved allogeneic skin: part I: clinical experiences and histological findings (overgraft and sandwich technique). Burns 20(Suppl 1):S23–S26
75. Lima Júnior EM, De Moraes Filho MO, Costa BA, Rohleder AVP, Sales Rocha MB, Fechine FV et al (2020) Innovative burn treatment using tilapia skin as a xenograft: a phase II randomized controlled trial. J Burn Care Res 41:585–592
76. Halim AS, Khoo TL, Mohd Yussof SJ (2010) Biologic and synthetic skin substitutes: an overview. Indian J Plast Surg 43:S23–S28
77. Schwarze H, Küntscher M, Uhlig C, Hierlemann H, Prantl L, Ottomann C et al (2008) Suprathel, a new skin substitute, in the management of partial-thickness burn wounds. Ann Plast Surg 60:181–185
78. Vogt PM, Kolokythas P, Niederbichler A, Knobloch K, Reimers K, Choi CY (2007) Innovative wound therapy and skin substitutes for burns. Chirurg 78:335–342
79. Greenwood JE, Dearman BL (2012) Comparison of a sealed, polymer foam biodegradable temporizing matrix against Integra® dermal regeneration template in a porcine wound model. J Burn Care Res 33:163–173
80. Greenwood JE, Dearman BL (2012) Split skin graft application over an integrating, biodegradable temporizing polymer matrix. J Burn Care Res 33:7–19
81. Schlottmann F, Obed D, Bingöl AS, März V, Vogt PM, Krezdorn N (2022) Treatment of complex wounds with NovoSorb® biodegradable Temporising matrix (BTM)—a retrospective analysis of clinical outcomes. J Pers Med 12:2002
82. Dickson K, Lee KC, Abdulsalam A, Amirize E, Kankam HKN, ter Horst B et al (2023) A histological and clinical study of MatriDerm® use in burn reconstruction. J Burn Care Res 44:1100–1109
83. De Vries HJ, Zeegelaar JE, Middelkoop E, Gijsbers G, Van Marle J, Wildevuur CH et al (1995) Reduced wound contraction and scar formation in punch biopsy wounds. Native collagen dermal substitutes. A clinical study. Br J Dermatol 132:690–697
84. Ryan CM, Schoenfeld DA, Malloy M, Schulz JT, Sheridan RL, Tompkins RG (2002) Use of Integra artificial skin is associated with decreased length of stay for severely injured adult burn survivors. J Burn Care Rehabil 23:311–317
85. Gravante G, Di Fede MC, Araco A, Grimaldi M, De Angelis B, Arpino A et al (2007) A randomized trial comparing ReCell system of epidermal cells delivery versus classic skin grafts for the treatment of deep partial thickness burns. Burns 33:966–972
86. Gallico GG, O'Connor NE, Compton CC, Kehinde O, Green H (1984) Permanent coverage of large burn wounds with autologous cultured human epithelium. N Engl J Med 311:448–451
87. Meuli M, Hartmann-Fritsch F, Hüging M, Marino D, Saglini M, Hynes S et al (2019) A cultured autologous Dermo-epidermal skin substitute for full-thickness skin defects: a phase I, open, prospective clinical trial in children. Plast Reconstr Surg 144:188–198
88. Seth AK, Friedstat JS, Orgill DP, Pribaz JJ, Halvorson EG (2017) Microsurgical burn reconstruction. Clin Plast Surg 44:823–832

89. Bingoel AS, Strauss S, Vogt PM (2022) Clinical application of wIRA irradiation in burn wounds. In: Water-filtered infrared a (wIRA) irradiation. Springer International Publishing, Cham, pp 189–194
90. Gładysz M, März V, Ruemke S, Rubalskii E, Vogt PM, Krezdorn N (2022) Limb salvage through intermediary wound coverage with acellular dermal matrix template after persistent pseudomonas aeruginosa infection in a burn patient. Eur Burn J 3:27–33
91. Oliver MA, Hussein LK, Molina EA, Keyloun JW, SM MK, Jimenez LM et al (2024) Cold atmospheric plasma is bactericidal to wound-relevant pathogens and is compatible with burn wound healing. Burns
92. Wu Y, Peng G, Wang Y, Chen J, Zhang B, Tang J et al (2024) Clinical efficacy of blood derivatives on wound healing: a systematic review and network meta-analysis. Int Wound J 21:e14622
93. Milewski MR, Schlottmann F, März V, Dieck T, Vogt PM (2024) The successful treatment of multi-resistant colonized burns with large-area atmospheric cold plasma therapy and dermis substitute matrix—a case report. Eur Burn J 5:271–282
94. Neubauer H, Stolle A, Ripper S, Klimitz F, Ziegenthaler H, Strupat M et al (2019) Evaluation of an International Classification of Functioning, Disability and Health-based rehabilitation for thermal burn injuries: a prospective non-randomized design. Trials 20:752
95. Schlottmann F, Lorbeer L (2024) Update burn surgery: overview of current multidisciplinary treatment concepts. Innov Surg Sci 9:181–190
96. Dastagir K, Obed D, Bucher F, Hofmann T, Koyro KI, Vogt PM (2021) Non-invasive and surgical modalities for scar management: a clinical algorithm. J Pers Med 11:1259
97. Farzan R, Ghorbani Vajargah P, Mollaei A, Karkhah S, Samidoust P, Takasi P et al (2023) A systematic review of social support and related factors among burns patients. Int Wound J 20:3349–3361
98. Klinger M, Giannasi S, Bandi V, Veronesi A, Maione L, Lisa A et al (2020) Update on "Fat injection for cases of severe burn outcomes: a new perspective of scar remodeling and reduction". Aesthet Plast Surg 44:1283–1285
99. Fergason JR, Blanck R (2011) Prosthetic management of the burn amputation. Phys Med Rehabil Clin N Am 22:277–299
100. Barrett LW, Fear VS, Waithman JC, Wood FM, Fear MW (2019) Understanding acute burn injury as a chronic disease. Burns Trauma 7
101. Hatch R, Young D, Barber V, Griffiths J, Harrison DA, Watkinson P (2018) Anxiety, depression and post traumatic stress disorder after critical illness: a UK-wide prospective cohort study. Crit Care 22:310
102. Davydow DS, Gifford JM, Desai SV, Bienvenu OJ, Needham DM (2009) Depression in general intensive care unit survivors: a systematic review. Intensive Care Med 35:796–809
103. Lerman SF, Sylvester S, Hultman CS, Caffrey JA (2021) Suicidality after burn injuries: a systematic review. J Burn Care Res 42:357–364
104. Katsu A, Mackenzie L, Tyack Z, Mackey M (2024) Understanding return-to-employment experiences after burns: qualitative scoping review findings. Aust Occup Ther J 71:113–131

Chapter 25
Wound Healing and Regenerative Medicine in Diabetes

Martin Kauke-Navarro, Felix J. Klimitz, and Henry C. Hsia

Abstract Diabetic wounds remain a significant source of morbidity and healthcare burden worldwide. Despite advances in standard wound care which include debridement, infection control, off-loading, revascularization, and adjunctive therapies, healing rates remain suboptimal, and the risk of severe complications such as infection, amputation, and mortality remains high. In light of the pathologic inflammatory response and impaired healing observed in diabetic wounds, stem cells, particularly those capable of suppressing pro-inflammatory cytokine production, expanding regulatory T cells, and supporting angiogenesis, have gained significant attention. Stem cell-based therapies, possibly in combination with smart biomaterials and cell-free approaches, hold the potential to transform the management of chronic diabetic wounds, bridging the gap between scientific innovation and clinical application. This chapter explores the most recent advances in stem cell-based therapies for diabetic wound healing, examining underlying mechanisms, delivery strategies, and current clinical evidence supporting their use.

Keywords Diabetic Wound Healing · Chronic Wounds · Advanced Wound Therapies · Stem Cells · Exosomes

Abbreviations

3D	Three dimensional
AAV	Adeno-associated virus
ABI	Ankle-brachial index
aFGF	Acidic fibroblast growth factor

M. Kauke-Navarro · F. J. Klimitz · H. C. Hsia (✉)
Department of Surgery, Division of Plastic Surgery, Yale School of Medicine, New Haven, CT, USA
e-mail: henry.hsia@yale.edu

F. De Francesco, K. Turksen (eds.), *Management and Strategies for Wound Healing*, Stem Cell Biology and Regenerative Medicine 79,
https://doi.org/10.1007/978-3-032-12442-5_25

AGEs	Advanced glycation end products
ASC	Adipose-derived stem cell
ASC-exo	Adipose-derived stem cell-derived exosome
bFGF	Basic fibroblast growth factor
BM-MSC	Bone marrow–derived mesenchymal stem cell
CCL2	Chemokine (C-C motif) ligand 2, or monocyte attractant protein 1
cGMP	Current Good Manufacturing Practice
CGRP	Calcitonin gene-related peptide
CLI	Critical limb ischemia
CVU	Chronic venous ulcer
CXCL1	Chemokine (C-X-C motif) ligand 1
CXCL8	Chemokine (C-X-C motif) ligand 8, or interleukin 8
DFO	Desferrioxamine
DFU	Diabetic foot ulcer
DM	Diabetes mellitus
DNA	Deoxyribonucleic acid
ECM	Extracellular matrix
EGF	Epidermal growth factor
ESC	Embryonic stem cell
EV	Extracellular vesicle
EXT	Embryonic stem cell extract
FDA	Food and Drug Administration
G-CSF	Granulocyte colony-stimulating factor
GelMA	Gelatin methacrylamide
GFP	Green fluorescent protein
HBOT	Hyperbaric oxygen therapy
HGF	Hepatocyte growth factor
HIF-1α	Hypoxia-inducible factor 1-alpha
HLA	Human leukocyte antigen
HUC-MSC	Human umbilical cord–derived mesenchymal stem cell
IDO	Indoleamine 2,3-dioxygenase
IGF-1	Insulin-like growth factor 1
iPSC	Induced pluripotent stem cell
KGF	Keratinocyte growth factor
MMP	Matrix metalloproteinases
MSC	Mesenchymal stem cell
NET	Neutrophil extracellular trap
NPWT	Negative pressure wound therapy
PAD	Peripheral arterial disease
PBMNC	Peripheral blood mononuclear cell
PDGF	Platelet-derived growth factor
PEG	Polyethylene glycol
PRP	Platelet-rich plasma
RAGE	Receptor for advanced glycation end products
RD	Risk difference

RNA	Ribonucleic acid
ROS	Reactive oxygen species
SDF-1α	Stromal cell-derived factor-1 alpha
SVF	Stromal vascular fraction
SVFC	Stromal vascular fraction cell
$TcPO_2$	Transcutaneous oxygen pressure
TGF-β	Transforming growth factor-beta
TIMP	Tissue inhibitors of metalloproteinases
UC-MSC	Umbilical cord–derived mesenchymal stem cell
VEGF	Vascular endothelial growth factor

Introduction

Diabetes mellitus (DM) is one of the most prevalent metabolic disorders worldwide, currently affecting an estimated 463 million individuals globally [1]. Beyond its metabolic consequences, DM significantly reduces average life expectancy by approximately 4 to 10 years and independently increases the risk of cardiovascular disease, renal dysfunction, and certain malignancies [1]. Furthermore, diabetes is a leading cause of non traumatic lower extremity amputations, with diabetic individuals facing a seven- to ten-fold higher risk than non-diabetic patients [1, 2].

Chronic wounds, particularly diabetic foot ulcers (DFUs), represent one of the most debilitating and costly complications of diabetes, often resulting from the interplay of neuropathy and impaired pressure sensation, vascular insufficiency, and dysregulated immune responses [3]. These factors contribute to a chronic, non-healing wound that current standard of care treatments often fail to address adequately [4].

Hyperglycemia plays a central role in these changes, for example, by priming neutrophils to undergo NETosis, a process by which neutrophils release neutrophil extracellular traps (NETs)—web-like structures composed of DNA, histones, and cytotoxic proteins—that trap pathogens but also cause collateral tissue damage [5]. In preclinical models, PAD4 gene knockout led to reduced NET formation and accelerated wound healing, underscoring the detrimental role of excessive neutrophil activity in diabetic wound pathology [5]. The pathogenesis of chronic diabetic wounds is, however, multifactorial, with neuropathic (sensory, motor, and autonomic) and microcirculatory changes playing major roles [6]. DFUs often develop due to a diminished ability to detect abnormal pressure, leading to repetitive microtraumas that can evolve into chronic wounds [7]. Globally, DFUs affect an estimated 18.6 million individuals annually, including 1.6 million in the US alone [7, 8]. Notably, up to 25% of DFUs progress to severe infections or limb-threatening complications requiring hospitalization or amputation [8, 9]. The consequences are profound—DFUs are associated with a 5-year mortality rate approaching 30%, particularly in patients requiring amputation [7].

While glycemic control is a cornerstone of diabetes management, its direct role in wound healing is nuanced. Although strict glycemic targets have not consistently shown benefit, moderate control—hemoglobin A1c between 7% and 8%—may be optimal in facilitating wound healing among patients with DFUs [9, 10].

Given the substantial morbidity, healthcare burden, and limited effectiveness of current wound care strategies, there is a critical need to explore novel regenerative therapies. Stem cells, with their immunomodulatory, pro-angiogenic, and tissue-regenerative properties, offer a promising avenue. These cells can replace damaged tissue, repopulate the wound bed with barrier-forming epithelial cells, and modulate the chronic inflammatory milieu of diabetic wounds [11].

Stem Cell Approaches to DFUs

In light of the pathologic inflammatory response and impaired healing observed in diabetic wounds, stem cells—particularly those capable of suppressing pro-inflammatory cytokine production, expanding regulatory T cells, and supporting angiogenesis—have gained significant attention [4, 11]. These cells hold the potential to reprogram the local immune response and stimulate tissue repair, and several strategies have been developed that leverage these properties through topical application, injection, or biomaterial delivery systems [12–14]. While the regenerative function of endogenous stem cells is often impaired in the diabetic wound environment, therapeutic stem cell delivery can reestablish healing trajectories.

This chapter explores the most recent advances in stem cell-based therapies for diabetic wound healing, examining the mechanisms, delivery strategies, and clinical evidence supporting their use.

Pathophysiology of Wound Healing in Diabetes

Physiological Wound Healing Process

The normal wound healing process progresses through four overlapping phases: hemostasis, inflammation, proliferation, and remodeling. Immediately after injury, platelets aggregate at the wound site, form a fibrin clot, and release growth factors such as platelet-derived growth factor (PDGF) and transforming growth factor-beta (TGF-β), which initiate immune cell recruitment and stabilize the clot [15–18]. During the inflammation phase, neutrophils are the first responders, clearing debris and pathogens. Macrophages follow, initially exhibiting a pro-inflammatory M1 phenotype and later transitioning to an anti-inflammatory M2 phenotype, which is essential for resolving inflammation and promoting tissue repair [19, 20]. This phenotypic shift is critical for the transition from inflammation to proliferation. In the proliferation phase, fibroblasts migrate into the wound and deposit extracellular matrix, primarily type III collagen. Fibroblasts, in concert with endothelial cells, drive angiogenesis under the influence of growth factors such as vascular endothelial growth factor (VEGF) and basic fibroblast growth factor (bFGF). Keratinocytes proliferate and migrate to reepithelialize the wound, forming granulation tissue that is rich in cells, matrix, and vasculature [21]. During the remodeling phase, type III collagen is gradually replaced by type I collagen. Myofibroblasts mediate wound contraction, while matrix metalloproteinases (MMPs) and tissue inhibitors of metalloproteinases (TIMPs) regulate extracellular matrix turnover. This leads to decreased vascularity and cellularity, resulting in a mature scar with restored tensile strength [15, 18, 22]. In summary, successful wound healing relies on the sequence of hemostasis, inflammation, proliferation, and remodeling, requiring the coordinated action of immune cells, fibroblasts, endothelial cells, keratinocytes, cytokines, growth factors, and extracellular matrix components.

Impairments in Diabetic Wound Healing

In diabetes mellitus, however, wound healing is often impaired, and wounds frequently stall in the inflammatory phase due to multiple hyperglycemia-driven pathologies.

Chronic hyperglycemia leads to the formation of advanced glycation end products (AGEs), which activate inflammatory pathways via the receptor for advanced glycation end products (RAGE) and contribute to extracellular matrix (ECM) stiffening, further impeding healing [23]. Hyperglycemia also increases oxidative stress through excess reactive oxygen species (ROS), damaging fibroblasts and keratinocytes, which impairs reepithelialization and tissue repair [24–26]. Persistent M1 macrophage and neutrophil activity, with insufficient transition to reparative M2 macrophages, results in chronic inflammation, impaired granulation tissue formation, and excessive protease activity [24, 27, 28].

Furthermore, impaired angiogenesis and tissue hypoxia are common due to both macrovascular and microvascular disease, with hyperglycemia inhibiting hypoxia-inducible factor 1-alpha (HIF-1α) and reducing VEGF expression, as well as causing endothelial progenitor cell dysfunction [29–31]. Diabetic neuropathy (sensory, motor, and autonomic) and neuropeptide dysregulation, including loss of substance P and CGRP, diminish neurogenic inflammation, angiogenesis, and keratinocyte function, further hindering healing [23]. Elevated MMPs with reduced TIMPs lead to excessive ECM and growth factor degradation, preventing stable granulation tissue and hindering healing progression [29].

Collectively, these factors—hyperglycemia-induced oxidative stress, chronic inflammation, poor angiogenesis, neuropathy, and ECM degradation—render diabetic wounds prone to chronicity and complications [28, 32, 33].

Current Management Strategies for Diabetic Wounds

Standard Clinical Care

The current standard of care for diabetic wound management is multifaceted and endorsed by major professional organizations, including the American Diabetes Association, the American Heart Association, and the Society for Vascular Surgery [34–36]. All advocate for a coordinated, multidisciplinary approach built on several core components. First, regular debridement of necrotic tissue and biofilm is essential to establish a healthy wound bed and promote healing. Sharp debridement is the preferred method, extending to the wound edges to remove nonviable tissue and disrupt biofilm, thereby enhancing cellular migration and the wound's responsiveness to growth factors [34]. Second, infection control is critical. Early diagnosis and targeted antibiotic therapy are vital, with surgical intervention warranted for deep infections, abscess formation, or necrotizing fasciitis. In cases of osteomyelitis, treatment may involve both surgical and medical management, tailored to the clinical context [37]. Third, pressure off-loading plays a central role, especially in the management of diabetic foot ulcers. The most effective off-loading device is a non-removable knee-high device, such as a total contact cast, though patient compliance can be a limiting factor. When necessary, removable cast walkers or specialized off-loading footwear serve as alternatives [7]. Fourth, revascularization is recommended for ischemic wounds to restore adequate perfusion. This may involve endovascular procedures or surgical bypass, always complemented by cardiovascular risk management [7, 34, 37]. Finally, moist wound care remains a cornerstone of treatment. Maintaining a moist wound environment is critical for cellular migration and granulation tissue formation. Hydrogels are preferred for dry or minimally exudative wounds, while alginates or hydrofibers are indicated for wounds with heavy exudate. Collagen and antimicrobial dressings may be considered based on wound bioburden and healing trajectory. Dressings should always be selected based on wound characteristics, including exudate level, infection risk, and location, to optimize healing and minimize complications [34, 38].

Advanced Therapies

In cases where these standard interventions have been optimized, and the wound remains non-healing, advanced therapies including topical growth factors, negative pressure wound therapy (NPWT), skin substitutes and bioengineered scaffolds, and hyperbaric oxygen therapy (HBOT) can be considered.

Topical growth factors, including recombinant PDGF (becaplermin), epidermal growth factor (EGF), and bFGF, have demonstrated efficacy in improving healing rates in diabetic foot ulcers, but their clinical utility is limited by rapid degradation in the protease-rich chronic wound environment and high cost. The American

Diabetes Association lists these agents as advanced wound therapies to consider when standard care fails to achieve adequate healing, typically defined as less than 50% reduction in wound size after 4 weeks of optimal management [36, 39]. NPWT is widely used for deep or complex diabetic wounds and is supported by multiple randomized clinical trials showing improved healing, granulation tissue formation, and wound contraction compared to standard moist wound care. It is particularly effective for wounds with significant depth or after surgical debridement [7]. Skin substitutes and bioengineered scaffolds, such as bilayered skin equivalents (e.g., Apligraf) and amniotic-derived matrices, provide a biological scaffold and deliver growth factors to promote closure in non-healing wounds. Meta-analyses and systematic reviews support their use in chronic, non-infected, non-ischemic diabetic foot ulcers that have not responded to standard care [7]. HBOT is best used as an adjunctive therapy for select advanced cases, particularly in patients with ischemic or infected ulcers unresponsive to standard care. The American Diabetes Association and the American Heart Association both recognize HBOT as a potential adjunct, though evidence is mixed, and it should be reserved for refractory cases [7, 34, 36].

Emerging Therapies

In addition to these standard and advanced therapies, there are several emerging therapies for diabetic wounds that aim to stimulate regenerative processes. They are generally considered when standard and advanced wound care fail to achieve adequate healing, and their use should be individualized based on wound characteristics and patient factors. Robust randomized controlled trial data are still limited for many of these modalities, and their adoption into routine practice is evolving [29, 36, 40, 41].

Platelet-rich plasma (PRP) is increasingly used as an adjunct for chronic diabetic wounds, leveraging its high concentration of growth factors to promote healing. The American Diabetes Association includes autologous blood products such as PRP among advanced wound therapies, and recent meta-analyses suggest PRP can reduce adverse events and improve healing rates in diabetic foot ulcers [36, 39].

Maggot debridement therapy, which uses sterile larvae to selectively debride necrotic tissue, is recognized as a safe and effective adjunct in selected chronic wounds, including diabetic ulcers, though it is less commonly used in the United States [42]. Low-level laser therapy (also called photobiomodulation) and extracorporeal shockwave therapy are noninvasive modalities that have demonstrated potential to enhance wound healing by stimulating cellular activity, angiogenesis, and tissue regeneration [29, 43].

Cell-based therapies, particularly those using mesenchymal stem cells (MSCs) derived from bone marrow or allogeneic sources, are also under active investigation [40, 44]. These therapies have shown promise in preclinical and early clinical studies by enhancing angiogenesis and modulating inflammation—key processes impaired in diabetic wound healing [29, 41]. These emerging strategies highlight the growing interest in regenerative approaches, setting the stage for stem cell-based therapies, which represent a frontier in diabetic wound management and will be discussed in the following sections.

Stem Cell-Based Regenerative Therapies

Overview of Stem Cell Types

Diabetic wounds are characterized by impaired healing due to chronic hyperglycemia and sustained inflammation, which disrupt extracellular matrix production, tissue remodeling, and the normal progression of wound repair [3, 45]. Prolonged exposure to elevated glucose levels contributes to cellular dysfunction, delays reepithelialization, and impairs angiogenesis. Stem cells have been explored as a therapeutic option for their intrinsic anti-inflammatory, paracrine, and pro-angiogenic properties, which can counteract these effects and promote tissue regeneration by replenishing damaged areas with lineage-specific cells [3, 4, 45]. Stem cells are defined by their ability to self-renew, proliferate, and differentiate into specialized cell types, allowing them to replace damaged or lost cells [46]. They are generally categorized into three types: embryonic stem cells (ESCs), adult (somatic) stem cells, and induced pluripotent stem cells (iPSCs) [3, 47].

Embryonic Stem Cells (ESCs)

Embryonic stem cells (ESCs) are derived from the inner cell mass of 4–5 day-old blastocysts and are characterized by their ability to maintain pluripotency [47]. Because ESCs originate from early-stage embryos with the potential to develop into functional human life, their use has generated substantial ethical debate, particularly in human research settings [46].

In addition to ethical concerns, ESC-based therapies face several scientific and clinical hurdles. A major limitation is their high immunogenicity when transplanted into genetically non-identical individuals, resulting in an increased risk of rejection [48, 49]. Another challenge is the potential for teratoma formation due to the possible uncontrolled differentiation of pluripotent cells into multiple cell types, including undesired tissues [46, 48].

Clinically, ESCs have been explored in trials for the regeneration of specific tissues. One example is the use of ESC-derived retinal pigment epithelial cells in patients with macular degeneration, which has shown promising results without evidence of abnormal growth or rejection when combined with chronic immunosuppression [49]. However, in the context of wound healing, ESCs remain limited to preclinical investigation. While their pluripotency makes them theoretically capable of regenerating complex skin structures composed of multiple cell types, their application in clinical wound care is hindered by both ethical and immunological barriers.

As a result, the field has shifted toward the use of alternative stem cell types, such as adult stem cells and iPSCs, which present fewer ethical challenges and reduced risk of rejection. In summary, ESCs possess the potential to contribute to skin regeneration due to their broad differentiation capacity, but their direct clinical use in wound healing remains unlikely in the near future.

Induced Pluripotent Stem Cells (IPSCs)

Induced pluripotent stem cells (iPSCs) were first described by Takahashi and Yamanaka, who successfully reprogrammed adult mouse fibroblasts into an embryonic stem cell-like state by introducing four transcription factors: Oct3/4, Sox2, c-Myc, and Klf4 [48]. One year later, the same group applied these factors to adult human dermal fibroblasts and generated the first human iPSCs [50].

iPSCs are somatic cells that have been genetically reprogrammed to revert to a pluripotent state, mimicking the characteristics of embryonic stem cells (ESCs), including the ability to self-renew and differentiate into any cell type of the three germ layers [48, 51]. Thus, given the aforementioned limitations with ESCs, iPSCs hold great promise in the field of regenerative medicine. Their unlimited ability to renew and pluripotent state make these cells desirable over adult somatic stem cells, especially in terms of renewing complex multicellular structures such as skin [3]. iPSCs have demonstrated therapeutic potential in various preclinical disease models involving the loss of specific cell types, including sickle cell anemia, diabetes, Parkinson's disease, and spinal cord injury [51–53].

In the context of wound healing, iPSCs can be generated from a patient's own cells, offering the advantage of genetic compatibility and reduced risk of rejection [51]. In a study by Wang et al., iPSCs were differentiated into skin, pancreatic islet, and cardiac cells and transplanted into mice. The iPSC-derived skin cells were able to engraft and contribute to the repair of wounded tissue [51]. In another preclinical study, Gorecka et al. evaluated the safety and efficacy of iPSC-derived smooth muscle cells embedded in a 3D collagen scaffold for diabetic wound healing. Application of these scaffolds to full-thickness back wounds in diabetic nude mice improved skin architecture, enhanced neovascularization, and reduced inflammation [54].

Despite their promise, iPSC-based therapies face important limitations. A major safety concern is the potential for teratoma formation if undifferentiated cells remain within the graft [50, 51]. Additionally, iPSCs are prone to chromosomal aberrations and may retain epigenetic memory from their reprogrammed state, increasing the risk of neoplastic transformation [55]. Thus, tumorigenicity remains one of the principal obstacles preventing widespread clinical use of iPSC-derived therapies. Ongoing research is focused on improving the safety and reliability of iPSC applications. Strategies to ensure full differentiation before transplantation and to monitor genetic stability are being explored. While iPSC-based wound healing therapies are still in the preclinical stage, large animal models are being developed to further assess their efficacy and safety in preparation for future clinical translation [56].

Mesenchymal Stem Cells (MSCs)

Mesenchymal stem cells (MSCs) represent one of the most relevant and widely studied classes of adult stem cells. They reside in various tissues, including bone marrow, adipose tissue, muscle, and many organs, where they serve as a local source of cell replacement and tissue repair [3]. MSCs are among the most commonly used stem cell types in regenerative therapies due to their accessibility, safety profile, and desirable immunomodulatory capabilities.

MSCs are defined by their multipotency, allowing them to differentiate into mesodermal lineage cells such as osteoblasts, chondrocytes, adipocytes, and myocytes, while retaining the ability to self-renew. Unlike pluripotent stem cells, MSCs have restricted differentiation potential but play a key role in modulating the immune environment through the paracrine secretion of a broad range of bioactive factors, including growth factors, cytokines, chemokines, extracellular matrix proteins, and metabolic molecules. These secreted factors can suppress lymphocyte activity and modulate both innate and adaptive immune responses [57].

Clinically, the immunomodulatory properties of MSCs have been applied in transplantation settings to promote immune tolerance. Studies have shown that infusion of bone marrow-derived MSCs can have dose-dependent effects in promoting tolerance in kidney and liver transplantation, with reduced maintenance immunosuppressive therapy requirements and a potentially lower incidence of biopsy-proven acute rejection [58, 59]. These findings also extend to vascularized composite allotransplantation.

In the context of wound healing, MSCs have shown therapeutic potential through their paracrine effects. Chen et al. demonstrated in a preclinical model that MSC-conditioned media can significantly enhance wound repair. In this study, bone marrow-derived MSCs and dermal fibroblasts were cultured, and their conditioned media applied to full-thickness skin wounds in mice. The MSC-conditioned media, enriched in pro-regenerative factors such as VEGF-α, IGF-1, EGF, and KGF, promoted keratinocyte and endothelial cell proliferation and accelerated wound closure compared to control groups [60]. These findings suggest that MSCs exert their effects primarily through paracrine signaling rather than direct differentiation or engraftment.

Despite promising preclinical results, the clinical efficacy of MSCs in diabetic wound healing remains limited, and their mechanisms of action are not fully understood. While stem cell therapies have shown benefits in tissue regeneration, including cardiac tissue repair, emerging data challenge the assumption that MSCs directly engraft and replace damaged tissue. For example, Vagnozzi et al. found that the beneficial effects of MSC injection in cardiac models were primarily due to an acute immune response, particularly macrophage recruitment, rather than durable engraftment or differentiation into cardiac cells [61]. This raises the possibility that MSC therapies in wound healing may similarly function by activating endogenous repair pathways rather than acting as direct cellular replacements.

Additionally, the route of MSC delivery—whether topical, intradermal, or systemic—may influence therapeutic outcomes and requires further investigation.

Optimizing delivery methods and understanding the exact mechanisms by which MSCs facilitate healing remain important areas for ongoing research [3].

Adipose-Derived Stem Cells (ASCs) and Stromal Vascular Fraction (SVF)

Adipose-derived stem cells (ASCs) are a subtype of mesenchymal stem cells (MSCs) that reside in adipose tissue and are isolated from the stromal vascular fraction (SVF) following liposuction and processing of lipoaspirate [3, 62].

ASCs share many of the regenerative properties of other MSCs—including immunomodulatory, proangiogenic, antifibrotic, and antiapoptotic effects—but offer distinct practical advantages due to their high abundance and ease of harvest. Compared to bone marrow, adipose tissue contains a significantly higher density of MSCs—approximately 5% in adipose tissue versus 0.01% in bone marrow [62, 63]. This accessibility and cell yield make ASCs an attractive option for regenerative applications, particularly in wound healing.

Fat grafting is a widely used technique in plastic and reconstructive surgery for tissue augmentation and restoration. Beyond volume replacement, autologous fat transfer has demonstrated beneficial effects on tissue quality, particularly in irradiated skin. For instance, in the context of breast reconstruction after mastectomy, fat grafting has been shown to improve skin pliability and reverse radiation-induced damage. The regenerative effects observed are attributed in part to the presence of ASCs within the graft, which are believed to contribute to tissue remodeling and healing [64–66].

In vitro studies have shown that ASCs produce a pro-regenerative secretome enriched with angiogenic cytokines such as CXCL8, CCL2, CXCL1, and G-CSF, along with growth factors like VEGF. These molecules collectively support enhanced vascularization and wound repair, highlighting the potential of ASCs in treating chronic and ischemic wounds [65].

SVF itself—obtained through enzymatic or mechanical processing of adipose tissue—contains not only ASCs but also endothelial cells, pericytes, T regulatory cells, and other regenerative stromal populations [67]. As such, it is increasingly used directly in clinical applications. In a phase 1 open-label single-arm clinical trial, Carstens et al. evaluated autologous SVF cell administration in 63 patients with type 2 diabetes and chronic (>3 cm) foot ulcers who were candidates for amputation. Patients received injections of 30 million SVF cells into the ulcer bed and pedal vasculature. The treatment was found to be safe with no serious adverse events and showed promising efficacy: 81% of patients achieved full wound closure by 6 months, accompanied by improved arterial perfusion on Doppler ultrasound—suggesting a dual benefit on both tissue repair and vascular remodeling [67].

In summary, ASCs combine the therapeutic strengths of MSCs with important logistical advantages. Their abundance, autologous origin, and biological activity—particularly in promoting angiogenesis and modulating inflammation—make them highly suitable for the development of advanced wound healing therapies.

Mechanisms of Action in Wound Healing

Several types of stem cells are currently being explored in the context of wound healing. These cell types share several similarities in their mechanisms of action, although the exact processes remain incompletely understood and are still under investigation. One of the primary pathways by which stem cells contribute to wound healing is through paracrine signaling and immunomodulation, which help create an environment conducive to tissue repair. Stem cells such as iPSCs, MSCs, and SVFCs secrete a rich milieu of growth factors and cytokines that promote wound healing, likely through paracrine effects rather than direct cellular integration [3, 60, 68]. Current evidence is promising and suggests that stem cells may help shift the balance in chronic, inflamed wounds toward a more regenerative and healing-prone state. Key effects on the wound microenvironment include the promotion of angiogenesis, which is fundamental to successful wound healing. Stem cell secretomes are highly angiogenic and contain factors such as VEGF, HGF, angiopoietin-1, and basic FGF [60, 68]. Angiogenesis improves local oxygen and nutrient delivery to the tissue, thereby enhancing the healing process. In vitro experiments have demonstrated that medium conditioned by MSCs significantly increases capillary formation in wounded skin and enhances perfusion [68]. In a mouse model, delivery of MSC-derived factors resulted in greater blood vessel density in wounds compared to controls, an effect largely mediated by VEGF family growth factors [68, 69].

In addition, the stem cell secretome contains chemokines—such as SDF-1α—that can recruit endothelial progenitor cells and other reparative progenitors to the wound site [60]. For example, in a study conducted by Chen et al., wounds treated with MSC-conditioned medium showed a higher influx of $CD34^+$ and $Flk\text{-}1^+$ endothelial progenitor cells and other pro-regenerative cells compared to untreated wounds [60].

Furthermore, paracrine signals from stem cells can directly stimulate the proliferation and migration of keratinocytes and dermal fibroblasts, thereby enhancing reepithelialization and tissue regeneration [60]. In vitro scratch assays have demonstrated that the addition of ASC- or MSC-derived secreted factors accelerates the closure of the wound gap by keratinocytes [57, 70]. In another study, factors such as keratinocyte growth factor (KGF) and TGF-β, secreted by stem cells, were shown to promote keratinocyte differentiation and ECM production [71]. Additionally, stem cell secretomes may contain microRNAs that post-transcriptionally regulate the translation of mRNAs in resident skin cells, thereby promoting regeneration [57, 65].

One of the most important mechanisms of stem cell action is the modulation of immunity and control of local inflammation. Chronic wounds are typically characterized by a persistent pro-inflammatory milieu, which hampers proper tissue repair. MSCs have demonstrated the ability to balance and resolve this inflammation. Stem cells aid in the resolution of chronic and excessive inflammation through the secretion of extracellular vesicles (EVs), such as exosomes, and anti-inflammatory cytokines, including prostaglandin E2, indoleamine 2,3-dioxygenase (IDO), and TGF-β.

Additionally, stem cells can directly interact with immune cells, for example, by inhibiting T lymphocyte proliferation [57, 72].

Extracellular vesicles released by MSCs contain various biologically active components—such as microRNAs, proteins, and lipids—which can fuse with the plasma membranes of target cells and modulate their gene expression and function [57]. For example, an early study showed that exosomal miR-125b-5p contributed to cardiac repair after myocardial infarction by preventing cardiomyocyte apoptosis [73].

In another study, exosomes enriched with miR-182 were isolated from mouse bone marrow–derived MSCs and administered following myocardial ischemia/reperfusion injury. These exosomes promoted macrophage polarization from a pro-inflammatory M1 phenotype to an anti-inflammatory M2 phenotype by targeting the TLR4/NF-κB/PI3K/Akt signaling pathway, ultimately reducing infarct size and improving cardiac function [74].

Another hallmark of chronic wounds is fibrosis and excessive scar formation due to unresolved inflammation. Stem cells not only reduce inflammation but also help mitigate fibrosis and modulate scar formation by influencing the quality of tissue repair. MSCs and ASCs secrete factors that downregulate pro-fibrotic signals and promote more organized collagen fiber deposition [68].

For instance, umbilical cord–derived MSC (UC-MSC) conditioned medium has been shown to shift dermal fibroblasts toward a regenerative phenotype. Treated fibroblasts demonstrated reduced myofibroblast differentiation, a decreased TGF-β1/β3 ratio, and an increased MMP/TIMP ratio. In a murine wound model, application of UC-MSC conditioned medium accelerated healing [75]. Similarly, an in vitro study showed that co-culturing ASCs with hypertrophic and keloid fibroblasts inhibited activation of the pro-fibrotic TGF-β1/Smad signaling pathway [76].

Delivery Methods of Stem Cells and Exosomes in Chronic Wound Healing

Considerable effort has been devoted to optimizing the delivery of stem cells for the treatment of chronic wounds. Stem cells can be administered either topically—via direct application to the wound surface or local injection—or systemically, through intravenous or intramuscular administration [3].

Topical delivery of stem cells can be achieved through various methods [77]. Although local injection or direct application is technically simple, it is often limited by low cell survival, short retention time, poor engraftment, and therefore limited efficacy in promoting healing [3, 78]. To address these challenges, researchers have developed biocompatible carriers such as hydrogels, collagen matrices, and porous 3D scaffolds that mimic the ECM. These carriers can be loaded with stem cells or their secretome to allow gradual release of regenerative factors while protecting them from early degradation [77]. Additionally, such scaffolds can be engineered to optimize the stem cell microenvironment, thereby enhancing the regenerative potential of the cells themselves [3]. For example, Martin et al. employed a 4-arm PEG-4MAL synthetic hydrogel as a 3D scaffold for MSC delivery [79]. In this preclinical model, the hydrogel significantly improved MSC persistence and retention within the wound, modulated the local immune response, and accelerated wound closure compared to saline controls.

Systemic delivery of stem cells has also shown potential benefits, especially in addressing systemic complications of diabetes, such as microvascular and macrovascular dysfunction. In patients with DFUs and critical limb ischemia (CLI), intramuscular injection of bone marrow–derived MSCs was associated with improved perfusion, symptom relief, and higher limb salvage rates [80].

More recently, cell-free approaches have emerged as an alternative to cell-based therapies. These strategies use the stem cell secretome, which includes exosomes, extracellular vesicles, soluble proteins, enzymes, and lipids [3, 77]. As previously described, exosomes—small membrane-bound vesicles containing bioactive molecules such as proteins, mRNAs, and microRNAs—can replicate many of the beneficial effects of MSCs without requiring the delivery of viable cells [3, 77]. Exosomes or conditioned media can be embedded into scaffolds or delivery systems in place of live cells. These cell-free approaches offer several advantages: they bypass ethical concerns related to the use of certain stem cells, reduce the risk of rejection associated with allogeneic cells, minimize tumorigenicity, and offer easier storage and handling compared to living cells [3].

Clinical Evidence in Chronic Diabetic and Other Chronic Wounds

Clinical trials over the past decade have demonstrated promising results for stem cell therapy—including bone marrow–derived mesenchymal stem cells (BM-MSCs), peripheral blood mononuclear cells (PBMNCs), and ASCs—particularly in the treatment of DFUs [81]. Most human clinical trials have focused on the use of MSCs, with other stem cell types less commonly employed in clinical practice.

A 2024 systematic review and meta-analysis of 20 randomized and non-randomized trials (n = 1304) found that stem cell therapy significantly improved ulcer healing rates (risk difference [RD] = +0.36), reduced the risk of amputation (RD = −0.19), and enhanced vascular parameters such as ankle-brachial index (ABI) and transcutaneous oxygen pressure ($TcPO_2$), all without an increase in adverse events compared to standard wound care or placebo [81]. In one cited study, Lu et al. demonstrated that autologous BM-MSCs accelerated wound healing and improved limb perfusion in 41 patients with DFUs when compared to controls receiving standard care [80]. Another trial by Dash et al. used PBMNCs in 45 patients and observed a marked reduction in ulcer size and increased capillary density after 6 months of treatment [82].

In other aspects of chronic wound management, allogeneic ABCB5$^+$ MSCs have been investigated for the treatment of chronic venous ulcers (CVUs). In a Phase 1/IIa single-arm clinical trial conducted in Germany (ClinicalTrials.gov Identifier: NCT03257098), Kerstan et al. evaluated the use of skin-derived ABCB5$^+$ MSCs as a topical treatment for CVUs (1×10^6 allogeneic ABCB5$^+$ MSCs per cm^2 of wound area) in addition to standard therapy [83]. The treatment was generally well tolerated, with only 3 of 83 adverse events considered cell-related and all categorized as mild. At 12 weeks of follow-up, the median ulcer area had decreased by approximately 80%. The same ABCB5$^+$ dermal MSC topical therapy was also evaluated for the management of diabetic neuropathic foot ulcers (ClinicalTrials.gov Identifier: NCT03267784) [84]. In this study, 23 adults with diabetic neuropathic foot ulcers received two topical applications of approximately 2×10^6 cells/cm^2 of ulcer surface and were followed for 3 months. By 12 weeks post-treatment, the median ulcer area was reduced by approximately 59–67%, with no treatment-related adverse events reported.

In another study, Schulman et al. investigated bone marrow–derived MSCs for the treatment of deep burn wounds and concluded that MSC therapy was safe and accelerated wound healing [85]. In this Phase 1 dose-escalation trial conducted in the United States (ClinicalTrials.gov Identifier: NCT02104713), 10 adult patients with second-degree burns involving less than 20% of total body surface area (TBSA) received allogeneic BM-MSCs either topically or by injection under a transparent dressing. Doses ranged from 2.5×10^3 to 5×10^3 cells/cm^2 of burn surface area. The authors reported that 100% of the burns healed with minimal scarring, and no adverse events were attributed to the cell product. Healing rates were higher in the high-dose group (10.47 vs. 3.64 cm^2/day), although the difference did not reach statistical significance.

The use of allogeneic umbilical cord lining MSCs (Corlicyte®) for the treatment of DFUs was evaluated by Wang et al. in a Phase 1 open-label safety study (ClinicalTrials.gov Identifier: NCT04104451) [86]. Nine patients were enrolled and received topical applications of cord lining MSCs across two distinct dosing cohorts. No treatment-related adverse events were observed. Sixty percent of patients in the low-dose group achieved complete ulcer closure by day 70, while the remaining patients demonstrated an average wound surface area reduction of approximately 60%. These findings support the safety of this cell product in the management of DFUs and warrant further clinical investigation [86].

Autologous bone marrow–derived MSCs and bone marrow–derived mononuclear cells (BM-MNCs) were evaluated by Lu et al. in the treatment of patients with CLI and DFUs [87]. In this study, 41 patients received intramuscular injections of either BM-MSCs, BM-MNCs, or saline (control) into one or both limbs and were followed for three years. The authors found that the BM-MSC group demonstrated superior outcomes, including a significantly lower amputation risk (hazard ratio 0.21 vs. control, $p = 0.04$) and faster ulcer healing with lower recurrence rates compared to both BM-MNC and saline groups. By 9 months, patients treated with BM-MSCs exhibited markedly improved limb perfusion and complete ulcer resolution, while BM-MNCs showed no significant effect. The study concluded that

BM-MSC therapy achieved superior limb salvage and wound healing compared to BM-MNCs and standard care.

In 2022, Zhang et al. published results from an open-label Phase 1 clinical trial in China (Chinese Clinical Trial Registry: ChiCTR2200055885) involving 14 patients with critical DFUs and peripheral arterial disease (PAD) who were not candidates for surgical revascularization [69]. Participants received allogeneic human umbilical cord–derived mesenchymal stem cells (HUC-MSCs) from Wharton's jelly, administered both topically in a hydrogel and intravenously at a dose of approximately 2×10^5 cells/kg (maximum ten million cells per dose), in addition to standard care. The treatment was well tolerated, with only two patients developing transient fevers postinfusion. All patients achieved more than 95% wound surface area closure within 6 weeks. In addition, symptoms of limb ischemia improved, and no amputations were reported during the 3-year follow-up period. Figure 25.1 provides an overview of available stem cell-based therapies for chronic wounds.

Emerging Concepts and Future Directions

In recent years, the general direction of stem cell-based wound care research has shifted from basic cell therapy to a precision-engineered multimodal approach that aims to further enhance the safety, efficacy, and clinical scalability and applicability of these futuristic wound care strategies.

Genetically Modified Stem Cells

One key area of advancement involves the genetic enhancement of stem cells and the development of cell-free therapies (e.g., exosomes) to augment wound healing. Genetic modification allows for the selective overexpression of genes known to promote regenerative processes, such as those involved in angiogenesis (e.g., VEGF, E-selectin) [3, 88]. The advent of CRISPR/Cas9 has made it possible to precisely edit the genomic sequence of stem cells—enabling targeted deletion or insertion of genetic material to optimize their therapeutic behavior [89].

In a preclinical study, bone marrow–derived MSCs were genetically modified using an adeno-associated virus (AAV) vector to overexpress E-selectin, with a GFP reporter for tracking [88]. These E-selectin–modified MSCs significantly accelerated wound closure compared to controls. Histologic and perfusion analyses revealed improved capillary ingrowth, enhanced collagen deposition, and superior muscle fiber preservation. The modified cells also demonstrated longer survival within the wound. The authors concluded that E-selectin overexpression augmented the MSCs' pro-angiogenic and regenerative effects.

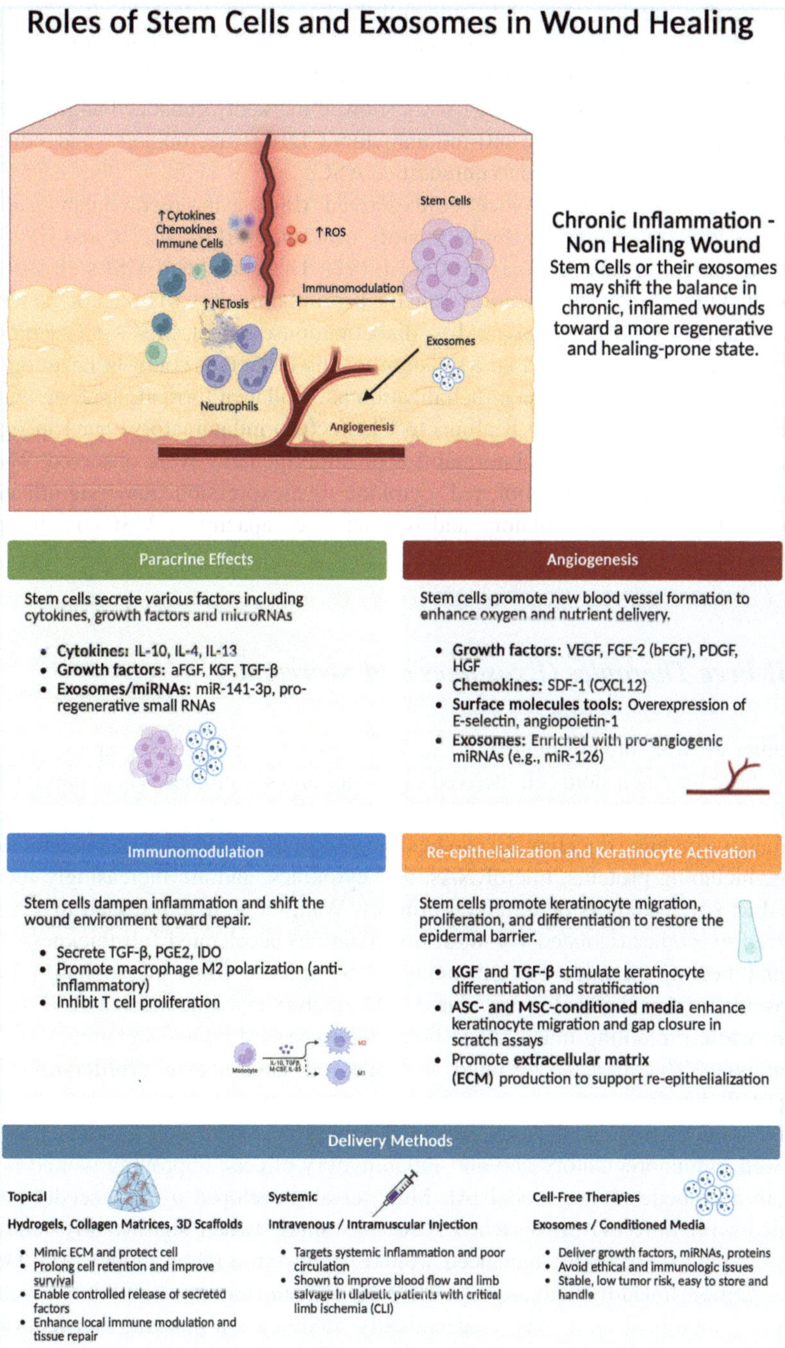

Fig. 25.1 Landscape of stem cell-based therapies for chronic wounds

In another preclinical model using full-thickness skin defects in diabetic rats, Zhu et al. employed ASCs that were genetically modified via lentiviral transduction to overexpress acidic fibroblast growth factor (aFGF) [90]. While healing improved across all treatment groups, aFGF-ASCs showed markedly enhanced angiogenesis (CD31 expression), increased anti-inflammatory CD163^{+} macrophages, and reduced local inflammation compared to unmodified ASCs.

Wu et al. engineered umbilical cord–derived MSCs using recombinant lentiviruses to co-express three anti-inflammatory cytokines—IL-4, IL-10, and IL-13—creating a cell line referred to as MSCs-3IL [91]. These modified MSCs effectively promoted macrophage polarization from the pro-inflammatory M1 phenotype to the anti-inflammatory M2 phenotype. In a diabetic mouse model, MSCs-3IL achieved >96% wound closure within 14 days—significantly faster than controls. Histological analysis showed enhanced reepithelialization and collagen formation, along with a higher ratio of CD206^{+} (pro-healing) to CD86^{+} (pro-inflammatory) macrophages. Importantly, no karyotypic abnormalities or tumorigenicity were observed. These findings suggest that multiplexed cytokine overexpression may significantly enhance the immunomodulatory and regenerative capacity of MSCs in diabetic wound repair.

Cell-Free Therapies (Exosomes and Secretome)

Another growing area of interest is the development of cell-free therapies, particularly those based on stem cell–derived exosomes and secretomes. These approaches harness the regenerative benefits of stem cells while avoiding risks such as tumorigenicity or immune rejection. Exosomes carry bioactive molecules secreted by stem cells, including proteins, microRNAs, and cytokines, and are increasingly recognized as key mediators of stem cell efficacy. Wang et al. demonstrated that ASC-derived exosomes (isolated via ultracentrifugation) accelerated full-thickness skin wound healing in mice by enhancing IL-33 release from macrophages [92]. Exosome treatment also downregulated inflammatory cytokines such as TNF-α and IL-6, while promoting macrophage polarization toward an anti-inflammatory M2 phenotype. This cascade resulted in improved keratinocyte proliferation and reepithelialization.

In another preclinical study, acellular embryonic stem cell extracts (EXTs) showed immunoregulatory and anti-inflammatory effects, improving wound healing in a diabetic mouse model [4]. Meng et al. developed a microneedle patch loaded with miR-141-3p–enriched ASC exosomes, which significantly reduced hypertrophic scarring and enhanced wound healing in a rabbit model [93]. Deng et al. demonstrated that conditioned medium derived from angiopoietin-1–overexpressing umbilical cord MSCs—genetically modified via plasmid transfection—significantly improved neovascularization and wound closure in diabetic mice. These findings illustrate the regenerative potential of engineered secretomes alone, highlighting how combining genetic modification with cell-free delivery strategies may further amplify the therapeutic effects of stem cell-based interventions.

Integration of Smart Biomaterials with Stem Cells

The integration of stem cells and exosomes with smart biomaterials has progressed substantially in recent years, enabling more precise, sustained, and responsive wound therapies. One example is the incorporation of ASC-derived exosomes (ASC-exo) into a smart hydrogel composed of MMP-degradable polyethylene glycol (PEG) [94]. This hydrogel was designed to degrade in response to elevated MMP activity, a hallmark of the wound microenvironment, thereby releasing ASC-exo on demand. In a diabetic mouse skin wound model, treatment with this hydrogel significantly accelerated wound closure compared to controls. In another study, Luo et al. developed an injectable gelatin methacrylamide (GelMA) hydrogel scaffold loaded with desferrioxamine (DFO)-releasing polydopamine nanoparticles to support ASCs in the form of 3D spheroids [95]. This smart scaffold reduced oxidative stress and hypoxia within the wound environment by scavenging ROS and stabilizing HIF-1α. Under these favorable conditions, 3D ASC spheroids secreted higher levels of growth factors. In a diabetic mouse model, this approach enhanced stem cell survival, increased angiogenesis and collagen deposition, and led to superior wound closure compared to controls.

Challenges and Considerations Regarding Stem Cell-Based Regenerative Therapies

Stem cell-based regenerative therapies for diabetic wounds have shown promise, but they face numerous challenges before routine clinical use. Key obstacles involve standardization of cell products, long-term safety, ethical and regulatory hurdles, high costs, and patient-specific factors [96–98].

Standardization of Cell Preparation and Dosing

There is currently no consensus on the optimal cell source, isolation or expansion method, or dosing regimen. Different studies utilize a variety of stem cell types—including bone marrow-derived, adipose-derived, umbilical cord-derived, and peripheral blood-derived stem cells—with disparate preparation techniques and cell doses, resulting in heterogeneous outcomes, making direct comparisons difficult [96, 99–101]. This heterogeneity is reflected in both preclinical and clinical studies, where protocols for cell isolation, expansion, delivery route (such as topical, injectable, or scaffold-based), and dosing vary widely. For example, some studies favor adipose-derived mesenchymal stem cells due to ease of harvest and expansion, while others use bone marrow or umbilical cord sources, each with different reported efficacies and safety profiles [96, 99–103]. The lack of standardized protocols for cell quality, viability, and dosing regimens further complicates the identification of best practices for therapeutic use in diabetic wound healing [97, 100, 104].

Long-Term Safety

Tumorigenesis is a major safety concern for stem cell-based regenerative therapies in diabetic wounds, particularly with pluripotent stem cells such as embryonic stem cells and induced pluripotent stem cells. Undifferentiated ESCs and iPSCs can form teratomas, and iPSC-derived cells may harbor genetic or epigenetic abnormalities that increase neoplastic risk. Ensuring complete differentiation and genomic stability of these cells before clinical use is essential to minimize tumorigenic potential [105–108]. Immune rejection is also a significant concern when using allogeneic (donor-derived) cells, as host immune responses can be provoked unless immunosuppression or HLA matching strategies are employed. This is particularly relevant for allogeneic pluripotent stem cell therapies, where immunogenicity can compromise graft survival and safety [108, 109]. Clinical trials of mesenchymal stem cells in chronic wounds have not reported significant adverse events compared to standard care, but prolonged follow-up is still required to confirm long-term safety and monitor for delayed complications. MSCs are considered to have a lower risk of tumorigenesis and immunogenicity compared to pluripotent stem cells, but the need for ongoing surveillance remains due to the potential for rare or late-onset adverse effects [110–113].

Ethical and Regulatory Barriers

The use of ESCs in regenerative therapies raises ethical issues because obtaining these cells involves the destruction of human embryos. This ethical dilemma has been a major driver of increased interest in alternative sources such as induced pluripotent stem cells, which do not require embryo use and thus circumvent these concerns. iPSCs are generated from adult somatic cells and offer similar pluripotency without the associated ethical controversy [98, 114]. In addition, regulatory approval is a major hurdle for stem cell therapies. These therapies must undergo rigorous evaluation for safety, efficacy, and quality and must meet strict manufacturing standards, which increases both development time and cost. The complexity of regulatory frameworks, particularly in the United States and the European Union, has contributed to the fact that few cell-based products have been approved for wound healing indications, as navigating these requirements is challenging. The U.S. Food and Drug Administration (FDA) requires that most stem cell-based products for wound healing be regulated as biologics, necessitating premarket approval and compliance with current good manufacturing practice (cGMP) standards [115, 116].

Cost and Scalability

Stem cell therapies are considerably more expensive than conventional wound treatments. The individualized and labor-intensive nature of autologous stem cell approaches, as well as the high manufacturing costs and stringent quality control requirements for allogeneic "off-the-shelf" products, contribute to significant economic and logistical challenges that have limited their widespread adoption in clinical practice [117–119]. Furthermore, these therapies often require specialized storage, have limited shelf lives and are associated with restricted indications and limited access, further increasing the overall cost of care, reflecting the need for careful patient selection and cost-effectiveness analysis [34, 120].

Individualized Therapy and Patient Factors

Patient heterogeneity influences the success of regenerative treatments. Diabetes itself impairs the function of endogenous stem cells and the wound-healing response through mechanisms such as increased oxidative stress, cellular senescence, and dysregulated cytokine and growth factor signaling, which collectively reduce the regenerative capacity of autologous cells and impair angiogenesis and tissue repair [29, 121]. Advanced age further diminishes regenerative potential as older individuals' stem cells exhibit attenuated activity due to intrinsic declines in stem cell function, reduced proliferative and differentiation capacity, and altered stem cell niches, as demonstrated by both molecular and single-cell analyses [122–124]. These factors can limit the efficacy of autologous cell therapy, as elderly diabetic patients' own cells may be insufficiently functional for optimal wound healing. While some preclinical data suggest that autologous diabetic adipose-derived stem cells may retain some therapeutic potential, the overall consensus is that diabetes and aging negatively impact autologous cell quality and function [44, 124, 125]. Allogeneic cells from healthy, younger donors could help overcome these limitations, provided that immune compatibility is managed, as allogeneic cells are less affected by the patient's comorbidities and may offer superior regenerative effects. However, immune rejection remains a challenge and requires appropriate immunomodulatory strategies [44, 126]. In summary, a personalized approach, tailoring regenerative therapy to each patient's specific condition, is likely required to achieve the best outcomes in this context, as supported by the need to address individual variability in wound pathophysiology, comorbidities, and stem cell function.

Conclusion

Diabetic wounds, particularly DFUs, remain a significant source of morbidity and healthcare burden worldwide. Despite advances in standard wound care—including debridement, infection control, off-loading, revascularization, and adjunctive therapies—healing rates remain suboptimal, and the risk of severe complications such as infection, amputation, and mortality remains high. These limitations highlight the need for innovative therapeutic approaches capable of addressing both the structural deficits and the dysregulated inflammatory environment characteristic of chronic diabetic wounds [7, 29, 41]. Stem cell-based therapies offer a promising avenue by leveraging their immunomodulatory, pro-angiogenic, and regenerative properties. These cells can influence the wound microenvironment, promote neovascularization, and modulate inflammation—key processes impaired in diabetic wound healing. Preclinical studies and early-phase clinical trials have demonstrated encouraging results, yet significant translational hurdles remain [96, 97, 100]. Continued translational research is essential to define optimal cell sources, dosing strategies, delivery methods, and patient selection criteria. Furthermore, rigorously designed clinical trials are needed to validate the efficacy and safety of stem cell-based therapies in diverse patient populations and to establish standardized treatment protocols [97, 100]. Looking ahead, the integration of regenerative medicine into mainstream diabetic wound care will likely depend on advancements in scalable manufacturing, regulatory approval pathways, cost-effectiveness, and personalization of therapy. As the field matures, stem cell-based therapies—possibly in combination with smart biomaterials and cell-free approaches—hold the potential to transform the management of chronic diabetic wounds, bridging the gap between scientific innovation and clinical application.

References

1. Chan JCN, Lim L-L, Wareham NJ et al (2020) The Lancet Commission on diabetes: using data to transform diabetes care and patient lives. Lancet 396(10267):2019–2082. https://doi.org/10.1016/S0140-6736(20)32374-6
2. Geiss LS, Li Y, Hora I, Albright A, Rolka D, Gregg EW (2019) Resurgence of diabetes-related nontraumatic lower-extremity amputation in the young and middle-aged adult U.S. population. Diabetes Care 42(1):50–54. https://doi.org/10.2337/dc18-1380
3. Ho J, Yue D, Cheema U, Hsia HC, Dardik A (2023) Innovations in stem cell therapy for diabetic wound healing. Adv Wound Care 12(11):626–643. https://doi.org/10.1089/wound.2021.0104
4. Loretelli C, Ben Nasr M, Giatsidis G et al (2020) Embryonic stem cell extracts improve wound healing in diabetic mice. Acta Diabetol 57(7):883–890. https://doi.org/10.1007/s00592-020-01500-0
5. Wong SL, Demers M, Martinod K et al (2015) Diabetes primes neutrophils to undergo NETosis, which impairs wound healing. Nat Med 21(7):815–819. https://doi.org/10.1038/nm.3887
6. Lavery LA, Armstrong DG, Wunderlich RP, Mohler MJ, Wendel CS, Lipsky BA (2006) Risk factors for foot infections in individuals with diabetes. Diabetes Care 29(6):1288–1293. https://doi.org/10.2337/dc05-2425

7. Armstrong DG, Tan T-W, Boulton AJM, Bus SA (2023) Diabetic foot ulcers: a review. JAMA 330(1):62. https://doi.org/10.1001/jama.2023.10578
8. Skrepnek GH, Mills JL, Armstrong DG (2015) A diabetic emergency one million feet long: disparities and burdens of illness among diabetic foot ulcer cases within emergency departments in the United States, 2006–2010. PLoS One 10(8):e0134914. https://doi.org/10.1371/journal.pone.0134914
9. Lane KL, Abusamaan MS, Voss BF et al (2020) Glycemic control and diabetic foot ulcer outcomes: a systematic review and meta-analysis of observational studies. J Diabetes Complicat 34(10):107638. https://doi.org/10.1016/j.jdiacomp.2020.107638
10. Xiang J, Wang S, He Y, Xu L, Zhang S, Tang Z (2019) Reasonable glycemic control would help wound healing during the treatment of diabetic foot ulcers. Diabetes Ther 10(1):95–105. https://doi.org/10.1007/s13300-018-0536-8
11. Leung TH, Cotsarelis G (2022) Cellular memories—more than skin deep. N Engl J Med 386(8):793–795. https://doi.org/10.1056/NEJMcibr2118516
12. Huelsboemer L, Knoedler L, Kochen A et al (2024) Cellular therapeutics and immunotherapies in wound healing -- on the pulse of time? Mil Med Res 11(1):23. https://doi.org/10.1186/s40779-024-00528-5
13. Knoedler S, Knoedler L, Kauke-Navarro M et al (2023) Regulatory T cells in skin regeneration and wound healing. Mil Med Res 10(1):49. https://doi.org/10.1186/s40779-023-00484-6
14. Knoedler S, Broichhausen S, Guo R et al (2023) Fibroblasts—the cellular choreographers of wound healing. Front Immunol 14:1233800. https://doi.org/10.3389/fimmu.2023.1233800
15. Diegelmann RF, Evans MC (2004) Wound healing: an overview of acute, fibrotic and delayed healing. Front Biosci 9:283–289. https://doi.org/10.2741/1184
16. Scopelliti F, Cattani C, Dimartino V, Mirisola C, Cavani A (2022) Platelet derivatives and the immunomodulation of wound healing. Int J Mol Sci 23(15):8370. https://doi.org/10.3390/ijms23158370
17. Barrientos S, Stojadinovic O, Golinko MS, Brem H, Tomic-Canic M (2008) Growth factors and cytokines in wound healing. Wound Repair Regen 16(5):585–601. https://doi.org/10.1111/j.1524-475X.2008.00410.x
18. Velnar T, Bailey T, Smrkolj V (2009) The wound healing process: an overview of the cellular and molecular mechanisms. J Int Med Res 37(5):1528–1542. https://doi.org/10.1177/147323000903700531
19. Hesketh M, Sahin KB, West ZE, Murray RZ (2017) Macrophage phenotypes regulate scar formation and chronic wound healing. Int J Mol Sci 18(7):1545. https://doi.org/10.3390/ijms18071545
20. Wang Z, Qi F, Luo H, Xu G, Wang D (2022) Inflammatory microenvironment of skin wounds. Front Immunol 13:789274. https://doi.org/10.3389/fimmu.2022.789274
21. Peña OA, Martin P (2024) Cellular and molecular mechanisms of skin wound healing. Nat Rev Mol Cell Biol 25(8):599–616. https://doi.org/10.1038/s41580-024-00715-1
22. Pensalfini M, Tepole AB (2023) Mechano-biological and bio-mechanical pathways in cutaneous wound healing. PLoS Comput Biol 19(3):e1010902. https://doi.org/10.1371/journal.pcbi.1010902
23. Peppa M, Stavroulakis P, Raptis SA (2009) Advanced glycoxidation products and impaired diabetic wound healing. Wound Repair Regen 17(4):461–472. https://doi.org/10.1111/j.1524-475X.2009.00518.x
24. Elajaili H, Lyttle BD, Lewis CV et al (2025) Increased ROS and persistent pro-inflammatory responses in a diabetic wound healing model (db/db): implications for delayed wound healing. Int J Mol Sci 26(10):4884. https://doi.org/10.3390/ijms26104884
25. Lan CC, Wu CS, Huang SM, Wu IH, Chen GS (2013) High-glucose environment enhanced oxidative stress and increased interleukin-8 secretion from keratinocytes: new insights into impaired diabetic wound healing. Diabetes 62(7):2530–2538. https://doi.org/10.2337/db12-1714

26. Hu SC, Lan CE (2016) High-glucose environment disturbs the physiologic functions of keratinocytes: focusing on diabetic wound healing. J Dermatol Sci 84(2):121–127. https://doi.org/10.1016/j.jdermsci.2016.07.008
27. Sharma S, Kishen A (2025) Dysfunctional crosstalk between macrophages and fibroblasts under LPS-infected and hyperglycemic environment in diabetic wounds. Sci Rep 15(1):17233. https://doi.org/10.1038/s41598-025-00673-4
28. den Dekker A, Davis FM, Kunkel SL, Gallagher KA (2019) Targeting epigenetic mechanisms in diabetic wound healing. Transl Res 204:39–50. https://doi.org/10.1016/j.trsl.2018.10.001
29. Mondal NS, Islam A, Gautam MK et al (2025) Unraveling the mechanisms of diabetic wounds: insights into pathogenesis and advanced treatment strategies. Curr Diabetes Rev. https://doi.org/10.2174/0115733998368222250509185511
30. Kamal R, Awasthi A, Pundir M, Thakur S (2024) Healing the diabetic wound: unlocking the secrets of genes and pathways. Eur J Pharmacol 975:176645. https://doi.org/10.1016/j.ejphar.2024.176645
31. Yadu N, Singh M, Singh D, Keshavkant S (2025) Mechanistic insights of diabetic wound: healing process, associated pathways and microRNA-based delivery systems. Int J Pharm 670:125117. https://doi.org/10.1016/j.ijpharm.2024.125117
32. Swoboda L, Held J (2022) Impaired wound healing in diabetes. J Wound Care 31(10):882–885. https://doi.org/10.12968/jowc.2022.31.10.882
33. Wan R, Weissman JP, Grundman K, Lang L, Grybowski DJ, Galiano RD (2021) Diabetic wound healing: the impact of diabetes on myofibroblast activity and its potential therapeutic treatments. Wound Repair Regen 29(4):573–581. https://doi.org/10.1111/wrr.12954
34. Gallagher KA, Mills JL, Armstrong DG et al (2024) Current status and principles for the treatment and prevention of diabetic foot ulcers in the cardiovascular patient population: a scientific statement from the American Heart Association. Circulation 149(4):e232–e253. https://doi.org/10.1161/cir.0000000000001192
35. Hingorani A, LaMuraglia GM, Henke P et al (2016) The management of diabetic foot: a clinical practice guideline by the Society for Vascular Surgery in collaboration with the American Podiatric Medical Association and the Society for Vascular Medicine. J Vasc Surg 63(2 Suppl):3s–21s. https://doi.org/10.1016/j.jvs.2015.10.003
36. Committee ADAPP (2025) 12. Retinopathy, neuropathy, and foot care: standards of care in diabetes-2025. Diabetes Care 48(1 Suppl 1):S252–s265. https://doi.org/10.2337/dc25-S012
37. Gornik HL, Aronow HD, Goodney PP et al (2024) ACC/AHA/AACVPR/APMA/ABC/SCAI/SVM/SVN/SVS/SIR/VESS guideline for the Management of Lower Extremity Peripheral Artery Disease: a report of the American College of Cardiology/American Heart Association Joint Committee on clinical practice guidelines. Circulation 149(24):e1313–e1410. https://doi.org/10.1161/cir.0000000000001251
38. Haxho F, Lytvyn Y, Shelley AJ, Ortega A, Melin M, Alavi A (2025) Wound Care - Part I: a practical approach to local wound care and dressing selection. J Am Acad Dermatol. https://doi.org/10.1016/j.jaad.2025.02.103
39. Wong AYW, Hooi NMF, Yeo BSY et al (2024) Improving diabetic wound-healing outcomes with topical growth factor therapies. J Clin Endocrinol Metab 109(8):e1642–e1651. https://doi.org/10.1210/clinem/dgae128
40. Kolimi P, Narala S, Nyavanandi D, Youssef AAA, Dudhipala N (2022) Innovative treatment strategies to accelerate wound healing: trajectory and recent advancements. Cells 11(15). https://doi.org/10.3390/cells11152439
41. Cho H, Blatchley MR, Duh EJ, Gerecht S (2019) Acellular and cellular approaches to improve diabetic wound healing. Adv Drug Deliv Rev 146:267–288. https://doi.org/10.1016/j.addr.2018.07.019
42. Singer AJ, Tassiopoulos A, Kirsner RS (2017) Evaluation and management of lower-extremity ulcers. N Engl J Med 377(16):1559–1567. https://doi.org/10.1056/NEJMra1615243

43. Oyebode OA, Jere SW, Houreld NN (2023) Current therapeutic modalities for the management of chronic diabetic wounds of the foot. J Diabetes Res 2023:1359537. https://doi.org/10.1155/2023/1359537
44. Mastrogiacomo M, Nardini M, Collina MC et al (2022) Innovative cell and platelet rich plasma therapies for diabetic foot ulcer treatment: the allogeneic approach. Front Bioeng Biotechnol 10:869408. https://doi.org/10.3389/fbioe.2022.869408
45. Blumberg SN, Berger A, Hwang L, Pastar I, Warren SM, Chen W (2012) The role of stem cells in the treatment of diabetic foot ulcers. Diabetes Res Clin Pract 96(1):1–9. https://doi.org/10.1016/j.diabres.2011.10.032
46. Lo B, Parham L (2009) Ethical issues in stem cell research. Endocr Rev 30(3):204–213. https://doi.org/10.1210/er.2008-0031
47. Xia H, Li X, Gao W et al (2018) Tissue repair and regeneration with endogenous stem cells. Nat Rev Mater 3(7):174–193. https://doi.org/10.1038/s41578-018-0027-6
48. Takahashi K, Yamanaka S (2006) Induction of pluripotent stem cells from mouse embryonic and adult fibroblast cultures by defined factors. Cell 126(4):663–676. https://doi.org/10.1016/j.cell.2006.07.024
49. Schwartz SD, Regillo CD, Lam BL et al (2015) Human embryonic stem cell-derived retinal pigment epithelium in patients with age-related macular degeneration and Stargardt's macular dystrophy: follow-up of two open-label phase 1/2 studies. Lancet 385(9967):509–516. https://doi.org/10.1016/S0140-6736(14)61376-3
50. Takahashi K, Tanabe K, Ohnuki M et al (2007) Induction of pluripotent stem cells from adult human fibroblasts by defined factors. Cell 131(5):861–872. https://doi.org/10.1016/j.cell.2007.11.019
51. Wang L, Cao J, Wang Y et al (2015) Immunogenicity and functional evaluation of iPSC-derived organs for transplantation. Cell Discov 1(1):15015. https://doi.org/10.1038/celldisc.2015.15
52. Hanna J, Wernig M, Markoulaki S et al (2007) Treatment of sickle cell anemia mouse model with iPS cells generated from autologous skin. Science 318(5858):1920–1923. https://doi.org/10.1126/science.1152092
53. Alipio Z, Liao W, Roemer EJ et al (2010) Reversal of hyperglycemia in diabetic mouse models using induced-pluripotent stem (iPS)-derived pancreatic β-like cells. Proc Natl Acad Sci U S A 107(30):13426–13431. https://doi.org/10.1073/pnas.1007884107
54. Gorecka J, Gao X, Fereydooni A et al (2020) Induced pluripotent stem cell-derived smooth muscle cells increase angiogenesis and accelerate diabetic wound healing. Regen Med 15(2):1277–1293. https://doi.org/10.2217/rme-2019-0086
55. Ben-David U, Benvenisty N (2011) The tumorigenicity of human embryonic and induced pluripotent stem cells. Nat Rev Cancer 11(4):268–277. https://doi.org/10.1038/nrc3034
56. Dash BC, Korutla L, Vallabhajosyula P, Hsia HC (2022) Unlocking the potential of induced pluripotent stem cells for wound healing: the next frontier of regenerative medicine. Adv Wound Care 11(11):622–638. https://doi.org/10.1089/wound.2021.0049
57. Han Y, Yang J, Fang J et al (2022) The secretion profile of mesenchymal stem cells and potential applications in treating human diseases. Signal Transduct Target Ther 7(1):92. https://doi.org/10.1038/s41392-022-00932-0
58. Casiraghi F, Perico N, Remuzzi G (2018) Mesenchymal stromal cells for tolerance induction in organ transplantation. Hum Immunol 79(5):304–313. https://doi.org/10.1016/j.humimm.2017.12.008
59. Elisseeff J, Badylak SF, Boeke JD (2021) Immune and genome engineering as the future of transplantable tissue. N Engl J Med 385(26):2451–2462. https://doi.org/10.1056/NEJMra1913421
60. Chen L, Tredget EE, Wu PYG, Wu Y (2008) Paracrine factors of mesenchymal stem cells recruit macrophages and endothelial lineage cells and enhance wound healing. PLoS One 3(4):e1886. https://doi.org/10.1371/journal.pone.0001886

61. Vagnozzi RJ, Maillet M, Sargent MA et al (2020) An acute immune response underlies the benefit of cardiac stem cell therapy. Nature 577(7790):405–409. https://doi.org/10.1038/s41586-019-1802-2
62. Gorecka A, Salemi S, Haralampieva D et al (2018) Autologous transplantation of adipose-derived stem cells improves functional recovery of skeletal muscle without direct participation in new myofiber formation. Stem Cell Res Ther 9(1):195. https://doi.org/10.1186/s13287-018-0922-1
63. Izadpanah R, Trygg C, Patel B et al (2006) Biologic properties of mesenchymal stem cells derived from bone marrow and adipose tissue. J Cell Biochem 99(5):1285–1297. https://doi.org/10.1002/jcb.20904
64. Ho Jeong J (2010) Adipose stem cells and skin repair. Curr Stem Cell Res Ther 5(2):137–140. https://doi.org/10.2174/157488810791268690
65. Gerini G, Mari E, Pontecorvi P et al (2025) 3D culturing as a promising strategy to enhance the angiogenic potential of adipose stem cell-derived secretome: insights into the role of miR-145-5p/ANGPT2 axis. Stem Cell Res Ther 16(1):153. https://doi.org/10.1186/s13287-025-04277-7
66. Kenny EM, Egro FM, Ejaz A, Coleman SR, Greenberger JS, Rubin JP (2021) Fat grafting in radiation-induced soft-tissue injury: a narrative review of the clinical evidence and implications for future studies. Plast Reconstr Surg 147(4):819–838. https://doi.org/10.1097/PRS.0000000000007705
67. Carstens MH, Quintana FJ, Calderwood ST et al (2021) Treatment of chronic diabetic foot ulcers with adipose-derived stromal vascular fraction cell injections: safety and evidence of efficacy at 1 year. Stem Cells Transl Med 10(8):1138–1147. https://doi.org/10.1002/sctm.20-0497
68. Zhu M, Chu Y, Shang Q et al (2020) Mesenchymal stromal cells pretreated with pro-inflammatory cytokines promote skin wound healing through VEGFC-mediated angiogenesis. Stem Cells Transl Med 9(10):1218–1232. https://doi.org/10.1002/sctm.19-0241
69. Zhang C, Huang L, Wang X et al (2022) Topical and intravenous administration of human umbilical cord mesenchymal stem cells in patients with diabetic foot ulcer and peripheral arterial disease: a phase I pilot study with a 3-year follow-up. Stem Cell Res Ther 13(1):451. https://doi.org/10.1186/s13287-022-03143-0
70. Walter MNM, Wright KT, Fuller HR, MacNeil S, Johnson WEB (2010) Mesenchymal stem cell-conditioned medium accelerates skin wound healing: an in vitro study of fibroblast and keratinocyte scratch assays. Exp Cell Res 316(7):1271–1281. https://doi.org/10.1016/j.yexcr.2010.02.026
71. Ceccarelli S, Nodale C, Vescarelli E et al (2018) Neuropilin 1 mediates keratinocyte growth factor signaling in adipose-derived stem cells: potential involvement in Adipogenesis. Stem Cells Int 2018:1–18. https://doi.org/10.1155/2018/1075156
72. Yang S-H, Park M-J, Yoon I-H et al (2009) Soluble mediators from mesenchymal stem cells suppress T cell proliferation by inducing IL-10. Exp Mol Med 41(5):315. https://doi.org/10.3858/emm.2009.41.5.035
73. Zhu L-P, Tian T, Wang J-Y et al (2018) Hypoxia-elicited mesenchymal stem cell-derived exosomes facilitates cardiac repair through miR-125b-mediated prevention of cell death in myocardial infarction. Theranostics 8(22):6163–6177. https://doi.org/10.7150/thno.28021
74. Zhao J, Li X, Hu J et al (2019) Mesenchymal stromal cell-derived exosomes attenuate myocardial ischaemia-reperfusion injury through miR-182-regulated macrophage polarization. Cardiovasc Res 115(7):1205–1216. https://doi.org/10.1093/cvr/cvz040
75. Li M, Luan F, Zhao Y et al (2017) Mesenchymal stem cellconditioned medium accelerates wound healing with fewer scars. Int Wound J 14(1):64–73. https://doi.org/10.1111/iwj.12551
76. Xie F, Teng L, Xu J et al (2020) Adiposederived mesenchymal stem cells inhibit cell proliferation and migration and suppress extracellular matrix synthesis in hypertrophicscar and keloid fibroblasts. Exp Ther Med 21(2):139. https://doi.org/10.3892/etm.2020.9571

77. Ghasempour A, Dehghan H, Mahmoudi M, Lavi Arab F (2024) Biomimetic scaffolds loaded with mesenchymal stem cells (MSCs) or MSC-derived exosomes for enhanced wound healing. Stem Cell Res Ther 15(1):406. https://doi.org/10.1186/s13287-024-04012-8
78. Sivaraj D, Chen K, Chattopadhyay A et al (2021) Hydrogel scaffolds to deliver cell therapies for wound healing. Front Bioeng Biotechnol 9:660145. https://doi.org/10.3389/fbioe.2021.660145
79. Martin KE, Hunckler MD, Chee E et al (2023) Hydrolytic hydrogels tune mesenchymal stem cell persistence and immunomodulation for enhanced diabetic cutaneous wound healing. Biomaterials 301:122256. https://doi.org/10.1016/j.biomaterials.2023.122256
80. Lu D, Chen B, Liang Z et al (2011) Comparison of bone marrow mesenchymal stem cells with bone marrow-derived mononuclear cells for treatment of diabetic critical limb ischemia and foot ulcer: a double-blind, randomized, controlled trial. Diabetes Res Clin Pract 92(1):26–36. https://doi.org/10.1016/j.diabres.2010.12.010
81. Mudgal SK, Kumar S, Gaur R et al (2024) Effectiveness of stem cell therapy for diabetic foot ulcers: a systematic review and GRADE compliant bootstrapped meta-analysis of randomized clinical trials. Int J Low Extrem Wounds:15347346241227530. https://doi.org/10.1177/15347346241227530
82. Dash NR, Dash SN, Routray P, Mohapatra S, Mohapatra PC (2009) Targeting nonhealing ulcers of lower extremity in human through autologous bone marrow-derived mesenchymal stem cells. Rejuvenation Res 12(5):359–366. https://doi.org/10.1089/rej.2009.0872
83. Kerstan A, Dieter K, Niebergall-Roth E et al (2022) Allogeneic ABCB5+ mesenchymal stem cells for treatment-refractory chronic venous ulcers: a phase I/IIa clinical trial. JID Innov 2(1):100067. https://doi.org/10.1016/j.xjidi.2021.100067
84. Kerstan A, Dieter K, Niebergall-Roth E et al (2022) Translational development of ABCB5+ dermal mesenchymal stem cells for therapeutic induction of angiogenesis in non-healing diabetic foot ulcers. Stem Cell Res Ther 13(1):455. https://doi.org/10.1186/s13287-022-03156-9
85. Schulman CI, Namias N, Pizano L et al (2022) The effect of mesenchymal stem cells improves the healing of burn wounds: a phase 1 dose-escalation clinical trial. Scars Burns Heal 8:20595131211070783. https://doi.org/10.1177/20595131211070783
86. Low Wang CC, Chong T, Moore G et al (2024) Results of the phase 1 open-label safety study of umbilical cord lining mesenchymal stromal/stem cells (Corlicyte®) to heal chronic diabetic foot ulcers. Biomedicine 12(6):1375. https://doi.org/10.3390/biomedicines12061375
87. Lu D, Jiang Y, Deng W et al (2019) Long term outcomes of BMMSC compared with BMMNC for treatment of critical limb ischemia and foot ulcer in patients with diabetes. Cell Transplant 28(5):645–652. https://doi.org/10.1177/0963689719835177
88. Huerta CT, Ortiz YY, Li Y et al (2023) Novel gene-modified mesenchymal stem cell therapy reverses impaired wound healing in ischemic limbs. Ann Surg 278(3):383–395. https://doi.org/10.1097/SLA.0000000000005949
89. Kauke-Navarro M, Noel OF, Knoedler L et al (2023) Novel strategies in transplantation: genetic engineering and vascularized composite allotransplantation. J Surg Res 291:176–186. https://doi.org/10.1016/j.jss.2023.04.028
90. Zhu Y, Chen P, Zhang Z et al (2025) aFGF gene-modified adipose-derived mesenchymal stem cells promote healing of full-thickness skin defects in diabetic rats. Stem Cell Res Ther 16(1):93. https://doi.org/10.1186/s13287-025-04241-5
91. Wu D, Liu R, Cen X et al (2025) Preclinical study of engineering MSCs promoting diabetic wound healing and other inflammatory diseases through M2 polarization. Stem Cell Res Ther 16(1):113. https://doi.org/10.1186/s13287-025-04248-y
92. Wang Y, Ding H, Bai R et al (2025) Exosomes from adipose-derived stem cells accelerate wound healing by increasing the release of IL 33 from macrophages. Stem Cell Res Ther 16(1):80. https://doi.org/10.1186/s13287-025-04203-x
93. Meng S, Wei Q, Chen S et al (2024) MiR1413pFunctionalized exosomes loaded in dissolvable microneedle arrays for hypertrophic scar treatment. Small 20(8):2305374. https://doi.org/10.1002/smll.202305374

94. Jiang T, Liu S, Wu Z et al (2022) ADSC-exo@MMP-PEG smart hydrogel promotes diabetic wound healing by optimizing cellular functions and relieving oxidative stress. Mater Today Bio 16:100365. https://doi.org/10.1016/j.mtbio.2022.100365
95. Luo T, Zhu P, Li S et al (2025) DFO-loaded PDA nanoparticles facilitated 3D stem cell spheroids for diabetic wound repair by normalizing the pathological microenvironment. Mater Today Bio 33:101973. https://doi.org/10.1016/j.mtbio.2025.101973
96. Farabi B, Roster K, Hirani R, Tepper K, Atak MF, Safai B (2024) The efficacy of stem cells in wound healing: a systematic review. Int J Mol Sci 25(5):3006. https://doi.org/10.3390/ijms25053006
97. Panda D, Nayak S (2024) Stem cell-based tissue engineering approaches for diabetic foot ulcer: a review from mechanism to clinical trial. Stem Cell Rev Rep 20(1):88–123. https://doi.org/10.1007/s12015-023-10640-z
98. Gorecka J, Kostiuk V, Fereydooni A et al (2019) The potential and limitations of induced pluripotent stem cells to achieve wound healing. Stem Cell Res Ther 10(1):87. https://doi.org/10.1186/s13287-019-1185-1
99. Tong L, Tang L, Tang B, Zhang J (2024) Impacts of stem cells from different sources on wound healing rate in diabetic foot ulcers: a systematic review and meta-analysis. Front Genet 15:1541992. https://doi.org/10.3389/fgene.2024.1541992
100. Krasilnikova OA, Baranovskii DS, Lyundup AV, Shegay PV, Kaprin AD, Klabukov ID (2022) Stem and somatic cell monotherapy for the treatment of diabetic foot ulcers: review of clinical studies and mechanisms of action. Stem Cell Rev Rep 18(6):1974–1985. https://doi.org/10.1007/s12015-022-10379-z
101. Cao Y, Yan J, Dong Z et al (2024) Adipose-derived mesenchymal stem cells are ideal for the cell-based treatment of refractory wounds: strong potential for angiogenesis. Stem Cell Rev Rep 20(1):313–328. https://doi.org/10.1007/s12015-023-10641-y
102. Gadelkarim M, Abushouk AI, Ghanem E, Hamaad AM, Saad AM, Abdel-Daim MM (2018) Adipose-derived stem cells: effectiveness and advances in delivery in diabetic wound healing. Biomed Pharmacother 107:625–633. https://doi.org/10.1016/j.biopha.2018.08.013
103. Mrozikiewicz-Rakowska B, Szabłowska-Gadomska I, Cysewski D et al (2023) Allogenic adipose-derived stem cells in diabetic foot ulcer treatment: clinical effectiveness, safety, survival in the wound site, and proteomic impact. Int J Mol Sci 24(2):1472. https://doi.org/10.3390/ijms24021472
104. Lopes L, Setia O, Aurshina A et al (2018) Stem cell therapy for diabetic foot ulcers: a review of preclinical and clinical research. Stem Cell Res Ther 9(1):188. https://doi.org/10.1186/s13287-018-0938-6
105. Yamanaka S (2020) Pluripotent stem cell-based cell therapy-promise and challenges. Cell Stem Cell 27(4):523–531. https://doi.org/10.1016/j.stem.2020.09.014
106. Lezmi E, Benvenisty N (2022) The tumorigenic potential of human pluripotent stem cells. Stem Cells Transl Med 11(8):791–796. https://doi.org/10.1093/stcltm/szac039
107. Fu X, Xu Y (2012) Challenges to the clinical application of pluripotent stem cells: towards genomic and functional stability. Genome Med 4(6):55. https://doi.org/10.1186/gm354
108. Rohani L, Johnson AA, Naghsh P, Rancourt DE, Ulrich H, Holland H (2018) Concise review: molecular cytogenetics and quality control: clinical guardians for pluripotent stem cells. Stem Cells Transl Med 7(12):867–875. https://doi.org/10.1002/sctm.18-0087
109. Simonson OE, Domogatskaya A, Volchkov P, Rodin S (2015) The safety of human pluripotent stem cells in clinical treatment. Ann Med 47(5):370–380. https://doi.org/10.3109/07853890.2015.1051579
110. Herberts CA, Kwa MS, Hermsen HP (2011) Risk factors in the development of stem cell therapy. J Transl Med 9:29. https://doi.org/10.1186/1479-5876-9-29
111. Gentile P, Garcovich S (2021) Systematic review: adipose-derived mesenchymal stem cells, platelet-rich plasma and biomaterials as new regenerative strategies in chronic skin wounds and soft tissue defects. Int J Mol Sci 22(4). https://doi.org/10.3390/ijms22041538

112. Huang YZ, Gou M, Da LC, Zhang WQ, Xie HQ (2020) Mesenchymal stem cells for chronic wound healing: current status of preclinical and clinical studies. Tissue Eng Part B Rev 26(6):555–570. https://doi.org/10.1089/ten.TEB.2019.0351
113. Nasadiuk K, Kolanowski T, Kowalewski C, Wozniak K, Oldak T, Rozwadowska N (2024) Harnessing mesenchymal stromal cells for advanced wound healing: a comprehensive review of mechanisms and applications. Int J Mol Sci 26(1). https://doi.org/10.3390/ijms26010199
114. Blau HM, Daley GQ (2019) Stem cells in the treatment of disease. N Engl J Med 380(18):1748–1760. https://doi.org/10.1056/NEJMra1716145
115. Song SJ, Nam Y, Rim YA, Ju JH, Sohn Y (2024) Comparative analysis of regulations and studies on stem cell therapies: focusing on induced pluripotent stem cell (iPSC)-based treatments. Stem Cell Res Ther 15(1):447. https://doi.org/10.1186/s13287-024-04065-9
116. Marks P, Gottlieb S (2018) Balancing safety and innovation for cell-based regenerative medicine. N Engl J Med 378(10):954–959. https://doi.org/10.1056/NEJMsr1715626
117. Brembilla NC, Vuagnat H, Boehncke WH, Krause KH, Preynat-Seauve O (2023) Adipose-derived stromal cells for chronic wounds: scientific evidence and roadmap toward clinical practice. Stem Cells Transl Med 12(1):17–25. https://doi.org/10.1093/stcltm/szac081
118. Talukdar S, Medhi J, Barbhuiya PA, Chakrabarti S (2025) Complementary approaches and advanced technologies in ameliorating rate of wound healing. Int J Pharm 682:125910. https://doi.org/10.1016/j.ijpharm.2025.125910
119. Deptuła M, Brzezicka A, Skoniecka A, Zieliński J, Pikuła M (2021) Adipose-derived stromal cells for nonhealing wounds: emerging opportunities and challenges. Med Res Rev 41(4):2130–2171. https://doi.org/10.1002/med.21789
120. Haxho F, Lytvyn Y, Shelley AJ, Ortega A, Melin M, Alavi A (2025) Wound care - part II: tissue-engineered skin substitutes & other advanced wound therapies. J Am Acad Dermatol. https://doi.org/10.1016/j.jaad.2025.06.045
121. Yan J, Tie G, Wang S et al (2018) Diabetes impairs wound healing by Dnmt1-dependent dysregulation of hematopoietic stem cells differentiation towards macrophages. Nat Commun 9(1):33. https://doi.org/10.1038/s41467-017-02425-z
122. Berlanga-Acosta JA, Guillén-Nieto GE, Rodríguez-Rodríguez N et al (2020) Cellular senescence as the pathogenic hub of diabetes-related wound chronicity. Front Endocrinol (Lausanne) 11:573032. https://doi.org/10.3389/fendo.2020.573032
123. Magenta A, Florio MC, Ruggeri M, Furgiuele S (2021) Autologous cell therapy in diabetes-associated critical limb ischemia: from basic studies to clinical outcomes (review). Int J Mol Med 48(3). https://doi.org/10.3892/ijmm.2021.5006
124. Brunet A, Goodell MA, Rando TA (2023) Ageing and rejuvenation of tissue stem cells and their niches. Nat Rev Mol Cell Biol 24(1):45–62. https://doi.org/10.1038/s41580-022-00510-w
125. Chen B, Wei Y, Cai J, Zitkovsky HS, Guo L (2024) Comparable effects on healing between autologous diabetic adipose-derived stem cells and allogeneic normal counterparts. Plast Reconstr Surg 153(2):379–388. https://doi.org/10.1097/prs.0000000000010647
126. Mardomi A, Mohammadi N, Ahmadzadeh F, Abediankenari S (2023) CTLA4-Ig alleviates the allogeneic immune responses against insulin-producing cells in a murine model of cell transplantation. Naunyn Schmiedeberg's Arch Pharmacol 396(11):3195–3206. https://doi.org/10.1007/s00210-023-02527-0

Chapter 26
Innovative Frontiers in Wound Healing: Integrating Bioengineering, Nanotechnology, and Regenerative Medicine

Sathish Muthu, Naveen Jeyaraman, A. S. P. V. S. Saketh, Sharun Khan, and Madhan Jeyaraman

Abstract Chronic wounds remain a pressing global health challenge, exacerbated by comorbid conditions such as diabetes and vascular disorders. Traditional wound care methods often fall short in managing non-healing wounds, leading to increased morbidity and healthcare costs. Recent advancements in biomedical research have paved the way for novel strategies to enhance and accelerate the wound healing process. Bioengineered dressings, such as hydrocolloids and scaffolds like Integra®, offer enhanced moisture retention and support tissue regeneration. Nanotechnology introduces antimicrobial nanoparticles and smart drug delivery systems that directly target the wound environment, reducing infection and promoting cell proliferation. Mesenchymal stem cells (MSCs) and extracellular vesicles (EVs) offer regenerative and immunomodulatory capabilities, enhancing angiogenesis and re-epithelialization while minimizing inflammation. Advanced materials like hydrogels and microneedle patches allow for sustained and localized release of bioactive agents, improving

S. Muthu (✉)
Orthopaedic Research Group, Coimbatore, Tamil Nadu, India

Central Research Laboratory, Meenakshi Medical College Hospital and Research Institute, Meenakshi Academy of Higher Education and Research, Chennai, Tamil Nadu, India

N. Jeyaraman · M. Jeyaraman
Orthopaedic Research Group, Coimbatore, Tamil Nadu, India

Department of Orthopaedics, ACS Medical College and Hospital, Dr MGR Educational and Research Institute, Chennai, Tamil Nadu, India

A. S. P. V. S. Saketh
Department of Orthopaedics & Sports Medicine, Yashoda Hospitals, Malakpet, Hyderabad, Telangana, India

S. Khan
Center for Regenerative Nanomedicine, Northwestern University, Chicago, IL, USA

Graduate Institute of Medicine, Yuan Ze University, Taoyuan, Taiwan

F. De Francesco, K. Turksen (eds.), *Management and Strategies for Wound Healing*, Stem Cell Biology and Regenerative Medicine 79,
https://doi.org/10.1007/978-3-032-12442-5_26

therapeutic outcomes. Moreover, 3D-printed scaffolds, designed to mimic the extracellular matrix, support cellular migration and tissue integration, while innovations in biomaterial biocompatibility reduce immune rejection. Artificial Intelligence (AI) further augments wound monitoring through predictive analytics and remote diagnostics, offering scalable solutions in resource-limited settings like rural India. Despite the promise of these technologies, challenges such as high production costs, regulatory bottlenecks, and infrastructure disparities limit widespread clinical adoption. Addressing these barriers through collaborative efforts, policy reform, and equitable healthcare strategies is essential. The integration of emerging biotechnologies with personalized medicine holds transformative potential for revolutionizing wound care and achieving scar-free, regenerative healing.

Keywords Chronic wounds · Nanotechnology · Wound healing · Bioengineering

Introduction

Chronic wounds present a significant and escalating global health burden. Affecting millions annually, these persistent lesions are often associated with underlying conditions such as diabetes, vascular diseases, and prolonged immobility. It is estimated that over 6.5 million individuals in the United States alone suffer from chronic wounds, with global costs projected in the billions due to long-term care and associated complications [1, 2].

Beyond the economic implications, the human toll is substantial. Chronic wounds can lead to reduced quality of life, increased risk of infection, amputations, and even death. The 5-year mortality rate for individuals with diabetic foot ulcers has been compared to that of many cancers [2]. Traditional wound care, while effective in some contexts, often falls short in addressing complex or non-healing wounds.

In recent years, a paradigm shift has emerged toward personalized medicine, leveraging individual patient profiles to tailor therapeutic strategies. This patient-centric approach is particularly promising in wound care, where genetic, metabolic, and lifestyle factors can dramatically affect healing outcomes [3]. As science advances, the integration of bioengineering, regenerative medicine, and smart biomaterials is paving the way for faster, more effective wound healing.

Novel Therapeutic Approaches

Bioengineered Dressings

The evolution of wound dressings from simple cotton gauze to sophisticated bioengineered materials represents a major advancement in wound care. Traditional dressings primarily provided mechanical protection, but they often failed to address the dynamic needs of the wound healing environment. Modern wound management now emphasizes the importance of maintaining a moist, controlled microenvironment that supports cell migration, tissue remodeling, and bacterial resistance—all of which are critical for accelerated wound healing [4].

Among the most widely used modern dressings are hydrocolloids, alginates, and foam-based materials. Hydrocolloid dressings contain gel-forming agents like carboxymethylcellulose, which interact with wound exudate to form a gel that maintains a moist environment while providing a barrier against bacteria. Alginates, derived from seaweed, are particularly effective for wounds with heavy exudate, as they can absorb large amounts of fluid while also facilitating autolytic debridement, a natural process that helps remove necrotic tissue without damaging healthy cells [5]. Foam dressings, on the other hand, offer excellent cushioning and absorbency, making them suitable for pressure ulcers or wounds in high-friction areas.

One notable example of a bioengineered dressing is the Integra® Dermal Regeneration Template, a bilayered construct comprising a matrix of bovine tendon collagen and glycosaminoglycans, covered by a silicone layer. This template serves as a scaffold that supports neodermis formation and vascularization, while the silicone acts as a temporary epidermal barrier. Integra has been particularly beneficial in managing full-thickness skin defects, including scalp wounds exposing the skull, where traditional skin grafts or flaps may not be viable options [6].

These advanced dressings not only reduce healing time but also lower the risk of complications such as infections or scarring. By combining mechanical protection with biological functionality, bioengineered dressings mark a critical step forward in wound management strategies for both acute and chronic injuries [4, 5]. Further, they are also capable of integrated drug release to aid in antimicrobial action during the healing phase of the wound, as shown in Fig. 26.1.

Nanotechnology

Nanotechnology has emerged as a transformative tool in the field of wound healing, offering innovative approaches to overcome some of the major limitations of conventional treatments. Chronic wounds, often characterized by prolonged inflammation, infection, and biofilm formation, are notoriously difficult to manage. In this context, nanomaterials—owing to their extremely small size (1–100 nanometers) and high surface-area-to-volume ratio—demonstrate unique physicochemical and biological properties that can significantly enhance wound care strategies [5].

One of the most valuable applications of nanotechnology in wound healing is the use of nanoparticles with intrinsic antimicrobial activity. Silver nanoparticles (AgNPs) are perhaps the most well-researched in this domain. Their ability to disrupt bacterial cell membranes, interfere with DNA replication, and generate reactive oxygen species makes them highly effective against a broad spectrum of bacteria, including antibiotic-resistant strains [7]. This makes silver-based nanomaterials particularly beneficial for preventing and treating infections in chronic wounds,

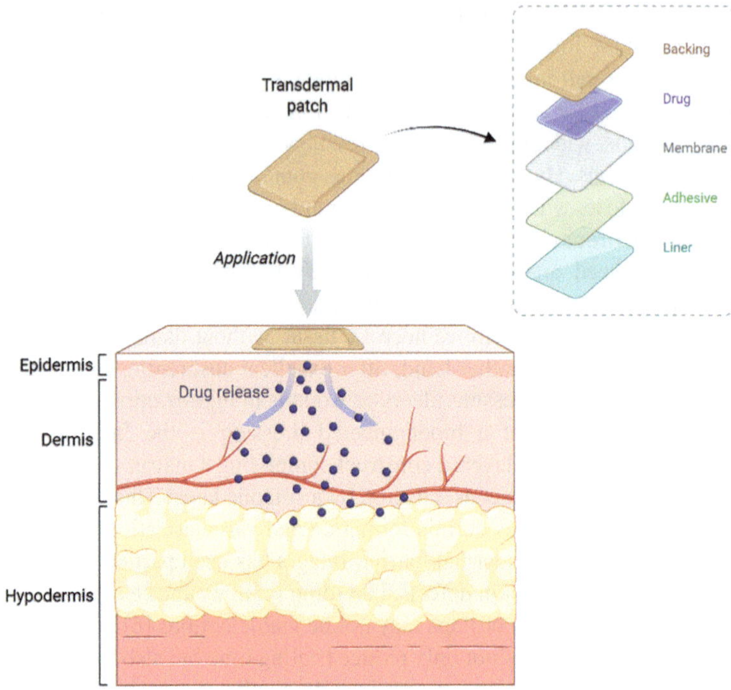

Fig. 26.1 Dermal patch with controlled drug delivery mechanism

where biofilm formation by bacteria like *Staphylococcus aureus* and *Pseudomonas aeruginosa* significantly delays healing.

Similarly, zinc oxide (ZnO) nanoparticles offer dual benefits of antimicrobial activity and wound repair. ZnO is known to promote epithelialization and modulate inflammation, both critical in the early stages of wound healing. Moreover, gold nanoparticles (AuNPs) have gained attention for their biocompatibility and ability to act as carriers for therapeutic molecules, including anti-inflammatory drugs and growth factors [5]. These metal-based nanoparticles not only eliminate infection risks but also create a microenvironment conducive to tissue regeneration.

Beyond antimicrobial functions, nanotechnology has paved the way for targeted drug delivery systems. Traditional wound therapies often involve topical or systemic administration of drugs, which can lead to off-target effects or insufficient drug concentration at the wound site. Nanoparticles can be engineered to encapsulate therapeutic agents such as growth factors (e.g., vascular endothelial growth factor (VEGF) and epidermal growth factor (EGF)), anti-inflammatory compounds, and antibiotics and release them in a sustained, localized manner [8]. This targeted delivery enhances bioavailability, minimizes systemic toxicity, and reduces the frequency of dressing changes, thereby improving patient compliance and comfort.

Moreover, polymeric nanoparticles and nanofibrous scaffolds are being developed to mimic the natural extracellular matrix, supporting cell adhesion and migration. These nanostructures can also be functionalized to respond to environmental stimuli such as pH or temperature, providing smart, on-demand drug release in response to changes in the wound environment [8].

The integration of nanotechnology into wound dressings and scaffolds represents a powerful convergence of materials science and biomedicine. With ongoing research and development, these nano-enabled systems are expected to transition from experimental models to routine clinical applications. Their ability to address multiple challenges, such as infection control, precise drug delivery, and promotion of tissue regeneration, positions nanotechnology at the forefront of next-generation wound care solutions [5, 7, 8].

Cell-Based Therapies

Stem Cells

In the realm of regenerative medicine, mesenchymal stem cells (MSCs) have garnered substantial attention for their potential to revolutionize wound healing strategies. These multipotent stromal cells, which can be isolated from various tissues including bone marrow, adipose tissue, and umbilical cord blood, exhibit a remarkable ability to differentiate into multiple cell types such as osteoblasts, adipocytes, and chondrocytes. Beyond their differentiation potential, MSCs exert powerful paracrine effects, releasing a wide array of bioactive molecules that orchestrate complex tissue repair processes [9, 10].

One of the key attributes that makes MSCs ideal for wound healing is their immunomodulatory function. Chronic wounds often remain in a prolonged inflammatory state, which hinders progression to the proliferative and remodeling phases of healing. MSCs can modulate immune responses by shifting macrophage phenotypes from the pro-inflammatory M1 to the anti-inflammatory M2 type. This transition is critical for reducing inflammation and promoting tissue regeneration. Moreover, MSCs secrete anti-inflammatory cytokines such as IL-10 and TGF-β, which help in resolving chronic inflammation, thereby accelerating the wound healing cascade [10].

Additionally, MSCs contribute significantly to angiogenesis, the formation of new blood vessels, which is essential for supplying oxygen and nutrients to regenerating tissues. They release angiogenic factors like VEGF and fibroblast growth factor (FGF), which promote endothelial cell proliferation and migration [11]. In ischemic and diabetic wounds, where blood supply is often compromised, this pro-angiogenic property becomes especially valuable.

MSCs also play a vital role in fibroblast activation and extracellular matrix (ECM) remodeling. Fibroblasts are responsible for producing collagen and other ECM proteins that provide structural support for new tissue formation. MSC-derived signals enhance fibroblast proliferation and migration while also promoting keratinocyte activity, which is essential for re-epithelialization—the final phase of wound closure [9].

A growing body of preclinical and clinical evidence supports the therapeutic application of MSCs in both acute and chronic wound models. In a study by L et al., transplantation of human umbilical cord-derived MSCs into burn-induced wounds in rats significantly accelerated re-epithelialization, enhanced neovascularization, and improved dermal architecture. These results demonstrated that MSC therapy not only speeds up wound closure but also leads to better quality skin regeneration [12].

Furthermore, advanced genetic engineering techniques have been employed to enhance the regenerative efficacy of MSCs. Chang et al. [11] developed a TALEN-mediated genetically modified MSC line capable of secreting hepatocyte growth factor (HGF), a potent mitogen and angiogenic factor. When applied to wound sites, these engineered MSCs substantially improved vascular density and tissue regeneration, outperforming their unmodified counterparts.

Another advantage of MSCs is their low immunogenicity, which allows for allogeneic transplantation without significant risk of immune rejection. This feature opens the door to off-the-shelf stem cell therapies that could be readily available in clinical settings. Umbilical cord-derived MSCs, in particular, are appealing due to their high proliferative capacity and non-invasive harvesting methods [12].

However, despite these promising outcomes, challenges remain in translating MSC therapy into widespread clinical practice. Issues such as optimal cell dosage, delivery method, and long-term safety need to be thoroughly addressed. Additionally, standardization in the isolation and expansion of MSCs is crucial to ensure consistency and reproducibility across studies [10].

Extracellular Vesicles (EVs)

In recent years, EVs have emerged as a promising alternative to traditional stem cell therapies in regenerative medicine, particularly in the context of wound healing. EVs are nano- to microscale, membrane-bound vesicles naturally secreted by a variety of cell types, including mesenchymal stem cells (MSCs). These vesicles serve as vehicles for transporting a rich cargo of bioactive molecules—such as proteins, lipids, mRNAs, and microRNAs—between cells, thus playing a pivotal role in intercellular communication and tissue homeostasis [3].

What makes EVs particularly attractive for therapeutic use is their ability to mimic many of the beneficial effects of MSCs without some of the limitations and risks associated with live-cell transplantation. MSC-derived EVs exert anti-inflammatory, angiogenic, and proliferative effects, all of which are crucial for promoting efficient and high-quality wound healing [13]. By delivering signaling molecules to target cells within the wound bed, EVs can modulate the immune response, enhance endothelial cell activity, and stimulate fibroblast migration and collagen synthesis—key steps in the healing process [3].

One of the greatest advantages of EV-based therapy is its favorable safety and regulatory profile. While MSC therapies require careful consideration of cell viability, engraftment, and potential tumorigenicity, EVs—being acellular and non-replicating—present a lower immunogenic risk and fewer ethical concerns. This makes them more amenable to standardized manufacturing, storage, and delivery, opening new avenues for clinical scalability and accessibility [14].

Moreover, EVs possess natural targeting abilities and can be engineered to enhance their therapeutic potential. Studies have shown that MSC-derived EVs can home in on inflamed or injured tissues, delivering their cargo precisely where it is needed. Their lipid bilayer protects the encapsulated molecules from enzymatic degradation, ensuring stability and bioactivity in vivo [14].

In experimental wound models, the application of MSC-derived EVs has been shown to accelerate re-epithelialization, promote angiogenesis, and reduce fibrotic tissue formation, thereby facilitating faster and more organized tissue repair. Given these promising results, EVs are now being actively explored as standalone biologics or as adjuncts to other advanced wound therapies [13]. EVs represent a cutting-edge, cell-free therapeutic platform that harnesses the regenerative potential of stem cells while avoiding many of their inherent risks. As research and bioengineering techniques advance, EVs are poised to become a cornerstone of next-generation therapies for chronic and complex wounds [3, 14].

Advanced Drug Delivery Systems

Hydrogels

Hydrogels are three-dimensional, water-swollen polymeric networks that have become a cornerstone in the development of advanced wound dressings. Due to their high water content and structural similarity to the extracellular matrix (ECM), hydrogels provide a moist and supportive environment conducive to cellular processes involved in wound healing, such as cell migration, proliferation, and differentiation [15]. Their physical and chemical properties can be finely tuned, allowing customization based on the type, severity, and stage of the wound.

One of the key advantages of hydrogels is their ability to serve as carriers for bioactive molecules. They can be loaded with various therapeutic agents, including growth factors like vascular endothelial growth factor (VEGF), fibroblast growth factor (FGF), and epidermal growth factor (EGF). These molecules play crucial roles in stimulating angiogenesis, granulation tissue formation, and epithelial regeneration—core components of the wound healing process. By enabling the sustained and localized release of these agents, hydrogels enhance their therapeutic efficacy while minimizing systemic side effects [15, 16].

Hydrogels also offer excellent biocompatibility and minimal immunogenicity, making them suitable for prolonged application on chronic wounds, such as diabetic ulcers or pressure sores. Their mechanical properties, such as stiffness and elasticity, can be modified to match different wound environments, improving integration with surrounding tissues and preventing mechanical disruption of the healing site [8, 16, 17].

Furthermore, some hydrogels can be engineered to be responsive to environmental stimuli, such as pH or temperature changes, enabling intelligent drug release in response to the wound's physiological state. This makes them highly versatile and adaptable for personalized wound care strategies. Hydrogels combine structural support with targeted therapy delivery, offering a multifaceted approach to chronic wound management and tissue regeneration [8, 15, 18].

Microneedles

Microneedle patches represent an innovative and minimally invasive approach for the delivery of bioactive molecules directly to the wound bed, a technique that has gained considerable attention in recent years for its potential to enhance wound healing. These patches are composed of arrays of microneedles—typically ranging from 50 to 900 micrometers in length—that are small enough to penetrate the outermost skin layer, the stratum corneum, without causing pain or significant discomfort [5, 19]. By bypassing this natural skin barrier, microneedles provide an efficient route for the local delivery of therapeutic agents, including peptides, proteins, and even stem cell-derived products, directly to the site of injury.

One of the primary advantages of microneedle-based systems is their ability to deliver molecules in a controlled and targeted manner, ensuring higher bioavailability at the wound site compared to conventional topical treatments or systemic drug administration. This targeted delivery can accelerate the healing process by promoting epithelialization and stimulating the formation of granulation tissue, key stages in the wound repair process [5, 20]. Furthermore, the microneedles themselves can be biodegradable, allowing them to dissolve once inserted into the skin, eliminating the need for removal and minimizing the risk of infection.

The versatility of microneedles is another attractive feature, as they can be engineered to release a wide range of therapeutic agents, including growth factors like epidermal growth factor (EGF) and fibroblast growth factor (FGF), which play essential roles in tissue regeneration and wound closure, as shown in Fig. 26.2. Additionally, microneedles can be used to enhance transdermal drug delivery for chronic wounds, offering an ideal solution for patients who require sustained, localized therapy. Microneedles offer a promising, low-risk platform for accelerating wound healing and improving patient outcomes, especially for complex or chronic wounds that do not respond well to traditional treatments [5, 21, 22].

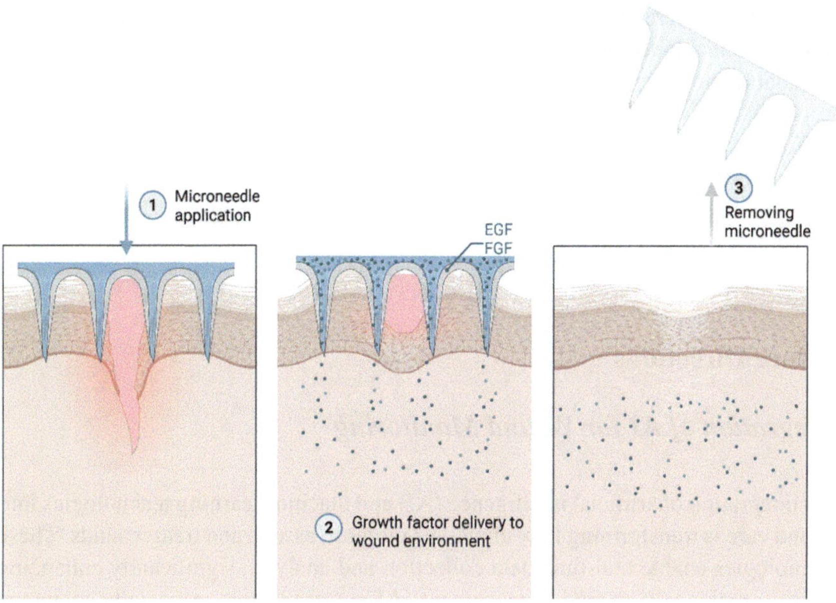

Fig. 26.2 Microneedles in wound healing

Emerging Biomaterials

Scaffolds

The advent of three-dimensional (3D) printed scaffolds has revolutionized tissue engineering, offering a new dimension in regenerative medicine. These scaffolds provide customizable, biologically relevant frameworks that are designed to closely mimic the structural and biochemical properties of native extracellular matrices (ECM). This allows for more effective and efficient wound repair, particularly in cases of complex or chronic wounds where traditional treatments may be insufficient. By mimicking the ECM, 3D scaffolds promote cell adhesion, migration, and proliferation, key processes essential for tissue regeneration [15].

One of the primary functions of 3D scaffolds is their ability to support cell infiltration, which is crucial for the formation of new tissue. Additionally, these scaffolds facilitate nutrient diffusion and neovascularization, enabling the growth of new blood vessels and providing essential oxygen and nutrients to the healing tissue. These properties are essential for effective wound healing, as they help to ensure that the newly forming tissue remains viable and functional [15].

Materials commonly used in the fabrication of 3D scaffolds include collagen, chitosan, and polylactic acid (PLA). Collagen, being a major component of the ECM, is often used because of its excellent biocompatibility and its ability to enhance cellular attachment and proliferation. Chitosan, derived from the exoskeletons of crustaceans, is another material widely used in scaffold construction due to its biodegradability and antimicrobial properties. PLA, a synthetic biodegradable polymer, is favored for its mechanical stability and regenerative properties, allowing for controlled degradation as the wound heals [15]. 3D-printed scaffolds represent a highly promising tool in wound healing, providing structural support and facilitating cellular and vascular growth essential for tissue repair. Their customization and versatility make them ideal for addressing a variety of chronic and complex wound conditions.

Future Directions

Integration of AI for Wound Monitoring

The integration of artificial intelligence (AI) and machine learning technologies into wound care is transforming how clinicians monitor, assess, and treat wounds. These technologies enable real-time data collection and analysis, significantly enhancing decision-making in wound management. AI-powered tools, especially in image analysis and predictive analytics, are proving instrumental in optimizing treatment and improving patient outcomes, even in complex wound cases.

One of the most promising AI applications in wound care is image analysis, where AI algorithms are used to assess wound size, depth, and healing progression. AI-based systems, utilizing deep learning techniques, can analyze high-resolution images of wounds with remarkable accuracy. These systems measure wound dimensions, detect changes over time, and assess factors like granulation tissue formation or infection, thereby assisting clinicians in tracking wound healing progress. This not only reduces human error but also speeds up the decision-making process [23]. For example, AI tools can provide automatic documentation, reducing the manual burden on healthcare professionals and ensuring consistent monitoring.

In the Indian context, the adoption of AI in healthcare is increasingly gaining traction, particularly in the realm of telemedicine and remote monitoring. Given the large rural population and limited access to healthcare professionals, AI-powered wound monitoring systems provide an essential tool for remote consultations. In regions where specialized care is scarce, these systems enable healthcare providers to monitor wounds from a distance and offer timely interventions. For instance, AI-based platforms could analyze images of chronic diabetic wounds, common in rural areas, and send alerts to healthcare providers if infection or other complications are detected. This type of remote support could help reduce the burden on healthcare systems by enabling timely intervention and preventing complications such as amputations [1].

Moreover, AI-driven predictive analytics is another area with significant potential in wound care. By analyzing historical patient data and real-time monitoring, predictive algorithms can forecast complications, such as infections or delayed wound healing, before they occur. These systems use factors like patient comorbidities, age, and wound characteristics to provide insights into potential risks and suggest preventive measures. In India, where diabetic foot ulcers are a major concern due to the high prevalence of diabetes [24], predictive analytics could play a crucial role in preventing infections and promoting faster healing. With the rising rates of chronic diseases, such as diabetes and hypertension, predictive models could also assist in identifying high-risk patients and prioritizing them for interventions, potentially reducing healthcare costs and improving quality of life for patients.

AI's ability to support clinicians with personalized treatment recommendations further enhances wound care. By combining data on wound type, healing stages, and patient history, AI systems can suggest tailored treatment plans, recommending specific dressings, therapies, or drug regimens. This type of precision medicine can be especially useful in complex wound cases, such as those resulting from burns or diabetic ulcers, where personalized interventions are necessary for optimal healing [2].

In India, however, there are still challenges to the widespread implementation of AI-based wound monitoring solutions. These include internet access in remote areas, technological infrastructure, and healthcare professional training to utilize AI tools effectively. Despite these hurdles, India's rapidly growing health tech sector and increasing digital literacy offer a promising foundation for integrating AI into healthcare systems. Investments in AI research and public private partnerships could help address these challenges, allowing AI-driven wound care tools to become accessible to both urban and rural populations.

The integration of AI into wound monitoring holds great potential in improving wound care, particularly in resource-constrained environments like India. With its ability to assist in real-time wound assessment, predictive analysis, and personalized treatment, AI can significantly enhance clinical outcomes, reduce complications, and contribute to more efficient healthcare delivery.

Challenges in Scaling Up Innovative Treatments

Despite significant progress in the development of innovative wound healing therapies, several key barriers continue to hinder their widespread adoption, particularly in low- and middle-income countries (LMICs) like India. The transition from laboratory discoveries to clinically applicable treatments often faces obstacles, including regulatory hurdles, high production costs, limited clinical data, and inequitable access to these technologies. A comprehensive summary of the recent advancements in the field of wound healing research and the considerations associated with each of the fronts is presented in Fig. 26.3.

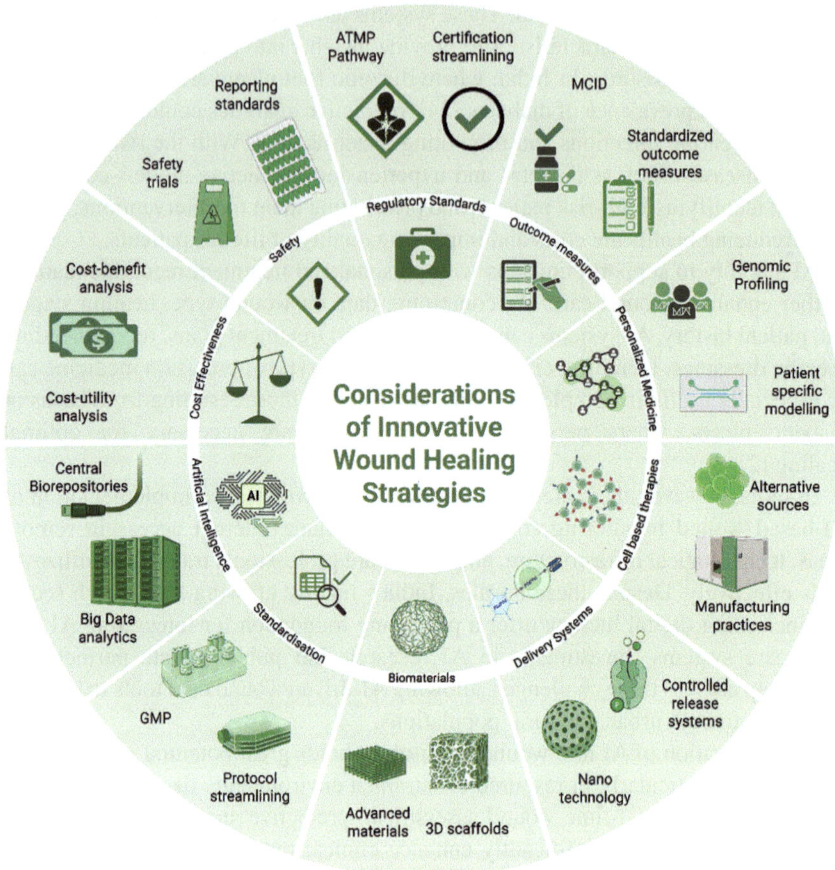

Fig. 26.3 Consideration of the recent advances in wound healing research

Regulatory Hurdles and Approval Processes

In India, as in other countries, regulatory approval for novel wound care technologies remains a major challenge. The approval process for new biomaterials, cell-based therapies, and bioengineered dressings is often complex and time-consuming. The Indian regulatory framework for medical devices and biologics is still evolving, which means that clear guidelines for the approval of these innovative treatments may be lacking or unclear [25]. Unlike pharmaceutical products, which have more established pathways for approval, biologics and regenerative therapies such as stem cell-based treatments and advanced dressings require extensive clinical trials to demonstrate both efficacy and safety, further complicating the approval process. This slow regulatory pace significantly delays the availability of innovative solutions to the broader population.

High Production Costs and Scalability

Another significant challenge in scaling up innovative wound care treatments is the high production cost of advanced therapies. Many bioengineered dressings, cell-based therapies, and nanotechnology-based solutions are costly to manufacture, particularly in terms of raw materials and the specialized equipment required for their production [26]. These costs are often passed on to the end users, making them unaffordable for a significant portion of the population in resource-constrained settings like India. Moreover, scalable manufacturing processes are critical to reduce these costs, but for many of these innovative treatments, the infrastructure to produce them at scale is not readily available. In India, manufacturing at scale for 3D-printed scaffolds or cell therapies requires significant investment in specialized facilities and equipment, which many healthcare systems cannot afford.

Biocompatibility

The biocompatibility of any implanted biomaterial is critical to its success in medical applications, particularly in the context of wound healing and tissue regeneration. Biocompatibility refers to the material's ability to integrate seamlessly with the surrounding tissue without causing harmful immune responses. A central challenge in biomaterial development is minimizing immune rejection while ensuring that the material supports proper integration with the host tissue to promote effective healing [27].

One of the key aspects of biocompatibility is controlling the immune response to the implanted material. The body's immune system must recognize the material as non-threatening while still activating appropriate defense mechanisms to prevent infection. However, prolonged or excessive immune activation can lead to chronic inflammation and fibrosis, hindering the healing process. To address this, researchers are focusing on innovative surface modifications of biomaterials that can influence the immune response. These modifications aim to promote favorable responses, such as the polarization of macrophages, which are critical cells in the immune system that play a key role in both inflammation and tissue repair [28].

Macrophage polarization is particularly important, as these cells can either promote inflammation and fibrosis (M1 phenotype) or facilitate healing and tissue regeneration (M2 phenotype). By using immune-instructive materials, it is possible to direct macrophages toward a pro-repair phenotype, thereby enhancing tissue regeneration and minimizing scar formation [3]. Achieving the right balance between immune activation for initial defense and immune resolution for tissue repair is crucial for successful biomaterial integration [27]. Ultimately, ensuring biocompatibility through these approaches not only reduces the risk of rejection and adverse effects but also enhances the regenerative potential of biomaterials in wound healing applications.

Limited Clinical Data and Long-Term Outcomes

In many cases, innovative wound care therapies, particularly those involving cell-based treatments like mesenchymal stem cells (MSCs) or extracellular vesicles (EVs), lack sufficient long-term clinical data to demonstrate their sustained efficacy over time. This absence of long-term studies and outcomes poses a barrier to their widespread acceptance by both clinicians and regulatory bodies. For India, where healthcare infrastructure varies significantly across regions, establishing the clinical data required to support these treatments is a difficult task. Clinical trials are expensive, and the lack of robust funding mechanisms for such studies means that large-scale trials are often not feasible. Without comprehensive evidence of the safety and efficacy of these treatments, healthcare providers may be reluctant to adopt them, especially when more traditional wound care methods are readily available [25].

Equity in Access and Infrastructure Disparities

Equity in access to advanced wound care technologies remains a crucial issue, particularly in low-resource settings. In India, there is a stark disparity between urban and rural areas in terms of access to innovative wound care treatments. Advanced technologies, such as hydrocolloid dressings, nanotechnology, and stem cell therapies, are often concentrated in urban centers with higher healthcare infrastructure. This concentration leaves rural populations, who bear a disproportionate burden of chronic wounds due to conditions such as diabetes and poor wound management practices, without access to cutting-edge treatments [24, 29]. Moreover, the lack of specialized healthcare professionals in rural India exacerbates the situation, as these innovations often require expert knowledge for proper application and monitoring.

In addition, India's healthcare system faces substantial gaps in infrastructure and training. Many hospitals and clinics, especially in rural areas, may lack the basic facilities needed for the effective delivery of advanced wound care. For instance, negative pressure wound therapy (NPWT) [30] and bioengineered dressings [31] require specialized equipment and trained staff, which are often unavailable in low-resource settings. Similarly, there is limited awareness among healthcare providers about the benefits of advanced treatments, which further restricts access.

Reimbursement Pathways and Cost Barriers

Another critical challenge in scaling up innovative treatments in India is the lack of clear reimbursement pathways. While advanced therapies offer great promise, the costs associated with them remain prohibitive for most patients, particularly those covered under government-sponsored healthcare schemes or lacking comprehensive health insurance. In many cases, patients are expected to pay out-of-pocket for treatments that are not covered by insurance or government health programs. This situation is particularly dire in a country like India, where the public health system is underfunded, and only a small portion of the population has access to private health insurance. Without reimbursement support or policy changes to include new wound care technologies in national health insurance plans, many patients in India will continue to rely on traditional, less effective treatments.

While India has made significant strides in healthcare and technology, the challenges of scaling up innovative wound care treatments are substantial. Addressing these barriers will require coordinated efforts between regulatory bodies, healthcare providers, research institutions, and policymakers. Additionally, creating awareness, establishing clear reimbursement pathways, and ensuring equitable access to advanced treatments for all populations, including rural and underserved communities, will be crucial to improving the nation's wound care outcomes. Through concerted action, India can overcome these challenges and harness the potential of innovative wound care therapies for better patient outcomes.

Conclusion

The field of wound healing is undergoing transformative changes, fueled by advancements in biotechnology, materials science, and clinical medicine. Cutting-edge solutions such as smart wound dressings, nanotechnology-based carriers, and stem cell-derived therapies hold the promise of revolutionizing chronic wound management and offering hope to patients facing persistent healing challenges. Among these innovations, the integration of artificial intelligence in wound diagnostics and monitoring is reshaping how clinicians assess, predict, and optimize treatment strategies with remarkable precision.

However, achieving the full potential of these technologies requires overcoming key barriers, including equitable access and robust clinical validation. In India, where healthcare disparities between urban and rural areas are profound, it is crucial to ensure that advanced wound care solutions are accessible to underserved populations. Regulatory reforms, infrastructure development, and policy initiatives must prioritize the adaptation of these treatments for resource-constrained settings, allowing for their widespread use and impact.

Looking ahead, the growing emphasis on personalized medicine and a deeper understanding of wound healing mechanisms brings us closer to achieving scar-free, regenerative healing for all patients. By fostering interdisciplinary collaborations among research institutions, healthcare providers, and policymakers, India has the opportunity to lead the way in integrating innovative treatments into mainstream healthcare. A focus on affordability, accessibility, and clinical validation will be instrumental in bridging the gap and transforming wound care for the better.

References

1. Nussbaum SR, Carter MJ, Fife CE, DaVanzo J, Haught R, Nusgart M et al (2018) An economic evaluation of the impact, cost, and medicare policy implications of chronic nonhealing wounds. Value Health 21(1):27–32. https://doi.org/10.1016/j.jval.2017.07.007
2. Armstrong DG, Swerdlow MA, Armstrong AA, Conte MS, Padula WV, Bus SA (2020) Five year mortality and direct costs of care for people with diabetic foot complications are comparable to cancer. J Foot Ankle Res 13(1):16. https://doi.org/10.1186/s13047-020-00383-2
3. Rodrigues M, Kosaric N, Bonham CA, Gurtner GC (2019) Wound healing: a cellular perspective. Physiol Rev 99(1):665–706. https://doi.org/10.1152/physrev.00067.2017
4. Korting HC, Schöllmann C, White RJ (2011) Management of minor acute cutaneous wounds: importance of wound healing in a moist environment. J Eur Acad Dermatol Venereol 25(2):130–137. https://doi.org/10.1111/j.1468-3083.2010.03775.x
5. Tottoli EM, Dorati R, Genta I, Chiesa E, Pisani S, Conti B (2020) Skin wound healing process and new emerging technologies for skin wound care and regeneration. Pharmaceutics 12(8):735. https://doi.org/10.3390/pharmaceutics12080735
6. Leclère FM (2017) The use of Integra® dermal regeneration template versus flaps for reconstruction of full-thickness scalp defects involving the calvaria: a cost-benefit analysis. Aesth Plast Surg 41(2):472–473. https://doi.org/10.1007/s00266-016-0765-z
7. Dreifke MB, Jayasuriya AA, Jayasuriya AC (2015) Current wound healing procedures and potential care. Mater Sci Eng C Mater Biol Appl 48:651–662. https://doi.org/10.1016/j.msec.2014.12.068
8. Okur ME, Karantas ID, Şenyiğit Z, Üstündağ ON, Siafaka PI (2020) Recent trends on wound management: new therapeutic choices based on polymeric carriers. Asian J Pharm Sci 15(6):661–684. https://doi.org/10.1016/j.ajps.2019.11.008
9. Díaz-García D, Filipová A, Garza-Veloz I, Martinez-Fierro ML (2021) A beginner's introduction to skin stem cells and wound healing. Int J Mol Sci 22(20):11030. https://doi.org/10.3390/ijms222011030
10. Leavitt T, Hu MS, Marshall CD, Barnes LA, Longaker MT, Lorenz HP (2016) Stem cells and chronic wound healing: state of the art. Chronic Wound Care Manag Res 3:7–27. https://doi.org/10.2147/CWCMR.S84369
11. Chang H-K, Kim P-H, Cho H-M, Yum S-Y, Choi Y-J, Son Y et al (2016) Inducible HGF-secreting human umbilical cord blood-derived MSCs produced via TALEN-mediated genome editing promoted angiogenesis. Mol Ther 24(9):1644–1654. https://doi.org/10.1038/mt.2016.120
12. Liu L, Yu Y, Hou Y, Chai J, Duan H, Chu W et al (2014) Human umbilical cord mesenchymal stem cells transplantation promotes cutaneous wound healing of severe burned rats. PLoS One 9(2):e88348. https://doi.org/10.1371/journal.pone.0088348
13. Nallakumarasamy A, Jeyaraman M, Maffulli N, Jeyaraman N, Suresh V, Ravichandran S et al (2022) Mesenchymal stromal cell-derived extracellular vesicles in wound healing. Life (Basel, Switzerland) 12(11):1733. https://doi.org/10.3390/life12111733
14. Monavarian M, Kader S, Moeinzadeh S, Jabbari E (2019) Regenerative scar-free skin wound healing. Tissue Eng Part B Rev 25(4):294–311. https://doi.org/10.1089/ten.TEB.2018.0350
15. Hosseini M, Shafiee A (2021) Engineering bioactive scaffolds for skin regeneration. Small (Weinheim an Der Bergstrasse, Germany) 17(41):e2101384. https://doi.org/10.1002/smll.202101384
16. Gounden V, Singh M (2024) Hydrogels and wound healing: current and future prospects. Gels 10(1):43. https://doi.org/10.3390/gels10010043
17. Lin X, Zhang X, Wang Y, Chen W, Zhu Z, Wang S (2025) Hydrogels and hydrogel-based drug delivery systems for promoting refractory wound healing: applications and prospects. Int J Biol Macromol 285:138098. https://doi.org/10.1016/j.ijbiomac.2024.138098
18. Oliva N, Conde J, Wang K, Artzi N (2017) Designing hydrogels for on-demand therapy. Acc Chem Res 50(4):669–679. https://doi.org/10.1021/acs.accounts.6b00536

19. Jose J, Khot KB, shastry P, Thomas SP, Chopra H, Gopan G et al (2023) Advances in microneedles-based drug delivery system on promoting wound healing. J Drug Deliv Sci Technol 90:105163. https://doi.org/10.1016/j.jddst.2023.105163
20. Ji M, Zhan F, Qiu X, Liu H, Liu X, Bu P et al (2024) Research progress of hydrogel microneedles in wound management. ACS Biomater Sci Eng 10(8):4771–4790. https://doi.org/10.1021/acsbiomaterials.4c00972
21. Sanshita P, Suman G, Vivek, Singh I (2023) Novel bio-inspired microneedles for wound healing applications. Expert Opin Drug Deliv 20(11):1463–1465. https://doi.org/10.1080/17425247.2023.2279116
22. Jian X, Deng Y, Xiao S, Qi F, Deng C (2025) Microneedles in diabetic wound care: multifunctional solutions for enhanced healing. Burns Trauma 13:tkae076. https://doi.org/10.1093/burnst/tkae076
23. Ennis WJ, Lee C, Gellada K, Corbiere TF, Koh TJ (2016) Advanced technologies to improve wound healing: electrical stimulation, vibration therapy, and ultrasound-what is the evidence? Plast Reconstr Surg 138(3 Suppl):94S–104S. https://doi.org/10.1097/PRS.0000000000002680
24. Margolis DJ, Hoffstad O, Nafash J, Leonard CE, Freeman CP, Hennessy S et al (2011) Location, location, location: geographic clustering of lower-extremity amputation among Medicare beneficiaries with diabetes. Diabetes Care 34(11):2363–2367. https://doi.org/10.2337/dc11-0807
25. Boyce ST, Lalley AL (2018) Tissue engineering of skin and regenerative medicine for wound care. Burns Trauma 6:4. https://doi.org/10.1186/s41038-017-0103-y
26. Han G, Ceilley R (2017) Chronic wound healing: a review of current management and treatments. Adv Ther 34(3):599–610. https://doi.org/10.1007/s12325-017-0478-y
27. Salthouse D, Novakovic K, Hilkens CMU, Ferreira AM (2023) Interplay between biomaterials and the immune system: challenges and opportunities in regenerative medicine. Acta Biomater 155:1–18. https://doi.org/10.1016/j.actbio.2022.11.003
28. Sinno H, Prakash S (2013) Complements and the wound healing cascade: an updated review. Plast Surg Int 2013:146764. https://doi.org/10.1155/2013/146764
29. Krisp C, Jacobsen F, McKay MJ, Molloy MP, Steinstraesser L, Wolters DA (2013) Proteome analysis reveals antiangiogenic environments in chronic wounds of diabetes mellitus type 2 patients. Proteomics 13(17):2670–2681. https://doi.org/10.1002/pmic.201200502
30. Saini R, Jeyaraman M, Jayakumar T, Iyengar KP, Jeyaraman N, Jain VK (2023) Evolving role of negative pressure wound therapy with instillation and dwell time (NPWTi-d-) in management of trauma and orthopaedic wounds: mechanism, applications and future perspectives. Indian J Orthop 57(12):1968–1983. https://doi.org/10.1007/s43465-023-01018-x
31. Tran HQ, Shahriar SMS, Yan Z, Xie J (2023) Recent advances in functional wound dressings. Adv Wound Care 12(7):399–427. https://doi.org/10.1089/wound.2022.0059

Chapter 27
Nanobiology of Cutaneous Wound Healing

Mohini Moulick, Aparajita Nandy, Pramod Bhasme, Shayani Mukherjee, Anita Yadav, and Subhadip Ghatak

Abstract The restoration of cutaneous tissue integrity following injury is a temporally orchestrated and biologically complex process, underpinned by a dynamic interplay among blood-borne immune cells, resident stromal and epithelial cells, soluble mediators, and extracellular matrix remodeling. Recent advances in nanobiology have highlighted the pivotal role of cell-specific endosome-originated small extracellular vesicles (sEV), also known as exosomes, as fundamental mediators of intercellular communication. These sEV facilitate the targeted transfer of bioactive cargos, including regulatory ribonucleic acid (RNAs), proteins, and lipids, which are instrumental in key reparative processes such as inflammation resolution, angiogenesis, and epithelial regeneration. In this chapter, we critically evaluate the emerging functions of sEV in cutaneous wound healing, elucidating their mechanistic roles and therapeutic potential. We delineate a series of bidirectional and tripartite signaling circuits between resident cells and infiltrating immune cells for effective tissue restitution. Disruption of these sEV-mediated exchanges leads to the pathophysiology of chronic, nonhealing wounds, characterized by persistent inflammation, fibroblast dysfunction, and epithelial staling. Concurrently, we investigate how bacterial outer membrane vesicles (OMVs) utilize the structural and functional mimicry of host exosomes to subvert regenerative signaling, amplify local inflammation, and facilitate biofilm formation, thereby making the bacteria recalcitrant to antimicrobial therapy. These microbial vesicles engage in sophisticated interkingdom communication, thereby reshaping the wound microenvironment to favor chronicity and immune evasion. Finally, we explore the translational trajectory of bioengineered exosomes and synthetic sEV mimetics as emerging nanotherapeutic platforms. These vesicles are being strategically engineered to exhibit enhanced cellular tropism, extended systemic persistence, and stimulus-responsive release kinetics. By incorporating precision cargos such as small interfering RNAs (siR-

M. Moulick · A. Nandy · P. Bhasme · S. Mukherjee · A. Yadav · S. Ghatak (✉)
McGowan Institute for Regenerative Medicine, Department of Surgery, University of Pittsburgh, Pittsburgh, PA, USA
e-mail: ghataks@pitt.edu

F. De Francesco, K. Turksen (eds.), *Management and Strategies for Wound Healing*, Stem Cell Biology and Regenerative Medicine 79,
https://doi.org/10.1007/978-3-032-12442-5_27

NAs), microRNAs (miRNAs), Clustered Regularly Interspaced Palindromic Repeats–associated protein (CRISPR-Cas9) gene-editing components, and immunomodulatory proteins, these next-generation vesicular systems hold significant promise for the targeted modulation of pathological signaling pathways in chronic wounds and for advancing cell-specific regenerative interventions.

Keywords Small extracellular vesicles · Cell-specific exosomes · Intercellular communication · Microbial vesicles · Chronic wounds · Wound healing · Tissue repair · Regenerative medicine

Abbreviations

Akt	Protein kinase B
AMP	Antimicrobial peptides
AP-1	Activator protein-1
CRISPR	Clustered Regularly Interspaced Short Palindromic Repeats
ADSC	Adipocyte-derived stem cells
CD	Cluster of Differentiation
CDC	Centers for Disease Control and Prevention
cGAS	Cyclic GMP-AMP synthase
GMP-AMP	Guanosine monophosphate-adenosine monophosphate
COVID-19	Coronavirus disease
DAMPs	Damage-associated molecular patterns
DNA	Deoxyribonucleic acid
ECM	Extracellular matrix
OMV	Outer membrane vesicles
eNOS	Endothelial nitric oxide synthase
EDC/NHS	1-ethyl-3-(3-dimethylaminopropyl)carbodiimide/*N*-hydroxysuccinimide
SDF-1	Stromal cell-derived factor-1
sEV	Small extracellular vesicles
GMP	Good Manufacturing Practices
hnRNPA2B1	Heterogeneous nuclear ribonucleoprotein A2B1
HIF-1α	Hypoxia-inducible factor 1-alpha
IL	Interleukin
ILVs	Intraluminal vesicles
LAMP-2B	Lysosome-associated membrane protein
LPS	Lipopolysaccharides
M1	Pro-inflammatory macrophage
M2	Pro-resolution macrophage
MASH	Metabolic dysfunction-associated steatohepatitis
MAPK	Mitogen-activated protein kinase
miRNA	MicroRNA
MISEV	Minimal Information for Studies of Extracellular Vesicles

MMP	Matrix metalloproteases
MPS	Mononuclear phagocyte systems
MSC	Mesenchymal stem cells
MVBs	Multivesicular bodies
NF-κB	Nuclear factor kappa-light-chain-enhancer of activated B cell
NOD	Nucleotide-binding oligomerization domain
NTA	Nanoparticle tracking analysis
PAMP	Pathogen-associated molecular patterns
PPARγ	Peroxisome proliferator-activated receptors
PRR	Pattern recognition receptors
PVAT	Perivascular adipose tissue
ROS	Reactive oxygen species
RGD	Arginylglycylaspartic acid
RNA	Ribonucleic acid
RVG	Rabies virus glycoprotein
siRNA	Small interfering RNA
siMDM2	Small interfering RNA targeting Murine double minute gene 2
SOX9	SRY-box transcription factor 9
STAT3	Signal transducer and activator of transcription 3
STING	Stimulator of interferon genes
TBK1	Tank binding kinase 1
TGF-β	Transforming growth factor beta
TLRs	Toll-like receptors
TNF-α	Tumor necrosis factor-alpha
TOMM70	Translocase outer mitochondrial membrane 70
TSG101	Tumor susceptibility gene 101
VEGF	Vascular endothelial growth factor
VSMC	Vascular smooth muscle cells
WNT/β	Wingless-related integration site/beta

Introduction

From the moment terrestrial life evolved a vascularized integument, the ability to rapidly seal breaches in that barrier became essential for survival [1–3]. In humans, the skin, which spans roughly 1.8 m^2 in adults, protects against dehydration, pathogens, and mechanical insults [4]. Nevertheless, it is persistently susceptible to damage induced by mechanical or thermal injury, iatrogenic trauma, and diverse systemic pathologies [5]. Modern health-care systems allocate billions of dollars annually to treat acute and chronic wounds [6]. The epidemiological burden is rising in parallel with aging populations, diabetes, obesity, and vascular disease [7–9]. Against this backdrop, a detailed knowledge of the wound-healing process is no longer merely academic; it is essential. Wound healing forms the foundation for a spectrum of emerging technologies, including stem cell-based regenerative therapeutics, bioengineered dermal scaffolds, epidermal biosensors, and individualized molecular interventions [10, 11].

Cutaneous repair is best conceptualized as an overlapping cascade with four interdependent phases: hemostasis, inflammation, proliferation, and remodeling [13, 14]. Each of these phases is orchestrated by a dynamic interplay of resident cells, infiltrating immune cells, extracellular matrix (ECM) components, and soluble mediators [15]. Seconds after injury, platelets aggregate and trigger the coagulation cascade, forming a fibrin-rich provisional matrix that both staunches blood loss and concentrates cytokines and growth factors [16]. Within hours, neutrophils and monocytes flood the wound, executing antimicrobial defense and clearing debris, while macrophages emerge as conductors of subsequent tissue renewal [17]. Days later, fibroblasts lay down nascent ECM, endothelial cells sprout new capillaries, and keratinocytes migrate to re-establish epidermal continuity [18]. Over weeks to months, myofibroblast-driven contraction, collagen remodeling, and scar maturation restore tensile strength albeit seldom matching that of uninjured skin [19].

Coagulation Cascade
The blood clotting process starts with a series of enzyme reactions, which result in clot formation to stop bleeding. The intrinsic and extrinsic pathways merge to activate thrombin through the common pathway. Thrombin transforms fibrinogen into fibrin which forms a stable blood clot [12].

Successful progression through these stages hinges on adequate perfusion, controlled inflammation, and synchronized cellular crosstalk [20]. When intrinsic factors such as diabetes-induced microangiopathy, immunosenescence, or genetic syndromes intersect with extrinsic challenges like bacterial biofilms or repeated pressure, the cascade stalls [9, 21]. The result is wound chronicity, a wound that lingers beyond 4 weeks, entrenching the patient in cycles of pain and infection [22]. Conversely, maladaptive acceleration, such as excess growth factor signaling or aberrant myofibroblast persistence, can yield hypertrophic scars or keloids, underscoring the need for balanced regulation [23].

At the molecular level, the old paradigm of single “magic-bullet” mediators has given way to network biology [24]. The miRNAs can fine-tune hundreds of target transcripts; reactive oxygen species (ROS) serve dual roles as bactericidal agents and redox signals; and extracellular vesicles (EVs) shuttle protein and RNA cargo across cell borders [25, 26]. Concurrently, emerging technologies such as CRISPR-engineered skin grafts, tissue nanotransfection, spatial transcriptomics, and AI-driven image analytics are redefining how we interrogate and manipulate wound biology in vivo [27]. Equally transformative is the growing emphasis on patient-centric metrics and not merely “time to closure,” but restoration of skin barrier function, recurrence-free intervals, and quality-of-life indices [28].

Note: For further reading, refer to (29) Chapter 14; Wound Healing by Sen C. K, and Roy S. in Plastic Surgery: Volume 1: Principles 3rd Edition by Geoffrey C Gurtner, Peter C. Neligan* [29]

This chapter revisits the pathophysiology of cutaneous wound healing through the emerging paradigm of nano-EV-mediated communication, with a focus on how exosomes and other small EV orchestrate intercellular signaling networks that regulate inflammation, angiogenesis, fibroblast activation, and epithelial regeneration [30–32]. By integrating recent advances in nanobiology and vesicle engineering, we elucidate how dysregulation of these communication pathways contributes to chronic wound persistence and how their therapeutic modulation offers novel avenues for regenerative intervention [30, 31].

The Biological Code for Translating Damage into Repair

Cutaneous wound healing unfolds not as a random cascade, but as a precisely timed molecular dialogue, a cellular consensus forged under stress [13, 33]. The choreography of soluble cytokines, matrix dynamics, and sequential cellular recruitment is, at its core, a manifestation of highly coordinated cellular crosstalk at the site of injury [34]. This intercellular dialogue is not incidental, but it is deterministic, context-sensitive, and evolutionarily optimized to restore tissue integrity. Each cell type involved, from resident keratinocytes and fibroblasts to endothelial cells and macrophages, operates as both a sender and a receiver, continuously updating its status based on the microenvironmental cues it detects [35]. These cell types contribute to a shifting molecular landscape, responding not only to gradients of diffusible signals like interleukin-1 beta (IL-1β), tumor necrosis factor-alpha (TNF-α), and transforming growth factor beta (TGF-β), but also to biomechanical cues from the ECM and metabolic regulators such as hypoxia-inducible factor 1-alpha (HIF-1α) [36]. But how do these cells decide when to act, what to express, and which pathways to engage? To answer this, we need to decode the cellular crosstalk executed through layered modalities such as juxtacrine contact, paracrine diffusion, and increasingly recognized, exosome-mediated signal transduction [37]. These nanosized vesicles serve as circuit-integrated messengers, capable of reprogramming recipient cells by transferring functional biomolecular cargo and organellar components such as mitochondrial proteins [38]. As a result, macrophages, keratinocytes, fibroblasts, and endothelial cells do not merely act in linear succession; they operate in a dynamic, bidirectional communication network [33]. It is this real-time, reciprocal exchange of biochemical information that determines the pace, quality, and outcome of wound repair [33].

Note: For further reading, refer to (33) Peña, O. A. & Martin, P. Cellular and Molecular Mechanisms of Skin Wound Healing. Nat Rev Mol Cell Biol 25, 599–616 (2024).*

Soluble Cytokines
The immune system uses soluble cytokines as small signaling proteins to control immune responses, inflammation, and cell communication. The proteins function by attaching to receptors on target cells to trigger growth. These exist in free form without attaching to cell surfaces. The cytokine family includes interleukins together with interferons and tumor necrosis factors as examples [39].

Keratinocyte-Macrophage Reciprocity

Immediately after injury, basal and suprabasal keratinocytes at the wound-edge(WE) SUMOylated the RNA-binding protein heterogeneous nuclear ribonucleoprotein A2B1 (hnRNPA2B1), responsible for packaging miRNA within the endosomal compartment [9, 40, 41]. This post-translational modification serves as a molecular barcode, guiding the selective packaging of pro-resolution miRNAs, such as miR-21, miR-146a, and miR-132, into budding multivesicular bodies (MVBs) [9, 38]. The resulting vesicles within MVB, hereafter Exo_{κ}, are released and traverse the epidermal-dermal junction [38, 41, 42]. These nanovesicles are taken up by the infiltrating blood-borne macrophages during the first 48 h following injury [41, 43, 44]. Once internalized, the miRNA payload gradually dampens the inflammatory signaling, lowers ROS formation, and accelerates the phenotypic switch from classically activated pro-inflammatory (M1) macrophages to their reparative pro-resolution (M2) counterparts [41]. Epidermal knockdown of hnRNPA2B1, by targeted lipid-nanoparticle siRNA, disrupts this export, prolongs inflammatory gene expression, and delays barrier restitution, thereby demonstrating the evidence that Exo_{κ} constitutes a nonredundant brake on innate immunity [41, 45].

SUMOylation
The post-translational modification process of SUMOylation involves the covalent attachment of Small Ubiquitin-like Modifier (SUMO) proteins to target proteins. The modification process affects how proteins exist in cells by altering their stability and localization patterns and interaction capabilities without leading to protein breakdown. The process serves as an essential mechanism for controlling DNA repair functions, as well as gene expression and cellular stress response mechanisms [46].

On the dermal front, the infiltrating blood-borne macrophages, once internalized the Exo_{κ}, initiate a reciprocal communication axis by selectively enriching their endosome-originated sEVs or exosomes, hereafter $Exo_{m\varphi}$, with a mitochondrial import receptor protein translocase of the outer mitochondrial membrane 70 (TOMM70) [41]. The TOMM70 is essential for protein translocation into the inner mitochondrial matrix [47]. This cargo selection within the $Exo_{m\varphi}$ occurs independently of mitophagy or mitochondrial fragmentation, suggesting the presence of a specialized vesicular trafficking mechanism responsible for sorting mitochondrial components [48]. These metabolically impaired epithelial cells exhibit a high avidity for $Exo_{m\varphi}$ uptake [49, 50]. Once internalized, TOMM70 rapidly integrates into the outer mitochondrial membrane, restoring import competency, enhancing oxidative phosphorylation, and reactivating glycolytic flux critical for keratinocyte proliferation and migration [49, 50]. This metabolic rescue, driven by macrophage altruism, reinstates actin cytoskeletal organization and promotes directed keratinocyte migration at the leading edge of the epithelial tongue [51]. Experimental blockade of exosome uptake or conditional deletion of TOMM70 in myeloid cells arrests WE keratinocyte migration and results in impaired re-epithelialization, effectively converting otherwise self-limited acute wounds into persistent, nonhealing ulcers [50]. Importantly, clinical specimens from chronic diabetic foot ulcers reveal a profound depletion of TOMM70-positive vesicles at the WE, supporting the translational relevance of this mechanism [50].

Taken together, these studies delineate a reciprocal regulatory circuit in which keratinocytes initiate the resolution of inflammation *via* the release of miRNA-enriched Exo_{κ} that suppresses macrophage pro-inflammatory activity [50]. In turn, macrophages return the molecular favor by exporting mitochondria-associated cargo such as TOMM70 within $Exo_{m\varphi}$, thereby restoring the metabolic competency and migratory potential of hypoxic keratinocytes at the WE [50]. The two axes, keratinocyte-to-macrophage and macrophage-to-keratinocyte, form a tightly regulated, bidirectional exosomal circuit critical for orchestrating sequential transitions in the wound healing cascade [41, 50]. The loop thus offers a molecular explanation for the persistent inflammation and metabolic exhaustion, the dual hallmarks of chronic wounds, and suggests that successful healing hinges on the timely hand-off between these nanoscopic couriers [50, 52].

Note: For further reading, refer to (50) Sharma, A. et al. *Mitochondrial Bioenergetics of Functional Wound Closure is Dependent on Macrophage-Keratinocyte Exosomal Crosstalk. ACS Nano 18, 30405–30420 (2024).***

Regeneration Dialogues Among Other Cell Types

Beyond the well-characterized interplay between keratinocytes and macrophages, tissue regeneration is orchestrated by a complex network of vesicle-mediated communications involving vascular cells, endothelial cells, adipocytes, and fibroblasts [53]. These interactions modulate cell fate and activity in processes fundamental to vascular restructuring, ECM remodeling, and epithelial restoration [54–57].

Bidirectional exosomal crosstalk between endothelial cells and vascular smooth muscle cells (VSMCs) influences vascular homeostasis [54–56]. Endothelial-derived exosomes enriched in TGF-β1 and miR-221-3p are internalized by VSMCs, where they drive a phenotypic shift toward a synthetic, ECM-producing and migration-competent state [57]. While this facilitates adaptive vascular remodeling, such as in revascularization of ischemic tissues, chronic activation of this loop as seen in atherosclerosis or persistent inflammation can tip the balance toward maladaptive changes [54]. Conversely, VSMC-derived exosomes, particularly those encapsulating miR-155, have deleterious effects on vascular integrity [54]. These vesicles, upon uptake by endothelial cells, impair endothelial junction maturity, increase permeability, and elevate the expression of adhesion molecules and inflammatory mediators, thereby amplifying vascular inflammation and promoting endothelial dysfunction [54]. This bidirectional exosomal traffic between endothelial and mural cells is thus a double-edged sword, supportive of physiological repair but detrimental when dysregulated in chronic wounds or metabolic disease [52, 57].

Perivascular adipose tissue (PVAT), which envelops the vasculature without a separating fascia, exosomes become particularly enriched with miRNA cargo, such as miR-221-3p, under inflammatory or obese conditions [58]. This exosomal miRNA has been shown to modulate the phenotype of VSMCs, promoting their proliferation, migration, and phenotypic switching from a contractile to a synthetic state [58]. This shift contributes to early vascular remodeling, a crucial yet tightly regulated process in wound angiogenesis [58]. However, in pathological settings such as diabetes or obesity, this signaling becomes dysregulated, leading to impaired insulin signaling through attenuation of the protein kinase Akt/eNOS (Protein kinase B/endothelial nitric oxide synthase) pathway, ultimately promoting vascular dysfunction and inflammation [59, 60].

Adipose-derived stem cell (ADSC)-derived exosomes have potent regenerative effects, particularly on dermal fibroblasts [61]. These exosomes modulate fibroblast behavior by delivering regulatory RNAs that suppress miR-19b, a repressor of SRY-box transcription factor 9 (SOX9), thereby enabling SOX9 activation [62]. Activated SOX9, in turn, promotes Wingless-related integration site/beta-catenin pathway (Wnt/β-catenin pathway), a critical axis for fibroblast proliferation, collagen synthesis, and matrix remodeling [62]. These ADSC exosomes are also enriched with trophic factors such as matrix metalloproteases (MMPs), stromal cell-derived factor-1 (SDF-1), TGF-β, vascular endothelial growth factor (VEGF), and interleukin-10 (IL-10), which collectively support fibroblast-mediated dermal regeneration while suppressing pro-inflammatory cytokines like interleukin-6 (IL-6), thereby tempering the inflammatory *milieu* [62]. In the later stages of wound healing, dermal fibroblasts secrete exosomes that direct keratinocyte behavior during re-epithelialization and epidermal stratification [61]. These vesicles carry bioactive components including miR-21, a miRNA known to enhance keratinocyte migration, proliferation, and survival [63]. Furthermore, surface expression of proteins such as cluster of differentiation 98 heavy chain (CD98hc), an integrin co-receptor, modulates keratinocyte adhesion dynamics and matrix interactions, facilitating epidermal closure [64]. These exosomes serve as integrative messengers, coordinating the cellular choreography required for efficient ECM remodeling, basement membrane deposition, and epidermal barrier restoration [38, 65].

Wnt/β-Catenin Pathway
It is a pathway that modulates essential cellular activities such as proliferation, differentiation, protein expression, migration, autophagy, and stem cell turnover in both embryonic development and adult tissues [66]. The Wnt signals preserve beta-catenin by blocking its degradation. It does so as Wnt ligands bind to their receptors, such as Frizzled and LRP5/6 [66]. Beta-catenin aggregation induces gene transcription [66].

Taken together, these findings underscore a complex, highly orchestrated network of exosome-mediated communications among adipocytes, vascular cells, fibroblasts, and keratinocytes. These dialogues are essential for maintaining tissue integrity and promoting successful wound resolution, and their dysregulation may contribute to the pathogenesis of chronic wounds and fibrotic disorders. Incorporating the cell-type-specific messages and their context-dependent impact on regenerative outcomes offers exciting therapeutic potential for engineering exosome-based interventions in wound healing and beyond.

Healing Trialogues

Within the wound microenvironment, while many cell–cell dialogues have been individually characterized, it is now increasingly evident that the healing trajectory involves multilayered communication among diverse cellular constituents each responding to and influencing others in a coordinated yet context-dependent manner. In this section, we focus on four critical tripartite interactions or "healing trialogues" where the coordinated action of three distinct cell types is essential for restoring tissue integrity. These triads exemplify how cellular communication orchestrated *via* exosomes and soluble factors determines the balance between regeneration and fibrosis, particularly under pathological conditions such as diabetes or chronic inflammation (Fig. 27.1).

Fibroblast-Macrophage-Keratinocyte Circuit: Linking Inflammation Control to Re-epithelialization

The interplay between fibroblasts, macrophages, and keratinocytes represents an essential healing trialogue, particularly at the interface of inflammation and re-epithelialization [41, 67–69]. Fibroblast-derived exosomes have been shown to influence macrophage polarization *via* the miRNA *miR-93-5p* [68]. Under diabetic conditions, elevated levels of miR-93-5p in fibroblast exosomes perpetuate M1 macrophage polarization by enhancing autophagic flux, thus sustaining a pro-inflammatory environment [68, 69]. This aberrant signaling delays the phenotypic

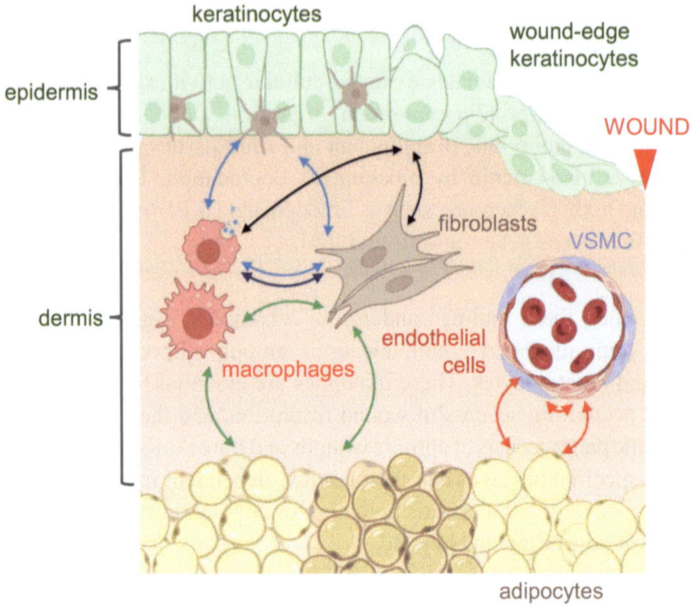

Fig 27.1 Healing Trialogues, multicellular communication network. Fibroblast, macrophage and keratinocyte circuit communication is essential for re-epithelialization. Adipocyte, macrophage, and fibroblast axis works to resolve inflammation and induce matrix remodeling. Endothelial cells, VSMC, and adipocyte triad that leads to vascular remodeling. Melanocyte, fibroblast-immune axis (macrophage or T-cells) communication leading to fibrotic scarring

shift to reparative M2 macrophages. Notably, silencing *miR-93-5p* in fibroblasts promotes the conversion of macrophages to the M2 subtype [68]. These reprogrammed macrophages then secrete exosomes enriched with reparative noncoding RNAs that act on basal keratinocytes, enhancing their proliferation, migration, and activation, thereby promoting re-epithelialization [41]. Disruption at any node in this fibroblast-macrophage-keratinocyte triad, whether through impaired exosomal release, altered cargo content, or dysfunctional uptake, can derail the healing process [41, 67–69]. The result may be persistent inflammation, inadequate barrier restoration, or even fibrotic remodeling [61, 67, 69]. Therapeutically targeting miRNA-regulated exosomal communication within this circuit offers a novel strategy to restore healing fidelity in chronic wounds [9, 61, 67].

Autophagic Flux

Cellular degradation and a cell's ability to recycle its constituents *via* autophagy are known as autophagic flux [70]. It can be measured by the rate of flow, Western blot, and transmission electron microscopy [70]. It is essential to understand the rate of autophagic flux as when it is disrupted it leads to neurodegenerative diseases, cardiovascular diseases, and tumorigenesis [70].

ADSC-Macrophage-Fibroblast Axis: A Circuit for Inflammation Resolution and Matrix Remodeling

The ADSCs play a pivotal role in cutaneous regeneration by modulating immune and stromal compartments through paracrine mechanisms [71]. Among these, ADSCs have been shown to promote the polarization of macrophages toward an anti-inflammatory M2 phenotype [71]. Mechanistically, this reprogramming occurs *via* activation of the peroxisome proliferator-activated receptors gamma/signal transducer and activator of transcription 3 (PPARγ/STAT3) signaling cascade [72]. Once polarized, M2 macrophages secrete TGF-β1, a potent cytokine known to stimulate fibroblast proliferation, migration, and ECM synthesis [72]. This coordinated ADSC-macrophage-fibroblast axis is especially relevant in diabetic wound environments, where chronic inflammation and impaired fibroblast function hinder proper tissue repair [71, 72]. By restoring the M2 macrophage phenotype and promoting fibroblast-driven matrix deposition, this triad effectively bridges the transition from the inflammatory to the proliferative phase of healing. Targeting this axis holds promise for resolving delayed or nonhealing wounds commonly observed in diabetic individuals [71, 72].

PPARγ/STAT3
Inhibition of STAT3 selectively is essential for cutaneous wound healing [73]. The event upregulated PPARs on activation lead to STAT3 inhibition by phosphorylation and impair inflammatory responses that lead to expedited tissue repair. This is especially true in osteoarthritis [73]. Therefore, it is essential for the orchestration of wound healing [73].

Endothelial-VSMC-Adipocyte Triad: Coordinating Vascular Remodeling in Wound Healing

Angiogenesis and vascular remodeling are indispensable for effective wound healing, and these processes are governed by a triad involving endothelial cells, VSMCs, and PVAT adipocytes [58–60, 74–76]. Following injury, endothelial cells initiate angiogenic sprouting, releasing exosomes that influence nearby stromal and immune cells [77]. Concurrently, PVAT adipocytes secrete exosomes carrying regulatory miRNAs such as miR-143/145 and miR-221-3p, which directly modulate VSMC phenotypes, balancing the switch between contractile and synthetic states that are crucial for vessel stabilization and remodeling [59, 60]. In the context of obesity or diabetic wounds, this triad becomes dysregulated. Adipocyte-derived exosomes may become pro-inflammatory or lose key regulatory cargo, leading to aberrant VSMC behavior and endothelial dysfunction [58–60]. This results in impaired perfusion, reduced vascular integrity, and delayed tissue regeneration [58–60]. The endothelial-VSMC-adipocyte trialogue underscores how tissue-specific metabolic cues intersect with vascular biology. Exploring how exosomal signals from metabolically active cells like adipocytes influence vascular remodeling could unveil new therapeutic targets to normalize angiogenesis in chronic or ischemic wounds [58–61, 74–76].

Melanocyte-Fibroblast-Immune Axis: A Hidden Triad in Fibrotic Scarring and Ethnic Variability

While melanocytes are traditionally viewed as pigment-producing cells, emerging data suggest their exosomal output may have functional consequences on fibroblast behavior and immune activation, particularly in the context of fibrotic wound healing [78]. Melanocyte-derived exosomes may be enriched in factors that potentiate TGF-β/Smad signaling, a central axis in fibroblast activation and ECM overproduction [78]. This could predispose certain individuals to hypertrophic scars or keloids, especially in melanin-rich skin types [78]. Immune cells, particularly macrophages and T cells, are also likely influenced by melanocyte signals, either directly or through fibroblast-mediated crosstalk [78]. The cumulative effect is a pro-fibrotic *milieu* characterized by chronic inflammation and unchecked matrix deposition [78]. Although understudied, this melanocyte-fibroblast-immune cell triad may explain racial disparities observed in scarring patterns and responses to wound healing [78]. Further investigation into this axis could not only deepen our knowledge of fibrotic remodeling but also guide the development of personalized wound therapies based on ethnic and cellular composition [78]. In essence, the concept of "healing trialogues" reveals that tissue repair is neither a linear cascade nor the product of single cell–cell contacts, but rather the emergent property of a highly coordinated network in which immune, stromal, vascular, and epithelial cells exchange a multitude of signals, both soluble and vesicular across space and time. However, the wound microenvironment is not solely sculpted by host-derived signals. In the presence of infection, particularly in chronic or nonhealing wounds, microbial inhabitants become active participants in the signaling landscape. Bacterial pathogens actively engage in interkingdom communication through the release of OMVs predominantly released by Gram-negative bacteria. These vesicles are structurally akin to host exosomes but carry distinct cargos that can profoundly alter the trajectory of wound healing. The following section explores how bacterial OMVs subvert host regenerative dialogues, rewire immune responses, and contribute to the pathophysiology of chronic, nonhealing wounds.

Nanovesicular Deception in Tissue Repair Signaling by Microbial Snoopers

Chronic wounds present a complex immunopathological landscape, where microbial colonization does not merely coexist with impaired healing but actively subverts regenerative processes. A growing body of evidence implicates nanosized bilayered proteoliposomal bacterial OMVs, released predominantly by Gram-negative pathogens as critical effectors in this sabotage. These OMVs mimic host-derived sEV in size, structure, and functional dynamics, enabling them to hijack endogenous intercellular communication networks that ordinarily coordinate wound repair.

The Fall of the Vesicular Sentinel as Host Defenses Overrun by Microbial Deception

Extracellular vesicles, particularly those secreted by epithelial cells, neutrophils, and macrophages, have emerged as critical mediators of innate immunity through their ability to encapsulate and deliver antimicrobial peptides (AMPs), including cathelicidins (e.g., LL-37), β-defensins, and lactoferrin [79]. Encapsulation of AMPs within sEV confers several functional advantages. It enhances their stability against proteolytic degradation, enables targeted delivery to infected or inflamed tissues, and facilitates synergistic immunomodulatory effects. Upon arrival at the site of infection, AMP-loaded sEV disrupt bacterial membranes, and recruit additional immune effectors, to extend the spatial range and functional precision of host defense mechanisms [80]. However, in chronic infected wounds such as those caused by *Pseudomonas aeruginosa* or *Staphylococcus aureus,* the pathogens have evolved sophisticated strategies to evade or counteract these vesicle-mediated antimicrobial responses [81]. One such strategy involves the secretion of bacterial OMVs, which act as decoys or molecular traps and neutralize host-derived AMPs [82] These OMVs contain surface-exposed lipopolysaccharides (LPS), proteases, and decoy lipoproteins that bind or degrade AMPs, effectively neutralizing their bactericidal activity [83]. In addition, some bacterial species remodel their membrane architecture by modifying lipid A acylation or increasing surface charge to reduce susceptibility to AMP-mediated membrane disruption [84]. Compounding this microbial evasion is the dysregulation of host sEV biogenesis during prolonged or pathologically inflamed conditions, such as diabetic or ischemic wounds. Under such conditions, the production, cargo sorting, and release of AMP-enriched sEV are compromised [85]. This leads to a decrease in vesicular AMP content and a shift toward pro-inflammatory, but non-antimicrobial, sEV phenotypes [86]. The loss of antimicrobial sEV functionality creates a permissive environment for microbial persistence and biofilm maturation, thereby accelerating the transition from a controlled to a chronic infection state [87].

Antimicrobial Peptides

Antimicrobial peptides (AMPs) are small, naturally occurring molecules produced by many organisms as part of their innate immune system [88]. They display broad-spectrum activity against bacteria, fungi, viruses, and even cancer cells by disrupting microbial cell membranes [88]. They are fast-acting and less prone to resistance development related to traditional antibiotics [88]. Their multifunctional roles also incorporate immune modulation and wound healing support [88].

Emerging evidence suggests that this dysfunction in AMP-EV-mediated defense likely precedes and facilitates the pathogenic dominance of bacterial OMVs. In the early stages of infection, AMP-loaded sEV act as a critical line of defense; however, their failure, due to either impaired host sEV production or active microbial neutralization, enables OMVs to exert control over the local microenvironment [89, 90]. These bacterial vesicles not only degrade residual AMPs but also interfere with sEV uptake pathways and modulate host immune responses to favor bacterial survival and tissue damage [91]. Collectively, this vesicular tug-of-war highlights a critical and dynamic axis of host-pathogen interaction. Thoroughly studying the kinetics, cargo specificity, and immunological impact of AMP-EVs and OMVs in various infection contexts may uncover new therapeutic opportunities aimed at restoring host antimicrobial defenses and interrupting vesicle-mediated pathogen sabotage in chronic wound environments.

Outer Membrane Vesicles

Outer membrane vesicles (OMVs) are bilayered structures from Gram-negative bacteria that enhance biofilms, resist antibiotics, and modulate immunity [92]. They transport enzymes, adhesins, and pathogen-associated molecular patterns (PAMPs), triggering Toll-like receptors (TLRs) and inflammatory pathways [92]. They disrupt epithelial repair, promote keratinocyte death, and impair immune responses by inhibiting chemokines and neutrophil recruitment, contributing to chronic infection and tissue damage [92].

Hijacked Healing

Unlike host sEV, particularly those secreted by resident cells and infiltrating immune cells, bacterial OMVs exploit this system to propagate infection, often leading to biofilm formation, suppress immune clearance, and delay wound closure,[93, 94] when OMVs gain access to host cells by co-opting canonical sEV uptake mechanisms such as clathrin-mediated endocytosis, caveolin-associated entry, and lipid raft-dependent internalization [95]. Once internalized, OMVs enriched with pathogen-associated molecular patterns (PAMPs), including LPS, peptidoglycan, and bacterial outer membrane proteins, activate pattern recognition receptors (PRRs) such as Toll-like receptors (TLRs) and nucleotide-binding oligomerization domain (NOD) receptors on host cells [91]. This initiates a robust cascade of downstream signaling through nuclear factor kappa-light-chain-enhancer of activated B cells (NF-κB), mitogen-activated protein kinase (MAPK), and activator protein-1 (AP-1) pathways, culminating in sustained production of pro-inflammatory cytokines such as IL-1β, IL-6, and TNF-α. These inflammatory cues not only override the immunoregulatory cargo typically delivered by anti-inflammatory host sEV (e.g., miR-146a, TGF-β), but also establish a microenvironment that favors chronicity and microbial persistence [96].

Clathrin-Mediated Endocytosis

Clathrin-mediated endocytosis is a cellular process where clathrin proteins form coated pits to internalize specific molecules [97]. Receptors on the plasma membrane capture cargo, and clathrin assembles to curve the membrane inward [97]. Dynamin helps pinch off the vesicle, which then sheds its coat and fuses with endosomes [97]. This pathway is vital for nutrient uptake, receptor recycling, and signal regulation [97].

In addition to initiating inflammatory cascades, bacterial OMVs also induce transcriptional and post-transcriptional reprogramming of host sEV biogenesis. For instance, in keratinocytes, OMV exposure leads to the release of host sEV aberrantly enriched with ROS, damage-associated molecular patterns (DAMPs), and pro-inflammatory miRNAs (e.g., miR-155) [82]. These vesicles act as amplifiers of sterile inflammation, thereby perpetuating local immune activation and impeding re-epithelialization. Simultaneously, OMV-stimulated macrophages exhibit suppressed production of reparative sEV and a polarization shift toward a sustained M1-like phenotype, further exacerbating local immune dysregulation [98]. This imbalance disrupts the coordinated crosstalk among immune and stromal cells that is essential for transitioning from the inflammatory to the proliferative phase of healing. Outer membrane vesicles also antagonize angiogenic signaling. In endothelial cells, OMVs activate the cyclic GMP-AMP synthase (cGAS) stimulator of interferon genes (STING)-TANK-binding kinase 1 (TBK1) axis, which inhibits the activity of pro-angiogenic host sEV carrying cargo such as vascular endothelial growth factor (VEGF) or miR-126 [99]. This inhibition of neovascularization compromises nutrient delivery and delays tissue regeneration. Beyond signaling interference, bacterial OMVs exert direct physical effects on the host vesicular landscape [100]. They may fuse with host sEV, interfere with their uptake at target cells, or degrade them *via* proteases and lipases, particularly those sEV that are embedded in the ECM. These disruptions impair the spatial and temporal signaling cues critical for directed cell migration and tissue patterning. Furthermore, OMVs serve as conduits for horizontal gene transfer, transporting nucleic acids such as small regulatory RNAs and deoxyribonucleic acid (DNA) fragments encoding antibiotic resistance or virulence factors. By modulating host transcriptional programs and reshaping the epigenetic landscape, OMVs exert long-range effects on cellular behavior [101]. Their cargo frequently includes TonB-dependent receptors and siderophores that enhance bacterial access to essential cofactors such as iron and zinc that are often sequestered by the host during infection, thereby supporting bacterial survival in the nutrient-deprived wound *milieu* [102].

TonB-Dependent Receptors
The TonB receptor proteins are outer membrane proteins in Gram-negative bacteria [103]. They transport iron siderophore complexes, vitamin B12, and heme [103]. Energy is supplied by the TonB-extracellular binding energy protein B (ExbB)-extracellular binding energy protein D (ExbD) complex from the inner membrane [103]. This enables active nutrient uptake across the outer membrane [103].

This nanovesicular deception fundamentally alters the vesicular equilibrium in chronic wounds. Instead of a regenerative vesicle network facilitating tissue repair, a pathogenic vesicle *milieu* emerges as one that is skewed toward inflammation, matrix degradation, and immunoevasion. This molecular sabotage by microbial OMVs represents a sophisticated strategy to corrupt host-derived signaling and disable intrinsic healing programs [104]. While OMVs have gained attention as novel virulence vectors, their structural and functional resemblance to host sEV also presents opportunities for therapeutic repurposing. Preclinical studies exploring engineered OMVs or hybrid EV-OMV systems as delivery vectors for immunomodulators or antimicrobial agents have shown promise. However, translational progress is hampered by concerns over the inherent immunogenicity and cytotoxicity of OMVs derived from pathogenic strains [105]. In conclusion, the interplay between bacterial OMVs and host sEV is not a passive consequence of infection but an active, competitive molecular dialogue. They act as microbial "snoopers" that intercept, mimic, and distort host vesicle-mediated signaling, thereby redefining the wound healing trajectory. A deeper investigation of this vesicle-level sabotage may enable the development of targeted strategies to restore regenerative signaling and counteract microbial deception in chronic wound environments (Fig. 27.2).

Nanotherapeutics: Bioengineered Extracellular Vesicles

Bioengineered exosomes represent a transformative innovation at the intersection of synthetic biology, nanomedicine, and regenerative therapeutics [106–108]. An expanding body of literature in PubMed over the past decade reflects a significant acceleration in efforts to reprogram these endogenous nanocarriers for the targeted delivery of therapeutic agents [108]. Endogenous endosomal-originated membrane-bound nanovesicles naturally secreted by most cell types possess intrinsic advantages such as biocompatibility, low immunogenicity, and the ability to cross biological barriers, including the blood-brain barrier [43, 106–108]. However, their clinical utility is constrained by cargo heterogeneity, inefficient payload loading, and limited targeting specificity [43, 106–108].

To circumvent these limitations, bioengineering strategies are broadly categorized into two modalities: producer cell engineering and post-isolation exosome

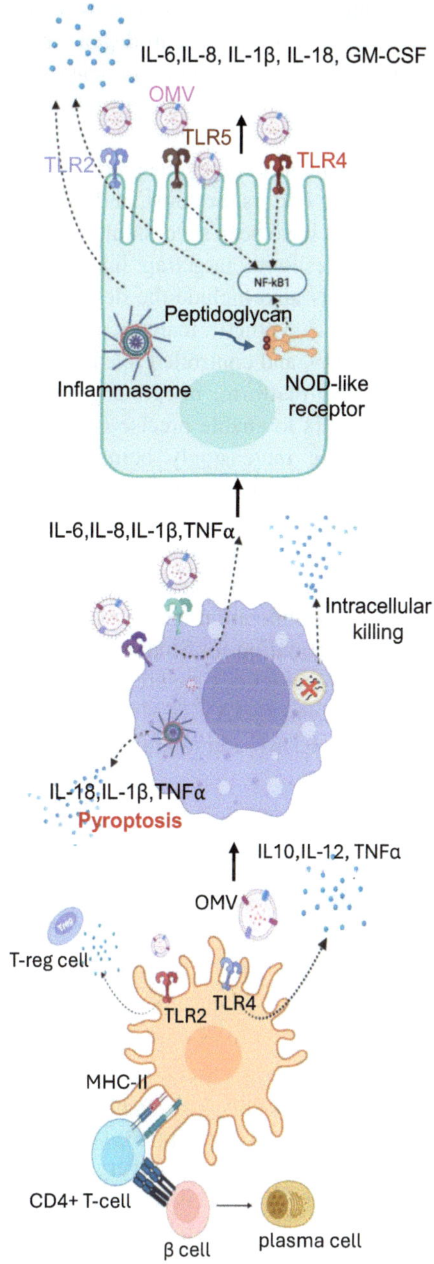

Fig 27.2 Innate immune response against OMV. OMV interaction with epithelial cell, macrophage and dendritic cells. OMVs with PAMPs, peptidoglycan, and bacterial outer membrane proteins, activate TLRs and nucleotide-binding oligomerization domain (NOD) receptors on host cells. This initiates a robust cascade of downstream signaling through Nuclear factor kappa-light-chain-enhancer of activated B cells (NF-κB), MAPK, and AP-1 pathways, culminating in sustained production of pro-inflammatory cytokines such as IL-1β, IL-6, and TNF-α

modification [108–111]. The former entails genetic manipulation of donor cells to incorporate therapeutic RNAs, proteins, or surface ligands into exosomes *via* fusion with membrane proteins such as cluster of differentiation 63 (CD63), cluster of differentiation 9 (CD9), cluster of differentiation 81 (CD81), or lysosome-associated membrane protein 2b (Lamp2b) [38, 112]. For example, Lamp2b fused with

RVG (rabies virus glycoprotein) peptide has facilitated efficient delivery of siRNAs to neurons in preclinical neurodegeneration models [38, 112]. Post-isolation engineering includes electroporation, sonication, and click chemistry to functionalize vesicles with cargos or surface ligands for enhanced targeting and extended circulation [113, 114].

Advancements in bioengineered EVs have enabled the encapsulation of a diverse array of gene editors, proteins, and even mitochondrial fragments [108, 115, 116]. These cargos have been successfully delivered to specific tissues by modifying exosome surfaces with targeting ligands or peptides [108, 117, 118]. Functional enhancements also include the development of hybrid nanovesicles that integrate exosomal and synthetic features, enabling better cargo loading, improved membrane stability, and controlled release [108, 119, 120]. Recent innovations, such as the GEMINI platform, integrate genetic modification with fusogens and RNA-loading motifs to enable precise CRISPR/Cas9 delivery to T cells. Small EVs or exosomes are increasingly being utilized for diagnostic purposes [121–123]. Engineered exosomes expressing fluorescent tags or reporter proteins have enabled real-time *in vivo* tracking, offering insights into biodistribution and therapeutic activity [124, 125]. To meet translational demands, scalable production systems such as hollow-fiber bioreactors and tangential flow filtration have been developed, alongside lyophilization protocols for long-term storage [126]. Adherence to Minimal Information for Studies of Extracellular Vesicles 2023 (MISEV2023) and EV-TRACK guidelines ensures rigorous characterization, including nanoparticle tracking analysis (NTA), TEM, and western blotting for canonical markers such as CD9, CD63, CD81, and tumor susceptibility gene 101 (TSG101) [127–130].

Cross-Domain Lessons: What We Learn from Other Therapeutic Areas

Bioengineered sEV have demonstrated remarkable versatility across diverse disease contexts, offering insights that inform and enrich strategies in regenerative medicine and beyond [131]. For example in sarcopenia, the decline in satellite cell (SC) function with age has been addressed through sEV loaded with antagomiRs against miR-467a-3p and miR-874-5p, improving SC stemness and myogenesis [132]. The sEV engineered to overexpress TSG101 for targeted SC uptake, combined with controlled hydrogel-based delivery, have restored muscle mass and function in preclinical models [131]. This illustrates the importance of both cell-specific targeting and temporal coordination in regenerative therapy.

In oncology, bioengineered sEV offer a means to circumvent chemoresistance and deliver gene-targeted interventions [133]. Murine double minute gene 2 (siMDM2)-loaded sEV have reversed drug resistance in colorectal cancer, while sEV functionalized with tumor-targeting peptides and imaging agents have facilitated both therapy and diagnostics in lung cancer [133]. These strategies emphasize the potential of sEV for combinatorial theranostics.

In liver disease, hepatocyte targeting *via* GalNAc ligands and cargo loading using RNA-binding protein fusions have enabled precision delivery of antifibrotic and anti-inflammatory agents [134]. The modularity and low immunogenicity of sEV provide advantages over conventional nanoparticle systems, especially in conditions like metabolic dysfunction-associated steatohepatitis (MASH) and hepatocellular carcinoma [135].

In the cardiovascular field, sEV have been used to promote angiogenesis, modulate inflammation, and resolve fibrosis [136]. The GPI-anchored proteins such as Cluster differential protein 14 (CD14) have been integrated into sEV surfaces to neutralize endotoxins and modulate post-infarction immune responses [137]. These findings support the use of sEV in both acute and chronic vascular injury.

Taken together, these diverse applications reveal a common blueprint for therapeutic success; rational engineering of sEV includes cell- or tissue-specific targeting motifs, cargo optimized for disease modulation, and delivery platforms that provide spatial and temporal control [137]. These cross-domain insights are especially instructive for advancing sEV-based therapies in cutaneous wound healing [137]. Lessons from muscle regeneration highlight the value of targeting resident progenitor cells, while oncology and liver disease models demonstrate the utility of combinatorial cargo design and immune modulation [138]. As infected and chronic wounds often involve biofilm formation and systemic inflammatory responses, strategies developed in sepsis and cardiovascular contexts can be leveraged to design sEV that simultaneously modulate the immune microenvironment, promote tissue regeneration, and counter microbial resistance [139]. The convergence of these insights reinforces the potential of bioengineered sEV as precision tools for orchestrating complex repair processes in the skin and beyond.

Bioengineered sEV in Diabetic Wound Healing

Diabetic wounds represent a pathological disruption of the conserved wound healing cascade, characterized by persistent inflammation, protease imbalance, and fibroblast senescence [140]. These aberrations impair the coordinated signaling among keratinocytes, endothelial cells, immune cells, and progenitor populations, leading to defective matrix remodeling and regenerative failure [141, 142]. Centers for Disease Control and Prevention (CDC) reports (2021–2022) highlight a 20% rise in antimicrobial resistance during the coronavirus (COVID-19) pandemic, further complicating chronic wound care [143]. Conventional topical treatments, though effective for some conditions (e.g., psoriasis, fungal infections), often fall short in delivering macromolecules like peptides or siRNAs to deeper dermal layers [142, 143].

Nanotechnology-based delivery platforms such as solid lipid nanoparticles, dendrimers, and electrospun nanofibers enable enhanced drug solubility, transdermal permeability, and controlled release kinetics [144]. Concurrently, hydrogel matrices and electroactive polymers embedded with nanozymes have been developed as bioresponsive dressings to facilitate wound closure and eradicate biofilm-forming pathogens [145]. In this context, stem cell-derived sEV, particularly from adipose-derived mesenchymal stem cells (MSCs), have shown regenerative capacity by modulating inflammation, promoting angiogenesis, and enhancing cellular proliferation [145]. Lentiviral engineering of donor MSCs to overexpress anti-apoptotic proteins like BCL-2 results in sEV with heightened reparative potential [146]. Embedding these engineered sEV into antimicrobial hydrogels composed of ε-poly-lysine and hyaluronic acid has led to accelerated wound closure, reduced cytokine expression, and enhanced neovascularization in diabetic mouse models [147].

Wound Infection, Inflammatory Syndromes, and Sepsis: Tailoring sEV for Precision Immunotherapy

Infected wounds are a major source of chronic inflammation and, if left unmanaged, can progress to systemic complications such as sepsis [148]. These wounds are often characterized by persistent microbial colonization, impaired tissue regeneration, and immune dysregulation [149]. One of the key pathological features of chronic infected wounds is the formation of biofilms embedded in an ECM that are recalcitrant to antibiotics and innate immune responses [150–153]. These microbial communities also release OMVs that mimic host sEV and can hijack local cellular signaling, exacerbating tissue damage and systemic inflammation [82]. In this context, bioengineered sEV offer a dual opportunity: local modulation of wound-associated inflammation and systemic immune recalibration to prevent or treat sepsis [109]. The sEV derived from immune cells or mesenchymal stem cells have been shown to carry anti-inflammatory miRNAs (e.g., miR-146a), cytokine modulators, or TLR antagonists that dampen cytokine storms and promote immune resolution [154]. When engineered with targeting ligands, these sEV can home to activate immune cells or inflamed endothelium, exerting site-specific effects [115].

Recent studies have embedded anti-inflammatory sEV within smart hydrogels or wound dressings to provide sustained, localized release in infected wounds, effectively reducing bacterial burden, disrupting biofilms, and promoting tissue regeneration [155]. In preclinical models, these hybrid systems have shown reduced systemic inflammatory markers, improved wound closure, and prevented transition to sepsis [156].

Furthermore, synthetic sEV mimetics, designed to replicate the membrane composition and bioactive content of natural sEV, are under development to address issues of batch variability and production scalability [157]. These mimetics hold potential for broader application in acute inflammatory syndromes, offering consistent therapeutic profiles and enhanced manufacturability [157].

Targeting Specificity: Natural Limitations and Bioengineered Solutions

Despite their biological origin and some degree of natural tissue tropism, unmodified sEV frequently demonstrate broad and nonspecific biodistribution when administered systemically [158]. This off-target accumulation poses challenges in achieving efficient therapeutic concentrations at the intended site of action and can reduce the efficacy of EV-based interventions [159]. The heterogeneity in native sEV surface proteins, lack of consistent receptor-ligand engagement, and clearance by mononuclear phagocyte systems (MPS) are major limiting factors that reduce their clinical reliability [160, 161].

To address these limitations, bioengineering strategies have been developed to endow sEV with enhanced targeting specificity [161]. One of the most widely used approaches involves the genetic fusion tetraspanins like CD9 and CD81 [125]. For example, Lamp2b fused with RVG enables neuron-specific delivery by engaging the nicotinic acetylcholine receptor on neuronal cells. Similarly, arginylglycylaspartic acid (RGD) peptides have been employed for integrin-targeted cancer delivery, while interleukin-3 receptor (IL-3R) ligands have facilitated targeting of leukemic stem cells [162]. Post-isolation modifications offer another avenue for fine-tuning sEV specificity. Surface functionalization using click chemistry, 1-ethyl-3-(3-dimethylaminopropyl) carbodiimide/N-hydroxysuccinimide (EDC/NHS) coupling, or lipid insertion techniques enables the attachment of antibodies, aptamers, or synthetic ligands that can recognize specific tissue or cellular markers [163]. These modifications can also include PEGylation to prolong circulation time and reduce opsonization and clearance [164].

Another innovative strategy involves hybridization of sEV with synthetic nanoparticles, forming chimeric vesicles that retain the bioactivity of natural sEV while gaining the precision targeting capabilities of engineered nanoparticles [165]. For instance, coating sEV with cell membrane fragments from specific cell types can help exploit homotypic targeting mechanisms, thereby increasing localization to target tissues such as tumors, inflamed vasculature, or fibrotic regions [166]. Additionally, spatiotemporal control over sEV localization and activity can be achieved by embedding bioengineered sEV into biomaterial scaffolds or injectable hydrogels [167]. These matrices not only localize sEV at the site of injury or disease but also offer sustained and programmable release, enhancing therapeutic persistence and bioavailability [168].

Advancements in imaging technologies have further enabled real-time tracking of sEV biodistribution, allowing researchers to correlate surface modifications with targeting efficiency and refine delivery strategies accordingly [169]. As sEV engineering progresses, an integrated approach combining genetic-, chemical-, and material-based modifications is likely to offer the highest degree of precision in achieving tissue-specific, disease-responsive therapeutic delivery [116].

Clinical Translation and Future Directions

Despite encouraging preclinical outcomes, the clinical translation of bioengineered sEV is hindered by donor variability, batch-to-batch inconsistency, and regulatory constraints [170]. Implementation of standardized protocols aligned with MISEV2023 guidelines, along with scalable Good Manufacturing Practices (GMP)-compliant manufacturing approaches such as tangential flow filtration and ultracentrifugation-free purification has been instrumental in mitigating translational barriers [130, 171]. Advanced functional assays (e.g., confocal microscopy, luciferase tracking, and live biodistribution studies) are critical for validating therapeutic impact [172]. Integration with transcriptomics and proteomics further supports cargo consistency and mechanistic understanding [173]. Bioengineered EVs have emerged as a cornerstone of next-generation therapeutics, integrating the fidelity of endogenous signaling with the precision of engineered nanocarriers [174]. From restoring satellite cell function in sarcopenia to overcoming chemoresistance and promoting hepatic or vascular repair, sEV offers a customizable, minimally immunogenic, and highly targeted solution to previously intractable medical challenges [175]. With the convergence of synthetic biology, materials science, and regenerative medicine, continued innovation in sEV engineering, delivery strategies, and GMP manufacturing will be essential. The path from bench to bedside is increasingly clear, as bioengineered sEV move closer to realizing their full potential in clinical medicine.

Acknowledgement This work was primarily supported by NIH R01 grant DK129592 to S.G.

Glossary

Angiogenesis Formation of new blood vessels through coordinated endothelial cell migration, proliferation, and differentiation, primarily driven by vascular endothelial growth factor (VEGF) signaling.

Barrier Restitution Rapid restoration of epithelial or endothelial barrier integrity in which pre-existing cells migrate to cover a defect without requiring cell proliferation.

Biofilm Structured, three-dimensional community of microorganisms (bacteria, fungi, or mixed consortia) encased within a self-produced matrix of proteins, polysaccharides, and nucleic acids that adheres to biotic or abiotic surfaces.

Chronic Wound Cutaneous injury that fails to progress through the normal phases of healing within 4–12 weeks and is characterized by persistent inflammation and associated comorbidities.

Collagen Predominant structural protein of skin and connective tissues that confers tensile strength and scaffolds new tissue during repair.

Clustered Regularly Interspaced Short Palindromic Repeats (CRISPR) Adaptive prokaryotic immune system repurposed as a genome-editing platform enabling sequence-specific DNA (and, with Cas variants, RNA) modification in living cells.

Cutaneous Wound Disruption of skin integrity that may involve superficial layers or extend into deeper tissues such as vasculature, muscle, or fat.

Cytokine Low-molecular-weight secreted protein (e.g., interleukins, interferons) that modulates immune responses, cell survival, and intercellular communication within the wound *milieu.*

Damage-Associated Molecular Patterns (DAMPs) Endogenous molecules released from stress, injured, or necrotic cells activate innate immune receptors and initiate sterile inflammation.

Diabetic Foot Ulcer Chronic open lesion on the plantar aspect of the foot arising from neuropathy and microvascular compromise in diabetes; highly prone to infection and amputation.

Endothelial Cells Specialized squamous cells lining the interior of blood and lymphatic vessels are crucial for angiogenesis, regulating vascular tone and maintaining barrier function.

Epithelial Cell Cell types that form continuous sheets covering body surfaces and lining cavities, providing protection, secretion, and absorption.

Exosome Nanosized (≈30–150 nm) extracellular vesicles are generated within multivesicular bodies and released upon fusion with the plasma membrane, carrying proteins, lipids, and nucleic acids that mediate intercellular signaling.

Extracellular Vesicle (EV) Generic terms for membrane-bound vesicles (exosomes, microvesicles, apoptotic bodies) secreted by most cell types; distinguished by size, biogenesis, and cargo but incapable of replication.

Fibrosis Excessive deposition and contraction of extracellular matrix—especially collagen—during wound repair, resulting in scar formation and potential loss of function.

Glycoprotein Protein covalently linked to oligosaccharide chains, abundant on exosomal and cell surfaces where they modulate recognition, adhesion, and signaling.

Homeostasis Dynamic process by which living organisms regulate internal conditions to maintain physiological equilibrium despite external fluctuations.

Hydrogel Three-dimensional, hydrophilic polymer network capable of absorbing large volumes of water or biological fluid; widely used for wound dressings, tissue engineering, and controlled drug delivery.

Iatrogenic Trauma Unintentional physical injury incurred as a consequence of medical intervention.

Immunosenescence Age-related decline in innate and adaptive immune functions, leading to impaired pathogen clearance, reduced vaccine efficacy, and increased incidence of chronic inflammatory diseases.

Intercellular Communication Exchange of information between cells via direct contact (e.g., gap junctions, juxtacrine signaling) or paracrine, endocrine, and EV-mediated pathways.

Juxtacrine Signaling Cell–cell communication mechanism requiring direct membrane contact, where ligand and receptor remain surface-bound.

Keloid Pathological tumor-like scar arising from excessive collagen synthesis that extends beyond the original wound margins.

Keratinocyte Predominant epidermal cells are responsible for barrier formation; it migrates, proliferates, and differentiates during re-epithelialization of cutaneous wounds.

Lipopolysaccharide (LPS) Endotoxin is located in the outer membrane of Gram-negative bacteria; it is essential for bacterial integrity and a potent activator of host immune responses.

Macrophage A professional phagocyte derived from monocytes that orchestrate inflammation, debris clearance, and tissue remodeling exhibits plastic phenotypes across the healing continuum.

Melanocyte Epidermal cell that synthesizes melanin pigments within melanosomes, thereby determining skin and hair coloration and providing UV protection.

Microangiopathy Pathology of small blood vessels leading to compromised microcirculation, commonly observed in diabetes and contributing to chronic wound etiology.

Mitophagy Selective autophagic degradation of damaged or superfluous mitochondria, safeguarding cellular energy homeostasis and limiting oxidative stress.

Mononuclear Phagocyte System (MPS) Network of circulating monocytes and tissue-resident macrophages and dendritic cells responsible for phagocytosis, antigen presentation, and cytokine production.

Multivesicular Body (MVB) Endosomal compartment containing numerous intraluminal vesicles; upon fusion with the plasma membrane, it releases exosomes.

Myofibroblast Contractile fibroblast phenotype expressing α-smooth muscle actin, pivotal for wound contraction and extracellular matrix remodeling; unchecked activity promotes fibrosis.

Nanobiology An interdisciplinary field applying nanoscale tools and concepts to elucidate and manipulate biological processes at the molecular and cellular levels.

Neovascularization Umbrella term encompassing angiogenesis and vasculogenesis that results in the formation of functional microvascular networks within ischemic or healing tissues.

Paracrine Signaling The mode of communication in which soluble factors are released by a cell acts on neighboring target cells within the same tissue microenvironment.

Perfusion Delivery of oxygenated blood to tissues through the vascular system, essential for nutrient supply, waste removal, and optimal healing.

Perivascular Adipose Tissue (PVAT) Adipose depot enveloping blood vessels that modulate vascular tone and inflammation through secretion of adipokines and EVs.

Protease An enzyme that hydrolyzes peptide bonds, facilitating extracellular matrix turnover and the regulation of bioactive molecules during tissue repair.

Reactive Oxygen Species (ROS) Highly reactive oxygen-containing molecules that function as signaling intermediates at low concentrations but cause oxidative damage when in excess.

Stromal Cell Nonparenchymal connective tissue cell (e.g., fibroblast, pericyte) that produces extracellular matrix and provides structural and biochemical support to organs.

Tetraspanin Four-pass transmembrane protein enriched on exosomal membranes that regulates vesicle biogenesis, cargo selection, and target-cell interaction.

Translocase of the Outer Mitochondrial Membrane 70 (TOMM70) Receptor subunit of the mitochondrial import machinery that guides precursor proteins into mitochondria and participates in stress signaling; mutations disrupt mitochondrial function.

Vascularized Integument Skin tissue contains an intrinsic vascular network that supplies oxygen and nutrients while removing metabolic waste.

Vesicular Trafficking Intracellular transport system that shuttles proteins, lipids, and nucleic acids between organelles and the plasma membrane *via* membrane-bound vesicles, underpinning secretion, endocytosis, and autophagy.

Chapter Questions

1. Describe cutaneous wound repair in your own words.
2. The four phases of wound healing: hemostasis, inflammation, proliferation, and remodeling. Why is it important for wound healing phases to follow this specific order? What would happen if the inflammation phase didn't take place?
3. Name the cell types involved in wound healing according to the different wound phases.
4. What is the result of keratinocyte and macrophage exosome-mediated crosstalk aberration during wound healing?
5. What would be the result of endothelial and VSMC exosome-mediated crosstalk aberration during wound healing?
6. What would be the result of ADSC not releasing any exosomes in a wounded environment?
7. What would be the result of PVAT not releasing any exosomes in a wounded environment?
8. What is the difference between PVAT and ADSC? What are some examples of cargo in PVAT-derived exosomes, and what are some examples of cargo in ADSC-derived exosomes?
9. What does a cellular trialogue signify?
10. What are some common exosome markers?
11. What are some examples of exosomal cargo in fibroblast, macrophage, and keratinocyte-derived exosomes? Why is this trialogue essential?
12. How do melanocytes impact wound scarring?
13. What is biofilm? What are extracellular vesicles released by bacterial cells called?
14. How do biofilms impact the wound environment?
15. What are some host defenses that work to inhibit the formation of biofilms in healthy individuals?
16. What are the benefits of having engineered sEV/exosomes as drug delivery systems?
17. List some challenges in the clinical trials with bioengineered extracellular vesicles.

References

1. Mehra R (2002) Historical survey of wound healing. Bull Indian Inst Hist Med Hyderabad 32:159–175
2. Broughton G 2nd, Janis JE, Attinger CE (2006) A brief history of wound care. Plast Reconstr Surg 117:6s–11s. https://doi.org/10.1097/01.prs.0000225429.76355.dd
3. Eming SA, Krieg T, Davidson JM (2007) Inflammation in wound repair: molecular and cellular mechanisms. J Invest Dermatol 127:514–525. https://doi.org/10.1038/sj.jid.5700701
4. Grice EA, Segre JA (2011) The skin microbiome. Nat Rev Microbiol 9:244–253. https://doi.org/10.1038/nrmicro2537
5. Simson N, Stonier T, Challacombe B, Wheatstone S (2019) When things go wrong: a surgeon's guide to iatrogenic injury (perspective). Int J Surg 72:93–95. https://doi.org/10.1016/j.ijsu.2019.10.029
6. Sen CK (2019) Human wounds and its burden: an updated compendium of estimates. Adv Wound Care (New Rochelle) 8:39–48. https://doi.org/10.1089/wound.2019.0946
7. Prince MJ et al (2015) The burden of disease in older people and implications for health policy and practice. Lancet 385:549–562. https://doi.org/10.1016/S0140-6736(14)61347-7
8. Atella V et al (2019) Trends in age-related disease burden and healthcare utilization. Aging Cell 18:e12861. https://doi.org/10.1111/acel.12861
9. Brown BA et al (2022) Analysis of Keratinocytic exosomes from diabetic and nondiabetic mice by charge detection mass spectrometry. Anal Chem 94:8909–8918. https://doi.org/10.1021/acs.analchem.2c00453
10. Bogadi S, Uddin ME, Rahman MH, Satyanarayana Reddy Karri VV, Begum R, Udeabor SE (2025) Wound healing in the modern era: emerging research, biomedical advances, and transformative clinical approaches. J Drug Delivery Sci Technol 110:107058. https://doi.org/10.1016/j.jddst.2025.107058
11. Tottoli EM, Dorati R, Genta I, Chiesa E, Pisani S, Conti B (2020) Skin wound healing process and new emerging technologies for skin wound care and regeneration. Pharmaceutics 12. https://doi.org/10.3390/pharmaceutics12080735
12. Smith SA, Travers RJ, Morrissey JH (2015) How it all starts: initiation of the clotting cascade. Crit Rev Biochem Mol Biol 50:326–336. https://doi.org/10.3109/10409238.2015.1050550
13. Gushiken LFS, Beserra FP, Bastos JK, Jackson CJ, Pellizzon CH (2021) Cutaneous wound healing: an update from physiopathology to current therapies. Life (Basel) 11. https://doi.org/10.3390/life11070665
14. Singer AJ (2022) Healing mechanisms in cutaneous wounds: tipping the balance. Tissue Eng Part B Rev 28:1151–1167. https://doi.org/10.1089/ten.TEB.2021.0114
15. Naba A (2024) Mechanisms of assembly and remodelling of the extracellular matrix. Nat Rev Mol Cell Biol 25:865–885. https://doi.org/10.1038/s41580-024-00767-3
16. Scridon A (2022) Platelets and their role in hemostasis and thrombosis-from physiology to pathophysiology and therapeutic implications. Int J Mol Sci 23. https://doi.org/10.3390/ijms232112772
17. Wilgus TA, Roy S, McDaniel JC (2013) Neutrophils and wound repair: positive actions and negative reactions. Adv Wound Care (New Rochelle) 2:379–388. https://doi.org/10.1089/wound.2012.0383
18. DeLeon-Pennell KY, Barker TH, Lindsey ML (2020) Fibroblasts: the arbiters of extracellular matrix remodeling. Matrix Biol 91-92:1–7. https://doi.org/10.1016/j.matbio.2020.05.006
19. Pakshir P, Hinz B (2018) The big five in fibrosis: macrophages, myofibroblasts, matrix, mechanics, and miscommunication. Matrix Biol 68-69:81–93. https://doi.org/10.1016/j.matbio.2018.01.019
20. Yadav A, Nandy A, Sharma A, Ghatak S (2024) In: Kloc M, Kubiak JZ, Halasa M (eds) Intercellular and interorganellar transfer and communication in biology and medicine. Springer International Publishing, pp 249–297

21. Camera A, Hopps E, Caimi G (2007) Diabetic microangiopathy: physiopathological, clinical and therapeutic aspects. Minerva Endocrinol 32:209–229
22. Darwin E, Tomic-Canic M (2018) Healing chronic wounds: current challenges and potential solutions. Curr Dermatol Rep 7:296–302. https://doi.org/10.1007/s13671-018-0239-4
23. Lian N, Li T (2016) Growth factor pathways in hypertrophic scars: molecular pathogenesis and therapeutic implications. Biomed Pharmacother 84:42–50. https://doi.org/10.1016/j.biopha.2016.09.010
24. Strebhardt K, Ullrich A (2008) Paul Ehrlich's magic bullet concept: 100 years of progress. Nat Rev Cancer 8:473–480. https://doi.org/10.1038/nrc2394
25. Shang R, Lee S, Senavirathne G, Lai EC (2023) microRNAs in action: biogenesis, function and regulation. Nat Rev Genet 24:816–833. https://doi.org/10.1038/s41576-023-00611-y
26. Sen CK, Ghatak S (2015) miRNA control of tissue repair and regeneration. Am J Pathol 185:2629–2640. https://doi.org/10.1016/j.ajpath.2015.04.001
27. Bhat P, Garibyan L (2022) The potential of CRISPR-guided therapies in the dermatology clinic. JID Innov 2:100103. https://doi.org/10.1016/j.xjidi.2022.100103
28. Tzelepis F, Sanson-Fisher RW, Zucca AC, Fradgley EA (2015) Measuring the quality of patient-centered care: why patient-reported measures are critical to reliable assessment. Patient Prefer Adherence 9:831–835. https://doi.org/10.2147/ppa.S81975
29. Chandan K, Sen SR (2012) Wound healing, vol 1, 3rd edn. Elsevier
30. Yadav A, Xuan Y, Sen CK, Ghatak S (2024) Standardized reporting of research on exosomes to ensure rigor and reproducibility. Adv Wound Care (New Rochelle) 13:584–599. https://doi.org/10.1089/wound.2024.0093
31. Yadav A, Nandy A, Sharma A, Ghatak S (2024) Exosome mediated cell-cell cross-talk in tissue injury and repair. Results Probl Cell Differ 73:249–297. https://doi.org/10.1007/978-3-031-62036-2_12
32. Vidal M (2019) Exosomes: revisiting their role as "garbage bags". Traffic 20:815–828. https://doi.org/10.1111/tra.12687
33. Peña OA, Martin P (2024) Cellular and molecular mechanisms of skin wound healing. Nat Rev Mol Cell Biol 25:599–616. https://doi.org/10.1038/s41580-024-00715-1
34. Bonaventura A, Montecucco F, Dallegri F (2016) Cellular recruitment in myocardial ischaemia/reperfusion injury. Eur J Clin Investig 46:590–601. https://doi.org/10.1111/eci.12633
35. Altan-Bonnet G, Mukherjee R (2019) Cytokine-mediated communication: a quantitative appraisal of immune complexity. Nat Rev Immunol 19:205–217. https://doi.org/10.1038/s41577-019-0131-x
36. Wang Y, Li N, Zhang X, Horng T (2021) Mitochondrial metabolism regulates macrophage biology. J Biol Chem 297. https://doi.org/10.1016/j.jbc.2021.100904
37. Driscoll J, Gondaliya P, Patel T (2022) Tunneling nanotube-mediated communication: a mechanism of intercellular nucleic acid transfer. Int J Mol Sci 23. https://doi.org/10.3390/ijms23105487
38. Zhang Y, Liu Y, Liu H, Tang WH (2019) Exosomes: biogenesis, biologic function and clinical potential. Cell Biosci 9:19. https://doi.org/10.1186/s13578-019-0282-2
39. Liu C, Chu D, Kalantar-Zadeh K, George J, Young HA, Liu G (2021) Cytokines: from clinical significance to quantification. Adv Sci (Weinh) 8:e2004433. https://doi.org/10.1002/advs.202004433
40. Patel GK, Wilson CH, Harding KG, Finlay AY, Bowden PE (2006) Numerous keratinocyte subtypes involved in wound re-epithelialization. J Invest Dermatol 126:497–502. https://doi.org/10.1038/sj.jid.5700101
41. Zhou X et al (2020) Exosome-mediated crosstalk between keratinocytes and macrophages in cutaneous wound healing. ACS Nano 14:12732–12748. https://doi.org/10.1021/acsnano.0c03064
42. Guda PR et al (2023) Nanoscopic and functional characterization of keratinocyte-originating exosomes in the wound fluid of non-diabetic and diabetic chronic wound patients. Nano Today 52. https://doi.org/10.1016/j.nantod.2023.101954

43. Chavda VP et al (2023) Exosome nanovesicles: a potential carrier for therapeutic delivery. Nano Today 49:101771. https://doi.org/10.1016/j.nantod.2023.101771
44. Prasai A, Jay JW, Jupiter D, Wolf SE, El Ayadi A (2022) Role of exosomes in dermal wound healing: a systematic review. J Invest Dermatol 142:662–678.e668. https://doi.org/10.1016/j.jid.2021.07.167
45. Liu Y, Shi SL (2021) The roles of hnRNP A2/B1 in RNA biology and disease. WIREs RNA 12:e1612. https://doi.org/10.1002/wrna.1612
46. Wilkinson KA, Henley JM (2010) Mechanisms, regulation and consequences of protein SUMOylation. Biochem J 428:133–145. https://doi.org/10.1042/bj20100158
47. Backes S et al (2018) Tom70 enhances mitochondrial preprotein import efficiency by binding to internal targeting sequences. J Cell Biol 217:1369–1382. https://doi.org/10.1083/jcb.201708044
48. Lechado Terradas A, Zittlau KI, Macek B, Fraiberg M, Elazar Z, Kahle PJ (2021) Regulation of mitochondrial cargo-selective autophagy by posttranslational modifications. J Biol Chem 297:101339. https://doi.org/10.1016/j.jbc.2021.101339
49. Yan T et al (2017) Hypoxia regulates mTORC1-mediated keratinocyte motility and migration via the AMPK pathway. PLoS One 12:e0169155. https://doi.org/10.1371/journal.pone.0169155
50. Sharma A et al (2024) Mitochondrial bioenergetics of functional wound closure is dependent on macrophage-keratinocyte Exosomal crosstalk. ACS Nano 18:30405–30420. https://doi.org/10.1021/acsnano.4c07610
51. DeWane G, Salvi AM, DeMali KA (2021) Fueling the cytoskeleton – links between cell metabolism and actin remodeling. J Cell Sci 134. https://doi.org/10.1242/jcs.248385
52. Yadav A, Sharma A, Moulick M, Ghatak S (2025) Nanomanaging chronic wounds with targeted exosome therapeutics. Pharmaceutics 17. https://doi.org/10.3390/pharmaceutics17030366
53. Villarreal-Ponce A et al (2020) Keratinocyte-macrophage crosstalk by the Nrf2/Ccl2/EGF signaling Axis orchestrates tissue repair. Cell Rep 33:108417. https://doi.org/10.1016/j.celrep.2020.108417
54. Zheng B et al (2017) Exosome-mediated miR-155 transfer from smooth muscle cells to endothelial cells induces endothelial injury and promotes atherosclerosis. Mol Ther 25:1279–1294. https://doi.org/10.1016/j.ymthe.2017.03.031
55. Heo J, Yang HC, Rhee WJ, Kang H (2020) Vascular smooth muscle cell-derived exosomal MicroRNAs regulate endothelial cell migration under PDGF stimulation. Cells 9. https://doi.org/10.3390/cells9030639
56. jiang y, Tan F, yu m, Yang W, Qian H-Y (2024) Abstract 4141302: endothelial cell-derived exosome regulates the proliferation and migration of smooth muscle cells and aggravate recurrent in-stent restenosis. Circulation 150:A4141302. https://doi.org/10.1161/circ.150.suppl_1.4141302
57. Xu J et al (2020) miRNA-221-3p in endothelial progenitor cell-derived exosomes accelerates skin wound healing in diabetic mice. Diabetes Metab Syndr Obes 13:1259–1270. https://doi.org/10.2147/dmso.S243549
58. Li X, Ballantyne LL, Yu Y, Funk CD (2019) Perivascular adipose tissue-derived extracellular vesicle miR-221-3p mediates vascular remodeling. FASEB J 33:12704–12722. https://doi.org/10.1096/fj.201901548R
59. Chen HH, Li HF, Tseng TL, Lin H (2023) Perivascular adipose tissue and adipocyte-derived exosomal miRNAs maintain vascular homeostasis. Heliyon 9:e22607. https://doi.org/10.1016/j.heliyon.2023.e22607
60. Blumensatt M et al (2014) Adipocyte-derived factors impair insulin signaling in differentiated human vascular smooth muscle cells via the upregulation of miR-143. Biochim Biophys Acta (BBA) – Mol Basis Dis 1842:275–283. https://doi.org/10.1016/j.bbadis.2013.12.001
61. Hu L et al (2016) Exosomes derived from human adipose mesenchymal stem cells accelerates cutaneous wound healing via optimizing the characteristics of fibroblasts. Sci Rep 6:32993. https://doi.org/10.1038/srep32993

62. Qian L, Pi L, Fang B-R, Meng X-X (2021) Adipose mesenchymal stem cell-derived exosomes accelerate skin wound healing via the lncRNA H19/miR-19b/SOX9 axis. Lab Investig 101:1254–1266. https://doi.org/10.1038/s41374-021-00611-8
63. Yang X et al (2011) miR-21 promotes keratinocyte migration and re-epithelialization during wound healing. Int J Biol Sci 7:685–690. https://doi.org/10.7150/ijbs.7.685
64. Tissot FS et al (2025) Functional transfer of integrin co-receptor CD98hc by small extracellular vesicles improves wound healing in vivo. Matrix Biol 135:99–105. https://doi.org/10.1016/j.matbio.2024.12.006
65. Jella KK, Nasti TH, Li Z, Malla SR, Buchwald ZS, Khan MK (2018) Exosomes, their biogenesis and role in inter-cellular communication, tumor microenvironment and cancer immunotherapy. Vaccines (Basel) 6. https://doi.org/10.3390/vaccines6040069
66. Liu J et al (2022) Wnt/β-catenin signalling: function, biological mechanisms, and therapeutic opportunities. Signal Transduct Target Ther 7:3. https://doi.org/10.1038/s41392-021-00762-6
67. Chen C et al (2025) Orchestration of macrophage polarization dynamics by fibroblast-secreted exosomes during skin wound healing. J Invest Dermatol 145:171–184.e176. https://doi.org/10.1016/j.jid.2024.05.007
68. Xu Z et al (2025) Exosomes derived from fibroblasts in DFUs delay wound healing by delivering miR-93-5p to target macrophage ATG16L1. Biochim Biophys Acta (BBA) – Mol Basis Dis 1871:167640. https://doi.org/10.1016/j.bbadis.2024.167640
69. Chen J, Zhou R, Liang Y, Fu X, Wang D, Wang C (2019) Blockade of lncRNA-ASLNCS5088-enriched exosome generation in M2 macrophages by GW4869 dampens the effect of M2 macrophages on orchestrating fibroblast activation. FASEB J 33:12200–12212. https://doi.org/10.1096/fj.201901610
70. Loos B, du Toit A, Hofmeyr JH (2014) Defining and measuring autophagosome flux—concept and reality. Autophagy 10:2087–2096. https://doi.org/10.4161/15548627.2014.973338
71. Hsu H-H, Wang AYL, Loh CYY, Pai AA, Kao H-K (2022) Therapeutic potential of exosomes derived from diabetic adipose stem cells in cutaneous wound healing of db/db mice. Pharmaceutics 14:1206
72. Li B, Qian L, Pi L, Meng X (2023) A therapeutic role of exosomal lncRNA H19 from adipose mesenchymal stem cells in cutaneous wound healing by triggering macrophage M2 polarization. Cytokine 165:156175. https://doi.org/10.1016/j.cyto.2023.156175
73. Kasembeli MM, Bharadwaj U, Robinson P, Tweardy DJ (2018) Contribution of STAT3 to inflammatory and fibrotic diseases and prospects for its targeting for treatment. Int J Mol Sci 19. https://doi.org/10.3390/ijms19082299
74. Agabiti-Rosei C et al (2018) Modulation of vascular reactivity by perivascular adipose tissue (PVAT). Curr Hypertens Rep 20:44. https://doi.org/10.1007/s11906-018-0835-5
75. Ouarné M, Pena A, Franco CA (2021) From remodeling to quiescence: the transformation of the vascular network. Cells Dev 168:203735. https://doi.org/10.1016/j.cdev.2021.203735
76. Gu W et al (2019) Single-cell RNA-sequencing and metabolomics analyses reveal the contribution of perivascular adipose tissue stem cells to vascular remodeling. Arterioscler Thromb Vasc Biol 39:2049–2066. https://doi.org/10.1161/atvbaha.119.312732
77. De Bock K, Georgiadou M, Carmeliet P (2013) Role of endothelial cell metabolism in vessel sprouting. Cell Metab 18:634–647. https://doi.org/10.1016/j.cmet.2013.08.001
78. Shen Z, Shao J, Sun J, Xu J (2022) Exosomes released by melanocytes modulate fibroblasts to promote keloid formation: a pilot study. J Zhejiang Univ Sci B 23:699–704. https://doi.org/10.1631/jzus.B2200036
79. Buzas EI (2023) The roles of extracellular vesicles in the immune system. Nat Rev Immunol 23:236–250. https://doi.org/10.1038/s41577-022-00763-8
80. Ciobanasu C (2025) Bacterial extracellular vesicles and antimicrobial peptides: a synergistic approach to overcome antimicrobial resistance. Antibiotics (Basel) 14. https://doi.org/10.3390/antibiotics14040414
81. Thammavongsa V, Kim HK, Missiakas D, Schneewind O (2015) Staphylococcal manipulation of host immune responses. Nat Rev Microbiol 13:529–543. https://doi.org/10.1038/nrmicro3521

82. Magaña G, Harvey C, Taggart CC, Rodgers AM (2023) Bacterial outer membrane vesicles: role in pathogenesis and host-cell interactions. Antibiotics (Basel) 13. https://doi.org/10.3390/antibiotics13010032
83. Michel LV, Gaborski T (2022) Outer membrane vesicles as molecular biomarkers for Gram-negative sepsis: taking advantage of nature's perfect packages. J Biol Chem 298:102483. https://doi.org/10.1016/j.jbc.2022.102483
84. Nawrocki KL, Crispell EK, McBride SM (2014) Antimicrobial peptide resistance mechanisms of gram-positive bacteria. Antibiotics (Basel) 3:461–492. https://doi.org/10.3390/antibiotics3040461
85. Lee J, Choi WG, Rhee M, Lee S-H (2025) Extracellular vesicle-mediated network in the pathogenesis of obesity, diabetes, Steatotic liver disease, and cardiovascular disease. Diabetes Metab J 49:348–367. https://doi.org/10.4093/dmj.2025.0184
86. Rima M, Dakramanji M, El Hayek E, El Khoury T, Fajloun Z, Rima M (2025) Unveiling the wonders of bacteria-derived extracellular vesicles: from fundamental functions to beneficial applications. Heliyon 11:e42509. https://doi.org/10.1016/j.heliyon.2025.e42509
87. Jiang B et al (2024) Microbial extracellular vesicles contribute to antimicrobial resistance. PLoS Pathog 20:e1012143. https://doi.org/10.1371/journal.ppat.1012143
88. Huan Y, Kong Q, Mou H, Yi H (2020) Antimicrobial peptides: classification, design, application and research Progress in multiple fields. Front Microbiol 11:582779. https://doi.org/10.3389/fmicb.2020.582779
89. Duarte-Mata DI, Salinas-Carmona MC (2023) Antimicrobial peptides' immune modulation role in intracellular bacterial infection. Front Immunol 14:1119574. https://doi.org/10.3389/fimmu.2023.1119574
90. Heimlich DR, Harrison A, Mason KM (2014) Host antimicrobial peptides in bacterial homeostasis and pathogenesis of disease. Antibiotics (Basel) 3:645–676. https://doi.org/10.3390/antibiotics3040645
91. Gan Y, Zhao G, Wang Z, Zhang X, Wu MX, Lu M (2023) Bacterial membrane vesicles: physiological roles, infection immunology, and applications. Adv Sci (Weinh) 10:e2301357. https://doi.org/10.1002/advs.202301357
92. Schwechheimer C, Kuehn MJ (2015) Outer-membrane vesicles from gram-negative bacteria: biogenesis and functions. Nat Rev Microbiol 13:605–619. https://doi.org/10.1038/nrmicro3525
93. Charpentier LA, Dolben EF, Hendricks MR, Hogan DA, Bomberger JM, Stanton BA (2023) Bacterial outer membrane vesicles and immune modulation of the host. Membranes (Basel) 13. https://doi.org/10.3390/membranes13090752
94. Jung AL, Schmeck B, Wiegand M, Bedenbender K, Benedikter BJ (2021) The clinical role of host and bacterial-derived extracellular vesicles in pneumonia. Adv Drug Deliv Rev 176:113811. https://doi.org/10.1016/j.addr.2021.05.021
95. O'Donoghue EJ, Krachler AM (2016) Mechanisms of outer membrane vesicle entry into host cells. Cell Microbiol 18:1508–1517. https://doi.org/10.1111/cmi.12655
96. Zhong Y et al (2024) NF-κB affected the serum levels of TNF-α and IL-1β via activation of the MAPK/NF-κB signaling pathway in rat model of acute pulmonary microthromboembolism. Pulm Circ 14:e12357. https://doi.org/10.1002/pul2.12357
97. Kaksonen M, Roux A (2018) Mechanisms of clathrin-mediated endocytosis. Nat Rev Mol Cell Biol 19:313–326. https://doi.org/10.1038/nrm.2017.132
98. Luo M, Zhao F, Cheng H, Su M, Wang Y (2024) Macrophage polarization: an important role in inflammatory diseases. Front Immunol 15:1352946. https://doi.org/10.3389/fimmu.2024.1352946
99. Lv H et al (2024) TET2-mediated tumor cGAS triggers endothelial STING activation to regulate vasculature remodeling and anti-tumor immunity in liver cancer. Nat Commun 15:6. https://doi.org/10.1038/s41467-023-43743-9
100. Zhou Z et al (2024) Exploring naturally tailored bacterial outer membrane vesicles for selective bacteriostatic implant coatings. Adv Sci (Weinh) 11:e2405764. https://doi.org/10.1002/advs.202405764

101. Patel NJ, Ashraf A, Chung EJ (2023) Extracellular vesicles as regulators of the extracellular matrix. Bioengineering (Basel) 10. https://doi.org/10.3390/bioengineering10020136
102. Ma L, Terwilliger A, Maresso AW (2015) Iron and zinc exploitation during bacterial pathogenesis. Metallomics 7:1541–1554. https://doi.org/10.1039/c5mt00170f
103. Fujita M, Mori K, Hara H, Hishiyama S, Kamimura N, Masai E (2019) A TonB-dependent receptor constitutes the outer membrane transport system for a lignin-derived aromatic compound. Commun Biol 2:432. https://doi.org/10.1038/s42003-019-0676-z
104. Li W et al (2025) Cell membrane-derived nanovesicles as extracellular vesicle-mimetics in wound healing. Mater Today Bio 31:101595. https://doi.org/10.1016/j.mtbio.2025.101595
105. Wang S, Gao J, Wang Z (2019) Outer membrane vesicles for vaccination and targeted drug delivery. Wiley Interdiscip Rev Nanomed Nanobiotechnol 11:e1523. https://doi.org/10.1002/wnan.1523
106. Hu W, Wang W, Chen Z, Chen Y, Wang Z (2024) Engineered exosomes and composite biomaterials for tissue regeneration. Theranostics 14:2099–2126. https://doi.org/10.7150/thno.93088
107. Mougenot MF et al (2022) Biomimetic Nanovesicles-sources, design, production methods, and applications. Pharmaceutics 14. https://doi.org/10.3390/pharmaceutics14102008
108. Sharma A, Yadav A, Nandy A, Ghatak S (2024) Insight into the functional dynamics and challenges of exosomes in pharmaceutical innovation and precision medicine. Pharmaceutics 16. https://doi.org/10.3390/pharmaceutics16060709
109. Ma Y et al (2025) Engineering therapeutical extracellular vesicles for clinical translation. Trends Biotechnol 43:61–82. https://doi.org/10.1016/j.tibtech.2024.08.007
110. Bashor CJ, Hilton IB, Bandukwala H, Smith DM, Veiseh O (2022) Engineering the next generation of cell-based therapeutics. Nat Rev Drug Discov 21:655–675. https://doi.org/10.1038/s41573-022-00476-6
111. Madl CM, Heilshorn SC, Blau HM (2018) Bioengineering strategies to accelerate stem cell therapeutics. Nature 557:335–342. https://doi.org/10.1038/s41586-018-0089-z
112. Luan X, Sansanaphongpricha K, Myers I, Chen H, Yuan H, Sun D (2017) Engineering exosomes as refined biological nanoplatforms for drug delivery. Acta Pharmacol Sin 38:754–763. https://doi.org/10.1038/aps.2017.12
113. Richter M, Vader P, Fuhrmann G (2021) Approaches to surface engineering of extracellular vesicles. Adv Drug Deliv Rev 173:416–426. https://doi.org/10.1016/j.addr.2021.03.020
114. Ciferri MC et al (2024) Standardized method to functionalize plasma-extracellular vesicles via copper-free click chemistry for targeted drug delivery strategies. ACS Appl Bio Mater 7:827–838. https://doi.org/10.1021/acsabm.3c00822
115. Gong Z, Cheng C, Sun C, Cheng X (2025) Harnessing engineered extracellular vesicles for enhanced therapeutic efficacy: advancements in cancer immunotherapy. J Exp Clin Cancer Res 44:138. https://doi.org/10.1186/s13046-025-03403-w
116. Liang X et al (2025) Engineering of extracellular vesicles for efficient intracellular delivery of multimodal therapeutics including genome editors. Nat Commun 16:4028. https://doi.org/10.1038/s41467-025-59377-y
117. Li L, Wang F, Zhu D, Hu S, Cheng K, Li Z (2025) Engineering exosomes and exosome-like nanovesicles for improving tissue targeting and retention. Fundam Res 5:851–867. https://doi.org/10.1016/j.fmre.2024.03.025
118. Kim HI, Park J, Zhu Y, Wang X, Han Y, Zhang D (2024) Recent advances in extracellular vesicles for therapeutic cargo delivery. Exp Mol Med 56:836–849. https://doi.org/10.1038/s12276-024-01201-6
119. Mondal J et al (2023) Hybrid exosomes, exosome-like nanovesicles and engineered exosomes for therapeutic applications. J Control Release 353:1127–1149. https://doi.org/10.1016/j.jconrel.2022.12.027
120. Peruzzi JA et al (2024) Enhancing extracellular vesicle cargo loading and functional delivery by engineering protein-lipid interactions. Nat Commun 15:5618. https://doi.org/10.1038/s41467-024-49678-z

121. Du Y, Liu Y, Hu J, Peng X, Liu Z (2023) CRISPR/Cas9 systems: delivery technologies and biomedical applications. Asian J Pharm Sci 18:100854. https://doi.org/10.1016/j.ajps.2023.100854
122. Pacesa M, Pelea O, Jinek M (2024) Past, present, and future of CRISPR genome editing technologies. Cell 187:1076–1100. https://doi.org/10.1016/j.cell.2024.01.042
123. Li T et al (2023) CRISPR/Cas9 therapeutics: progress and prospects. Signal Transduct Target Ther 8:36. https://doi.org/10.1038/s41392-023-01309-7
124. Han J et al (2024) Engineered exosomes with a photoinducible protein delivery system enable CRISPR-Cas–based epigenome editing in Alzheimer's disease. Sci Transl Med 16:eadi4830. https://doi.org/10.1126/scitranslmed.adi4830
125. Salunkhe S, Dheeraj, Basak M, Chitkara D, Mittal A (2020) Surface functionalization of exosomes for target-specific delivery and in vivo imaging & tracking: Strategies and significance. J Control Release 326:599–614. https://doi.org/10.1016/j.jconrel.2020.07.042
126. Nie M, Shima A, Yamamoto M, Takeuchi S (2025) Scalable tissue biofabrication via perfusable hollow fiber arrays for cultured meat applications. Trends Biotechnol. https://doi.org/10.1016/j.tibtech.2025.02.022
127. Zhang Y, Lan M, Chen Y (2024) Minimal information for studies of extracellular vesicles (MISEV): ten-year evolution (2014–2023). Pharmaceutics 16:1394
128. Vaiaki EM, Falasca M (2024) Comparative analysis of the minimal information for studies of extracellular vesicles guidelines: advancements and implications for extracellular vesicle research. Semin Cancer Biol 101:12–24. https://doi.org/10.1016/j.semcancer.2024.04.002
129. Upadhya D, Shetty AK (2024) MISEV2023 provides an updated and key reference for researchers studying the basic biology and applications of extracellular vesicles. Stem Cells Transl Med 13:848–850. https://doi.org/10.1093/stcltm/szae052
130. Welsh JA et al (2024) Minimal information for studies of extracellular vesicles (MISEV2023): from basic to advanced approaches. J Extracell Vesicles 13:e12404. https://doi.org/10.1002/jev2.12404
131. Mittal A, Jakhmola VR, Baweja S (2025) Bioengineered extracellular vesicles: the path to precision medicine in liver diseases. Liver Res 9:17–28. https://doi.org/10.1016/j.livres.2025.02.002
132. Alway SE, Myers MJ, Mohamed JS (2014) Regulation of satellite cell function in sarcopenia. Front Aging Neurosci 6:246. https://doi.org/10.3389/fnagi.2014.00246
133. Singh V, Khan N, Jayandharan GR (2022) Vector engineering, strategies and targets in cancer gene therapy. Cancer Gene Ther 29:402–417. https://doi.org/10.1038/s41417-021-00331-7
134. Debacker AJ, Voutila J, Catley M, Blakey D, Habib N (2020) Delivery of oligonucleotides to the liver with GalNAc: from research to registered therapeutic drug. Mol Ther 28:1759–1771. https://doi.org/10.1016/j.ymthe.2020.06.015
135. Du S et al (2023) Extracellular vesicles: a rising star for therapeutics and drug delivery. J Nanobiotechnol 21:231. https://doi.org/10.1186/s12951-023-01973-5
136. Xiao J, Sluijter JPG (2025) Extracellular vesicles in cardiovascular homeostasis and disease: potential role in diagnosis and therapy. Nat Rev Cardiol. https://doi.org/10.1038/s41569-025-01141-2
137. Pugin J, Kravchenko VV, Lee JD, Kline L, Ulevitch RJ, Tobias PS (1998) Cell activation mediated by glycosylphosphatidylinositol-anchored or transmembrane forms of CD14. Infect Immun 66:1174–1180. https://doi.org/10.1128/iai.66.3.1174-1180.1998
138. Dey D, Goldhamer DJ, Yu PB (2015) Contributions of muscle-resident progenitor cells to homeostasis and disease. Curr Mol Biol Rep 1:175–188. https://doi.org/10.1007/s40610-015-0025-z
139. Zhao G et al (2013) Biofilms and inflammation in chronic wounds. Adv Wound Care (New Rochelle) 2:389–399. https://doi.org/10.1089/wound.2012.0381
140. Dasari N et al (2021) Updates in diabetic wound healing, inflammation, and scarring. Semin Plast Surg 35:153–158. https://doi.org/10.1055/s-0041-1731460

141. Klicznik MM, Szenes-Nagy AB, Campbell DJ, Gratz IK (2018) Taking the lead – how keratinocytes orchestrate skin T cell immunity. Immunol Lett 200:43–51. https://doi.org/10.1016/j.imlet.2018.06.009
142. Benhadou F, Mintoff D, Del Marmol V (2019) Psoriasis: keratinocytes or immune cells - which is the trigger? Dermatology 235:91–100. https://doi.org/10.1159/000495291
143. COVID-19: U.S. impact on antimicrobial resistance, special report 2022; (2022): https://www.cdc.gov/antimicrobial-resistance/media/pdfs/covid19-impact-report-508.pdf
144. Ghasemiyeh P, Mohammadi-Samani S (2018) Solid lipid nanoparticles and nanostructured lipid carriers as novel drug delivery systems: applications, advantages and disadvantages. Res Pharm Sci 13:288–303. https://doi.org/10.4103/1735-5362.235156
145. Zhang F et al (2025) A dynamically phase-adaptive regulating hydrogel promotes ultrafast antifibrotic wound healing. Nat Commun 16:3738. https://doi.org/10.1038/s41467-025-58987-w
146. Ni X et al (2017) Lentiviral vector-mediated co-overexpression of VEGF and Bcl-2 improves mesenchymal stem cell survival and enhances paracrine effects in vitro. Int J Mol Med 40:418–426. https://doi.org/10.3892/ijmm.2017.3019
147. Ju Y et al (2025) Antimicrobial dual-crosslinked hydrogel synergizes bioengineered extracellular vesicles for enhanced diabetic wound healing. Mater Today Bio 32:101870. https://doi.org/10.1016/j.mtbio.2025.101870
148. Ousey K et al (2008) Wound infection in clinical practice. An international consensus. Int Wound J 5(Suppl 3):iii–11. https://doi.org/10.1111/j.1742-481X.2008.00488.x
149. Cavallo I et al (2024) Bacterial biofilm in chronic wounds and possible therapeutic approaches. Biology (Basel) 13. https://doi.org/10.3390/biology13020109
150. Wu Y-K, Cheng N-C, Cheng C-M (2019) Biofilms in chronic wounds: pathogenesis and diagnosis. Trends Biotechnol 37:505–517. https://doi.org/10.1016/j.tibtech.2018.10.011
151. Raziyeva K, Kim Y, Zharkinbekov Z, Kassymbek K, Jimi S, Saparov A (2021) Immunology of acute and chronic wound healing. Biomolecules 11. https://doi.org/10.3390/biom11050700
152. Uberoi A, McCready-Vangi A, Grice EA (2024) The wound microbiota: microbial mechanisms of impaired wound healing and infection. Nat Rev Microbiol 22:507–521. https://doi.org/10.1038/s41579-024-01035-z
153. Aswathanarayan JB, Rao P, Hm S, Gs S, Rai RV (2023) In: Donelli G (ed) Advances in microbiology, infectious diseases and public health: volume 17. Springer International Publishing, pp 55–75
154. Xiao X et al (2021) Mesenchymal stem cell-derived small extracellular vesicles mitigate oxidative stress-induced senescence in endothelial cells via regulation of miR-146a/Src. Signal Transduct Target Ther 6:354. https://doi.org/10.1038/s41392-021-00765-3
155. Li Z, Dong H, Yang S, Wang X, Li Z (2025) An injectable ibuprofen sustained-release composite hydrogel system effectively accelerates diabetic wound healing via anti-inflammatory effects and angiogenesis. Int J Nanomedicine 20:4535–4550. https://doi.org/10.2147/ijn.S504924
156. Pastar I et al (2018) In: Marques AP, Pirraco RP, Cerqueira MT, Reis RL (eds) Skin tissue models. Academic, pp 223–253
157. Chen R, Bhavsar C, Lourie R, Li S, Wu SY (2025) Development of an innovative extracellular vesicle mimetic delivery platform for efficient miRNA delivery to tumours. Biomaterials 321:123282. https://doi.org/10.1016/j.biomaterials.2025.123282
158. Kooijmans SAA, Schiffelers RM, Zarovni N, Vago R (2016) Modulation of tissue tropism and biological activity of exosomes and other extracellular vesicles: new nanotools for cancer treatment. Pharmacol Res 111:487–500. https://doi.org/10.1016/j.phrs.2016.07.006
159. Ezike TC et al (2023) Advances in drug delivery systems, challenges and future directions. Heliyon 9:e17488. https://doi.org/10.1016/j.heliyon.2023.e17488
160. Wang S et al (2023) Nanoarray enabled size-dependent isolation and proteomics profiling of small extracellular vesicle subpopulations toward accurate cancer diagnosis and prognosis. Anal Chem 95:15276–15285. https://doi.org/10.1021/acs.analchem.3c02594

161. Di Bella MA (2022) Overview and update on extracellular vesicles: considerations on exosomes and their application in modern medicine. Biology (Basel) 11. https://doi.org/10.3390/biology11060804
162. Zhang R et al (2022) sEVs(RVG) selectively delivers antiviral siRNA to fetus brain, inhibits ZIKV infection and mitigates ZIKV-induced microcephaly in mouse model. Mol Ther 30:2078–2091. https://doi.org/10.1016/j.ymthe.2021.10.009
163. Smyth T et al (2014) Surface functionalization of exosomes using click chemistry. Bioconjug Chem 25:1777–1784. https://doi.org/10.1021/bc500291r
164. Li C et al (2024) Research progress on the PEGylation of therapeutic proteins and peptides (TPPs). Front Pharmacol 15:1353626. https://doi.org/10.3389/fphar.2024.1353626
165. Shao M et al (2024) Revolutionizing cancer treatment via bioengineered extracellular vesicles: exploring nanovesicles to fully synthetic solutions. Appl Mater Today 40:102395. https://doi.org/10.1016/j.apmt.2024.102395
166. Yang LY et al (2025) Enhanced tumor self-targeting of lemon-derived extracellular vesicles by embedding homotypic cancer cell membranes for efficient drug delivery. J Nanobiotechnol 23:74. https://doi.org/10.1186/s12951-025-03161-z
167. Giebel B, Lim SK (2025) Overcoming challenges in MSC-sEV therapeutics: insights and advances after a decade of research. Cytotherapy. https://doi.org/10.1016/j.jcyt.2025.03.505
168. Deng M et al (2025) Decellularized tissue matrices hydrogels functionalized with extracellular vesicles promote macrophage reprogramming and neural stem cell differentiation for spinal cord injury repair. J Nanobiotechnol 23:139. https://doi.org/10.1186/s12951-025-03152-0
169. Deepa A, Mastan A, Buddolla V (2025) Advances in imaging techniques for real-time microbial visualization in wastewater treatment reactors: challenges, applications, and process optimization. TrAC Trends Anal Chem 188:118227. https://doi.org/10.1016/j.trac.2025.118227
170. Younis MA, Tawfeek HM, Abdellatif AAH, Abdel-Aleem JA, Harashima H (2022) Clinical translation of nanomedicines: challenges, opportunities, and keys. Adv Drug Deliv Rev 181:114083. https://doi.org/10.1016/j.addr.2021.114083
171. Saint-Pol J, Culot M (2025) Minimum information for studies of extracellular vesicles (MISEV) as toolbox for rigorous, reproducible and homogeneous studies on extracellular vesicles. Toxicol In Vitro 106:106049. https://doi.org/10.1016/j.tiv.2025.106049
172. Krmpot AJ et al (2019) Functional fluorescence microscopy imaging: quantitative scanning-free confocal fluorescence microscopy for the characterization of fast dynamic processes in live cells. Anal Chem 91:11129–11137. https://doi.org/10.1021/acs.analchem.9b01813
173. Sanches PHG, de Melo NC, Porcari AM, de Carvalho LM (2024) Integrating molecular perspectives: strategies for comprehensive multi-omics integrative data analysis and machine learning applications in Transcriptomics, proteomics, and metabolomics. Biology (Basel) 13. https://doi.org/10.3390/biology13110848
174. Rai A, Claridge B, Lozano J, Greening DW (2024) The discovery of extracellular vesicles and their emergence as a next-generation therapy. Circ Res 135:198–221. https://doi.org/10.1161/circresaha.123.323054
175. Marzetti E, Lozanoska-Ochser B, Calvani R, Landi F, Coelho-Júnior HJ, Picca A (2024) Restoring mitochondrial function and muscle satellite cell signaling: remedies against age-related sarcopenia. Biomolecules 14. https://doi.org/10.3390/biom14040415

Chapter 28
Adjunctive Therapy for Wound Healing

Mohamed Abdelhakim and Rei Ogawa

Abstract Chronic and nonhealing wounds represent a major clinical and economic burden, particularly in patients with complex comorbidities such as diabetes, vascular disease, and autoimmune disorders. Conventional wound care alone is often insufficient in these cases, driving the development and adoption of adjunctive therapies that enhance the body's intrinsic healing mechanisms. This chapter reviews key adjunctive modalities used in complex wound management, focusing on their mechanisms, clinical applications, and emerging evidence. Negative pressure wound therapy (NPWT) has become a mainstay in the treatment of acute and chronic wounds, improving drainage, reducing bacterial burden, enhancing perfusion, and stimulating granulation tissue formation, with demonstrated benefits in wound closure rates, infection control, and cost effectiveness. Extracorporeal shock wave therapy (ESWT), as a form of mechanotherapy, promotes angiogenesis, modulates fibroblast activity, remodels extracellular matrix, and shows promise in refractory wounds and scar management. Oxygen therapy interventions, including hyperbaric oxygen therapy (HBOT) and topical oxygen therapy (TOT), improve tissue oxygenation, support collagen synthesis and angiogenesis, and reduce infection in difficult-to-heal ulcers and burns. Additional adjunctive options such as recombinant growth factors, stem cell therapies (including MSC- and ASC-based approaches), electrical stimulation, and photobiomodulation therapy(PBMT) are discussed, alongside bioengineered scaffolds, skin substitutes, and nanotechnology-enabled delivery systems that further optimize the wound microenvironment. The chapter also addresses practical challenges, including high costs, limited reimbursement, stringent regulatory frameworks, the need for careful patient selection, and adherence to often intensive treatment protocols. Finally, we highlight future directions, such as portable devices, combined-therapy strategies, and the integration of artificial intelligence for personalized treatment planning and outcome prediction.

M. Abdelhakim (✉) · R. Ogawa
Department of Plastic, Reconstructive, and Aesthetic Surgery, Nippon Medical School Hospital, Tokyo, Japan
e-mail: r.ogawa@nms.ac.jp

F. De Francesco, K. Turksen (eds.), *Management and Strategies for Wound Healing*, Stem Cell Biology and Regenerative Medicine 79,
https://doi.org/10.1007/978-3-032-12442-5_28

Collectively, these adjunctive therapies are reshaping modern wound care and hold substantial potential to improve outcomes in patients with the most challenging wounds

Keywords Chronic wounds · Adjunctive therapies · Wound healing · Diabetes and vascular disease · Complex wound management

Abbreviations

NPWT	negative pressure wound therapy
OTI	oxygen therapy interventions
SWT	shockwave therapies
HBOT	hyperbaric oxygen therapy
TOT	topical oxygen therapy
PDGF	platelet-derived growth factor
VEGF	vascular endothelial growth factor
FGF	fibroblast growth factor
EGF	epidermal growth factor
KGF	keratinocyte growth factor
TGF-β	transforming growth factor beta
GM-CSF	granulocyte-macrophage colony-stimulating factor
MSCs	mesenchymal stem cells
ASCs	adipose-derived stem cells
iPSCs	induced pluripotent stem cells
ESWT	extracorporeal shock wave therapy
ISMST	International Society for Medical Shockwave Treatment
PBMT	photobiomodulation therapy
LLLT	low-level laser therapy
AI	artificial intelligence

Introduction

Chronic or nonhealing wounds can arise in the context of various medical conditions, leading to significant morbidity and substantial healthcare costs. These challenging wounds may occur following trauma, infection, cancer, radiation therapy, frostbite, animal bites, or prolonged immobility. Their complexity increases when patients also have conditions such as diabetes mellitus, peripheral vascular disease, autoimmune disorders, neuropathy, steroid dependence, or venous stasis. Despite extensive laboratory and clinical research, no single treatment has proven universally effective for all wound types. However, numerous devices that promote wound healing have established valuable roles as adjunctive therapies. These therapies have become essential in managing complex wounds, improving outcomes by enhancing the body's natural healing mechanisms.

These therapies provide solutions for wounds that fail to heal with conventional treatments, leveraging advanced technologies and biological mechanisms to accelerate recovery and improve quality of life. Herein, in this chapter we discuss the most prominent adjunctive therapies including: negative pressure wound therapy (NPWT), oxygen therapy interventions (OTI), and shockwave therapies (SWT). Additionally, we explore other emerging modalities, discussing their mechanisms of action, clinical indications, and the evidence supporting their use. These interventions hold promise for transforming wound care practices globally [1, 2].

Negative Pressure Wound Therapy (NPWT)

Known as vacuum-assisted closure (VAC), this therapy has rapidly become a mainstay in chronic wound management. Developed for clinical use at the Bowman Gray School of Medicine, the V.A.C. TM uses a subatmospheric pressure dressing to convert an open wound to a controlled closed wound. With the development of negative pressure wound therapy (NPWT), the outcomes of pathological scar treatment have significantly improved. The application of NPWT in burn surgery has garnered much attention [3]. Its role in promoting wound repair has become widely recognized. However, its specific function in pathological scar treatment has received less focus. In recent years, NPWT has been recommended as a significant approach for pathological scar treatment [4].

The technique is straightforward and yields favorable prognoses. Since published studies on pathological scar treatment rarely isolate the effects of NPWT, these studies also refer to wounds with similar characteristics to pathological scars. NPWT consists of a foam sponge, drainage tube, and semipermeable membrane. Two main types of foam are available: a polyurethane foam or a denser polyvinyl alcohol foam. Suction in negative pressure creates a clean and one-way sealed environment around the wound. The characteristics of foam sponges have been widely reported [5]. The semipermeable membrane is also related to effectiveness and comfort [6]. It is a one-way, breathable, transparent film, and its main components are polyurethane and acrylic. Oxygen can enter through the dressing, and water vapor and carbon dioxide leave the wound. Consequently, little infiltration occurs around the wound.

Several technical advances in NPWT devices and dressings have been made. The feasibility of home NPWT and single-use NPWT has been verified [7].

Additionally, other dressings may be considered for use in combination with NPWT. The gauze is placed over the wound to prevent the granulation from growing into the negative pressure material. Silver-ion dressings may also increase the antibacterial effect of NPWT [8].

In addition, the foam dressing is fixed around the negative pressure material to avoid tension blisters. Numerous clinical studies have validated the efficacy of

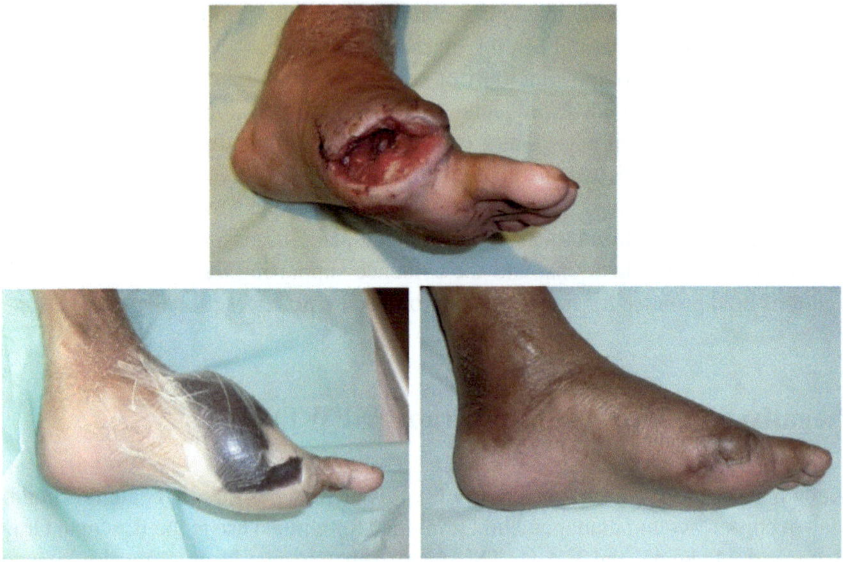

Fig. 28.1 Progressive wound healing of a big toe using negative pressure wound therapy (NPWT), with significant recovery evident after 8 weeks

NPWT. A systematic review of randomized controlled trials found significant improvements in wound closure rates, reduced time to healing, and lower incidences of infection in patients treated with NPWT compared to conventional therapies [8]. Early clinical results of case reports and small series were impressive, prompting surgeons to apply this technology to a wide range of clinical problems, commonly used to treat pressure sores, open abdomens, sternal wounds, traumatic wounds, diabetic foot infections, second-degree burns, and skin graft recipient sites. Moreover, NPWT has been shown to reduce hospital stays and overall treatment costs, making it a cost-effective option for managing complex wounds [5] (Fig. 28.1).

Mechanism of Action of NPWT

Negative pressure therapy exerts many effects on both the gross and microscopic levels. It operates by applying a controlled, continuous, or intermittent negative pressure to the wound bed. This subatmospheric pressure promotes wound healing through several mechanisms [8].

Drain Secretions and Remove Necrotic Tissue

NPWT effectively enhances wound drainage and removes bacterial products [5].

It also prevents cross-infection, as this process is carried out in a closed environment, minimizing the risk of cross-infection [9]. NPWT plays a role in promoting skin graft success by providing a vascularized bed with a low degree of bacterial colonization which increases the likelihood of successful skin grafting [10]. A study has reported higher CD34 and CD68 levels in the NPWT group compared to traditional methods which indicates its significant role in the inflammatory response during wound healing [11]. Finally, NPWT effectively drains secretions and removes necrotic tissues, playing a crucial role in wound healing.

Promote Circulation and Lymphatic Return, Reduce Hematoma and Edema

The mechanical traction from negative pressure enhances the pressure difference across capillaries and widens the endothelial intercellular space in lymphatic capillaries [12].

Subsequently, blood supply and lymphatic return flow are significantly improved.

Moreover, tissue edema is decreased through the NPWT process. NPWT also results in vascular endothelial growth factor (VEGF) and angiopoietin-2 (Ang-2) to reach higher levels due to the closed and hypoxic environment of NPWT [14]. The exerted negative pressure continuously supports local circulation, reducing seroma and hematoma compared to traditional dressings [13].

Stimulate Granulation Tissue Growth and Improve Vascular Bed Conditions

Proper preparation of the wound bed is critical to preventing graft loss [14]. NPWT supports healing through local immune modulation, hypoxia-mediated signaling, and activation of mechanoreceptors. NPWT promotes the expression of cellular fibronectin (cFN) and transforming growth factor-β1 (TGF-β1), thus leading to enhanced granulation tissue growth [14]. Continuous negative pressure reduces healing time before skin grafting by drawing interstitial fluid, encouraging capillary and granulation tissue growth, and improving blood circulation [15–17].

Indications, Contraindications, and Precautions

NPWT is widely indicated for various types of wounds, including acute wounds such as surgical wounds, traumatic wounds, and dehisced incisions which can benefit from NPWT by reducing the risk of infection and enhancing healing speed. Chronic wounds including pressure ulcers, diabetic foot ulcers, and venous leg ulcers are effectively managed with NPWT, particularly when standard care fails [18, 19]. Particularly in complex wounds, it is useful for wounds with large tissue defects, exposed bones or tendons, and those requiring skin grafts or flaps.

In the treatment of diabetic foot ulcers, NPWT promotes wound area reduction, wound bed granulation, and microbial clearance [18], therefore enabling a higher rate of limb salvage, especially in Wagner grade 3 and grade 4 ulcers [19].

NPWT prevents digit amputation in the nonoperative management of scleroderma ulcers [20–23].

To take in consideration, NPWT should be used cautiously or avoided in patients with untreated osteomyelitis, malignancy in the wound bed, or wounds with exposed blood vessels. It is also contraindicated in the presence of necrotic tissue unless adequate debridement is performed. Careful patient selection and device management are essential to prevent complications such as maceration or bleeding.

Extracorporeal Shock Wave Therapy (ESWT)

ESWT is a standard treatment indicated for urinary stones. In orthopedics, specifically in Japan, ESWT is covered by insurance only for refractory plantar fasciitis. Meanwhile, the Consensus Statement of the International Society for Medical Shockwave Treatment (ISMST) has indicated that ESWT is a feasible and noninvasive treatment for intractable skin wounds and skin ulcers; these findings are congruent with those of a previous systematic review and meta-analysis. Some recent studies have examined ESWT for hypertrophic and postburn scars [24–26].

In refractory skin wounds, the peripheral limbs are prone to decreases in blood flow due to peripheral circulatory insufficiency, and wound healing is likely to be delayed owing to the lack of soft tissue. In addition, motor functional structures, such as tendons and bones, are susceptible to infection and are therefore prone to becoming dysfunctional [27].

Mechanism of Action of ESWT

The mechanism of action of ESWT on skin wounds is thought to be angiogenic, induced by cavitation [28]. Cavitation is a phenomenon characterized by intracellular microbubble generation that disappears in the ischemic area of the vascular endothelium that has undergone low-power ESWT. This leads to a type of shear stress on the vascular endothelial cells, followed by increased expression of vascular endothelial growth factor, stromal cell-derived factors, and angiogenesis-promoting factors such as nitric oxide. These factors are thought to stimulate the secretion of vasodilators [28, 29].

In addition, these effects are reportedly maximized at approximately 0.1 mJ/mm^2, as reported by an in vitro study performed using cultured human vascular endothelial cells [28].

Low-output ESWT is reported to improve incision wound healing due to increased expression of transforming growth factor-β1 levels, including the release of cytokines, which stimulates collagen production to increase skin tensile strength [29].

On the other hand, when diabetic skin ulcers, pressure ulcers, and skin ulcers associated with vasculitis become chronic, the wound-healing process may stagnate and be accompanied by bacterial infection. The use of ESWT is effective in combination with antibacterial agents. Shock waves are reported to be effective in destroying drug-resistant biofilms both in vitro and in vivo; thus, it is suggested that ESWT may be useful for infected wounds [28].

ESWT has been reported to provide pain relief by destroying free nerve endings [29, 30].

The exact clinical efficacy of this technology in specific wound types and the precise mechanisms of action are now beginning to be understood. Shock wave therapy may be perceived by cell surface receptors through extracellular matrix (ECM) and fluid effects. Mechanoreceptors, including integrins, ion channels, connexins, and/or the lipid component of the plasma membrane activation, could all possibly be affected by shock wave therapy. Akt-mediated mechanotransduction in fibroblasts has been shown to play a role in hypertrophic scar formation in response to mechanical forces, suggesting that Akt and other upstream components such as focal adhesion kinase would be important candidates to study in the future for shock wave therapy. We believe that additional basic science studies along with randomized controlled trials and registry studies powered to detect clinically relevant outcomes will be necessary to increase our understanding of this technology.

Contraindications for SWT include malignancies, active infections, bone fractures, pregnancy, and application near growth plates in children. Patients with clotting disorders or those on anticoagulant therapy require careful consideration due to the mechanical impact on tissues.

Role of Mechanotherapy in SWT and Scar Management

SWT is classified as mechanotherapy, which uses mechanical forces to enhance tissue healing and regeneration. Like NPWT, SWT activates cellular repair mechanisms through different pathways. It converts external mechanical stimuli into biochemical cell reactions (e.g., gene transcription leading to collagen remodeling) in living tissue, supported by mechanotransduction pathways, which result in the activation of numerous cellular events, responsible for the positive effects of SWT on cell metabolism and cell cycle [31].

Moreover, as in some other mechanotherapies applied in clinical practice, the main action of SWT seems to focus on inducing tissue regeneration and matrix remodeling in vivo [32].

Fibroblasts are the major mechanoresponsive cells in connective tissue and are therefore the main target of SWT [33].

Fibroblasts play a crucial role in remodeling of the extracellular matrix (ECM) by synthesizing and organizing connective tissue components, and there is a strong belief that SWT has a modulating effect on these actions and could therefore be of interest in scar prevention and management.

While NPWT focuses on maintaining a moist environment and removing exudate, SWT targets deeper cellular responses, making them complementary in managing difficult wounds [8, 19, 20].

Fibrous tissue can be reduced by SWT (at the origin) during wound-healing processes, but it can also be remodeled in a second phase of scar formation. Primary data show that height, pliability, vascularity, and pigmentation, all relevant scar parameters, show improvement after SWT. The change in these physical and physiological parameters will probably lead to amelioration in function, which is, for example, demonstrated by an increase of passive ROM reported in a study on retracting hand scars. Subjective pain measures also show a decrease in pain at the scar site after SWT [34, 35].

On a histopathological level, the effects of SWT on fibrosis are plural. A down-regulation of alpha-SMA expression, myofibroblast phenotype, TGF-β1 expression, fibronectin, and collagen type I is measured. A significant increase in dermal fibroblast-like phenotype with low contractility and high migratory ability, small vessel density, and precursors of extracellular matrix components probably leads to new and thinner collagen fascicles and parallel orientation to the dermo-epidermal junction. Synergistic alterations in pro- and anti-fibrotic proteins (TGF-β1 and MMP-2, respectively) suggest a reduced capsule formation after silicone implantation [35–37].

Oxygen Therapy Interventions (OTI)

Hyperbaric Oxygen Therapy (HBOT)

In plastic surgery, HBOT is regarded as a successful adjunctive therapy for promoting wound healing, reducing inflammatory reactions, and improving flap survival [38].

Keloid incisions are often vulnerable to infection after radiotherapy. HBOT can improve local blood circulation and wound healing by increasing the blood oxygen capacity and tension of the body. Based on this capacity, we innovatively applied adjunctive HBOT to keloid treatment and achieved satisfactory results. Oxygen is necessary for hydroxylation of proline and lysine, the polymerization and crosslinking of procollagen strands, collagen transport, fibroblast and endothelial cell replication, effective leukocyte killing, angiogenesis, and many other processes [39].

HBOT involves administering 100% oxygen at elevated atmospheric pressures in a specialized hyperbaric chamber. The elevated oxygen levels dissolve more oxygen in the blood plasma, significantly increasing oxygen delivery to hypoxic tissues.

HBOT stimulates new blood vessel formation by enhancing VEGF expression and endothelial cell proliferation. Oxygen is a critical substrate for collagen production, ensuring fibroblasts synthesize strong, well-organized fibers, which subsequently enhances collagen synthesis.

conduits for mobilization of MSC. Local application, topically or via injection, of MSC would place these progenitor cells at the site of injury, assisting in homing and delivery. Human studies with recent small case-series transplanting autologous cultured marrow-derived MSC and pure marrow aspirate have shown rapid improvement in granulation and healing in chronic wounds [49–53].

Pioneering Research and Clinical Applications

Japan has been at the forefront of stem cell research and applications, with numerous studies showcasing the potential of MSCs in wound healing, for instance, demonstrating the efficacy of adipose-derived stem cells (ASCs) in improving healing outcomes in chronic wounds, such as diabetic foot ulcers. These studies highlighted how ASCs significantly enhanced re-epithelialization and reduced inflammation through their paracrine effects [54, 55].

Another notable study investigated the use of bone marrow-derived MSCs in treating radiation-induced skin injuries. The findings revealed accelerated wound closure and improved dermal thickness, attributed to the MSCs' ability to promote angiogenesis and reduce oxidative stress. This research underscores the potential of stem cells in managing complex wounds that are resistant to conventional therapies [56].

Integration of Bioengineering with Stem Cell Therapy

To maximize the therapeutic potential of stem cells, Japanese researchers have explored combining MSCs with bioengineered scaffolds and hydrogel matrices. A study introduced a novel hydrogel containing MSCs, which provided structural support while facilitating cell survival and proliferation. This approach significantly enhanced wound contraction and collagen deposition in animal models, paving the way for its clinical application [57, 58].

Challenges and Future Directions

Despite the promising results, several challenges remain. Ethical concerns, particularly surrounding the use of embryonic stem cells, have prompted researchers to focus more on adult-derived stem cells such as ASCs and induced pluripotent stem cells (iPSCs). The high cost of cell-based therapies and the stringent regulatory landscape in Japan further complicate the path to widespread clinical adoption. Moreover, advancements in 3D bioprinting and nanotechnology offer exciting prospects for the integration of stem cells with customizable scaffolds, enabling precise delivery and localized effects. Future research in Japan is likely to focus on optimizing these combinations to enhance safety, efficacy, and scalability for wound care applications [59, 60].

Electrical Stimulation Therapy

Electrical stimulation therapy employs low-voltage currents to influence cellular processes, thereby accelerating wound healing. Electrical currents are known to stimulate cell migration (galvanotaxis), enhance blood flow, and promote the release of endogenous growth factors. These effects contribute to faster re-epithelialization and wound contraction [61].

Studies demonstrate that electrical stimulation can be synergistically combined with negative pressure wound therapy (NPWT) or hyperbaric oxygen therapy (HBOT) to enhance outcomes in chronic wounds. For instance, diabetic ulcers and pressure injuries respond particularly well to this adjunctive treatment [61, 62].

Practical considerations include the need for precise parameter settings, patient adherence, and potential contraindications such as the presence of pacemakers. Innovations in wearable electrical stimulation devices aim to simplify application and improve patient compliance.

Photobiomodulation Therapy (PBMT)

Photobiomodulation therapy (PBMT), also known as low-level laser therapy (LLLT), utilizes specific wavelengths of light to promote tissue repair and reduce inflammation. By enhancing ATP production in mitochondria, PBMT stimulates cellular proliferation, collagen synthesis, and angiogenesis. The anti-inflammatory properties of PBMT also mitigate oxidative stress and cytokine activity, contributing to a favorable wound-healing environment.

Emerging evidence supports PBMT's application in treating diabetic foot ulcers, venous leg ulcers, and radiation-induced wounds. Additionally, portable and user-friendly PBMT devices have expanded its accessibility in outpatient and home settings. Current research focuses on optimizing light parameters such as wavelength, intensity, and duration to maximize therapeutic efficacy [63, 64].

Emerging Modalities

Advancements in biomaterials and nanotechnology have led to the development of cutting-edge solutions for wound management. Bioengineered skin substitutes, such as allografts and xenografts, mimic the structural and functional properties of native skin, providing a scaffold for cellular growth and tissue regeneration [65].

Similarly, nanotechnology-based drug delivery systems enable the controlled release of antimicrobial agents, growth factors, or stem cells directly to the wound site.

Examples include nanoparticle-coated dressings and hydrogels embedded with therapeutic agents [66]. These innovations not only accelerate healing but also reduce the risk of infection and minimize the need for frequent dressing changes. As research progresses, integrating these technologies with adjunctive therapies like PBMT or electrical stimulation could revolutionize wound care [66].

Challenges and Considerations in Adjunctive Therapy

Firstly, high costs remain a significant barrier to the widespread adoption of advanced adjunctive therapies, including HBOT, stem cell therapy, and bioengineered skin substitutes. Insurance reimbursement policies often do not cover these treatments, limiting accessibility for many patients. Policymakers and healthcare providers must collaborate to develop cost-effective solutions and expand coverage, ensuring equitable access [67].

Secondly, the success of adjunctive therapies hinges on careful patient selection. Factors such as wound type, comorbidities, and individual tolerance to treatment must be thoroughly evaluated. For instance, not all patients can tolerate HBOT due to risks of oxygen toxicity or barotrauma. Similarly, certain conditions may contraindicate the use of electrical stimulation devices. Personalized treatment plans based on clinical evaluation are essential to maximize efficacy and minimize risks [51]. Finally, adherence to treatment protocols is critical for achieving optimal outcomes. Adjunctive therapies like HBOT require consistent session attendance, while NPWT devices demand proper maintenance and usage. Noncompliance can compromise treatment efficacy and lead to complications, such as infections or device malfunction. Regular monitoring by healthcare professionals, combined with patient education, is vital to address these challenges.

Future Directions

The field of adjunctive wound therapies is rapidly advancing, driven by innovations in technology and materials science. Notable areas of progress include portable hyperbaric oxygen therapy (HBOT) devices designed to reduce costs and enhance accessibility, particularly in outpatient settings; the development of cost-effective recombinant growth factor formulations using plant-based or microbial systems, and synergistic combination therapies, such as integrating negative pressure wound therapy (NPWT) with advanced dressings or localized drug delivery systems. Additionally, artificial intelligence (AI) is emerging as a transformative tool, offering personalized treatment plans, predictive analytics, and optimized therapy protocols to enhance decision-making, improve patient outcomes, and minimize complications [68, 69].

Conclusion

Adjunctive therapies, including NPWT, HBOT, and emerging modalities like stem cell therapy and PBMT, have transformed the landscape of wound management. By leveraging advanced technologies and multidisciplinary approaches, clinicians can accelerate wound healing, reduce complications, and improve patient quality of life. Despite challenges such as cost and patient compliance, ongoing research and innovation are paving the way for more accessible and effective treatments. The integration of AI, biomaterials, and nanotechnology heralds a new era in wound care, empowering healthcare providers to achieve superior results in even the most challenging cases.

References

1. Normandin S, Safran T, Winocour S, Chu CK, Vorstenbosch J, Murphy AM, Davison PG (2021) Negative pressure wound therapy: mechanism of action and clinical applications. Semin Plast Surg 35(3):164–170. https://doi.org/10.1055/s-0041-1731792. Epub 2021 Sep 10. PMID: 34526864; PMCID: PMC8432996
2. Bishop A (2008) Role of oxygen in wound healing. J Wound Care 17(9):399–402. https://doi.org/10.12968/jowc.2008.17.9.30937. PMID: 18833899
3. Chai J, Shen C (2015) Lay emphasis on the application of negative pressure wound therapy technique in burn surgery. Chin J Burns 31(2):81–83. https://doi.org/10.3760/cma.j.issn.1009-2587.2015.02.001
4. Cai JH, Deng HP, Shen ZA, Sun TJ, Li DJ, Li DW et al (2017) Effects of scar excision combined with negative-pressure on repair of hypertrophic scar in burn children. Zhonghua Shao Shang Za Zhi 33(7):410–414. https://doi.org/10.3760/cma.j.issn.1009-2587.2017.07.003
5. Agarwal P, Kukrele R, Sharma D (2019) Vacuum assisted closure (VAC)/negative pressure wound therapy (NPWT) for difficult wounds: a review. J Clin Orthop Trauma 10(5):845–848. https://doi.org/10.1016/j.jcot.2019.06.015
6. Dooley TP, Ellis AL, Belousova M, Petersen D, DeCarlo AA (2012) Dense chitosan surgical membranes produced by a coincident compression-dehydration process. J Biomater Sci Polym Ed 24(5):621–643. https://doi.org/10.1080/09205063.2012.701549
7. Lim X, Zhang L, Hong Q, Yong E, Neo S, Chandrasekar S et al (2021) Novel home use of mechanical negative pressure wound therapy in diabetic foot ulcers. J Wound Care 30(12):1006–1010. https://doi.org/10.12968/jowc.2021.30.12.1006
8. Huang C, Leavitt T, Bayer LR, Orgill DP (2014) Effect of negative pressure wound therapy on wound healing. Curr Probl Surg 51(7):301–331. https://doi.org/10.1067/j.cpsurg.2014.04.001
9. Huang Q, Huang K, Xue J (2021) Vacuum sealing drainage combined with free anterolateral femoral skin flap grafting in 16 cases of pediatric soft tissue damage to the foot and ankle. Transl Pediatr 10(10):2489–2495. https://doi.org/10.21037/tp-21-399

10. Kantak NA, Mistry R, Halvorson EG (2016) A review of negative-pressure wound therapy in the management of burn wounds. Burns 42(8):1623–1633. https://doi.org/10.1016/j.burns.2016.06.011
11. Yang H, Liu L, Li G, Chen Y, Jiang D, Wang W et al (2021) Growth promoting effect of vacuum sealing drainage in the healing processes of diabetic foot ulcers. Ther Clin Risk Manag 17:65–71. https://doi.org/10.2147/TCRM.S28284
12. Cagney D, Simmons L, O'Leary DP, Corrigan M, Kelly L, O'Sullivan MJ et al (2020) The efficacy of prophylactic negative pressure wound therapy for closed incisions in breast surgery: a systematic review and meta-analysis. World J Surg 44(5):1526–1537. https://doi.org/10.1007/s00268-019-05335-x
13. Nagata T, Miura K, Homma Y, Fukamizu H (2018) Comparison between negative-pressure fixation and film dressing in wound management after tissue expansion: a randomized controlled trial. Plast Reconstr Surg 142(1):37–41. https://doi.org/10.1097/PRS.0000000000004470
14. Yang SL, Zhu LY, Han R, Sun LL, Dou JT (2017) Effect of negative pressure wound therapy on cellular fibronectin and transforming growth factor-β1 expression in diabetic foot wounds. Foot Ankle Int 38(8):893–900. https://doi.org/10.1177/1071100717704940
15. Hsiao SF, Ma H, Wang YH, Wang TH (2017) Occlusive drainage system for split-thickness skin graft: a prospective randomized controlled trial. Burns 43(2):379–387. https://doi.org/10.1016/j.burns.2016.08.025
16. Glass GE, Murphy GF, Esmaeili A, Lai LM, Nanchahal J (2014) Systematic review of molecular mechanism of action of negative-pressure wound therapy. Br J Surg 101(13):1627–1636. https://doi.org/10.1002/bjs.9636
17. Sun T, Ying W, Wang S, Chen C, Sun P, Tan J (2021) Clinical application of vacuum sealing drainage for the treatment of deep burn wounds. Am Surgeon™ 89(4):1018–1023. https://doi.org/10.1177/00031348211054527
18. Eid J (2016) Consensus statement on ESWT indications and contraindications [Internet]. International Society for Medical Shockwave Treatment; [cited 2021 May 29]. Available from: https://www.shockwavetherapy.org/fileadmin/user_upload/dokumente/PDFs/Formulare/ISMST_consensus_statement_on_indications_and_contraindications_20161012_final.pdf
19. Mittermayr R, Antonic V, Hartinger J et al (2012) Extracorporeal shock wave therapy (ESWT) for wound healing: technology, mechanisms, and clinical efficacy. Wound Repair Regen 20:456–465
20. Zhang L, Weng C, Zhao Z et al (2017) Extracorporeal shock wave therapy for chronic wounds: a systematic review and meta-analysis of randomized controlled trials. Wound Repair Regen 25:697–706
21. Nather A, Hong NY, Lin WK, Sakharam JA (2011) Effectiveness of bridge V.A.C. dressings in the treatment of diabetic foot ulcers. Diabet Foot Ankle 2:5893
22. Ulusal AE, Sahin MS, Ulusal B, Cakmak G, Tuncay C (2011) Negative pressure wound therapy in patients with diabetic foot. Acta Orthop Traumatol Turc 45(4):254–260
23. Patel RM, Nagle DJ (2012) Nonoperative management of scleroderma of the hand with tadalafil and subatmospheric pressure wound therapy: case report. J Hand Surg 37(4):803–806
24. Zhao JC, Zhang BR, Shi K et al (2018) Lower energy radial shock wave therapy improves characteristics of hypertrophic scar in a rabbit ear model. Exp Ther Med 15:933–939
25. Nishida T, Shimokawa H, Oi K et al (2004) Extracorporeal cardiac shock wave therapy markedly ameliorates ischemia-induced myocardial dysfunction in pigs in vivo. Circulation 110:3055–3061
26. Yang G, Luo C, Yan X et al (2011) Extracorporeal shock wave treatment improves incisional wound healing in diabetic rats. Tohoku J Exp Med 225:285–292
27. Datey A, Thaha CSA, Patil SR et al (2019) Shockwave therapy efficiently cures multispecies chronic periodontitis in a humanized rat model. Front Bioeng Biotechnol 7:382
28. Gnanadhas DP, Elango M, Janardhanraj S et al (2015) Successful treatment of biofilm infections using shock waves combined with antibiotic therapy. Sci Rep 5:17440

29. Ohtori S, Inoue G, Mannoji C et al (2001) Shock wave application to rat skin induces degeneration and reinnervation of sensory nerve fibres. Neurosci Lett 315:57–60
30. Takahashi N, Wada Y, Ohtori S et al (2003) Application of shock waves to rat skin decreases calcitonin gene-related peptide immunoreactivity in dorsal root ganglion neurons. Auton Neurosci 107:81–84
31. Huang C, Holfeld J, Schaden W, Orgill D, Ogawa R (2013) Mechanotherapy: revisiting physical therapy and recruiting mechanobiology for a new era in medicine. Trends Mol Med 19:555–564
32. d'Agostino MC, Craig K, Tibalt E, Respizzi S (2015) Shock wave as biological therapeutic tool: from mechanical stimulation to recovery and healing, through mechanotransduction. Int J Surg 24:147–153
33. Wang JH-C, Thampatty BP, Lin J-S, Im H-J (2007) Mechanoregulation of gene expression in fibroblasts. Gene 391:1–15
34. Saggini R, Saggini A, Spagnoli AM, Dodaj I, Cigna E, Maruccia M, Soda G, Bellomo RG, Scuderi N (2016) Extracorporeal shock wave therapy: an emerging treatment modality for retracting scars of the hands. Ultrasound Med Biol 42:185–195. [PubMed] [CrossRef]
35. Rinella L, Marano F, Berta L, Bosco O, Fraccalvieri M, Fortunati N, Frairia R, Catalano MG (2016) Extracorporeal shock waves modulate myofibroblast differentiation of adipose-derived stem cells. Wound Repair Regen 24:275–286. [PubMed] [CrossRef]
36. Cui HS, Hong AR, Kim J-B, Yu JH, Cho YS, Joo SY, Seo CH (2018) Extracorporeal shock wave therapy alters the expression of fibrosis-related molecules in fibroblast derived from human hypertrophic scar. Int J Mol Sci. [PMC free article] [PubMed][https://doi.org/10.3390/ijms19010124]
37. Fischer S, Mueller W, Schulte M, Kiefer J, Hirche C, Heimer S, Köllensperger E, Germann G, Reichenberger MA (2015) Multiple extracorporeal shock wave therapy degrades capsular fibrosis after insertion of silicone implants. Ultrasound Med Biol 41:781–789
38. Zhang T, Gong W, Li Z et al (2007) Efficacy of hyperbaric oxygen on survival of random pattern skin flap in diabetic rats. Undersea Hyperb Med 34(5):335–339
39. Stone JA, Cianci P (1997) The adjunctive role of hyperbaric oxygen therapy in the treatment of lower extremity wounds in patients with diabetes. Diabetes Spectr 10(2):1–11
40. Bhutani S, Vishwanath G (2012) Hyperbaric oxygen and wound healing. Indian J Plast Surg 45(2):316–324. https://doi.org/10.4103/0970-0358.101309. PMID: 23162231; PMCID: PMC3495382
41. Oropallo A, Andersen CA (2024) Topical oxygen. [Updated 2023 Aug 28]. In: StatPearls [Internet]. StatPearls Publishing, Treasure Island
42. Chan BC, Campbell KE (2020 Dec) An economic evaluation examining the cost-effectiveness of continuous diffusion of oxygen therapy for individuals with diabetic foot ulcers. Int Wound J 17(6):1791–1808
43. Almeleh R. Spontaneous accelerated epithelialization in deep dermal burns using an oxygen-delivering hydrogel: a report of two cases, WOUNDS. 2013. https://www.woundsresearch.com/article/online-exclusive-spontaneous-accelerated-epithelialization-deep-dermal-burns-using-oxygen-de 25, 18–25
44. Park JW, Hwang SR, Yoon IS (2017) Advanced growth factor delivery systems in wound management and skin regeneration. Molecules 22:E1259. https://doi.org/10.3390/molecules22081259
45. Trengove NJ, Bielefeldt-Ohmann H, Stacey MC (2000) Mitogenic activity and cytokine levels in non-healing and healing chronic leg ulcers. Wound Repair Regen 8:13–25. https://doi.org/10.1046/j.1524-475x.2000.00013.x
46. Brown LF, Yeo KT, Berse B et al (1992) Expression of vascular permeability factor (vascular endothelial growth factor) by epidermal keratinocytes during wound healing. J Exp Med 176:1375–1379. https://doi.org/10.1084/jem.176.5.1375
47. Powers CJ, McLeskey SW, Wellstein A (2000) Fibroblast growth factors, their receptors and signaling. Endocr Relat Cancer 7:165–197. https://doi.org/10.1677/erc.0.0070165

48. Uchi H, Igarashi A, Urabe K et al (2009) Clinical efficacy of basic fibroblast growth factor (bFGF) for diabetic ulcer. Eur J Dermatol 19(5):461–468. https://doi.org/10.1684/ejd.2009.0750
49. Fathke C, Wilson L, Hutter J, Kapoor V, Smith A, Hocking A (2004) contribution of bone marrow-derived cells to skin: collagen deposition and wound repair. Stem Cells 22:812–822
50. McFarlin K, Gao X, Liu YB, Dulchavsky DS, Kwon D, Arbab AS et al (2006) Bone marrow-derived mesenchymal stromal cells accelerate wound healing in the rat. Wound Repair Regen 14:471–478
51. Bauer SM, Goldstein LJ, Bauer RJ, Chen H, Putt M, Velazquez OC (2006) The bone marrow-derived endothelial progenitor cell response is impaired in delayed wound healing from ischemia. J Vasc Surg 43:134–141
52. Badiavas EV, Ford D, Liu P, Kouttab N, Morgan J, Richards A (2007) Long-term bone marrow culture and its clinical potential in chronic wound healing. Wound Repair Regen 15:856–865
53. Falanga V, Iwamoto S, Chartier M, Yufit T, Butmarc J, Kouttab N et al (2007) Autologous bone marrow-derived cultured mesenchymal stem cells delivered in a fibrin spray accelerate healing in murine and human cutaneous wounds. Tissue Eng 13:1299–1312
54. Lin Z, Shibuya Y, Imai Y, Oshima J, Sasaki M, Sasaki K, Aihara Y, Khanh VC, Sekido M (2023) Therapeutic potential of adipose-derived stem cell-conditioned medium and extracellular vesicles in an in vitro radiation-induced skin injury model. Int J Mol Sci 24(24):17214. https://doi.org/10.3390/ijms242417214. PMID: 38139042; PMCID: PMC10743562
55. Shingyochi Y, Orbay H, Mizuno H (2015) Adipose-derived stem cells for wound repair and regeneration. Expert Opin Biol Ther 15(9):1285–1292. https://doi.org/10.1517/14712598.2015.1053867
56. Wu SH, Shirado T, Mashiko T, Feng J, Asahi R, Kanayama K, Mori M, Chi D, Sunaga A, Sarukawa S et al (2018) Therapeutic effects of human adipose-derived products on impaired wound healing in irradiated tissue. Plast Reconstr Surg 142:383–391
57. Zhang K, Chen X, Li H, Feng G, Nie Y, Wei Y et al (2020b) A nitric oxide-releasing hydrogel for enhancing the therapeutic effects of mesenchymal stem cell therapy for Hindlimb ischemia. Acta Biomater 113:289–304. https://doi.org/10.1016/j.actbio.2020.07.011
58. Nagano H, Suematsu Y, Takuma M, Aoki S, Satoh A, Takayama E et al (2021) Enhanced cellular engraftment of adipose-derived mesenchymal stem cell spheroids by using Nanosheets as scaffolds. Sci Rep 11:14500. https://doi.org/10.1038/s41598-021-93642-6
59. Mehrban N, Teoh GZ, Birchall MA (2016) 3D bioprinting for tissue engineering: stem cells in hydrogels. Int J Bioprinting 2(1):6–19. https://doi.org/10.18063/IJB.2016.01.006
60. Murphy S, Atala A (2014) 3D bioprinting of tissues and organs. Nat Biotechnol 32:773–785. https://doi.org/10.1038/nbt.2958
61. Cheah YJ, Buyong MR, Mohd Yunus MH (2021) Wound healing with electrical stimulation technologies: a review. Polymers (Basel) 13(21):3790. https://doi.org/10.3390/polym13213790. PMID: 34771347; PMCID: PMC8588136
62. Konstantinou E, Zagoriti Z, Pyriochou A, Poulas K (2020) Microcurrent stimulation triggers MAPK signaling and TGF-β1 release in fibroblast and osteoblast-like cell lines. Cells 9:1924. https://doi.org/10.3390/cells9091924
63. Bayat M, Albright R, Hamblin MR, Chien S (2022) Impact of blue light therapy on wound healing in preclinical and clinical subjects: a systematic review. J Lasers Med Sci 13:e69. https://doi.org/10.34172/jlms.2022.69. PMID: 37041783; PMCID: PMC10082909
64. Nadene Houreld N, Winfield Jere S (2020) Impact of photobiomodulation therapy on chronic wound healing. In: A closer look at wound infections and healing. Nova Science Publishers, Inc., pp 129–155
65. Dai C, Shih S, Khachemoune A (2020) Skin substitutes for acute and chronic wound healing: an updated review. J Dermatolog Treat 31(6):639–648. https://doi.org/10.1080/09546634.2018.1530443
66. Sharda D, Attri K, Choudhury D (2024) Greener healing: sustainable nanotechnology for advanced wound care. Discov Nano 19(1):127. Published 2024 Aug 13. https://doi.org/10.1186/s11671-024-04061-1

67. Carter MJ (2013) Health economics information in wound care: the elephant in the room. Adv Wound Care (New Rochelle) 2(10):563–570. https://doi.org/10.1089/wound.2013.0479. PMID: 24527322; PMCID: PMC3865616
68. Weigelt MA, Lev-Tov HA, Tomic-Canic M, Lee WD, Williams R, Strasfeld D, Kirsner RS, Herman IM (2022) Advanced Wound Diagnostics: Toward Transforming Wound Care into Precision Medicine. Adv Wound Care (New Rochelle) 11(6):330–359. https://doi.org/10.1089/wound.2020.1319. Epub 2021 Jul 21. PMID: 34128387; PMCID: PMC8982127
69. Chemello G, Salvatori B, Morettini M, Tura A (2022) Artificial intelligence methodologies applied to technologies for screening, diagnosis and care of the diabetic foot: a narrative review. Biosensors 12:985. https://doi.org/10.3390/bios12110985

Chapter 29
Nutrition and Wound Healing: Mechanisms and Clinical Integration

Scott A. Sylvester, Abigail Peoples, Samuel Pendergraft, and Joseph A. Molnar

Keywords Nutrition · wound healing · Tissue regeneration · Macronutrients · Micronutrients

Introduction

Nutrition has long been suspected to influence wound healing and tissue regeneration, with historical practices offering early insights into its therapeutic potential. Twelfth-century convents prescribed pottage, a legume and grain based stew high in proteins and vitamins, for individuals with wounds, believing it could strengthen both flesh and spirit [1, 2]. In his memoirs, Dominique Jean Larrey, Napoleon's chief surgeon, makes numerous observations on the impact of diet and the importance of "broth." He later credited such nourishing broths with reducing gangrene by enhancing patient resilience [3]. Observations during World Wars I and II further underscored the importance of supporting the body's healing capacity, including through nutrition [4]. The development of parenteral nutrition by Dudrick [5] in the

S. A. Sylvester (✉) · A. Peoples
Department of Plastic and Reconstructive Surgery, Atrium Health Wake Forest Baptist Hospital, Medical Center Boulevard, Winston-Salem, NC, USA
e-mail: scott.sylvester@advocatehealth.org

S. Pendergraft
Wake Forest Institute for Regenerative Medicine, Wake Forest University School of Medicine, Winston-Salem, NC, USA

J. A. Molnar
Department of Plastic and Reconstructive Surgery, Atrium Health Wake Forest Baptist Hospital, Medical Center Boulevard, Winston-Salem, NC, USA

Wake Forest Institute for Regenerative Medicine, Wake Forest University School of Medicine, Winston-Salem, NC, USA

F. De Francesco, K. Turksen (eds.), *Management and Strategies for Wound Healing*, Stem Cell Biology and Regenerative Medicine 79,
https://doi.org/10.1007/978-3-032-12442-5_29

1970s marked a turning point, making it possible to deliver essential nutrients to critically ill patients and reinforcing the idea that nutrition is integral to recovery.

Today, wound healing is understood as a complex, multi-phase biological process with many distinct phases. While surgical techniques, pharmacological agents, and biomaterials have traditionally dominated wound care strategies, nutrition is now recognized as a central, modifiable factor that actively shapes the trajectory and quality of healing [6–8]. Far from being a passive adjunct, nutrition influences immune responses, extracellular matrix (ECM) remodeling, angiogenesis, and cell signaling—each of which is essential for effective tissue repair and regeneration. This evolving understanding bridges centuries of empirical observation with modern biomedical science, positioning nutrition not only as supportive care but also as a dynamic co-therapy in both conventional and regenerative medicine.

The burden of chronic wounds such as diabetic ulcers, pressure injuries, and venous leg ulcers is substantial, affecting millions globally and costing healthcare systems billions of dollars per annum [9, 10]. These wounds are characterized by stalled healing, persistent inflammation, and poor tissue regeneration. In this context, nutritional deficiencies or imbalances can exacerbate dysfunction, delay recovery, and increase susceptibility to infection. Conversely, targeted nutritional interventions can restore immune competence, enhance collagen synthesis, and promote vascularization, thereby accelerating healing and improving outcomes [11]. This may become increasingly important in an increasingly older global health population in which malnutrition and vitamin deficiency may often go unnoticed [12].

In recent years, the field of regenerative medicine has expanded rapidly, offering novel therapies such as stem cell transplantation, bioengineered tissue scaffolds, and growth factor delivery systems. However, the role of nutrition in optimizing these interventions remains underappreciated. Emerging evidence suggests that nutritional status can influence the efficacy of regenerative therapies by modulating the micro- and macroenvironment in which cells and biomaterials operate [13, 14]. This evolving understanding positions nutrition not merely as supportive care but as potential co-therapy, a synergistic partner for wound healing. Nutrients such as arginine, glutamine, vitamin C, zinc, and omega-3 fatty acids have demonstrated specific roles in enhancing wound healing and regenerative outcomes [15, 16]. These nutrients, and others, can be strategically applied to complement regenerative interventions, improving tissue integration, reducing inflammation, and accelerating functional recovery [17, 18].

The key themes explored in this chapter revolve around how nutrition supports wound healing and regeneration by influencing inflammation, tissue remodeling, and cell function. We will examine the molecular and cellular mechanisms through which nutrients exert their effects and how this knowledge can be translated into clinical practice and integrated into emerging regenerative therapies. By reframing nutrition as an active modulator of healing rather than a passive background factor, we open new avenues for personalized, multi-modal approaches to wound care that are not only more effective but also more sustainable and accessible.

Wound Healing

Wound healing is a dynamic, multi-phase biological process comprised of four stages: *hemostasis, inflammation, proliferation, and remodeling*, each with distinct nutritional demands. (Fig. 29.1)

- In the *hemostasis stage*, the body initiates clot formation and vasoconstriction to prevent blood loss. Although this phase is brief, adequate intake of vitamin K and calcium is essential for the coagulation cascade function.
- The *inflammatory stage* follows, characterized by immune cell infiltration and pathogen clearance. Nutrients such as vitamins A, C, and E, zinc, and protein are critical here, supporting immune cell activation, antioxidant defense, and tissue debridement.
- In the *proliferative stage*, fibroblasts, keratinocytes, and endothelial cells drive granulation tissue formation, angiogenesis, and re-epithelialization. This stage demands amino acids (arginine and glutamine), vitamins, iron, and lipids, which fuel cell proliferation and collagen synthesis.
- Finally, the *remodeling stage* involves collagen maturation and ECM reorganization, requiring sustained intake of vitamin C, zinc, and water to support tensile strength and skin integrity [18, 19] (Fig. 29.1).

The metabolic demands of acute and chronic wound healing are substantial [6, 20]. Energy expenditure can increase by more than 100% of basal energy requirements depending on wound acuity and severity, with protein requirements rising by up to 250% in some cases [21, 22]. Chronic wounds, such as pressure ulcers and diabetic foot ulcers, are often accompanied by catabolic stress, driven by persistent inflammation, oxidative damage, and impaired anabolic signaling. Malnutrition exacerbates these challenges, impairing fibroblast function, collagen synthesis, angiogenesis, and immune regulation [23]. Deficiencies in key nutrients like zinc, vitamin C, and protein can lead to delayed wound closure, increased infection risk, and poor tissue remodeling. Moreover, there is ongoing recognition of a positive feedback loop between malnutrition and wound healing, as chronic wounds often

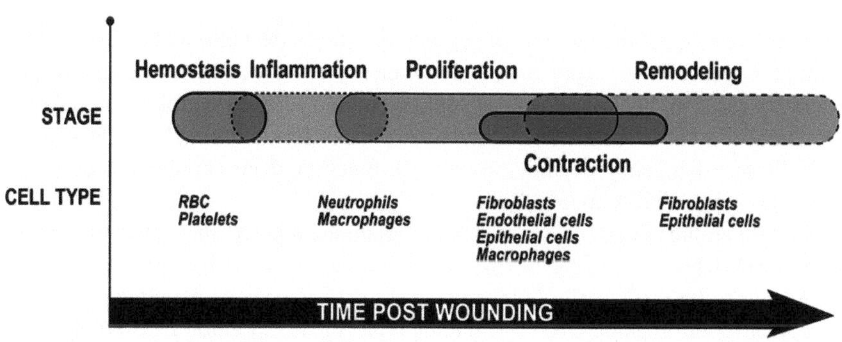

Fig. 29.1 Stages of wound healing. (Molnar J)

arise in the malnourished, and the increase in metabolic demands to attempt wound healing further exacerbates nutrient depletion, further worsening malnutrition [24].

Mechanistically, nutrition influences wound healing at the cellular level by modulating redox balance, macrophage polarization, and keratinocyte migration. For example, arginine and glutamine support nitric oxide synthesis and antioxidant defense, while omega-3 fatty acids promote anti-inflammatory macrophage (M2) polarization. These nutrients also regulate keratinocyte behavior, enhancing migration and re-epithelialization [19, 25]. Functional amino acids like aspartate and tryptophan have been shown to influence immune cell metabolism and cytokine production, further shaping the wound microenvironment. This interplay between nutrition and cellular function lays the foundation for the next section, which explores the role of various macro- and micronutrients in wound healing [26].

Macronutrients in Wound and Tissue Repair

Wound healing is a biologically intricate process that demands precise coordination of cellular and molecular events. Research on wound healing within the last half century has elucidated the importance of one's nutritional status, particularly the availability and balance of macronutrients (proteins, fats, and carbohydrates). These nutrients serve not only as energy sources but also as structural and functional components that directly influence the complex processes involved in tissue repair.

Understanding the role of macronutrients in wound healing has allowed clinicians and researchers to design more effective nutritional strategies that complement regenerative therapies. This chapter explores how each macronutrient class contributes to the healing process and how targeted nutritional support can optimize outcomes in both acute and chronic wounds.

Proteins and Amino Acids

Proteins are foundational to wound healing, serving as the building blocks for granulation tissue, collagen deposition, and immune system modulation. During the proliferative phase of healing, fibroblasts rely on amino acids to synthesize collagen, a structural protein essential for extracellular matrix formation and tensile strength [27]. Protein deficiency impairs immune cell function, delays epithelialization, and increases susceptibility to infection.

Certain amino acids, such as arginine, glutamine, and leucine, can also function as nutraceuticals.

Arginine plays a pivotal role in wound healing and tissue remodeling [28]. It is a precursor for nitric oxide (NO), a signaling molecule that promotes vasodilation and enhances tissue perfusion. Arginine also contributes to proline synthesis, a key component of collagen, and supports macrophage function, particularly in transitioning from pro-inflammatory (M1) to pro-resolving (M2) phenotypes. Clinical trials have shown that arginine-enriched formulas improve wound closure rates and reduce pressure ulcer size [29].

Glutamine, a conditionally essential amino acid, is vital for maintaining epithelial integrity and modulating oxidative stress. Among its roles, glutamine serves as a fuel source for rapidly dividing cells and supports glutathione synthesis, a major antioxidant defense mechanism. Glutamine supplementation has been associated with improved immune responses and reduced infection rates in patients with chronic wounds [30].

Leucine, a branched-chain amino acid, activates the mTOR signaling pathway, which regulates protein synthesis and cellular growth [31]. In wound healing, leucine helps preserve lean body mass and supports anabolic processes necessary for tissue regeneration. Its role is particularly important in patients with sarcopenia and those recovering from major surgery or trauma [32].

Fats

Fats contribute to wound healing in many ways. Cholesterol, phospholipids, prostaglandins, and fatty acids consumed via diet play a role in wound healing [33]. They comprise cell membranes and maintain stromal integrity. Lipids transport micronutrients and provide necessary energy. Pro-resolving lipid mediators (PMLMs) such as resolvins and protectins help resolve inflammation, promote angiogenesis, and support fibroblast activity critical to wound healing. Omega-3s also modulate immune responses, reducing excessive inflammation that can impair healing.

The balance between omega-6 and omega-3 fatty acids is critical. Western diets often contain disproportionately high levels of omega-6 fatty acids, which can promote chronic inflammation. A more balanced ratio supports anti-inflammatory pathways and improves outcomes in chronic wounds [34].

Carbohydrates

Carbohydrates provide the primary energy substrate for wound healing, fueling cellular processes such as collagen synthesis and immune activation. However, glycemic control is essential, especially in patients with diabetes. Hyperglycemia impairs endothelial function, reduces oxygen delivery, and increases the risk of infection.

Elevated blood glucose levels also promote the formation of advanced glycation end products (AGEs), which disrupt collagen structure and impair wound remodeling [35]. AGEs contribute to chronic inflammation and oxidative stress, further delaying healing. Diets emphasizing complex carbohydrates and low glycemic index foods can mitigate these effects.

Together, these macronutrients shape the cellular environment essential for tissue regeneration. Proteins and amino acids fuel anabolic processes and immune modulation; fats regulate inflammation and vascularization; and carbohydrates provide energy while influencing metabolic stress. Understanding their roles allows clinicians to tailor nutritional interventions that complement regenerative therapies and optimize healing outcomes.

Micronutrients and Bioactive Compounds in Wound Healing

- Micronutrients and other bioactive compounds play a pivotal role in the orchestration of wound healing, influencing cellular behavior, enzymatic activity, immune modulation, and tissue regeneration. While macronutrients provide the structural and energetic foundation for repair, micronutrients act as cofactors and regulators of key biological processes. Emerging evidence also highlights the therapeutic potential of bioactive compounds such as polyphenols and plant-derived antioxidants in modulating inflammation and enhancing healing outcomes. This section explores the mechanistic roles of essential micronutrients and bioactive compounds in wound physiology.
- *Zinc* is indispensable for wound healing due to its involvement in matrix metalloproteinase (MMP) regulation, epithelial cell migration, and immune function. MMPs are zinc-dependent enzymes that remodel the extracellular matrix (ECM), facilitating tissue regeneration and angiogenesis [36]. Zinc also supports keratinocyte proliferation and migration, essential for re-epithelialization. Zinc deficiency impairs neutrophil and macrophage function, leading to prolonged inflammation and delayed healing [37].
- *Vitamin C* (ascorbic acid) is a critical cofactor for prolyl and lysyl hydroxylase, enzymes required for collagen cross-linking and stabilization. Collagen provides tensile strength to healing tissues, and its proper synthesis is essential for wound integrity. Vitamin C also acts as a potent antioxidant, neutralizing reactive oxygen species (ROS) generated during inflammation and protecting cells from oxidative damage [38].
- *Vitamin A* regulates epithelial differentiation and maintains mucosal barrier integrity, both of which are vital for wound closure and defense against pathogens. It enhances fibroblast activity and supports the formation of granulation tissue. Retinoic acid, the active metabolite of vitamin A, influences gene expression related to cell proliferation and immune modulation [39].
- *Vitamin D* plays a vital role in innate cellular immunity, modulating the activity of macrophages and antimicrobial peptides. It enhances the expression of cathelicidin and defensins, proteins that help to protect against microbial invasion in open wounds. Vitamin D also influences inflammatory cytokine profiles, promoting a shift toward resolution and tissue repair [40].
- *Iron and copper* are essential for oxygen transport and serve as cofactors for enzymes involved in collagen synthesis and angiogenesis. Iron is a component of

hemoglobin and cytochromes, facilitating oxygen delivery to hypoxic wound tissues, and is a cofactor in the formation of collagen hydroxyproline [41]. Copper is required for lysyl oxidase activity, which cross-links collagen and elastin in the ECM.

- *Emerging Bioactive Compounds* such as curcumin, resveratrol, and polyphenols are gaining attention for their anti-inflammatory and pro-healing properties. Curcumin, derived from turmeric, inhibits NF-κB signaling and reduces pro-inflammatory cytokine production, thereby promoting resolution of inflammation [42]. Resveratrol, commonly found in fruits (grapes & berries), enhances angiogenesis and fibroblast proliferation while protecting against oxidative stress and decreasing circulating cytokines [43]. Polyphenols from plant sources modulate immune responses and support epithelial regeneration. While much is still unknown about their mechanisms, these compounds seem to act synergistically with micronutrients, enhancing bioavailability and therapeutic efficacy. Their integration into wound care, whether through topical formulations or dietary supplementation, represents a promising frontier in regenerative medicine.

This mechanistic understanding of micronutrients and bioactive compounds lays the foundation for personalized nutritional strategies in wound care and regenerative medicine. By targeting specific deficiencies and leveraging the healing potential of bioactive compounds, clinicians can optimize and individualize tissue repair to improve patient outcomes.

Nutritional Influence on Regenerative Mechanisms

Regenerative medicine seeks to restore or replace damaged tissues through biological processes that mimic natural healing. Nutrition plays a critical role in modulating regenerative mechanisms, influencing ECM synthesis and remodeling, stem cell behavior, and intracellular signaling pathways. This section explores how specific nutrients—vitamins, amino acids, trace minerals, and polyunsaturated fats—affect tissue regeneration at the molecular and cellular levels, with emphasis on their roles in ECM dynamics and stem cell modulation.

Extracellular Matrix (ECM) Synthesis and Remodeling

The ECM is composed of structural proteins such as collagen, elastin, and fibronectin, which provide mechanical support and regulate cell behavior. ECM remodeling is essential for wound healing and tissue regeneration, involving the coordinated degradation and synthesis of matrix components. Nutritional status directly influences ECM integrity. For example, vitamin C is a cofactor for prolyl and lysyl hydroxylases, enzymes critical for collagen cross-linking and stabilization [38].

Zinc is another key nutrient that regulates matrix metalloproteinases (MMPs), enzymes responsible for ECM turnover. Zinc-dependent MMPs facilitate the removal of damaged matrix and promote angiogenesis and cell migration. Additionally, copper supports lysyl oxidase activity, which cross-links collagen and elastin, enhancing ECM strength and elasticity. Iron, through its role in oxygen transport and enzymatic function, supports fibroblast activity and collagen synthesis, particularly in hypoxic wound environments [41].

Nutrients and Extracellular Matrix (ECM)

Nutrients influence the expression of ECM proteins through multiple mechanisms and pathways. Amino acids such as glycine, proline, and lysine are direct precursors for collagen synthesis. Vitamin A regulates epithelial differentiation and mucosal barrier integrity, indirectly affecting ECM composition by promoting keratinocyte proliferation and stratification [39]. Polyunsaturated fatty acids (PUFAs), particularly omega-3s, modulate inflammatory responses and reduce fibrotic ECM deposition, promoting a regenerative rather than scarring phenotype [33].

Nutritional Influence on Stem Cell Behavior and Regeneration

Stem cells are central to tissue regeneration, capable of differentiating into specialized cell types and replenishing damaged tissues. Their behavior is tightly regulated by the ECM and nutrient availability. Nutrients act as signaling molecules within the stem cell environment, influencing proliferation, differentiation, and epigenetic programming. For instance, vitamin D enhances mesenchymal stem cell (MSC) proliferation and differentiation by modulating innate immune responses and cytokine profiles. Vitamin A, through retinoic acid signaling, promotes lineage commitment and epithelial regeneration [44].

Amino acids such as leucine and arginine activate nutrient-sensing pathways, such as the mTOR pathway, that regulate stem cell fate. These pathways are interconnected with others that respond to cellular energy status and mediate longevity and stress resistance [45]. The interplay between nutrients and intracellular signaling pathways is crucial for regenerative outcomes. The mTOR pathway integrates signals from amino acids, insulin, and growth factors to regulate protein synthesis and cell proliferation [46].

These pathways are sensitive to dietary inputs. Caloric restriction and nutrient composition can shift the balance between anabolic and catabolic states, influencing stem cell function and regenerative capacity.

Nutrition is a powerful modulator of regenerative mechanisms, influencing ECM composition, stem cell behavior, and intracellular signaling. Vitamins, amino acids, trace minerals, and bioactive compounds work synergistically to support tissue repair and regeneration. Understanding these interactions provides a foundation for integrating nutritional strategies into regenerative medicine, optimizing outcomes in wound healing and tissue engineering. Future studies on these interactions between nutrition and regenerative medicine may be individualized to patients and scenarios by using knowledge of the impact of certain nutrients on wound healing [47]. The next section will focus on applying this knowledge in the clinical setting.

Clinical Nutrition Strategies

Determining patient nutrition status can be challenging for any clinician. Nutrition status is impacted by multiple physiologic and pathologic conditions. Methods for evaluating nutrition status are also variable based on patients' presentation at the time of evaluation. For example, evaluation of nutritional status in the critically ill patient in the intensive care unit is quite different from evaluating the nutritional status of a patient in the outpatient ambulatory setting. There are multiple different tools that can be used to determine a patient's nutritional status. Each uses different combinations of patient-reported history, anthropomorphic measurements, and lab data. A list of screening tools, their indications, and dates of creation are included in Table 29.1, created by Tabernara et al. in their scoping review of current tools for evaluation of nutritional status [48].

Multiple lab values can be helpful when evaluating both macro- and micronutritional status. Commonly obtained are albumin, prealbumin, and transferrin. These plasma proteins are synthesized in the liver and respond to metabolic stress and nutritional deficiency. While in isolation, they may not provide a complete picture, they may help provide a gestalt for the patient's overall metabolic state.

- *Albumin:* Plasma protein with a long half-life (19–21 days). It is known to decrease with age and during inflammatory states. Lab values can be falsely low in hypocalcemia, so lab values should be interpreted after correcting for the patient calcium level. Reference range of 3.4–5.4 g/dL [49]. Studies have shown an inverse correlation between serum albumin levels and incidence of complications, including wound healing issues [50, 51].
- *Prealbumin:* Plasma protein with a much shorter half-life than albumin (2–3 days) [52] making it more useful for determining trends in nutritional status over a period of weeks compared to the months needed to trend albumin. Prealbumin can also be decreased during periods of inflammation and elevated in renal failure. Reference range of 16–30 mg/dL, with below 10 mg/dL indicating malnutrition [53].
- *Transferrin:* Plasma protein for iron transport with a half-life of 10 days. Like albumin and prealbumin, serum concentration is affected by inflammation (decreases) and renal failure (increases). Reference range 0.16–0.36 g/dL [52, 53].

Table 29.1 Current nutritional status assessment tools

Tool (acronym, release year)	Features	Applications (A)/ populations (P)/ diseases where used (D)	Recent article [ref.]
Instant nutritional assessment (INA, 1979)	Computation of serum albumin levels and total lymphocyte counts within an equation	(A) Warning for an increase in potential complications and deaths in acute diseases	Catarci et al.
		(P) All community	
		(D) Major cancer surgery, liver disease, pancreatic fistula, postoperative sepsis, chronic pancreatitis	
Patient-generated subjective global assessment (PG-SGA, 1987)	Two parts including history examination (weight change, dietary intake change, gastrointestinal symptoms that have persisted for greater than 2 weeks, and functional capacity) and physical examination (loss of subcutaneous fat, muscle wasting, ankle/sacral edema, and ascites) compiled from 13 questions	(A) Identification of cancer patients with malnutrition	Sealy et al.
		(P) All communities (self-reported)	
		(D) Cirrhosis, pediatric patients, and so forth	
Prognostic inflammatory and nutritional index (PINI, 1990)	Integration of C-reactive protein, orosomucoid, albumin, and transthyretin in an assessment formula	(A) Prediction of an inflammatory and nutritional process	Kirov et al.
		(P) Geriatric and oncological population	
		(D) Fibrosis, various types of cancer, (pancreatic, cervical, colon). Cirrhosis, pediatric patients	
Nutritional screening initiative checklist (DETERMINE, 1998)	Ten Yes/No subjective questions with different quantitative values associated with the nutritional well-being of older people	(A) Evaluation of low-nutrient intake risk and subsequent health problems	Chye et al.
		(P) Elderly population (self-reported)	
Mini-nutritional assessment (MNA, 1999)	Eighteen questions categorized in blocks using anthropometric measures, questions about clinical history (medication, mobility, nutritional assessment), and a subjective assessment of nutritional status	(A) Detection of undernutrition	Miller et al.
		(P) Elderly at home, nursing homes, or hospitalized community	
		(D) Cancer, inflammatory bowel disease, Crohn's disease, ulcerative colitis, and so forth	

(continued)

Table 29.1 (continued)

Tool (acronym, release year)	Features	Applications (A)/ populations (P)/ diseases where used (D)	Recent article [ref.]
Malnutrition screening tool (MST, 1999)	Two items related to recent appetite status and weight loss	(A) Early estimation of malnutrition	Porter et al.
		(P) Residential aged care community and adults in the inpatient, outpatient hospital community.	
		(D) Cancer patients	
Nutritional risk screening—2002 (NRS—2002, 2003)	Nutritional information about BMI, weight loss, and acute disease score (no nutritional intake for >5 days)	(A) Indication of nutritional assessment recommended by ESPEN for hospitalized patients	Müller et al.
		(P) Hospitalized patients	
		(D) Sarcopenia, various cancers, ICU patients, and so forth	
Glasgow prognostic score (GPS, 2004)	Estimation of serum levels of albumin and C-reactive protein (CRP)	(A) Warning factor for patients with advanced diseases	Mitani et al.
		(P) All inpatients	
		(D) Advanced gastric cancer, colorectal cancer, inflammatory bowel disease, pancreatitis, and so forth	
Malnutrition Universal Screening Tool (MUST, 2004)	Three clinical makers: BMI, weight loss, and illness in relation with food intake over 5 days	(A) Identification of nutritional status	Hiller et al.
		(P) Ambulatory hospitalized adults	
		(D) Cancer (colorectal, gastrointestinal), cachexia, obesity, cirrhosis, sclerosis, and so forth	
Controlling nutritional status (CONUT, 2005)	Hospital data based on the clinical history and laboratory data (serum albumin, cholesterol, total lymphocytes, and hematocrit)	(A) Automatic determination of daily nutritional status of all inpatients	Wang et al.
		(P) All inpatients.	
		(D) Cancer (all types), cirrhosis, pancreatitis, pulmonary disease	

(continued)

Table 29.1 (continued)

Tool (acronym, release year)	Features	Applications (A)/ populations (P)/ diseases where used (D)	Recent article [ref.]
Nutritional risk index (NRI, 2004)	Measurements of serum albumin, current/usual body weight ratio within an equation	(A) Early detection of potential complications during hospitalization	Cho et al.
		(P) All inpatients	
		(D) Cancer, Crohn's disease, diabetes, liver disease, sepsis, and so forth	
Simplified Nutritional Appetite Questionnaire (SNAQ, 2005)	Four items related to appetite, food timing during day, food preferences, and daily number of meals	(A) Verification of malnutrition in hospital elderly inpatients	Ilhan et al.
		(P) Hospital elderly inpatients	
		(D) Cancer (gastrointestinal, gynecologic), cirrhosis, kidney disease, and so forth	
Geriatric Nutritional Risk Index (GNRI, 2009)	Evaluation of serum albumin and the relationships between the current weight and the theoretical ideal weight according to the Lorentz formula, within an equation	(A) Determination of the risk of nutrition-related morbidity and mortality at admission into hospital	Nishi et al.
		(P) Elderly community	
		(D) Kidney failure (hemodialysis), cancer, and so forth	
Mini Nutritional Assessment—short form (MNA-SF, 2009)	Six items (instead of 18 from the full MNA) emphasizing anthropometric measurement, global assessment, dietary questions, and health assessment items	(A) Detection of undernutrition	Lin et al.
		(P) Elderly community patients at home, nursing homes, or hospitals	
		(D) Cancer, inflammatory bowel disease, Crohn's disease, ulcerative colitis, and so forth	

(continued)

Table 29.1 (continued)

Tool (acronym, release year)	Features	Applications (A)/ populations (P)/ diseases where used (D)	Recent article [ref.]
Nutrition risk in critically ill (NUTRIC, 2011)	Evaluation of age, APACHE II score, SOFA score, comorbidities, days in the hospital before admission to the ICU and interleukin-6	(A) Identification of nutritional risk in critically ill patients	Hung et al. and Rahman et al.
		(P) ICU inpatient	
		(D) Cardiac disease, respiratory disease, polytraumatism, poisoning, and so forth	
Nutritional perception screening questionnaire—9 (NPSQ9, 2017)	Nine-item questionnaire aggregated into two groups (management and perception and habits) related to the self-perception of a healthy diet	(A) Early estimation of nutritional status	San-Cristobal et al.
		(P) All community (self-reported)	

In addition to these serum proteins, which can be used to assess and trend nutritional status over time, micronutrients should be considered as well. Key vitamins and minerals that are components of tissue regeneration and wound healing should be assessed in patients with risk factors. A patient history of an unvaried diet, lack of consumption or access to fruits and vegetables, and diet dependence on processed foods can indicate the need for micronutrient screening labs. Important vitamins include B vitamins (thiamine, riboflavin, niacin, pyridoxine, folic acid, and cobalamin), fat-soluble vitamins A, D, E, and K, as well as vitamin C, which is a key component in collagen synthesis. In addition to vitamins, a lack of essential minerals (iron, zinc, selenium) can also impair wound healing.

Critical to wound healing is protein intake. In a practice guideline published by the American College of Gastroenterology, 1.5–2.0 g/kg per day is recommended for all hospitalized patients, noting that protein intake needs may even be higher in patients with trauma or wounds [54]. In meta-analysis, supplementation specifically with arginine and glutamine, two key amino acids in wound healing pathways, has been shown to be beneficial for wound healing [29].

After the initial assessment of patient nutritional status via history and laboratory tests comes the more challenging task of improving patient nutrition. In ideal situations, patients would have access to fresh foods and the skill and time needed to prepare balanced and nutritional meals. Food insecurity is a known sociomedical issue in the United States, with the United States Department of Agriculture identifying 13.5% of U.S. households as food insecure in 2023 [55]. In addition to food security, the concept of nutritional security has been introduced to capture that simply accessing calories is not sufficient; access to food of high nutritional value is critical to overall health [56]. In the context of food insecurity and nutritional insecurity, the task of assisting patients in improving their nutritional status becomes more complex. However, we will attempt to outline some initial steps clinicians can take to encourage improved nutritional status, specifically protein intake, for wounded patients.

For wounded patients admitted to the hospital, clinicians have a higher degree of control over nutritional offerings for patients than they do in the outpatient setting. The simplest intervention is at the level of diet orders. Clinicians should be in active conversation with their patients about the importance of protein intake for wound healing. Modifications to the standard hospital diet offerings, like orders for double protein or protein supplement shakes delivered with meals, can allow the clinician to support the patient in increasing their protein intake. For patients who are critically ill, with suppressed appetites, or who are otherwise unable to meet their caloric and protein diet goals with oral intake alone, enteral feeding supplementation should be considered. Choosing to pursue enteral nutrition supplementation via nasogastric feeding tube or surgically placed gastric/jejunal feeding tube depends on the anticipated chronicity of the patient's nutritional deficit. As a last resort, total parenteral nutrition is an option for patients who are not able to be fed enterally. For all patients with acute nutritional needs, but especially those requiring tube feeds or parenteral nutrition, a registered dietitian should be consulted. Indirect calorimetry may be used to determine the energy needs of ill patients and help to prevent over- or undernutrition [57].

Providing nutritional support in the outpatient setting becomes more difficult as interventions rely on more active participation by patients and their caretakers. Cost and access become more salient factors in the outpatient environment. The multitude of personal and systemic factors that lead to poor nutritional status follow patients into the wound care clinic and are unlikely to be overcome by the care of a single clinician. Still, wound clinics should have basic nutritional educational materials to provide patients, and specifically materials geared toward increasing protein intake. Protein supplementation can be framed as a prescription for wound healing. Patients should be encouraged to add protein in ways that work best for their palate and financial constraints. Some institutions may have access to ancillary health programs for nutrition, weight loss, and diabetes management that may be helpful in providing dedicated nutritional support to patients.

Special Populations and Considerations

While each patient has their own unique nutritional needs and challenges, some populations frequently cared for in wound care centers have comorbidities to consider.

Diabetes Hyperglycemia is known to impair the wound healing process through prolongation of the inflammatory phase of healing [58]. Diabetic patients may need additional support from a primary care provider, endocrinologist, or diabetes educator to meet their glucose control goals through a combination of lifestyle changes and medications. The wound care clinician should be aware of and track the average blood glucose over the course of wound treatment episodes. There should be a low threshold to refer to other specialists as needed for glycemic management assistance.

Chronic Kidney Disease Renal disease should be considered when counseling patients on increasing dietary protein. While the evidence remains mixed, some studies have demonstrated renal injury from high-protein diets [59].

Vascular Insufficiency Patients with lower extremity wounds often have some component of vascular pathology in their lower extremities. Peripheral arterial disease and chronic venous insufficiency both impair blood flow to and from wounds, which in turn impairs the delivery of critical nutrients as well as oxygen. Addressing vascular disease while addressing nutritional needs is mutually beneficial.

Advanced Age Geriatric patients experience physiological changes that impair wound healing, like sarcopenia, anabolic resistance, and frailty. Importantly from a nutritional perspective, this patient population may be affected by decreased appetite or social factors that prevent them from having full agency over their food sources and preparation. Clinicians may need to work in conjunction with caretakers and care facilities to address nutritional needs for the elderly wound patient.

Obesity Excess adiposity has been associated with chronic inflammation and altered micronutrient status [60, 61]. This patient population is often counseled on nutrition and weight loss by clinicians they encounter. Counseling on weight loss, however, is counterintuitive to wound healing. Weight loss requires a catabolic state, while wound healing requires an anabolic physiologic state. The wound care clinician must be cautious in maintaining adequate protein intake while controlling carbohydrate and caloric intake. Additionally, the wound care clinician should not assume that a patient's excess weight indicates adequate nutrition. Protein-energy malnutrition can affect this patient population and is not limited to patients who are underweight [62].

Immunosuppression Patients with autoimmune disease or transplant recipients on chronic steroids or biologic treatments can have wound healing challenges due to the impact of their medications on the inflammatory response needed for wound healing. Nutritional supplementation with vitamin A can be of use for this patient population, as it has been shown to mitigate the impact of corticosteroids on wound healing [39].

Future Directions and Research Gaps

While there is a significant and growing body of literature focused on nutrition and wound healing, gaps in the literature remain. More work needs to be done to understand how nutrition delivery can be optimized for patients in various clinical settings. The needs of inpatient and outpatient wound patients are drastically different, and more studies analyzing the nutritional needs of both should be undertaken to understand their effect on modern regenerative medicine. Additionally, future research is needed into the nutritional needs of specific populations based on their unique single and combined comorbidities. This is particularly true of the growing obese diabetic population. While we have laboratory values and clinical tools for assessing patient nutritional status, there is room for improvement in these tools. In particular, we are lacking in our ability to quantify and account for systemic inflammation's impact on a patient's clinical presentation. Wounds are often grouped together as a monolith. However, there are many unique types of wounds that we treat. For example, a patient with a diabetic foot ulcer likely has different nutritional needs than an otherwise healthy spinal cord patient who has a pressure injury. Additional research should be done to understand the role of patient nutritional status for these various types of wounds. Finally, additional research is always warranted to elucidate the social determinants of health that impact our patients' abilities to obtain adequate nutrition. Understanding patients' lives inside and outside the medical system allows us to build interventions in the pursuit of nutritional security for all patients.

References

1. Brocke T, Barr J (2020) The history of wound healing. Surg Clin N Am 100(4):787–806. https://doi.org/10.1016/j.suc.2020.04.004
2. Broughton G, Janis JE, Attinger CE (2006) A brief history of wound care. Plast Reconstr Surg 117(7 Suppl):6S–11S. https://doi.org/10.1097/01.prs.0000225429.76355.dd
3. Larrey DJ, Hall RW (1814) Memoirs of a military surgeon, vol 1, 1st edn, Baltimore
4. Unknown. Injuries in World War I
5. Dudrick SJ, Wilmore DW, Vars HM, Rhoads JE (1969) Can intravenous feeding as the sole means of nutrition support growth in the child and restore weight loss in an adult? An affirmative answer. Ann Surg 169(6):974–984. https://doi.org/10.1097/00000658-196906000-00018
6. Shields BA, Nakakura AM (2023) Nutrition considerations for burn patients: optimizing recovery and healing. Eur Burn J 4(4):537–547. https://doi.org/10.3390/ebj4040035
7. Reintam Blaser A, Starkopf J, Alhazzani W et al (2017) Early enteral nutrition in critically ill patients: ESICM clinical practice guidelines. Intensive Care Med 43(3):380–398. https://doi.org/10.1007/s00134-016-4665-0
8. Chourdakis M, Bouras E, Shields BA, Stoppe C, Rousseau AF, Heyland DK (2020) Nutritional therapy among burn injured patients in the critical care setting: an international multicenter observational study on "best achievable" practices. Clin Nutr 39(12):3813–3820. https://doi.org/10.1016/j.clnu.2020.04.023
9. Fact_Sheet-Economic_Burden_of_Wounds_10.5.17
10. Nussbaum SR, Carter MJ, Fife CE et al (2018) An economic evaluation of the impact, cost, and Medicare policy implications of chronic nonhealing wounds. Value Health 21(1):27–32. https://doi.org/10.1016/j.jval.2017.07.007
11. Melo PG, Mota JF, Nunes CA d B et al (2022) Effects of oral nutritional supplementation on patients with venous ulcers: a clinical trial. J Clin Med 11(19). https://doi.org/10.3390/jcm11195683
12. van der Pols-Vijlbrief R, Wijnhoven HAH, Schaap LA, Terwee CB, Visser M (2014) Determinants of protein-energy malnutrition in community-dwelling older adults: a system-

atic review of observational studies. Ageing Res Rev 18:112–131. https://doi.org/10.1016/j.arr.2014.09.001
13. McMillan DC, Maguire D, Talwar D (2019) Relationship between nutritional status and the systemic inflammatory response: micronutrients. Proc Nutr Soc 78(1):56–67. https://doi.org/10.1017/S0029665118002501
14. Mao AS, Mooney DJ (2015) Regenerative medicine: current therapies and future directions. Proc Natl Acad Sci USA 112(47):14452–14459. https://doi.org/10.1073/pnas.1508520112
15. Newsholme P, Diniz VLS, Dodd GT, Cruzat V (2023) Glutamine metabolism and optimal immune and CNS function. Proc Nutr Soc 82(1):22–31. https://doi.org/10.1017/S0029665122002749
16. Jin D, Wei X, He Y et al (2024) The nutritional roles of zinc for immune system and COVID-19 patients. Front Nutr 11:1385591. https://doi.org/10.3389/fnut.2024.1385591
17. Martin ND, Schott LL, Miranowski MK et al (2024) Exploring the impact of arginine-supplemented immunonutrition on length of stay in the intensive care unit: a retrospective cross-sectional analysis. PLoS One 19(4):e0302074. https://doi.org/10.1371/journal.pone.0302074
18. Ju M, Kim Y, Seo KW (2023) Role of nutrition in wound healing and nutritional recommendations for promotion of wound healing: a narrative review. Ann Clin Nutr Metab 15(3):67–71. https://doi.org/10.15747/acnm.2023.15.3.67
19. Grada A, Phillips TJ (2022) Nutrition and cutaneous wound healing. Clin Dermatol 40(2):103–113. https://doi.org/10.1016/j.clindermatol.2021.10.002
20. Demling RH (2009) Nutrition, anabolism, and the wound healing process: an overview. Eplasty 9:e9
21. Frankenfield D (2006) Energy expenditure and protein requirements after traumatic injury. Nutr Clin Pract 21(5):430–437. https://doi.org/10.1177/0115426506021005430
22. Miles JM (2006) Energy expenditure in hospitalized patients: implications for nutritional support. Mayo Clin Proc 81(6):809–816. https://doi.org/10.4065/81.6.809
23. Hasselgren PO (2000) Catabolic response to stress and injury: implications for regulation. World J Surg 24(12):1452–1459. https://doi.org/10.1007/s002680010262
24. Leininger SM (2002) The role of nutrition in wound healing. Crit Care Nurs Q 25(1):13–21
25. Molnar JA, Vlad LG, Gumus T (2016) Nutrition and chronic wounds: improving clinical outcomes. Plast Reconstr Surg 138(3 Suppl):71S–81S. https://doi.org/10.1097/PRS.0000000000002676
26. Ghaly P, Iliopoulos J, Ahmad M (2021) The role of nutrition in wound healing: an overview. Br J Nurs 30(5):S38–S42. https://doi.org/10.12968/bjon.2021.30.5.S38
27. Bainbridge P (2013) Wound healing and the role of fibroblasts. J Wound Care 22(8):407–408, 410–412. https://doi.org/10.12968/jowc.2013.22.8.407
28. Wittmann F, Prix N, Mayr S et al (2005) L-arginine improves wound healing after trauma-hemorrhage by increasing collagen synthesis. J Trauma 59(1):162–168. https://doi.org/10.1097/01.ta.0000171529.06625.a8
29. Arribas-López E, Zand N, Ojo O, Snowden MJ, Kochhar T (2021) The effect of amino acids on wound healing: a systematic review and meta-analysis on arginine and glutamine. Nutrients 13(8). https://doi.org/10.3390/nu13082498
30. Torsy T, Tency I, Beeckman D, Isoherranen K, Litchford M, De Vylder F (2025) The role of glutamine and arginine in wound healing of pressure ulcers: a systematic review. Wound Repair Regen 33(4):e70077. https://doi.org/10.1111/wrr.70077
31. Chen S, Li G, Jiang Z, Xu Y, Aipire A, Li J (2025) Regulation of dendritic cell biology by amino acids and their transporters. Front Immunol 16:1626973. https://doi.org/10.3389/fimmu.2025.1626973
32. Martínez-Arnau FM, Fonfría-Vivas R, Buigues C et al (2020) Effects of leucine Administration in Sarcopenia: a randomized and placebo-controlled clinical trial. Nutrients 12(4). https://doi.org/10.3390/nu12040932

33. Choudhary V, Choudhary M, Bollag WB (2024) Exploring skin wound healing models and the impact of natural lipids on the healing process. Int J Mol Sci 25(7). https://doi.org/10.3390/ijms25073790
34. McDaniel JC, Belury M, Ahijevych K, Blakely W (2008) Omega-3 fatty acids effect on wound healing. Wound Repair Regen 16(3):337–345. https://doi.org/10.1111/j.1524-475X.2008.00388.x
35. Van Putte L, De Schrijver S, Moortgat P (2016) The effects of advanced glycation end products (AGEs) on dermal wound healing and scar formation: a systematic review. Scars Burn Heal 2:2059513116676828. https://doi.org/10.1177/2059513116676828
36. Cabral-Pacheco GA, Garza-Veloz I, Castruita-De la Rosa C et al (2020) The roles of matrix Metalloproteinases and their inhibitors in human diseases. Int J Mol Sci 21(24). https://doi.org/10.3390/ijms21249739
37. Kogan S, Sood A, Garnick MS (2017) Zinc and wound healing: a review of zinc physiology and clinical applications. Wounds 29(4):102–106
38. Moores J (2013) Vitamin C: a wound healing perspective. Br J Community Nurs Suppl:S6:S8–S11. https://doi.org/10.12968/bjcn.2013.18.sup12.s6
39. Zinder R, Cooley R, Vlad LG, Molnar JA (2019) Vitamin A and wound healing. Nutr Clin Pract 34(6):839–849. https://doi.org/10.1002/ncp.10420
40. Wu Y, Gong Y, Ma Y et al (2024) Effects of vitamin D status on cutaneous wound healing through modulation of EMT and ECM. J Nutr Biochem 134:109733. https://doi.org/10.1016/j.jnutbio.2024.109733
41. Jomova K, Makova M, Alomar SY et al (2022) Essential metals in health and disease. Chem Biol Interact 367:110173. https://doi.org/10.1016/j.cbi.2022.110173
42. Barchitta M, Maugeri A, Favara G et al (2019) Nutrition and wound healing: an overview focusing on the beneficial effects of curcumin. Int J Mol Sci 20(5). https://doi.org/10.3390/ijms20051119
43. Pignet AL, Schellnegger M, Hecker A, Kohlhauser M, Kotzbeck P, Kamolz LP (2021) Resveratrol-induced signal transduction in wound healing. Int J Mol Sci 22(23). https://doi.org/10.3390/ijms222312614
44. Godoy-Parejo C, Deng C, Zhang Y, Liu W, Chen G (2020) Roles of vitamins in stem cells. Cell Mol Life Sci 77(9):1771–1791. https://doi.org/10.1007/s00018-019-03352-6
45. Meng D, Frank AR, Jewell JL (2018) mTOR signaling in stem and progenitor cells. Development 145(1). https://doi.org/10.1242/dev.152595
46. Sadria M, Layton AT (2021) Interactions among mTORC, AMPK and SIRT: a computational model for cell energy balance and metabolism. Cell Commun Signal 19(1):57. https://doi.org/10.1186/s12964-021-00706-1
47. McKinley KL, Longaker MT, Naik S (2023) Emerging frontiers in regenerative medicine. Science 380(6647):796–798. https://doi.org/10.1126/science.add6492
48. Taberna DJ, Navas-Carretero S, Martinez JA (2019) Current nutritional status assessment tools for metabolic care and clinical nutrition. Curr Opin Clin Nutr Metab Care 22(5):323–328. https://doi.org/10.1097/MCO.0000000000000581
49. Albumin blood (serum) test. https://www.ucsfhealth.org/medical-tests/albumin-blood-(serum)-test
50. Griffin FS, Stead TS, Zeyl VG, Mehrzad R, King VA, Kalliainen LK (2024) Low preoperative albumin levels significantly associated with increased risk of wound infection and bleeding after panniculectomy. Plast Surg (Oakv). Published online October 25, 2024:22925503241292350. https://doi.org/10.1177/22925503241292350
51. Garwe T, Albrecht RM, Stoner JA, Mitchell S, Motghare P (2016) Hypoalbuminemia at admission is associated with increased incidence of in-hospital complications in geriatric trauma patients. Am J Surg 212(1):109–115. https://doi.org/10.1016/j.amjsurg.2015.06.013
52. Haldeman-Englert C, Turley Jr R, Novick T. Health encyclopedia. University of Rochester Medical Center

53. Keller U (2019) Nutritional laboratory markers in malnutrition. J Clin Med 8(6). https://doi.org/10.3390/jcm8060775
54. McClave SA, Dibaise JK, Mullin GE, Martindale RG (2016) ACG clinical guideline: nutrition therapy in the adult hospitalized patient. Am J Gastroenterol 111(3):315–334. https://doi.org/10.1038/ajg.2016.28
55. Rabiitt MP, Hales LJ, Reed-Jones M. Food Security in the U.S.- Key statistics & graphics. https://www.ers.usda.gov/topics/food-nutrition-assistance/food-security-in-the-us/key-statistics-graphics
56. Odoms-Young A, Brown AGM, Agurs-Collins T, Glanz K (2024) Food insecurity, neighborhood food environment, and health disparities: state of the science, research gaps and opportunities. Am J Clin Nutr 119(3):850–861. https://doi.org/10.1016/j.ajcnut.2023.12.019
57. Singer P, Singer J (2016) Clinical guide for the use of metabolic carts: indirect calorimetry–no longer the orphan of energy estimation. Nutr Clin Pract 31(1):30–38. https://doi.org/10.1177/0884533615622536
58. Roth J, Sommerfeld O, Birkenfeld AL et al (2021) Blood sugar targets in surgical intensive care—management and special considerations in patients with diabetes. Dtsch Arztebl Int 118(38):629–636. https://doi.org/10.3238/arztebl.m2021.0221
59. Ko GJ, Rhee CM, Kalantar-Zadeh K, Joshi S (2020) The effects of high-protein diets on kidney health and longevity. J Am Soc Nephrol 31(8):1667–1679. https://doi.org/10.1681/ASN.2020010028
60. Kawai T, Autieri MV, Scalia R (2021) Adipose tissue inflammation and metabolic dysfunction in obesity. Am J Physiol Cell Physiol 320(3):C375–C391. https://doi.org/10.1152/ajpcell.00379.2020
61. Saltiel AR, Olefsky JM (2017) Inflammatory mechanisms linking obesity and metabolic disease. J Clin Invest 127(1):1–4. https://doi.org/10.1172/JCI92035
62. Mwala NN, Borkent JW, Van Der Meij BS, De Van Der Schueren MAE (2024) Challenges in identifying malnutrition in obesity; an overview of the state of the art and directions for future research. Nutr Res Rev. Published online 2024. https://doi.org/10.1017/S095442242400012X

Chapter 30
Scar Problems After Wound Healing

Mohamed Abdelhakim and Rei Ogawa

Abstract The emotional and social implications of visible scars, especially those that are large and disfiguring, can contribute to reduced self-esteem and social withdrawal. As a result, effective management of keloids and hypertrophic scars remains a clinical challenge. Understanding the mechanisms behind their development has been a focus of research, with recent advancements providing new insights into the molecular and cellular processes that contribute to excessive scar formation.

Keywords Hypertrophic scars · Keloid scars · Scar management · Assessment of scars · Protein kinase C

Abbreviations

YAP/TAZ	Yes-associated protein/transcriptional coactivator with PDZ-binding motif
JSS	Japan Scar Scale
JSW	Japan Scar Scale (original name)
POSAS	Patient and Observer Scar Assessment Scale
TGF-β	Transforming Growth Factor-Beta
VEGF	Vascular Endothelial Growth Factor
PDGF	Platelet-Derived Growth Factor
ECM	Extracellular Matrix
MMPs	Matrix Metalloproteinases
IL-6	Interleukin-6
IL-8	Interleukin-8

M. Abdelhakim (✉) · R. Ogawa
Department of Plastic, Reconstructive, and Aesthetic Surgery, Nippon Medical School Hospital, Tokyo, Japan
e-mail: mhmedhakim7@gmail.com; r.ogawa@nms.ac.jp

F. De Francesco, K. Turksen (eds.), *Management and Strategies for Wound Healing*, Stem Cell Biology and Regenerative Medicine 79,
https://doi.org/10.1007/978-3-032-12442-5_30

HS	Hypertrophic Scar
SNPs	Single-Nucleotide Polymorphisms
COL1A1	Collagen Type I Alpha 1 Chain Gene
EGF	Epidermal Growth Factor
RT	Radiation Therapy
Gy	Gray (unit of radiation dose)
CRISPR-Cas9	Clustered Regularly Interspaced Short Palindromic Repeats-Associated Protein 9

Introduction

Scarring is a critical aspect of the wound-healing process, involving a complex interplay of cellular and molecular mechanisms. While scars are a natural outcome of healing, their excessive formation can lead to hypertrophic scars and keloids, which pose significant clinical challenges [1, 2].

While pathological scars, including both keloids and hypertrophic scars, result from similar biological processes, there are important distinctions between them. These scars usually form as a result of an injury, surgery, or skin trauma and are characterized by excessive collagen deposition and fibrosis, leading to raised, often discolored, and thickened skin [3].

Not only do these scars affect the physical appearance of individuals but also can significantly impact their psychological well-being. The emotional and social implications of visible scars, especially those that are large and disfiguring, can contribute to reduced self-esteem and social withdrawal. As a result, effective management of keloids and hypertrophic scars remains a clinical challenge [4].

Understanding the mechanisms behind their development has been a focus of research, with recent advancements providing new insights into the molecular and cellular processes that contribute to excessive scar formation.

This growing understanding has paved the way for innovative therapies aimed at improving treatment outcomes and reducing the recurrence of these scars after interventions.

Herein, in this chapter, we discuss the types of pathological scars, their etiology, and advancements in their treatment based on our accumulated experiences over the years from Japan.

Types of Scars

Scars are a natural outcome of the body's wound-healing process, and their manifestations can vary widely due to etiological factors, patient-specific characteristics, and wound care. Scars are broadly categorized into physiological and pathological types, with pathological scars including hypertrophic scars, keloids, atrophic scars, contracture scars, stretch marks (striae distensae), pitted scars, pigmented scars, and mixed scars. Among these, hypertrophic scars and keloids are the most recognized, characterized by distinct features and underlying mechanisms [5].

These scars often share overlapping characteristics in their clinical features, making their diagnosis and classification complex for clinicians.

Hypertrophic Scars

Hypertrophic scars are characterized by their raised and thickened appearance, yet they remain confined to the boundaries of the original wound. These scars typically form within weeks of an injury, often growing progressively for several months before stabilizing. Unlike keloids, hypertrophic scars have a tendency to regress over time, sometimes even without intervention [6].

The collagen fibers within hypertrophic scars are organized in parallel bundles, contributing to their firm texture [6]. Initially, these scars may appear red or pink but often fade as they mature.

Hypertrophic scars are commonly associated with wounds that experience prolonged inflammation, infection, or mechanical tension during the healing process.

Keloid Scars

Keloids, in contrast to hypertrophic scars, are pathological scars that grow beyond the boundaries of the original wound. They often appear months after the initial injury and are marked by their firm, rubbery texture. Keloids can vary in color, ranging from pink or red to dark brown, depending on the individual's skin type and the scar's location [7, 8].

What distinguishes keloids apart is their aggressive growth pattern and resistance to regression. While hypertrophic scars may stabilize and shrink over time, keloids tend to persist and, in some cases, continue to grow indefinitely [7].

This persistent growth is a result of an exaggerated wound-healing response, where excessive collagen production and abnormal fibroblast activity lead to the formation of disorganized and dense scar tissue [4, 5] (Fig. 30.1).

Studies conducted at the Mechanobiology Laboratory of Nippon Medical School have demonstrated that areas of high skin tension and stretching, such as the chest, shoulders, and knees, are particularly susceptible to keloid formation, suggesting that mechanical force may play a role in localized inflammation and fibroblast activation [9].

Additionally, recent research has shown that mechanosensitive pathways, particularly those involving YAP/TAZ signaling, are upregulated in areas of high mechanical stress. These pathways activate fibroblasts, which produce excess collagen and contribute to the exaggerated appearance of hypertrophic scars [10].

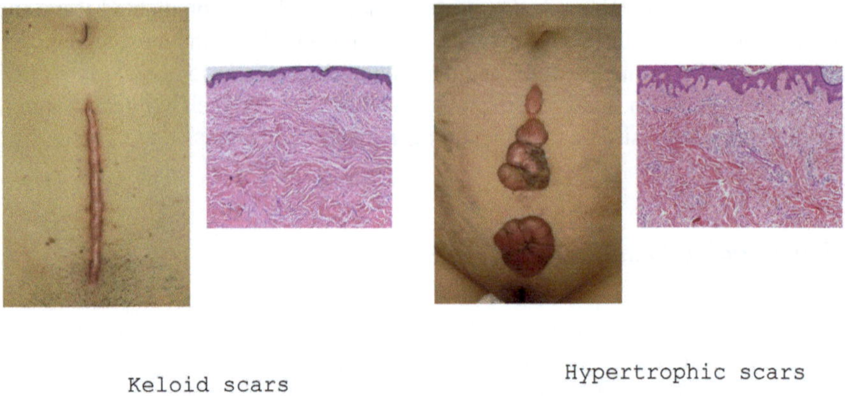

Fig. 30.1 Clinical and histopathological images illustrating key differences between keloids and hypertrophic scars, highlighting their unique morphological and tissue characteristics

Keloids not only pose aesthetic concerns but also have a significant psychological impact. Patients often report feelings of frustration, embarrassment, and social withdrawal due to the conspicuous nature of these scars. This highlights the importance of addressing both the physical and emotional aspects of keloid management in clinical practice [4].

Intermediate lesions and varying spectra of scarring dynamically exist, with many scars displaying characteristics of both hypertrophic and keloid scars. These intermediate lesions blur the boundaries between traditional classifications, complicating the diagnostic process. Their formation is influenced by factors such as the intensity and duration of the inflammatory response, genetic predisposition, wound care practices, and environmental conditions. This continuous spectrum underscores the importance of adopting a nuanced approach to scar assessment and management.

Assessment of Scars and Classification

In Japan, the Japan Scar Scale (JSS) has been developed as a standardized tool to enhance the assessment and classification of scars [11]. The first version of this scar assessment tool was named the JSW scar scale (JSS), and it was reported in 2011. The revised second version was reported in 2015. The JSS consists of two tables; one is a scar classification table that is used to determine whether the scar is a normal mature scar, a hypertrophic scar, or a keloid. This grading system helps the user to select the most appropriate treatment method for the scar; the other table in the JSS is an evaluation table that is used to judge the response to treatment and for follow-up. Both tables contain sample images of each subjective keloid/hypertrophic scar item that allow the user to evaluate each item without hesitation.

The JSS offers a comprehensive approach by incorporating both clinical evaluations and patient-reported outcomes. Unlike traditional classification systems that primarily rely on physical characteristics, such as the Vancouver Scar Scale, the Manchester Scar Scale, and the Patient and Observer Scar Assessment Scale (POSAS), which are all very well-known scar evaluation methods based on a number of scar variables, including color, height, and pliability, the JSS provides a multidimensional perspective on scarring [12]. The JSS evaluates a range of factors, including the scar's appearance, symptoms such as itching or pain, and its impact on daily activities [12].

We believe that this patient-centered approach ensures both subjective experiences and objective clinical findings are considered when planning treatment or monitoring progress.

Additionally, the JSS is flexible enough to accommodate the wide spectrum of scar types, from hypertrophic scars and keloids to intermediate lesions. By doing so, it facilitates a more precise understanding of scar presentations and guides targeted therapeutic interventions.

Advancing Scar Management

The study of pathological scars has evolved to recognize the continuous and multifaceted nature of scar formation. While traditional classifications focus on clinical and physical characteristics, there is an increasing emphasis on integrating molecular and histological markers to improve diagnostic accuracy. Tools like the JSS represented a shift toward more personalized approaches, allowing clinicians to address not only the aesthetic but also the functional and emotional challenges posed by abnormal scars.

This comprehensive understanding is essential for tailoring treatments to the specific needs of each patient, ensuring better outcomes and improved quality of life.

The aesthetic and psychological impact of keloids is significant. Patients often report feelings of frustration, embarrassment, and social withdrawal due to the conspicuous nature of these scars. This highlights the importance of addressing not only the physical but also the emotional aspects of keloid management in clinical practice.

Pathophysiology of Scar Formation

In normal wounds, a balance is achieved between new tissue biosynthesis and degradation mediated by apoptosis and remodeling of the extracellular matrix [13].

The complex orchestration of cellular and molecular events ensures that the healing process proceeds in a controlled manner. However, when this balance is disrupted, it can result in excessive or insufficient healing responses. In the case of hypertrophic scars and keloids, the dysregulated processes favor the excessive accumulation of extracellular matrix components, leading to the formation of problematic scars [13].

The distinct phases of wound healing—hemostasis, inflammation, proliferation, and remodeling are—marked by the interplay of several growth factors such as TGF-β, VEGF, and PDGF. Dysregulation at any stage, particularly during the inflammatory and proliferative phases, may predispose individuals to pathological scarring [14].

During excessive scar formation, a dysfunction of the underlying regulatory mechanisms may lead to persistent inflammation, excessive collagen synthesis, or deficient matrix degradation and remodeling, as elaborated in the following phases:

Inflammatory Phase

The inflammatory phase marks the beginning of wound healing; after an injury, the body initiates an inflammatory response aimed at cleaning the wound site and promoting tissue repair. In this phase, immune cells, such as macrophages and neutrophils, are recruited to the injury site, releasing inflammatory mediators like cytokines and growth factors. In keloid formation, this inflammatory response is prolonged and excessive, leading to the activation of fibroblasts. Fibroblasts are the primary cells responsible for collagen synthesis in the extracellular matrix (ECM) [15].

In hypertrophic scars, this inflammatory phase is more regulated, but fibroblast activation still leads to excessive ECM production. The primary difference between hypertrophic scars and keloids is that while hypertrophic scars often regress with time, keloids continue to grow due to ongoing fibroblast activity and the persistence of collagen deposition.

Proliferative Phase

During the proliferative phase, the body forms new tissue to cover the wound. Fibroblasts are activated and proliferate to synthesize collagen, elastin, and other extracellular matrix components. In this phase, capillaries form to restore blood supply to the healing tissue, a process known as angiogenesis. The accumulation of collagen provides a structural foundation for the wound to heal.

However, in pathological scars such as hypertrophic scars and keloids, this phase can become exaggerated. An overproduction of collagen, mediated by fibroblasts, occurs under the influence of growth factors like TGF-β, which plays a central role in fibroblast differentiation and collagen deposition. This leads to thickening and raised scars [16].

The imbalance between collagen synthesis and degradation during this phase is a key factor in excessive scarring. While the normal wound-healing process includes the remodeling of collagen, in hypertrophic scars and keloids, collagen remains disorganized and accumulates in large quantities. Furthermore, the matrix metalloproteinases (MMPs), which are responsible for breaking down collagen, may be inhibited in these conditions, leading to the persistence of excessive scar tissue [16].

Maturation and Remodeling Phase

The final phase of wound healing is the remodeling phase, where the newly formed tissue is reorganized to increase its strength and elasticity. This phase can last for months to years. During this time, collagen is continuously remodeled, and the wound site is increasingly restored to a state similar to its original tissue.

However, in hypertrophic scars and keloids, the remodeling phase is impaired. The collagen fibers in these scars are disorganized, more densely packed, and not properly aligned, leading to the raised appearance and rigidity. In keloids, the continued overproduction of collagen during the remodeling phase may result from persistent fibroblast activation. A study has shown that these fibroblasts exhibit an abnormal response to growth factors, leading to their prolonged survival and activity [17]. This abnormal collagen deposition and the failure to achieve normal tissue remodeling are characteristic of pathological scarring. In hypertrophic scars, collagen remodeling is arrested early, and in keloids, the process is even more severe, leading to the continuous deposition of extracellular matrix material [18].

For example, wounds on the major joints tend to develop hypertrophic scars because the joint movements place strong cyclical tension on the wound. This tension provokes chronic inflammation of the dermis, namely, the unceasing influx and activation of inflammatory cells, the persistent generation of blood vessels and nerve fibers, and the constant production of collagen by the activated fibroblasts. This chronic inflammation blocks the conversion of the granulation tissue into dermis-like tissue by the remodeling process and results in an immature hypertrophic scar that is red, elevated, hard, and painful.

The above observations suggest that, to prevent pathological scarring after surgery, it is necessary to ensure that the sutures cause the wound edges to adhere to each other without any tension, even when strong extrinsic forces are placed on the wound. This will allow the granulation tissue to convert smoothly into dermis-like tissue, thereby yielding minimal scarring.

Etiology and Contributing Factors

Multiple episodes of trauma to the same area are also a risk factor for hypertrophic scarring and keloid scars [19–21]. One example is the use of earrings. Each time a person puts on or takes off the earring, there can be some amount of trauma to the area. The site of ear piercings is a common area for the development of keloids, and there is an unclear association with nickel or other metals commonly used in ear jewelry. Burn wounds are prone to hypertrophic scarring, especially if they affect the deeper dermis (second- and third-degree burns).

The etiology of hypertrophic scars and keloids is multifactorial, with genetic predisposition, local wound factors, and systemic conditions playing key roles. Advancements in genetic research have revealed the involvement of specific single-nucleotide polymorphisms (SNPs) that predispose certain individuals to abnormal scarring. Moreover, lifestyle factors, including smoking, poor nutrition, and inadequate wound care, can exacerbate the risk [19].

A variety of risk factors are linked to both hypertrophic scars and keloids. Multiple lines of evidence suggest that the intensity/duration of dermal inflammation in keloids/HSs is dictated by genetic, systemic, local, and lifestyle risk factors.

Genetic Factors

Both hypertrophic scars and keloids have a genetic component. Keloids, in particular, are more common in individuals of African, Asian, and Hispanic descent. Genetics can play a role in hypertrophic scars or keloids. Particularly, certain single-nucleotide polymorphisms are reported to be associated with keloid scars in persons of Japanese descent [7]. In addition to studies suggesting that certain genetic mutations may lead to abnormal wound-healing responses, predisposing individuals to keloid formation [20]. Research has identified several genes involved in collagen production and fibroblast activity that may contribute to this predisposition [21].

Additionally, recent genetic research has shown that polymorphisms in the TGF-β receptor gene may contribute to the excessive fibrosis seen in keloids. Furthermore, mutations in the COL1A1 gene, which encodes collagen type I, have been associated with abnormal collagen formation in hypertrophic scars and keloids [22].

Moreover, these emerging studies have also identified epigenetic modifications, such as DNA methylation and histone acetylation, as contributors to the pathophysiology of hypertrophic scars and keloids. These epigenetic changes may influence the expression of key genes involved in inflammation, fibroblast activity, and extracellular matrix production [20–22].

Ethnicity, Age, and Sex

Darker skin tones are more susceptible to keloid formation due to higher levels of melanin and greater reactivity of melanocytes and fibroblasts [23].

Additionally, younger individuals, particularly those in their second and third decades of life, are more prone to developing keloids and hypertrophic scars, likely due to the more active cellular and immune responses in younger skin, possibly female sex [24].

Role of Melanin in Pathological Scarring

Melanin is thought to contribute to the increased risk and severity of pathological scarring, particularly hypertrophic scars and keloids, in individuals with darker skin tones.

This relationship is driven by melanin's influence on inflammation, fibroblast activity, and growth factor production, especially transforming growth factor-beta (TGF-β), which stimulates collagen overproduction.

Elevated melanin levels correlate with prolonged inflammation, intensifying fibroblast proliferation and excessive extracellular matrix (ECM) deposition, key factors in scar formation. Additionally, melanin can lead to pigmentation changes in scars, such as hyperpigmentation or hypopigmentation, making these scars more visible and thus more impactful aesthetically. Genetic predispositions linked to melanin production further increase susceptibility to abnormal scarring, highlighting the importance of tailored interventions for individuals with higher melanin levels to prevent and treat pathological scars effectively [25].

Wound Site and Mechanical Factors (e.g., Tension, Strain, Stretch)

Scars that form over areas of high tension are more likely to become keloidal. Our previous studies showed that keloids are particularly prevalent on body regions with strong skin-stretching force due to body movements, such as the joints, anterior neck, chest wall, scapula/upper arm, and suprapubic region (Fig. 30.2).

Mechanical tension at the wound site is a critical factor that can lead to scarring. Studies have shown that mechanical forces can activate cellular pathways such as the Rho/ROCK pathway, which leads to fibroblast contraction and increased collagen deposition [26–28].

At Nippon Medical School's Mechanobiology laboratory, a study was conducted using finite-element analysis; it was suggested that keloid shape is largely determined by the direction of the tension on the skin around the wound/scar site [29].

The mechanical forces exerted on the healing wound can stimulate fibroblast activity and collagen production, leading to excessive scarring. *Ogawa* et al. have emphasized the role of skin tension and mechanical stress in the pathophysiology of these scars [30].

Moreover, our studies showed that changes in human position are strongly correlated with mechanical loading of the predilection sites, which leads to increased mechanical strain in the peripheral tissue surrounding keloids. Furthermore, increased mechanical strain in the peripheral tissue, which is the site of keloid progression.

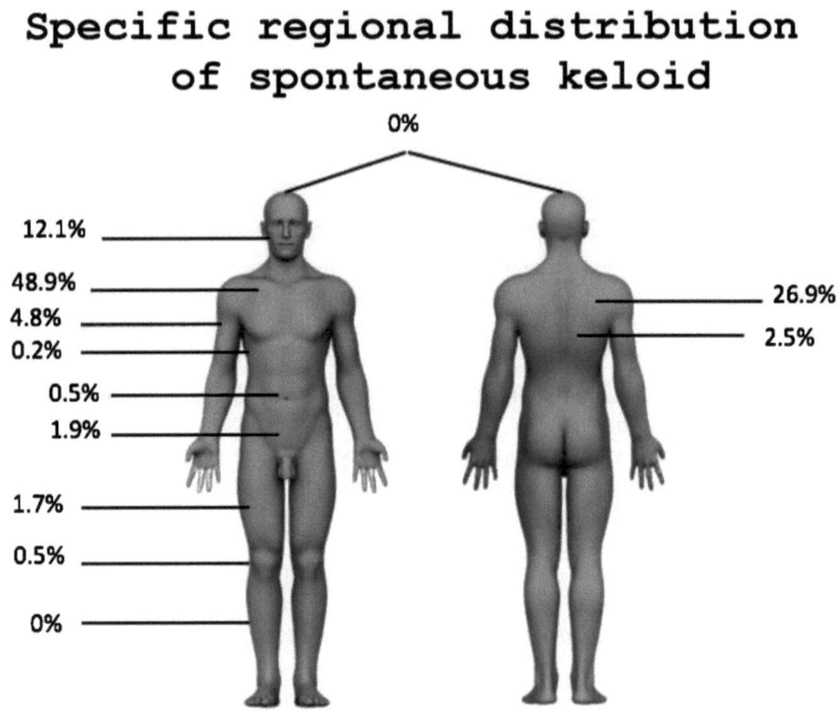

Fig. 30.2 Body map showing the specific regional distribution of keloids, with areas of higher prevalence indicated

Chronic Inflammation

The abnormal growth of these scars is amplified by chronic inflammation in the reticular dermis that is triggered by dermal injury or irritation, including trauma, burns, surgery, vaccination, skin piercing, acne, and herpes zoster. In general, the scars that spontaneously lose this dermal inflammation over time are defined as HSs, whereas the scars with enduring inflammation are defined as keloids. The persistence of inflammatory cytokines, particularly IL-6 and IL-8, promotes fibroblast activation and the overproduction of extracellular matrix components [31].

This difference in inflammation durability explains the disparate growth habits of HSs and keloids. Thus, the temporary, waning inflammation in HSs means that they do not grow over the borders of the original wound. By contrast, the continuous, escalating inflammation of keloids causes them to grow into the adjacent normal skin.

In terms of histopathology, the inflammation and the resulting accumulation of extracellular matrix cause dermal nodules and thick eosinophilic (hyalinizing) collagen bundles called "keloidal collagen" to form over time [32].

If the dermal nodule is the main histological finding, it is diagnosed as an HS. This suggests that keloids/HSs are manifestations of the same inflammatory fibroproliferative condition and simply differ in the intensity and duration of inflammation. Our research highlights the importance of controlling inflammation during the early stages of wound healing to prevent pathological scarring [31].

Clinical Challenges in Managing Hypertrophic Scars and Keloids

Despite advancements in therapeutic approaches, managing hypertrophic scars and keloids remains a clinical challenge. The unpredictable nature of these scars, coupled with their high recurrence rates, necessitates a comprehensive and personalized treatment strategy.

Patient education plays a crucial role in management. Informing patients about the risk factors, potential outcomes, and the importance of adhering to prescribed treatments can significantly improve compliance and overall outcomes. Additionally, involving patients in shared decision-making fosters trust and enhances their satisfaction with the treatment process.

Aesthetic and Functional Concerns

Scars, particularly those on the face or over joints, can significantly impact a patient's appearance and physical function. Contractures over joints can restrict movement, leading to functional impairment and discomfort [33].

Recurrence and Resistance to Treatment

Keloids are notoriously challenging to manage due to their high recurrence rates, which can reach up to 70% or higher depending on the treatment modality used. After surgical excision or other therapies, keloids can recur, often larger than before. This makes them particularly challenging to manage, necessitating a combination of treatment modalities and long-term follow-up (Fig. 30.3). Moreover, long-term follow-up is essential in keloid management to monitor for signs of recurrence and implement timely interventions. Furthermore, patient education plays a pivotal role in reducing recurrence risks, emphasizing the importance of adherence to post-treatment care, including the use of silicone gel sheets, pressure garments, and sun protection.

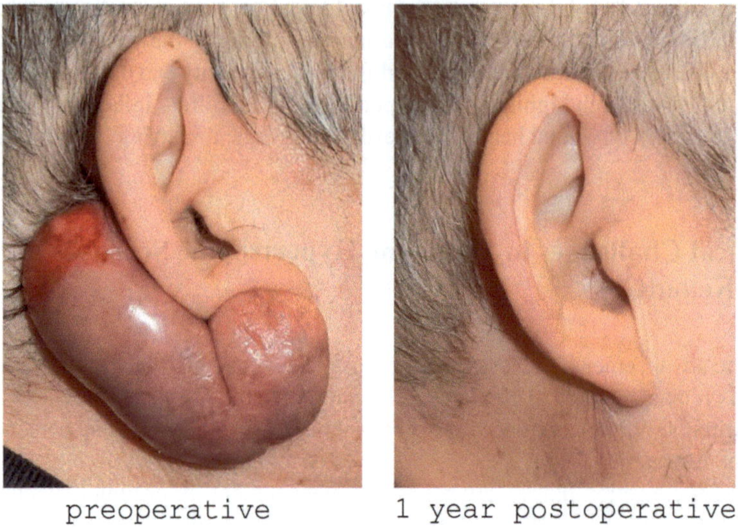

Fig. 30.3 Pre- and post-operative images of an ear keloid caused by prolonged mask use during the COVID-19 pandemic, demonstrating significant improvement 1 year after surgical intervention. (These figures are cited brom Lumintang LM, Dohi T, Ogawa R. Ear Keloid as an Unusual Complication of Prolonged Mask Use during the COVID-19 Pandemic. Plast Reconstr Surg Glob Open. 2024 Jan 22;12(1):e5541)

Psychological Impact

The disfigurement caused by hypertrophic scars and keloids can lead to significant psychological distress, affecting self-esteem and quality of life. Patients may experience anxiety, depression, and social withdrawal due to the visibility and chronic nature of these scars [34].

At the Scar Clinic of Nippon Medical School, rehabilitation makeup technique has been introduced as an additional complementary therapy addressing the patient's distress and self-esteem, which is applied to camouflage physical damage, such as that caused by skin diseases, traumatic injuries, or post-operative scars that may remain cosmetically unpleasant. The goal is to enhance the patient's quality of life by facilitating reentry to society and providing a new option for solving mental and social problems that are not covered by present medical services. We believe that this concept will be accepted as a means of improving the QOL of patients who desire to return to society [35].

Current and Emerging Approaches to Scar Management

The management of hypertrophic scars and keloids has evolved significantly over the years, with an increasing emphasis on evidence-based practices and multidisciplinary approaches. Combining multiple treatment modalities, including pharmacological, surgical, and noninvasive therapies, often yields the best outcomes.

Emerging therapies, such as nanotechnology-based drug delivery systems and biomimetic scaffolds, hold promise for revolutionizing scar management. These innovative approaches aim to deliver therapeutic agents directly to the scar tissue, minimizing systemic side effects and enhancing efficacy. Additionally, advancements in regenerative medicine, such as the use of stem cells and tissue-engineered constructs, offer potential for scarless wound healing.

Topical Therapies

Silicone Gel Sheets

Silicone sheets have long been used as a first-line treatment for hypertrophic scars and keloids. These sheets create a hydrated environment that helps to soften the scar tissue, reduce tension, and improve scar appearance. Studies have shown that silicone gel sheets can significantly improve the appearance and texture of scars when used consistently over several months. Silicone therapy works by reducing collagen synthesis through its influence on local epidermal growth factor (EGF) and transforming growth factor-beta (TGF-β) [36].

Steroid Tapes and Intralesional Injections

Corticosteroids, such as triamcinolone, are commonly injected into hypertrophic scars and keloids to reduce inflammation and collagen production. However, repeated treatments are often necessary, and the response can be variable.

In Japan, recognized for its advancements in scar treatment, steroid tape has become a standard approach, particularly at facilities like the Scar and Keloid-specialized Clinic at Nippon Medical School Hospital. Unlike traditional methods relying on frequent corticosteroid injections such as triamcinolone to reduce inflammation and collagen production, [37] the clinic now minimizes the use of intralesional injections, favoring steroid tapes. This method provides a controlled release of corticosteroids directly onto the scar, reducing the need for repeated invasive treatments and offering a more consistent response [38] (Fig. 30.4).

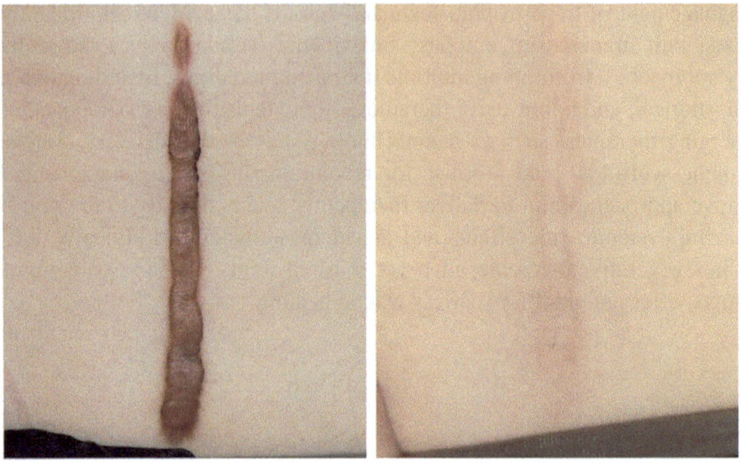

Fig. 30.4 Positive treatment outcome observed 36 months after using steroid tape (Deprodone propionate plaster) on a hypertrophic scar, showing reduced scar size and improved texture. (These figures are cited from Ogawa R. The Most Current Algorithms for the Treatment and Prevention of Hypertrophic Scars and Keloids: A 2020 Update of the Algorithms Published 10 Years Ago. Plast Reconstr Surg. 2022 Jan 1;149(1):79e-94e)

Laser Therapy

Fractional CO_2 laser and pulsed dye laser (PDL) are extensively utilized in the management of scars to improve texture and color. These laser modalities work through distinct yet complementary mechanisms. The fractional CO_2 laser creates microthermal zones in the scar tissue, which stimulate controlled dermal remodeling, enhance collagen turnover, and reduce skin irregularities [39]. On the other hand, the PDL specifically targets blood vessels within the scar tissue by utilizing selective photothermolysis, effectively reducing redness and vascular prominence commonly associated with scars [40].

Evidence also highlights the efficacy of these lasers when used as standalone treatments and in combination with adjunctive therapies. The study underscores the role of the fractional CO_2 laser in breaking down excessive collagen deposits and encouraging a more organized collagen matrix. This aligns with its ability to soften scar texture and improve pliability. Similarly, PDL has demonstrated significant benefits in reducing erythema and modulating inflammation, which are critical factors in minimizing scar hypertrophy and improving aesthetic outcomes [41].

At our clinic we emphasize the synergistic effects of combining laser therapy with other modalities, such as corticosteroid injections or topical silicone therapy.

The integration of lasers enhances the penetration and efficacy of these treatments by disrupting the rigid scar tissue barrier. For instance, using a fractional CO_2 laser before applying silicone gel or administering intralesional corticosteroids increases the absorption of therapeutic agents, thereby enhancing their anti-inflammatory and anti-fibrotic effects [41].

Furthermore, it is important to tailor laser settings to individual patient needs. Factors such as skin type, scar maturity, and underlying vascularity influence the choice of laser parameters, ensuring optimized outcomes with minimal adverse effects.

Combining fractional CO_2 laser and PDL in sequential sessions can offer complementary benefits, addressing both the vascular and structural components of scar pathology.

Our scar outpatient clinic has actively contributed to the growing body of evidence supporting the combined use of laser therapy and adjunctive treatments.

The clinic's data align with Ogawa's findings, demonstrating significant improvements in scar appearance and patient satisfaction when lasers are incorporated into a multimodal therapeutic approach. In conclusion, the fractional CO_2 laser and PDL represent powerful tools in scar management, offering targeted approaches to improving texture and color. When combined with other therapies, these lasers enhance treatment efficacy, addressing the multifaceted nature of scar pathology. Continuous advancements in laser technology and protocol optimization are expected to further refine these outcomes, benefiting a broader range of patients.

Radiation Therapy

Low-dose radiation therapy (RT) has been increasingly employed as an adjuvant treatment to prevent the recurrence of keloids following surgical excision. Its mechanism of action involves the inhibition of fibroblast proliferation and the suppression of excessive collagen deposition, both of which are central to keloid formation. By targeting these cellular processes, RT effectively mitigates the overactive wound-healing response characteristic of keloids.

According to Ogawa et al., the timing and dose of radiation therapy are critical factors influencing its efficacy and safety. The study emphasizes that RT is most effective when administered within 24 h of surgical excision. This timing aligns with the period when fibroblast activity is highest, allowing RT to directly interfere with the cellular and molecular events that drive keloid recurrence. The recommended dose typically ranges between 10 and 15 Gy, fractionated over 2–3 sessions, which balances efficacy with the minimization of adverse effects [42]. *Ogawa et al's* research also highlights the importance of fibroblast radiosensitivity. Keloid-derived fibroblasts exhibit a heightened response to radiation compared to normal fibroblasts, making RT particularly effective in targeting the pathological processes involved in keloid formation. This selective impact on abnormal fibroblasts reduces the likelihood of damaging surrounding healthy tissue, which is a critical consideration in preserving skin integrity [43].

Despite its advantages, radiation therapy is not without risks; the potential for long-term side effects, including the induction of skin malignancies, must be taken into consideration.

While the absolute risk remains low, it is a significant concern, particularly for younger patients and those receiving radiation in areas with high sun exposure. As a result, RT is generally reserved for patients with severe or recurrent keloids that have not responded to other treatments. The study underscores the importance of patient selection and informed consent, with a thorough discussion of the benefits and risks before initiating therapy.

Emerging Therapies

Emerging therapies in scar management are paving the way for more effective and targeted treatments by focusing on the molecular mechanisms underlying excessive scarring. Gene therapy, with tools like CRISPR-Cas9, is being explored to suppress genes such as TGF-β1, which play a key role in fibrosis and collagen overproduction. Anti-fibrotic agents, including pirfenidone and tranilast, are being adapted from systemic fibrosis treatments to dermatological applications, offering new avenues to inhibit fibroblast activation and extracellular matrix deposition. Growth factor inhibitors, such as bevacizumab targeting VEGF, show promise in reducing scar vascularity and erythema. Stem cell-based approaches, particularly using adipose-derived and mesenchymal stem cells, leverage their anti-inflammatory and regenerative properties to improve dermal remodeling, with even greater potential when combined with advanced laser techniques [44]. Additionally, biologics like monoclonal antibodies targeting TGF-β are demonstrating efficacy in reducing fibroblast-driven collagen synthesis.

Nanotechnology-driven therapies are also emerging, offering precise and controlled delivery of drugs and growth factor inhibitors directly to scar tissue. These innovative approaches, often integrated with existing modalities such as lasers and corticosteroids, aim to enhance treatment outcomes and reduce scarring severity. Despite challenges in translating these advances to widespread clinical practice, the future of scar management is poised for significant improvement with the continued evolution of gene editing, bioengineering, and regenerative medicine [44, 45].

Conclusion

The management of hypertrophic scars and keloids remains a significant challenge in clinical medicine, posing both diagnostic and therapeutic complexities. These scars not only result in physical discomfort and functional limitations but also contribute significantly to psychological distress and social challenges, profoundly affecting patients' quality of life. Despite advances in both noninvasive and invasive therapies, including laser treatments, corticosteroid injections, surgical excisions, and emerging biologics, achieving consistent and satisfactory outcomes remains elusive for many individuals. The heterogeneity of scar types, combined with individual differences in healing responses, underscores the need for a more personalized approach to treatment. Continued research into the molecular mechanisms underlying scar formation, particularly the role of inflammation, fibroblast activity, and extracellular matrix remodeling, offers promising avenues for innovation. Identifying key pathways and genetic markers involved in abnormal scar development could lead to the development of targeted therapies that minimize scar formation while preserving the integrity of the healing process.

As we gain a deeper understanding of the molecular and cellular mechanisms that govern scar formation, the potential for more effective and tailored treatments becomes increasingly viable. The integration of gene therapy, advanced wound care products, and biologic agents, such as monoclonal antibodies targeting fibrotic pathways, is poised to revolutionize the field of scar management in the coming years. Additionally, technologies such as 3D bioprinting for skin grafts and the application of artificial intelligence in predicting treatment outcomes could further enhance personalized care. The development and refinement of comprehensive classification systems, like the Japan Scar Scale, provide an essential framework for evaluating scars and guiding treatment. Such systems enable clinicians to assess scars holistically, incorporating both clinical observations and patient-reported outcomes. This approach ensures that management strategies address not only the physical aspects of scars but also their broader psychosocial impacts. Ultimately, the future of hypertrophic scar and keloid management lies in multidisciplinary collaboration, combining insights from molecular biology, clinical dermatology, and patient-centered care. By advancing our understanding and embracing innovative therapeutic options, we can move closer to achieving more effective, predictable, and patient-specific solutions for scar management.

References

1. Macarak EJ, Wermuth PJ, Rosenbloom J, Uitto J (2021) Keloid disorder: fibroblast differentiation and gene expression profile in fibrotic skin diseases. Exp Dermatol 30(1):132–145
2. Pereira D, Sequeira I (2021) A scarless healing tale: comparing homeostasis and wound healing of Oral mucosa with skin and Oesophagus. Front Cell Dev Biol 9:682143
3. Matsumoto NM, Peng WX, Aoki M et al (2017) Histological analysis of hyalinised keloidal collagen formation in ear- lobe keloids over time: collagen hyalinisation starts in the perivascular area. Int Wound J 14(6):1088–1093
4. Mathew M et al (2024) Biopsychosocial impact of keloids on quality of life. JAAD Rev 2:13–19
5. Martin P, Nunan R (2015) Cellular and molecular mechanisms of repair in acute and chronic wound healing. Br J Dermatol 173(2):370–378

6. Schmieder SJ, Ferrer-Bruker SJ (2024) Hypertrophic scarring. [Updated 2023 Sep 4]. In: StatPearls [Internet]. StatPearls Publishing, Treasure Island. Available from: https://www.ncbi.nlm.nih.gov/books/NBK470176/
7. Huang C, Murphy GF, Akaishi S, Ogawa R (2013) Keloids and hypertrophic scars: update and future directions. Plast Reconstr Surg Glob Open 1:e25
8. Dunkin CS, Pleat JM, Gillespie PH, Tyler MP, Roberts AH, McGrouther DA (2007) Scarring occurs at a critical depth of skin injury: precise measurement in a graduated dermal scratch in human volunteers. Plast Reconstr Surg 119:1722–1732
9. Ogawa R, Okai K, Tokumura F et al (2012) The relationship between skin stretching/contraction and pathologic scarring: the important role of mechanical forces in keloid generation. Wound Repair Regen 20(2):149–157. https://doi.org/10.1111/j.1524-475X.2012.00766.x
10. Ross R, Guo Y, Walker RN, Bergamaschi D, Shaw TJ, Connelly JT (2024) Biomechanical activation of keloid fibroblasts promotes lysosomal remodeling and exocytosis. J Invest Dermatol 144(12):2730–2741. https://doi.org/10.1016/j.jid.2024.04.015
11. Ogawa R (2020) Chapter 15: Japan Scar Workshop (JSW) Scar Scale (JSS) for Assessing Keloids and Hypertrophic Scars. 2020 Dec 8. In: Téot L, Mustoe TA, Middelkoop E et al (eds) Textbook on scar management: state of the art management and emerging technologies [Internet]. Springer, Cham; Available from: https://www.ncbi.nlm.nih.gov/books/NBK586118/, https://doi.org/10.1007/978-3-030-44766-3_15
12. Baryza MJ, Baryza GA (1995) The Vancouver scar scale: an administration tool and its interrater reliability. J Burn Care Rehabil 16:535–538
13. Gurtner GC, Werner S, Barrandon Y et al (2008) Wound repair and regeneration. Nature 453(7193):314–321. In-depth analysis of wound repair stages and scar tissue formation
14. Sinno H, Prakash S (2013) Complements and the wound healing cascade: an updated review. Plast Surg Int 2013:146764
15. Landen NX, Li D, Stahle M (2016) Transition from inflammation to proliferation: a critical step during wound healing. Cell Mol Life Sci 73:3861–3885
16. Aarabi S, Longaker MT, Gurtner GC (2007) Hypertrophic scar formation following burns and trauma: new approaches to treatment. PLoS Med 4(9):e234. https://doi.org/10.1371/journal.pmed.0040234
17. Tredget EE, Nedelec B, Scott PG, Ghahary A (1997) Hypertrophic scars, keloids, and contractures: the cellular and molecular basis for therapy. Surg Clin N Am 77(3):701–730. https://doi.org/10.1016/S0039-6109(05)70572-2
18. Verhaegen PD, van Zuijlen PP, Pennings NM et al (2009) Differences in collagen architecture between keloid, hypertrophic scar, normotrophic scar, and normal skin: an objective histopathological analysis. Wound Repair Regen 17(5):649–656. https://doi.org/10.1111/j.1524-475X.2009.00533.x
19. Xia G, Dohi T, Abdelhakim M, Tosa M, Ogawa R (2024) The effects of systemic diseases, genetic disorders and lifestyle on keloids. Int Wound J 21(4):e14865. https://doi.org/10.1111/iwj.14865. PMID: 38584345; PMCID: PMC10999570
20. Ogawa R, Watanabe A, Than Naing B et al (2014 Jul) Associations between keloid severity and single-nucleotide polymorphisms: importance of rs8032158 as a biomarker of keloid severity. J Invest Dermatol 134(7):2041–2043
21. Yagi Y, Kuwatsuka Y, Asai M, Honda M, Utani A (2018) Coexistence of keloids and pilomatricoma in a patient with Rubinstein-Taybi syndrome. Dermatol Online J 24(1):13030/qt4rq2k5fr
22. Zhu YO, MacDonnell S, Kaplan T et al (2023) Defining a unique gene expression profile in mature and developing keloids. JID Innovations 3(5):100211. Published 2023 May 19. https://doi.org/10.1016/j.xjidi.2023.100211
23. LeFlore IC (1980) Misconceptions regarding elective plastic surgery in the black patient. J Natl Med Assoc 72:947–948
24. Noishiki C, Hayasaka Y, Ogawa R (2019) Sex differences in keloidogenesis: an analysis of 1659 keloid patients in Japan. Dermatol Ther (Heidelb) 9(4):747–754

25. Gao FL, Jin R, Zhang L, Zhang YG (2013) The contribution of melanocytes to pathological scar formation during wound healing. Int J Clin Exp Med 6(7):609–613. PMID: 23936604; PMCID: PMC3731197
26. Dohi T, Padmanabhan J, Akaishi S, Than P, Terashima M, Matsumoto N, Ogawa R, Gurtner G (2019) The interplay of mechanical stress, strain, and stiffness at the keloid periphery correlates with increased Caveolin-1/ROCK signaling and scar progression. Plast Reconstr Surg 144(1):58e–67e. https://doi.org/10.1097/PRS.0000000000005717
27. Berry CE, Downer M Jr, Morgan AG, Griffin M, Liang NE, Kameni L, Laufey Parker JB, Guo J, Longaker MT, Wan DC (2023) The effects of mechanical force on fibroblast behavior in cutaneous injury. Front Surg 10:1167067. https://doi.org/10.3389/fsurg.2023.1167067. PMID: 37143767; PMCID: PMC10151708
28. Kuehlmann B, Bonham CA, Zucal I, Prantl L, Gurtner GC (2020) Mechanotransduction in wound healing and fibrosis. J Clin Med 9(5):1423. https://doi.org/10.3390/jcm9051423. PMID: 32403382; PMCID: PMC7290354
29. Akaishi S, Akimoto M, Ogawa R, Hyakusoku H (2008) The relationship between keloid growth pattern and stretching tension: visual analysis using the finite element method. Ann Plast Surg 60(4):445–451
30. Ogawa R, Akaishi S, Huang C et al (2011) Clinical applications of basic research that shows reducing skin tension could prevent and treat abnormal scarring: the importance of fascial/subcutaneous tensile reduction sutures and flap surgery for keloid and hypertrophic scar reconstruction. J Nippon Med Sch 78(2):68–76
31. Ogawa R (2017) Keloid and hypertrophic scars are the result of chronic inflammation in the reticular dermis. Int J Mol Sci 18(3):606. https://doi.org/10.3390/ijms18030606
32. Matsumoto NM, Peng WX, Aoki M et al (2017) Histological analysis of hyalinised keloidal collagen formation in earlobe keloids over time: collagen hyalinisation starts in the perivascular area. Int Wound J 14(6):1088–1093
33. Trace AP, Enos CW, Mantel A, Harvey VM (2016) Keloids and hypertrophic scars: a spectrum of clinical challenges. Am J Clin Dermatol 17:201–223. https://doi.org/10.1007/s40257-016-0175-7
34. Olaitan PB (2009) Keloids: assessment of effects and psychosocial- impacts on subjects in a black African population. Indian J Dermatol Venereol Leprol 75:368–372
35. Reiko K (2019) Rehabilitation makeup. J Japan Lingual Orthod Assoc 2019(29):22–26
36. O'Brien L, Jones DJ (2013) Silicone gel sheeting for preventing and treating hypertrophic and keloid scars. Cochrane Database Syst Rev 2013(9):CD003826. Published 2013 Sep 12. https://doi.org/10.1002/14651858.CD003826.pub3
37. Morelli Coppola M, Salzillo R, Segreto F, Persichetti P (2018) Triamcinolone acetonide intralesional injection for the treatment of keloid scars: patient selection and perspectives. Clin Cosmet Investig Dermatol 11:387–396. https://doi.org/10.2147/CCID.S133672. PMID: 30087573; PMCID: PMC6063260
38. Ogawa R, Quong WL (2024) Effective treatment of an aggressive chest wall keloid in a woman using deprodone propionate plaster without surgery, radiotherapy, or injection. Plast Reconstr Surg Glob Open 12(9):e6117. Published 2024 Sep 3. https://doi.org/10.1097/GOX.0000000000006117
39. Xi W, Xie Y, Zhang Z et al (2021) 3D mesh releasing method: a retrospective analysis of fractional CO2 treatment on contracture scars. Lasers Surg Med 53:227–235
40. Gauglitz GG, Poetschke J (2020) Strategies for keloid and hypertrophic scars with lasers. In: Shokrollahi K (ed) Laser Management of Scars. Springer, Cham
41. Ogawa R (2022) The Most current algorithms for the treatment and prevention of hypertrophic scars and keloids: a 2020 update of the algorithms published 10 years ago. Plast Reconstr Surg 149(1):79e–94e. https://doi.org/10.1097/PRS.0000000000008667
42. Dohi T, Kuribayashi S, Tosa M et al (2019) Z-plasty and postoperative radiotherapy for upper-arm keloids: an analysis of 38 patients. Plast Reconstr Surg Glob Open 7:e2496

43. Ogawa R, Tosa M, Dohi T, Akaishi S, Kuribayashi S (2019) Surgical excision and postoperative radiotherapy for keloids. Scars Burn Heal 5:2059513119891113. Published 2019 Dec 10. https://doi.org/10.1177/2059513119891113
44. Mizuno H, Hyakusoku H (2018) Adipose-derived stem cells for scar remodeling: current evidence and future directions. Regenerative Med Jpn 13(4):242–251
45. Liu Y, Ren J (2018) Targeting TGF-β pathway for scar prevention and treatment: potential applications of gene therapy. Mol Ther 26(6):1316–1332

Chapter 31
Therapeutic Approaches to Treat Hypertrophic Scars

Vanessa C. Smith, Steve Smith, and Tugba Ozdemir

Abstract The purpose of this paper is to compare the similarities and differences between the types of hypertrophic scarring, a form of pathological scarring. Hypertrophic scarring is caused by abnormal wound healing after cutaneous trauma, and it can present in several ways due to the patient's genetic predisposition. We will explore hypertrophic scars that are burn-related, keloids, and scleroderma. This paper will examine the extracellular matrix components, such as fibrillar proteins, non-fibrillar proteins, and glycosaminoglycans, and cellular components of the condition, to understand the key differences in expression among the types of hypertrophic scarring, as well as highlight some potential underlying factors for future research. Therapeutic solutions like pharmacotherapy, biomaterials, and invasive treatments are also compared in relation to each presentation type of hypertrophic scarring. This allows for greater understanding in the differences between preferred treatment options and for which type of pathological scarring each method should be implemented.

Keywords Keloids · Scleroderma · Hypertrophic scars · Burns · Wound healing

Abbreviations

5-FU	5-Fluorouracil
BM	Basement Membrane
BMP	Bone Morphogenic Protein
Botox	Botulinum Toxin A
CTGF	Connective Tissue Growth Factor

V. C. Smith · S. Smith · T. Ozdemir (✉)
Department of Nanoscience & Biomedical Engineering, South Dakota School of Mines and Technology, Rapid City, SD, USA
e-mail: Tugba.Ozdemir@sdsmt.edu

F. De Francesco, K. Turksen (eds.), *Management and Strategies for Wound Healing*, Stem Cell Biology and Regenerative Medicine 79,
https://doi.org/10.1007/978-3-032-12442-5_31

DIEP	Deep Inferior Epigastric Artery
DLQI	Dermatology Life Quality Index
ECM	Extracellular Matrix
EGC	Enteric Glial Cell
FGF	Fibroblast Growth Factors
GPI	Glycosylphosphatidylinositol
HA	Hyaluronic Acid
HS	Heparan Sulfate
HSPG	Heparan Sulfate Proteoglycan
MMP	Matrix Metalloproteinases
Nd:YAG	Neodymium-Doped Yttrium Aluminum Garnet
NK	Natural Killer
OEC	Olfactory Ensheathing Cell
PDGF	Platelet-Derived Growth Factor
PDL	Pulsed-Dye Laser
PRG4	Proteoglycan 4
SGC	Satellite Glial Cell
SLRP	Small Leucine-Rich Proteoglycans
TGF-β	Transforming Growth Factor-Beta
TIMP	Tissue Inhibitors of Metalloproteinases
TNC	Tenascin-C
VEGF	Vascular Endothelial Growth Factor

Introduction

Abnormal wound healing can occur with cutaneous trauma, including surgery, deep cuts, and burns. Areas with higher tension and greater stretching are more likely to form hypertrophic scars [1]. According to the Cleveland Clinic, hypertrophic scars are defined as raised, visible scars that are the result of abnormal wound healing responses. Hypertrophic scars form over a dermal injury site and develop approximately 4 to 8 weeks after the wound closure [2]. The visible scars are created through the body producing excessive collagen in an attempt to heal an injury. The resulting collagen deposition has a rigid structure and surface within the boundaries of the original wound site. Different types of hypertrophic scars can be listed as:

- Burns
- Keloids
- Scleroderma

In this section, we will give an overview of each type of hypertrophic scar.

Burns

A common trauma that leads to the development of cutaneous scars is burn injuries. In 2003, a study evaluated the prevalence of hypertrophic scarring post-burn injuries. The results of that study indicated a prevalence of 67% of the patients experiencing hypertrophic scars [3]. However, a comprehensive literature review conducted in 2012 found a prevalence range of 32 to 72% using research articles published in English between 1965 and the present time of the review [4]. Hypertrophic scarring leads to a variety of physical and psychosocial adverse effects that impact burn victims following the inciting incident and recovery process.

Physical impacts on the patient's quality of life include contractures, pruritus (itching sensation), pain, loss of function, and restricted movement. The potential dyschromia of the injury as well as the other recognizable features of the hypertrophic scarring could lead to an impact on psychosocial well-being during and after recovery. These physical and psychosocial challenges decrease the quality of life for patients and delay the recovery and reintegration periods.

Several factors control the depth and severity of the burn injury. The first is the temperature of the burns' inciting event. The second is the total duration of contact with the inciting event. This includes any absorption into the skin by the inciting material post the incident. The third factor is the thickness of the epidermis skin layer. Eyelids, arms, hands, and genital regions are notable areas with thin skin that are especially at risk for burn injuries. The fourth factor is the blood supply of the skin site. More blood at the location of the injury leads to a dissipation of the heat. Burns can be classified into three degrees of clinical significance: first-degree (superficial burns), second-degree (partial-thickness burns), and third-degree (full-thickness burns) [5]. Superficial burns only affect the epidermis layer, while partial-thickness burns affect the epidermis and any point in the dermal layer. The most severe full-thickness burns permeate the epidermis, dermis, and subcutaneous tissue. The majority of burn wounds are superficial first-degree or second-degree burns which may result in non-hypertrophic scars on the more severe side. Deep dermal second-degree and third-degree burns are at a greater risk of developing into hypertrophic scars [1].

The depth of the burn injury and the wound healing are both factors in determining the likelihood of the development of hypertrophic scars. According to a study conducted by Deitch et al., the optimal time to predict the development of hypertrophic scars on burn injuries is 21 days post-injury. It was determined that if a burn wound healed after that 3-week time frame, 78% of the burn sites would develop hypertrophic scars [6]. The location of the burn site on the body does affect this period as facial burns' critical point is 10 days post-injury due to fragility of the face [7].

Keloids

Hypertrophic scars and keloids are diagnosed separately both clinically and pathologically, despite the similar features and characteristics they share. However, it has been suggested that keloids and hypertrophic scars could be successive stages of the same skin condition [8]. Keloids are an abnormal fibroproliferative wound healing response that does not evolve in the same pattern of stability as hypertrophic scars. In Fig. 31.1, it can be demonstrated that the hypertrophic scar grows along the edges of the original normotrophic scar site, whereas the keloid scar excessively grows in all directions to create a new edge line. Keloids have a much longer period of recovery before developing. They may develop from as early as 3 months post-injury to several years after the inciting incident. Keloids, like hypertrophic scars, follow abnormal wound healing of cutaneous trauma. The most defining characteristic of keloid that differs from hypertrophic scars is the invasive growth from the original wound site boundaries [8]. Keloids continue to grow after development and rarely regress [9]. Due to the proliferation of keloid tissue, keloids are benign tumors.

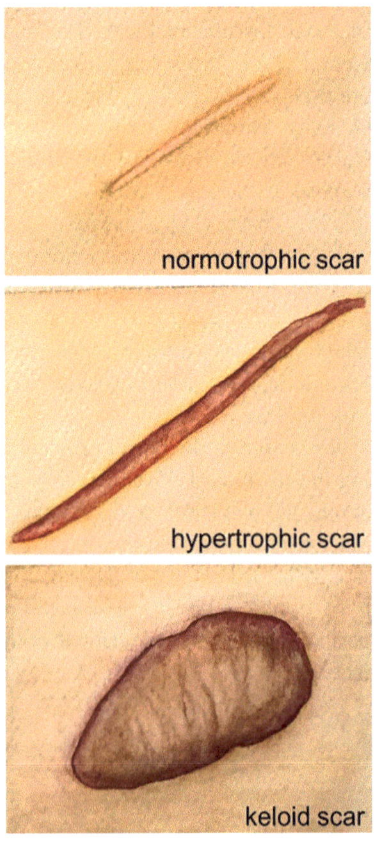

Fig. 31.1 This figure depicts a watercolor illustration of a normotrophic, a hypertrophic, and a keloid scar [2]

There are two distinct keloid phenotypes that are characterized by their elevation level on the skin: flat keloids and raised keloids.

Literature reviews suggest that keloid formation has a strong genetic predisposition component. The risk of keloid development increases in relation to an increase in skin pigmentation. The predominantly affected racial types are those of African and Asian descent [10]. Genetically, keloids have a higher incidence in people of Afro-Caribbean ethnicities [11]. People of the Hispanic population also have a higher incidence of keloids. However, pigmentation itself is not the only factor in ethnic differentiation of incidence rates.

Keloids have pruritus, pain, loss of function, and restricted movement. A Turkish case study conducted in 2009 resulted in no significant difference in the Dermatology Life Quality Index (DLQI) scores of patients with keloids and hypertrophic scars and patients with psoriasis [12].

Keloid-inducing events include trauma in surgical and non-surgical wound healing sites. Some examples of non-surgical wound-healing trauma are insect bites, acne, injections, vaccinations, body piercings, lacerations, abrasions, hair removal, tattooing, excoriations, and blunt trauma. Burns can lead to keloids, but it is more common for those to develop into hypertrophic scars [13].

Scleroderma

Scleroderma, a chronic autoimmune disease, affects the body's connective tissue and causes excessive deposition of collagen in the skin and internal organs. The cause of scleroderma is not known, but it is believed to be caused by the body mistakenly attacking its own tissues.

Scleroderma can present itself either localized or in a systemic form, with the former being more commonly referred to as morphea. Morphea primarily affects skin tissue, with some cases affecting fat tissue, muscle, and fascia, and it is characterized by inflamed fibrotic patches that appear anywhere on the dermal layer [14]. A primary distinction between hypertrophic scarring, keloids, and scleroderma is the scar site, or in the case of scleroderma scar-like site, the origin of the excessive collagen deposition differs. Hypertrophic scarring and keloids begin at a cutaneous trauma, whereas scleroderma can occur anywhere.

There are two forms of systemic scleroderma: limited systemic scleroderma and diffuse systemic scleroderma [15]. Systemic scleroderma impacts pulmonary, renal, and cardiac systems as well as connective tissue. One of the key symptoms of systemic scleroderma is the narrowing of blood vessels as an example of Raynaud's phenomenon. Skin lesions can be found all over the body, specifically on the extremities and the facial region. The skin lesions lead to abnormalities in the structure of the microvascular area of the same location. According to a study conducted in 2021, systemic scleroderma has a prevalence of 17.6 per 100,000 in a population with an incidence rate of 1.4 per 100,000 people [16].

Morphea is a rare disorder with an estimated prevalence of 220 per 100,000 by 80 years old [14]. There is also a sex predominance in morphea, appearing approximately 2.6 times more in male patients than female patients. Systemic scleroderma has a much lower incidence rate in the United States of 7 to 489 cases per million.

There are several subtypes of morphea. The first is plaque-like morphea which is defined by either a raised or depressed plaque surrounded by erythema. Sebaceous glands atrophy at the site as well as hair loss to accompany the hyper/hypopigmentation. The second subtype is linear scleroderma which can be characterized by linear bands of sclerosis causing deformities. Generalized morphea causes disseminated skin sclerosis leading to contractures and deformities in the limbs. Bullous morphea is a rare form of morphea with a blister-like appearance [17]. The fifth subtype is deep morphea which affects the deep reticular dermis and subcutaneous layers and sometimes the underlying fascia as well. Disabling pansclerotic morphea is a rare childhood disease characterized by rapid progression of cutaneous fibrosis with contractures and musculoskeletal atrophy. Eosinophilic fasciitis, also known as Shulman syndrome, is considered a variant of scleroderma as a sudden onset of symmetrical hardening of limbs following strenuous exercise.

ECM Characteristics

The expression and organization of extracellular matrix (ECM) proteins in normal and hypertrophic scars show differences. In this section, we will summarize the main characteristics of ECM proteins in hypertrophic scars. Figure 31.2 illustrates some of the differences between the ECM of burn, keloid, scleroderma, and normal wound sites.

Fibrillar Proteins

(a) Collagen

Collagen is the main structural protein in the ECM or the connective tissue. Collagen type I and type III are essential in the wound healing process. Type III is deposited onto a wound site first to support the formation of granulation tissue. Collagen type I then replaces type III as the dominant collagen in skin [18]. Cross-linking of collagen type I leads to tensile strength in the scar tissue.

Hypertrophic scars, keloids, and scleroderma all exhibit an excessive deposition of collagen type I on wound healing sites, leading to cutaneous scars. Collagens' organization being more straight and parallel leads to a stronger representation of scarring. The ratio of collagen type I to collagen type III at hypertrophic wound sites has been altered from normal wound healing processes [19]. Keloid fibroblasts have been shown to produce a significant increase in collagen type I in parallel bundles

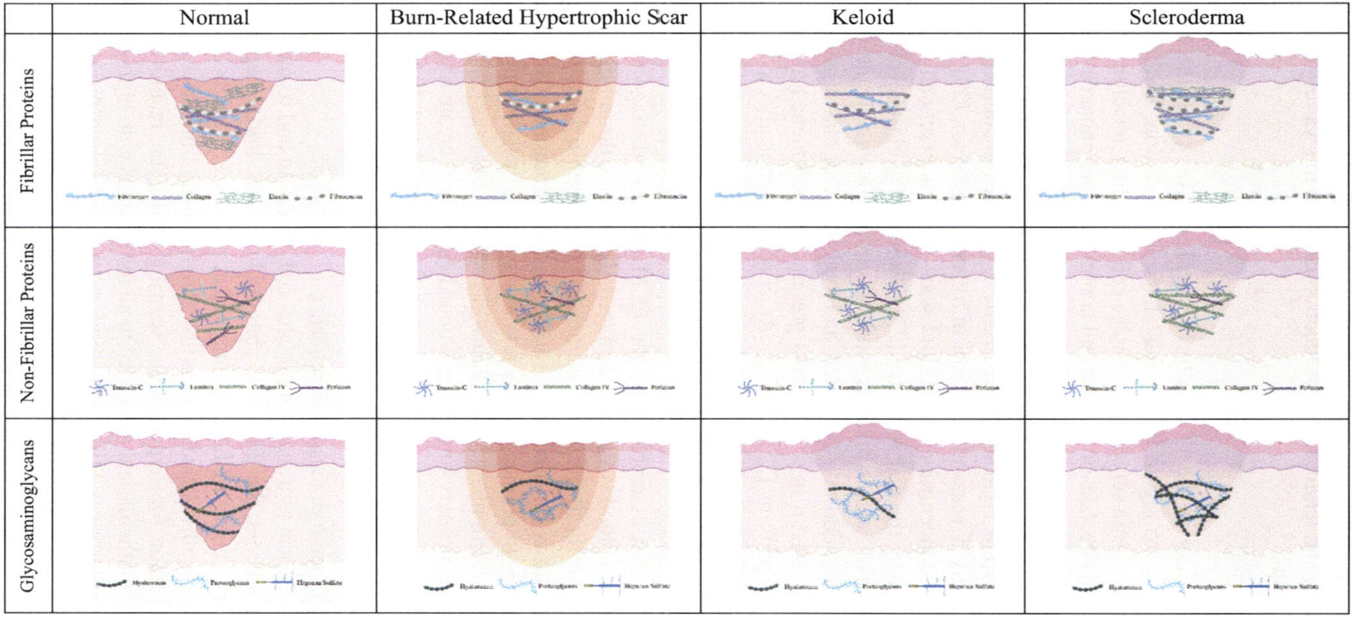

Fig. 31.2 Table illustrating the general differences in molecular components under the categories fibrillar proteins, non-fibrillar proteins, and glycosaminoglycans in normal, burn, keloid, and scleroderma wound sites. The fibrillar proteins shown in the image are fibrinogen, collagen, elastin, and fibronectin. It can be noted that elastin is absent in burn-related hypertrophic scars and keloids. The elastin in scleroderma is parallel and straightened in comparison to the normal wound site. The non-fibrillar proteins shown in the image are tenascin-C, laminin, collagen IV, and perlecan. All three types of pathological scarring have an increase in the presence of tenascin-C; however, burn-related hypertrophic scars have an increased expression of laminin and scleroderma has an increased expression of collagen IV. The glycosaminoglycans shown in the image are hyaluronan, proteoglycans, and heparan sulfate. It is illustrated that there is a decrease of expression of hyaluronan in burn-related hypertrophic scars and keloids in contrast to the increase of expression in scleroderma

[20]. The Mayo Clinic attributes scleroderma to being caused by too much collagen production in body tissue. Since collagen strengthens and supports connective tissues such as skin, bones, and tendons, overproduction can lead to a variety of complications in patient health.

(b) Elastin

Elastin is a fibrillar protein in the ECM which provides elasticity to the newly formed tissue, allowing it to regain its original shape and function. In the early stages of wound healing, elastin serves to encourage cell spreading, attachment, and signaling through wound site. Elastin lessens cell contraction and enhances dermal regeneration [21].

The elasticity which provides stretching and flexibility in the skin is visibly absent in scar tissue due to its rigid structure. Elasticity has been documented to be missing for approximately 5 years after a burn in hypertrophic scars [22]. Elastin has reportedly such low concentrations in keloid scars that it is practically absent from the dermal layer where the scar is located [23]. However, the elastin volume in the deeper dermal layers of keloids presents more elastin than normal healthy skin tissue [24]. Elastin in scleroderma is disrupted, and the balance of elastin and elastase is altered [25]. The elastin in scleroderma presents in straightened and compressed fibers in parallel patterns [25]. The rigidity of scar tissue is affected by the imbalance of elastin and the organization that it presents in.

(c) Fibrin/Fibrinogen

Fibrinogen is a protein that assembles into the ECM at tissue damage sites for wound healing [26]. When blood vessels are damaged, the enzyme thrombin cleaves the protein fibrinogen to form fibrin. The concentration of thrombin determines the structure of the fibrin that is formed as a result. Low concentrations of thrombin produce thick fibrin fibers in permeable network arrangements, whereas high concentrations produce thin fibrin fibers in dense, low-permeable networks [27]. Fibrin is an insoluble protein. Its primary function in wound healing is to create a structure of fibrin threads cross-linking to create a mesh with the design to trap blood cells and platelets. This forms a clot to stop bleeding and initiates the wound healing process. After the provisional ECM created by the fibrin is no longer needed, the fibrinolysis process will begin to dissolve the fibrin clot.

As the production of fibrin as a scaffold initiates cell migration and tissue repair in the wound healing process, the effects of abnormality in the initial deposition would lead to a cascade of issues in the remaining repair processes. It has been theorized through multiscale mathematical models that a high-density fibrin clot deposition could be an indicator for hypertrophic scarring [28]. The results of the experimentation of this mathematical model led to all the samples with high-density fibrin clots leading to raised scars [29]. This indicates that the relationship between ECM density is affected by the fibrin density in deposition. Another factor that plays a role in scar formation is the degradation of fibrin after wound healing. It was discovered using a fibrin gel model system, that in keloids there is a lack of microscopically detectable fibrin degradation [30]. Fibrin in scleroderma is abnormally deposited in tissues, which contributes to fibrosis and vascular damage [31]. It has also been theorized that an increased deposition of fibrin may cause an increased collagen deposition; such claims are currently being investigated experimentally.

(d) Fibronectin

Fibronectin is a glycoprotein which has a variety of functions in wound healing. One of the first exhibits of fibronectin is the binding of fibrin to fibronectin to create a clot as part of the platelet plug to prevent blood flow from the wound [32]. The cytokines transforming growth factor-beta (TGF-β) and connective tissue growth factor (CTGF) stimulate the synthesis of fibronectin and fibrogenesis [33]. The deposition of fibronectin into the ECM leads to a deposition of collagen and thus wound contraction.

Abnormally high levels of fibronectin are found in hypertrophic scarring to the point that it is considered a hallmark feature [34]. Fibronectin synthesis is accelerated nearly fourfold in keloids, leading to an overproduction [35]. A fivefold increase in fibronectin was observed in scleroderma patients compared to control subjects [36].

Non-fibrillar Proteins

(a) Tenascin-C

Tenascin-C (TNC) is a glycoprotein in the ECM that is essential for angiogenesis, which is the formation of new blood vessels, during wound healing [37]. TNC is also involved in the modulation of the ECM integrity [38]. A study was conducted which determined that decreased TNC levels inhibit collagen production and ECM deposition [39]. The results of this study lead to the hypothesis that TNC overexpression is present in hypertrophic scars. The same expected increased expression of TNC has been documented in keloids [40]. Elevated levels of TNC have been observed in scleroderma and as it is involved in the development of fibrosis [41, 42]. While the direct cause of the correlation between scar formation and TNC levels is yet to be understood, the repeatedly high expression of TNC in conditions that consist of prevalent scar formation confirms the relationship.

(b) Laminin

Laminins are the most abundant non-collagenous basement membrane proteins that assist in wound healing by promoting re-epithelialization and angiogenesis [43]. In the re-epithelialization process, laminins provide a surface for keratinocytes to migrate and adhere to the wound which establishes an intact epithelial barrier [44]. During wound angiogenesis, laminin stimulates the formation of blood vessels by regulating adhesion of endothelial cells [45]. In addition to those processes, laminin serves as regulators of cell adhesion, migration, differentiation, and proliferation within the ECM [46].

Hypertrophic scars have increased levels of laminin which leads to an excessive ECM deposition. This is theorized to be that an increase in laminin expression levels leads to reconstruction of the ECM which then leads to hypertrophic scarring [47]. Endothelial cells in keloid tissue present a downregulation of various types of laminins which leads to an influx of inflammation at the wound site [48]. The downregulation of laminin types is theorized to potentially promote the chronic inflammation associated with keloid wound healing. In scleroderma, peripheral blood mononuclear cells displayed cellular immunity to laminin [49].

(c) Perlecan

Perlecan is a large basement membrane-specific heparan sulfate proteoglycan which acts as a scaffolding protein in the ECM with its ability to bind to fibroblast growth factors (FGFs), platelet-derived growth factor (PDGF), vascular endothelial growth factor (VEGF), and bone morphogenetic proteins (BMPs) [50]. FGFs and PDGF impact the cell proliferation and differentiation of fibroblasts. VEGF is a signaling protein that stimulates the growth and development of new blood vessels. BMPs promote bone formation and differentiation of mesenchymal stem cells into osteoblasts. These growth factors and proteins regulate cell behavior and promote healing. Hypertrophic scars have defective regeneration of basement membranes (BM) such as perlecan during fibrosis [51]. This is presented in scars developed from burns, keloids, and scleroderma. This indicates that there is dysfunction in the regulation of cell behavior in cutaneous scar sites.

(d) Collagen IV

Collagen type IV is a protein in the BM that supports new blood vessel growth and cell migration during wound healing [18]. Levels of BM proteins are drastically reduced in hypertrophic scars and keloids [52]. This impacts the ability for vital cells to come to the wound site to further facilitate the healing process. In contrast, the levels of collagen type IV present in scleroderma are significantly elevated in comparison to normal skin levels [53]. The difference in expression of collagen type IV between the cutaneous scar conditions is potentially related to the vascular aspects of scleroderma. As collagen type IV is impacted by constant vascular damage such as Raynaud's phenomenon which presents in scleroderma [54].

Glycosaminoglycans

(a) Hyaluronic Acid

Hyaluronic acid (HA), or hyaluronan, is a glycosaminoglycan with a multitude of functions that are present in wound healing including creating pathways for cell migration [55] and regulating normal ECM structural integrity and development [56]. Hyaluronan contributes to wound healing in the inflammatory, proliferative, and remodeling phases. In the inflammatory phase, the synthesis of HA increases rapidly around the wound margin [57] and HA binds with fibrinogen to begin the clotting pathway [55]. During the proliferative phase, fibroblasts are drawn to the HA at the wound site and begin to create collagen and glycosaminoglycans, such as HA, in the newly formed ECM [55]. This process of HA stimulating the fibroblasts at the wound site to produce collagen continues into the remodeling phase.

Hypertrophic scars have decreased levels of HA in comparison to normal skin and mature scar tissue [58]. In matured scars, levels of HA were comparable to normal skin, but the HA was distributed as a thinner layer; in hypertrophic scars the HA is primarily concentrated in the papillary dermis [59]. Keloid tissue, in a similarity to hypertrophic scars, has significantly decreased levels of HA compared to normal skin [60]. Levels of HA are elevated during early stages of scleroderma and as the disease matures, the levels decrease [61]. The cause for elevated levels of HA in scleroderma is theorized to be related to abnormal stimulation by connective tissue [62].

(b) Proteoglycans

(i) Aggrecan

Aggrecan is an extracellular proteoglycan that can be found in scar tissue [63]. Normal adult skin is absent of aggrecan, where it can be found in cartilage, connective tissue sheath, and under hair follicles [64]. Aggrecan aggregates with HA and is responsible for high water retention [65]. The accumulation of aggrecans hinders migration of cells to wound sites and inhibits the maturation of progenitor cells to fibroblasts [66]. Keloids were experimentally shown to have a cartilage-like composition partially due to the overexpression of aggrecan [52]. In scleroderma, the aggrecan protein has been identified to be significantly elevated [52].

(ii) Small Leucine-Rich Proteoglycans

Small Leucine-Rich Proteoglycans (SLRPs) are a family of proteoglycans whose functions can be split into either of two purposes: skin development or wound healing [67]. SLRPs contribute a vital role in collagen fibril growth, fibril organization, and ECM assembly in normal healthy skin during wound healing [63]. SLRPs are also responsible for linking and interacting with collagen fibrils to regulate tendon structure and biomechanics [68]. SLRPs impact fibroblasts during wound healing which could lead to an effect on multiple critical fibroblast functions. If SLRP is abnormally expressed, it could lead to a disruption of the wound healing process and could result in pathological scarring, such as keloids and hypertrophic scars [68]. Altered expression of SLRP has been observed in both localized and systemic scleroderma [69].

(iii) Proteoglycan 4

Proteoglycan 4 (PRG4) is a glycoprotein that is a factor in lubricating and protecting the joints. PRG4 is most found in synovial fluid of joints and is very minimally expressed in skin. The relationship between PRG4 and wound healing is still unclear. However, it has been experimentally shown that mice lacking PRG4 expression showed a complete loss of wound healing ability [70]. It has been indicated that the presence of PRG4 expression switches the healing response to pro-regenerative [71]. There has been preliminary research which suggests that PRG4 has the potential to suppress scar formation [70]. The connections between PRG4 expression in hypertrophic scars, keloids, and scleroderma have yet to be properly explored.

(iv) Versican

Versican is a large proteoglycan with hyaluronan-binding domains and is located in the ECM, which functions to facilitate TGFβ-signaling, promote fibroblast proliferation and differentiation, and accelerate wound closure [72, 73]. Versican levels in normal skin are considered undetectable, but due to elevated expression of versican, it can be easily identified in hypertrophic scars [74]. Excessive deposition of versican can also be seen in keloid scars, to the extent that versican was suggested as a marker for keloids [75]. A study was conducted to understand the relationship between versican expression and systemic sclerosis; it demonstrated that versican is upregulated in a subset of scleroderma patients [76]. The high versican levels detected were associated with fibrotic characteristics of scleroderma such as diffuse cutaneous involvement, widespread skin thickening, and tightening.

(c) Heparan Sulfate

Heparan Sulfate (HS) is a glycosaminoglycan that is important in angiogenesis, cell growth, migration, and differentiation during wound healing. When a protein has multiple HS polysaccharide chains covalently bond to it, it becomes a Heparan Sulfate Proteoglycan (HSPG) [77]. Syndecans and glypicans are the two main HSPG families on the cell surface [78]. Syndecans play a crucial role in wound healing by acting as receptors that mediate cell-matrix interactions, cell migration, and angiogenesis, and their dysregulation can lead to delayed wound healing [79]. Glypicans are glycosylphosphatidylinositol (GPI)-anchored HSPGs that play a role in wound healing by modulating signaling pathways, cell migration, and ECM remodeling, impacting processes like angiogenesis and tissue repair [80]. HS also contributes a crucial role in interactions between the cells and ECM, such as stimulating the cellular adhesion to ECM by binding the cells to matrix macromolecules, for instance, fibronectin or laminin [78].

Scanning for the presence of HS in hypertrophic scars indicates that basal membrane restoration was incomplete at 3 months after wound incident, but no significant difference in HSPG expression was observed [81]. There has been no research to indicate a relationship between HS and cutaneous scarring such as hypertrophic scars, keloids, or scleroderma. However, TGF-β, such as TGF-β1 and TGF-β2, binds to HS and they interact strongly with one another [82]. TGF-β plays a crucial role in the development of scars by promoting the synthesis of the ECM [82]. This indicates that while there is no evidence of a direct correlation between HS and cutaneous scarring, there is reason to believe that the interactions between HS and TGF-β may impact the development of scar formation.

Cellular Components

There are many cell varieties which play pivotal roles in the wound healing process. In this section, we will summarize the main characteristics of some of these relevant cell types and the way they are present in hypertrophic scars.

Fibroblasts/Myofibroblasts

Fibroblasts are a type of cell that synthesizes and maintains the ECM, which forms the structural framework of tissues. Consequently, fibroblasts impact wound healing and tissue repair. Fibroblasts contribute to wound healing during the inflammatory, proliferative, and remodeling phase. Myofibroblasts are a specialized, contractile form of fibroblasts that also play a role in wound healing [83].

During the inflammatory phase, fibroblasts assist in modulating the recruitment of immune cells as well as regulating the behavior, retention, and survival of those immune cells at the wound site [84]. The crosstalk between macrophages and fibroblasts plays a role in instigating the transition from inflammatory to proliferative phases.

In the proliferative phase, fibroblasts contribute to angiogenesis and the granulation tissue formation. Fibroblasts proliferate and migrate to the wound site, and synthesize ECM components like collagen, proteoglycans, and other components [85]. In the granulation tissue, fibroblasts stimulated by injury or stress differentiate into myofibroblast, which can be identified by the marker of α-smooth muscle actin expression.

The myofibroblast synthesizes and replaces the provisional matrix with a more permanent extracellular matrix structure. The contractile properties of myofibroblast are essential in the contraction and maturation of granulation tissue [83]. During the remodeling process of wound healing, collagen is reorganized and reoriented by myofibroblasts as the contraction of the wound occurs.

The healing process of cutaneous wounds that leads to hypertrophic scarring involves an aberrant form of the wound healing process. One of the abnormalities that can occur is through uncontrolled proliferation of fibroblasts which leads to excessive deposition of ECM, contributing to the scar formation. Thus, it can be determined that the levels of growth factors and collagen secreted by fibroblasts during wound healing are directly related to the formation of scar tissue. The specific growth factors and collagen types relating to scar formation are TGF-β1, VEGF, collagen I, and collagen III [86].

TGF-β has been linked to the development of hypertrophic scarring and keloids though its strong and persistent expression in fibroblasts of post-burn hypertrophic scars and keloid-derived fibroblasts [87, 88]. Due to the increased presence of growth factors such as TGF-β and Platelet-Derived Growth Factors (PDGF), keloidal fibroblasts have an increase in proliferation, and have lower rates of apoptosis compared to typical fibroblasts in wound healing [89].

Fibroblasts regulate matrix overproduction through the expression of the enzyme matrix metalloproteinases (MMPs) which degrade several components of the ECM such as collagen, elastin, gelatin, and casein [90]. The expression of MMPs is regulated by fibroblast secretion of tissue inhibitors of metalloproteinases (TIMPs) to balance the ECM turnover. The levels of TIMPs are significantly raised in patients with scleroderma, supporting the understanding of one of the means in which the ECM undergoes unregulated synthesis in scleroderma [91].

Keratinocytes

Keratinocytes are the primary cells of the epidermis, and they play a multitude of roles during cutaneous wound repair. The expression of the keratinocytes depends on where in the epidermal layer it resides. The top layer of the epidermis is the stratum corneum, where anucleated keratinocytes terminally differentiate into corneocytes, commonly called squames, that serve as a protective barrier against the external environment [92]. The next layer down in the epidermis is the stratum lucidum which has thin layers of keratinocytes. Following is the stratum granulosum, and this layer has keratohyalin granules found within keratinocytes [93]. These keratohyalin granules function to play a key role in keratinization as well as serving as a waterproof barrier of the skin. The layer below is the stratum spinosum in which the keratinocytes are adhered together through the protein desmosomes. The keratinocytes in this layer are called prickle cells or spinous cells. Stratum basale is the deepest layer of the epidermis. This is the area in which new skin cells are derived from epidermal stem cells in the basal layer.

In wound healing, during the proliferative phase, re-epithelialization occurs. Keratinocytes at the edges of wound sites proliferate and migrate to cover the wound area which in turn forms a new epithelial layer [94]. It has been shown that the contraction of wound healing sites is related to keratinocytes as well as fibroblasts [95]. This is due to the early wound closure caused by keratinocytes before the migration of fibroblasts and myofibroblasts to the wound site [96].

Keratinocytes lead to epithelialization and restoration of the vascular network of burn wounds [97]. Studies have shown that keratinocytes have a role in pathological fibrosis development in hypertrophic scar tissue by regulating fibroblast and myofibroblasts [98]. Another study has shown that keloid keratinocytes co-cultured with normal dermal fibroblasts and keloid-derived fibroblasts modulated the collagen production enough to suggest that another potential cause of keloids is due to the signaling pathways between cell types [99]. In scleroderma, keratinocytes promote the activation of fibroblasts independent of TGF-β [100].

Hair Follicles

Hair follicles are small tube-like structures that surround the root and strand of a hair throughout the epidermis and dermis layers of the skin. The functions of hair follicles are to help repair skin after a wound, angiogenesis, and neurogenesis. Hair follicles have three main structural segments: infundibulum, isthmus, and inferior. The infundibulum is the region from the tubular invagination in the epidermal surface to the opening of the sebaceous gland, which secretes sebum [101]. The isthmus extends from the opening of the sebaceous duct and the bulge, which is connected to the arrector pili muscle [102]. The inferior segment is the segment from the bulge to the base of the hair follicle. The bulb is surrounded by the follicular matrix, where keratinocytes and melanocytes proliferate to form the hair shaft, and contains the dermal papilla, the mesenchymal components of a hair follicle [102].

Due to the proximity of hair follicles to wound sites, epidermal stem cells from the bulge migrate toward the center of the wound [103, 104]. Hair follicle transplant has been considered as a form of wound healing therapy based on the migration through the wound site [105]. Hair follicles are often damaged and/or destroyed during the wound healing process; this leads to a lack of hair follicles on scarred tissue such as hypertrophic scars and keloids. The areas affected by scleroderma include hair follicles, which again would lead to scarring alopecia [106].

Immune Cells

According to the Immune Deficiency Foundation, immune cells, both innate (leukocytes and macrophages) or adaptive (B cells and T cells), are a pivotal part of the body's defense system to fight off infections and defend against foreign invaders. The most common types of immune cells are categorized into lymphocytes, neutrophils, and monocytes. Lymphocytes can further be classified into T cells, B cells, and Natural killer (NK) cells. T cells control the immune system response and attack and kill infected cells. B cells make antibodies to target foreign invaders. NK cells recognize and kill infected cells without prior sensitization. Neutrophils are the most prominent leukocyte, and they engulf and destroy pathogens. Monocytes circulate in the bloodstream to combat infections and diseases. Once monocytes enter the tissue, they differentiate into either macrophages or dendritic cells. Macrophages functionality is to engulf and destroy pathogens and dead cells. Dendritic cells functionality is to capture and process the antigens and then signal to T cells and B cells to initiate and regulate the adaptive immune response [107].

During the hemostasis phase, immune cells, primarily neutrophils, migrate to the site of injury to initiate the inflammatory phase of wound healing [108]. Neutrophils are documented to regulate the chemokines that recruit macrophages, T cells, and additional neutrophils to the wound site [109]. In the inflammatory phase, the neutrophils and macrophages clear debris, prevent tissue damage, and combat any pathogens at the wound site [110]. Neutrophils undergo apoptosis and macrophages clear the apoptotic neutrophils through efferocytosis; this process occurs in the transition of inflammatory to proliferative phases [111]. Macrophages also secrete VEGF which triggers angiogenesis [112]. During wound healing remodeling, macrophages secrete MMPs to control leukocyte migration and ECM degradation [113, 114]. Immune cells continue to protect the wound site from infection during the entirety of the wound healing process.

In hypertrophic and keloid wound healing sites, there is a dysregulation of the immune response. The most prominent example would be the overabundance of inflammatory cells leading to the remodeling phase which leads to inflammation overlapping with proliferative and remodeling and then the formation of pathological scars [115]. This expressive inflammation response is considered to have a direct correlation with the severity of scar formation. Another aspect that influences the formation of pathological scars is the dysregulation of immune cells and cytokine secretion to stimulate fibroblasts [116]. Scleroderma is also characterized by a dysregulation of immune cells such as increase in proliferation of macrophages which leads to inflammation [117, 118]. Macrophages secrete TGF-β and its activation during wound healing; an overexpression of TGF-β has been linked to fibrosis in scleroderma [119, 120].

Blood Vessels

Blood vessels are the channels that form a closed loop carrying blood throughout the body from and to the heart. There are five types of blood vessels each with different functions. The first type is arteries which function by carrying oxygen-rich blood from the heart to the rest of the body. The second is arterioles which are smaller branches that extend from the arteries to tissue; they dilate and constrict to control blood pressure. Third are veins which carry oxygen-poor blood back to the heart. The fourth is capillaries which connect arteries and veins. Capillaries exchange oxygen, nutrients, and waste products between the blood and tissue cells. The fifth type of blood vessel is venules whose function is to receive blood from capillaries and deliver it to veins. All types of blood vessels play a crucial role in wound healing.

During the hemostasis phase, blood vessels in the area immediately constrict to stop bleeding; this process is called vasoconstriction. Then in the inflammatory phase, blood vessels begin vasodilation in which the vessels dilate to increase blood flow, delivering essential oxygen and nutrients to the injured area, which is crucial for tissue repair and regeneration. In the proliferative phase, angiogenesis starts, and new blood vessels are formed from the existing blood vessels. The process of angiogenesis relies on the endothelial cells that line the inner layer of the blood vessels to proliferate, migrate, and differentiate [121].

Blood vessels have been shown to dilate in keloid and hypertrophic scar [122]. Keloids and hypertrophic scars have an abundant increase in vascularization since it has been demonstrated that there is a higher density of blood vessels at pathological scar sites in comparison to normal skin [123]. This increased vascularization leads to an increase of blood flow in the area as well as a pink or red hue appearance to the wound site. Scleroderma is a disease that can often lead to microvascular damage. One example commonly associated with scleroderma is Raynaud's phenomenon which is a condition that involves vasospasm, sudden involuntary constriction of blood vessels, on extremities in response to cold or emotional stress. Another condition associated with scleroderma is telangiectasias which is the appearance of flat, red marks on the skin caused by dilated blood vessels or capillaries becoming noticeably prominent.

Neural Cells

Neural cells are also known as neurons or nerve cells. They are the fundamental units of the brain and nervous system. Some of the functions that neural cells have throughout the body are receiving, processing, and transmitting information by means of electrical and chemical signaling. A study was conducted on rat skin, and it was found that wound contraction was delayed on the samples that underwent experimentally derived denervation [124]. This alludes to some connection between neural cells and wound healing processes.

Peripheral glial cells are non-neuronal cells that are a part of the peripheral nervous system, and they are essential for nerve function, including myelination, support, and regeneration. There are four major types of glial cells: Schwann cells, satellite glial cells (SGCs), enteric glial cells (EGCs), and olfactory ensheathing cells (OECs) [125]. Schwann cells are involved in a pivotal role in tissue repair and regeneration by undergoing dedifferentiation, proliferation, and releasing paracrine factors that promote healing such as TGF-β [126]. In keloids, it has been demonstrated that the skin surrounding the keloid has an increase in Schwann cells; however, there are fewer Schwann cells in scar tissue once it has matured [127]. Since the Schwann cells do not remain in the tissue after wound healing, this could indicate that the cells serve a tissue repair function that is not currently understood. The areas of peripheral nervous system affected by scleroderma include symptoms such as trigeminal neuropathy, peripheral sensorimotor polyneuropathy, and carpal tunnel syndrome. In those cases, the damage of the neural cells could play a role in the wound healing process.

Therapeutic Solutions to Treat Dermal Hypertrophy

A multitude of methods exist to treat and prevent the formation of pathological scarring. This section serves to explore some of the most commonly available therapeutic solutions.

Pharmacotherapy

Pharmacotherapy is the method of pharmaceutical products, more commonly referred to as drugs, for medication purposes. One common treatment and prevention method for hypertrophic scars and keloids is the usage of corticosteroid injections. The steroid reduces the size of the scarred skin by breaking the bonds of the collagen in the scar tissue. Triamcinolone acetonide is the most used corticosteroid medicine for the hypertrophic and keloid scars [128]. Studies have shown that triamcinolone acetonide injections reduce pain in patients with scleroderma as well as showing a reduction in Raynaud's phenomenon [129]. Corticosteroids suppress the body's inflammatory response and serve as vasoconstrictors to reduce blood flow to the injection site [128]. In some cases, chemotherapeutic drugs, such as 5-fluorouracil (5-FU), can be used alongside the intralesional steroid injection treatment. 5-FU is a drug that slows or stops tumor cells in the body; this is due to a hindering of TGF-β1. It is also suspected that 5-FU hinders the collagen type I gene expression [130].

It has been suggested that botulinum toxin A (Botox) could be used to treat hypertrophic scarring and keloids since the neurotoxin suppresses scar tension and fibroblast activity [131]. There have been promising results for preliminary clinical trials of Botox injections on keloids which have led to more experiments being conducted to explore the treatment method [132]. Research has shown evidence of Botox being used to treat Raynaud's phenomenon and microstomia, the reduction of the oral aperture, in patients with scleroderma; however, the results indicate that it is not a universal solution for all scleroderma patients [133].

Verapamil is a phenylalkylamine calcium channel blocker, which means it has the ability to decrease ECM collagen production and induce collagenase synthesis [134]. These properties have led to intralesional verapamil injections being used as an alternative to corticosteroid injections for hypertrophic scars and keloid management [135]. Calcium channel blockers, such as verapamil, have been shown to reduce the frequency, severity, and pain of Raynaud's phenomenon attacks in patients with scleroderma [136].

There are topical drug treatments for cutaneous scarring such as retinoids. Retinoids are a class of chemical compounds derived from vitamin A that reduce fibroblast proliferation and inhibit collagen type I gene expression induced by TGF-β1 [134]. Various types of retinoids have been shown to be effective in the treatment of scleroderma [137].

Biomaterials

(a) *Silicone Gel Sheets*
Silicone gel sheets are flexible, self-adhesive dressings. The benefits of silicone gel sheets are hydration of the wound site, a preventive barrier from bacteria, modulation of growth factors, increased collagenase level to break down excess collagen, and reduction of itching feeling on the skin at the wound site [138]. This treatment option is considered the first-line prevention and treatment for hypertrophic scars and keloids due to the efficacy and lack of invasive nature [139, 140]. Silicone gel sheets have been implemented as treatment for scleroderma-like diseases and have been shown to improve joint range and mobility [141]. One danger of this treatment method is that the prolonged usage can lead to maceration and breakdown of the skin tissue on the affected area under the sheet [142].

(b) *Hydrogels*
Hydrogels are characterized by a three-dimensional, hydrophilic polymer network that is cross-linked together to absorb water and maintain structural integrity. Hydrogels have been shown to have strong biocompatibility. There are three types of hydrogels classified based on their source: natural, synthetic, or hybrid. Natural hydrogels are derived from naturally occurring polymers such as gelatin and collagen [143]. Since natural hydrogels contain components naturally found in the ECM such as hyaluronic acid, laminin, fibrin, growth factors, and MMPs, they are the most biocompatible form of hydrogel [144]. HA hydrogels have been widely used in wound healing applications due to the tunable cell behavior responses [145]. There have been promising results of treating burn injuries with HA hydrogels to deliver adipose stem cells to the wound site prior to scar formation [146]. Synthetic hydrogels are created using synthetic polymers such as polyamides and polyethylene glycol. These polymers can be tailored to achieve certain mechanical properties [147]. Hybrid hydrogels are able to combine the benefits of natural and synthetic hydrogels through chemical alteration of natural polymers or a combination of natural and synthetic polymers [144]. Hydrogels have been shown to be an effective treatment method for hypertrophic scars and keloids by reducing the scar's appearance and minimizing symptoms such as itchiness and pain [144]. Hydrogels loaded with mesenchymal stem cells have shown experimental promise as a treatment method for systemic scleroderma with the capabilities of inhibiting fibrosis and ameliorating disease progression [148].

Invasive Treatments

(a) *Laser therapy*
The first type of laser therapy used to treat scars is pulsed-dye laser (PDL). PDL uses a concentrated beam of yellow light to target blood vessel abnormalities in the skin. It has been shown to reduce keloid scars with high patient satisfaction

[149]. PDL is commonly used to treat hypertrophic scars, specifically caused by burns, before scar maturation [150]. Telangiectasias is a condition of small red blood vessels being visible on the surface of the skin; it is a common symptom of systemic scleroderma. Based on previously conducted studies, PDL has shown promise in treating telangiectasias [151]. There is also evidence to suggest that PDL could be beneficial in reducing the skin hardening associated with morphea [152].

Neodymium-doped yttrium aluminum garnet (Nd:YAG) lasers are a type of solid-state laser that uses a Nd:YAG crystal to emit infrared light. This type of laser is used in a dermatological application to treat various skin conditions. Nd:YAG lasers have been shown to be significantly more effective in the treatment of hypertrophic scars rather than keloids [153]. Inactive sclerotic and atrophic lesions of morphea have been shown to be effectively treated by Nd:YAG lasers when dealing with extreme contractures [154].

(b) *Cryotherapy*

Cryotherapy can be applied in an external or intralesional manner. In the safer treatment method, an intralesional cryosurgery cryoneedle is inserted into the hypertrophic scar or keloid, and the needle is connected to liquid nitrogen which freezes the scar from the inside out [155]. This method of therapy reduces the volume of scar tissue with minimal adverse effects [156]. Cryotherapy is pro-inflammatory, so it has been advised to be avoided on patients with scleroderma even though it has been shown to reduce the severity of Raynaud's phenomenon [157].

(c) *Surgery*

Surgical intervention is often only considered as a treatment option when pharmacotherapy treatments have been unsuccessful. Scar revision surgery aims to minimize the appearance of scar tissue and have it blend in with the surrounding dermal tissue [158]. This is often done with less severe forms of scar tissue, whereas scar removal surgery is the method of cutting out scar tissue. A lot of care and attention is paid to the reduction of skin tension applied when removing keloid scar tissue since keloids reform on trauma zones. Hypertrophic scars can be excised if the scar is raised. Surgical methods are often accompanied by injections simultaneously during the healing process to reduce the risk of scars forming on the new trauma site. There is no precedent of scar removal of scleroderma. However, the surgical method of deep inferior epigastric artery perforator (DIEP) free flaps is used to repair or reconstruct skin areas affected by morphea [159].

Another method of surgical intervention is through the usage of grafting. Skin grafting is done through full layers of skin, full-thickness skin grafts, or partial layers of skin, split-thickness skin grafts, being taken from the donor site and transplanted to the scar site. For smaller scars, punch grafts can be used to take a circular area of skin and bring it to the affected area. Autologous fat grafts, also known as lipofilling, are a surgical procedure where fat tissue is harvested from one area of the body and transplanted to a different area. Often when being used in therapeutic settings, the fat tissue gathered is mixed with patient's stem cells. Fat grafting has been shown to be beneficial to have a positive effect on scars, though there is not a lot of literature relating this treatment method to hypertrophic scars and keloids [160]. Autologous fat grafts have been shown to have effects in preventing irreversible deformities in patients with scleroderma as well as improving patient skin quality [161, 162].

Conclusions and Future Directions

Hypertrophic scars, keloids, and scleroderma are all conditions which lead to pathological scarring. Current research understands some of the methodology in scar development, but there is room for further characterization and understanding of interactions of the cellular and the ECM components. Figure 31.3 shows an

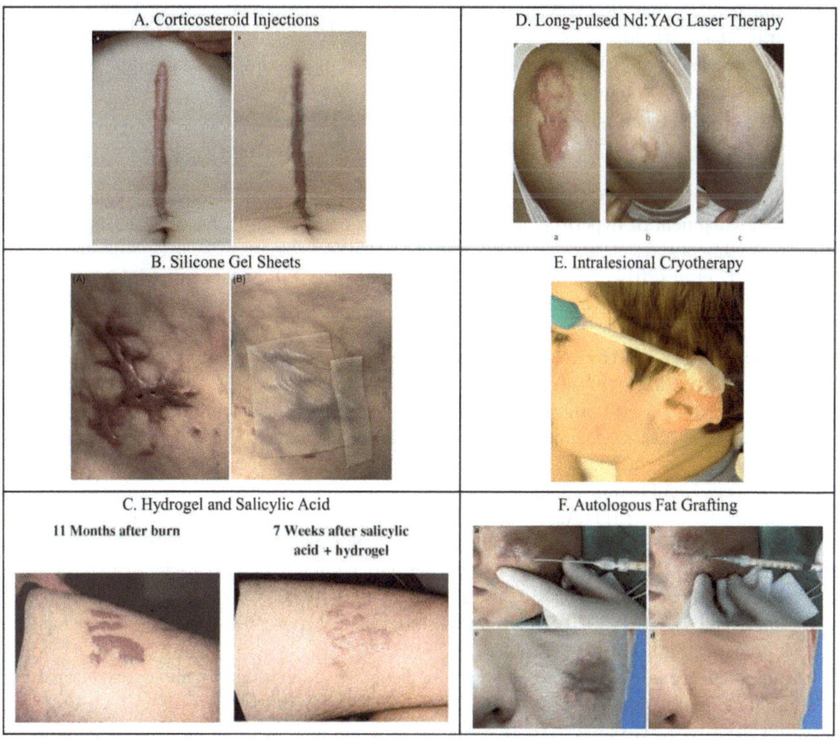

Fig. 31.3 This table shows therapeutic applications or the results of specific therapy methods. (**a**) This image shows the wound progression of a patient receiving corticosteroid injections in 4-week intervals [163]. (**b**) Shows the method of applying silicone gel sheets to the pathological scar site [164]. (**c**) Shows the effect of a hydrogel therapy method application after 7 weeks post treatment [165]. (**d**) Shows the results of Nd:YAG laser therapy after 30 months of treatment [166]. (**e**) Shows the application of a cryoneedle into a keloid [156]. (**f**) Illustrates the method of autologous fat grafting application into scar tissue [167]

overview of some of the common treatment methods of pathological scars. In the future, once the molecular mechanisms of cutaneous scarring are understood, the cause and prevention of these conditions can be explored. There is a lot of research being done currently to investigate other innovative avenues of treating hypertrophic scarring, including the exploration of nanomaterial applications and inhibitors of growth factors. Progress toward alternative treatment therapies and eventual preventive measures is continuously made across modern research.

References

1. Finnerty CC et al (2016) Hypertrophic scarring: the greatest unmet challenge after burn injury. Lancet 388(10052):1427-1436
2. Limandjaja GC et al (2021) Hypertrophic scars and keloids: overview of the evidence and practical guide for differentiating between these abnormal scars. Exp Dermatol 30(1):146–161
3. Bombaro KM et al (2003) What is the prevalence of hypertrophic scarring following burns? Burns 29(4):299–302
4. Lawrence JW et al (2012) Epidemiology and impact of scarring after burn injury: a systematic review of the literature. J Burn Care Res 33(1):136–146
5. Warby R, Maani CV (2023) Burn Classification. StatPearls [Internet]
6. Deitch EA et al (1983) Hypertrophic burn scars: analysis of variables. J Trauma 23(10):895–898
7. Cubison TCS, Pape SA, Parkhouse N (2006) Evidence for the link between healing time and the development of hypertrophic scars (HTS) in paediatric burns due to scald injury. Burns 32(8):992–999
8. Huang C et al (2014) Are keloid and hypertrophic scar different forms of the same disorder? A fibroproliferative skin disorder hypothesis based on keloid findings. Int Wound J 11(5):517–522
9. Andrews JP et al (2016) Keloids: the paradigm of skin fibrosis—Pathomechanisms and treatment. Matrix Biol 51:37–46
10. Sadiq A, Khumalo NP, Bayat A (2020) Genetics of Keloid Scarring. In: Téot L et al (eds) Textbook on scar management: state of the art management and emerging technologies. Springer Copyright 2020, The Editor(s) (if applicable) and The Author(s). This book is an open access publication, Cham (CH)
11. Ashcroft KJ et al (2011) Assessment of the influence of HLA class I and class II loci on the prevalence of keloid disease in Jamaican Afro-Caribbeans. Tissue Antigens 78(5):390–396
12. Balci DD et al (2009) DLQI scores in patients with keloids and hypertrophic scars: a prospective case control study. JDDG: J Dtsch Dermatol Ges 7(8):688–691
13. Limandjaja GC et al (2020) The Keloid disorder: heterogeneity, histopathology, mechanisms and models. Front Cell Dev Biol 8:360
14. Rongioletti F et al (2018) Scleroderma with an update about clinico-pathological correlation. G Ital Dermatol Venereol 153(2):208–215
15. Kowalska-Kępczyńska A (2022) Systemic Scleroderma—definition, clinical picture and laboratory diagnostics. J Clin Med 11. https://doi.org/10.3390/jcm11092299
16. Bairkdar M et al (2021) Incidence and prevalence of systemic sclerosis globally: a comprehensive systematic review and meta-analysis. Rheumatology 60(7):3121–3133
17. Sánchez-Pérez S et al (2017) Bullous Morphea: description of a new case and discussion of etiologic and pathogenic factors in bulla formation. Actas Dermosifiliogr 108(1):75–76
18. Mathew-Steiner SS, Roy S, Sen CK (2021) Collagen in wound healing. Bioengineering 8. https://doi.org/10.3390/bioengineering8050063

19. Chiang RS et al (2016) Current concepts related to hypertrophic scarring in burn injuries. Wound Repair Regen 24(3):466–477
20. Muthusubramaniam L et al (2014) Effect of collagen nanotopography on keloid fibroblast proliferation and matrix synthesis: implications for dermal wound healing. Tissue Eng A 20(19–20):2728–2736
21. Gardeazabal L, Izeta A (2024) Elastin and collagen fibres in cutaneous wound healing. Exp Dermatol 33(3):e15052
22. Finnerty CC et al (2016) Hypertrophic scarring: the greatest unmet challenge after burn injury. Lancet 388(10052):1427–1436
23. Bhangoo KS, Quinlivan JK, Connelly JR (1976) Elastin fibers in scar tissue. Plast Reconstr Surg 57(3):308–313
24. Amadeu TP et al (2004) Fibrillin-1 and elastin are differentially expressed in hypertrophic scars and keloids. Wound Repair Regen 12(2):169–174
25. Walters R, Pulitzer M, Kamino H (2009) Elastic fiber pattern in scleroderma/morphea. J Cutan Pathol 36(9):952–957
26. Laurens N, Koolwijk P, de Maat MP (2006) Fibrin structure and wound healing. J Thromb Haemost 4(5):932–939
27. Wolberg AS (2023) Fibrinogen and fibrin: synthesis, structure, and function in health and disease. J Thromb Haemost 21(11):3005–3015
28. Mony MP et al (2023) An updated review of hypertrophic scarring. Cells 12. https://doi.org/10.3390/cells12050678
29. Wang Y et al (2019) A multiscale hybrid mathematical model of epidermal-dermal interactions during skin wound healing. Exp Dermatol 28(4):493–502
30. Tuan TL et al (1996) Elevated levels of plasminogen activator inhibitor-1 may account for the altered fibrinolysis by keloid fibroblasts. J Invest Dermatol 106(5):1007–1011
31. Falanga V, Kruskal JB, Franks JJ (1991) Fibrin and fibrinogen-related antigens in systemic sclerosis (scleroderma). J Am Acad Dermatol 25(5 Pt 1):771–775
32. Clark RAF (2003) Fibrin is a many splendored thing. J Invest Dermatol 121(5):xxi–xxii
33. Lenselink EA (2015) Role of fibronectin in normal wound healing. Int Wound J 12(3):313–316
34. Kischer CW, Hendrix MJ (1983) Fibronectin (FN) in hypertrophic scars and keloids. Cell Tissue Res 231(1):29–37
35. Babu M, Diegelmann R, Oliver N (1989) Fibronectin is overproduced by keloid fibroblasts during abnormal wound healing. Mol Cell Biol 9(4):1642–1650
36. Bhattacharyya S et al (2014) FibronectinEDA promotes chronic cutaneous fibrosis through toll-like receptor signaling. Sci Transl Med 6(232):232ra50
37. Wang Y, Wang G, Liu H (2022) Tenascin-C: a key regulator in angiogenesis during wound healing. Biomolecules 12(11):1689
38. Choi YE et al (2020) Effects of tenascin C on the integrity of extracellular matrix and skin aging. Int J Mol Sci 21. https://doi.org/10.3390/ijms21228693
39. Xiang W et al (2024) The role of tenascin-C in hypertrophic scar formation: insights from cell and animal experiments. Clin Cosmet Investig Dermatol 17:1637–1648
40. Dalkowski A et al (1999) Increased expression of tenascin C by keloids in vivo and in vitro. Br J Dermatol 141(1):50–56
41. Bhattacharyya S, Midwood KS, Varga J (2022) Tenascin-C in fibrosis in multiple organs: translational implications. Semin Cell Dev Biol 128:130–136
42. Bhattacharyya S et al (2016) Tenascin-C drives persistence of organ fibrosis. Nat Commun 7:11703
43. Iorio V, Troughton LD, Hamill KJ (2014) Laminins: roles and utility in wound repair. Adv Wound Care 4(4):250–263
44. Michopoulou A et al (2020) A novel mechanism in wound healing: Laminin 332 drives MMP9/14 activity by recruiting syndecan-1 and CD44. Matrix Biol 94:1–17
45. Robinson CJ et al (2024) Release of miR-29 target Laminin C2 improves skin repair. Am J Pathol 194(2):195–208

46. Hamill KJ et al (2009) Laminin deposition in the extracellular matrix: a complex picture emerges. J Cell Sci 122(24):4409–4417
47. Xie Q et al (2017) Homeobox B9 facilitates hypertrophic scar formation via activating the mitogen-activated protein kinase signaling pathway. Mol Med Rep 16(2):1669–1676
48. Saijo Y et al (2024) Vascular basement membrane fragmentation in keloids and the expression of key basement membrane component genes. Plast Reconstr Surg Glob Open 12(12):e6366
49. Huffstutter JE, Delustro FA, LeRoy EC (1985) Cellular immunity to collagen and laminin in scleroderma. Arthritis Rheum 28(7):775–780
50. Hayes AJ et al (2022) Perlecan, a multi-functional, cell-instructive, matrix-stabilizing proteoglycan with roles in tissue development has relevance to connective tissue repair and regeneration. Front Cell Dev Biol 10:856261
51. Wilson SE (2022) Defective perlecan-associated basement membrane regeneration and altered modulation of transforming growth factor beta in corneal fibrosis. Cell Mol Life Sci 79(3):144
52. Barallobre-Barreiro J et al (2019) Cartilage-like composition of keloid scar extracellular matrix suggests fibroblast mis-differentiation in disease. Matrix Biology Plus 4:100016
53. Motegi SI et al (2017) Possible association of elevated serum collagen type IV level with skin sclerosis in systemic sclerosis. J Dermatol 44(2):167–172
54. Juhl P et al (2020) Type III, IV, and VI collagens turnover in systemic sclerosis—a longitudinal study. Sci Rep 10(1):7145
55. Frenkel JS (2014) The role of hyaluronan in wound healing. Int Wound J 11(2):159–163
56. Garantziotis S, Savani RC (2019) Hyaluronan biology: a complex balancing act of structure, function, location and context. Matrix Biol 78-79:1–10
57. Oksala O et al (1995) Expression of proteoglycans and hyaluronan during wound healing. J Histochem Cytochem 43(2):125–135
58. Cassuto D, Vinshtok Y (2020) Treatment of scar contracture with intralesional jet-assisted injection of hyaluronic acid. J Dermatol Res Ther 6:94
59. Bertheim U, Hellstrom S (1994) The distribution of hyaluronan in human skin and mature, hypertrophic and keloid scars. Br J Plast Surg 47(7):483–489
60. Meyer LJ et al (2000) Reduced hyaluronan in keloid tissue and cultured keloid fibroblasts. J Invest Dermatol 114(5):953–959
61. Neudecker BA, Stern R, Connolly MK (2004) Aberrant serum hyaluronan and hyaluronidase levels in scleroderma. Br J Dermatol 150(3):469–476
62. Engström-Laurent A et al (1985) Raised serum hyaluronate levels in scleroderma: an effect of growth factor induced activation of connective tissue cells? Ann Rheum Dis 44(9):614–620
63. Pang X, Dong N, Zheng Z (2020) Small leucine-rich proteoglycans in skin wound healing. Front Pharmacol 10:1649
64. Smith MM, Melrose J (2013) Proteoglycans in normal and healing skin. Adv Wound Care 4(3):152–173
65. Melrose J et al (2016) The cartilage extracellular matrix as a transient developmental scaffold for growth plate maturation. Matrix Biol 52-54:363–383
66. Velasco J et al (2011) *Adamts5* deletion blocks murine dermal repair through CD44-mediated aggrecan accumulation and modulation of transforming growth factor β1 (TGFβ1) signaling*. J Biol Chem 286(29):26016–26027
67. Merline R, Schaefer RM, Schaefer L (2009) The matricellular functions of small leucine-rich proteoglycans (SLRPs). J Cell Commun Signal 3(3):323–335
68. Xu X et al (2022) Small leucine-rich proteoglycans in tendon wound healing. Adv Wound Care (New Rochelle) 11(4):202–214
69. Chen S, Birk DE (2013) The regulatory roles of small leucine-rich proteoglycans in extracellular matrix assembly. FEBS J 280(10):2120–2137
70. Krawetz RJ et al (2022) Proteoglycan 4 (PRG4) treatment enhances wound closure and tissue regeneration. npj Regen Med 7(1):32

71. Ninkovic N et al (2024) Proteoglycan 4 (PRG4) treatment improves skin wound healing in a porcine model. FASEB J 38(6):e23547
72. Islam S et al (2020) Accumulation of versican facilitates wound healing: implication of its initial ADAMTS-cleavage site. Matrix Biol 87:77–93
73. Yang W, Yee AJM (2014) Versican 3′-untranslated region (3′UTR) promotes dermal wound repair and fibroblast migration by regulating miRNA activity. Biochimica et Biophysica Acta (BBA) – Mol Cell Res 1843(7):1373–1385
74. Scott PG et al (1995) Immunohistochemical localization of the proteoglycans decorin, biglycan and versican and transforming growth factor-beta in human post-burn hypertrophic and mature scars. Histopathology 26(5):423–431
75. Yagi Y et al (2013) An ex vivo model employing keloid-derived cell–seeded collagen sponges for therapy development. J Invest Dermatol 133(2):386–393
76. Masuda A et al (2013) Versican is upregulated in circulating monocytes in patients with systemic sclerosis and amplifies a CCL2-mediated pathogenic loop. Arthritis Res Ther 15(4):R74
77. Cagno V et al (2019) Heparan sulfate proteoglycans and viral attachment: true receptors or adaptation bias? Viruses 11. https://doi.org/10.3390/v11070596
78. Olczyk P, Mencner Ł, Komosinska-Vassev K (2015) Diverse roles of heparan sulfate and heparin in wound repair. Biomed Res Int 2015(1):549417
79. Fears CY, Woods A (2006) The role of syndecans in disease and wound healing. Matrix Biol 25(7):443–456
80. Thota LNR, Chignalia AZ (2022) The role of the glypican and syndecan families of heparan sulfate proteoglycans in cardiovascular function and disease. Am J Phys Cell Phys 323(4):C1052–C1060
81. Andriessen MP et al (1998) Hypertrophic scarring is associated with epidermal abnormalities: an immunohistochemical study. J Pathol 186(2):192–200
82. Rider CC, Mulloy B (2017) Heparin, heparan sulphate and the TGF-β cytokine superfamily. Molecules 22. https://doi.org/10.3390/molecules22050713
83. Darby IA et al (2014) Fibroblasts and myofibroblasts in wound healing. Clin Cosmet Investig Dermatol 7:301–311
84. Cialdai F, Risaliti C, Monici M (2022) Role of fibroblasts in wound healing and tissue remodeling on earth and in space. Front Bioeng Biotechnol 10:958381
85. Schultz GS et al (2011) Principles of wound healing. In: Fitridge R, Thompson M (eds) Mechanisms of vascular disease: a reference book for vascular specialists. University of Adelaide Press. © The Contributors 2011, Adelaide (AU)
86. Chun Q et al (2016) Dynamic biological changes in fibroblasts during hypertrophic scar formation and regression. Int Wound J 13(2):257–262
87. Schmid P et al (1998) Enhanced expression of transforming growth factor-beta type I and type II receptors in wound granulation tissue and hypertrophic scar. Am J Pathol 152(2):485–493
88. Lee TY et al (1999) Expression of transforming growth factor beta 1, 2, and 3 proteins in keloids. Ann Plast Surg 43(2):179–184
89. McGinty S, Siddiqui WJ (2025) Keloid. In: StatPearls. StatPearls Publishing. Copyright © 2025, StatPearls Publishing LLC, Treasure Island (FL)
90. Zitka O et al (2010) Matrix metalloproteinases. Curr Med Chem 17(31):3751–3768
91. Gilbane AJ, Denton CP, Holmes AM (2013) Scleroderma pathogenesis: a pivotal role for fibroblasts as effector cells. Arthritis Res Ther 15(3):215
92. Murphrey MB, Miao JH, Zito PM (2025) Histology, Stratum Corneum. In: StatPearls. StatPearls Publishing. Copyright © 2025, StatPearls Publishing LLC, Treasure Island (FL)
93. Freeman SC, Sonthalia S (2025) Histology, Keratohyalin Granules. In: StatPearls. StatPearls Publishing. Copyright © 2025, StatPearls Publishing LLC, Treasure Island (FL).
94. Pastar I et al (2014) Epithelialization in wound healing: a comprehensive review. Adv Wound Care 3(7):445–464
95. Souren JM, Ponec M, van Wijk R (1989) Contraction of collagen by human fibroblasts and keratinocytes. In Vitro Cell Dev Biol 25(11):1039–1045

96. Rakar J, Krammer MP, Kratz G (2015) Human melanocytes mitigate keratinocyte-dependent contraction in an in vitro collagen contraction assay. Burns 41(5):1035–1042
97. ter Horst B et al (2018) Advances in keratinocyte delivery in burn wound care. Adv Drug Deliv Rev 123:18–32
98. Lee JS et al (2019) Effect of keratinocytes on myofibroblasts in hypertrophic scars. Aesth Plast Surg 43(5):1371–1380
99. Alghamdi MA et al (2020) Secreted factors from keloid keratinocytes modulate collagen deposition by fibroblasts from normal and fibrotic tissue: a pilot study. Biomedicine 8. https://doi.org/10.3390/biomedicines8070200
100. McCoy SS et al (2017) Scleroderma keratinocytes promote fibroblast activation independent of transforming growth factor beta. Rheumatology 56(11):1970–1981
101. Brown TM, Krishnamurthy K (2025) Histology, hair and follicle. In: StatPearls. StatPearls Publishing. Copyright © 2025, StatPearls Publishing LLC., Treasure Island (FL)
102. Martel JL et al (2025) Anatomy, Hair Follicle. In: StatPearls. StatPearls Publishing. Copyright © 2025, StatPearls Publishing LLC, Treasure Island (FL)
103. Joulai Veijouye S et al (2017) Bulge region as a putative hair follicle stem cells niche: a brief review. Iran J Public Health 46(9):1167–1175
104. Ito M et al (2005) Stem cells in the hair follicle bulge contribute to wound repair but not to homeostasis of the epidermis. Nat Med 11(12):1351–1354
105. Nuutila K (2020) Hair follicle transplantation for wound repair. Adv Wound Care 10(3):153–163
106. Adigun R, Goyal A, Hariz A (2025) Systemic Sclerosis (Scleroderma). In: StatPearls. StatPearls Publishing. Copyright © 2025, StatPearls Publishing LLC, Treasure Island (FL)
107. Song L et al (2018) The function of dendritic cells in modulating the host response. Mol Oral Microbiol 33(1):13–21
108. Engelhardt E et al (1998) Chemokines IL-8, GROα, MCP-1, IP-10, and Mig are sequentially and differentially expressed during phase-specific infiltration of leukocyte subsets in human wound healing. Am J Pathol 153(6):1849–1860
109. Ellis S, Lin EJ, Tartar D (2018) Immunology of wound healing. Curr Derm Rep 7(4):350–358
110. Dovedytis M, Liu ZJ, Bartlett S (2020) Hyaluronic acid and its biomedical applications: a review. Eng Regen 1:102–113
111. Landén NX, Li D, Ståhle M (2016) Transition from inflammation to proliferation: a critical step during wound healing. Cell Mol Life Sci 73(20):3861–3885
112. Azari Z et al (2023) The role of immune cells in therapeutic angiogenesis: concepts in tissue engineering. Curr Opin Biomed Eng 28:100470
113. Raziyeva K et al (2021) Immunology of acute and chronic wound healing. Biomolecules 11. https://doi.org/10.3390/biom11050700
114. Elkington PT, Green JA, Friedland JS (2009) Analysis of matrix metalloproteinase secretion by macrophages. Methods Mol Biol 531:253–265
115. Wang Z-C et al (2020) The roles of inflammation in keloid and hypertrophic scars. Front Immunol 11:603187
116. Zhang X, Wu X, Li D (2023) The communication from immune cells to the fibroblasts in keloids: implications for immunotherapy. Int J Mol Sci 24. https://doi.org/10.3390/ijms242015475
117. Brown M, O'Reilly S (2019) The immunopathogenesis of fibrosis in systemic sclerosis. Clin Exp Immunol 195(3):310–321
118. Dai B et al (2022) Contributions of immune cells and stromal cells to the pathogenesis of systemic sclerosis: recent insights. Front Pharmacol 13:826839
119. Toledo DM, Pioli PA (2019) Macrophages in systemic sclerosis: novel insights and therapeutic implications. Curr Rheumatol Rep 21(7):31
120. Morris E et al (2011) Endoglin promotes TGF-β/Smad1 signaling in scleroderma fibroblasts. J Cell Physiol 226(12):3340–3348

121. Velazquez OC (2007) Angiogenesis and vasculogenesis: inducing the growth of new blood vessels and wound healing by stimulation of bone marrow-derived progenitor cell mobilization and homing. J Vasc Surg 45(6):A39–A47
122. Yoo MG, Kim I-H (2014) Keloids and hypertrophic scars: characteristic vascular structures visualized by using dermoscopy. Ann Dermatol 26(5):603–609
123. Amadeu T et al (2003) Vascularization pattern in hypertrophic scars and keloids: a stereological analysis. Pathol Res Pract 199(7):469–473
124. Engin C et al (1996) Delayed effect of denervation on wound contraction in rat skin. Plast Reconstr Surg 98(6):1063
125. Verkhratsky A, Butt A (2013) Peripheral glial cells. In: Glial physiology and pathophysiology, pp 381–430
126. Jiang D, Machens H-G, Rinkevich Y (2024) Sensory neurons: unveiling the symphony of wound healing. Signal Transduct Target Ther 9(1):177
127. Direder M et al (2022) Schwann cells contribute to keloid formation. Matrix Biol 108:55–76
128. Morelli Coppola M et al (2018) Triamcinolone acetonide intralesional injection for the treatment of keloid scars: patient selection and perspectives. Clin Cosmet Investig Dermatol 11:387–396
129. DeLea SL et al (2011) Sonographically guided hydrodissection and corticosteroid injection for scleroderma hand. Clin Rheumatol 30(6):805–813
130. Shah VV et al (2016) 5-fluorouracil in the treatment of keloids and hypertrophic scars: a comprehensive review of the literature. Dermatol Ther 6(2):169–183
131. Sohrabi C, Goutos I (2020) The use of botulinum toxin in keloid scar management: a literature review. Scars Burns Heal 6:2059513120926628
132. Zhibo X, Miaobo Z (2009) Intralesional botulinum toxin type a injection as a new treatment measure for keloids. Plast Reconstr Surg 124(5):275e
133. Bello RJ et al (2017) The therapeutic efficacy of botulinum toxin in treating scleroderma-associated Raynaud's phenomenon: a randomized, double-blind, Placebo-controlled clinical trial. Arthritis Rheumatol 69(8):1661–1669
134. Viera MH et al (2010) Innovative therapies in the treatment of keloids and hypertrophic scars. J Clin Aesthet Dermatol 3(5):20–26
135. Wang R et al (2016) Role of verapamil in preventing and treating hypertrophic scars and keloids. Int Wound J 13(4):461–468
136. Rirash F et al (2017) Calcium channel blockers for primary and secondary Raynaud's phenomenon. Cochrane Database Syst Rev 12
137. Yamada E et al (1999) Tocoretinate inhibited the contraction of collagen gel matrices by human dermal fibroblasts with tenascin-C expression. J Dermatol Sci 22(1):45–53
138. Puri N, Talwar A (2009) The efficacy of silicone gel for the treatment of hypertrophic scars and keloids. J Cutan Aesthet Surg 2(2):104–106
139. Gold MH et al (2001) Prevention of hypertrophic scars and keloids by the prophylactic use of topical silicone gel sheets following a surgical procedure in an office setting. Dermatologic Surg 27(7):641–644
140. Chan KY et al (2005) A randomized, placebo-controlled, double-blind, prospective clinical trial of silicone gel in prevention of hypertrophic scar development in median sternotomy wound. Plast Reconstr Surg 116(4):1013–1020; discussion 1021–2
141. Dıncer S et al (2004) Silicone gel sheet dressing for sclerodermatous type chronic graft-versus-host-disease (cGVHD). Turkish J Hematol 21:107–110
142. De Decker I et al (2022) The use of fluid silicone gels in the prevention and treatment of hypertrophic scars: a systematic review and meta-analysis. Burns 48(3):491–509
143. Gyles DA et al (2017) A review of the designs and prominent biomedical advances of natural and synthetic hydrogel formulations. Eur Polym J 88:373–392
144. Zhong Y et al (2024) Hydrogel loaded with components for therapeutic applications in hypertrophic scars and keloids. Int J Nanomedicine 19:883–899

145. Lam J, Truong NF, Segura T (2014) Design of cell–matrix interactions in hyaluronic acid hydrogel scaffolds. Acta Biomater 10(4):1571–1580
146. Dong Y et al (2020) Conformable hyaluronic acid hydrogel delivers adipose-derived stem cells and promotes regeneration of burn injury. Acta Biomater 108:56–66
147. Madduma-Bandarage USK, Madihally SV (2021) Synthetic hydrogels: synthesis, novel trends, and applications. J Appl Polym Sci 138(19):50376
148. Nie M et al (2022) MSCs-laden injectable self-healing hydrogel for systemic sclerosis treatment. Bioact Mater 17:369–378
149. S. S et al (2011) Treatment of refractory keloids with pulsed dye laser alone and with rotational pulsed dye laser and intralesional corticosteroids. A retrospective case series. Laser Therapy 20(4):279–286
150. Kim YH et al (2024) Very early pulsed dye laser intervention for optimal cosmetic outcome in post-thyroidectomy scars. Clin Cosmet Investig Dermatol 17:877–884
151. Kottler D et al (2023) Efficacy, tolerance and acceptability of pulsed dye laser on facial and neckline telangiectasias in systemic scleroderma: a prospective open-label monocentric study in 21 patients. J Cosmet Laser Ther 25(5–8):77–85
152. Tawfik AA et al (2013) Pulsed dye laser in the treatment of localized scleroderma and its effects on CD34+ and factor XIIIa+ cells: an immunohistochemical study. Am J Clin Dermatol 14(3):235–241
153. Koike S et al (2014) Nd: YAG laser treatment for keloids and hypertrophic scars: an analysis of 102 cases. Plast Reconstr Surg Glob Open 2(12):e272
154. Szczepanik-Kułak P, Michalska-Jakubus M, Krasowska D (2021) Laser therapy for the treatment of Morphea: a systematic review of literature. J Clin Med 10. https://doi.org/10.3390/jcm10153409
155. Gauglitz GG (2013) Management of keloids and hypertrophic scars: current and emerging options. Clin Cosmet Investig Dermatol 6:103–114
156. O'Boyle CP, Shayan-Arani H, Hamada MW (2017) Intralesional cryotherapy for hypertrophic scars and keloids: a review. Scars, Burns & Healing 3:2059513117702162
157. Guillot X et al (2013) Cryotherapy in inflammatory rheumatic diseases: a systematic review. Expert Rev Clin Immunol 10:281
158. Tullington, J.E. and R. Gemma, Scar Revision, in StatPearls. 2025, StatPearls Publishing. Copyright © 2025, StatPearls Publishing LLC.: Treasure Island (FL).
159. Lee JH et al (2017) Surgical management of localized scleroderma. Arch Craniofac Surg 18(3):166–171
160. Lee G, Hunter-Smith DJ, Rozen WM (2017) Autologous fat grafting in keloids and hypertrophic scars: a review. Scars Burns Heal 3:2059513117700157
161. Strong AL et al (2021) Fat grafting subjectively improves facial skin elasticity and hand function of scleroderma patients. Plast Reconstr Surg Glob Open 9(1):e3373
162. Strong AL et al (2019) Fat grafting for the treatment of scleroderma. Plast Reconstr Surg 144(6):1498–1507
163. Lee J, Kim J (2020) Minimal-invasive technologies for treatment of HTS and keloids: corticosteroids. In: Téot L et al (eds) Textbook on scar management: state of the art management and emerging technologies. Springer International Publishing, Cham, pp 243–250
164. Oliveira G, Gold M (2020) Silicone sheets and new gels to treat hypertrophic scars and keloids: a short review. Dermatol Ther 33:e13705
165. Danielson JR, Walter RJ (2005) Case studies: use of salicylic acid (Avosil) and hydrogel (Avogel) in limiting scar formation. J Burns Wounds 4:e6
166. Ogawa R, Akaishi S (2016) Endothelial dysfunction may play a key role in keloid and hypertrophic scar pathogenesis—keloids and hypertrophic scars may be vascular disorders. Med Hypotheses 96:51–60
167. Dai Q et al (2021) Early intervention with autologous fat grafting to reduce hypertrophic scar formation: a preliminary study. Aesth Plast Surg 45(5):2409–2416

Chapter 32
AI and Personalized Medicine in Wound Care

Necip Gurler, Raphael A. Yaakov, H. Samet Bulbul, Patrick Cheng, Kyle Wu, and Ozgur Guler

Abstract Chronic wounds pose significant global health challenges due to their prevalence, economic burden, and substantial impact on patient quality of life. Artificial intelligence (AI) presents transformative opportunities in personalized wound care by leveraging diverse datasets, including medical imaging, electronic health records, wearable sensors, and laboratory biomarkers. This chapter reviews the key data sources for AI applications and explores current AI-driven assessment techniques, such as predictive analytics, risk stratification, computer vision tools, and multimodal data fusion, along with rule-based and data-driven therapeutic strategies. Additionally, a multimodal agentic AI framework is proposed to autonomously integrate diverse patient data, perform comprehensive wound assessments, and dynamically customize treatment plans based on continuous clinical feedback. Lastly, the ethical, regulatory, economic, and policy dimensions of deploying AI technologies, emphasizing data privacy, bias mitigation, and interoperability, are also discussed.

Keywords Artificial intelligence · AI-driven wound care · Personalized medicine · Precision medicine · Wound healing · Chronic wounds · Predictive analytics · Multimodal data fusion · Agentic AI · Ethical considerations · Regulatory frameworks

Abbreviations

AI	Artificial Intelligence
AUC	Area Under the Curve
BERT	Bidirectional Encoder Representations from Transformers
CDSS	Clinical Decision Support System

N. Gurler (✉) · R. A. Yaakov · H. S. Bulbul · P. Cheng · K. Wu · O. Guler
eKare Inc, Fairfax, VA, USA
e-mail: ngurler@ekareinc.com

F. De Francesco, K. Turksen (eds.), *Management and Strategies for Wound Healing*, Stem Cell Biology and Regenerative Medicine 79,
https://doi.org/10.1007/978-3-032-12442-5_32

CHMP3	Charged Multivesicular Body Protein 3
CMS	Centers for Medicare and Medicaid Services
CNN	Convolutional Neural Network
CONSORT-AI	Consolidated Standards of Reporting Trials—Artificial Intelligence
CPT	Current Procedural Terminology
CTP	Cellular and Tissue-Based Products
CV	Computer Vision
DiGA	Digitale Gesundheitsanwendungen
DFU	Diabetic Foot Ulcer
EHR	Electronic Health Record
FHIR	Fast Healthcare Interoperability Resources
FDA	U.S. Food and Drug Administration
FUTURE-AI	Fair, Universal, Traceable, Usable, Robust, Explainable—Artificial Intelligence
GAN	Generative Adversarial Network
GEO	Gene Expression Omnibus
GPT	Generative Pre-Trained Transformer
H&E	Hematoxylin and Eosin
HTAR	Health Technology Assessment Regulation
HITECH Act	Health Information Technology for Economic and Clinical Health Act
HL7	Health Level Seven
HSI	Hyperspectral Imaging
LLM	Large Language Model
LOINC	Logical Observation Identifiers Names and Codes
LSTM	Long Short-Term Memory
MHRA	UK Medicines and Healthcare Products Regulatory Agency
MCP	Model Context Protocol
ML	Machine Learning
MLOps	Machine Learning Operations
NICE	National Institute for Health and Care Excellence
NLP	Natural Language Processing
NPWT	Negative Pressure Wound Therapy
NTAP	New Technology Add-On Payment
PCCP	Predetermined Change Control Plan
PHI	Protected Health Information
RAG	Retrieval-Augmented Generation
R-CNN	Region-Based Convolutional Neural Network
RGB	Red, Green, and Blue
RL	Reinforcement Learning
RNF103	Ring Finger Protein 103
SaMD	Software as a Medical Device
SCUBE1	Signal Peptide, CUB Domain, EGF-Like Domain-Containing Protein 1

SNOMED CT	Systematized Nomenclature of Medicine—Clinical Terms
SPIRIT-AI	Standard Protocol Items: Recommendations for Interventional Trials—Artificial Intelligence
U-Net	U-Shaped Convolutional Neural Network
W2GAN	Wasserstein-2 Generative Adversarial Network
WHO	World Health Organization
2D	Two-Dimensional
3D	Three-Dimensional

Introduction

Acute and chronic wounds pose significant global health challenges, affecting millions of individuals worldwide and placing substantial economic and clinical burdens on the healthcare system. A recent study reported that, in the United States alone, chronic wounds affected approximately 14.5% of Medicare beneficiaries in 2014, and this prevalence increased to 16.3% in 2019 [17]. Another comprehensive analysis indicated that chronic wounds impact approximately 2.5% of the total U.S. population, with a disproportionate effect on the elderly individuals with chronic conditions such as diabetes, obesity, and vascular insufficiencies [81]. These studies highlight a growing public health concern in managing wounds, contributing to $22.5 billion in Medicare costs [17]. Furthermore, chronic wounds continue to be associated with significant morbidity and mortality. For instance, diabetic foot ulcers (DFUs) are one of the examples of this severity, as their five-year mortality rates are comparable to those observed in certain cancers [81].

Despite the progress in advanced dressings, growth factors, and adjunctive devices in wound care, considerable gaps persist between clinical needs and actual healing outcomes. Current clinical practices often rely heavily on empirical methods and individual judgment rather than standardized diagnostic procedures, resulting in variable treatment effectiveness [32, 102]. Furthermore, chronic wounds typically involve complex interactions among pathophysiological factors such as ischemia, biofilm-related infections, persistent inflammation, and underlying systemic conditions, which conventional assessments often do not fully capture [30]. Consequently, the widening gap between subjective clinical practices and objective diagnostic needs strongly motivates the adoption of personalized medicine approaches. These approaches aim to precisely align diagnostics, assessments, and therapeutic interventions closely with each patient's unique biological characteristics [44].

AI coupled with personalized medicine promises to transform wound management by enabling patient-specific interventions that surpass traditional wound care. Additionally, personalized medicine in wound care individualizes therapy by aligning treatment with each patient's risk profile, wound biology, and comorbidities [94, 102]. Achieving this level of personalization in wound treatment requires multidimensional datasets, the interpretation of which exceeds human analytic capabilities.

This is where *AI* acts as a catalyst for personalized medicine, helping clinicians transition toward evidence-based, individualized wound assessment and treatment decisions that were previously unattainable with conventional methods. For example, modern machine-learning algorithms digest electronic health records (EHRs), imaging, and physiologic data to forecast 12-week healing trajectories with the area under the curve (AUC) $\geq$0.85, which enables the early identification of non-healing wounds [11, 23]. Another recent study shows that deep-learning classifiers integrating wound images and anatomic location reach accuracies above 90% across ulcer subtypes [68]. Additionally, emerging smart biosensor dressings now stream real-time pH, temperature, and moisture metrics, creating continuous feedback loops for AI-driven decision support [97]. Collectively, these AI-driven modalities integrate mixed and multimodal datasets and generate clinically actionable insights, enabling evidence-based assessments and supporting patient-specific therapeutic interventions in wound care.

This chapter presents a comprehensive overview of AI-driven personalized medicine within wound care. It begins by describing essential data sources, including medical imaging, electronic health records (EHR), wearable sensor data, and biomarkers. Subsequently, the chapter explores the AI-driven assessments, predictive analytics methods, patient risk stratification, and computer vision applications for precise wound characterization. Then, the chapter explores AI-enabled therapeutic strategies customized for individual patient needs. Building upon this, it introduces an agentic AI framework based on generative AI and multimodal systems, which is designed to seamlessly integrate and manage data, assessment, and treatment modalities from one framework. The chapter continues by critically analyzing the broader economic, ethical, and regulatory implications associated with deploying these sophisticated technologies in clinical settings. Finally, it synthesizes key insights and identifies directions for future research aimed at enhancing personalized clinical outcomes in wound management.

Data Sources in Wound Care

Medical Imaging of Wounds

Medical imaging is a primary data source for assessing wounds. Standard 2D digital photography is widely used to document wound appearance and size [53]. However, traditional 2D photography lacks essential depth information. To overcome this limitation, 3D imaging techniques such as stereophotogrammetry and 3D laser scanning provide precise measurements of wound depth and volume, which are critical for tracking healing in deep or irregular wounds [45]. AI algorithms analyze these images to measure wound dimensions and identify tissue types or pathologic signs. For example, a deep-learning model differentiated rare conditions like pyoderma gangrenosum from common leg ulcers with higher sensitivity than expert dermatologists [13].

Beyond dimensional assessments, advanced optical modalities significantly enhance wound characterization. Fluorescence imaging, for example, can identify bacterial contamination, enabling targeted interventions [50]. Furthermore, hyperspectral imaging modalities provide detailed insights into tissue oxygenation and perfusion, facilitating early detection of ischemia or compromised blood flow [76]. Similarly, thermal imaging detects skin temperature patterns, which can be used to identify inflammation, ischemia, or burn depth [75]. The integration of these advanced imaging techniques expands AI's ability to comprehensively characterize wounds, enabling more timely and personalized interventions.

Electronic Health Records and Clinical Data

EHRs contain vital contextual data for personalized wound care. Patient demographics, medical history, medications, lab results, and prior wound treatments all reside in EHRs. AI models trained on large EHR datasets can identify patterns and risk factors that influence wound healing. For example, machine-learning models using EHR data from over one million wounds could predict the likelihood of chronic wounds healing by 12 weeks with an AUC around 0.85 [11], highlighting how clinical variables, such as wound size, depth, duration, or comorbidities, contribute to healing outcome predictions. Moreover, EHRs store clinical notes and assessments that become an additional data source. By incorporating structured numerical data along with unstructured clinician observations obtained through natural language processing, AI systems can achieve a comprehensive understanding of each patient's wound condition, which is essential for personalized medicine.

Real-Time Monitoring: Wearable Devices and Smart Dressings

Wearable health technology is increasingly recognized as a valuable source of real-time data for personalized medicine [10]. These devices continuously monitor patient activity and physiological parameters, providing valuable insights into wound healing [67]. For example, pressure-sensing wearables have been demonstrated in experimental settings to effectively monitor the duration a patient remains in one position, providing timely alerts to caregivers for repositioning patients at risk of developing pressure ulcers [37]. Smart wound dressings embedded with biosensors offer another advanced data source. These dressings extend continuous monitoring directly to the wound interface, converting passive wound coverings into interactive diagnostic platforms. For instance, bandages embedded with pH sensors can indicate bacterial infection by changing color or transmitting signals when the wound environment becomes alkaline. Other flexible sensors embedded in dressings can track wound temperature, moisture levels, and exudate composition, providing comprehensive real-time monitoring [97].

Through real-time data collection and continuous monitoring, wearable devices and smart dressings significantly enhance AI-driven personalized medicine. AI algorithms utilize this comprehensive data to promptly detect subtle changes in wound conditions, allowing dynamic adjustment of therapies tailored precisely to each patient's physiological responses and healing trajectory.

Biobanks and Laboratory Data

Beyond bedside measurements, biobanks and laboratory analyses contribute valuable data for AI-driven personalized wound care, especially in understanding the biological mechanisms of wound healing. Chronic wound tissues and fluids stored in biobanks offer extensive genomic, proteomic, and microbiological data, invaluable for advanced computational analysis. For example, transcriptomic profiling of diabetic foot ulcers (DFU) using GEO datasets revealed that machine-learning-identified feature genes, SCUBE1 and RNF103-CHMP3, whose downregulation correlated with healing and immune signaling changes [104]. Additionally, deep-learning algorithms applied to Hematoxylin and Eosin (H&E) stained biopsies obtained from the ComplexWounds@Manchester biobank enabled automated segmentation of tissue types and immune cell infiltration, reaching up to 97% accuracy in human chronic wound histology [74]. These multidimensional datasets, spanning gene expression to histopathology, enable AI-driven personalized wound management to pinpoint precise molecular targets and guide tailored therapeutic interventions.

Data Quality and Interoperability Challenges

While diverse data hold the promise of generating powerful AI-driven insights, there are significant challenges related to data quality and interoperability that must be addressed to fully realize AI-driven personalized wound care. Variability in wound imaging, including inconsistent lighting, angles, and device calibration, limits the generalizability and reliability of AI models [36]. Clinical documentation often suffers from incomplete or inconsistent entries; for instance, a national audit revealed widespread gaps and insufficient detail in wound descriptors within electronic health records [21]. Furthermore, biosensors introduce additional reliability issues, such as measurement drift and sensor durability problems, potentially compromising the integrity of continuous monitoring data [97]. Interoperability challenges, on the other hand, including non-standardized data formats, diverse terminologies, and fragmented data storage across healthcare institutions, significantly impede data integration.

Addressing the above challenges requires a multipronged approach. Recent literature and industry efforts propose several solutions to enhance data quality and interoperability, thus supporting effective AI applications in wound care. These solutions are discussed in detail below.

Data-Curation Pipelines

A fundamental step involves implementing robust data-curation pipelines that systematically clean and standardize data from multiple sources. These pipelines typically handle tasks such as imputing missing values, reconciling inconsistent units, and flagging potential outliers for further review. AI can also help on these tasks; for example, natural language processing (NLP) tools can transform free-text wound descriptions into structured, coded data, while validation rules ensure entries like wound size or date of onset remain within acceptable ranges. Curating data in this way reduces noise and bias, directly improving the training datasets for machine learning [89].

Interoperability and Standardized Formats

Another key solution involves adopting common data standards for wound care information. Standard terminologies such as SNOMED CT for clinical terms and LOINC for laboratory tests and measurements offer a unified vocabulary, ensuring consistent encoding of wound characteristics across diverse systems. On the technical side, adopting data standards like HL7 FHIR (Fast Healthcare Interoperability Resources) can greatly improve structural interoperability. FHIR defines consistent data models, called resources, for health information exchange. For example, an FHIR observation resource could represent a wound measurement with standard fields for value, units, and timestamp. By using FHIR APIs, an AI-driven wound care application can easily pull data from an EHR or send new observations back without using the custom interfaces for each system. The integration of FHIR with standard vocabularies was shown to achieve both semantic and structural interoperability in personal health records [19].

Model Context Protocol for Health Information Exchange

The Model Context Protocol (MCP) provides an open and standardized method for AI models to securely communicate and exchange information with diverse data sources [8]. As of the authors' current knowledge, there are no peer-reviewed publications specifically addressing MCP in healthcare settings; however, while MCP was not initially designed for healthcare, the protocol holds significant promise for simplifying and standardizing secure data interactions in clinical settings.

Within an MCP workflow, an AI client, such as a Large Language Model (LLM)-driven chatbot, communicates with an MCP gateway, which coordinates interactions with backend data repositories like FHIR servers, imaging archives, or wearable sensor platforms. The gateway manages authentication, audit logging, and context-specific filtering, providing the AI model only the necessary data to perform specific tasks. Therefore, this architecture presents several practical benefits. Firstly, MCP simplifies model training by enabling AI clients to retrieve diverse datasets from multiple sources through one standardized interface, facilitating the creation of multimodal wound care AI models. Secondly, MCP functions as an interoperability layer by translating AI-generated queries into uniform FHIR-compatible requests, harmonizing data access across different systems. Thirdly, MCP supports real-time decision-making by securely retrieving current clinical data, such as wound photographs, laboratory results, and physiological parameters, thereby enabling automated documentation and timely clinical interventions at the bedside.

In summary, diverse and robust data sources form a comprehensive foundation for AI-driven personalized wound care. However, fully realizing the potential of these data sources requires systematically addressing key challenges related to data quality and interoperability. Emerging technological approaches, such as structured data-curation pipelines, standardized frameworks like FHIR, and innovative tools such as MCP, offer promising solutions to overcome these barriers. By adopting these advancements, healthcare institutions can reliably integrate multimodal data, empowering AI models to generate timely, accurate, and personalized insights that significantly enhance patient outcomes and clinical efficiency in wound management.

AI-Driven Assessment for Personalized Wound Care

AI technologies have rapidly advanced the assessment of wounds, enabling more objective and comprehensive evaluations that can be tailored to individual patients. Traditional wound assessment, including estimating wound size, depth, tissue composition, and healing prognosis, is often subjective and variable between clinicians [85]. AI-driven approaches address this variability by learning from vast datasets of wound images and clinical data, essentially providing an expert eye trained on thousands of wound cases. In the past decade, researchers have developed machine-learning and deep-learning models that can automatically measure wound dimensions, classify tissue types, predict healing trajectories, and stratify patient risk across all major wound types [32]. This section reviews key developments in AI for wound assessment, including computer vision methods for wound imaging, predictive analytics for healing outcomes and risk stratification, and emerging multimodal and generative AI techniques that together contribute to a holistic understanding of each patient's wound.

AI Models for Wound Size, Depth, and Tissue Classification

Computer vision has become a cornerstone of AI-driven wound assessment. Modern deep-learning algorithms can interpret wound photographs to quantify wound size (area, depth, volume) and identify the composition of tissues in the wound bed (such as granulation tissue, slough, and eschar) with high accuracy [87]. Accurate measurement of wound dimensions is critical for tracking healing progress and making treatment decisions, yet manual measurements using rulers or tracings are often imprecise. AI models, particularly convolutional neural networks (CNNs), have demonstrated the ability to segment wound boundaries and calculate wound area automatically [16, 64, 98, 99].

Beyond overall size, AI vision models can assess wound depth and severity, which is an important task, especially in burns and diabetic foot ulcers, where depth correlates with treatment needs. Deep networks can be trained to recognize subtle color and texture cues that indicate how deep a wound extends. In acute burn care, for example, Boissin et al. [14] developed a deep-learning algorithm to analyze burn images; it could correctly delineate 87% of burn wound areas and predict whether a burn would require surgical intervention with an AUC of 0.885. This level of performance approaches that of experienced burn specialists, demonstrating that AI can assist in burn depth classification and triage.

AI-driven image analysis also excels at tissue type classification within wounds, which is crucial for monitoring healing. Wound healing is typically marked by increasing granulation and decreasing necrotic tissue, but visual estimation of these tissue proportions is subjective. Deep-learning algorithms allow objective quantification on tissue type classification. For instance, Lien et al. [54] employed an active-learning CNN model that could detect the growth of granulation tissue in diabetic foot ulcers over time. Similarly, Liu et al. [56] leveraged a deep EfficientNet model to create color-coded maps of ischemic vs. infected regions in diabetic foot ulcers, based on real patient images. Taken together, these advances mean that an AI system can take a wound photograph and output rich information: the wound's exact area and depth and the composition of tissue within it (e.g., "50% granulation, 30% slough, 20% eschar"), all with consistency beyond what most humans could achieve.

Overall, AI models for wound sizing, depth estimation, and tissue classification are making assessments more precise. These tools turn images into objective data that can be trended over time to see if a wound is truly improving. This lays the foundation for personalized care because accurate assessment is the first step to deciding the right intervention for the right patient.

Predictive Models for Wound Healing Trajectories

In addition to static assessment, AI is increasingly used to predict how a wound will progress over time. Chronic wounds are influenced by many factors, such as patient comorbidities (diabetes, vascular disease, nutrition), wound attributes (size, duration, infection), and treatments applied. Machine-learning algorithms can integrate these variables to forecast outcomes such as healing time. Predictive models provide clinicians and patients with foresight: for example, identifying a wound that is unlikely to heal with standard care so that advanced therapies can be introduced earlier.

Over the past decade, researchers have built numerous prognostic models for wounds, often using techniques like logistic regression, random forests, or neural networks on clinical datasets. Cho et al. [20] developed logistic regression and classification tree models using electronic medical records of over 620,000 wounds from 532 clinics, achieving an AUC of approximately 0.71 when predicting wound closure within 12 weeks. They found wound size, depth, location, and cause to be more influential predictors than demographic or comorbid factors. Extending this research, Berezo et al. [11] utilized gradient-boosted decision trees trained on data from 1.22 million wounds from commercial electronic health records, attaining even stronger predictive accuracy (AUC 0.85) for non-healing at 4, 8, and 12 weeks. Their analysis highlighted modifiable risk factors such as prolonged treatment duration, wound depth, and surface area using Shapley values. Moving beyond structured datasets, He et al. [40] introduced an interpretable deep-learning approach using histological images to assess collagen fiber organization. Their model accurately classified healing stages with 85% accuracy, and visualization techniques clearly indicated the collagen features influencing predictions, providing clinicians with valuable insight into the biological reasons for delayed wound healing. Collectively, these studies illustrate how varied retrospective data, from clinical records to microscopic tissue images, can effectively predict wound healing trajectories and highlight intervention opportunities.

Risk Stratification and Patient Profiling

Traditional methods of risk stratification in wound care commonly rely on standardized assessment scales or clinician judgment. Instruments such as the Braden Scale provide broad categories of risk based on factors like sensory perception, moisture, activity, mobility, nutrition, and friction or shear [12]. While these tools are valuable, their static nature and limited scope often fail to capture the dynamic and individual-specific factors that influence wound healing outcomes, thereby limiting personalized management strategies.

Recent advancements in AI and machine learning have the potential to address these limitations by integrating extensive patient data to produce personalized and dynamic risk profiles. For example, Gabison et al. [31] developed an AI-driven, non-contact monitoring system that used load cells under hospital beds to accurately classify patient posture (94% accuracy) that allows clinicians to identify and mitigate pressure ulcer risks proactively. Similarly, Danilovich et al. [24] employed computer vision algorithms applied to continuous video monitoring systems, classifying patient positions with 95% accuracy, further enabling real-time risk assessment and preventive repositioning strategies.

In DFU management, AI-based prediction models have also demonstrated particular effectiveness in stratifying patient populations into meaningful subgroups tailored by wound and patient characteristics. For instance, Wang et al. [100] developed several machine-learning models to predict the risk of hard-to-heal DFUs using clinical data from 362 patients. Among these, a Naïve Bayes model exhibited high predictive performance (AUC = 0.864, recall = 0.907), demonstrating the capability of AI-driven methods to accurately stratify patient risk even with relatively small datasets. Such models offer feasible, precise risk assessment tools suitable for use in resource-constrained clinical environments.

Beyond structured clinical data, NLP significantly enhances personalized wound care by extracting nuanced patient information directly from clinical narratives. For instance, Topaz et al. [89] developed and validated an NLP-based application capable of accurately identifying wound details such as wound type, pressure ulcer stage, wound size, anatomical location, and treatment interventions from clinical notes, achieving high accuracy. Expanding further on this paradigm, recent advancements in generative AI show potential to further enhance risk profiling by summarizing complex patient information clearly and consistently. LLMs, for example, can distill extensive clinical documentation into concise, structured risk summaries. Integrated reasoning models can provide transparent explanations about how specific risk factors influence predicted outcomes, increasing clinician confidence and interpretability.

In summary, AI-driven risk stratification transforms large amounts of patient data into actionable risk levels. Each patient gets a more nuanced risk profile than conventional methods provide, allowing clinicians to prioritize interventions for those who need them most. High-risk patients can be funneled into more intensive monitoring or advanced therapies early, while low-risk patients might be managed with standard care, thus optimizing resource use.

Computer Vision in Wound Imaging and Histopathology

Computer vision (CV) techniques form the backbone of many AI assessment tools, extending beyond just standard photographs to various imaging modalities used in wound care. High-resolution digital photography is the most common input for CV models as described in section "AI Models for Wound Size, Depth, and Tissue Classification", but AI is now applied to hyperspectral imaging, fluorescence imaging, and even microscopic histopathology images to collect information that complements what clinicians see with the naked eye.

One impactful application is using fluorescence imaging to detect wound infection. Specialized autofluorescence devices are used to identify bacteria directly within wounds. Under specific lighting conditions, typically violet or ultraviolet light, certain bacteria naturally emit a distinct fluorescent signal, known as autofluorescence. These devices capture and visualize this fluorescence, allowing clinicians to detect and localize bacterial presence without needing to culture the bacteria first [50]. A clinical study by Viswanathan et al. [95] involving 178 patients demonstrated that the autofluorescence device used in this study achieved an accuracy of 89.54% in identifying bacterial Gram types compared to traditional microbial culture. This early identification of bacterial presence supports timely initiation of antimicrobials or surgical debridement, which is crucial for personalizing wound management by proactively addressing infections.

Another impactful CV application in wound care is hyperspectral imaging (HSI), which uses a broader range of wavelengths beyond standard visible imaging. HSI captures continuous spectral data across hundreds of narrow wavelength bands, creating detailed reflectance spectra for each pixel. This rich spectral information reveals biochemical changes such as tissue perfusion and oxygenation levels, thereby addressing critical needs in wound assessment and monitoring [76]. In burn wound assessment, for example, Wang et al. [98, 99] combined near-infrared HSI with CNNs and deep transfer learning to assess burn wounds, achieving over 96% accuracy in classifying burn depth. In DFU, Hornberger et al. [42] utilized HSI in fluorescence mode combined with principal component analysis, achieving 83% sensitivity and 100% specificity in detecting significant bacterial colonization. Further demonstrating HSI's clinical utility, Zhao et al. [105] reconstructed hyperspectral images from standard diabetic foot ulcer photographs and applied deep-learning methods to measure oxygen saturation, effectively distinguishing healing (54–64% saturation) from non-healing ulcers (32–48%). Collectively, these studies highlight how AI-enabled HSI supports non-invasive, real-time decision-making, significantly enhancing personalized therapeutic strategies in wound care.

In the realm of wound histopathology, AI is beginning to analyze microscopic images of wound tissue and biopsies. Although not routine for all wounds, biopsies are done in cases of atypical wounds or suspected malignancy. AI-based image analysis can assist pathologists by identifying cellular patterns associated with chronic inflammation or impaired healing. One study presented a deep-learning approach that can objectively analyze H&E-stained wound biopsy sections, extracting features like extent of angiogenesis (new blood vessel formation), inflammation levels, and collagen deposition [74]. By quantifying these histological markers, the AI can provide an assessment of wound healing at the tissue level as a "virtual pathologist" for wound biopsies. Another promising development is *virtual staining*, where AI models transform imaging data from unstained tissues or other modalities into images resembling traditional histologically stained pathology slides. Deng et al. [25] developed a deep learning-based virtual staining technique specifically tailored for chronic wound tissue pathology. Pathologist evaluations revealed that virtual staining images closely resembled traditional manual staining, achieving accuracy scores of 71.1% for cellular features, 75.4% for tissue structures, and 77.8% for overall assessment.

In summary, computer vision technologies in wound care span from macro-level assessments, such as wound size and tissue composition, to microscopic analysis of cellular characteristics. By extending human visual capabilities beyond the visible spectrum, using hyperspectral imaging to evaluate perfusion and fluorescence imaging to detect bacterial presence, and surpassing the processing limits of the human eye with deep-learning methods, AI significantly enhances clinicians' ability to perform comprehensive wound assessments. This enhanced diagnostic insight enables earlier and more precise identification of wound complications, forming a solid foundation for personalized care. Consequently, interventions can be better tailored, both in timing and method, to address each patient's specific clinical needs effectively.

Multimodal Data Fusion for Comprehensive Wound Understanding

Thus far in section "AI Driven Assessment for Personalized Wound Care", we have examined various AI-driven techniques leveraging individual modalities, such as text-based EHRs, lab results, clinical notes, or diverse imaging methods ranging from standard RGB images to advanced hyperspectral imaging. However, wound assessment inherently requires integration of multiple complementary data types, as relying on a single modality alone fails to capture the comprehensive biological complexity involved. Multimodal data fusion, therefore, offers significant advantages by synthesizing structured and unstructured data streams to provide richer, context-aware assessments.

Recent studies demonstrate that combining structured and unstructured inputs can notably enhance wound assessments and personalized treatment planning. For example, Anisuzzaman et al. [7] developed a deep neural network-based multimodal classifier that combines wound images with corresponding anatomical location data to classify wounds into diabetic, pressure, surgical, and venous ulcers. They created a body map to efficiently tag wound locations, integrating this spatial context with visual data. Their multimodal approach achieved maximum accuracies ranging from 72.95% to 97.12% across various wound-class experiments, significantly outperforming previous single-modality classification methods. This study demonstrates how combining visual imagery with spatial location can substantially enhance the accuracy and reliability of wound classification.

Another study involves postoperative wound monitoring. McLean et al. [59] developed a multimodal AI framework that combines wound photographs with patient-reported outcome measures, such as pain scores and symptom descriptions, to predict early surgical site infections. Conducted across two clinical cohorts following gastrointestinal surgeries, this integrated model achieved diagnostic accuracy comparable to expert clinicians (AUC 0.76 vs. clinicians' 0.78) and demonstrated excellent generalizability in external validations. Notably, the AI-based approach reduced clinician review time by approximately 80%, underscoring how integrating visual and subjective patient data can streamline workflows and enable proactive, remote wound monitoring.

Expanding beyond images and textual data, multimodal AI approaches are also applied via integrated multi-sensor devices. Chang et al. [18] introduced a portable assessment system combining RGB imaging, 3D depth scanning, thermal imaging, multispectral imaging, and chemical vapor sensing for pressure ulcer evaluation. In a clinical study involving 23 patients, the multimodal device accurately identified subtle wound status changes undetectable by standard methods alone. For instance, thermal imaging detected inflammation and potential infection, while multispectral imaging differentiated tissue types effectively. The comprehensive fusion of these data streams generated precise metrics and intuitive 3D visualizations of wound progression, significantly enhancing clinical assessments and facilitating real-time, informed treatment decisions.

Looking ahead, multimodal data integration will likely become fundamental to holistic wound care management. Given the complexity of wound healing dynamics influenced by biological, environmental, and behavioral factors, AI systems integrating diverse modalities, including imaging, clinical narratives, sensor data, and even genomic information, will drive more precise and tailored interventions. Moreover, the emergence of mobile health technologies, telemedicine, and wearable sensor-equipped smart bandages will further accelerate multimodal AI applications, enabling real-time wound assessment and automated clinical decision support. In the next section, AI-driven treatment will be explored in depth, highlighting how intelligent systems can guide precise, personalized interventions and enhance therapeutic outcomes in wound care.

AI-Driven Treatment for Personalized Wound Care

Effective wound care requires not only accurate assessment but also the selection and implementation of appropriate treatments tailored to the individual. Building on data sources and AI-driven assessment as detailed in previous sections, this section discusses key developments in AI-driven wound treatment, highlighting how machine learning and intelligent devices can support clinicians in delivering precise, timely, and patient-specific interventions. This lays a foundation for the subsequent chapter on an agentic AI framework for personalized wound care, where we propose more autonomous and interactive AI systems.

Rule-Based and Knowledge-Driven Systems in Wound Treatment

Early efforts to apply AI in wound care took the form of rule-based or expert systems [48, 69]. These systems encode clinical knowledge and guidelines as logical rules, which the computer uses to recommend treatments. In the first epoch of medical AI, rule-based expert systems were developed to assist clinicians in decision-making, for example, by suggesting an appropriate wound dressing based on wound characteristics (size, exudate level, and infection status) and established protocols [80]. *This* approach was useful but inherently limited, because the systems could not learn from new cases or adapt beyond their programmed rules. In wound care, a rule-based program might, for instance, prompt a provider to consider offloading and debridement for a diabetic foot ulcer if certain criteria are met or recommend a moisture-retentive dressing for a superficial surgical wound based on healing principles. Such logic mimics clinical practice guidelines. However, the rigidity of rule-based AI means it cannot easily handle the variability of wound patients; nuances like unusual comorbidities or atypical wound presentations may fall outside the programmed rules. Consequently, purely rule-based wound care decision aids have largely been replaced by data-driven models in recent years. They remain an important historical stepping-stone and are sometimes embedded within clinical decision support tools as default care pathways or alerts. These provide consistency with guidelines, but modern AI aims to be far more flexible and personalized.

Data-Driven Systems for Personalized Therapy

The advent of machine learning, especially supervised learning, has significantly enhanced the capacity for adaptive and data-driven support in wound treatment planning. Techniques such as healing prediction, wound classification, and tissue analysis, which were already examined in section "AI Driven Assessment for Personalized Wound Care", primarily inform clinicians about wound prognosis and the effectiveness of ongoing treatment strategies. Although valuable for real-time assessment and feedback on current therapeutic interventions, these approaches typically serve as supportive tools rather than direct providers of treatment plans.

One of the most immediate and transformative ways AI contributes directly to treatment strategies is through Clinical Decision Support Systems (CDSS). These AI-driven recommendation engines utilize a patient's wound characteristics, severity levels, comorbid conditions, and risk factors, which are often initially assessed by AI tools discussed previously. The CDSS then integrates these data points, matching them to personalized treatment pathways that have demonstrated the highest likelihood of success in clinically analogous cases.

Despite considerable progress in AI-driven assessments, AI-based treatment suggestions remain relatively rare in the literature. Nevertheless, promising applications are emerging. A prominent example is in the management of DFUs, where selecting an appropriate offloading device is critical. Researchers recently proposed an AI-powered CDSS that provides personalized offloading prescriptions for patients with diabetes at risk of plantar foot ulcers [1]. By analyzing patient-specific factors, such as foot shape, ulcer location, walking patterns, and lifestyle needs, the system recommends customized solutions like orthopedic shoes, insoles tailored for pressure redistribution, or total contact casting. This personalized approach aims to improve patient adherence and efficacy by targeting each individual's unique pressure points and lifestyle requirements, representing a shift toward precision offloading strategies.

AI decision support is also influential in dressing selection and topical therapies. Various dressing types must be chosen based on wound characteristics such as exudate level, infection status, and tissue type. As mentioned previously, knowledge-based systems have been developed to encode best-practice algorithms and suggest appropriate dressings based on wound inputs [80]. More sophisticated ML models, trained on extensive datasets of wound outcomes, can identify which dressing types accelerate healing in specific patient scenarios. For instance, an AI might learn that for venous leg ulcers with heavy exudate, a particular multilayer compression dressing yields better outcomes than alternatives in similar patients. Additionally, AI-powered smart dressings integrated with sensor technologies are promising for personalized wound treatment by enabling continuous, real-time monitoring and adaptive dressing recommendations. These systems can measure critical wound parameters such as temperature, pH, moisture level, and exudate production patterns, which are then analyzed by AI algorithms to predict wound healing trajectories. For example, ML algorithms embedded in smart dressings can recommend switching to antimicrobial dressings if sensor readings indicate a heightened risk of infection, thus optimizing the timing and type of intervention [33].

Another area is the use of AI for recommending advanced therapies, such as hyperbaric oxygen therapy, negative pressure wound therapy (NPWT), and cellular and tissue-based products (CTPs). These treatments, typically reserved for wounds unresponsive to standard care, involve substantial costs and clinical resources; therefore, determining the appropriate timing and patient eligibility is crucial. AI-driven predictive models, leveraging extensive historical clinical datasets, can provide precise forecasts of wound healing trajectories and facilitate timely escalation to these advanced interventions. For instance, AI systems might suggest initiating NPWT based on wound stagnation, size, and exudate characteristics, or identify candidates suitable for bioengineered skin substitutes or CTPs by evaluating wound chronicity and predicted healing likelihood [49].

Another emerging data-driven paradigm for precision therapy is the digital twin, which is a virtual counterpart of an individual wound that evolves in near real-time as new clinical data arrive [46]. Digital twin technology integrates multimodal data sources, including sensor-embedded dressings, 3D imaging, laboratory values, and detailed patient histories, creating dynamic computational models that closely simulate the actual biological state and progression of wounds. Recent proof-of-concept studies highlight the potential of digital twins in personalized chronic wound management; for instance, Sarp et al. [79] demonstrated a digital twin framework that accurately predicted diabetic foot ulcer trajectories and recommended individualized dressing strategies, leading to earlier detection of stalled healing and significantly reducing healing times compared to conventional care approaches. Although digital twin adoption faces challenges related to data accuracy, integration complexity, high initial costs, and data privacy concerns, ongoing advancements position this technology as a key future component in personalized wound care, offering clinicians a robust tool for proactive decision-making and significantly enhancing therapeutic precision.

Reinforcement Learning for Optimizing Wound Therapy

While supervised learning techniques primarily provide static predictions to support clinical decisions, reinforcement learning (RL) presents an opportunity for dynamic, adaptive decision-making in personalized wound care. RL algorithms are capable of identifying optimal therapeutic sequences and adjustments through continuous interaction with the treatment environment, leveraging trial-and-error learning in either simulated or controlled real-world scenarios.

Current research in this area remains exploratory but promising. For instance, Lu et al. [58] presented a novel approach involving an adaptive closed-loop control system integrating deep learning for wound state estimation with deep RL algorithms for optimizing treatment interventions. Their methodology modeled wound healing as a nonlinear dynamic process, enabling the RL agent to dynamically learn intervention strategies aimed at minimizing healing duration. In their simulated experiments, the RL-derived policy significantly reduced healing times by approximately 45.6% compared to control scenarios without algorithmic intervention, suggesting considerable potential for RL-driven therapeutic optimization in clinical wound care scenarios.

While these AI methodologies hold significant promise, it is important to note that current implementations remain largely investigational or in the proof-of-concept phase, without widespread clinical adoption. Rigorous validation in clinical trials, regulatory approvals, and integration into routine clinical workflows are still required before such AI-driven treatment recommendations become standard practice. Nonetheless, these research initiatives represent an important step toward truly personalized wound care, potentially allowing clinicians in the future to make more informed and timely decisions regarding advanced therapeutic interventions.

Generative AI in Wound Care: From Synthetic Data to Treatment Suggestions

Generative AI refers to computational models capable of generating novel data or content, such as images, texts, or designs, by learning inherent patterns from existing datasets. Within wound care, generative models are emerging as valuable tools in multiple capacities, notably in data augmentation, image analysis, treatment planning, and personalized therapy design.

A primary application of generative AI in wound care involves image generation and data augmentation. Chronic wound datasets are typically limited due to privacy concerns, data-collection challenges, and variability in wound presentations. Generative adversarial networks (GANs), capable of synthesizing highly realistic wound images, help overcome these limitations. Sarp et al. [77, 78], for instance, successfully developed the W2GAN model, a GAN-based system generating synthetic yet clinically realistic wound images to augment datasets, thereby enhancing the training and generalization of wound assessment algorithms. Similarly, conditional GAN architectures have been effectively applied to automated wound segmentation tasks. Jishnu et al. [43] developed a model (automatic foot ulcer segmentation generative adversarial network), achieving a Dice coefficient over 93%, demonstrating GANs' potential in managing the complexity and variability inherent in wound imaging, such as diverse lighting and tissue types.

Another significant aspect of generative AI involves the use of LLMs, such as GPT-4, to generate clinical documentation, patient educational content, and preliminary treatment recommendations based on learned medical guidelines and best practices [66]. Gurler et al. [38] introduced a novel conversational AI framework specifically designed for personalized wound management. Utilizing advanced LLM techniques and retrieval-augmented generation (RAG), the proposed system dynamically synthesizes treatment suggestions adapted to guidelines and institution-specific protocols, facilitating standardized care planning.

Another significant frontier of generative AI in wound care is the development of reasoning-centric LLM agents, which integrate foundational generative models with capabilities such as planning, memory management, external tool interaction, and dynamic information retrieval [101]. Unlike traditional generative AI models that focus primarily on single-task outputs, these advanced AI agents orchestrate multiple, interconnected tasks by autonomously decomposing clinical queries into manageable subtasks, retrieving pertinent data from diverse medical databases or protocols, and iteratively refining clinical recommendations or care strategies. These advanced architectures envision AI systems that continuously analyze multimodal patient data, dynamically reason through evolving clinical contexts, and autonomously suggest or even implement therapeutic interventions within carefully defined safety boundaries. This emerging paradigm is positioned to substantially enhance precision and adaptability in wound care, forming the core topic discussed in detail in the subsequent section.

An Agentic, Multimodal, Multi-Agent AI Framework for Personalized Wound Care

Vision and Structure

The preceding sections of this chapter have detailed the wide-ranging data sources, AI-driven assessments, and personalized treatment paradigms that currently characterize modern wound care. However, the practical clinical implementation of these advancements necessitates a structured framework capable of effectively integrating diverse data streams, analytic models, and therapeutic decisions into a seamless, holistic system. Therefore, in this section, we propose a comprehensive multimodal, multi-agent framework for personalized wound care (illustrated in Fig. 32.1) that effectively integrates these elements into a single coherent structure.

The framework is structured through a hierarchical design wherein clinical settings, including hospitals, outpatient facilities, and home-based care, serve as interaction points for data collection and the implementation of treatment strategies, linking clinical interventions with automated analytical and decision-making agents. This structure ensures that clinical actions and AI-driven recommendations remain aligned with patient-specific health contexts, continuously enhancing patient outcomes through evidence-based personalization. This framework establishes a

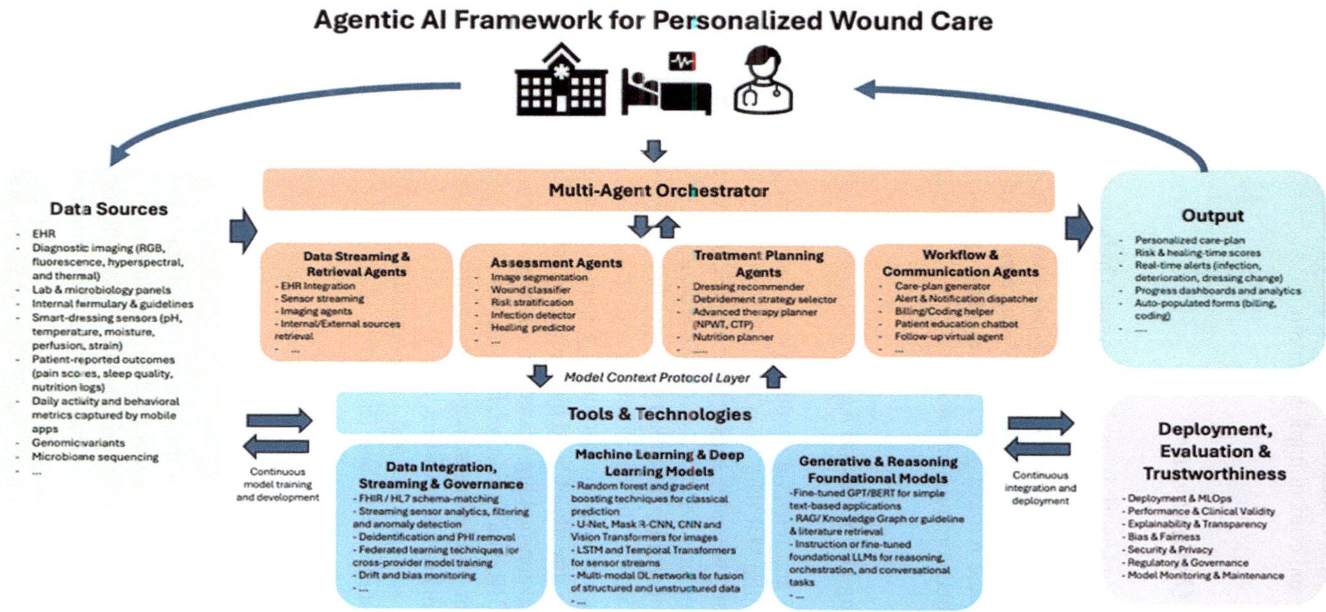

Fig. 32.1 Multimodal multi-agent framework for personalized wound care

continuous loop starting with the clinician at the clinical setting, progressing through analytical and decision-making processes, and ultimately returning to the clinician for final review and clinical implementation.

Multi-agent Orchestrator

Central to this framework is the Multi-Agent Orchestrator, which functions as the central coordinating component for the entire system. This orchestrator dynamically manages a variety of specialized agents, assigning each of them precise tasks tailored to their expertise. These agents are categorized into four primary groups: data streaming and retrieval agents, assessment agents, treatment planning agents, and workflow and communication agents.

Data Streaming and Retrieval Agents

Data streaming and retrieval agents serve as the initial gateway for capturing and integrating multimodal data from diverse clinical inputs such as EHRs, medical imaging systems, wearable sensor technologies, and various other clinical sources described comprehensively in section "Data Sources in Wound Care". The data collected at clinical settings have dual purposes: first, they continuously feed into training loops to enhance the development and refinement of AI-driven models, thus contributing to the adaptive learning capabilities of the system. Second, real-time data streams are directly utilized by specialized retrieval agents to support immediate clinical decision-making. Examples of these agents include those responsible for real-time EHR data extraction, high-resolution imaging data retrieval for wound characterization, and sensor management agents that stream live physiological metrics such as temperature, moisture, and pH from smart dressing technologies. These agents ensure standardized and compliant data handling, safeguarding privacy, maintaining data integrity, and facilitating seamless integration within the overall clinical workflow.

Assessment Agents

This agent group performs the critical role of transforming diverse and complex multimodal data, comprehensively described in section "AI Driven Assessment for Personalized Wound Care", into clinically actionable insights. Leveraging sophisticated machine-learning and deep-learning algorithms, these agents systematically interpret collected information from medical imaging, clinical records, and real-time physiological sensor data. Image segmentation agents, for example, accurately delineate wound boundaries and quantify tissue types, facilitating precise wound assessments. Risk stratification agents integrate extensive clinical datasets to generate predictive analyses on wound healing trajectories, factoring in individual patient variables such as comorbidities and wound characteristics. Additionally, agents specialized in infection detection rapidly analyze fluorescence imaging data and physiological sensor outputs, such as pH and temperature fluctuations, to identify early signs of microbial colonization, enabling immediate clinical interventions. Collectively, these assessment agents utilize advanced AI-driven tools and technologies, ensuring comprehensive wound understanding, proactive management, and optimized patient outcomes.

Treatment Planning Agents

Treatment Planning Agents translate analytical insights derived by assessment agents into detailed, patient-centric therapeutic strategies, as discussed in detail in section "AI-Driven Treatment for Personalized Wound Care". For instance, agents in this category may suggest specific wound dressing types tailored to manage varying levels of exudate and infection risk, based on real-time data. Other agents might evaluate and recommend appropriate debridement techniques, determining whether autolytic, enzymatic, surgical, or mechanical methods would optimally accelerate healing for specific wound conditions. Additionally, treatment planning agents assess patient eligibility for advanced therapeutic interventions, such as NPWT or the application of CTPs, by carefully analyzing the complexity and healing trajectory of individual wounds. Nutrition-focused agents further enrich patient care by proposing tailored dietary adjustments to address specific micronutrient deficiencies highlighted in clinical assessments, thus enhancing overall recovery and therapeutic effectiveness.

Workflow and Communication Agents

This agent group plays an essential role by effectively translating treatment recommendations into concrete clinical actions, thereby improving the efficiency and consistency of wound management practices. While Treatment Planning Agents primarily focus on generating therapeutic strategies based on data insights, Workflow and Communication Agents take these strategies and operationalize them into practical and implementable tasks. These agents create specific outputs formatted for user interfaces such as comprehensive dashboards or interactive platforms, enabling clinicians and healthcare providers to interact seamlessly with generated data and recommendations. They also encompass various automated systems, such as care-plan generators that systematically create detailed and actionable clinical instructions, and alert and notification systems that proactively escalate critical clinical events. Additionally, patient-education chatbots provide tailored and accessible information directly to patients, and AI-powered voice chatbots communicate effectively with outpatients, conducting routine follow-ups and gathering updates on their condition. Further enhancing operational efficiency, automated billing and coding agents manage administrative tasks by generating accurate, compliant documentation essential for reimbursement and reporting quality metrics.

It is important to emphasize that these agents represent envisioned capabilities within this framework and are not limited to the examples provided. The architecture is intentionally designed to accommodate additional agents and functionalities as technological advancements and clinical needs evolve.

Tools and Technologies

Within the agentic AI framework concept, agents execute specific operations by utilizing various tools accessible to them. The interactions between agents and these tools can be standardized through the MCP [8]. MCP is an open, model-agnostic protocol that standardizes interactions between AI agents and various computational tools or data sources, providing secure and controlled data exchanges with role-based access and clear audit trails. By serving as a universal interface, MCP ensures transparency, accountability, and seamless integration across diverse agents, computational resources, and clinical data repositories.

The tools utilized by agents can range broadly from rule-based algorithms to sophisticated AI-driven models, facilitating numerous clinical tasks as described comprehensively in the preceding sections. The comprehensive suite of technological components includes various tools such as data integration mechanisms, machine-learning and deep-learning models, and generative and reasoning AI frameworks. These tools handle diverse analytical tasks, from managing data governance processes like de-identification and schema harmonization to performing advanced predictive and diagnostic analytics using neural network architectures. Additionally, generative AI facilitates tasks such as automated summarization of clinical notes, retrieval of evidence-based guidelines, and the generation of dynamic dialogues to enhance clinician-patient interactions.

System Outputs and Clinical Integration

The outputs generated by this integrated multi-agent framework are diverse clinical artifacts specifically designed for practical application in real-world healthcare environments. These outputs may take various forms, such as personalized care plans detailing specific wound care strategies, structured clinical reports, detailed documentation, automatically rescheduled patient appointments, real-time alerts for urgent clinical interventions, and analytical insights presented through interactive dashboards. These dashboards can be accessed by clinicians via desktop or mobile interfaces, facilitating convenient and flexible visualization and management of wound care strategies and patient progress.

The cycle initiates with patient registration and clinician-driven data collection. The system subsequently provides real-time insights, predictive recommendations, and alerts back to clinicians for immediate review and potential adjustments. This continuous feedback loop ensures ongoing refinement and adaptation of treatment strategies, enabling clinicians to respond promptly to evolving patient conditions, real-time physiological feedback, and emerging therapeutic opportunities. Consequently, this iterative process ensures constant improvement in clinical decision-making and patient outcomes.

Deployment, Evaluation, Trustworthiness

Successful deployment and sustained operation of a multi-agent AI framework in clinical wound care require robust infrastructure supporting continuous delivery, ongoing performance monitoring, and rigorous quality assurance. Prior to clinical adoption, it is essential to validate the predictive and therapeutic capabilities of the framework through prospective trials adhering to established standards such as the SPIRIT-AI and CONSORT-AI extensions [22, 55]. Once operational, maintaining system reliability demands sophisticated machine-learning operations (MLOps) pipelines to monitor and promptly address issues such as data drift, bias, and performance degradation, a practice already validated in clinical settings [70].

Explainability and transparency are equally critical components. For clinicians to confidently integrate AI-driven recommendations into clinical decision-making, each recommendation should be accompanied by clear, interpretable rationales. Periodic bias-and-fairness audits further support equitable healthcare delivery by systematically assessing algorithm performance across diverse patient demographics [72]. Rigorous security and privacy protocols must safeguard patient data, facilitated by frameworks such as the MCP, which provides structured and secure interactions among AI agents and clinical data repositories, and the HL7 FAST Security Implementation Guide, which defines standardized procedures for secure data exchanges [41].

From a regulatory standpoint, adherence to recognized principles such as the Good Machine-Learning Practice (GMLP) guidelines, jointly developed by regulatory bodies including the FDA, Health Canada, and MHRA, is necessary to ensure safety, effectiveness, and compliance with postmarket surveillance expectations [92]. Complementary ethical guidance from the World Health Organization (WHO) further underscores the importance of transparency, human oversight, and equitable distribution of AI-driven healthcare benefits [103].

Finally, realizing the benefits of personalized wound care requires more than isolated technological solutions. Integrating multimodal data sources, specialized analytical agents, and standardized operational protocols, as highlighted in the FUTURE AI consensus guidelines [52], forms the comprehensive foundation necessary for meaningful clinical outcomes, workflow efficiency, and continuous learning capabilities. This integrated perspective establishes the basis for subsequent discussions on broader policy, regulatory, economic, and ethical considerations outlined in the next section.

Policy, Regulatory, Economic, and Ethical Considerations for Artificial Intelligence in Wound Care

The rapid advancement and clinical potential of AI in personalized wound care, as outlined in previous sections, bring with them critical considerations beyond technical and clinical innovations. While the transformative promise of AI-driven assessment, tailored treatments, and agentic frameworks has been established, realizing their full potential necessitates navigating an intricate landscape of regulatory requirements, ethical principles, economic realities, and healthcare policy implications.

Historically, wound care has been marginalized within healthcare, characterized by economic invisibility and underfunded research. Unlike primary care providers, wound care facilities, particularly independent outpatient centers, did not receive direct financial incentives from policies such as the HITECH Act [34, 35, 47, 84]. This lack of financial motivation, coupled with the significant implementation costs, hindered widespread adoption of EHR within these settings [2, 34, 47]. Furthermore, wound care centers frequently operated independently from broader healthcare systems, limiting integration, data sharing, and continuity of care.

Given these challenges, there is now an emerging consensus on the need for supportive healthcare policies, transparent regulatory frameworks, and sustainable economic models. Such frameworks aim to facilitate equitable access to advanced wound care technologies while ensuring robust ethical standards. This section examines these critical dimensions by highlighting emerging global regulatory structures, policy considerations affecting reimbursement and economic viability, and ethical dilemmas posed by AI applications in wound care.

Emerging Regulatory Frameworks Across the Globe

The regulatory landscape for AI in healthcare is still evolving. In the United States, the FDA regulates AI under its broader oversight of Software as a Medical Device (SaMD). The proposed approach supports a total product lifecycle, where regulatory controls adapt to continuous learning [90]. It promotes transparency and consistency in premarket and postmarket review to ensure AI-based applications maintain clinical relevance over time [88]. Predetermined Change Control Plans (PCCPs) allow a mechanism for developers to submit pre-approved plans for algorithm modifications learning [91]. PCCPs define what can change, how it can change, and metrics by which changes are assessed.

Further reinforcing these efforts, FDA's AI/ML Action Plan outlined a commitment to fostering innovation while ensuring safety and effectiveness. In collaboration with international regulators, including Health Canada and MHRA, GMLP was developed [92]. It sets foundational principles for development, validation, deployment, and lifecycle management of AI/ML-based technologies.

In the European Union, the AI Act (Regulation EU 2024/1689) introduced a uniform risk-based framework to ensure AI systems are safe, ethical, transparent, and aligned with fundamental rights. AI systems are categorized into four risk levels based on harm to people or society [15, 27, 29]. Manipulative systems that exploit vulnerabilities, socially score individuals, or rate them unfairly are banned. High-risk systems have strict rules for risk assessment, data quality, cybersecurity, and human oversight. Chatbots or emotional recognition systems fall under limited risk and are subject to transparency obligations. Users must be informed that they are interacting with an AI system and certain outputs are artificially generated. Systems that do not significantly impact people's rights or safety are considered minimal or no risk. These include AI-based agents for scheduling meetings, spam filters, and recommendation engines.

Globally, AI regulatory approaches vary considerably. China enforces stringent regulations emphasizing algorithm transparency and national security [82, 86]. In contrast, Japan, Singapore, and South Korea adopt softer guidelines encouraging voluntary compliance focused on transparency and human-centric design [6, 51, 83]. India is developing comprehensive AI legislation within broader digital strategies [61, 65].

Policy, Reimbursement, and Economic Considerations

Reimbursement policies significantly influence the transition of AI technologies from experimental prototypes to routine clinical practice, particularly within wound care. Despite promising developments, current reimbursement frameworks in the United States for AI-driven wound management remain fragmented, presenting both challenges and opportunities. In the context of Medicare fee-for-service reimbursement, the Centers for Medicare & Medicaid Services (CMS) have yet to establish dedicated wound-specific reimbursement codes explicitly for AI-enabled clinical decision support systems. However, several existing mechanisms offer potential pathways to adoption:

- Incremental progress includes the introduction of Category III CPT codes for tracking clinical use of emerging technologies and reimbursement for select AI tools, as seen in radiology and ophthalmology [3, 5, 26, 71].
- The New Technology Add-On Payment (NTAP) program further supports technologies demonstrating clinical improvement, providing potential models for wound care AI reimbursement [62, 96].

Private insurers, like UnitedHealthcare, Anthem, and Cigna, explore reimbursement models contingent upon clinical validation, rigorous evidence, and economic outcomes, highlighting the importance of transparency and explainability in algorithmic decisions [4, 60, 93].

Internationally, several models offer important lessons:

- The European Union's Health Technology Assessment Regulation (HTAR) harmonizes evaluation criteria, including cost-effectiveness analyses for high-risk digital health technologies, providing a streamlined market entry process for AI-based wound management solutions across member states [28].
- In Germany, the fast track process for Digital Health Applications (DiGA) enables evidence-based solutions, including wound care applications, to enter the market quickly with reimbursement coverage while long-term outcomes are evaluated [73].
- The United Kingdom's NICE Evidence Standards Framework specifies economic and clinical evidence standards for digital health technologies, thus establishing clear guidelines for integrating AI tools within the National Health Service [63].

Challenges and Opportunities

Integrating AI tools into routine clinical practice for personalized wound care remains challenging, despite significant advancements in algorithmic performance. Central among these challenges is the issue of data fragmentation. Clinical data, including wound images, physiological measurements, and narrative clinical notes, are frequently captured and stored in isolated, institution-specific systems across hospitals, outpatient centers, long-term care facilities, and patients' homes. Such decentralized data repositories hinder the development of large, diverse, and representative datasets, which are critical for training robust and generalizable AI models. Emerging approaches, such as federated learning and differential privacy techniques, allow for multi-institutional collaborative model training without compromising data privacy or ownership. Additionally, standardized data exchange protocols, such as HL7 FHIR, facilitate interoperability and data integration [32, 57].

Beyond technical limitations, effective integration of AI into established clinical workflows is crucial. Interoperability challenges, such as additional authentication steps, redundant documentation requirements, and slow data transfers, negatively impact clinician acceptance and overall user experience. To streamline these workflows, innovative approaches involve embedding AI-driven microservices directly within existing EHRs, mobile wound assessment applications, and intuitive voice-activated interfaces.

Furthermore, the inherent lack of transparency in many AI systems, referred to as "black-box" models, significantly reduces clinicians' trust in their recommendations. To address this concern, advanced explainability techniques, including visual aids, such as saliency maps and comprehensible natural language explanations, must be developed to clarify the rationale behind AI decisions [9, 77, 78]. Such explainability methods are pivotal for enhancing transparency, building clinical confidence, and facilitating the identification and correction of potential biases.

Ensuring fairness and equity represents another critical challenge. AI models trained on datasets under-representing specific populations risk producing biased or inaccurate clinical assessments, potentially leading to inequitable healthcare outcomes. Systematic strategies, including equity audits, synthetic data augmentation, and counterfactual analyses, are essential mechanisms for detecting and mitigating algorithmic biases. Complementary outreach and data-collection initiatives targeting underrepresented patient groups further ensure that AI performance maintains uniform efficacy across diverse patient demographics and care environments [39].

Sustainability of AI implementation relies significantly on available skilled personnel and robust technological infrastructures, both of which are often limited in many wound care facilities. Cloud-based MLOps platforms and "AI-as-a-service" models mitigate this gap by providing automated version control, performance monitoring, drift detection, and cybersecurity capabilities, thus reducing resource demands for ongoing maintenance.

Effectively addressing these interconnected challenges and fully realizing the potential opportunities of AI-driven wound care necessitate sustained interdisciplinary collaboration. Clinicians, engineers, data scientists, ethicists, regulators, payers, industry representatives, and patients must actively participate in shaping the development, implementation, and governance of these technologies. Such collaboration facilitates the establishment of robust regulatory and ethical frameworks, ensuring transparency in AI decision-making processes, equitable access across diverse patient populations, and demonstrable clinical value in real-world practice. Building a socio-technical ecosystem that integrates these diverse perspectives promotes not only technical innovation but also fosters trust, accountability, and long-term sustainability in personalized wound care delivery.

Conclusion

Artificial intelligence stands poised to redefine wound management through precise, personalized, and adaptive care strategies. By integrating diverse multimodal data, including high-resolution medical imaging, real-time physiological metrics, and structured clinical records, advanced AI-driven frameworks can accurately assess wound status, predict healing trajectories, and recommend evidence-based interventions. The multi-agent AI framework presented in this chapter illustrates how AI systems can effectively interface across clinical, analytical, therapeutic, and operational domains, enabling enhanced clinical decision-making, optimized patient outcomes, and streamlined workflows. However, translating this technical potential into routine clinical practice demands rigorous attention to explainability, transparency, data interoperability, and robust cybersecurity measures, alongside sustainable economic models and supportive regulatory frameworks.

Ultimately, fulfilling the transformative potential of AI-driven personalized wound care requires more than technological innovation alone. It necessitates meaningful collaboration among diverse stakeholders to the various challenges and opportunities highlighted throughout this chapter. As these collaborations progress, AI-driven wound care is positioned not only to improve clinical outcomes significantly but also to transform wound management into an exemplar of equitable, efficient, and patient-centered precision medicine.

References

1. Ahmed S, Kabir MA, Chowdhury MEH, Nancarrow S (2024) AI-driven personalised offloading device prescriptions: a cutting-edge approach to preventing diabetes-related plantar forefoot ulcers and complications. IntechOpen. https://doi.org/10.5772/intechopen.1003960
2. Ajami S, Bagheri-Tadi T (2013) Barriers for adopting electronic health records (EHRs) by physicians. Acta Inform Med 21(2):129–134. https://doi.org/10.5455/aim.2013.21.129-134
3. American Medical Association (2023a) Oversight needed for payers' use of AI in prior authorization. Retrieved June 10, 2025, from https://www.ama-assn.org/practice-management/prior-authorization/oversight-needed-payers-use-ai-prior-authorization
4. American Medical Association (2023b) CPT code set facilitates development of useful AI-powered devices. Retrieved June 9, 2025, from https://www.ama-assn.org/practice-management/cpt/cpt-code-set-facilitates-development-useful-ai-powered-devices
5. American Medical Association (2025) Augmented intelligence in medicine. Retrieved June 9, 2025, from https://www.ama-assn.org/practice-management/digital-health/augmented-intelligence-medicine
6. Andrews C (2025) Japan passes innovation-focused AI governance bill. International Association of Privacy Professionals. Retrieved June 9, 2025, from https://iapp.org/news/a/japan-passes-innovation-focused-ai-governance-bill
7. Anisuzzaman DM, Patel Y, Rostami B, Niezgoda J, Gopalakrishnan S, Yu Z (2022) Multimodal wound classification using wound image and location by deep neural network. Sci Rep 12:Article 20057. https://doi.org/10.1038/s41598-022-21813-0
8. Anthropic (2024) Introducing the Model Context Protocol (MCP) [White paper]. Anthropic. https://www.anthropic.com/news/model-context-protocol
9. Aziz NA, Manzoor A, Qureshi MDM, Qureshi MA, Rashwan W (2024) Unveiling explainable AI in healthcare: current trends, challenges, and future directions (Version 1). medRxiv. https://doi.org/10.1101/2024.08.10.24311735
10. Babu M, Lautman Z, Lin X, Sobota MHB, Snyder MP (2024) Wearable devices: implications for precision medicine and the future of health care. Annu Rev Med 75:401–415. https://doi.org/10.1146/annurev-med-052422-020437
11. Berezo M, Budman J, Deutscher D, Hess CT, Smith K, Hayes D (2022) Predicting chronic wound healing time using machine learning. Adv Wound Care 11(6):281–296. https://doi.org/10.1089/wound.2021.0073
12. Bergstrom N, Braden BJ, Laguzza A, Holman V (1987) The Braden scale for predicting pressure sore risk. Nurs Res 36(4):205–210
13. Birkner M, Schalk J, von den Driesch P, Schultz ES (2022) Computer-assisted differential diagnosis of pyoderma gangrenosum and venous ulcers with deep neural networks. J Clin Med 11(23):7103. https://doi.org/10.3390/jcm11237103
14. Boissin C, Laflamme L, Fransén J, Lundin M, Huss F, Wallis L, Allorto N, Lundin J (2023) Development and evaluation of deep learning algorithms for assessment of acute burns and the need for surgery. Sci Rep 13:Article 1794. https://doi.org/10.1038/s41598-023-28164-4
15. Busch F, Kather JN, Johner C, Moser M, Truhn D, Adams LC, Bressem KK (2024) Navigating the European Union artificial intelligence act for healthcare. NPJ Digit Med 7:Article 210. https://doi.org/10.1038/s41746-024-01094-7

16. Carrión H, Jafari M, Bagood MD, Yang H-Y, Isseroff RR, Gomez M (2022) Automatic wound detection and size estimation using deep learning algorithms. PLoS Comput Biol 18(3):e1009852. https://doi.org/10.1371/journal.pcbi.1009852
17. Carter MJ, DaVanzo J, Haught R, Nusgart M, Cartwright D, Fife CE (2023) Chronic wound prevalence and the associated cost of treatment in Medicare beneficiaries: changes between 2014 and 2019. J Med Econ 26(1):894–901. https://doi.org/10.1080/13696998.2023.2232256
18. Chang M-C, Yu T, Luo J, Duan K, Tu P, Zhao Y, Nagraj N, Rajiv V, Priebe M, Wood EA, Stachura M (2018) Multimodal sensor system for pressure ulcer wound assessment and care. IEEE Trans Industr Inform 14(3):1186–1196. https://doi.org/10.1109/TII.2017.2782213
19. Chatterjee A, Pahari N, Prinz A (2022) HL7 FHIR with SNOMED-CT to achieve semantic and structural interoperability in personal health data: a proof-of-concept study. Sensors 22(10):3756. https://doi.org/10.3390/s22103756
20. Cho SK, Mattke S, Gordon H, Sheridan M, Ennis W (2020) Development of a model to predict healing of chronic wounds within 12 weeks. Adv Wound Care 9(9):516–524. https://doi.org/10.1089/wound.2019.1091
21. Coleman S, Nelson EA, Vowden P, Vowden K, Adderley U, Sunderland L, Harker J, Conroy T, Fiori S, Bezer N, Holding E, Atkin L, Stables E, Dumville J, Gavelle S, Sandoz H, Moore K, Chambers T, Napper S, Nixon J (2017) Development of a generic wound care assessment minimum data set. J Tissue Viability 26(4):226–240. https://doi.org/10.1016/j.jtv.2017.09.007
22. Cruz Rivera S, Liu X, Chan A-W, Denniston AK, Calvert MJ, SPIRIT-AI and CONSORT-AI Working Group, SPIRIT-AI and CONSORT-AI Steering Group, SPIRIT-AI and CONSORT-AI Consensus Group (2020) Guidelines for clinical trial protocols for interventions involving artificial intelligence: the SPIRIT-AI extension. Nat Med 26:1351–1363. https://doi.org/10.1038/s41591-020-1037-7
23. Dallmann AC, Sheridan M, Mattke S, Ennis W (2024) Prediction of healing trajectory of chronic wounds using a machine learning approach. Adv Wound Care. https://doi.org/10.1089/wound.2024.0095
24. Danilovich I, Moshkin V, Reimche A, Tevelevich M, Mikhaylovskiy N (2021) Video monitoring over anti-decubitus protocol execution with a deep neural network to prevent pressure ulcer. In: 2021 IEEE engineering in medicine & Biology Society Annual International Conference (EMBC). IEEE, pp 1384–1387. https://doi.org/10.1109/EMBC46164.2021.9630830
25. Deng J, Shi G, Ye Z, Xiao Q, Zhang X, Ren L, Yang F, Wang M (2025) Unveiling and swift diagnosing chronic wound healing with artificial intelligence assistance. Chin Chem Lett 36(3):Article 110496. https://doi.org/10.1016/j.cclet.2024.110496
26. Digital Diagnostics (2019) New CPT code for automated point-of-care retinal imaging. Retrieved June 9, 2025, from https://www.digitaldiagnostics.com/new-cpt-code-for-automated-point-of-care-retinal-imaging/
27. EU Artificial Intelligence Act (2024) High-level summary of the AI Act. Retrieved June 9 2025, from https://artificialintelligenceact.eu/high-level-summary/
28. European Medicines Agency (2024) The use of artificial intelligence (AI) in the medicinal product lifecycle (EMA/CHMP/CVMP/83833/2023). https://www.ema.europa.eu/en/use-artificial-intelligence-ai-medicinal-product-lifecycle
29. European Parliament (2023) EU AI act: first regulation on artificial intelligence. https://www.europarl.europa.eu/topics/en/article/20230601STO93804/eu-ai-act-first-regulation-on-artificial-intelligence
30. Frykberg RG, Banks J (2015) Challenges in the treatment of chronic wounds. Adv Wound Care 4(9):560–582. https://doi.org/10.1089/wound.2015.0635
31. Gabison S, Pupic N, Evans G, Dolatabadi E, Dutta T (2022) Measuring repositioning in home care for pressure injury prevention and management. Sensors 22(18):7013. https://doi.org/10.3390/s22187013
32. Ganesan O, Morris MX, Guo L, Orgill D (2024) A review of artificial intelligence in wound care. Artificial intelligence surgery 4(4):364–375. https://doi.org/10.20517/ais.2024.68

33. Gefen A (2025) Promoting wound healing through artificial intelligence-powered dressing development. Wounds Int 16(1):30–33. Retrieved from https://woundsinternational.com/journal-articles/promoting-wound-healing-through-artificial-intelligence-powered-dressing-development/
34. Gettinger A, Zayas-Cabán T (2021) HITECH to 21st century cures: clinician burden and evolving health IT policy. J Am Med Inform Assoc 28(5):1022–1025. https://doi.org/10.1093/jamia/ocaa330
35. Gold M, McLaughlin C (2016) Assessing HITECH implementation and lessons: five years later. Milbank Q 94(3):654–687. https://doi.org/10.1111/1468-0009.12214
36. Griffa D, Natale A, Merli Y, Starace M, Curti N, Mussi M, Castellani G, Melandri D, Piraccini BM, Zengarini C (2024) Artificial intelligence in wound care: a narrative review of the currently available mobile apps for automatic ulcer segmentation. BioMedInformatics 4(4):2321–2337. https://doi.org/10.3390/biomedinformatics4040126
37. Gruenerbel L, Heinrich F, Böhlhoff-Martin J, Röper L, Machens H-G, Gruenerbel A, Schillinger M, Kist A, Wenninger F, Richter M, Steinbacher L (2023) Wearable prophylaxis tool for AI-driven identification of early warning patterns of pressure ulcers. Bioengineering 10(10):1125. https://doi.org/10.3390/bioengineering10101125
38. Gurler N, Bulbul HS, Cheng P, Kyle WL, Guler O (2024) Novel conversational artificial intelligence framework for customization and standardization of wound management [Conference poster]. In: 2024 Annual meeting of the wound healing society (SAWC spring/WHS joint meeting), Orlando, FL, United States
39. Hasanzadeh F, Josephson CB, Waters G, Adedinsewo D, Azizi Z, White JA (2025) Bias recognition and mitigation strategies in artificial intelligence healthcare applications. npj Digit Med 8:Article 154. https://doi.org/10.1038/s41746-025-01503-7
40. He J, Wang X, Wang Z, Xie R, Zhang Z, Liu T-M, Cai Y, Chen L (2025) Interpretable deep learning method to predict wound healing progress based on collagen fibers in wound tissue. Comput Biol Med 191:Article 110110. https://doi.org/10.1016/j.compbiomed.2025.110110
41. HL7 International (2024) Security for scalable registration, authentication, and authorization (HL7 FHIR® FAST security implementation guide, Release 1.0.0 – STU 1). https://hl7.org/fhir/us/udap-security/
42. Hornberger C, Herrmann B, Daeschlein G, Podewils S, Sicher C, Kuhn J, Masur K, Meister M, Wahl P (2020) Detecting bacteria on wounds with hyperspectral imaging in fluorescence mode. Curr Direct Biomed Eng 6(3):264–267. https://doi.org/10.1515/cdbme-2020-3067
43. Jishnu P, Shreyamsha Kumar BK, Jayaraman S (2023) Automatic foot ulcer segmentation using conditional generative adversarial network (AFSegGAN): a wound management system. PLoS Digit Health 2(11):e0000344. https://doi.org/10.1371/journal.pdig.0000344
44. Johnson KB, Wei W-Q, Weeraratne D, Frisse ME, Misulis K, Rhee K, Zhao J, Snowdon JL (2021) Precision medicine, AI, and the future of personalized health care. Clin Transl Sci 14(1):86–93. https://doi.org/10.1111/cts.12884
45. Jørgensen LB, Sørensen JA, Jemec GBE, Yderstræde KB (2016) Methods to assess area and volume of wounds: a systematic review. Int Wound J 13(4):540–553. https://doi.org/10.1111/iwj.12472
46. Katsoulakis E, Wang Q, Wu H, Shahriyari L, Fletcher R, Liu J, Achenie L, Liu H, Jackson P, Xiao Y, Syeda-Mahmood T, Tuli R, Deng J (2024) Digital twins for health: a scoping review. NPJ Digit Med 7:Article 77. https://doi.org/10.1038/s41746-024-01073-0
47. Kruse CS, Kristof C, Jones B, Mitchell E, Martinez A (2016) Barriers to electronic health record adoption: a systematic literature review. J Med Syst 40(12):Article 252. https://doi.org/10.1007/s10916-016-0628-9
48. Larsen RA, Evans RS, Burke JP, Pestotnik SL, Gardner RM, Classen DC (1989) Improved perioperative antibiotic use and reduced surgical wound infections through use of computer decision analysis. Infect Control Hosp Epidemiol 10(7):316–320. https://doi.org/10.1086/646035
49. Larson BJ, Roakes A, Yurick S, Netravali NA (2024) Precision in prevention: tailoring single-use negative pressure wound therapy utilization through artificial intelligence-based

surgical site complications risk and cost modeling. Surg Infect 25(4):315–321. https://doi.org/10.1089/sur.2023.274
50. Le L, Baer M, Briggs P, Bullock N, Cole W, DiMarco D, Hamil R, Harrell K, Kasper M, Li W, Patel K, Sabo M, Thibodeaux K, Serena TE (2021) Diagnostic accuracy of point-of-care fluorescence imaging for the detection of bacterial burden in wounds: results from the 350-patient fluorescence imaging assessment and guidance trial. Adv Wound Care 10(3):123–136. https://doi.org/10.1089/wound.2020.1272
51. Legal Business Online (2025) Explainer: Korea's AI Basic Act. Retrieved June 9, 2025, from https://www.legalbusinessonline.com/features/explainer-korea%E2%80%99s-ai-basic-act
52. Lekadir K, Frangi AF, Porras AR, Glocker B, Cintas C, Langlotz CP, Weicken E, Asselbergs FW, Prior F, Collins GS, Kaissis G, Tsakou G, Buvat I, Kalpathy-Cramer J, Mongan J, Schnabel JA, Kushibar K, Riklund K, Marias K et al (2025) FUTURE-AI: international consensus guideline for trustworthy and deployable artificial intelligence in healthcare. BMJ 388:e081554. https://doi.org/10.1136/bmj-2024-081554
53. Li S, Mohamedi AH, Senkowsky J, Nair A, Tang L (2020) Imaging in chronic wound diagnostics. Adv Wound Care 9(5):245–263. https://doi.org/10.1089/wound.2019.0967
54. Lien A-S-Y, Lai C-Y, Wei J-D, Yang H-M, Yeh J-T, Tai H-C (2022) A granulation tissue detection model to track chronic wound healing in DM foot ulcers. Electronics 11(16):2617. https://doi.org/10.3390/electronics11162617
55. Liu X, Cruz Rivera S, Moher D, Calvert MJ, Denniston AK, SPIRIT-AI and CONSORT-AI Working Group (2020) Reporting guidelines for clinical trial reports for interventions involving artificial intelligence: the CONSORT-AI extension. Nat Med 26(9):1364–1374. https://doi.org/10.1038/s41591-020-1034-x
56. Liu Z, John J, Agu E (2022) Diabetic foot ulcer ischemia and infection classification using EfficientNet deep learning models. IEEE Open J Eng Med Biol 3:189–201. https://doi.org/10.1109/OJEMB.2022.3219725
57. Liu M, Ning Y, Teixayavong S, Liu X, Mertens M, Shang Y, Li X, Miao D, Liao J, Xu J, Ting DSW, Cheng LT-E, Ong JCL, Teo ZL, Tan TF, RaviChandran N, Wang F, Celi LA, Ong MEH, Liu N (2025) A scoping review and evidence gap analysis of clinical AI fairness. NPJ Digit Med 8:Article 360. https://doi.org/10.1038/s41746-025-01667-2
58. Lu F, Zlobina K, Rondoni NA, Teymoori S, Gomez M (2024) Enhancing wound healing through deep reinforcement learning for optimal therapeutics. R Soc Open Sci 11(7):240228. https://doi.org/10.1098/rsos.240228
59. McLean KA, Sgrò A, Brown LR, Buijs LF, Mountain KE, Shaw CA, Drake TM, Pius R, Knight SR, Fairfield CJ, Skipworth RJE, Tsaftaris SA, Wigmore SJ, Potter MA, Bouamrane M-M, Harrison EM, TWIST Collaborators (2025) Multimodal machine learning to predict surgical site infection with healthcare workload impact assessment. NPJ Digit Med 8:Article 121. https://doi.org/10.1038/s41746-024-01419-8
60. Minemyer P (2024) Why UnitedHealthcare is putting a focus on simplicity in its AI strategy. Fierce Healthcare. Retrieved June 10, 2025, from https://www.fiercehealthcare.com/payers/why-unitedhealthcare-putting-focus-simplicity-its-ai-strategy
61. Mohanty A, Sahu S (2024) India's advance on AI regulation. Carnegie Endowment for International Peace. Retrieved June 9, 2025, from https://carnegieendowment.org/research/2024/11/indias-advance-on-ai-regulation?lang=en
62. Morey JR, Katana M, Li L, Hohmann S, Weber E, Ferket B (2021) Adoption and trends in the Medicare new technology add-on payment program. J Gen Intern Med 36(7):2174–2176. https://doi.org/10.1007/s11606-020-05930-y
63. National Institute for Health and Care Excellence (2022) Evidence standards framework for digital health technologies (Corporate Document ECD7). https://www.nice.org.uk/corporate/ecd7
64. Niri R, Zahia S, Stefanelli A, Sharma K, Probst S, Pichon S, Chanel G (2025) Wound segmentation with U-Net using a dual attention mechanism and transfer learning. J Imaging Inform Med. https://doi.org/10.1007/s10278-025-01386-w

65. NITI Aayog (2023) National strategy for artificial intelligence. Government of India. https://www.niti.gov.in/sites/default/files/2023-03/National-Strategy-for-Artificial-Intelligence.pdf
66. OpenAI (2024) GPT-4 technical report. arXiv preprint arXiv:2303.08774. arXiv. https://doi.org/10.48550/arXiv.2303.08774
67. Patel S, Ershad F, Zhao M, Isseroff RR, Duan B, Zhou Y, Wang Y, Yu C (2022) Wearable electronics for skin wound monitoring and healing. Soft Sci 2:Article 9. https://doi.org/10.20517/ss.2022.13
68. Patel Y, Shah T, Dhar MK, Zhang T, Niezgoda J, Gopalakrishnan S, Yu Z (2024) Integrated image and location analysis for wound classification: a deep learning approach. Sci Rep 14:Article 7043. https://doi.org/10.1038/s41598-024-56626-w
69. Peleg M, Shachak A, Wang D, Karnieli E (2009) Using multi-perspective methodologies to study users' interactions with the prototype front end of a guideline-based decision support system for diabetic foot care. Int J Med Inform 78(7):482–493. https://doi.org/10.1016/j.ijmedinf.2009.02.008
70. Rajagopal A, Ayanian S, Ryu AJ, Qian R, Legler SR, Peeler EA, Issa M, Coons TJ, Kawamoto K (2024) Machine learning operations in health care: a scoping review. Mayo Clin Proc Digit Health 2(3):421–437. https://doi.org/10.1016/j.mcpdig.2024.06.009
71. Rajesh AE, Lee AY (2025) AI for DR screening: where are we in 2025? Retina Specialist. Retrieved June 9, 2025, from https://www.retina-specialist.com/article/ai-for-dr-screening-where-are-we-in-2025
72. Rajkomar A, Hardt M, Howell MD, Corrado G, Chin MH (2018) Ensuring fairness in machine learning to advance health equity. Ann Intern Med 169(12):866–872. https://doi.org/10.7326/M18-1990/
73. Rammos T, Schwarz T (2023) Is 2024 the breakthrough year for digital health across Europe? Taylor Wessing. https://www.taylorwessing.com/en/interface/2023/predictions-2024/is-2024-the-breakthrough-year-for-digital-health-across-europe
74. Robinson CJ, Dickie B, Lindner C, Herrera J, Dingle L, Reid AJ, Wong JKF, Hiebert P, Cootes TF, Kurinna S (2025) Complex wound analysis using AI. Comput Biol Med 190:Article 109945. https://doi.org/10.1016/j.compbiomed.2025.109945
75. Rumiński J, Kaczmarek M, Renkielska A, Nowakowski A (2007) Thermal parametric imaging in the evaluation of skin burn depth. IEEE Trans Biomed Eng 54(2):303–312. https://doi.org/10.1109/TBME.2006.886607
76. Saiko G, Lombardi P, Au Y, Queen D, Armstrong D, Harding K (2020) Hyperspectral imaging in wound care: a systematic review. Int Wound J 17(6):1840–1856. https://doi.org/10.1111/iwj.13474
77. Sarp S, Kuzlu M, Wilson E, Cali U, Guler O (2021a) The enlightening role of explainable artificial intelligence in chronic wound classification. Electronics 10(12):Article 1406. https://doi.org/10.3390/electronics10121406
78. Sarp S, Kuzlu M, Wilson E, Guler O (2021b) WG2AN: synthetic wound image generation using generative adversarial network. J Eng 2021:286–294. https://doi.org/10.1049/tje2.12033
79. Sarp S, Kuzlu M, Zhao Y, Gueler O (2023) Digital twin in healthcare: a study for chronic wound management. IEEE J Biomed Health Inform 27(11):5634–5643. https://doi.org/10.1109/JBHI.2023.3299028
80. Sellmer D, Carvalho CM, Carvalho DR, Malucelli A (2013) Sistema especialista para apoiar a decisão na terapia tópica de úlceras venosas [Expert system to support the decision in topical therapy for venous ulcers]. Rev Gaucha Enferm 34(2):154–162. https://doi.org/10.1590/S1983-14472013000200020
81. Sen CK (2023) Human wound and its burden: updated 2022 compendium of estimates. Adv Wound Care 12(12):657–670. https://doi.org/10.1089/wound.2023.0150
82. Sheehan M (2023) China's AI regulations and how they get made (Working paper). Carnegie Endowment for International Peace. https://carnegie-production-assets.s3.amazonaws.com/static/files/202307-Sheehan_Chinese%20AI%20gov-1.pdf

83. Shivhare S, Park KB (2025) South Korea's new AI framework act: a balancing act between innovation and regulation. Future of Privacy Forum. Retrieved June 9, 2025, from https://fpf.org/blog/south-koreas-new-ai-framework-act-a-balancing-act-between-innovation-and-regulation/
84. Stark P (2010) Congressional intent for the HITECH act. Am J Manag Care 16:SP24–SP28
85. Stremitzer S, Wild T, Hoelzenbein T (2007) How precise is the evaluation of chronic wounds by health care professionals? Int Wound J 4:156–161. https://doi.org/10.1111/j.1742-481X.2007.00334.x
86. Su G (2025) Could one strategy fit all? A comparison of regulatory guidance from China, Europe, and the USA on medical device clinical evaluation throughout the total product lifecycle. Expert Rev Med Devices 22(1):5–13. https://doi.org/10.1080/17434440.2024.2448848
87. Swerdlow M, Lo J, Armstrong DG (2024) Reliability of an AI-powered application across different mobile devices for assessment of chronic wounds. Adv Wound Care 13(1):14–21. https://doi.org/10.1089/wound.2022.0095
88. Taib BG, Karwath A, Wensley K, Minku L, Gkoutos GV, Moiemen N (2023) Artificial intelligence in the management and treatment of burns: a systematic review and meta-analyses. J Plast Reconstr Aesthet Surg 77:133–161. https://doi.org/10.1016/j.bjps.2022.11.049
89. Topaz M, Lai K, Dowding D, Lei VJ, Zisberg A, Bowles KH, Zhou L (2016) Automated identification of wound information in clinical notes of patients with heart diseases: developing and validating a natural language processing application. Int J Nurs Stud 64:25–31. https://doi.org/10.1016/j.ijnurstu.2016.09.013
90. U.S. Food and Drug Administration (2023) Total product life cycle for medical devices. CDRH Transparency. https://www.fda.gov/about-fda/cdrh-transparency/total-product-life-cycle-medical-devices
91. U.S. Food and Drug Administration (2024) Predetermined change control plans for medical devices: Draft guidance for industry and FDA staff (Docket No. FDA-2024-D-2338) [Draft guidance]. https://www.fda.gov/regulatory-information/search-fda-guidance-documents/predetermined-change-control-plans-medical-devices
92. U.S. Food and Drug Administration, Health Canada, & Medicines and Healthcare products Regulatory Agency (2025) Good machine learning practice for medical device development: Guiding principles (Release 1.0). https://www.fda.gov/medical-devices/software-medical-device-samd/good-machine-learning-practice-medical-device-development-guiding-principles
93. UnitedHealth Group (2025) Making health care work smarter. Retrieved June 10, 2025, from https://www.unitedhealthgroup.com/ns/making-health-care-work-smarter.html
94. Veličković VM, Spelman T, Clark M, Probst S, Armstrong DG, Steyerberg E (2023) Individualized risk prediction for improved chronic wound management. Adv Wound Care 12(7):387–398. https://doi.org/10.1089/wound.2022.0017
95. Viswanathan V, Govindan S, Selvaraj B, Rupert S, Kumar R (2021) A clinical study to evaluate autofluorescence imaging of diabetic foot ulcers using a novel artificial intelligence enabled noninvasive device. Int J Low Extrem Wounds 23(1):169–176. https://doi.org/10.1177/15347346211047098
96. Viz.ai (2021) Viz.ai receives New Technology Add-on Payment renewal for stroke AI software from CMS. Retrieved June 9, 2025, from https://www.viz.ai/news/viz-ai-receives-new-technology-add-on-payment-renewal-for-stroke-ai-software-from-cms
97. Vo D-K, Trinh KTL (2025) Advances in wearable biosensors for wound healing and infection monitoring. Biosensors 15(3):139. https://doi.org/10.3390/bios15030139
98. Wang C, Anisuzzaman DM, Williamson V, Dhar MK, Rostami B, Niezgoda J, Gopalakrishnan S, Yu Z (2020a) Fully automatic wound segmentation with deep convolutional neural networks. Sci Rep 10:Article 21897. https://doi.org/10.1038/s41598-020-78799-w
99. Wang P, Li P, Yin M, Li Y, Wu J (2020b) Burn wound assessment system using near-infrared hyperspectral imaging and deep transfer features. Infrared Phys Technol 111:103558. https://doi.org/10.1016/j.infrared.2020.103558

100. Wang S, Xia C, Zheng Q, Wang A, Tan Q (2022) Machine learning models for predicting the risk of hard-to-heal diabetic foot ulcers in a Chinese population. Diabetes Metab Syndr Obes Targets Ther 15:3347–3359. https://doi.org/10.2147/DMSO.S383960
101. Wang L, Ma C, Feng X, Zhang Z, Yang H, Zhang J, Chen Z, Tang J, Chen X, Lin Y, Zhao WX, Wei Z, Wen J-R (2025) A survey on large language model based autonomous agents. arXiv preprint arXiv:2308.11432. arXiv. https://doi.org/10.48550/arXiv.2308.11432
102. Weigelt MA, Lev-Tov HA, Tomic-Canic M, Lee WD, Williams R, Strasfeld D, Kirsner RS, Herman IM (2022) Advanced wound diagnostics: toward transforming wound care into precision medicine. Adv Wound Care 11(6):330–359. https://doi.org/10.1089/wound.2020.1319
103. World Health Organization (2025) Ethics and governance of artificial intelligence for health: guidance on large multi-modal models. https://www.who.int/publications/i/item/9789240084759
104. Yu X, Wu Z, Zhang N (2024) Machine learning–driven discovery of novel therapeutic targets in diabetic foot ulcers. Mol Med 30:Article 215. https://doi.org/10.1186/s10020-024-00955-z
105. Zhao L, Chu X, Kong L, Dong L, Liu M (2023) Prediction of diabetic foot healing based on hyperspectral reconstruction. In: Optics in health care and biomedical optics XIII (Proceedings of SPIE 12770, 127701B). SPIE. https://doi.org/10.1117/12.2688641

Index

A

Accelerated wound healing, 157, 231, 269, 317, 329, 340, 455, 573–592, 667, 678, 679, 697
Adipose-derived stem cells (ASCs), 115–117, 146, 224, 252, 409–428, 587, 600, 612, 613, 675 678, 682, 683, 685, 720, 757
Adjunctive wound therapies, 755, 759
Advanced therapy, 120, 365, 567, 616, 670–671, 707, 708, 841, 843, 848
Advanced wound dressings, 76, 290, 371, 441, 483, 548–552, 568, 602, 701
Advanced wound therapies, 671, 701
Agentic AI, 836, 846, 854
AI-driven wound care, 705, 839, 859, 860
Allograft, 17, 426, 456, 458, 473–475, 502, 504–508, 510–511, 599, 630, 631, 637, 653, 654, 758
Angiogenesis, 6, 38, 48, 63, 86, 127, 202, 221, 238, 269, 295, 314, 357, 391, 414, 441, 469, 501, 545, 574, 601, 643, 668, 699, 717, 752, 766, 790, 813, 844
Animal model, 17, 156, 160–161, 163, 233, 251, 252, 262–268, 270, 273–279, 290, 291, 299–300, 303, 305, 337–341, 372, 373, 375, 400, 413, 426, 469, 474, 606–609, 673, 757
Artificial intelligence (AI), 63, 71, 78, 119, 120, 303–305, 374, 567, 658, 704–705, 709, 716, 759, 760, 801, 833–860
Autograft, 456, 458, 472, 473, 502, 503, 511, 610, 630, 632–637, 656

B

Bioactive scaffolds, 589
Bioengineered scaffolds, 86, 94, 102, 103, 114, 120, 483–485, 670, 671, 757
Bioengineering, 103, 106, 111–114, 118–120, 483–485, 544, 695–709, 728, 733, 757, 800
Biomaterials, 65, 98, 109, 111, 112, 114, 118–120, 143, 169, 231, 273–277, 291, 292, 303, 359, 371, 374, 376, 410, 413, 418, 421, 428, 457, 463–486, 495–533, 544, 559, 574, 581–588, 598, 599, 603, 604, 610, 615–617, 654, 668, 683, 686, 696, 703–704, 706, 707, 733, 756, 758, 760, 766, 823
Burns, 17, 115, 132, 202, 230, 264, 289, 384, 425, 452, 467, 497, 546, 585, 598, 628, 642, 679, 700, 749, 792, 806, 837
Burn units, 561, 644

C

Cadherin, 181–195, 393
Cell adhesion, 26, 183–185, 187, 189, 193–195, 223, 249, 357, 410, 416, 418, 480, 483, 511–513, 515, 519, 521, 523, 527–529, 583, 584, 589, 604, 618, 698, 703, 813

F. De Francesco, K. Turksen (eds.), *Management and Strategies for Wound Healing*, Stem Cell Biology and Regenerative Medicine 79,
https://doi.org/10.1007/978-3-032-12442-5

Cell-based therapy, 332, 548, 611
Cell proliferation, 17, 78, 94, 95, 97, 98, 100, 101, 106, 107, 143, 158, 211, 224, 227, 231, 251, 317, 321, 327, 333–335, 372, 375, 420, 457, 466, 469, 505, 506, 514, 515, 524, 549, 552, 674, 699, 754, 755, 767, 770, 772, 814
Cellular energetics, 125–169
Chronic dermal inflammation, 794
Chronic inflammation, 50, 51, 54–56, 63, 93, 107, 117, 120, 144, 225, 239–240, 242–244, 246, 247, 249, 251–253, 394, 417, 449, 553, 669, 699, 707, 721, 723, 724, 732, 769, 779, 791, 794–795, 814, 844
Chronic wounds, 46, 62, 86, 130, 182, 201, 221, 238, 270, 295, 314, 355, 393, 411, 439, 467, 498, 543, 574, 605, 667, 696, 715, 749, 766, 835
Clinical care, 670
Critical limb ischemia (CLI), 359–364, 678, 679
Cultured epithelial autografts (CEA), 627–638
Cytokines, 30, 47, 64, 86, 127, 182, 202, 221, 240, 265, 290, 314, 356, 387, 415, 465, 500, 547, 574, 601, 636, 648, 668, 699, 716, 753, 768, 790, 813

D
Debridement, 6, 76, 77, 106, 112, 203, 356, 364, 365, 397, 441, 443, 444, 446–449, 451, 454, 456, 513, 521, 544, 550, 565, 577, 599, 643, 651–652, 658, 670, 671, 686, 697, 752, 767, 843, 846, 853
Decellularization, 503, 511
Dermis, 17, 22, 24, 25, 29–35, 37–38, 40, 49, 68, 182, 200–202, 270, 271, 275, 297, 298, 358, 384–388, 397, 399, 424, 474, 475, 498, 499, 503–506, 510, 517, 527, 547, 607, 608, 615, 616, 637, 643, 654, 791, 792, 794, 807, 810, 815, 819
Diabetic foot ulcers (DFUs), 14, 50–52, 62, 77, 102, 103, 106, 150, 169, 210, 237–253, 272, 290, 300, 304, 356, 359–365, 368, 370, 395, 413, 420, 425, 428, 457, 458, 467, 476, 477, 479, 485, 497, 499, 520, 546, 548, 550, 553, 555, 558, 561, 565, 575–579, 612, 613, 615, 667, 668, 670, 671, 678–680, 686, 696, 705, 719, 751, 752, 755–758, 767, 780, 835, 838, 841, 842, 844, 846–848
Diabetic wound healing, 159, 246, 252, 317, 323, 336–341, 367, 368, 478, 525, 668, 669, 671, 673, 674, 684, 686, 731–732
Diagnostic tools, 73, 74, 239
Digital wound care, 71

E
ECM remodeling, 30, 87, 90, 91, 93–95, 98, 100, 101, 103, 106, 117, 118, 144–145, 192, 204, 223, 226, 228, 229, 233, 242, 245, 247, 248, 250–253, 269, 270, 296, 333, 335, 359, 367, 417, 421–423, 457, 470, 471, 502, 522, 550, 556, 600, 602, 616, 699, 719, 720, 756, 766, 771, 801, 816
Electrospinning, 514, 523, 525, 529–530, 583, 584, 589, 604
Epidermal stem cells/holoclones, 34, 117, 200, 424, 600, 628–630, 634, 636, 638, 643, 818, 819
Epidermis, 17, 22, 24–31, 34–39, 49, 54, 98, 182, 183, 185–195, 200, 201, 210, 267, 270, 271, 297, 384, 386–388, 390, 394, 397, 399, 473–475, 498, 499, 501, 503, 504, 518, 527, 582, 599, 600, 607, 608, 615, 634, 636, 643, 807, 818, 819
Ethical considerations, 112, 277–278, 292, 299–301, 607, 609, 651, 855–859
Evaluation, 9, 63–67, 69–71, 73, 77, 161, 270, 274, 291, 301, 305, 364, 369, 370, 452, 610, 614, 616, 617, 646, 647, 684, 759, 773, 774, 776, 777, 788, 789, 840, 844, 845, 855, 858
Exosomes, 47, 119, 140, 227, 251, 315, 366, 413, 468, 545, 584, 600, 676, 717
Extracellular matrix (ECM), 30, 49, 86, 136, 182, 202, 222, 239, 294, 314, 357, 393, 416, 465, 499, 546, 574, 600, 669, 699, 716, 753, 766, 790, 810
Extracellular vesicles (EVs), 146, 169, 228, 252, 311–343, 353–376, 417, 421, 422, 424, 468, 479, 480, 545, 546, 561, 562, 568, 584, 585, 589, 602, 606, 610, 614, 676–678, 700–701, 708, 716, 725, 728–733, 756
Extracorporeal shock wave therapy (ESWT), 671, 752–754

F

Fibrin matrix carrier, 638
Fibroblast activation, 86, 89, 92, 97, 142, 324, 325, 417, 420–422, 557, 562, 699, 717, 724, 787, 790, 791, 794, 800
Fibrosis, 50, 55, 68, 69, 73, 87, 89, 96, 101, 105, 107, 119, 137, 142, 145, 168, 207, 211, 223, 228, 229, 231, 232, 237–253, 264, 271, 272, 274, 275, 277, 306, 338, 357, 359, 395, 417, 422, 423, 427, 470, 508, 513, 602, 608, 609, 677, 707, 721, 731, 754, 774, 786, 792, 800, 810, 813, 814, 818, 820, 823
4D bioprinting, 533

G

Gene therapy, 86, 95, 98, 102, 103, 106, 108, 111–114, 118–120, 552, 574, 588–592, 800, 801
Growth factors, 30, 47, 65, 86, 128, 182, 203, 223, 241, 265, 290, 314, 356, 388, 413, 441, 465, 500, 546, 574, 601, 628, 656, 668, 698, 716, 752, 766, 790, 814, 835
Growth factor therapy, 232, 552–553, 756

H

Hair follicle, 22, 24, 29–34, 38–40, 200, 225, 267, 275, 276, 290, 302, 328, 330, 384, 471, 484, 499, 560, 580, 582, 584, 589–591, 600, 607, 615, 634, 636, 643, 815, 819
Hard-to-heal wounds, 61–78, 103, 118, 120
Hemostasis, 47, 48, 56, 86–88, 90, 91, 98, 113, 120, 224, 228, 245, 268, 269, 288, 314, 356, 389, 390, 465, 466, 500, 514, 516, 519, 525, 543, 544, 549, 550, 574, 583, 589, 592, 642, 650, 668, 669, 716, 767, 790, 819, 820
Histology, 21–40, 53, 291, 292, 581, 582, 838
Hypertrophic scar model, 272, 590
Hypertrophic scars, 107, 221, 250, 265, 272–275, 277, 279, 317, 325, 335, 423, 471, 582, 607, 609, 612, 643, 716, 724, 753, 786–793, 795–798, 800, 801, 805–826
Hypodermis, 22, 25, 34–40, 200, 201, 485, 499, 657

I

Imaging, 54, 63, 65, 66, 71–73, 76, 78, 140, 161–163, 183, 195, 270, 273, 275, 277, 291, 297–299, 303–305, 342, 374, 531, 561, 647, 730, 733, 836–840, 843–846, 848, 849, 852, 853, 859
Immune cells, 30, 47, 64, 86, 127, 192, 202, 221, 239, 263, 290, 326, 358, 386, 414, 468, 520, 561, 574, 668, 716, 767, 790, 817, 838
Immune response, 7, 17, 29, 33, 48, 63, 64, 89, 90, 92, 93, 97, 100, 105, 108, 110, 112–114, 118, 138, 141, 144, 145, 148, 150, 155, 204, 207, 221, 233, 237–253, 264, 269, 291, 300, 305, 323, 327, 329, 375, 411, 412, 414, 417, 426, 484, 498, 503, 511, 523, 552, 563, 601, 602, 606–608, 667, 668, 674, 678, 684, 699, 700, 707, 718, 724, 726, 729, 731, 732, 766, 769, 771, 772, 792, 819, 820
Immunomodulation, 86, 332, 333, 364, 416, 420, 421, 425, 545, 550, 562, 601, 676
Infection control, 52, 88, 271, 401, 441, 443, 504, 528, 544, 546, 548, 552, 555, 670, 686, 699
Inflammation, 5, 33, 46, 62, 86, 127, 182, 203, 221, 238, 261, 288, 314, 356, 387, 411, 449, 465, 500, 543, 574, 601, 637, 643, 668, 697, 716, 756, 766, 787, 814, 835
Integrin, 24, 25, 49, 96, 152, 153, 183, 184, 189–195, 223, 249, 358, 361, 472, 516, 520, 577, 720, 753
Intensive care units (ICUs), 644, 646–650, 658, 773, 775, 777
Intercellular communication, 252, 314, 315, 323, 324, 329, 342, 366, 602, 700, 725
In vitro model, 32, 160, 262, 269, 277, 278, 304, 605, 618
In vivo model, 261, 262, 272, 289, 423, 579

K

Keloids, 107, 221, 264, 302, 305, 359, 379, 423, 607, 677, 716, 724, 754, 786–801, 806, 808–825

M
Mechanobiology, 16
Mechanotransduction, 153, 154, 418, 580–582, 753
Medical texts, 2–4, 18
Mesenchymal stromal cells (MSCs), 47, 94, 146, 251, 305, 314, 356, 400, 411, 476, 544, 589, 600, 671, 699, 732, 756, 772
Metabolic reprogramming, 128, 143, 145–146
Microbial vesicles, 728
Micrografting–CEA hybrid technique, 633, 636
Micropatterning, 530
Mitochondrial dynamics, 129–132, 137, 146, 147, 158, 168
Moisture balance, 64, 356, 439, 441, 443, 548
Multidisciplinary care, 61–78
Multimodal data fusion, 845–846
Multimodal scar management, 657, 799
Murine wound healing, 267, 677

N
Nanopatterning, 530
Nanotechnology, 109, 110, 119, 166–167, 520, 552, 695–709, 731, 756–760, 797, 800
Natural polymers, 230, 232, 511–513, 520, 522, 527, 529, 531, 559, 615, 823
Negative pressure wound therapy (NPWT), 16–18, 77, 203, 264, 305, 442, 455–456, 554–555, 568, 670, 671, 708, 749–754, 758–760, 848, 853

O
Oxygen therapy (HBOT/TOT), 14, 153, 557–559, 568, 648, 670, 671, 754–755, 758–760

P
Paracrine signaling, 86, 118, 343, 366, 371, 419, 476, 545, 550, 568, 602, 604, 606, 636, 674, 676
Pathological scarring, 423, 790, 791, 793, 795, 811, 815, 822, 825
Peripheral artery disease (PAD), 50, 55, 62, 64, 238, 243, 356, 359, 360, 363–364, 558, 680, 779
Personalized medicine, 103, 119, 301, 305, 618, 658, 696, 709, 833–860
Plastic surgery procedures, 11, 17, 754
Platelet derivatives, 475–480, 486
Precision medicine, 113, 119, 164, 374, 705, 860
Preclinical model, 259–279, 295, 296, 302, 304–306, 667, 674, 678, 682, 730, 732
Predictive analytics, 704, 705, 759, 836, 840
Proliferation, 14, 33, 46, 63, 86, 127, 192, 202, 221, 239, 265, 289, 314, 356, 389, 416, 448, 465, 500, 543, 574, 602, 630, 643, 668, 699, 716, 754, 767, 790, 808

R
Regeneration, 6, 29, 47, 62, 86, 128, 192, 205, 223, 241, 263, 290, 314, 365, 384, 411, 441, 467, 498, 547, 580, 600, 636, 643, 671, 697, 717, 753, 765, 812
Regenerative medicine, 18, 78, 109, 111, 113, 114, 116–121, 169, 211, 277, 289, 302, 358, 368, 375, 410–412, 414–416, 418, 426, 428, 468, 475–480, 485, 486, 526, 544, 559, 561–563, 592, 599, 617, 628, 665–686, 695–709, 730, 734, 756, 766, 771, 773, 780, 797, 800
Regulatory frameworks, 113, 166, 374, 427, 617, 684, 706, 856–857, 859
Rehabilitation, 239, 656–658, 796
Remodeling, 30, 46, 86, 127, 185, 202, 222, 238, 269, 289, 314, 356, 389, 413, 457, 466, 500, 543, 574, 600, 643, 668, 697, 716, 753, 766, 789, 814
Risk assessment, 67–70, 646, 842, 843, 857

S
Scaffolds, 24, 49, 86, 127, 191, 223, 336, 356, 390, 410, 456, 467, 499, 547, 581, 603, 638, 654, 670, 697, 716, 757, 766, 797, 812
Scar formation, 86, 128, 144, 187, 202, 224, 248, 264, 270, 272, 277, 296, 334, 335, 339, 420, 422, 428, 476, 481, 503, 507, 590, 601, 604, 608–610, 616, 643, 677, 707, 753, 754, 786, 789, 790, 793, 801, 812, 813, 816, 820, 823
Scar management, 369, 611, 657, 753–754, 789, 797–801
Scleroderma, 752, 806, 809–816, 818–825
Skin, 2, 22, 46, 63, 99, 129, 182, 200, 221, 238, 261, 288, 314, 355, 384, 421, 440, 465, 497, 543, 579, 598, 628,

642, 670, 697, 715, 750, 767, 786, 807, 837
Skin glands, 31, 32, 471, 560, 600, 615
Skin regeneration, 6, 141, 209, 325, 330, 365, 467, 476–478, 480, 481, 483, 484, 498, 503, 512, 513, 515, 516, 524, 527, 533, 559, 600–603, 605, 611, 615, 656, 673, 700
Skin substitutes, 17, 18, 78, 108–109, 111, 119, 263, 275, 291, 292, 301, 305, 365, 456, 467, 472–478, 484–486, 495–533, 560, 568, 599, 606, 615, 616, 618, 654–656, 670, 671, 758, 759, 848
Skin transplantation, 473, 673, 824
Small extracellular vesicles (sEVs), 52, 584, 589, 718, 725–728, 730–734
Small molecule inhibitors, 96, 98, 574, 577–582
Small-molecule therapeutics, 98, 576
Smart skin substitutes, 532
Stem cell–based regenerative approaches, 598, 672
Stem cells, 26, 48, 78, 86, 146, 194, 200, 224, 252, 273, 290, 358, 384, 411, 465, 517, 544, 584, 598, 628, 643, 668, 699, 715, 756, 766, 787, 814
Stromal vascular fraction (SVF), 412–415, 421, 425, 427, 477, 478, 486, 611, 614, 675
Synthetic polymers, 482, 483, 511, 512, 522–524, 526, 530, 531, 553, 592, 618, 654, 823

T
Therapeutic innovation, 86, 125–169, 205, 219–233, 261
3D bioprinting, 109, 113, 114, 119, 301, 411, 483, 485, 486, 531–532, 559–561, 568, 603, 604, 615, 618, 652, 757, 801
Tissue engineering, 116, 289, 301, 302, 410, 412, 418, 463–486, 499, 559–561, 568, 585, 605, 615, 617, 652, 703, 773
Tissue repair, 48, 54, 56, 63, 78, 86, 87, 89, 91–95, 99, 100, 103, 113–115, 117–119, 127–130, 132, 133, 136–141, 143, 144, 146, 147, 149, 151, 153, 155–157, 160, 161, 169, 189, 203, 210, 211, 221, 222, 226, 227, 229, 240, 245, 247, 248, 250, 251, 261, 266, 293, 295, 306, 330, 332, 342, 357–359, 366, 368, 373, 384, 390, 392, 393, 410, 413, 414, 416, 417, 457, 467, 473, 476, 504, 506, 507, 511, 518–520, 552, 565, 566, 603, 668, 669, 674–677, 685, 699, 701, 704, 707, 723–728, 758, 766, 768–771, 790, 812, 816, 817, 820, 821

V
Venous leg ulcer (VLU), 50, 54–55, 150, 203, 359–361, 365, 420, 470, 475, 548, 550, 556, 561, 751, 756, 758, 766, 847

W
Wound assessment, 63–66, 69–73, 77, 78, 301, 303, 305, 442, 452, 705, 836, 840, 844–846, 849, 852, 858
Wound care, 2, 56, 63, 90, 155, 203, 233, 239, 264, 304, 364, 413, 440, 480, 509, 544, 574, 643, 668, 696, 731, 749, 766, 786, 835
Wound epithelialization, 181–195, 372
Wound healing, 3, 30, 46, 64, 86, 127, 182, 201, 221, 239, 260, 289, 314, 356, 384, 411, 439, 465, 500, 543, 574, 598, 635, 642, 667, 696, 715, 748, 765, 786, 806, 837
Wound management, 12, 47, 65, 70, 71, 73, 76–78, 86, 91, 95, 104, 109, 113, 118, 120, 159, 358–371, 413, 425, 439, 443, 452, 478, 482, 497, 509, 516, 520, 533, 599, 601, 615, 618, 644, 670, 672, 679, 697, 702, 704, 708, 709, 749, 756, 758, 760, 835, 836, 838, 840, 843, 848, 850, 853, 857–860
Wound repair, 64, 97, 98, 101, 103–105, 113, 114, 117, 119, 120, 129, 130, 132–133, 137, 139, 140, 156, 161, 182, 187, 194, 201, 202, 204, 208, 224, 225, 227, 244, 246, 269, 279, 290, 292, 314, 329, 330, 334, 336, 340–342, 367, 371, 421, 466, 480, 511, 522, 545, 568, 588, 600, 601, 611, 672, 674, 675, 682, 698, 702, 703, 717, 725, 749, 756, 818

X
Xenograft, 17, 473, 498, 502, 511, 581, 582, 599, 654, 758

MIX
Papier aus verantwortungsvollen Quellen
Paper from responsible sources
FSC® C105338
FSC
www.fsc.org

If you have any concerns about our products,
you can contact us on
ProductSafety@springernature.com

In case Publisher is established outside the EU,
the EU authorized representative is:
Springer Nature Customer Service Center GmbH
Europaplatz 3, 69115 Heidelberg, Germany

Printed by Libri Plureos GmbH
in Hamburg, Germany